Lecture Notes in Artificial Intelligence   5590

Edited by R. Goebel, J. Siekmann, and W. Wahlster

Subseries of Lecture Notes in Computer Science

Claudio Sossai   Gaetano Chemello (Eds.)

Symbolic and Quantitative Approaches to Reasoning with Uncertainty

10th European Conference, ECSQARU 2009
Verona, Italy, July 1-3, 2009
Proceedings

 Springer

Series Editors

Randy Goebel, University of Alberta, Edmonton, Canada
Jörg Siekmann, University of Saarland, Saarbrücken, Germany
Wolfgang Wahlster, DFKI and University of Saarland, Saarbrücken, Germany

Volume Editors

Claudio Sossai
ISIB-CNR
Corso Stati Uniti 4
35127 Padova, Italy
E-mail: claudio.sossai@isib.cnr.it

Gaetano Chemello
ISIB-CNR
Corso Stati Uniti, 4
35127 Padova, Italy
E-mail: gaetano.chemello@isib.cnr.it

Library of Congress Control Number: Applied for

CR Subject Classification (1998): I.2, F.4.1, F.3, F.4, I.2.3, I.2.4

LNCS Sublibrary: SL 7 – Artificial Intelligence

ISSN   0302-9743
ISBN-10   3-642-02905-1 Springer Berlin Heidelberg New York
ISBN-13   978-3-642-02905-9 Springer Berlin Heidelberg New York

springer.com

© Springer-Verlag Berlin Heidelberg 2009
Printed in Germany

Typesetting: Camera-ready by author, data conversion by Scientific Publishing Services, Chennai, India
Printed on acid-free paper   SPIN: 12702138   06/3180   5 4 3 2 1 0

Preface

These are the proceedings of the 10th European Conference on Symbolic and Quantitative Approaches to Reasoning with Uncertainty, ECSQARU 2009, held in Verona (Italy), July 1–3, 2009.

The biennial ECSQARU conferences are a major forum for advances in the theory and practice of reasoning under uncertainty. The first ECSQARU conference was held in Marseille (1991), and since then it has been held in Granada (1993), Fribourg (1995), Bonn (1997), London (1999), Toulouse (2001), Aalborg (2003), Barcelona (2005) and Hammamet (2007).

The 76 papers gathered in this volume were selected out of 118 submissions from 34 countries, after a rigorous review process. In addition, the conference included invited lectures by three outstanding researchers in the area: Isabelle Bloch ("Fuzzy and bipolar mathematical morphology, applications in spatial reasoning"), Petr Cintula ("From (deductive) fuzzy logic to (logic-based) fuzzy mathematics"), and Daniele Mundici ("Conditionals and independence in many-valued logics").

Two special sessions were presented during the conference: "Conditioning, independence, inference" (organized by Giulianella Coletti and Barbara Vantaggi) and "Mathematical fuzzy logic" (organized by Stefano Aguzzoli, Brunella Gerla, Lluís Godo, Vincenzo Marra, Franco Montagna)

On the whole, the program of the conference provided a broad, rich and up-to-date perspective of the current high-level research in the area which is reflected in the contents of this volume.

We would like to thank the members of the Program Committee, the additional reviewers, the invited speakers, the organizers of the special sessions and the members of the Organizing Committee for their support in making this conference successful. We also thank the creators and maintainers of the free conference management system EasyChair (http://www.easychair.org).

Finally, we gratefully acknowledge the sponsoring institutions for their support.

April 2009

Claudio Sossai
Gaetano Chemello

Organization

ECSQARU 2009 was organized by the Institute of Biomedical Engineering of the Italian National Research Council (ISIB-CNR, Padova) and by the Department of Computer Science of the University of Verona.

Executive Committee

Conference Chair   Claudio Sossai (ISIB-CNR, Italy)
Organizing Committee   Paolo Bison (ISIB-CNR, Italy)
     Claudio Cavaggion (ISIB-CNR, Italy)
     Gaetano Chemello (ISIB-CNR, Italy)
     Matteo Cristani (University of Verona, Italy)
     Elisabetta Di Prisco (Ecsqaru Verona 2009, Italy)
     Paolo Fiorini (University of Verona, Italy)
     Massimiliano Giacomin (University of Brescia, Italy)
     Gianni Perbellini (Ecsqaru Verona 2009, Italy)
     Silvia Zoletto (ISIB-CNR, Italy)

Program Committee

Stefano Aguzzoli (Italy)
Alain Appriou (France)
Nahla Ben Amor (Tunisia)
Boutheina Ben Yaghlane (Tunisia)
Salem Benferhat (France)
Philippe Besnard (France)
Isabelle Bloch (France)
Gerd Brewka (Germany)
Claudette Cayrol (France)
Gaetano Chemello (Italy)
Carlos Chesñevar (Argentina)
Laurence Cholvy (France)
Agata Ciabattoni (Austria)
Giulianella Coletti (Italy)
Fabio Cozman (Brazil)
Fabio Cuzzolin (UK)
Luís Miguel de Campos (Spain)
Gert de Cooman (Belgium)
James P. Delgrande (Canada)
Thierry Denoeux (France)
Antonio di Nola (Italy)

Marek Druzdzel (USA)
Didier Dubois (France)
Zied Elouedi (Tunisia)
Francesc Esteva (Spain)
Paolo Fiorini (Italy)
Hector Geffner (Spain)
Brunella Gerla (Italy)
Massimiliano Giacomin (Italy)
Angelo Gilio (Italy)
Lluís Godo (Spain)
Michel Grabisch (France)
Petr Hájek (Czech Republic)
Andreas Herzig (France)
Eyke Hüllermeier (Germany)
Anthony Hunter (UK)
Katsumi Inoue (Japan)
Gabriele Kern-Isberner (Germany)
Ivan Kramosil (Czech Republic)
Rudolf Kruse (Germany)
Jérôme Lang (France)
Pedro Larrañaga (Spain)

Jonathan Lawry (UK)

Churn-Jung Liau (Taiwan)

Paolo Liberatore (Italy)

Weiru Liu (UK)

Thomas Lukasiewicz (Italy)

Pierre Marquis (France)

Vincenzo Marra (Italy)

Khaled Mellouli (Tunisia)

Luís Moniz Pereira (Portugal)

Franco Montagna (Italy)

Serafín Moral (Spain)

Thomas Dyhre Nielsen (Denmark)

Kristian G. Olesen (Denmark)

Ewa Orlowska (Poland)

Odile Papini (France)

Simon Parsons (USA)

Ramón Pino-Pérez (Venezuela)

David Poole (Canada)

Henri Prade (France)

Maria Rifqi (France)

Alessandro Saffiotti (Sweden)

Sandra Sandri (Spain)

Torsten Schaub (Germany)

Romano Scozzafava (Italy)

Prakash P. Shenoy (USA)

Guillermo Simari (Argentina)

Enric Trillas (Spain)

Linda van der Gaag (The Netherlands)

Leon van der Torre (The Netherlands)

Barbara Vantaggi (Italy)

Emil Weydert (Luxembourg)

Mary-Anne Williams (Australia)

Nevin L. Zhang (Hong Kong, China)

Additional Reviewers

Teresa Alsinet

Mohamed Anis Bach Tobji

Marco Baioletti

Philippe Balbiani

Matteo Bianchi

Martin Caminada

Andrea Capotorti

Walter Carnielli

Federico Cerutti

Tao Chen

Sylvie Coste-Marquis

Andrei Doncescu

Phan Minh Dung

Tommaso Flaminio

Alfredo Gabaldon

Fabio Gadducci

Cipriano Galindo

Pere García-Calvés

Martin Gebser

Daniele Genito

Hervé Glotin

Joanna Golinska-Pilarek

Stijn Heymans

Andrzej Hildebrandt

Szymon Jaroszewicz

Reinhard Kahle

Petri Kontkanen

Sylvain Lagrue

Ada Lettieri

Jianbing Ma

Enrico Marchioni

Christophe Marsala

Marie-Hélène Masson

Christian Moewes

Daniele Mundici

Pascal Nicolas

David Picado Muiño

Gabriella Pigozzi

Valentina Poggioni

Benjamin Quost

Thorsteinn Rognvaldsson

Pavel Rusnok

Georg Ruß

Chiaki Sakama

Matthias Steinbrecher

Umberto Straccia

Carlos Uzcátegui

Yi Wang

Gregory Wheeler

Roland Winkler

Eric Würbel

Anbu Yue

Sponsoring Institutions

Institute of Biomedical Engineering (ISIB-CNR), Padova
Department of Computer Science, University of Verona
Fondazione Arena di Verona

Table of Contents

Bayesian Networks

Belief Functions

Belief Revision and Inconsistency Handling

Classification and Clustering

Conditioning, Independence, Inference

Default Reasoning

Foundations of Reasoning and Decision Making under Uncertainty

Fuzzy Sets and Fuzzy Logic

Implementation and Applications of Uncertain Systems

Logics for Reasoning under Uncertainty

Markov Decision Processes

Mathematical Fuzzy Logic

Fuzzy and Bipolar Mathematical Morphology, Applications in Spatial Reasoning

Isabelle Bloch

Télécom ParisTech (ENST), CNRS UMR 5141 LTCI, Paris, France
isabelle.bloch@enst.fr

Abstract. Mathematical morphology is based on the algebraic framework of complete lattices and adjunctions, which endows it with strong properties and allows for multiple extensions. In particular, extensions to fuzzy sets of the main morphological operators, such as dilation and erosion, can be done while preserving all properties of these operators. Another, more recent, extension, concerns bipolar fuzzy sets. These extensions have numerous applications, two of each being presented here. The first one concerns the definition of spatial relations, for applications in spatial reasoning and model-based recognition of structures in images. The second one concerns the handling of the bipolarity feature of spatial information.

Keywords: Fuzzy mathematical morphology, bipolar mathematical morphology, spatial relations, bipolar spatial information, spatial reasoning.

1 Algebraic Framework of Mathematical Morphology

Mathematical morphology [1] requires the algebraic framework of complete lattices [2]. Let $(\mathcal{T}, \leq)$ be a complete lattice, $\vee$ the supremum and $\wedge$ the infimum. A dilation is an operator δ on $\mathcal{T}$ which commutes with the supremum: $\forall(x_i) \in \mathcal{T}$, $\delta(\vee_i x_i) = \vee_i \delta(x_i)$. An erosion is an operator ε on $\mathcal{T}$ which commutes with the infimum: $\forall(x_i) \in \mathcal{T}$, $\varepsilon(\wedge_i x_i) = \wedge_i \varepsilon(x_i)$ [3]. Such operators are called algebraic dilation and erosion. An important property is that they are increasing with respect to $\leq$.

An adjunction on $(\mathcal{T}, \leq)$ is a pair of operators (ε, δ) such that $\forall(x,y) \in \mathcal{T}^2$, $\delta(x) \leq y \Leftrightarrow x \leq \varepsilon(y)$. If (ε, δ) is an adjunction, then ε is an algebraic erosion and δ an algebraic dilation. Additionally, the following properties hold: $\varepsilon\delta \geq Id$, where Id denotes the identity mapping on $\mathcal{T}$, $\delta\varepsilon \leq Id$, $\varepsilon\delta\varepsilon = \varepsilon$, $\delta\varepsilon\delta = \delta$, $\varepsilon\delta\varepsilon\delta = \varepsilon\delta$ et $\delta\varepsilon\delta\varepsilon = \delta\varepsilon$ (the compositions $\delta\varepsilon$ and $\varepsilon\delta$ are known as morphological opening and closing, respectively, and can also be formalized in the framework of Moore families [4]).

In the particular case of the lattice of subparts of $\mathbb{R}^n$ or $\mathbb{Z}^n$, denoted by $\mathcal{S}$ in the following, endowed with inclusion as partial inclusion, adding a property of invariance under translation leads to the particular following forms (called morphological dilation and erosion):

$$\forall X \subseteq \mathcal{S}, \delta_B(X) = \{x \in \mathcal{S} \mid \check{B}_x \cap X \neq \emptyset\}, \quad \varepsilon_B(X) = \{x \in \mathcal{S} \mid B_x \subseteq X\},$$

C. Sossai and G. Chemello (Eds.): ECSQARU 2009, LNAI 5590, pp. 1–13, 2009.

where B is a subset of $\mathcal{S}$ called structuring element, B_x denotes its translation at point x and $\check{B}$ its symmetrical with respect to the origin of space. Opening and closing are defined by composition (using the same structuring element).

These definitions are general and apply to any complete lattice. In the following, we focus on the lattice of fuzzy sets defined on $\mathcal{S}$ and on the lattice of bipolar fuzzy sets. Other works have been done on the lattice of logical formulas in propositional logics [5,6,7,8], with applications to fusion, revision, abduction, mediation, or in modal logics [9], with applications including qualitative spatial reasoning.

Mathematical morphology can therefore be considered as a unifying framework for spatial reasoning, leading to knowledge representation models and reasoning tools in quantitative, semi-quantitative (or fuzzy) and qualitative settings [10].

2 Fuzzy Mathematical Morphology

Extending mathematical morphology to fuzzy sets was proposed in the early 90's, by several teams independently [11,12,13,14,15], and was then largely developed (see e.g. [16,17,18,19,20,21]). An earlier extension of Minkowski's addition (which is directly linked to dilation) was defined in [22].

Let $\mathcal{F}$ be the set of fuzzy subsets of $\mathcal{S}$. For the usual partial ordering $\leq$ ($\mu \leq \nu \Leftrightarrow \forall x \in \mathcal{S}, \mu(x) \leq \nu(x)$), $(\mathcal{F}, \leq)$ is a complete lattice, on which algebraic operations can be defined, as described in Section 1. Adding a property of invariance under translation leads to the following general forms of fuzzy dilation and erosion [12,16]:

$$\forall x \in \mathcal{S}, \ \delta_\nu(\mu)(x) = \sup_{y \in \mathcal{S}} T[\nu(x-y), \mu(y)], \quad \varepsilon_\nu(\mu)(x) = \inf_{y \in \mathcal{S}} S[c(\nu(y-x)), \mu(y)],$$

where ν denotes a fuzzy structuring element in $\mathcal{F}$, μ a fuzzy set, c an involutive negation (or complementation), T a t-norm and S a t-conorm. The adjunction property imposes that S be the t-conorm derived from the residual implication I of T: $\forall(\alpha, \beta) \in [0,1]^2, S(\alpha, \beta) = I(c(\alpha), \beta)$, with $I(\alpha, \beta) = \sup\{\gamma \in [0,1], T(\alpha, \gamma) \leq \beta\}$. The erosion represents the degree to which the translation of the structuring element at point x intersects μ, while the dilation represents the degree to which it is included in μ.

For applications dealing with spatial objects for instance, it is often important to also have a duality property between dilation and erosion, with respect to the complementation. Then T and S have to be dual operators with respect to c. This property, along with the adjunction property, limits the choice of T and S to generalized Lukasiewicz operators [23,24]: $T(\alpha, \beta) = \max(0, \varphi^{-1}(\varphi(\alpha) + \varphi(\beta) - 1))$ and $S(\alpha, \beta) = \min(1, \varphi^{-1}(\varphi(\alpha) + \varphi(\beta)))$ where φ is a continuous strictly increasing function on $[0,1]$ with $\varphi(0) = 0$ and $\varphi(1) = 1$.

The links between definitions obtained for various forms of conjunctions and disjunctions have been presented from different perspectives in [16,20,23,24,25].

Opening and closing are defined by composition, as in the general case. The adjunction property guarantees that these operators are idempotent, and that opening (resp. closing) is anti-extensive (resp. extensive) [16,17,24].

3 Bipolar Fuzzy Mathematical Morphology

Bipolarity is important to distinguish between (i) positive information, which represents what is guaranteed to be possible, for instance because it has already been observed or experienced, and (ii) negative information, which represents what is impossible or forbidden, or surely false [26].

A bipolar fuzzy set on $\mathcal{S}$ is defined by a pair of functions (μ, ν) such that $\forall x \in \mathcal{S}, \mu(x) + \nu(x) \leq 1$. For each point x, $\mu(x)$ defines the membership degree of x (positive information) and $\nu(x)$ the non-membership degree (negative information). This formalism allows representing both bipolarity and fuzziness.

Let us consider the set $\mathcal{L}$ of pairs of numbers (a, b) in $[0, 1]$ such that $a + b \leq 1$. It is a complete lattice, for the partial order defined as [27]: $(a_1, b_1) \preceq (a_2, b_2)$ iff $a_1 \leq a_2$ and $b_1 \geq b_2$. The greatest element is $(1, 0)$ and the smallest element is $(0, 1)$. The supremum and infimum are respectively defined as: $(a_1, b_1) \vee (a_2, b_2) = (\max(a_1, a_2), \min(b_1, b_2))$, $(a_1, b_1) \wedge (a_2, b_2) = (\min(a_1, a_2), \max(b_1, b_2))$. The partial order $\preceq$ induces a partial order on the set of bipolar fuzzy sets: $(\mu_1, \nu_1) \preceq (\mu_2, \nu_2)$ iff $\forall x \in \mathcal{S}, \mu_1(x) \leq \mu_2(x)$ and $\nu_1(x) \geq \nu_2(x)$, and infimum and supremum are defined accordingly. It follows that, if $\mathcal{B}$ denotes the set of bipolar fuzzy sets on $\mathcal{S}$, $(\mathcal{B}, \preceq)$ is a complete lattice.

Mathematical morphology on bipolar fuzzy sets has been first introduced in [28]. Once we have a complete lattice, it is easy to define algebraic dilations and erosions on this lattice, as described in Section 1, as operators that commute with the supremum and the infimum, respectively. Their properties are derived from general properties of lattice operators.

Let us now consider morphological operations based on a structuring element. A degree of inclusion of a bipolar fuzzy set (μ', ν') in another bipolar fuzzy set (μ, ν) is defined as: $\inf_{x \in \mathcal{S}} I((\mu'(x), \nu'(x)), (\mu(x), \nu(x)))$, where I is an implication operator. Two types of implication can be defined [29], one derived from a bipolar t-conorm $\perp$[1]: $I_N((a_1, b_1), (a_2, b_2)) = \perp((b_1, a_1), (a_2, b_2))$, and one derived from a residuation principle from a bipolar t-norm $\top$[2]: $I_R((a_1, b_1), (a_2, b_2)) = \sup\{(a_3, b_3) \in \mathcal{L} \mid \top((a_1, b_1), (a_3, b_3)) \preceq (a_2, b_2)\}$, where $(a_i, b_i) \in \mathcal{L}$ and (b_i, a_i) is the standard negation of (a_i, b_i). Two types of t-norms and t-conorms are considered in [29]: operators called t-representable t-norms and t-conorms, which can be expressed using usual t-norms t and t-conorms T, and Lukasiewicz operators, which are not t-representable. A similar approach has been used for intuitionistic fuzzy sets in [30], but with weaker properties (in particular an important property such as the commutativity of erosion with the conjunction may be lost).

---

[1] A bipolar disjunction is an operator D from $\mathcal{L} \times \mathcal{L}$ into $\mathcal{L}$ such that $D((1, 0), (1, 0)) = D((0, 1), (1, 0)) = D((1, 0), (0, 1)) = (1, 0)$, $D((0, 1), (0, 1)) = (0, 1)$ and that is increasing in both arguments. A bipolar t-conorm is a commutative and associative bipolar disjunction such that the smallest element of $\mathcal{L}$ is the unit element.

[2] A bipolar conjunction is an operator C from $\mathcal{L} \times \mathcal{L}$ into $\mathcal{L}$ such that $C((0, 1), (0, 1)) = C((0, 1), (1, 0)) = C((1, 0), (0, 1)) = (0, 1)$, $C((1, 0), (1, 0)) = (1, 0)$ and that is increasing in both arguments. A bipolar t-norm is a commutative and associative bipolar conjunction such that the largest element of $\mathcal{L}$ is the unit element.

Based on these concepts, the morphological erosion of $(\mu, \nu) \in \mathcal{B}$ by a bipolar fuzzy structuring element $(\mu_B, \nu_B) \in \mathcal{B}$ is defined as:

$$\forall x \in \mathcal{S}, \varepsilon_{(\mu_B, \nu_B)}((\mu, \nu))(x) = \inf_{y \in \mathcal{S}} I((\mu_B(y - x), \nu_B(y - x)), (\mu(y), \nu(y))).$$

Dilation can be defined based on a duality principle or based on the adjunction property. Applying the duality principle to bipolar fuzzy sets using a complementation c (typically the standard negation $c((a, b)) = (b, a)$) leads to the following definition of morphological bipolar dilation:

$$\delta_{(\mu_B, \nu_B)}((\mu, \nu)) = c[\varepsilon_{(\mu_B, \nu_B)}(c((\mu, \nu)))].$$

Let us now consider the adjunction principle, as in the general algebraic case. The bipolar fuzzy dilation, adjoint of the erosion, is defined as:

$$\delta_{(\mu_B, \nu_B)}((\mu, \nu))(x) = \inf\{(\mu', \nu')(x) \mid (\mu, \nu)(x) \preceq \varepsilon_{(\mu_B, \nu_B)}((\mu', \nu'))(x)\}$$
$$= \sup_{y \in \mathcal{S}} \top((\mu_B(x - y), \nu_B(x - y)), (\mu(y), \nu(y))).$$

It has been shown that the adjoint operators are all derived from the Lukasiewicz operator, using a continuous bijective permutation on $[0, 1]$ [29]. Hence equivalence between both approaches can be achieved only for this class of operators.

Properties of these operations are consistent with the the ones holding for sets and for fuzzy sets, and are detailed in [28,31,32,33,34]. Interpretations of these definitions as well as some illustrative examples can also be found in these references.

4   Spatial Relations and Spatial Reasoning

Mathematical morphology provides tools for spatial reasoning at several levels [10]. The notion of structuring element captures the local spatial context, in a fuzzy and bipolar way here, which endows dilation and erosion with a low level spatial reasoning feature. At a more global level, several spatial relations between spatial entities can be expressed as morphological operations, in particular using dilations [35,10], leading to large scale spatial reasoning.

The interest of relationships between objects has been highlighted in very different types of works: in vision, for identifying shapes and objects, in database system management, for supporting spatial data and queries, in artificial intelligence, for planning and reasoning about spatial properties of objects, in cognitive and perceptual psychology, in geography, for geographic information systems. In all these domains, objects, relations, knowledge and questions to be answered may suffer from imprecision, and benefit from a fuzzy modeling, as stated in the 75's [36]. Spatial relations can be intrinsically fuzzy (for instance *close to*, *between*...) or have to be fuzzified in order to cope with imprecisely defined objects.

Fuzzy mathematical morphology has then naturally led to the definition of fuzzy spatial relations (see [35] for a review on fuzzy spatial relations, including morphological approaches). In our previous work, we proposed original definitions for both topological and metric relations (according to the classification of [37]): adjacency, distances, directional relations, and more complex relations such as *between* and *along*. Here we just discuss a few important features (the reader can refer to [35] and the references cited therein for the mathematical developments).

In spatial reasoning, two important questions arise: (i) to which degree is a relation between two objects satisfied? (ii) which is the spatial region in which a relation to a reference object is satisfied (to some degree)? Fuzzy models allow answering these two types of questions. Let us consider the directional relation *to the right of* [38]. Two objects are displayed in Figure 1. Object B is, to some degree, to the right of R. The region of space to the right of R (c) is defined as the dilation of R with a fuzzy structuring element providing the semantics of the relation (b). The membership degree of each point provides the degree to which the relation is satisfied at that point. The definition of the relation as a dilation is a generic model, but the structuring element can be adapted to the context. This type of representation deals with the first type of question. As for the second type, the adequation between B and the fuzzy dilated region can be evaluated. Other fuzzy approaches to this type of relation are reviewed in [39].

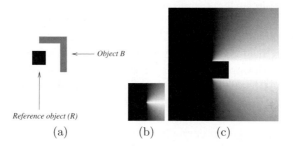

Fig. 1. (a) Two example objects. (b) Fuzzy structuring element defining, in the spatial domain, the semantics of *to the right of*. (c) Fuzzy Dilation of R (black square). Membership values range from 0 (black) to 1 (white).

This example also highlights another important issue, which concerns the representation, for which several forms can be interesting, depending on the raised question. While in the crisp case, a relation between two objects is usually represented by a number (either 0/1 for an all-or-nothing relation, or a numerical value for a distance for instance), in the fuzzy case, several representations are possible. They can typically be intervals, for instance representing necessity and possibility degrees, fuzzy numbers, distributions [40,41], representing actual measurements or the semantics of some linguistic variables. Details can be found in [42,43] in the case of distances. These representations are adequate to answer questions of type 1, since they rely on some computation procedure between two known objects. As for the second type of question, spatial representations are more appropriate, as fuzzy sets in the spatial domain.

Fuzzy representations are also interesting in terms of robustness. For instance set relationships and adjacency are highly sensitive, since in the binary case, the result can depend on one point only [44]. The satisfaction of a relation can be drastically modified depending on the digitization of the space, on the way objects are defined, on errors due to some segmentation process, etc. This is clearly a limitation of binary (all or nothing) definitions. In the fuzzy case, the problem is much less crucial. Indeed, there is no more strict membership, the fuzziness allows dealing with some gradual transition between objects or between object and background, and relations become then a matter of degree. In this respect, the fuzziness, even on digital images, could be interpreted as a partial recovering of the continuity lost during the digitization process.

Finally, some relations depend not only on the applicative context, but also on the shape of the considered objects. This is the case for the *between* relation, where the semantics changes depending on whether the objects have similar spatial extensions or very different ones [45]. Here again, fuzzy mathematical morphology leads to models adapted to each situation [46].

Let us now illustrate how these relations can be used in spatial reasoning, in particular for guiding structure recognition and segmentation in medical images. For instance in brain imaging, anatomical knowledge is often expressed as linguistic descriptions of the structures and their spatial arrangement. Spatial relations play a major role in such descriptions. Moreover, they are more stable than shape or size information and are less prone to inter-individual variations, even in the presence of pathologies. Recently, this knowledge was formalized, in particular using ontologies such as the Foundational Model of Anatomy (FMA) [47], just to mention one. However these models do not yet incorporate much structural descriptions. In [48], we proposed an ontology of spatial relations, which has been integrated in the part of the FMA dedicated to brain structures. This ontology has been further enriched by fuzzy models of the spatial relations (defining their semantics). This formalism partially solves the semantic gap issue, by establishing links between symbolic concepts of the ontology and their representation in the image domain (and hence with percepts that can be extracted from images). These links allow using concretely the ontology to help in image interpretation and object recognition. Mathematical morphology is directly involved in these fuzzy representations, but also at the reasoning level, since tools from morphologics can be integrated in the description logics [49].

In our group, we developed two different types of approaches for recognition, working either sequentially or globally. In the sequential approach [50,51], structures are recognized successively according to some order, and the recognition of each structure relies on its relations to previously detected structures. This allows reducing the search space, as in a process of focus of attention. For instance, anatomical knowledge includes statements such as *the right caudate nucleus is to the right of and close to the right lateral ventricle*. The search space for the right caudate nucleus is then defined as the fuzzy region resulting from the conjunctive fusion of the dilations of the right lateral ventricle using fuzzy structuring elements expressing the semantics of each of these relations. The application domain is here very important, since this semantics highly depends on it. It is clear for instance that the semantics of *close to* is not the same for brain structures or for stars. This is actually encoded in the parameters of the membership functions that define the relations, which can be learned from a data base of images [52]. Within the obtained restricted search region, a precise segmentation can be performed, for instance using deformable models integrating spatial relations in the energy functional [51]. The order according to which the structures are processed can also be learned, as proposed in [53,54].

In global approaches [55], several objects are extracted from the image using any segmentation method, generally providing an over-segmentation, and recognition is then based on the relations existing between these segmented regions, in comparison to those expressed in the knowledge base or ontology. Graph-based approaches [55], or constraint satisfaction problems approaches [56] have been developed, implementing these ideas. The ontological modeling allows, using classification tools for instance, filtering the knowledge base so as to keep only the objects that share some given relations. This leads to a reduced combinatorics in the search for possible associations between image regions and structures of the model.

Segmentation results for a few internal brain structures obtained with the sequential approach are illustrated in Figure 2 for a normal case and in Figure 3 for two pathological cases. The original images are 3D magnetic resonance images (MRI). In the pathological cases, the tumors strongly deform the normal structures. In such situations, methods based on shape and size fail, while using spatial relations (with only slight adaptations) leads to correct results.

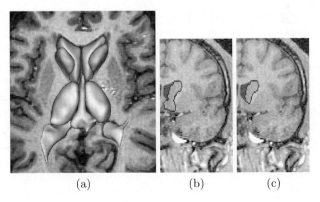

(a)   (b)   (c)

Fig. 2. Segmentation results for a few internal structures in a normal case [51]. (a) Results are superimposed to a part of an axial slice of the original 3D MRI image. (b) Segmentation of the caudate nucleus (shown on one slice) without using the spatial relations: the contour does not match the anatomical constraints and leak outside the structure. (c) Result using the spatial relations: anatomical knowledge is respected and the final segmentation is now correct.

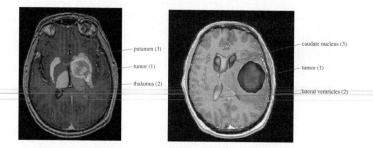

Fig. 3. Segmentation and recognition results in two pathological cases [52]. The order in which structures are segmented is indicated into parentheses.

5   Application of Bipolar Morphology to Spatial Reasoning

Let us provide a few examples where bipolarity occurs when dealing with spatial information, in image processing or for spatial reasoning applications: when assessing the position of an object in space, we may have positive information expressed as a set of possible places, and negative information expressed as a set of impossible or forbidden places (for instance because they are occupied by other objects). As another example, let us consider spatial relations. Human beings consider "left" and "right" as opposite relations. But this does not mean that one of them is the negation of the other one. The semantics of "opposite" captures a notion of symmetry (with respect to some axis or plane) rather than a strict complementation. In particular, there may be positions which are considered neither to the right nor to the left of some reference object, thus leaving room for some indifference or neutrality. This corresponds to the idea that the union of positive and negative information does not cover all the space. Concerning semantics, it should be noted that a bipolar fuzzy set does not necessarily represent one physical object or spatial entity, but rather more complex information, potentially issued from different sources.

In this section, we illustrate a typical scenario showing the interest of bipolar representations of spatial relations and of morphological operations on these representations for spatial reasoning. An example of a brain image is shown in Figure 4, with a few labeled structures of interest.

Let us first consider the right hemisphere (i.e. the non-pathological one). We consider the problem of defining a region of interest for the RPU, based on a known segmentation of RLV and RTH. An anatomical knowledge base or ontology provides some information about the relative position of these structures: (i) *directional information*: the RPU is exterior (left on the image) of the union of RLV and RTH (positive information) and cannot be interior (negative information); (ii) *distance information*: the RPU is quite close of the union of RLV and RTH (positive information) and cannot be very far (negative information). These pieces of information are represented in the image space based on morphological dilations using appropriate structuring elements (representing the semantics of

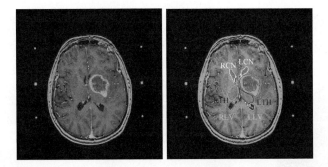

Fig. 4. A slice of a 3D MRI brain image, with a few structures: left and right lateral ventricles (LLV and RLV), caudate nuclei (LCN and RCN), putamen (LPU and RPU) and thalamus (LTH and RTH). A ring-shaped tumor is present in the left hemisphere (the usual "left is right" convention is adopted for the visualization).

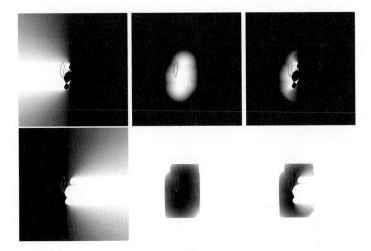

Fig. 5. Bipolar fuzzy representations of spatial relations with respect to RLV and RTH. Top: positive information, bottom: negative information. From left to right: directional relation, distance relation, conjunctive fusion. The contours of the RPU are displayed to show the position of this structure with respect to the region of interest.

the relations) and are illustrated in Figure 5. The neutral area between positive and negative information allows accounting for potential anatomical variability. The conjunctive fusion of the two types of relations is computed as a conjunction of the positive parts and a disjunction of the negative parts. As shown in the illustrated example, the RPU is well included in the bipolar fuzzy region of interest which is obtained using this procedure. This region can then be efficiently used to drive a segmentation and recognition technique of the RPU.

Let us now consider the left hemisphere, where a ring-shaped tumor is present. The tumor induces a deformation effect which strongly changes the shape of the normal structures, but also their spatial relations, to a less extent. In particular

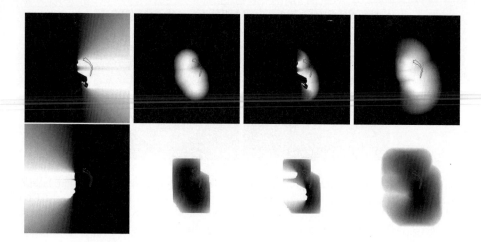

Fig. 6. Bipolar fuzzy representations of spatial relations with respect to LLV and LTH. From left to right: directional relation, distance relation, conjunctive fusion, Bipolar fuzzy dilation. First line: positive parts, second line: negative parts. The contours of the LPU are displayed to show the position of this structure.

the LPU is pushed away from the inter-hemispheric plane, and the LTH is pushed towards the posterior part of the brain and compressed. Applying the same procedure as for the right hemisphere does not lead to very satisfactory results in this case (see Figure 6). The default relations are here too strict and the resulting region of interest is not adequate: the LPU only satisfies with low degrees the positive part of the information, while it also slightly overlaps the negative part. In such cases, some relations (in particular metric ones) should be considered with care. This means that they should be more permissive, so as to include a larger area in the possible region, accounting for the deformation induced by the tumor. This can be easily modeled by a bipolar fuzzy dilation of the region of interest, as shown in the last column of Figure 6. Now the obtained region is larger but includes the right area. This bipolar dilation amounts to dilate the positive part and to erode the negative part.

Other examples are provided in [34]. Exploring further these ideas is planned for future work.

References

1. Serra, J.: Image Analysis and Mathematical Morphology. Academic Press, London (1982)
2. Ronse, C.: Why Mathematical Morphology Needs Complete Lattices. Signal Processing 21(2), 129–154 (1990)
3. Heijmans, H.J.A.M., Ronse, C.: The Algebraic Basis of Mathematical Morphology – Part I: Dilations and Erosions. Computer Vision, Graphics and Image Processing 50, 245–295 (1990)

4. Ronse, C., Heijmans, H.J.A.M.: The Algebraic Basis of Mathematical Morphology – Part II: Openings and Closings. Computer Vision, Graphics and Image Processing 54, 74–97 (1991)
5. Bloch, I., Lang, J.: Towards Mathematical Morpho-Logics. In: 8th International Conference on Information Processing and Management of Uncertainty in Knowledge based Systems IPMU 2000, Madrid, Spain, vol. III, pp. 1405–1412 (2000)
6. Bloch, I., Pino-Pérez, R., Uzcátegui, C.: Explanatory Relations based on Mathematical Morphology. In: Benferhat, S., Besnard, P. (eds.) ECSQARU 2001. LNCS, vol. 2143, pp. 736–747. Springer, Heidelberg (2001)
7. Bloch, I., Pino-Pérez, R., Uzcategui, C.: A Unified Treatment of Knowledge Dynamics. In: International Conference on the Principles of Knowledge Representation and Reasoning, KR 2004, Canada, pp. 329–337 (2004)
8. Bloch, I., Pino-Pérez, R., Uzcategui, C.: Mediation in the Framework of Morphologic. In: European Conference on Artificial Intelligence ECAI 2006, Riva del Garda, Italy, pp. 190–194 (2006)
9. Bloch, I.: Modal Logics based on Mathematical Morphology for Spatial Reasoning. Journal of Applied Non Classical Logics 12(3-4), 399–424 (2002)
10. Bloch, I., Heijmans, H., Ronse, C.: Mathematical Morphology. In: Aiello, M., Pratt-Hartman, I., van Benthem, J. (eds.) The Logic of Space, pp. 857–947. Kluwer, Dordrecht (2007)
11. Bandemer, H., Näther, W.: Fuzzy Data Analysis. Theory and Decision Library. Serie B: Mathematical and Statistical Methods. Kluwer Academic Publisher, Dordrecht (1992)
12. Bloch, I.: Triangular Norms as a Tool for Constructing Fuzzy Mathematical Morphologies. In: Int. Workshop on "Mathematical Morphology and its Applications to Signal Processing", Barcelona, Spain, May 1993, pp. 157–161 (1993)
13. De Baets, B., Kerre, E., Gupta, M.: The Fundamentals of Fuzzy Mathematical Morphology Part 1: Basic Concepts. International Journal of General Systems 23(2), 155–171 (1995)
14. De Baets, B., Kerre, E., Gupta, M.: The Fundamentals of Fuzzy Mathematical Morphology Part 2: Idempotence, Convexity and Decomposition. International Journal of General Systems 23(4), 307–322 (1995)
15. Sinha, D., Dougherty, E.R.: Fuzzification of Set Inclusion: Theory and Applications. Fuzzy Sets and Systems 55, 15–42 (1993)
16. Bloch, I., Maître, H.: Fuzzy Mathematical Morphologies: A Comparative Study. Pattern Recognition 28(9), 1341–1387 (1995)
17. De Baets, B.: Generalized Idempotence in Fuzzy Mathematical Morphology. In: Kerre, E., Nachtegael, M. (eds.) Fuzzy Techniques in Image Processing. Studies in Fuzziness and Soft Computing, vol. 52, pp. 58–75. Physica Verlag/Springer, Heidelberg (2000)
18. Deng, T.Q., Heijmans, H.: Grey-Scale Morphology Based on Fuzzy Logic. Journal of Mathematical Imaging and Vision 16, 155–171 (2002)
19. Maragos, P.: Lattice Image Processing: A Unification of Morphological and Fuzzy Algebraic Systems. Journal of Mathematical Imaging and Vision 22, 333–353 (2005)
20. Nachtegael, M., Kerre, E.E.: Classical and Fuzzy Approaches towards Mathematical Morphology. In: Kerre, E.E., Nachtegael, M. (eds.) Fuzzy Techniques in Image Processing. Studies in Fuzziness and Soft Computing, pp. 3–57. Physica-Verlag/Springer, Heidelberg (2000)
21. Popov, A.T.: Morphological Operations on Fuzzy Sets. In: IEE Image Processing and its Applications, Edinburgh, UK, July 1995, pp. 837–840 (1995)

22. Dubois, D., Prade, H.: Inverse Operations for Fuzzy Numbers. In: Sanchez, E., Gupta, M. (eds.) Fuzzy Information, Knowledge Representation and Decision Analysis, IFAC Symposium, Marseille, France, July 1983, pp. 391–396 (1983)

23. Bloch, I.: Duality vs Adjunction and General Form for Fuzzy Mathematical Morphology. In: Bloch, I., Petrosino, A., Tettamanzi, A.G.B. (eds.) WILF 2005. LNCS, vol. 3849, pp. 354–361. Springer, Heidelberg (2006)

24. Bloch, I.: Duality vs. Adjunction for Fuzzy Mathematical Morphology and General Form of Fuzzy Erosions and Dilations. Fuzzy Sets and Systems 160, 1858–1867 (2009)

25. Sussner, P., Valle, M.: Classification of Fuzzy Mathematical Morphologies based on Concepts of Inclusion Measure and Duality. Journal of Mathematical Imaging and Vision 21, 139–159 (2008)

26. Dubois, D., Kaci, S., Prade, H.: Bipolarity in Reasoning and Decision, an Introduction. In: International Conference on Information Processing and Management of Uncertainty, IPMU 2004, Perugia, Italy, pp. 959–966 (2004)

27. Cornelis, C., Kerre, E.: Inclusion Measures in Intuitionistic Fuzzy Sets. In: Nielsen, T.D., Zhang, N.L. (eds.) ECSQARU 2003. LNCS (LNAI), vol. 2711, pp. 345–356. Springer, Heidelberg (2003)

28. Bloch, I.: Dilation and Erosion of Spatial Bipolar Fuzzy Sets. In: Masulli, F., Mitra, S., Pasi, G. (eds.) WILF 2007. LNCS (LNAI), vol. 4578, pp. 385–393. Springer, Heidelberg (2007)

29. Deschrijver, G., Cornelis, C., Kerre, E.: On the Representation of Intuitionistic Fuzzy t-Norms and t-Conorms. IEEE Transactions on Fuzzy Systems 12(1), 45–61 (2004)

30. Nachtegael, M., Sussner, P., Mélange, T., Kerre, E.: Some Aspects of Interval-Valued and Intuitionistic Fuzzy Mathematical Morphology. In: IPCV 2008 (2008)

31. Bloch, I.: A Contribution to the Representation and Manipulation of Fuzzy Bipolar Spatial Information: Geometry and Morphology. In: Workshop on Soft Methods in Statistical and Fuzzy Spatial Information, Toulouse, France, September 2008, pp. 7–25 (2008)

32. Bloch, I.: Bipolar Fuzzy Spatial Information: First Operations in the Mathematical Morphology Setting. In: De, R.K., Mandal, D.P., Ghosh, A. (eds.) Machine Interpretation of Patterns: Image Analysis, Data Mining and Bioinformatics. World Scientific Press, Singapore (2009)

33. Bloch, I.: Geometry of Spatial Bipolar Fuzzy Sets based on Bipolar Fuzzy Numbers and Mathematical Morphology. In: International Workshop on Fuzzy Logic and Applications WILF, Palermo, Italy (June 2009)

34. Bloch, I.: Bipolar Fuzzy Mathematical Morphology for Spatial Reasoning. In: International Symposium on Mathematical Morphology ISMM 2009, Groningen, The Netherlands (August 2009)

35. Bloch, I.: Fuzzy Spatial Relationships for Image Processing and Interpretation: A Review. Image and Vision Computing 23(2), 89–110 (2005)

36. Freeman, J.: The Modelling of Spatial Relations. Computer Graphics and Image Processing 4(2), 156–171 (1975)

37. Kuipers, B.J., Levitt, T.S.: Navigation and Mapping in Large-Scale Space. AI Magazine 9(2), 25–43 (1988)

38. Bloch, I.: Fuzzy Relative Position between Objects in Image Processing: a Morphological Approach. IEEE Transactions on Pattern Analysis and Machine Intelligence 21(7), 657–664 (1999)

39. Bloch, I., Ralescu, A.: Directional Relative Position between Objects in Image Processing: A Comparison between Fuzzy Approaches. Pattern Recognition 36, 1563–1582 (2003)
40. Dubois, D., Prade, H.: On Distance between Fuzzy Points and their Use for Plausible Reasoning. In: Int. Conf. Systems, Man, and Cybernetics, pp. 300–303 (1983)
41. Rosenfeld, A.: Distances between Fuzzy Sets. Pattern Recognition Letters 3, 229–233 (1985)
42. Bloch, I.: On Fuzzy Distances and their Use in Image Processing under Imprecision. Pattern Recognition 32(11), 1873–1895 (1999)
43. Bloch, I.: On Fuzzy Spatial Distances. In: Hawkes, P. (ed.) Advances in Imaging and Electron Physics, vol. 128, pp. 51–122. Elsevier, Amsterdam (2003)
44. Bloch, I., Maître, H., Anvari, M.: Fuzzy Adjacency between Image Objects. International Journal of Uncertainty, Fuzziness and Knowledge-Based Systems 5(6), 615–653 (1997)
45. Mathet, Y.: Etude de l'expression en langue de l'espace et du déplacement : analyse linguistique, modélisation cognitive, et leur expérimentation informatique. PhD thesis, Université de Caen, France (December 2000)
46. Bloch, I., Colliot, O., Cesar, R.: On the Ternary Spatial Relation Between. IEEE Transactions on Systems, Man, and Cybernetics SMC-B 36(2), 312–327 (2006)
47. Rosse, C., Mejino, J.L.V.: A Reference Ontology for Bioinformatics: The Foundational Model of Anatomy. Journal of Biomedical Informatics 36, 478–500 (2003)
48. Hudelot, C., Atif, J., Bloch, I.: Fuzzy Spatial Relation Ontology for Image Interpretation. Fuzzy Sets and Systems 159, 1929–1951 (2008)
49. Hudelot, C., Atif, J., Bloch, I.: A Spatial Relation Ontology Using Mathematical Morphology and Description Logics for Spatial Reasoning. In: ECAI 2008 Workshop on Spatial and Temporal Reasoning, Patras, Greece, July 2008, pp. 21–25 (2008)
50. Bloch, I., Géraud, T., Maître, H.: Representation and Fusion of Heterogeneous Fuzzy Information in the 3D Space for Model-Based Structural Recognition - Application to 3D Brain Imaging. Artificial Intelligence 148, 141–175 (2003)
51. Colliot, O., Camara, O., Bloch, I.: Integration of Fuzzy Spatial Relations in Deformable Models - Application to Brain MRI Segmentation. Pattern Recognition 39, 1401–1414 (2006)
52. Atif, J., Hudelot, C., Fouquier, G., Bloch, I., Angelini, E.: From Generic Knowledge to Specific Reasoning for Medical Image Interpretation using Graph-based Representations. In: International Joint Conference on Artificial Intelligence IJCAI 2007, Hyderabad, India, January 2007, pp. 224–229 (2007)
53. Fouquier, G., Atif, J., Bloch, I.: Local Reasoning in Fuzzy Attributes Graphs for Optimizing Sequential Segmentation. In: Escolano, F., Vento, M. (eds.) GbRPR. LNCS, vol. 4538, pp. 138–147. Springer, Heidelberg (2007)
54. Fouquier, G., Atif, J., Bloch, I.: Sequential Spatial Reasoning in Images based on Pre-Attention Mechanisms and Fuzzy Attribute Graphs. In: European Conference on Artificial Intelligence ECAI, Patras, Greece, July 2008, pp. 611–615 (2008)
55. Bengoetxea, E., Larranaga, P., Bloch, I., Perchant, A., Boeres, C.: Inexact Graph Matching by Means of Estimation of Distribution Algorithms. Pattern Recognition 35, 2867–2880 (2002)
56. Nempont, O., Atif, J., Angelini, E., Bloch, I.: Structure Segmentation and Recognition in Images Guided by Structural Constraint Propagation. In: European Conference on Artificial Intelligence ECAI, Patras, Greece, July 2008, pp. 621–625 (2008)

From (Deductive) Fuzzy Logic to (Logic-Based) Fuzzy Mathematics

Petr Cintula

Institute of Computer Science, Academy of Sciences of the Czech Republic
Pod Vodárenskou věží 2, 182 07 Prague, Czech Republic
cintula@cs.cas.cz

It is indisputable that mathematical structures arising around vague/ fuzzy/ non-bivalent concepts have a broad range of applications; therefore they have been intensively investigated during the last five decades. The discipline studying these structures is, maybe unfortunately, called *Fuzzy Mathematics*.

This discipline started by Zadeh's Fuzzy Set Theory [7] (although there are several almost forgotten predecessors) and already from its early days the role of logic have been noticed, stressed, and studied. However, fuzzy logic as a formal symbolic system (I will use the term 'deductive fuzzy logic' in this text) in the spirit of other non-classical logics has been thoroughly developed only recently. The paper [1] describes the deference between traditional and deductive fuzzy logic (refining Zadeh's original distinction between broad and narrow fuzzy logic). Very roughly speaking: deductive fuzzy logic deals with degrees of truth only, whereas traditional fuzzy logic speaks about also about degrees of belief, preference, entropy, necessity, etc. The consequences of this restriction narrow down the agenda of deductive fuzzy logic but give methodological clarity, determine the applicability scope and provide a research *focus* which leads to rapid development of the theory (and hopefully of some applications soon).

There is an ongoing project of the Prague research group in fuzzy logic, directed towards developing the *logic-based* fuzzy mathematics, i.e., an 'alternative' mathematics built in a formal analogy with classical mathematics, but using *deductive* fuzzy logic instead of the classical logic. The core of the project is a formulation of certain formalistic methodology (see [4]), a proposed foundational theory (see [3,5]), and development of the particular disciplines of fuzzy mathematics within the foundational theory using the formalistic methodology. The proposed foundational theory is called Fuzzy Class Theory (FCT) and it is a first-order theory over multi-sorted predicate fuzzy logic, with a very natural axiomatic system which approximates nicely Zadeh's original notion of fuzzy set. In paper [4] the authors claim that the whole enterprize of Fuzzy Mathematics can be formalized in FCT. This is still true as classical logic is formally interpretable inside deductive fuzzy logic, however in the parts of fuzzy mathematics 'incompatible' with the requirements of deductive fuzzy logic our approach provides (very) little added value.

An important feature of the theory is the gradedness of all defined concepts, which makes it more genuinely fuzzy than traditional approaches. Indeed, e.g. in the theory of fuzzy relations the majority of traditional characterizing properties, such as reflexivity, symmetry, transitivity, and so forth, are defined in a strictly

C. Sossai and G. Chemello (Eds.): ECSQARU 2009, LNAI 5590, pp. 14–15, 2009.

crisp way, i.e., as properties that either hold fully or do not hold at all (the notion of fuzzy inclusion is a notable exception; graded properties of fuzzy relations were originally studied by Siegfried Gottwald in [6]). One may be tempted to argue that it is somewhat peculiar to fuzzify relations by allowing intermediate degrees of relationships, but, at the same time, to still enforce strictly crisp properties on fuzzy relations. This particularly implies that all results are effective only if some assumptions are fully satisfied, but say nothing at all if the assumptions are only fulfilled to a certain degree (even if they are *almost* fulfilled).

In this talk I start by formulating and explaining the restrictions of the deductive fuzzy logic and presenting advantages (and disadvantages) of such restrictions. Then I sketch the methodology and formalism of FCT and illustrate it using simple examples from the theory of fuzzy relation (from the paper [2]). Finally I put this approach in the context of other nonclassical-logic-based mathematics (intuitionistic, relevant, substructural, etc.); compare logic-based, categorial, and traditional fuzzy mathematics; and address the possible outlooks of FCT and its role in future fuzzy mathematics.

References

1. Běhounek, L.: On the difference between traditional and deductive fuzzy logic. Fuzzy Sets and Systems 159(10), 1153–1164 (2008)
2. Běhounek, L., Bodenhofer, U., Cintula, P.: Relations in Fuzzy Class Theory: Initial steps. Fuzzy Sets and Systems 159(14), 1729–1772 (2008)
3. Běhounek, L., Cintula, P.: Fuzzy class theory. Fuzzy Sets and Systems 154(1), 34–55 (2005)
4. Běhounek, L., Cintula, P.: From fuzzy logic to fuzzy mathematics: A methodological manifesto. Fuzzy Sets and Systems 157(5), 642–646 (2006)
5. Běhounek, L., Cintula, P.: Fuzzy Class Theory: A primer v1.0. Technical Report V-939, Institute of Computer Science, Academy of Sciences of the Czech Republic, Prague (2006), www.cs.cas.cz/research/library/reports_900.shtml
6. Gottwald, S.: Fuzzy Sets and Fuzzy Logic: Foundations of Application—from a Mathematical Point of View. Vieweg, Wiesbaden (1993)
7. Zadeh, L.A.: Fuzzy sets. Information and Control 8(3), 338–353 (1965)

Conditionals and Independence in Many-Valued Logics

Daniele Mundici

Department of Mathematics, University of Florence
Viale Morgagni 67/a, 50134 Florence, Italy
mundici@math.unifi.it

Formulas and valuations in boolean logic are a traditional source of examples of "events" and "possible worlds". However, many events of interest in everyday life are more general than yes-no events, as described in boolean logic. Their possible outcomes typically range over a continuous spectrum, which after a suitable normalization can be restricted within the unit real interval $[0,1]$.

Events and Possible Worlds from Physical Systems. States and observables of physical systems provide a very general source of continuously valued events and possible worlds. Let SYST be a "physical system". Following [6, pp.362–369], a rigorous account of SYST is given by its associated C^*-algebra A, with the set $\mathsf{A_{sa}} \subseteq \mathsf{A}$ of self-adjoint elements, and the set $\mathsf{S}^* \subseteq \mathbb{R}^{\mathsf{A_{sa}}}$ of real-valued normalized positive linear functionals on $\mathsf{A_{sa}}$.[1] For any $W \in \mathsf{S}^*$ and $X \in \mathsf{A_{sa}}$ the real number $W(X)$ is *the expectation value of the observable X when SYST is prepared in mode W.*

Given a set $E = \{X_1,\ldots,X_m\}$ of nonzero positive elements of $\mathsf{A_{sa}}$, W determines, by normalization, the map $w\colon E \to [0,1]$ given by $w(X_i) = w(X_i)/||X_i||$, where $||X_i||$ is the norm of X_i. Intuitively, the *event* X_i says "the observable X_i has a high value," and w evaluates how true X_i is. The set $\mathsf{W} \subseteq [0,1]^E$ of *possible worlds* is defined by $\mathsf{W} = \{w \mid W \in \mathsf{S}^*\}$. W is a closed nonempty set in the cube $[0,1]^E = [0,1]^n$.

Coherent Bets on E and W. Having thus presented a sufficiently general framework for the notions of "event" and "possible world", we will now consider two *abstract* sets $E = \{X_1,\ldots,X_m\}$ and $\mathsf{W} \subseteq [0,1]^{\{X_1,\ldots,X_m\}} = [0,1]^n$, without any reference to observables and states of physical systems.

Suppose two players Ada (the bookmaker) and Blaise (the bettor) wager money on the outcome of events $X_1,\ldots,X_m$ within a prescribed set W of possible worlds. By definition, Ada's book is a map $\beta\colon E \to [0,1]$, containing a "betting odd" $\beta(X_i)$ for each event. Blaise, who knows β, chooses a "stake" $\sigma_i \in \mathbb{R}$ for each $i = 1,\ldots,m$: by definition, σ_i is the amount of money (measured in euro for definiteness) to be paid to the bettor if event X_i occurs. Money transfers are oriented in such a way that "positive" means Blaise-to-Ada. For each $i = 1,\ldots,m$, $\sigma_i \cdot \beta(X_i)$ euro are paid, with the proviso that $-\sigma_i \cdot v(X_i)$ euro will be paid back, in the possible world $v \in \mathsf{W}$. Any stake $\sigma_i < 0$ results in a sort of "reverse bet", where the bookmaker-bettor roles are interchanged: Ada first pays Blaise $|\sigma_i| \cdot \beta(X_i)$ euro, and Blaise will pay back $|\sigma_i| \cdot v(X_i)$ in the possible world v.

Ada's book β would lead her to financial disaster if Blaise could choose stakes $\sigma_1,\ldots,\sigma_m$ ensuring him to win at least one million euro in every possible world. Replacing the word "disaster" by "incoherence", we have the following definition:

---

[1] SYST is said to be classical if its associated C^*-algebra is commutative.

C. Sossai and G. Chemello (Eds.): ECSQARU 2009, LNAI 5590, pp. 16–21, 2009.
© Springer-Verlag Berlin Heidelberg 2009

A map $\beta\colon E \to [0,1]$ is *W-incoherent* if for some $\sigma_1,\ldots,\sigma_m \in \mathbb{R}$ we have $\sum_{i=1}^{m} \sigma_i \cdot (\beta(X_i) - v(X_i)) < 0$ for all $v \in W$. Otherwise, β is *W-coherent*.

In the particular case when $W \subseteq \{0,1\}^E$, we obtain De Finetti's no-Dutch-Book criterion for coherent probability assessments of yes-no events (see [3, §7, p. 308], [4, pp. 6-7], [5, p. 87]).

The Role of Łukasiewicz Logic and MV-Algebras in $[0,1]$-valued probability. We refer to [2] for background on Łukasiewicz (always propositional and infinite-valued) logic Ł$_\infty$, and MV-algebras. We will denote by $\odot, \oplus, \neg$ the connectives of conjunction, disjunction and negation. F_m is the set of all formulas whose variables are in the set $\{X_1,\ldots,X_m\}$. A *(Łukasiewicz) valuation of* F_m is a function $v\colon \mathsf{F}_m \to [0,1]$ such that

$$v(\neg\phi) = 1 - v(\phi), \quad v(\phi \oplus \psi) = \min(1, v(\phi) + v(\psi)), \quad v(\phi \odot \psi) = \max(0, v(\phi) + v(\psi) - 1)$$

for all $\phi, \psi \in \mathsf{F}_m$. A formula is a *tautology* if it is satisfied by every valuation[2]. We say that v *satisfies a set of formulas* $\Psi \subseteq \mathsf{F}_m$ if $v(\theta) = 1$ for all $\theta \in \Psi$. Ψ is *consistent* if it is satisfied by at least one valuation. A formula θ is *consistent* if so is $\{\theta\}$. Two formulas $\phi, \psi \in \mathsf{F}_m$ are Ψ-*equivalent* if from Ψ one obtains $\phi \leftrightarrow \psi$ (i.e., $(\neg\phi \oplus \psi) \odot (\neg\psi \oplus \phi)$) using all tautologies and modus ponens. We denote by $\frac{\psi}{\equiv_\Psi}$ the Ψ-equivalence class of formula ψ.

As is well known, MV-algebras stand to Ł$_\infty$ as boolean algebras stand to classical two-valued propositional logic. Thus for instance, the set of Ψ-equivalence classes of formulas forms the MV-algebra

$$\mathsf{L}(\Psi) = \frac{\mathsf{F}_m}{\equiv_\Psi} = \left\{ \frac{\phi}{\equiv_\Psi} \mid \phi \in \mathsf{F}_m \right\}.$$

Part of the proof of the following theorem is in [11]. The rest will appear elsewhere.

Theorem 1. *For any set $E = \{X_1,\ldots,X_m\}$ and closed nonempty set $W \subseteq [0,1]^E = [0,1]^{\{1,\ldots,m\}} = [0,1]^m$, there is a set Θ of formulas in the variables $X_1,\ldots,X_m$ such that W coincides with the set of restrictions to E of all valuations satisfying Θ. Further, a map $\beta\colon E \to [0,1]$ is W-coherent iff it can be extended to a convex combination of valuations satisfying Θ iff there is a state s of $\mathsf{L}(\Theta)$ such that $\beta(X_i) = s(X_i/\equiv_\Theta)$, for all $i = 1,\ldots,m$.*

As the reader will recall, a *state* of an MV-algebra B is a map $s\colon B \to [0,1]$ such that $s(1) = 1$ and $s(x \oplus y) = s(x) + s(y)$ whenever $x \odot y = 0$. We say that s is *faithful* if $s(x) = 0$ implies $x = 0$. We say that s is *invariant* if $s(\alpha(x)) = s(x)$ for every automorphism α of B and element $x \in B$.

Under the restrictive hypothesis $W \subseteq \{0,1\}^E$ the above theorem boils down to De Finetti's well known characterization of coherent assessments of yes-no events (see [3, pp.311-312], [4, Chapter 1], [5, pp.85-90]).

De Finetti's theorem was extended by Paris [15] to several modal logics, by Kühr et al., [9] to all $[0,1]$-valued logics whose connectives are *continuous*, including all

---

[2] It is tacitly understood that all valuations and formulas are of F_m.

finite-valued logics. In their paper [1], Aguzzoli, Gerla and Marra further extend De Finetti's criterion to Gödel logic [7], a logic with a *discontinuous* implication connective. By Theorem 1, the various kinds of "events", "possible worlds" and "coherent probability assessments" arising in all these logic contexts can be re-interpreted in Łukasiewicz logic.

Conditionals and their Invariance. Given the universal role of Łukasiewicz logic and MV-algebraic states for the treatment of coherent probability assessments, one is naturally led to develop a theory of conditionals in this logic. In [12, 3.1-3.2], the present author gave the following definition:

> A *conditional* is a map $P: \theta \mapsto P_\theta$ such that, for every $m = 1, 2, \ldots$ and every consistent formula $\theta \in F_m$, P_θ is a state of the MV-algebra $L(\{\theta\})$. We say that P is *invariant* if for any two consistent formulas $\phi \in F_m$, $\psi \in F_n$, and isomorphism η of $L(\{\phi\})$ onto $L(\{\psi\})$, we have $P_\phi = P_\psi \circ \eta$, where $\circ$ denotes composition. P is said to be *faithful* if so is every state P_θ.

The main result of [12] is

Theorem 2. *Łukasiewicz logic $Ł_\infty$ has a faithful invariant conditional P^*.*

It follows that P^* is invariant under equivalent reformulations of the same event. In more detail:

Corollary (a). *For every formula $\psi \in F_m$ let us write $P^*_\theta(\psi)$ instead of $P^*_\theta\left(\frac{\psi}{\equiv_{\{\theta\}}}\right)$, and say that $P^*_\theta(\psi)$ is the probability of ψ given θ. We then have*

$$P^*_{\psi \leftrightarrow \psi}(\psi) = P^*_{\psi \leftrightarrow X_{m+1}}(X_{m+1}). \tag{1}$$

Proof. We assume familiarity with [12] and [2]. A *rational polyhedron* in $[0,1]^n$ is a finite union of simplexes in $[0,1]^n$, such that the coordinates of the vertices of each simplex are rational.

Given rational polyhedra $P \subseteq \mathbb{R}^n$ and $Q \subseteq \mathbb{R}^m$ by a $\mathbb{Z}$-*homeomorphism* we understand a piecewise linear homeomorphism η of P onto Q such that each linear piece of both η and η^{-1} has integer coefficients.

We denote by f_ψ the McNaughton function of ψ, (see e.g., [2, p.221]). Let the rational polyhedron $D \subseteq [0,1]^{m+1}$ be defined by

$$D = \{(x_1, \ldots, x_{m+1}) \in [0,1]^{m+1} \mid x_{m+1} = f_\psi(x_1, \ldots, x_m)\}.$$

Up to isomorphism, the MV-algebra $L(\{\psi \leftrightarrow \psi\})$ of the tautology $\psi \leftrightarrow \psi$ is the free m-generator MV-algebra $Free_m$, i.e., (by McNaughton theorem [2, 9.1.5]) the MV-algebra $M([0,1]^m)$ of all McNaughton functions $f: [0,1]^m \to [0,1]$.

By a routine variant of [12, 2.3], the MV-algebra $L(\{\psi \leftrightarrow X_{m+1}\})$ is the MV-algebra

$$M(D) = \{l \upharpoonright D \mid l \in M([0,1]^{m+1})\}$$

obtained by restricting to D the McNaughton functions of $Free_{m+1} = M([0,1]^{m+1})$.

We can safely use the identifications

$$\frac{\psi}{\equiv_{\{\psi \leftrightarrow \psi\}}} = f_\psi \qquad \text{and} \qquad \frac{X_{m+1}}{\equiv_{\{X_{m+1} \leftrightarrow \psi\}}} = \pi_{m+1} \restriction D, \qquad (2)$$

where $\pi_{m+1} \colon [0,1]^{m+1} \to [0,1]$ is the $(m+1)$th coordinate function

$$\pi_{m+1}(x_1, \ldots, x_{m+1}) = x_{m+1}.$$

In view of Theorem 2, to conclude the proof, we must only construct an isomorphism of $M([0,1]^m)$ onto $M(D)$ sending f_ψ to the coordinate function π_{m+1}. The map

$$\eta \colon (x_1, \ldots, x_m) \mapsto (x_1, \ldots, x_m, f_\psi(x_1, \ldots, x_m))$$

is promptly seen to be a $\mathbb{Z}$-homeomorphism of $[0,1]^m$ onto D. The inverse map projects D onto the face of $[0,1]^{m+1}$ given by $x_{m+1} = 0$. In symbols,

$$\eta^{-1} = (\pi_1, \ldots, \pi_m) \restriction D.$$

The map

$$\Omega \colon g \in M(D) \mapsto g \circ \eta \in M([0,1]^m)$$

is a one-one homomorphism of $M(D)$ into $M([0,1]^m)$. The map

$$\mho \colon h \in M([0,1]^m) \mapsto h \circ \eta^{-1} \in M(D)$$

is a one-one homomorphism of $M([0,1]^m)$ into $M(D)$. Trivially, these two maps are inverse of each other, and in view of (2) we can write

$$\mho \colon \mathsf{L}(\{\psi \leftrightarrow \psi\}) = M([0,1]^m) \cong M(D) = \mathsf{L}(\{\psi \leftrightarrow X_{m+1}\}).$$

The two elements $\dfrac{\psi}{\equiv_{\{\psi \leftrightarrow \psi\}}}$ and $\dfrac{X_{m+1}}{\equiv_{\{X_{m+1} \leftrightarrow \psi\}}}$ correspond under the isomorphism $\mho$. This completes the proof. $\square$

A supplementary analysis of the proof of the main theorem of [12] shows that

$$P^*_{\psi \leftrightarrow X_{m+1}}(X_{m+1}) = P^*_{\psi \leftrightarrow \psi}(\psi) = \int_{[0,1]^m} f_\psi. \qquad (3)$$

More generally, a similar argument proves:

Corollary (b). *For any formula ψ and consistent formula θ we have*

$$P^*_\theta(\psi) = P^*_{\theta \odot (\psi \leftrightarrow X)}(X), \qquad (4)$$

provided the variable X does not occur in θ and ψ.

We are now in a position to introduce a reasonable notion of independence, by saying that a formula α is P^*-*independent of* (a consistent formula) θ if the probability of α given θ coincides with the unconditional probability of α. In view of Corollary (a) we can equivalently write

$$P^*_\theta(\alpha) = P^*_{\theta \leftrightarrow \theta}(\alpha) = P^*_{\alpha \leftrightarrow \alpha}(\alpha) = P^*_{X \leftrightarrow \alpha}(X), \qquad (5)$$

where X is a fresh variable. As a consequence of Corollary (b) we have

Corollary (c). *If α and θ are two formulas in disjoint sets of variables $\{Y_1, \ldots, Y_m\}$ and $\{Z_1, \ldots, Z_n\}$, and θ is consistent, then α is P^*-independent of θ.*

Concluding Remarks. For any finite set $E = \{X_1, \ldots, X_n\}$ whose elements are called "events", and closed set $W \subseteq [0,1]^E$ whose elements are called "possible worlds", following De Finetti we have defined a bookmaker's map $b \colon E \to [0,1]$ to be W-incoherent if a bettor can fix (positive or negative) stakes $s_1, \ldots, s_n$ ensuring him a profit of least one million euro (equivalently, a profit > 0) in any possible world of W.

No matter the physical or logical nature of E and W, Theorem 1 shows that there is a theory Θ in Łukasiewicz logic such that W-coherent maps coincide with restrictions to E of states of the MV-algebra $L(\Theta)$. In particular, when W is a set of valuations in *any* $[0,1]$-valued logic L, and E is a set of formulas in L, W-coherent maps on E can always be interpreted in Łukasiewicz logic.

It is often claimed that De Finetti's coherence criterion yields an axiomatic approach to finitely additive probability measures, missing the full strength of Kolomgorov axioms. We beg to dissent: by the Kroupa-Panti theorem [8,13], in every MV-algebra A—whence in particular in every boolean algebra— the set of (*finitely additive*) states of A is in canonical one-one correspondence with the set of (*countably additive*) regular Borel probability measures on the maximal spectrum of A. Thus the theory of finitely additive measures (i.e., states) on boolean algebras has the same degree of generality as the theory of regular Borel measures on their Stone spaces. Passing to the much larger class of MV-algebras, Theorem 1 in combination with the Kroupa-Panti theorem shows that De Finetti's coherence criterion has the same degree of generality as the theory of regular probability Borel measures on *any compact space*.

Theorem 2 shows that Łukasiewicz logic has a faithful invariant conditional P^*. In Corollary (a)-(b) a new result is proved, to the effect that P^* *does not make any distinction between (i) the probability of ψ given θ, and (ii) the probability of (the event described by) a fresh variable X given θ together with the information that X is equivalent to ψ.* A novel notion of independence is built on P^*, having various desirable properties, some of which are summarized in (5) and in Corollary (c).

For further information on MV-algebraic probability theory, including other approaches to conditional probability and independence, see [14, Chapters 20-22].

References

1. Aguzzoli, S., Gerla, B., Marra, V.: De Finetti's no-Dutch-Book criterion for Gödel logic. Studia Logica 90, 25–41 (2008); Special issue on Many-valued Logic and Cognition. Ju, S., et al. (eds.)
2. Cignoli, R., D'Ottaviano, I.M.L., Mundici, D.: Algebraic Foundations of Many-Valued Reasoning. In: Trends in Logic, vol. 7. Kluwer/Springer, Dordrecht, New York (2000)
3. De Finetti, B.: Sul significato soggettivo della probabilitá. Fundamenta Mathematicae 17, 298–329 (1931); Translated into English as On the Subjective Meaning of Probability. In: Monari, P., Cocchi, D. (eds.) Probabilitá e Induzione, Clueb, Bologna, pp. 291–321 (1993)
4. De Finetti, B.: La prévision: ses lois logiques, ses sources subjectives. Annales de l'Institut H. Poincaré 7, 1–68 (1937); Translated into English by Kyburg Jr., H.E.: Foresight: Its Logical Laws, its Subjective Sources. In: Kyburg Jr., H.E., Smokler, H.E. (eds.) Studies in Subjective Probability, 7, 53–118. Wiley, New York (1980); Second edition published by Krieger, New York, pp. 53–118, 1980.
5. De Finetti, B.: Theory of Probability, vol. 1. John Wiley and Sons, Chichester (1974)

6. Emch, G.G.: Mathematical and Conceptual Foundations of 20th Century Physics. North-Holland, Amsterdam (1984)
7. Hájek, P.: Metamathematics of fuzzy logic. Trends in Logic, vol. 4. Kluwer/ Springer, Dordrecht, New York (1998)
8. Kroupa, T.: Every state on semisimple MV-algebra is integral. Fuzzy Sets and Systems 157, 2771–2782 (2006)
9. Kühr, J., Mundici, D.: De Finetti theorem and Borel states in [0,1]-valued algebraic logic. International Journal of Approximate Reasoning 46, 605–616 (2007)
10. Łukasiewicz, J., Tarski, A.: Investigations into the Sentential Calculus. In: [16], pp. 38–59
11. Mundici, D.: Bookmaking over infinite-valued events. International J. Approximate Reasoning 43, 223–240 (2006)
12. Mundici, D.: Faithful and invariant conditional probability in Łukasiewicz logic. In: Makinson, D., Malinowski, J., Wansing, H. (eds.) Proceedings of the conference Trends in Logic IV, Torun, Poland. Trends in Logic, vol. 28, pp. 213–232. Springer, New York (2008)
13. Panti, G.: Invariant measures in free MV-algebras. Communications in Algebra 36, 2849–2861 (2008)
14. Pap, E. (ed.): Handbook of Measure Theory, I,II. North-Holland, Amsterdam (2002)
15. Paris, J.: A note on the Dutch Book method. In: De Cooman, G., Fine, T., Seidenfeld, T. (eds.) Proceedings of the Second International Symposium on Imprecise Probabilities and their Applications, ISIPTA 2001, Ithaca, NY, USA, pp. 301–306. Shaker Publishing Company (2001), http://www.maths.man.ac.uk/DeptWeb/Homepages/jbp/
16. Tarski, A.: Logic, Semantics, Metamathematics. Clarendon Press, Oxford (1956); reprinted, Hackett, Indianapolis (1983)

Inference from Multinomial Data Based on a MLE-Dominance Criterion

Alessio Benavoli and Cassio P. de Campos

Dalle Molle Institute for Artificial Intelligence
Manno, Switzerland
{alessio,cassio}@idsia.ch

Abstract. We consider the problem of inference from multinomial data with chances $\boldsymbol{\theta}$, subject to the a-priori information that the true parameter vector $\boldsymbol{\theta}$ belongs to a known convex polytope $\boldsymbol{\Theta}$. The proposed estimator has the parametrized structure of the conditional-mean estimator with a prior Dirichlet distribution, whose parameters $(s, \mathbf{t})$ are suitably designed via a *dominance* criterion so as to guarantee, for any $\boldsymbol{\theta} \in \boldsymbol{\Theta}$, an improvement of the *Mean Squared Error* over the *Maximum Likelihood Estimator* (MLE). The solution of this MLE-dominance problem allows us to give a different interpretation of: (1) the several Bayesian estimators proposed in the literature for the problem of inference from multinomial data; (2) the *Imprecise Dirichlet Model* (IDM) developed by Walley [13].

1 Introduction

An important estimation problem that has been treated extensively in the literature is the problem of inference from multinomial data, as the number of potential applications is huge. Accuracy of results relies on the quality of model parameters. Ideally, with enough data, it is possible to learn by standard statistical analysis like maximum likelihood estimation. However, the amount of training data may be small, for example, because of the cost of acquisition or natural conditions. In spite of that, domain knowledge through constraints is available in many real applications and can improve estimations.

The problem is as follows. Consider an infinite population which can be categorized in k categories or types from the set $C = \{c_1, \ldots, c_k\}$. The proportion of units of type c_j is denoted θ_j and called the chance of c_j. The population is thus characterized by the vector of chances $\boldsymbol{\theta} = [\theta_1, \ldots, \theta_k]' \in \mathcal{S}_{\boldsymbol{\theta}}$, where $\mathcal{S}_{\boldsymbol{\theta}} = \{\theta_j : 0 \le \theta_j \le 1$ for all j and $\boldsymbol{\theta}'\mathbf{1} = 1\}$. The observed data consist in a sample of size N from the population, summarized by the counts $\mathbf{n} = [n_1, n_2, \ldots, n_k]'$, where n_j is the number of units of type c_j and $\mathbf{n}'\mathbf{1} = N$. The chances $\boldsymbol{\theta}$ are unknown parameters and the goal is to construct an estimator $\hat{\boldsymbol{\theta}}$ of the true chances $\boldsymbol{\theta}$, from the observations $\mathbf{n}$, that is close to $\boldsymbol{\theta}$ in some sense.

A popular measure of estimator's performance is the expected value of the quadratic loss function which is also called Mean-Squared Error (MSE) and is defined as

$$E_{\mathbf{n}}[(\hat{\boldsymbol{\theta}} - \boldsymbol{\theta})(\hat{\boldsymbol{\theta}} - \boldsymbol{\theta})'] = (E_{\mathbf{n}}[\hat{\boldsymbol{\theta}}] - \boldsymbol{\theta})(E_{\mathbf{n}}[\hat{\boldsymbol{\theta}}] - \boldsymbol{\theta})' + (\hat{\boldsymbol{\theta}} - E_{\mathbf{n}}[\hat{\boldsymbol{\theta}}])(\hat{\boldsymbol{\theta}} - E_{\mathbf{n}}[\hat{\boldsymbol{\theta}}])' \quad (1)$$

C. Sossai and G. Chemello (Eds.): ECSQARU 2009, LNAI 5590, pp. 22–33, 2009.

where the first term of the summation is the squared-bias of the estimator and the second term is its variance matrix. The unknown parameter vector $\boldsymbol{\theta}$ is assumed to be deterministic and, thus, the expectation is only over the data.

With respect to the problem of inference from multinomial data, the MLE estimate $\hat{\boldsymbol{\theta}}_{MLE}$ has two properties: (1) it is unbiased, which means that $E_{\mathbf{n}}[\hat{\boldsymbol{\theta}}] = \boldsymbol{\theta}$; (2) it achieves the *Cramer-Rao Lower Bound* (CRLB) for unbiased estimators, i.e. $E_{\mathbf{n}}[(\hat{\boldsymbol{\theta}}_{MLE} - \boldsymbol{\theta})(\hat{\boldsymbol{\theta}}_{MLE} - \boldsymbol{\theta})'] = \boldsymbol{\Sigma}_{MLE}$, where $\boldsymbol{\Sigma}_{MLE}$ is the inverse of the Fisher information matrix. Minimum variance and unbiasedness are suitable properties, but this does not imply that MLE always provides a small MSE, especially for small data samples. In fact, exploiting the relationship "MSE=variance + squared bias" and trading-off bias for variance, estimators may exist which provide a MSE lower than the CRLB for unbiased estimators.

Ranking the estimators in terms of MSE is not obvious (the MSE depends on the unknown $\boldsymbol{\theta}$) and an important practical question is how to decide which estimator to use. Although in general this question is hard to answer, some estimators may be uniformly better than others in terms of MSE. An estimator $\hat{\boldsymbol{\theta}}$ is said to *dominate* a given estimator $\hat{\boldsymbol{\theta}}_0$ on a set $\boldsymbol{\Theta}$ if its MSE is never greater than that of $\hat{\boldsymbol{\theta}}_0$ for all values of $\boldsymbol{\theta}$ in $\boldsymbol{\Theta}$, and is strictly smaller for some $\boldsymbol{\theta}$ in $\boldsymbol{\Theta}$. An estimator that is not dominated by any other estimator is said to be *admissible* on $\boldsymbol{\Theta}$ [5]. Hence, a desirable property of an estimator is to be admissible: otherwise it is dominated by some other estimator that have smaller MSE for all choices of $\boldsymbol{\theta}$. It can be proved that the MLE is admissible w.r.t. the MSE criterion [10]. However, estimators that dominate MLE may exist if a subregion $\boldsymbol{\Theta} \subseteq \mathcal{S}_{\boldsymbol{\theta}}$ of the parameters space is considered.

In this paper, we derive a procedure for determining estimators of a particular structure which dominates MLE on a polytopic membership set $\boldsymbol{\Theta}$. We consider an estimator $\hat{\boldsymbol{\theta}} = (\mathbf{n} + s\mathbf{t})/(N + s)$, which has the shape of the conditional-mean estimator obtained by assuming a prior Dirichlet distribution with parameters s and $\mathbf{t}$ for $\boldsymbol{\theta}$ and we design parameters s and $\mathbf{t}$, based on the knowledge $\boldsymbol{\Theta}$, so as to guarantee the MLE-dominance for the MSE. This proposal may be somehow interpreted as an intermediate approach between Bayesian and frequentist views. We assume that the unknown parameter θ is deterministic, which is in fact a frequentist approach. However our estimator would yield the optimal MMSE estimate under a Bayesian approach in the case the unknown vector $\boldsymbol{\theta}$ is actually Dirichlet distributed. This and the fact that the MLE is admissible w.r.t. the MSE are our motivations for choosing the MSE as risk function.

The idea of estimators that dominate MLE is not new (e.g. [1,7,9,12]). A similar approach has also been followed in [1] for the problem of estimating an unknown parameter vector in an additive Gaussian-noise linear model by designing an estimator which dominates the least-square estimator. To our knowledge, the idea of designing the parameters s and $\mathbf{t}$, so as to guarantee the MLE-dominance inside $\boldsymbol{\Theta}$ and analysing other approaches under this perspective have not been explored in the literature.

2   Inference from Multinomial Data

Consider the problem of inference from multinomial data discussed at the beginning of Section 1. The objective is to compute an estimate of the parameter vector $\boldsymbol{\theta}$ based on the vector of observations $\mathbf{n}$. The probability of observing $\mathbf{n}$, conditionally on $\boldsymbol{\theta}$, is given by the multinomial distribution: $P(\boldsymbol{\theta}, \mathbf{n}) = \binom{N}{\mathbf{n}} \prod_{j=1}^{k} \theta_j^{n_j}$. The MLE can be obtained by maximizing the likelihood $L(\boldsymbol{\theta}, \mathbf{n}) \propto \prod_{j=1}^{k} \theta_j^{n_j}$ w.r.t. $\boldsymbol{\theta}$ subject to the constraint $\boldsymbol{\theta}'\mathbf{1} = 1$, which gives: $\hat{\boldsymbol{\theta}}_{MLE} = \mathbf{n}/N$.

Another approach is to assume a Dirichlet model over $\boldsymbol{\theta}$, it generates a Dirichlet posterior density function:

$$p(\boldsymbol{\theta}|\mathbf{n}) \propto L(\boldsymbol{\theta}, \mathbf{n})D(s, \mathbf{t}, \boldsymbol{\theta}) \propto \prod_{j=1}^{k} \theta_j^{n_j + st_j - 1} \qquad (2)$$

where $D(s, \mathbf{t}, \boldsymbol{\theta}) \propto \prod_{j=1}^{k} \theta_j^{st_j - 1}$ is the prior, with $s > 0$, $\mathbf{t} = [t_1, t_2, \ldots, t_k]'$, $\mathbf{0} < \mathbf{t} < 1$ and $\mathbf{t}'\mathbf{1} = 1$. Using the posterior expectation of $\boldsymbol{\theta}$ given $\mathbf{n}$ as estimator, one gets:

$$\hat{\boldsymbol{\theta}} := E[\boldsymbol{\theta}|\mathbf{n}] = \frac{\mathbf{n} + s\mathbf{t}}{N + s} \qquad (3)$$

The parameters s and $\mathbf{t}$ represent the a-priori information. In case no prior information is available, the common approach is to select these parameters to represent a non-informative prior. The most used non-informative priors select $t_j = 1/k$ for $j = 1, 2, \ldots, k$ but differ in the choice of the value for s. Bayes and Laplace suggest to use a uniform prior $s = k$, Perks [11] suggests $s = 1$, Jeffreys $s = k/2$, and Haldane $s = 0$ [8].

3   MLE-Dominance

We derive a procedure for choosing the values of the parameters s and $\mathbf{t}$ by using the MLE-dominance criterion. The idea is to design an estimator of structure as Equation (3) which dominates MLE on a polytopic membership set $\boldsymbol{\Theta}$. We choose the free parameters s and $\mathbf{t}$ so as to guarantee that:

$$E_{\mathbf{n}}[(\boldsymbol{\theta} - \hat{\boldsymbol{\theta}})(\boldsymbol{\theta} - \hat{\boldsymbol{\theta}})'] \leq E_{\mathbf{n}}[(\boldsymbol{\theta} - \hat{\boldsymbol{\theta}}_{MLE})(\boldsymbol{\theta} - \hat{\boldsymbol{\theta}}_{MLE})'] := \boldsymbol{\Sigma}_{MLE} \qquad (4)$$

for each vector $\boldsymbol{\theta}$ in the convex polytope $\boldsymbol{\Theta}$ of vertices $\boldsymbol{\theta}_{v_1}, \boldsymbol{\theta}_{v_2}, \ldots, \boldsymbol{\theta}_{v_m}$, i.e. $\boldsymbol{\Theta} = Co\{\boldsymbol{\theta}_{v_1}, \boldsymbol{\theta}_{v_2}, \ldots, \boldsymbol{\theta}_{v_m}\}$ where $Co\{\cdots\}$ stands for *convex hull*. $\boldsymbol{\Sigma}_{MLE} = (\sigma_{ij})$ represents the covariance matrix of the MLE whose elements are $\sigma_{ii} = \theta_i(1 - \theta_i)/N$ and $\sigma_{ij} = -\theta_i\theta_j/N$ for $i, j = 1, 2, \ldots, k$ and $i \neq j$. In order to compute the expectaction on the left-side of (4), it is convenient to rewrite the observation vector as $\mathbf{n} = N\boldsymbol{\theta} + \mathbf{v}$, where $\mathbf{v}$ is a random vector such that

$$E_{\mathbf{n}}[\mathbf{v}] = \mathbf{0}, \qquad E_{\mathbf{n}}[\mathbf{v}\mathbf{v}'] = N^2 E[(\mathbf{n}/N - \boldsymbol{\theta})(\mathbf{n}/N - \boldsymbol{\theta})'] = N^2\boldsymbol{\Sigma}_{MLE} \qquad (5)$$

where the vector of observations $\mathbf{n}$ is assumed to be unbiased, i.e. $E_{\mathbf{n}}[\mathbf{n}/N - \boldsymbol{\theta}] = \mathbf{0}$. The inequality (4) can be rewritten in the following way:

$$
\begin{aligned}
E_{\mathbf{n}}[(\boldsymbol{\theta} - \hat{\boldsymbol{\theta}})(\boldsymbol{\theta} - \hat{\boldsymbol{\theta}})'] &= E_{\mathbf{v}}\left[\left(\boldsymbol{\theta} - \frac{N\boldsymbol{\theta} + \mathbf{v} + s\mathbf{t}}{N + s}\right)\left(\boldsymbol{\theta} - \frac{N\boldsymbol{\theta} + \mathbf{v} + s\mathbf{t}}{N + s}\right)'\right] \\
&= \frac{s^2}{(N + s)^2}(\boldsymbol{\theta} - \mathbf{t})(\boldsymbol{\theta} - \mathbf{t})' + \frac{N^2}{(N + s)^2}\boldsymbol{\Sigma}_{MLE} \leq \boldsymbol{\Sigma}_{MLE} \\
&\iff (\boldsymbol{\theta} - \mathbf{t})(\boldsymbol{\theta} - \mathbf{t})' \leq (\tfrac{2}{s} + \tfrac{1}{N})N\boldsymbol{\Sigma}_{MLE}
\end{aligned}
\tag{6}
$$

The estimator $\hat{\boldsymbol{\theta}}$ has a MSE lower than that of MLE for each $\boldsymbol{\theta} \in \boldsymbol{\Theta}$ if s and $\mathbf{t}$ are chosen to guarantee that (6) is satisfied.

If the vertices of the polytope $\boldsymbol{\Theta}$ satisfy $0 < \theta_{v_i} < 1$ for $i = 1, 2, \ldots, m$, then we can derive an alternative expression for (6)[1]. Since $N\boldsymbol{\Sigma}_{MLE} = \boldsymbol{\Lambda_\theta} - \boldsymbol{\theta\theta}'$, where $\boldsymbol{\Lambda_\theta} = diag[\theta_1, \theta_2, \ldots, \theta_k]$, by the matrix inversion lemma one gets

$$
(\boldsymbol{\Lambda_\theta} - \boldsymbol{\theta\theta}')^{-1} = \boldsymbol{\Lambda_\theta}^{-1} + \boldsymbol{\Lambda_\theta}^{-1}\boldsymbol{\theta}(1 + \boldsymbol{\theta}'\boldsymbol{\Lambda_\theta}^{-1}\boldsymbol{\theta})\boldsymbol{\theta}'\boldsymbol{\Lambda_\theta}^{-1} = \boldsymbol{\Lambda_\theta}^{-1} + \frac{1}{2}\mathbf{1}\mathbf{1}'
$$

By a property of the Schur-complement, the inequality (6) becomes

$$
(\boldsymbol{\theta} - \mathbf{t})'(N\boldsymbol{\Sigma}_{MLE})^{-1}(\boldsymbol{\theta} - \mathbf{t}) \leq (\tfrac{2}{s} + \tfrac{1}{N})
\tag{7}
$$

Manipulating the left-side, it follows

$$
\begin{aligned}
(\boldsymbol{\theta} - \mathbf{t})'(N\boldsymbol{\Sigma}_{MLE})^{-1}(\boldsymbol{\theta} - \mathbf{t}) &= (\boldsymbol{\theta} - \mathbf{t})'(\boldsymbol{\Lambda_\theta}^{-1} + \tfrac{1}{2}\mathbf{1}\mathbf{1}')(\boldsymbol{\theta} - \mathbf{t}) \\
&= (\boldsymbol{\theta} - \mathbf{t})'\boldsymbol{\Lambda_\theta}^{-1}(\boldsymbol{\theta} - \mathbf{t}) \\
&= \sum_{i=1}^{k} \frac{(\theta_i - t_i)^2}{\theta_i} \leq (\tfrac{2}{s} + \tfrac{1}{N})
\end{aligned}
\tag{8}
$$

Notice that to derive the previous expression it has been exploited the fact that $\boldsymbol{\theta}'\mathbf{1} = 1$ and $\mathbf{t}'\mathbf{1} = 1$. By calculating the Hessian of the left-side of (8) and exploiting the fact that $\theta_k = 1 - \sum_{i=1}^{k-1} \theta_i$, it can be verified that such function is convex on $\boldsymbol{\theta}$ (and also on $\mathbf{t}$). Therefore, (8) is satisfied for each $\boldsymbol{\theta} \in \boldsymbol{\Theta}$ if the inequality (8) holds on the vertices of the polytope $\boldsymbol{\Theta}$. Thus, the MLE-dominance is guaranteed if s and $\mathbf{t}$ are chosen to satisfy the following m-inequalities

$$
\sum_{i=1}^{k} \frac{(\theta^i_{v_j} - t_i)^2}{\theta^i_{v_j}} \leq (\tfrac{2}{s} + \tfrac{1}{N}), \quad \text{for } j = 1, 2, \ldots, m
\tag{9}
$$

where $\theta^i_{v_j}$ denotes the i-th component of the j-th vertices.

The previous inequalities define all the values of s and $\mathbf{t}$ which guarantee the MLE-dominance. However, since the MSE of the admissible estimators depends

---

[1] The problem of having vertices on the border of the probability simplex can be managed using (6) to define the constraints directly.

on the true value of $\boldsymbol{\theta}$, there is no way to decide which admissible estimator is preferable in terms of MSE and, thus, to select one value for s and $\mathbf{t}$. Nevertheless, we can define an ad-hoc criterion to choose a single value for s and $\mathbf{t}$ [1]:

$$\max_{s,\mathbf{t}} s$$

subject to:

$$\begin{cases} \sum_{i=1}^{k} \frac{(\theta_{v_j}^i - t_i)^2}{\theta_{v_j}^i} \leq (\frac{2}{s} + \frac{1}{N}), \quad \text{for } j = 1, 2, \ldots, m \\ \mathbf{t} \in \boldsymbol{\Theta}, \quad \mathbf{t}'\mathbf{1} = 1, \quad s > 0 \end{cases} \tag{10}$$

Maximizing s means minimizing the prior uncertainty on $\boldsymbol{\theta}$. Notice that, since we know that $\boldsymbol{\theta} \in \boldsymbol{\Theta}$ it is also natural to constrain $\mathbf{t}$ to be inside $\boldsymbol{\Theta}$. By denoting the solution of (10) with $(s_0, \mathbf{t}_0)$, it can be noticed that any pair $(s, \mathbf{t}_0)$ with $0 < s < s_0$ is also a feasible solution of (10). Furthermore, notice that the optimal solution $(s_0, \mathbf{t}_0)$ of (10) depends on N. Given two values N_1 and N_2, such that $N_1 \leq N_2$, from the previous remarks we have $s_0(N_2) \leq s_0(N_1)$. Therefore, $s_0(N_2)$ is a feasible value for s for any $N \leq N_2$.

The dependence of s on N violates the coherence principle, as it makes the prior information depending on the size of the data. However, taking the limit $N \to \infty$, the dependency on N can be removed from (10) and the resulting value for s will still be a feasible solution for any finite value of N. Another approach is to fix s and consider all the prior distributions defined in (9) by the inequalities in $\mathbf{t}$ and deal with a set of distributions like in the IDM [13]. Coherence and set of distributions will be discussed in Section 5.

4   Binomial Data

Consider the case where only two categories ($k = 2$) are distinguished. Since in this case $\boldsymbol{\theta} = [\theta_1, \theta_2]$ and $\theta_2 = 1 - \theta_1$, there is only one degree of freedom, i.e. only one parameter to be estimated. It can easily be verified that the matrix-inequality (6) is satisfied if and only if:

$$\theta_1^2(1 + \tfrac{2}{s} + \tfrac{1}{N}) + \theta_1(-2t_1 - \tfrac{2}{s} - \tfrac{1}{N}) + t_1^2 \leq 0 \tag{11}$$

Given t_1 and s, this inequality must be satisfied for each $\theta_1 \in \boldsymbol{\Theta} = [\theta_a, \theta_b]$ with $0 \leq \theta_a < \theta_b \leq 1$, i.e. in the binomial case the polytope is just an interval. Because of the convexity of the left side of (11), we can guarantee that the inequality is satisfied for all $\theta_1 \in [\theta_a, \theta_b]$ if it is satisfied in the extremes of the interval:

$$\begin{cases} \theta_a^2(1 + \tfrac{2}{s} + \tfrac{1}{N}) + \theta_a(-2t_1 - \tfrac{2}{s} - \tfrac{1}{N}) + t_1^2 \leq 0 \\ \theta_b^2(1 + \tfrac{2}{s} + \tfrac{1}{N}) + \theta_b(-2t_1 - \tfrac{2}{s} - \tfrac{1}{N}) + t_1^2 \leq 0 \end{cases} \tag{12}$$

The values of the parameters $\theta_a \leq t_1 \leq \theta_b$ and $s > 0$ which satisfy both inequalities in (12) define all the admissible estimators (3) which dominate MLE in $[\theta_a, \theta_b]$. Solving the previous inequalities w.r.t t_1, one gets:

$$\begin{cases} \theta_a \leq t_1 \leq \theta_a + \sqrt{\alpha_s \theta_a(1 - \theta_a)} \\ \theta_b - \sqrt{\alpha_s \theta_b(1 - \theta_b)} \leq t_1 \leq \theta_b \end{cases} \tag{13}$$

where $\alpha_s = \frac{2}{s} + \frac{1}{N}$. If $\theta_a = 0$ and $\theta_b = 1$ then (13) yields that it does not exist any value of t_1 which satisfies all the inequalities and no estimator exists which dominates MLE (apart from the obvious case $s = 0$ in which (3) reduces to MLE). In general there is a feasible solution if

$$\theta_b - \sqrt{\alpha_s \theta_b(1 - \theta_b)} \leq \theta_a + \sqrt{\alpha_s \theta_a(1 - \theta_a)} \tag{14}$$

By solving (14) w.r.t to s one gets:

$$s \leq \frac{2N\left(\sqrt{\theta_a(1 - \theta_a)} + \sqrt{\theta_b(1 - \theta_b)}\right)^2}{N(\theta_b - \theta_a)^2 - \left(\sqrt{\theta_a(1 - \theta_a)} + \sqrt{\theta_b(1 - \theta_b)}\right)^2} \tag{15}$$

Notice that when the denominator is not greater than zero, any value of $s > 0$ satisfies (14). It can be seen that both s and the inequalities on t_1 depend on N. However, since in (15) the upper bound of s is a monotone decreasing function of N, the dependence on N can be dropped by taking the limit $N \to \infty$ and, thus, considering the most conservative bound

$$s \leq \frac{2\left(\sqrt{\theta_a(1 - \theta_a)} + \sqrt{\theta_b(1 - \theta_b)}\right)^2}{(\theta_b - \theta_a)^2} \tag{16}$$

For instance in the case $\theta_a = 0.1$ and $\theta_b = 0.9$, the previous bound states that $s \leq 1.125$. In this case, for large values of N, the right-side member of (16) and $t_1 = 1/2$ is the optimal solution of the problem (10).

An interesting result can be found by interpreting the dominance conditions (12) under the point of view of the Bayesian approach in the case of the non informative priors discussed in Section 2. Consider for the example the case $\theta_a = \epsilon$, $\theta_b = 1 - \epsilon$ with $\epsilon < 0.5$. In this case, by selecting $t_1 = 1/2$, the inequalities in (12) are satisfied if

$$0.5\left(1 - \sqrt{1 - \frac{1}{1 + \frac{2}{s} + \frac{1}{N}}}\right) \leq \epsilon < 0.5 \tag{17}$$

The values of the true θ_1 for which the MLE-dominance condition is satisfied when $N \to \infty$ are: Haldane ($s = 0$) needs $0 \leq \theta_1 \leq 1$; Jeffreys ($s = 0.5$) needs $0.05 \leq \theta_1 \leq 0.95$; Perks ($s = 1$) needs $0.1 \leq \theta_1 \leq 0.9$; Bayes/Laplace ($s = 2$) needs $0.15 \leq \theta_1 \leq 0.85$.

A remark is that the length of the interval where the Bayesian estimators dominate MLE decreases at the increasing of s. Thus, under the MLE-dominance point of view, only the Haldane's prior is really non-informative. The other Bayesian estimators express preferences among subregions of the parameter space and these preferences are as stronger as higher is the value of s.

5   Imprecise Dirichlet Model

An important argument against the Bayesian approach is that, at least without a large amount of samples, inferences depend on the value of $\mathbf{t}$ to be fixed in advance, typically without having sufficient information to guide the choice. This problem is addressed by the imprecise Dirichlet model proposed by Walley [13,14] as a model for prior ignorance about the chances $\boldsymbol{\theta}$. It avoids unjustifiable prior assumption by relying on the set of all Dirichlet distributions $D(s, \mathbf{t}, \boldsymbol{\theta})$ that can be obtained by varying the values of the vector of parameters $\mathbf{t}$. In the IDM, prior information is defined as the set $\mathcal{M}_0$ of all Dirichlet distributions on $\boldsymbol{\theta}$ with a fixed parameter $s > 0$,

$$\mathcal{M}_0 = \{D(s, \mathbf{t}, \boldsymbol{\theta}) \ \forall \ \mathbf{t} = [t_1, t_2, \ldots, t_k]' \text{ s.t. } 0 < t_j < 1, \ j = 1, \ldots, k, \ \mathbf{t}'\mathbf{1} = 1\}$$

After observing the data $\mathbf{n}$, each Dirichlet distribution in the set $\mathcal{M}_0$ is updated by Bayes' theorem as in (2). Under the IDM, the posterior lower and upper expectations are obtained by the minimization or maximization of $E[\boldsymbol{\theta}|\mathbf{n}]$ w.r.t. $\mathbf{t}$, which gives:

$$\underline{E}[\boldsymbol{\theta}|\mathbf{n}] = \inf_{0 < t < 1} \frac{\mathbf{n} + s\mathbf{t}}{N + s} = \frac{\mathbf{n}}{N + s}, \qquad \overline{E}[\boldsymbol{\theta}|\mathbf{n}] = \sup_{0 < t < 1} \frac{\mathbf{n} + s\mathbf{t}}{N + s} = \frac{\mathbf{n} + s\mathbf{1}}{N + s} \quad (18)$$

Notice that when no data are available, the lower and upper expectations reduce to $\underline{E}[\boldsymbol{\theta}] = \mathbf{0}$ and $\overline{E}[\boldsymbol{\theta}] = \mathbf{1}$ which is the vacuous probability model used to encode the initial lack of information on $\boldsymbol{\theta}$.

In the IDM, the parameter s determines how quickly upper and lower expectations converge as statistical data accumulate. The value of s must not depend on k to guarantee the *representation invariance principle* or the number of observations to guarantee the *coherence* [13,14]. An important criterion for the choice of s is the requirement that the IDM should be cautious enough to encompass frequentist or Bayesian alternatives, but not too cautious to avoid too weak inferences. Several convincing arguments [3,4,14] lead to choosing $1 \leq s \leq 2$. Notice in fact that in the binomial case, Haldane ($s = 0$), Perks ($s = 1$) and uniform ($s = 2$) models are encompassed by the IDM with $s = 2$. Similar relationships have also been proved for other statistical tools such as frequentist p-values and Bayesian significance levels [3,4,14].

The degree of imprecision in the IDM is defined as $\overline{E}[\boldsymbol{\theta}|\mathbf{n}] - \underline{E}[\boldsymbol{\theta}|\mathbf{n}] = s/(N + s)$. This is precisely the weight of $(\boldsymbol{\theta} - \mathbf{t})(\boldsymbol{\theta} - \mathbf{t})'$ in the MSE see Equation (6). Hence, the degree of imprecision in the IDM specifies the trade-off between bias and variance in the MSE.

The IDM, like the mentioned Bayesian approaches, expresses some preference among subregions of the parameter space. Since $\mathbf{0} < \mathbf{t} < \mathbf{1}$, this preference is due to the choice of s. In the binomial case, for the IDM MLE-dominance is guaranteed for each value of t_1 if it holds for the extreme values $t_1 = 0$ and $t_1 = 1$ (because of the convexity of (11) w.r.t t_1):

$$\begin{cases} \theta_1^2(1 + \frac{2}{s} + \frac{1}{N}) + \theta_1(-\frac{2}{s} - \frac{1}{N}) \leq 0 \\ \theta_1^2(1 + \frac{2}{s} + \frac{1}{N}) + \theta_1(-2 - \frac{2}{s} - \frac{1}{N}) + 1 \leq 0 \end{cases} \quad (19)$$

A value of θ_1 which satisfies both the inequalities exists if:

$$\frac{1}{1 + \frac{2}{s} + \frac{1}{N}} \leq \frac{\frac{2}{s} + \frac{1}{N}}{1 + \frac{2}{s} + \frac{1}{N}} \tag{20}$$

or, equivalently, if $1 \leq \frac{2}{s} + \frac{1}{N}$. Dropping the dependence of N by assuming that $N \to \infty$, it follows that $s \leq 2$. Therefore, in the IDM, $s \leq 2$ is a necessary and sufficient condition to the existence of one value of θ_1 for which all the IDM estimators dominate MLE. Still, when $s = 2$ only if the true value θ_1 is $1/2$ the IDM dominates MLE, while for $s = 1$ the true value θ_1 must be in $[1/3, 2/3]$ to have the dominance. It can be proved that, in the multinomial case, $s \leq 2$ is a necessary condition. When s is fixed, the MLE-dominance criterion proposed on Section 3 can also be interpreted in the imprecise probability formalism. However, IDM and the proposed approach are different. The set of distributions considered in this paper includes all the estimators that dominate MLE. Conversely, the IDM considers all the estimators consistent with the *near-ignorance* set of priors used to model the prior uncertainty on $\boldsymbol{\theta}$.

6   Numerical Examples

Consider a multinomial experiment with three categories. We assume that the available information is in the form of the convex polytope Θ, shown in Figure 1 and defined by the following vertices: $\theta_{v_1} = [0.15, 0.15, 0.7]'$, $\theta_{v_2} = [0.5, 0.15, 0.35]'$, $\theta_{v_3} = [0.5, 0.4, 0.1]'$, $\theta_{v_4} = [0.4, 0.5, 0.1]'$, and $\theta_{v_5} = [0.15, 0.5, 0.35]'$. The following estimators are compared: $\hat{\boldsymbol{\theta}}_{MLE} = \frac{\mathbf{n}}{N}$, $\hat{\boldsymbol{\theta}}_{MMSE_s} = \frac{\mathbf{n} + s\mathbf{t}}{N + s}$,

$$\hat{\boldsymbol{\theta}}_{MLE_h} = \arg \max_{\substack{\boldsymbol{\theta} \in \Theta \\ \boldsymbol{\theta}'\mathbf{1} = 1}} \sum_{i=1}^{k} n_i \log(\theta_i), \quad \hat{\boldsymbol{\theta}}_{MMSE_h} = \frac{\int_{\boldsymbol{\theta}} \boldsymbol{\theta}\, p(\boldsymbol{\theta}|\mathbf{n})\,\mathcal{U}(\Theta)\,d\boldsymbol{\theta}}{\int_{\boldsymbol{\theta}} p(\boldsymbol{\theta}|\mathbf{n})\,\mathcal{U}(\Theta)\,d\boldsymbol{\theta}},$$

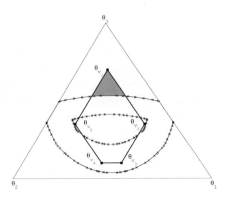

Fig. 1. Graphical representation of the a-priori information on $\boldsymbol{\theta}$. The polytope Θ (square-marked line) is plotted inside the probability simplex $\mathcal{S}_{\boldsymbol{\theta}}$. The star-marked and plus-marked lines delimitate the set of values of $\mathbf{t}$ which guarantee the MLE-dominance over Θ for $s = 2$ and, respectively, $s = 1$.

where $p(\boldsymbol{\theta}|\mathbf{n})$ is defined by Eq. (2), and $\mathcal{U}(\boldsymbol{\Theta})$ denotes the uniform distribution over $\boldsymbol{\Theta}$. These estimators are respectively the MLE, the MLE-dominance estimator (MMSE$_s$), the constrainted MLE (MLE$_h$), and the constrained Bayesian MMSE with $s = 1$ (MMSE$_h$), where the last two explicitly impose the hard constraint $\boldsymbol{\theta} \in \boldsymbol{\Theta}$. The parameters of the MLE-dominance estimator are $\mathbf{t} = [0.332, 0.332, 0.336]'$ and $s = 6.77$ for $N = 3$, and $s = 3.78$ for $N = 10$, obtained using the available a-priori information, as specified in (10). The estimators are compared by evaluating the MSE via Monte Carlo simulations. More specifically, 300 vectors of parameters $\boldsymbol{\theta}$ are randomly generated, uniformly over $\boldsymbol{\Theta}$, and for each vector 1000 independent trials are run by varying the realizations of $\mathbf{n}$.

The simulation results are reported in Table 1. More precisely, Table 1 reports the MSE of each component of the parameter vector of the MLE estimator as well as the % MSE reductions, w.r.t. the MLE estimator. Furthermore, the best-case ($\boldsymbol{\theta} = \mathbf{t}$) and the worst-case (average over the vertices $\boldsymbol{\theta}_{v_i}$ of the polytope) are also reported; these results are labelled as *best* and, respectively, *worst* in the table while the label *rand* refers to averages w.r.t randomly generated $\boldsymbol{\theta}$.

From Table 1, as expected, the MMSE$_h$ estimator provides the best results for *rand* and *best* cases, because it exploits the constraints in a hard way [2] (the same as the MLE$_h$ does, but it does not force the solution to be in the border as the latter). The drawbacks are: (i) it overweights distributions that are central to the polytope and its performance quickly degradates towards the border; (ii) it is computationally expensive, since it requires the solution of an integration that has no closed form, which is therefore tackled by numerical methods. Conversely, the estimator we propose, which is the second best overall, is much less time consuming since we have a convex quadratic programming problem that can be solved in polynomial time. Furthermore, the solution of this programming problem does not depend on the measurements but only on $\boldsymbol{\theta}$ and, therefore, can be calculated just once. Actually, MMSE$_s$ depends on the size of the data N, but it could be evaluated off-line for all desired N. Moreover, this dependency can be completely removed by taking the value of s corresponding to $N \to \infty$ (i.e. $s = 3.18$ in the example), which we call MMSE$_{s_\infty}$. Despite of that, its performance is still good as it can be seen in Table 1. Finally, the performance gain of MMSE$_s$ is specially relevant when size of the constrained region is small compared to the variance of the MLE (this is usually the case when the data set is small, i.e. $N = 3$ in the example). Figure 1 shows the set of all values of $\mathbf{t}$ that satisfy the MLE-dominance criterion for $N = 10$ in the $s = 1$ and $s = 2$ cases, i.e. the convex sets on $\mathbf{t}$ defined by (9) and $\mathbf{0} < \mathbf{t} < \mathbf{1}$. As discussed in Section 5, given s, this convex set defines the set of all estimators which dominate MLE on $\boldsymbol{\Theta}$. The upper and lower expectations of (3) w.r.t. the values of the MLE-dominating $\mathbf{t}$ are

$$\underline{E}[\boldsymbol{\theta}|\mathbf{n}] = \frac{\mathbf{n} + s\mathbf{t_0}}{N + s} \qquad \overline{E}[\boldsymbol{\theta}|\mathbf{n}] = \frac{\mathbf{n} + s\mathbf{t_1}}{N + s} \qquad (21)$$

where $\mathbf{t_1} \approx [0.6625, 0.6625, 0.5325]'$ and $\mathbf{t_0} \approx [0, 0, 0.0375]'$ for the $s = 1$ case and $\mathbf{t_1} \approx [0.5125, 0.5125, 0.4125]'$ and $\mathbf{t_0} \approx [0.1225, 0.1225, 0.22]'$ for the $s = 2$ case. These estimators can be compared with those defined for the IDM in (18). In

Table 1. Simulation results

	MSE	MLE N=3	MLE N=10	MLE$_h$ N=3	MLE$_h$ N=10	MMSE$_s$ N=3	MMSE$_s$ N=10	MMSE$_{s_\infty}$ N=3	MMSE$_{s_\infty}$ N=10	MMSE$_h$ N=3	MMSE$_h$ N=10
rand	$\hat{\theta}_1$	0.0746	0.0221	-77%	-43%	-89%	-47%	-76%	-42%	-95%	-89%
	$\hat{\theta}_2$	0.0743	0.0224	-77%	-43%	-89%	-47%	-76%	-42%	-95%	-89%
	$\hat{\theta}_3$	0.0702	0.0213	-55%	-23%	-87%	-47%	-75%	-42%	-93%	-83%
best	$\hat{\theta}_1$	0.0740	0.0226	-77%	-41%	-91%	-48%	-77%	-43%	-99%	-95%
	$\hat{\theta}_2$	0.0735	0.0219	-77%	-41%	-91%	-48%	-77%	-43%	-99%	-95%
	$\hat{\theta}_3$	0.0721	0.0222	-55%	-23%	-91%	-49%	-77%	-43%	-96%	-93%
worst	$\hat{\theta}_1$	0.0660	0.0199	-72%	-56%	-70%	-38%	-67%	-35%	-26%	+5%
	$\hat{\theta}_2$	0.0654	0.0200	-72%	-56%	-70%	-38%	-67%	-35%	-26%	+5%
	$\hat{\theta}_3$	0.0563	0.0169	-50%	-35%	-47%	-27%	-54%	-26%	-24%	+90%

the case **t** is constrained to be inside the polytope, the set of MLE-dominating **t** is given by the intersection of the previous set with the polytope. For instance, when $s = 1$ the resulting set will be the polytope excluding the shadowed area (which is exactly where IDM does not dominate MLE).

6.1   Learning Bayesian Networks

Bayesian networks encode joint probability distributions using a compact representation based on a graph with nodes associated to random variables and conditional distributions specified for variables given parents in the graph. It can be defined by a triple $(\mathcal{G}, \mathcal{X}, \mathcal{P})$, where $\mathcal{G}$ is a directed acyclic graph with nodes associated to variables $\mathcal{X} = \{X_1, \ldots, X_n\}$ (which we assume to be discrete), and $\mathcal{P}$ is a collection of parameters $p(x_{ik}|\pi_{ij})$, with $\sum_k p(x_{ik}|\pi_{ij}) = 1$, where $x_{ik} \in \Omega_{X_i}$ is a category or state of X_i and $\pi_{ij} \in \times_{Y \in \pi_i} \Omega_Y$ a complete instantiation for the parents π_i of X_i in $\mathcal{G}$ (j is viewed as an index for each parent configuration). In a Bayesian network every variable is conditionally independent of its non-descendants given its parents. Hence, the joint probability distribution is obtained by $p(\mathcal{X}) = \prod_i p(X_i|\pi_i)$. We perform parameter learning in a Bayesian network where $\mathcal{G}$ is known in advance. Because of the decomposition properties of Bayesian networks, the local distributions $p(X_i|\pi_{ij}) \in \mathcal{P}$, for each X_i and configuration π_{ij} can be learned separately using the ideas previously discussed. Using three well known Bayesian network graphs (*Asia, Insurance* and *Alarm* networks), we generate true parameters for the distributions in $\mathcal{P}$. Using these parameters, datasets are randomly created with distinct sizes (10, 50 and 100 observations). Furthermore, constraints are generated such that true values certainly lie inside the constrained set (one interval constraint for each parameter). To compare the results, we work with five estimators: (unconstrained) MLE, hard constrained MLE, constrained maximum entropy (as described in [6]), Bayesian Dirichlet model with $s = 1$ and uniform **t** (named MMSE), and the MLE-dominant MMSE$_s$ estimator. Note that MMSE$_h$ is not included in the experiment because it is computationally too expensive, since there are hundreds of distributions to be learned. The bars in Fig. 2 represent the average Kullback–Leibler (KL) divergence for 30 runs of these methods. Size of data and nodes in the networks are presented in the labels. The number of local distributions is much higher, as it depends on the number of categories and states of

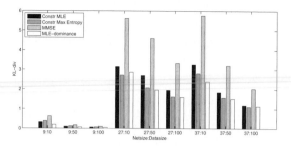

Fig. 2. Comparison between Bayesian network learning ideas

the parents of each node: *Asia, Insurance* and *Alarm* have 21, 411 and 243 local distributions, respectively. The same set of constraints and data are applied to each method in each run. The results of MLE are not displayed because they are more than 5 times worse than the best estimator. MMSE_s achieves the best results. It is clearly superior to MMSE and constrained MLE, and in general better than constrained maximum entropy.

7   Conclusion

This paper addresses the problem of inference from multinomial data under polytopic constraints on the parameter vector to be estimated, following a MLE-dominance approach. This approach consists of designing free parameters of the estimator so as to guarantee, for any admissible value of the unknown parameter vector to be estimated, an improvement of the MSE w.r.t. the standard MLE estimator. This allows us to define an objective method to choose the values of parameters s and $\mathbf{t}$ of the Dirichlet in contrast to ad-hoc practices. Using such method, we derive an estimator that is compared with existing estimators for constrained parameters, obtaining good performance. In fact, if we consider a trade-off between accuracy and computational time, our proposal surpasses the other analyzed methods. It is indeed inferior to constrained minimum mean squared error estimator (MMSE_h), but the latter cannot be run in many practical situations because of its computational cost. Besides that, the proposed estimator uses the constraints in a soft way, which makes it more robust than hard constrained estimators (like the MMSE_h) in case constraints are incorrect.

The relationship between the proposed method and the Imprecise Dirichlet Model (IDM) are briefly discussed, but some interesting conclusions already appear. For instance, according to the MLE-dominance criterion, s shall be less than 2 in the IDM. As future work, we intend to explore deeper connections of our approach with the IDM, for example, the design of a minimum value of s that guarantees MLE-dominance. As results look promising, we also intend to apply the method to large domains where data scarceness is one of the challenges.

Acknowledgements. This work has been partially supported by the Swiss NSF grant n. 200020-121785/1 (first author) and the "Ticino in Rete" project (second author). The authors are grateful to Marco Zaffalon for his suggestions.

References

1. Benavoli, A., Chisci, L., Farina, A.: Estimation of constrained parameters with guaranteed MSE improvement. IEEE T. on Signal Processing 55, 1264–1274 (2007)
2. Benavoli, A., Chisci, L., Farina, A., Ortenzi, L., Zappa, G.: Hard-constrained vs. soft-constrained parameter estimation. IEEE Trans. on Aerospace and Electronic Systems 42, 1224–1239 (2006)
3. Bernard, J.M.: Bayesian interpretation of frequentist procedures for a bernoulli process. Amer. Statist. 50, 7–13 (1996)
4. Bernard, J.M.: An introduction to the imprecise dirichlet model for multinomial data. Int. J. of Approximate Reasoning, 123–150 (2005)
5. Casella, G., Lehmann, E.L.: Theory of Point Estimation. Springer Series in Statistics, New York (1999)
6. de Campos, C.P., Ji, Q.: Improving bayesian network parameter learning using constraints. In: Int. Conference on Pattern Recognition (2008)
7. Ghosh, M., Hwang, J.T., Tsui, K.W.: Construction of improved estimators in multiparameter estimation for discrete exponential families. Ann. Statist., 351–376 (1983)
8. Haldane, J.B.S.: The precision of observed values of small frequencies. Biometrika 35, 297–300 (1948)
9. Hwang, J.T.: Improving upon standard estimators in discrete exponential families with applications to poisson and negative binomial. Ann. Statist., 857–867 (1982)
10. Johnson, B.M.: On admissible estimators for certain fixed sample binomial problems. Ann. Math. Statist. 42, 1579–1587 (1971)
11. Perks, W.: Some observations on inverse probability including a new indifference rule. J. Inst. Actuaries 73, 285–334 (1947)
12. Stein, C.: Inadmissibility of the usual estimator for the mean of a multivariate normal distribution, pp. 197–206 (1956)
13. Walley, P.: Statistical Reasoning with Imprecise Probabilities. Chapman and Hall, New York (1991)
14. Walley, P.: Inferences from multinomial data: learning about a bag of marbles. J. R. Statist. Soc. B, 3–57 (1996)

Local Computation Schemes with Partially Ordered Preferences

Hélène Fargier[1] and Nic Wilson[2]

[1] IRIT, 118 route de Narbonne 31062 Toulouse Cedex, France
[2] Cork Constraint Computation Centre, University College Cork, Ireland
fargier@irit.fr, n.wilson@4c.ucc.ie

Abstract. Many computational problems linked to uncertainty and preference management can be expressed in terms of computing the marginal(s) of a combination of a collection of valuation functions. Shenoy and Shafer showed how such a computation can be performed using a local computation scheme. A major strength of this work is that it is based on an algebraic description: what is proved is the correctness of the local computation algorithm under a few axioms on the algebraic structure. The instantiations of the framework in practice make use of totally ordered scales. The present paper focuses on the use of partially ordered scales and examines how such scales can be cast in the Shafer-Shenoy framework and thus benefit from local computation algorithms. It also provides many examples of such scales, thus showing that each of the algebraic structures explored here is of interest.

Keywords: Soft CSP, Dynamical programming, Valuation networks/ algebra.

1 Introduction

Many computational problems linked to reasoning under uncertainty can be expressed in terms of computing the marginal(s) of the combination of a collection of (local) valuation functions. Shenoy and Shafer [16,15] showed how such a computation can be performed using only local computation (see also, in particular, [9]). A major strength of this work, is that it is based on an algebraic description: what is proved is the correctness of the local computation algorithm under a few axioms on the algebraic structure. Hence, the same algorithm may be used for computing the projection on a given variable of a joint probability distribution described by a Bayesian net, for making the fusion of several basic probability assignments with Dempster's rule of combination, for computing the degree of consistency of a possibilistic knowledge base. The scope of Shenoy and Shafer's framework also encompasses several optimization problems, like the MAX CSP problem [5] or the VCSP problem [14].

But in practice, the all the actual applications of the Shenoy-Shafer framework rely on totally ordered scales of scoring. On the other hand, AI has witnessed the emergence of frameworks based on partial orders. Let us for instance cite semiring

C. Sossai and G. Chemello (Eds.): ECSQARU 2009, LNAI 5590, pp. 34–45, 2009.
© Springer-Verlag Berlin Heidelberg 2009

constraint satisfaction problems [1], order of magnitude reasoning [18], or belief revision [2,6]; other examples are obviously provided by multicriteria decision making. The purpose of this paper is to show whether and how such partially ordered problems can be cast in Shenoy and Shafer's framework, so as to provide them with local computation algorithms. We also give examples of preference relations in order to show that the algebraic structures explored here are of interest.

2 Axioms for Local Computation

We recall here some basics of the Shenoy-Shafer framework [16,15,9]. Consider a finite set $X = \{x_1, \dots, x_n\}$ of variables, each x_i ranging over a finite state space (or "domain") D_i. D_S will denote the cartesian product of the domains of variables in S. For the sake of simplicity, considering a function f over D_S we shall extend the notation $f(d)$ to tuples d assigning a superset of S, i.e., $f(d) = f(d')$ where $d' = proj(d, S)$, the projection/restriction of d to S. We also adopt the convention that the state space for the empty set $\emptyset$ consists of a single object $\Diamond$: $D_\emptyset = \{\Diamond\}$.

Given a set $S \subseteq X$ of variables there is a set V_S. The elements of V_S are called valuations and S is the scope of each $\sigma \in V_S$—let us denote it $scope(\sigma) = S$. $V = \bigcup_{S \subseteq X} V_S$ the set of valuations. Valuations are primitives in the Shenoy-Shafer framework and as such require no definition. They are simply entities that can be *combined* and *marginalized*:

- The combination of two valuations σ and τ, denoted $\sigma \boxtimes \tau$ is a valuation whose scope is $scope(\sigma) \cup scope(\tau)$.
- The marginalisation of one valuation σ over a set of variables $T \subseteq scope(\sigma)$ is a valuation whose scope is T. Let us denote it $\sigma^{\downarrow T}$.

Call $(V, \boxtimes, \downarrow)$ a *valuation algebra*. A valuation network (VN) is a finite set $\Sigma = \{\tau_1, \dots, \tau_m\} \subseteq V$. The marginal of Σ over a subset T of X is:

$$(\boxtimes\Sigma)^{\downarrow T} = (\tau_1 \boxtimes \dots \boxtimes \tau_m)^{\downarrow T}$$

Bayesian nets are instances of VNs, where valuations are conditional probability distributions, combined by the product and marginalized using summation. These instances, among many others, satisfy the Shenoy-Shafer axioms for local computation:

Axiom A1: If $S \subseteq T \subseteq scope(\sigma)$, then $((\sigma)^{\downarrow T})^{\downarrow S} = \sigma^{\downarrow S}$

Axiom A2: $\boxtimes$ is associative and commutative

Axiom A3 (distributivity of $\downarrow$ over $\boxtimes$):
If $scope(\sigma) \subseteq T \subseteq scope(\sigma) \cup scope(\tau)$, then $(\sigma \boxtimes \tau)^{\downarrow T} = \sigma \boxtimes (\tau)^{\downarrow T \cap scope(\tau)}$

It is then shown in [16,15,9] that if the valuation algebra satisfies Axioms A1, A2, A3, then for any valuation network over X and for any $Y \subseteq X$, the marginal of the network over Y can be computed by successive variable eliminations. More technically, given a VN network Σ, the basic procedure can be defined as:

$$Elim(\Sigma, T) = \Sigma_{\neg T} \cup \{(\boxtimes\Sigma_T)^{\downarrow X\backslash T}\}$$

where $\Sigma_{\neg T} = \{\sigma \in \Sigma, scope(\sigma) \cap T = \emptyset\}$ is the subset of valuations in Σ that do not bear on any variable in T and $\Sigma_T = \Sigma \backslash \Sigma_{\neg T}$ is the subset of valuations that do. If Axioms A1, A2, A3, hold, then it can be proved that:

$$(\boxtimes\Sigma)^{\downarrow X\backslash T} = \boxtimes Elim_T(\Sigma)$$

So, we can go from Σ to a new set of valuations, not bearing on T, by combining all the valuation that bear on T, computing its marginal over $X \backslash T$ and adding it to the set of valuations that do not bear on T. Applying principle iteratively w.r.t. a sequence of variables $Y = (x_{p1}x_{p2}\ldots x_{pk})$, the algorithm computes the marginal of the VN over $X \backslash Y$:

$$(\boxtimes\Sigma)^{\downarrow X\backslash\{x_{p1}\ldots x_{pk}\}} = \boxtimes Elim_{\{x_{pk}\}}(Elim_{\{x_{p_{k-1}}\}}(\ldots Elim_{\{x_{p1}\}}(\Sigma)\ldots))$$

Axioms for local computation are sufficient conditions for the correctness of the sequential elimination procedure. They also ensure the correctness of algorithms of message passing in a join tree decomposition of the VN. What is important for the purpose of the present paper, is that it is granted that when axioms A1, A2 and A3 hold such algorithms are available.

Applications of local computation focus on the case when optimization is made w.r.t. a total order (though see [12,10]). We will show that it applies to many other situations, which involve only partially ordered scales.

3   Optimization in Utility Structures

3.1   Utility Structures

Let L be a scale on which alternatives, state of the world, possible choices (the interpretation depends on the application) are scored[1] and let $\preceq$ denote the preference relation over scores. We use notation $\prec$ for the associated strict preference ($a \prec b$ iff $a \preceq b$ and not($b \preceq a$)) and $\sim$ for the corresponding indifference relation. We adopt the convention that $a \preceq b$ means that the score a is at least as good as the score b, i.e., we are oriented toward minimization. Each alternative d receives a collection $\langle c_1(d), ..., c_m(d)\rangle$ of scores; the c_i can be criteria, soft constraints, etc. The global score of d is the aggregation of all the $c_i(d)$ according to $\otimes$.

Definition 1. *A utility structure is a triplet $\langle L, \preceq, \otimes\rangle$ which forms an ordered commutative monoid. Its neutral element will be denoted denoted $\mathbf{1}$.*

That is to say, $\preceq$ is a partial order: a reflexive, anti-symmetric and transitive relation over L (hence $a \sim b$ iff $a = b$), and L is equipped with an internal operation $\otimes$ which is associative, commutative and monotonic w.r.t. $\preceq$ ($a \preceq b \implies a \otimes c \preceq b \otimes c$) and such that $a \otimes \mathbf{1} = a$ for all a.

Before going in more details about the possible property of utility structures, let us present a large class of examples that can be captured by the framework:

---

[1] We cautiously avoid the term "(e)valuation" because of the potential confusion with the notion of valuation used in VNs; in VNs, a valuation is not an element of a scale L, but (often) a function taking its values in L.

• *MAX CSP and VCSP.* In the MAX CSP [5] and (resp. VCSPs [14]) framework, the aim is to find a d that minimizes the number of violated constraints (resp. a combination, generally the sum, of the weight of the violated constraints). We shall use $L = \mathbb{N} \cup \{+\infty\}$. $\otimes$ is the addition of numbers and $\preceq \, = \, \leq$. In these examples, L is totally ordered, $\otimes$ admits a neutral element (0) which is the best score is L.

• *Cumulative prospect theory (CPT)* is an old attempt to take into account the positive and negative aspects of decision making [17]. In CPT, each c_i evaluates each possible decision d with a score may be either a positive real (i is in favor of d) or a negative real (i is against decision d). The global score of d is the sum of the positive and negative scores and should be maximized. Here, $L = \mathbb{R} \cup \{-\infty\}$, $\otimes$ is the addition of numbers and $\preceq$ is follows the classical comparison of reals ($a \preceq b$ iff $a \geq b$), since our convention minimizes while CPT maximizes. Notice that L is totally ordered, that $\otimes$ admits an annihilator $(-\infty)$ and a neutral element (0). The main difference with MAX CSP is that the neutral element does not need to be the optimal element in L.

• *Bi-attribute Pareto decision making.* In many multicriteria problems one has to simultaneously optimize several non commensurable quantities, like cost, time, security, etc. In the problem of bi-scaled shortest path for instance [7], each edge in a graph is labeled by a cost and a duration. The cost (resp. the duration) of a path is the sum of the costs (resp. durations) of its edges. For these problems, we can use $L = (\mathbb{N} \cup \{+\infty\}) \times (\mathbb{N} \cup \{+\infty\})$, $\boxtimes$ being the pointwise addition $(a, b) \otimes (a', b') = (a + a', b + b')$. Pairs are compared according to Pareto's rule: $(a, b) \preceq (a, b)$ iff $a \leq a'$ and $b \leq b'$. $\preceq$ is a partial order, e.g., $(3, 2)$ and $(2, 3)$ are incomparable.

• *Order Of Magnitude (OOM) Reasoning.* In the system of order of magnitude reasoning described in [18], the elements of L are pairs $\langle s, r \rangle$ where $s \in \{+, -, \pm\}$, and $r \in \mathbb{Z} \cup \{\infty\}$. The system is interpreted in terms of "order of magnitude" values of utility, so, for example, $\langle -, r \rangle$ represents something which is negative and has order of magnitude K^r (for a large number K). Element $\langle \pm, r \rangle$ arises from the sum of $\langle +, r \rangle$ and $\langle -, r \rangle$. $\langle \pm, r \rangle$ can be thought of as the interval between $\langle -, r \rangle$ and $\langle +, r \rangle$, since the sum of a positive quantity of order K^r and a negative quantity of order K^r can be either positive or negative and of any order less than or equal to r. Let $A_{oom} = \{\langle \pm, -\infty \rangle\} \cup \{\langle s, r \rangle \{s \in \{+, -, \pm\}, r \in \mathbb{Z} \cup \{+\infty\}\}$.

The interpretation leads to define $\otimes$ by: $\langle s, r \rangle \otimes \langle s', r' \rangle = \langle s, r \rangle$ if $r > r'$; it is equal to $\langle s', r' \rangle$ if $r < r'$; and is equal to $\langle s \vee s', r \rangle$ if $r = r'$, where $\vee$ is given by: $+ \vee + = +$ and $- \vee - = -$, and otherwise, $s \vee s' = \pm$. Operation $\otimes$ is commutative and associative with neutral element $\langle \pm, \infty \rangle$. $\preceq$ is defined by the following instances:[2] (i) for all r and s, $\langle +, r \rangle \preceq \langle -, s \rangle$; (ii) for all $s \in \{+, -, \pm\}$, and all r, r' with $r \geq r'$: $\langle +, r \rangle \preceq \langle s, r' \rangle \preceq \langle -, r \rangle$. $\preceq$ is a partial order. However, there are incomparable elements, e.g. $\langle \pm, r \rangle$ and $\langle \pm, s \rangle$ when $r \neq s$.

---

[2] This definition is slightly stronger than the original one, which doesn't allow $\langle +, r \rangle \preceq \langle \pm, r \rangle \preceq \langle -, r \rangle$; either order can be justified, but our choice is more discriminating.

- *Discrimax comparison.* In the application described by [11] one has to satisfy n agents, each of them expressing her preferences by weighted formulas of propositional logic. The "disutilty" of an agent is then the combination, normally the sum (resp. the max), of those of her formulae that are not satisfied: in other terms, $L = (\mathbb{N} \cup \{+\infty\})^n$ and $\otimes$ is the pointwise addition (resp. maximum) of the vectors. In this application, decision making must be fair and egalitarist. So, the disutility of the most unsatisfied agent is minimized: for two vectors a and b, $a \preceq b$ iff $max_{j=1,n,a_j \neq b_j} a_j < max_{j=1,n,a_j \neq b_j} b_j$ or $a = b$. The restriction $a_j \neq b_j$ allows one not to consider the agents that are equally satisfied by a and b (the comparison is made Ceteris Paribus). Hence the name "Discrimax". Other uses of discrimax comparison include belief revision [2] and multi criteria optimisation [8]. The preference relation is only a partial order. For instance, with two agents, $\langle 0, 5 \rangle$ and $\langle 5, 0 \rangle$ are incomparable vectors. Both $\otimes = +$ and $\otimes = \max$ are associative, commutative, with a neutral element $\langle 0, \ldots, 0 \rangle$, but $\otimes = +$ is not monotonic. With $\otimes = \max$, monotonicity holds.

- *Tolerant Pareto.* The problem with a Pareto-based comparison is that the preference provided is often not decisive enough. For instance the two pairs $a = (a_{cost}, a_{time})$ and $b = (b_{cost}, b_{time})$ are incomparable as soon as $a_{cost} < b_{cost}$ and $b_{time} < a_{time}$, and this even if the difference between a_{cost} and b_{cost} is much greater than difference between b_{time} and a_{time}.

Consider our time/cost pair. The idea is to use indifference thresholds, say α_{cost} for the first dimension, and α_{time} for the second one. If $a_{cost} + \alpha_{cost} < b_{cost}$, we shall say that the cost dimension has a strong preference for a over b, and opposes a veto to the opposite preference. Then we decide that an alternative is better than the other iff it Pareto dominates, but with respect to the thresholds of tolerance. Formally decide:

$$a \prec b \text{ iff either} \begin{cases} b_{cost} - a_{cost} > \alpha_{cost} \text{ and} \\ b_{time} - a_{time} \geq -\alpha_{time}; \text{ or} \\ b_{time} - a_{time} > \alpha_{time} \text{ and} \\ b_{cost} - a_{cost} \geq -\alpha_{cost} \end{cases}$$

So, when one dimension strongly prefers alternative a while the other does not oppose a veto we do not get an incomparability, like in the classical Pareto case, but a strict preference $a \prec b$. This decision rule is related to the Electre method (see e.g. [13]). It yields a preference relation that is not complete nor transitive: it may happen that $a \prec b$ and $b \prec c$ while a and c are not comparable (e.g. because the time dimension that does not oppose a veto to $a \prec b$ nor to $b \prec c$ is a vetoer for $a \prec c$). Nevertheless, $\prec$ is acyclic.

This example cannot be cast as a utility network stricto sensu, but its closure by transitivity can be, using pointwise addition as the combination. Let $\prec^*$ be the transitive closure of $\prec$. It can be shown that $a \prec^* b$ holds if and only if either (i) $b_{cost} - a_{cost} > 0$ and $b_{time} - a_{time} > 0$, or (ii) there exists $k \in \{1, 2, \ldots\}$ such that either (a) $b_{cost} - a_{cost} > k\alpha_{cost}$ and $b_{time} - a_{time} \geq -k\alpha_{time}$ or (b) $b_{time} - a_{time} > k\alpha_{time}$ and $b_{cost} - a_{cost}c \geq -k\alpha_{cost}$.

In this rule, the thresholds are considered as elementary units of strong preference. So, a is better than b when, going from b to a, the enhancement on one dimension (e.g. the cost dimension) is greater than the degradation in the other dimension, this enhancement (resp. degradation) being evaluated on a scale whose unit is α_{cost} (resp. α_{time}).

Let us return to the algebraic framework, $\langle L, \preceq \otimes \rangle$. Remark that in all the problems, the worst score annihilates $\otimes$. Indeed, for any ordered monoid, we can suppose without loss of generality that L contains a unique maximal (worst) element $\top$ and a unique minimal (best) element $\bot$, and that $\top$ annihilates $\otimes$.

- If $\bot$ is the neutral element, then it holds that $a \preceq a \otimes b$. $\langle L, \preceq, \otimes \rangle$ is then said to be negative.
- If there exists an associative and commutative operator $\oplus$ such that $a \preceq b \iff a \oplus b = a$, then we say that $\oplus$ represents $\preceq$. It is well known that such a $\oplus$ exists iff $\langle L, \preceq \rangle$ forms a meet semilattice.
- If $\preceq$ is a total order this operator necessarily exists ($\oplus = \min$).
- If $\forall a, b, \forall c \neq \bot, \top, a \prec b \Rightarrow a \otimes c \prec b \otimes c$ then $\langle L, \preceq, \otimes \rangle$ is said to be strictly monotonic.

Negative structures are well known in flexible constraint satisfaction. In semiring CSP [1], the first two properties are assumed (semiring CSP are utility structures where $\langle L, \otimes, \oplus \rangle$ is a negative commutative semiring). If the completeness of $\preceq$ is moreover assumed, the network is a soft CSP in the sense of [3]. Max CSPs and VCSPs are instances of soft CSPs (and thus of semiring CSPs). Pure Pareto Cost/Time problems are semiring CSPs (just set $(a, b) \oplus (c, d) = (\min(a, c), \min(b, d))$). Both are based on a negative structure. But there are utility structures that cannot be captured by soft CSPs nor semiring CSPs: in the CPT and OOM examples, $\bot$ is not the neutral element; in the Tolerant Pareto example, there exist no operator $\oplus$ encoding $\preceq$. The reason of the last assertion is that in these two cases, $\preceq$ is not a meet semilattice. Intuitively, in these cases, there may be several candidates for $a \oplus b$.

3.2   Optimisation in Utility Networks

Let us now use utility structure in combinatorial optimisation problems, thus defining utility networks:

Definition 2. *Given a utility structure $\langle L, \preceq, \otimes \rangle$ and a set X of variables:*
- *A local function is a function from the domain D_S of some $S \subseteq X$ into L.*
- *A utility network $\mathcal{C}$ is a set of local functions.*

Definition 3. *Given a utility network $\mathcal{C}$ on $\langle L, \preceq, \otimes \rangle$, the global score of d is $score_\mathcal{C}(d) = \bigotimes_{c_i \in \mathcal{C}} c_i(d)$.*
We shall also write $Scores(\mathcal{C}) = \{score_\mathcal{C}(d) : d \in D_X\}$.

When the scale is totally ordered, as for CPT or VCSP, the usual optimization request is to compute the minimal value for $score_\mathcal{C}(d)$ (generally, together with the d leading to this score). When $\preceq$ is partial, there may be several optimal scores that are pairwise incomparable.

Definition 4. $d \in D_X$ *is an optimal solution for* (C) *if there is no* d' *in* D_X *such that* $score_C(d') \prec score_C(d)$.

a is an optimal score for C *if* $a = score_C(d)$ *for some optimal solution* d.

For partial order $\preceq$ and any $A \subseteq L$, let us denote $Kernel_\preceq(A)$ (the kernel of A) as the set of $\preceq$-minimal elements of A, i.e., the set of elements $a \in A$ such that there exists no $b \in A$ with $b \prec a$. It is easy to see that the set of optimal scores is the Kernel of $Scores(C)$ w.r.t. $\preceq$:

Proposition 1. $a \in Kernel_\preceq(Scores(C))$ *iff* a *is an optimal score for* C.

So, if $\preceq$ is a total order, $Kernel_\preceq(Scores(C))$ is the singleton set containing the optimal score for C.

When compared to soft CSPs (resp. semiring CSPs), our utility networks relax the assumption of $\preceq$ being a total order (resp. a semilattice) as well as the requirement about the neutral element. However, this does not increase the complexity of the problem. Let $\mathcal{L} = \langle L, \preceq, \otimes \rangle$ be a utility structure. We consider the following two problems:

[OPT$_\mathcal{L}$]: *Given a network* C *built on utility structure* $\mathcal{L}$ *and* $a \in L$, *does there exist an assignment* d *such that* $score_C(d) \prec a$.

[FULLOPT$_\mathcal{L}$]: *Given a network* C *built on utility structure* $\mathcal{L}$, *and given* $H \subseteq L$, *does there exist an assignment* d *such that* $\exists a \in H, score_C(d) \prec a$.

There problems are easily seen to be in NP. Furthermore they are NP-hard under very weak assumptions, as shown by the following result which is similar to Proposition 5 of [4].

Proposition 2. *Let* $\mathcal{L} = \langle L, \preceq, \otimes \rangle$ *be a utility structure. Suppose that testing* $a \preceq b$ *is polynomial, that computing the combination of a multiset of elements of* L *is polynomial, and that* L *contains some element* a *such that* $a \succ 1$. *Then* $OPT_\mathcal{L}$ *and* $FULLOPT_\mathcal{L}$ *are* NP-*complete*.

So, the optimization problem in its simple version (find an element of the Kernel) or its full version (find the Kernel) is not harder in the case of a partially ordered scale than in the case of a totally ordered one. Branch and Bound algorithms can always be used for computing a single optimal solution or even for computing the Kernel. But this analysis is somewhat biased, since the size of the Kernel is theoretically large. In the worst case, it is equal to the width of $\preceq$. The width of the Pareto comparison, for instance, is exponential, hence the weakness of the rule; the width of the OOM rule, on the contrary, is limited by the number of levels in the scale.

4   Casting Utility Networks in the Local Computation Scheme

In the following, we focus on the ways of embedding utility networks into Shenoy and Shafer's framework in order to benefit from the local computation machinery.

First, we show that a direct encoding of the utility structure is inadequate. Two alternative ways are then investigated: the use of a a refinement of the original preference order (this provides one of the optimal scores, provided that such a refinement exists) and the use of a set encoding of the utility structure (this is always possible and provides all the the optimal scores).

4.1   Direct Encoding

Utility networks can be simply cast as a problem of combination of valuations, letting $\mathcal{V} = \bigcup_{S \subseteq X} \{f : D_S \mapsto L\}$ and defining $\boxtimes$ in a pointwise fashion:

Definition 5. *Let $\langle L, \preceq, \otimes \rangle$ be a utility structure and σ, τ two functions for a subset of X to L. For any $d \in D_{scope(\sigma) \cup scope(\tau)}$, define $(\tau \boxtimes \sigma)(d) = \tau(d) \otimes \sigma(d)$*

Then the global score function is simply the combination of the c_i in $\mathcal{C}$.

Proposition 3. *For any utility network $\mathcal{C}$ over $\langle L, \preceq, \otimes \rangle$, $Score = \boxtimes_{c_i \in \mathcal{C}} c_i$.*

Also, $\boxtimes$ satisfies axiom $A2$ iff $\otimes$ is associative and commutative, which gives a fundamental justification for having $\otimes$ associative and commutative in preference structures. Now, the difficulties arise with the marginalisation operator. The only trivial case is when $\preceq$ is totally ordered. Then the min operator is well defined and we can set

$$\sigma^{\downarrow T}(d) = min_{d' \in D_S, d=proj(d',T)} \sigma(d')$$

This definition ensures the satisfaction of A1 and A3, and that for any $\mathcal{C}$ built on $\langle L, \preceq, \otimes \rangle$, $(\boxtimes_{c \in \mathcal{C}} c)^{\downarrow \emptyset}$ is the optimal score for $\mathcal{C}$. We can consider using the same technique there exists an operator $\oplus$ such that $a \preceq b \iff a \oplus b = a$ and $\langle L, \otimes, \oplus \rangle$ is a semiring. It is always possible to define the marginalisation operator:

Definition 6. *If there exists an operator $\oplus$ such as $a \preceq b \iff a \oplus b = a$ and $\langle L, \otimes, \oplus \rangle$ is a semiring, let us define $\downarrow$ as:*

$$\forall \sigma, d : \sigma^{\downarrow T}(d) = \bigoplus_{d' \in D_S, d=proj(d',T)} \sigma(d').$$

Proposition 4. *Axioms A1, A2 and A3 are satisfied by $\boxtimes$ and $\oplus$ as defined in Definitions 5 and 6.*

See [10], Theorem 2. The problem is that $\oplus$ and $\downarrow$ are not faithful to the notion of optimality in L. First of all because there may be more than one score in the kernel. Secondly, and maybe more importantly, because it may happen the score computed by this marginalisation is not achievable: $(\boxtimes_{c \in \mathcal{C}} c)^{\downarrow \emptyset}$ *does not necessarily belong to the kernel at all.* More precisely, it holds that:

Proposition 5. *Given a utility structure $\langle L, \preceq, \otimes \rangle$, and if $\boxtimes$ and $\downarrow$ are defined according to Definitions 5, the following assertions are equivalent:*

- $\forall \mathcal{C}, (\boxtimes_{c \in \mathcal{C}} c)^{\downarrow \emptyset} \in Kernel_{\preceq}(Scores(\mathcal{C}))$
- $\preceq$ *is a total order.*

What local computation computes with this direct encoding is actually a (greatest) lower bound of the Kernel:

Proposition 6. *If $\boxtimes$ and $\downarrow$ are defined according to Definitions 5 and 6, then $\forall a \in Kernel_{\preceq}(Scores(\mathcal{C}))$, $(\boxtimes_{c_i \in \mathcal{C}} c_i)^{\downarrow \emptyset} \preceq a$.*

But once again, it may happen that this score does not belong to the kernel. Proposition 5 is a rather negative result. A variable elimination approach is indeed potentially exponential in time and space. It may be worthwhile using it if it were providing the optimal score. But the computational cost is too high for just an approximation of the result. We shall circumvent the difficulty, by working with another comparator. The first solution is to simply refine $\preceq$.

4.2   Refining $\preceq$

A classical approach in Pareto-based multicriteria optimisation problems is to optimize a linear combination of the criteria. The important idea here is that one optimizes according to a new comparator, say $\trianglelefteq$, such that $a \preceq b$ implies $a \trianglelefteq b$: if a is preferred to b according to the original relation, then it is still the case with the new one. But $\trianglelefteq$ can rank scores that are incomparable w.r.t. $\preceq$. Such a relation is called a refinement of the original relation.

Definition 7. *$\trianglelefteq$ refines $\preceq$ if and only if $a \preceq b$ implies $a \trianglelefteq b$.*

In the Cost/Time Pareto case, we shall decide $a \trianglelefteq b$ iff $a_{cost} + \beta \cdot a_{time} \leq b_{cost} + \beta \cdot b_{time}$. $\trianglelefteq$ is complete and if β is high enough, there are no ties, i.e. $\trianglelefteq$ is a total order.

Optimizing with respect to a refinement leads to solutions that are optimal with respect to the original relation. More precisely:

Proposition 7. *If $\trianglelefteq$ refines $\preceq$, then whatever A, $Kernel_{\trianglelefteq}(A) \subseteq Kernel_{\preceq}(A)$.*

Now, if there exists a totally ordered refinement $\trianglelefteq$ of $\preceq$ such that $< L, \trianglelefteq, \otimes >$ is a monotonic utility structure, it is then possible to define $\oplus$ as the min of two scores according to $\trianglelefteq$. Then Definitions 5 and 6 can be applied from $< L, \trianglelefteq, \otimes >$ and Axioms A1, A2 and A3 are satisfied, thanks to Proposition 4.

Like the approach described in Section 4.1, the present one provides the user with a unique score among the optimal ones, but this one has the advantage of being reached by one of the optimal solutions.

Unfortunately, such a totally ordered refinement does not necessarily exist. Consider for instance the case where L is the set of integers, $\otimes = \times$ and any $\preceq$ making $\langle L, \preceq, \otimes \rangle$ a utility structure (e.g., $a \preceq b \iff a = b$). Since $\trianglerighteq$ is total, we have either $1 \trianglerighteq -1$ or $-1 \trianglerighteq 1$. $1 \trianglerighteq -1$ implies by monotonicity $1 \otimes -1 \trianglerighteq -1 \otimes -1$, and so $-1 \trianglerighteq 1$. Similarly, $-1 \trianglerighteq 1$ implies $1 \trianglerighteq -1$, so in either case we have $1 \trianglerighteq -1 \trianglerighteq 1$, contradicting antisymmetry. The following result gives sufficient conditions for an appropriate refinement to exist. (a^1 is defined to be a, and, for $k \geq 1$, $a^{k+1} = a^k \otimes a$.)

Theorem 1. *Let $\langle L, \preceq, \otimes \rangle$ be a utility structure with unit element $\mathbf{1}$, which also satisfies the following two properties:*

(i) for all $a, b \in L$ with $a \neq b$ and all $k > 0$ we have $a^k \neq b^k$;
(ii) $a \otimes c \preceq b \otimes c \Rightarrow a \preceq b$ for all $a, b, c \in L$.

Then there exists a total order $\trianglelefteq$ on L extending $\preceq$ and such that for all $a, b, c \in L$, $a \otimes c \trianglelefteq b \otimes c \iff a \trianglelefteq b$, and so, in particular, $\langle L, \trianglelefteq, \otimes \rangle$ is a utility structure.

4.3   Set Encoding

There is a definitive way of using Shafer and Shenoy's framework to optimize over a utility structure. The idea is to move from L to 2^L, the set of subsets of L. A score can then be a set of scores. Each c_i provides a singleton, nothing is really changed from this point of view. What changes, is the ability of computing a "min": when a and b are not comparable, we keep both when marginalizing. This transformation has been used by Rollon and Larrosa [12] in problems of multiobjective optimization problems based on a Pareto comparison. We show here that algebraic utility networks are rich enough to use this kind of transformation.

More formally, let $\mathbf{L} = \{A \subseteq L, A \neq \emptyset, \text{ s.t. } A = Kernel_\prec(A)\}$. Notice that a singleton is its own kernel, thus belongs to $\mathbf{L}$ and that $\mathbf{L}$ is stable w.r.t. the kernel based union : for any $A, B \in \mathbf{L}, Kernel_\prec(A \cup B) \in \mathbf{L}$.

For any constraint c, let $\mathbf{c}$ be the constraint taking it scores in $\mathbf{L}$ defined by: $\mathbf{c}(d) = \{c(d)\}$ and denote $\mathbf{C} = \{\mathbf{c} : c \in \mathcal{C}\}$ the transformation of $\mathcal{C}$ by this "singletonization". Let us now define an operator $\oplus_s$ between sets of scores:

Definition 8. *For all non-empty subsets A and B of L, define: $A \oplus_s B = Kernel_\prec(A \cup B)$.*

The operation of aggregation now has to be able to handle sets of scores.

Definition 9. *$\forall A, B \subseteq L, A \otimes_s B = Kernel(\{a \otimes b, a \in A, b \in B\})$.*

Proposition 8. *$\langle \mathbf{L}, \otimes_s, \oplus_s \rangle$ is a (commutative) semiring.*

Proposition 4 then implies:

Proposition 9. *Axioms A1, A2 and A3 are satisfied by $\boxtimes$ and $\downarrow$ as defined in Definitions 5 and 6 from the set operations $\otimes_s$ and $\oplus_s$ provided by Definitions 8 and 9.*

The following is the key result that shows that the set of optimal elements $Kernel_\prec(Scores(\mathcal{C}))$ can be expressed as the projection of a combination, which can be computed using local computation because of Proposition 9.

Proposition 10. *$(\boxtimes_{\mathbf{c} \in \mathbf{C}} \mathbf{c})^{\downarrow \emptyset} = Kernel_\prec(Scores(\mathcal{C}))$.*

A direct consequence of these propositions is that local computation can be used to compute the set of optimal values of any utility network, i.e., variable elimination is possible for *any* utility network.

Now, the theoretical application of local computation must not overshadow its practical range of application. It is known that variable elimination is in the worst case exponential w.r.t. the treewidth of the constraint graph. This is the case if we consider that size of the score sets is 1. Depending on how discriminating $\preceq$ is, we may get a larger score set at some point in the computation. The worst case complexity of variable elimination, in time and space, must thus be multiplied by the size of the largest subset of L that contains elements that are pointwisely incomparable with respect to $\preceq$. Mathematically, this number is known as the width of $\preceq$. It is relatively small for some of our examples:

- Its value is 1, obviously, for the total orders (Max CSP and CPT);
- For Pareto comparison on n criteria, the width is exponentially large in the number of criteria. This is an additional reason to prefer refinements when meaningful.
- For the OOM case, the largest kernel is $\{\alpha_1^{\pm}, \ldots, \alpha_k^{\pm}\}$, $\{\alpha_1, \ldots, \alpha_k\}$ being the set of possible values for the order of magnitude — typically, reduced to a small selection of qualitative values: "null", negligible", "weak", "significant", "high", "very high".

In practice the width of the order can be a minor issue in comparison to the original complexity of the variable elimination procedure, which is exponential in the treewidth of the elimination sequence. If variable elimination is affordable, it may work well over partially ordered scales.

5   Conclusion and Perspectives

This paper mainly focused on the ways of embedding utility networks into Shenoy and Shafer's framework. More precisely, we have shown that (i) a direct encoding is not always sound w.r.t. optimality, (ii) the definition of a refinement, widely used in multi criteria optimization, can be applied in certain cases (though not always); (iii) it always possible to benefit from the local computation machinery by using a set encoding.

But as it is the case for the tolerant Pareto example, there are meaningful structures of preferences that are not captured by utility networks. Other examples include preorders and semiorders, that allow richer indifference relations. Further research will be developed around the algebraic study of such structures.

Acknowledgements

The work of the second author has been supported by the Science Foundation Ireland under Grant No. 00/PI.1/C075 and Grant No. 08/PI/I1912.

References

1. Bistarelli, S., Montanari, U., Rossi, F.: Constraint solving over semirings. In: IJCAI 1995, pp. 624–630 (1995)
2. Brewka, G.: Preferred subtheories: An extended logical framework for default reasoning. In: IJCAI 1989, pp. 1043–1048 (1989)
3. Cooper, M., Schiex, T.: Arc consistency for soft constraints. Artif. Intelligence 154(1-2), 199–227 (2004)
4. Fargier, H., Wilson, N.: Algebraic structures for bipolar constraint-based reasoning. In: Mellouli, K. (ed.) ECSQARU 2007. LNCS, vol. 4724, pp. 623–634. Springer, Heidelberg (2007)
5. Freuder, E., Wallace, R.: Partial constraint satisfaction. Artificial Intelligence 58(1-3), 21–70 (1992)
6. Friedman, N., Halpern, J.: Plausibility measures and default reasoning. In: AAAI 1996, pp. 1297–1304 (1996)
7. Henig, M.I.: The shortest path problem with two objective functions. EJOR 25, 281–291 (1985)
8. Junker, U.: Preference-based search and multi-criteria optimization. In: AAAI 2002, pp. 34–40 (2002)
9. Kohlas, J.: Information Algebras: Generic Structures for Inference. Springer, Heidelberg (2003)
10. Kohlas, J., Wilson, N.: Semiring induced valuation algebras: Exact and approximate local computation algorithms. Artificial Intelligence 172, 1360–1399 (2008)
11. Lafage, C., Lang, J.: Logical representation of preferences for group decision making. In: KR 2000, pp. 457–468 (2000)
12. Rollon, E., Larrosa, J.: Bucket elimination for multiobjective optimization problems. J. Heuristics 12(4-5), 307–328 (2006)
13. Roy, B.: The outranking approach and the foundations of ELECTRE methods. Theory and Decision 31(1), 49–73 (1991)
14. Schiex, T., Fargier, H., Verfaillie, G.: Valued constraint satisfaction problems: Hard and easy problems. In: IJCAI 1995, Montreal, pp. 631–637 (1995)
15. Shenoy, P.P.: Valuation-based systems for discrete optimisation. In: UAI 1990, pp. 385–400 (1990)
16. Shenoy, P.P., Shafer, G.: Axioms for probability and belief-function proagation. In: UAI, pp. 169–198 (1988)
17. Tversky, A., Kahneman, D.: Advances in prospect theory: Cumulative representation of uncertainty. Journal of Risk and Uncertainty 5, 297–323 (1992)
18. Wilson, N.: An order of magnitude calculus. In: UAI 1995, pp. 548–555 (1995)

Inference in Hybrid Bayesian Networks with Deterministic Variables

Prakash P. Shenoy and James C. West

University of Kansas School of Business, 1300 Sunnyside Ave.,
Summerfield Hall, Lawrence, KS 66045-7585 USA
{pshenoy,cully}@ku.edu

Abstract. The main goal of this paper is to describe an architecture for solving large general hybrid Bayesian networks (BNs) with deterministic variables. In the presence of deterministic variables, we have to deal with non-existence of joint densities. We represent deterministic conditional distributions using Dirac delta functions. Using the properties of Dirac delta functions, we can deal with a large class of deterministic functions. The architecture we develop is an extension of the Shenoy-Shafer architecture for discrete BNs. We illustrate the architecture with some small illustrative examples.

Keywords: Hybrid Bayesian networks, deterministic variables, Dirac delta functions, Shenoy-Shafer architecture.

1 Introduction

Bayesian networks (BNs) and influence diagrams (IDs) were invented in the mid 1980s (see e.g., [19, 9]) to represent and reason with large multivariate discrete probability models and decision problems, respectively. Several efficient algorithms exist to compute exact marginals of posterior distributions for discrete BNs (see e.g., [13, 27]), and to solve discrete influence diagrams exactly (see e.g., [18, 22, 24]).

The state of the art exact algorithm for mixtures of Gaussians hybrid BNs is the Lauritzen-Jensen [14] algorithm. This requires the conditional distributions of continuous variables to be conditional linear Gaussians (CLG), and that discrete variables do not have continuous parents.

If a BN has discrete variables with continuous parents, Murphy [17] uses a variational approach to approximate the product of the potentials associated with a discrete variable and its parents with a CLG. Lerner [15] uses a numerical integration technique called Gaussian quadrature to approximate non-CLG distributions with CLG, and this same technique can be used to approximate the product of potentials associated with a discrete variable and its continuous parents. Murphy's and Lerner's approach is then embedded in the Lauritzen-Jensen [14] algorithm to solve the resulting mixtures of Gaussians BN. Shenoy [26] proposes approximating non-CLG distributions by mixtures of Gaussians using a nonlinear optimization technique, and using arc reversals to ensure discrete variables do not have continuous parents. The resulting mixture of Gaussians BN is then solved using Lauritzen-Jensen [14] algorithm.

C. Sossai and G. Chemello (Eds.): ECSQARU 2009, LNAI 5590, pp. 46–58, 2009.

Moral *et al.* [16] proposes approximating probability density functions (PDFs) by mixtures of truncated exponentials (MTE), which are easy to integrate in closed form. Since the family of MTE is closed under combination and marginalization, the Shenoy-Shafer [27] architecture can be used to solve the MTE BN. Cobb *et al.* [5] proposes using a non-linear optimization technique for finding MTE approximation for the many commonly used PDFs. Cobb and Shenoy [2, 3] extend this approach to BNs with linear and non-linear deterministic variables. In the latter case, they approximate non-linear deterministic functions by piecewise linear ones.

Shenoy and West [29] propose mixtures of polynomials (MOP) to approximate PDFs. Like MTE, MOP are easy to integrate, and are closed under combination and marginalization. Unlike MTE, they can be easily found using the Taylor series expansion of differentiable functions, and they are closed under a larger family of deterministic functions than MTE, which are closed only for linear functions.

For Bayesian decision problems, Kenley [11] (see also [23]) describes the representation and solution of Gaussian IDs that include continuous chance variables with CLG distributions. Poland [20] extends Gaussian IDs to mixtures of Gaussians IDs. Thus, continuous chance variables can have any distributions, and these are approximated by mixtures of Gaussians. Cobb and Shenoy [4] extend MTE BNs to MTE IDs for the special case where all decision variables are discrete.

In this paper, we describe a generalization of the Shenoy-Shafer architecture for discrete BNs so that it applies to hybrid BNs with deterministic variables. The functions associated with deterministic variables do not have to be linear (as in the CLG case) or even invertible. We use Dirac delta functions to represent such functions and also for observations of continuous variables. We use mixed potentials to keep track of the nature of potentials (discrete and continuous). We define combination and marginalization of mixed potentials. Finally, we illustrate our architecture by solving some small examples that include non-linear, non-invertible deterministic variables.

An outline of the remainder of the paper is as follows. In Section 2, we describe our architecture for making inferences in hybrid BNs with deterministic variables. This is the main contribution of this paper. In Section 3, we solve three small examples to illustrate the architecture. In Section 4, we end with a summary and discussion.

2 The Extended Shenoy-Shafer Architecture

In this section, we describe the extended Shenoy-Shafer architecture for representing and solving hybrid BNs with deterministic variables. The architecture and notation is adapted from Cinicioglu and Shenoy [1], and Cobb and Shenoy [2].

Variables and States. We are concerned with a finite set V of *variables*. Each variable $X \in V$ is associated with a set Ω_X of its possible *states*. If Ω_X is a finite set or countably infinite, we say X is *discrete*, otherwise X is *continuous*. We will assume that the state space of continuous variables is the set of real numbers (or some subset of it), and that the state space of discrete variables is a set of symbols (not necessarily real numbers). If $r \subseteq V$, $r \neq \varnothing$, then $\Omega_r = \times\{\Omega_X \mid X \in r\}$. If $r = \varnothing$, we will adopt the convention that $\Omega_\varnothing = \{\blacklozenge\}$. If $\mathbf{r} \in \Omega_r$, $\mathbf{s} \in \Omega_s$, and $r \cap s = \varnothing$, then $(\mathbf{r}, \mathbf{s}) \in \Omega_{r \cup s}$. Therefore, $(\mathbf{r}, \blacklozenge) = \mathbf{r}$.

Projection of States. Suppose $r \in \Omega_r$, and suppose $s \subseteq r$. Then the *projection* of r to s, denoted by $r^{\downarrow s}$, is the state of s obtained from r by dropping states of $r \setminus s$. Thus, $(w, x, y, z)^{\downarrow \{W, X\}} = (w, x)$, where $w \in \Omega_W$, and $x \in \Omega_X$. If $s = r$, then $r^{\downarrow s} = r$. If $s = \varnothing$, then $r^{\downarrow s} = \blacklozenge$.

In a BN, each variable has a conditional distribution function for each state of its parents. A conditional distribution function associated with a continuous variable is said to be *deterministic* if the variances (for each state of its parents) are zeros. For simplicity, henceforth, we will refer to continuous variables with non-deterministic conditionals as *continuous*, and continuous variables with deterministic conditionals as *deterministic*. In a BN, discrete variables are denoted by rectangular shaped nodes, continuous variables by oval shaped nodes, and deterministic variables by oval nodes with a double border.

Discrete Potentials. In a BN, the conditional probability functions associated with the variable are represented by functions called *potentials*. If A is discrete, it is associated with conditional probability mass functions, one for each state of its parents. The conditional probability mass functions are represented by functions called *discrete potentials*. Suppose $r \subseteq V$ is such that it contains a discrete variable. A discrete potential α for r is a function $\alpha \colon \Omega_r \to [0, 1]$. The values of discrete potentials are probabilities. We will sometimes write the range of α as $[0, 1](m)$ to denote that the values in $[0, 1]$ are probability masses.

Although the domain of the function α is Ω_r, for simplicity, we will refer to r as the *domain* of α. Thus, the domain of a potential representing the conditional probability function associated with some variable X in a BN is always the set $\{X\} \cup pa(X)$, where $pa(X)$ denotes the set of parents of X in the BN graph.

Density Potentials. If Z is continuous, then it is associated with a *density* potential. Suppose $r \subseteq V$ is such that it contains a continuous variable. A density potential ζ for r is a function $\zeta \colon \Omega_r \to \mathbb{R}^+$, where $\mathbb{R}^+$ is the set of non-negative real numbers. The values of density potentials are probability densities. We will sometimes write the range of ζ as $\mathbb{R}^+(d)$ to denote that the values in $\mathbb{R}^+$ are densities.

Dirac Delta Functions. $\delta \colon \mathbb{R} \to \mathbb{R}^+(d)$ is called a *Dirac delta function* if $\delta(x) = 0$ if $x \neq 0$, and $\int \delta(x)\, dx = 1$. Whenever the limits of integration of an integral are not specified, the entire range $(-\infty, \infty)$ is to be understood. δ is not a proper function since the value of the function at 0 doesn't exist, i.e., $\delta(0)$ is not finite. It can be regarded as a limit of a certain sequence of functions (such as, e.g., the Gaussian density function with mean 0 and variance σ^2 in the limit as $\sigma \to 0$). However, it can be used as if it were a proper function for practically all our purposes without getting incorrect results. It was first defined by Dirac [6].

As defined above, the value $\delta(0)$ is undefined, i.e., $\delta(0) = \infty$, when considered as density. We argue that we can *interpret* the value $\delta(0)$ as probability 1. Consider the normal PDF with mean 0 and variance σ^2. Its moment generating function (MGF) is $M(t) = e^{\sigma^2 t}$. In the limit as $\sigma \to 0$, $M(t) = 1$. Now, $M(t) = 1$ is the MGF of the distribution $X = 0$

with probability 1. Therefore, we can interpret the value $\delta(0)$ as probability 1. This is strictly for interpretation only.

Some basic properties of the Dirac delta function are as follows [6, 7, 8, 10, 21, 12].

(i) If $f(x)$ is any function that is continuous in the neighborhood of a, then $f(x)\,\delta(x-a) = f(a)\,\delta(x-a)$, and $\int f(x)\,\delta(x-a)\,dx = f(a)$.

(ii) $\int \delta(x-h(u,v))\,\delta(y-g(v,w,x))\,dx = \delta(y-g(v,w,h(u,v)))$. This follows from (i).

(iii) If $g(x)$ has real (non-complex) zeros at $a_1, \ldots, a_n$, and is differentiable at these points, and $g'(a_i)\neq0$ for $i = 1, \ldots, n$, then $\delta(g(x)) = \Sigma_i \delta(x - a_i)/|g'(a_i)|$. For example, $\delta(ax) = \delta(x)/|a|$, if $a \neq 0$. Therefore, $\delta(-x) = \delta(x)$.

(iv) Suppose continuous variable X has PDF $f_X(x)$ and $Y=g(X)$. Then Y has PDF $f_Y(y) = \int f_X(x)\,\delta(y-g(x))\,dx$. The function g does not have to be invertible.

A more extensive list of properties of the Dirac delta function that is relevant for uncertain reasoning is stated in [1].

Dirac Potentials. Deterministic variables have conditional distributions containing functions. We will represent such functions by *Dirac* potentials. Suppose $x = r \cup s$ is a set of variables containing some discrete variables r and some continuous variables s. We assume $s \neq \varnothing$. A Dirac potential ξ for x is a function $\xi\colon \Omega_x \to \mathbb{R}^+(d)$ such that $\xi(r, s)$ is of the form $\Sigma\{p_{r,i}\,\delta(z - g_{r,i}(s^{\downarrow(s\setminus\{Z\})})) \mid i = 1, \ldots, n,$ and $r \in \Omega_r\}$, where $s \in \Omega_s$, $Z \in s$ is a continuous or deterministic variable, $z \in \Omega_Z$, $\delta(z - g_{r,i}(s^{\downarrow(s\setminus\{Z\})}))$ are Dirac delta functions and $p_{r,i}$ are probabilities for all $i = 1, \ldots, n$, and $r \in \Omega_r$, and n is a positive integer. Here, we are assuming that Z is a weighted sum of functions $g_{r,i}(s^{\downarrow(s\setminus\{Z\})})$ of the other continuous variables in s, weighted by $p_{r,i}$, and that the nature of the functions and weights may depend on $r \in \Omega_r$, and/or on some latent index i.

Suppose X is a deterministic variable with continuous parent Z, and suppose that the deterministic relationship is $X = Z^2$. This conditional distribution is represented by the Dirac potential $\xi(z, x) = \delta(x - z^2)$ for $\{Z, X\}$. In this case, $n = 1$, and $r = \varnothing$.

A more general example of a Dirac potential for $\{Z, X\}$ is $\xi(z, x) = (\frac{1}{2})\,\delta(x - z) + (\frac{1}{2})\,\delta(x - 1)$. Here, X is a continuous variable with continuous parent Z. As argued before, we can interpret the value $\xi(x, x)$ as $\frac{1}{2}(m)$, and the value $\xi(1, x)$ as $\frac{1}{2}(m)$. All other values are equal to zero. The conditional distribution of X is as follows: $X = Z$ with probability $\frac{1}{2}$, and $X = 1$ with probability $\frac{1}{2}$. Notice that X is not deterministic since the variances of its conditional distributions are not zeros.

Continuous Potentials. Both density and Dirac potentials are special instances of a broader class of potentials called *continuous* potentials. Suppose $x \subseteq V$ is such that it contains a continuous or deterministic variable. Then a *continuous potential* ξ for x is a function $\xi\colon \Omega_x \to \mathbb{R}^+(d)$. For example, consider a continuous variable X with a mixed distribution: a probability of 0.5 at $X = 1$, and a probability density of 0.5 f, where f is a PDF. This mixed distribution can be represented by a continuous potential ξ for $\{X\}$ as follows: $\xi(x) = 0.5\,\delta(x-1) + 0.5\,f(x)$. Notice that $\int \xi(x)\,dx = 0.5\int \delta(x-1)\,dx + 0.5\int f(x)\,dx = 0.5 + 0.5 = 1$.

Consider the BN in Fig. 1. A is discrete (with two states, a_1 and a_2), Z is continuous, and X is deterministic. Let α denote the discrete potential for $\{A\}$. Then $\alpha(a_1) = 0.5$, $\alpha(a_2) = 0.5$. Let ζ denote the density potential for $\{Z\}$. Then $\zeta(z) = f(z)$. Let ξ denote the Dirac potential for $\{A, Z, X\}$. Then $\xi(a_1, z, x) = \delta(x - z)$, and $\xi(a_2, z, x) = \delta(x - 1)$.

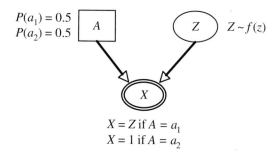

$$X = Z \text{ if } A = a_1$$
$$X = 1 \text{ if } A = a_2$$

Fig. 1. A hybrid BN with a discrete, a continuous, and a deterministic variable

Mixed Potentials. In reasoning with hybrid models, we need to define mixed potentials. A mixed potential has two parts, the first part is a discrete potential and the second part is a continuous potential. Formally, suppose α is a discrete potential for r. Then a mixed potential representation of α is $\mu_1 = (\alpha, \iota)$ for r, where ι denotes the *identity* potential for the empty set, $\iota(\blacklozenge) = 1$. Suppose ζ is a continuous potential for s. Then, a mixed potential representation of ζ is $\mu_2 = (\iota, \zeta)$ for s. Mixed potentials can have non-vacuous discrete and continuous parts. Thus $\mu_3 = (\alpha, \zeta)$ is a mixed potential for $r \cup s$. Such a mixed potential would be the result of combining μ_1 and μ_2, which we will define next. The main idea behind mixed potentials is to represent the nature (discrete or continuous) of potentials.

Combination of Potentials. Suppose α is a discrete or continuous potential for some subset a of variables and β is a discrete or continuous potential for b. Then the *combination* of α and β, denoted by $\alpha \otimes \beta$, is the potential for $a \cup b$ obtained from α and β by pointwise multiplication, i.e.,

$$(\alpha \otimes \beta)(x) = \alpha(x^{\downarrow a}) \, \beta(x^{\downarrow b}) \text{ for all } x \in \Omega_{a \cup b}. \tag{2.1}$$

If α and β are both discrete potentials, then $\alpha \otimes \beta$ is a discrete potential, and if α and β are both continuous potentials, then $\alpha \otimes \beta$ is a continuous potential. The definition of combination in (2.1) is valid also if α is discrete and β is continuous and vice-versa, and will be used when we define marginalization of mixed potentials. However, the nature of the potential $\alpha \otimes \beta$ when α is discrete and β is continuous (or vice-versa) will not arise in the combination operation since we will used mixed potentials to represent the potentials, and as we will see, combination of mixed potentials avoids such combinations.

The identity potential ι_r for r has the property that given any potential ξ for $s \supseteq r$, $\xi \otimes \iota_r = \xi$. If $r = \varnothing$, then we will let ι denote $\iota_\varnothing$.

Combination of Mixed Potentials. Suppose $\mu_1 = (\alpha_1, \zeta_1)$, and $\mu_2 = (\alpha_2, \zeta_2)$ are two mixed potentials with discrete parts α_1 for r_1 and α_2 for r_2, respectively, and continuous parts ζ_1 for s_1 and ζ_2 for s_2, respectively. Then, the combination $\mu_1 \otimes \mu_2$ is a mixed potential for $r_1 \cup s_1 \cup r_2 \cup s_2$ given by

$$\mu_1 \otimes \mu_2 = (\alpha_1 \otimes \alpha_2, \zeta_1 \otimes \zeta_2). \tag{2.2}$$

Since $\alpha_1 \otimes \alpha_2$ is a discrete potential and $\zeta_1 \otimes \zeta_2$ is a continuous potential, the definition of combination of mixed potentials in (2.2) is consistent with the definition of mixed potentials.

If $\mu_1 = (\alpha, \iota)$ represents the discrete potential α for r, and $\mu_2 = (\iota, \zeta)$ represents the continuous potential for s, then $\mu_1 \otimes \mu_2 = (\alpha, \zeta)$ is a mixed potential for $r \cup s$.

Since combination is pointwise multiplication, and multiplication is commutative, combination of potentials (discrete or continuous) is commutative ($\alpha \otimes \beta = \beta \otimes \alpha$) and associative (($\alpha \otimes \beta) \otimes \gamma = \alpha \otimes (\beta \otimes \gamma)$). Since the combination of mixed potentials is defined in terms of combination of discrete and continuous potentials, each of which is commutative and associative, combination of mixed potentials is also commutative and associative.

Marginalization of Potentials. The definition of marginalization depends on whether the variable being marginalized is discrete or continuous. We marginalize discrete variables by addition, and continuous variables by integration. Integration of potentials containing Dirac delta functions is done using the properties of Dirac delta functions. Also, after marginalization, the nature of a potential could change, e.g., from continuous to discrete (if the domain of the marginalized potential contains only discrete variables) and from discrete to continuous (if the domain of the marginalized potential contains only continuous variables). We will make this more precise when we define marginalization of mixed potentials.

Suppose α is a discrete or continuous potential for a, and suppose X is a discrete variable in a. Then the *marginal* of α by deleting X, denoted by α^{-X}, is the potential for $a \setminus \{X\}$ obtained from α by addition over the states of X, i.e.,

$$\alpha^{-X}(\mathbf{y}) = \Sigma\{\alpha(x, \mathbf{y}) \mid x \in \Omega_X\} \text{ for all } \mathbf{y} \in \Omega_{a \setminus \{X\}}. \tag{2.3}$$

If X is a continuous variable in a, then the marginal of α by deleting X is obtained by integration over the state space of X, i.e.,

$$\alpha^{-X}(\mathbf{y}) = \int \alpha(x, \mathbf{y}) \, dx \text{ for all } \mathbf{y} \in \Omega_{a \setminus \{X\}}. \tag{2.4}$$

If α contains Dirac delta functions, then we have to use the properties of Dirac delta functions in doing the integration.

If ξ is a discrete or continuous potential for $\{X\} \cup pa(X)$ representing the conditional distribution for X in a BN, then ξ^{-X} is an identity potential for $pa(A)$.

If we marginalize a discrete or continuous potential by deleting two (or more) variables from its domain, then the order in which the variables are deleted does not matter, i.e., $(\alpha^{-A})^{-B} = (\alpha^{-B})^{-A} = \alpha^{-\{A, B\}}$.

If α is a discrete or continuous potential for a, β is a discrete or continuous potential for b, $A \in a$, and $A \notin b$, then $(\alpha \otimes \beta)^{-A} = (\alpha^{-A}) \otimes \beta$. This is a key property of combination and marginalization that allows local computation [Shenoy and Shafer 1990]. We call this property *local computation*.

Marginalization of Mixed Potentials. Mixed potentials allow us to represent the nature of potentials, and marginalization of mixed potentials allows us to represent the nature of the marginal. Suppose $\mu = (\alpha, \zeta)$ is a mixed potential for $r \cup s$ with discrete part α for r, and continuous part ζ for s. Let C denote the set of continuous variables, and let D denote the set of discrete variables. The marginal of μ by deleting $X \in r \cup s$, denoted by μ^{-X}, is defined as follows.

$$
\mu^{-X} = \begin{cases}
(\alpha^{-X}, \zeta), & \text{if } X \in r,\ X \notin s,\ \text{and } r\backslash\{X\} \not\subset C, & (2.5) \\
(\iota, \alpha^{-X} \otimes \zeta), & \text{if } X \in r,\ X \notin s,\ \text{and } r\backslash\{X\} \subset C, & (2.6) \\
(\alpha, \zeta^{-X}), & \text{if } X \notin r,\ X \in s,\ \text{and } s\backslash\{X\} \not\subset D, & (2.7) \\
(\alpha \otimes \zeta^{-X}, \iota), & \text{if } X \notin r,\ X \in s,\ \text{and } s\backslash\{X\} \subset D, & (2.8) \\
((\alpha \otimes \zeta)^{-X}, \iota), & \text{if } X \in r,\ X \in s,\ \text{and } (r \cup s)\backslash\{X\} \subset D,\ \text{and} & (2.9) \\
(\iota, (\alpha \otimes \zeta)^{-X}), & \text{if } X \in r,\ X \in s,\ \text{and } (r \cup s)\backslash\{X\} \not\subset D. & (2.10)
\end{cases}
$$

Some comments about the definition of marginalization of mixed potentials are as follows. First, if the variable being deleted belongs only to one part (discrete or continuous, as in cases (2.5)–(2.8)), then the local computation property allow us to delete the variable from that part only leaving the other part unchanged. If the variable being deleted belongs to both parts (as in cases (2.9)–(2.10)), then we first need to combine the two parts before deleting the variable. Second, when we have only continuous variables left in a discrete potential after marginalization, we move the potential to the continuous part (2.6), and when we only have discrete variables left, we move the potential to the discrete part (2.8), otherwise we don't change the nature of the marginalized potentials ((2.5) and (2.7)). In cases (2.9)–(2.10), when we have to combine the discrete and continuous potentials before marginalizing X, if only discrete variables are left, then we have to classify it as a discrete potential (2.9), and if we have only continuous variables left, then we have to classify it as a continuous potential (2.10). However, if it has discrete and continuous variables, it could be classified as either discrete or continuous, and the definition above has chosen to classify it as continuous (2.10). It makes no difference one way or the other.

Division of Potentials. The Shenoy-Shafer [27] architecture requires only the combination and marginalization operations. However, at the end of the propagation, we need to normalize the potentials, and this involves division. Divisions are also involved in doing arc reversals [1].

Suppose ρ is a discrete or continuous potential for r, and suppose $X \in r$. Then the *division* of ρ by ρ^{-X}, denoted by $\rho \oslash (\rho^{-X})$, is the potential for r obtained by pointwise division of ρ by ρ^{-X}, i.e.,

$$(\rho \oslash (\rho^{-X}))(x, y) = \rho(x, y)/\rho^{-X}(y) \tag{2.11}$$

for all $y \in \Omega_{r \setminus \{X\}}$ and $x \in \Omega_X$. Notice that if $\rho^{-X}(y) = 0$, then $\rho(x, y) = 0$. In this case, we will simply define 0/0 as 0. Also, notice that $(\rho \oslash (\rho^{-X})) \otimes \rho^{-X} = \rho$.

Suppose ξ is a discrete or continuous potential for $\{X\}$ representing the unnormalized posterior marginal for X. To normalize ξ, we divide ξ by ξ^{-X}. Thus the normalized posterior marginal for X is $\xi \oslash (\xi^{-X})$. The value $\xi^{-X}(\blacklozenge)$ represents the probability of the evidence, and is the same regardless of variable X for which we are computing the marginal.

3 Some Illustrative Examples

In this section, we illustrate the extended Shenoy-Shafer architecture using several small examples. More examples can be found in [28].

Example 1 (*Transformation of variables*). Consider a BN with continuous variable Y and deterministic variable Z as shown in Fig. 2. Notice that the function defining the deterministic variable is not invertible.

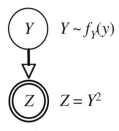

Fig. 2. A BN with a deterministic variable with a non-invertible, non-linear function

Let ψ and ζ_1 denote the mixed potentials for $\{Y\}$ and $\{Y, Z\}$, respectively. Then,

$$\psi(y) = (1, f_Y(y)) \tag{4.1}$$

$$\zeta_1(y, z) = (1, \delta(z - y^2)) \tag{4.2}$$

To find the prior marginal distribution of Z, first we combine ψ and ζ_1, and then we marginalize Y from the combination.

$$(\psi \otimes \zeta_1)(y, z) = (1, f_Y(y)\, \delta(z - y^2)) \tag{4.3}$$

$$((\psi \otimes \zeta_1)^{-Y})(z) = (1, \int f_Y(y)\, \delta(z - y^2)\, dy) = (1, (1/(2\sqrt{z}))(f_Y(\sqrt{z}) + f_Y(-\sqrt{z})))$$
$$= (1, f_Z(z)) \text{ for } z > 0, \text{ where } f_Z(z) = (1/(2\sqrt{z}))(f_Y(\sqrt{z}) + f_Y(-\sqrt{z})) \tag{4.4}$$

The result in (4.4) follows from (2.7) and properties (iii) and (iv) of Dirac delta functions. Now suppose we observe $Z = c$, where c is a constant such that $f_Z(c) > 0$, i.e., $c > 0$ and $f_Y(\sqrt{c}) > 0$ or $f_Y(-\sqrt{c}) > 0$ or both. This observation is represented by the mixed potential for Z, $\zeta_2(z) = (1, \delta(z - c))$. Then, the un-normalized posterior marginal distribution of Y is computed as follows:

$$((\zeta_1 \otimes \zeta_2)^{-Z})(y) = (1, \int \delta(z - y^2)\, \delta(z - c))\, dz) = (1, \delta(c - y^2)) = (1, \delta(y^2 - c)) \quad (4.5)$$

$$(\psi \otimes (\zeta_1 \otimes \zeta_2)^{-Z})(y) = (1, f_Y(y)\, \delta(y^2 - c))$$

$$= (1, f_Y(y)\, (\delta(y - \sqrt{c}) + \delta(y + \sqrt{c}))/(2\sqrt{c}))$$

$$= (1, (f_Y(\sqrt{c})\, \delta(y - \sqrt{c}) + f_Y(-\sqrt{c})\, \delta(y + \sqrt{c}))/(2\sqrt{c})) \quad (4.6)$$

The normalization constant is $(f_Y(\sqrt{c}) + f_Y(-\sqrt{c}))/(2\sqrt{c})$. Therefore the normalized posterior distribution of Y is $(f_Y(\sqrt{c})\delta(y - \sqrt{c}) + f_Y(-\sqrt{c})\, \delta(y + \sqrt{c}))/ (f_Y(\sqrt{c}) + f_Y(-\sqrt{c}))$, i.e., $Y = \sqrt{c}$ with probability $f_Y(\sqrt{c})/(f_Y(\sqrt{c}) + f_Y(-\sqrt{c}))$, and $Y = -\sqrt{c}$ with probability $f_Y(-\sqrt{c})/(f_Y(\sqrt{c}) + f_Y(-\sqrt{c}))$.

Example 2 (*Mixed distributions*). Consider the hybrid BN shown in Fig. 1 with three variables. A is discrete with state space $\Omega_Y = \{a_1, a_2\}$, Z is continuous, and X is deterministic. What is the prior marginal distribution of X? Suppose we observe $X = 1$. What is the posterior marginal distribution of A?

Let α, ζ, and ξ_1 denote the mixed potentials for $\{A\}$, $\{Z\}$, and $\{A, Z, X\}$, respectively. Then:

$$\alpha(a_1) = (0.5, 1), \ \alpha(a_2) = (0.5, 1); \quad (4.7)$$

$$\zeta(z) = (1, f_Z(z)); \quad (4.8)$$

$$\xi_1(a_1, z, x) = (1, \delta(x - z)), \ \xi_1(a_2, z, x) = (1, \delta(x - 1)). \quad (4.9)$$

The prior marginal distribution of X is given by $(\alpha \otimes \zeta \otimes \xi_1)^{-\{A, Z\}} = ((\alpha \otimes \xi_1)^{-A} \otimes \zeta)^{-Z}$.

$$(\alpha \otimes \xi_1)(a_1, z, x) = (0.5, \delta(x - z)), \ (\alpha \otimes \xi_1)(a_2, z, x) = (0.5, \delta(x - 1)); \quad (4.10)$$

$$((\alpha \otimes \xi_1)^{-A})(z, x) = (1, 0.5\, \delta(x - z) + 0.5\, \delta(x - 1)); \quad (4.11)$$

$$((\alpha \otimes \xi_1)^{-A} \otimes \zeta)(z, x) = (1, 0.5\, \delta(x - z)\, f_Z(z) + 0.5\, \delta(x - 1)\, f_Z(z)); \quad (4.12)$$

$$(((\alpha \otimes \xi_1)^{-A} \otimes \zeta)^{-Z})(x) = (1, \int 0.5\, \delta(x - z)\, f_Z(z)\, dz + 0.5\, \delta(x - 1) \int f(z)\, dz)$$
$$= (1, 0.5\, f_Z(x) + 0.5\, \delta(x - 1)). \quad (4.13)$$

Thus the prior marginal distribution of X is mixed with PDF $0.5\, f_Z(x)$ and a mass of 0.5 at $X = 1$. (4.11) results from use of (2.9) since Y is in the domain of discrete and continuous parts. (4.13) follows from (2.7).

Let ξ_2 denote the observation $X = 1$. Thus, $\xi_2(x) = (1, \delta(x - 1))$. The (unnormalized) posterior marginal of A is given by $(\alpha \otimes \zeta \otimes \xi_1 \otimes \xi_2)^{-\{Z, X\}} = \alpha \otimes (\zeta \otimes (\xi_1 \otimes \xi_2)^{-X})^{-Z}$.

$$(\xi_1 \otimes \xi_2)(a_1, z, x) = (1, \delta(x - z)\,\delta(x - 1)),$$

$$(\xi_1 \otimes \xi_2)(a_2, z, x) = (1, \delta(x - 1)\,\delta(x - 1)) = (1, \delta(x - 1)); \qquad (4.14)$$

$$(\xi_1 \otimes \xi_2)^{-X}(a_1, z) = (1, \textstyle\int \delta(x - z)\,\delta(x - 1)\,dx) = (1, \delta(1 - z)) = (1, \delta(z - 1)),$$

$$(\xi_1 \otimes \xi_2)^{-X}(a_2, z) = (1, \textstyle\int \delta(x - 1)\,dx) = (1, 1); \qquad (4.15)$$

$$(\zeta \otimes (\xi_1 \otimes \xi_2)^{-X})(a_1, z) = (1, \delta(z - 1)\,f_Z(z)),$$

$$((\zeta \otimes (\xi_1 \otimes \xi_2)^{-X})(a_2, z) = (1, f_Z(z)); \qquad (4.16)$$

$$(\zeta \otimes (\xi_1 \otimes \xi_2)^{-X})^{-Z}(a_1) = (\textstyle\int \delta(z - 1)\,f_Z(z)\,dz, 1) = (f_Z(1), 1),$$

$$((\zeta \otimes (\xi_1 \otimes \xi_2)^{-X})^{-Z}(a_2) = (\textstyle\int f(z)\,dz, 1) = (1, 1); \qquad (4.17)$$

$$(\alpha \otimes (\zeta \otimes (\xi_1 \otimes \xi_2)^{-X})^{-Z})(a_1) = (0.5\,f_Z(1), 1),$$

$$(\alpha \otimes (\zeta \otimes (\xi_1 \otimes \xi_2)^{-X})^{-Z})(a_2) = (0.5, 1). \qquad (4.18)$$

Notice that the un-normalized posterior marginal for A is in units of density for $A = a_1$, and in units of probability for $A = a_2$. Thus, after normalization, the posterior probability of a_1 is 0, and the posterior probability of a_2 is 1.

Example 3 (*Discrete variable with a continuous parent*). Consider the hybrid BN consisting of a continuous variable Y, a discrete variable A, and a deterministic variable X as shown in Fig. 3. A is an indicator variable with states $\{a_1, a_2\}$ such that $A = a_1$ if $0 < Y \le 0.5$, and $A = a_2$ if $0.5 < Y < 1$. What is the prior marginal distribution of X? If we observe $X = 0.25$, what is the posterior marginal distribution of Y?

Let ψ, α, and ξ_1 denote the mixed potentials for $\{Y\}$, $\{Y, A\}$, and $\{Y, A, X\}$, respectively. The *Heaviside* function $H(\cdot)$ is: $H(y) = 0$ if $y < 0$, and $= 1$ if $y > 0$.

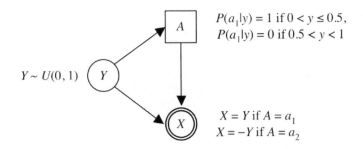

Fig. 3. A hybrid BN with a discrete variable that has a continuous parent

$$\psi(y) = (1, f_Y(y)), \text{ where } f_Y(y) = 1 \text{ if } 0 < y < 1, = 0 \text{ otherwise}; \qquad (4.19)$$

$$\alpha(a_1, y) = (H(y) - H(y - 0.5), 1),$$

$$\alpha(a_2, y) = (H(y - 0.5) - H(y - 1), 1); \qquad (4.20)$$

$$\xi_1(a_1, y, x) = (1, \delta(x - y)),$$

$$\xi_1(a_2, y, x) = (1, \delta(x + y)). \tag{4.21}$$

To find the marginal distribution of X, first we combine α and ξ_1 and marginalize A from the combination, next we combine the result with ψ and marginalize Y from the combination.

$$(\alpha \otimes \xi_1)(a_1, y, x) = (H(y) - H(y - 0.5), \delta(x - y)),$$

$$(\alpha \otimes \xi_1)(a_2, y, x) = (H(y - 0.5) - H(y - 1), \delta(x + y)); \tag{4.22}$$

$$(\alpha \otimes \xi_1)^{-A}(y, x) = (1, (H(y) - H(y - 0.5)) \, \delta(x - y) +$$
$$(H(y - 0.5) - H(y - 1)) \, \delta(x + y)); \tag{4.23}$$

$$(((\alpha \otimes \xi_1)^{-A}) \otimes \psi)(y, x) = (1, f_Y(y)[((H(y) - H(y - 0.5)) \, \delta(x - y) +$$
$$(H(y - 0.5) - H(y - 1)) \, \delta(x + y))]); \tag{4.24}$$

$$(((\alpha \otimes \xi_1)^{-A}) \otimes \psi)^{-Y}(x) = (1, f_Y(x)((H(x) - H(x - 0.5))$$
$$+ f_Y(-x)(H(-x - 0.5) - H(-x - 1))$$
$$= (1, H(x) - H(x - 0.5) + H(-x - 0.5) - H(-x - 1)). \tag{4.25}$$

Thus, the prior marginal distribution of X is uniform in the interval $(-1, -0.5) \cup (0, 0.5)$. Let ξ_2 be the mixed potential denoting the observation that $X = 0.25$. Thus, $\xi_2(x) = (1, \delta(x - 0.25))$. The (unnormalized) posterior marginal of Y is given by $(\xi_1 \otimes (\xi_2 \otimes \alpha))^{-\{A, X\}}) \otimes \psi$.

$$(\xi_2 \otimes \alpha)(a_1, y, x) = (H(y) - H(y - 0.5), \delta(x - 0.25)),$$

$$(\xi_2 \otimes \alpha)(a_2, y, x) = (H(y - 0.5) - H(y - 1), \delta(x - 0.25)); \tag{4.26}$$

$$(\xi_1 \otimes (\xi_2 \otimes \alpha))(a_1, y, x) = (H(y) - H(y - 0.5), \delta(x - 0.25) \, \delta(x - y)),$$

$$(\xi_1 \otimes (\xi_2 \otimes \alpha))(a_2, y, x) = (H(y - 0.5) - H(y - 1), \delta(x - 0.25) \, \delta(x + y)); \tag{4.27}$$

$$((\xi_1 \otimes (\xi_2 \otimes \alpha))^{-\{A, X\}})(y) = (1, [(H(y) - H(y - 0.5)) \int \delta(x - 0.25) \, \delta(x - y) \, dx]$$
$$+ [(H(y - 0.5) - H(y - 1)) \int \delta(x - 0.25) \, \delta(x + y) \, dx]$$
$$= (1, (H(y) - H(y - 0.5)) \, \delta(y - 0.25)); \tag{4.28}$$

$$(((\xi_1 \otimes (\xi_2 \otimes \alpha))^{-\{A, X\}}) \otimes \psi)(y) = (1, f_Y(y)([(H(y) - H(y - 0.5)) \, \delta(y - 0.25)]))$$
$$= (1, \delta(y - 0.25)). \tag{4.29}$$

The posterior marginal for Y is $Y = 0.25$ with probability 1.

4 Summary and Discussion

We have described a generalization of the Shenoy-Shafer architecture for discrete BNs so it applies to hybrid BNs with deterministic variables. We use Dirac delta functions to represent conditionals of deterministic variables, and observations of continuous variables. We use mixed potentials to keep track of the discrete and continuous nature of potentials. Marginalization of discrete variables is using addition and marginalization of continuous variables is by integration. We define marginalization of mixed potentials to keep track of the nature of marginalized potentials.

We have ignored the computational problem of integration of density potentials. In some cases, e.g., Gaussian density functions, there does not exist a closed form solution of the integral of the Gaussian density. We assume that we can somehow work around such problems by approximating such density functions by mixtures of truncated exponentials [16] or mixtures of polynomials [29]. In any case, this needs further investigation.

Acknowledgments. We are grateful to Barry Cobb for many discussions.

References

1. Cinicioglu, E.N., Shenoy, P.P.: Arc Reversals in Hybrid Bayesian Networks with Deterministic Variables. Int. J. Approx. Reas. 50(5), 763–777 (2009)
2. Cobb, B.R., Shenoy, P.P.: Hybrid Bayesian Networks with Linear Deterministic Variables. In: Bacchus, F., Jaakkola, T. (eds.) UAI 2005, pp. 136–144. AUAI Publishers, Corvallis (2005)
3. Cobb, B.R., Shenoy, P.P.: Nonlinear Deterministic Relationships in Bayesian networks. In: Godo, L. (ed.) ECSQARU 2005. LNCS, vol. 3571, pp. 27–38. Springer, Heidelberg (2005)
4. Cobb, B.R., Shenoy, P.P.: Decision Making with Hybrid Influence Diagrams Using Mixtures of Truncated Exponentials. Eur. J. Oper. Res. 186(1), 261–275 (2008)
5. Cobb, B.R., Shenoy, P.P., Rumi, R.: Approximating Probability Density Functions in Hybrid Bayesian Networks with Mixtures of Truncated Exponentials. Stat. & Comp. 16(3), 293–308 (2006)
6. Dirac, P.A.M.: The Physical Interpretation of the Quantum Dynamics. Proc. Royal Soc. London. Series A 113(765), 621–641 (1927)
7. Dirac, P.A.M.: The Principles of Quantum Mechanics, 4th edn. Oxford Univ. Press, London (1958)
8. Hoskins, R.F.: Generalised Functions. Ellis Horwood, Chichester (1979)
9. Howard, R.A., Matheson, J.E.: Influence Diagrams. In: Howard, R.A., Matheson, J.E. (eds.) The Principles and Applications of Decision Analysis, vol. 2, pp. 719–762. Strategic Decisions Group, Menlo Park (1984)
10. Kanwal, R.P.: Generalized Functions: Theory and Technique, 2nd edn. Birkhäuser, Boston (1998)
11. Kenley, C.R.: Influence Diagram Models with Continuous Variables. PhD thesis, Dept. of Engg.-Econ. Sys. Stanford University, Stanford (1986)
12. Khuri, A.I.: Applications of Dirac's Delta Function in Statistics. Int. J. Math. Edu. in Sci. & Tech. 32(2), 185–195 (2004)
13. Lauritzen, S.L., Spiegelhalter, D.J.: Local Computations with Probabilities on Graphical Structures and their Application to Expert Systems. J. Royal Stat. Soc. B 50(2), 157–224 (1988)
14. Lauritzen, S.L., Jensen, F.: Stable Local Computation with Conditional Gaussian Distributions. Stat. & Comp. 11, 191–203 (2001)
15. Lerner, U.N.: Hybrid Bayesian Networks for Reasoning About Complex Systems. PhD thesis, Dept. of Comp. Sci. Stanford University, Stanford (2002)

16. Moral, S., Rumi, R., Salmeron, A.: Mixtures of Truncated Exponentials in Hybrid Bayesian Networks. In: Benferhat, S., Besnard, P. (eds.) ECSQARU 2001. LNCS (LNAI), vol. 2143, pp. 156–167. Springer, Heidelberg (2001)
17. Murphy, K.: A Variational Approximation for Bayesian Networks with Discrete and Continuous Latent Variables. In: Laskey, K.B., Prade, H. (eds.) UAI 1999, pp. 457–466. Morgan Kaufmann, San Francisco (1999)
18. Olmsted, S.M.: On Representing and Solving Decision Problems. PhD thesis, Dept. of Engg.-Econ. Sys. Stanford University, Stanford (1983)
19. Pearl, J.: Fusion, Propagation and Structuring in Belief Networks. Art. Int. 29, 241–288 (1986)
20. Poland III, W.B.: Decision Analysis with Continuous and Discrete Variables: A Mixture Distribution Approach. PhD thesis, Dept. of Engg.-Econ. Sys., Stanford University, Stanford (1994)
21. Saichev, A.I., Woyczyński, W.A.: Distributions in the Physical and Engineering Sciences, vol. 1. Birkhäuser, Boston (1997)
22. Shachter, R.D.: Evaluating Influence Diagrams. Oper. Res. 34(6), 871–882 (1986)
23. Shachter, R.D., Kenley, C.R.: Gaussian Influence Diagrams. Mgmt. Sci. 35(5), 527–550 (1989)
24. Shenoy, P.P.: Valuation-Based Systems for Bayesian Decision Analysis. Oper. Res. 40(3), 463–484 (1992)
25. Shenoy, P.P.: A New Method for Representing and Solving Bayesian Decision Problems. In: Hand, D.J. (ed.) Artificial Intelligence Frontiers in Statistics: AI and Statistics III, pp. 119–138. Chapman & Hall, London (1993)
26. Shenoy, P.P.: Inference in Hybrid Bayesian Networks Using Mixtures of Gaussians. In: Dechter, R., Richardson, T. (eds.) UAI 2006, pp. 428–436. AUAI Press, Corvallis (2006)
27. Shenoy, P.P., Shafer, G.: Axioms for Probability and Belief-Function Propagation. In: Shachter, R.D., Levitt, T.S., Lemmer, J.F., Kanal, L.N. (eds.) UAI 1988, vol. 4, pp. 169–198. North-Holland, Amsterdam (1990)
28. Shenoy, P.P., West, J.C.: Inference in Hybrid Bayesian Networks with Deterministic Variables. Working Paper No. 318. School of Business, University of Kansas (2009)
29. Shenoy, P.P., West, J.C.: Inference in Hybrid Bayesian Networks with Mixtures of Polynomials. Working Paper No. 321. School of Business, University of Kansas (2009)

Extracting the Core of a Persuasion Dialog to Evaluate Its Quality

Leila Amgoud and Florence Dupin de Saint-Cyr

IRIT-CNRS, 118, route de Narbonne,
31062 Toulouse Cedex 4 France
{amgoud,bannay}@irit.fr

Abstract. In persuasion dialogs, agents exchange arguments on a subject on which they disagree. Thus, each agent tries to persuade the others to change their minds. Several systems, grounded on argumentation theory, have been proposed in the literature for modeling persuasion dialogs. It is important to be able to analyze the quality of these dialogs. Hence, *quality criteria* have to be defined in order to perform this analysis.

This paper tackles this important problem and proposes one criterion that concerns the conciseness of a dialog. A dialog is concise if all its moves are relevant and useful in order to reach the same outcome as the original dialog. From a given persuasion dialog, in this paper we compute its corresponding "ideal" dialog. This ideal dialog is concise. A persuasion dialog is thus interesting if it is close to its ideal dialog.

1 Introduction

Persuasion is one of the main types of dialogs encountered in everyday life. A persuasion dialog concerns two (or more) agents who disagree on a state of affairs, and each of them tries to persuade the others to change their minds. For that purpose, agents exchange arguments of different strengths. Several systems have been proposed in the literature for allowing agents to engage in persuasion dialogs (e.g. [1,2,3,4,5,6,7]). A dialog system is built around three main components: i) a *communication language* specifying the locutions that will be used by agents during a dialog for exchanging information, arguments, etc., ii) a *protocol* specifying the set of rules governing the well-definition of dialogs such as who is allowed to say what and when? and iii) agents' strategies which are the different tactics used by agents for selecting their moves at each step in a dialog. All the existing systems allow agents to engage in dialogs that obey to the rules of the protocol. Thus, the only properties that are guaranteed for a generated dialog are those related to the protocol. For instance, one can show that a dialog terminates, the turn shifts equally between agents in that dialog (if such rule is specified by the protocol), agents can refer only to the previous move or are allowed to answer to an early move in the dialog, etc. The properties inherited from a protocol are related to the way the dialog is generated. However, the protocol is not concerned by the *quality* of that dialog. Moreover, it is well-known that under the same protocol, different dialogs on the same subject may be generated. It is important to be able to

C. Sossai and G. Chemello (Eds.): ECSQARU 2009, LNAI 5590, pp. 59–70, 2009.

compare them w.r.t. their quality. Such a comparison may help to refine the protocols and to have more efficient ones.

While there are a lot of works on dialog protocols (eg. [8]), no work is done on defining criteria for evaluating the persuasion dialogs generated under those protocols, except a very preliminary proposal in [9]. The basic idea of that paper is, given a finite persuasion dialog, it can be analyzed w.r.t. three families of criteria. The first family concerns the quality of arguments exchanged in this dialog. The second family checks the behavior of the agents involved in this dialog. The third family concerns the dialog as a whole. In this paper, we are more interested by investigating this third family of quality criteria. We propose a criterion based on the conciseness of the generated dialog. A dialog is concise if all its moves (i.e. the exchanged arguments) are both *relevant* to the subject (i.e. they don't deviate from the subject of the dialog) and *useful* (i.e. they are important to determine the outcome of the dialog). Inspired from works on proof procedures that have been proposed in argumentation theory in order to check whether an argument is accepted or not [10], we compute and characterize a sub-dialog of the original one that is concise. This sub-dialog is considered as *ideal*. The closer the original dialog to its ideal sub-dialog, the better is its quality. All the proofs are in [11].

The paper is organized as follows: Section 2 recalls the basics of argumentation theory. Section 3 presents the basic concepts of a persuasion dialog. Section 4 defines the notions of relevance and usefulness in a dialog. Section 5 presents the concept of ideal dialog founded on an ideal argumentation tree built from the initial dialog.

2   Basics of Argumentation Systems

Argumentation is a reasoning model based on the construction and the comparison of arguments. Arguments are reasons for believing in statements, or for performing actions. In this paper, the origins of arguments are supposed to be unknown. They are denoted by lowercase Greek letters. In [12], an argumentation system is defined by:

Definition 1 (Argumentation system). *An argumentation system is a pair* $\mathsf{AS} = \langle \mathcal{A}, \mathcal{R} \rangle$, *where* $\mathcal{A}$ *is a set of arguments and* $\mathcal{R} \subseteq \mathcal{A} \times \mathcal{A}$ *is an attack relation. We say that an argument* α *attacks an argument* β *iff* $(\alpha, \beta) \in \mathcal{R}$.

Note that to each argumentation system is associated a directed graph whose nodes are the different arguments, and the arcs represent the attack relation between them.

Since arguments are conflicting, it is important to know which arguments are acceptable. For that purpose, in [12], different *acceptability semantics* have been proposed. In this paper, we consider the case of *grounded* semantics. Remaining semantics are left for future research.

Definition 2 (Defense–Grounded extension). *Let* $\mathsf{AS} = \langle \mathcal{A}, \mathcal{R} \rangle$ *and* $\mathcal{B} \subseteq \mathcal{A}$.

- $\mathcal{B}$ *defends an argument* $\alpha \in \mathcal{A}$ *iff* $\forall \, \beta \in \mathcal{A}$, *if* $(\beta, \alpha) \in \mathcal{R}$, *then* $\exists \delta \in \mathcal{B}$ *s.t.* $(\delta, \beta) \in \mathcal{R}$.
- *The* grounded *extension of* AS, *denoted by* $\mathcal{E}$, *is the least fixed point of a function* $\mathcal{F}$ *where* $\mathcal{F}(\mathcal{B}) = \{\alpha \in \mathcal{A} \mid \mathcal{B} \text{ defends } \alpha\}$.

When the argumentation system is finite in the sense that each argument is attacked by a finite number of arguments, $\mathcal{E} = \bigcup_{i>0} \mathcal{F}^i(\emptyset)$.

Now that the acceptability semantics is defined, we can define the status of any argument. As we will see, an argument may have two possible statuses: *accepted* or *rejected*.

Definition 3 (Argument status). *Let* AS $= \langle \mathcal{A}, \mathcal{R} \rangle$ *be an argumentation system, and* $\mathcal{E}$ *its grounded extension. An argument* $\alpha \in \mathcal{A}$ *is* accepted *iff* $\alpha \in \mathcal{E}$, *it is* rejected *otherwise. We denote by* Status(α, AS) *the status of* α *in* AS.

Property 1 ([10]). Let AS $= \langle \mathcal{A}, \mathcal{R} \rangle$, $\mathcal{E}$ *its grounded extension, and* $\alpha \in \mathcal{A}$. *If* $\alpha \in \mathcal{E}$, *then* α *is indirectly defended[1] by non-attacked arguments against all its attackers.*

3   Persuasion Dialogs

This section defines persuasion dialogs in the same spirit as in [1]. A persuasion dialog consists mainly of an exchange of arguments between different agents of the set Ag $=$ $\{a_1, \dots, a_m\}$. The subject of such a dialog is an argument, and its aim is to provide the status of that argument. At the end of the dialog, the argument may be either "accepted" or "rejected", this status is the output of the dialog. In what follows, we assume that agents are *only* allowed to exchange arguments.

Each participating agent is supposed to be able to recognize all elements of $\arg(\mathcal{L})$ and $\mathcal{R}_{\mathcal{L}}$, where $\arg(\mathcal{L})$ is the set of all arguments that may be built from a logical language $\mathcal{L}$ and $\mathcal{R}_{\mathcal{L}}$ is a binary relation that captures all the conflicts that may exist among arguments of $\arg(\mathcal{L})$. Thus, $\mathcal{R}_{\mathcal{L}} \subseteq \arg(\mathcal{L}) \times \arg(\mathcal{L})$. For two arguments $\alpha, \beta \in \arg(\mathcal{L})$, the pair $(\alpha, \beta) \in \mathcal{R}_{\mathcal{L}}$ means that the argument α attacks the argument β. Note that this assumption does not mean at all that an agent is aware of all the arguments. But, it means that agents use the same logical language and the same definitions of arguments and conflict relation.

Definition 4 (Moves). *A move* m *is a triple* $\langle S, H, \alpha \rangle$ *such that:*

- $S \in$ Ag *is the agent that utters the move,* Speaker$(m) = S$
- $H \subseteq$ Ag *is the set of agents to which the move is addressed,* Hearer$(m) = H$
- $\alpha \in \arg(\mathcal{L})$ *is the content of the move,* Content$(m) = \alpha$.

During a dialog several moves may be uttered. Those moves constitute a sequence denoted by $\langle m_1, \dots, m_n \rangle$, where m_1 is the initial move whereas m_n is the final one. The empty sequence is denoted by $\langle \rangle$. These sequences are built under a given protocol. A protocol amounts to define a function that associates to each sequence of moves, a set of valid moves. Several protocols have been proposed in the literature, like for instance [1,6]. In what follows, we don't focus on particular protocols.

Definition 5 (Persuasion dialog). *A* persuasion dialog D *is a non-empty and finite sequence of moves* $\langle m_1, \dots, m_n \rangle$ *s.t. the subject of D is* Subject$(D) =$ Content(m_1), *and the* length *of D, denoted $|D|$, is the number of moves: n. Each sub-sequence* $\langle m_1, \dots, m_i \rangle$ *is a* sub-dialog D^i *of D. We will write also* $D^i \sqsubseteq D$.

---

[1] An argument α is *indirectly defended* by β iff there exists a finite sequence $a_1, \dots, a_{2n+1}$ such that $\alpha = a_1$, $\beta = a_{2n+1}$, and $\forall i \in [\![1, 2n]\!]$, $(a_{i+1}, a_i) \in \mathcal{R}, n \in \mathbf{N}^*$.

To each persuasion dialog, one may associate an argumentation system that will be used to evaluate the status of each argument uttered during it and to compute its output.

Definition 6 (AS of a pers. dialog). *Let $D = \langle m_1, \ldots, m_n \rangle$ be a persuasion dialog. The* argumentation system *of D is the pair* $\mathsf{AS}_D = \langle \mathsf{Args}(D), \mathsf{Confs}(D) \rangle$ *such that:*

- $\mathsf{Args}(D) = \{\mathsf{Content}(m_i) \mid i \in [\![1, n]\!]\}$
- $\mathsf{Confs}(D) = \{(\alpha, \beta) \mid \alpha, \beta \in \mathsf{Args}(D) \text{ and } (\alpha, \beta) \in \mathcal{R}_\mathcal{L}\}$

In other words, $\mathsf{Args}(D)$ and $\mathsf{Confs}(D)$ return respectively, the set of arguments exchanged during the dialog and the different conflicts among those arguments.

Example 1. *Let D_1 be the following persuasion dialog between two agents a_1 and a_2.*
$D_1 = \langle \langle a_1, \{a_2\}, \alpha_1 \rangle, \langle a_2, \{a_1\}, \alpha_2 \rangle, \langle a_1, \{a_2\}, \alpha_3 \rangle, \langle a_1, \{a_2\}, \alpha_4 \rangle, \langle a_2, \{a_1\}, \alpha_1 \rangle \rangle.$
Let us assume that there exist conflicts in $\mathcal{R}_\mathcal{L}$ among some of these arguments. Those conflicts are summarized in the figure below.

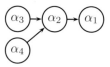

Here, $\mathsf{Args}(D_1) = \{\alpha_1, \alpha_2, \alpha_3, \alpha_4\}$ *and* $\mathsf{Confs}(D_1) = \{(\alpha_2, \alpha_1), (\alpha_3, \alpha_2), (\alpha_4, \alpha_2)\}.$

Property 2. Let $D = \langle m_1, \ldots, m_n \rangle$ be a persuasion dialog. $\forall D^j \sqsubseteq D$, it holds that $\mathsf{Args}(D^j) \subseteq \mathsf{Args}(D)$, and $\mathsf{Confs}(D^j) \subseteq \mathsf{Confs}(D)$.

The output of a dialog is the status of the argument under discussion (i.e., the subject):

Definition 7 (Output of a persuasion dialog). *Let D be a persuasion dialog. The output of D, denoted by $\mathsf{Output}(D)$, is $\mathsf{Status}(\mathsf{Subject}(D), \mathsf{AS}_D)$.*

4   Criteria for Dialog Quality

In this paper, we are interested in evaluating the conciseness of a dialog D which is already generated under a given protocol. This dialog is assumed to be *finite*. Note that this assumption is not too strong since a main property of any protocol is the termination of the dialogs it generates [13]. A consequence of this assumption is that the argumentation system AS_D associated to D is finite as well. In what follows, we propose two criteria that evaluate the importance of the moves that are exchanged in D, then we propose a way to compute the "ideal" dialog that reaches the same outcome as D.

In everyday life, it is very common that agents deviate from the subject of the dialog. The first criterion evaluates to what extent the moves uttered are in relation with the subject of the dialog. This amounts to check whether there exists a path from a move to the subject in the graph of the argumentation system associated to the dialog.

Definition 8 (Relevant and useful move)
Let $D = \langle m_1, \ldots, m_n \rangle$ be a persuasion dialog. A move m_i, $i \in [\![1, n]\!]$, is relevant *to D iff there exists a path (not necessarily directed) from $\mathsf{Content}(m_i)$ to $\mathsf{Subject}(D)$ in the directed graph associated with AS_D. m_i is* useful *iff there exists a directed path from $\mathsf{Content}(m_i)$ to $\mathsf{Subject}(D)$ in this graph.*

Example 2. *Let D_2 be a persuasion dialog. Let* $\texttt{Args}(D_2) = \{\alpha_1, \alpha_3, \beta_1, \beta_2\}$. *The conflicts among the four arguments are depicted in the figure below.*

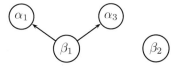

Suppose that $\texttt{Subject}(D_2) = \alpha_1$. *It is clear that the arguments* α_3, β_1 *are relevant, while* β_2 *is irrelevant. Here* β_1 *is useful, but* α_3 *is not.*

Property 3. If a move m is useful in a dialog D, then m is relevant to D.

On the basis of the notion of relevance, one can define a measure that computes the percentage of moves that are relevant in a dialog D. In Example 2, $\texttt{Relevance}(D_2) = $ 3/4. It is clear that the greater this degree is, the better the dialog. When the relevance degree of a dialog is equal to 1, this means that agents did not deviate from the subject of the dialog. The useful moves are moves that have a more direct influence on the status of the subject. However, this does not mean that their presence has an impact on the result of the dialog, i.e., on the status of the subject. The moves that have a real impact on the status of the subject are said "decisive".

Definition 9 (Decisive move). *Let $D = \langle m_1, \ldots, m_n \rangle$ be a persuasion dialog and $\textsf{AS}_D$ its argumentation system. A move m_i ($i = 1, \ldots, n$) is decisive in D iff*

$$\texttt{Status}(\texttt{Subject}(D), \textsf{AS}_D) \neq \texttt{Status}(\texttt{Subject}(D), \textsf{AS}_D \ominus \texttt{Content}(m_i))$$

where $\textsf{AS}_D \ominus \texttt{Content}(m_i) = \langle A', R' \rangle$ *s.t.* $A' = \texttt{Args}(D) \setminus \{\texttt{Content}(m_i)\}$ *and* $R' = \texttt{Confs}(D) \setminus \{(x, \texttt{Content}(m_i)), (\texttt{Content}(m_i), x) \mid x \in \texttt{Args}(D)\}$.

Property 4. If a move m is decisive in a persuasion dialog D then m is useful in D.

From the above property, it follows that each decisive move is also relevant. Note that the converse is not true as shown in the following example.

Example 3. *Let D_3 be a dialog whose subject is α_1 and whose graph is the following:*

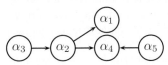

The set $\{\alpha_1, \alpha_3, \alpha_5\}$ is the only grounded extension of $\textsf{AS}_{D_3}$. It is clear that the argument α_4 is relevant to α_1, but it is not decisive for D_3. Indeed, the removal of α_4 will not change the status of α_1 which is accepted.

Example 4. *Let D_4 be a dialog whose subject is α_1, and whose graph is the following:*

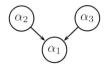

In this example, neither α_2 nor α_3 is decisive in D_4. However, this does not mean that the two arguments should be removed since the status of α_1 depends on at least one of them (they are both useful).

On the basis of the above notion of decisiveness of moves, we can define the degree of decisiveness of the entire dialog as the percentage of moves that are decisive.

5   Computing the Ideal Dialog

As already said, it is very common that dialogs contain redundancies in the sense that some moves are uttered but these are useless for the subject, or have no impact on the output of the dialog. Only a subset of the arguments is necessary to determine the status of the subject. Our aim is to compute the subset that returns exactly the same status for the subject of the dialogue as the whole set of arguments, and that is sufficient to convince that this result holds against any attack available in the initial dialog. That subset will form the "ideal" dialog. In what follows, we will provide a procedure for finding this subset and thus the ideal dialog.

A subset of arguments that will be convenient for our purpose contains those arguments that belong to a proof tree leading to the status of the subject. This is due to the fact that a proof tree contains every necessary argument for obtaining the status of the subject. When the subject is accepted, the proof tree contains defenders of the subject against any attack. When the subject is rejected, the proof tree contains at least every non attacked attacker. Hence, proof trees seem adequate to summarize perfectly the dialog. However, it is important to say that not any proof theory that exists in the literature will lead to the ideal dialog. This is due to the fact that some of them are not concise. In [10], a comparison of proof theories for grounded semantics shows that the one used here is the most concise.

5.1   Canonical Dialogs

Let us define a sub-dialog of a given persuasion dialog D that reaches the same output as D. In [10], a proof procedure that tests the membership of an argument to a grounded extension has been proposed. The basic notions of this procedure are revisited and adapted for the purpose of characterizing canonical dialogs.

Definition 10 (Dialog branch). *Let D be a persuasion dialog and $\mathsf{AS}_D = \langle \mathrm{Args}(D), \mathrm{Confs}(D) \rangle$ its argumentation system. A dialog branch for D is a sequence $\langle \alpha_0, \ldots, \alpha_p \rangle$ of arguments s. t. $\forall i, j \in [\![0, p]\!]$*

1. $\alpha_i \in \mathrm{Args}(D)$
2. $\alpha_0 = \mathrm{Subject}(D)$
3. *if $i \neq 0$ then $(\alpha_i, \alpha_{i-1}) \in \mathrm{Confs}(D)$*
4. *if i and j are even and $i \neq j$ then $\alpha_i \neq \alpha_j$*
5. *if i is even and $i \neq 0$ then $(\alpha_{i-1}, \alpha_i) \notin \mathrm{Confs}(D)$*
6. $\forall \beta \in \mathrm{Args}(D), \langle \alpha_0, \ldots, \alpha_p, \beta \rangle$ *is not a dialog branch.*

Intuitively, a dialog branch is a kind of partial sub-graph of AS_D in which the nodes contains arguments and the arcs represents inverted conflicts. Note that arguments that appear at even levels are not allowed to be repeated. Moreover, these arguments should strictly attack[2] the preceding argument. The last point requires that a branch is maximal. Let us illustrate this notion on examples.

Example 5. *The only dialog branch that can be built from dialog D_2 is depicted below:*

Example 6. *Let D_5 be a persuasion dialog with subject α whose graph is the following:*

The only possible dialog branch associated to this dialog is the following:

Property 5. A dialog branch is non-empty and finite.

This result comes from the definitions of a dialog branch and of a persuasion dialog. Moreover, it is easy to check the following result:

Property 6. For each dialog branch $\langle \alpha_0, ..., \alpha_k \rangle$ of a persuasion dialog D there exists a unique directed path $\langle \alpha_k, \alpha_{k-1}, ..., \alpha_0 \rangle$ of same length[3] (k) in the directed graph associated to AS_D.

In what follows, we will show that when a dialog branch is of even-length, then its leaf is not attacked in the original dialog.

Theorem 1. *Let D be a persuasion dialog and $\langle \alpha_0, ... \alpha_p \rangle$ be a given dialog branch of D. If p is even, then $\nexists \beta \in \mathtt{Args}(D)$ such that $(\beta, \alpha_p) \in \mathtt{Confs}(D)$.*

Let us now introduce the notion of a dialog tree.

Definition 11 (Dialog tree). *Let D be a persuasion dialog and $\mathsf{AS}_D = \langle \mathtt{Args}(D), \mathtt{Confs}(D) \rangle$ its argumentation system. A dialog tree of D, denoted by D^t, is a finite tree whose branches are all the possible dialog branches that can be built from D.*
We denote by AS_{D^t} the argumentation system associated to D^t, $\mathsf{AS}_{D^t} = \langle A^t, C^t \rangle$ s.t. $A^t = \{ \alpha \in \mathtt{Args}(D)$ s.t. α appears in a node of $D^t \}$ and $C^t = \{ (\alpha, \beta) \in \mathtt{Confs}(D)$ s.t. (β, α) is an arc of $D^t \}$.

Hence, a dialog tree is a tree whose root is the subject of the persuasion dialog.

Example 7. *Let us consider D_6 whose subject is α_1 and whose graph is the following:*

---

[2] An argument α strictly attacks an argument β in a argumentation system $\langle A, R \rangle$ iff $(\alpha, \beta) \in R$ and $(\beta, \alpha) \notin R$.

[3] The length of a path is defined by its number of arcs.

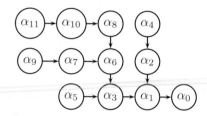

The dialog tree associated to this dialog is depicted below:

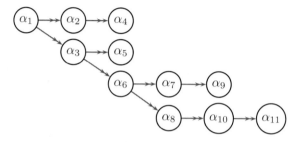

Note that the argument α_0 does not belong to the dialog tree.

Property 7. Each persuasion dialog has exactly one corresponding dialog tree.

An important result states that the status of the subject of the original persuasion dialog D is exactly the same in both argumentation systems AS_D and AS_{D^t} (where AS_{D^t} is the argumentation system whose arguments are all the arguments that appear in the dialog tree D^t and whose attacks are obtained by inverting the arcs between those arguments in D^t).

Theorem 2. *Let D be a persuasion dialog and AS_D its argumentation system. It holds that* $\mathtt{Status}(\mathtt{Subject}(D), \mathsf{AS}_D) = \mathtt{Status}(\mathtt{Subject}(D), \mathsf{AS}_{D^t})$.

In order to compute the status of the subject of a dialog, we can consider the dialog tree as an And/Or tree. A node of an even level is an And node, whereas a node of odd level is an Or one. This distinction between nodes is due to the fact that an argument is accepted if it can be defended against all its attackers. A dialog tree can be decomposed into one or several trees called canonical trees.

Definition 12 (Canonical tree). *Let D be a persuasion dialog, and let D^t its dialog tree. A canonical tree is a subtree of D^t whose root is $\mathtt{Subject}(D)$ and which contains all the arcs starting from an even node and exactly one arc starting from an odd node.*

It is worth noticing that from a dialog tree one may extract at least one canonical tree. Let $D_1^c, \ldots, D_m^c$ denote those canonical trees. We will denote by $\mathsf{AS}_1^c, \ldots, \mathsf{AS}_m^c$ their corresponding argumentation systems. It can be checked that the status of $\mathtt{Subject}(D)$ is not necessarily the same in these different systems.

Example 8. *From the dialog tree of D_6, two canonical trees can be extracted:*

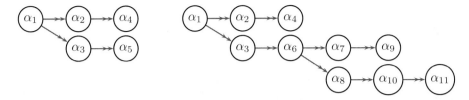

It can be checked that the argument α_1 is accepted in the argumentation system of the canonical tree on the left while it is rejected in the one of the right.

The following result characterizes the status of $\texttt{Subject}(D)$ in the argumentation system $\textsf{AS}_i^c$ associated to a canonical tree D_i^c.

Theorem 3. *Let D be a persuasion dialog, D_i^c a canonical tree and $\textsf{AS}_i^c$ its corresponding argumentation system.*

- *$\texttt{Subject}(D)$ is accepted in $\textsf{AS}_i^c$ iff all the branches of D_i^c are of even-length.*
- *$\texttt{Subject}(D)$ is rejected in $\textsf{AS}_i^c$ iff there exists a branch of D_i^c of odd-length.*

The following result follows immediately from this Theorem and Theorem 1.

Corollary 1. *Let D be a persuasion dialog, D_i^c a canonical tree and $\textsf{AS}_i^c$ its corresponding argumentation system.*
If $\texttt{Subject}(D)$ is accepted in $\textsf{AS}_i^c$, then all the leaves of D_i^c are not attacked in D.

An important result shows the link between the outcome of a dialog D and the outcomes of the different canonical trees.

Theorem 4. *Let D be a persuasion dialog, D_1^c, ..., D_m^c its different canonical trees and $\textsf{AS}_1^c, \ldots, \textsf{AS}_m^c$ their corresponding argumentation systems.*

- *$\texttt{Output}(D)^4$ is accepted iff $\exists\, i \in [\![1, m]\!]$ s.t. $\texttt{Status}(\texttt{Subject}(D), \textsf{AS}_i^c)$ is accepted.*
- *$\texttt{Output}(D)$ is rejected iff $\forall j \in [\![1, m]\!]$, $\texttt{Status}(\texttt{Subject}(D), \textsf{AS}_j^c)$ is rejected.*

This result is of great importance since it shows that a canonical tree whose branches are all of even-length is sufficient to reach the same outcome as the original dialog in case the subject is accepted. When the subject is rejected, the whole dialog tree is necessary to ensure the outcome.

Example 9. *In Example 7, the subject α_1 of dialog D_6 is accepted since there is a canonical tree whose branches are of even length (it is the canonical tree on the left in Example 8). It can also be checked that α_1 is in the grounded extension $\{\alpha_1, \alpha_4, \alpha_5, \alpha_8, \alpha_9, \alpha_{11}\}$ of $\textsf{AS}_D$.*

So far, we have shown how to extract from a graph associated with a dialog its canonical trees. These canonical trees contain only useful (hence relevant) moves:

Theorem 5. *Let D_i^c be a canonical tree of a persuasion dialog D. Any move built on an argument of D_i^c is useful in the dialog D.*

---

[4] Recall that $\texttt{Output}(D) = \texttt{Status}(\texttt{Subject}(D), \textsf{AS}_D)$.

The previous theorem gives an upper bound of the set of moves that can be used to build a canonical dialog, a lower bound is the set of decisive moves.

Theorem 6. *Every argument of a decisive move belongs to the dialog tree and to each canonical dialog.*

The converse is false since many arguments are not decisive, as shown in Example 4. Indeed, there are two attackers that are not decisive but the dialog tree contains both of them (as does the only canonical dialog for this example).

5.2   The Ideal Dialog

In the previous section, we have shown that from each dialog, a dialog tree can be built. This dialog tree contains direct and indirect attackers and defenders of the subject. From this dialog tree, interesting subtrees can be extracted and are called canonical trees. A canonical tree is a subtree containing only particular entire branches of the dialog tree (only one argument in favor of the subject is chosen for attacking an attacker while each argument against a defender is selected). In case the subject of the dialog is accepted it has been proved that there exists at least one canonical tree such that the subject is accepted in its argumentation system. This canonical tree is a candidate for being an ideal tree since it is sufficient to justify the acceptance of the subject against any attack available in the initial dialog. Among all these candidate we define the ideal tree as the smallest one. In the case the subject is rejected in the initial dialog, then the dialog tree contains all the reasons to reject it, hence we propose to consider the dialog tree itself as the only ideal tree.

Definition 13 (ideal trees and dialogs). *If a dialog D has an accepted output*

- *then an ideal tree associated to D is a canonical tree of D in which Subject(D) is accepted and having a minimal number of nodes among all the canonical graphs that also accept Subject(D)*
- *else the ideal tree is the dialog tree of D.*

A dialog using once each argument of an ideal graph is called an ideal dialog.

Example 10.   *An ideal Dialog for Dialog D6 (on the left) has the following graph (on the right):*

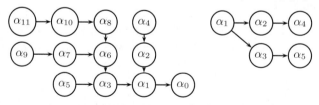

Given the above definition, an ideal dialog contains exactly the same number of moves that the number of nodes of the ideal graph.

Property 8. Given a dialog D whose subject is accepted. An ideal dialog ID for D is the shortest dialog with the same output, and s.t. every argument in favor of the subject in ID (including $Subject(D)$ itself) is defended against any attack (existing in D).

This property ensures that, when the subject is accepted in the initial dialog D, an ideal dialog ID is the more concise dialog that entails an acceptation. In other words, we require that the ideal dialog should contain a set of arguments that sumarize D.

Note that the ideal dialog exists but is not always unique. Here is an example of an argumentation system of a dialog which leads to two ideal trees (hence it will lead to at least two ideal dialogs).

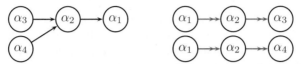

So far, we have formally defined the notion of ideal dialog, and have shown how it is extracted from a persuasion dialog. It is clear that the closer (it terms of set-inclusion of the exchanged arguments) to its ideal version the dialog is, the better the dialog.

6 Conclusion

In this paper, we have proposed three criteria for evaluating the moves of a persuasion dialog with respect to its subject: relevance, usefulness and decisiveness. Relevance only expresses that the argument of the move has a link with the subject (this link is based on the attack relation of the argumentation system). Usefulness is a more stronger relevance since it requires a directed link from the argument of the move to the subject. Decisive moves have a heavier impact on the dialog, since their omission changes the output of the dialog.

Inspired from works on proof theories for grounded semantics in argumentation, we have defined a notion of "ideal dialog". More precisely, we have first defined a dialog tree associated to a given dialog as the graph that contains every possible direct and indirect attackers and defenders of the subject. From this dialog tree, it is then possible to extract sub-trees called "ideal trees" that are sufficient to prove that the subject is accepted or rejected in the original dialog and this, against any possible argument taken from the initial dialog. A dialog is good if it is close to that ideal tree. Ideal dialogs have nice properties with respect to conciseness, namely they contain only useful and relevant arguments for the subject of the dialog. Moreover for every decisive move its argument belongs to all ideal trees.

From the results of this paper, it seems natural that a protocol generates dialogs of good quality if (1) irrelevant and not useful moves are penalized until there is a set of arguments that relate them to the subject (2) adding arguments in favor of the subject that are attacked by already present arguments has no interest (since they do not belong to any ideal tree). By doing so, the generated dialogs are more *concise* (i.e. all the uttered arguments have an impact on the result of the dialog), and more *efficient* (i.e. they are the minimal dialogs that can be built from the information exchanged and that reach the goal of the persuasion).

Note that in our proposal, the order of the arguments has not to be constrained since the generated graph does not take it into account. The only thing that matters in order to obtain a conclusion is the final set of interactions between the exchanged arguments. But the criteria of being relevant to the previous move or at least to a move not too far in the dialog sequence could be taken into account for analyzing dialog quality. Moreover, all the measures already defined in the literature and cited in the introduction could also be used to refine the proposed preference relation on dialogs and finally could help to formalize general properties of protocols in order to generate good dialogs.

Furthermore, it may be the case that from the set of formulas involved in a set of arguments, new arguments may be built. This give birth to a new set of arguments and to a new set of attack relations called complete argumentation system associated to a dialog. Hence, it could be interesting to define dialog trees on the basis of the complete argumentation system then more efficient dialogs could be obtained (but this is not guaranteed). However, some arguments of the complete argumentation system may require the cooperation of the agents. It would mean that in an ideal but practicable dialog, the order of the utterance of the arguments would be constrained by the fact that each agent should be able to build each argument at each step.

References

1. Amgoud, L., Maudet, N., Parsons, S.: Modelling dialogues using argumentation. In: Proc. of the International Conference on Multi-Agent Systems, Boston, MA, pp. 31–38 (2000)
2. Bench-Capon, T.: Persuasion in practical argument using value-based argumentation frameworks. J. of Logic and Computation 13(3), 429–448 (2003)
3. Dunne, P., Bench-Capon, T.: Two party immediate response disputes: Properties and efficiency. Artificial Intelligence 149, 221–250 (2003)
4. Gordon, T.F.: The pleadings game. Artificial Intelligence and Law 2, 239–292 (1993)
5. Parsons, S., McBurney, P.: Games that agents play: A formal framework for dialogues between autonomous agents. J. of Logic, Language and Information 11(3), 315–334 (2002)
6. Prakken, H.: Coherence and flexibility in dialogue games for argumentation. Journal of Logic and Computation 15, 1009–1040 (2005)
7. Zabala, S., Lara, I., Geffner, H.: Beliefs, reasons and moves in a model for argumentative dialogues. In: Proc. 25th Latino-American Conf. on Computer Science (1999)
8. Johnson, M., McBurney, P., Parsons, S.: When are two protocols the same? In: Huget, M.-P. (ed.) Communication in Multiagent Systems. LNCS (LNAI), vol. 2650, pp. 253–268. Springer, Heidelberg (2003)
9. Amgoud, L., Dupin de Saint-Cyr, F.: Measures for persuasion dialogs: a preliminary investigation. In: 2nd Int. Conf. on Computational Models of Argument, pp. 13–24 (2008)
10. Amgoud, L., Cayrol, C.: A reasoning model based on the production of acceptable arguments. Annals of Mathematics and Artificial Intelligence 34, 197–216 (2002)
11. Amgoud, L., Dupin de Saint-Cyr, F.: Extracting the core of a persuasion dialog to evaluate its quality. Technical Report IRIT/RR–2009-10–FR, Toulouse (2009), http://www.irit.fr/recherches/RPDMP/persos/Bannay/publis/ecsqaruwithproof.pdf
12. Dung, P.M.: On the acceptability of arguments and its fundamental role in nonmonotonic reasoning, logic programming and n-person games. Artificial Intelligence Journal 77, 321–357 (1995)
13. Torroni, P.: A study on the termination of negotiation dialogues. In: Proceedings of the first international joint conference on Autonomous agents and multiagent systems, pp. 1223–1230. ACM, New York (2002)

On Revising Argumentation-Based Decision Systems

Leila Amgoud and Srdjan Vesic

Institut de Recherche en Informatique de Toulouse
118, route de Narbonne,
31062 Toulouse Cedex 9 France
{amgoud,vesic}@irit.fr

Abstract. Decision making amounts to define a preorder (usually a complete one) on a set of options. Argumentation has been introduced in decision making analysis. In particular, an argument-based decision system has been proposed recently by Amgoud et al. The system is a variant of Dung's abstract framework. It takes as input a set of options, different arguments and a defeat relation among them, and returns as outputs a status for each option, and a total preorder on the set of options. The status is defined on the basis of the acceptability of their supporting arguments.

The aim of this paper is to study the revision of this decision system in light of a new argument. We will study under which conditions an option may change its status when a new argument is received and under which conditions this new argument is useless. This amounts to study how the acceptability of arguments evolves when the decision system is extended by new arguments.

1 Introduction

Decision making, often viewed as a form of reasoning toward action, has raised the interest of many scholars including economists, psychologists, and computer scientists for a long time. A decision problem amounts to selecting the "best" or sufficiently "good" action(s) that are feasible among different options, given some available information about the current state of the world and the consequences of potential actions. Available information may be incomplete or pervaded with uncertainty. Besides, the goodness of an action is judged by estimating how much its possible consequences fit the preferences of the decision maker.

Argumentation has been introduced in decision making analysis by several researchers only in the last few years (e.g. [2,4,7]). Indeed, in everyday life, decision is often based on arguments and counter-arguments. Argumentation can also be useful for explaining a choice already made. Recently, in [1], a decision model in which the pessimistic decision criterion was articulated in terms of an argumentation process has been proposed. The model is an instantiation of Dung's abstract framework ([6]). It takes as input a set of options, a set of arguments and a defeat relation among arguments. It assigns a status for each

C. Sossai and G. Chemello (Eds.): ECSQARU 2009, LNAI 5590, pp. 71–82, 2009.

option on the basis of the acceptability of its supporting arguments. This paper studies deeply the revision of option status in light of a new argument. This amounts to study how the acceptability of arguments evolves when the decision system is extended by new arguments without computing the whole extensions. All the proofs are in [3].

This paper is organized as follows: Section 2 recalls briefly the decision model proposed in [1]. Section 3 studies the revision of option status when a new argument is received. In section 4 we study the revision of option status under some assumptions on the decision model. The last section concludes.

2   An Argumentation Framework for Decision Making

This section recalls briefly the argument-based framework for decision making that has been proposed in [1].

Let $\mathcal{L}$ denote a logical language. From $\mathcal{L}$, a finite set $\mathcal{O}$ of n distinct *options* is identified. Two kinds of arguments are distinguished: arguments supporting options, called *practical arguments* and arguments supporting beliefs, called *epistemic arguments*. Arguments supporting options are collected in a set $\mathcal{A}_o$ and arguments supporting beliefs are collected in a set $\mathcal{A}_b$ such that $\mathcal{A}_o \cap \mathcal{A}_b = \emptyset$ and $\mathcal{A} = \mathcal{A}_b \cup \mathcal{A}_o$. Note that the structure of arguments is assumed not known. Moreover, arguments in $\mathcal{A}_o$ highlight positive features of their conclusions, i.e., they are *in favor* of their conclusions. Practical arguments are linked to the options they support by a function $\mathcal{H}$ defined as follows:

$$\mathcal{H}: \mathcal{O} \to 2^{\mathcal{A}_o} \text{ s.t. } \forall i, j \text{ if } i \neq j \text{ then } \mathcal{H}(o_i) \cap \mathcal{H}(o_j) = \emptyset \text{ and } \mathcal{A}_o = \bigcup_{i=1}^n \mathcal{H}(o_i)$$

Each practical argument a supports only one option o. We say also that o is the conclusion of the practical argument a, and we write $\texttt{Conc}(a) = o$. Note that there may exist options that are not supported by arguments (i.e., $\mathcal{H}(o) = \emptyset$).

Example 1. Let us assume a set $\mathcal{O} = \{o_1, o_2, o_3\}$ of three options, a set $\mathcal{A}_b = \{b_1, b_2, b_3\}$ of three epistemic arguments, and finally a set $\mathcal{A}_o = \{a_1, a_2, a_3\}$ of three practical arguments. The arguments supporting the different options are summarized in table below.

$$\mathcal{H}(o_1) = \{a_1\}$$
$$\mathcal{H}(o_2) = \{a_2, a_3\}$$
$$\mathcal{H}(o_3) = \emptyset$$

Three binary relations between arguments have been defined. They express the fact that arguments may not have the same strength. The first preference relation, denoted by $\geq_b$, is a partial preorder[1] on the set $\mathcal{A}_b$. The second relation, denoted by $\geq_o$, is a partial preorder on the set $\mathcal{A}_o$. Finally, a third preorder, denoted by $\geq_m$ (m for *mixed* relation), captures the idea that any epistemic argument is stronger then any practical argument. The role of epistemic arguments in a decision problem is to validate or to undermine the beliefs on which practical

---

[1] Recall that a relation is a preorder iff it is *reflexive* and *transitive*.

arguments are built. Indeed, decisions should be made under certain information. Thus, $(\forall a \in \mathcal{A}_b)(\forall a' \in \mathcal{A}_o)\ (a, a') \in \geq_m \wedge (a', a) \notin \geq_m$. Note that $(a, a') \in \geq_x$ with $x \in \{b, o, m\}$ means that a is *at least as good as* a'. In what follows, $>_x$ denotes the strict relation associated with $\geq_x$. It is defined as follows: $(a, a') \in >_x$ iff $(a, a') \in \geq_x$ and $(a', a) \notin \geq_x$. We will sometimes write $(a, a') \in \odot$ to refer to one of the four possible situations: $(a, a') \in \geq_x \wedge (a', a) \in \geq_x$, meaning that the two arguments a and a' are *indifferent* for the decision maker, $(a, a') \in >_x$, meaning that a is *strictly preferred* to a', $(a', a) \in >_x$, meaning that a' is strictly preferred to a, $(a, a') \notin \geq_x \wedge (a', a) \notin \geq_x$, meaning that the two arguments are *incomparable*.

Generally arguments may be conflicting. These conflicts are captured by a binary relation on the set of arguments. Three such relations are distinguished. The first one, denoted by $\mathcal{R}_b$ captures the different conflicts between epistemic arguments. The second relation, denoted $\mathcal{R}_o$ captures the conflicts among practical arguments. Two practical arguments are conflicting if they support different options. Formally, $(\forall a, b \in \mathcal{A}_o)\ (a, b) \in \mathcal{R}_o$ iff $\text{Conc}(a) \neq \text{Conc}(b)$. Finally, practical arguments may be attacked by epistemic ones. The idea is that an epistemic argument may undermine the belief part of a practical argument. However, practical arguments are not allowed to attack epistemic ones. This avoids wishful thinking, i.e., avoids making decisions according to what might be pleasing to imagine instead of by appealing to evidence or rationality. This relation, denoted by $\mathcal{R}_m$, contains pairs (a, a') where $a \in \mathcal{A}_b$ and $a' \in \mathcal{A}_o$. Before introducing the framework, we need first to combine each preference relation $\geq_x$ (with $x \in \{b, o, m\}$) with the conflict relation $\mathcal{R}_x$ into a unique relation between arguments, denoted Def_x, and called *defeat* relation.

Definition 1. *(Defeat relation) Let $a, b \in \mathcal{A}$. $(a, b) \in \text{Def}_x$ iff $(a, b) \in \mathcal{R}_x$ and $(b, a) \notin \geq_x$.*

Let Def_b, Def_o and Def_m denote the three defeat relations corresponding to three attack relations. Since arguments in favor of beliefs are always preferred (in the sense of $\geq_m$) to arguments in favor of options, it holds that $\mathcal{R}_m = \text{Def}_m$.

Example 2. (Example 1 cont.) The graph on the left depicts different attacks among arguments. Let us assume the following preferences: $(b_2, b_3) \in \geq_b$, $(a_2, a_1) \in \geq_o$ and $(a_1, a_3) \in \geq_o$. The defeats are depicted on the right of figure below.

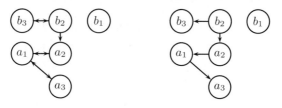

The different arguments of $\mathcal{A}_b \cup \mathcal{A}_o$ are evaluated within the system $\mathcal{AF} = \langle \mathcal{A} = \mathcal{A}_b \cup \mathcal{A}_o, \text{Def} = \text{Def}_b \cup \text{Def}_o \cup \text{Def}_m \rangle$ using any Dung's acceptability semantics.

Definition 2. *(Conflict-free, Defense) Let* $\langle \mathcal{A}, \mathtt{Def} \rangle$ *be an argumentation sys-tem[2], $\mathcal{B} \subseteq \mathcal{A}$, and $a \in \mathcal{A}$.*

- $\mathcal{B}$ *is* conflict-free *iff* $\nexists$ $a, b \in \mathcal{B}$ *s.t.* $(a, b) \in \mathtt{Def}$.
- $\mathcal{B}$ defends *a iff* $\forall$ $b \in \mathcal{A}$, *if* $(b, a) \in \mathtt{Def}$, *then* $\exists$ $c \in \mathcal{B}$ *s.t.* $(c, b) \in \mathtt{Def}$.

The main semantics introduced by Dung are recalled in the following definition.

Definition 3. *(Acceptability semantics) Let* $\mathcal{AF} = \langle \mathcal{A}, \mathtt{Def} \rangle$ *be an argumenta-tion system, and $\mathcal{E}$ be a conflict-free set of arguments.*

- $\mathcal{E}$ *is a* preferred extension *iff* $\mathcal{E}$ *is a maximal (w.r.t set* $\subseteq$*) set that defends any element in* $\mathcal{E}$.
- $\mathcal{E}$ *is a* grounded extension, *denoted* GE, *iff* $\mathcal{E}$ *is the least fixpoint of function $\mathcal{F}$ where* $\mathcal{F}(S) = \{a \in \mathcal{A} \mid S \text{ defends } a\}$, *for* $S \subseteq \mathcal{A}$.

Using these acceptability semantics, the status of each argument can be defined.

Definition 4. *(Argument status) Let* $\mathcal{AF} = \langle \mathcal{A}, \mathtt{Def} \rangle$ *be an argumentation sys-tem,* $\mathcal{E}_1, \ldots, \mathcal{E}_x$ *its extensions under a given semantics and let $a \in \mathcal{A}$.*

- *a is* skeptically accepted *iff exists at least one extension and* $(\forall \mathcal{E}_i)$ $a \in \mathcal{E}_i$.
- *a is* credulously accepted *iff* $(\exists \mathcal{E}_i)$ *s.t.* $a \in \mathcal{E}_i$ *and* $(\exists \mathcal{E}_j)$ *s.t.* $a \notin \mathcal{E}_j$.
- *a is* rejected *iff* $(\nexists \mathcal{E}_i)$ *s.t.* $a \in \mathcal{E}_i$.

Example 3. (Example 1 cont.) There is one preferred extension, which is also the grounded one, $\{a_1, b_1, b_2\}$. It is clear that a_1, b_1 and b_2 are skeptically accepted while other arguments are rejected.

Let $\mathcal{AF} = \langle \mathcal{A}, \mathtt{Def} \rangle$ be an argumentation system. $\mathtt{Sc}(\mathcal{AF})$, $\mathtt{Cr}(\mathcal{AF})$ and $\mathtt{Rej}(\mathcal{AF})$ denote respectively the sets of skeptically accepted arguments, credulously ac-cepted arguments and rejected arguments of the system $\mathcal{AF}$. It can be shown that these three sets are disjoint. Moreover, their union is the set $\mathcal{A}$ of arguments.

Proposition 1. *Let* $\mathcal{AF} = \langle \mathcal{A}, \mathcal{R} \rangle$ *be an argumentation system and* $\mathtt{Sc}(\mathcal{AF})$, $\mathtt{Cr}(\mathcal{AF})$, $\mathtt{Rej}(\mathcal{AF})$, *its sets of arguments.*

1. $\mathtt{Sc}(\mathcal{AF}) \cap \mathtt{Cr}(\mathcal{AF}) = \emptyset$, $\mathtt{Sc}(\mathcal{AF}) \cap \mathtt{Rej}(\mathcal{AF}) = \emptyset$, $\mathtt{Cr}(\mathcal{AF}) \cap \mathtt{Rej}(\mathcal{AF}) = \emptyset$
2. $\mathtt{Sc}(\mathcal{AF}) \cup \mathtt{Cr}(\mathcal{AF}) \cup \mathtt{Rej}(\mathcal{AF}) = \mathcal{A}$.

The status of an option is defined from the status of its arguments.

Definition 5. *(Option status) Let* $o \in \mathcal{O}$.

- *o is* acceptable *iff* $\exists a \in \mathcal{H}(o)$ *s.t.* $a \in \mathtt{Sc}(\mathcal{AF})$.
- *o is* rejected *iff* $\mathcal{H}(o) \neq \emptyset$ *and* $\forall a \in \mathcal{H}(o)$, $a \in \mathtt{Rej}(\mathcal{AF})$.
- *o is* negotiable *iff* $(\nexists a \in \mathcal{H}(o))$ $(a \in \mathtt{Sc}(\mathcal{AF})) \wedge (\exists a' \in \mathcal{H}(o))$ $(a' \in \mathtt{Cr}(\mathcal{AF}))$.
- *o is* non-supported *iff it is neither acceptable, nor rejected nor negotiable.*

---

[2] At some places, it will be referred to as a decision system.

Let $\mathcal{O}_a$ (resp. $\mathcal{O}_n$, $\mathcal{O}_{ns}$, $\mathcal{O}_r$) be the set of acceptable (resp. negotiable, non-supported, rejected) options.

Example 4. (Example 1 cont.) Option o_1 is acceptable, o_2 is rejected and o_3 is non-supported.

It can be checked that an option has only one status. This status may change in light of new arguments as we will show in next sections.

Proposition 2. *Let $o \in \mathcal{O}$. o has exactly one status.*

The choice of a semantics has an impact on the acceptability of arguments and, consequently, on the status of options. We have studied the impact of several semantics on the status of options. However, due to lack of space, we present only the results related to preferred and grounded semantics. Let $\mathcal{O}_y^x$ denote the set of options having status y under semantics x.

Proposition 3. *It holds that: $O_a^g \subseteq O_a^p$, $O_r^p \subseteq O_r^g$, $O_{ns}^p = O_{ns}^g$ and $O_n^g = \emptyset$.*

In [1], the status of options makes it possible to compare them, thus to define a preference relation $\succeq$ on O. The basic idea is the following: acceptable options are preferred to negotiable ones. Negotiable options are themselves preferred to non-supported options, which in turn are better than rejected options.

3   Revising Option Status

Given a decision system $\mathcal{AF} = \langle \mathcal{A} = \mathcal{A}_b \cup \mathcal{A}_o, \mathtt{Def} = \mathtt{Def}_b \cup \mathtt{Def}_o \cup \mathtt{Def}_m \rangle$ that defines a preorder on a set $\mathcal{O}$ of options, we study how the status of each option in $\mathcal{O}$ may change when a new argument is added to the set $\mathcal{A}$ of arguments. In this paper, we investigate the case where the new argument, say e, is practical. Let $\mathcal{AF} \oplus e = \langle \mathcal{A}', \mathtt{Def}' \rangle$ denote the new decision system. When $e \in \mathcal{A}$, $\mathcal{A}' = \mathcal{A}$ and $\mathtt{Def}' = \mathtt{Def}$, all the arguments and all the options keep their original status (i.e., the one computed with $\mathcal{AF}$). Things are different when $e \notin \mathcal{A}$. In this case, $\mathcal{A}' = \mathcal{A} \cup \{e\}$ and $\mathtt{Def}' = \mathtt{Def} \cup \{(x,e) \mid x \in \mathcal{A}_b$ and $(x,e) \in \mathcal{R}_m^{\mathcal{L}}\}^3 \cup \{(e,y) \mid y \in \mathcal{A}_o$ and $\mathtt{Conc}(y) \neq \mathtt{Conc}(e)$ and $(y,e) \notin \geq_o\} \cup \{(y,e) \mid y \in \mathcal{A}_o$ and $\mathtt{Conc}(y) \neq \mathtt{Conc}(e)$ and $(e,y) \notin \geq_o\}$. Throughout the paper, we assume that $e \notin \mathcal{A}_o$.

In this section we will use *grounded semantics* to compute acceptability of arguments. We will denote by $\mathcal{O}_x(\mathcal{AF})$, with $x \in \{a, r, ns\}$, the set of accept-able (resp. rejected and non-supported) options of the original system $\mathcal{AF}$ and $\mathcal{O}_x(\mathcal{AF} \oplus e)$ the corresponding sets of the new system. For example, $\mathcal{O}_r(\mathcal{AF} \oplus e)$ is the set of rejected options when argument e is added to the system $\mathcal{AF}$.

In this section, we will study the properties of an argument that can change the status of an option. For that purpose, we start by studying when an accepted argument in the system $\mathcal{AF}$ remains accepted (resp. becomes rejected) in $\mathcal{AF} \oplus$

---

^{3} $\mathcal{R}_m^{\mathcal{L}}$ contains all the attacks from epistemic arguments to practical arguments of a logical language $\mathcal{L}$.

e. Then, we show under which conditions an option in $\mathcal{O}_x(\mathcal{AF})$ will move to $\mathcal{O}_y(\mathcal{AF} \oplus e)$ with $x \neq y$.

The first results states that a new practical arguments e will have no impact on existing epistemic arguments. This is due to the fact that a practical argument is not allowed to attack an epistemic one. Formally:

Proposition 4. *Let e be a new practical argument. It holds that $\mathrm{Sc}(\mathcal{AF} \oplus e) \cap \mathcal{A}_b = \mathrm{Sc}(\mathcal{AF}) \cap \mathcal{A}_b$.*

This result in not necessarily true for the practical arguments of the set $\mathcal{A}_o$. However, this can be the case when the new argument is defeated by a skeptically accepted epistemic argument. In this case, the argument e is clearly useless.

Proposition 5. *Let e be a new practical argument. If $(\exists a \in \mathcal{A}_b \cap \mathrm{Sc}(\mathcal{AF}))$ such that $(a, e) \in \mathrm{Def}$ then $\mathrm{Sc}(\mathcal{A} \oplus e) \cap \mathcal{A}_o = \mathrm{Sc}(\mathcal{AF}) \cap \mathcal{A}_o$.*

From the two above propositions, the following trivial result holds:

Proposition 6. *Let e be a new practical argument. If $(\exists a \in \mathcal{A}_b \cap \mathrm{Sc}(\mathcal{AF}))$ such that $(a, e) \in \mathrm{Def}$ then $\mathrm{Sc}(\mathcal{A} \oplus e) = \mathrm{Sc}(\mathcal{AF})$.*

It can be shown that each skeptically accepted practical argument can be defended either by an epistemic argument or by another practical argument that supports the same option. Before presenting formally this result, let us first introduce a notation. Recall that $\mathrm{Sc}(\mathcal{AF}) = \bigcup_{i=1}^{\infty} \mathcal{F}^{(i)}(\emptyset)$. Let $\mathrm{Sc}^1(\mathcal{AF}) = \mathcal{F}(\emptyset)$ and let $(\forall i \in \{2, 3, \ldots\})$ $\mathrm{Sc}^i(\mathcal{AF})$ denote $\mathcal{F}^{(i)}(\emptyset) \setminus \mathcal{F}^{(i-1)}(\emptyset)$, i.e., the arguments reinstated at step i.

Proposition 7. *Let $o \in \mathcal{O}$, $a_i \in \mathcal{H}(o)$, $a_i \in \mathrm{Sc}^i(\mathcal{AF})$ and $x \in \mathcal{A}$ such that $(x, a_i) \in \mathrm{Def}$.*

1. *If $x \in \mathcal{A}_b$ then $(\exists j \geq 1)$ $(j < i) \wedge (\exists a_j \in \mathcal{A}_b \cap \mathrm{Sc}^j(\mathcal{AF}))$ $(a_j, x) \in \mathrm{Def}$,*
2. *If $x \in \mathcal{A}_o$ then $(\exists j \geq 1)$ $(j < i) \wedge (\exists a_j \in (\mathcal{A}_b \cup \mathcal{H}(o)) \cap \mathrm{Sc}^j(\mathcal{AF}))$ $(a_j, x) \in \mathrm{Def}$.*

The following result states that a new practical argument will never influence the accepted arguments supporting the same option as the new argument e.

Theorem 1. *Let e be a new argument such that $\mathrm{Conc}(e) = o$. Then, $(\forall a \in \mathcal{H}(o))$ $a \in \mathrm{Sc}(\mathcal{AF}) \Rightarrow a \in \mathrm{Sc}(\mathcal{AF} \oplus e)$.*

We can also show that if the new practical argument e induces a change in the status of a given practical argument from rejection to acceptance, then this argument supports the same option as e. This means that a new practical argument can improve the status of arguments supporting its own conclusion, thus it can improve the status of option it supports. However, it can never improve the status of other options.

Theorem 2. *Let $o \in \mathcal{O}$, and $a \in \mathcal{H}(o)$. If $a \in \mathrm{Rej}(\mathcal{AF})$ and $a \in \mathrm{Sc}(\mathcal{AF} \oplus e)$, then $e \in \mathcal{H}(o)$.*

Before continuing with the results on the revision of the status of options, let us define the set of arguments defended by epistemic arguments in $\mathcal{AF}$.

Definition 6. *(Defense by epistemic arguments) Let $\mathcal{AF} = \langle \mathcal{A}, \mathrm{Def} \rangle$ be an argumentation system and $a \in \mathcal{A}$. We say that a is defended by epistemic arguments in $\mathcal{AF}$ and we write $a \in \mathrm{Dbe}(\mathcal{AF})$ iff $(\forall x \in \mathcal{AF})$ $(x, a) \in \mathrm{Def} \Rightarrow (\exists \alpha \in \mathrm{Sc}(\mathcal{AF}) \cap \mathcal{A}_b)$ $(\alpha, x) \in \mathrm{Def}$.*

Note that, since elements of $\mathrm{Sc}^1(\mathcal{AF})$ are not attacked at all, they are also defended by epistemic arguments, i.e., $\mathrm{Sc}^1(\mathcal{AF}) \subseteq \mathrm{Dbe}(\mathcal{AF})$. We can prove that the set of arguments defended by epistemic arguments is skeptically accepted.

Proposition 8. *It holds that $\mathrm{Dbe}(\mathcal{AF}) \subseteq \mathrm{Sc}(\mathcal{AF})$.*

Given an option which is accepted in the system $\mathcal{AF}$, it becomes rejected in $\mathcal{AF} \oplus e$ if three conditions are satisfied: e is not in favor of the option o, there is no skeptically accepted epistemic argument that defeats e, and e defeats all the arguments in favor of option o that are defended by epistemic arguments.

Theorem 3. *Let $o \in \mathcal{O}_a(\mathcal{AF})$ and let agent receive new practical argument e. Then: $o \in \mathcal{O}_r(\mathcal{AF} \oplus e)$ iff $e \notin \mathcal{H}(o) \wedge (\nexists x \in \mathcal{A}_b \cap \mathrm{Sc}(\mathcal{AF}))$ $(x, e) \in \mathrm{Def} \wedge (\forall a \in \mathrm{Dbe}(\mathcal{AF}) \cap \mathcal{H}(o))$ $(e, a) \in \mathrm{Def}$.*

This result is important in a negotiation. It shows the properties of a good argument that may kill an option that is not desirable for an agent.

Similarly, we can show that it is possible for an option to move from a rejection to an acceptance. The idea is to send a practical argument that supports this option and that is accepted in the new system. Formally:

Theorem 4. *Let $o \in \mathcal{O}_r(\mathcal{AF})$ and let agent receive new practical argument e. Then: $o \in \mathcal{O}_a(\mathcal{AF} \oplus e)$ iff $e \in \mathcal{H}(o) \wedge e \in \mathrm{Sc}(\mathcal{AF} \oplus e)$.*

4   Revising Complete Decision Systems

So far, we have analyzed how an argument may change its status when a new practical argument is received, and similarly how an option may change its status without computing the new grounded extension. The decision system that is used assumes that an option may be supported by several arguments, each of them pointing out to a particular goal satisfied by the option. In some works on argument-based decision making, an argument in favor of an option refers to all the goals satisfied by that option. Thus, there is one argument per option. A consequence of having one argument per option is that all the practical arguments are conflicting. In this section, we will use this particular system, but we will allow multiple arguments in favor of an option under the condition that they all attack each other in the sense of $\mathcal{R}$. We assume also that the set of epistemic arguments is empty. The argumentation system that is used is then $\mathcal{AF}_o = \langle \mathcal{A}_o, \mathrm{Def}_o \rangle$, and we will use preferred semantics for computing the acceptability of arguments.

In this system, the status of each argument can be characterized as follows:

Proposition 9. *Let $\mathcal{AF}_o = \langle \mathcal{A}_o, \text{Def}_o \rangle$ be a complete argumentation framework for decision making, and a be an arbitrary argument. Then:*

1. *a is skeptically accepted iff $(\forall x \in \mathcal{A}_o)\ (a, x) \in \geq_o$.*
2. *a is rejected iff $(\exists x \in \mathcal{A})\ (x, a) \in >_o$.*
3. *a is credulously accepted iff*
   $((\exists x' \in \mathcal{A})\ (a, x') \notin \geq_o) \wedge ((\forall x \in \mathcal{A})\ ((a, x) \notin \geq_o) \Rightarrow (x, a) \notin \geq_o)).$

It can be checked that all skeptically accepted arguments in this system are equally preferred.

Proposition 13. *Let $a, b \in \text{Sc}(\mathcal{AF}_o)$. Then $(a, b) \in \geq_o$ and $(b, a) \in \geq_o$.*

We will now prove that in this particular system, there are two possible cases: the case where there exists at least one skeptically accepted argument but there are no credulously accepted arguments, and the case where there are no skeptically accepted arguments but there is "at least" one credulously accepted argument. This means that one cannot have a state with both skeptically accepted and credulously accepted arguments. Moreover, it cannot be the case that all the arguments are rejected. Formally:

Theorem 5. *Let $\mathcal{AF}_o = \langle \mathcal{A}_o, \text{Def}_o \rangle$ be an argumentation system. The following implications hold:*

1. *If $\text{Sc}(\mathcal{AF}_o) \neq \emptyset$ then $\text{Cr}(\mathcal{AF}_o) = \emptyset$.*
2. *If $\text{Cr}(\mathcal{AF}_o) = \emptyset$ then $\text{Sc}(\mathcal{AF}_o) \neq \emptyset$.*

We will now show that an arbitrary argument e is in the same relation with all accepted arguments. Recall that we use the notation $(e, a) \in \odot$ to refer to one particular relation between the arguments e and a.

Proposition 14. *Let e be an arbitrary argument.*
If $(\exists a \in \text{Sc}(\mathcal{AF}_o))$ such that $(a, e) \in \odot$ then $(\forall a' \in \text{Sc}(\mathcal{AF}_o))\ (a', e) \in \odot$.

Let us now have a look at credulously accepted arguments. While all the skeptically accepted arguments are in the same class with respect to the preference relation $\geq_o$, this is not always the case with credulously accepted arguments. The next proposition shows that credulously accepted arguments are either incomparable or indifferent with respect to $\geq_o$.

Proposition 15. *$\mathcal{AF}_o = \langle \mathcal{A}_o, \text{Def}_o \rangle$ be an argumentation system and $\text{Cr}(\mathcal{AF}_o)$ its set of credulously accepted arguments. Then $(\forall a, b \in \text{Cr}(\mathcal{AF}_o))$ it holds that*

$$((a, b) \in \geq_o \wedge (b, a) \in \geq_o) \vee ((a, b) \notin \geq_o \wedge (b, a) \notin \geq_o).$$

The next proposition shows that if a' is credulously accepted then there exists another credulously accepted argument a'' such that they are incomparable in the sense of preference relation.

Proposition 16. *Let $\mathcal{AF}_o = \langle \mathcal{A}_o, \text{Def}_o \rangle$ be an argumentation system for decision making, and $\text{Cr}(\mathcal{AF}_o) \neq \emptyset$. Then it holds that: $(\forall a' \in \text{Cr}(\mathcal{AF}_o))\ (\exists a'' \in \text{Cr}(\mathcal{AF}_o))\ (a', a'') \notin \geq_o \wedge (a'', a') \notin \geq_o$.*

The next proposition will make some reasoning easier, because it shows that, in this particular framework, the definition of negotiable options can be simplified.

Proposition 17. *Let $o \in \mathcal{O}$. The option o is negotiable iff there is at least one credulously accepted argument in its favor.*

As a consequence of the above propositions, the following result shows that negotiable options and acceptable ones cannot exist at the same time.

Theorem 6. *Let $\mathcal{AF}_o = \langle \mathcal{A}_o, \mathtt{Def}_o \rangle$ be a complete argumentation framework for decision making. The following holds: $\mathcal{O}_a \neq \emptyset \Leftrightarrow \mathcal{O}_n = \emptyset$.*

4.1   Revising the Status of an Argument

Like in the previous section, we assume that an agent receives a new practical argument e. The question is, how the status of an argument given by the system $\mathcal{AF}_o$ may change in the system $\mathcal{AF} \oplus e$ without having to compute the preferred extensions of $\mathcal{AF}_o \oplus e$.

The first result states that rejected arguments in $\mathcal{AF}_o$ remain rejected in the new system $\mathcal{AF}_o \oplus e$. This means that rejected arguments cannot be "saved".

Proposition 18. *Let $\mathcal{AF}_o = \langle \mathcal{A}_o, \mathtt{Def}_o \rangle$ be an argumentation system. If $a \in \mathtt{Rej}(\mathcal{AF}_o)$, then $a \in \mathtt{Rej}(\mathcal{AF}_o \oplus e)$.*

We can also show that an argument that was credulously accepted in $\mathcal{AF}_o$ can never become skeptically accepted in $\mathcal{AF}_o \oplus e$. It can either remain credulously accepted, either become rejected.

Proposition 19. *Let $\mathcal{AF}_o = \langle \mathcal{A}_o, \mathtt{Def}_o \rangle$ be an argumentation system. If $a \in \mathtt{Cr}(\mathcal{AF}_o)$, then $a \notin \mathtt{Sc}(\mathcal{AF}_o \oplus e)$.*

The next proposition is simple but will be very useful later in this section.

Proposition 20. *Let $\mathcal{AF}_o = \langle \mathcal{A}_o, \mathtt{Def}_o \rangle$ be a decision system.*

1. *If $a \in \mathtt{Sc}(\mathcal{AF}_o)$ then $a \in \mathtt{Sc}(\mathcal{AF}_o \oplus e)$ iff $(a, e) \in \geq_o$.*
2. *If $a \notin \mathtt{Rej}(\mathcal{AF}_o)$ then $a \in \mathtt{Rej}(\mathcal{AF}_o \oplus e)$ iff $(e, a) \in >_o$.*

The next proposition shows that all the skeptically accepted arguments will have the "same destiny" when a new argument is recieved.

Proposition 21. *Let $\mathcal{AF}_o = \langle \mathcal{A}_o, \mathtt{Def}_o \rangle$ be an argumentation system and $a, b \in \mathtt{Sc}(\mathcal{AF}_o)$. Let $e \notin \mathcal{A}_o$.*

1. *If $a \in \mathtt{Sc}(\mathcal{AF}_o \oplus e)$ then $b \in \mathtt{Sc}(\mathcal{AF}_o \oplus e)$.*
2. *If $a \in \mathtt{Cr}(\mathcal{AF}_o \oplus e)$ then $b \in \mathtt{Cr}(\mathcal{AF}_o \oplus e)$.*
3. *If $a \in \mathtt{Rej}(\mathcal{AF}_o \oplus e)$ then $b \in \mathtt{Rej}(\mathcal{AF}_o \oplus e)$.*

The next theorem analyzes the status of all skeptically accepted arguments after a new argument has arrived.

Theorem 7. *Let $\mathcal{AF}_o = \langle \mathcal{A}_o, \text{Def}_o \rangle$ be a complete argumentation framework for decision making, $a \in \text{Sc}(\mathcal{AF}_o)$ and $e \notin \mathcal{A}_o$. The following holds:*

1. $a \in \text{Sc}(\mathcal{AF}_o \oplus e) \wedge e \in \text{Sc}(\mathcal{AF}_o \oplus e)$ iff $((a, e) \in \geq_o) \wedge ((e, a) \in \geq_o)$
2. $a \in \text{Rej}(\mathcal{AF}_o \oplus e) \wedge e \in \text{Sc}(\mathcal{AF}_o \oplus e)$ iff $(e, a) \in >_o$
3. $a \in \text{Sc}(\mathcal{AF}_o \oplus e) \wedge e \in \text{Rej}(\mathcal{AF}_o \oplus e)$ iff $(a, e) \in >_o$
4. $a \in \text{Cr}(\mathcal{AF}_o \oplus e) \wedge e \in \text{Cr}(\mathcal{AF}_o \oplus e)$ iff $((a, e) \notin \geq_o) \wedge ((a, e) \notin \geq_o)$

Note that, according to Proposition 14, all skeptically accepted arguments are in the same relation with e as a is. Formally, if a and e are in a particular relation i.e., $(a, e) \in \odot$, then $(\forall b \in \mathcal{A}_o) ((b \in \text{Sc}(\mathcal{AF}_o)) \Rightarrow (b, e) \in \odot)$. Hence, the condition "let $a \in \text{Sc}(\mathcal{AF}_o)$ and $(a, e) \in \odot$" in the previous theorem is equivalent to the condition $(\forall a \in \mathcal{A}_o) ((a \in \text{Sc}(\mathcal{AF}_o)) \Rightarrow (a, e) \in \odot)$.

Theorem 7 stands as a basic tool for reasoning about the status of new arguments as well as about the changes in the status of other arguments. Once the argument status is known, it is much easier to determine the status of options.

We will now analyze the relation between credulously accepted arguments and new arguments.

The next result shows that if there are credulously accepted arguments in $\mathcal{AF}_o$ and the new argument e is preferred to all of them, then it is strictly preferred to all of them.

Proposition 22. *Let $\mathcal{AF}_o = \langle \mathcal{A}_o, \text{Def}_o \rangle$ s.t. $\text{Cr}(\mathcal{AF}_o) \neq \emptyset$. The following result holds: $((\forall a \in \text{Cr}(\mathcal{AF}_o)) (e, a) \in >_o)$ iff $((\forall a \in \text{Cr}(\mathcal{AF}_o)) (e, a) \in \geq_o)$.*

Proposition 23. *Let $\mathcal{AF}_o = \langle \mathcal{A}_o, \text{Def}_o \rangle$ s.t. $\text{Cr}(\mathcal{AF}_o) \neq \emptyset$. The following holds: $((\forall a \in \text{Cr}(\mathcal{A}_o)) a \in \text{Rej}(\mathcal{A}_o \oplus e))$ iff $((\forall a \in \text{Cr}(\mathcal{A}_o)) (e, a) \in >_o)$.*

The next theorem analyzes the case when there are no skeptically accepted arguments in $\mathcal{AF}_o$.

Theorem 8. *Let $\mathcal{AF}_o = \langle \mathcal{A}_o, \text{Def}_o \rangle$ be an argumentation framework such that $\text{Cr}(\mathcal{AF}_o) \neq \emptyset$. Then, the following holds:*

1. $(\forall a \in \text{Cr}(\mathcal{AF}_o)) (e, a) \in >_o$ iff $e \in \text{Sc}(\mathcal{AF}_o \oplus e) \wedge \mathcal{A}_o = \text{Rej}(\mathcal{AF}_o \oplus e)$.
2. $(\exists a \in \text{Cr}(\mathcal{AF}_o)) (e, a) \notin >_o \wedge (\nexists a' \in \text{Cr}(\mathcal{AF}_o))$
   $(a', e) \in >_o$ iff $e \in \text{Cr}(\mathcal{AF}_o \oplus e)$
3. $(\exists a \in \text{Cr}(\mathcal{AF}_o)) (a, e) \in >_o$ iff $e \in \text{Rej}(\mathcal{AF}_o \oplus e) \wedge \mathcal{A}_o = \text{Cr}(\mathcal{AF}_o \oplus e)$.

Recall that, according to Proposition 22, the condition $(\forall a \in \text{Cr}(\mathcal{AF}_o)) (e, a) \in >_o$ in the previous theorem is equivalent to the condition $(\forall a \in \text{Cr}(\mathcal{AF}_o)) (e, a) \in \geq_o$.

While all the skeptically accepted arguments have the "same destiny" after a new argument arrives, this is not the case with credulously accepted arguments. Some of them may remain credulously accepted while the others may become rejected.

4.2 Revising the Status of an Option

We will now show under which conditions an option can change its status. We start by studying acceptable options.

Theorem 9. *Let $\mathcal{AF}_o = \langle \mathcal{A}_o, \mathtt{Def}_o \rangle$ be an argumentation system and $o \in \mathcal{O}_a(\mathcal{AF}_o)$. Suppose that $a \in \mathtt{Sc}(\mathcal{AF}_o)$ is an arbitrary skeptically accepted argument. Then:*

1. *$o \in \mathcal{O}_a(\mathcal{AF}_o \oplus e)$ iff $((a,e) \in \geq_o) \vee (e \in \mathcal{H}(o)) \wedge ((e,a) \in >_o)$*
2. *$o \in \mathcal{O}_n(\mathcal{AF}_o \oplus e)$ iff $((a,e) \not\in \geq_o) \wedge ((e,a) \not\in \geq_o))$*
3. *$o \in \mathcal{O}_r(\mathcal{AF}_o \oplus e)$ iff $(e \notin \mathcal{H}(o)) \wedge (e,a) \in >_o$*

Recall that, according to Proposition 14, all skeptically accepted arguments are in the same relation with an arbitrary argument. Hence, the condition $(\exists a \in \mathtt{Sc}(\mathcal{AF}_o))$ $(a,e) \in \odot)$ in the previous theorem is equivalent to the condition $(\forall a \in \mathtt{Sc}(\mathcal{AF}_o))$ $(a,e) \in \odot)$.

A similar characterization is given bellow for negotiable options.

Theorem 10. *Let $\mathcal{AF}_o = \langle \mathcal{A}_o, \mathtt{Def}_o \rangle$ be an argumentation system and $o \in \mathcal{O}_n\mathcal{AF}$. Then:*

1. *$o \in \mathcal{O}_a(\mathcal{AF}_o \oplus e)$ iff $(e \in \mathcal{H}(o)) \wedge ((\forall a \in \mathtt{Cr}(\mathcal{A}_o))$ $(e,a) \in >)$*
2. *$o \in \mathcal{O}_n(\mathcal{AF}_o \oplus e)$ iff $((e \in \mathcal{H}(o)) \wedge (\exists a' \in \mathtt{Cr}(\mathcal{AF}_o))$ $(e,a') \not\in >_o \wedge (\nexists a'' \in \mathtt{Cr}(\mathcal{AF}_o))$ $(a'',e) \in >_o) \vee ((\exists a' \in \mathtt{Cr}(\mathcal{AF}_o))$ $(a' \in \mathcal{H}(o) \wedge (e,a') \not\in >_o))$*
3. *$o \in \mathcal{O}_r(\mathcal{AF}_o \oplus e)$ iff $((e \notin \mathcal{H}(o)) \wedge ((\forall a \in \mathtt{Cr}(\mathcal{AF}_o))$ $(a \in \mathcal{H}(o)) \Rightarrow (e,a) \in >_o))$.*

Note that, according to Proposition 22, the condition $(\forall a \in \mathtt{Cr}(\mathcal{A}_o))$ $(e,a) \in >$ in the previous theorem is equivalent to condition $(\forall a \in \mathtt{Cr}(\mathcal{AF}_o))$ $(e,a) \in \geq_o$.

Let us now analyze when a rejected option in $\mathcal{AF}_o$ may change its status in $\mathcal{AF} \oplus e$.

Theorem 11. *Let $\mathcal{AF}_o = \langle \mathcal{A}_o, \mathtt{Def}_o \rangle$ be an argumentation system and $o \in \mathcal{O}_r(\mathcal{AF})$. Then:*

1. *$o \in \mathcal{O}_a(\mathcal{AF}_o \oplus e)$ iff $(e \in \mathcal{H}(o)) \wedge ((\forall a \in \mathcal{A}_o)$ $(e,a) \in \geq_o)$*
2. *$o \in \mathcal{O}_n(\mathcal{AF}_o \oplus e)$ iff $(e \in \mathcal{H}(o)) \wedge ((\forall a \in \mathcal{A}_o)$ $(a,e) \not\in >_o) \wedge ((\exists a \in \mathcal{A}_o)$ $(e,a) \not\in >_o)$*
3. *$o \in \mathcal{O}_r(\mathcal{AF}_o \oplus e)$ iff $(e \notin \mathcal{H}(o)) \vee ((e \in \mathcal{H}(o)) \wedge (\exists a \in \mathcal{A}_o)(a,e) \in >)$*

5 Conclusion

This paper has tackled the problem of revising argument-based decision models. To the best of our knowledge, in this paper we have proposed the first investigation on the impact of a new argument on the outcome of a decision system. The basic idea is to check when the status of an option may shift when a new argument is received without having to compute the whole new ordering on options. For that purpose, we have considered a decision model that has recently been proposed in the literature. This model computes a status for each option on the basis of the status of their supporting arguments. We have studied two cases: the case where an option may be supported by several arguments and the case where an option is supported by only one argument. In both cases, we assumed that the new argument is practical, i.e., it supports an option. We have provided a full characterization of acceptable options that become rejected, negotiable or remain accepted. Similarly, we have characterized any shift from one status to another. These results are based on a characterization of a shift of the status of arguments themselves.

These results may be used to determine strategies for negotiation, since at a given step of a dialog an agent has to choose an argument to send to another agent in order to change the status of an option. Moreover, they may help to understand which arguments are useful and which arguments are useless in a given situation, which allows us to understand the role of argumentation in a negotiation.

Note that a recent work has been done on revision in argumentation systems in [5]. That paper addresses the problem of revising the set of extensions of an abstract argumentation system. It studies how the extensions of an argumentation system may evolve when a new argument is received. Nothing is said on the revision of a particular argument. In our paper, we are more interested by the evolution of the status of a given argument without having to compute the extensions of the new argumentation system. We have also studied how the status of an option changes when a new argument is received. Another main difference with this work is that in [5] only the case of adding an argument having only one interaction with an argument of the initial argumentation system is studied. In our paper we have studied the more general case, i.e., the new argument may attack and be attacked by an arbitrary number of arguments of the initial argumentation system.

References

1. Amgoud, L., Dimopoulos, Y., Moraitis, P.: Making decisions through preference-based argumentation. In: Proc. of Int. Conf. on Principles of Knowledge Representation and Reasoning, pp. 113–123. AAAI Press, Menlo Park (2008)
2. Amgoud, L., Prade, H.: Using arguments for making and explaining decisions. Artificial Intelligence Journal 173, 413–436 (2009)
3. Amgoud, L., Vesic, S.: On revising offer status in argument-based negotiations. IRIT/RR–2009-09–FR (2009), http://www.irit.fr/~Srdjan.Vesic
4. Bonet, B., Geffner, H.: Arguing for decisions: A qualitative model of decision making. In: Proceedings of the 12th Conference on Uncertainty in Artificial Intelligence (UAI 1996), pp. 98–105 (1996)
5. Cayrol, C., Bannay, F., Lagasquie, M.: Revision of an argumentation system. In: Int. Conf. on Principles of Knowledge Representation and Reasoning, pp. 124–134. AAAI Press, Menlo Park (2008)
6. Dung, P.M.: On the acceptability of arguments and its fundamental role in non-monotonic reasoning, logic programming and n-person games. Artificial Intelligence Journal 77, 321–357 (1995)
7. Fox, J., Das, S.: Safe and Sound. Artificial Intelligence in Hazardous Applications. AAAI Press/ MIT Press (2000)

Encompassing Attacks to Attacks in Abstract Argumentation Frameworks

Pietro Baroni, Federico Cerutti, Massimiliano Giacomin, and Giovanni Guida

Dipartimento di Elettronica per l'Automazione, Università di Brescia,
Via Branze 38, I-25123 Brescia, Italy

Abstract. In the traditional definition of Dung's abstract argumentation framework (AF), the notion of attack is understood as a relation between arguments, thus bounding attacks to start from and be directed to arguments. This paper introduces a generalized definition of abstract argumentation framework called $AFRA$ (Argumentation Framework with Recursive Attacks), where an attack is allowed to be directed towards another attack. From a conceptual point of view, we claim that this generalization supports a straightforward representation of reasoning situations which are not easily accommodated within the traditional framework. From the technical side, we first investigate the extension to the generalized framework of the basic notions of conflict-free set, acceptable argument, admissible set and of Dung's fundamental lemma. Then we propose a correspondence from the $AFRA$ to the AF formalism, showing that it satisfies some basic desirable properties. Finally we analyze the relationships between $AFRA$ and a similar extension of Dung's abstract argumentation framework, called $EAF+$ and derived from the recently proposed formalism EAF.

1 Introduction

An argumentation framework (AF in the following), as introduced in a seminal paper by Dung [1], is an abstract entity consisting of a set of elements, called *arguments*, whose origin, nature and possible internal structure is not specified and by a binary relation of *attack* on the set of arguments, whose meaning is not specified either. This abstract formalism has been shown to be able to encompass a large variety of more specific formalisms in areas ranging from nonmonotonic reasoning to logic programming and game theory, and, as such, is widely regarded as a powerful tool for theoretical analysis. Several variations of the original AF formalism have been proposed in the literature. On one hand, some approaches introduce new elements in the basic scheme in order to encompass explicitly some additional conceptual notions, useful for a "natural" representation of some reasoning situations. This is the case for instance of value-based argumentation frameworks [2], where a notion of value is associated to arguments, and of bipolar argumentation frameworks [3], where a relation of support between arguments is considered besides the one of attack. On the other hand, one may investigate generalized versions of the original AF definition (in particular, of the notion

C. Sossai and G. Chemello (Eds.): ECSQARU 2009, LNAI 5590, pp. 83–94, 2009.

of attack) while not introducing additional concepts within the basic scheme, as in [4,5,6]. This paper lies in the latter line of investigation and pursues the goal of generalizing the AF notion of attack by allowing an attack, starting from an argument, to be directed not just towards an argument but also towards any other attack. This will be achieved by a recursive definition of the attack relation leading to the introduction and preliminary investigation of a formalism called $AFRA$ (Argumentation Framework with Recursive Attacks).

The paper is organised as follows. In Section 2 motivations for a recursive notion of attack will be discussed, leading in Section 3 to the formal definition of $AFRA$ and of the necessary companion notions. Section 4 proposes a translation procedure from $AFRA$ to AF, able to ensure a full correspondence between the notions of conflict-free set, acceptable argument and admissible set. In section 5 we compare $AFRA$ with an alternative way to encompass attacks to attacks, called $EAF+$ (in turn based on the EAF formalism, proposed in [4,5,6]). Section 6 concludes the paper.

2   Background and Motivations

In Dung's theory an argumentation framework $AF = \langle \mathcal{A}, \mathcal{R} \rangle$ is a pair where $\mathcal{A}$ is a set of arguments (whatever this may mean) and $\mathcal{R} \subseteq \mathcal{A} \times \mathcal{A}$ a binary relation on it. The terse intuition behind this formalism is that arguments may attack each other and useful formal definitions and theoretical investigations may be built on this simple basis. In particular, the related fundamental notions of conflict-free set, acceptable argument and admissible set are recalled in Definition 1.

Definition 1. *Given an argumentation framework* $AF = \langle \mathcal{A}, \mathcal{R} \rangle$:

- *a set* $\mathcal{S} \subseteq \mathcal{A}$ *is* conflict-free *if* $\nexists A, B \in \mathcal{S}$ *s.t.* $(A, B) \in \mathcal{R}$;
- *an argument* $A \in \mathcal{A}$ *is* acceptable *with respect to a set* $\mathcal{S} \subseteq \mathcal{A}$ *if* $\forall B \in \mathcal{A}$ *s.t.* $(B, A) \in \mathcal{R}$, $\exists C \in \mathcal{S}$ *s.t.* $(C, B) \in \mathcal{R}$;
- *a set* $\mathcal{S} \subseteq \mathcal{A}$ *is* admissible *if* $\mathcal{S}$ *is conflict-free and every element of* $\mathcal{S}$ *is* acceptable *with respect to* $\mathcal{S}$.

The notions recalled in Definition 1 lie at the heart of the definitions of Dung's argumentation semantics, a topic which is only marginally covered by this paper. For our purposes it is sufficient to recall that an argumentation semantics identifies for an argumentation framework, a set of *extensions*, namely sets of arguments which are "collectively acceptable", or, in other words, are able to survive together the conflict represented by the attack relation.

Even from this quick review it emerges that the main role of the notion of attack in Dung's theory is supporting the identification of "surviving" arguments, on which the definition of extensions is exclusively focused. On this basis, one might say that attacks are necessary but accessory in this theory.

At a merely abstract level, one might then envisage an alternative approach where attacks are ascribed an extended (in a sense, empowered) role. A simple way to achieve this is allowing an attack to be directed also towards another attack. From a general point of view, this amounts to conceive an attack as an entity able to affect any other entity (be it an argument or an attack) rather than just a by-product of how arguments relate each other. As a further consequence, this opens the way to include attacks as first-class elements in the definitions of all the basic notions we have seen before, from conflict-free sets to extensions.

However, before proceeding with what could be regarded as a sort of technical exercise, one might wonder whether there are practical motivations and concrete intuitions backing this kind of investigation. We provide an affirmative answer by means of an example in the area of modeling decision processes. Suppose Bob is deciding about his Christmas holidays and, as a general rule of thumb, he always buys cheap last minute offers. Suppose two such offers are available, one for a week in Gstaad, another for a week in Cuba. Then, using his behavioral rule, Bob can build two arguments, one, let say G, whose premise is "There is a last minute offer for Gstaad" and whose conclusion is "I should go to Gstaad", the other, let say C, whose premise is "There is a last minute offer for Cuba" and whose conclusion is "I should go to Cuba" (note that if more last minute offers were available, more arguments of the same kind would be constructed). As the two choices are incompatible, G and C attack each other, a situation giving rise to an undetermined choice. Suppose however that Bob has a preference P for skiing and knows that Gstaad is a ski resort, how can we represent this fact?

P might be represented implicitly by suppressing the attack from C to G, but this is unsatisfactory, since it would prevent, in particular, further reasoning on P, as described below. So let us consider P as an argument whose premise is "Bob likes skiing" and whose conclusion is "When it is possible, Bob prefers to go to a ski resort". P might attack C, but this does not seem sound since P is not actually in contrast with the existence of a good last minute offer for Cuba and the fact that, according to Bob's general behavioral rule, this gives him a reason for going to Cuba. Thus, it seems more reasonable to represent P as attacking the attack from C to G, causing G to prevail. Note that the attack from C to G is not suppressed, but only made ineffective, in the specific situation at hand, due to the attack of P.

Assume now that Bob learns that there have been no snowfalls in Gstaad since one month and from this fact he derives that it might not be possible to ski in Gstaad. This argument (N), whose premise is "The weather report informs that in Gstaad there were no snowfalls since one month" and whose conclusion is "It is not possible to ski in Gstaad", does not affect neither the existence of last minute offers for Gstaad nor Bob's general preference for ski, rather it affects the ability of this preference to affect the choice between Gstaad and Cuba. Thus argument N attacks the attack originated from P.

Suppose finally that Bob is informed that in Gstaad it is anyway possible to ski, thanks to a good amount of artificial snow. This allows to build an argument, let say A, which attacks N, thus in turn reinstating the attack originated from P and intuitively supporting the choice of Gstaad. A graphical illustration of this example is provided in Figure 1.

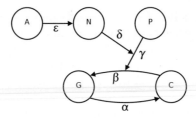

Fig. 1. Bob's last minute dilemma

While the quick formalization adopted for this largely informal example is clearly not the only possible one and might even be questionable, we believe that the example anyway provides the kind of intuitive backing we sought for the notion of attacks towards attacks in abstract argumentation and the necessity of extending concepts like acceptability, admissibility and reinstatement to attacks too. A formal counterpart to this intuition will be provided in the next section.

3   *AFRA*: Argumentation Framework with Recursive Attacks

An Argumentation Framework with Recursive Attacks ($AFRA$) is defined, similarly to Dung's argumentation framework, as a pair consisting of a set of arguments and a set of attacks. Unlike the original definition, every attack is defined recursively as a pair where the first member is an argument and the second is another attack or an argument (base case).

Definition 2 (*AFRA*). *An Argumentation Framework with Recursive Attacks (AFRA) is a pair $\langle \mathcal{A}, \mathcal{R} \rangle$ where:*

- *$\mathcal{A}$ is a set of arguments;*
- *$\mathcal{R}$ is a set of attacks, namely pairs $(A, \mathcal{X})$ s.t. $A \in \mathcal{A}$ and $(\mathcal{X} \in \mathcal{R}$ or $\mathcal{X} \in \mathcal{A})$.*

Given an attack $\alpha = (A, \mathcal{X}) \in \mathcal{R}$, we will say that A is the source of α, denoted as $src(\alpha) = A$ and $\mathcal{X}$ is the target of α, denoted as $trg(\alpha) = \mathcal{X}$.

We start substantiating the role played by attacks by introducing a notion of defeat which regards attacks, rather than their source arguments, as the subjects able to defeat arguments or other attacks, as encompassed by Definition 3.

Definition 3 (Direct Defeat). *Let $\langle \mathcal{A}, \mathcal{R} \rangle$ be an AFRA, $\mathcal{V} \in \mathcal{R}$, $\mathcal{W} \in \mathcal{A} \cup \mathcal{R}$, then $\mathcal{V}$ directly defeats $\mathcal{W}$ iff $\mathcal{W} = trg(\mathcal{V})$.*

Moreover, as we are interested also in how attacks are affected by other attacks, we introduce a notion of indirect defeat for an attack, corresponding to the situation where its source receives a direct defeat.

Definition 4 (Indirect Defeat). *Let* $\langle \mathcal{A}, \mathcal{R} \rangle$ *be an AFRA,* $\mathcal{V} \in \mathcal{R}$, $\mathcal{W} \in \mathcal{A}$, *if* $\mathcal{V}$ *directly defeats* $\mathcal{W}$ *then* $\forall \alpha \in \mathcal{R}$ *s.t.* $src(\alpha) = \mathcal{W}$, $\mathcal{V}$ *indirectly defeats* α.

Therefore an element $\mathcal{V}$ of a $AFRA$ defeats another element $\mathcal{W}$ if there is a direct or an indirect defeat from $\mathcal{V}$ to $\mathcal{W}$.

Definition 5 (Defeat). *Let* $\langle \mathcal{A}, \mathcal{R} \rangle$ *be an AFRA,* $\mathcal{V} \in \mathcal{R}$, $\mathcal{W} \in \mathcal{A} \cup \mathcal{R}$, *then* $\mathcal{V}$ *defeats* $\mathcal{W}$, *denoted as* $\mathcal{V} \rightarrow_R \mathcal{W}$, *iff* $\mathcal{V}$ *directly or indirectly defeats* $\mathcal{W}$.

To exemplify, the case of Fig. 1 can be represented by an $AFRA$ $\Gamma = \langle \mathcal{A}, \mathcal{R} \rangle$ where $\mathcal{A} = \{C, G, P, N, A\}$ and $\mathcal{R} = \{\alpha, \beta, \gamma, \delta, \epsilon\}$ with $\alpha = (G, C)$, $\beta = (C, G)$, $\gamma = (P, \beta)$, $\delta = (N, \gamma)$, $\epsilon = (A, N)$. There are five direct defeats, namely $\epsilon \rightarrow_R N$, $\delta \rightarrow_R \gamma$, $\gamma \rightarrow_R \beta$, $\beta \rightarrow_R G$, $\alpha \rightarrow_R C$, and three indirect defeats: $\epsilon \rightarrow_R \delta$, $\beta \rightarrow_R \alpha$, $\alpha \rightarrow_R \beta$.

The definition of conflict-free set follows directly.

Definition 6 (Conflict–free). *Let* $\langle \mathcal{A}, \mathcal{R} \rangle$ *be an AFRA,* $\mathcal{S} \subseteq \mathcal{A} \cup \mathcal{R}$ *is conflict–free iff* $\nexists \mathcal{V}, \mathcal{W} \in \mathcal{S}$ *s.t.* $\mathcal{V} \rightarrow_R \mathcal{W}$.

Note that, while the definition of conflict-free set for $AFRA$ is formally almost identical to the corresponding one in AF, actually they feature substantial differences, related to the underlying notion of defeat. In fact in $AFRA$ every set of arguments $\mathcal{S} \subseteq \mathcal{A}$ is conflict-free, since only the explicit consideration of attacks gives rise to conflict in this approach. For instance if $A, B \in \mathcal{A}$ and $\alpha = (A, B) \in \mathcal{R}$, the set $\{A, B\}$ is conflict-free, while the set $\{A, B, \alpha\}$ is not.

Continuing the above example, the maximal, w.r.t. inclusion, conflict free sets of Γ are $\{C, P, A, \epsilon, \beta\}$, $\{C, G, P, N, A, \delta\}$, $\{C, G, P, N, A, \gamma\}$, $\{C, G, A, P, \epsilon, \gamma\}$, $\{C, P, N, A, \delta, \beta\}$, $\{G, P, N, A, \delta, \alpha\}$, $\{G, P, N, A, \alpha, \gamma\}$, $\{G, A, P, \epsilon, \gamma, \alpha\}$.

Also the definition of acceptability is formally very similar to the traditional one, but now it is applied to both arguments and attacks.

Definition 7 (Acceptability). *Let* $\langle \mathcal{A}, \mathcal{R} \rangle$ *be an AFRA,* $\mathcal{S} \subseteq \mathcal{A} \cup \mathcal{R}$, $\mathcal{W} \in \mathcal{A} \cup \mathcal{R}$, $\mathcal{W}$ *is acceptable w.r.t.* $\mathcal{S}$ *iff* $\forall \mathcal{Z} \in \mathcal{R}$ *s.t.* $\mathcal{Z} \rightarrow_R \mathcal{W}$ $\exists \mathcal{V} \in \mathcal{S}$ *s.t.* $\mathcal{V} \rightarrow_R \mathcal{Z}$.

On this basis, the definition of admissibility is formally identical to the one proposed by Dung. As a consequence, it is also possible to directly introduce the notion of *preferred extension* (which is simply a maximal admissible set) lying at the basis of Dung's well-known *preferred semantics* [1].

Definition 8 (Admissibility). *Let* $\langle \mathcal{A}, \mathcal{R} \rangle$ *be an AFRA,* $\mathcal{S} \subseteq \mathcal{A} \cup \mathcal{R}$ *is admissible iff it is conflict–free and each element of* $\mathcal{S}$ *is acceptable w.r.t.* $\mathcal{S}$.

Definition 9 (Preferred extension). *A preferred extension of an AFRA is a maximal (w.r.t. set inclusion) admissible set.*

Referring again to the example of Fig. 1, the only preferred extension of Γ is $\{A, P, G, \epsilon, \gamma, \alpha\}$.

It is also possible to prove that a straigthforward transposition of Dung's fundamental lemma holds in the context of $AFRA$.

Lemma 1 (Fundamental lemma). *Let $\langle A, \mathcal{R} \rangle$ be an AFRA, $\mathcal{S} \subseteq A \cup \mathcal{R}$ an admissible set and $\mathscr{A}, \mathscr{A}' \in A \cup \mathcal{R}$ elements acceptable w.r.t. $\mathcal{S}$. Then:*

1. *$\mathcal{S}' = \mathcal{S} \cup \{\mathscr{A}\}$ is admissible; and*
2. *$\mathscr{A}'$ is acceptable w.r.t. $\mathcal{S}'$.*

Proof.

1. $\mathscr{A}$ is acceptable w.r.t. $\mathcal{S}$ so each element of $\mathcal{S}'$ is acceptable w.r.t. $\mathcal{S}'$. Suppose $\mathcal{S}'$ is not conflict–free; therefore there exists an element $\mathscr{B} \in \mathcal{S}$ such that either $\mathscr{A} \to_R \mathscr{B}$ or $\mathscr{B} \to_R \mathscr{A}$. From the admissibility of $\mathcal{S}$ and the acceptability of $\mathscr{A}$ there exists an element $\bar{\mathscr{B}} \in \mathcal{S}$ such that $\bar{\mathscr{B}} \to_R \mathscr{B}$ or $\bar{\mathscr{B}} \to_R \mathscr{A}$. Since $\mathcal{S}$ is conflict–free it follows that $\bar{\mathscr{B}} \to_R \mathscr{A}$. But then there must exist an element $\hat{\mathscr{B}} \in \mathcal{S}$ such that $\hat{\mathscr{B}} \to_R \bar{\mathscr{B}}$. Contradiction.
2. Obvious.

4   A Correspondence between *AFRA* and *AF*

We consider now the issue of expressing an $AFRA$ in terms of a traditional AF and drawing the relevant correspondences between the basic notions introduced in Section 3. This kind of correspondence provides a very useful basis for further investigations as it allows one to reuse or adapt, in the context of $AFRA$, the many theoretical results available in Dung's framework.

Definition 10. *Let $\Gamma = \langle A, \mathcal{R} \rangle$ be an AFRA, the corresponding AF $\widetilde{\Gamma} = \langle \widetilde{A}, \widetilde{\mathcal{R}} \rangle$ is defined as follows:*

- *$\widetilde{A} = A \cup \mathcal{R}$;*
- *$\widetilde{\mathcal{R}} = \{(\mathscr{V}, \mathscr{W}) | \mathscr{V}, \mathscr{W} \in A \cup \mathcal{R} \text{ and } \mathscr{V} \to_R \mathscr{W}\}$.*

In words both arguments and attacks of the original $AFRA$ Γ become arguments of its corresponding AF version $\widetilde{\Gamma}$ while the defeat relations in AF correspond to all direct and indirect defeats in the original $AFRA$. We can now examine the relationships between the relevant basic notions in Γ and $\widetilde{\Gamma}$.

Lemma 2. *Let $\Gamma = \langle A, \mathcal{R} \rangle$ an AFRA and $\widetilde{\Gamma} = \langle \widetilde{A}, \widetilde{\mathcal{R}} \rangle$ its corresponding AF:*

1. *$\mathcal{S}$ is a conflict–free set for Γ iff $\mathcal{S}$ is a conflict–free set for $\widetilde{\Gamma}$;*
2. *$\mathscr{A}$ is acceptable w.r.t. $\mathcal{S} \subseteq A \cup \mathcal{R}$ in Γ iff $\mathscr{A}$ is acceptable w.r.t. $\mathcal{S}$ in $\widetilde{\Gamma}$;*
3. *$\mathcal{S}$ is an admissible set for Γ iff $\mathcal{S}$ is an admissible set for $\widetilde{\Gamma}$.*

Proof

1. The conclusion follows directly from Def. 10.
2. *Right to left half:* Let $\mathscr{A} \in A \cup \mathcal{R}$ be acceptable w.r.t. $\mathcal{S} \subseteq A \cup \mathcal{R}$ in Γ and suppose $\mathscr{A}$ is not acceptable w.r.t. $\mathcal{S}$ of $\widetilde{\Gamma}$. So, there exists $\mathscr{B} \in \widetilde{A} = A \cup \mathcal{R}$ s.t. $(\mathscr{B}, \mathscr{A}) \in \widetilde{\mathcal{R}}$ and $\nexists \mathscr{C} \in \mathcal{S}$ s.t. $(\mathscr{C}, \mathscr{B}) \in \widetilde{\mathcal{R}}$. From Def. 10, $(\mathscr{B}, \mathscr{A}) \in \widetilde{\mathcal{R}}$ iff $\mathscr{B} \to_R \mathscr{A}$ and $(\mathscr{C}, \mathscr{B}) \in \widetilde{\mathcal{R}}$ iff $\mathscr{C} \to_R \mathscr{B}$. Then $\exists \mathscr{A} \in \mathcal{S}$, $\exists \mathscr{B} \in A \cup \mathcal{R}$ s.t. $\mathscr{B} \to_R \mathscr{A}$ and $\nexists \mathscr{C} \in \mathcal{S}$ s.t. $\mathscr{C} \to_R \mathscr{B}$. Therefore $\mathscr{A}$ is not acceptable w.r.t. $\mathcal{S}$ in Γ. Contradiction.
   Left to right half: Follows the same reasoning line with obvious modifications.
3. Follows directly from 1 and 2.

5   Comparing *AFRA* with *EAF+*

AFRA provides a formally quite simple way to generalize the notion of attack of Dung's framework, such simplicity partially hiding some substantial underlying differences, concerning in particular the notion of conflict-free set. A more restricted, but similar in spirit, generalization of Dung's framework has recently been proposed in [4,5,6], with the name of Extended Argumentation Framework (*EAF*). This approach is motivated by the need to express preferences between arguments and supports a very interesting form of meta–level argumentation about the values that arguments promote [2]. In *EAF* a limited notion of attacks to attacks is encompassed: only attacks whose target is an argument (i.e. the "traditional" ones) can be attacked, while attacks whose target is another attack can not be in turn attacked. Referring to Figure 1, only the attack originated from P could be represented, while the one originated from N could not. On the other hand, the notion of conflict–free set introduced in [5] for *EAF* is somehow closer to Dung's original one. To compare this kind of approach with *AFRA* we investigate in this section an extension of *EAF* (called *EAF+*) which allows for recursive attacks, while attempting to follow as close as possible the original *EAF* definitions provided in [5].

The definition of *EAF+* keeps the original elements of Dung's definition, adding, as a separate entity, a relation of attack between attacks.

Definition 11 (*EAF+*). *An EAF Plus (EAF+) is a tuple $\langle \mathcal{A}, \mathcal{R}, \mathcal{D}+ \rangle$ s.t.:*

1. *$\mathcal{A}$ is a set of arguments;*
2. *$\mathcal{R} \subseteq \mathcal{A} \times \mathcal{A}$;*
3. *$\mathcal{D}+$ is a set of pairs (A, δ) s.t. $A \in \mathcal{A}$ and ($\delta \in \mathcal{R}$ or $\delta \in \mathcal{D}+$).*

We extend in the obvious way the notions of source and target of an attack to the pairs both in $\mathcal{R}$ and in $\mathcal{D}+$. The notion of defeat for *EAF+* turns out to be articulated into four cases as it has to encompass the roles of both arguments and attacks.

Definition 12 (Defeat). *Let $\langle \mathcal{A}, \mathcal{R}, \mathcal{D}+ \rangle$ be an EAF+, $\mathcal{E} = \mathcal{R} \cup \mathcal{D}+$, $\mathcal{V}, \mathcal{W} \in \mathcal{A} \cup \mathcal{E}$, $\mathcal{V}$ defeats $\mathcal{W}$ (denoted $\mathcal{V} \rightarrow_E \mathcal{W}$) if any of the following conditions holds:*

1. *$\mathcal{V}, \mathcal{W} \in \mathcal{A}, (\mathcal{V}, \mathcal{W}) \in \mathcal{R}$;*
2. *$\mathcal{V} \in \mathcal{A}, \mathcal{W} \in \mathcal{E}$ and $(\mathcal{V}, \mathcal{W}) \in \mathcal{E}$;*
3. *$\mathcal{V}, \mathcal{W} \in \mathcal{E}$ and $\mathcal{W} = trg(\mathcal{V})$;*
4. *$\mathcal{V} \in \mathcal{E}, \mathcal{W} \in \mathcal{A}$ and $\mathcal{W} = trg(\mathcal{V})$.*

Keeping again distinct the treatment of arguments and attacks, the property of being conflict-free has to be introduced for pairs consisting of a set of arguments and a set of attacks. Moreover we need to constrain the sets of arguments and attacks in the pair, by restricting our attention on *self-contained* pairs, where the presence of an attack implies also the presence of its source.

Definition 13 (Self–contained pair). *Let $\langle \mathcal{A}, \mathcal{R}, \mathcal{D}+ \rangle$ be an EAF+, $\mathcal{E} = \mathcal{R} \cup \mathcal{D}+$, $\mathcal{S} \subseteq \mathcal{A}$, $\mathcal{T} \subseteq \mathcal{E}$, a pair $\mathcal{W} = (\mathcal{S}, \mathcal{T})$ is self–contained iff $\forall \alpha \in \mathcal{T}$, $src(\alpha) \in \mathcal{S}$. We will denote a self–contained pair as $\widehat{\mathcal{W}} = \widehat{(\mathcal{S}, \mathcal{T})}$.*

Definition 14 (Conflict–free). *Let $\langle \mathcal{A}, \mathcal{R}, \mathcal{D}+ \rangle$ be an $EAF+$, a self–contained pair $\widehat{W} = \widehat{(\mathcal{S}, \mathcal{T})}$ is conflict–free iff $\forall A, B \in \mathcal{S}$ s.t. $A \rightarrow_E B$ (so there exists $\alpha = (A, B) \in \mathcal{R}$), $\exists \beta \in \mathcal{T}$ s.t. $\beta \rightarrow_E \alpha$ and $\nexists \gamma \in \mathcal{T}, \nexists \mathcal{D} \in \mathcal{S} \cup \mathcal{T}$ s.t. $\mathcal{D} = trg(\gamma)$.*

In words, any attack between the arguments in $\mathcal{S}$ has to be attacked in turn by $\mathcal{T}$ (and thus made ineffective), while attacks in $\mathcal{T}$ directed against any other element in the pair are simply not allowed. Note that, differently from the definition provided for AF, there may be a conflict even in absence of attacks in $\mathcal{T}$, so a pair $\widehat{W} = \widehat{(\mathcal{S}, \emptyset)}$ may not be conflict-free.

In $EAF+$, like in Dung's AF, an element (argument or attack) is acceptable with respect to a self–contained pair when it is defended against any attack it receives. Defense may consist of a defeat against the attack or against its source.

Definition 15 (Acceptability). *Let $\langle \mathcal{A}, \mathcal{R}, \mathcal{D}+ \rangle$ be an $EAF+$, $\mathcal{E} = \mathcal{R} \cup \mathcal{D}+$, and $\widehat{W} = \widehat{(\mathcal{S}, \mathcal{T})}$ a self–contained pair:*

- *$A \in \mathcal{A}$ is acceptable w.r.t. $\widehat{W} = \widehat{(\mathcal{S}, \mathcal{T})}$ iff $\forall \beta \in \mathcal{R}$ s.t. $A = trg(\beta)$, $\exists \alpha \in \mathcal{T}$ s.t. $\alpha \rightarrow_E src(\beta)$ or $\alpha \rightarrow_E \beta$;*
- *$\alpha \in \mathcal{E}$ is acceptable w.r.t. $\widehat{W} = \widehat{(\mathcal{S}, \mathcal{T})}$ iff $src(\alpha)$ is acceptable w.r.t. $\widehat{W}$ and $\forall \beta \in \mathcal{D}+$ s.t. $\beta \rightarrow_E \alpha$, $\exists \gamma \in \mathcal{T}$ s.t. $\gamma \rightarrow_E src(\beta)$ or $\gamma \rightarrow_E \beta$.*

Following [1,5], we consider a self–contained pair admissible if and only if it is conflict–free and any of its elements is acceptable with respect to it.

Definition 16 (Admissibility). *Let $\langle \mathcal{A}, \mathcal{R}, \mathcal{D}+ \rangle$ be an $EAF+$, $\mathcal{E} = \mathcal{R} \cup \mathcal{D}+$, a self–contained pair $\widehat{W} = \widehat{(\mathcal{S}, \mathcal{T})}$ is admissible iff it is conflict–free, $\forall A \in \mathcal{S}$, A is acceptable w.r.t. $\widehat{W} = \widehat{(\mathcal{S}, \mathcal{T})}$ and $\forall \alpha \in \mathcal{T}$, α is acceptable w.r.t. $\widehat{W} = \widehat{(\mathcal{S}, \mathcal{T})}$.*

In order to introduce the notion of preferred extension for $EAF+$ we have to define an inclusion relation between self–contained pairs.

Definition 17 (Inclusion of self–contained pair). *Let $\langle \mathcal{A}, \mathcal{R}, \mathcal{D}+ \rangle$ be an $EAF+$, $\mathcal{E} = \mathcal{R} \cup \mathcal{D}+$, $\mathcal{S} \subseteq \mathcal{A}$, $\mathcal{T} \subseteq \mathcal{E}$, $\widehat{W'} = \widehat{(\mathcal{S'}, \mathcal{T'})}$ is included in $\widehat{W} = \widehat{(\mathcal{S}, \mathcal{T})}$ iff $\mathcal{S'} \subseteq \mathcal{S}$ and $\mathcal{T'} \subseteq \mathcal{T}$.*

Definition 18 (Preferred extension). *A preferred extension of an $EAF+$ is a maximal (according to the inclusion relation introduced in Definition 17) admissible self–contained pair of $EAF+$.*

It is also possible to prove Dung's fundamental Lemma for the case of $EAF+$.

Lemma 3 (Fundamental lemma). *Let $\langle \mathcal{A}, \mathcal{R}, \mathcal{D}+ \rangle$ be an $EAF+$, $\mathcal{E} = \mathcal{R} \cup \mathcal{D}+$, $\widehat{W} = \widehat{(\mathcal{S}, \mathcal{T})}$ an admissible self–contained pair and $\mathcal{A}, \mathcal{A'} \in \mathcal{A} \cup \mathcal{E}$ acceptable w.r.t. $\widehat{W}$. Then:*

1. *$\widehat{W'} = \begin{cases} \widehat{(\mathcal{S'}, \mathcal{T})}, \mathcal{S'} = \mathcal{S} \cup \{\mathcal{A}\} \text{ if } \mathcal{A} \in \mathcal{A} \\ \widehat{(\mathcal{S}, \mathcal{T'})}, \mathcal{T'} = \mathcal{T} \cup \{\mathcal{A}\} \text{ if } \mathcal{A} \in \mathcal{E} \end{cases}$ is admissible; and*
2. *$\mathcal{A'}$ is acceptable w.r.t. $\widehat{W'}$.*

Proof. To prove this lemma, we have to consider four different cases:

A. $\mathscr{A}, \mathscr{A}' \in \mathcal{A}$;
B. $\mathscr{A} \in \mathcal{A}$, $\mathscr{A}' \in \mathcal{E}$;
C. $\mathscr{A} \in \mathcal{E}$, $\mathscr{A}' \in \mathcal{A}$;
D. $\mathscr{A}, \mathscr{A}' \in \mathcal{E}$.

Case A.
1. $\mathscr{A}$ is acceptable w.r.t. $\widehat{\mathcal{W}} = \widehat{(\mathcal{S}, \mathcal{T})}$ therefore in $\widehat{\mathcal{W}'} = \widehat{(\mathcal{S}', \mathcal{T})}$, with $\mathcal{S}' = \mathcal{S} \cup \{\mathscr{A}\}$, $\forall A \in \mathcal{S}'$, A is acceptable w.r.t. $\widehat{\mathcal{W}'}$ and $\forall \alpha \in \mathcal{T}$, α is acceptable w.r.t. $\widehat{\mathcal{W}'}$. Suppose $\widehat{\mathcal{W}'}$ is not conflict-free. There are two possible cases: (a) $\exists A, B \in \mathcal{S}'$ s.t. $\alpha = (A, B) \in \mathcal{R}$ and $\nexists \beta \in \mathcal{T}$ s.t. $\beta \to_E \alpha$; (b) $\exists \alpha \in \mathcal{T}, \exists \mathscr{D} \in \mathcal{S}' \cup \mathcal{T}$ s.t. $\mathscr{D} = trg(\alpha)$.

   Considering case (b), from the admissibility of $\widehat{\mathcal{W}}$ we have $\mathscr{D} = \mathscr{A} = trg(\alpha)$; from the acceptability of $\mathscr{A}$ w.r.t. $\widehat{\mathcal{W}}$ and Def. 15 two cases are possible but $\alpha \in \mathcal{T}$, so both cases imply that $\widehat{\mathcal{W}}$ is not conflict–free: contradiction.

   Considering case (a), from the admissibility of $\widehat{\mathcal{W}}$, $A = \mathscr{A}$ or $B = \mathscr{A}$.

   Suppose $A = \mathscr{A}$, i.e. $\alpha = \mathscr{A} \to_E B$. Since B is acceptable w.r.t. $\widehat{\mathcal{W}}$ from Def. 15 two cases are in turn possible: (i) $\exists \beta \in \mathcal{T}$ s.t. $\beta \to_E \mathscr{A}$ and $src(\beta) \in \mathcal{S}$; (ii) $\exists \beta \in \mathcal{T}$ s.t. $\beta \to_E \alpha$.

   In case (i) from the acceptability of $\mathscr{A}$ with respect to $\widehat{\mathcal{W}}$ we would have that β or its source (both belonging to $\widehat{\mathcal{W}}$) should be defeated by an element of $\widehat{\mathcal{W}}$ thus contradicting the fact that $\widehat{\mathcal{W}}$ is conflict free. In case (ii) we contradict the assumption (a).

   Suppose now $B = \mathscr{A}$; so $trg(\alpha) = \mathscr{A}$. From the acceptability of $\mathscr{A}$ w.r.t. $\widehat{\mathcal{W}}$ there exists $\beta \in \mathcal{T}$ s.t. $\beta \to_E \alpha$ or $\beta \to_E src(\alpha)$, but again in both cases we contradict the fact that $\widehat{\mathcal{W}}$ is conflict free.
2. Obvious.

Case B.
1. Same as Case A, item 1.
2. Suppose $\mathscr{A}'$ is acceptable w.r.t. $\widehat{\mathcal{W}} = \widehat{(\mathcal{S}, \mathcal{T})}$ but not w.r.t. $\widehat{\mathcal{W}'} = \widehat{(\mathcal{S}', \mathcal{T})}$; then (i) $src(\mathscr{A}')$ is not acceptable w.r.t. $\widehat{\mathcal{W}'}$ or (ii) $\exists \beta \in \mathcal{D}+$ s.t. $\beta \to_E \mathscr{A}'$ and $\nexists \gamma \in \mathcal{T}$ s.t. $\gamma \to_E src(\beta)$ or $\gamma \to_E \beta$. In case (i), noting that $src(\mathscr{A}')$ is acceptable w.r.t. $\widehat{\mathcal{W}}$, we contradict what we have proved in Case A, item 1. Case (ii) contradicts the fact that $\mathscr{A}'$ is acceptable w.r.t. $\widehat{\mathcal{W}}$.

Case C.
1. $\mathscr{A}$ is acceptable w.r.t. $\widehat{\mathcal{W}} = \widehat{(\mathcal{S}, \mathcal{T})}$ so in $\widehat{\mathcal{W}'} = \widehat{(\mathcal{S}, \mathcal{T}')}$, with $\mathcal{T}' = \mathcal{T} \cup \{\mathscr{A}\}$, $\forall A \in \mathcal{S}$, A is acceptable w.r.t. $\widehat{\mathcal{W}'}$ and $\forall \alpha \in \mathcal{T}'$, α is acceptable w.r.t. $\widehat{\mathcal{W}'}$. Suppose $\widehat{\mathcal{W}'}$ is not conflict–free. There are two possible cases: (a) $\exists A, B \in \mathcal{S}$ s.t. $\alpha = (A, B) \in \mathcal{R}$ and $\nexists \beta \in \mathcal{T}'$ s.t. $\beta \to_E \alpha$; (b) $\exists \alpha \in \mathcal{T}', \exists \mathscr{D} \in \mathcal{S} \cup \mathcal{T}'$ s.t. $\mathscr{D} = trg(\alpha)$.

   Case (a) is impossible because $\mathcal{T}' \supseteq \mathcal{T}$ and $\widehat{\mathcal{W}}$ is admissible.

   Let us consider case (b): from the admissibility of $\widehat{\mathcal{W}}$, $\alpha = \mathscr{A}$ or $\mathscr{D} = \mathscr{A}$.

Suppose $\alpha = \mathscr{A}$; so $\mathscr{A} \rightarrow_E \mathscr{D}$ and $src(\mathscr{A})$ is acceptable w.r.t. $\widehat{\mathcal{W}}$ (because $\mathscr{A}$ is acceptable w.r.t. $\widehat{\mathcal{W}}$). From the acceptability of $\mathscr{D}$ w.r.t. $\widehat{\mathcal{W}}$, there exists $\gamma \in \mathcal{T}$, $src(\gamma) \in \mathcal{S}$ s.t. (i) $\gamma \rightarrow_E src(\mathscr{A})$ or (ii) $\gamma \rightarrow_E \mathscr{A}$. In both cases, from the acceptability of $src(\mathscr{A})$ w.r.t. $\widehat{\mathcal{W}}$, there exists $\delta \in \mathcal{T}$ s.t. $\delta \rightarrow_E src(\gamma)$ or $\delta \rightarrow_E \gamma$. Either case implies that $\widehat{\mathcal{W}}$ is not conflict–free: contradiction.

Suppose $\mathscr{D} = \mathscr{A}$; so $\alpha \rightarrow_E \mathscr{A}$. Using the acceptability of $\mathscr{A}$ w.r.t. $\widehat{\mathcal{W}}$, we can then apply the same reasoning as above leading to contradict the fact that $\widehat{\mathcal{W}}$ is conflict–free.

2. Obvious.

Case D.

1. Same of Case C, item 1.
2. Same of Case B, item 2.

Let us now analyse the relation between $EAF+$ and $AFRA$: first it is possible to draw a direct correspondence between the two formalisms.

Definition 19 ($AFRA$–$EAF+$ **correspondence**). *For any $EAF+$ $\Delta = \langle \mathcal{A}, \mathcal{R}, \mathcal{D}+ \rangle$ we define the corresponding $AFRA$ as $\Delta_R = \langle \mathcal{A}, \mathcal{R} \cup \mathcal{D}+ \rangle$. For any $AFRA$ $\Gamma = \langle \mathcal{A}, \mathcal{R}' \rangle$ we define the corresponding $EAF+$ as $\Gamma_E = \langle \mathcal{A}, \mathcal{R}' \cap (\mathcal{A} \times \mathcal{A}), \mathcal{R}' \setminus (\mathcal{A} \times \mathcal{A}) \rangle$.*

The correspondence between the notions of defeat is drawn in Lemma 4.

Lemma 4. *Let $\Delta = \langle \mathcal{A}, \mathcal{R}, \mathcal{D}+ \rangle$ be an $EAF+$ and Δ_R its corresponding $AFRA$. For any $\mathscr{V}, \mathscr{W} \in (\mathcal{A} \cup \mathcal{R} \cup \mathcal{D}+)$, $\mathscr{V} \rightarrow_E \mathscr{W}$ in Δ iff the disjunction of the following conditions holds:*

(a) $\mathscr{W}$ is directly defeated by $\mathscr{V}$ in Δ_R;
(b) $\exists \mathscr{Z} \in (\mathcal{R} \cup \mathcal{D}+)$ s.t. $\mathscr{V} = src(\mathscr{Z})$ and $\mathscr{Z}$ directly defeats $\mathscr{W}$ in Δ_R.

Proof. Consider the four cases of defeat for $EAF+$ of Def. 12. The disjunction of cases 1 and 2 is equivalent to condition (b). In fact, in these cases $\exists \mathscr{Z} \in (\mathcal{R} \cup \mathcal{D}+)$ s.t. $\mathscr{V} = src(\mathscr{Z})$ and $\mathscr{W} = trg(\mathscr{Z})$. It follows that $\mathscr{Z}$ directly defeats $\mathscr{W}$ in Δ_R. Conversely, if $\exists \mathscr{Z} \in (\mathcal{R} \cup \mathcal{D}+)$ s.t. $\mathscr{V} = src(\mathscr{Z})$ and $\mathscr{Z}$ directly defeats $\mathscr{W}$ in Δ_R it follows that $\mathscr{W} = trg(\mathscr{Z})$ which implies that case 1 or 2 holds in Δ. The disjunction of cases 3 and 4 is equivalent to condition (a): the fact that either 3 or 4 implies (a) follows directly from Definition 3, the converse implication is immediate too.

From the proof of Lemma 4 we can observe that the four-cases definition of defeat in $EAF+$ is somewhat redundant and might be summarized referring to the more synthetic formulation used in $AFRA$. We can now analyse the relationship involving the notions of conflict-free set, acceptability and admissibility: those defined for $EAF+$ turn out to be a specialization of the ones in $AFRA$.

Lemma 5. *Let* $\Delta = \langle \mathcal{A}, \mathcal{R}, \mathcal{D}+ \rangle$ *an EAF+ and* $\Gamma = \langle \mathcal{A}, \mathcal{R}' \rangle = \Delta_R$:

1. *if* $\widehat{\mathcal{W}} = \widehat{(\mathcal{S}, \mathcal{T})}$ *is conflict–free in* Δ *then* $\mathcal{S} \cup \mathcal{T}$ *is a conflict–free set for* Γ;
2. *if* $\mathcal{V}$ *is acceptable w.r.t.* $\widehat{\mathcal{W}} = \widehat{(\mathcal{S}, \mathcal{T})}$ *in* Δ *then* $\mathcal{V}$ *is acceptable w.r.t.* $\mathcal{S} \cup \mathcal{T}$ *in* Γ;
3. *if* $\widehat{\mathcal{W}} = \widehat{(\mathcal{S}, \mathcal{T})}$ *is admissible for* Δ *then* $\mathcal{S} \cup \mathcal{T}$ *is an admissible set for* Γ.

Proof

1. Let $\widehat{\mathcal{W}} = \widehat{(\mathcal{S}, \mathcal{T})}$ a conflict–free pair for Δ. From Def. 14 we have: (a) $\forall A, B \in \mathcal{S}, \alpha = (A, B) \in \mathcal{R}, \exists \beta \in \mathcal{T}$ s.t. $\beta \to_E \alpha$; and (b) $\nexists \gamma \in \mathcal{T}, \nexists \mathcal{D} \in \mathcal{S} \cup \mathcal{T}$ s.t. $\mathcal{D} = trg(\gamma)$.

   Let $\mathcal{U} = \mathcal{S} \cup \mathcal{T} \subseteq \mathcal{A} \cup \mathcal{R}'$. According to Def. 6 we have to show that $\nexists \mathcal{V}, \mathcal{W} \in \mathcal{U}$ s.t. $\mathcal{V} \to_R \mathcal{W}$ which in turn amounts to require: (i) $\nexists \mathcal{V} \in \mathcal{U} \cap \mathcal{R}'$, $\nexists \mathcal{W} \in \mathcal{U}$ s.t. $\mathcal{W} = trg(\mathcal{V})$; (ii) $\nexists \mathcal{V}, \mathcal{Z} \in \mathcal{U} \cap \mathcal{R}'$, s.t. $src(\mathcal{Z}) = trg(\mathcal{V})$. From (b) it follows that $\nexists \mathcal{V} \in \mathcal{U} \cap \mathcal{R}'$, $\nexists \mathcal{W} \in \mathcal{U}$ s.t. $\mathcal{W} = trg(\mathcal{V})$, i.e. condition (i). To prove (ii), assume that $\exists \mathcal{V}, \mathcal{Z} \in \mathcal{U} \cap \mathcal{R}'$, s.t. $src(\mathcal{Z}) = trg(\mathcal{V})$. To avoid contradiction with (b) we have to assume $src(\mathcal{Z}) \notin \mathcal{S}$, but this contradicts in turn the fact that $\widehat{\mathcal{W}}$ is self-contained.
2. Let $\mathcal{V}$ be acceptable w.r.t. $\widehat{\mathcal{W}} = \widehat{(\mathcal{S}, \mathcal{T})}$. By Def. 7, we have to show that $\forall \mathcal{Z} \in \mathcal{R}'$ s.t. $\mathcal{Z} \to_R \mathcal{V}$ (i.e. (i) $\mathcal{V} = trg(\mathcal{Z})$ or, if $\mathcal{V} \in \mathcal{R}'$, (ii) $src(\mathcal{V}) = trg(\mathcal{Z})$), $\exists \mathcal{W} \in \mathcal{U} = \mathcal{S} \cup \mathcal{T}$ s.t. $\mathcal{W} \to_R \mathcal{Z}$ (i.e. (iii) $\mathcal{Z} = trg(\mathcal{W})$ or (iv) $src(\mathcal{Z}) = trg(\mathcal{W})$). Assume first (i) and $\mathcal{V} \in \mathcal{A}$: then (iii) or (iv) follows directly from the first part of Def. 15. Assuming (i) and $\mathcal{V} \in \mathcal{R}'$, according to the second part of Def. 15 we obtain again that (iii) or (iv) holds. Assuming (ii), we have $\mathcal{V} \in \mathcal{R}'$ and, by Def. 15, $src(\mathcal{V})$ must be acceptable w.r.t. $\widehat{\mathcal{W}} = \widehat{(\mathcal{S}, \mathcal{T})}$ and (iii) or (iv) follows again from the first part of Def. 15.
3. Follows directly from 1 and 2.

6 Discussion and Conclusions

We have proposed a preliminary investigation about $AFRA$, a generalization of Dung's argumentation framework where attacks to attacks are recursively encompassed without restriction. An intuitive justification for this kind of formalism has been provided in relation with the representation of decision processes.

The idea of encompassing attacks to attacks in abstract argumentation framework has been first considered in [7], in the context of an extended framework encompassing argument strengths and their propagation. In this quite different context, deserving further development, Dung style semantics issues have not been considered.

Focusing on approaches closer to "traditional" Dung's framework, attacks to attacks have been considered in the context of reasoning about preferences [4,5,6] and reasoning about coalitions [8]. In both cases only attacks to attacks between arguments are covered, i.e. only one level of recursion is allowed. A detailed conceptual analysis motivating EAF with respect to a variety of reference domains is provided in particular in [6]. While developing this kind of analysis for $AFRA$ and the relevant comparison with EAF is beyond the scope of this paper, one can note that EAF adopts some specific assumptions, for instance a limited level of recursion and a constraint on some attacks to be symmetric, when the involved arguments represent conflicting preferences. These assumptions are fully justified in the context of reasoning about preferences but may not be necessary in general. The $AFRA$ formalism, though originally conceived to support some intuitive forms of reasoning in the context of decision making, addresses the general need to reason about conflicts which may be themselves defeasible. In order to satisfy this need, it seems reasonable to provide attacks with an ontological status encompassing defeasibility, differently than in AF.

To complete the comparison from a more technical point of view, we have considered in Section 5 the $EAF+$ formalism, namely a possible extension of EAF aimed at overcoming these restrictive assumptions, and we have drawn correspondences between $EAF+$ and $AFRA$. It turns out that the $AFRA$ formalism supports more compact (and in some cases also more general) definitions of the fundamental notions of defeat, conflict–free set, acceptability and admissibility. The "translation" from $AFRA$ to Dung's AF proposed in Section 4 opens the way to one of the main future work directions, namely enlarging the theoretical bases of $AFRA$ and investigating the definition of argumentation semantics in this context, possibly exploiting the rich corpus of results available for the traditional framework.

References

1. Dung, P.M.: On the acceptability of arguments and its fundamental role in non-monotonic reasoning, logic programming, and n-person games. Artificial Intelligence 77(2), 321–357 (1995)
2. Bench-Capon, T.J.M.: Persuasion in practical argument using value based argumentation frameworks. Journal of Logic and Computation 13(3), 429–448 (2003)
3. Amgoud, L., Cayrol, C., Lagasquie-Schiex, M.C., Livet, P.: On bipolarity in argumentation frameworks. Int. Journal of Intelligent Systems 23, 1062–1093 (2008)
4. Modgil, S.: An abstract theory of argumentation that accommodates defeasible reasoning about preferences. In: Mellouli, K. (ed.) ECSQARU 2007. LNCS, vol. 4724, pp. 648–659. Springer, Heidelberg (2007)
5. Bench-Capon, T.J.M., Mogdil, S.: Integrating object and meta-level value based argumentation. In: Proc. 2nd Int. Conf. on Computational Models of Argument (COMMA 2008), Toulouse, F, pp. 240–251 (2008)
6. Modgil, S.: Reasoning about preferences in argumentation frameworks. Artificial Intelligenc (in press, 2009)
7. Barringer, H., Gabbay, D.M., Woods, J.: Temporal dynamics of support and attack networks: From argumentation to zoology. In: Hutter, D., Stephan, W. (eds.) Mechanizing Mathematical Reasoning. LNCS, vol. 2605, pp. 59–98. Springer, Heidelberg (2005)
8. Boella, G., van der Torre, L., Villata, S.: Social viewpoints for arguing about coalitions. In: Proc. 11th Pacific Rim Int. Conf. on Multi-Agents (PRIMA 2008), pp. 66–77 (2008)

Social Argument Justification: Some Mechanisms and Conditions for Their Coincidence[*]

Gustavo Adrián Bodanza[1,2,3] and Marcelo Roberto Auday[2]

[1] Artificial Intelligence Research and Development Laboratory (LIDIA)
[2] Logic and Philosophy of Science Research Center (CILF),
Universidad Nacional del Sur, (8000) Bahía Blanca, Argentina
[3] Consejo Nacional de Investigaciones Científicas y Técnicas (CONICET), Argentina
{ccbodanz,ccauday}@criba.edu.ar

Abstract. In this paper we analyze the problem of aggregating different individual argumentation frameworks over a common set of arguments in order to obtain a unique socially justified set of arguments. This can be done in two different ways: a social attack relation is built up from the individual ones, and then is used to produce a set of justified arguments, or this set is directly obtained from the sets of individually justified arguments. Our main concern here is whether these two procedures can coincide or under what conditions this could happen. To deal with this, we consider different voting by quota mechanisms, and the aggregation mechanisms by decisive sets.

1 Introduction

The work by P.M. Dung ([6]) has paved the way for the study of argument justification in a highly abstract level, offering different ways of selecting arguments according to a given attack relation among them. An attack relation represents a criterion by which arguments pose threats of defeat on other arguments, and justified arguments are those sanctioned by some "extension" semantics, based on a notion of acceptability. Roughly speaking, an agent can accept an argument a if she has some arguments to reply any possible attack on a. A structure composed of a set of arguments and an attack relation among them constitutes an *argumentation framework*.

Bench-Capon [3] and Bench-Capon and Modgil [4] have studied meta-level argumentation where some arguments express preferences about the social values that other arguments promote, so that different audiences having different preferences can yield different defeat relations among arguments. In the same way, we consider that different individuals can yield different defeat relations on the basis of their own preferences. In this paper we think of a society of agents, each one supporting a particular criterion of attack, yielding her own individual argumentation framework and, as a consequence, her own justified arguments. Given a pair of arguments a and b, each agent can individually express her criterion by choosing one of four alternatives: both arguments are perfectly compatible, a attacks b, b attacks a, or they attack each other (expressing that

---

[*] Partially supported by SeCyT - Universidad Nacional del Sur, CONICET and ANPCYT.

C. Sossai and G. Chemello (Eds.): ECSQARU 2009, LNAI 5590, pp. 95–106, 2009.

they are in conflict but indifferent). The main issue we address is how the society can yield argument justifications on the basis of the individual criteria of its members.

A *social* justification of arguments can be thought through two different kinds of procedures:

P1: by aggregating all the individual attack relations into one social relation, and then obtaining the socially justified arguments w.r.t. this relation through some specific extension semantics;

P2: by aggregating all the sets of individually justified arguments into one set of socially justified arguments.

Social Choice theory (SCT) has studied several mechanisms to characterize "fair" aggregations, in accordance with general democratic principles ([1]). Taking into account that argumentation is at the very heart of deliberative democracy, we are interested in applying aggregation mechanisms for characterizing socially justified arguments in both of the above mentioned ways. In particular, we will explore the instantiation of P1 and P2 through the majority voting mechanism and mechanisms requiring arbitrary quotas of votes; on the other hand, we will also consider mechanisms that can be represented by the sets of individuals which are decisive to impose their will. Tohmé et al. ([13]) have studied procedure P1 (aggregation of attack relations), showing how some well-known arrovian principles of SCT are accomplished when the algebraic structure of the class of decisive sets is that of a proper prefilter. Coste-Marquis et al. ([5]) instantiate the P1 procedure by merging individual argumentation frameworks that are expanded into a partial system over the set of all arguments considered by the group of agents. Procedure P2 is rejected by these last authors arguing that it only would make sense if all agents consider the same set of arguments (e.g. Rahwan and Larson do this in [12]). We agree at this point, but then the authors point out another drawback: the attack relations (from which extensions are characterized) are not taken into consideration any more once extensions have been computed (see [5], p. 615). On the contrary, we think that analogously to SCT, where choice sets often say something about the preference involved, extensions —in certain circumstances— can say something about the attack relation involved. We will show, in fact, that there are sensible conditions under which P1 and P2 mechanisms lead to logically related outcomes.

A concrete question we ask is whether procedures P1 and P2 will coincide in general or not. This problem replicates in some way the *discursive dilemma* or *doctrinal paradox*, where a collective judgment reached by voting on the set of premises of an argument can yield a different outcome to that reached by voting on the conclusion ([8], [11], [7], etc.). Though our approach is concerned with the justification of arguments instead of judgments (i.e. sentences), the problem seems to be the same at the underlying level in which a deliberation on some input information (premises of an argument/attacks criteria) can yield a collective result that is different to that of the deliberation on the possible outputs (conclusion of the argument/justified arguments). The following example illustrates the problem.

Example 1. Assume that a medical team of three M.D.'s, say 1, 2 and 3, deliberate about which therapy should be applied to some patient. After a few rounds of discussion the deliberation focuses on three arguments, a, b and c, each one concluding that a given therapy must be applied on basis of some observed symptoms. But each doctor has a particular opinion about the importance of the observed symptoms and about the compatibility or incompatibility of the therapies, which leads to three different criteria of attack among the arguments. Say that each individual i proposes an attack criterion $\rightharpoonup_i$ over $\{a, b, c\}$ as follows:

$$a \rightharpoonup_1 b \rightharpoonup_1 c,$$
$$c \rightharpoonup_2 b \rightharpoonup_2 a,$$
$$b \rightharpoonup_3 a, b \rightharpoonup_3 c.$$

(For example, individual 1 thinks that the therapy suggested by argument b is incompatible with those suggested by a and c, while these ones are compatible between them; moreover, she prefers a to b and b to c.) Assume now that this little society formed by the team has two alternative ways to reach a decision on which are the justified arguments, consisting in respective implementations of P1 and P2 through majority voting, say $P1_m$ and $P2_m$. Then we have:

- $(P1_m)$ The team obtains the social attack relation $\rightharpoonup$ as follows:
  - $b \rightharpoonup c$, since b attacks c under $\rightharpoonup_1$ and $\rightharpoonup_3$;
  - $b \rightharpoonup a$, since b attacks a under $\rightharpoonup_2$ and $\rightharpoonup_3$.

  Thus $\rightharpoonup$ (which coincides with $\rightharpoonup_3$) yields the only socially justified argument b (b is the only non attacked argument and it attacks the other two arguments).
- $(P2_m)$ The society chooses the justified arguments by aggregating all the sets of individually justified arguments. Assuming that each individual follows any of the Dung's extension semantics as the justification criterion, then:
  - individual 1 can justify the choice set $\{a, c\}$ according to $\rightharpoonup_1$;
  - individual 2 can justify the choice set $\{a, c\}$ according to $\rightharpoonup_2$;
  - individual 3 can justify the choice set $\{b\}$ according to $\rightharpoonup_3$.

Thus, according to majority voting on these individual choice sets, the socially justified arguments should be a and c. Hence, clearly $P1_m$ and $P2_m$ yield different outcomes, resembling the doctrinal paradox.

As this example shows, procedures P1 and P2 will not coincide in general. On the other hand, most of the paper is devoted to propose different kinds of restrictions which suffice for the coincidence. The restrictions will mainly concern limitations on the number of arguments, the number of individuals, or impositions on the individual attack relations.

The paper is organized as follows: Section 2 offers preliminary definitions, introducing social argumentation frameworks formally. Section 3 refers to voting mechanisms by quotas showing how P1 and P2 mechanisms can match under special but sensible conditions. Section 4 shows how decisive groups of individuals voting on the arguments they choose will sanction a preferred extension of the framework obtained by an aggregate attack relation. Conclusions are offered in section 5.

2   Preliminaries

Dung defines an argumentation framework as a pair $AF = \langle AR, \; \rightharpoonup \rangle$, where AR is a set of abstract entities called 'arguments' and $\rightharpoonup \; \subseteq AR \times AR$ denotes an attack relation among arguments. This relation determines which sets of arguments become "defended" from attacks. Different characterizations of the notion of defense yield alternative sets called *extensions* of AF. These extensions are seen as the semantics of the argumentation framework, i.e. the classes of arguments that can be deemed as the outcomes of the whole process of argumentation. Dung introduces the notions of *preferred*, *stable*, *complete*, and *grounded* extensions, each corresponding to different requirements on the attack relation.

Definition 1. (Dung ([6])). *In any argumentation framework AF, an argument a is said* acceptable *w.r.t. a subset S of arguments of AR, in case that for every argument b such that $b \; \rightharpoonup \; a$, there exists some argument $c \in S$ such that $c \; \rightharpoonup \; b$. A set of arguments S is said* admissible *if each $a \in S$ is acceptable w.r.t. S, and is conflict-free, i.e., the attack relation does not hold for any pair of arguments belonging to S. A* preferred extension *is any maximally admissible set of arguments of AF. A* complete extension *of AF is any conflict-free subset of arguments which is a fixed point of $\Phi(\cdot)$, where $\Phi(S) = \{a : a$ is acceptable w.r.t. $S\}$, while the* grounded extension *is the least (w.r.t. $\subseteq$) complete extension. Moreover, a* stable extension *is a conflict-free set S of arguments which attacks every argument not belonging to S.*

Dung defines *well-founded* argumentation frameworks as those in which there exists no infinite sequence $a_0, a_1, \ldots, a_n, \ldots$ such that for each i, $a_{i+1} \; \rightharpoonup \; a_i$. An important result is that well-founded argumentation frameworks have only one extension that is grounded, preferred, stable and complete (*cf.* [6], theorem 30, p. 331).

We propose *social argumentation frameworks* to model situations of social debate (e.g. arguments pro and con sanctioning a law, deliberative group decisions, etc.). Given the set AR of all the arguments involved in the debate, each individual establishes her own attack criterion on them, which can be based on particular values of preference, intention, desire, etc. (these values will be abstracted in the model).

Definition 2. *A* social argumentation framework *is a structure $SAF_M = \langle N, \; AR, \; \{AF_i\}_{i \in N}, \; \rightharpoonup_M \rangle$, where: $N = \{1, \ldots, n\}$ is a set of individuals; AR is a finite set of arguments; $AF_i = \langle AR, \rightharpoonup_i \rangle$ is the argumentation framework of individual i, built up from her own attack criterion $\rightharpoonup_i \; \subseteq AR \times AR$; and $\rightharpoonup_M \; \subseteq AR \times AR$ is the social attack relation of SAF_M, obtained through the aggregation of $\rightharpoonup_1, \ldots, \rightharpoonup_n$ according to some specified mechanism M.*

3   Aggregation Mechanisms by Quotas

For the sake of simplicity, we will assume that each individual chooses her set of justified arguments according to the Dung's grounded semantics. In this way we will have to compute only one set per individual. Furthermore, since it is well known that in all argumentation frameworks the grounded extension is contained in the intersection of all the complete extensions, we will always have a minimum of justified arguments per individual. The set of socially justified arguments obtained through a P2 procedure should arise as an aggregation of all the sets of individually justified arguments. The aggregation can be done according to various criteria. A *quota* criterion establishes a minimum support among the individuals, measured in terms of a given number of them.

3.1   Conditions for the Coincidence between Quota Mechanisms

A natural way of defining a quota-based mechanism for the election of an attack criterion, for any given quota q, is as follows:

$$\rightarrow_q^{\geq} =_{def} \{(a,b) : |\{i \in N : a \rightarrow_i b\}| \geq q\}. \tag{1}$$

That is, $a \rightarrow_q^{\geq} b$ if at least q individuals support the attack of a on b. On the other hand, the aggregation of sets of arguments can be done in the same fashion. Let G_i be the grounded extension of AF_i (*i.e.*, the set of justified arguments for the individual i). The set of socially justified arguments can be defined as follows:

$$G_q^{\geq} =_{def} \{a : |\{i \in N : a \in G_i\}| \geq q\}. \tag{2}$$

Our first question is: will $G_q^{\geq}$ coincide with the grounded extension of $\langle AR, \rightarrow_q^{\geq} \rangle$ in general? The answer is no; example 1 shows a counterexample where $q = 2$. Moreover, it is not granted that $G_q^{\geq}$ will be conflict-free in any setting, so this mechanism can get out of rationality. A more general question now is: which conditions suffice for the coincidence of some P1 and P2 mechanisms? (Note that if the coincidence is granted, the conflict-freeness condition is granted too.) While it is difficult to find sensible conditions for such coincidence in presence of four or more arguments, we will show next some arguably reasonable restrictions under which the goal is accomplished in presence of two or three arguments. Restrictions will vary the quota, the number of individuals or the properties of the attack relations.

Variations on two arguments. The most simple setting that deserves analysis consists of two individuals ($N = \{1, 2\}$) deliberating about only two arguments ($AR = \{a, b\}$). In fact, such a setting is the most basic test for any aggregation procedure. In addition, it is well known from SCT that paradoxes arise when three (or more) people and three (or more) options are at stake (see [2], chapter 3). A total agreement about the attack criterion should obviously be reflected in the resulting social choice[1], but to impose that the outcome must be established *only* under unanimous decision is too strong. Instead, a quota $q = 1$ seems to be not so restrictive and reasonable at the same time. Note that, using this quota, $a \rightarrow_q^{\geq} b$ if for at least one individual i, $a \rightarrow_i b$ (in case that $a \rightarrow_1 b$ and $b \rightarrow_2 a$, we will have $a \rightarrow_q^{\geq} b$ and $b \rightarrow_q^{\geq} a$, a well known situation in the argument systems literature).

For the aggregation of sets of individually justified arguments we have to consider the conflict-freeness problem. If $SAF_q^{\geq}$ is such that $a \rightarrow_q^{\geq} b$ or $b \rightarrow_q^{\geq} a$ and we obtain $a, b \in G_q^{\geq}$, then the mechanism to obtain $G_q^{\geq}$ via $q = 1$ would be sanctioning a not

---

[1] This is known as the Weak Pareto Condition in SCT.

conflict-free set of socially justified arguments. To avoid this situation (in this setting with only two arguments), the requirement of an unanimous choice of arguments seems to be right. For $q = 1$, this amounts to ask that:

$$G_q^> =_{def} \{a : |\{i \in N : a \in G_i\}| > q\}. \tag{3}$$

Now we are in conditions to offer a first result of coincidence.

Notation. '$G(\rightarrow_M)$' will denote the grounded extension of SAF_M.

Proposition 1. *If* $|AR| = 2$, $|N| = 2$ *and* $q = 1$, *then* $G_q^> = G(\rightarrow_q^{\geq})$.

Proof.

$a \in G_q^>$,

iff $a \in G_1 \wedge a \in G_2$,

iff $(a \not\rightarrow_1 a \wedge b \not\rightarrow_1 a) \wedge (a \not\rightarrow_2 a \wedge b \not\rightarrow_2 a)$,

iff $a \not\rightarrow_q^{\geq} a \wedge b \not\rightarrow_q^{\geq} a$,

iff $a \in G(\rightarrow_q^{\geq})$.   □

This result cannot be generalized for any $|N|$ or any q, even imposing sensible restrictions on the domain of individual attack relations. Though we will not offer a proof stricto sensu, an example will be enough to show the problem.

Example 2. Assume the individual attack relations are irreflexive, asymmetric and complete according to the following definitions:

Irreflexive: $\forall x \in AR : \neg(x \rightarrow x)$
Complete: $\forall x, y \in AR, x \neq y : (x \rightarrow y) \vee (y \rightarrow x)$
Asymmetric: $\forall x, y \in AR, x \neq y : (x \rightarrow y) \Rightarrow \neg(y \rightarrow x)$

Let $N = \{1, 2, 3, 4, 5\}$ and the individual attack relations be as follows:

$\rightarrow_1 = \rightarrow_2 = \rightarrow_3 = \{(a, b)\}$;
$\rightarrow_4 = \rightarrow_5 = \{(b, a)\}$.

The individual grounded extensions are $G_1 = G_2 = G_3 = \{a\}$, and $G_4 = G_5 = \{b\}$. Then we have:

- for $q = 1$: $G_q^> = \{a, b\}$; $\rightarrow_q^{\geq} = \{(a, b), (b, a)\}$; $G(\rightarrow_q^{\geq}) = \emptyset$;
- for $q = 2$: $G_q^> = \{a\}$; $\rightarrow_q^{\geq} = \{(a, b), (b, a)\}$; $G(\rightarrow_q^{\geq}) = \emptyset$;
- for $q = 3$: $G_q^> = \emptyset$; $\rightarrow_q^{\geq} = \{(a, b)\}$; $G(\rightarrow_q^{\geq}) = \{a\}$;
- for $q = 4$: $G_q^> = \emptyset$; $\rightarrow_q^{\geq} = \emptyset$; $G(\rightarrow_q^{\geq}) = \{a, b\}$.

In the next step, a slight variation in the use of the quota for aggregating the individual attack relations will lead us to new conditions for the coincidence. Here the attack of an argument on another will be socially sanctioned if there exist *at most* q individuals that *do not support* that attack:

$$\rightarrow_q^{\leq} =_{def} \{(a, b) : |\{i \in N : a \not\rightarrow_i b\}| \leq q\}. \tag{4}$$

The coincidence can be found for any number n of individuals and for any quota q, $1 \leq q < n$. The only condition in the domain of individual attacks is irreflexivity, i.e. no individual is allowed to support self-attacks (clearly, this condition can be justified as a rationality requirement).

Proposition 2. *If $|AR| = 2$, $1 \leq q < n$ for an arbitrary number n of individuals, and for every individual i, $\rightarrowtail_i$ is irreflexive, then $G_q^{>} = G(\rightarrowtail_{\frac{q}{q}}^{\leq})$.*

Proof.
$\qquad a \in G_q^{>}$,
iff $|\{i \in N : a \in G_i| > q$,
iff $|\{i \in N : b \not\rightarrowtail_i a\}| > q$,
iff $b \not\rightarrowtail_{\frac{q}{q}}^{\leq} a$,
iff $a \in G(\rightarrowtail_{\frac{q}{q}}^{\leq})$. $\square$

A similar result can be obtained by maintaining the original aggregation of the individual attacks given in (1), and varying the use of the quota in the aggregation of the sets of individually justified arguments as follows:

$$G_q^{<} =_{def} \{a : |\{i \in N : a \notin G_i\}| < q\}. \tag{5}$$

Proposition 3. *If $|AR| = 2$, $1 < q < n$ for an arbitrary number n of individuals, and for every individual i, $\rightarrowtail_i$ is irreflexive, then $G_q^{<} = G(\rightarrowtail_{\frac{q}{q}}^{\geq})$.*

Proof.
$\qquad a \in G_q^{<}$,
iff $|\{i \in N : a \notin G_i| < q$,
iff $|\{i \in N : b \rightarrowtail_i a\}| < q$,
iff $b \not\rightarrowtail_{\frac{q}{q}}^{\geq} a$,
iff $a \in G(\rightarrowtail_{\frac{q}{q}}^{\geq})$. $\square$

The results given in propositions 2 and 3 can be extended in an obvious way to settings with more than two arguments, assuming that the individuals unanimously agree in the justification of at least $|AR| - 2$ arguments, i.e. $\exists S \subseteq AR : (|S| \geq |AR| - 2 \land \forall i \forall a, b \in S : a \not\rightarrowtail_i b)$.

Three arguments: coincidence of absolute majority mechanisms. In social argumentation frameworks with at most three arguments, a coincidence of absolute majority mechanisms can be found by imposing some reasonable restrictions on the profile of individual attack relations.

We require the profile being such that all individuals avoid supporting cycles of attack (i.e., $\langle AF_i, \rightarrowtail_i \rangle$ must be well-founded for every i); all individuals agree about the attack on at least one pair of arguments; and all individuals agree about all the conflicts within AR, in the sense of the following definition of 'conflictually definite':

Definition 3. *Let N be the set of individuals of a social argumentation framework SAF_M. We say that two arguments $a, b \in AR$ are socially conflicting (in SAF_M), in symbols, '$a \otimes b$', iff for each individual $i \in N$, $a \rightarrowtail_i b$ or $b \rightarrowtail_i a$; and that they are* socially coherent (in SAF_M), in symbols, '$a \odot b$', *iff for each individual* $i \in N$, $a \nrightarrow_i b$ *and* $b \nrightarrow_i a$. *We say that* SAF_M *is* conflictually definite *iff for every pair of arguments* $a, b \in AR$, $a \otimes b$ *or* $a \odot b$.

With respect to the implementation of the absolute majority mechanisms, we will require an odd number of individuals (in order to avoid ties) and a quota $q = (|N| - 1)/2$. So the desired mechanism for the aggregation of sets of individually justified arguments is obtained with $G_q^>$, while absolute majority for the aggregation of individual attack relations is obtained with the operator:

$$\rightarrow_q^> =_{def} \{(a, b) : |\{i \in N : a \rightarrow_i b\}| > q\}. \tag{6}$$

Proposition 4. *If*

1. $|AR| = 3$,
2. $|N|$ *is odd*,
3. $q = (|N| - 1)/2$ *(quota for absolute majority)*,
4. $\rightarrow_i$ *is acyclic for every* $i \in N$ *(i.e. individual argumentation frameworks are well-founded)*,
5. $SAF_q^>$ *is conflictually definite, and*
6. *there exists a pair of arguments* a *and* b *such that either* $\forall i: a \rightarrow_i b$ *or* $\forall i: b \rightarrow_i a$,

then $G_q^> = G(\rightarrow_q^>)$.

Proof. Let us assume condition 6 supposing, without lost of generality, that $\forall i : a \rightarrow_i b$. We have to consider three cases with respect to the third argument c: (i) all the individuals agree in that a and c are in conflict, or (ii) all the individuals agree in that b and c are in conflict or (iii) all the individuals agree in that neither a and c nor b and c are in conflict.

Case (i): By hypothesis, $\forall i : (a \rightarrow_i c \lor c \rightarrow_i a)$. By conditions 2 and 3, either (a) $|\{i \in N : a \rightarrow_i c\}| > q$ or (b) $|\{i \in N : c \rightarrow_i a\}| > q$ (but not both cases). If (a) is the case, then there exist at least $q + 1$ individuals whose attack relations contain the pairs (a, b) and (a, c) and, hence, by condition 3, these pairs are also in the aggregate social attack relation. Therefore, $G_q^> = \{a\} = G(\rightarrow_q^>)$, since there are at least $q + 1$ individuals i such that $G_i = \{a\}$. If (b) is the case we have either $b \odot c$ or $|\{i \in N : c \rightarrow_i b\}| > q$ ($b \rightarrow_i c$ cannot be majoritarily supported given condition 4), therefore we will have either $G_q^> = \{b, c\} = G(\rightarrow_q^>)$ or $G_q^> = \{c\} = G(\rightarrow_q^>)$.

Case (ii): By hypothesis, $\forall i : (b \rightarrow_i c \lor c \rightarrow_i b)$. By conditions 2 and 3, we have two possible cases: (a) $|\{i \in N : b \rightarrow_i c\}| > q$; and (b) $|\{i \in N : c \rightarrow_i b\}| > q$; If (a) is the case, then either $a \odot c$ or there exists at least $q + 1$ individuals whose attack relations contain the pair (a, c) (the pair (c, a) is inhibited by condition 4), therefore we will have either $G_q^> = \{a, c\} = G(\rightarrow_q^>)$ or $G_q^> = \{a\} = G(\rightarrow_q^>)$. If (b) is the case, then either (b.1) there exists at least $q + 1$ individuals which attack relations contain the pair (a, c), in which case clearly $G_q^> = \{a\} = G(\rightarrow_q^>)$; or (b.2) there exists at least $q + 1$ individuals whose attack relations contain the pair (c, a), in which case clearly $G_q^> = \{c\} = G(\rightarrow_q^>)$; or (b.3) $a \odot c$, in which case clearly $G_q^> = \{a, c\} = G(\rightarrow_q^>)$.

Case (iii): Given Condition 6 all the individuals have the same preferences, that is, $\forall i : \rightarrow_i = \{(a, b)\}$, so the aggregation problem is trivial. $\quad\square$

The requirements established in this result (especially condition 6) are aimed to avoid undesirable cycles in the aggregate attack relation and the well-known problem in SCT called *The Condorcet's Paradox* ([10]).

The next example shows that the above result cannot be extended to four arguments.

Example 3. Assume:

$$\rightarrow_1 = \{(a,b),(b,c),(c,d)\},$$
$$\rightarrow_2 = \{(b,a),(b,c),(c,d)\},$$
$$\rightarrow_3 = \{(b,a),(c,b),(c,d)\}.$$

So $G_1 = G_3 = \{a,c\}$ and $G_2 = \{b,d\}$, then $G_q^> = \{a,c\}$. On the other hand, $\rightarrow_q^> = \{(b,a),(b,c),(c,d)\}$, then $G(\rightarrow_q^>) = \{b,d\}$. Therefore, $G(\rightarrow_q^>) \neq G_q^>$.

4 Aggregation Mechanisms by Decisive Sets

A set $\Omega \subseteq N$ of individuals is said to be *decisive* to impose an alternative x if whenever that alternative is voted by all $i \in \Omega$ then x is socially chosen. For example, in absolute majority voting, any set containing more than fifty percent of the voters is decisive. We are interested in characterizing the aggregation of both individual attack relations and sets of individually justified arguments via decisive groups, and in finding some correspondence between them. To do this we will make use of the notion of 'resolution' of an argumentation framework, introduced by Modgil ([9]).

An argumentation framework $\langle AR, \mathcal{R}' \rangle$ is a *resolution* of $\langle AR, \mathcal{R} \rangle$ iff every pair of arguments belonging to $\mathcal{R}'$ also belongs to $\mathcal{R}$, if $(a,b) \in \mathcal{R}$ and $(b,a) \notin \mathcal{R}$, then $(a,b) \in \mathcal{R}'$ and if $(a,b) \in \mathcal{R}$ and $(b,a) \in \mathcal{R}$, then only one of these pairs is in $\mathcal{R}'$. Modgil shows that a set of arguments E is admissible in AF iff there exists a resolution AF' of AF such that E is admissible in AF' ([9], Lemma 1). If we interpret AF as (a partial structure of) a social argumentation framework SAF and the set of all the resolutions AF' as the set of individual argumentation frameworks of SAF, then we would say that E is admissible for that society iff there exist some individuals for which E is admissible. Now, in conflictually definite social argumentation frameworks, if those individual argumentation frameworks are well founded (and consequently their attack relations are acyclic), then their respective extensions —which are grounded, preferred, complete and stable— will be preferred extensions of AF.

Lemma 1. *Let $SAF = \langle N, AR, \{AF_i\}_{i \in N} \rightarrow \rangle$ be conflictually definite. For every $i \in N$, if AF_i is a well-founded resolution of $\langle AR, \rightarrow \rangle$, then the only extension G_i of AF_i is a preferred extension of AF.*

Proof

1. Assuming AF_i is well-founded, the only extension G_i of AF_i is obviously conflict-free in AF_i.
2. To prove that G_i is conflict-free in AF, let us assume the contrary; then there exist at least two arguments $a, b \in G_i$ such that $(a \rightarrow b)$ or $(b \rightarrow a)$. Since SAF is conflictually definite, we have $a \rightarrow_i b$ or $b \rightarrow_i a$. Therefore, G_i is not conflict-free in AF_i. Contradiction.

3. To see that every argument belonging to G_i is acceptable w.r.t. G_i in AF, assume that there exists $b \in AR$ such that $b \rightharpoonup a$ and there not exist $c \in G_i$ such that $c \rightharpoonup b$. $a \rightharpoonup_i b$ cannot be the case since $\rightharpoonup_i \subseteq \rightharpoonup$ and, since the system is conflictually definite, we have that $b \rightharpoonup_i a$. Moreover, there not exist $c \in G_i$ such that $c \rightharpoonup_i b$. Therefore, a is not acceptable w.r.t. G_i in AF_i, which is absurd.

4. Finally, let us prove the maximal admissibility showing by contraposition that if a is acceptable w.r.t. G_i in AF then $a \in G_i$. Assume $a \notin G_i$. Now, since AF_i is well-founded G_i is stable, hence, for any $a \notin G_i$ there exists $b \in G_i$ such that $b \rightharpoonup_i a$. Since $\rightharpoonup_i \subseteq \rightharpoonup$, $b \rightharpoonup a$. Therefore, a is not acceptable w.r.t. G_i in AF.   □

This lemma shows that if an individual proposes a "normal" attack criterion (i.e. without cycles of attack) for every socially recognized conflicting pair of arguments, then the arguments she supports will not be rejected in the social choice[2].

Note that asking for *every* individual of a society to be "normal" seems too demanding. Moreover, the lemma does not tell us how an aggregation mechanism can be such that each individual argumentation framework is a resolution of the social argumentation framework. The notion of decisive sets will help us to solve these problems: on the one hand, asking for all the members *of a decisive set* to be "normal" is not too demanding and, on the other hand, the class of all decisive sets of individuals can easily be put in correspondence with a specific aggregation mechanism. So let us define, first, how a social attack relation can be obtained according to a decisive set.

Assume that Ω is a decisive set according to some given mechanism and that $a \rightharpoonup_i b$ for every $i \in \Omega$; then the attack of a on b should be socially accepted (even though the attack of b on a could also be accepted, for instance, if it is imposed by another decisive set). Formally,

$$\rightharpoonup_d =_{def} \{(a,b) : \exists \Omega \forall i \in \Omega : a \rightharpoonup_i b\}. \tag{7}$$

Now, the aggregation of individually justified arguments can also be characterized by the decisive sets of a mechanism. Intuitively, the society should elect the arguments that are individually justified by all the members of a decisive group.

Definition 4. *A set of arguments* $E \subseteq AR$ *is socially eligible iff* $E = \bigcap_{i \in \Omega} G_i$ *for some decisive set* Ω.

Of course, different decisive groups can lead to different socially eligible sets of arguments that could not be combined into one set without the chance of yielding conflicts. Indeed, conflicts could occur even inside socially eligible sets, so the mechanism at stake should define some requirements to avoid that. Another way of preventing conflicts not depending on the mechanism is to establish some general, sufficient conditions for that. For example, if all the members of a decisive set Ω agree with respect to the attack relation, then all of them will justify the same arguments, and so $\bigcap_{i \in \Omega} G_i$ will be conflict-free. This kind of agreement will suffice for finding the result we are looking for:

---

[2] The term 'rejected' is also taken from Modgil ([9]) and means that an argument does not belong to any extension sanctioned by the semantics at stake.

Proposition 5. *Let $SAF = \langle N, AR, \{AF_i\}_{i \in N}, \rightarrow_d \rangle$ be conflictually definite. For any decisive set $\Omega \subseteq N$, if every AF_i, $i \in \Omega$, is well founded and $\forall i, j \in \Omega$, $\rightarrow_i = \rightarrow_j$ then every socially eligible set of arguments is a preferred extension of $\langle AR, \rightarrow_d \rangle$.*

Proof

1. For every socially eligible set E there exists a decisive set Ω such that $E = \bigcap_{i \in \Omega} G_i$.
2. Given $\forall i, j \in \Omega$, $\rightarrow_i = \rightarrow_j$, it follows that $E = G_i$ for all $i \in \Omega$.
3. Since Ω is decisive, by (7) we have $\rightarrow_i \subseteq \rightarrow_d$.
4. Since AF_i is well-founded, $\rightarrow_i$ is acyclic;
5. SAF is conflictually definite, hence $a \rightarrow_d b$ implies $a \rightarrow_i b$ or $b \rightarrow_i a$, for all $i \in \Omega$. This fact together with 3 and 4 implies that AF_i is a resolution of $\langle AR, \rightarrow_d \rangle$.
6. By 5 and lemma 1 we have that G_i is a preferred extension of $\langle AR, \rightarrow_d \rangle$. □

In sum, this result says that in social argumentation frameworks which are conflictually definite, an agreement reached in a decisive set of individuals will suffice for leading to a set of arguments that is socially preferred according to the aggregate attack criterion.

On the other hand, the following example shows that not every preferred extension yielded by $\rightarrow_d$ is an eligible set.

Example 4. Assume $AR = \{a, b, c\}$; $N = \{1, 2\}$; the individual attack relations and the corresponding extensions are, respectively:

$a \rightarrow_1 b \rightarrow_1 c$; $\{a, c\}$;
$c \rightarrow_2 b \rightarrow_2 a$; $\{a, c\}$.

Let the decisive sets be $\{1\}$ and $\{2\}$; then the social attack relation is:

$a \rightarrow_d b, b \rightarrow_d a, b \rightarrow_d c, c \rightarrow_d b,$

which yields the preferred extensions $\{a, c\}$ and $\{b\}$. Clearly, $\{b\}$ is not socially eligible. (Note that we could have naturally assumed that $\{1, 2\}$ were also decisive, but the outcome would also have been one of no coincidence.)

5   Conclusions

In this paper, we have addressed the problem of how a society can yield argument justifications on the basis of the individual criteria of its members. A social justified set of arguments can be obtained through two different procedures: we can first aggregate the individual attack relations into a social one, and then generate the set of arguments from such relation, or we can get such set by directly aggregating the sets of individually justified arguments. Using very common voting mechanisms (as majority rule and quota rules) we have shown that these two procedures do not coincide in general, but they do under sensible conditions when there are at most two arguments. In addition, we have also proven a possibility result for three arguments, but stronger conditions are required. Finally, we have established a more general result showing the conditions under which aggregation procedures based on decisive sets can coincide.

Aside from its intrinsic theoretical interest, the problem of the coincidence or not of both aggregation procedures has a deep practical relevance: when the procedures do not generate the same set of social justified arguments, the society should then also decide which mechanism to rely on. In order to address this, more work on the relationship between both ways of aggregating argumentation frameworks is needed.

Acknowledgments

We thank three anonymous reviewers for constructive criticisms that improved the paper.

References

1. Arrow, K.J., Sen, A.K., Suzumura, K.: Handbook of Social Choice and Welfare. Elsevier, Amsterdam (2002)
2. Austen-Smith, D., Banks, J.S.: Positive Political Theory I: Collective Preference. Michigan Studies in Political Analysis. University of Michigan Press, Ann Arbor (2000)
3. Bench-Capon, T.: Persuasion in practical argument using value-based argumentation frameworks. Journal of Logic and Computation 13(3), 429–448 (2003)
4. Modgil, S., Bench-Capon, T.: Integrating object and meta-level value based argumentation. In: Besnard, P., Doutre, S., Hunter, A. (eds.) Proc. of Computational Models of Argument, COMMA 2008, Toulouse, France, May 28-30, 2008. Frontiers in Artificial Intelligence and Applications, vol. 172, pp. 240–251. IOS Press, Amsterdam (2008)
5. Coste-Marquis, S., Devred, C., Konieczny, S., Lagasquie-Schiex, M.C., Marquis, P.: On the merging of Dung's argumentation systems. Artificial Intelligence 171, 740–753 (2007)
6. Dung, P.M.: On the acceptability of arguments and its fundamental role in nonmonotonic reasoning, logic programming and n-person games. Artificial Intelligence 77, 321–358 (1995)
7. Konieczny, S., Pino-Pérez: Propositional belief base merging or how to merge beliefs/goals coming from several sources and some links with social choice theory. European Journal of Operational Research 160, 785–802 (2005)
8. List, C., Pettit, P.: Aggregating sets of judgments. Two impossibility results compared. Synthese 140, 207–235 (2004)
9. Modgil, S.: Value based argumentation in hierarchical argumentation frameworks. In: Dunne, P., Bench-Capon, T. (eds.) Proc. of Computational Models of Argument, COMMA 2006, Liverpool, UK, September 11-12. Frontiers in Artificial Intelligence and Applications, vol. 144, pp. 297–308. IOS Press, Amsterdam (2006)
10. Nurmi, H.: Voting Paradoxes and How to Deal with Them. Springer, Heidelberg (1999)
11. Pigozzi, G.: Belief merging and the discursive dilemma: An argument-based account to paradoxes of judgment aggregation. Synthese 152, 285–298 (2006)
12. Rahwan, I., Larson, K.: Mechanism design for abstract argumentation. In: Padgham, L., Parkes, D., Müller, J., Parsons, S. (eds.) Proc. of the 7th International Joint Conference on Autonomous Agents and Multiagent Systems, AAMAS 2008. Estoril, Portugal, May 12-16, vol. 2, pp. 1031–1038. IFAAMAS (2008)
13. Tohmé, F., Bodanza, G., Simari, G.: Aggregation of attack relations: a Social-Choice theoretical analysis of defeasibility criteria. In: Hartmann, S., Kern-Isberner, G. (eds.) FoIKS 2008. LNCS, vol. 4932, pp. 8–23. Springer, Heidelberg (2008)

Dynamics in Argumentation with Single Extensions: Abstraction Principles and the Grounded Extension

Guido Boella[1], Souhila Kaci[2], and Leendert van der Torre[3]

[1] Department of Computer Science, University of Torino, Italy
guido@di.unito.it
[2] Université Lille-Nord de France, Artois, CRIL,
CNRS UMR 8188 – IUT de Lens, F-62307, France
kaci@cril.fr
[3] Computer Science and Communication, University of Luxembourg, Luxembourg
leendert@vandertorre.com

Abstract. In this paper we consider the dynamics of abstract argumentation in Baroni and Giacomin's framework for the evaluation of extension based argumentation semantics. Following Baroni and Giacomin, we do not consider individual approaches, but we define general principles or postulates that individual approaches may satisfy. In particular, we define abstraction principles for the attack relation, and for the arguments in the framework. We illustrate the principles on the grounded extension. In this paper we consider only principles for the single extension case, and leave the multiple extension case to further research.

1 Introduction

Argumentation is a suitable framework for modeling interaction among agents. Dung introduced a framework for abstract argumentation with various kinds of so-called semantics. Baroni and Giacomin introduced a more general framework to study general principles of sets of semantics [1]. This is a very promising approach, since due to the increase of different semantics we need abstract principles to study the proposals, compare them, and select them for applications. So far Dung's argumentation framework has been mainly considered as static, in the sense that the argumentation framework is fixed. The dynamics of argumentation framework has attracted a recent interest where the problem of revising an argumentation framework has been addressed [5,7]. In this paper, we address complementary problems and study how the semantics of an argumentation framework remains unchanged when we change the set of arguments or the attack relations between them. In particular, we consider the case in which arguments or attack relations are removed, for example when agents retract arguments in a dialogue. More precisely, we address the following questions:

1. Which principles can be defined for abstracting (i.e., removing) an attack relation?
2. Which principles can be defined for abstracting (i.e., removing) an argument?
3. Which of these principles are satisfied by the grounded semantics?

We use the general framework of Baroni and Giacomin for arbitrary argumentation semantics, but we consider only semantics that give precisely one extension, like the

C. Sossai and G. Chemello (Eds.): ECSQARU 2009, LNAI 5590, pp. 107–118, 2009.

grounded extension or the skeptical preferred semantics. Baroni and Giacomin [1] define the so-called directionality and resolution principles, which may be considered as argument and attack abstraction principles respectively. However, whereas directionality only considers abstraction from disconnected arguments, we also consider abstraction from arguments which are connected. To define the principles, we use Caminada's distinction between accepted, rejected and undecided arguments [4]. We find some results for the most popular semantics used in argumentation, namely the grounded extension.

In this paper we consider only principles for the single extension case, and leave the multiple extension case to further research. Moreover, we consider only abstractions which differ only one attack or one argument.

The layout of this paper is as follows. In Section 2 we give a recall of Dung's argumentation framework, the framework of Baroni and Giacomin, Caminada labeling, and we introduce the notion of abstraction. In Section 3 we consider the abstraction of attack relations and in Section 4 we consider the abstraction of arguments.

2   Formal Framework for Abstraction Principles

2.1   Dung's Argumentation Framework

Argumentation is a reasoning model based on constructing arguments, determining potential conflicts between arguments and determining acceptable arguments. Dung's framework [6] is based on a binary *attack* relation. In Dung's framework, an argument is an abstract entity whose role is determined only by its relation to other arguments. Its structure and its origin are not known. We restrict ourselves to *finite* argumentation frameworks, i.e., those frameworks in which the set of arguments is *finite*.

Definition 1 (Argumentation framework). *An argumentation framework is a tuple* $\langle \mathcal{B}, \rightarrow \rangle$ *where* $\mathcal{B}$ *is a finite set (of arguments) and* $\rightarrow$ *is a binary (attack) relation defined on* $\mathcal{B} \times \mathcal{B}$.

The output of $\langle \mathcal{B}, \rightarrow \rangle$ is derived from the set of selected acceptable arguments, called extensions, with respect to some acceptability semantics. We need the following definitions before we recall the most widely used acceptability semantics of arguments given in the literature.

Definition 2. *Let* $\langle \mathcal{B}, \rightarrow \rangle$ *be an argumentation framework. Let* $\mathcal{S} \subseteq \mathcal{B}$.

– $\mathcal{S}$ *defends* a *if* $\forall b \in \mathcal{B}$ *such that* $b \rightarrow a$, $\exists c \in \mathcal{S}$ *such that* $c \rightarrow b$.
– $\mathcal{S} \subseteq \mathcal{B}$ *is conflict-free if and only if there are no* $a, b \in \mathcal{S}$ *such that* $a \rightarrow b$.

The following definition summarizes the well-known acceptability semantics.

Definition 3 (Acceptability semantics). *Let* $AF = \langle \mathcal{B}, \rightarrow \rangle$ *be an argumentation framework. Let* $\mathcal{S} \subseteq \mathcal{B}$.

– $\mathcal{S}$ *is an* admissible *extension if and only if it is conflict-free and defends all its elements.*
– $\mathcal{S}$ *is a* complete extension *if and only if it is conflict-free and* $\mathcal{S} = \{a \mid \mathcal{S}$ *defends* $a\}$.

- S is *a* grounded extension *of AF if and only if S is the smallest (for set inclusion) complete extension of AF.*
- S is *a* preferred extension *of AF if and only if S is maximal (for set inclusion) among admissible extensions of AF.*
- S is *the* skeptical preferred extension *of AF if and only if S is the intersection of all preferred extensions of AF.*
- S is *a* stable extension *of AF if and only if S is conflict-free and attacks all arguments of $\mathcal{B} \backslash S$.*

Which semantics is most appropriate in which circumstances depends on the application domain of the argumentation theory. The grounded extension is the most basic one, in the sense that its conclusions are not controversial, each argumentation framework has a grounded extension (it may be the empty set), and this extension is unique. The grounded extension therefore plays an important role in the remainder of this paper. The preferred semantics is more credulous than the grounded extension. There always exists at least one preferred extension but it does not have to be unique. Stable extensions have an intuitive appeal, but their drawbacks are that extensions do not have to be unique and do not have to exist. Stable extensions are used, for example, in answer set programming, where it makes sense that some programs do not have a solution.

2.2 Baroni and Giacomin's Framework

In this paper we use the recently introduced formal framework for argumentation of Baroni and Giacomin [1]. They assume that the set $\mathcal{B}$ represents the set of arguments produced by a reasoner at a given instant of time, and they therefore assume that $\mathcal{B}$ is finite, independently of the fact that the underlying mechanism of argument generation admits the existence of infinite sets of arguments. Like in Dung's original framework, they consider an argumentation framework as a pair $\langle \mathcal{B}, \rightarrow \rangle$ where $\mathcal{B}$ is a set and $\rightarrow \subseteq (\mathcal{B} \times \mathcal{B})$ is a binary relation on $\mathcal{B}$, called the attack relation. In the following it will be useful to explicitly refer to the set of all arguments which can be generated, which we call $\mathcal{N}$ for the universe of arguments.

The generalization of Baroni and Giacomin is based on a function $\mathcal{E}$ that maps an argumentation framework $\langle \mathcal{B}, \rightarrow \rangle$ to its set of extensions, i.e., to a set of sets of arguments. However, this function is not formally defined. To be precise, they say: "An extension-based argumentation semantics is defined by specifying the criteria for deriving, for a generic argumentation framework, a set of extensions, where each extension represents a set of arguments considered to be acceptable together. Given a generic argumentation semantics S, the set of extensions prescribed by S for a given argumentation framework AF is denoted as $\mathcal{E}_S(AF)$." The following definition captures the above informal meaning of the function $\mathcal{E}$. Since Baroni and Giacomin do not give a name to the function $\mathcal{E}$, and it maps argumentation frameworks to the set of accepted arguments, we call $\mathcal{E}$ the *acceptance function*.

Definition 4. *Let $\mathcal{N}$ be the universe of arguments. A multiple extensions acceptance function $\mathcal{E} : \mathcal{N} \times 2^{\mathcal{N} \times \mathcal{N}} \rightarrow 2^{2^{\mathcal{N}}}$ is*

1. *a partial function which is defined for each argumentation framework $\langle \mathcal{B}, \rightarrow \rangle$ with finite $\mathcal{B} \subseteq \mathcal{N}$ and $\rightarrow \subseteq \mathcal{B} \times \mathcal{B}$, and*

2. which maps an argumentation framework $\langle \mathcal{B}, \rightarrow \rangle$ to sets of subsets of $\mathcal{B}$: $\mathcal{E}(\langle \mathcal{B}, \rightarrow \rangle) \subseteq 2^{\mathcal{B}}$.

The generality of the framework of Baroni and Giacomin follows from the fact that they have to define various principles which are built-in in Dung's framework. For example, Baroni and Giacomin identify the following two fundamental principles underlying the definition of extension-based semantics in Dung's framework, the *language independent* principle and the *conflict free* principle (see [1] for a discussion on these principles). In the following, we assume that these two principles are satisfied.

Definition 5 (Language independence). *Two argumentation frameworks* $\mathcal{AF}_1 = \langle \mathcal{B}_1, \rightarrow_1 \rangle$ *and* $\mathcal{AF}_2 = \langle \mathcal{B}_2, \rightarrow_2 \rangle$ *are isomorphic if and only if there is a bijective mapping* $m : \mathcal{B}_1 \rightarrow \mathcal{B}_2$, *such that* $(\alpha, \beta) \in \rightarrow_1$ *if and only if* $(m(\alpha), m(\beta)) \in \rightarrow_2$. *This is denoted as* $\mathcal{AF}_1 \doteq_m \mathcal{AF}_2$.

A semantics $\mathcal{S}$ *satisfies the* language independence *principle if and only if* $\forall AF_1 = \langle \mathcal{B}_1, \rightarrow_1 \rangle$, $\forall AF_2 = \langle \mathcal{B}_2, \rightarrow_2 \rangle$ *such that* $AF_1 \doteq_m AF_2$ *we have* $\mathcal{E}_{\mathcal{S}}(AF_2) = \{M(E) \mid E \in \mathcal{E}_{\mathcal{S}}(AF_1)\}$, *where* $M(E) = \{\beta \in \mathcal{B}_2 \mid \exists \alpha \in E, \beta = m(\alpha)\}$.

Definition 6 (Conflict free). *Given an argumentation framework* $AF = \langle \mathcal{B}, \rightarrow \rangle$, *a set* $S \subseteq \mathcal{B}$ *is* conflict free, *denoted as* $cf(S)$, *iff* $\nexists a, b \in S$ *such that* $a \rightarrow b$. *A semantics* $\mathcal{S}$ *satisfies the conflict free principle if and only if* $\forall AF, \forall E \in \mathcal{E}_{\mathcal{S}}(AF)$, E *is conflict free.*

2.3   The Single Extension Case

In this paper we consider only the case in which the semantics of an argumentation framework contains precisely one extension. Examples are the grounded and the skeptical preferred extension.

Definition 7. *Let* $\mathcal{N}$ *be the universe of arguments. A single extension acceptance function* $\mathcal{A} : \mathcal{N} \times 2^{\mathcal{N} \times \mathcal{N}} \rightarrow 2^{\mathcal{N}}$ *is*

1. *a total function which is defined for each argumentation framework* $\langle \mathcal{B}, \rightarrow \rangle$ *with finite* $\mathcal{B} \subseteq \mathcal{N}$ *and* $\rightarrow \subseteq \mathcal{B} \times \mathcal{B}$, *and*
2. *which maps an argumentation framework* $\langle \mathcal{B}, \rightarrow \rangle$ *to a subset of* $\mathcal{B}$: $\mathcal{A}(\langle \mathcal{B}, \rightarrow \rangle) \subseteq \mathcal{B}$.

Principles of Baroni and Giacomin defined for multiple acceptance functions such as directionality and conflict free are defined also for the single extension case, because the set of all single extension acceptance functions is a subset of the set of all multiple extensions acceptance functions. For example, a semantics $\mathcal{S}$ satisfies the conflict free principle when the unique extension is conflict free: $\forall AF, A_{\mathcal{S}}(AF)$ is conflict free.

2.4   Abstraction

We now define abstraction relations between argumentation frameworks.

Definition 8 (Abstraction). *Let* $\langle \mathcal{B}, \mathcal{R} \rangle$ *and* $\langle \mathcal{B}', \mathcal{S} \rangle$ *be two argumentation frameworks.*

– $\langle \mathcal{B}, \mathcal{R} \rangle$ *is an argument abstraction from* $\langle \mathcal{B}', \mathcal{S} \rangle$ *iff* $\mathcal{B} \subseteq \mathcal{B}'$ *and* $\forall a, b \in \mathcal{B}$, $a\mathcal{R}b$ *if and only if* $a\mathcal{S}b$.

- $\langle \mathcal{B}, \mathcal{R} \rangle$ is an attack *abstraction from* $\langle \mathcal{B}', \mathcal{S} \rangle$ *iff* $\mathcal{B} = \mathcal{B}'$ *and* $\mathcal{R} \subseteq \mathcal{S}$.
- $\langle \mathcal{B}, \mathcal{R} \rangle$ is an argument-attack *abstraction from* $\langle \mathcal{B}', \mathcal{S} \rangle$ *iff* $\mathcal{B} \subseteq \mathcal{B}'$ *and* $\mathcal{R} \subseteq \mathcal{S}$.

Baroni and Giacomin also introduce two principles which may be interpreted as argument abstraction or attack abstraction principles, the so-called directionality and resolution principles. Directionality says that unattacked sets are unaffected by the remaining part of the argumentation framework as far as extensions are concerned (a principle which, as they show, does not hold for stable semantics, but it does hold for most other semantics). This may be seen as an argument abstraction principle, in the sense that when we abstract away all arguments not affecting a part of the argumentation framework, then the extensions of this part of the framework are not affected either (see their paper for the details).

2.5 Caminada Labeling

In the definition of principles in the following section, it is useful to distinguish between rejected and undecided arguments. The following definition gives Caminada's [4] translation from extensions to three valued labelling functions. Caminada uses this translation only for complete extensions, such as the grounded extension, such that an argument is accepted if and only if all its attackers are rejected, and an argument is rejected if and only if it has at least one attacker that is accepted. We use it also for extensions which are not complete, such as the skeptical preferred extension, such that Caminada's labelling principles may not hold in general. We assume only that extensions are conflict free, i.e., an accepted argument cannot attack another accepted argument.

Definition 9 (Rejected and undecided arguments). *Let* $\mathcal{A}(AF)$ *be a conflict free extension of an argumentation framework* $AF = \langle \mathcal{B}, \rightarrow \rangle$, *then* $\mathcal{B}$ *is partitioned into* $\mathcal{A}(AF)$, $\mathcal{R}(AF)$ *and* $\mathcal{U}(AF)$, *where:*

- $\mathcal{A}(AF)$ *is the set of accepted arguments,*
- $\mathcal{R}(AF) = \{a \in \mathcal{B} \mid \exists x \in \mathcal{A}(AF) : x \rightarrow a\}$ *is the set of rejected arguments, and*
- $\mathcal{U}(AF) = \mathcal{B} \setminus (\mathcal{A}(AF) \cup \mathcal{R}(AF))$ *is the set of undecided arguments.*

3 Attack Abstraction Principles

In this section, we consider the situation where the set of arguments remains the same, but the attack relation may shrink (abstraction). Our framework considers principles where we remove a single attack relation $a \rightarrow b$ from an argumentation framework. We distinguish whether arguments a and b are accepted, rejected or undecided.

Principle 1 (Attack abstraction). *An acceptance function* $\mathcal{A}$ *satisfies the* $\mathcal{X}\mathcal{Y}$ *attack abstraction principle, where* $\mathcal{X}, \mathcal{Y} \in \{\mathcal{A}, \mathcal{R}, \mathcal{U}\}$, *if for all argumentation frameworks* $AF = \langle \mathcal{B}, \rightarrow \rangle$, $\forall a \in \mathcal{X}(AF) \forall b \in \mathcal{Y}(AF) : \mathcal{A}(\langle \mathcal{B}, \rightarrow \setminus \{a \rightarrow b\} \rangle) = \mathcal{A}(AF)$.

We start with two useful lemmas. The first says that there cannot be an attack relation from an accepted argument to another accepted or an undecided argument, denoted $\mathcal{A}\mathcal{A}$ and $\mathcal{A}\mathcal{U}$ respectively.

Lemma 1. *Each semantics that satisfies the consistency principle (and thus each semantics defined by Dung) satisfies the following principle.*

$\mathcal{AA}$; $\mathcal{AU}$: *There is no attack from an accepted argument to an accepted or undecided argument.*

Proof (sketch). $\mathcal{AA}$ is a reformulation of the consistency principle. $\mathcal{AU}$ follows directly from the notion of undecided in Definition 9.

The second lemma gives a new characterization of the distinction between grounded and skeptical preferred semantics. It says that there cannot be an attack relation from an undecided argument to an accepted argument, denoted $\mathcal{UA}$. In other words, if an accepted argument is attacked, then its attacker is itself attacked by an accepted argument and thus rejected.

Lemma 2. *The grounded semantics satisfies the following principle, whereas the skeptical preferred semantics does not.*

$\mathcal{UA}$: *There is no attack from an undecided argument to an accepted argument.*

Proof (sketch). A counterexample for the skeptical preferred semantics is a well known example distinguishing the two semantics, that contains four arguments $\{a, b, c, d\}$ where a and b attack each other, both a and b attack c, and c attacks d. The grounded extension is empty, whereas the skeptical preferred extension contains only d. All the other arguments are undecided, there are no rejected arguments. Thus, in the skeptical preferred semantics there is an undecided argument that attacks an accepted argument, whereas in the grounded semantics, there is no such argument. The fact that in the grounded semantics there are no undecided arguments attacking accepted arguments follows by structural induction on the construction of the grounded extension.

The following proposition shows that the grounded extension satisfies seven of the nine abstraction principles.

Proposition 1. *The grounded semantics satisfies the $\mathcal{AA}$, $\mathcal{AU}$, $\mathcal{UA}$, $\mathcal{UR}$, $\mathcal{RA}$, $\mathcal{RU}$ and $\mathcal{RR}$ attack abstraction principles, and it does not satisfy the $\mathcal{AR}$ and $\mathcal{UU}$ attack abstraction principles. Intuitively the satisfied principles reflect the following ideas:*

$\mathcal{AA}$; $\mathcal{AU}$; $\mathcal{UA}$: *hold vacuously, since there is nothing to remove (Lemma 1 and 2).*

$\mathcal{RA}$; $\mathcal{RR}$; $\mathcal{RU}$: *the attacks from a rejected argument do not influence the extension. This principle holds for any attacked argument b.*

$\mathcal{UR}$: *the attacks on a rejected argument b by an undecided argument a do not influence the extension. Intuitively, this means that an argument is rejected only when it is attacked by an accepted argument.*

Proof (sketch). The satisfied principles can be proven by induction. Take the argumentation framework and the abstracted one, and show that in each step of the construction of the grounded extension, the two remain the same. Counterexamples for $\mathcal{UU}$ and $\mathcal{AR}$ attack abstraction are given below.

$\mathcal{UU}$: *Consider an argumentation framework composed of two arguments a and b attacking each other so both will be undecided. If we remove the attack from a to b, then b attacks a, so b will be accepted and a is rejected by the grounded extension, and all other reasonable acceptability semantics.*

$\mathcal{AR}$: *For the latter, consider again an argumentation framework composed of two arguments a and b, where a attacks b. a is accepted while b is rejected. If we remove this attack relation, then both are accepted.*

Proposition 1 leaves two interesting cases for further principles: the removal of an attack relation from an undecided argument to another undecided argument, i.e., $\mathcal{UU}$, and the removal of an attack relation from an argument that is accepted to an argument that is rejected, i.e., $\mathcal{AR}$. In both cases, the extension can stay the same only under conditions: the challenge is therefore to define suitable conditions. We first define conditional attack abstraction. The idea is that if we remove an attack from a to b, then in those two cases there must be another reason why b does not become accepted.

Principle 2 (Conditional attack abstraction). *An acceptance function $\mathcal{A}$ satisfies the $\mathcal{XY}(\mathcal{Z})$ attack abstraction principle, where $\mathcal{X}, \mathcal{Y}, \mathcal{Z} \in \{\mathcal{A}, \mathcal{R}, \mathcal{U}\}$, if for all argumentation frameworks $AF = \langle \mathcal{B}, \rightarrow \rangle$,*

$$\forall a \in \mathcal{X}(AF) \forall b \mathcal{Y}(AF): \text{if } \exists c \in \mathcal{Z}(AF) \text{ such that } c \neq a, c \rightarrow b \text{ then}$$
$$\mathcal{A}(\langle \mathcal{B}, \rightarrow \setminus \{a \rightarrow b\} \rangle) = \mathcal{A}(AF),$$

Proposition 2. *The grounded semantics satisfies the $\mathcal{UU}(\mathcal{A})$ and $\mathcal{UU}(\mathcal{U})$ attack abstraction principles. It does not satisfy the $\mathcal{AR}(\mathcal{A})$, $\mathcal{AR}(\mathcal{U})$, $\mathcal{AR}(\mathcal{R})$, $\mathcal{UU}(\mathcal{R})$ attack abstraction principles.*

$\mathcal{UU}(\mathcal{A})$: *holds vacuously since this situation never occurs.*

$\mathcal{UU}(\mathcal{U})$: *represents that for attacks among undecided arguments, it is only important that there is at least one of such attacks, additional attacks to the same undecided argument do not change the extension.*

Proof (sketch). $\mathcal{UU}(\mathcal{U})$ can be proven by induction. Take the argumentation framework and the abstracted one, and show that in each step of the construction of the grounded extension, the two remain the same.

$\mathcal{AR}(\mathcal{A})$: *Consider three arguments $\{a, b, c\}$ with a attacks b, b attacks c and c attacks b. The grounded extension is $\{a, c\}$, and b is rejected. If we remove the attack from a to b, then the grounded extension is $\{a\}$.*

$\mathcal{AR}(\mathcal{U})$: *Consider five arguments $\{a, b, c, d, e\}$ with a attacks b, b attacks c, c attacks b, c attacks d and d attacks c, b attacks e. The grounded extension is $\{a, e\}$, b is rejected, c and d are undecided. If we remove the attack from a to b, then the grounded extension is $\{a\}$, all others are undecided.*

$\mathcal{AR}(\mathcal{R})$: *Follows from Proposition 1. Due to $\mathcal{RR}$ abstraction principle, we can remove the attack among the rejected arguments without changing the extension. Do this for all attacks among rejected arguments. Then due to the fact that grounded extension does not satisfy $\mathcal{AR}$, it also does not satisfy $\mathcal{AR}(\mathcal{R})$.*

$\mathcal{UU}(\mathcal{R})$: *Follows from Proposition 1. Due to $\mathcal{RU}$ abstraction principle, we can remove the attack from the rejected arguments to the undecided arguments without changing the extension. Do this for all attacks among rejected arguments. Then due to the fact that grounded extension does not satisfy $\mathcal{UU}$, it also does not satisfy $\mathcal{UU}(\mathcal{R})$.*

Note that the skeptical extension does not satisfy $\mathcal{UU}(\mathcal{U})$. Consider again the example given in the proof of Lemma 2. The preferred extensions are $\{a, d\}$ and $\{b, d\}$. So the skeptical extension is $\{d\}$. All other arguments are undecided. Let us now remove the attack from a to c. Then the preferred extensions are $\{a, c\}$ and $\{b, d\}$. Thus the skeptical extension is empty. In other words, the $\mathcal{UU}(\mathcal{U})$ abstraction principle characterizes another distinction between the grounded and skeptical preferred semantics.

Summarizing, for argument b to remain undecided when we remove the attack from a to b, there must be another reason besides a why b is undecided. The other reason cannot be an accepted argument attacking b, since in that case b would be rejected. Consequently the extension may change. And the other reason cannot be a rejected argument attacking b since an attack from a rejected argument doesn't necessarily make an argument undecided. Therefore, we ask that there is another undecided argument c attacking b. This motivates the $\mathcal{UU}(\mathcal{U})$ principle.

Proposition 2 leaves two interesting cases for further development, $\mathcal{AR}(\mathcal{A})$ and $\mathcal{AR}(\mathcal{U})$ abstraction. We start with $\mathcal{AR}(\mathcal{A})$. Due to the removal of the attack, the status of argument b may remain rejected or change from rejected to accepted or undecided. When the argument becomes accepted then it should belong to the extension following all reasonable acceptability semantics. Therefore we consider the cases where b remains rejected or becomes undecided. When b is still rejected, this means that there is another accepted argument c unequal to a, which attacks b. This motivates $\mathcal{AR}(\mathcal{A})$ principle. However, the counterexample shows that the reason that c is accepted, may be b itself! The counterexample indicates ways in which conditional abstraction can be further developed: the rejection of b should not be the reason for the acceptance of c.

One simple way to prevent the possibility that the rejection of b is the cause of the acceptance of c, is to prevent any paths from b to c. In fact, we can do a little better, because only odd paths can lead to acceptance. This motivates the following principle.

Principle 3 (Acyclic conditional attack abstraction). *An acceptance function $\mathcal{A}$ satisfies the acyclic $\mathcal{XY}(\mathcal{Z})$ attack abstraction principle, where $\mathcal{X}, \mathcal{Y}, \mathcal{Z} \in \{\mathcal{A}, \mathcal{R}, \mathcal{U}\}$, if for all argumentation frameworks $AF = \langle \mathcal{B}, \rightarrow \rangle$,*

$\forall a \in \mathcal{X}(AF) \forall b \mathcal{Y}(AF)$: *if $\exists c \in \mathcal{Z}(AF)$ such that $c \neq a, c \rightarrow b$, and there is no odd length sequence of attacks from b to c, then $\mathcal{A}(\langle \mathcal{B}, \rightarrow \setminus \{a \rightarrow b\} \rangle) = \mathcal{A}(AF)$,*

Proposition 3. *The grounded semantics satisfies the acyclic $\mathcal{AR}(\mathcal{A})$ abstraction principle.*

We now consider $\mathcal{AR}(\mathcal{U})$. Thus there is an accepted argument a attacking rejected argument b, which is also attacked by undecided argument c. In this case, the argument b may change into undecided. In that case, argument b is still not in the extension, but there may be implications in other parts of the argumentation framework. As the counterexample in Proposition 2 shows, b should not be the cause of acceptance of another argument.

Principle 4 (Stronger conditional attack abstraction). *An acceptance function $\mathcal{A}$ satisfies the $\mathcal{XY}(\mathcal{Z},\mathcal{W})$ attack abstraction principle, if for all argumentation frameworks*
$AF = \langle \mathcal{B}, \rightarrow \rangle,$

> $\forall a \in \mathcal{X}(AF) \forall b \mathcal{Y}(AF)$: *if* $\exists c \in \mathcal{Z}(AF)$ *such that* $c \neq a, c \rightarrow b$ *and* $\forall d \in \mathcal{W}(AF)$,
> *we do not have* $b \rightarrow d$, *then* $\mathcal{A}(\langle \mathcal{B}, \rightarrow \setminus \{a \rightarrow b\}\rangle) = \mathcal{A}(AF)$,

where $\mathcal{X}, \mathcal{Y}, \mathcal{Z}, \mathcal{W} \in \{\mathcal{A}, \mathcal{R}, \mathcal{U}\}$.

Proposition 4. *The grounded semantics satisfies the $\mathcal{AR}(\mathcal{U}, \mathcal{A})$ abstraction principle.*

4 Argument Abstraction Principles

In this section we consider the abstraction of an argument, including the attack relations involving this argument. It builds on the abstraction principles for attack relations in the previous section. As before, we distinguish between removal of an argument which is accepted, which is rejected and which is undecided. In the first case the extension should be the extension without the accepted argument, in the other two cases the extension should stay the same. So we have to consider three cases. Let $\rightarrow_a$ denote the set of attack relations related to an argument a. Formally, we have the following principles:

Principle 5 (Argument abstraction). *An acceptance function $\mathcal{A}$ satisfies the $\mathcal{X} \in \{\mathcal{R}, \mathcal{U}, A\}$ argument abstraction principle if for all argumentation frameworks $AF = \langle \mathcal{B}, \rightarrow \rangle$, if $a \in \mathcal{X}(AF)$, then $\mathcal{A}(\langle \mathcal{B} \setminus \{a\}, \rightarrow \setminus \rightarrow_a\rangle) = \mathcal{A}(AF) \setminus \{a\}$.*

The following proposition shows that only $\mathcal{R}$ abstraction is satisfied. For example, if we remove an accepted argument, then arguments attacked by the removed accepted argument may become accepted.

Proposition 5. *The grounded extension satisfies $\mathcal{R}$ argument abstraction, and it does not satisfy $\mathcal{A}$ and U argument abstraction. This represents the following idea.*

$\mathcal{R}$ *Rejected arguments do not play a role in the argumentation and can be removed.*

Proof (sketch). Follows from the abstraction principles of the attack relation or their counterexamples in Proposition 1. For $\mathcal{U}$ abstraction take two arguments a and b attacking each other, and for $\mathcal{A}$ abstraction take two arguments a and b where a attacks b.

This leave two interesting cases, $\mathcal{U}$ abstraction and $\mathcal{A}$ abstraction. We start with $\mathcal{U}$ abstraction. Unlike rejected arguments whose attacks are inoffensive, an attack from an undecided argment may "block" an argument, i.e., prevents the argument to be accepted. Therefore the removal of an undecided argument may change the extension. One way to keep the extension unchanged is that the removed undecided argument only attacks arguments which are in the extension (so the attacks have not been successful) or out of the extension but due to other arguments. Formally, we have the following principles:

Principle 6 ($\mathcal{U}$ argument abstraction 1). *$\mathcal{A}$ satisfies the undecided argument abstraction 1 principle if for all argumentation frameworks $AF = \langle \mathcal{B}, \rightarrow \rangle$, if $a \in \mathcal{U}(AF)$ is an undecided argument and a attacks only accepted arguments in $\mathcal{A}(AF)$, then $\mathcal{A}(\langle \mathcal{B} \setminus \{a\}, \rightarrow \setminus \rightarrow_a\rangle) = \mathcal{A}(AF).$*

Principle 7 (*$\mathcal{U}$* **argument abstraction 2**). *A satisfies the* undecided argument abstraction 2 principle *if for all argumentation frameworks $AF = \langle \mathcal{B}, \rightarrow \rangle$, if $a \in \mathcal{U}(AF)$ is an undecided argument and a attacks only rejected arguments in $\mathcal{R}(AF)$, then $\mathcal{A}(\langle \mathcal{B} \setminus \{a\}, \rightarrow \setminus \rightarrow_a \rangle) = \mathcal{A}(AF)$.*

Moreover, inspired by $\mathcal{U}\mathcal{U}(\mathcal{U})$ attack abstraction principle, we define the following $\mathcal{U}$ argument abstraction principle.

Principle 8 (*$\mathcal{U}$* **argument abstraction 3**). *A satisfies the* undecided argument abstraction 3 principle *if for all argumentation frameworks $AF = \langle \mathcal{B}, \rightarrow \rangle$, if $a \in \mathcal{U}(AF)$ is an undecided argument and for each undecided argument b attacked by a, there is another undecided argument $c \neq b$ that attacks b, then $\mathcal{A}(\langle \mathcal{B} \setminus \{a\}, \rightarrow \setminus \rightarrow_a \rangle) = \mathcal{A}(AF)$.*

Proposition 6. *The grounded extension satisfies $\mathcal{U}$ argument abstraction principle 1, 2 and 3.*

Proof (sketch). $\mathcal{U}$ argument abstraction 1 principle holds vacuously (Lemma 2). The others follow from Proposition 1 and 2.

Finally, we consider weakened $\mathcal{A}$ abstraction principles. The first principle of accepted argument abstraction is Baroni and Giacomin's principle of directionality restricted to single arguments. If we remove an argument that does not attack another argument besides possibly itself, then the extensions will remain the same.

Principle 9 (*$\mathcal{A}$* **abstraction 1**). *A satisfies the* accepted argument abstraction 1 principle *if for all argumentation frameworks $AF = \langle \mathcal{B}, \rightarrow \rangle$, if $a \in \mathcal{A}(AF)$ is an accepted argument, and there is no argument $b \in \mathcal{B}$ unequal to a such that a attacks b, then $\mathcal{A}(\langle \mathcal{B} \setminus \{a\}, \rightarrow \setminus \rightarrow_a \rangle) = \mathcal{A}(AF) \setminus \{a\}$.*

The second principle of accepted argument abstraction is inspired by Proposition 3. It says that if we remove an argument $b \in \mathcal{B}$ which attacks only arguments which are also attacked by other arguments, then the extensions remain the same.

Principle 10 (*$\mathcal{A}$* **abstraction 2**). *A satisfies the* accepted argument abstraction 2 principle *if for all argumentation frameworks $AF = \langle \mathcal{B}, \rightarrow \rangle$, if $a \in \mathcal{A}(AF)$ is an accepted argument, and for each rejected argument $b \in \mathcal{R}(AF)$ such that $a \rightarrow b$, there is another argument $c \in \mathcal{A}(AF)$ such that $c \rightarrow b$, and there is no odd attack sequence from b to c, then $\mathcal{A}(\langle \mathcal{B} \setminus \{a\}, \rightarrow \setminus \rightarrow_a \rangle) = \mathcal{A}(AF) \setminus \{a\}$*

Proposition 7. *The grounded extension satisfies $\mathcal{A}$ abstraction 1 and 2 principles.*

5   Related Research

Besides the work of Baroni and Giacomin on principles for the evaluation of argumentation semantics, there is various work on dialogue and a few very recent approaches on the dynamics of argumentation. Researchers in the multi-agent systems area have been looking at this problem under various names. Cayrol et al. [5] define a typology of refinement (the dual of abstraction) (called revision in their paper), i.e. adding an argument. Then they define principles and condition so that each type of refinement becomes a revision (called classical revision in their paper), i.e., the new argument is accepted. Refinement and revision is different from abstraction considered in this paper, and they do not define general principles as we do. Rotstein et al. [7] introduce the notion of dynamics into the concept of abstract argumentation frameworks, by considering arguments built from evidence and claims. They do not consider abstract arguments and general principles as we do in this paper. Barringer et al. [2] consider internal dynamics by extending Dung's theory in various ways, but without considering general principles.

6 Summary and Further Research

Motivated by situations where the argumentation framework may change, for example by dialogues in which arguments can be retracted or added, we study principles where the extension does not change when we consider the dynamics of the argumentation framework. The most interesting attack abstraction principles are listed in Table 1. This table should be read as follows. Each line represents a principle. For example, the first line says that $\mathcal{A}$ satisfies the $\mathcal{RA};\mathcal{UA}$ principle if for all argumentation frameworks $AF = \langle \mathcal{B}, \rightarrow \rangle$, if $b \in \mathcal{A}(AF)$ is an accepted argument, then for all arguments $a \in \mathcal{R}(AF) \cup \mathcal{U}(AF)$, $\mathcal{A}(\langle \mathcal{B}, \rightarrow \setminus \{a \rightarrow b\}\rangle) = \mathcal{A}(AF)$. For each principle, we state whether it is satisfied by the grounded extension or not. Vacuously means that the situation does not occur for the grounded extension.

Table 1. Attack abstraction: if $\forall AF = \langle \mathcal{B}, \rightarrow \rangle$ condition, then $\mathcal{A}(\langle \mathcal{B}, \rightarrow \setminus \{a \rightarrow b\}\rangle) = \mathcal{A}(AF)$

Principle	Condition	Grounded extension
$\mathcal{AA}, \mathcal{AU}, \mathcal{UA}$	$a \in \mathcal{A}(AF) \cup \mathcal{U}(AF), b \in \mathcal{A}(AF)$ or $a \in \mathcal{A}(AF), b \in \mathcal{A}(AF) \cup \mathcal{U}(AF)$	yes (vacuously)
$\mathcal{RA}, \mathcal{RR}, \mathcal{RU}$	$a \in \mathcal{R}(AF)$	yes
$\mathcal{UR}$	$a \in \mathcal{U}(AF), b \in \mathcal{R}(AF)$	yes
$\mathcal{UU}(\mathcal{A})$	$a \in \mathcal{U}(AF), b \in \mathcal{U}(AF), \exists c \in \mathcal{A}(AF), c \rightarrow b$	yes (vacuously)
$\mathcal{UU}(\mathcal{U})$	$a \in \mathcal{U}(AF), b \in \mathcal{U}(AF), \exists c \in \mathcal{U}(AF), c \neq a, c \rightarrow b$	yes
$\mathcal{AR}(\mathcal{U}), \mathcal{AR}(\mathcal{R}), \mathcal{UU}(\mathcal{R})$		no
$\mathcal{AR}(\mathcal{A})$	$a \in \mathcal{A}(AF), b \in \mathcal{R}(AF), \exists c \in \mathcal{A}(AF), c \neq a, c \rightarrow b$ no odd length sequence of attacks from b to c	yes
$\mathcal{AR}(\mathcal{U}, \mathcal{A})$	$a \in \mathcal{A}(AF), b \in \mathcal{R}(AF), \exists c \in \mathcal{U}(AF), c \rightarrow b,$ $\forall d \in \mathcal{A}(AF)$, we do not have $b \rightarrow d$	yes

Moreover, the most interesting argument abstraction principles are listed in Table 2.

There are several directions of future research. We would like to extend our dynamic analysis to a wider range of principles. For example, we are interested in refinement of argumentation frameworks [3], and their relation to argument abstraction. Moreover, besides considering situations in which the extension stays the same, we are interested in questions about the change needed to change an argument from being accepted to rejected, or vice versa. Also, we would like to study principles for the multiple extension

Table 2. Argument abstraction: if $\forall AF = \langle \mathcal{B}, \rightarrow \rangle$ condition, then $\mathcal{A}(\langle \mathcal{B}\backslash\{a\}, \rightarrow \backslash \rightarrow_a\rangle) = \mathcal{A}(AF)\backslash\{a\}$

Principle	Condition	Grounded extension
$\mathcal{R}$	$a \in \mathcal{R}(AF)$	yes
$\mathcal{U}$	$a \in \mathcal{U}(AF)$	no
$\mathcal{U}\,1$	$a \in \mathcal{U}(AF)$, if $a \rightarrow c$ then $c \in \mathcal{A}(AF)$	yes (vacuously)
$\mathcal{U}\,2$	$a \in \mathcal{U}(AF)$, if $a \rightarrow c$ then $c \in \mathcal{R}(AF)$	yes
$\mathcal{U}\,3$	$a \in \mathcal{U}(AF)$, if $a \rightarrow c$ and $c \in \mathcal{U}(AF)$ then ...	yes
$\mathcal{A}$	$a \in \mathcal{A}(AF)$	no
$\mathcal{A}\,1$	$a \in \mathcal{A}(AF)$, $\nexists b \neq a, a \rightarrow b$	yes
$\mathcal{A}\,2$	$a \in \mathcal{A}(AF), \forall b \in \mathcal{R}(AF), a \rightarrow b, \exists c \in \mathcal{A}(AF), c \rightarrow b$	yes

case. Also, we would like to apply our theory to practical problems such as evaluation of argumentation semantics, or sensitivity analysis of a dispute.

References

1. Baroni, P., Giacomin, M.: On principle-based evaluation of extension-based argumentation semantics. Artificial Intelligence 171(10-15), 675–700 (2007)
2. Barringer, H., Gabbay, D.M., Woods, J.: Temporal dynamics of support and attack networks: From argumentation to zoology. In: Hutter, D., Stephan, W. (eds.) Mechanizing Mathematical Reasoning. LNCS, vol. 2605, pp. 59–98. Springer, Heidelberg (2005)
3. Boella, G., Kaci, S., van der Torre, L.: Dynamics in argumentation with single extensions: Attack refinement and the grounded extension (short paper). In: Proc. of 8th Int. Conf. on Autonomous Agents and Multiagent Systems (AAMAS 2009) (2009)
4. Caminada, M.: On the issue of reinstatement in argumentation. In: Fisher, M., van der Hoek, W., Konev, B., Lisitsa, A. (eds.) JELIA 2006. LNCS, vol. 4160, pp. 111–123. Springer, Heidelberg (2006)
5. Cayrol, C., de Saint Cyr Bannay, F.D., Lagasquie-Schiex, M.C.: Revision of an argumentation system. In: 11th International Conference on Principles of Knowledge Representation and Reasoning (KR 2008), pp. 124–134 (2008)
6. Dung, P.M.: On the acceptability of arguments and its fundamental role in nonmonotonic reasoning, logic programming and n-person games. Artificial Intelligence 77(2), 321–357 (1995)
7. Rotstein, N.D., Moguillansky, M.O., Garcia, A.J., Simari, G.R.: An abstract argumentation framework for handling dynamics. In: Proceedings of the Argument, Dialogue and Decision Workshop in NMR 2008, Sydney, Australia, pp. 131–139 (2008)

An Algorithm for Generating Arguments in Classical Predicate Logic

Vasiliki Efstathiou and Anthony Hunter

Department of Computer Science,
University College London,
Gower Street, London WC1E 6BT, UK
{v.efstathiou,a.hunter}@cs.ucl.ac.uk

Abstract. There are a number of frameworks for modelling argumentation in logic. They incorporate a formal representation of individual arguments and techniques for comparing conflicting arguments. A common assumption for logic-based argumentation is that an argument is a pair $\langle \Phi, \alpha \rangle$ where Φ is a minimal subset of the knowledgebase such that Φ is consistent and Φ entails the claim α. Different logics provide different definitions for consistency and entailment and hence give us different options for argumentation. An appealing option is classical first-order logic which can express much more complex knowledge than possible with defeasible or classical propositional logics. However the computational viability of using classical first-order logic is an issue. Here we address this issue by using the notion of a connection graph and resolution with unification. We provide a theoretical framework and algorithm for this, together with some theoretical results.

1 Introduction

Argumentation is a vital aspect of intelligent behaviour by humans used to deal with conflicting information. There are a number of proposals for logic-based formalisations of argumentation (for reviews see [6,14,5]). These proposals allow for the representation of arguments and for counterargument relationships between arguments. In a number of key examples of argumentation systems, an argument is a pair where the first item in the pair is a minimal consistent set of formulae that proves the second item which is a formula (see for example [2,9,3,1,10,4,7,13]). Algorithms have been developed for finding arguments from a knowledgebase using defeasible logic. However, there is a lack of viable algorithms for finding arguments for first-order classical logic.

We propose an approach to this problem by extending an existing proposal for propositional logic [8] based on the connection graph proof procedure [11,12]. This extension is based on the idea of resolution with unification [15]. We use a connection graph structure where nodes are clauses from the knowledgebase and arcs link contradictory literals and then apply heuristic search strategies to follow the arcs of the graph and create partial instances of the visited clauses based on the unification of atoms that appear at either end of an arc. The

C. Sossai and G. Chemello (Eds.): ECSQARU 2009, LNAI 5590, pp. 119–130, 2009.

aim of the search is to retrieve an unsatisfiable grounded version of a subset of the knowledgebase and use the refutation completeness of the resolution rule and Herbrand's theorem to obtain a proof for the claim. The minimality and consistency of this proof is achieved according to some restrictions applied during the search.

2   Argumentation for a Language of Quantified Clauses

In this section we review an existing proposal for argumentation based on classical logic [3] and in particular an extension of this [4] dealing with first-order logic. For a first-order language $\mathcal{F}$, the set of formulae that can be formed is given by the usual inductive definitions of classical logic.

In this paper we use a restricted function-free first-order language of quantified clauses $\mathcal{F}$ consisting of n-ary predicates $(n \geq 1)$ where we allow both existential and universal quantifiers and we consider arguments whose claims consist of one disjunct (i.e. unit clauses). This language is composed of the set of n-ary $(n \geq 1)$ predicates $\mathcal{P}$, a set of constant symbols $\mathcal{C}$, a set of variables $\mathcal{V}$, the quantifiers $\forall$ and $\exists$, the connectives $\neg$ and $\vee$ and the bracket symbols ( ). The clauses of $\mathcal{F}$ are in prenex normal form, consisting of a quantification string followed by a disjunction of literals. Literals are trivially defined as positive or negative atoms where an atom is an n-ary predicate. The quantification part consists of a sequence of quantified variables that appear as parameters of the predicates of the clause. These need not follow some ordering, that is any type of quantifier (existential or universal) can preceed any type of quantifier. Deduction in classical logic is denoted by the symbol $\vdash$.

Example 1. If $\{a, b, c, d, e\} \subset \mathcal{C}$ and $\{x, y, z, w\} \subset \mathcal{V}$, then each of the elements of Φ is a clause in $\mathcal{F}$ where $\Phi = \{\forall x \exists z (P(x) \vee \neg Q(z, a)), \exists x \exists z (P(x) \vee \neg Q(z, a), \forall w \exists x \exists z (P(x) \vee \neg Q(z, a) \vee P(b, w, x, z)), \forall w (\neg Q(w, b, a)), \neg Q(e, b, a) \vee R(d), \neg P(a, d)\}$. In addition $\forall w (\neg Q(w, b, a))$ is a unit clause, $\neg Q(e, b, a) \vee R(d)$ is a ground clause and $\neg P(a, d)$ is a ground unit clause.

Given a set Δ of first-order clauses, we can define an argument as follows.

Definition 1. *An **argument** is a pair $\langle \Phi, \psi \rangle$ such that (1) $\Phi \subseteq \Delta$, (2) $\Phi \vdash \psi$, (3) $\Phi \nvdash \perp$ and (4) there is no $\Phi' \subset \Phi$ such that $\Phi' \vdash \psi$.*

Example 2. Let $\Delta = \{\forall x (\neg P(x) \vee Q(x)), P(a), \forall x \forall y (P(x, y) \vee \neg P(x)), R(a, b), \exists x (R(x, b)), \exists x (\neg S(x, b))\}$. Some arguments are:

$\langle \{\forall x (\neg P(x) \vee Q(x)), P(a)\}, Q(a) \rangle$ $\langle \{R(a, b)\}, \exists x (R(x, b)) \rangle$
$\langle \{\forall x \forall y (P(x, y) \vee \neg P(x)), P(a)\}, \forall y (P(a, y)) \rangle$ $\langle \{P(a)\}, \exists y (P(y)) \rangle$

3   Relations on Clauses

In this section we define some relations on $\mathcal{F}$ that we use throughout this paper. We use the terms 'term', 'variable' and 'constant' in the usual way. We define functions $\mathsf{Variables}(X)$ and $\mathsf{Constants}(X)$ to return the set of all the variables and constants respectively that appear in a literal or a clause or a set X.

Definition 2. *For a language $\mathcal{L}$, with variables $\mathcal{V}$ and constant symbols $\mathcal{C}$, the set of bindings $\mathcal{B}$ is $\{x/t \mid x \in \mathcal{V} \text{ and } t \in \mathcal{V} \cup \mathcal{C}\}$.*

Definition 3. *For a clause ϕ and a set of bindings $B \subseteq \mathcal{B}$, $\mathsf{Assign}(\phi, B)$ returns clause ϕ with the values of B assigned to the terms of ϕ. So, for each x/t, if x is a variable in ϕ, then x is replaced by t and the quantifier of x is removed.*

Example 3. Let $\phi = \exists x \forall y \exists z \forall w (P(c, x) \vee Q(x, y, z) \vee R(y, c, w))$. Some assignments for ϕ with the corresponding values assigned to ϕ are:

$$B_1 = \{x/a, z/b\}, \ \mathsf{Assign}(\phi, B_1) = \forall y \forall w (P(c, a) \vee Q(a, y, b) \vee R(y, c, w))$$
$$B_2 = \{x/a, y/b, z/b\}, \ \mathsf{Assign}(\phi, B_2) = \forall w (P(c, a) \vee Q(a, b, b) \vee R(b, c, w))$$
$$B_3 = \{x/a, z/b, w/b\}, \ \mathsf{Assign}(\phi, B_3) = \forall y (P(c, a) \vee Q(a, y, b) \vee R(y, c, b))$$
$$B_4 = \{x/a, y/a, z/b, w/b\}, \ \mathsf{Assign}(\phi, B_4) = P(c, a) \vee Q(a, a, b) \vee R(a, c, b)$$
$$B_5 = \{w/z\}, \ \mathsf{Assign}(\phi, B_5) = \exists x \forall y \exists z (P(c, x) \vee Q(x, y, z) \vee R(y, c, z))$$

Function $\mathsf{Assign}(\phi, B)$ gives a specific instance of ϕ, indicated by the bindings in B. We define next the function that returns all the possible instances for a clause ϕ and the function that returns all the possible instances for all the elements of a set of clauses Ψ.

Definition 4. *For a clause ϕ, $\mathsf{Assignments}(\phi)$ returns the set of all the possible instances of ϕ: $\mathsf{Assignments}(\phi) = \{\mathsf{Assign}(\phi, B_i) \mid B_i \in \wp(\mathcal{B})\}$. For a set of clauses Ψ, $\mathsf{SetAssignments}(\Psi) = \bigcup_{\phi \in \Psi} \{\mathsf{Assignments}(\phi)\}$.*

We use the assignment functions to create partial instances of the clauses from the knowledgebase during the search for arguments. As no restrictions apply to the order of the quantifiers in the quantification of a clause from $\mathcal{F}$, the order of interchanging universal and existential quantifiers in a clause ϕ is taken into account when a partial instance of ϕ is created. For this, we define function $\mathsf{Prohibited}(\phi)$ to return the sets of bindings that are not allowed for ϕ.

Definition 5. *Let ϕ be a clause. Then, $\mathsf{Prohibited}(\phi) \subseteq \wp(\mathcal{B})$ returns the set of sets of bindings such that for each $B \in \mathsf{Prohibited}(\phi)$ there is at least one $y_i/t_i \in B$ such that y_i is a universally quantified variable which is in the scope of an existentially quantified variable x_i for which either $x_i = t_i$ or $x_i/t_i \in B$.*

Example 4. For the sets of bindings of example 3, $B_3, B_4, B_5 \in \mathsf{Prohibited}(\phi)$ and $B_1, B_2 \notin \mathsf{Prohibited}(\phi)$.

We now define a function that gives a partial instance of a clause ϕ where each of the existentially quantified variables is replaced by a distinct arbitrary constant from $\mathcal{C} \setminus \mathsf{Constants}(\phi)$. This is a form of Skolemization.

Definition 6. *For a clause ϕ, $\mathsf{ExistentialGrounding}(\phi, B) = \mathsf{Assign}(\phi, B)$ where $B \in \wp(\mathcal{B})$ is such that: (1) $x_i/t_i \in B$ iff $x_i \in \mathsf{Variables}(\phi)$ and x_i is existentially quantified (2) $t_i \in \mathcal{C} \setminus \mathsf{Constants}(\phi)$ and (3) for all $x_j/t_j \in B$, if $x_j \neq x_i$ then $t_j \neq t_i$. If $\phi' = \mathsf{ExistentialGrounding}(\phi, B)$ for some $\phi \in \mathcal{F}$ and $B \in \wp(\mathcal{B})$, we say that ϕ' is an* **existential instance** *of ϕ.*

Example 5. For $\phi = \exists x \forall y \exists z \forall w (P(c,x) \vee Q(x,y,z) \vee R(y,c,w))$, $\mathsf{Constants}(\phi) = \{c\}$ and so each of the elements of the set of constants $I = \{a,b\} \subset \mathcal{C} \setminus \mathsf{Constants}(\phi)$ can be used for the substitution of each of the existentially bound variables x, z. For $B = \{x/a, z/b\}$, $\mathsf{ExistentialGrounding}(\phi, B) = \forall y \forall w (P(c,a) \vee Q(a,y,b) \vee R(y,c,w))$.

Definition 7. *For a clause* $\phi = Q_1 x_1, \ldots, Q_m x_m (p_1 \vee \ldots \vee p_k)$, $\mathsf{Disjuncts}(\phi)$ *returns the set of disjuncts of* ϕ. $\mathsf{Disjuncts}(\phi) = \{p_1 \vee \ldots \vee p_k\}$. *For each* $p_i \in \mathsf{Disjuncts}(\phi)$, $\mathsf{Unit}(\phi, p_i)$ *returns the unit clause that consists of* p_i *as its unique disjunct and the part of the quantification* $Q_1 x_1, \ldots, Q_m x_m$ *of* ϕ *that involves the variables that occur in* p_i *as its quantification:* $\mathsf{Unit}(\phi, p_i) = Q_j x_j \ldots Q_l x_l (p_i)$ *where* $\{Q_j x_j, \ldots, Q_l x_l\} \subseteq \{Q_1 x_1, \ldots, Q_m x_m\}$ *and* $\{x_j, \ldots, x_l\} = \mathsf{Variables}(p_i)$.

Example 6. Let $\phi = \forall x \forall y \exists z (P(x) \vee Q(a) \vee \neg R(x,y,z,b) \vee S(a,b,c))$ and let $p = P(x), q = Q(a), r = \neg R(x,y,z,b)$ and $s = S(a,b,c)$. Then, $\mathsf{Disjuncts}(\phi) = \{p,q,r,s\}$ and

$$\mathsf{Unit}(\phi, p) = \forall x (P(x)) \qquad \mathsf{Unit}(\phi, r) = \forall x \forall y \exists z (\neg R(x,y,z,b))$$
$$\mathsf{Unit}(\phi, q) = Q(a) \qquad \mathsf{Unit}(\phi, s) = S(a,b,c)$$

Definition 8. *For a clause* ϕ, $\mathsf{Units}(\phi) = \{\mathsf{Unit}(\phi, p_i) \mid p_i \in \mathsf{Disjuncts}(\phi)\}$.

Example 7. Continuing example 6, for $\phi = \forall x \forall y \exists z (P(x) \vee Q(a) \vee \neg R(x,y,z,b) \vee S(a,b,c))$, $\mathsf{Units}(\phi) = \{\forall x (P(x)), Q(a), \forall x \forall y \exists z (\neg R(x,y,z,b)), S(a,b,c)\}$

We now define some binary relations that express contradiction between clauses. For this we define contradiction between unit clauses ϕ and ψ as follows: ϕ and ψ **contradict** each other iff $\phi \vdash \neg \psi$. Then we say that ψ is a complement of ϕ and we write $\phi = \overline{\psi}$. Using the contradiction relation between the units of a pair of clauses, we define the following relations of attack.

Definition 9. *Let* ϕ, ψ *be clauses. Then,* $\mathsf{Preattacks}(\phi, \psi) = \{a_i \in \mathsf{Units}(\phi) \mid \exists a_j \in \mathsf{Units}(\psi) \text{ s.t. } a_i = \overline{a}_j\}$.

Example 8. According to definition 9, the following relations hold.

8.1) $\mathsf{Preattacks}(\forall x (\neg N(x) \vee R(x)), N(a) \vee \neg R(b)) = \{\forall x (\neg N(x)), \forall x (R(x))\}$
8.2) $\mathsf{Preattacks}(\forall x (\neg N(x) \vee R(x)), N(a) \vee \neg R(a)) = \{\forall x (\neg N(x)), \forall x (R(x))\}$
8.3) $\mathsf{Preattacks}(P(a) \vee \neg Q(b), \neg P(a) \vee Q(b)) = \{P(a), \neg Q(b)\}$
8.4) $\mathsf{Preattacks}(\forall x (P(x) \vee \neg Q(a,x)), \exists x (\neg P(a) \vee Q(x,b))) = \{\forall x (P(x))\}$
8.5) $\mathsf{Preattacks}(\exists x (\neg P(a) \vee Q(x,b)), \forall x (P(x) \vee \neg Q(a,x))) = \{\neg P(a)\}$

We now define a special case of the preattacks relation which we use to define arcs for trees in the next section.

Definition 10. *For clauses* ϕ, ψ, *if* $|\mathsf{Preattacks}(\phi, \psi)| = 1 = |\mathsf{Preattacks}(\psi, \phi)|$ *then* $\mathsf{Attacks}(\phi, \psi) = \alpha$, *where* $\alpha \in \mathsf{Preattacks}(\phi, \psi)$, *otherwise* $\mathsf{Attacks}(\phi, \psi) = \mathsf{Attacks}(\psi, \phi) = null$.

Example 9. For examples 8.1, 8.2 and 8.3, $\mathsf{Attacks}(\phi, \psi) = null$. For examples 8.4-8.5, $\mathsf{Attacks}(\phi, \psi) = \mathsf{Preattacks}(\phi, \psi)$.

Although the Attacks relation might be *null* for a pair of clauses ϕ, ψ, it can sometimes hold for instances of ϕ and ψ.

Example 10. In example 8.1, let $\phi = \forall x (\neg N(x) \vee R(x))$ and $\psi = N(a) \vee \neg R(b)$. Then $|\mathsf{Preattacks}(\phi, \psi)| > 1$ and so, $\mathsf{Attacks}(\phi, \psi) = null$. There are instances ϕ' of ϕ though for which $\mathsf{Attacks}(\phi', \psi) \neq null$. Let $B_1 = \{x/a\}$, and $B_2 = \{x/b\}$. Then for $\phi_1 = \mathsf{Assign}(\phi, B_1) = \neg N(a) \vee R(a)$ and $\phi_2 = \mathsf{Assign}(\phi, B_2) = \neg N(b) \vee R(b)$, $\mathsf{Attacks}(\phi_1, \psi) = \neg N(a)$ and $\mathsf{Attacks}(\phi_2, \psi) = R(b)$. For all other instances ϕ' of ϕ $\mathsf{Attacks}(\phi', \phi) = null$.

Example 11. In example 8.2, let $\gamma = \forall x (\neg N(x) \vee R(x))$ and $\delta = N(a) \vee \neg R(a)$. Then, for all the instances γ' of γ, $|\mathsf{Preattacks}(\gamma', \delta)| \neq 1$ and so there is no instance γ' of γ for which $\mathsf{Attacks}(\gamma', \delta) \neq null$.

4   Assignment Trees

Using the attack relations defined in section 3, we define in this section the notion of an assignment tree which represents a tentative proof of an argument. The definition is designed for use with the algorithm we introduce in section 5 for searching for arguments.

Definition 11. *Let Δ be a clause knowledgebase and α be a unit clause and let $\Delta' = \Delta \cup \{\neg \alpha\}$. An **assignment tree** for Δ and α is tuple (N, A, e, f, g, h) where N is a set of nodes and A is a set of arcs such that (N, A) is a tree and e, f, g, h are functions such that: $e : N \mapsto \Delta'$, $f : N \mapsto \mathsf{SetAssignments}(\Delta')$, $g : N \mapsto \wp(\mathcal{B})$, $h : N \mapsto \mathsf{SetAssignments}(\Delta')$ and*

(1) *if p is the root of the tree, then $e(p) = \neg \alpha$*
(2) *$f(p)$ is an existential instance of $e(p)$ s.t. $\mathsf{Constants}(f(p)) \subseteq \mathsf{Constants}(g(p))$*
(3) *for any nodes p, q in the same branch, if $e(p) = e(q)$ then $g(p) \neq g(q)$*
(4) *for all $p \in N, g(p) \cap \mathsf{Prohibited}(e(p)) = \emptyset$*
(5) *for all $p \in N, h(p) = \mathsf{Assign}(f(p), g(p))$*
(6) *for all $p, q \in N$, if p is the parent of q, then $\mathsf{Attacks}(h(q), h(p)) \neq null$*
(7) *for all $p, q \in N$, $(\mathsf{Constants}(f(p)) \setminus \mathsf{Constants}(e(p))) \cap \mathsf{Constants}(\Delta') = \emptyset$, & $(\mathsf{Constants}(f(p)) \setminus \mathsf{Constants}(e(p))) \cap (\mathsf{Constants}(f(q)) \setminus \mathsf{Constants}(e(q))) = \emptyset$*

Each of the functions e, f, g, h for a node p gives the state of the tentative proof for an argument for α. Function $e(p)$ identifies for p the clause ϕ from $\Delta \cup \{\neg \alpha\}$ and $f(p)$ is an existential instance of $e(p)$. $g(p)$ is a set of bindings that when assigned to $e(p)$ creates the instance $h(p)$ of $e(p)$. Hence, $g(p)$ contains the set of bindings that create the existential instance $f(p)$ of $e(p)$ together with the bindings that unify atoms of contradictory literals connected with arcs on the tree as condition 6 indicates. Condition 7 ensures that the existential instances used in the proof are created by assigning to the existentially quantified variables of a clause $e(p)$ constants that do not appear anywhere else in $\Delta \cup \{\neg \alpha\}$ or the other instances of the clauses of the tentative proof. Finally, condition 3 ensures that an infinite sequence of identical nodes on a branch will be avoided.

In all the examples that follow, assignment trees are represented by the value $h(p)$ for each node p. Hence, all the variables that appear in a tree representation are universally quantified and so universal quantifiers are omitted for simplicity.

Example 12. Let $\Delta = \{\forall y(\neg P(y) \vee Q(b,y)), \forall y\exists x(P(d) \vee P(a) \vee M(x,y)), R(c),$ $\forall x\forall y(\neg M(x,y)), \exists x\forall y(Q(x,y) \vee R(x,y)), Q(a,b) \vee \neg N(a,b), \forall x\forall y(L(x,y,a)),$ $\forall x(\neg R(x,x) \vee S(x,y)), \neg Q(a,b) \vee N(a,b), \forall x\forall y(\neg S(x,y)), \neg L(c,d,a), \forall x(P(x)),$ $\neg R(c,a)\}$. The following is an assignment tree for Δ and $\alpha = \exists x\exists y(Q(x,y))$

$e(p_0) = f(p_0) = \forall x\forall y(\neg Q(x,y)), g(p_0) = \{x/b, y/d\}$

$e(p_1) = f(p_1) = \forall y(\neg P(y) \vee Q(b,y)), g(p_1) = \{y/d\}$

$e(p_2) = \forall y\exists x(P(d) \vee P(a) \vee M(x,y)), g(p_2) = \{x/f\}$

$f(p_2) = \forall y(P(d) \vee P(a) \vee M(f,y))$

$e(p_3) = f(p_3) = \forall y(\neg P(y) \vee Q(b,y)), g(p_3) = \{y/a\}$

$e(p_4) = f(p_4) = \forall x\forall y(\neg M(x,y)), g(p_4) = \{x/f\}$

Definition 12. *A **complete assignment tree** (N, A, e, f, g, h) is an assignment tree such that for any $x \in N$ if y a child of x then there is a $\bar{b}_i \in \mathsf{Units}(h(x))$ such that $\mathsf{Attacks}(h(y), h(x)) = b_i$ and for each $b_j \in \mathsf{Units}(h(y)) \setminus \{b_i\}$*

(1) either there is exactly one child z of y s.t. $\mathsf{Attacks}(h(z), h(y)) = \bar{b}_j$
(2) or there is a node w in the branch containing y s.t. $b_j = \mathsf{Attacks}(h(y), h(w))$

Definition 13. *A **grounded assignment tree** (N, A, e, f, g, h) is an assignment tree such that for any $x \in N$, $h(x)$ is a ground clause.*

Example 13. The assignment tree of example 12 is neither complete nor grounded. It is not a complete assignment tree because for $Q(b,a) \in \mathsf{Units}(h(p_3))$ the conditions of definition 12 do not hold. Adding a node p_5 as a child of p_3 with $e(p_5) = f(p_5) = \forall x\forall y(\neg Q(x,y)), g(p_5) = \{x/b, y/a\}$ for which $h(p_5) = \neg Q(b,a)$ gives a complete assignment tree. It is not a grounded assignment tree because for nodes p_2 and p_4 $h(p_2) = P(d) \vee P(a) \vee M(f,y)$ and $h(p_4) = \neg M(f,y)$ are non-ground clauses. If we substitute the non-ground term y in $h(p_2)$ and $h(p_4)$ with the same arbitrary constant value ($e \in \mathcal{C}$ for instance), the resulting tree still satisfies the conditions for an assignment tree and it is also a grounded assignment tree.

assignment tree 1
(grounded)

assignment tree 2
(complete)

assignment tree 3
(grounded & complete)

For a complete grounded assignment tree we have the following result on the entailment of a claim α for an argument.

Proposition 1. *If* (N, A, e, f, g, h) *is a complete grounded assignment tree for* Δ *and* α, *then* $\{e(p) \mid p \in N\} \setminus \{\neg\alpha\} \vdash \alpha$.

Example 14. For the complete and grounded assignment tree of example 13, $\{e(p) \mid p \in N\} \setminus \{\neg\alpha\} = \{\forall y(\neg P(y) \vee Q(b, y)), \forall y \exists x (P(d) \vee P(a) \vee M(x, y)), \forall x \forall y (\neg Q(x, y)), \forall x \forall y (\neg M(x, y))\} \vdash \exists x \exists y (Q(x, y))$.

Although all the assignment trees in example 13 correspond to the same subset of clauses $e(p)$ from Δ, it is not always the case that a non-grounded or non-complete assignment tree is sufficient to indicate a proof for α.

The following definitions introduce additional constraints on the definition of a complete assignment tree for Δ and α that give properties related to the minimality and the consistency of the proof for α indicated by the set of nodes in the assignment tree.

Definition 14. (N, A, e, f, g, h) *is* **a minimal assignment tree** *for* Δ *and* α *if for any arcs* (p, q), (p', q') *such that* $\mathsf{Attacks}(h(q), h(p)) = \mathsf{Assign}(\beta, g(q))$ *for some* $\beta \in \mathsf{Units}(e(q))$, *and* $\mathsf{Attacks}(h(q'), h(p')) = \mathsf{Assign}(\beta', g(q'))$ *for some* $\beta' \in \mathsf{Units}(e(q'))$, $\beta \vdash \beta'$ *holds iff* $e(q) = e(q')$.

Example 15. The following (N, A, e, f, g, h) is a complete assignment tree for a knowledgebase Δ and $\alpha = \exists x (\neg M(x))$, with $\{e(p) \mid p \in N\} = \{\forall x (M(x)), \forall x (\neg S(a) \vee \neg M(x) \vee \neg T(x)), \forall x (S(a) \vee N(x)), \forall x (T(x) \vee N(x)), \forall x (\neg N(x)), \forall x (\neg N(x) \vee R(x)), \forall x (\neg R(x))\}$. (N, A, e, f, g, h) is not minimal because of $\beta = \forall x (\neg N(x)) \in \mathsf{Units}(e(q))$ and $\beta' = \forall x (\neg N(x)) \in \mathsf{Units}(e(q'))$. If a copy of the subtree rooted at p in (N, A, e, f, g, h) is substituted by the subtree rooted at p', a minimal assignment tree (N', A', e', f', g', h') with $\{e(p) \mid p \in N'\} = \{\forall x (M(x)), \forall x (\neg R(x)), \forall x (\neg S(a) \vee \neg M(x) \vee \neg T(x)), \forall x (S(a) \vee N(x)), \forall x (T(x) \vee N(x)), \forall x (\neg N(x) \vee R(x))\}$ is obtained. Similarly, if a copy of the subtree rooted at p' is substituted by the subtree rooted at p, another minimal assignment tree $(N'', A'', e'', f'', g'', h'')$ is obtained, with $\{e(p) \mid p \in N''\} = \{\forall x (M(x)), \forall x (\neg S(a) \vee \neg M(x) \vee \neg T(x)), \forall x (S(a) \vee N(x)), \forall x (T(x) \vee N(x)), \forall x (\neg N(x))\}$.

$$M(x)$$
$$|$$
$$\neg S(a) \vee \neg M(x) \vee \neg T(x)$$

$$S(a) \vee N(x)_p \qquad T(x) \vee N(x)_{p'}$$
$$| \qquad\qquad\qquad |$$
$$\neg N(x)_q \qquad\qquad \neg N(x) \vee R(x)_{q'}$$
$$|$$
$$\neg R(x)$$

Definition 15. *Let* (N, A, e, f, g, h) *be a minimal assignment tree for* Δ *and* α. *Then,* (N, A, e, f, g, h) *is* **a consistent assignment tree** *if for any arcs* (p, q),

(p', q') *where* $\mathsf{Attacks}(h(q), h(p)) = \mathsf{Assign}(\beta, g(q))$ *for some* $\beta \in \mathsf{Units}(e(q))$ *and* $\mathsf{Attacks}(h(q'), h(p')) = \mathsf{Assign}(\beta', g(q'))$ *for some* $\beta' \in \mathsf{Units}(e(q'))$, $\beta \vdash \overline{\beta'}$ *holds iff* $e(q) = e(p')$.

Example 16. The following minimal assignment tree (N, A, e, f, g, h) with $\{e(p) \mid p \in N\} = \{\forall x \forall y (\neg Q(x, y)), \forall x \forall y (P(a) \vee \neg P(b) \vee Q(x, y)), \forall x (\neg P(x)), \forall x (P(x))\}$ is not consistent because for q, q', $\beta = \forall x (\neg P(x)) \in \mathsf{Units}(e(q))$, $\beta' = \forall x (P(x)) \in \mathsf{Units}(e(q'))$, $\beta \vdash \overline{\beta'}$ but $e(q) \neq e(p')$.

$$\neg Q(x, y)$$
$$|$$
$$P(a) \vee \neg P(b) \vee Q(x, y)_{p\,=\,p'}$$
$$\diagup \qquad \diagdown$$
$$\neg P(a)_q \qquad\qquad P(b)_{q'}$$

An assignment tree (N', A', e', f', g', h') with the same tree structure as above can be formed from the set of clauses $\{e(p) \mid p \in N'\} = \{\forall x \forall y (\neg Q(x, y)), \forall x \forall y (P(a) \vee \neg P(b) \vee Q(x, y)), \neg P(a), P(b)\}$. In this case, (N', A', e', f', g', h') satisfies the conditions of definition 15.

Using the definitions for minimality and consistency for an assignment tree we have the following result.

Proposition 2. *Let* (N, A, e, f, g, h) *be a complete, consistent grounded assignment tree. Then* $\langle \Phi, \alpha \rangle$ *with* $\Phi = \{e(p) \mid p \in N\} \setminus \{\neg \alpha\}$ *is an argument.*

5   Algorithms

In this section we present an algorithm to search for all the minimal and consistent complete assignment trees for a unit clause α from a given knowledgebase Δ. If a grounded version of a complete assignment trees exists, then according to proposition 2 this gives an argument for α.

Algorithm 1 builds a depth-first search tree T that represents the steps of the search for arguments for a claim α from a knowledgebase Δ. Every node in T is an assignment tree, every node is an extension of the assignment tree in its parent node. The leaf node of every complete accepted branch is a complete consistent assignment tree.

$\mathsf{Reject}(T)$, and $\mathsf{Accept}(T)$ are boolean functions which, given the current state T of the search tree, test whether the leaf node of the currently built branch can be expanded further. $\mathsf{Reject}(T)$ rejects the current branch of the search tree if the assignment tree in its leaf node does not satisfy the conditions for an assignment tree. $\mathsf{Accept}(T)$ checks whether a solution has been found. Hence, $\mathsf{Accept}(T)$ tests whether the assignment tree in the leaf node of the currently built branch is a complete assignment tree. When either of these functions returns true, the algorithm rejects or outputs the current branch accordingly and the algorithm backtracks and continues to the next node of tree T to be expanded. NextChild(T) adds to T one of the next possible nodes for its current leaf. A next possible node for the current branch can be any extension of the assignment tree contained in its current leaf which satisfies the conditions of definition 11.

Algorithm 1. $Build(T)$
___
if Reject(T) **then**
   return $T = null$
end if
if Accept(T) **then**
   return T
   $S =$ NextChild(T)
end if
while $S \neq null$ **do**
   $Build(S)$
   $S =$ NextChild(T)
end while
___

The search is based on a graph structure whose vertices are represented by clauses from $\Delta \cup \{\neg \alpha\}$ and arcs link clauses ϕ, ψ for which Preattacks(ϕ, ψ) $\neq \emptyset$. In fact, the algorithm works by visiting a subgraph of this graph which we call the **query graph** of α in Δ. The query graph is the component (N, A) of the graph where for each node $\phi \in N$: (1) ϕ is linked to $\neg \alpha$ through a path in A and (2) $\forall a_i \in$ Units(ϕ) there is a $\psi \in N$ with $a_i \in$ Preattacks(ϕ, ψ). Hence, each unit in each clause of the search space has a link in the query graph associated to it. Figure 1 illustrates the structure of the query graph of $\alpha = \exists x \exists y (Q(x, y))$ in Δ from example 12. The idea in building an assignment tree by using the structure of the query graph, is to start from the negation of the claim and walk over the graph by following the links and unifying the atoms of pairs of contradictory

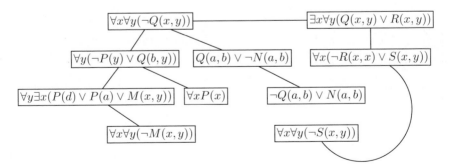

Fig. 1. The query graph of $\alpha = \exists x \exists y (Q(x, y))$ in Δ. The negation of the claim $\neg \alpha = \forall x \forall y (\neg Q(x, y))$ on the top left of the graph is the starting point for the search for arguments for α.

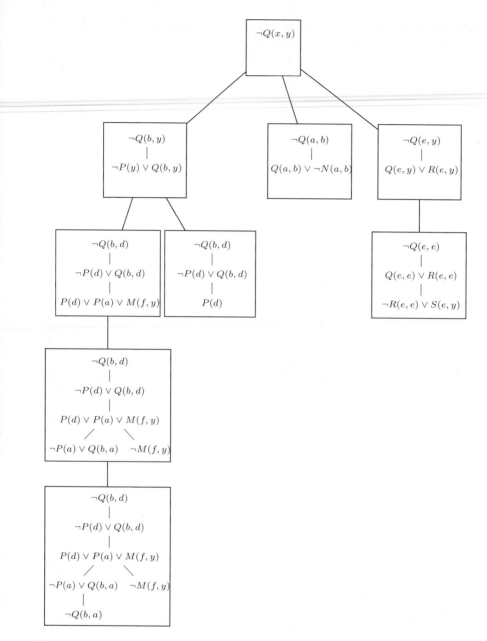

Fig. 2. A search tree generated using algorithm 1 by exploring the graph in figure 1. Each node of this search tree represents an assignment tree which extends the assignment tree contained in its parent node by one level. For this the algorithm adds clauses each of which preattacks their parent clause on a different unit. The atoms of the contradictory units between a parent and a child clause are unified and the assignments of the unification are passed on to any other clauses that can be affected in the assignment tree of this node.

literals connected with arcs. Hence, the algorithm at the same time follows the arcs of the graph and also produces partial instances of the clauses it visits as the unification of atoms indicates. The partial instances produced while walking over the graph are generated with respect to the conditions of definitions 11, 12, 14 and 15. Every time a clause ϕ on the graph is visited, a node q for an assignment tree is created with $e(q) = \phi$. For this node, an existential-free instance of $e(q)$ is generated by substituting each of its existentially quantified variables with an arbitrary constant that does not appear anywhere in $\Delta \cup \{\neg \alpha\}$ or in the instances already created during the search. This instantiation initializes the value $g(q)$ and sets the value $f(q)$ for q: $f(q) = \mathsf{Assign}(e(q), g(q))$. The value of $h(q)$ is also initialized at this stage to be equal to $f(q)$. After node q has been initialized as an assignment tree node, another instantiation process follows, which is based on unifying the atoms of the contradictory units in $h(q)$ and its parent with their most general unifier. This updates values $g(q)$ and $h(q)$. Let p be the parent of q in an assignment tree. If $\theta \subset \wp(\mathcal{B})$ is the the most general unifier of the atoms of a pair of contradictory units from $h(q)$ and $h(p)$, then $g(q) = g(q) \cup \theta$ and $h(q) = \mathsf{Assign}(h(q), g(q))$. Apart from node q, the values for g and h for any other node in the assignment tree are also updated. Every time such a unification binding is retrieved, its values are forwarded to the rest of the nodes in the assignment tree. These values are assigned to any of the corresponding clauses that can be associated through a sequence of arcs in the assignment tree to the variables of $e(q)$ and can therefore be affected by the bindings in θ.

Figure 2 represents the result of searching for arguments for $\alpha = \exists x \exists y (Q(x, y))$ using the query graph of figure 1. The result of the first branch of the search tree (at the leaf) is a complete consistent assignment tree which by substituting variable y in disjuncts $M(f, y)$ and $\neg M(f, y)$ by the same arbitrary constant gives a complete grounded assignment tree. The leaf of the second branch corresponds to a complete grounded assignment tree while the third branch is rejected because for node p with $e(p) = f(p) = h(p) = Q(a, b) \vee \neg N(a, b)$ there is only one arc in the graph that connects $e(p)$ with a clause that contains a complement of $\neg N(a, b)$. This is clause $\neg Q(a, b) \vee N(a, b)$ but a child q of p with $e(q) = \neg Q(a, b) \vee N(a, b)$ cannot be created because there is no assignment $g(q)$ for which $\mathsf{Attacks}(h(q), h(p)) \neq null$. The last branch of the search tree is rejected because adding node s with $e(s) = \forall x (\neg R(x, x) \vee S(x, y))$ as a child of r with $h(r) = Q(e, y) \vee R(e, y)$ requires unifying $R(x, x)$ with $R(e, y)$ which updates the value of $g(r)$ to $g(r) = \{x/e, y/e\} \in \mathsf{Prohibited}(e(r))$ and so condition 4 of the definition for an assignment tree is violated.

6   Discussion

Classical first-order logic has many advantages for representing and reasoning with knowledge. However, in general it is computationally challenging to generate arguments from a knowledgebase using classical logic. In this paper we propose a method for retrieving arguments in a rich first-order language. We have provided a theoretical framework, algorithms and theoretical results for this proposal.

References

1. Amgoud, L., Cayrol, C.: A model of reasoning based on the production of acceptable arguments. Annals of Math. and A.I. 34, 197–216 (2002)
2. Benferhat, S., Dubois, D., Prade, H.: Argumentative inference in uncertain and inconsistent knowledge bases. In: Proceedings of the 9th Annual Conference on Uncertainty in Artificial Intelligence (UAI 1993), pp. 1449–1445 (1993)
3. Besnard, Ph., Hunter, A.: A logic-based theory of deductive arguments. Artificial Intelligence 128, 203–235 (2001)
4. Besnard, P., Hunter, A.: Practical first-order argumentation. In: Proceedings of the 20th American National Conference on Artificial Intelligence (AAAI 2005), pp. 590–595. MIT Press, Cambridge (2005)
5. Besnard, Ph., Hunter, A.: Elements of Argumentation. MIT Press, Cambridge (2008)
6. Chesñevar, C., Maguitman, A., Loui, R.: Logical models of argument. ACM Computing Surveys 32, 337–383 (2000)
7. Dung, P., Kowalski, R., Toni, F.: Dialectical proof procedures for assumption-based admissible argumentation. Artificial Intelligence 170, 114–159 (2006)
8. Efstathiou, V., Hunter, A.: Algorithms for effective argumentation in classical propositional logic: A connection graph approach. In: Hartmann, S., Kern-Isberner, G. (eds.) FoIKS 2008. LNCS, vol. 4932, pp. 272–290. Springer, Heidelberg (2008)
9. Elvang-Gøransson, M., Krause, P., Fox, J.: Dialectic reasoning with classically inconsistent information. In: Proceedings of the 9th Conference on Uncertainty in Artificial Intelligence (UAI 1993), pp. 114–121. Morgan Kaufmann, San Francisco (1993)
10. García, A., Simari, G.: Defeasible logic programming: An argumentative approach. Theory and Practice of Logic Programming 4(1), 95–138 (2004)
11. Kowalski, R.: A proof procedure using connection graphs. Journal of the ACM 22, 572–595 (1975)
12. Kowalski, R.: Logic for problem solving. North-Holland Publishing, Amsterdam (1979)
13. Prakken, H., Sartor, G.: Argument-based extended logic programming with defeasible priorities. Journal of Applied Non-Classical Logics 7, 25–75 (1997)
14. Prakken, H., Vreeswijk, G.: Logical systems for defeasible argumentation. In: Gabbay, D. (ed.) Handbook of Philosophical Logic. Kluwer, Dordrecht (2000)
15. Robinson, J.A.: A machine-oriented logic based on the resolution principle. J. ACM 12(1), 23–41 (1965)

Modelling Argument Accrual in Possibilistic Defeasible Logic Programming

Mauro J. Gómez Lucero, Carlos I. Chesñevar, and Guillermo R. Simari

National Council of Scientific and Technical Research (CONICET)
Artificial Intelligence Research & Development Laboratory (LIDIA)
Universidad Nacional del Sur (UNS), Bahía Blanca, Argentina
{mjg,cic,grs}@cs.uns.edu.ar

Abstract. Argumentation frameworks have proven to be a successful approach to formalizing commonsense reasoning. Recently, some argumentation frameworks have been extended to deal with possibilistic uncertainty, notably Possibilistic Defeasible Logic Programming (P-DeLP). At the same time, modelling *argument accrual* has gained attention from the argumentation community. Even though some preliminary formalizations have been advanced, they do not take into account possibilistic uncertainty when accruing arguments. In this paper we present a novel approach to model argument accrual in the context of P-DeLP in a constructive way.

1   Introduction

Argumentation frameworks have proven to be a succesful approach to formalizing qualitative, commonsense reasoning. Recently, some argumentation frameworks have emerged which incorporate the treatment of possibilistic uncertainty (e.g. those proposed by Amgoud et al. with application in decision making [1] and merging conflicting databases [2]). Possibilistic Defeasible Logic Programming (P-DeLP) [3,4] is an argumentation framework based on logic programming which incorporates the treatment of possibilistic uncertainty at the object-language level.

At the same time, the notion of argument accrual has received some attention from the argumentation community [5,6,7]. This notion is based on the intuitive idea that having more reasons or arguments for a given conclusion makes such a conclusion more credible. However, none of the existing approaches to model argument accrual deals explicitly with possibilistic uncertainty.

In this paper we propose an approach based on P-DeLP to model argument accrual in a possibilistic setting. Our proposal is partly based on previous work in a workshop paper [7] where possibilistic uncertainty was not taken into account. We show that accrued arguments can be conceptualized as weighted structures which, as is the case with P-DeLP arguments, can be subject to a dialectical analysis in order to determine if their conclusions are warranted. As we will see, this is not a simple task. On the one hand, in our formalization we want to combine the propagation of necessity degrees when performing rule-based

C. Sossai and G. Chemello (Eds.): ECSQARU 2009, LNAI 5590, pp. 131–143, 2009.
© Springer-Verlag Berlin Heidelberg 2009

inference with a way of accumulating necessity values coming from different rules with the same conclusion. On the other hand, we do not want to commit ourselves to a specific way of aggregating necessity degrees; this will be abstracted away in terms of a user-defined function.

The rest of this paper is structured as follows. The next section briefly describes P-DeLP. Next we present the notion of *accrued structure*, which plays a central role in our proposal. Based on this notion, we then formalize the notions of attack and defeat among accrued structures. We show then how to perform a dialectical analysis on accrued structures, formalizing the notion of warranted literal. Finally, we present the main conclusions that have been obtained.

2   Argumentation in P-DeLP: An Overview

In order to make this paper self-contained, we will present next some of the main definitions that characterize the P-DeLP framework (for details the reader is referred to [3]). The language of P-DeLP is inherited from the language of logic programming, including the usual notions of atom, literal, rule and fact, but defined over an extended set of atoms where a new atom $\sim a$ is added for each original atom a. Therefore, a *literal* in P-DeLP is either an atom a or a (negated) atom of the form $\sim a$.

A *weighted* clause is a pair (φ, α), where φ is a rule $q \leftarrow p_1 \wedge \ldots \wedge p_k$ or a fact q (i.e., a rule with empty antecedent), where $q, p_1, \ldots, p_k$ are literals, and $\alpha \in [0, 1]$ expresses a lower bound for the necessity degree of φ. We distinguish between *certain* and *uncertain* clauses. A clause (φ, α) is referred as certain if $\alpha = 1$ and uncertain, otherwise. A set of P-DeLP clauses Γ will be deemed as *contradictory*, denoted $\Gamma \vdash \perp$, if , for some atom a, $\Gamma \vdash (a, \alpha)$ and $\Gamma \vdash (\sim a, \beta)$, with $\alpha > 0$ and $\beta > 0$, where $\vdash$ stands for deduction by means of the following particular instance of the *generalized modus ponens rule*:

$$\frac{(q \leftarrow p_1 \wedge \cdots \wedge p_k, \ \alpha)}{(q, \ \min(\alpha, \beta_1, \ldots, \beta_k))} \ [GMP]$$

A P-DeLP *program* $\mathcal{P}$ (or just program $\mathcal{P}$) is a pair (Π, Δ), where Π is a non-contradictory finite set of certain clauses, and Δ is a finite set of uncertain clauses. Formally, given a program $\mathcal{P} = (\Pi, \Delta)$, we say that a set $\mathcal{A} \subseteq \Pi \cup \Delta$ (of clauses) is an *argument* for a literal h with necessity degree $\alpha > 0$, denoted $\langle \mathcal{A}, h, \alpha \rangle$, iff:

1. $\mathcal{A} \vdash (h, \ \alpha)$,
2. $\Pi \cup \mathcal{A}$ is non-contradictory,
3. $\mathcal{A}$ is minimal w.r.t. set inclusion, i.e. there is no $\mathcal{A}_1 \subset \mathcal{A}$ s.t. $\mathcal{A}_1 \vdash (h, \ \alpha)$. [1]

---

[1] The definition of argument adopted here differs slightly from the one used in [3] in order to make simpler the formalization of the notion of accrued structure.

Moreover, if $\langle \mathcal{A}, h, \alpha \rangle$ and $\langle \mathcal{S}, k, \beta \rangle$ are arguments w.r.t. a program $\mathcal{P} = (\Pi, \Delta)$, we say that $\langle \mathcal{S}, k, \beta \rangle$ is a *subargument* of $\langle \mathcal{A}, h, \alpha \rangle$ whenever $\mathcal{S} \subseteq \mathcal{A}$.

Let $\mathcal{P}$ be a P-DeLP program, and let $\langle \mathcal{A}_1, h_1, \alpha_1 \rangle$ and $\langle \mathcal{A}_2, h_2, \alpha_2 \rangle$ be two arguments w.r.t. $\mathcal{P}$. We say that $\langle \mathcal{A}_1, h_1, \alpha_1 \rangle$ counterargues (or attacks) $\langle \mathcal{A}_2, h_2, \alpha_2 \rangle$ [2] iff there exists a subargument (called *disagreement subargument*) $\langle \mathcal{S}, k, \beta \rangle$ of $\langle \mathcal{A}_2, h_2, \alpha_2 \rangle$ such that $h_1 = \overline{k}$. In such a case, we say that $\langle \mathcal{A}_1, h_1, \alpha_1 \rangle$ is a *defeater* for $\langle \mathcal{A}_2, h_2, \alpha_2 \rangle$ when $\alpha_1 \geq \beta$.

3   Modelling Argument Accrual with Possibilistic Uncertainty

As stated in the introduction, our goal is to model argument accrual in a possibilistic setting taking into account several issues. In P-DeLP, the *GMP* inference rule allows us to propagate necessity degrees; however, given different arguments supporting the same conclusion, we want to be able to accumulate their strength in terms of possibilistic values. To do this we will define the notion of *accrued structure*, which will account for several arguments supporting the same conclusion, and whose necessity degree is defined in terms of two mutually recursive functions: $f_\Phi^+(\cdot)$ (the accruing function) and $f_\Phi^{MP}(\cdot)$ (which propagates necessity degrees as *GMP*). As we do not want to commit ourselves to a specific way of aggregating necessity degrees, we will assume that $f_\Phi^+(\cdot)$ is parameterized w.r.t. a user-specified function ACC. Additionally, we identify two properties that we believe reasonable to hold for any candidate instantiation of ACC: [**Non-depreciation**] $ACC(\alpha_1, \ldots, \alpha_n) \geq max(\alpha_1, \ldots, \alpha_n)$ (i.e., accruing arguments results in a necessity degree not lower than any single argument involved in the accrual). [**Maximality**] $ACC(\alpha_1, \ldots, \alpha_n) = 1$ only if $\alpha_i = 1$ for some i, $1 \leq i \leq n$ (i.e., accrual means total certainty only if there is an argument with necessity degree 1).

Definition 1 (Accrued Structure). *Let $\mathcal{P}$ be a P-DeLP program, and let Ω be a set of arguments in $\mathcal{P}$ supporting the same conclusion h, i.e., $\Omega = \{\langle \mathcal{A}_1, h, \alpha_1 \rangle, \ldots, \langle \mathcal{A}_n, h, \alpha_n \rangle\}$. We define the accrued structure for h (or just a-structure) from the set Ω (denoted $Accrual(\Omega)$) as a 3-uple $[\Phi, h, \alpha]$, where $\Phi = \mathcal{A}_1 \cup \ldots \cup \mathcal{A}_n$ and α is obtained using two mutually recursive functions, $f_\Phi^+(\cdot)$ and $f_\Phi^{MP}(\cdot)$, defined as follows. Let q be a literal appearing in Φ and let $(\varphi_1, \beta_1), \ldots, (\varphi_n, \beta_n)$ be all the weighted clauses in Φ with head q. Then*

$$f_\Phi^+(q) =_{def} ACC(f_\Phi^{MP}(\varphi_1), \ldots, f_\Phi^{MP}(\varphi_n))$$

Let (φ, β) be a weighted clause in Φ. Then

$$f_\Phi^{MP}(\varphi) =_{def} \begin{cases} \beta & \text{if } \varphi \text{ is a fact } q; \\ \\ min(f_\Phi^+(p_1), \ldots, f_\Phi^+(p_n), \beta) & \text{if } \varphi = q \leftarrow p_1, \ldots, p_n \end{cases}$$

---

[2] In what follows, for a given literal h, we will write $\overline{h}$ to denote "$\sim a$" if $h \equiv a$, and "a" if $h \equiv \sim a$.

Finally, $\alpha = f_\Phi^+(h)$. When $\Omega = \emptyset$ we get the special accrued structure $[\emptyset, \epsilon, 0]$, representing the accrual of no argument.

Next we will present the function $ACC_{1'}$ (one-complement accrual) as a possible instantiation for ACC, which will be used in the examples that follow. Formally:

$$ACC_{1'}(\alpha_1, \ldots, \alpha_n) = 1 - \prod_{i=1}^{n}(1 - \alpha_i)$$

It can be shown that $ACC_{1'}$ satisfies non-depreciation and maximality.[3]

Example 1. Consider a P-DeLP program $\mathcal{P}$ where:

$$\mathcal{P} = \left\{ \begin{array}{llll} (x \leftarrow z, \; 0.7) & (z \leftarrow v, \; 0.5) & (y \leftarrow u, \; 0.3) & (s \leftarrow p, \; 0.7) \quad (t, \; 1) \quad (w, \; 1) \\ (x \leftarrow y, \; 1) & (\sim z \leftarrow w, \; 0.4) & (\sim y \leftarrow p, \; 0.4) & (\sim s \leftarrow t, \; 0.9) \; (u, \; 1) \; (p, \; 1) \\ (z \leftarrow t, \; 0.6) & (\sim z \leftarrow s, \; 0.8) & (\sim x \leftarrow q, \; 0.45) \; (q, \; 1) & \qquad\qquad\qquad (v, \; 1) \end{array} \right\}$$

Let $\langle \mathcal{A}_1, x, 0.6 \rangle = \langle \{(x \leftarrow z, \; 0.7), (z \leftarrow t, \; 0.6), (t, \; 1)\}, x, 0.6 \rangle$,
$\langle \mathcal{A}_2, x, 0.5 \rangle = \langle \{(x \leftarrow z, \; 0.7), (z \leftarrow v, \; 0.5), (v, \; 1)\}, x, 0.5 \rangle$ and
$\langle \mathcal{A}_3, x, 0.3 \rangle = \langle \{(x \leftarrow y, \; 1), (y \leftarrow u, \; 0.3), (u, \; 1)\}, x, 0.3 \rangle$ be arguments in $\mathcal{P}$. Then
$Accrual(\{\langle \mathcal{A}_1, x, 0.6 \rangle, \langle \mathcal{A}_3, x, 0.3 \rangle\}) = [\Phi_1, z, 0.72]$ where
$\Phi_1 = \{(x \leftarrow z, \; 0.7), (z \leftarrow t, \; 0.6), (t, \; 1), (x \leftarrow y, \; 1), (y \leftarrow u, \; 0.3), (u, \; 1)\}$ (Fig. 1a)
$Accrual(\{\langle \mathcal{A}_1, x, 0.6 \rangle, \langle \mathcal{A}_2, x, 0.5 \rangle\}) = [\Phi_2, x, 0.7]$ where
$\Phi_2 = \{(x \leftarrow z, \; 0.7), (z \leftarrow t, \; 0.6), (t, \; 1), (z \leftarrow v, \; 0.5), (v, \; 1)\}$ (Fig. 1b)
$Accrual(\{\langle \mathcal{A}_1, x, 0.6 \rangle, \langle \mathcal{A}_2, x, 0.5 \rangle, \langle \mathcal{A}_3, x, 0.3 \rangle\}) = [\Phi_3, x, 0.79]$ where
$\Phi_3 = \{(x \leftarrow z, \; 0.7), (z \leftarrow t, \; 0.6), (t, \; 1), (z \leftarrow v, \; 0.5), (v, \; 1), (x \leftarrow y, \; 1), (y \leftarrow u, \; 0.3),$
$(u, \; 1)\}$ (Fig. 1c).

An a-structure for a conclusion h can be seen as a special kind of argument which subsumes different chains of reasoning which provide support for h. For instance, the a-structure $[\Phi_1, x, 0.72]$ (see Fig. 1a) provides two alternative chains of reasoning supporting x, both coming from each of the arguments accrued. The case of $[\Phi_2, x, 0.7]$ in Ex. 1 (see Fig. 1b) illustrates a situation similar to the previous one, but in this case the arguments involved share their topmost parts (more precisely the weighted clause $(x \leftarrow z, \; 0.7)$), differing in the reasons supporting the (shared) intermediate conclusion z. Figure 1 also shows how the possibilistic values associated with the depicted a-structures are obtained from the weighted clauses conforming them, using the functions $f_\Phi^+(\cdot)$ and $f_\Phi^{MP}(\cdot)$. Notice that weighted clauses were represented as black arrows labeled with their associated necessity measures. The values in gray ovals are computed using the mutually recursive functions.

An important question that naturally emerges when considering the way we accrue arguments is what happens if we accrue two arguments that are in conflict (for instance because they have contradictory intermediate conclusions.) We will come back to this issue later.

Definition 2. *Let $[\Phi, h, \alpha]$ be an a-structure. Then the set of arguments in $[\Phi, h, \alpha]$, denoted as $Args([\Phi, h, \alpha])$, is the set of all arguments $\langle \mathcal{A}_i, h, \alpha_i \rangle$ s.t. $\mathcal{A}_i \subseteq \Phi$. Note that $Args([\emptyset, \epsilon, 0]) = \emptyset$.*

---

[3] Proofs are not included for space reasons.

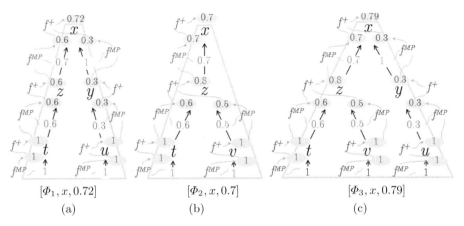

$$[\Phi_1, x, 0.72]$$ $$[\Phi_2, x, 0.7]$$ $$[\Phi_3, x, 0.79]$$

(a)   (b)   (c)

Fig. 1. Accrued Structures

Example 2. Consider the arguments and a-structures presented in Ex. 1. Then $Args([\Phi_1, x, 0.72]) = \{\langle A_1, x, 0.6\rangle, \langle A_3, x, 0.3\rangle\}$ and $Args([\Phi_3, x, 0.79]) = \{\langle A_1, x, 0.6\rangle, \langle A_2, x, 0.5\rangle, \langle A_3, x, 0.3\rangle\}$.

Definition 3 (Maximal a-structure). *Let $\mathcal{P}$ be a P-DeLP program. We say that an a-structure $[\Phi, h, \alpha]$ is maximal iff $Args([\Phi, h, \alpha])$ contains all arguments in $\mathcal{P}$ with conclusion h.*

Example 3. Consider the P-DeLP program $\mathcal{P}$ and the a-structures in Ex. 1. Then $[\Phi_3, x, 0.79]$ is a maximal a-structure in $\mathcal{P}$, whereas $[\Phi_1, z, 0.72]$ and $[\Phi_2, x, 0.7]$ are not.

Property 1 (uniqueness of maximal a-structures). Let $\mathcal{P}$ be a P-DeLP program and let h be a literal. If $[\Phi, h, \alpha]$ and $[\Phi_1, h, \alpha_1]$ are two maximal a-structures w.r.t. $\mathcal{P}$, then $[\Phi, h, \alpha] = [\Phi_1, h, \alpha_1]$.

Next we will introduce the notion of *narrowing* of an a-structure, which is analogous to the notion of narrowing in [5]. Intuitively, a narrowing of an a-structure $[\Phi, h, \alpha]$ is an a-structure $[\Theta, h, \beta]$ accounting for a subset of $Args([\Phi, h, \alpha])$.

Definition 4 (Narrowing of an a-structure). *Let $[\Phi, h, \alpha]$ and $[\Theta, h, \beta]$ be two a-structures. We say that $[\Theta, h, \beta]$ is a narrowing of $[\Phi, h, \alpha]$ iff $Args([\Theta, h, \beta]) \subseteq Args([\Phi, h, \alpha])$.*

Example 4. Consider the a-structures in Ex. 1. Then $[\Phi_1, x, 0.72]$, $[\Phi_2, x, 0.7]$ and $[\Phi_3, x, 0.79]$ itself are narrowings of $[\Phi_3, x, 0.79]$.

Next we will introduce the notion of accrued sub-structure, that is analogous to the notion of subargument but for a-structures. Intuitively, an accrued sub-structure of an a-structure $[\Phi, h, \alpha]$ is an a-structure supporting an intermediate conclusion k of $[\Phi, h, \alpha]$ and accounting for a subset of the reasons that support k in $[\Phi, h, \alpha]$. The one that accounts for all the reasons supporting k in $[\Phi, h, \alpha]$ is called *complete*.

Definition 5 (a-substructure and complete a-substructure). *Let* $[\Phi, h, \alpha]$ *and* $[\Theta, k, \gamma]$ *be two a-structures. Then we say that* $[\Theta, k, \gamma]$ *is an accrued substructure (o just a-substructure) of* $[\Phi, h, \alpha]$ *iff* $\Theta \subseteq \Phi$. *We also say that* $[\Theta, k, \gamma]$ *is a* complete a-substructure *of* $[\Phi, h, \alpha]$ *iff for any other a-substructure* $[\Theta', k, \gamma']$ *of* $[\Phi, h, \alpha]$ *it holds that* $\Theta' \subset \Theta$.

Example 5. Consider the a-structure $[\Phi_2, x, 0.7]$ in Ex. 1. Then the a-structures $[\{(z \leftarrow t,\ 0.6), (t,\ 1)\}, z, 0.6]$, $[\{(z \leftarrow t,\ 0.6), (t,\ 1), (z \leftarrow v,\ 0.5), (v,\ 1)\}, z, 0.8]$ and $[\Phi_2, x, 0.7]$ itself are a-substructures of $[\Phi_2, x, 0.7]$. Moreover, the two latter are complete.

4   Modelling Conflict and Defeat in Accrued Structures

Next we will formalize the notion of attack between a-structures, which differs from the notion of attack in argumentation frameworks in several respects. First, an a-structure $[\Phi, h, \alpha]$ stands for (possibly) several chains of reasoning (arguments) supporting the conclusion h. Besides, some intermediate conclusions in $[\Phi, h, \alpha]$ could be shared by some, but not necessarily all the arguments in $[\Phi, h, \alpha]$. Thus, given two a-structures $[\Phi, h, \alpha]$ and $[\Psi, k, \beta]$, if the conclusion k of $[\Psi, k, \beta]$ contradicts some intermediate conclusion h' in $[\Phi, h, \alpha]$, then only those arguments in $Args([\Phi, h, \alpha])$ involving h' will be affected by the conflict.

Next we will define the notion of *partial attack*, where the attacking a-structure generally affects only a narrowing of the attacked one (that one containing exactly the arguments in the attacked a-structure affected by the conflict), and we will refer to this narrowing as the *attacked narrowing*.

Definition 6 (Partial Attack and Attacked Narrowing). *Let* $[\Phi, h, \alpha]$ *and* $[\Psi, k, \beta]$ *be two a-structures. We say that* $[\Psi, k, \beta]$ *partially attacks* $[\Phi, h, \alpha]$ *at literal* h', *iff there exists a complete a-substructure* $[\Phi', h', \alpha']$ *of* $[\Phi, h, \alpha]$ *such that* $k = \overline{h'}$. *The a-substructure* $[\Phi', h', \alpha']$ *will be called the* disagreement a-substructure. *We will also say that* $[\Lambda, h, \gamma]$ *is the* attacked narrowing *of* $[\Phi, h, \alpha]$ *associated with the attack iff* $[\Lambda, h, \gamma]$ *is the minimal narrowing of* $[\Phi, h, \alpha]$ *that has* $[\Phi', h', \alpha']$ *as an a-substructure.*

Example 6. Consider the a-structures $[\Phi_3, x, 0.79]$ and $[\Psi_1, \sim z, 0.82]$ in Fig. 2. Then $[\Psi_1, \sim z, 0.82]$ partially attacks $[\Phi_3, x, 0.79]$ with disagreement a-substructure $[\Phi', z, 0.8] = [\{(z \leftarrow t,\ 0.6), (t,\ 1), (z \leftarrow v,\ 0.5), (v,\ 1)\}, z, 0.8]$. The attacked narrowing of $[\Phi_3, x, 0.79]$ is $[\{(x \leftarrow z,\ 0.7), (z \leftarrow t,\ 0.6), (t,\ 1), (z \leftarrow v,\ 0.5), (v,\ 1)\}, x, 0.7]$. Graphically, this attack relation will be depicted with a dotted arrow (see Fig. 2).

4.1   Accrued Structures: Evaluation and Defeat

As in P-DeLP, we will use the necessity measures associated with a-structures in order to decide if a partial attack really succeeds and constitutes a defeat.

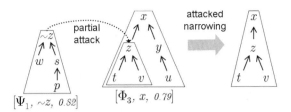

Fig. 2. Partial Attack

Definition 7 (Partial Defeater). *Let* $[\Phi, h, \alpha]$ *and* $[\Psi, k, \beta]$ *be two a-structures. Then we say that* $[\Psi, k, \beta]$ *is a* partial defeater *of* $[\Phi, h, \alpha]$ *(or equivalently that* $[\Psi, k, \beta]$ *is a* successful attack *on* $[\Phi, h, \alpha]$*) iff 1)* $[\Psi, k, \beta]$ *attacks* $[\Phi, h, \alpha]$ *at literal* h'*, where* $[\Phi', h', \alpha']$ *is the disagreement a-substructure, and 2)* $\beta \geq \alpha'$.

Example 7. Consider the attack from $[\Psi_1, \sim z, 0.82]$ against $[\Phi_3, x, 0.79]$ with disagreement a-substructure $[\Phi', z, 0.8]$ in Ex. 6 (Fig. 2). As the necessity measure associated with the attacking a-structure (0.82) is greater than the one associated with the disagreement a-substructure (0.8), then the attack succeeds, constituting a defeat. Graphically, this defeat relation will be depicted with a continuous arrow (see Fig. 3).

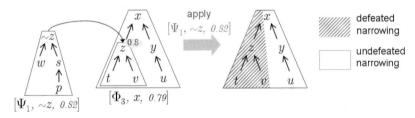

Fig. 3. Defeated and Undefeated Narrowings

Given an attack relation, we will identify two complementary narrowings associated with the attacked a-structure: the narrowing that becomes defeated as a consequence of the attack, and the narrowing that remains undefeated.

Definition 8 (Undefeated and Defeated Narrowings). *Let* $[\Phi, h, \alpha]$ *and* $[\Psi, k, \beta]$ *be two a-structures such that* $[\Psi, k, \beta]$ *attacks* $[\Phi, h, \alpha]$*. Let* $[\Lambda, h, \gamma]$ *be the attacked narrowing of* $[\Phi, h, \alpha]$*. Then the* defeated narrowing *of* $[\Phi, h, \alpha]$ *associated with the attack, denoted as* $\mathsf{N}_w^{\mathsf{D}}([\Phi, h, \alpha], [\Psi, k, \beta])$*, is defined by cases as follows: 1)* $\mathsf{N}_w^{\mathsf{D}}([\Phi, h, \alpha], [\Psi, k, \beta]) =_{def} [\Lambda, h, \gamma]$*, if* $[\Psi, k, \beta]$ *is a partial defeater of* $[\Phi, h, \alpha]$*, or 2)* $\mathsf{N}_w^{\mathsf{D}}([\Phi, h, \alpha], [\Psi, k, \beta]) =_{def} [\emptyset, \epsilon, 0]$*, otherwise. The* undefeated narrowing *of* $[\Phi, h, \alpha]$ *associated with the attack, denoted as* $\mathsf{N}_w^{\mathsf{U}}([\Phi, h, \alpha], [\Psi, k, \beta])$*, is the a-structure* $Accrual(Args([\Phi, h, \alpha]) \setminus Args(\mathsf{N}_w^{\mathsf{D}}([\Phi, h, \alpha], [\Psi, k, \beta])))$*.*

Example 8. Fig. 3 illustrates a successful attack from $[\Psi_1, \sim z, 0.82]$ against $[\Phi_3, x, 0.79]$, as well as the associated defeated and undefeated narrowings of

$[\Phi_3, x, 0.79]$. As another example, consider the attack from $[\Psi_2, \sim x, 0.45] = [\{(\sim x \leftarrow q,\ 0.45), (q,\ 1)\}, \sim x, 0.45]$ against $[\Phi_3, x, 0.79]$, with $[\Phi_3, x, 0.79]$ itself as disagreement a-substructure. In this case the attack does not succeed, and then $[\emptyset, \epsilon, 0]$ is the defeated narrowing and $[\Phi_3, x, 0.79]$ is the undefeated narrowing.

4.2   Combined Attack

Until now we have considered only *single* attacks. When a single attack succeeds, a nonempty narrowing of the attacked a-structure becomes defeated. But two or more a-structures could simultaneously attack another, possibly affecting different narrowings of the target a-structure, and thus causing a bigger narrowing to become defeated (compared with the defeated narrowings associated with the individual attacks). Fig. 4a illustrates a *combined* attack from the a-structures $[\Psi_1, \sim z, 0.82]$ and $[\Psi_3, \sim y, 0.4]$ against $[\Phi_3, x, 0.79]$. Even though each attacking a-structure defeats only a proper narrowing of $[\Phi_3, x, 0.79]$, the whole $[\Phi_3, x, 0.79]$ becomes defeated after applying *both* attacks.

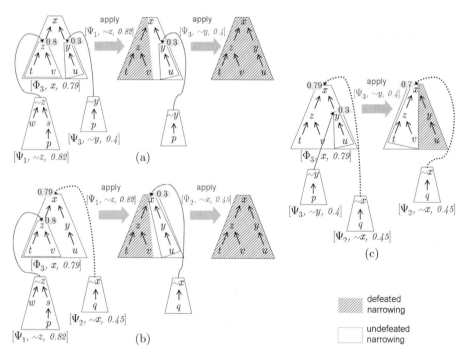

Fig. 4. Combined Defeat

Consider now the combined attack against $[\Phi_3, x, 0.79]$ shown in Fig. 4b. One of the attacking a-structures ($[\Psi_1, \sim z, 0.82]$) defeats a narrowing of $[\Phi_3, x, 0.79]$ on its own, whereas the other ($[\Psi_2, \sim x, 0.45]$) only attacks $[\Phi_3, x, 0.79]$. Note also that, although $[\Phi_3, x, 0.79]$ is stronger than $[\Psi_2, \sim x, 0.45]$, $[\Psi_2, \sim x, 0.45]$ is stronger than $[\Phi', x, 0.3] = [\{(x \leftarrow y, \ 1), (y \leftarrow u, \ 0.3), (u, \ 1)\}, x, 0.3]$, a proper narrowing of $[\Phi_3, x, 0.79]$. Then, as shown in Fig. 4b, when the a-structures $[\Psi_1, \sim z, 0.82]$ and $[\Psi_2, \sim x, 0.45]$ combine their attacks, they cause the whole $[\Phi_3, x, 0.79]$ to become defeated. The reason is that the successful attack of $[\Psi_1, \sim z, 0.82]$ weakens the target a-structure, allowing the attack of $[\Psi_2, \sim x, 0.45]$ to succeed. Figure 4c illustrates a combined attack from $[\Psi_2, \sim x, 0.45]$ and $[\Psi_3, \sim y, 0.4]$ against $[\Phi_3, x, 0.79]$. In this case, a nonempty narrowing of the attacked a-structure remains undefeated.

The purpose of the following definitions is to formally capture the notions of defeated and undefeated narrowings associated with a given combined attack from a set Σ of attacking a-structures against an a-structure $[\Phi, h, \alpha]$. In particular the first definition is a formalization of the procedure suggested by Figs. 4a, 4b and 4c, and described as follows. (1) Pick a defeater in Σ of $[\Phi, h, \alpha]$ (if any) and apply it, obtaining an undefeated narrowing $[\Theta, h, \beta]$ of $[\Phi, h, \alpha]$. (2) Repeat step 1 taking the resulting a-structure $[\Theta, h, \beta]$ as the new target for defeaters, until there is no more defeaters for $[\Theta, h, \beta]$ in Σ.

Definition 9 (Sequential Degradation). *Let $[\Phi, h, \alpha]$ be an a-structure and let Σ be a set of a-structures attacking $[\Phi, h, \alpha]$. A sequential degradation of $[\Phi, h, \alpha]$ associated with the combined attack of the a-structures in Σ, consists of a finite sequence of narrowings of $[\Phi, h, \alpha]$:*

$$[\Phi_1, h, \alpha_1], \ [\Phi_2, h, \alpha_2], \ \ldots, [\Phi_{m+1}, h, \alpha_{m+1}]$$

provided there exists a finite sequence of a-structures in Σ:

$$[\Psi_1, k_1, \beta_1], \ [\Psi_2, k_2, \beta_2], \ \ldots, [\Psi_m, k_m, \beta_m]$$

where $[\Phi_1, h, \alpha_1] = [\Phi, h, \alpha]$, for each i, $1 \leq i \leq m$, $[\Psi_i, k_i, \beta_i]$ partially defeats $[\Phi_i, h, \alpha_i]$ with associated undefeated narrowing $[\Phi_{i+1}, h, \alpha_{i+1}]$ and $[\Phi_{m+1}, h, \alpha_{m+1}]$ has no defeaters in Σ.

Given a combined attack against an a-structure $[\Phi, h, \alpha]$, there could exist several possible orders of defeater applications, and hence, more than one sequential degradation associated with the combined attack. Interestingly, it can be shown that all sequential degradations associated with a given combined attack converge to the same a-structure, provided that the function ACC satisfies non-depreciation.

Property 2 (Convergence). Let $[\Phi, h, \alpha]$ be an a-structure and let Σ be a set of a-structures attacking $[\Phi, h, \alpha]$. Let $[\Phi_1, h, \alpha_1], \ldots, [\Phi_m, h, \alpha_m]$ and $[\Phi'_1, h, \alpha'_1], \ldots, [\Phi'_n, h, \alpha'_n]$ be two sequential degradations of $[\Phi, h, \alpha]$ associated with the combined attack of the a-structures in Σ. Then $[\Phi_m, h, \alpha_m] = [\Phi'_n, h, \alpha'_n]$, provided that the ACC function satisfies non-depreciation.

Convergence can also be achieved without requiring non-depreciation for ACC function by slightly refining the notion of sequential degradation in the same way it was done in [7]. We do not present such a refinement here due to space limitations.

Definition 10 (Narrowings associated with a Combined Attack). *Let* $[\Phi, h, \alpha]$ *be an a-structure and let* Σ *be a set of a-structures attacking* $[\Phi, h, \alpha]$. *Let* $[\Phi_1, h, \alpha_1], ..., [\Phi_{m+1}, h, \alpha_{m+1}]$ *be a sequential degradation of* $[\Phi, h, \alpha]$ *associated with the combined attack of the a-structures in* Σ. *Then* $[\Phi_{m+1}, h, \alpha_{m+1}]$ *is the undefeated narrowing of* $[\Phi, h, \alpha]$ *associated with the combined attack, and* $Accrual(Args([\Phi, h, \alpha]) \setminus Args([\Phi_{m+1}, h, \alpha_{m+1}]))$ *is its defeated narrowing.*

Example 9. Consider the combined attack of $[\Psi_1, \sim z, 0.82]$ and $[\Psi_2, \sim x, 0.45]$ against $[\Phi_3, x, 0.79]$ (Fig. 4b). The associated undefeated narrowing of $[\Phi_3, x, 0.79]$ is $[\emptyset, \epsilon, 0]$, *i.e.*, the whole $[\Phi_3, x, 0.79]$ results defeated. On the other hand, when $[\Psi_2, \sim x, 0.45]$ and $[\Psi_3, \sim y, 0.4]$ attack $[\Phi_3, x, 0.79]$ (Fig. 4c), its associated undefeated narrowing is $[\{(x \leftarrow z, \ 0.7), (z \leftarrow t, \ 0.6), (t, \ 1), (z \leftarrow v, \ 0.5), (v, \ 1)\}, x, 0.7]$.

5 Dialectical Analysis for Accrued Structures

Given a P-DeLP program $\mathcal{P}$ and a literal h, we are interested in determining if h is ultimately accepted (or *warranted*), and if so, with which necessity degree. In order to determine this, we will consider the maximal a-structure $[\Phi, h, \alpha]$ supporting h, and we will analyze which is the final undefeated narrowing of $[\Phi, h, \alpha]$ after considering all possible a-structures attacking it. As those attacking a-structures may also have other a-structures attacking them, this strategy prompts a recursive dialectical analysis formalized as discussed below.

It must be remarked that this dialectical analysis can be seen as a generalization of the notion of dialectical tree used in P-DeLP [3], DeLP [8] and other argumentation frameworks (e.g. Besnard & Hunter's argument tree [9]). In such frameworks, nodes in dialectical trees stand for *individual* arguments, whereas in our case nodes correspond to accrued structures, each of them standing for many arguments supporting a given conclusion.

Definition 11 (Accrued Dialectical Tree). *Let* $\mathcal{P}$ *be a P-DeLP program and let* h *be a literal. Let* $[\Phi, h, \alpha]$ *be the maximal a-structure for* h *in* $\mathcal{P}$. *The accrued dialectical tree for* h, *denoted* $\mathcal{T}_h$, *is defined as follows:*

1. *The root of the tree is labeled with* $[\Phi, h, \alpha]$.
2. *Let* N *be an internal node labelled with* $[\Theta, k, \beta]$. *Let* Σ *be the set of all disagreement a-substructures associated with the attacks in the path from the root to* N. *Let* $[\Theta_i, k_i, \beta_i]$ *be a maximal a-structure attacking* $[\Theta, k, \beta]$ *s.t.* $[\Theta_i, k_i, \beta_i]$ *has no a-substructures in* Σ. *Then the node* N *has a child node* N_i *labelled with* $[\Theta_i, k_i, \beta_i]$. *If there is no a-structure attacking* $[\Theta, k, \beta]$ *satisfying the above condition, then* N *is a leaf.*

The condition involving the set Σ avoids the introduction of a new a-structure as a child of a node N if it is already present in the path from the root to N (resulting in a circularity). This requirement is needed in order to avoid *fallacious* reasoning, as discussed in [10].

Once the dialectical tree has been constructed, each combined attack is analyzed, from the deepest ones to the one against the root, in order to determine the undefeated narrowing of each node in the tree.

Definition 12 (Undefeated narrowing of a Node). *Let $\mathfrak{T}_h$ be the accrued dialectical tree for a given literal h. Let N be a node of $\mathfrak{T}_h$ labelled with $[\Theta, k, \beta]$. Then the* undefeated narrowing *of N is defined as follows:*

1. *If N is a leaf, then the undefeated narrowing of N is its own label $[\Theta, k, \beta]$.*
2. *Otherwise, let M_1, ..., M_n be the childs of N and let $[\Lambda_i, k, \gamma_i]$ be the undefeated narrowing of the a-structure labelling the child node M_i, $1 \leq i \leq n$. Then the undefeated narrowing of N is the undefeated narrowing of $[\Theta, k, \beta]$ associated with the combined attack involving all the $[\Lambda_i, k, \gamma_i]$, $1 \leq i \leq n$.*

Example 10. Fig. 5a shows the accrued dialectical tree for x w.r.t. program $\mathcal{P}$ (Ex. 1). Fig. 5b shows the accrued dialectical tree for x, where the undefeated narrowings of each node are highlighted.

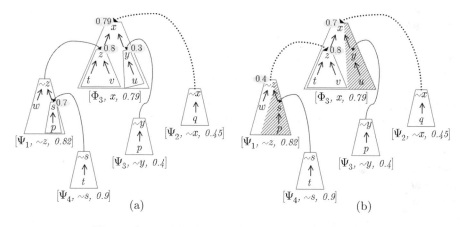

Fig. 5. Accrued Dialectical Tree and Warrant analysis

Definition 13 (Warrant). *Let $\mathcal{P}$ be a P-DeLP program and let h be a literal. Let $[\Phi, h, \alpha]$ be the maximal a-structure for h such that its undefeated narrowing in $\mathfrak{T}_h$ is a nonempty a-structure $[\Phi', h, \alpha']$. Then we say that h is* warranted *w.r.t. $\mathcal{P}$ with necessity α' and that $[\Phi', h, \alpha']$ is a* warranted a-structure.

According to the dialectical tree in Fig. 5b, the literal x is warranted w.r.t. $\mathcal{P}$ with necessity 0.7 and $[\{(x \leftarrow z,\ 0.7), (z \leftarrow t,\ 0.6), (t,\ 1), (z \leftarrow v,\ 0.5), (v,\ 1)\}, x, 0.7]$ is a warranted a-structure.

The following property establishes that the a-structure emerging as a result of the above dialectical process cannot involve contradictory literals.

Property 3. Let $\mathcal{P}$ be a P-DeLP program, and let $[\Phi, h, \alpha]$ be a warranted a-structure w.r.t. $\mathcal{P}$. Then there exist no intermediate conclusions k and r in $[\Phi, h, \alpha]$ s.t. $k = \bar{r}$.

6 Conclusions

In this paper we have proposed a novel approach based on P-DeLP to model argument accrual in a possibilistic setting. This approach is based on the notion of *accrued structure*, which accounts for different P-DeLP arguments supporting a given conclusion and associates a necessity measure to its conclusion obtained as an aggregation of the necessity measures of the individual arguments it accounts for. We have shown how accrued structures can be in conflict in terms of the notion of partial attack, and how possibilistic information is used to determine if a given attack succeeds, becoming a defeat. The notions of combined attack and sequential degradation were also defined, allowing us to characterize a dialectical process in which all accrued structures in favor and against a given conclusion h are taken into account in order to determine if h is warranted, and if so, with which necessity. Finally, an interesting property (Prop. 3) of our approach was stated, which ensures that accrued structures which are ultimately accepted as warranted will never involve conflicting arguments.

There exist other argument-based approaches modeling the notion of accrual, mainly the ones of Prakken [6] and Verheij [5]. However, none of them deals explicitly with possibilistic uncertainty. Associated with his approach, Prakken has proposed a set of desirable principles for a sound modeling of accrual, and it can be shown that our formalization verifies them all. [4] A complete analysis of Prakken's principles in the context of our approach, together with a comparison with other existing formalizations of accrual can be found in [7] (we do not include it here for space reasons).

In order to test the applicability of our proposal we are developing an implementation of our formalization using the DeLP system [5] as a basis. We are studying different theoretical results emerging from our proposal which could help to speed up the computation of accrued dialectical trees. Research in this direction is currently being pursued.

Acknowledgements. This research was partially supported by CONICET (Project PIP 112-200801-02798), UNS (Projects 24/ZN10 and 24/N023), and Projects TIN2007-68005-C04-01 and TIN2006-15662-C02-02 (MEC, Spain).

References

1. Amgoud, L., Prade, H.: Explaining qualitative decision under uncertainty by argumentation. In: AAAI (2006)
2. Amgoud, L., Kaci, S.: An argumentation framework for merging conflicting knowledge bases. Int. J. Approx. Reasoning 45(2), 321–340 (2007)

---

[4] One of Prakken's principles states that "accruals are sometimes weaker than their elements". In order to satisfy it, the notion of sequential degradation must be slightly refined in the same way it was done in [7], ensuring thus convergence of sequential degradation even in case of a "depreciating" *ACC* function. As mentioned in section 4.2 we do not present such a refinement in this work due to space limitations.

[5] See http://lidia.cs.uns.edu.ar/delp

3. Alsinet, T., Chesñevar, C.I., Godo, L., Simari, G.R.: A logic programming framework for possibilistic argumentation: Formalization and logical properties. Fuzzy Sets and Systems 159(10), 1208–1228 (2008)
4. Alsinet, T., Chesñevar, C., Godo, L., Sandri, S., Simari, G.: Formalizing argumentative reasoning in a possibilistic logic programming setting with fuzzy unification. International Journal of Approximate Reasoning 48(3), 711–729 (2008)
5. Verheij, B.: Rules, Reasons, Arguments: Formal studies of argumentation and defeat. Doctoral dissertation, University of Maastricht (1996)
6. Prakken, H.: A study of accrual of arguments, with applications to evidential reasoning. In: ICAIL 2005: Proceedings of the 10th international conference on Artificial intelligence and law, pp. 85–94. ACM, New York (2005)
7. Lucero, M.J.G., Chesñevar, C.I., Simari, G.R.: Formalizing accrual in defeasible logic programming. In: Proc. of the 12th Intl. Workshop on Nonmonotonic Reasoning (NMR 2008), pp. 122–130 (2008)
8. García, A., Simari, G.: Defeasible logic programming: An argumentative approach. Theory Practice of Logic Programming 4(1), 95–138 (2004)
9. Besnard, P., Hunter, A.: A logic-based theory of deductive arguments. Artif. Intell. 128(1-2), 203–235 (2001)
10. Chesñevar, C., Simari, G., Alsinet, T., Godo, L.: A Logic Programming Framework for Possibilistic Argumentation with Vague Knowledge. In: Proc. of the Intl. Conf. in Uncertainty in Artificial Intelligence (UAI 2004), Canada, pp. 76–84 (2004)

Generalized Abstract Argumentation: Handling Arguments in FOL Fragments

Martín O. Moguillansky, Nicolás D. Rotstein,
Marcelo A. Falappa, and Guillermo R. Simari

Consejo Nacional de Investigaciones Científicas y Técnicas (CONICET)
Artificial Intelligence Research and Development Laboratory (LIDIA)
Department of Computer Science and Engineering (DCIC)
Universidad Nacional del Sur (UNS) – Bahía Blanca, Argentina
{mom,ndr,maf,grs}@cs.uns.edu.ar

Abstract. Generalized argumentation frameworks relate formulae in classical logic to arguments based on the Dung's classic framework. The main purpose of the generalization is to provide a theory capable of reasoning (following argumentation technics) about inconsistent knowledge bases (KB) expressed in FOL fragments. Consequently, the notion of argument is related to a single formula in the KB. This allows to share the same primitive elements from both, the framework (arguments) and, the KB (formulae). A framework with such features would not only allow to manage a wide range of knowledge representation languages, but also to cope with the dynamics of knowledge in a straightforward manner.

1 Introduction

The formalism studied in this work is based on the widely accepted Dung's argumentation framework (AF) [1]. An AF is deemed as abstract since the language used to define arguments remains unspecified, thus, arguments in an AF are treated as "black boxes" of knowledge. In this work we go one step further into a not-so-abstract form of argumentation by proposing an argument language $\mathbb{A}\text{rgs}$ in order to provide some structure to the notion of arguments while keeping them abstract. Intuitively, an *argument* may be seen as *an indivisible piece of knowledge inferring a claim from a set of premises*. Since claims and premises are distinguishable entities of any argument, we will allow both to be expressed through different sublanguages. The proposed *argument language* $\mathbb{A}\text{rgs}$ is thus characterized through the interrelation between its inner components. Assuming arguments specified through $\mathbb{A}\text{rgs}$ would bring about a highly versatile framework given that different knowledge representation languages could be handled through it. But consequently, some basic elements of the argumentation machinery should be accommodated, giving rise to a new kind of abstract argumentation frameworks identified as generalized (GenAF). The first approach to a GenAF in [2] was inspired by [3,4].

The GenAF here proposed aims at reasoning about inconsistent knowledge bases (KB) expressed through some fragment of first order logic (FOL). Consequently, $\mathbb{A}\text{rgs}$ will be reified to the restriction imposed to the FOL KB. Thus, the maximum expressive power of a GenAF is imposed by restricting the inner components of $\mathbb{A}\text{rgs}$ to be

C. Sossai and G. Chemello (Eds.): ECSQARU 2009, LNAI 5590, pp. 144–155, 2009.
© Springer-Verlag Berlin Heidelberg 2009

bounded to some logic $\mathcal{L}^\kappa$, with $\kappa \in \mathbb{N}_0$[1]. Formulae in $\mathcal{L}^\kappa$ are those of FOL that can be built with the help of predicate symbols with arity $\leq \kappa$, including equality and constant symbols, but without function symbols. An example of an $\mathcal{L}^2$-compliant logic is the $\mathcal{ALC}$ DL used to describe basic ontologies. The interested reader is referred to [5,6].

A normal form for a $\mathcal{L}^\kappa$ KB is presented to reorganize the knowledge in the KB through sentences conforming some minimal pattern, which will be interpreted as single arguments in the GenAF. Therefore, a GenAF may be straightforwardly adapted to deal with dynamics of knowledge as done in [4]: deleting an argument from the framework would mean deleting a statement from the KB. Argumentation frameworks were also related to FOL in [7], however, since there was no intention to cope with dynamics of arguments, no particular structure was provided to manage statements in the KB through single arguments. In this sense, our proposal is more similar to that in [8], although we relate the notions of deduction and conflict to FOL interpretations.

Specifying Args could bring about some problems: the language for claims may consider conjunctive and/or disjunctive formulae. For the former case, the easiest option is to trigger a different claim for each conjunctive term. For the case of disjunctive formulae for claims, the problem seems to be more complicated. To that matter we introduce the notion of *coalition*, which is a structure capable of grouping several arguments with the intention to support an argument's premise, identify conflictive sources of knowledge, or even to infer new knowledge beyond the one specified through the arguments considered in it. In argumentation theory, an argument's premises are satisfied in order for that argument to reach its claim. This is usually referred as *support relation* [3], handled in this work through coalitions.

Usually, an abstract argument is treated as an indivisible entity that suffices to support a claim; here arguments are also indivisible but they play a smaller role: they are aggregated in structures which can be thought as if they were arguments in the usual sense [7]. However, we will see that they do not always guarantee the achievement of the claim. The idea behind the aggregation of arguments within a structure is similar to that of sub-arguments [9]. Besides, classic argumentation frameworks consider ground arguments, that is, a claim is directly inferred if the set of premises are conformed. In our framework, we consider two different kinds of arguments: *ground* and *schematic*. In this sense, a set of premises might consider free variables, meaning that the claim, and therefore the inference, will depend on them. Thus, when an argument $\mathcal{B}$ counts with free variables in its claim or premises, it will be called schematic; whereas $\mathcal{B}$ is referred as ground, when its variables are instantiated. Instantiation of variables within a schematic argument may occur as a consequence of its premises being supported.

Finally, a basic acceptability semantics is proposed, inspired in the grounded semantics [10]. These semantics ensure the obtention of a consistent set of arguments, from which the accepted knowledge (warranted formulae) can be identified.

2 Foundations for a Generalized AF

For $\mathcal{L}^\kappa$, we use $p, p_1, p_2, \ldots$ and $q, q_1, q_2, \ldots$ to denote monadic predicate letters, $r, r_1, r_2, \ldots$ for dyadic predicate letters, x, y for free variable objects, and a, b, c, d for

---

[1] Natural numbers are enclosed in the sets $\mathbb{N}_0 = \{0, 1, \ldots\}$ and $\mathbb{N}_1 = \{1, 2, \ldots\}$.

constants (individual names). Besides, the logic $\mathcal{L}_A \subset \mathcal{L}^\kappa$ identifies the fragment of $\mathcal{L}^\kappa$ describing *assertional formulae* (ground atoms and their negations). Recall that ground atoms are atomic formulae which do not consider variable objects. The logic $\mathcal{L}^\kappa$ is interpreted as usual through interpretations $\mathcal{I} = \langle \Delta^{\mathcal{I}}, p^{\mathcal{I}}, p_1^{\mathcal{I}}, \ldots, q^{\mathcal{I}}, q_1^{\mathcal{I}}, \ldots, r^{\mathcal{I}}, r_1^{\mathcal{I}}, \ldots \rangle$, where $\Delta^{\mathcal{I}}$ is the interpretation domain, and $p^{\mathcal{I}}, p_1^{\mathcal{I}}, \ldots, q^{\mathcal{I}}, q_1^{\mathcal{I}}, \ldots, r^{\mathcal{I}}, r_1^{\mathcal{I}}, \ldots$ interpret $p, p_1, \ldots, q, q_1, \ldots, r, r_1, \ldots$, respectively. For an interpretation $\mathcal{I}$, some $a \in \Delta^{\mathcal{I}}$, and a formula $\varphi(x)$, we write $\mathcal{I} \models \varphi(a)$ if $\mathcal{I}, v \models \varphi(x)$, for the assignment v mapping x to a. For simplicity we omit universal quantifiers writing $\varphi(x)$ to refer to $(\forall x)(\varphi(x))$.

As mentioned before, we will rely on a (abstract) language $\mathbb{A}\mathrm{rgs}$ (for arguments) composed by two (unspecified) inner sub-languages: $\mathcal{L}_{\mathrm{pr}}$ (for premises) and $\mathcal{L}_{\mathrm{cl}}$ (claims).

Definition 1 (Argument Language). *Given the logic $\mathcal{L}^\kappa$, an **argument language** $\mathbb{A}\mathrm{rgs}$ is defined as $2^{\mathcal{L}_{\mathrm{pr}}} \times \mathcal{L}_{\mathrm{cl}}$, where $\mathcal{L}_{\mathrm{cl}} \subseteq \mathcal{L}^\kappa$ and $\mathcal{L}_{\mathrm{pr}} \subseteq \mathcal{L}^\kappa$ are recognized as the respective languages for claims and premises in $\mathbb{A}\mathrm{rgs}$.*

Since a premise is supported through the claim of other argument/s, the expressivity of both languages $\mathcal{L}_{\mathrm{pr}}$ and $\mathcal{L}_{\mathrm{cl}}$ should be controlled in order to allow every describable premise to be supported by formulae from the language for claims. Therefore, to handle the language $\mathbb{A}\mathrm{rgs}$ at an abstract level, we will characterize it by relating $\mathcal{L}_{\mathrm{pr}}$ and $\mathcal{L}_{\mathrm{cl}}$.

Definition 2 (Legal Argument Language). *An argument language $2^{\mathcal{L}_{\mathrm{pr}}} \times \mathcal{L}_{\mathrm{cl}}$ is **legal** iff for every $\rho \in \mathcal{L}_{\mathrm{pr}}$ there is a set $\Phi \subseteq \mathcal{L}_{\mathrm{cl}}$ such that $\Phi \models \rho$ **(support)**.*

In the sequel any argument language used will be assumed to be legal. Argumentation frameworks are a tool to reason about potentially inconsistent knowledge bases. Due to complexity matters, it would be interesting to interpret any $\mathcal{L}^\kappa$ KB directly as an argumentation framework with no need to transform the KB to a GenAF. Intuitively, an argument poses a reason to believe in a claim if it is the case that its premises are supported. This intuition is similar to the notion of material conditionals (implications "$\rightarrow$") in classical logic. Hence, statements from a KB could give rise to a single argument. To this end, we propose a normal form for $\mathcal{L}^\kappa$ KBs.

Definition 3 (pANF). *Given a knowledge base $\Sigma \subseteq \mathcal{L}^\kappa$, and an argument language $\mathbb{A}\mathrm{rgs}$, Σ conforms to the **pre-argumental normal form** (pANF) iff every formula $\varphi \in \Sigma$ is an assertion in $\mathcal{L}_A$, or it corresponds to the form $\rho_1 \wedge \ldots \wedge \rho_n \rightarrow \alpha$, where $\alpha \in \mathcal{L}_{\mathrm{cl}}$ and $\rho_i \in \mathcal{L}_{\mathrm{pr}}$ $(1 \leq i \leq n)$. Hence, each formula $\varphi \in \Sigma$ is said to be in pANF.*

Example 1. [2] Suppose $\mathcal{L}_{\mathrm{cl}}$ and $\mathcal{L}_{\mathrm{pr}}$ are concretized as follows: $\mathcal{L}_{\mathrm{cl}}$ allows disjunctions but prohibits conjunctions; whereas $\mathcal{L}_{\mathrm{pr}}$ avoids both conjunctions and disjunctions. This would require for a formula like $(p_1(x) \wedge p_2(x)) \vee (p_3(x) \wedge p_4(x)) \rightarrow q_1(x) \wedge (q_2(x) \vee q_3(x))$ to be reformatted into the pANF formulae $p_1(x) \wedge p_2(x) \rightarrow q_1(x)$, $p_1(x) \wedge p_2(x) \rightarrow q_2(x) \vee q_3(x)$, $p_3(x) \wedge p_4(x) \rightarrow q_1(x)$ and $p_3(x) \wedge p_4(x) \rightarrow q_2(x) \vee q_3(x)$.

Next we formalize the generalized notion of argument independently from a KB. The relation between premises and claims wrt. a KB could be referred to Remark 1.

---

[2] For simplicity, examples are enclosed within $\mathcal{L}^2$ to consider only predicates of arity ≤ 2.

Definition 4 (Argument). *An **argument** $\mathcal{B} \in \mathbb{A}\text{rgs}$ is a pair $\langle \Gamma, \alpha \rangle$, where $\Gamma \subseteq \mathcal{L}_{\text{pr}}$ is a finite set of finite premises, $\alpha \in \mathcal{L}_{\text{cl}}$, its finite claim, and $\Gamma \cup \{\alpha\} \not\models \bot$ (**consistency**).*

Usually, *evidence* is considered a basic irrefutable piece of knowledge. This means that evidence does not need to be supported given that it is self-justified by definition. Thus, two options appear to specify evidence: as a separate entity in the framework, or as arguments with no premises to be satisfied. In this article we assume the latter posture, referring to them as *evidential arguments*.

Definition 5 (Evidence). *Given an argument $\mathcal{B} \in \mathbb{A}\text{rgs}$, $\mathcal{B}$ is referred as **evidential argument** (or just evidence) iff $\mathcal{B} = \langle \{\}, \alpha \rangle$ with $\alpha \in \mathcal{L}_{\text{A}}$ (assertional formulae).*

Given $\mathcal{B} \in \mathbb{A}\text{rgs}$, its claim and set of premises are identified by the functions $\mathfrak{cl}$: $\mathbb{A}\text{rgs} \longrightarrow \mathcal{L}_{\text{cl}}$, and $\mathfrak{pr}$: $\mathbb{A}\text{rgs} \longrightarrow 2^{\mathcal{L}_{\text{pr}}}$, respectively. For instance, given $\mathcal{B} = \langle \{\rho_1, \rho_2\}, \alpha \rangle$, its premises are $\mathfrak{pr}(\mathcal{B}) = \{\rho_1, \rho_2\}$, and its claim, $\mathfrak{cl}(\mathcal{B}) = \alpha$. Arguments will be obtained from pANF formulae through an *argument translation function* $\mathfrak{arg}$: $\mathcal{L}^{\kappa} \longrightarrow \mathbb{A}\text{rgs}$ such that $\mathfrak{arg}(\varphi) = \langle \{\rho_1, \ldots, \rho_n\}, \alpha \rangle$ iff $\varphi \in \mathcal{L}^{\kappa}$ is a pANF formula $\rho_1 \wedge \ldots \wedge \rho_n \rightarrow \alpha$ and $\mathfrak{arg}(\varphi)$ verifies the conditions in Def. 4. Otherwise, $\mathfrak{arg}(\varphi) = \langle \emptyset, \bot \rangle$. An evidential argument $\mathfrak{arg}(\varphi) = \langle \emptyset, \alpha \rangle$ appears if φ is $\rightarrow \alpha$.

Example 2 (Continued from Ex. 1). For the formulae given in Ex. 1, the arguments $\langle \{p_1(x), p_2(x)\}, q_1(x) \rangle$, $\langle \{p_1(x), p_2(x)\}, q_2(x) \vee q_3(x) \rangle$, $\langle \{p_3(x), p_4(x)\}, q_1(x) \rangle$ and $\langle \{p_3(x), p_4(x)\}, q_2(x) \vee q_3(x) \rangle$, are triggered by effect of the function "$\mathfrak{arg}$".

As mentioned before, it is important to recall that the notion of argument adopted in this work differs from its usual usage. This is made clear in the following remark.

Remark 1. Given a pANF KB $\Sigma \subseteq \mathcal{L}^{\kappa}$, a formula $\varphi \in \Sigma$, and its associated argument $\mathfrak{arg}(\varphi) = \langle \Gamma, \alpha \rangle$; it follows $\Sigma \models (\bigwedge \Gamma) \rightarrow \alpha$, but $\Gamma \models \alpha$ does not necessarily hold.

A more restrictive definition of argument could consider conditions like $\Gamma \not\models \alpha$, and/or $\Gamma \setminus \{\rho\} \not\models \rho$, with $\rho \in \Gamma$. However, its appropriate discussion exceeds the scope of this article. For the usual notion of argument see *argumental structures* in Def. 15.

The formalization of the GenAF will rely on *normality conditions*: user defined constraints in behalf of the appropriate construction of the argumentation framework.

Definition 6 (GenAF). *A **generalized abstract argumentation framework** (GenAF) is a pair $\langle \mathbf{A}, \mathbf{N} \rangle$, where $\mathbf{A} \subseteq \mathbb{A}\text{rgs}$ is a finite set of arguments, and $\mathbf{N} \subseteq \mathbb{N}\text{orm}$, a finite set of normality condition functions $\mathfrak{nc}$: $2^{\mathbb{A}\text{rgs}} \longrightarrow \{true, false\}$. The domain of functions $\mathfrak{nc}$ is identified through $\mathbb{N}\text{orm}$, and $\mathbb{G}$ identifies the class of every GenAF. The set $\mathbf{E} \subseteq \mathbf{A}$ encloses every evidential argument from $\mathbf{A}$.*

A normality condition required through $\mathbf{N}$, could be to require evidence to be consistent, that is no pair of contradictory evidential arguments should be available in the framework. Other conditions could be to restrict arguments from being non-minimal justifications for the claim, or from including the claim itself as a premise.

(evidence coherency) there is no pair $\langle \{\}, \alpha \rangle \in \mathbf{A}$ and $\langle \{\}, \neg \alpha \rangle \in \mathbf{A}$.
(minimality) there is no pair $\langle \Gamma, \alpha \rangle \in \mathbf{A}$ and $\langle \Gamma', \alpha \rangle \in \mathbf{A}$ such that $\Gamma' \subset \Gamma$.
(relevance) there is no $\langle \Gamma, \alpha \rangle \in \mathbf{A}$ such that $\alpha \in \Gamma$.

Other normality conditions may appear depending on the concretization of the logic for arguments and the environment the framework is set to model. The complete study of these features falls out of the scope of this article. Given a GenAF $T = \langle \mathbf{A}, \mathbf{N} \rangle$, we will say that T is a *theory* *iff* for every normality condition $\mathfrak{nc} \in \mathbf{N}$ it follows $\mathfrak{nc}(\mathbf{A}) = true$. From now on we will work only with theories, thus unless the contrary is stated, every framework will be assumed to conform a theory. Moreover, the framework specification is done in such a way that its correctness does not rely on the normality conditions required. Thus, a GenAF $\langle \mathbf{A}, \mathbf{N} \rangle$, with $\mathbf{N} = \emptyset$, will be trivially a theory.

In order to univocally determine a single GenAF from a given KB and a set of normality conditions, it is necessary to assume a comparison criterion among formulae in the KB. Such criterion could be defined for instance, upon entrenchment of knowledge, *i.e.*, levels of importance are related to formulae in the KB. As will be seen later, this criterion will determine the *argument comparison criterion* from which the attack relation is usually specified in the classic argumentation literature. Next, we define a *theory function* to identify the GenAF associated to a KB.

Definition 7 (Theory Function). *Given a* pANF *knowledge base* $\Sigma \subseteq \mathcal{L}^\kappa$, *and a set* $\mathbf{N} \subseteq \mathrm{Norm}$ *of normality condition functions* $\mathfrak{nc}$, *a **theory function** $\mathfrak{genaf} : 2^{\mathcal{L}^\kappa} \times 2^{\mathrm{Norm}} \longrightarrow \mathbb{G}$ identifies the GenAF* $\mathfrak{genaf}(\Sigma, \mathbf{N}) = \langle \mathbf{A}, \mathbf{N} \rangle$, *where* $\mathbf{A} \subseteq \{\mathfrak{arg}(\varphi) | \varphi \in \Sigma$ *and* $\mathfrak{arg}(\varphi)$ *is an argument*$\} \cup \{\mathfrak{arg}(\varphi \to \varphi') | (\neg \varphi' \to \neg \varphi) \in \Sigma$ *and* $\mathfrak{arg}(\varphi \to \varphi')$ *is an argument*$\}$ *and* $\mathbf{A}$ *is the maximal set (wrt. set inclusion and the comparison criterion in* Σ*) such that for every* $\mathfrak{nc} \in \mathbf{N}$ *it holds* $\mathfrak{nc}(\mathbf{A}) = true$.

The GenAF obtained by the function "$\mathfrak{genaf}$" will consider a maximal subset of the KB Σ such that the resulting set of arguments (triggered by "$\mathfrak{arg}$") is compliant with the normality conditions. Note that also the counterpositive formula of each one considered is assumed to conform an argument in the resulting GenAF. This is natural since counterpositive formulae from the statements in a KB are implicitly considered to reason in classical logic. In a GenAF, this issue is done by considering both explicitly.

3   The GenAF Argumentation Machinery

The purpose of generalizing an abstract argumentation framework comes from the need of managing different argument languages specified through some FOL fragment. Given the specification of Args, different possibilities may arise, for instance, the language for claims may accept disjunction of formulae. Thus, it is possible to infer a formula in $\mathcal{L}_{cl}$ from several arguments in the GenAF through their claims. Consider for example, two arguments $\langle \{p_1(x)\}, q_1(x) \vee q_2(x) \rangle$ and $\langle \{p_2(x)\}, \neg q_2(x) \rangle$, the claim $q_1(x)$ may be inferred. This kind of constructions are similar to arguments themselves, but are implicitly obtained from the GenAF at issue. To such matter, the notion of *claiming-coalition* is introduced as *a coalition required to infer a new claim*.

In general, a *coalition* might be interpreted as *a minimal and consistent set of arguments guaranteeing certain requirement*. We say that a coalition $\widehat{\mathcal{C}} \subseteq \mathrm{Args}$ is consistent *iff* $\mathfrak{prset}(\widehat{\mathcal{C}}) \cup \mathfrak{clset}(\widehat{\mathcal{C}}) \not\models \bot$, while minimality ensures that $\widehat{\mathcal{C}}$ guarantees a requirement θ *iff* there is no proper subset of $\widehat{\mathcal{C}}$ guaranteeing θ. The functions $\mathfrak{clset} : 2^{\mathrm{Args}} \longrightarrow 2^{\mathcal{L}_{cl}}$ and $\mathfrak{prset} : 2^{\mathrm{Args}} \longrightarrow 2^{\mathcal{L}_{pr}}$ are defined as $\mathfrak{clset}(\widehat{\mathcal{C}}) = \{\mathfrak{cl}(\mathcal{B}) | \mathcal{B} \in \widehat{\mathcal{C}}\}$, and $\mathfrak{prset}(\widehat{\mathcal{C}}) =$

$\bigcup_{\mathcal{B} \in \widehat{\mathcal{C}}} \mathfrak{pr}(\mathcal{B})$, to respectively identify the set of claims and premises from $\widehat{\mathcal{C}}$. In this article, three types of coalitions will be considered. Regarding claiming-coalitions, the requirement θ is a new inference in $\mathcal{L}_{\mathtt{cl}}$ from the arguments considered by the coalition.

Definition 8 (Claiming-Coalition). *Given a* GenAF $\langle \mathbf{A}, \mathbf{N} \rangle \in \mathbb{G}$, *and a formula* $\alpha \in \mathcal{L}_{\mathtt{cl}}$, *a set of arguments* $\widehat{\mathcal{C}} \subseteq \mathbf{A}$ *is a claiming-coalition, or just a **claimer**, of* α *iff* $\widehat{\mathcal{C}}$ *is the minimal coalition guaranteeing* $\mathfrak{clset}(\widehat{\mathcal{C}}) \models \alpha$ *and* $\widehat{\mathcal{C}}$ *is consistent.*

Note that a claiming-coalition containing a single argument $\mathcal{B}$ is a primitive coalition for the claim of $\mathcal{B}$. As said before, an argument needs to find its premises supported as a functional part of the reasoning process to reach its claim. In this framework, due to the characterization of $\mathbb{Args}$, sometimes a formula from $\mathcal{L}_{\mathtt{pr}}$ could be satisfied only through several formulae from $\mathcal{L}_{\mathtt{cl}}$. This means that a single argument is not always enough to support a premise of another argument. Thus, we will extend the usual definition of supporter [3] by introducing the notion of *supporting-coalition*.

Definition 9 (Supporting-Coalition). *Given a* GenAF $\langle \mathbf{A}, \mathbf{N} \rangle \in \mathbb{G}$, *an argument* $\mathcal{B} \in \mathbf{A}$, *and a premise* $\rho \in \mathfrak{pr}(\mathcal{B})$. *A set of arguments* $\widehat{\mathcal{C}} \subseteq \mathbf{A}$ *is a supporting-coalition, or just a **supporter**, of* $\mathcal{B}$ *through* ρ *iff* $\widehat{\mathcal{C}}$ *is the minimal coalition guaranteeing* $\mathfrak{clset}(\widehat{\mathcal{C}}) \models \rho$ *and* $\widehat{\mathcal{C}} \cup \{\mathcal{B}\}$ *is consistent.*

Example 3. Assume $\mathbf{A} = \{\mathcal{B}_1, \mathcal{B}_2, \mathcal{B}_3, \mathcal{B}_4\}$, where $\mathcal{B}_1 = \langle \{p_1(x)\}, q_1(x) \rangle$, $\mathcal{B}_2 = \langle \{p_1(x)\}, q_2(x) \rangle$, $\mathcal{B}_3 = \langle \{p_2(x)\}, p_1(x) \vee q_1(x) \rangle$, and $\mathcal{B}_4 = \langle \{p_3(x)\}, \neg q_1(x) \rangle$. The set $\widehat{\mathcal{C}} = \{\mathcal{B}_3, \mathcal{B}_4\}$ is a supporter of $\mathcal{B}_2$. Note that $\widehat{\mathcal{C}}$ cannot be a supporting-coalition of $\mathcal{B}_1$ since it violates (supporter) consistency.

When not every necessary argument to conform the supporting-coalition is present in $\mathbf{A}$, the (unsupported) premise is referred as free.

Definition 10 (Free Premise). *Given a* GenAF $\langle \mathbf{A}, \mathbf{N} \rangle \in \mathbb{G}$ *and an argument* $\mathcal{B} \in \mathbf{A}$, *a premise* $\rho \in \mathfrak{pr}(\mathcal{B})$ *is **free** wrt.* $\mathbf{A}$ *iff there is no supporter* $\widehat{\mathcal{C}} \subseteq \mathbf{A}$ *of* $\mathcal{B}$ *through* ρ.

From Ex. 3, premises $p_2(x) \in \mathfrak{pr}(\mathcal{B}_3)$, $p_3(x) \in \mathfrak{pr}(\mathcal{B}_4)$, and $p_1(x) \in \mathfrak{pr}(\mathcal{B}_1)$ are free wrt. $\mathbf{A}$; whereas $p_1(x) \in \mathfrak{pr}(\mathcal{B}_2)$ is not.

When a schematic argument is fully supported from evidence ($\widehat{\mathcal{C}} \subseteq \mathbf{E}$), its claim is ultimately instantiated ending up as a ground formula. Therefore, an argument $\mathcal{B}$ may be included in a supporting coalition $\widehat{\mathcal{C}}$ of $\mathcal{B}$ itself due to the substitution of variables. This situation is made clearer later and may be referred to Ex. 5. The quest for a supporter $\widehat{\mathcal{C}}$ of some argument $\mathcal{B}$ through a premise ρ in it, describes a recursive supporting process given that each premise in $\widehat{\mathcal{C}}$ needs to be also supported. When this process does ultimately end in a supporter containing only evidential arguments, we will distinguish $\rho \in \mathfrak{pr}(\mathcal{B})$ not only as non-free but also as *closed*.

Definition 11 (Closed Premise). *Given a* GenAF $\langle \mathbf{A}, \mathbf{N} \rangle \in \mathbb{G}$, *and an argument* $\mathcal{B} \in \mathbf{A}$, *a premise* $\rho \in \mathfrak{pr}(\mathcal{B})$ *is **closed** wrt.* $\mathbf{A}$ *iff there exists a supporter* $\widehat{\mathcal{C}} \subseteq \mathbf{A}$ *of* $\mathcal{B}$ *through* ρ *such that either* $\mathfrak{prset}(\widehat{\mathcal{C}}) = \emptyset$, *or every premise in* $\mathfrak{prset}(\widehat{\mathcal{C}})$ *is closed.*

The idea behind closing premises is to identify those arguments that effectively state a reason from the GenAF to believe in their claims. Such arguments will be those for which the support of each of its premises does ultimately end in a set of evidential arguments –and therefore no more premises are required to be supported. Thus, every premise in an argument is closed *iff* the claim is *inferrable*. This is natural since inferrable claims can be effectively reached from evidence. Finally, when the claiming-coalition of an inferrable claim passes the acceptability analysis, the claim ends up *warranted*. Acceptability analysis and warranted claims will be detailed later, in Sect. 5.

Definition 12 (Inferrable Formula). *Given a* GenAF $\langle \mathbf{A}, \mathbf{N} \rangle \in \mathbb{G}$, *a formula* $\alpha \in \mathcal{L}_{\mathrm{cl}}$ *is* **inferrable** *from* $\mathbf{A}$ *iff there exists a claiming-coalition* $\widehat{\mathcal{C}} \subseteq \mathbf{A}$ *for* α *such that either* $\mathrm{prset}(\widehat{\mathcal{C}}) = \emptyset$, *or every premise in* $\mathrm{prset}(\widehat{\mathcal{C}})$ *is closed.*

The supporting process closing every premise in a claiming-coalition $\widehat{\mathcal{C}}$ to verify whether the claim is inferrable, clearly conforms a tree rooted in $\widehat{\mathcal{C}}$. We will refer to such tree as *supporting-tree*, and to each branch in it as *supporting-chain*.

Definition 13 (Supporting-Chain). *Given a* GenAF $\langle \mathbf{A}, \mathbf{N} \rangle \in \mathbb{G}$, *a formula* $\alpha \in \mathcal{L}_{\mathrm{cl}}$, *and a sequence* $\lambda \in (2^{\mathbf{A}})^n$ *such that* $\lambda = \widehat{\mathcal{C}}_1 \ldots \widehat{\mathcal{C}}_n$, *where* $n \in \mathbb{N}_1$, $\widehat{\mathcal{C}}_1$ *is a claiming-coalition for* α, *and for every* $i \in \mathbb{N}_1$ *it follows* $\widehat{\mathcal{C}}_i \subseteq \mathbf{A}$, *and* $\widehat{\mathcal{C}}_{i+1}$ *is a supporting-coalition through some* $\rho_i \in \mathrm{prset}(\widehat{\mathcal{C}}_i)$. *The notations* $|\lambda| = n$ *and* $\lambda[i]$ *are used to respectively identify the* **length** *of* λ *and the* **node** $\widehat{\mathcal{C}}_i$ *in it. The last supporting-coalition in* λ *(referred as* **leaf***) is identified through the function* $\mathrm{leaf}(\lambda) = \lambda[|\lambda|]$. *The function* $\overleftarrow{\lambda} : (2^{\mathbf{A}})^n \times \mathbb{N}_0 \longrightarrow \mathcal{L}_{\mathrm{cl}} \cup \mathcal{L}_{\mathrm{pr}} \cup \{\perp\}$ *identifies the* **link** $\overleftarrow{\lambda}[0] = \alpha$; *or* $\overleftarrow{\lambda}[i] = \rho_i$ $(0 < i < |\lambda|)$, *where* $\rho_i \in \mathrm{prset}(\lambda[i])$ *is supported by* $\lambda[i+1]$; *or* $\overleftarrow{\lambda}[i] = \perp$ $(i \geq |\lambda|)$. *The set* $\lambda^* = \bigcup_i \lambda[i]$ *(with* $0 < i \leq |\lambda|$) *identifies the set of arguments included in* λ. *Finally,* λ *is a* **supporting-chain for** α *wrt.* $\mathbf{A}$ *iff it guarantees:*

(minimality) $\widehat{\mathcal{C}} \subseteq \lambda^*$ *is a supporter (claimer if* $i = 0$*) of* $\overleftarrow{\lambda}[i]$ *iff* $\widehat{\mathcal{C}} = \lambda[i+1]$ $(0 \leq i < |\lambda|)$.
(exhaustivity) *every* $\rho \in \mathrm{prset}(\mathrm{leaf}(\lambda))$ *is free wrt.* λ^*.
(acyclicity) $\overleftarrow{\lambda}[i] = \overleftarrow{\lambda}[j]$ *iff* $i = j$, *with* $\{i, j\} \subseteq \{0, \ldots, |\lambda| - 1\}$.
(consistency) $\mathrm{prset}(\lambda^*) \cup \mathrm{clset}(\lambda^*) \not\models \perp$.

From the definition above, a supporting-chain is a finite sequence of interrelated supporting-coalitions $\widehat{\mathcal{C}}_i$ through a link $\rho_i \in \mathrm{prset}(\widehat{\mathcal{C}}_i)$ supported by $\widehat{\mathcal{C}}_{i+1}$. It is finite indeed, given that the set $\mathbf{A}$ is also finite, and that no link could be repeated in the chain (acyclicity). The minimality condition (wrt. set inclusion over λ^*) stands to consider as less arguments from $\mathbf{A}$ as it is possible in order to obtain the same chain, whereas the exhaustivity condition (wrt. the length $|\lambda|$) ensures that the chain is as long as it is possible wrt. λ^* (without cycles), that is, λ has all the possible links that can appear from the arguments considered to build it. Note that from minimality no pair of arguments for a same claim could be simultaneously considered in any supporting-chain. Finally, consistency is required given that the intention of the supporting-chain is to provide a tool to close a premise from the claiming-coalition. Next, supporting-trees are formalized upon the definition of supporting-chains.

Definition 14 (Supporting-Tree). *Given a* GenAF $\langle \mathbf{A}, \mathbf{N} \rangle \in \mathbb{G}$, *a formula* $\alpha \in \mathcal{L}_{\mathrm{cl}}$, *and a tree* $\mathcal{T}$ *of coalitions* $\widehat{\mathcal{C}} \subseteq \mathbf{A}$ *such that each node* $\widehat{\mathcal{C}}$ *is either:*

- **the root** iff $\widehat{\mathcal{C}}$ is a claiming-coalition for α; or
- **an inner node** iff $\widehat{\mathcal{C}}$ is a supporting-coalition through $\rho \in \mathfrak{prset}(\widehat{\mathcal{C}'})$, where $\widehat{\mathcal{C}'} \subseteq$ **A** is either an inner node or the root.

  The membership relation will be overloaded by writing $\lambda \in \mathcal{T}$ and $\widehat{\mathcal{C}} \in \mathcal{T}$ to respectively identify the branch λ and the node $\widehat{\mathcal{C}}$ from $\mathcal{T}$. The set $\mathcal{T}^* = \bigcup_{\widehat{\mathcal{C}} \in \mathcal{T}} \widehat{\mathcal{C}}$ identifies the set of arguments included in $\mathcal{T}$. Hence, $\mathcal{T}$ is a **supporting-tree** iff it guarantees:

(completeness) every $\lambda \in \mathcal{T}$ is a supporting-chain of α wrt. **A**.

(minimality) for every $\lambda \in \mathcal{T}$, $\widehat{\mathcal{C}} \subseteq \mathcal{T}^*$ is a supporting-coalition (claimer if $i = 0$) through $\overleftarrow{\lambda}[i]$ iff $\widehat{\mathcal{C}} = \lambda[i+1]$ $(0 \le i < |\lambda|)$.

(exhaustivity) for every $\rho \in \mathfrak{prset}(\mathcal{T}^*)$, if there is no $\lambda \in \mathcal{T}$ such that $\overleftarrow{\lambda}[i] = \rho$ $(0 < i < |\lambda|)$ then ρ is free wrt. $\mathcal{T}^*$.

(consistency) $\mathfrak{prset}(\mathcal{T}^*) \cup \mathfrak{clset}(\mathcal{T}^*) \not\models \bot$.

Finally, the notation $\mathfrak{Trees}_{\mathbf{A}}(\alpha)$ *identifies the set of all supporting-trees for α from* **A**.

The completeness condition is required in order to restrict the supporting-tree to consider only supporting-chains as their branches. Similar to supporting-chain, minimality is required to avoid considering extra arguments to build the tree, while exhaustivity stands to ensure that every possible supporting-coalition $\widehat{\mathcal{C}} \subseteq \mathcal{T}^*$ through a premise in $\mathfrak{prset}(\mathcal{T}^*)$ is an inner node in the tree. Finally, consistency ensures that the whole supporting process of the premises in the claiming-coalition will end being non-contradictory, even among branches. It is important to note that a supporting-tree for $\alpha \in \mathcal{L}_{\mathrm{cl}}$ determines the set of arguments used in the (possibly inconclusive)[3] supporting process of some claiming-coalition of α. Such set will be referred as *structure*.

Definition 15 (Structure). *Given a* GenAF $\langle \mathbf{A}, \mathbf{N} \rangle \in \mathbb{G}$, *and a formula* $\alpha \in \mathcal{L}_{\mathrm{cl}}$, *a set* $\mathbb{S} \subseteq \mathbf{A}$ *identifies a* **structure** *for* α *iff there is a supporting-tree* $\mathcal{T} \in \mathfrak{Trees}_{\mathbf{A}}(\alpha)$ *for* α *such that* $\mathbb{S} = \mathcal{T}^*$. *The claim and premises of* $\mathbb{S}$ *can be respectively determined through the functions* $\mathfrak{cl} : 2^{\mathrm{Args}} \longrightarrow \mathcal{L}_{\mathrm{cl}}$ *and* $\mathfrak{pr} : 2^{\mathrm{Args}} \longrightarrow 2^{\mathcal{L}_{\mathrm{pr}}}$, *such that* $\mathfrak{cl}(\mathbb{S}) = \alpha$ *and* $\mathfrak{pr}(\mathbb{S}) = \{\rho \in \mathfrak{prset}(\mathbb{S}) \mid \rho$ *is a free premise wrt.* $\mathbb{S}\}$. *Finally, the structure* $\mathbb{S}$ *is* **argumental** *iff* $\mathfrak{pr}(\mathbb{S}) = \emptyset$, *otherwise* $\mathbb{S}$ *is* **schematic**.

Note that functions "$\mathfrak{pr}$" and "$\mathfrak{cl}$" are overloaded and can be applied both to arguments and structures. This is not going to be problematic since either usage will be rather explicit. Besides, a structure $\mathbb{S}$ formed by a single argument is referred as *primitive* iff $|\mathbb{S}| = 1$. Thus, if $\mathbb{S} = \{\mathcal{B}\}$ then $\mathfrak{pr}(\mathcal{B}) = \mathfrak{pr}(\mathbb{S})$ and $\mathfrak{cl}(\mathcal{B}) = \mathfrak{cl}(\mathbb{S})$. However, not every single argument has an associated primitive structure. For instance, unless relevance would be required as a framework's normality condition, no structure could contain an argument $\langle \{p(x)\}, p(x) \rangle$ given that it would violate (supporting-chain) acyclicity. Finally, when no distinction is needed, we refer to primitive, schematic, or argumental structures, simply as structures.

Example 4. Given two arguments $\mathcal{B}_1 = \langle \{p(x)\}, q(x) \rangle$ and $\mathcal{B}_2 = \langle \{q(x)\}, p(x) \rangle$. The set $\{\mathcal{B}_1, \mathcal{B}_2\}$ cannot be a structure for $q(x)$ since $\{\mathcal{B}_1\}\{\mathcal{B}_2\}\{\mathcal{B}_1\} \ldots$ is a supporting-chain violating acyclicity. Similarly, $\{\mathcal{B}_1, \mathcal{B}_2\}$ could neither be a structure for $p(x)$.

---

[3] Inconclusive supporting processes lead to schematic structures with non-free premises wrt. **A**.

Given two structures $\mathbb{S} \subseteq \mathbf{Args}$ for $\alpha \in \mathcal{L}_{cl}$, and $\mathbb{S}' \subseteq \mathbf{Args}$ for $\alpha' \in \mathcal{L}_{cl}$, $\mathbb{S}'$ is a sub-structure of $\mathbb{S}$ (noted as $\mathbb{S}' \sqsubseteq \mathbb{S}$) *iff* $\mathbb{S}' \subseteq \mathbb{S}$. Besides, $\mathbb{S}' \sqsubset \mathbb{S}$ *iff* $\mathbb{S}' \subset \mathbb{S}$.

Proposition 1. [4] *Given a* GenAF $\langle \mathbf{A}, \mathbf{N} \rangle \in \mathbb{G}$, *a formula* $\alpha \in \mathcal{L}_{cl}$, *and two structures* $\mathbb{S} \subseteq \mathbf{A}$ *for* α *and* $\mathbb{S}' \subseteq \mathbf{A}$ *for* α,

- *if* $\mathbb{S}' \sqsubset \mathbb{S}$ *then* $\mathfrak{pr}(\mathbb{S}) \neq \mathfrak{pr}(\mathbb{S}')$.
- *if* $\mathbb{S}$ *is argumental then* $\mathfrak{leaf}(\lambda) \subseteq \mathbf{E}$, *for every* $\lambda \in \mathcal{T}$ *where* $\mathcal{T} \in \mathfrak{Trees}_{\mathbb{S}}(\alpha)$.

Lemma 1. *Given a* GenAF $\langle \mathbf{A}, \mathbf{N} \rangle \in \mathbb{G}$, *and a formula* $\alpha \in \mathcal{L}_{cl}$, *a structure* $\mathbb{S} \subseteq \mathbf{A}$ *for* α *is argumental iff* α *is inferrable.*

If a formula $\varphi(x) \in \mathcal{L}_{cl}$ (where x is a free variable) is inferrable then there exists an argumental structure $\mathbb{S}$ for $\varphi(x)$. Note now that since every argumental structure contains an empty set of premises, its supporting-tree $\mathcal{T}$ has only evidential arguments in their leaves. Thus, since the claim of evidential arguments are expressed in the language $\mathcal{L}_A$–it considers no free variables– the inner supporting process of $\mathbb{S}$ performed through $\mathcal{T}$ ends up applying a *variable substitution*, for instance mapping x to a, such that $\mathfrak{cl}(\mathbb{S}) = \varphi(a)$. Finally, if a structure states a property about some element of the world through a claim considering only free variables then it is schematic.

Lemma 2. *Given a* GenAF $\langle \mathbf{A}, \mathbf{N} \rangle \in \mathbb{G}$, *and a formula* $\varphi(x) \in \mathcal{L}_{cl}$, *a structure* $\mathbb{S} \subseteq \mathbf{A}$ *for* $\varphi(x)$ *is argumental iff* $\mathfrak{cl}(\mathbb{S}) = \varphi(a)$ *and* $\varphi(a), v \models \varphi(x)$, *where* v *maps* x *to* a.

Example 5. Assume the GenAF $\langle \mathbf{A}, \mathbf{N} \rangle$ such that $\{\mathcal{B}_1, \mathcal{B}_2, \mathcal{B}_3\} \subseteq \mathbf{A}$ where $\mathcal{B}_1 = \langle \{p(x)\}, (\exists y)(\neg r(x,y) \vee p(y)) \rangle$, $\mathcal{B}_2 = \langle \{\}, r(a,b) \rangle$, and $\mathcal{B}_3 = \langle \{\}, p(a) \rangle$.

The argumental structure $\mathbb{S}_1 = \{\mathcal{B}_1, \mathcal{B}_3\}$ for $(\exists y)(\neg r(a,y) \vee p(y))$ appears. Moreover, $\widehat{\mathcal{C}}_1 = \{\mathcal{B}_1, \mathcal{B}_2\}$ is a supporter of $\mathcal{B}_1$ through $p(x)$, where the free variables x and y are mapped to a and b, respectively. Note that as a result of such variables substitutions, we have $\mathfrak{pr}(\widehat{\mathcal{C}}_1) = \{p(a)\}$, which in turn will be supported through the primitive coalition $\{\mathcal{B}_3\}$. Hence, the schematic structure $\mathbb{S}_2 = \{\mathcal{B}_1, \mathcal{B}_2, \mathcal{B}_3\}$ for $p(b)$ appears, where $\mathbb{S}_1 \sqsubset \mathbb{S}_2$. Note that, $\mathcal{T} \in \mathfrak{Trees}_{\mathbb{S}_2}(p(b))$ has a unique supporting-chain $\{\mathcal{B}_1\}\{\mathcal{B}_1, \mathcal{B}_2\}\{\mathcal{B}_3\}$.

4 Conflict Identification

As will be formalized in Def. 16, two argumental structures are in conflict whenever their claims cannot be assumed together. Schematic structures may also be conflictive if it is the case that a claim of one of them could support a premise of the other, but a supporting-coalition does not exist given consistency would be violated. A second option of conflict between schematic structures appears when the premises of one of them infer the premises of the other, and either claim is in conflict with some premise from the other, or both claims cannot be assumed together. The intuition for this may

---

[4] In this work, proofs were omitted due to space reasons.

be seen as a framework lacking of evidence to close every premise in each structure, but a hypothetical addition of the lacking evidence of one of them would be enough to include in the new framework two different argumental structures containing each original schematic structure. In such a case, the conflict conforms to the first case given.

This discussion may be made extensive to coalitional sets of structures. Analogous to coalitions of arguments, a *coalition of structures* might be interpreted as *a minimal and consistent set of structures guaranteeing certain requirement*. To go one step further into the formalization of a coalition $\widehat{\mathbb{C}} \subseteq 2^{\mathrm{Args}}$ of structures $\mathbb{S} \subseteq \mathbf{A}$, we will rely on the set $\mathbb{C}^* = \bigcup_{\mathbb{S} \in \widehat{\mathbb{C}}} \mathbb{S}$ of arguments from $\widehat{\mathbb{C}}$. Therefore, we say that a coalition $\widehat{\mathbb{C}}$ of structures $\mathbb{S}$, is consistent *iff* $\mathrm{prset}(\mathbb{C}^*) \cup \mathrm{clset}(\mathbb{C}^*) \not\models \bot$, while minimality ensures $\widehat{\mathbb{C}}$ guarantees a requirement θ *iff* there is no proper subset of $\widehat{\mathbb{C}}$ guaranteeing θ, and there is no $\widehat{\mathbb{C}'} \subseteq 2^{\mathrm{Args}}$ guaranteeing θ such that $\mathbb{C}'^* \subset \mathbb{C}^*$. Note that minimality not only looks for the smallest set of structures, but also for the smallest structures.

Coalition of structures are sets grouping structures to guarantee certain requirement θ: conflict. For the formalization of the notion of conflict, we will rely on the functions $\mathrm{clset} : 2^{2^{\mathbf{A}}} \longrightarrow 2^{\mathcal{L}_{\mathrm{cl}}}$ and $\mathrm{prset} : 2^{2^{\mathbf{A}}} \longrightarrow 2^{\mathcal{L}_{\mathrm{pr}}}$, which are respectively defined as $\mathrm{clset}(\widehat{\mathbb{C}}) = \{\mathrm{cl}(\mathbb{S}) | \mathbb{S} \in \widehat{\mathbb{C}}\}$, and $\mathrm{prset}(\widehat{\mathbb{C}}) = \bigcup_{\mathbb{S} \in \widehat{\mathbb{C}}} \mathrm{pr}(\mathbb{S})$. Note that functions "clset" and "prset" are overloaded and can be applied both to sets of arguments (for instance coalitions $\widehat{\mathcal{C}}$) and to coalitions $\widehat{\mathbb{C}}$ of structures. For this latter case, the functions' outcomes are the claims and premises of the structures included by the coalition $\widehat{\mathbb{C}}$. Next, we specify the notion of conflict between pairs of coalition of structures.

Definition 16 (Conflicting Coalitions). *Given a* GenAF $\langle \mathbf{A}, \mathbf{N} \rangle \in \mathbb{G}$, *two coalitions* $\widehat{\mathbb{C}} \subseteq 2^{\mathbf{A}}$ *and* $\widehat{\mathbb{C}'} \subseteq 2^{\mathbf{A}}$ *of structures are in **conflict** iff it follows:*

- *Both coalitions are related either through dependency or support:*
  (dependency) $\mathrm{prset}(\widehat{\mathbb{C}}) \models \mathrm{prset}(\widehat{\mathbb{C}'})$.
  (support) $\mathrm{clset}(\widehat{\mathbb{C}}) \models \mathrm{prset}(\widehat{\mathbb{C}'})$.
- *The conflict appears either from claim-clash or premise-clash:*
  (claim-clash) $\mathrm{clset}(\widehat{\mathbb{C}}) \cup \mathrm{clset}(\widehat{\mathbb{C}'}) \models \bot$.
  (premise-clash) $\mathrm{clset}(\widehat{\mathbb{C}}) \cup \mathrm{prset}(\widehat{\mathbb{C}'}) \models \bot$, *or* $\mathrm{clset}(\widehat{\mathbb{C}'}) \cup \mathrm{prset}(\widehat{\mathbb{C}}) \models \bot$.

It is important to note that for any conflicting pair, each involved coalition of structures guarantees minimality and consistency. Later on we will see how acceptability of arguments benefits from these requirements. Next we exemplify the four different types of conflict that may be recognized from a GenAF following Def. 16.

Example 6. Let $\{\mathcal{B}_1, \mathcal{B}_2, \mathcal{B}_3, \mathcal{B}_4, \mathcal{B}_5, \mathcal{B}_6, \mathcal{B}_7\} \subseteq \mathbf{A}$ where $\mathcal{B}_1 = \langle\{p_1(x)\}, p_2(x)\rangle$, $\mathcal{B}_2 = \langle\{p_2(x)\}, p_3(x)\rangle$, $\mathcal{B}_3 = \langle\{p_1(x)\}, \neg p_3(x)\rangle$, $\mathcal{B}_4 = \langle\{\neg p_3(x)\}, p_1(x)\rangle$, $\mathcal{B}_5 = \langle\{p_1(x), \neg p_2(x)\}, p_3(x)\rangle$, $\mathcal{B}_6 = \langle\{p_4(x)\}, \neg p_3(x) \vee \neg p_1(x)\rangle$, $\mathcal{B}_7 = \langle\{p_5(x)\}, p_1(x)\rangle$.

(dependency & claim-clash) $\widehat{\mathbb{C}}_1 = \{\{\mathcal{B}_1, \mathcal{B}_2\}\}$ and $\widehat{\mathbb{C}}_2 = \{\{\mathcal{B}_3\}\}$.
(dependency & premise-clash) $\widehat{\mathbb{C}}_3 = \{\{\mathcal{B}_1\}\}$ and $\widehat{\mathbb{C}}_4 = \{\{\mathcal{B}_5\}\}$.
(support & claim-clash) $\widehat{\mathbb{C}}_1 = \{\{\mathcal{B}_1, \mathcal{B}_2\}\}$ and $\widehat{\mathbb{C}}_5 = \{\{\mathcal{B}_6\}, \{\mathcal{B}_7\}\}$.
(support & premise-clash) $\widehat{\mathbb{C}}_1 = \{\{\mathcal{B}_1, \mathcal{B}_2\}\}$ and $\widehat{\mathbb{C}}_6 = \{\{\mathcal{B}_4\}\}$.

In order to decide which coalition of structures succeeds from a conflicting pair, an *argument comparison criterion* "$\succcurlyeq$" is assumed to be determined from the comparison criterion among formulae in the KB (see Sect. 2). Afterwards, two conflicting coalitions of structures $\widehat{\mathbb{C}}_1$ and $\widehat{\mathbb{C}}_2$ are assumed to be ordered by a function "$\mathfrak{pref}$" relying on "$\succcurlyeq$", where $\mathfrak{pref}(\widehat{\mathbb{C}}_1, \widehat{\mathbb{C}}_2) = (\widehat{\mathbb{C}}_1, \widehat{\mathbb{C}}_2)$ implies the attack relation $\widehat{\mathbb{C}}_1 \mathbf{R_A} \widehat{\mathbb{C}}_2$, *i.e.*, $\widehat{\mathbb{C}}_1$ is a *defeater of* (or it defeats) $\widehat{\mathbb{C}}_2$. In such a case, $\widehat{\mathbb{C}}_2$ is said to be *defeated*. Moreover, if there is no defeater of $\widehat{\mathbb{C}}_1$ then it is said to be *undefeated*. Note that when no pair of arguments is related by "$\succcurlyeq$", both $\widehat{\mathbb{C}}_1 \mathbf{R_A} \widehat{\mathbb{C}}_2$ and $\widehat{\mathbb{C}}_2 \mathbf{R_A} \widehat{\mathbb{C}}_1$ appear from any conflicting pair $\widehat{\mathbb{C}}_1$ and $\widehat{\mathbb{C}}_2$. Finally, the set $\mathbf{R_A} = \{(\widehat{\mathbb{C}}_1, \widehat{\mathbb{C}}_2) \mid \widehat{\mathbb{C}}_1$ and $\widehat{\mathbb{C}}_2$ are in conflict and $\mathfrak{pref}(\widehat{\mathbb{C}}_1, \widehat{\mathbb{C}}_2) = (\widehat{\mathbb{C}}_1, \widehat{\mathbb{C}}_2)\}$ identifies the *attack relations* from a GenAF $\langle \mathbf{A}, \mathbf{N} \rangle \in \mathbb{G}$.

Theorem 1. *Given a* GenAF $\langle \mathbf{A}, \mathbf{N} \rangle \in \mathbb{G}$, $\mathcal{L}_{cl} = \mathcal{L}_{pr} = \mathcal{L}_A$ *iff* $\langle A, \hookrightarrow \rangle$ *is a Dung's AF, where* $A = \{\mathbb{S} \subseteq \mathbf{A} \mid \mathbb{S}$ *is an argumental structure* $\}$ *and* $\hookrightarrow = \{(\mathbb{S}_1, \mathbb{S}_2) \subseteq \mathbf{A} \times \mathbf{A} \mid (\{\mathbb{S}_1\}, \{\mathbb{S}_2\}) \in \mathbf{R_A}\}$.

5   Acceptability Analysis

Assuming a set of normality conditions $\mathbf{N}$, an inconsistent KB Σ leads to conflicting arguments within the associated $\mathfrak{genaf}(\Sigma, \mathbf{N}) = \langle \mathbf{A}, \mathbf{N} \rangle$. Thus, each minimal source of inconsistency within Σ is reflected as an attack in the resulting GenAF. Since the objective of a GenAF is to reason about a KB under uncertainty, there is a need for a mechanism that allows us to obtain those arguments that prevail over the rest. That is, those arguments that can be consistently assumed together, following some policy. For instance, structures with no defeaters should always prevail, since there is nothing strong enough to be posed against them. The tool we need to resolve inconsistency is the notion of *acceptability of arguments*, which is defined on top of an *argumentation semantics* [10]. There are several well-known argumentation semantics, such as the grounded, the stable, and the preferred semantics [1]. These semantics ensure the obtention of consistent sets of arguments, namely *extensions*. That is, the set of accepted arguments calculated following any of these semantics is such that no pair of conflicting arguments appears in that same extension. Finally, an extension determines a maximal consistent subset of the KB Σ.

It is important to notice that some problems like multiple extensions may arise from semantics like both the *stable* and the *preferred*. This would require to make a choice among them. On the other hand, the outcome of the *grounded semantics* is always a single extension, which could be empty. Finally, since dealing with multiple extensions is a problem that falls outside the scope of this article, we will choose the grounded semantics, which can be implemented with a simple algorithm. Consequently, we define a mapping $\mathfrak{sem} : \mathbb{G} \longrightarrow 2^{\text{Args}}$, that intuitively behaves as follows. The set $X \subseteq \mathbf{A}$ is the minimal set verifying $X \subseteq \bigcup_{(\widehat{\mathbb{C}}', \widehat{\mathbb{C}}) \in \mathbf{R_A}} \mathbb{C}^*$ for every undefeated $\widehat{\mathbb{C}}'$ defeating $\widehat{\mathbb{C}}$, and for each $\widehat{\mathbb{C}}$ it follows $\mathbb{C}^* \cap X \neq \emptyset$. As a result, other coalition of structures defeated by $\widehat{\mathbb{C}}$ could appear undefeated. Thus, this process is iteratively applied over the set of arguments $\mathbf{A} \setminus X$ until no conflicting pair is identified. Finally, the extension of the GenAF is determined.

As stated before, the outcome of a grounded semantics could be an empty extension. Such an issue arises when there is a loop in the structures attack graph, that is $(\widehat{\mathbb{C}}', \widehat{\mathbb{C}}) \in$

$\mathbf{R_A}$ and $(\widehat{\mathbb{C}}, \widehat{\mathbb{C}'}) \in \mathbf{R_A}$. To overcome this, some argument from either $\widehat{\mathbb{C}}$ or $\widehat{\mathbb{C}'}$ could be included in X, and therefore the loop would be broken, and the process determined by applying "$\mathfrak{sem}$" can be reconsidered.

Given a (potentially inconsistent) pANF knowledge base $\Sigma \subseteq \mathcal{L}^\kappa$, and a set of normality conditions $\mathbf{N} \subseteq \mathtt{Norm}$, it is possible to redefine the notion of entailment "$\models$" from Σ by reasoning about it over its associated GenAF $\mathfrak{genaf}(\Sigma, \mathbf{N})$, such that $\Sigma \models_G \alpha$ *iff* there exists an argumental structure $\mathbb{S}$ for α such that $\mathbb{S} \subseteq \mathfrak{sem}(\mathfrak{genaf}(\Sigma, \mathbf{N}))$. In such a case, the inferrable claim α is said to be *warranted* and therefore, $\Sigma \models_G \alpha$. Note that if Σ is consistent and $\alpha \in \mathcal{L}_{\mathtt{cl}}$, "$\models_G$" equals the classical entailment "$\models$".

Theorem 2. *Given a consistent* pANF *knowledge base* $\Sigma \subseteq \mathcal{L}^\kappa$, *a set of normality conditions* $\mathbf{N} \subseteq \mathtt{Norm}$, *and a formula* $\alpha \in \mathcal{L}_{\mathtt{cl}}$, $\Sigma \models \alpha$ *iff* $\Sigma \models_G \alpha$.

6   Concluding Remarks

A novel argumentation framework was presented as a generalization of the classical Dung's AF named GenAF. A GenAF aims at providing a straightforward reification tool to reason about inconsistent knowledge bases specified through FOL fragments.

In the last few years, a great effort has been put to the area of ontology change. For instance, ontology evolution intends to restore consistency to inconsistent ontologies. Description logics are probably the most important ontological representation language. Part of our current investigations is done on the research of possible reifications of the here presented GenAF into highly expressible DLs. Consequently, not only ontology evolution could be resolved but also, reasoning about inconsistent ontologies. Some previous work may be referred to [2], where a preliminary investigation on these matters have been done. There, a dynamic version of the GenAF is presented to apply change in a consistent manner to (potentially inconsistent) ontologies.

Finally, since the grounded semantics [1] could return empty extensions, the usage of different semantics [10] is part of the ongoing work to overcome this issue.

References

1. Dung, P.: On the Acceptability of Arguments and its Fundamental Role in Nonmonotonic Reasoning and Logic Programming and n-person Games. Artif. Intell. 77, 321–357 (1995)
2. Moguillansky, M., Rotstein, N., Falappa, M.: A Theoretical Model to Handle Ontology Debugging & Change Through Argumentation. In: IWOD (2008)
3. Rotstein, N., Moguillansky, M., García, A., Simari, G.: An Abstract Argumentation Framework for Handling Dynamics. In: NMR, pp. 131–139 (2008)
4. Rotstein, N., Moguillansky, M., Falappa, M., García, A., Simari, G.: Argument Theory Change: Revision Upon Warrant. In: COMMA, pp. 336–347 (2008)
5. Borgida, A.: On the Relative Expressiveness of Description Logics and Predicate Logics. Artif. Intell. 82(1-2), 353–367 (1996)
6. Baader, F.: Logic-Based Knowledge Representation. Artif. Intell., Today 13–41 (1999)
7. Besnard, P., Hunter, A.: Practical First-Order Argumentation. In: AAAI, pp. 590–595 (2005)
8. Vreeswijk, G.: Abstract Argumentation Systems. Artif. Intell. 90(1-2), 225–279 (1997)
9. Martínez, D.C., García, A.J., Simari, G.R.: Modelling Well-Structured Argumentation Lines. In: IJCAI, pp. 465–470 (2007)
10. Baroni, P., Giacomin, M.: On Principle-Based Evaluation of Extension-Based Argumentation Semantics. Artificial Intelligence 171(10-15), 675–700 (2007)

Probability Density Estimation by Perturbing and Combining Tree Structured Markov Networks

Sourour Ammar[1], Philippe Leray[1], Boris Defourny[2], and Louis Wehenkel[2]

[1] Knowledge and Decision Team,
Laboratoire d'Informatique de Nantes Atlantique (LINA) UMR 6241,
Ecole Polytechnique de l'Université de Nantes, France
sourour.ammar@etu.univ-nantes.fr, philippe.leray@univ-nantes.fr
[2] Department of Electrical Engineering and Computer Science & GIGA-Research,
University of Liège, Belgium
boris.defourny@ulg.ac.be, L.Wehenkel@ulg.ac.be

Abstract. To explore the Perturb and Combine idea for estimating probability densities, we study mixtures of tree structured Markov networks derived by bagging combined with the Chow and Liu maximum weight spanning tree algorithm, or by pure random sampling. We empirically assess the performances of these methods in terms of accuracy, with respect to mixture models derived by EM-based learning of Naive Bayes models, and EM-based learning of mixtures of trees. We find that the bagged ensembles outperform all other methods while the random ones perform also very well. Since the computational complexity of the former is quadratic and that of the latter is linear in the number of variables of interest, this paves the way towards the design of efficient density estimation methods that may be applied to problems with very large numbers of variables and comparatively very small sample sizes.

1 Introduction

Learning of graphical probabilistic models essentially aims at discovering a maximal factorization of the joint density of a set of random variables according to a graph structure, based on a random sample of joint observations of these variables [1]. Such a graphical probabilistic model may be used for elucidating the conditional independencies holding in the data-generating distribution, for automatic reasoning under uncertainties, and for Monte-Carlo simulations. Unfortunately, currently available optimization algorithms for graphical model structure learning are either restrictive in the kind of distributions they search for, or of too high computational complexity to be applicable in very high dimensional spaces [2]. Moreover, not much is known about the behavior of these methods in small sample conditions and, as a matter of fact, one may suspect that they will suffer from overfitting when the number of variables is very large and the sample size is comparatively very small.

In the context of supervised learning, a generic framework which has led to many fruitful innovations is called "Perturb and Combine". Its main idea is to

C. Sossai and G. Chemello (Eds.): ECSQARU 2009, LNAI 5590, pp. 156–167, 2009.

on the one hand perturb in different ways the optimization algorithm used to derive a predictor from a dataset and on the other hand to combine in some appropriate fashion a set of predictors obtained by multiple iterations of the perturbed algorithm over the dataset. In this framework, ensembles of weakly fitted randomized models have been studied intensively and used successfully during the last two decades. Among the advantages of these methods, let us quote the improved predictive accuracy of their models, and the potentially improved scalability of their learning algorithms. For example, ensembles of bagged (derived from bootstrap copies of the dataset) or extremely randomized decision or regression trees, as well as random forests, have been applied successfully in complex high-dimensional tasks, as image and sequence classification [3].

In the context of density estimation, bagging (and boosting) of normal distributions has been proposed by Ridgeway [4]. In [5] the Perturb and Combine idea for probability density estimation with probabilistic graphical models was first explored by comparing large ensembles of randomly generated (directed) poly-trees and randomly generated undirected trees. One of the main findings of that work is that poly-trees, although more expressive, do not yield more accurate ensemble models in this context than undirected trees.

Thus, in the present paper we focus on ensembles of tree structured undirected probabilistic graphical networks (we call them Markov tree mixtures) and we study various randomization and averaging schemes for generating such models. We consider two simple and in some sense extreme instances of this class of methods, namely ensembles of optimal trees derived from bootstrap copies of the dataset by the Chow and Liu algorithm [6], which is of quadratic complexity with respect to the number of variables (we call this bagging of Markov trees), and mixtures of tree structures generated in a totally randomized fashion with linear complexity in the number of variables (we call them totally randomized Markov tree mixtures). We assess the accuracy of these two methods empirically on a set of synthetic test problems in comparison to EM-based state of the art methods building respectively Naive Bayes models and mixtures of trees, as well as a golden standard which uses the structure of the target distribution.

The rest of this paper is organized as follows. Section 2 recalls the classical Bayesian framework for learning mixtures of models and Section 3 describes the proposed algorithms. Section 4 collects our simulation results, Section 5 discusses the main findings of our work, and Section 6 briefly concludes and highlights some directions for further research.

2   Bayesian Modeling Framework

Let $X = \{X_1, \ldots, X_n\}$ be a finite set of discrete random variables, and $D = (x^1, \cdots, x^d)$ be a dataset (sample) of joint observations $x^i = (x_1^i, \cdots, x_n^i)$ independently drawn from some data-generating density $\mathbb{P}_G(X)$.

In the full Bayesian approach, one assumes that $\mathbb{P}_G(X)$ belongs to some space of densities $\mathcal{D}$ described by a model-*structure* $M \in \mathcal{M}$ and model-*parameters* $\theta_M \in \Theta_M$, and one infers from the dataset a mixture of models described by the following equation:

$$\mathbb{P}_{\mathcal{D}}(X|D) = \sum_{M \in \mathcal{M}} \mathbb{P}(M|D)\ \mathbb{P}(X|M, D), \tag{1}$$

where $\mathbb{P}(M|D)$ is the posterior probability over the model-space $\mathcal{M}$ conditionally to the data D, and where $\mathbb{P}(X|M, D)$ is the integral:

$$\mathbb{P}(X|M, D) = \int_{\Theta_M} \mathbb{P}(X|\theta_M, M)\ d\mathbb{P}(\theta_M|M, D). \tag{2}$$

So $\mathbb{P}_{\mathcal{D}}(X|D)$ is computed by:

$$\mathbb{P}_{\mathcal{D}}(X|D) = \sum_{M \in \mathcal{M}} \mathbb{P}(M|D) \int_{\Theta_M} \mathbb{P}(X|\theta_M, M)\ d\mathbb{P}(\theta_M|M, D), \tag{3}$$

where $d\mathbb{P}(\theta_M|M, D)$ is the posterior model-parameter density and $\mathbb{P}(X|\theta_M, M)$ is the likelihood of observation X for the structure M with parameters θ_M.

When the space of model-structures $\mathcal{M}$ and corresponding model-parameter Θ_M is the space of Bayesian networks or the space of Markov networks over X, approximations have to be done in order to make tractable the computation of equation (3). For Bayesian networks for example, it is shown in [7] that equation (2) can be simplified by the likelihood estimated with the parameters of maximum a posteriori probability $\tilde{\theta}_M = \arg\max_{\theta_M} \mathbb{P}(\theta_M|M, D)$, under the assumption of a Dirichlet distribution (parametrized by its coefficients α_i) for the prior distribution of the parameters $\mathbb{P}(\theta_M|M)$.

Another approximation to consider is simplifying the summation over all the possible model-structures M. As the size of the set of possible graphical model structures is super-exponential in the number of variables [8], the summation of equation (1) must in practice be performed over a strongly constrained subspace $\hat{\mathcal{M}}$ obtained for instance by sampling methods [9,10,11], yielding the approximation

$$\mathbb{P}_{\hat{\mathcal{M}}}(X|D) = \sum_{M \in \hat{\mathcal{M}}} \mathbb{P}(M|D)\mathbb{P}(X|\tilde{\theta}_M, M). \tag{4}$$

Let's note here that this equation is simplified once more when using classical structure learning methods, by keeping only the model $M = \tilde{M}$ maximizing $\mathbb{P}(M|D)$ over $\mathcal{M}$:

$$\mathbb{P}_{\tilde{M}}(X|D) = \mathbb{P}(X|\tilde{\theta}_{\tilde{M}}, \tilde{M}). \tag{5}$$

3   Randomized Markov Tree Mixtures

In this work, we propose to choose as set $\hat{\mathcal{M}}$ in equation (4) a randomly generated subset of pre-specified cardinality of Markov tree models.

3.1   Poly-Tree Models

A poly-tree model for the density over X is defined by a Directed Acyclic Graph (DAG) structure P which skeleton is acyclic and connected, and the set of vertices of which is in bijection with $X = \{X_1, \dots, X_n\}$, together with a set of conditional densities $\mathbb{P}_P(X_i|pa_P(X_i))$, where $pa_P(X_i)$ denotes the set of variables in bijection with the parents of X_i in P. Like more general DAGs, this structure P represents graphically the density factorization

$$\mathbb{P}_P(X) = \prod_{i=1}^{n} \mathbb{P}_P(X_i|pa_P(X_i)). \tag{6}$$

The model parameters are thus here specified by the conditional distributions:

$$\theta_P = (\mathbb{P}_P(X_i|pa_P(X_i)))_{i=1}^{n}. \tag{7}$$

The structure P can be exploited for probabilistic inference over $\mathbb{P}_P(X)$ with a computational complexity linear in the number of variables n [12].

One can define nested subclasses $\mathcal{P}^p$ of poly-tree structures by imposing constraints on the maximum number p of parents of any node. In these subclasses, not only inference but also parameter learning is of linear complexity in the number of variables.

3.2 Markov Tree Models

The smallest subclass of poly-tree structures is called the Markov tree subspace, in which nodes have exactly one parent ($p = 1$). Markov tree models have the essential property of having no v-structures [1], in addition to the fact that their skeleton is a tree, and their dependency model may be read-off without taking into account the direction of their arcs. In other words, a poly-tree model without v-structures is a Markov tree and is essentially defined by its skeleton. These are the kind of models that we will consider subsequently in this paper. Importantly, Markov tree models may be learned efficiently by the Chow and Liu algorithm which is only quadratic in the number of vertices (variables) [6]. Given the skeleton of the Markov tree, one can derive an equivalent directed acyclic (poly-tree) graph from it by arbitrarily choosing a root node and by orienting the arcs outwards from this node in a depth-first fashion.

3.3 Mixtures of Markov Trees

A mixture distribution $\mathbb{P}_{\hat{\mathcal{T}}}(X_1, \ldots, X_n)$ over a set $\hat{\mathcal{T}} = \{T_1, \ldots, T_m\}$ of m Markov trees is defined as a convex combination of elementary Markov tree densities, ie.

$$\mathbb{P}_{\hat{\mathcal{T}}}(X) = \sum_{i=1}^{m} \mu_i \mathbb{P}_{T_i}(X), \tag{8}$$

where $\mu_i \in [0, 1]$ and $\sum_{i=1}^{m} \mu_i = 1$, and where we leave for the sake of simplicity implicit the values of the parameter sets $\tilde{\theta}_i$ of the individual Markov tree densities.

While single Markov tree models impose strong restrictions on the kind of densities they can faithfully represent, mixtures of Markov trees, as well as mixtures of empty graphs (i.e. Naive Bayes with hidden class), are universal approximators (see, e.g., [13]).

3.4   Generic Randomized Markov Tree Mixture Learning Algorithm

Our generic procedure for learning a random Markov tree mixture distribution from a dataset D is described by Algorithm 1; it receives as inputs X, D, m, and three procedures *DrawMarkovtree*, *LearnPars*, *CompWeights*.

Algorithm 1 (Learning a Markov tree mixture)

1. *Repeat for $i = 1, \cdots, m$:*
    (a) *Draw random number ρ_i,*
    (b) *$T_i = DrawMarkovtree(D, \rho_i)$,*
    (c) *$\tilde{\theta}_{T_i} = LearnPars(T_i, D, \rho_i)$*
2. *$(\mu)_{i=1}^m = CompWeights((T_i, \tilde{\theta}_{T_i}, \rho_i)_{i=1}^m, D)$*
3. *Return $\left(\mu_i, T_i, \tilde{\theta}_{T_i} \right)_{i=1}^m$.*

Line (a) of this algorithm draws a random number in $\rho_i \in [0; 1)$ according to a uniform disribution, which may be used as a random seed for *DrawMarkovtree* which builds a tree structure T_i possibly depending on the dataset D and ρ_i and *LearnPars* which estimates the parameters of T_i. Versions of these two procedures used in our experiments are further discussed in the next section. The algorithm returns the m tree-models, along with their parameters θ_{T_i} and the weights of the trees μ_i.

3.5   Specific Variants

In our investigations reported below, we have decided to compare various versions of the above generic algorithm.

In particular, we consider two variants of the *DrawMarkovtree* function: one that randomly generates unconstrained Markov trees (by sampling from a uniform density over the set $\mathcal{P}^1$ of all Markov tree models), and one that builds optimal tree structures by applying the MWST (Maximum Weight Spanning Tree) structure learning algorithm published in the late sixties by Chow and Liu [6] on a random bootstrap [14] replica of the initial learning set D. The random sampling procedure of the first variant is described in [5]. The second variant reminds the Bagging idea of [15].

Concerning the parameter estimation by *LearnPars*, we use the BDeu score maximization for each tree structure individually, which is tantamount to selecting the estimates using Dirichlet priors. More specifically, in our experiments which are limited to binary random variables, we used non-informative priors, which then amounts to using $\alpha = 1/2$, i.e. $p(\theta, 1 - \theta) \propto \theta^{-1/2}(1 - \theta)^{-1/2}$ for the prior density of the parameters characterizing the conditional densities attached the Markov tree nodes, once this tree is oriented in an arbitrary fashion. Notice that in the case of tree-bagging, these parameters are estimated from the bootstrap sample used to generate the corresponding tree structure.

Finally, we consider two variants for the *CompWeights* function, namely uniform weighting (where coefficients are defined by $\mu_i = \frac{1}{m}, \forall i = 1, \ldots, m$) and Bayesian averaging (where coefficients μ_i are proportional to the posterior probability of the Markov tree structure T_i, derived from its BDeu score [1] computed from the full dataset D).

4   Empirical Simulations

4.1   Protocol

In order to evaluate the four different variants of our algorithm, we carried out repetitive experiments for different data-generating (or target) densities, by proceeding in the following way.

Choice of target density. All our experiments were carried out with models for a set of 8 and 16 binary random variables. We chose to start our investigations in such a simple setting in order to be able to compute accuracies exactly (see Section 4.1), and so that we can easily analyze the graphical structures of the target densities and of the inferred set of trees.

To choose a target density $\mathbb{P}_G(X)$, we first decide whether it will factorize according to a poly-tree or to a more general directed acyclic graph structure. Then we use the appropriate random structure generation algorithm described in [5] to draw a structure and, we choose the parameters of the target density by selecting for each conditional density of the structure (they are all related to binary variables) two random numbers in the interval $[0, 1]$ and by normalizing.

Generation of datasets. For each target density and dataset size, we generated 10 different datasets by sampling values of the random variables using the Monte-Carlo method with the target structure and parameter values.

We carried out simulations with dataset sizes of N = 250 elements for models with 8 or 16 variables and for N= 2000 for the models with 16 variables. Given the total number of 2^n possible configurations of our n random variables, we thus look at rather small datasets.

Learning of mixtures. For a given dataset and for a given variant of the mixture learning algorithm we generate ensemble models of growing sizes, respectively $m = 1$, $m = 10$, and then up to $m = 500$ by increments of 10. This allows us to appraise the effect of the ensemble size on the quality of the resulting model.

Accuracy evaluation. The quality of any density inferred from a dataset is evaluated by the (asymmetric) Kullback-Leibler divergence [16] between this density and the data-generating density $\mathbb{P}_G(X)$ used to generate the dataset. This is exactly computed by

$$KL(\mathbb{P}_G, \mathbb{P}_M) = \sum_{X \in \mathcal{X}} \mathbb{P}_G(X) \ln \frac{\mathbb{P}_G(X)}{\mathbb{P}_M(X)}, \qquad (9)$$

where $\mathbb{P}_M(X)$ denotes the density that is evaluated, and $\mathcal{X}$ denotes the set of all possible configurations of the random variables in X.

Reference methods. We also provide comparative accuracy values obtained in the same fashion with five different reference methods, namely (i) a *golden standard* denoted by GO which is obtained by using the data-generating structure and reestimating its parameters from the dataset D, (ii) a series of *Naive Bayes* models with growing number of hidden classes denoted by NBE^* and built according to the Expectation-Maximization (EM) algorithm [17] as proposed in [18] but without pruning, (iii) a series of *Optimal Tree Mixtures* with growing number of terms denoted by $MixTree$ and built according to the algorithm proposed by Meila-Predoviciu which combines the Chow and Liu MWST algorithm with the EM algorithm for parameter estimation [13], (iv) a baseline method denoted by BL which uses a complete (fully connected) DAG structure whose parameters are estimated from the dataset D, and (v) a single Markov tree built using the Chow and Liu algorithm on the whole dataset (denoted by CL, below).

Software implementation. Our various algorithms were implemented in C++ with the Boost library (http://www.boost.org/) and APIs provided by the ProBT© platform (http://bayesian-programming.org).

4.2   Results

Fig. 1 (resp. Fig. 2, Fig. 3 and Fig. 4) provides a representative set of learning curves for a target density corresponding to a poly-tree distribution (resp. DAG distribution). The horizontal axis corresponds to the number m of mixture terms, whereas the vertical axis corresponds to the KL measures with respect to the target density. All the curves represent average results obtained over ten different datasets of 250 learning samples (2000 in Fig. 4) and five target distributions (only four target distributions in Fig. 3 and Fig. 4). Before analyzing these curves, let us first remind that in our first experiments reported in [5], which compared mixtures of fully randomly generated poly-trees with mixtures of fully randomly generated Markov trees, we found that general poly-tree mixtures were not significantly different from Markov tree mixtures in terms of their accuracies. Thus we have decided to report in the present paper only results obtained with our Markov tree mixtures (MTU, $MTBDeu$, $MBTU$, $MBTBDeu$) and a broader set of reference methods (BL, GO, $MixTree$, CL, NBE^*).

MTU corresponds to uniform mixtures of totally randomly generated trees, while $MTBDeu$ corresponds to the same mixtures when the terms are weighted according to their posterior probabilities given the dataset. $MBTU$ and $MBTBDeu$ correspond to mixtures of bagged trees with respectively uniform and posterior probability weighting.

We thus observe in Fig. 1 that our four random Markov tree mixture methods are clearly outperforming the baseline BL in terms of accuracy, and some of them are already quite close to the golden standard GO. For this reason, BL results are not reported in all other figures. All four variants also nicely behave in a monotonic fashion: the more terms in the mixture the more accurate the resulting model.

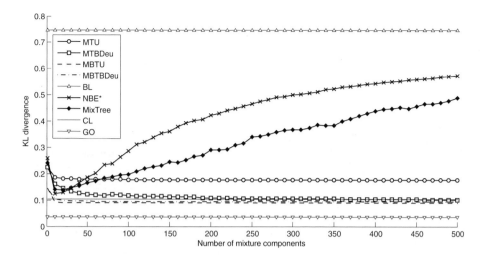

Fig. 1. Mixtures of trees for density estimation with a poly-tree target distribution. 10 experiments with a sample size of 250 for 5 random target distributions of 8 variables. (lower is better).

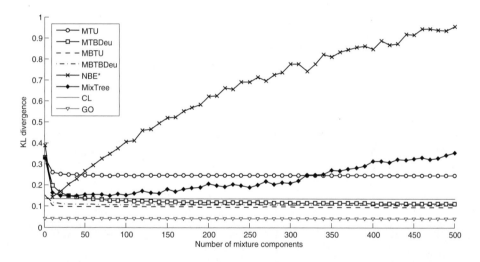

Fig. 2. Mixtures of trees for density estimation with a DAG target distribution. 10 experiments with a sample size of 250 for 5 random target distributions of 8 variables. (lower is better).

Concerning the totally randomly generated tree mixtures, we also observe that when they are weighted proportionally to their posterior probability given the dataset they provide much better performances as compared to a uniform weighting procedure. Concerning the mixtures of bagged trees we observe from all figures that they both outperform the mixtures of randomly generated trees in terms of asymptotic (with respect to the number of mixture terms) accuracy

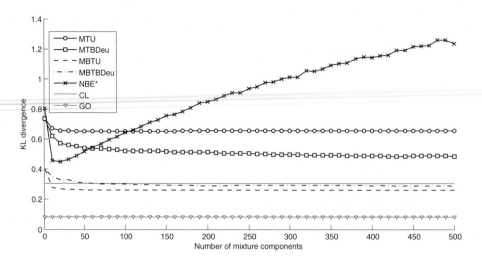

Fig. 3. Mixtures of trees for density estimation with a DAG target distribution. 10 experiments with a sample size of 250 for 4 random target distributions of 16 variables. (lower is better).

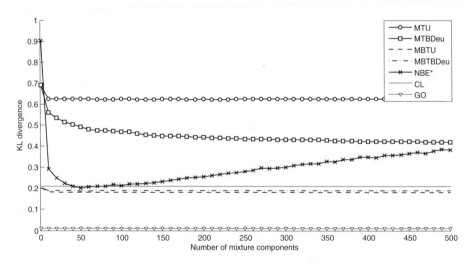

Fig. 4. Mixtures of trees for density estimation with a DAG target distribution. 10 experiments with a sample size of 2000 for 4 random target distributions of 16 variables. (lower is better).

and even more in terms of speed of convergence. With this bagging approach, we also notice that the uniform weighting procedure is actually slightly better than the one using weights based on the posterior probabilities given the dataset. We believe that non-uniform weighting is counterproductive in the case of bagged ensembles because it increases the variance of the ensemble model without decreasing its bias. Finally, we note that both bagging methods provide slightly better results than single CL trees built on the whole dataset, as soon as the number mixture terms is larger than about ten.

The NBE^* algorithm obtains results very close but slightly less good than those of bagged tree mixtures for a very small number of components (hidden classes). However, contrary to random trees or bagged trees, in the Naive Bayes method one clearly observes the fact that adding new components in the mixture eventually, and rather quickly, leads to overfitting which is stronger when the sample size is smaller (Fig. 1, Fig. 2, Fig. 3 and Fig. 4).

All in all, the consistently best method in these trials is the method which uses bagging of tree structures combined with a uniform weighting scheme.

To further illustrate our results, we plot in Fig. 1 and Fig. 2 curves corresponding to the $MixTree$ algorithm. We observe that the $MixTree$ algorithm provides similar results (although better) than those of Naive Bayes, namely for a small number of terms it yields accuracies close but slightly less good to those of the bagged tree mixtures, but then when the number mixture terms increases it also leads to overfitting.

5 Discussion

Our simulation results showed that in small sample conditions the mixtures of Markov trees turned out to be in general largely superior to the complete structure baseline BL.

Bagged ensembles of Markov trees significantly outperform totally randomized ensembles of Markov trees, both in terms of accuracy and in terms of speed of convergence when the number of mixture components is increased. Contrary to the more sophisticated EM-based Naive Bayes and $Mixtree$ methods, our methods do not lead to overfitting when the number of mixture terms is increased.

From a computational point of view, bagging which uses the Chow Liu MWST algorithm as baselearner is quadratic in the number of variables while the generation of random tree structures may be done in linear time (see [5]). Bagged ensembles of Markov trees slightly outperform the CL method, which is also quadratic in the number of variables, in terms of accuracy.

In case a linear complexity is needed, random mixtures of Markov trees, namely $MTBDeu$, give acceptable results. When a quadratic complexity can be accepted, CL remains slightly preferable to our methods with bagging.

In between these two extreme randomization schemes, one can imagine a whole range of methods based on the combination of bootstrap resampling and more or less strong randomizations of the generation of the Markov trees leading to different computational trade-offs. Our methods with bagging can be improved to supply algorithms with same accuracy and complexity better than quadratic. Also, out-of-bag estimates may be exploited to compute unbiased accuracies of the ensemble models [4].

6   Summary and Future Works

We have investigated in this paper the transposition of the "Perturb and Combine" idea celebrated in supervised learning to the context of unsupervised density estimation with graphical probabilistic models. We have presented a generic framework for doing this, based on randomized mixtures of tree structured Markov networks, where the perturbation was done by generating in a totally random fashion the structure component, or by bootstrapping data before optimizing this structure component.

The first results obtained in the context of a simple test protocol are already very promising, while they also highlight a certain number of immediate future research directions.

Thus, a first line of research will be to apply our experimental protocol to a larger set of problems including high-dimensional ones and a larger range of sample sizes. We believe also that a more in depth analysis of the accuracy results with respect to the basic properties of the target distributions as well as sample sizes would be of interest, in particular with the aim of characterizing more precisely under which conditions our methods are more effective than state-of-the-art ones. Of course, these investigations should also aim at systematically comparing all these algorithm variants from the point of view of their computational complexity.

Another more generic direction of research, is to adapt importance sampling approaches (e.g. the cross-entropy method [19]) in order to generate randomized ensembles of simple structures (chains, trees, poly-trees, etc.) that fit well the given dataset.

A more simple direction is to improve our methods of bagged ensembles of Markov trees by forcing the complexity of the optimization level in the Chow Liu MWST algorithm to come down below the quadratic.

While the class of methods investigated in this paper is based on generating an ensemble by drawing its terms from a same distribution (which could be done in parallel), we believe that the combination of these methods with sequential methods such as Boosting or Markov-Chain Monte-Carlo which have already been applied in the context of graphical probabilistic models (see e.g. [20]), might provide a very rich avenue for the design of novel density estimation algorithms.

Acknowledgments

This paper presents research results of the Belgian Network BIOMAGNET, funded by the Interuniversity Attraction Poles Programme, initiated by the Belgian State, Science Policy Office. The authors thank Pierre Geurts and François Schnitzler for useful discussions and suggestions to improve the manuscript.

References

1. Cowell, R., Dawid, A., Lauritzen, S., Spiegelhalter, D.: Probabilistic Networks and Expert Systems. Springer, Heidelberg (1999)
2. Auvray, V., Wehenkel, L.: Learning inclusion-optimal chordal graphs. In: Proceedings of the 24th Conference on Uncertainty in Artificial Intelligence (UAI 2008), pp. 18–25. Morgan Kaufmann, San Francisco (2008)
3. Geurts, P., Ernst, D., Wehenkel, L.: Extremely randomized trees. Machine Learning 63(1), 3–42 (2006)
4. Ridgeway, G.: Looking for lumps: boosting and bagging for density estimation. Computational Statistics & Data Analysis 38(4), 379–392 (2002)
5. Ammar, S., Leray, P., Defourny, B., Wehenkel, L.: High-dimensional probability density estimation with randomized ensembles of tree structured bayesian networks. In: Proceedings of the fourth European Workshop on Probabilistic Graphical Models (PGM 2008), pp. 9–16 (2008)
6. Chow, C., Liu, C.: Approximating discrete probability distributions with dependence trees. IEEE Transactions on Information Theory 14(3), 462–467 (1968)
7. Chickering, D., Heckerman, D.: Efficient approximations for the marginal likelihood of bayesian networks with hidden variables. Machine Learning 29(2-3), 181–212 (1997)
8. Robinson, R.: Counting unlabeled acyclic digraphs. In: Little, C.H.C. (ed.) Combinatorial Mathematics V. Lecture Notes in Mathematics, vol. 622, pp. 28–43. Springer, Berlin (1977)
9. Madigan, D., Raftery, A.: Model selection and accounting for model uncertainty in graphical models using occam's window. The American Statistical Association 89, 1535–1546 (1994)
10. Madigan, D., York, J.: Bayesian graphical models for discrete data. International Statistical Review 63, 215–232 (1995)
11. Friedman, N., Koller, D.: Being bayesian about network structure. In: Boutilier, C., Goldszmidt, M. (eds.) Proceedings of the 16th Conference on Uncertainty in Artificial Intelligence (UAI 2000), pp. 201–210. Morgan Kaufmann Publishers, San Francisco (2000)
12. Pearl, J.: Fusion, propagation, and structuring in belief networks. Artificial Intelligence 29, 241–288 (1986)
13. Meila-Predoviciu, M.: Learning with Mixtures of Trees. PhD thesis, MIT (1999)
14. Efron, B., Tibshirani, R.J.: An introduction to the Bootstrap. Monographs on Statistics and Applied Probability, vol. 57. Chapman and Hall, Boca Raton (1993)
15. Breiman, L.: Bagging predictors. Machine Learning 24(2), 123–140 (1996)
16. Kullback, S., Leibler, R.: On information and sufficiency. Annals of Mathematical Statistics 22(1), 79–86 (1951)
17. Dempster, A., Laird, N., Rubin, D.: Maximum likelihood from incomplete data via the EM algorithm. Journal of the Royal Statistical Society. Series B (Methodological) 39(1), 1–38 (1977)
18. Lowd, D., Domingos, P.: Naive bayes models for probability estimation. In (ICML 2005) Proceedings of the 22nd international conference on Machine Learning, pp. 529–536. ACM Press, New York (2005)
19. Rubinstein, R., Kroese, D.: The Cross-Entropy Method. A Unified Approach to Combinatorial Optimization, Monte-Carlo Simulation, and Machine Learning. In: Information Science and Statistics. Springer, Heidelberg (2004)
20. Rosset, S., Segal, E.: Boosting density estimation. In: Proceedings of the 16th International Conference on Neural Information Processing Systems (NIPS), Vancouver, British Columbia, Canada, pp. 267–281 (2002)

Integrating Ontological Knowledge for Iterative Causal Discovery and Visualization

Montassar Ben Messaoud[1], Philippe Leray[2], and Nahla Ben Amor[1]

[1] LARODEC, Institut Supérieur de Gestion Tunis,
41, Avenue de la liberté, 2000 Le Bardo, Tunisie
benmessaoud.montassar@hotmail.fr, nahla.benamor@gmx.fr
[2] Knowledge and Decision Team,
Laboratoire d'Informatique de Nantes Atlantique (LINA) UMR 6241,
Ecole Polytechnique de l'Université de Nantes, France
philippe.leray@univ-nantes.fr

Abstract. Bayesian networks (BN) have been used for prediction or classification tasks in various domains. In the first applications, the BN structure was causally defined by expert knowledge. Then, algorithms were proposed in order to learn the BN structure from observational data. Generally, these algorithms can only find a structure encoding the right conditional independencies but not all the causal relationships. Some new domains appear where the model will only be learnt in order to discover these causal relationships. To this end, we will focus on discovering causal relations in order to get Causal Bayesian Networks (CBN). To learn such models, interventional data (i.e. samples conditioned on the particular values of one or more variables that have been experimentally manipulated) are required. These interventions are usually very expensive to perform, therefore the choice of variables to experiment on can be vital when the number of experimentations is restricted. In many cases, available ontologies provide high level knowledge for the same domain under study. Consequently, using this semantical knowledge can turn out of a big utility to improve causal discovery. This article proposes a new method for learning CBNs from observational data and interventions. We first extend the greedy approach for perfect observational and experimental data proposed in [13], by adding a new step based on the integration of ontological knowledge, which will allow us to choose efficiently the interventions to perform in order to obtain the complete CBN. Then, we propose an enriched visualization for better understanding of the causal graphs.

1 Introduction

Over the last few years, the use of ontologies is becoming increasingly widespread by the computer science community. The main advantage of ontologies is essentially that they try to capture the semantics of domain expertise by deploying knowledge representation primitives enabling a machine to understand the relationships between concepts in a domain [3]. A lot of solutions based on ontologies

C. Sossai and G. Chemello (Eds.): ECSQARU 2009, LNAI 5590, pp. 168–179, 2009.

have been implemented in several real applications in different areas as natural language translation, medicine, electronic commerce and bioinformatics.

In the other hand, causal discovery remains a challenging and important task [12]. One of the most common techniques for causal modeling are *Causal Bayesian Networks* (CBNs), which are probabilistic graphical models, where causal relationships are expressed by directed edges [18]. Like classical Bayesian Networks, CBNs need *observational data* to learn a partially directed model but also they require *interventional data*, i.e. samples conditioned on the particular values of one or more variables that have been experimentally manipulated, in order to fully orient the model.

Several researches have proposed algorithms to learn the structure of a CBN from observational and experimental data.

This paper goes farther by integrating ontological knowledge for more efficient causal discovery and proposes, henceforth, an enriched visualization of causal graphs based on semantical knowledge.

Less works has been done on this aspect of combining the power of Bayesian networks and ontologies. And most of them focus on developing probabilistic frameworks for combining different ontology mapping methods [16,22].

The remainder of this paper is organized as follows: Section 2 provides some notations and basic definitions on Bayesian networks, Causal Bayesian Networks as well as ontologies. Section 3 describes the new approach for causal discovery based on ontological knowledge. Finally, section 5 proposes different tools to obtain an enriched visualization of CBN.

2   Background and Notations

This section gives basic definitions on Bayesian networks, causal Bayesian networks and ontologies. The following notations and syntactical conventions are used: Let $X = \{X_1, X_2, ..., X_N\}$ be a set of variables. By x we denote any instance of X. For any node $X_i \in X$ we denote by $\mathrm{Pa}(X_i)$ (resp. $\mathrm{Ne}(X_i)$) the parents (resp. neighbors) of the variable X_i. $\mathrm{P}(X_i)$ is used to denote the probability distribution over all possible values of variable X_i, while $\mathrm{P}(X_i{=}x_k)$ is used to denote the probability that the variable X_i is equal to x_k.

2.1   Bayesian Networks

A *Bayesian Network* (BN) [17] consists of a Directed Acyclic Graph (DAG) and a set of conditional probability tables (CPTs) such that each node in the DAG corresponds to a variable, and the associated CPT contains the probability of each state of the variable given every possible combination of its parents states i.e. $\mathrm{P}(X_i|\mathrm{Pa}(X_i))$. Bayesian networks are very suited for probabilistic inference, since they satisfy an important property known as *Markov property*, which states that each node is independent of its non-descendants given its parents and leads to a direct factorization of the joint distribution into the product of the conditional distribution of each variable X_i given its parents $\mathrm{Pa}(X_i)$. Therefore, the probability distribution relative to X=$(X_1, X_2,..., X_n)$ can be computed by the following chain rule:

$$P(X_1, X_2, ..., X_n) = \prod_{i=1..n} P(x_i \mid Pa(X_i)). \qquad (1)$$

Note that several BNs can model the same probability distribution. Such networks are called equivalent or Markov equivalent [24].

Definition 1. *A **Complete Partially Directed Acyclic Graph** (CPDAG) is a representation of all equivalent BNs. The CPDAG contains the same skeleton as the original DAG, but possesses both directed and undirected edges. Every directed edge $X_i \rightarrow X_j$ of a CPDAG denotes that all DAGs of this class contain this edge, while every undirected edge $X_i—X_j$ in this CPDAG-representation denotes that some DAGs contain the directed edge $X_i \rightarrow X_j$, while others contain the oppositely orientated edge $X_i \leftarrow X_j$.*

Under Causal sufficiency assumption (i.e. there are no latent variables that influence the system under study), many structure learning techniques using perfect *observational data* are available and can be used to learn CPDAG and then choose a possible complete instantiation in the space of equivalent graphs defined by this CPDAG. These techniques can be classified into two groups, namely *score-based* and *constraint-based* algorithms.

Score-based algorithms [4,5] attempt to identify the network that maximizes a scoring function evaluating how well the network fits the data while *constraint-based* algorithms [12,21] look for (in)dependencies in the data and try to model that information directly into the graphical structure.

2.2   Causal Bayesian Networks

A *Causal Bayesian Network* (CBN) [18] is a Bayesian network with the added property that all edges connecting variables represent autonomous causal relations. Given a CBN, we can go further than probabilistic inference to perform causal inference. Pearl has introduced the *do operator* as standard notification for external intervention on causal models. In fact, the effect of an action "do$(X_j=x_k)$" in a causal model corresponds to a minimal perturbation of the existing system that forces the variable X_j to the value x_k. In other terms, causal inference is the process of calculating the effect of manipulating some set of variables X_i on the probability distribution of some other set of variables X_j, this is denoted as P$(X_i|$do$(X_j=x_k))$.

Several researches in the literature have proposed algorithms to learn CBN's structure. We can, in particular, mention Tong and Koller [23] and Cooper and Yoo [6], which developed a score-based techniques to learn a CBN from a mixture of experimental and observational data. Eberhardt et al. [9] performed a theoretical study on the lower bound of the worst case for the number of experiments to perform to recover the causal structure.

Recently, Meganck et al. [13] proposed a greedy contraint-based learning algorithm for CBNs using experiments. The MyCaDo (My Causal DiscOvery) algorithm, which represents the main application of their approach, is a structure learning algorithm able to select appropriate interventions or experiments (i.e. randomly assigning values to a single variable and measuring some other possible response variables) needed to built a CBN. As input, the MyCaDo algorithm needs a CPDAG. Meganck et al. proposed to use the PC algorithm proposed by Spirtes et al. [21] with modified orientation rules to learn the initial CPDAG, but other structure learning algorithms can be taken into consideration.

When applying PC algorithm, we start with a complete undirected network. Then, we remove edges when independencies are found; we call such step Skeleton discovery. The edge orientation, in such algorithm, is based on v-structure discovery and edge propagation. Based on the already oriented edges, an inferred edges step will apply some orientation rules until no more edges can be oriented.

In this step, we need also experimental data to perform interventions on the system. To learn a CBN from interventional data, three major parts can be distinguished in the MyCaDo algorithm:

- First of all, it tries to maximize an utility function based on three variables: $gain(exp)$, $cost(exp)$, $cost(meas)$, respectively, the gained information, the cost of performing an experiment and the cost of measuring other variables, to decide which experiment should be performed and hence also which variables will be measured.

  If we denote performing an experiment on X_i by A_{X_i}, and measuring the neighboring variables by M_{X_i}, then the utility function will be as follows:

  $$U(A_{X_i}) = \frac{\alpha gain(A_{X_i})}{\beta cost(A_{X_i}) + \gamma cost(M_{X_i})} \qquad (2)$$

  where α, β and γ are measures of importance for every term.

  In this formula, $gain(A_{X_i})$ takes into consideration the number of undirected neighbors $Ne_U(X_i)$ (e.g. nodes that are connected to X_i by an undirected edge) and the amount of possible inferred edges, after performing an experiment on X_i.

  Three decision criteria were proposed *Maximax*, *Maximin* and *Expected Utility* depending on the type of situation in which to perform the experiments it might be advantageous to choose a specific criterion.
- Secondly, the selected experiment will be performed.
- Finally, the results of this experiment, will be analyzed in order to direct a number of edges in the CPDAG.

Note that the amount of edges of which the direction can be inferred after performing an experiment is entirely based on the instantiation (i.e. assignation of a direction) of the undirected edges connected to the one being experimented on.

This process will be iterated until we obtain the correct structure of the CBN. Note that we can have a non-complete causal graph (i.e. not all edges are oriented) as final output of MyCaDo algorithm.

Borchani et al. [2] have also described another approach for learning CBNs from incomplete observational data and interventions.

Furthermore, other approaches were proposed to learn graphical models that can handle latent variables. For this kind of hidden variable modeling, Meganck et al. [14,15] and Maes et al [11] studied which experiments were needed to learn CBN with latent variables under the assumption of faithful distribution (i.e. the observed samples come from a distribution which independence properties are exactly matched by those present in the causal structure of a CBN). Their solution offers a new representation based on two paradigms that model the latent variables implicitly, namely *Maximal Ancestral Graphs* and *Semi-Markovian Causal Models*, in order to perform probabilistic as well as causal inference.

2.3   Ontology

An **ontology** [10] is defined as a formal explicit specification of a shared conceptualization. In other terms, it is a formal representation of a set of concepts within a domain and the relations between these concepts. The relations in an ontology are either taxonomic (e.g. is-a, part-of) or non-taxonomic (i.e. user-defined). The taxonomic ones are the commonly used relations. In the simple case, the ontology takes the form of a tree or hierarchy representing concept taxonomy. Formally:

Definition 2. *A **Concept Taxonomy** $H = (C, E, Rt)$ is a directed acyclic graph where, $C = \{c_1, c_2, ..., c_n\}$ is the set of all concepts in the hierarchy, E is the set of all subsumption links (is-a) and Rt is the unique root of this DAG. The primary structuring element of concept taxonomy is the subsumption relationship, which supposed that if the concept c_i is a child of concept c_j then all properties of c_j are also properties of c_i and we say that c_j subsumes c_i .*

Regarding any concept taxonomy hierarchy, we can give the following notations:

- $pths(c_i, c_j)$: the set of paths between the concepts c_i and c_j, where $i \neq j$,
- $len_e(e)$: the length in number of edges of the path e,
- $mscs(c_i, c_j)$: the most specific common subsumer of c_i and c_j, where $i \neq j$.

The major contribution of using concept taxonomies is, essentially, to present the domain knowledge in a declarative formalism.

For instance, the Gene Ontology (GO) [25] is one of the principal knowledge resource repository in the bioinformatic field and represents an important tool for the representation and processing of information about genes and functions. It provides controlled vocabularies for the designations of cellular components, molecular functions and biological processes.

Example 1. Figure 1 shows an ontology toy example or more precisely, an is-a tree, where the leaf nodes (X1, X2...,X5) are genes. Their direct subsumers

(F1, F2, F3) represent the biological functions i.e. every set of genes in a GO share a very specified function. Also, the biological functions are subsumed by their biological super-functions (SF1, SF2). Note that we can use the term of generalization as shown in figure 1 to refer the subsuming process. For this case of is-a tree, only one path can be find between two different concepts. To more illustrate this purpose, let us consider the two concepts X1 and SF2. Here, pths(X1, SF2) represents the set of the edges X1-F1, F1-SF1, SF1-RT and RT-SF2. Consequently the length in number of edges of the corresponding path will be equal to four.

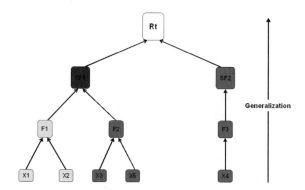

Fig. 1. Example of hierarchy representing a taxonomy of concepts

Recently, several works highlighted the importance of evaluating the strength of the semantic links inside domain ontologies. We can distinguish three major classes of semantic measures, namely *semantic relatedness, semantic similarity* and *semantic distance*, evaluating, respectively, the closeness, the resemblance and the disaffection between two concepts.

The *semantic similarity* represents a special case of *semantic relatedness*. For instance, if we consider the two concepts *wind turbine* and *wind*, they would be more closely related than, for example the pair *wind turbine* and *solar panel*. However the latter concepts are more similar. Therefore, all pairs of concepts with a high semantic similarity value (i.e. high resemblance) have a high semantic relatedness value whereas the inverse is not necessarily true. In the other hand, the semantic distance is an inverse notion to the semantic relatedness.

The major approaches of measuring semantic distance are *Rada et al.'s distance, Sussna's distance* and *Jiang and Conrath's distance*. For the semantic similarity, we find *Leacock and Chodorow's similarity, Wu and Palmer's similarity* and *Lin similarity*, while for semantic relatedness, the most used one is *Hirst and St Onge's relatedness*. See [1] for a comparative study of these measures.

In what follows, we will focus on semantic distances and in particular on the classical *Rada et al.'s distance* [19], which can be replaced by any other semantic distance.

This distance is based on the shortest path between the nodes corresponding to the items being compared such that the shorter the path from one node to another, the more similar they are. Thus, given multiple paths between two concepts, we should take the length of the shortest one. Formally, given two concepts c_i and c_j the Rada et al.'s distance is defined by:

$$dist_{rmbb}(c_i, c_j) = \min_{p \in pths(c_i, c_j)} len_e(p) \tag{3}$$

Consequently, we will compute a distance matrix, giving us all distances between all pairs of concepts.

Table 1. Rada et al.'s distance matrix between the nodes given in Fig.1

	Rt	SF1	SF2	F1	F2	F3	X1	X2	X3	X4	X5
Rt	0	1	1	2	2	2	3	3	3	3	3
SF1	1	0	2	1	1	3	2	2	2	4	2
SF2	1	2	0	3	3	1	4	4	4	2	4
F1	2	1	3	0	2	4	1	1	3	5	3
F2	2	1	3	2	0	4	3	3	1	5	1
F3	2	3	1	4	4	0	5	5	5	1	5
X1	3	2	4	1	3	5	0	2	4	6	4
X2	3	2	4	1	3	5	2	0	4	6	4
X3	3	2	4	3	1	5	4	4	0	6	2
X4	3	4	2	5	5	1	6	6	6	0	6
X5	3	2	4	3	1	5	4	4	2	6	0

Example 2. In figure 1, every term has at least one generalization's path back to the top node. The full Rada et al.'s distance matrix concerning the concept taxonomy is presented in Table 1. For instance, $dist_{rmbb}(SF1, X2) = 2$ since the shortest path between the two concepts SF1 and X2 is composed of the two edges SF1-F1 and F1-X2.

3   Our Approach for Causal Discovery and Visualization Based on Ontological Knowledge

Causal Bayesian networks are used to model domains under uncertainty [17]. In many cases, available ontologies provide consensual representation of the same domain and a full description of the knowledge model. Basically, ontologies attempt to find the complete set of concepts covering any domain as well as the relationships between them. Such domain knowledge can be efficiently used to enrich the learning process of Causal Bayesian networks and optimize the causal discovery.

In order, to illustrate this idea we propose to extend the MyCaDo (My Causal DiscOvery) algorithm [13] used to learn causal structures with experiments by integrating the ontological knowledge, extracted using Rada et al. semantic distance calculation, for causal discovery and visualization.

We make the assumption that we consider a concept taxonomy hierarchy, in which the leaves are also the nodes of the Bayesian network. However, the causal relations we want to discover and model with our CBN did not exist in the corresponding ontology, therefore, each model representation complete the other.

3.1   Integrating Ontological Knowledge for Causal Discovery

In what follows let us assume that after performing an experiment on a variable X_i, we can measure all neighboring variables $Ne_U(X_i)$.
Our utility function $U(.)$ is an extension of the one proposed in [13] (see subsection 2.2) by generalizing the first term $Ne_U(Xi)$ and replacing it by the semantical inertia, denoted by $SemIn(Ne_U(X_i))$.

This notion will enable us to integrate our ontological knowledge extracted by calculating semantic distances from ontologies and guide the choice of the best experiment. In many situations, we can have a node X_i with a high number of neighbors but all those neighbors are very close semantically. This implies that we will have, exactly, two alternatives (i.e. X_i is the direct cause of all his neighbors or the inverse).

For instance, in bio-informatics, biologists would discover more causal relations between biological functions. Considering a neighborhood which is very close semantically will reduce the number of functions under study and consequently the informative contribution of the experiment would be very low. However, in the case of distant neighbors, there will be more cause-to-effect relations between biological functions to find. Such causal discoveries have an important scientific contribution in the bio-informatic field.

The semantical inertia will be as follows:

$$SemIn(M) = \frac{\sum_{X_i \in M} \min_{p \in pths(X_i,\ mscs(M))} (len_e(p))}{card(M)} \quad (4)$$

Let us consider the CPDAG in figure 2 in order to illustrate our approach and compare it to the original MyCaDo. We will also use the concept taxonomy hierarchy presented in figure 1. Suppose that we will perform an action on X_2. Figure 3 summarizes all possible instantiations of the edges $X_i\text{-}Ne(Xi)$.

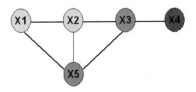

Fig. 2. An example of CPDAG

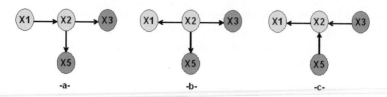

Fig. 3. All possible instantiations for $X_2 - Ne_U(X_2)$

It is clear that the following edges $X_2 - X_3$ and $X_2 - X_5$ have the same direction because the two neighbors X_3 and X_5 share the same function in the corresponding ontology. In this case, we will have three instantiations whereas, by applying MyCaDo we will find exactly six. Consequently, we reduced considerably the set of possible instantiations in the particular PDAG (i.e. CPDAG with some oriented edges).

Here the most specific commun subsumer of $Ne_U(X_2)=\{X_1, X_3, X_5\}$ is the concept SF1. According to the ontology in figure 1 and table 1, the semantical inertia of the nodes X_1, X_3 and X_5 is as follows:

$$
SemIn(X_1, X_3, X_5) = \frac{\sum_{X_j \in Ne_U(X_2)} \min_{p \in pths(X_j, \, mscs(Ne_U(X_2)))} (len_e(p))}{\#Ne_U}
$$

$$
= \frac{2+2+2}{3} = 2
$$

The semantical inertia presents three major characteristics. At the first glance, where all undirected neighboring nodes belong to the same biological function, the semantical inertia of the neighborhood will be equal to one. Secondly, the semantical inertia depends on the number of undirected neighbors. For example, if we eliminate X_1 from the neighborhood of X_2, SemIn will automatically decrease.

$$
SemIn(X_3, X_5) = \frac{1+1}{2} = 1
$$

And, finally, the more the neighboring variables are distant according to the ontology, the more the semantical inertia will be important and the utility maximized. Here, if we replace the node X_3 by X_4, which is more distant from the rest of neighbors of X_2, SemIn will increase considerably.

$$
SemIn(X_1, X_4, X_5) = \frac{3+3+3}{3} = 3
$$

It is clear now that the semantical inertia represent a generalization of $\#Ne_U(Xi)$ and introducing it in the utility function will allow a better choice of experiments, based on ontological knowledge. With such method, we can focus the causal discovery on relations between distant concepts.

3.2   Integrating Ontological Knowledge for Causal Graph Visualization

Visualizing large networks of hundreds or thousands nodes is a real challenging task. It is essentially due to the limitation of the screen, the huge number of nodes and edges and the limitations of human visual abilities. In this context, our approach can turn out of big utility to improve actual visual tools. The main purpose was to propose an enriched visualization of causal models by integrating the power of semantical knowledge. The Rada et al.'s distance matrix can implement different graph drawing algorithms among which MultiDimentional Scalling (MDS) and Force Directed Placement [7,8]. More precisely, we will adjust the node's position in the screen, referring to the matrix distance calculated above. Similarly, in [20], Ranwez et al. used the ontological distance measures to propose an alternative information visualization on conceptual maps.

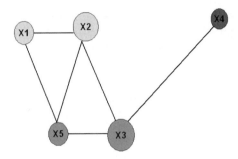

Fig. 4. Enriched visualization corresponding to the CPDAG in Figure 2

In figure 4, we show an enriched visualization of the CPDAG used in the previous subsection. The biological functions are more distinguished and the causal relations between those functions are, visually, more revealed.

In the other hand, we propose to adapt the size of the nodes to the utility function. This method allows to biologists to determine the ideal node to make interventions.

While judicious use of semantical distances and node's size can help considerably the understanding of the causal bayesian network, we can go farther in the visualization by adding a zoom-in/zoom-out function that allows one to visualize either the global structure of the graph or just, smaller components reduced to more general concepts in the ontology. For example, we can pass from visualizing the genes, in our example, to an abstraction reduced to only biological functions or even to the biological super-functions.

All those visualization attributes can lead to different valuable informations about the causal bayesian network. Moreover, we will be able to investigate further into the adapted structure of the network. Since the enriched visualization can offer assistance to the domain experts, we can change the MyCaDo algorithm into semi-automatic method, combining the use of optimal solution of utility and visualization of causal graph. Consequently, we can ease considerably the depiction of causality.

4   Conclusion

In this article, we discussed how ontological knowledge can be useful for iterative causal discovery. More precisely, we extend the MyCaDo algorithm via introducing the notion of the semantical inertia. By supporting the assignment of costs to experiments and measurements, our approach permit us to guide the choice of the best experiment in order to obtain the complete causal bayesian network. We then proposed an enriched visualization of causal models, using the power of semantical knowledge extracted from ontologies.

To this end, we considered a toy example ontology as a starting point to develop our approach but this choice does not exclude the application of our works to more realistic ontology's domains. We emphasize that the work described here represents a major step in a longer-term project focusing on the knowledge integration for causal bayesian networks learning.

Directions for future work include studying how ontological knowledge can be integrated to learn causal graphical models with latent variables, or other links between causal bayesian network learning and ontologies construction.

References

1. Blanchard, E., Harzallah, M., Briand, H., Kuntz, P.: A typology of ontology-based semantic measures. In: 2nd INTEROP-EMOI Open Workshop on Enterprise Models and Ontologies for Interoperability at the 17th Conference on Advanced Information Systems Engineering (CAISE 2005), vol. 160, pp. 407–412. CEUR-WS (2005)
2. Borchani, H., Chaouachi, M., Ben Amor, N.: Learning causal bayesian networks from incomplete observational data and interventions. In: Proceedings of Symbolic and Quantitative Approaches to Reasoning with Uncertainty, pp. 17–29 (2007)
3. Cannataro, M., Massara, A., Veltri, P.: The OnBrowser ontology manager: Managing ontologies on the Grid. In: International Workshop on Semantic Intelligent Middleware for the Web and the Grid, Valencia, Spain (2004)
4. Chickering, D.M.: Optimal Structure Identification With Greedy Search. Journal of Machine Learning Research, 507–554 (2002)
5. Cooper, G.F., Herskovits, E.: A Bayesian Method for the Induction of Probabilistic Networks from Data. Machine Learning, 309–347 (1992)
6. Cooper, G.F., Yoo, C.: Causal discovery from a mixture of experimental and observational data. In: Proceedings of Uncertainty in Artificial Intelligence, pp. 116–125 (1999)
7. Crampes, M., Ranwez, S., Villerd, J., Velickovski, F., Mooney, C., Emery, A., Mille, N.: Concept Maps for Designing Adaptive Knowledge Maps. In: Tergan, S.-O., Keller, T., Burkhard, R. (Guest eds.) Concept Maps, A Special Issue of Information Visualization, vol. 5(3). Palgrave - Macmillan (2006)

8. Crampes, M., Ranwez, S., Velickovski, F., Mooney, C., Mille, N.: An integrated visual approach for music indexing and dynamic playlist composition. In: Proceedings of 13th Annual Multimedia Computing and Networking (MMCN 2006), San Jose, CA, US (2006)

9. Eberhardt, F., Glymour, C., Scheines, R.: N-1 experiments suffice to determine the causal relations among n variables. Technical report, Carnegie Mellon University (2005)

10. Gruber, T.R.: Toward Principles for the Design of Ontologies Used for Knowledge Sharing. International Journal Human-Computer Studies 43, 907–928 (1993)

11. Maes, S., Meganck, S., Leray, P.: An integral approach to causal inference with latent variables. In: Russo, F., Williamson, J. (eds.) Causality and Probability in the Sciences (Texts in Philosophy), pp. 17–41. College Publications (2007)

12. Mani, S., Cooper, G.F.: Causal discovery using a Bayesian local causal discovery algorithm. In: Proceedings of MedInfo, pp. 731–735. IOS Press, Amsterdam (2004)

13. Meganck, S., Leray, P., Manderick, B.: Learning causal bayesian networks from observations and experiments: A decision theoritic approach. In: Torra, V., Narukawa, Y., Valls, A., Domingo-Ferrer, J. (eds.) MDAI 2006. LNCS, vol. 3885, pp. 58–69. Springer, Heidelberg (2006)

14. Meganck, S., Maes, S., Leray, P., Manderick, B.: Learning semi-markovian causal models using experiments. In: Proceedings of The third European Workshop on Probabilistic Graphical Models, PGM 2006, pp. 195–206 (2006)

15. Meganck, S., Leray, P., Manderick, B.: Causal graphical models with latent variables: Learning and inference. In: Mellouli, K. (ed.) ECSQARU 2007. LNCS, vol. 4724, pp. 5–16. Springer, Heidelberg (2007)

16. Pan, R., Ding, Z., Yu, Y., Peng, Y.: A Bayesian Network Approach to Ontology Mapping. In: Gil, Y., Motta, E., Benjamins, V.R., Musen, M.A. (eds.) ISWC 2005. LNCS, vol. 3729, pp. 563–577. Springer, Heidelberg (2005)

17. Pearl, J.: Probabilistic reasoning in intelligent systems: networks of plausible inference. Morgan Kaufmann Publishers, San Francisco (1988)

18. Pearl, J.: Causality: Models, reasoning and inference. MIT Press, Cambridge (2000)

19. Rada, R., Mili, H., Bicknell, E., Blettner, M.: Development and application of a metric on semantic nets. IEEE Transactions on Systems, Man and Cybernetics 19, 17–30 (1989)

20. Ranwez, S., Ranwez, V., Villerd, J., Crampes, M.: Ontological Distance Measures for Information Visualization on Conceptual Maps. In: Meersman, R., Tari, Z., Herrero, P. (eds.) OTM 2006 Workshops. LNCS, vol. 4278, pp. 1050–1061. Springer, Heidelberg (2006)

21. Spirtes, P., Glymour, C., Scheines, R.: Causation, Prediction and Search. MIT Press, Cambridge (2000)

22. Sváb, O., Svátek, V.: Combining Ontology Mapping Methods Using Bayesian Networks. In: Workshop on Ontology Matching at ISWC (2006)

23. Tong, S., Koller, D.: Active learning for structure in bayesian networks. In: Proceedings of the Seventeenth International Joint Conference on Artificial Intelligence (IJCAI), pp. 863–869 (2001)

24. Verma, T., Pearl, J.: Equivalence and Synthesis of Causal Models. In: Proceedings of the Sixth Conference on Uncertainty in Artificial Intelligence (1990)

25. The Gene Ontology Consortium. Gene Ontology: tool for the unification of biology, Nature Genet., 25–29 (2000)

Binary Probability Trees for Bayesian Networks Inference

Andrés Cano, Manuel Gómez-Olmedo, and Serafín Moral

Dept. Computer Science and Artificial Intelligence,
University of Granada, Granada, 18071, Spain
{acu,mgomez,smc}@decsai.ugr.es

Abstract. The present paper introduces a new kind of representation for the potentials in a Bayesian network: Binary Probability Trees. They allow to represent finer grain context-specific independences than those which can be encoded with probability trees. This enhanced capability leads to more efficient inference algorithms in some types of Bayesian networks. The paper explains how to build a binary tree from a given potential with a similar procedure to the one employed for probability trees. It also offers a way of pruning a binary tree if exact inference cannot be performed with exact trees, and provides detailed algorithms for performing directly with binary trees the basic operations on potentials (restriction, combination and marginalization). Finally, some experiments are shown that use binary trees with the variable elimination algorithm to compare the performance with standard probability trees.

Keywords: Bayesian networks inference, approximate computation, variable elimination algorithm, deterministic algorithms, probability trees.

1 Introduction

Bayesian networks are graphical models which can be used to handle uncertainty in probabilistic expert systems. They enable efficient representation of joint probability distributions. It is known that exact computation [1] of the posterior probabilities, given certain evidence, may become unfeasible in large networks. This has led to the proposal of different approximate algorithms. They provide results in shorter time (albeit inexact). Some of the methods are based on Monte Carlo simulation, and others rely on deterministic procedures. Among the deterministic methods, one can find those that use alternative representations for potentials, such as *probability trees* [2,3,4]. This representation offers the chance to take advantage of *context-specific independences*. Probability trees can be pruned and converted into smaller ones when potentials are too large, thus providing approximate algorithms. In the present paper, we introduce a new kind of probability trees in which the internal nodes always have two children. They will be called *binary probability trees*. These trees are capable of specifying context-specific independences with finer grain than those which can be

C. Sossai and G. Chemello (Eds.): ECSQARU 2009, LNAI 5590, pp. 180–191, 2009.

represented with standard trees, and should work better than standard probability trees when the Bayesian networks contain variables with a large number of states.

The remainder of the paper is organized as follows: In Section 2 we describe the problem of probabilities propagation in Bayesian networks. Section 3 studies the use of probability trees to represent potentials compactly and presents the related notation. In Section 4, we introduce binary probability trees and describe how they can be built from a potential, and how to approximate them by pruning terminal trees; we also show the algorithms for direct application of the basic operations with potentials to binary probability trees. Section 5 provides details of the experimental work. And finally, Section 6 gives the conclusions and future work.

2   Probability Propagation in Bayesian Networks

Let $\mathbf{X} = \{X_1, \ldots, X_n\}$ be a set of variables. Let us assume that each variable X_i takes values on a finite set of states Ω_{X_i} (the domain of X_i). We shall use x_i to denote one of the values of X_i, $x_i \in \Omega_{X_i}$. If I is a set of indices, we shall write $\mathbf{X}_I$ for the set $\{X_i | i \in I\}$. $N = \{1, \ldots, n\}$ will denote the set of indices of all the variables in the network; thus $\mathbf{X} = \mathbf{X}_N$. The Cartesian product $\times_{i \in I}\Omega_{X_i}$ will be denoted by $\Omega_{\mathbf{X}_I}$. The elements of $\Omega_{\mathbf{X}_I}$ are called configurations of $\mathbf{X}_I$ and will be written with $\mathbf{x}$ or $\mathbf{x}_I$. We denote $\mathbf{x}_I^{\downarrow \mathbf{X}_J}$ to the projection of the configuration $\mathbf{x}_I$ to the set of variables $\mathbf{X}_J$, $\mathbf{X}_J \subseteq \mathbf{X}_I$.

A mapping from a set $\Omega_{\mathbf{X}_I}$ into $\mathbb{R}_0^+$ will be called a potential p for $\mathbf{X}_I$. Given a potential p, we denote with $s(p)$ to the set of variables for which p is defined. The process of inference in probabilistic graphical models requires the definition of two operations on potentials: combination $p_1 \otimes p_2$ (multiplication) and marginalization $p^{\downarrow J}$ (by summing out all the variables not in $\mathbf{X}_J$).

A Bayesian network is a directed acyclic graph, where each node represents a random event X_i, and the topology of the graph shows the independence relations between variables according to the d-separation criterion [5]. Each node X_i also has a conditional probability distribution $p_i(X_i | \Pi(X_i))$ for that variable, given its parents $\Pi(X_i)$. A Bayesian network determines a joint probability distribution:

$$p(\mathbf{X} = \mathbf{x}) = \prod_{i \in N} p_i(x_i | \Pi(x_i)) \quad \forall \mathbf{x} \in \Omega_{\mathbf{X}} \tag{1}$$

Let $\mathbf{E} \subset \mathbf{X}_N$ be the set of observed variables and $\mathbf{e} \in \Omega_{\mathbf{E}}$ the instantiated value. An algorithm that computes the posterior distributions $p(x_i | \mathbf{e})$ for each $x_i \in \Omega_{X_i}$, $X_i \in \mathbf{X}_N \setminus \mathbf{E}$ is called propagation algorithm or inference algorithm.

3   Probability Trees

Probability trees [6] have been used as a flexible data structure that enables the specification of context-specific independences (see [4]) as well as using exact or approximate representations of probability potentials. A *probability tree* $\mathcal{T}$ is a directed labelled tree, in which each internal node represents a variable and each leaf represents a non-negative real number. Each internal node has one outgoing arc for each state of the variable that labels that node; each state labels one arc.

A probability tree $\mathcal{T}$ on variables $\mathbf{X}_I = \{X_i | i \in I\}$ represents a potential $p : \Omega_{\mathbf{X}_I} \to \mathbb{R}_0^+$ if for each $\mathbf{x}_I \in \Omega_{\mathbf{X}_I}$ the value $p(\mathbf{x}_I)$ is the number stored in the leaf node that is reached by starting from the root node and selecting the child corresponding to coordinate x_i for each internal node labelled X_i. We use L_t to denote the *label of node* t (a variable for an internal node, and a real number for a leaf). We say that a subtree of $\mathcal{T}$ is a *terminal tree* if it contains only one node labelled with a variable, and all the children are numbers (leaf nodes).

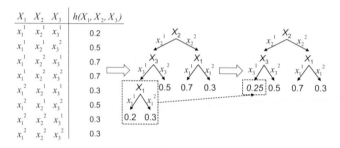

Fig. 1. Potential p, its representation as a probability tree and its approximation after pruning various branches

A probability tree is usually a more compact representation of a potential than a table. This is illustrated in Fig. 1, which displays a potential p and its representation, using a probability tree. The tree contains the same information as the table, but only requires five values rather than eight. Furthermore, trees enable even more compact representations in exchange for loss of accuracy. This is achieved by pruning certain leaves and replacing them with the average value, as shown in the right tree in Fig. 1.

If $\mathcal{T}$ is a probability tree on $\mathbf{X}_I$ and $\mathbf{X}_J \subseteq \mathbf{X}_I$, we use $\mathcal{T}^{R(\mathbf{x}_J)}$ (probability tree restricted to the configuration $\mathbf{x}_J$) to denote the *restriction operation* which consists of returning the part of the tree which is consistent with the values of the configuration $\mathbf{x}_J \in \Omega_{\mathbf{X}_J}$. For example, in the left probability tree in Figure 1, $\mathcal{T}^{R(X_2=x_2^1, X_3=x_3^1)}$ represents the terminal tree enclosed by the dashed line square. This operation is used to define combination and marginalization operations as well as for conditioning.

We use LP_t to denote the *labelling of the branches* from root to another node t (not necessarily to a leaf). A labelling LP_t defines a configuration $\mathbf{x}^t$ for the variables in $\mathbf{X}_I^t, \mathbf{X}_I^t \subseteq \mathbf{X}_I$, where $\mathbf{X}_I$ is the set of variables of the potential represented by the probability tree, and $\mathbf{X}_I^t$ is the set of variables that labels the internal nodes contained in the path from the root node to descendant node t (excluding node t). We say that $\mathbf{x}^t$ is the *associated configuration* for node t.

The basic operations (*combination, marginalization*) in potentials can be performed directly in probability trees ([6]).

4   Binary Probability Trees

A *binary probability tree* $\mathcal{BT}$ is similar to a probability tree. It is also a directed labelled tree, where each internal node is labelled with a variable, and each leaf is labelled with a non-negative real number. It also allows a potential for a set of variables $\mathbf{X}_I$ to be represented. But now, each internal node has always two outgoing arcs, and a variable can appear more than once labelling the nodes in the path from the root to a leaf node. Another difference is that, for an internal node labelled with X_i, the outgoing arcs can generally be labelled with more than one state of the domain of X_i, Ω_{X_i}.

At a given node t of $\mathcal{BT}$, labelled with variable X_i, we denote with $\Omega^t_{X_i}$, $\Omega^t_{X_i} \subseteq \Omega_{X_i}$, the set of *available states* of X_i at node t. In general, this set is a subset of Ω_{X_i}. The available states of X_i at node t will be distributed between two subsets, in order to label its two outgoing arcs. We denote with $L_{lb(t)}$ and $L_{rb(t)}$ the labels (two subsets of $\Omega^t_{X_i}$) of the left and right branches of t. We denote with t_l and t_r the two children of t.

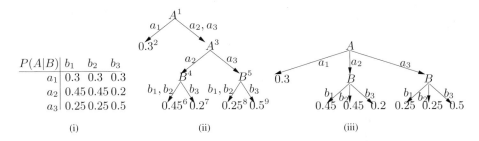

Fig. 2. Potential $P(A|B)$ as a table, as binary probability tree and as probability tree

For example, Fig. 2 (ii) shows a binary probability tree for the table in (i). In the figure we use a superscript number at each node of the tree, in order to easily identify them. The domain of A, Ω_A, is the set of states $\{a_1, a_2, a_3\}$. At root node (node 1, labelled with A), we have $\Omega^1_A = \{a_1, a_2, a_3\}$. That is, the available states of A at the root node coincides with its domain. The left branch of node 1 is labelled with $\{a_1\}$ and the right branch with $\{a_2, a_3\}$. At node 3 (also labelled with A) $\Omega^3_A = \{a_2, a_3\}$. This potential can also be represented with the probability tree in Fig. 2 (iii). It can be seen that the binary probability tree contains only five leaves, whereas the probability tree contains seven.

Another difference with probability trees is that the labelling LP_t of a path from the root to a descendant node t now determines an *extended configuration* $\mathbf{A}_{\mathbf{X}^t_I}$ for the variables in $\mathbf{X}^t_I, \mathbf{X}^t_I \subseteq \mathbf{X}_I$, rather than a standard configuration $\mathbf{x}^t$. This new concept is required in binary probability trees in order to express that a variable X_i in $\mathbf{X}_I$ belongs to a subset of Ω_{X_i}, instead of stating that $X_i = x_i$. Extended configurations will be denoted with $\mathbf{A}$ or $\mathbf{A_{X_I}}$. Thus, an extended configuration $\mathbf{A_{X_I}}$ defines a set of configurations $S_{\mathbf{A_{X_I}}}$ for $\mathbf{X}_I$, which is obtained with the Cartesian product of the subsets of states in $\mathbf{A_{X_I}}$. For example, an extended configuration for the set of variables $\{A, B\}$ could be $\{\{a_3\}, \{b_1, b_2\}\}$. This means that A is a_3 and B can be b_1 or b_2. Therefore, it determines the set of configurations $\{\{a_3, b_1\}, \{a_3, b_2\}\}$. This extended configuration corresponds with the labelling of the path from the root to node 8 in Fig. 2 (ii). $\mathbf{A}^t$ denotes the *associated extended configuration* for node t.

4.1   Constructing a Binary Probability Tree

Cano, Moral and Salmerón [7,6] proposed a methodology for constructing a probability tree $\mathcal{T}$ from a potential p. It was inspired by the methods for inducing classification trees, such as Quinlan's ID3 algorithm [8], which builds a *decision tree* from a set of examples. A decision tree represents a sequential procedure for deciding the class membership of a given instance of the attributes of the problem. That is, the leaves of decision trees provide the class for given instances of the attributes. However a leaf in a probability tree contains a probability value. This means that the measure used as the *splitting criterion* in probability trees was particularly adapted to probabilities. For binary probability trees, we follow a similar methodology, although the *splitting criterion* will be adapted to them.

Let p be a potential for a set of variables $\mathbf{X}_I$. In general, it is possible to obtain several binary probability trees representing potential p, depending on the order we place the variables of $\mathbf{X}_I$ in the internal nodes of the tree, and how we distribute at each internal node, the available states of the variable between its outgoing arcs.

There is a need to extend the definition of *restriction* operation in probability trees to an arbitrary potential p: If p is a potential for $\mathbf{X}_I$ and $\mathbf{x}_J$ is a configuration for $\mathbf{X}_J$, $\mathbf{X}_J \subseteq \mathbf{X}_I$, we denote with $p^{R(\mathbf{x}_J)}$ the potential p *restricted* to configuration $\mathbf{x}_J$, which consists of returning the part of the potential consistent with $\mathbf{x}_J$. Furthermore, we denote with $p^{R(\mathbf{A_{x_J}})}$ the potential p *restricted* to the extended configuration $\mathbf{A_{X_J}}$, which consists of returning the part of the potential consistent with one of the configurations in $S_{\mathbf{A_{x_J}}}$. For example, restricting the potential in Fig. 2 (i) to $A = a_1$ produces a potential with only the first row in that table; restricting to the extended configuration $\{\{a_2\}, \{b_1, b_2\}\}$ produces a potential with the first two numbers in the second row. We denote with $|p^{R(\mathbf{x}_J)}|$ or $|p^{R(\mathbf{A_{x_J}})}|$ the number of values in the restricted potential.

The proposed methodology for constructing a binary probability tree $\mathcal{BT}$ from a given potential p for the set of variables $\mathbf{X}_I$ is very similar to the one used to build a probability tree (see [7,6]). The process begins with a binary probability tree $\mathcal{BT}_0$ with only one node (a leaf node) labelled with the average of values in the potential: $L_t = \sum_{\mathbf{x}_I \in \Omega_{\mathbf{X}_I}} p(\mathbf{x}_I)/|\Omega_{\mathbf{X}_I}|$.

A greedy step is then applied successively until we obtain an exact binary probability tree, or until a given *stop criterion* is satisfied. At each step, a new binary tree $\mathcal{BT}_{j+1}$ is obtained from the previous one $\mathcal{BT}_j$. The greedy step requires the definition of a *splitting criterion*. It consists of expanding one of the leaf nodes t in $\mathcal{BT}_j$ with a terminal tree (with t rooting the terminal tree, and t_l and t_r as children of t). Node t will be labelled with one of the *candidate variables*. The set of available states $\Omega^t_{X_i}$ of the chosen candidate variable X_i at node t, will be partitioned into two subsets $\Omega^{t_l}_{X_i}$ and $\Omega^{t_r}_{X_i}$, to label the two outgoing arcs (left and right) of t. The two leaf nodes t_l and t_r in the new terminal tree will be labelled with the average of values of p consistent with the extended configurations $\mathbf{A}^{t_l}$ and $\mathbf{A}^{t_r}$ respectively (the associated extended configurations for nodes t_l and t_r).

After applying the previous process, we say that the leaf node t has been expanded with variable X_i and the sets of states $\Omega^{t_l}_{X_i}$ and $\Omega^{t_r}_{X_i}$. The result of applying previous splitting to $\mathcal{BT}$ will be denoted with $\mathcal{BT}(t, X_i, \Omega^{t_l}_{X_i}, \Omega^{t_r}_{X_i})$. For example, the binary probability tree in Fig. 2 (i) was built by selecting A in the first splitting (at root node), and the sets of states $\Omega^{t_l}_A = \{a_1\}$ and $\Omega^{t_r}_A = \{a_2, a_3\}$ to label the left and right outgoing arcs. A variable $X_i \in \mathbf{X}_I$ can be a candidate variable to expand a leaf node t, if it contains more than one state in its set $\Omega^t_{X_i}$ of available states at node t.

The definition of the splitting criterion, requires a distance to measure the goodness of the approximation of a binary probability tree $\mathcal{BT}$ to a given potential p. If we denote by $d_{\mathcal{BT}}$ and d_p the probability distributions proportional to $\mathcal{BT}$ and p, respectively, then the *distance* from a binary tree $\mathcal{BT}$ to a potential p is measured with the Kullback-Leibler's divergence [9]:

$$D(p, \mathcal{BT}) = \sum_{\mathbf{x}_I \in \Omega_{\mathbf{x}_I}} d_p(\mathbf{x}_I) \log \frac{d_p(\mathbf{x}_I)}{d_{\mathcal{BT}}(\mathbf{x}_I)}. \qquad (2)$$

Kullback-Leibler's divergence is always positive or zero. It is equal to zero if $\mathcal{BT}$ represents exactly potential p. In the definition of the splitting criterion, we propose following the same methodology to construct probability trees (see [6]), but adapting it to binary probability trees.

Definition 1 (Splitting criterion). *Let p be the potential we are constructing and $\mathcal{BT}_j$ the binary tree in step j of the greedy algorithm and t a leaf node, then node t can be expanded with the candidate variable X_i and a partition of its available states $\Omega^t_{X_i}$ into sets $\Omega^{t_l}_{X_i}$ and $\Omega^{t_r}_{X_i}$ if X_i and the partition of $\Omega^t_{X_i}$ maximizes the following expression:*

$$I(t, X_i, \Omega^{t_l}_{X_i}, \Omega^{t_r}_{X_i}) = D(p, \mathcal{BT}_j) - D(p, \mathcal{BT}_j(t, X_i, \Omega^{t_l}_{X_i}, \Omega^{t_r}_{X_i})) \qquad (3)$$

This expression represents the information gain obtained in the current binary tree $\mathcal{BT}_j$ after performing the mentioned expansion on leaf node t. It is clear to see that $I(t, X_i, \Omega^{t_l}_{X_i}, \Omega^{t_r}_{X_i}) \geq 0$. By maximizing $I(t, X_i, \Omega^{t_l}_{X_i}, \Omega^{t_r}_{X_i})$, we manage to minimize Kullback-Leibler's distance to potential p.

In our experiments (Section 5) we will not check every possible partition of $\Omega^t_{X_i}$ into $\Omega^{t_l}_{X_i}$ and $\Omega^{t_r}_{X_i}$, because this would be a very time-consuming task. Assuming that the set of available states for X_i at node t are ordered,

$\Omega^t_{X_i} = \{x_1, \dots, x_n\}$ we will only check partitions of $\Omega^t_{X_i}$ into subsets with consecutive states, $\Omega^{t_l}_{X_i} = \{x_1, \dots, x_j\}$ and $\Omega^{t_r}_{X_i} = \{x_{j+1}, \dots, x_n\}$, for each $j \in [1, n-1]$.

Proposition 1. *The information gain (expression (3)) obtained by expanding node t with variable X_i and partition of its set of available states into $\Omega^{t_l}_{X_i}$ and $\Omega^{t_r}_{X_i}$ can be calculated in the following way:*

$$I(t, X_i, \Omega^{t_l}_{X_i}, \Omega^{t_r}_{X_i}) = sum(p^{R(\mathbf{A}^t)}) \cdot \log(|\Omega^t_{X_i}| / sum(p^{R(\mathbf{A}^t)}))$$
$$+ sum(p^{R(\mathbf{A}^{t_l})}) \cdot \log(sum(p^{R(\mathbf{A}^{t_l})}) / |\Omega^{t_l}_{X_i}|)$$
$$+ sum(p^{R(\mathbf{A}^{t_r})}) \cdot \log(sum(p^{R(\mathbf{A}^{t_r})}) / |\Omega^{t_r}_{X_i}|) \qquad (4)$$

where $sum(q)$ is the addition of all the values of potential q.

Due to space restrictions we do not include the proof of this proposition. It should be noted that the information only depends on the values of the potential consistent with node t, and it can therefore be locally computed. The methodology explained in this section for building a binary probability tree can also be used to reorder the variables or the split sets of a binary tree resulting from an operation of combination or marginalization.

4.2   Pruning Binary Probability Trees

If we need to reduce the size of a binary probability tree, we can apply the same process *pruning* as in the case of probability trees. Again, a *pruning* in a binary probability tree consists of replacing a terminal tree by the average of values that the terminal tree represents. For example, if we wish to prune the terminal tree rooted by node 4 in the binary tree of Fig. 2 (ii), we must replace it by $(0.45 + 0.45 + 0.2)/3$. The following definition shows when a terminal tree can be pruned.

Definition 2 (Pruning of a terminal tree). *Let $\mathcal{BT}$ be a binary probability tree, t the root of a terminal tree labelled with variable X_i, t_l and t_r its child nodes, $\Omega^{t_l}_{X_i}$ and $\Omega^{t_r}_{X_i}$ the sets of states that label left and right branches, respectively, and Δ a given threshold, $\Delta \geq 0$, then that terminal tree can be pruned if:*

$$I(t, X_i, \Omega^{t_l}_{X_i}, \Omega^{t_r}_{X_i}) \leq \Delta \qquad (5)$$

In previous definition, I is calculated using expression (4). In this case, potential p in expression (4) is the binary tree $\mathcal{BT}$ to be pruned. Again, the information can be locally computed at node t in the current binary tree. The goal of the pruning involves detecting leaves that can be replaced by one value without a big increment in Kullback-Leibler's divergence of the potential that $\mathcal{BT}$ represents before and after pruning. Here, I is considered as the *information loss* produced in the current binary tree if the terminal tree rooted by node t is pruned. The pruning process would finish when there are no more terminal trees in $\mathcal{BT}$ verifying condition (5). If $\Delta = 0$, an exact binary probability tree will be obtained: a terminal tree t will be pruned only if $L_{t_l} = L_{t_r}$.

4.3   Operations with Binary Probability Trees

Inference algorithms require three operations with potentials: *restriction, combination* and *marginalization*. Herein we describe these for binary probability trees.

The *restriction* operation is trivial. If $\mathcal{BT}$ is a binary probability tree and $\mathbf{x}_J$ a configuration for the set of variables $\mathbf{X}_J$, $\mathcal{BT}^{R(\mathbf{x}_J)}$ ($\mathcal{BT}$ restricted to configuration $\mathbf{x}_J$) is obtained with the same algorithm as the one for probability trees: Replace in $\mathcal{BT}$ every node t labelled with X_k, $X_k \in X_J$, by subtree $\mathcal{BT}_k$, children of t, corresponding to the value of X_k in $\mathbf{x}_J$. For binary trees, we need to extend the definition of *restriction* operation, to an extended configurations $\mathbf{A}_{\mathbf{X}_J}$, which means returning the part of the tree consistent with one of the configurations in set $S_{\mathbf{A}_{\mathbf{X}_J}}$. This operation is easy to specify if we first define the restriction of $\mathcal{BT}$ to a set of states S_{X_j} ($S_{X_j} \subseteq \Omega_{X_j}$) of variable X_j, denoted by $\mathcal{BT}^{R(S_{X_j})}$. It can be obtained with Algorithm 1. As an example, Fig. 3 shows the tree of Fig. 2 (ii) restricted to $A \in \{a_1, a_2\}$.

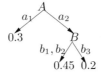

Fig. 3. Restriction of tree of Fig. 2 (ii) to $A \in \{a_1, a_2\}$

Input   : t (root node of $\mathcal{BT}$); X_j (variable to restrict); S_{X_j} (set of states of X_j to restrict)
Output: The root of $\mathcal{BT}^{R(S_{X_j})}$

if t *is not a leaf node* **then**
  if $L_t == X_j$ **then**
    Set $S_{X_j}^l = L_{lb(t)} \cap S_{X_j}$ and $S_{X_j}^r = L_{rb(t)} \cap S_{X_j}$
    if $S_{X_j}^l == \emptyset$ **then**
      return $Restrict(t_r, X_j, S_{X_j}^r)$
    else if $S_{X_j}^r == \emptyset$ **then**
      return $Restrict(t_l, X_j, S_{X_j}^l)$
    else
      Set $L_{lb(t)} = S_{X_j}^l$, $L_{rb(t)} = S_{X_j}^r$ the new labels of the branches of t
      Set $Restrict(t_l, X_j, S_{X_j}^l)$ as the new left child of t
      Set $Restrict(t_r, X_j, S_{X_j}^r)$ as the new right child of t
  else
    Set $Restrict(t_l, X_j, S_{X_j})$ as the new left child of t
    Set $Restrict(t_r, X_j, S_{X_j})$ as the new right child of t
return t

Algorithm 1. Restrict

The restriction of a binary tree to an extended configuration $\mathbf{A_{X_J}}$, $\mathcal{BT}^{R(\mathbf{A_{X_J}})}$, can be performed by repeating Algorithm 1 for each one of the variables in $\mathbf{X}_J$. The combination of two probability trees $\mathcal{BT}1$ and $\mathcal{BT}2$, $\mathcal{BT}1 \otimes \mathcal{BT}2$, can be achieved with Algorithm 2. The combination process is illustrated in Fig. 4.

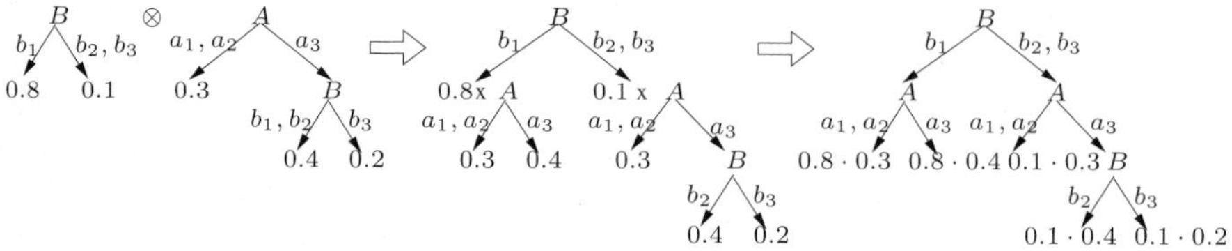

Fig. 4. Combination of two binary trees

Input : $t1$ and $t2$ (root nodes of $\mathcal{BT}1$ and $\mathcal{BT}2$)
Output: The root of $\mathcal{BT} = \mathcal{BT}1 \otimes \mathcal{BT}2$

Build a new node t
if $t1$ *is a leaf node* **then**
   if $t2$ *is a leaf node* **then**
       $L_t = L_{t1} \cdot L_{t2}$
   else
      Set $L_t = L_{t2}$ the label of t
      Set $L_{lb(t)} = L_{lb(t2)}$ and $L_{rb(t)} = L_{rb(t2)}$ labels of the two branches of t
      Set Combine($t1,t2_l$) the left child of t
      Set Combine($t1,t2_r$) the right child of t

else
    $X_i = L_{t1}$
   Set $L_t = L_{t1}$ the label of t
   Set $L_{lb(t)} = L_{lb(t1)}$ and $L_{rb(t)} = L_{rb(t1)}$ labels of the two branches of t
   Set Combine($t1_l, \mathcal{BT}2^{R(X_i,L_{lb(t1)})}$) the left child of t
   Set Combine($t1_r, \mathcal{BT}2^{R(X_i,L_{rb(t1)})}$) the right child of t
return t

Algorithm 2. Combine

Given a binary tree $\mathcal{BT}$ representing a potential p defined for a set of variables $\mathbf{X}_I$, Algorithm 3 obtains a binary tree $\mathcal{BT}^{\downarrow \mathbf{X}_I \setminus \{X_j\}}$ for potential $p^{\downarrow \mathbf{X}_I \setminus \{X_j\}}$. This algorithm must be called using $|\Omega_{X_j}|$ as the input parameter f. In recursive calls to the algorithm, f will be set to the number of available states of X_j at current node of the tree. This algorithm uses the $Sum(\mathcal{BT}1, \mathcal{BT}2)$ algorithm, which is not included here, but which contains the same steps as the *Combine* algorithm, replacing product by addition. The process of marginalization is illustrated in Fig. 5.

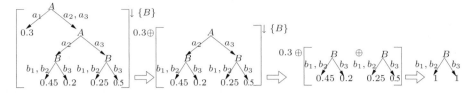

Fig. 5. Marginalizing out variable A

Input : t(root node of $\mathcal{BT}$); X_j (variable to remove); f (a factor for multiplying the labels of leaf nodes)
Output: The root of $\mathcal{BT}^{\downarrow \mathbf{X}_I \setminus X_j}$

if t *is a leaf node* **then**
   Build a new node tn
   Set $L_{tn} = L_t \cdot f$ the label of tn
else
   if $L_t == X_j$ **then**
      $t1$=Marginalize(t_l,X_j,$|L_{lb(t)}|$)
      $t2$=Marginalize(t_r,X_j,$|L_{rb(t)}|$)
      tn=Sum($t1$,$t2$)
   else
      Build a new node tn
      Set $L_{tn} = L_t$ the label of tn
      Set $L_{lb(tn)} = L_{lb(t)}$; $L_{rb(tn)} = L_{rb(t)}$ the labels of the branches of tn
      Set Marginalize(t_l,X_j,f) the left child of tn
      Set Marginalize(t_r,X_j,f) the right child of tn
return tn

Algorithm 3. Marginalize

5 Experiments

In order to compare standard probability trees with binary trees, we conducted some experiments using the Alarm network. In the case of binary trees, we obtained the posterior distribution for each variable in the network using the adapted version of Variable Elimination displayed in Algorithm 4. For standard probability trees, an equivalent algorithm was used, but this was based upon a different way of computing the *information gain* (expression (5)) when reordering and pruning a tree (see [6]). Each run of Algorithm 4 obtains an approximate posterior distribution for one variable. The accuracy of the approximation is controlled by parameter Δ (see expression (5)). We repeated it for each variable in the network and for each value of $\Delta \in [0.0, 0.1]$ with increments of 0.002, using the two versions (standard probability trees and binary trees). In each trial, we calculated computing time, error and largest tree size. The error for one variable X_i was computed with Kullback-Leibler's divergence (expression 2) of the exact posterior distribution respect to the approximate one. The global error for all the variables in the network is measured with the average of Kullback-Leibler's

Input : $\mathbf{P} = \{p_i : i = 1, \ldots, n\}$ the set of potentials of a Bayesian network; $\mathbf{e}$ the set of observed values, $\mathbf{e} \in \Omega_{\mathbf{E}}$; a variable of interest Z, $Z \in \mathbf{X}_N \setminus \mathbf{E}$; and Δ the threshold for pruning
Output: $p(z_i|\mathbf{e})$ for each $z_i \in \Omega_Z$, $Z \in \mathbf{X}_N \setminus \mathbf{E}$

1 Get the set $S_{\mathcal{BT}}$ of binary trees transforming each p_i into $\mathcal{BT}_i$ (Section 4.1)
2 Transform each $\mathcal{BT}_i$ into $\mathcal{BT}_i^{R(\mathbf{e})}$ (restrict to evidence): Algorithm 1
3 Prune each $\mathcal{BT}_i$ with Δ threshold (Section 4.2)
4 **foreach** $Y \in \mathbf{X}_N \setminus (\mathbf{E} \cup Z)$ **do**
5 $\quad$ Let be $\mathbf{S}_Y = \{\mathcal{BT}_i | Y \in s(\mathcal{BT}_i)\}$
6 $\quad$ Calculate $\mathcal{BT}_{prod} = \prod_{\mathcal{BT}_i \in \mathbf{S}_Y} \mathcal{BT}_i$: Algorithm 2
7 $\quad$ Calculate $\mathcal{BT}_{sum} = \mathcal{BT}_{prod}^{\downarrow s(\mathcal{BT}_{prod}) \setminus Y}$: Algorithm 3
8 $\quad$ Reorder variables and split sets in $\mathcal{BT}_{sum}$: (Section 4.1)
9 $\quad$ Prune $\mathcal{BT}_{sum}$ (Section 4.2)
10 $\quad$ $S_{\mathcal{BT}} = \{(S_{\mathcal{BT}} \setminus \mathbf{S}_Y)\} \cup \mathcal{BT}_{sum}$
11 Calculate $\mathcal{BT} = \prod_{\mathcal{BT}_i \in S_{\mathcal{BT}}} \mathcal{BT}_i$
12 Normalize $\mathcal{BT}$
13 **return** $\mathcal{BT}$

Algorithm 4. Variable Elimination

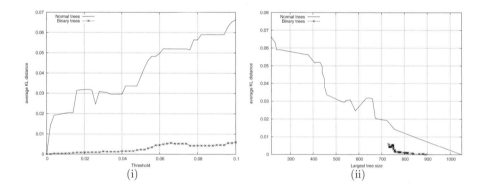

Fig. 6. Average KL errors versus pruning threshold and maximum tree size

divergences. Fig. 6 reports the comparison of the error of the approximations, using standard probability trees and binary trees for each value of Δ, as well as the comparison of error versus largest tree size. Computing times are not shown due to space requirements, but they are very similar in both cases. From such figures, we can conclude that we need to use smaller trees to obtain better approximations if we propagate with binary trees. This makes that propagation algorithms that use binary trees are more efficient than those that use standard trees.

6   Conclusions and Future Work

In the present paper, we have introduced a new type of probability trees for representing the potentials in a Bayesian network: binary probability trees. Experiments show that binary trees are a promising tool, because they can achieve better approximations of posterior probabilities than probability trees, in some Bayesian networks with a similar computing time. The accuracy of the approximation can be controlled with a Δ parameter, which is used in the process of pruning the trees. Propagation with binary trees (instead of standard probability trees) requires smaller trees and offers smaller errors (KL divergence). This confirm that binary trees are able to capture finer grain context-specific independences than probability trees, making possible to use them during propagation.

As regards future directions of research, we intend to study the behaviour of binary trees in other Bayesian networks with more variables, and more complex structures. Moreover, we will adapt other propagation algorithms for Bayesian networks to use binary trees. We also believe that binary trees can be applied to other problems, such as influence diagrams evaluation, propagation in credal networks and supervised classification.

Acknowledgments. This research was jointly supported by the Spanish Ministry of Education and Science under the project TIN2007-67418-C03-03, by the European Regional Development Fund (FEDER), and by the Spanish research programme Consolider Ingenio 2010: MIPRCV (CSD2007-00018).

References

1. Cooper, G.F.: The computational complexity of probabilistic inference using Bayesian belief networks. Artificial Intelligence 42, 393–405 (1990)
2. Cano, A., Moral, S., Salmerón, A.: Penniless propagation in join trees. International Journal of Intelligent Systems 15(11), 1027–1059 (2000)
3. Kozlov, D., Koller, D.: Nonuniform dynamic discretization in hybrid networks. In: Geiger, D., Shenoy, P. (eds.) Proceedings of the 13th Conference on Uncertainty in Artificial Intelligence, pp. 302–313. Morgan & Kaufmann, San Francisco (1997)
4. Boutilier, C., Friedman, N., Goldszmidt, M., Koller, D.: Context-specific independence in Bayesian networks. In: Proceedings of the Twelfth Annual Conference on Uncertainty in Artificial Intelligence (UAI 1996), Portland, Oregon, pp. 115–123 (1996)
5. Pearl, J.: Probabilistic Reasoning with Intelligent Systems. Morgan & Kaufman, San Mateo (1988)
6. Salmerón, A., Cano, A., Moral, S.: Importance sampling in Bayesian networks using probability trees. Computational Statistics and Data Analysis 34, 387–413 (2000)
7. Cano, A., Moral, S.: Propagación exacta y aproximada mediante árboles de probabilidad en redes causales. In: Actas de la VII Conferencia de la Asociación Española para la Inteligencia Artificial, Málaga, pp. 635–644 (1997)
8. Quinlan, J.R.: Induction of decision trees. Machine Learning 1, 81–105 (1986)
9. Kullback, S., Leibler, R.A.: On information and sufficiency. Annals of Mathematical Statistics 22, 76–86 (1951)

Marginals of DAG-Isomorphic Independence Models

Peter R. de Waal

Department of Information and Computing Sciences, Faculty of Sciences,
Universiteit Utrecht, P.O. Box 80089, 3508TB Utrecht, The Netherlands
waal@cs.uu.nl

Abstract. Probabilistic and graphical independence models both satisfy the semi-graphoid axioms, but their respective modelling powers are not equal. For every graphical independence model that is represented by d-separation in a directed acyclic graph, there exists an isomorphic probabilistic independence model, i.e. it has exactly the same independence statements. The reverse does not hold, as there exist probability distributions for which there are no perfect maps. We investigate if a given probabilistic independence model can be augmented with latent variables to a new independence model that is isomorphic with a graphical independence model of a directed acyclic graph. The original independence model can then be viewed as the marginal of the model with latent variables. We show that for some independence models we need infinitely many latent variables to accomplish this.

1 Introduction

Probabilistic models in artificial intelligence are typically built on the semi-graphoids axioms of independence. These axioms are exploited explicitly in graphical models, where independence is captured by topological properties, such as separation of vertices in an undirected graph or d-separation in a directed graph. A graphical representation with directed graphs for use in a decision support system has the advantage that it allows an intuitive interpretation by domain experts in terms of influences between the variables.

Ideally a probabilistic model is represented as a graphical model in a one-to-one way, that is, independence in the one representation implies independence in the other representation. The probabilistic model then is said to be isomorphic with the graphical model, and vice versa. Pearl and Paz [5] established a set of sufficient and necessary conditions under which a probabilistic model is isomorphic with an undirected graph. In this paper we shall not consider representations of independence with undirected graphs, but focus on directed representations. Contrary to undirected graphs directed graphs allow the representation of induced dependencies: if a specific independence has been established given some evidence, it is possible that this independence becomes invalid if more evidence is obtained. Pearl gave a set of necessary conditions for directed graph isomorphism in [6]. To the best of our knowledge there is no known set of sufficient conditions.

C. Sossai and G. Chemello (Eds.): ECSQARU 2009, LNAI 5590, pp. 192–203, 2009.

Pearl [6] also provides an example of an independence model not isomorphic with a directed graphical model, that can be made isomorphic by the introduction of an auxiliary variable. In [6] the isomorphism is then established by conditioning on the auxiliary variable. In this paper we choose a different approach. We extend a model with auxiliary variables to a directed graph isomorph and we then take the marginal over the original variables of this extended model. For this we introduce the concept of the marginal of an independence model. The model with auxiliary variables can then be considered as a latent perfect map. We show that it is possible to establish isomorphism in this manner, but that we may need an infinite number of auxiliary variables to accomplish this. We also show that there exists a probabilistic independence model that needs infinitely many latent variables.

This paper is organised as follows. In Sect. 2 we briefly review probabilistic and graphical independence models, and the semi-graphoid properties of these models. In Sect. 3 we introduce the concept of marginals of an independence model and latent perfect maps. In Sect. 4 we discuss the existence of latent perfect maps, and in Sect. 5 we wrap up with conclusions and recommendations.

2 Preliminaries

In this section, we provide some preliminaries on probabilistic independence models as defined by conditional independence for probability distributions, graphical independence models as defined by d-separation in directed acyclic graphs, and algebraic independence models that capture the properties that probabilistic and graphical models have in common.

2.1 Conditional Independence Models

We consider a finite set of distinct symbols $V = \{V_1, \ldots, V_N\}$, called the *attributes* or *variable names*. With each variable V_i we associate a finite domain set $\mathcal{V}_i$, which is the set of possible values the variable can take. We define the domain of V as $\mathcal{V} = \mathcal{V}_1 \times \cdots \times \mathcal{V}_N$, the Cartesian product of the domains of the individual variables.

A *probability measure* over V is defined by the domains $\mathcal{V}_i$, $i = 1, \ldots, N$, and a probability mapping $P : \mathcal{V} \to [0, 1]$ that satisfies the three basic axioms of probability theory [4].

For any subset $X = \{V_{i_1}, \ldots, V_{i_k}\} \subset V$, for some $k \geq 1$, we define the domain $\mathcal{X}$ of X as $\mathcal{X} = \mathcal{V}_{i_1} \times \cdots \mathcal{V}_{i_k}$. For a probability mapping P on V we define its *marginal* mapping over X, denoted by P^X, as the probability measure P^X over $\mathcal{X}$, defined by

$$P^X(x) = \sum \left\{ P(x, y) \,\middle|\, y \in \underset{\{i \mid V_i \notin X\}}{\times} \mathcal{V}_i \right\}$$

for $x \in \mathcal{X}$. By definition $P^V \equiv P$, $P^\varnothing \equiv 1$, and $(P^X)^Y = (P^Y)^X = P^{X \cap Y}$, for $X, Y \subset V$.

We denote the set of ordered triplets $(X, Y|Z)$ for disjoint subsets X, Y and Z of V as $\mathcal{T}(V)$. For any ternary relation $\mathcal{I}$ on V we shall use the notation

$\mathcal{I}(X,Y|Z)$ to indicate $(X,Y|Z) \in \mathcal{I}$. For simplicity of notation we will often write XY to denote the union $X \cup Y$, for $X, Y \subset V$. To avoid complicated notation we also allow Xy to denote $X \cup \{y\}$, for $X \subset V$ and $y \in V$.

Definition 1 (Conditional independence). *Let X, Y and Z be disjoint subsets of V, with domains $\mathcal{X}$, $\mathcal{Y}$, and $\mathcal{Z}$, respectively. The sets X and Y are defined to be conditionally independent under P given Z, if for every $x \in \mathcal{X}$, $y \in \mathcal{Y}$ and $z \in \mathcal{Z}$, we have*

$$P^{XYZ}(x,y,z) \cdot P^Z(z) = P^{XZ}(x,z) \cdot P^{YZ}(y,z)$$

Definition 2. *Let V be a set of variables and P a probability measure over V. The probabilistic independence model $\mathcal{I}_P$ of P is defined as the ternary relation $\mathcal{I}_P$ on V for which $\mathcal{I}_P(X,Y|Z)$ if and only if X and Y are conditionally independent under P given Z.*

If no ambiguity can arise we may omit the reference to the probability measure and just refer to the probabilistic independence model.

2.2   Graphical Independence Models in Directed Acyclic Graphs

We first introduce the standard concepts of blocking and d-separation in directed graphs. We consider a directed acyclic graph (DAG) $G = (V,A)$, with V the set of vertices and A the set of arcs. A path s in G of length $k-1$ from a vertex V_{i_1} to V_{i_2} is a k-tuple $s = (W_1, W_2, \ldots, W_k)$ with $W_i \in V$ for $i = 1, \ldots, k$, $W_1 = V_{i_1}$, $W_k = V_{i_2}$ and for each $i = 1, \ldots, k-1$ either $(W_i, W_{i+1}) \in A$ or $(W_{i+1}, W_i) \in A$. Without loss of generality we assume that a path has no loops, so there are no duplicates in $\{W_1, \ldots, W_k\}$. We define a path s to be *unidirectional* if all the arcs in s point in the same direction. More specifically, we define the unidirectional $s = (W_1, W_2, \ldots, W_k)$ to be a *descending* path if $(W_i, W_{i+1}) \in A$, for all $i = 1, \ldots, k-1$. A vertex Y is called a descendant of a vertex X if there is a descending path from X to Y.

Definition 3. *Let Z be a subset of V. We say that a path s is blocked in G by Z, if s contains three consecutive vertices W_{i-1}, W_i, and W_{i+1} for which one of the following conditions hold:*

- $W_{i-1} \leftarrow W_i \rightarrow W_{i+1}$, *and* $W_i \in Z$,
- $W_{i-1} \rightarrow W_i \rightarrow W_{i+1}$, *and* $W_i \in Z$,
- $W_{i-1} \leftarrow W_i \leftarrow W_{i+1}$, *and* $W_i \in Z$,
- $W_{i-1} \rightarrow W_i \leftarrow W_{i+1}$, *and* $\sigma(W_i) \cap Z = \varnothing$, *where* $\sigma(W_i)$ *consists of* W_i *and all its descendants.*

We refer to the first three conditions as blocking by presence, *and the last condition as* blocking by absence. *We refer to node* W_i *in the last condition as a* converging *or* colliding node *on the path.*

While the concept of blocking is defined for a single path, the d-separation criterion applies to the set of all paths in G.

Definition 4. *Let $G = (V, A)$ be a DAG, and let X, Y and Z be disjoint subsets of V. The set Z is said to d-separate X and Y in G, if every path s from any variable $x \in X$ to any variable $y \in Y$ is blocked in G by Z.*

Based on the d-separation criterion we can define the notion of a graphical independence model.

Definition 5. *Let $G = (V, A)$ be a DAG. The* graphical independence model *$\mathcal{I}_G$ defined by G is a ternary relation on V such that $\mathcal{I}_G(X, Y|Z)$ if and only if Z d-separates X and Y in G.*

2.3   Algebraic Independence Models

Both a probabilistic independence model on a set of variables V and a graphical independence model on a DAG $G = (V, A)$ define a ternary relation on V. In fact we can capture this in an algebraic construct of an independence model.

Definition 6. *An* algebraic independence model *on a set V is a ternary relation on V.*

Probabilistic and graphical independence models satisfy a set of axioms of independence. We use these axioms to define a special class within the set of algebraic independence models.

Definition 7. *A* ternary relation $\mathcal{I}$ on V is a *semi-graphoid independence model, or semi-graphoid for short, if it satisfies the following four axioms:*

A1: $\mathcal{I}(X, Y|Z) \Rightarrow \mathcal{I}(Y, X|Z)$,   *(symmetry),*
A2: $\mathcal{I}(X, YW|Z) \Rightarrow \mathcal{I}(X, Y|Z) \wedge \mathcal{I}(X, W|Z)$,   *(decomposition),*
A3: $\mathcal{I}(X, YW|Z) \Rightarrow \mathcal{I}(X, Y|ZW)$,   *(weak union),*
A4: $\mathcal{I}(X, Y|Z) \wedge \mathcal{I}(X, W|ZY) \Rightarrow \mathcal{I}(X, YW|Z)$,   *(contraction).*

for all disjoint sets of variables $W, X, Y, Z \subset V$.

The axioms convey the idea that learning irrelevant information does not alter the relevance relationships among the other variables discerned. They were first introduced in [1] for probabilistic conditional independence. The properties were later recognised in artificial intelligence as properties of separation in graphs [5,6], and are since known as the *semi-graphoid* axioms.

In the formulation that we have used so far we can allow X and Y to be empty, which leads to the so-called *trivial independence* axiom:

A0: $\mathcal{I}_P(X, \varnothing|Z)$,

This axiom trivially holds for both probabilistic independence and graphical independence.

An axiomatic representation allows us to derive qualitative statements about conditional independence that may not be immediate from a numerical representation of probabilities. It also enables a parsimonious specification of an independence model, since it is sufficient to enumerate the so-called dominating independence statements, from which all other statements can be derived by application of the axioms [9].

2.4   Graph-Isomorph

Probabilistic independence models and graphical independence models both satisfy the semi-graphoid axioms, so it is interesting to investigate whether they have equal modelling power. Can any probabilistic independence model be represented by a graphical model, and vice versa? For this we introduce the notions of I-maps and P-maps.

Definition 8. *Let* $\mathcal{I}$ *be an algebraic independence model on* V, *and* $G = (V, A)$ *a DAG that defines a graphical independence model* $\mathcal{I}_G$ *through d-separation.*

1. *The graph* G *is called an* independence map, *or* I-map *for short, for* $\mathcal{I}$, *if for all disjoint* X, Y, $Z \subset V$ *we have:* $\mathcal{I}_G(X, Y | Z) \Rightarrow \mathcal{I}(X, Y | Z)$. *If* G *is an I-map for* $\mathcal{I}$, *and deleting any arc makes* G *cease to be an I-map for* $\mathcal{I}$, *then* G *is called a* minimal I-map *for* $\mathcal{I}$.
2. *The graph* G *is called a* perfect map, *or* P-map *for short, for* $\mathcal{I}$, *if for all disjoint* X, Y, $Z \subset V$ *we have:* $\mathcal{I}_G(X, Y | Z) \Leftrightarrow \mathcal{I}(X, Y | Z)$,

Definition 9 (DAG-isomorph). *An independence model* $\mathcal{I}$ *on* V *is said to be a* DAG-isomorph, *if there exists a graph* $G = (V, A)$ *that is a perfect map for* $\mathcal{I}$.

Since a graphical independence model satisfies the semi-graphoid axioms, a DAG-isomorph has to be a semi-graphoid itself. Being a semi-graphoid is not a sufficient condition for DAG-isomorphism, however. To the best of our knowledge there does not exists a sufficient set of conditions, although Pearl presents a set of necessary conditions in [6].

Some results from literature describe the modelling powers of the types independence model that we presented in the previous sections. Geiger and Pearl show that for every DAG graphical model there exists a probability model for which that particular DAG is a perfect map [3]. The reverse does not hold, there exist probability models for which there is no DAG perfect map [6].

Studený shows in [7] that the semi-graphoid axioms are not complete for probabilistic independence models. He derives a new axiom for probabilistic independence models that is not implied by the semi-graphoid axioms. He also shows in [8] that probabilistic independence models cannot be characterised by a finite set of inference rules.

3   Marginal of an Independence Model

A set of necessary conditions for an algebraic independence model to be a DAG-isomorph is known from [6]. These conditions are derived from properties of d-separation in DAG's. One of the conditions that is not already implied by the semi-graphoid axioms is the so-called *chordality condition*:

$$\mathcal{I}(x, y | zw) \wedge \mathcal{I}(z, w | xy) \Rightarrow \mathcal{I}(x, y | z) \vee \mathcal{I}(x, y | w)$$

for all $x, y, z, w \in V$. Pearl shows in [6, Sect. 3.3.3] by example how conditioning on an auxiliary variable can be used to dispose of this chordality condition. In his example the independence model is not DAG-isomorph, but there exists a DAG with one extra variable, that, when conditioned on the auxiliary variable, is isomorphic with the independence model.

In this paper we take a different approach as we introduce an auxiliary variable without conditioning to create a DAG that is a P-map for an independence model. We formulate this in the following definition.

Definition 10. *Let $\mathcal{I}$ be an independence model on a set of variables V, and let A be a subset of V. We define the* marginal of $\mathcal{I}$ on A, *denoted by $\mathcal{I}^A$, as $\mathcal{I}^A = \mathcal{I} \cap \mathcal{T}(A)$.*

The following lemma follows immediately from this definition. It implies that taking the marginal of a probabilistic model is equivalent to taking the probabilistic model of the marginal of a probability measure. As such it justifies our use of the phrase "marginal of an independence model".

Lemma 1. *Let V be a set of variables, P a probability measure on V with probabilistic independence model $\mathcal{I}_P$. For every subset $A \subseteq V$ we have $\mathcal{I}_P^A = \mathcal{I}_{P^A}$.*

Proof. Let X, Y and Z be disjoint subsets of A, then the following holds:

$$\mathcal{I}_{P^A}(X, Y | Z)$$

$$\Leftrightarrow (P^A)^{XYZ}(x, y, z)(P^A)^Z(z) = (P^A)^{XZ}(x, z)(P^A)^{YZ}(y, z)$$

$$\Leftrightarrow P^{XYZ}(x, y, z)P^Z(z) = P^{XZ}(x, z)P^{YZ}(y, z)$$

$$\Leftrightarrow \mathcal{I}_P(X, Y | Z)$$

$$\Leftrightarrow \mathcal{I}_P^A(X, Y, Z)$$

The first and third steps follow from Definition 2, the second step from the observation that $(P^A)^W = P^W$ for any $W \subseteq A$, and the final step follows from $(X, Y | Z) \in \mathcal{T}(A)$ and Definition 10.   □

We can now extend the definition of DAG-isomorphism.

Definition 11 (DAG-isomorph marginal). *Let V be a set of variables, and $\mathcal{I}$ an independence model on V. We say that $\mathcal{I}$ is a* DAG-isomorph marginal, *if there exists a finite set of variables $\overline{V} \supseteq V$, an independence model $\overline{\mathcal{I}}$ on $\overline{V}$ and a DAG $\overline{G} = (\overline{V}, \overline{A})$, such that $\overline{G}$ is a P-map for $\overline{\mathcal{I}}$ and $\overline{\mathcal{I}}^V = \mathcal{I}$. We then say that $\overline{G}$ is a* latent P-map *of $\mathcal{I}$.*

A latent P-map would be used primarily for modelling purposes and specifically for representing all the independencies in a compact graphical form. For this reason we require $\overline{V}$ to be a finite set.

Note that if $\mathcal{I}$ is a DAG-isomorph, then it is by Definition 11 also a DAG-isomorph marginal.

Example 1. As an example we present the variable set $V = \{V_1, V_2, V_3, V_4\}$ and the algebraic independence model $\mathcal{I}$ on V defined by the following non-trivial independence statements (and their symmetric equivalents):

$$(S1) : \mathcal{I}(V_1, V_2 | \varnothing) \qquad (S2) : \mathcal{I}(V_1, V_2 | V_3)$$
$$(S3) : \mathcal{I}(V_2, V_3 | \varnothing) \qquad (S4) : \mathcal{I}(V_1, V_4 | \varnothing)$$
$$(S5) : \mathcal{I}(V_1, V_2 | V_4) \qquad (S6) : \mathcal{I}(V_2, V_3 | V_1)$$
$$(S7) : \mathcal{I}(V_1, V_4 | V_2)$$

The DAG $G_1 = (V, A)$ defined on the variables V as depicted on the left-hand side in Fig. 1, is a minimal I-map for $\mathcal{I}$, since the non-trivial graphical independence statement that can be derived from the DAG correspond to the statements (S1), (S2), (S3), and (S6). It is not a P-map for $\mathcal{I}$, since the statements (S4), (S5), and (S7) are not reflected as graphical independence statements in G_1. An alternative minimal I-map is G_2, as depicted on the right-hand side in Fig. 1. According to [2, Lemma 5.1] there does not exist a P-map for $\mathcal{I}$ on V, although $\mathcal{I}$ satisfies the necessary conditions for DAG-isomorphism of [6].

We can, however, construct a DAG $\overline{G}$ on a superset $\overline{V}$ of V for which the corresponding graphical independence model $\mathcal{I}_{\overline{G}}$ satisfies all the independence statements (S1)–(S7). This DAG is depicted in Fig. 2. It has an extra, latent, variable V_0. The graphical independence model $\mathcal{I}_{\overline{G}}$ satisfies more independence statements than (S1)–(S7), like for instance $\mathcal{I}_{\overline{G}}(V_1, V_2 | V_0)$. There are, however, no new independence statements $\mathcal{I}_{\overline{G}}(X, Y | Z)$ in $\mathcal{I}_{\overline{G}}$ for subsets X, Y, $Z \subset V$, other than (S1)–(S7). All new independence statements involve the latent variable V_0 in one of the arguments. By Definition 10 $\mathcal{I}$ in Example 1 is the marginal of $\mathcal{I}_{\overline{G}}$ on V, and $\overline{G}$ is a latent P-map of $\mathcal{I}$.

For Example 1 we have from [3] that there exists a probability distribution $\overline{P}$ on $\overline{V}$ that has $\overline{G}$ of Fig. 2 as a perfect map. The structure of $\overline{G}$ implies that $\overline{P}$ factorises as:

Fig. 1. G_1 (left) and G_2 (right), minimal I-maps for Example 1

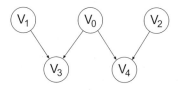

Fig. 2. $\overline{G}$, a latent P-map for Example 1

$$\overline{P}(v_0, v_1, v_2, v_3, v_4) =$$
$$p_0(v_0)\, p_1(v_1)\, p_2(v_2)\, p_3(v_3|v_0 v_1) p_4(v_4|v_0 v_2)$$

for some functions $p_1, \ldots, p_4$. It can be shown that the DAG's G_1 and G_2 are minimal I-maps for the marginal distribution of $\overline{P}$ on V. G_1 corresponds to a factorisation of P as:

$$P(v_1, v_2, v_3, v_4) = \tag{1}$$
$$p_1(v_1)\, p_2(v_2)\, p_3'(v_3|v_1)\, p_4'(v_4|v_1 v_2 v_3)$$

and G_2 corresponds to a factorisation of P as:

$$P(v_1, v_2, v_3, v_4) = \tag{2}$$
$$p_1(v_1)\, p_2(v_2)\, p_3''(v_3|v_1 v_2 v_4)\, p_4''(v_4|v_2)$$

In the example we thus have a probability distribution P and the corresponding independence model $\mathcal{I}_P$ on V that is not DAG-isomorphic, but it is the marginal of a distribution $\overline{P}$ that corresponds to a DAG-isomorphic probabilistic independence model.

For a probability measure we can now present a refined definition of DAG-isomorph marginal based on the probabilistic notion of a marginal.

Definition 12 (P-DAG-isomorph marginal). *Let V be a set of variables and P a probability measure on V. We say that P is a P-DAG-isomorph marginal, if there exists a finite set of variables $\overline{V} = \{\overline{V}_1, \ldots, \overline{V}_{\overline{N}}\} \supseteq V$ with domains $\overline{\mathcal{V}}_i$, $i = 1, \ldots, \overline{N}$, a DAG $\overline{G} = (\overline{V}, \overline{A})$, and a probability measure $\overline{P}$ on $\overline{V}$, such that*

- *The domains of the variables V_i in V for $\overline{P}$ are the same as for P ,*
- *The marginal distribution $\overline{P}^V$ of $\overline{P}$ over V is equal to P,*
- *$\overline{G}$ is a perfect map for $\overline{P}$.*

4   Existence of a Latent Perfect Map

Weak transitivity and chordality are necessary conditions for DAG-isomorphism. Assume that we have an independence model $\mathcal{I}$ on the set of variables V that does not satisfy either of these two conditions. For any independence model $\overline{\mathcal{I}}$ on a superset $\overline{V} \supseteq V$ the conflicting conditions remain unsatisfied, since the independence statements in $\mathcal{I}$ that violated the conditions will also be in $\overline{\mathcal{I}}$. This implies that any independence model that does not satisfy any of these two properties, is not a DAG-isomorph marginal.

For Example 1, which satisfies weak transitivity and chordality, there does not exist a P-map, but we were able to construct a latent perfect map. In this section we show that a latent perfect map does not always exists, even if the independence model satisfies all the necessary conditions for a DAG-isomorph. The main result is captured in the following theorem.

Theorem 1. *There exists an independence model satisfying the necessary conditions for DAG-isomorphism that neither has a P-map nor a latent P-map.*

We shall prove Theorem 1 by showing that there is no latent perfect map for the following independence model.

Example 2. Let $V = \{B, C, D, E\}$ and let $\mathcal{I}^*$ be the independence model on V, that consists of the following three non-trivial independent statements (and their symmetric equivalents):

$$(T1) : \mathcal{I}^*(B, E|CD)$$
$$(T2) : \mathcal{I}^*(C, E|\varnothing)$$
$$(T3) : \mathcal{I}^*(C, D|B)$$

It is a straightforward exercise to verify that $\mathcal{I}^*$ is indeed a semi-graphoid. Application of the semi-graphoid axioms on (T1)–(T3) does not yield any new non-trivial independence statements. Moreover, $\mathcal{I}^*$ satisfies Pearl's necessary conditions for DAG-isomorphism.

We prove by contradiction that $\mathcal{I}^*$ is not a DAG-isomorph marginal. The steps in the proof are summarised in the following four lemmas.

Lemma 2. *Assume that $\mathcal{I}^*$, as defined in Example 2, is a DAG-isomorph marginal and $\overline{G}$ is a latent P-map for $\mathcal{I}^*$, then there exists at least one path in $\overline{G}$ from C to E that is neither blocked by B nor by D.*

Proof. By contradiction: assume that there are no paths in $\overline{G}$ between C and E. C and E are then d-separated by any subset of $\overline{V}$, which contradicts, for instance, $\neg\mathcal{I}^*(C, E|BD)$.

Assume that all paths in $\overline{G}$ between C and E are blocked by B or D. Since there is at least one path in $\overline{G}$ from C to E, this contradicts $\neg\mathcal{I}^*(C, E|BD)$. □

Lemma 3. *Assume that $\mathcal{I}^*$, as defined in Example 2, is a DAG-isomorph marginal, $\overline{G}$ is a latent P-map for $\mathcal{I}^*$, and s is a path in $\overline{G}$ from C to E, then s has at least one converging node.*

Proof. Let s be a path from C to E. Due to $\mathcal{I}^*(C, E|\varnothing)$ s must be blocked by $\varnothing$, which implies that s has a converging node. □

Lemma 4. *Assume that $\mathcal{I}^*$, as defined in Example 2, is a DAG-isomorph marginal, $\overline{G}$ is a latent P-map of $\mathcal{I}^*$, s a path in $\overline{G}$ from C to E that is neither blocked by B nor by D, and let F be a converging node on s, then $D \in \sigma(F)$ and $B \in \sigma(F)$. Moreover every descending path from F to D is blocked by B.*

Proof. If there exists a converging node F on s for which $B \notin \sigma(F)$ or $D \notin \sigma(F)$, then the path s would be blocked by B or D, which is in contradiction with the definition of s.

Let F be a converging node on s. Since $D \in \sigma(F)$, there exists a descending path s_1 from F to D. We now construct a new path s_2 from C to D by concatenating the subpath of s between C and F with s_1 (see Fig. 4). Due to $\mathcal{I}^*(C, D|B)$ this path must be blocked by B. It cannot be blocked by B on the segment between C and F, since then also the original path s would be blocked by B. Therefore s_2 must be blocked by B on the subpath s_1. Since s_1 is descending, it is unidirectional. Hence B must lie on s_1 and s_1 is blocked by B. □

Lemma 5. *Assume that $\mathcal{I}^*$, as defined in Example 2, is a DAG-isomorph marginal, $\overline{G}$ is a latent P-map of $\mathcal{I}^*$, s is a path in $\overline{G}$ from C to E that is neither blocked by B nor by D. For any converging node F on s there is also a second converging node on the subpath of s between F and E.*

Proof. Let F be a converging node on s, which exists due to Lemma 3. From Lemma 4 we have that any descending path from F to D has B on it. At least one such path, say s_1, must exist, since $B \neq D$, and thus D cannot be equal to the converging node F. We now construct a path s_3 from B to E by concatenating the reverse of the part of subpath s_1 between B and F with the subpath of s between F and E (see also Fig. 4).

Now s_3 is a path from B to E via F. Due to $\mathcal{I}^*(B, E|CD)$, this path s_3 must be blocked by CD. Since s_1 is descending and thus unidirectional, the first part of s_3 between B and F is unidirectional. D is not on this subpath, so this subpath cannot be blocked by D. The second part of s_3 between F and E cannot be blocked by D, since it is part of the original path s and s is not blocked by D. In path s_3 the node F, where the two subpaths join, is not a converging node, so we conclude that s_3 cannot be blocked by D. This implies that s_3 must be blocked by C.

There are two possibilities for C to block s_3. The first possibility is that C blocks s_3 by presence on the (unidirectional) subpath of s_3 between B and F. If this is the case, then we can construct a new path s_4 from C to E, by dropping from s_3 the first part between B and C. This new path s_4 consists of a unidirectional path between C and F, that has neither B nor D on it. The second part of the path, between F and E, is the segment of the original path s. Since F is not a converging node on s_4 and s is not blocked by B nor D, we conclude that s_4 is also not blocked by B nor by D. From Lemma 3 we conclude that s_4 must have a converging node, which can lie only between F and E. Therefore this converging node must also lie on the original path s.

The second possibility for C to block s_3 is through absence, if there is a converging node on s_3 that does not have C as a descendant. Since the first part of s_3 between B and F is unidirectional, and F is not a converging node on s_3, this converging node must lie on the segment of s_3 strictly between F and E and therefore also on s. □

Proof (Of Theorem 1). Let $\mathcal{I}^*$ be as defined in Example 2. Due to Lemma 2 we know that there is at least one path s between C and E that is not blocked by B nor by D. According to Lemma 3 this path s must have at least one converging node and due to Lemma 5 we can conclude that s must have an infinite number of converging nodes. Therefore $\overline{V}$ cannot be finite, and $\mathcal{I}^*$ is not a DAG-isomorph marginal. □

The next theorem shows that there is also a probabilistic independence model without a latent perfect map.

Theorem 2. *There exists a set of variables V and a probability distribution on V that is not a P-DAG-isomorph marginal.*

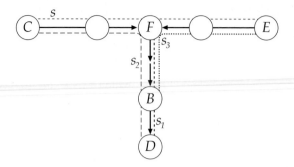

Fig. 3. The paths used in the proofs of Lemmas 4 and 5

Proof. Consider the set of binary variables $V = \{B, C, D, E\}$. Define the probability measure P^* on V as follows:

B C D E	$P^*(B,C,D,E)$	B C D E	$P^*(B,C,D,E)$
0 0 0 0	48/1357	1 0 0 0	96/1357
0 0 0 1	48/1357	1 0 0 1	96/1357
0 0 1 0	144/1357	1 0 1 0	192/1357
0 0 1 1	48/1357	1 0 1 1	64/1357
0 1 0 0	48/1357	1 1 0 0	27/1357
0 1 0 1	96/1357	1 1 0 1	54/1357
0 1 1 0	240/1357	1 1 1 0	90/1357
0 1 1 1	48/1357	1 1 1 1	18/1357

It can be verified that the probabilistic independence model $\mathcal{I}_{P^*}$ of P^* has exactly the same independence statements as $\mathcal{I}^*$ as defined in Example 2.   □

5   Conclusions

In this paper we have introduced the concept of the marginal of an algebraic independence model. We have shown that some independence models are in fact the marginals of models that are DAG-isomorphs, while the marginals themselves are not DAG-isomorphs. We have also proved that there exist some independence models for which we need to introduce an infinite number of auxiliary variables to obtain a latent perfect map. In examples for both cases the marginal independence models satisfy the sufficient conditions of [6] for DAG-isomorphism. It is an interesting topic for future research to investigate if necessary and sufficient conditions can be established to guarantee the existence of a latent perfect map.

It is also worthwhile to investigate if existence results for latent P-maps can be established for other types of graphical model. We can show that this is not true for undirected graphs, and we plan to investigate if it is possible for chain graphs.

Acknowledgements

We would like to thank Richard D. Gill for suggesting this problem in the context of probabilistic graphical models.

References

1. Dawid, A.P.: Conditional Independence in Statistical Theory. Journal of the Royal Statistical Society B 41(1), 1–31 (1979)
2. Dawid, A.P., Studený, M.: Conditional Products: An Alternative Approach to Conditional Independence. In: Whittaker, J., Heckerman, D. (eds.) Artificial Intelligence and Statistics 1999, Proceedings of the 7th workshop, pp. 32–40. Morgan Kaufmann, San Francisco (1999)
3. Geiger, D., Pearl, J.: On the Logic of Causal Models. In: Shachter, R.D., Levitt, T.S., Kanal, L.N., Lemmer, J.F. (eds.) Uncertainty in Artificial Intelligence, vol. 4, pp. 3–14. Elsevier Science Publishers, Amsterdam (1990)
4. Kolmogorov, A.N.: Foundation of the Theory of Probability. Chelsea Publishing, New York (1950)
5. Pearl, J., Paz, A.: Graphoids, Graph-based Logic for Reasoning About Relevance Relations. In: du Boulay, B., Hogg, D., Steele, L. (eds.) Advances in Artificial Intelligence II, pp. 357–363 (1987)
6. Pearl, J.: Probabilistic Reasoning in Intelligent Systems - Networks of Plausible Inference. Morgan Kaufman, San Francisco (1988)
7. Studený, M.: Multi-information and the Problem of Characterization of Conditional Independence Relations. Problems of Control and Information Theory 18(1), 3–16 (1989)
8. Studený, M.: Conditional Independence Relations Have No Finite Complete Characterization. In: Kubik, S., Visek, J.A. (eds.) Proc. of the 11th Prague conf. on Information Theory, Statistical Decision Functions and Random Processes, pp. 377–396 (1992)
9. Studený, M.: Complexity of structural models. In: Proceedings of the Joint Session of the 6th Prague Conference on Asymptotic Statistics and the 13th Prague Conference on Information Theory, Statistical Decision Functions and Random Processes, Prague, vol. II, pp. 521–528 (1998)

The Probabilistic Interpretation of Model-Based Diagnosis

Ildikó Flesch[1] and Peter J.F. Lucas[2]

[1] Tilburg Centre for Creative Computing
Tilburg University, The Netherland
i.flesch@uvt.nl
[2] Institute for Computing and Information Sciences
Radboud University Nijmegen, Nijmegen, The Netherlands
peterl@cs.ru.nl

Abstract. Model-based diagnosis is the field of research concerned with the problem of finding faults in systems by reasoning with abstract models of the systems. Typically, such models offer a description of the structure of the system in terms of a collection of interacting components. For each of these components it is described how they are expected to behave when functioning normally or abnormally. The model can then be used to determine which combination of components is possibly faulty in the face of observations derived from the actual system. There have been various proposals in literature to incorporate uncertainty into the diagnostic reasoning process about the structure and behaviour of systems, since much of what goes on in a system cannot be observed. This paper proposes a method for decomposing the probability distribution underlying probabilistic model-based diagnosis in two parts: (i) a part that offers a description of uncertain abnormal behaviour in terms of the Poisson-binomial probability distribution, and (ii) a part describing the deterministic, normal behaviour of system components.

1 Introduction

Almost from the inception of the field of probabilistic graphical models, Bayesian networks have been popular as formalisms to built *model-based*, diagnostic systems [1]. An alternative theory of model-based diagnosis was developed at approximately the same time, founded on techniques from logical reasoning [2,3]. The General Diagnostic Engine, GDE for short, is a well-known implementation of the logical theory; however, it also includes a restricted form of uncertainty reasoning to focus the diagnostic reasoning process [4]. Previous research by Geffner and Pearl showed that the GDE approach to model-based diagnosis can be equally well dealt with by Bayesian networks [5,1]. Geffner and Pearl's result is basically a mapping from the logical representation as traditionally used within the model-based diagnosis community to a specific Bayesian-network representation. The theory of model-based diagnosis supports multiple-fault diagnoses, which are similar to maximum a posteriori hypotheses, MAP hypotheses for

C. Sossai and G. Chemello (Eds.): ECSQARU 2009, LNAI 5590, pp. 204–215, 2009.

short, in Bayesian networks [6]. Thus, although the logical and the probabilistic theory of model-based diagnosis have different origins, they are closely related. In fact, in his research Darwich has extensively explored this relationship, although ignoring uncertainty [7]. However, whereas the traditional theory of model-based diagnosis is strong in providing models that are easily understood in relationship to the actual, real-world systems, it is weak on dealing with uncertain information. With Bayesian networks taken as representations of models of systems, it is the other way around. Thus, developing ways to combine both approaches can be advantageous.

In logical model-based diagnosis, it is clear that a diagnosis should be interpreted as behaviour assumptions of particular components that are compatible with, and possibly explain, the observations; however, probabilistic diagnosis defies giving similar straightforward interpretations. This is because the logical reasoning, implemented by deterministic probability distributions, and uncertainty reasoning (nondeterministic probability distributions) are mingled. To tackle this problem, this paper proposes a new way to look at model-based diagnosis, taking the Bayesian-network representation by Geffner and Pearl as the point of departure [5,1]. It is shown that after adding probabilistic information to a model of a system, the predictions that can be made by the model can be naturally decomposed into a logical and a probabilistic part. The logical specifications are determined by the system components that are assumed to behave normally, constituting part of the system behaviour. This is complemented by uncertainty about behaviour for components that are assumed to behave abnormally. It is shown that the Poisson-binomial distribution plays a central role in governing this uncertain behaviour. The results of this paper establish new links between traditional logic-based diagnosis, Bayesian networks and probability theory.

2   Poisson-Binomial Distribution

First, we begin by summarising some of the relevant theory of discrete probability distributions (cf. [8,9]).

Let $s = (s_1, \ldots, s_n)$ be a Boolean vector with elements $s_k \in \{0, 1\}$, $k = 1, \ldots, n$, where s_k is a Bernoulli discrete random variable that expresses that the outcome of trial k is either success (1) or failure (0). Let the probability of success of trial k be indicated by $p_k \in [0, 1]$ and, thus, the probability of failure is set to $1 - p_k$. Then, the probability of obtaining vector s as outcome is equal to

$$P(s) = \prod_{k=1}^{n} p_k^{s_k} (1 - p_k)^{1-s_k}. \tag{1}$$

This probability distribution acts as the basis for the *Poisson-binomial distribution*. The Poisson binomial distribution is employed to describe the outcomes of n independent Bernoulli distributed random variables, where only the number of success and failure are counted. The probability that there are m, $m \leq n$, successful outcomes amongst the n trials performed is then defined as:

$$f(m;n) = \sum_{s_1 + \cdots + s_n = m} \prod_{k=1}^{n} p_k^{s_k} (1 - p_k)^{1-s_k}, \tag{2}$$

where f is a probability function. Here, the summation means that we sum over all the possible values of elements of the vector s, where the sum of the values of the elements must be equal to m.

It is easy to check that when all probabilities p_k are equal, i.e. $p_1 = \cdots = p_n = p$, where p denotes this identical probability, then the probability function $f(m;n)$ becomes that of the well-known *binomial distribution*:

$$g(m;n) = \binom{n}{m} p^m (1 - p)^{n-m}. \tag{3}$$

Finally, suppose that we model interactions between the outcomes of the trials by means of a Boolean function b. This means that we have an oracle that is able to observe the outcomes, and then gives a verdict whether the overall outcome is successful. The *expectation* or *mean* of this Boolean function is then equal to:

$$\mathcal{E}_P(b(S)) = \sum_{s} b(s)P(s). \tag{4}$$

with P defined according to Equation (1). This expectation also acts as the basis for the theory of causal independence, where a causal process is modelled in terms of interacting independent processes (cf. [10]). Note that for $b(s) = b_m(s) \equiv s_1 + \cdots + s_n = m$ (i.e., the Boolean function that checks whether the number of successful trials is equal to m), we have that $\mathcal{E}_P(b_m(S)) = f(m;n)$. Thus, Equation (4) can be looked on as a generic way to combine the outcome of independent trials.

In the theory of model-based diagnosis, it is common to represent models of systems by means of logical specifications, which are equivalent to Boolean functions. Below, it will become clear that if we interpret the success probabilities p_k as the probability of observing the expected output of a system's component under the assumption that the component is faulty, then the theory of Poisson-binomial distributions can be used to describe part of the probabilistic model-based diagnostic process. However, first the necessary background to model-based diagnosis research is reviewed.

3   Uncertainty in Model-Based Diagnosis

3.1   Model-Based Diagnosis

In the theory of model-based diagnosis [2], the structure and behaviour of a system is represented by a *logical diagnostic system* $\mathcal{S}_L = (\text{SD}, \text{COMPS})$, where

- SD denotes the *system description*, which is a finite set of logical formulae, specifying structure and behaviour;
- COMPS is a finite set of constants, corresponding to the *components* of the system; these components can be faulty.

The system description consists of *behaviour descriptions* and *connections*. A behavioural description is a formula specifying *normal* and *abnormal* (faulty) functionalities of the components. A connection is a formula of the form $i_c = o_{c'}$, where i_c and $o_{c'}$ denote the input and output of components c and c', respectively. This way an equivalence relation on the inputs and outputs is defined, denoted by $IO_{\backslash \equiv}$. The class representatives from this set are denoted by $[r]$.

A *logical diagnostic problem* is defined as a pair $\mathcal{P}_L = (\mathcal{S}_L, OBS)$, where $\mathcal{S}_L$ is a logical diagnostic system and OBS is a finite set of logical formulae, representing *observations*.

Adopting the definition from [3], a diagnosis in the theory of consistency-based diagnosis is defined as follows. Let Δ be the assignment of either a normal or an abnormal behavioural assumption to *each* component. Then, Δ is a *consistency-based diagnosis* of the logical diagnostic problem $\mathcal{P}_L$ iff the observations are consistent with both the system description and the diagnosis:

$$SD \cup \Delta \cup OBS \nvDash \bot. \tag{5}$$

Here, $\nvDash$ stands for the negation of the logical entailment relation, and $\bot$ represents a contradiction.

Example 1. Consider the logical circuit depicted in Figure 1, which represents a full adder, i.e. a circuit that can be used for the addition of two bits with carry-in and carry-out bits. It is an example frequently used to illustrate concepts from model-based diagnosis. This circuit consists of two AND gates ($A1$ and $A2$), one OR gate ($R1$) and two exclusive-OR (XOR) gates ($X1$ and $X2$). These are the components that can be either faulty (abnormal) or normal.

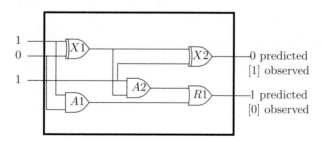

Fig. 1. Full adder with inputs $\{i_1, \bar{i}_2, i_3\}$, and observed ($\{o_{X2}, \neg o_{R1}\}$) and predicted outputs ($\{\neg o_{X2}, o_{R1}\}$)

3.2 Probabilistic Model-Based Diagnosis

In this section, we will map logical diagnostic problems onto probabilistic representations, called *Bayesian diagnostic problems*, using the Bayesian-network representation proposed by Flesch et al. [11], which was inspired by previous work by Geffner and Pearl [5,1]. As will become clear below, a Bayesian diagnostic problem is defined as (*i*) a Bayesian diagnostic system representing the components, including their behaviour and interaction, based on information from the logical diagnostic system of concern, and (*ii*) a set of observations.

Graphical Representation. First the graphical structure used to represent the structural information from a logical diagnostic system is defined. It has the form of an acyclic directed graph $G = (V, E)$, where V is the set of *vertices* and $E \subseteq (V \times V)$ is the set of *arcs*.

Definition 1 (*diagnostic mapping*). *Let* $\mathcal{S}_L$ = (SD, COMPS) *be a logical diagnostic system. The* diagnostic mapping m_d *maps* $\mathcal{S}_L$ *onto an acyclic directed graph* $G = m_d(\mathcal{S}_L)$, *as follows (see Figure 2):*

- *The vertices V of graph G are created according to the following rules:*
  - *Each component $c \in$ COMPS yields a vertex A_c used to represent its normal and abnormal behaviour;*
  - *Each class representative of an input or output $[r] \in \mathrm{IO}_{\backslash\equiv}$ yields an associated vertex $[r]$.*
  The set of all abnormality vertices A_c is denoted by Δ, i.e. $\Delta = \{A_c \mid c \in$ COMPS$\}$. The vertices of graph G are, thus, obtained as follows:

  $$V = \Delta \cup \mathrm{IO}_{\backslash\equiv},$$

  where $\mathrm{IO}_{\backslash\equiv} = I \cup O$, *with disjoint sets of* input vertices I *and* output vertices O.
- *The arcs E of G are constructed as follows:*
  - *There is an arc from each each input of a component c to each output of the component;*
  - *There is an arc for each component c from $A_c \in V$ to the corresponding output of the component.*

An example of using the diagnostic mapping is given below.

Example 2. Figure 3 shows the graphical representation of the full-adder circuit from Figure 1. The set V of vertices is:

$$V = \Delta \cup O \cup I$$
$$= \{A_{X1}, A_{X2}, A_{A1}, A_{A2}, A_{R1}\} \cup \{O_{X1}, O_{X2}, O_{A1}, O_{A2}, O_{R1}\}$$
$$\cup \{I_1, I_2, I_3\}.$$

The arcs from E connect (*i*) outputs of two components such as $O_{X1} \rightarrow O_{X2}$, (*ii*) an abnormality vertex with an output vertex such as $A_{A2} \rightarrow O_{A2}$ and (*iii*) an input vertex with an output vertex such as $I_3 \rightarrow O_{X2}$.

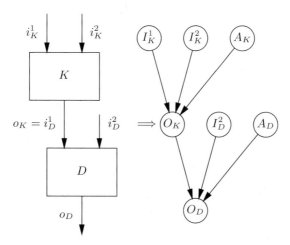

Fig. 2. The diagnostic mapping

Bayesian Diagnostic Problems. Recall that Bayesian networks that act as the basis for diagnostic Bayesian networks consist of two parts: a joint probability distribution and a graphical representation of the relations among the random variables defined by the joint probability distribution. Based on the definition of Bayesian networks, particular parts of a logical diagnostic system will be related to the graphical structure of a diagnostic Bayesian network, whereas other parts will have a bearing on the content of the probability table of the Bayesian network.

Having introduced the mapping of a logical diagnostic system to its associated graph structure, we next introduce the full concept of a Bayesian diagnostic system.

Definition 2 *(Bayesian diagnostic system)*. *Let* $\mathcal{S}_L = (\text{SD}, \text{COMPS})$ *be a logical diagnostic system, and* $G = m_d(\mathcal{S}_L)$ *be obtained by applying the diagnostic mapping. Let* P *be a joint probability distribution of the vertices of* G, *interpreted as random variables. Then,* $\mathcal{S}_B = (G, P)$ *is the associated* Bayesian diagnostic system.

Recall that by the definition of a Bayesian network, the joint probability distribution P of a Bayesian diagnostic system can be factorised as follows:

$$P(I, O, \Delta) = \prod_c P(O_c \mid \pi(O_c)) P(I) P(\Delta), \tag{6}$$

where O_c is an output variable associated to component $c \in \text{COMPS}$, and $\pi(O_c)$ are the random variables corresponding to the *parents* of O_c. The parents will normally consist of inputs I_c and an abnormality variable A_c.

To simplify notation, in the following, (sets of) random variables of a Bayesian diagnostic problem have the same names as the corresponding vertices. By a_c is indicated that abnormality variable A_c takes the value 'true', whereas by $\bar{a}_c$ it is

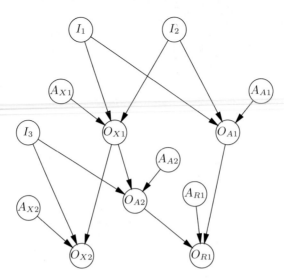

Fig. 3. A Bayesian diagnostic system corresponding to the circuit in Figure 1

indicated that A_c takes the value 'false'. A similar notation will be used for the other random variables. Finally, a specific abnormality assumption concerning all abnormality variables is denoted by δ_C, which is defined as follows:

$$\delta_C = \{a_c \mid c \in C\} \cup \{\bar{a}_c \mid c \in \text{COMPS} - C\},$$

with $C \subseteq \text{COMPS}$. There are some sensible constraints on the joint probability distribution P of a Bayesian diagnosis system that can be derived from the specification of the logical diagnostic system. These will be discussed later.

As with logical diagnostic problems, we need to add observations to Bayesian diagnostic systems in order to be able to solve diagnostic problems. In logical diagnostic systems, observations are the inputs and outputs of a system. It is generally not the case that the entire set of inputs and outputs of a system is observed. The set of input and output variables that have been observed, are referred to by I_ω and O_ω, respectively. The unobserved input and output variables will be referred to as I_u and O_u, respectively. We will use the notation i_ω to denote the values of the observed inputs, and o_ω for the observed output values. The set of *observations* is then denoted as $\omega = i_\omega \cup o_\omega$.

Now, we are ready to define the notion of Bayesian diagnostic problem, which is a Bayesian diagnostic system augmented by a set of observations.

Definition 3 (Bayesian diagnostic problem). *A Bayesian diagnostic problem, denoted by $\mathcal{P}_B$, is defined as the pair $\mathcal{P}_B = (\mathcal{S}_B, \omega)$, where $\mathcal{S}_B$ is a Bayesian diagnostic system and ω the set of observations of this system.*

Determining the diagnoses of a Bayesian diagnostic problem amounts to computing $P(\delta_C \mid \omega)$, and then finding the δ_C which maximises $P(\delta_C \mid \omega)$, i.e.

$$\delta_C^* = \arg\max_{\delta_C} P(\delta_C \mid \omega).$$

This problem is NP-hard; however, many special methods to make probabilistic inference feasible are known [6]. The probability $P(\delta_C \mid \omega)$ can be computed by Bayes' rule, using the probabilities from the specification of a Bayesian diagnostic system:

$$P(\delta_C \mid \omega) = \frac{P(\omega \mid \delta_C)P(\delta_C)}{P(\omega)}. \tag{7}$$

As a consequence of the independences that hold for a Bayesian diagnostic system, it is possible to simplify the computation of the conditional probability distribution $P(\omega \mid \delta_C)$. According to the definition of a Bayesian diagnostic system it holds that

$$P(i \mid \delta_C) = P(i),$$

for each $i \subseteq (i_\omega \cup i_u)$, as the input variables and abnormality variables are independent. In addition, it is assumed that the input variables are independent.

Using these results, basic probability theory and the definition of a Bayesian diagnostic problem yields the following derivation:

$$
\begin{aligned}
P(\omega \mid \delta_C) &= P(i_\omega, o_\omega \mid \delta_C) \\
&= \sum_{i_u} P(i_u)P(i_\omega, o_\omega \mid i_u, \delta_C) \\
&= P(i_\omega) \sum_{i_u} P(i_u) \sum_{o_u} \prod_c P(O_c \mid \pi(O_c)),
\end{aligned} \tag{8}
$$

since it holds by the axioms of probability theory that

$$P(i_\omega, o_\omega \mid i_u, \delta_C) = \sum_{o_u} P(i_\omega) \prod_c P(O_c \mid \pi(O_c)) \ .$$

To emphasise that the set of parents $\pi(O_c)$ includes an abnormality variable that is assumed to be true, i.e. the component is assumed to behave abnormally, the following notation is used $P(O_c \mid \pi(O_c) : a_c)$; similar, for the situation where the component c is assumed to behave normally the notation $P(O_c \mid \pi(O_c) : \bar{a}_c)$ is employed. Finally, the following assumptions are made and will be used in the remainder of this paper:

– $P(O_c \mid \pi(O_c) : a_c) = P(O_c \mid a_c)$, i.e. the probabilistic behaviour of a component that is faulty is independent of its inputs;
– $P(O_c \mid \pi(O_c) : \bar{a}_c) \in \{0, 1\}$, i.e. normal components behave deterministically.

The probability $P(o_c \mid a_c)$ will be abbreviated in the following section as p_c; thus $P(\bar{o}_c \mid a_c) = 1 - p_c$ These are realistic assumptions, as it is unlikely that detailed functional behaviour will be known for a component that is faulty, whereas when the component is not faulty, it is certain it will behave as intended. Note that the latter assumption is identical to that used in traditional, logical model-based diagnosis.

4   Decomposition of Probability Distribution

To establish that probabilistic model-based diagnosis can be partly interpreted in terms of a Poisson-binomial distribution, it is necessary to decompose Equation (8) into various parts. The first part will represent the probabilities that components c produce the right, o_c, or wrong, $\bar{o}_c$, output, which correspond to the success and failure probabilities, respectively, of a Poisson-binomial distribution. The second part represents a normally functioning system fragment, which will be represented by a Boolean function. There is also a third part, which concerns the observed and unobserved inputs. We start by distinguishing between various types of components, inputs and outputs, in order to make the necessary distinction:

- The sets of components assumed to function *normally* and *abnormally* will be denoted by $C^{\bar{a}}$ and C^a, respectively, with $C^{\bar{a}}, C^a \subseteq \text{COMPS}$;
- The sets $C^{\bar{a}}$ and C^a are partitioned into sets of components, for *observed* and *unobserved* outputs, indicated by the sets $C_{\omega}^{\bar{a}}$, $C_u^{\bar{a}}$, C_{ω}^a and C_u^a, respectively.

Thus, $C^{\bar{a}} = C_{\omega}^{\bar{a}} \cup C_u^{\bar{a}}$ and $C^a = C_{\omega}^a \cup C_u^a$. In addition, we will sometimes make a distinction between components c for which o_c has been observed, and components c for which $\bar{o}_c$ has been observed. These sets will be denoted by C_{ω}^o and $C_{\omega}^{\bar{o}}$, respectively. It holds that C_{ω}^o and $C_{\omega}^{\bar{o}}$ constitute a partition of C_{ω}. The notations can also be combined, e.g., as $C_{\omega}^{a,o}$ and $C_{\omega}^{a,\bar{o}}$. Furthermore, we will sometimes use a similar notation for sets of output variables, e.g., $O_u^{\bar{a}} = \{O_c \mid c \in C_u^{\bar{a}}\}$ and $O_{\omega}^{\bar{a}} = \{O_c \mid c \in C_{\omega}^{\bar{a}}\}$, and input variables, e.g., $I_u^{\bar{a}} = \bigcup_{c \in C_u^{\bar{a}}} I_c$ indicates unobserved inputs of components that are assumed to behave normally and $I_{\omega}^{\bar{a}} = \bigcup_{c \in C_{\omega}^{\bar{a}}} I_c$ are observed inputs of components that are assumed to behave normally, with I_c the set of input variables of component $c \in \text{COMPS}$ and $I^{\bar{a}} = I_{\omega}^{\bar{a}} \cup I_u^{\bar{a}}$.

The following lemma shows that it is possible to decompose part of the joint probability distribution of Equation (6) using the component sets defined above.

Lemma 1. *The following statements hold:*

- *The joint probability distribution of the outputs of the set of assumed normally functioning components $C^{\bar{a}}$, can be decomposed into two products as follows:*

$$\prod_{c \in C^{\bar{a}}} P(O_c \mid \pi(O_c) : \bar{a}_c)$$

$$= \prod_{c \in C_u^{\bar{a}}} P(O_c \mid \pi(O_c) : \bar{a}_c) \prod_{c \in C_{\omega}^{\bar{a}}} P(O_c \mid \pi(O_c) : \bar{a}_c).$$

- *Similarly, the joint probability distribution of the outputs of the set of assumed abnormally functioning components C^a, can be decomposed into two products as follows:*

$$\prod_{c \in C^a} P(O_c \mid \pi(O_c) : a_c) = \prod_{c \in C_u^a} P(O_c \mid a_c) \prod_{c \in C_{\omega}^a} P(O_c \mid a_c).$$

Proof: The decompositions follows from the definitions of the sets C^a, C^a_ω, C^a_u, $C^{\bar{a}}_u$ and $C^{\bar{a}}_\omega$, and the independence assumptions underlying the distribution P. □

Now, based on Lemma 1, we can also decompose the product of the *entire* set of components, as follows:

$$\prod_c P(O_c \mid \pi(O_c))$$

$$= \prod_{c \in C^{\bar{a}}_u} P(O_c \mid \pi(O_c) : \bar{a}_c) \prod_{c \in C^{\bar{a}}_\omega} P(O_c \mid \pi(O_c) : \bar{a}_c)$$

$$\times \prod_{c \in C^a_u} P(O_c \mid a_c) \prod_{c \in C^a_\omega} P(O_c \mid a_c).$$

Next, we show that the outputs of the set of observed abnormal components C^a_ω only depend on probabilities $p_c = P(o_c \mid a_c)$, $c \in C^a_\omega$.

Lemma 2. *The joint probability of observed outputs of the abnormally assumed components can be written as:*

$$\prod_{c \in C^a_\omega} P(O_c \mid \pi(O_c) : a_c) = \prod_{c \in C^{a,o}_\omega} p_c \prod_{c \in C^{a,\bar{o}}_\omega} (1 - p_c).$$

Proof: This follows straight from the definitions of C^a_ω, $C^{a,o}_\omega$ and $C^{a,\bar{o}}_\omega$.   □

Recall that the probability of an output of a normally functioning component was assumed to be either 0 or 1, i.e. $P(O_c \mid \pi(O_c) : \bar{a}_c) \in \{0, 1\}$. Clearly, these probabilities yield, when multiplied, Boolean functions. One of these Boolean functions, denoted by φ, is defined as follows: $\varphi(o^{\bar{a}}_u, o^a_u, i^{\bar{a}}) = \prod_{c \in C^{\bar{a}}_u} P(O_c \mid \pi(O_c) : \bar{a}_c)$, where the set of parents $\pi(O_c)$ may, but need not, contain variables from the sets of variables O^a_u and $I^{\bar{a}}$. However, $\pi(O_c)$ does not contain variables from the set I^a, as these only condition variables that are assumed to behave abnormally and are then ignored, as mentioned at the end of the previous section. Similarly, we define Boolean functions $\psi(o_u, o^{\bar{a}}_\omega, i^{\bar{a}}) = \prod_{c \in C^{\bar{a}}_\omega} P(O_c \mid \pi(O_c) : \bar{a}_c)$.

Lemma 3. *For each value o^a_u and $i^{\bar{a}}$, there exists exactly one value $o^{\bar{a}}_u$ of the set of variables $O^{\bar{a}}_u = \{O_c \mid c \in C^{\bar{a}}_u\}$ for which it holds that $\varphi(o^a_u, o^{\bar{a}}_u, i^{\bar{a}}) = 1$; similarly, for each value o_u and $i^{\bar{a}}$ there exists one value $o^{\bar{a}}_\omega$ of the set of variables $O^{\bar{a}}_\omega = \{O_c \mid c \in C^{\bar{a}}_\omega\}$ for which it holds that $\psi(o_u, o^{\bar{a}}_\omega, i^{\bar{a}}) = 1$.*

Proof: As both the functions φ and ψ are defined as products of conditional probability distributions $P(O_c \mid \pi(O_c) : \bar{a}_c)$, for which we have that $P(o_c \mid \pi(O_c) : \bar{a}_c) \in \{0, 1\}$, there is, due to the axioms of probability theory, for any value of the variables corresponding to the parents of the variables O_c at most one value for each O_c for which the joint probability $\prod_c P(O_c \mid \pi(O_c) : \bar{a}_c) = 1$.   □

The following lemma, which is used later, is a consequence of the definition of these Boolean functions.

Lemma 4. *Let the Boolean functions φ and ψ be as defined above, then:*

$$\sum_{o_u} \varphi(o_u^a, o_u^{\bar{a}}, i^{\bar{a}}) \psi(o_u, o_w^{\bar{a}}, i^{\bar{a}}) \prod_{c \in C^a} P(O_c \mid a_c) =$$

$$\sum_{o_u^a} b(o_u^a, i^{\bar{a}}) \prod_{c \in C^{a,o}} p_c \prod_{c \in C^{a,\bar{o}}} (1 - p_c),$$

with Boolean function b and $p_c = P(o_c \mid a_c)$.

Proof: First, the Boolean function b is defined for a given set of observed outputs o_w: $b(o_u, i^{\bar{a}}) = \varphi(o_u^a, o_u^{\bar{a}}, i^{\bar{a}}) \psi(o_u, o_w^{\bar{a}}, i^{\bar{a}})$, then,

$$\sum_{o_u} \varphi(o_u^a, o_u^{\bar{a}}, i^{\bar{a}}) \psi(o_u, o_w^{\bar{a}}, i^{\bar{a}}) \prod_{c \in C^a} P(O_c \mid a_c) = \sum_{o_u} b(o_u, i^{\bar{a}}) \prod_{c \in C^a} P(O_c \mid a_c).$$

Furthermore, due to Lemma 3, it suffices to only consider the restriction of the function b to the variables O_u^a and $I^{\bar{a}}$, as for given values o_u^a and $i^{\bar{a}}$, $b(o_u^a, o_u^{\bar{a}}, i^{\bar{a}}) = 0$ for all but one value of $O_u^{\bar{a}}$. This function is denoted by $b(o_u^a, i^{\bar{a}})$. The product term results from application of a slight generalisation of Lemma 2. $\square$

We are now ready to establish that $P(\omega \mid \delta_C)$ can be written as the sum of weighted products of the form $\prod_c p_c \prod_{c'} (1 - p_{c'})$, i.e. Equation (1).

Theorem 1. *Let $\mathcal{P}_B = (\mathcal{S}_B, \omega)$ be a Bayesian diagnostic problem. Then, $P(\omega \mid \delta_C)$ can be expressed as follows:*

$$P(\omega \mid \delta_C) = P(i_\omega) \sum_{i_u^{\bar{a}}} P(i_u^{\bar{a}}) \sum_{o_u^a} b(o_u^a, i^{\bar{a}}) \prod_{c \in C^{a,o}} p_c \prod_{c \in C^{a,\bar{o}}} (1 - p_c),$$

where $b(o_u^a, i^{\bar{a}}) \in \{0, 1\}$ and $p_c = P(o_c \mid a_c)$.

Proof: The result follows from the above lemmas and the fact that we sum over (part of) the input variables I. Note that only the variables $I^{\bar{a}}$ are used as conditioning variables, which follows from the assumption that $P(O_c \mid \pi(O_c) : a_c) = P(O_c \mid a_c)$. As only the input variables $i_u^{\bar{a}}$ are assumed to be dependent of output variables, we obtain: $\sum_{i_u, o_u^a} P(i_u) \cdots = \sum_{i_u^{\bar{a}}, o_u^a} P(i_u^{\bar{a}}) \cdots$. The Boolean function $b(o_u^a, i^{\bar{a}})$ is as above. $\square$

An alternative version of the theorem can be obtained in terms of expectations using Equation (4) for the Poisson-binomial distribution:

$$P(i_\omega) \sum_{i_u^{\bar{a}}} P(i_u^{\bar{a}}) \sum_{o_u^a} b(o_u^a, i^{\bar{a}}) \prod_{c \in C^{a,o}} p_c \prod_{c \in C^{a,\bar{o}}} (1 - p_c)$$

$$= P(i_\omega) \prod_{c \in C_\omega^a} P(O_c \mid a_c) \sum_{i_u^{\bar{a}}} P(i_u^{\bar{a}}) \mathcal{E}_P(b_{i^{\bar{a}}}(O_u^a)),$$

i.e. the sum of the mean of the Boolean functions $b_{i^{\bar{a}}}$, which are functions of the unobserved inputs $i_u^{\bar{a}}$, in terms of the probability function P (Equation (4)), weighed by the prior probability of unobserved inputs $i_u^{\bar{a}}$. Combining this with Equation (7) yields $P(\delta_C \mid \omega)$. Thus, to probabilistically rank diagnoses δ_C it is necessary to compute: (i) $\mathcal{E}_P(b_{i^{\bar{a}}}(O_u^a))$, the Poisson-binomial distribution mean of the behaviour of the normally assumed, unknown components, (ii) $P(i_u^{\bar{a}})$, (iii) $\prod_{c \in C_\omega^a} P(O_c \mid a_c)$, the observed abnormal components, and (iv) the prior $P(\delta_c)$. Note that $P(i_\omega)$ can be cancelled by $P(\omega)$ in Equation (7) and both probabilities are irrelevant for ranking.

5   Conclusions

We have shown that probabilistic model-based diagnosis, which is an extension of traditional GDE-like model-based diagnosis, can be decomposed into computation of various probabilities, in which a central role is played by the Poisson-binomial distribution. When all probabilities $p_c = P(o_c \mid a_c)$ are assumed to be equal, a common simplifying assumption in model-based diagnosis, the analysis reduces to the use of the standard binomial distribution.

So far, most other research on integrating probabilistic reasoning with logic-based model-based diagnosis took probabilistic reasoning as adding some sort of uncertain, abductive reasoning to logical reasoning. No attempts were made in related research to look inside what happens in the diagnostic process, as was done in this paper. We expect that it becomes thus possible to investigate further variations in probabilistic model-based diagnosis, for example, by adopting assumptions different from those in this paper with regard to fault behaviour in systems.

References

1. Pearl, J.: Probabilistic Reasoning in Intelligent Systems: Networks of Plausible Inference. Morgan Kauffman, San Francisco (1988)
2. Reiter, R.: A theory of diagnosis from first principles. Artificial Intelligence 32, 57–95 (1987)
3. de Kleer, J., Mackworth, A.K., Reiter, R.: Characterizing diagnoses and systems. Artificial Intelligence 52, 197–222 (1992)
4. de Kleer, J., Williams, B.C.: Diagnosing multiple faults. Artificial Intelligence 32, 97–130 (1987)
5. Geffner, H., Pearl, J.: Distributed diagnosis of systems with multiple faults. In: Proc. of the 3rd IEEE Conference on AI Applications, pp. 156–162. IEEE, Los Alamitos (1987)
6. Gámez, J.: Abductive inference in Bayesian Networks: a review, pp. 101–120. Springer, Heidelberg (2004)
7. Darwiche, A.: Model-based diagnosis using structured system descriptions. Artificial Intelligence Research, 165–222 (1998)
8. Cam, L.L.: An approximation theorem for the poisson binomial distribution. Pacific Journal of Mathematics 10, 1181–1197 (1960)
9. Darroch, J.: On the distribution of the number of successes in independent trials. The Annals of Mathematical Statistics 35, 1317–1321 (1964)
10. Lucas, P.: Bayesian network modelling through qualitative patterns. Artificial Intelligence 163, 233–263 (2005)
11. Flesch, I., Lucas, P., van der Weide, T.: Conflict-based diagnosis: adding uncertainty to model-based diagnosis. In: Proceedings of IJCAI 2007, pp. 380–388. Morgan Kauffman, NJ (2007)

Simplifying Learning in Non-repetitive Dynamic Bayesian Networks

Ildikó Flesch and Eric O. Postma

Tilburg centre for Creative Computing, Tilburg University
{ildikoflesch,eric.postma}@gmail.com

Abstract. Dynamic Bayesian networks (DBNs) are increasingly adopted as tools for the modeling of dynamic domains involving uncertainty. Due to their ease of modeling, repetitive DBNs have become the standard. However, repetition does not allow the independence relations to vary over time. *Non-repetitive* DBNs do allow for modeling time-varying relations, but are hard to apply to dynamic domains.

This paper presents a novel method that facilitates the use of non-repetitive DBNs and simplifies learning DBNs in general. This is achieved by learning disjoint sets of independence relations of separate parts of a DBN, and, subsequently, joining these relations together to obtain the complete set of independence relations of the DBN. Our simplified learning method improves previous methods by removing redundant operations which yields computational savings in the learning process of the network. Experimental results show that the simplified learning method facilitates the use of non-repetitive DNBs and enables us to build them in a seamless fashion.

1 Introduction

In recent years, Dynamic Bayesian networks (DBNs) became a popular tool for representing time-related uncertain processes. They are explored in various fields of applications, such as speech recognition (e.g. [2]) and gene-expression analysis [6]. The reason for their succes is their ease of representation of independence relations of random variables in dynamic processes. DBNs are distinguished into two main classes: repetitive and non-repetitive networks [4]. In repetitive DBNs the (in)dependence relations do not vary beyond the repetition cycle. In contrast, non-repetitive DBNs may change their structure over time.

Repetitive DBNs have become a standard due to their simple structure for modeling and analysis [6]. Since the appropriate modeling of many real-life dynamic processes requires taking the changing indepence relations into account, *non-repetitive* DBNs should be the method of choice for real-life applications. As a case in point, Tucker and Liu [8] established that non-repetitive DBNs are practically useful. More recently, experimental results have shown that for a real-life problem domain with finite time interval, non-repetitive DBNs perform better than repetitive DBNs [3]. In this earlier work, atemporal and temporal independence relations were distinguished, and subsequently joined together to

C. Sossai and G. Chemello (Eds.): ECSQARU 2009, LNAI 5590, pp. 216–227, 2009.

obtain all the independences of a DBN. This way of modeling simplifies the learning procedure of non-repetitive DBNs.

This paper builds on the earlier work by identifying and removing redundancies in the joining of atemporal and temporal relations. We propose a simplified learning method to further facilitate the analysis and application of non-repetitive DBNs. To avoid redundancy, our approach makes use of a more detailed distinction of atemporal relations in the graphical representation of DBNs, in which the separate parts of the atemporal relations include only *disjoint* sets of dependence information. These disjoint sets, which are the inputs for the simplified learning method, can then be joined to determine all relations of a DBN. Since our simplified learning method considers disjoint independence relations only, our approach provides computational savings as compared to the original method. The performance of the proposed method is shown by experiments on a real-life problem domain with infinite discrete time represented by a non-repetitive DBN. As far as the authors know this is the first time that a non-repetitive DBN is succesfully applied to a real-life problem with an *infinite* time interval.

The separation of repetitive and non-repetitive parts of a DBN has been proposed in work on multi-network models [1]. The simplified learning method differs from the earlier work that it enables to deal with time-related processes.

2   Motivating Example

A real-world data set of vessel traffic at the Dutch coast is used as an example throughout this paper. In this data set, we will focus on two vessels: the Sirius and the Anton (the vessels have been renamed for privacy and security reasons). Fig. 1 shows the movements of the two vessels, whose routes intersect at some point. In our data, each vessel has dynamic features, such as speed, longitude, and latitude; the values of these features may vary in time.

Initially, we learned a DBN, which represents the movements of the Sirius and the Anton, *without* considering any interaction between the two vessels. Fig. 2 shows the graphical representation of the learned model in three consecutive time intervals. In this figure, Sirius and Anton are denoted by S and A, and the longitude and latitude are abbreviated to 'long' and 'lat', respectively. The three rectangular boxes, labelled 1, 2, and 3, represent consecutive time steps. In this network, at each time step, for both vessels it holds that the longitude and latitude are dependent on the vessel-speed and the longitude and the latitude are dependent on each other (solid arrows), moreover, each variable also influences its own value at the consecutive time step (dotted arrows). This is in accordance with our expectations. Note that as the structure of this network does not change over time it is *repetitive*.

In general, not considering intersections between vessels is unrealistic. Therefore, the question arises if the structure remains repetitive if we take into account that the routes of vessels Sirius and Anton intersect each other at a certain time step. In our data, at time step 1, Sirius and Anton are approaching the same water area. Subsequently, at time step 2 they are passing each others route,

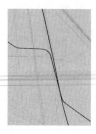

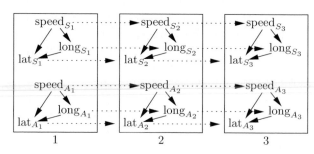

Fig. 1. Two inter-
secting vessel routes
(solid thick lines)

Fig. 2. A repetitive DBN representing the movements of
two *non-intersecting* vessels. Solid arcs represent atemporal dependences, whereas dashed arrows temporal dependences.

and at time step 3 they are not at the same sea area again. In this case, at
time steps 1 and 3 the speed and position of these two vessels are independent
of each other, while at time step 2 their variables might depend on each other
preventing a possible collision (cf. Fig. 1). This means that we have different
dependences at different time steps, and, therefore, non-repetitive DBNs are
preferred to represent our vessel traffic domain. This is our motivation for the
study of non-repetitive DBNs. Our example is further examined in Section 7.

3   Preliminaries

3.1   Dynamic Bayesian Networks

Let X_V be a set of *discrete* random variables with index set V, let X_v with
$v \in V$ denote a random variable, and let X_W with $W \subseteq V$ denote a set of
random variables. Furthermore, let P denote a joint probability distribution
(JPD) on X_V. The set X_U is said to be *conditionally independent* of X_W *given*
X_Z, with $U, W, Z \subseteq V$, if

$$P(X_U \mid X_W, X_Z) = P(X_U \mid X_Z) , \qquad (1)$$

denoted by $X_U \perp\!\!\!\perp_P X_W \mid X_Z$. If the set X_Z is empty, then it is abbreviated to
$X_U \perp\!\!\!\perp_P X_W$. These independence statements in the joint probability distribution
can also be graphically represented by an acyclic directed graph $G = (V, A)$ with
the set of vertices V (representing random variables) and the set of arcs A, having
no directed path $V_1 \to V_2 \to \ldots \to V_n$ s.t. $V_1 = V_n$. In an acyclic directed graph
G, the *independence relation* (i.e. the entire set of independence statements in G)
is denoted by $\perp\!\!\!\perp_G$ and the *dependence relation* (i.e. the entire set of dependence
statements in G) by $\not\perp\!\!\!\perp_G$. In this paper, the derived (in)dependence statements
from an example graph are derived from the d-separation criterion [5].

A *Bayesian network* is defined as a pair $\mathcal{B} = (G, P)$, where G is an acyclic
directed graph representing relations of random variables X_V, P is the joint
probability distribution on X_V, and each independence statement represented

in G is also a valid independence statement in P. Bayesian networks that include the dimension of time are called Dynamic Bayesian networks (DBNs); time is denoted by T. The graphical representation of a DBN consists of two parts: an atemporal (time-independent) part, and a temporal (time-dependent) part.

An atemporal relation of time step $t \in T$ is represented by a *timeslice* acyclic directed graph $G_t = (V_t, A_t^a)$ with set of vertices V_t and set of *atemporal arcs* $A_t^a \subseteq V_t \times V_t$. All the timeslices together form the *atemporal network* G:

$$G = \{G_t \mid t \in T\} = \{(V_t, A_t^a) \mid t \in T\} . \tag{2}$$

Related to the time-dependent relations, a *temporal arc* connects two vertices in different timeslices; it directs *always* from the past to the future. In this paper it is drawn as a dotted arrow. The set of temporal arcs is denoted by A^t. A *temporal network* $N = (V_T, A)$ consists of the set of vertices of the timeslices, i.e. V_T, and the union of the atemporal and temporal arcs, i.e. $A = A^a \cup A^t$.

A *Dynamic Bayesian network* (DBN) is formally defined as a pair $\mathcal{DBN} = (N, P)$, where P is the JPD on the entire set of random variables. Clearly, a temporal network N is an acyclic directed graph and it is the graphical representation of a DBN.

For example, Fig. 2 represents three timeslices. In this network, speed$_{A_1}$ -> lat$_{A_1}$ is an atemporal arc, since it connects vertices from the same timeslice G_1, with $t = 1$, whereas speed$_{A_1}$ $\cdots$> lat$_{A_2}$ is a temporal arc.

3.2   Atemporal and Temporal Relations and Their Join

The atemporal and temporal independence relations can be separated in DBNs based on the concept of trails [3]. A *trail* in a graph is a sequence of unique vertices $v_1, v_2, \ldots, v_m$, where consecutive vertices are connected by an arc pointing either forward or backward; each arc occurs only once. The set of trails is denoted by Θ.

In a temporal network, an atemporal independence can be represented by means of an *atemporal trail* θ^a containing no temporal arcs, whereas a temporal relation is defined by a *temporal trail* θ^t consisting of *at least one* temporal arc. The sets of all atemporal and temporal trails are denoted by Θ^a and Θ^t, respectively. With regards to the temporal relationships we only need to consider temporal trails resulting in a *reduced temporal network* $N_{|\Theta^t} = (V_T, A_{\Theta^t})$, with its set of arcs $A_{\Theta^t} \subseteq A$ consisting of all the arcs included on the temporal trails in Θ^t. Since temporal trails may consist of *both* atemporal and temporal trails a further partitioning is defined. The *atemporal part of the reduced temporal network* is denoted by $N_{|\Theta^t}^a = (V_T, A_{\Theta^t}^a)$, where $A_{\Theta^t}^a \subseteq A^a$ consists of all *atemporal* arcs in the reduced temporal network. The *temporal part of the reduced temporal network* is denoted by $N_{|\Theta^t}^t = (V_T, A_{\Theta^t}^t)$, where $A_{\Theta^t}^t \subseteq A^t$ consists of all *temporal* arcs in the reduced temporal network. Fig. 3(a) summarises the independence relations of these above-defined networks and Fig. 4 shows an example.

It has been shown that for the sound and complete composition of the above-defined temporal and atemporal independence relations of DBNs, the dependence preservation and independence concatenation properties has to be satisfied [3].

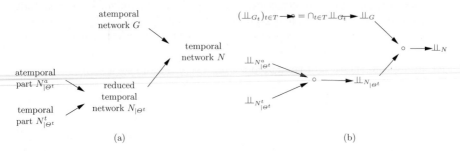

(a)   (b)

Fig. 3. (a) Various parts of a DBN, (b) an earlier presented join method, where initially the atemporal and temporal parts of the reduced temporal network are joined, and subsequently the atemporal network is joined with the reduced temporal network

The *dependence preservation* property establishes that a dependence statement dominates an independence statement, therefore, joining either two dependence or one dependence and one independence statements will result in a dependence statement. For example, suppose we want to join independence relations $\perp\!\!\!\perp$ and $\perp\!\!\!\perp'$ of two graphs defined on the same set of vertices. Then, when we join dependence statement $U \!\not\!\perp\!\!\!\perp W \mid Z$ with either dependence statement $U \!\not\!\perp\!\!\!\perp' W \mid Z$ or with independence statement $U \perp\!\!\!\perp' W \mid Z$, then in the joined relation $\perp\!\!\!\perp''$ dependence statement $U \!\not\!\perp\!\!\!\perp'' W \mid Z$ will hold. On the other hand, when joining two independence statements, we either apply the d-separation criterion to the composite graph, or apply the *independence concatenation* procedure, which is a special version of d-separation exploiting some special properties regarding the temporal composition. For example, depending on the graphical representation of independence relation $\perp\!\!\!\perp''$, the join of $U \perp\!\!\!\perp W \mid Z$ with $U \perp\!\!\!\perp' W \mid Z$ can remain independence $U \perp\!\!\!\perp'' W \mid Z$ or can turn into dependence $U \!\not\!\perp\!\!\!\perp'' W \mid Z$.

Then, satisfying both the dependence preservation and the independence concatenation properties, the *join operator*, denoted by ○, joins two independence relations $\perp\!\!\!\perp$ and $\perp\!\!\!\perp'$ (both defined on the same vertex set V) resulting in independence relation $\perp\!\!\!\perp''$. An overview of the various join operations defined for the different parts of a temporal network N is given in Fig. 3(b).

4   Demonstration of Redundancy of Join Operations

As was mentioned in the introduction, the theoretical separation of the independence relations of a DBN is not complete yet, since there are some join operations which are redundant. In this section, we discuss this redundancy and we also demonstrate it by example.

Consider the atemporal network G and the reduced temporal network $N_{|\Theta^t}$ of the graphical representation of a DBN. By definition, the relations in the reduced temporal network are constructed by temporal trails consisting of both atemporal and temporal arcs. However, these atemporal arcs are also included in the atemporal network G, since they also define time-independent relations.

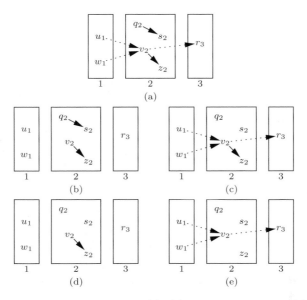

Fig. 4. Separation of a temporal network (a): (b) atemporal network, (c) reduced temporal network, (d) atemporal part of the reduced temporal network, and (e) temporal part of the reduced temporal network

Therefore, when a dependence statement in the atemporal network G is joined by a dependence statement in the reduced temporal network $N_{|\Theta^t}$, these dependence statements might be defined by the *same* set of atemporal arcs. This implies that joining independence relations $\perp\!\!\!\perp_G$ and $\perp\!\!\!\perp_{N_{|\Theta^t}}$ (see Fig. 3(b)) includes some unnecessary join operations. This issue is illustrated by the following example.

Example 1. Consider the temporal network in Fig. 4(a). Suppose that we want to join its atemporal network G in Fig. 4(b) with its reduced temporal network $N_{|\Theta^t}$ in Fig. 4(c). Then, we need to join the two dependence statements $v_2 \not\perp\!\!\!\perp_G z_2$ and $v_2 \not\perp\!\!\!\perp_{N_{|\Theta^t}} z_2$ with each other, which results into dependence $v_2 \not\perp\!\!\!\perp_N z_2$ (here, dependence preservation is applied). Now observe that the two dependence statements in the atemporal network G and the reduced temporal network $N_{|\Theta^t}$ are defined by the *same* atemporal arc, namely $v_2 \to z_2$. This can happen, since the reduced temporal network includes atemporal arcs.

The above-mentioned redundant join operations impose a further partitioning of some independence relations, which will be shown in the next section.

5   The Atemporal Partitioning of DBNs

In the previous section it was discussed that the original join operation includes redundant work suggesting that for a theoretically complete separation of the independence relations, a more detailed partitioning of these relations is still

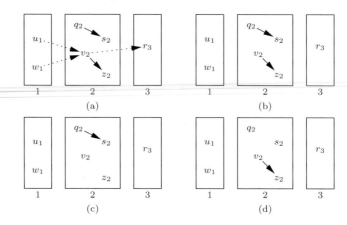

Fig. 5. The atemporal parts of the atemporal networks: (a) temporal network, (b) atemporal network, (c) atemporal part of the atemporal network, and (d) temporal part of the atemporal network

needed. In this section, to start, this separation is defined, and, subsequently, it is shown by a theorem, why this separation leads to the solution for preventing unnecessary join operations.

Recall that an atemporal network is defined as the set of vertices and the set of atemporal arcs of a temporal network. Since atemporal arcs might still have relations with temporal parts by being included in a temporal trail, as we have seen in the previous section, it imposes that this set has to be further partitioned into two disjoint subsets implying the following definitions.

Definition 1. (atemporal part of the atemporal network). *Let* $N = (V, A)$ *be a temporal network with associated atemporal network* $G = (V, A^a)$. *Then,* $G^a = (V, A^{a^a})$ *is called the* atemporal part of the atemporal network *with set of vertices* V, *and with set of arcs* $A^{a^a} \subseteq A^a$ *equal to the set of atemporal arcs in the atemporal network that are* not *included on any temporal trail of* N.

Definition 2. (temporal part of the atemporal network). *Let* $N = (V, A)$ *be a temporal network with associated atemporal network* $G = (V, A^a)$. *Then,* $G^t = (V, A^{a^t})$ *is called the* temporal part of the atemporal network, *with set of vertices* V, *and with set of arcs* $A^{a^t} \subseteq A^a$ *equal to the set of atemporal arcs in the atemporal network that are* included on *at least one* temporal trail of N.

Note that definitions 1 and 2 establish that the sets of atemporal arcs A^{a^a} and A^{a^t} are *disjoint* sets and their union is equal to the set of atemporal arcs, i.e. $A^{a^a} \cap A^{a^t} = \emptyset$ and $A^a = A^{a^a} \cup A^{a^t}$. The definitions above are illustrated by means of the following example.

Example 2. Fig. 5(a) shows a temporal network and Fig. 5(b) shows its atemporal network, whereas the figures 5(c), and 5(d) give the various parts of the atemporal network defined in the definitions above. As the arc $q_2 \rightarrow s_2$ is not included

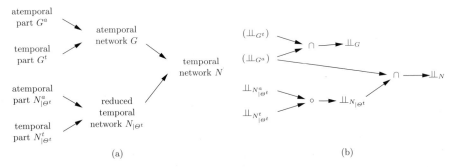

Fig. 6. (a) The various parts of the temporal network of a DBN, (b) the simplified learning method

on any temporal trail of the temporal network 5(a) it belongs to the atemporal part of the atemporal network, i.e. $A^{a^a} = \{(q_2, s_2)\}$. However, the atemporal arc $v_2 \rightarrow z_2$ is a member of several temporal trails, such as $w_1 \rightarrow v_2 \rightarrow z_2$, and thus it is a member of the temporal part of the atemporal network. Clearly, it holds that $A^{a^t} = \{(v_2, z_2)\}$.

Fig. 6(a) offers a summary of the various graphical parts of the temporal network including the new separation of the atemporal network.

The reason that these definitions form the basis to leave out needless join operations is established in the following theorem. Here, special properties related to independence relations in atemporal networks are derived.

Theorem 1. *Let $N = (V, A)$ be a temporal network with associated atemporal network $G = (V, A^a)$ and reduced temporal network $N_{|\Theta^t} = (V, A_{|\Theta^t})$. Then,*

- *the dependence relation of the atemporal part G^a of the atemporal network and the dependence relation of the reduced temporal network $N_{|\Theta^t}$ are disjoint, i.e. $\not\perp\!\!\!\perp_{G^a} \cap \not\perp\!\!\!\perp_{N_{|\Theta^t}} = \emptyset$;*
- *the dependence relation of the temporal part G^t of the atemporal network is a subset of the dependence relation of the reduced temporal network $N_{|\Theta^t}$, i.e. $\not\perp\!\!\!\perp_{G^t} \subseteq \not\perp\!\!\!\perp_{N_{|\Theta^t}}$.*

Proof. To start, the first item is proved. The independence relation $\perp\!\!\!\perp_{G^a}$ is defined on the set of atemporal arcs A^{a^a} that are not included in any temporal trail of the temporal network. In addition, the reduced temporal network $N_{|\Theta^t}$ consists of a set of dependence statements that are *only* related to the set of atemporal arcs that are included in at least one temporal trail; this set was defined by A^{a^t}. Since the sets A^{a^a} and A^{a^t} are disjoint, the set of dependence statements of the relations $\not\perp\!\!\!\perp_{G^a}$ and $\not\perp\!\!\!\perp_{N_{|\Theta^t}}$ are also disjoint.

For the proof of the second item note that the set of atemporal arcs A^{a^t} in G^t is included in the set of arcs of the temporal trails and, therefore, in the set of arcs of the reduced temporal network. Therefore, the dependence statements that are represented in the graph G^t are a subset of the dependence statements of the reduced temporal network, which completes the proof.   □

The previous theorem shows that when independence relations $\perp\!\!\!\perp_G$ and $\perp\!\!\!\perp_{N_{|\Theta^t}}$ are joined, it is redundant to join the subset $\perp\!\!\!\perp_{G^t} \subseteq \perp\!\!\!\perp_G$ with relation $\perp\!\!\!\perp_{N_{|\Theta^t}}$. This issue will be exploited in our simplified learning method as one of the main contribution of this paper.

6   The Simplified Learning Method

In this section, the simplified learning method to join independence relations is presented, which is based on the further separation of atemporal networks introduced in the previous section. More specifically, in Section 6.1, the join of the atemporal and temporal parts of the atemporal network G is studied. Subsequently, in Section 6.2 the simplified learning method for joining all the separate parts of the entire graphical representation of a DBN is presented.

6.1   Joining Atemporal Relations

In this section, we show that in the context of the atemporal network the join operator $\circ$ can be interpreted as the intersection of the independence relations. This is established in Proposition 1 followed by an example.

Proposition 1. *Let $N = (V, A)$ be a temporal network of a DBN with atemporal network $G = (V, A^a)$. Then, it holds that*

- $\not\!\perp\!\!\!\perp_G = \not\!\perp\!\!\!\perp_{G^a} \cup \not\!\perp\!\!\!\perp_{G^t}$, *and*
- $\perp\!\!\!\perp_G = \perp\!\!\!\perp_{G^a} \cap \perp\!\!\!\perp_{G^t}$.

Proof. The first equality holds due to the following reasons. Recall that the set of atemporal arcs A^a in the atemporal network G is split up into two disjoint sets, namely into sets A^{a^a} and A^{a^t} (see Section 5). Since these sets are disjoint, they do not share any vertex, their associated sets of dependences are also disjoint, and, therefore, the union of dependence relations $\not\!\perp\!\!\!\perp_{G^a}$ and $\not\!\perp\!\!\!\perp_{G^t}$ can be taken. The second equality is proved by applying logical rule $a \to b \equiv \neg b \to \neg a$ on the first equality. $\qquad\square$

Example 3. We reconsider our example of an atemporal network shown in Fig. 5(b). In graph 5(c), the independence relation consists of the dependence $q_2 \not\!\perp\!\!\!\perp_{G^a} s_2 \mid \emptyset$, and the temporal part in graph 5(d) includes the dependence statement $v_2 \not\!\perp\!\!\!\perp_{G^t} z_2 \mid \emptyset$. Thus, the dependence relation $\not\!\perp\!\!\!\perp_G$ of the atemporal network G consists of the union of these two dependence statements. The independence relation $\perp\!\!\!\perp_G$ is simply the complement of this dependence relation.

It is worth to mention that the intersection derived in Proposition 1 will not be included in the simplified learning method, however, it is necessary for the sketch of the entire connections of the independence relations of the network.

Based on Proposition 1, the following theorem defines the necessary machinery to be able to avoid unnecessary join operations.

Theorem 2. *Let* $G = (V, A^a)$ *be the atemporal network of temporal network* $N = (V, A)$, *and let* $G^a = (V, A^{a^a})$ *and* $G^t = (V, A^{a^t})$ *be the atemporal and temporal parts of the atemporal network* G. *Furthermore, let* $N_{|\Theta^t}$ *be the reduced temporal network. Then, the following two conditions hold:*

- $\not\perp\!\!\!\perp_{G^a} \cup \not\perp\!\!\!\perp_{N_{|\Theta^t}} = \not\perp\!\!\!\perp_N$, *and*
- $\perp\!\!\!\perp_{G^a} \circ \perp\!\!\!\perp_{N_{|\Theta^t}} = \perp\!\!\!\perp_{G^a} \cap \perp\!\!\!\perp_{N_{|\Theta^t}} = \perp\!\!\!\perp_N$.

Proof. First the proof for the first statement is provided. Recall that the sets of arcs that define the dependence relations in G^a and $N_{|\Theta^t}$ are disjoint sets and the associate vertices of these two sets of arcs are disjoint. Therefore, since dependence is represented by arcs, the dependence statements $\not\perp\!\!\!\perp_{G^a}$ and $\not\perp\!\!\!\perp_{N_{|\Theta^t}}$ are also disjoint explaining the opportunity to take the union over these sets to obtain the dependence relation $\not\perp\!\!\!\perp_N$ of temporal network N.

The second item shows that the join operator can be replaced by the intersection, which is a consequence of the two items in Theorem 1 and the fact that $\perp\!\!\!\perp_{G^a} \cap \perp\!\!\!\perp_{G^t} = \perp\!\!\!\perp_G$, shown in Proposition 1. The second equality of this item can be proven according to the logical rule $a \to b \equiv \neg b \to \neg a$ which replaces the independence into dependence and the intersection into a union. $\qquad\square$

6.2   The Simplified Learning Method

In this section, the simplified learning method discussed, which is based on theorems 1 and 2. The method is summarised in Fig. 6(b).

The method comprises two steps. First, the atemporal and temporal parts of the reduced temporal network are joined to obtain the independence relation of the reduced temporal network. It should be noted that this join process does not involve any redundancy in joining two statements, because their associates sets of arcs are disjoint. In the second step, the intersection of the previously obtained independence relation of the reduced temporal network and of the independence relation $\perp\!\!\!\perp_{G^a}$ are used to determine the independence relation of the temporal network.

The difference between the previous and this simplified join procedure is illustrated by figures 3(b) and 6(b), and can be summarised as follows:

- the simplified learning method does not include redundant join operations to obtain the relation $\perp\!\!\!\perp_N$ due to the fact that the set of arcs of G^a and $N_{|\Theta^t}$ are disjoint;
- in the simplified approach, the independence relation $\perp\!\!\!\perp_G$ does not have to be computed; only the intersection of the independence relations $\perp\!\!\!\perp_{G^a}$ and $\perp\!\!\!\perp_{N_{|\Theta^t}}$ has to be taken.

The main advantage of the simplified learning method is that the join operation has to be applied only once, instead of twice. The following example illustrates the simplified learning method.

Example 4. Applying the join procedure introduced in this section, first, the independence relation of the reduced temporal network is computed, as it was done

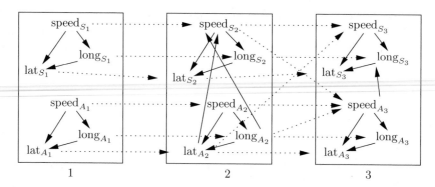

Fig. 7. A non-repetitive DBN representing the movements of two *intersecting* vessels

Table 1. The experimental results for the repetitive and non-repetitive DBNs

	repetitive DBN	non-repetitive DBN
AIC	-1038.16	-1063.91
BIC	-1325.96	-1404.94
Log-likelihood	-692.16	-653.908

in the original method. However, the independence relation $\perp\!\!\!\perp_G$ does *not* have to be computed. Given the independence relation $\perp\!\!\!\perp_{G^a}$ represented in Fig. 5(d), the union of the two independence relations $\perp\!\!\!\perp_{G^a}$ and $\perp\!\!\!\perp_{N_{|\Theta^t}}$ can be taken to obtain the independence relation $\perp\!\!\!\perp_N$ of the reduced temporal network. In doing so, the redundant join of the statements $v_2 \not\perp\!\!\!\perp_G z_2$ and $v_2 \not\perp\!\!\!\perp_{N_{|\Theta^t}} z_2$ does not have to be considered, since the dependence statement $v_2 \not\perp\!\!\!\perp_G z_2$ is not included in relation $\perp\!\!\!\perp_{G^a}$, which is the case for the original join method.

7   Experimental Results

In this section, the effectiveness of the simplified learning method is discussed, followed by a comparison of the repetitive and non-repetitive models.

Fig. 7 shows the learned non-repetitive DBN for the vessel traffic domain introduced in Section 2. In this model, the simplified learning method considers the atemporal and temporal parts G^a and G^t of the atemporal network G separately. According to definitions 1 and 2, at each time step, all three atemporal arcs speed $\rightarrow$ long, speed $\rightarrow$ lat, and long $\rightarrow$ lat belong to the temporal part G^t. In contrast to the temporal part, the atemporal part G^a consists of *no* arcs. According to theorems 1 and 2, the relations in G^t are already included in another independence relation. Thus, to avoid redundancy they are not considered in the simplified learning method when we build our DBN from the separate parts.

As a measure of the quality of the learned repetitive (Fig. 2) and non-repetitive networks (Fig. 7) we used the AIC score, BIC score, and log-likelihood function [1], [4]. For both the AIC and BIC scores it holds that given two models, the model with the lower score is preferred. For the log-likelihood function a higher value indicates a better fit of the distribution to the data. Table 1 summarises our experimental results, from which we can conclude that for all three performance measures the non-repetitive network performs better.

8 Conclusion

In this paper, we proposed a new simplified learning method for DBNs and we showed that non-repetitive DBNs provide a more profound graphical representation for certain dynamic processes. This simplified learning method makes use of a detailed partitioning of atemporal independence relations. The main contribution of the simplified learning method is that it avoids redundancy providing computational savings in the learning process. Using a real-life example, we also showed that non-repetitive DBNs yield better results than repetitive DBNs.

Acknowledgements

This work has been carried out as part of the Poseidon project under the responsibility of the Embedded Systems Institute (ESI), Eindhoven, The Netherlands. This project is partially supported by the Dutch Ministry of Economic Affairs under the BSIK03021 program.

References

1. Castillo, E., Gutiérrez, J.M., Hadi, A.S.: Expert Systems and Probabilistic Network Models. Springer, New York (1997)
2. Deviren, M., Daoudi, K.: Continuous speech recognition using dynamic Bayesian networks: a fast decoding algorithm. In: Proc PGM 2002, Spain, pp. 54–60 (2002)
3. Flesch, I., Lucas, P.J.F.: Independence Decomposition in Dynamic Bayesian Networks. In: Proc. ECSQARU, pp. 560–571 (2007)
4. Jensen, F.V., Nielsen, T.D.: Bayesian Networks and Decision Graphs. Springer, Heidelberg (2007)
5. Cowell, R.G., Philip Dawid, A., Lauritzen, S.L., Spiegelhalter, D.J.: Probabilistic Networks and Expert Systems. Springer, New York (1999)
6. Murphy, K.P.: Dynamic Bayesian Networks: Representation, Inference and Learning. PhD Thesis, UC Berkeley (2002)
7. Kjaerulff, U.: A computational scheme for reasoning in dynamic probabilistic networks. In: Proc. UAI 1992, pp. 121–129 (1992)
8. Tucker, A., Liu, X.: Learning Dynamic Bayesian Networks from Multivariate Time Series with Changing Dependencies. IDA (2003)

Surprise-Based Qualitative Probabilistic Networks

Zina M. Ibrahim, Ahmed Y. Tawfik, and Alioune Ngom

School of Computer Science, University of Windsor,
401 Sunset Ave., Windsor, Ontario, Canada N9B 3P4
{ibrahim,atawfik,angom}@uwindsor.ca

Abstract. This paper discusses a modification of the kappa measure of surprise and uses it to build semi-qualitative probabilistic networks. The new measure is designed to enable the definition of partial-order relations on its conditional values and is hence used to define qualitative influences over the edges of the network, similarly to Qualitative Probabilistic Networks. The resulting networks combine the advantages of kappa calculus of robustness and ease of assessment and the efficiency of Qualitative Probabilistic Networks. The measure also enables a built-in tradeoff resolution mechanism for the proposed network.

1 Introduction

Qualitative probabilistic networks (QPNs) [6,13] abstract Bayesian Networks (BNs) by replacing the numerical relations defined on the arcs of a BN (i.e. the conditional probability tables) by relations that describe how evidence given for one node influences other nodes in the network [13]. These influences are qualitative in nature, in the sense that the only information they capture is the direction of the influence (i.e. whether the evidence makes a node more or less likely) and is hence represented by its sign, being positive, negative, zero (constant) or unknown.

Despite the efficiency in reasoning with QPNs [3] (propagating influences along the arcs is efficient compared to the NP-hard reasoning in BNs), QPNs may suffer from over-abstraction because its reasoning mechanism is only concerned with finding the effect of new evidence on each node in terms of the sign of the change in belief (increase or decrease) [3] which may lead to problems when a node receives two influences of conflicting signs and a tradeoff must be resolved in order to continue reasoning. This issue has been addressed by several and solutions have been found using various tradeoff resolution strategies [9,10,14].

Apart from QPNs, there exist other formalisms for qualitatively reasoning about uncertain beliefs. One such formalism is the κ calculus, which uses the order of magnitude of probability to establish a measure, called the κ measure, where integers capture the relative degree of surprise associated with the event occurring.

C. Sossai and G. Chemello (Eds.): ECSQARU 2009, LNAI 5590, pp. 228–239, 2009.

The κ calculus has been used to create what is known as κ networks [4] to perform reasoning similarly to BNs [2]. κ networks have the advantage of being more robust and easier create than regular BNs as κ values are less easily affected by change and are easier to estimate compared to numerical probabilities [2]. However, κ networks still suffer from the inefficiency resulting from the polynomial size of the conditional tables associated with the nodes of the network, and as a result remain NP-hard in terms of reasoning.

To combine the efficiency of QPNs and the robustness and ease of estimation of the κ calculus, [9] proposes the use of κ values as measures of strength of the influences of a QPN. The approach retains the efficiency of arc-based reasoning of QPNs while reducing the unwanted coarseness in the representation by using κ values as measures of strength of QPN influences and resorting to them for tradeoff resolution. Despite the advantages of this work, it is only capable of capturing situations where it is possible to categorize the influences on the edges (i.e. as being positive, negative or constant). When it is not possible to do so (i.e. the influence is unknown), one must resort to quantitative probabilities to complete the reasonings task, which is contrary to the advantages of using QPNs in the first place.

Inspired by the ideas presented in [9] and the problem of unknown influences, we aim at creating a qualitative network in which the nodes represent an order of magnitude abstraction of probabilities and (i.e. as in κ networks) for which reasoning is possible on the arc-level (via qualitative influences) when the types of influences are known and in which it is possible to resort to node-based reasoning on a qualitative level when it is not possible to decide the type of the influences, only losing out on efficiency and retaining the robustness of qualitative calculi.

For this purpose, we define a new abstraction of probability, namely κ^{++}, which is based on the same concepts of κ but has additional semantics that enable its use to establish qualitative influences. For example, in the κ calculus, the rules governing the relation between $\kappa(g)$ and $\kappa(\neg g)$ are not rich enough to numerically deal with compliments in a manner that enables the propagation of their values through conditioning [5], in contrast to κ^{++} as the paper will show.

This paper is structured as follows. After providing preliminary concepts relating to qualitative probabilistic networks in section 2 and the κ calculus in section 3, we introduce the new qualitative measure, κ^{++}, in section 4. We present the details of constructing κ^{++}-based networks and using them to perform influence-based reasoning in section 5 and illustrate how reasoning can be performed if there is no information available about qualitative influences in section 6. We conclude in section 7 and outline some of our future research.

2   Qualitative Probabilistic Networks

Qualitative Probabilistic Networks (QPNs) are directed acyclic graphs that represent a qualitative abstraction of Bayesian Networks. The structure of a QPN captures the conditional independence among the variables the network represents and the arcs encode the relationships that exist between the nodes. Unlike in BNs however, the arcs of a QPN capture qualitative relations instead of conditional probabilities [6,13].

Formally, a QPN is given by a pair $G = (V, Q)$, where V is the set of variables it represents and Q is the set of qualitative relations among the variables. There are two types of qualitative relations in Q, qualitative influences and synergies.

Influences describe how the change of the value of one variable effects that of another. There are essentially four types of influences, positive, negative, constant and unknown.

A positive influence exists between two variable X and Y (X is said to positively influence Y) if observing higher values for X makes higher values of Y more probable regardless of the value of any other variable which may directly influence Y. The inequality given below describes the notion of a positive influence more formally. The inequality assumes that the variables X and Y are binary and places a partial order on their values such that for a variable X with two values x and $\neg x$, $x > \neg x$.

$$I^+(X, Y) \text{ iff } Pr(y|x, W) \geq Pr(y|\neg x, W)$$

$I^+(X, Y)$ reads: X positively influences Y, and W denotes all the other variables other than X that may directly influence Y. Negative, constant and unknown influences are defined analogously.

An example of a QPN is given in figure 1. In the figure, $V = \{A,B,C,D\}$ and $Q = \{(B,C),(A,C),(C,D)\}$. The only information encoded in the arcs are the signs of the influences from one node to another. For instance, the figure shows that node A positively influences node C, while B on the other hand has a negative influence on C.

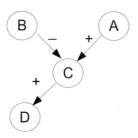

Fig. 1. A Qualitative Probabilistic Network

Observed evidence is propagated through the network via qualitative operators that combine influences and produce their net effect. Essentially, there are two such operators, each is used for a specific topology of arcs. When evaluating the net effect of two influences in a chain (for example, in order to obtain the effect of A on D, we have to examine a chain of two influences, that of A on C and of C on D), the sign multiplication operator, given in the left portion of table 1, is used. On the other hand, parallel connections (for example, two influences in parallel are required to establish the net effect of nodes A and B on node C, that of B on C and of A on C) are evaluated using the sign addition operator given in the right portion of the table. The signs propagate through the network until the net effect of the evidence is observed on the required node or all the nodes are known to be visited twice by the sign-propagation algorithm given in [3].

Table 1. Sign multiplication ($\otimes$) and sign addition ($\oplus$) Operators [13]

$\otimes$	+	−	0	?		$\oplus$	+	−	0	?
+	+	−	0	?		+	+	?	+	?
−	−	+	0	?		−	?	−	−	?
0	0	0	0	0		0	+	−	0	?
?	?	?	0	?		?	?	?	?	?

3   Overview of the Kappa Calculus

The kappa calculus [4,11] is a system that abstracts probability theory by using order of magnitude of probabilities as an approximation of probability values. It does so by capturing the degree of disbelief in a proposition g, or the degree of incremental surprise or abnormality associated with finding g to be true [4], labeled $\kappa(g)$. The value of $\kappa(g)$ is assigned so that probabilities having the same order of magnitude belong to the same κ class, and that $\kappa(g)$ grows inversely to the order of magnitude of the probability value $p(g)$.

The abstraction is achieved via a procedure which begins by representing the probability of a proposition g, $p(g)$, by a polynomial function of one unknown, ϵ, an infinitesimally small positive number $(0 < \epsilon < 1)$. The rank κ of a proposition g is represented by the power of the most significant ϵ-term in the polynomial representing $p(g)$ (the lowest power of ϵ in the polynomial). Accordingly, the relation between probability and κ values is that $p(g)$ is of the same order as ϵ^k, where $k = \kappa(g)$ [11], that is:

$$\epsilon < \frac{p(g)}{\epsilon^k} \leq 1 \text{ or equivalently: } \epsilon^{k+1} < p(g) \leq \epsilon^k$$

Where ϵ^k is the most significant ϵ-term of the polynomial representing $p(g)$.

The κ-calculus is useful because it provides an abstraction that only requires specifying the κ values of propositions, which is an easier task than specifying the exact probabilities associated with the specific value of the proposition. The κ values are in turn representative of the interval in which the probability falls [2].

$$\kappa(g) = \begin{cases} \min\{k \ such \ that \ \lim\limits_{\epsilon \to \infty} \dfrac{p(g)}{\epsilon^k} \neq 0\} & iff \ p(g) > 0 \\ \infty & iff \ p(g) = 0 \end{cases}$$

The above defines $\kappa(g)$ as the power of the most significant term (the term with the least power of ϵ, since $\epsilon < 1$) in the polynomial representing $p(g)$. In other words, $\kappa(g) = k$ if $p(g)$ has the same order of magnitude as ϵ^k [4].

A direct consequence of how $\kappa(g)$ is obtained is that since the most significant term is that with the smallest k it corresponds to the inverse of the likelihood of g, and is therefore representative of the degree of surprise associated with believing g. This can be seen in a more intuitive manner in the table below (obtained from [4]), which shows an example of how kappas can be mapped to linguistic quantifiers of beliefs.

$p(g) = \epsilon^0$	g and $\neg g$ are possible	$\kappa(g) = 0$
$p(g) = \epsilon^1$	$\neg g$ is believed	$\kappa(g) = 1$
$p(g) = \epsilon^2$	$\neg g$ is strongly believed	$\kappa(g) = 2$

The above abstraction yields an integer-based calculus which enables combining κ's via rules that are derived from those of probability theory by replacing multiplication by addition and addition by minimum [11]. The resulting properties are given below, along with their probability-theory equivalents.

$$\kappa(g_1) = \min_{\omega \models g_1} \qquad p(g_1) = \sum_{\omega \models g_1} p(\omega)$$

$$\kappa(g_1) \vee \kappa(\neg g_1) = 0 \qquad p(g_1) + p(\neg g_1) = 1$$

$$\kappa(g_2|g_1) = \kappa(g_2 \wedge g_1) - \kappa(g_1) \qquad p(g_2|g_1) = p(g_2 \wedge g_1)/p(g_1)$$

4   A New Ranking Function, $\kappa^{++}(.)$

We propose a ranking function that is based on taking the order of magnitude of the surprise associated with observing a certain event. The function is based on the surprise measure proposed by Weaver [12] which calculates the surprise associated with an event G having specific value g_r (where G has a total of I possible values) by dividing the expected value of probability of G by the probability $p(G = g_r)$ as given by equation 1 below.

$$\mathcal{W}(g_r) = \frac{\sum_{i=1}^{I} p(g_i)^2}{p(g_r)} \qquad (1)$$

Given a distribution ζ, the abstraction we form here is similar to that of the $\kappa(.)$ function in that it depends on the idea of writing the probability of every event $g \in \zeta$ as a polynomial χ of infinitesimally small numbers, ϵ (for example, $p(g) = 1 - c1\epsilon^2 + c2\epsilon^4$) and obtaining the most significant term whose power is representative of the order of magnitude of $p(g)$. Moreover, since we are only interested in the most significant term of the polynomial, we adapt the notation χ_g^n to denote the polynomial representing $p(g)$, where n is the lowest exponent of ϵ in the polynomial.

Given the above, a ranking function is formulated to reflect the order of magnitude to which $\mathcal{W}(g_r)$, the surprise associated with the event g_r, belongs. Hence, we redefine the κ-measure [4] by associating it with the surprise measure of an event instead of the probability. The result is a qualitative measure of the surprise associated with an event g_r. The new measure, we call it κ^{++} to distinguish it from the original κ measure of [4], is derived below.

Let $\chi^n_{g_r}$ be the polynomial representing $p(g_r)$, and for every other value g_i of g, let $\chi^{\beta_i}_{g_i}$ denote the polynomial corresponding to $p(g_i)$, with β_i representing the minimum power of ϵ in the polynomial. According to equation 1, the surprise associated with $p(g_r)$, namely $\mathcal{W}(g_r)$, is:

$$\mathcal{W}(g_r) = \frac{\sum_{i=1}^{I} p(g_i)^2}{p(g_r)} = \frac{\sum_{i=1}^{I} (\chi^{\beta_i}_{g_i})^2}{\chi^n_{g_r}}$$

Where I, as given previously, is the number of possible values of G. Since all the polynomials χ are to the base ϵ, it is possible to add the terms that have equal exponents. This makes the summation:

$$\frac{\sum_{i=1}^{I} (\chi^{\beta_i}_{g_i})^2}{\chi_{g_r}{}^n} = \frac{\alpha_1 \epsilon^{2\beta_1} + ... + \alpha_I \epsilon^{2\beta_I} + \alpha_{I+1} \epsilon^{2\alpha_1} + ... + \alpha_l \epsilon^{2\phi_k}}{\chi^n_{g_r}}$$

$\forall \beta_i, 1 \leq i \leq I$, $\alpha_i \epsilon^{2\beta_i}$ is a term whose power is a candidate to be the minimum power of the polynomial representing $\sum_{i=1}^{I} (\chi^{\beta_i}_{g_i})^2$ as each $2\beta_i$ is the minimum power of $(\chi^{\beta_i}_{g_i})^2$. The α terms in the equation above are non-minimum terms and therefore, their number ($k = 1\text{-}(I+1)$) and values are irrelevant for our purpose.

Let m be such term, i.e. $m = \beta_i$ is the minimum of the minimum powers of the polynomials $\chi^{\beta_i}_{g_i}$. The surprise measure $\mathcal{W}(g_r)$ can now be represented only in terms of polynomials as:

$$\mathcal{W}(g_r) = \frac{\chi^{2m}_{g_i}}{\chi_{g_r}{}^n} \tag{2}$$

Having represented $\mathcal{W}(g_r)$ as a fraction of two polynomials, we now construct an abstraction of $\mathcal{W}(g_r)$ which maybe regarded as the order of magnitude class to which $\mathcal{W}(g_r)$ belongs. The integer value resulting from the abstraction is denoted by $\kappa^{++}(g_r)$ and can be used to rank the belief g_r. $\kappa^{++}(g)$ is given below.

Definition 1. *For an event g_r whose probability is given by the polynomial $\chi^n_{g_r}$, the qualitative degree of surprise associated with g_r, namely $\kappa^{++}(g_r)$ is the power of the most significant ϵ-term in the polynomial representing the numerical surprise associated with g_r, $\mathcal{W}(g_r)$. In other words:*

$$\kappa(g) = \begin{cases} \min\{k \text{ such that } \lim_{\epsilon \to \infty} \frac{\mathcal{W}(g)}{\epsilon^k} \neq 0\} & \text{iff } \mathcal{W}(g) > 0 \\ \infty & \text{iff } \mathcal{W}(g) = 0. \end{cases}$$

According to the above, $\kappa^{++}(g_r) = \log \mathcal{W}(g_r) = 2m - n$, where m is the minimum of all minimum powers in the polynomials $p(g_i)$, $1 \leq i \leq I$, and n is the minimum power in $p(g_r)$.

4.1   Semantics of κ^{++}

The $\kappa^{++}(.)$ function can now be understood as a function which ranks events according to the surprise associated with finding that the event has occurred. $\kappa^{++}(g_r) = 2m - n$ returns a signed integer whose value and sign are representative of the degree of surprise. Accordingly, the larger the value of $\kappa^{++}(g_r)$, the greater the difference between its constituent quantities ($2m$ and n), and as a result the more surprising the event in question, g_r, is. Therefore, the signed integer produced by $\kappa^{++}(.)$ carries the semantics defined by three possible classes for its value.

Positive: $(\kappa^{++}(g_r) = 2m - n) > 0$ implies that the event g_r is a lot less likely than the other events g_i ($1 \leq i \leq I$) of the distribution, i.e. $2m > n$. Hence, the occurrence of g_r indicates a surprise. Moreover, the larger the value of $\kappa^{++}(g_r)$ (the greater the difference is between $2m$ and n), the more surprising the event g_r is.

Zero: $\kappa^{++}(g_r) = 0$ represents the normal world where both g_r and $\neg g_r$ are likely to occur as the order of magnitude of the probability of the variable g_r is comparable to that of the distribution, i.e. $2m = n$.

Negative: $\kappa^{++}(g_r) < 0$ refers to the case in which having the event g_r to be false is surprising as g_r becomes more likely than unlikely compared to other events in the distribution (because $n > 2m$), which implies that $\neg g_r$ is unlikely and its $\kappa^{++}(.)$ should indicate a surprise. In this case, the smaller the value of $\kappa^{++}(.)$, the more surprising $\neg g_r$ is.

Because $\kappa^{++}(.)$ is obtained through order-of-magnitude abstraction, its rules can be derived from those of probability theory by replacing multiplication by addition and addition by minimum, and can be summarized below.

1. $\kappa^{++}(G_1|G_2) = \kappa^{++}(G_1 \wedge G_2) - (G_2)$
2. $\kappa^{++}(G_1 \wedge G_2) = \kappa^{++}(G_1) + \kappa^{++}(G_2)$
   Given that G_1 and G_2 are two independent variables.
3. $\kappa^{++}(G_1) + \kappa^{++}(G_2) = 0$
4. $\kappa^{++}(G_1 \vee G_2) = \min(\kappa^{++}(G_1), \kappa^{++}(G_2))$

It is worth noting that the κ^{++} semantics introduced earlier permit the creation of a correspondence between κ^{++} and linguistic quantifiers such as the one given below.

g is strongly believed	$\kappa^{++}(g) = -2$
g is believed	$\kappa^{++}(g) = -1$
g and $\neg g$ are possible	$\kappa^{++}(g) = 0$
$\neg g$ is believed	$\kappa^{++}(g) = 1$
$\neg g$ is strongly believed	$\kappa^{++}(g) = 2$

5   Efficient Reasoning in κ^{++} Networks

Similarly to the κ calculus, κ^{++} can be used to abstract probabilistic networks to construct a semi-qualitative equivalent whose nodes have κ^{++}s as values instead of probabilities. Despite the added robustness of the resulting network, inference will remain NP-hard, i.e. of the same complexity as its quantitative equivalent [2].

Alternatively, we use κ^{++} to perform Qualitative Probabilistic Network-like inference by utilizing the sign and magnitude of κ^{++} to define partial order relations governing the conditional κ^{++} values of pairs of nodes in the κ^{++} network. In other words, we define notions of influences [13] on the arcs of the network by, when possible, ordering the conditional probabilities of the nodes connecting the arc.

The influences defined using κ^{++} values will not only be identified by their signs, which designates the type of influence, but also by a signed integer that can be used to evaluate their relative strength and to propagate them across the network. Tradeoff resolution comes natural in this case because conflicting signs can be resolved by assessing the magnitude of the influences in conflict. The result is a κ^{++} network capturing the semantics of conditional independence that can be used to propagate beliefs qualitatively and has a built-in conflict-resolution mechanism. In what follows, we define the notion of κ^{++}-based influences.

5.1 κ^{++}-Based Influences

We define four types of influences analogous to those defined over QPNs, positive, negative, zero and unknown. In this section, we restrict our discussion to the first three types of influences and delay the discussion of unknown influences to section 6.

Positive Influences: A binary variable X is said to positively influence another binary variable Y if the degree of conditional surprise associated with Y being true given X is observed, $\kappa^{++}(y|x)$, is lower than that of Y being true given that X is not observed $\kappa^{++}(y|\neg x)$ regardless of the value of any other variable which may directly influence Y. Definition 2 formally states this notion.

Definition 2. $I^+_{\kappa^{++}}(X,Y) \text{ iff } \kappa^{++}(y|\neg x, W) - \kappa^{++}(y|x, W) > 0.$

W represents any other variable other than X that directly influences Y, which maybe written as $\pi(Y) \backslash X$ (where there is more than one such variable, W is thought of as the conjunction of the possible values of such variables [7]). We denote the influence by a subscript κ^{++} to enforce the idea that they are defined over κ^{++} values and not probability values as in QPNs, and will follow the nomenclature for negative, zero and unknown influences.

It is important to see that the semantics of κ^{++} guarantee that the constraint given by the definition holds, which is what we show in proposition 1.

Proposition 1. *For two binary variables X and Y:*

$$\kappa^{++}(y|x, W) < \kappa^{++}(y|\neg x, W) \rightarrow \kappa^{++}(y|\neg x, W) - \kappa^{++}(y|x, W) \in \mathbb{Z}^+.$$

Proof
There are essentially two cases that result from the inequality $\kappa^{++}(y|x, W) < \kappa^{++}(y|\neg x, W)$:

- Case 1: $\kappa^{++}(y|x, W) \in \mathbb{Z}^-$ and $\kappa^{++}(y|\neg x, W) \in \mathbb{Z}^+$
  In this case, the fact that $\kappa^{++}(y|\neg x, W) - \kappa^{++}(y|x, W) \in \mathbb{Z}^+$ is intuitive.
- Case 2: Both $\kappa^{++}(y|x, W)$ and $\kappa^{++}(y|\neg x, W) \in \mathbb{Z}^+$
  In this case, the semantics of κ^{++} enforces that for $\kappa^{++}(y|x, W)$ to be less surprising than $\kappa^{++}(y|\neg x, W)$, it must possess a higher magnitude, which will guarantee the result.

Negative Influences. Similarly to positive influences, a binary variable X negatively influences another binary variable Y if the degree of conditional surprise associated with Y being true given X is observed, $\kappa^{++}(y|x)$, is higher than that of Y being true given that X is not observed $\kappa^{++}(y|\neg x)$ regardless of the value of any other variable which may directly influence Y as given in definition 3 below.

Definition 3. $I^-_{\kappa^{++}(X,Y)}$ *iff* $\kappa^{++}(y|\neg x, W) - \kappa^{++}(y|x; W) < 0.$

Zero Influences are defined in the same manner and is given in definition 4.

Definition 4. $I^0_{\kappa^{++}(X,Y)}$ *iff* $\kappa^{++}(y|x, W) - \kappa^{++}(y|\neg x, W) = 0.$

Although the influences given in this work are defined over binary variables, the definitions can be naturally extended to multi-valued variables as we have adopted the order of $x > \neg x$ to denote that a true value has a higher value than a false one.

5.2   Influence Propagation

To combine influences, we redefine the $\oplus$ and $\otimes$ operators to accommodate the sign and magnitude properties of the κ^{++}-based influences.

Chained Influences. As done in [3,9,13], we propagate influences along chains using the order of magnitude multiplication operator. Since our influences include sign and magnitude components, these components are handled separately to obtain the net effect on the variables.

The sign portion of the influence is dealt with using sign multiplication as in [13] while the magnitude portion is handled in accordance with the rules of order of magnitude multiplication by adding the corresponding values (since the magnitude represent the difference between two κ^{++} values, which are in essence order of magnitude abstractions of the numerical surprise associated with the variable). The result is presented in the table below.

$\otimes$	$+ve$	$-ve$	0	?
$+u$	$+(u+v)$	$-(u+v)$	0	?
$-u$	$-(u+v)$	$+(u+v)$	0	?
0	0	0	0	0
?	?	?	?	?

Parallel Influences. For influences in parallel chains, the net effect is decided by that of the strongest influence incident on the node. Accordingly, the effect is achieved via the $\oplus$ operator, presented in the table given below.

$\oplus$	$+ve$	$-ve$	0	?
$+u$	$+\min\{u,v\}$	a)	$+u$	?
$-u$	b)	$-\min\{u,v\}$	$-u$	?
0	$+v$	$-ve$	0	?
?	?	?	?	?

a) $= +u$, if $u < v$
$\quad = -v$, otherwise
b) $= -u$, if $u < v$
$\quad = +ve$, otherwise

Combining influences in chains and in parallel can be illustrated via an example such as the one given in the network of figure 2. In the figure, when nodes A and C are received as evidence, the discovery of the influences in the network propagates as follows. The net influence of node A on node E through B is given by -1 $\otimes$ +2 = -3 because this influence consists of two influences in a chain whose effect is obtained via the $\otimes$ operator. Similarly, Node D receives evidence from both A and C with the net influence being evaluated as +4 $\oplus$ -5 = +4 because node D has two arcs incident on it, which implies that the net effect on D is obtained through the discovery of the combined influences in parallel, which is achieved through the $\oplus$ operator. Similarly, the net influence of A and C on E through D is given by +4 $\otimes$ +5 = +9. Finally, node E receives as a net influence -3 $\oplus$ +9 = -3. As a result, the net influence of observing A and C on E is a negative one.

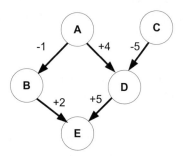

Fig. 2. Reasoning with a κ^{++}-based Qualitative Probabilistic Network

6   The Case of Unknown Influences

Because influences only exist when one is able to establish a partial order on the conditional κ^{++} of two variables [7], it is a weak concept that may not be defined when such order does not exist. In this case, it is imperative to resolve to methods at a finer level of resolution. In our approach, since our network are based on κ^{++} values, it is not necessary to go to probabilities and is sufficient to go back to node-based inference on the κ^{++}-level. Although this reduces the efficiency of the inference, it is a necessary last resort when orders are not definable. Moreover, the network retains its qualitative nature as we are still dealing with κ^{++}'s, which are easier to assess than numerical probabilities.

7   Conclusions and Future Work

We presented κ^{++}, a qualitative measure that uses sign and magnitude to designate the degree of surprise associated with an event and used it to construct a qualitative system with two levels of resolution. The resulting system enables the definition of qualitative influences and when not possible, can be used to reason on the node level. The system has its built-in conflict resolution mechanism and is as efficient as previous QPN systems when used with known influences. When the influences are unknown, the system we presented has the advantage of providing the option of reasoning (although on the node, and not arc level) qualitatively without having to resort back to numerical probabilities. Our current work involves an empirical analysis of the κ^{++} calculus and its possible application to study qualitative patterns in gene regulatory networks [8].

References

1. Bolt, Y., Van Der Gaag, L., Renooij, S.: Introducing situational signs in qualitative probabilistic networks. International Journal of Approximate Reasoning 38(3), 333–354 (2005)
2. Darwiche, A., Goldszmidt, M.: On the relation between kappa calculus and probabilistic reasoning. In: Proceedings of the International Conference on Uncertainty in Artificial Intelligence, pp. 136–144 (1994)
3. Drudzel, M.J., Henrion, M.: Efficient Reasoning in Qualitative Probabilistic Networks. In: Proceedings of the International Conference on Uncertainty in Artificial Intelligence, pp. 548–553 (1993)
4. Goldszmidt, M., Pearl, J.: Qualitative Probabilities for Default reasoning, belief revision, and causal modeling. Artificial Intelligence 84, 145–151 (1996)
5. Huber, F.: Ranking functions and rankings on languages. Artificial Intelligence 170(4-5), 462–471 (2006)
6. Neufeld, E.: Defaults and probabilities; Extensions and Coherence. In: Proceedings of the International Conference on Principles of Knowledge Representation and Reasoning, pp. 312–323 (1989)
7. Parsons, S.: Qualitative Methods for Reasoning Under Uncertainty. MIT Press, Cambridge (2001)
8. Pisabarro, A.G., Perez, G., Lavin, J.L., Ramirez, L.: Genetic Networks for the Functional Study of Genomes. Briefings in Functional Genomics and Protemics 7(4), 249–263 (2008)
9. Renooij, S., Parsons, S., Pardeick, P.: Using kappas as indicators of strength in Qualitative Probabilistic Networks. In: Proceedings of the European Conference on Symbolic and Quantitative Approaches to Reasoning and Uncertainty, pp. 87–99 (2003)
10. Renooij, S., Van Der Gaag, L.C., Parsons, S., Green, S.: Pivotal Pruning of Trade-offs in Qualitative Probabilistic Networks. In: Proceedings of the International conference on Uncertainty in Artificial Intelligence, pp. 515–522 (2000)
11. Spohn, W.: A General Non-Probabilistic Theory of Inductive Reasoning. In: Proceedings of the International Conference on Uncertainty in Artificial Intelligence, vol. 4, pp. 149–158 (1988)

12. Weaver, W.: Probability, Rarity, Interest and Surprise. Scientific Monthly 67, 390–392 (1948)
13. Wellman, M.: Fundamental Concepts of Qualitative Probabilistic Networks. Artificial Intelligence 44, 357–303 (1990)
14. Wellman, M., Liu, C.L.: Incremental Tradeoff Resolution in Qualitative Probabilistic Networks. In: Proceedings of the International Conference on Uncertainty in Artificial Inteligence, pp. 338–345 (1998)

Maximum Likelihood Learning of Conditional MTE Distributions

Helge Langseth[1], Thomas D. Nielsen[2], Rafael Rumí[3], and Antonio Salmerón[3]

[1] Department of Computer and Information Science,
The Norwegian University of Science and Technology, Norway
[2] Department of Computer Science, Aalborg University, Denmark
[3] Department of Statistics and Applied Mathematics University of Almería, Spain

Abstract. We describe a procedure for inducing conditional densities within the mixtures of truncated exponentials (MTE) framework. We analyse possible conditional MTE specifications and propose a model selection scheme, based on the BIC score, for partitioning the domain of the conditioning variables. Finally, experimental results demonstrate the applicability of the learning procedure as well as the expressive power of the conditional MTE distribution.

1 Introduction

The main difficulty when modelling hybrid domains (i.e., domains containing both discrete and continuous variables) using Bayesian networks, is to find a representation of the joint distribution that is compatible with the operations used by existing inference algorithms: Algorithms for exact inference based on local computations, like the Shenoy-Shafer scheme [1], require that the joint distribution over the variables in the network are closed under marginalization and multiplication.

This can be achieved by discretizing the domain of the continuous variables [2,3], which is a simple (but sometimes inaccurate) solution. A more elaborate approach is based on the use of mixtures of truncated exponentials (MTE) [4]. One of the advantages of this representation is that MTE distributions allow discrete and continuous variables to be treated in a uniform fashion, and since the family of MTEs is closed under marginalization and multiplication, inference in an MTE network can be performed efficiently using the Shenoy-Shafer architecture [1]. Also, the expressive power of MTEs was demonstrated in [5], where the most commonly used distributions were accurately approximated by MTEs.

The task of learning MTEs from data was initially approached using least squares estimation [6,7]. However, this technique does not combine well with more general model selection problems, as many standard score functions for model selection, including the Bayesian information criterion (BIC) [8], assume Maximum likelihood (ML) parameter estimates to be available.

Two kinds of distributions can be found in a Bayesian network: univariate distributions (for nodes with no parents), and conditional distributions (for nodes

C. Sossai and G. Chemello (Eds.): ECSQARU 2009, LNAI 5590, pp. 240–251, 2009.

with parents). ML learning of univariate distributions was introduced in [9]. However, the problem of learning conditional densities has so far only been described using least squares estimation [10]. In this paper, we study ML-based learning of conditional densities from data.

2   Preliminaries

Consider the problem of estimating a conditional density $f(x|\boldsymbol{y})$ from data. In this paper we will concentrate on the case in which X and $\boldsymbol{Y} = \{Y_1, \dots Y_r\}$ are continuous, and use $\Omega_{X,\boldsymbol{Y}} \subseteq \mathbb{R}^{r+1}$ to represent the support of the distribution function $f(x, \boldsymbol{y})$. Furthermore, we let $\{\boldsymbol{I}_1, \dots, \boldsymbol{I}_K\}$ be a partition of $\Omega_{X,\boldsymbol{Y}}$. An *MTE potential* [4] for the random vector $\{X, Y_1, \dots, Y_r\}$ is a function that, for each $k \in \{1, \dots, K\}$, can be written as

$$f(x, \boldsymbol{y}) = a_0 + \sum_{j=1}^{m} a_j \exp\left(b_j x + \boldsymbol{c}_j^{\mathrm{T}} \boldsymbol{y}\right), (x, \boldsymbol{y}) \in \boldsymbol{I}_k. \tag{1}$$

The main problems to solve when inducing MTE potentials from data are *i)* determining the partition $\{\boldsymbol{I}_1, \dots, \boldsymbol{I}_K\}$, *ii)* determining m (the number of exponential terms) for each $\boldsymbol{I}_k$, and *iii)* estimating the parameters. Throughout the paper we will consider a training data set $\mathcal{D}$ with n records, and each record containing observations of all $r + 1$ variables without missing values. We will write $\mathcal{D}(\boldsymbol{R})$ to denote the subset of $\mathcal{D}$ where the restriction $\boldsymbol{R}$ is fulfilled. For example $\mathcal{D}(y_1 \leq \alpha)$ selects all records for which the variable $y_1 \leq \alpha$.

3   Conditional Distributions and MTEs

Before we investigate methods for learning conditional distributions from data, we will consider how to define conditional MTE distributions. Unfortunately, since the class of MTE potentials is not closed under division, we do not know the most general form of the conditional distribution function. However, for a function $g(x, \boldsymbol{y})$ to be a conditional MTE distribution, there are three assumptions we will require to be fulfilled:

1. **Generating joint:** $g(x, \boldsymbol{y}) \cdot f(\boldsymbol{y})$ should be an MTE potential over $(x, \boldsymbol{y})$, and the result should be equal to the joint density $f(x, \boldsymbol{y})$, where $f(y)$ is the marginal distribution for Y.
2. **Conditioning:** $g(x, \boldsymbol{y}_0)$ should be an MTE density over X for any fixed $\boldsymbol{y}_0$.
3. **Closed under marginalization:** For any BN structure and specification of conditionals, the product $\prod_{i=1}^{n} g(x_i, \mathrm{pa}\,(x_i))$ must support marginalization of any variable in closed form, and the result should be an MTE potential.

The first two conditions are local in nature, whilst the third is global. Attending only to the local conditions, the natural way to define a conditional density would be as $f(x|\boldsymbol{y}) = f(x, \boldsymbol{y})/f(\boldsymbol{y})$, where $f(x, \boldsymbol{y})$ and $f(\boldsymbol{y})$ are MTEs. Formally, a conditional MTE density under these assumptions would be of the form

$$f(x|\boldsymbol{y}) = \frac{f(x, \boldsymbol{y})}{f(\boldsymbol{y})} = \frac{a_0}{f(\boldsymbol{y})} + \sum_{i=1}^{m} \frac{a_i \exp(\boldsymbol{c}_i^{\mathrm{T}} \boldsymbol{y})}{f(\boldsymbol{y})} \cdot \exp(b_i x), (x, \boldsymbol{y}) \in \boldsymbol{I}_k, \quad (2)$$

where we have assumed that $f(x, \boldsymbol{y}) = a_0 + \sum_{i=1}^{m} a_i \exp\left(b_i x + \boldsymbol{c}_i^{\mathrm{T}} \boldsymbol{y}\right)$, and $f(\boldsymbol{y}) = \int_x f(x, \boldsymbol{y}) dx$. Note that for any $\boldsymbol{y}_0$, $f(x|\boldsymbol{y}_0)$ specifies an MTE potential for x.

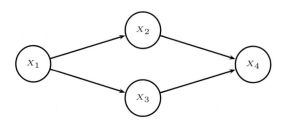

Fig. 1. A example of Bayesian network

However, the problems come when the defined conditional densities are considered globally in a Bayesian network, in which the marginalization operation is necessary to perform inference. To illustrate the problem, consider the network structure in Fig. 1. Observe that the joint distribution is

$$\begin{aligned} f(x_1, x_2, x_3, x_4) &= f(x_1)f(x_2|x_1)f(x_3|x_1)f(x_4|x_2, x_3) \\ &= f(x_1) \frac{f(x_2, x_1)}{f(x_1)} \frac{f(x_3, x_1)}{f(x_1)} \frac{f(x_4, x_2, x_3)}{f(x_2, x_3)} \\ &= f(x_1, x_2) \frac{f(x_3, x_1)}{f(x_1)} \frac{f(x_4, x_2, x_3)}{f(x_2, x_3)}, \end{aligned} \quad (3)$$

but if the original conditional distributions are as in Equation (2), we find that the joint distribution in the network, shown in Equation (3), is not an MTE. Furthermore, standard inference algorithms, such as variable elimination [11], cannot be directly applied. For instance, if the first variable to eliminate is X_2, the operation to carry out would be

$$\int_{x_2} f(x_1, x_2) \frac{f(x_4, x_2, x_3)}{f(x_2, x_3)} dx_2,$$

which cannot be calculated in closed form if the potentials are as in Equation (2); the variable to integrate out appears in both the numerator and in the denominator.

When MTEs were first introduced [4], Moral *et al.* avoided these problems by defining the conditional MTE distribution as follows:

Definition 1. *Let* $\boldsymbol{X}_1 = (\boldsymbol{Y}_1, \boldsymbol{Z}_1)$ *and* $\boldsymbol{X}_2 = (\boldsymbol{Y}_2, \boldsymbol{Z}_2)$ *be two mixed random variables. We say that an MTE potential* ϕ *defined over* $\Omega_{\boldsymbol{X}_1 \cup \boldsymbol{X}_2}$ *is a conditional MTE density if for each* $\boldsymbol{x}_2 \in \Omega_{\boldsymbol{X}_2}$, *it holds that the restriction of* ϕ *to* $\boldsymbol{x}_2$, $\phi^{R(\boldsymbol{X}_2 = \boldsymbol{x}_2)}$, *is an MTE density for* $\boldsymbol{X}_1$.

In this paper we focus on conditional distributions of continuous variables with continuous parents. In our notation, Definition 1 is therefore equivalent to requiring that the conditional distribution must have the functional form

$$f(x|\boldsymbol{y}) = \alpha_0 + \sum_{j=1}^{m} \alpha_j \exp\left(\beta_j x + \boldsymbol{\gamma}_j^{\mathrm{T}} \boldsymbol{y}\right), (x, \boldsymbol{y}) \in \boldsymbol{I}_k, \qquad (4)$$

where we will assume that $m < \infty$ in the following.

Moral *et al.* [4] noted that if one adopts the structural form of Equation (4), then specific requirements are in play to ensure that $f(x|\boldsymbol{y})$ is a conditional distribution. We will investigate one of these requirements in the following, namely that $\sum_k \int_{x:(x,\boldsymbol{y})\in\boldsymbol{I}_k} f(x|\boldsymbol{y})\,dx = 1$, for all $\boldsymbol{y}$. As an example, consider Fig. 2, where the support for $f(x|y)$ is divided into 4 hypercubes $\boldsymbol{I}_1, \ldots, \boldsymbol{I}_4$, such that a specific MTE potential MTE_k is defined for each $\boldsymbol{I}_k$. In this example, the requirement above implies that, e.g., $f(x|y = 0)$ ties the two MTE potentials MTE_1 and MTE_2 together, with the consequence that we cannot learn the MTE potentials separately.

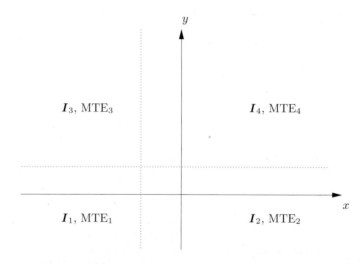

Fig. 2. The support for $f(x|y)$ is depicted, and divided into 4 hypercubes $\boldsymbol{I}_1$, $\boldsymbol{I}_2$, $\boldsymbol{I}_3$, and $\boldsymbol{I}_4$. An MTE potential MTE_k is connected to each $\boldsymbol{I}_k$.

The effect of tying the parameters of different MTE potentials is a dramatic increase in the computational burden of learning conditional MTE distributions. In this paper we will therefore assume *parameter independence* for simplicity. One consequence of this assumption is that $\int_x f(x|\boldsymbol{y})\,dx$ must be a constant w.r.t. $\boldsymbol{y}$, which corresponds to:

$$\frac{\partial}{\partial y_\ell} \int_x f(x|\boldsymbol{y}) \, dx = \frac{\partial}{\partial y_\ell} \int_x \sum_{j=1}^m \alpha_j \exp\left(\beta_j x + \boldsymbol{\gamma}_j^{\mathrm{T}} \boldsymbol{y}\right) \, dx$$

$$= \sum_{j=1}^m \alpha_j \gamma_{j\ell} \exp\left(\boldsymbol{\gamma}_j^{\mathrm{T}} \boldsymbol{y}\right) \int_x \exp\left(\beta_j x\right) \, dx$$

$$= 0.$$

Thus, for all $(x_0, \boldsymbol{y}_0) \in \boldsymbol{I}_k$ where $f(\boldsymbol{y}_0) > 0$, we should have

$$\sum_{j=1}^m \alpha_j \gamma_{j\ell} \exp\left(\boldsymbol{\gamma}_j^{\mathrm{T}} \boldsymbol{y}_0\right) \int_x \exp\left(\beta_j x\right) \, dx = 0. \tag{5}$$

Now, fixate an ϵ-ball around $(x_0, \boldsymbol{y}_0)$ s.t. the ball is in $\boldsymbol{I}_k$ and in the support of $f(x, \boldsymbol{y})$. We are interested in varying $\boldsymbol{y}$ inside this ball. Then, Equation (5) gives rise to uncountably many constraints (one for each $\boldsymbol{y}$ in the ball), but where we only have $\mathcal{O}(m)$ parameters that can be used to adhere to the constraints. This over-specified system of equations can only be solved if $\boldsymbol{\gamma}_j = \boldsymbol{0}$ for all j (remember that $\alpha_j = 0$ is not a viable solution, since we need the density to have some mass allocated). Thus, if the conditional distribution functions are to follow Definition 1, and at the same time adhere to parameter independence, we must constrain the functional form of the conditional distribution to

$$f(x|\boldsymbol{y}) = \sum_{j=1}^m \alpha_j \exp\left(\beta_j x\right), (x, \boldsymbol{y}) \in \boldsymbol{I}_k. \tag{6}$$

Thus, the conditional MTE potential $f(x|\boldsymbol{y})$ is constant in $\boldsymbol{y}$ inside each hypercube $\boldsymbol{I}_1, \ldots, \boldsymbol{I}_K$, and the only effect of the conditioning variables $\boldsymbol{y}$ on x is through the definition of the hypercubes. This may at first glance seem like a serious limitation on the expressiveness of conditional MTE distributions, however, as we show in Section 5 this restricted form can still capture complex conditional distributions.

In summary, there are some conditions that apply to the specification of conditional MTE potentials in order to use MTEs with standard inference algorithms. Moral *et al.* [4] therefore defined that the conditional MTE distributions must be of the functional form given in Equation (4). However, this general form implies parameter dependence, making automatic learning intractable. One approach to solve this problem is to assume parameter independence, which restricts conditional MTE distributions to the form given in Equation (6). In this case, learning conditional distributions reduces to the following two tasks:

1. Finding the *split points*/hybercubes for the conditioning variables.
2. Learning the parameters of Equation (6) for each hybercube.

The latter item can be solved by algorithms for learning univariate MTE potentials [9]. We will turn to this issue shortly, and thereafter look at a method for learning the definition of the hypercubes from data.

4 Learning Maximum Likelihood Distributions

4.1 The Univariate MTE Potentials

As already mentioned, MTE learning can be seen as a model selection problem where the number of exponential terms and the split points must be determined. In the following we briefly describe a learning procedure for the univariate case, the interested reader is referred to [9] for details.

When determining the number of exponential terms for a fixed interval I_k, we iteratively add exponential terms (starting with the MTE potential having only a constant term) as long as the BIC score improves or until some other termination criterion is met. The learning algorithm regards the parameter learning (with fixed structure) as a constrained optimisation problem, and uses Lagrange multipliers to find the maximum likelihood parameters.

To determine the split points of the domain of the variable, a set of candidate split points is chosen. Since the BIC score will be the same for any split points defining the same partitioning of the data, it is not required to look at a set of possible splits that is larger than the set of midpoints between every two consecutive observations of Y. However, to reduce the computational complexity of the learning algorithm we consider a smaller set of potential split points in the current implementation: Each lth consecutive midpoint is selected, where l is chosen so that we get a total of 10 candidate split points. We use a myopic approach to select among the candidate split points, so that the one offering the highest gain in BIC score is selected at each iteration. This is repeated until no candidate split point increases the BIC score.

4.2 Learning Conditional Distributions

After having defined how to learn the parameters of the marginal distribution of a variable X from data, we will now consider learning the hybercubes (i.e., the split points) that define the conditioning part of the distribution (cf. Section 3). We will again use the BIC-score for model selection, and for simplicity we will start the discussion assuming that X has only one continuous parent Y. Recall that learning the conditional distribution $f(x|y)$ consists of two parts: i) Find the split points for Y, and ii) learn the parameters of the marginal distribution for X inside each interval.

The previous subsection reviewed how we can learn the marginal distribution for X, and we will now turn to finding split points for Y. As was the case when we learned the split points of the marginal distributions, we will also now learn the split points of the conditioning variable using a myopic strategy: When evaluating a candidate split point for a given interval I_k, we compare the BIC score when partitioning I_k into two (convex) sub-intervals with the score obtained with no partitioning, i.e., we employ a one step look-a-head that does not consider possible further refinements of the two sub-intervals.

Recall that we use $\mathcal{D} = [\boldsymbol{y}, \boldsymbol{x}]$ to describe the data, and the notation $\mathcal{D}(R)$ to denote the subset of data for which a restriction R is true. The skeleton of the learning algorithm can then be described as in Algorithm 1.

```
1 Function LearnConditionalsplit points
  Data. 𝒟, ωₛ, ωₑ
  Result. splits
2 currentScore = Score(𝒟, ωₛ, ωₑ);
3 newScore = −∞;
4 for each potential split point sᵢ do
5      tmpScore = Score(𝒟(y ≤ sᵢ), ωₛ, sᵢ) + Score(𝒟(y > sᵢ), sᵢ, ωₑ) ;
6      if tmpScore > newScore then
7           newScore = tmpScore;
8           bestSplit = sᵢ;
9      end
10 end
11 if newScore > currentScore then
12      splits = [
13        LearnConditionalsplit points(𝒟(y ≤ bestSplit), ωₛ, bestSplit),
14        bestSplit,
15        LearnConditionalsplit points(𝒟(y > bestSplit), bestSplit, ωₑ)];
16 else
17      splits = ∅;
18 end
19 return splits;
```

Algorithm 1. Skeleton for the algorithm that learns the split points for the conditioning variable y

Algorithm 1 calls the external function Score to evaluate the different configurations of split points, both the current setting (in Line 2), and the one after adding a potential split point (in Line 5); note that the score function takes three or four parameters depending on whether we split the interval. One way of defining this score could be to fit a marginal MTE potential for each interval (looking only at data defined for the corresponding intervals), and afterwards calculate the BIC score for each of the intervals.

In Line 4, all potential split points for the conditioning variable are considered. Obviously, it suffices to only consider the observed values of the conditioning variable as potential split points. Note, however, that if t different split points are considered in Line 4, we will have to calculate the score $2t$ times in Line 5, and if we let t be equal to the number of observations in our database, the computational complexity of the algorithm will be intractable. We solve this in the same way as we did when learning marginal distributions, and select a fixed number of candidate split points. In the current implementation, we select the candidate splits by performing equal frequency-binning (with 10 bins) for each of the conditioning variables, and using the boundaries as candidate split points during learning.

The computational cost of calling the score-function in Line 5 may still be substantial though, and we have therefore considered alternatives ways of evaluating a split point. First of all, recall that the intuition behind defining a split point s for the conditioning variable y is that $f(x|y \leq s)$ and $f(x|y > s)$ are different (otherwise, we would reduce the BIC-score by introducing additional parameters for the new hypercube). By following this line of argument, we drop the calculations of the BIC-score in Line 5, and rather try to find good split points based on the Kolmogorov-Smirnov test [12] for determining whether two sets of data come from the same distribution. This modification is immediately accommodated in Algorithm 1 by simply replacing lines 2 and 5 with `CurrentScore`$= -\infty$ and `tmpScore`$= 1 - $ `kstest`$(\mathcal{D}_j(y \leq s_i), \mathcal{D}_j(y > s_i))$, respectively; `kstest`$(\mathcal{D}_1, \mathcal{D}_2)$ returns the p-value of the test that $\mathcal{D}_1$ and $\mathcal{D}_2$ come from the same distribution.

When working with more than one conditioning variable (i.e., $r > 1$) we need to select both a split variable and a split point. As before we do the selection greedily: iterate over all the conditioning variables, and for each variable find the best split point. After that select the conditioning variable having the best scoring split point. The recursive nature of the algorithm defines a binary tree in which each internal node is a conditioning variable and the arcs emanating from a node defines a partitioning of the associated interval (a so-called *probability tree*, [4]). Each leaf is associated with a univariate MTE distribution over x conditioned on the hybercube defined by the path from the root to the leaf. The final algorithm is similar to Algorithm 1 except that for each conditioning variable we also need to iterate over lines 4–10 and pick the best scoring variable to split on; the full specification has been left out due to space restrictions.

5 Examples

Our first example shows how our Algorithms learn a conditional Gaussian distribution, and in particular we examine the effect of the size of the dataset we learn from. We generated datasets of size 30, 50, 100, 250, and 1000 from the distribution

$$\begin{bmatrix} x \\ y \end{bmatrix} \sim \mathcal{N}\left(\begin{bmatrix} 0 \\ 0 \end{bmatrix}, \begin{bmatrix} 5 & 2 \\ 2 & 1 \end{bmatrix} \right), \tag{7}$$

and focused our attention on learning the conditional distribution $f(x|y)$. The potential split points for the conditioning variable y were defined by using the split points in an 11-bin equal-frequency histogram, meaning that 10 candidate split points were considered in each learning situation. As previously described, we used the results of the Kolmogorov-Smirnov tests to prioritise these candidate split points, and the BIC-score to determine whether or not a given candidate split point is to be accepted. The distribution over x for each y-interval was selected in order to maximise the BIC score.

The obtained results for all datasets are given in Fig. 3 part (a) to (e) respectively, and should be compared to the exact conditional density given in Fig. 3 (f); for further comparison, a scatter plot of the dataset with 250 cases in shown in Fig. 5. The results are promising: We see a strong resemblance between the target distribution and the gold standard distribution for all data sizes. We can

248   H. Langseth et al.

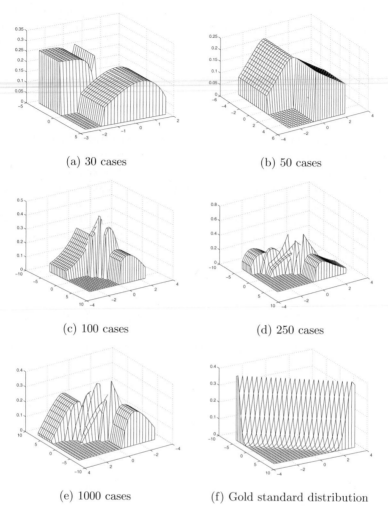

(a) 30 cases

(b) 50 cases

(c) 100 cases

(d) 250 cases

(e) 1000 cases

(f) Gold standard distribution

Fig. 3. The plots show the results of learning from 30, 50, 100 and 250 cases respectively. The gold-standard distribution, shown in part (f), is the conditional Gaussian distribution $f(x|y)$ derived from the joint distribution in Equation (7). The Kolmogorov-Smirnov tests were used to find the split points of the conditioning variable (axis on the left-hand side), whereas the marginal distributions were obtained by maximisation of the BIC score.

also see the effect of using the BIC-score to determine whether or not a candidate split point for y should be accepted: When only a few splits are employed for the smaller datasets, all candidate split points for y were used when learning from the largest dataset. Finally, it is also worth noticing that the support of x is never divided into subintervals in these runs. This is to be expected, considering that these MTE potentials are typically learned from only about 25 observations each.

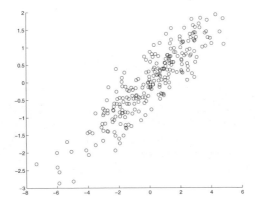

Fig. 4. A scatter of the data set with 250 cases sampled from the joint distribution in Equation (7). Note, in particular, the few cases sampled from the tails of the distribution.

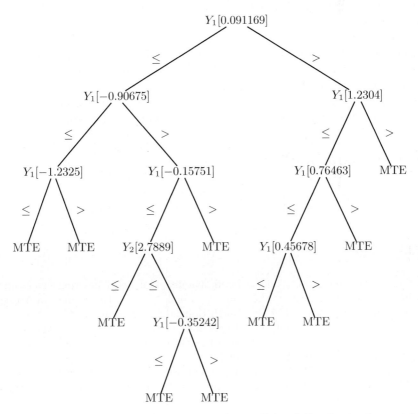

Fig. 5. A binary tree structure representing the conditional distribution learned from 250 cases sampled from the distribution in Equation (8)

For illustration, we also ran the algorithm using a data set containing 250 configurations sampled from the distribution

$$
\begin{bmatrix} x \\ y_1 \\ y_2 \end{bmatrix} \sim \mathcal{N} \left(\begin{bmatrix} 0 \\ 0 \\ 3 \end{bmatrix}, \begin{bmatrix} 5 & 2 & 2 \\ 2 & 1 & 1 \\ 2 & 1 & 2 \end{bmatrix} \right). \tag{8}
$$

The result of the learning can be seen in Fig. 5, where we have skipped the specification of the marginal MTE distributions in the leaves. Observe that the tree is more fine-grained around the mean of Y_1 compared to the parts of the interval with smaller support in the distribution (i.e., more data is available to capture the refinement). In particular, note also that the algorithm conditions on Y_2 is this interval.

6 Conclusions

In this paper we have investigated two alternatives for the definition of conditional MTE densities. We have shown that only the most restrictive one is compatible with standard efficient algorithms for inference in Bayesian networks.

We have also shown how the induction of this kind of conditional densities can be approached from the point of view of maximum likelihood estimation, including model selection for determining the partitioning of the domain of the conditioning variables based on the BIC score.

Our future work on this subject will include the definition of a structural learning algorithm based on the tools proposed on this paper and in [9].

Acknowledgments

This work has been partially supported by the Spanish Ministry of Science and Innovation under project TIN2007-67418-C03-02 and by FEDER funds.

References

1. Shenoy, P., Shafer, G.: Axioms for probability and belief function propagation. In: Shachter, R., Levitt, T., Lemmer, J., Kanal, L. (eds.) Uncertainty in Artificial Intelligence, vol. 4, pp. 169–198. North Holland, Amsterdam (1990)
2. Friedman, N., Goldszmidt, M.: Discretizing continuous attributes while learning bayesian networks. In: Proceedings of the Thirteenth International Conference on Machine Learning, pp. 157–165 (1996)
3. Kozlov, D., Koller, D.: Nonuniform dynamic discretization in hybrid networks. In: Geiger, D., Shenoy, P. (eds.) Proceedings of the 13th Conference on Uncertainty in Artificial Intelligence, pp. 302–313. Morgan & Kaufmann, San Francisco (1997)
4. Moral, S., Rumí, R., Salmerón, A.: Mixtures of truncated exponentials in hybrid Bayesian networks. In: Benferhat, S., Besnard, P. (eds.) ECSQARU 2001. LNCS, vol. 2143, pp. 156–167. Springer, Heidelberg (2001)

5. Cobb, B., Shenoy, P., Rumí, R.: Approximating probability density functions with mixtures of truncated exponentials. Statistics and Computing 16, 293–308 (2006)
6. Rumí, R., Salmerón, A., Moral, S.: Estimating mixtures of truncated exponentials in hybrid Bayesian network. Test 15, 397–421 (2006)
7. Romero, V., Rumí, R., Salmerón, A.: Learning hybrid Bayesian networks using mixtures of truncated exponentials. International Journal of Approximate Reasoning 42, 54–68 (2006)
8. Schwarz, G.: Estimating the dimension of a model. Annals of Statistics 6, 461–464 (1978)
9. Langseth, H., Nielsen, T., Rumí, R., Salmerón, A.: Parameter estimation in mixtures of truncated exponentials. In: Jaeger, M., Nielsen, T. (eds.) Proceedings of the 4th European Workshop on Probabilistic Graphical Models (PGM 2008), pp. 169–176 (2008)
10. Moral, S., Rumí, R., Salmerón, A.: Approximating conditional MTE distributions by means of mixed trees. In: Nielsen, T.D., Zhang, N.L. (eds.) ECSQARU 2003. LNCS, vol. 2711, pp. 173–183. Springer, Heidelberg (2003)
11. Zhang, N., Poole, D.: Exploiting causal independence in Bayesian network inference. Journal of Artificial Intelligence Research 5, 301–328 (1996)
12. DeGroot, M.H.: Optimal Statistical Decisions. McGraw-Hill, New York (1970)

A Generalization of the Pignistic Transform for Partial Bet

Thomas Burger[1] and Alice Caplier[2]

[1] Université Européenne de Bretagne, Université de Bretagne-Sud, CNRS,
Lab-STICC, Centre de Recherche Yves Coppens BP 573,
F-56017 Vannes cedex, France
thomas.burger@univ-ubs.fr
http://www-labsticc.univ-ubs.fr/~burger/
[2] Gipsa-Lab, 961 rue de la Houille Blanche, Domaine universitaire, BP 46, 38402
Saint Martin d'Hères cedex, France
alice.caplier@gips-lab.inpg.fr
http://www.lis.inpg.fr/pages_perso/caplier/

Abstract. The Transferable Belief Model is a powerful interpretation of belief function theory where decision making is based on the pignistic transform. Smets has proposed a generalization of the pignistic transform which appears to be equivalent to the Shapley value in the transferable utility model. It corresponds to the situation where the decision maker bets on several hypotheses by associating a subjective probability to non-singleton subsets of hypotheses. Naturally, the larger the set of hypotheses is, the higher the Shapley value is. As a consequence, it is impossible to make a decision based on the comparison of two sets of hypotheses of different size, because the larger set would be promoted. This behaviour is natural in a game theory approach of decision making, but, in the TBM framework, it could be useful to model other kinds of decision processes. Hence, in this article, we propose another generalization of the pignistic transform where the belief in too large focal elements is normalized in a different manner prior to its redistribution.

1 Introduction

The Transferable Belief Model [1] (TBM) is based on the decomposition of the problem into two stages: the **credal level**, in which the pieces of knowledge are aggregated under the formalism of belief functions, and the **pignistic level**, where the decision is made by applying the Pignistic Transform (PT): It converts the final belief function (resulting from the fusions of the credal level) into a probability function. Then, a classical probabilistic decision is made.

The manner in which belief functions allow to deal with compound hypotheses (i.e. set of several singleton hypotheses) is one of the main interests of the TBM. On the other hand, the decision making in the TBM only allows betting on singletons. Hence, at the decision making level, part of the belief function flexibility is lost. Of course, it is made on purpose, as, betting on a compound

C. Sossai and G. Chemello (Eds.): ECSQARU 2009, LNAI 5590, pp. 252–263, 2009.
© Springer-Verlag Berlin Heidelberg 2009

hypothesis is equivalent to remain hesitant among several singletons. It would mean no real decision is made, or equivalently, that no bet is booked, which seems curious, as the PT is based on betting ("pignistic" is derived from the Latin word for "bet").

Nevertheless, there are situations in which it could be interesting to bet on compound hypotheses. From the TBM point of view, it means generalising the PT so that it can handles compound bets. Smets has already presented such a generalisation [2], and it appears [3] to corresponds to the situation of a "n-person games" [4] presented by Shapley in the Transferable Utility Model in 1953. This work on game theory considers the case of a coalition of gamblers who wants to share fairly the gain with respect to the involvement of each. Once the formula is transposed to the TBM, the purpose is to share a global belief between several compound hypotheses. Obviously, one expects the transform to promote the hypotheses the cardinality of which is the greatest... Roughly, it means that, if for the same book, it is possible to bet on the singleton hypothesis $\{h_1\}$ or on the compound hypothesis $\{h_1, h_2\}$, then, this latter must be preferred (even if the chances for h_1 are far more interesting than for h_2). Practically, this intuitive behaviour looks perfectly accurate, and of course, the generalization proposed by Smets behaves so.

On the other hand, there are yet other situations, where it should be encouraged to bet on singleton hypotheses when possible, whereas it should remain allowed to bet on compound hypotheses when it is impossible to be more accurate. Hence, we depict a "progressive" decision process, where it is possible to remain slightly hesitant, and to manually tune the level between hesitation and bet. Let us imagine such a situation: the position of a robot is modelled by a state-machine, and its trajectory along a discrete time scale is modelled by a lattice. At each iteration of the discrete time, the sensors provide information to the robot, and these pieces of information are processed in the TBM framework: they are fused together (the credal level) and the state of the robot is inferred by a decision process (the pignistic level). At this point several stances are possible:

- the classical PT is used. Unfortunately, as the sensors are error-prone, the inferred state is not always the right one. Finally, the inferred trajectory is made of right and wrong states with respect to the ground-truth (Fig. 1). Of course, the TBM provides several tools to filter such trajectories [5,6,7], and, in spite of a relative computational cost, they are really efficient.
- Instead of betting on a single state at each iteration of the time, it is safer to bet on a compound hypothesis (i.e. on a group of several states, knowing that, the more numerous, the less chance to make a mistake). Unfortunately, the risk is now to face a situation where no real decision is made and the inferred trajectory is too imprecise (Fig. 2).
- The balance between these two extreme stances would be to automatically tune the level of hesitation in the bet: When the decision is difficult to make, a compound hypothesis is assessed to avoid a mistake, and otherwise, a singleton hypothesis is assessed, to remain accurate (Fig. 3).

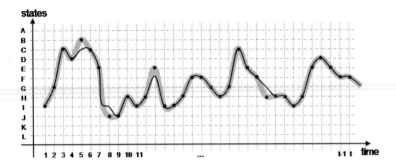

Fig. 1. The space-time lattice: the horizontal axis represents the time iterations, and the vertical axis, the states. The real trajectory (ground truth) is represented by the black line, and the inferred states are presented by black dots linked by the grey line. The real and inferred trajectories differ, as few mistakes are made in the decision process.

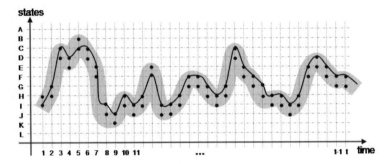

Fig. 2. In a similar manner to figure 1, the real trajectory (ground truth) is compared to the inferred one. As a matter of fact, no mistake is made on the inferred trajectory, but, as a drawback, it is really imprecise.

The first stance corresponds to classical decision making. The second and third stances both correspond to situations where it is possible to bet on compound hypotheses, but in a different manner. The second stance is rather classical from belief functions point of view, and several types of decision based on non-additive measures [8] achieve efficient results (such as [4,9]). Nevertheless, in spite of an adapted mathematical structure, they do not model the problem in a manner that corresponds to the kind of decision we expect in the third stance (as shown in 3.1). By now, the only way to perform a decision according to the third stance is to set an ad-hoc method. For instance, it is possible to consider compound hypotheses, and practise hypothesis testing, such as in classical statistical theory. With such a method, the size of the selected compound hypothesis is related to the p-value desired. In a similar way, but in a more subjective state of mind, it is also possible to associate a cost to each decision and to minimise the cost function. Finally, it is possible to simply assess a threshold T to the probability of each decision, to sort them in descending order, and to select the first n hypotheses so that their probabilities add up to a value superior to T. For all these methods, we do not provide bibliographical links, as they are based on very basic scholar knowledge.

This paper aims at defining a decision process according to the third stance in the context of the TBM. In section 2, we briefly present the TBM. In section 3 we analyse related works and we focus on the Shapley value and the corresponding PT generalization. We show that minor modifications lead to the expected results. In section 4, we present our new method to generalize the PT, and give some interesting properties. Finally, section 5 illustrates it with real examples.

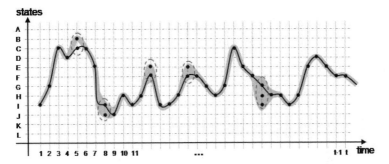

Fig. 3. In a similar manner to figure 1, the real trajectory (ground truth) is compared to the inferred one. A trade-off between risky bets (a singleton state is assessed) and imprecise decisions (circled by a dot line) allows limiting the number of mistake while remaining quite precise

2   Transferable Belief Model

In this section we rapidly cover the basis of the TBM [1] and of the belief function theory [10], in order to set the notations. We assume the reader to be familiar with belief functions.

Let Ω be the set of N exclusive hypotheses $\Omega = \{h_1, \ldots h_N\}$ for a variable X. Ω is called the **frame of discernment**. Let 2^Ω, called the **powerset** of Ω, be the set of all the subsets A of Ω, including the empty set (it is the sigma-algebra of Ω): $2^\Omega = \{A/A \subseteq \Omega\}$.

A **belief function**, or a **basic belief assignment**(BBA) $m(.)$ is a set of scores defined on 2^Ω that adds up to 1:

$$m : 2^\Omega \to [0, 1]$$
$$A \mapsto m(A) \text{ with } \sum_{A \subseteq \Omega} m(A) = 1$$

A **focal element** is an element of the powerset to which a non-zero belief is assigned. We call as the **cardinality** of a focal element, noted $|.|$, the number of elements of Ω it contains. For sake of simplicity, we say that a hypothesis or a focal element is larger (or wider) than another when its cardinality is greater. Hence, a BBA represents a subjective belief in the propositions that correspond to the elements of 2^Ω and nothing wider or smaller.

The **conjunctive combination** is a N-ary symmetrical and associative operator that models the fusion of the pieces of information coming from N independent sources (it is the core of the **credal level**):

$$\odot : \mathfrak{B}^\Omega \times \mathfrak{B}^\Omega \times \ldots \times \mathfrak{B}^\Omega \to \mathfrak{B}^\Omega$$

$$m_1 \odot m_2 \odot \ldots \odot m_N \mapsto m_\odot$$

with, $\mathfrak{B}^\Omega$ corresponding to the set of the BBA defined on Ω, and

$$m_\odot(A) = \sum_{\cap_{i=1}^N A_i = A} \left(\prod_{n=1}^N m_n(A_n) \right) \ \forall A \subseteq 2^\Omega$$

The pignistic probability measure (Bet$\mathbb{P}$) is defined by the use of the **pignistic transform** (in the **pignistic level**):

$$\mathrm{Bet}\mathbb{P}(X = h) = \frac{1}{1 - m(\emptyset)} \sum_{h \in A, \ A \subset \Omega} \frac{m(A)}{|A|} \ \forall h \in \Omega$$

Then, the pignistic probability distribution is computed: $p(h) = \mathrm{Bet}\mathbb{P}(X = h) \ \forall h \in \Omega$, or, in other words: $p = \mathrm{PT}(m)$. Finally, the hypothesis of maximum pignistic probability $\tilde{h}$ is selected: $\tilde{h} = \mathrm{argmax}_{h_i \in \Omega}(p(.))$

3   The Shapley Value and the Pignistic Transform

3.1   Related Work

Several generalizations/alternatives to the PT exist in the literature. When proposing such a work, authors do not have the same objective, which explains this manifold. In [11,12], the point is to find a conversion method between probabilistic models and evidential ones. Then, the PT is compared to the plausibility transform, and this later is assessed to be more adapted. In [13], the point is to face the computational complexity of BF by finding an adequate probabilistic approximation. In [16] the Generalized Pignistic Transformation is defined. In spite of its name, is not related to the TBM framework: it is the counterpart of the PT in a framework which is an attempt of generalization of the TBM. This framework, its potential applications and the way it generalizes the PT or propose alternative transforms have nothing in common with this work. In [14,15], several transforms are proposed as alternatives to derive a bet-like decision from a BF. Finally, in each of these works, the alternatives to the BFs (i.e. the alternative mathematical structures that support the information prior to the decision) are either out of our interests [16], either probabilistic. Consequently, as interesting these works remain, they are not in the scope of this paper, in the meaning that they do not propose any generalizations to partial bets, such as targeted here.

Nevertheless, there are several works in which alternative structures (i.e. neither BF, neither probabilities) are proposed and may fit our need here. These structures belong to the general class of fuzzy measures [8] (also called capacities, or non-additive measures [19]). Unfortunately, these works aim at defining alternative structures which are computationally more efficient than BF; and the definition of a non-additive measure adapted to partial bet is not investigated. For instance, the works of Cuzzolin [17] prove that the space of the BFs defined on a dedicated frame is a simplex, and it provides a framework to analyse the various transforms of the DST from a geometrical point of view. For us, it brings new insights with respect to decision making as this geometrical work stresses the link between structures on which decision is classically made (e.g. a bayesian BF, a plausibility function), and the geometrical transforms that make the conversion amongst these structures (e.g. the PT, the Mbius transforms). In [18,19], Gravisch introduces k-order additive fuzzy measures, which can be seen as an intermediate type of structure between probabilities (1-order measures) and BF: it corresponds to BF for which the cardinality of the largest focal element is k. Once again, the main objective is to define more efficient structure for computational aspects.

These papers do not investigate the consequences of the use of these structures in decision making. To our knowledge, the only work in which decision making scenarios with compound hypotheses are considered was carried out by Shapley [4], and then re-investigated by Smets [2,3], in the context of the TBM. Hence, we mainly base our work on this latter.

3.2   The Shapley Value

In [3], Smets summarizes his work of [2] and explains how to derive the PT in case of non-singleton bets. As it is explained altogether with the assessment of the result, it concurs with the work of Shapley [4]:

$$\mathrm{Bet}\mathbb{P}'(X = B) = \frac{1}{1 - m(\emptyset)} \sum_{A \subseteq \Omega} \frac{m(A) \cdot |A \cap B|}{|A|} \quad \forall B \subseteq \Omega$$

In this equation, the value associated to B is the sum of (1) all the hypotheses of strictly smaller cardinality which are nested in B, (2) $m(B)$ itself (3) an "inherited" mass from wider hypotheses in which B is nested, and (4) all other hypotheses for which there is no inclusion relation with B, but the intersection of which with B is non-empty:

$$\mathrm{Bet}\mathbb{P}'(X = B) = \frac{1}{1 - m(\emptyset)} \cdot \left[\sum_{A \subset B} m(A) + m(B) \right.$$
$$+ \sum_{B \subset A \subseteq \Omega} \frac{m(A) \cdot |B|}{|A|}$$
$$\left. + \sum_{\text{other } A \subseteq \Omega} \frac{m(A) \cdot |A \cap B|}{|A|} \right] \quad \forall B \subseteq \Omega$$

Of course, in case of B being a singleton hypothesis, it corresponds to the classical PT: The first and the fourth terms are zero-valued, and $|B| = 1$, so that:

$$\text{Bet}\mathbb{P}'(X = B) = \frac{1}{1 - m(\emptyset)} \cdot \left[m(B) + \sum_{B \neq A, B \in A \subseteq \Omega} \frac{m(A)}{|A|} \right] = \text{Bet}\mathbb{P}(X = B) \; \forall B \in \Omega$$

3.3   Application of Shapley Work to Partial Bets

Now, let us forget the original interest of this formula, and let us consider it through our own aim: Is it sensible to consider the Shapley value as a generalization of the PT which allows comparing hypotheses of different cardinality? Basically, the first term in the previous equation means, that the value associated to a compound hypothesis $\{h_1, h_2\}$ is increased by the belief of all the hypotheses which are nested in it : $\{h_1\}$ and $\{h_2\}$. Moreover, as all the considered hypotheses inherit belief of wider hypotheses in a manner proportional to their cardinality, it is impossible to assign a pignistic probability to $\{h_1\}$ or $\{h_2\}$ which is greater to the one assigned to $\{h_1, h_2\}$. As a consequence, larger hypotheses are always promoted in the decision making, which leads to situations such as the one illustrated in figure 2.

   In addition, the fourth term is also problematic with respect to partial bets. Because of it, an important belief in a compound hypothesis $\{h_1, h_2\}$ increases the Shapley value for another compound hypothesis $\{h_1, h_3\}$ as their intersection is none empty. In our situation, $\{h_1, h_2\}$ and $\{h_1, h_3\}$ are different and exclusive choices for the decision. The value assess to a compound hypothesis must keep an evidential interpretation, as we deal with hypothesis of different cardinality simultaneously (as in the credal level). Then, it must not yet be understood as a probabilistic sigma-algebra, and we should stick to an interpretation similar to Shafer's concerning the belief assignment [10], reading that, *the BBA in a focal element models the belief that can be placed in it, and nothing smaller or larger.*

   The transform leading to the Shapley value is really interesting as a natural generalization of the PT. Unfortunately, it does not lead to an acceptable solution when a partial bet is expected. On the other hand, as we root in the TBM, the decision process we aim at must also remain related to the PT. As a consequence, we propose to start from the Shapley value, and to modify it, so that it fulfils our requirements. The first natural step is to remove terms 1 and 4, as they appear to be problematic. On the contrary, the terms 2 (which represents the belief in the considered hypothesis) and 3 (which represents inherited beliefs) are perfectly natural, except for normalization considerations: As some of the redistributions of the belief have been discarded, it is natural to consider to renormalize their values so that the total mass is conservative.

3.4   Axiomatic Justification of the PT

By now, let us recall the axiomatic of the PT, in order to make sure that we respect it. In contradiction to what is often read, the PT is not justified by the principle of insufficient reason [20], nor it is justified by the proof that it is impossible to built a Dutch book against the PT. As a matter of fact, Smets is rather explicit on these two points:

- An intuitive generalization of the principle of insufficient reason is a cue of the interest of the PT, but it does not justify it [21].
- In spite of a particular Dutch Book discarded in [3], the proof that it resists to all Diachronic Dutch Books in general is not given.

Then, the only justifications of the PT rely in five axioms (linearity, projectivity, efficiency, anonymity, false event). These axioms are not always accepted beyond the TBM interpretation of the Belief function theory, and consequently, the PT is also discussed by supporters of other interpretations [11,12]. Within the TBM, it is nonetheless the only decision process accepted. As this work roots in the TBM, **our single concern is to remain coherent with this framework and with these five axioms**, to which we add a sixth, the conservation principle.

4   Generalization to Partial Bets

As explained in our introductive example, the point is to allow hesitation, but to control it, so that, when it is not necessary, no hesitation occurs. A very simple way to control the hesitation is that the decision maker defined a maximum amount of authorized hesitation. Thus, let $\gamma \in \mathbb{N} \cap [1, |\Omega|]$ a threshold that models this amount. Let L_i the set of hypotheses of cardinality i (L stands for "level"). The decision is made within $\Delta_\gamma = \{L_1, \ldots, L_\gamma\}$.

Then, our purpose is to define a probability measure $\mathbb{B}$ on Δ_γ, so that a decision can be made by selecting the element of Δ_γ the value of which is the greatest. The corresponding probability space $(\Delta_\gamma, \mathcal{F}, \mathbb{B})$, where $\mathcal{F}$ is the canonical sigma-algebra of Δ_γ, must be derived from the measured space $(\Omega, 2^\Omega)$ in a manner similar to the probability space $(\Omega, 2^\Omega, \mathrm{Bet}\mathbb{P})$, i.e. by the definition of an appropriate transform. Intuitively, $\mathbb{B}$ looks like a γ-additive BF [18], but as a matter of fact, its interpretation as such is problematic.

Δ_γ must be understood as a decision space in itself, in which a variable D (which stands for "Decision") takes its value, and which is not related to 2^Ω, in which the variable X takes its value. This may appear as strange, as the elements of Δ_γ corresponds to elements of 2^Ω, but from the decision point of view, $\{h_1, h_2\}$ and $\{h_1\}$ are two different elements of Δ_γ, and they are decisions which are exclusive one another. On the contrary, for X, $\{h_1, h_2\}$ and $\{h_1\}$ are nested. That is why $\mathbb{B}$ is not a BF. As a consequence, in spite of a similar mathematical structure, it can not be interpreted as a γ-additive BF. In [18], Gravisch stresses that the interpretation of a non-additive measure is rather difficult, and the interpretation of $\mathbb{B}$ perfectly illustrates this fact. Equivalently, $\mathrm{Bet}\mathbb{P}$, the result of the PT, which has a structure equivalent to a bayesian BF, can not be interpreted as such [11,12]; otherwise its combination with a BF thanks to Dempster's rule would be significant. As we consider the TBM as the frame of this work, it is obvious that the interpretation of $\mathbb{B}$ as a BF is as problematic as the interpretation of BetP as a BF.

Let $A \notin \Delta_\gamma$. As explained, the point is to "take" the belief $m(A)$, to "replace" it by a zero value, and to "redistribute" it to hypotheses within Δ_γ. Let B such a hypothesis within Δ_γ. In a way similar to the PT, the redistribution must be linear, that is, proportional to the size of the hypothesis that inherits it. Moreover, the redistribution must remain conservative, so that, making a decision on Ω by selecting an element of Δ_{γ_1}, then, making a more precise decision by selecting an element of Δ_{γ_2} with $\gamma_2 < \gamma_1$ is equivalent to directly make a decision on Ω by selecting an element of Δ_{γ_2}.

As a consequence, all the hypotheses within a level L_i must inherit the same amount of belief, and each level L_i must globally inherit a belief proportional to $|L_i| \times i$. Hence, we propose to share $m(A)$ into N parts, so that, all the elements of $L_i, \forall i \leq \gamma$ inherits i parts of $m(A)$. N depends on the number of hypotheses of D_γ which are nested in B. This number depends in turn of γ, and the size of A. An elementary enumeration leads to the following formula:

Definition 1

$$N(|A|, \gamma) = \sum_{k=1}^{\gamma} C_k^{|A|} \cdot k$$

where $C_p^n = \frac{n!}{p!(n-p)!}$ is the number of combinations of p elements among n.

Now that the redistribution pattern is defined, let us derive the transform itself:

Definition 2. *The probability measure $\mathbb{B}_\gamma$ is derived from the BBA $m(.)$ by the following transform:*

$$\mathbb{B}_\gamma(D = B) = \frac{1}{1 - m(\emptyset)} \cdot \left[m(B) + \sum_{B \subset A \subseteq \Omega, A \notin \Delta_\gamma} \frac{m(A) \cdot |B|}{N(|A|, \gamma)} \right] \quad \forall B \in \Delta_\gamma$$

Proposition 1. *The pignistic transform is a particular case of definition 2, as we have $\mathbb{B}_1(.) = \mathrm{BetP}(.)$.*

Proof. If $\gamma = 1$, several simplifications occur: $\Delta_\gamma \equiv \Omega$, $X \equiv D$, $N(|A|, \gamma) = |A|$ and $|B| = 1$. As B is a singleton, let us note it h. One has $\forall h \in \Omega$:

$$\mathbb{B}_1(D = h) = \frac{1}{1 - m(\emptyset)} \cdot \left[m(h) + \sum_{h \in A \subseteq \Omega, |A| \neq 1} \frac{m(A)}{|A|} \right]$$

$$= \frac{1}{1 - m(\emptyset)} \cdot \sum_{h \in A \subseteq \Omega} \frac{m(A)}{|A|} = \mathrm{BetP}(D = h) = \mathrm{BetP}(X = h) \quad \square$$

Another interesting property is derived from the conservation principle:

Proposition 2. *Making a decision by selecting $\delta_1 \in \Delta_{\gamma_1}$ with the use of $\mathbb{B}_{\gamma_1}$, and then, making a decision on δ_1 by the use of $\mathbb{B}_{\gamma_2}, \gamma_2 < \gamma_1$, is equivalent to directly make a decision by the use of the probability measure $\mathbb{B}_{\gamma_2}$.*

From a applicative point of view, this last result is really interesting, as it means, it is possible to make a decision on $\Delta_{|\Omega|-1}$, by redistributing the belief from $L_{|\Omega|}$, in order to discard a single element of Ω, then to make a decision on $\Delta_{|\Omega|-2}$ by redistributing only the belief from $L_{|\Omega|-1}$, and so on, until Δ_1. This set of operations has a computational cost similar to the one necessary to make a decision over Δ_1: the belief at each level L_i is redistributed one time to compute $\mathbb{B}_1$ or $\mathbb{B}_\gamma, \forall \gamma \in [1, |\Omega|]$. Then, it is possible for the decision maker to rapidly analyse the capability of the decision process to focus on a compound hypothesis of restricted cardinality, prior to the definition of γ.

Now the transform is explicit, the removal of two of the four terms in the original Shapley value may appear as arbitrary: We mainly explain it from a "functional" point of view. On the other hand, here are strong evidences that a more mathematical construction is also achievable : (1) the γ-additive structure, with $\gamma = 1$ being equivalent to the PT and with $\gamma = |\Omega|$ being equivalent to the original BF, and (2) our formula has strong similarities with the orthogonal projection of a BF on the probability simplex [17]. Hence, identifying the geometrical transform that justify it is an interesting future works to focus on.

5   Applications to American Sign Language Recognition

In this section, we briefly summarize a previous work of ours [22], in which the transform has been used for gesture recognition: We proposed to recognize an American Sign Language gesture performed in front of a video camera, among a set 19 possible gestures. For any gesture G^i, a dedicate Hidden Markov Model HMM^i is trained. For any new occurrence $G^?$ to recognize, the system computes the probability of the observed gesture to be the observation sequence produced by each HMM^i. A classical method is to recognize the new gesture as an occurrence of the gesture G^* which HMM^* produces the highest likelihood.

Amongst the 19 gestures, few pairs of them are so closed to each others than the system does not discriminate. Consequently, the overall accuracy for the recognition task is reasonably good (75.88% on 228 items), but several mistakes occur between the similar pairs or triplets. This is why, in this article, we have first proposed to set a decision method which allows to produce a single decision when it is possible and to produce an incomplete decision otherwise, in order to complete it in a second step. The use of the $\mathbb{B}_2$ and $\mathbb{B}_3$ provides far better result than $\mathbb{B}_1$ or equivalently $\text{Bet}\mathbb{P}$. On the 228 items in the test set, there are 189 examples for which a complete decision is made. For this singleton decision, the accuracy is 79.37%: the fewer remaining singletons are less error-prone. For the other examples the decision is imprecise. If we consider the decision as a right one when one of the elements of the compound hypothesis is the good one, then, the overall accuracy is 82.02%. From the applicative point of view, it shows that the concept of such a decision process is accurate as it allows focusing the imprecision of a decision only when it is necessary.

In a second step, we fuse the information of the manual gestures with additional non-manual gestures (face/shoulders motions, facial expression, which are very important in ASL) in order to help do discriminate among a cluster of similar gestures. These non-manual gestures are completely inefficient to discriminate if they are used in the first step altogether with the manual ones, as their variability is hidden by variability of the manual feature. Hence, the progressive nature of the decision process is helpful for a hierarchical data fusion.

Finally, when compared to classical Bayesian methods, it appears that our complete system is both more accurate and more robust: For instance, 31.1% of the mistakes are avoided with respect to a situation where the second step is systematically used: In such a use, the second step put back into question some good decisions of the first step (see [22] for a comprehensive evaluation).

6   Conclusion

In this article, we have presented a generalisation of the pignistic transform that differs from Smets' one, as it does not corresponds to the Shapley value. It provides an alternative when it is necessary to control the trade-of between hesitation and bet in decision making. Moreover, the classical pignistic transform from Smets appears to be a particular case of this generalisation. From a theoretical point of view, the main change with respect to Shapley's work relies in (1) the manner the belief in too large focal elements is normalized prior to its redistribution, and (2) the restriction to focal elements of cardinality $\leq \gamma$, as with k-additive belief functions. From a practical point of view, its application to Sign Language recognition stresses its interest on real problems. Tree directions are considered for future works: (1) application to other real problems in pattern recognition, (2) its application to credal time-state models [5], and (3) the geometrical explanation of the transform in the probability simplex [17].

References

1. Smets, P., Kennes, R.: The transferable belief model. Art. Int. 66(2), 191–234 (1994)
2. Smets, P.: Constructing the pignistic probability function in a context of uncertainty. Uncertainty in Artificial Intelligence 5, 29–39 (1990)
3. Smets, P.: No Dutch book can be built against the TBM even though update is not obtained by bayes rule of conditioning. In: Workshop on probabilistic expert systems, Societa Italiana di Statistica, Roma, pp. 181–204 (1993)
4. Shapley, L.: A Value for n-person Games. Contributions to the Theory of Games. In: Kuhn, H.W., Tucker, A.W. (eds.) Annals of Mathematical Studies, vol. 2(28), pp. 307–317. Princeton University Press, Princeton (1953)
5. Ramasso, E., Rombaut, M., Pellerin, D.: Forward-Backward-Viterbi procedures in the Transferable Belief Model for state sequence analysis using belief functions, ECSQARU, Hammamet, Tunisia (2007)
6. Xu, H., Smets, P.: Reasoning in Evidential Networks with Conditional Belief Functions. International Journal of Approximate Reasoning 14, 155–185 (1996)
7. Ristic, B., Smets, P.: Kalman filters for tracking and classification and the transferable belief model. In: FUSION 2004, pp. 4–46 (2004)

8. Dennenberg, D.: Non-Additive Measure and Integral. Kluwer Academic Publishers, Dordrecht (1994)
9. Dubois, D., Prade, H.: Possibility theory: an approach to computerized processing of uncertainty. Plenum Press (1988)
10. Shafer, G.: A Mathematical Theory of Evidence. Princeton University Press, Princeton (1976)
11. Cobb, B., Shenoy, P.: On the plausibility transformation method for translating belief function models to probability models. Int. J. of Approximate Reasoning (2005)
12. Cobb, B., Shenoy, P.: A Comparison of Methods for Transforming Belief Functions Models to Probability Models. In: Nielsen, T.D., Zhang, N.L. (eds.) ECSQARU 2003. LNCS (LNAI), vol. 2711, pp. 255–266. Springer, Heidelberg (2003)
13. Voorbraak, F.: A computationally efficient approximation of Dempster-Shafer theory. International Journal on Man-Machine Studies 30, 525–536 (1989)
14. Daniel, M.: Probabilistic Transformations of Belief Functions. In: Godo, L. (ed.) ECSQARU 2005. LNCS, vol. 3571, pp. 539–551. Springer, Heidelberg (2005)
15. Sudano, J.: Pignistic Probability Transforms for Mixes of Low- and High- Probability Events. In: Int. Conf. on Information Fusion, Montreal, Canada (2001)
16. Dezert, J., Smarandache, F., Daniel, M.: The Generalized Pignistic Transformation. In: 7th International Conference on Information Fusion Stockholm, Sweden (2004)
17. Cuzzolin, F.: Two new Bayesian approximations of belief functions based on convex geometry. IEEE Trans. on Systems, Man, and Cybernetics - B 37(4), 993–1008 (2007)
18. Grabisch, M.: K-order additive discrete fuzzy measures and their representation. Fuzzy sets and systems 92, 167–189 (1997)
19. Miranda, P., Grabisch, M., Gil, P.: Dominance of capacities by k-additive belief functions: EJOR 175, 912–930 (2006)
20. Keynes, J.: Fundamental Ideas. A Treatise on Probability. Macmillan, Basingstoke (1921)
21. Smets, P.: Decision Making in the TBM: the Necessity of the Pignistic Transformation. Int. J. Approximate Reasoning 38, 133–147 (2005)
22. Aran, O., Burger, T., Caplier, A., Akarun, L.: A Belief-Based Sequential Fusion Approach for Fusing Manual and Non-Manual Signs. Pattern Recognition (2009)

Using Logic to Understand Relations between DSmT and Dempster-Shafer Theory

Laurence Cholvy

ONERA Centre de Toulouse,
2 avenue Edouard Belin,
31055 Toulouse, France
cholvy@cert.fr

Abstract. In this paper[1,2], we study the relations that exist between Dempster-Shafer Theory and one of its extensions named DSmT. In particular we show, by using propositional logic, that DSmT can be reformulated in the classical framework of Dempster-Shafer theory and that any combination rule defined in the DSmT framework corresponds to a rule in the classical framework. The interest of DSmT rather concerns the compacity of expression it manipulates.

Keywords: Dempster-Shafer Theory, DSmT, Propositional Logic.

1 Introduction

Dempster-Shafer Theory [14] is one of the main theories which deal with uncertainty representation. Because of the combinaison rule it provides, it has widely been used in information fusion context. This theory has motivated and still motivates many studies: applicative studies whose aim is to apply this theory on concrete cases [1], [6], [12], [10] or more fundamental works whose aim is to extend this theory or to propose new combination rules, or to give interpretations to this theory [11], [15], [16],[17], [2], [3].

Among the different extensions of Dempster-Shafer Theory, one can cite the so-called DSmT, whoses basis are described in [7] then extended in [13]. The authors of DSmT, aware of the fact that their new theory is an extension of Dempster-Shafer Theory, affirm that it allows one to solve problems that the original Dempster-Shafer Theory fails to solve: *The Dezert-Smarandache Theory (DSmT) of plausible and paradoxical reasoning is a natural extension of the classical Dempster-Shafer Theory (DST) but includes fundamental differences with the DST. (...) DSmT is able to solve complex, static or dynamic fusion problems beyond the limits of the DST framework, specially when conflicts between sources become large and when the refinement of the frame of the problem under consideration becomes inaccessible because of vague, relative and imprecise nature of elements of it.* ([13]).

---

[1] The french version of this paper has been presented at LFA'08 (Rencontres francophones sur la Logique Floue et ses Applications) in Lens, october 16th 2008.
[2] This work has been supported by ONERA, grant number 13631.01.

C. Sossai and G. Chemello (Eds.): ECSQARU 2009, LNAI 5590, pp. 264–274, 2009.
© Springer-Verlag Berlin Heidelberg 2009

In this present paper, we aim at formally comparing the DSmT and Dempster-Shafer Theory. More precisely, we show that DSmT can be reformulated in Dempster-Shafer Theory and thus, from expressivity point of view, DSmT is equivalent to Dempster-Shafer Theory.

For establishing this correspondance, we apply one of the major properties of propositional logic. Given a propositional language, this property ensures the existence of a bijection between the quotient set defined by the equivalence relation on the the set of formulas and the power set of interpretations. More precisely, any formula of the language is associated with the unique set of its models (called its truth-set). Moreover, all the formulas which are equivalent are associated with the same truth-set.

DSmT considers that hypothesis of a discernment frame can be non-exclusive. We show that such a frame can be considered as a propositional language and that expressions which are assigned a mass are formulas of this language. By the previous bijection, this comes to assign masses to sets of interpretations which are, by definition, exclusive. Thus, we are projected in the classical Dempster-Shafer framework in which hypothesis are exclusive.

This paper is organized as follows. Section 2 reminds the notions of logic needed for the comprehension of the paper. Section 3 analyses DSmT in the light of logic. We first study the case of free models then we study the more general case of hybrid models. Finally, section 4 concludes the paper.

2   Propositional Logic

A propositional language Θ is defined by an alphabet whose vocabulary is composed of a set of propositional letters, connectors $\neg, \wedge, \vee, \rightarrow, \leftrightarrow$ and parenthesis.

The set of propositional formulas, denoted $FORM$, is the smallest set of words built on this alphabet such that: if A is a letter, then A is a formula; $\neg A$ is a formula if A is a formula; $A \wedge B$ is a formula if A and B are formulas. Other formulas are defined by abbreviation. More precisely, $A \vee B$ denotes $\neg(\neg A \wedge \neg B)$; $A \rightarrow B$ denotes $\neg A \vee B$; $A \leftrightarrow B$ denotes $(A \rightarrow B) \wedge (B \rightarrow A)$.

In this paper, it is sufficient to consider a finite language i.e a language whose finite set of letters. We will note σ_0 the formula which is the disjonction of all these letters.

A clause is a particular formula which is a disjunction of literals, a literal being a letter or the negation of a letter. One says that a clause subsumes another clause if the literals of the first one are literals of the second one too.

An interpretation i is an application from the set of letters to the set of truth value $\{0, 1\}$. An interpretation i can be extended to the set of formulas by: $i(\neg A) = 1$ iff $i(A) = 0$; $i(A \wedge B) = 1$ iff $i(A) = 1$ and $i(B) = 1$. Consequently, $i(A \vee B) = 1$ iff $i(A) = 1$ or $i(B) = 1$, and $i(A \rightarrow B) = 1$ iff $i(A) = 0$ or $i(B) = 1$; $i(A \leftrightarrow B) = 1$ iff $i(A) = i(B)$.

The set of interpretations of Θ will be denoted I_Θ.

Frequently, an interpretation i is represented by a set made of the propositional letters a such that $i(a) = 1$ and negation of propositional letters b such that $i(b) = 0$. For instance, if a, b, c are the three letters of the languae, then the interpretation i defined by: $i(a) = 1$ and $i(b) = i(c) = 0$ is denoted by the set $\{a, \neg b, \neg c\}$. This notation will be used in the following.

The interpretation i is *a model* of formula A iff $i(A) = 1$. We say that i satisfies A.

The *"truth set"* (or set of models) of formula A is the set of interpretations which are models of A. It is denoted $mod(A)$. We have: $mod(A) \in 2^{I_\Theta}$.

A is *satisfiable* iff there exists an interpretation i such that $i(A) = 1$.

Let A and B two formulas. We have: $A \models B$ iff $mod(A) \subseteq mod(B)$ and $\models A \leftrightarrow B$ iff $mod(A) = mod(B)$. Thus, two equivalent formulas have the same truth-set.

Definition 1. *Let σ be a satisfiable formula. We define an equivalence relation denoted $r^\sigma_\leftrightarrow$ on $FORM$ as follows: let A and B two formulas, we have A $r^\sigma_\leftrightarrow$ B iff $\sigma \models A \leftrightarrow B$.*

Definition 2. *Let σ be a satisfiable formula. Let S be a set of formulas. We note $S/r^\sigma_\leftrightarrow$ the quotient set of S given the equivalence relation $r_\leftrightarrow$.*

Definition 3. *Let σ be a satisfiable formula. We define a function Mod^σ from $FORM/r^\sigma_\leftrightarrow$ to $2^{mod(\sigma)}$ by: if c belongs to $FORM/r^\sigma_\leftrightarrow$, then $Mod^\sigma(c) = mod(A \wedge \sigma)$ where $A \in c$.*

Proposition 1. *Mod^σ is a bijection from $FORM/r^\sigma_\leftrightarrow$ to $2^{mod(\sigma)}$*

This proposition states that equivalent formulas given relation $r^\sigma_\leftrightarrow$ do have the same truth set and that any subset of $mod(\sigma)$ is the truth-set of an unique set of formulas which are equivalent given $r^\sigma_\leftrightarrow$.

Example 1. Consider a language whose letters are a, b, c. I_Θ has got 8 elements which are the following interpretations:

$i_1 = \{a, b, c\}$, $i_2 = \{a, b, \neg c\}$, $i_3 = \{a, \neg b, c\}$, $i_4 = \{a, \neg b, \neg c\}$,
$i_5 = \{\neg a, b, c\}$, $i_6 = \{\neg a, b, \neg c\}$, $i_7 = \{\neg a, \neg b, c\}$, $i_8 = \{\neg a, \neg b, \neg c\}$.
σ_0 is the formula $a \vee b \vee c$, et $mod(\sigma_0) = \{i_1, ... i_7\}$.

Let c_1 be the equivalence class of $a \wedge (b \vee c)$. $Mod^{\sigma_0}(c_1) = \{i_1, i_2, i_3\}$.

Let $I_1 = \{i_2, i_4\}$. $Mod^{\sigma_0^{-1}}(I_1)$ is the equivalence class of the formula $a \wedge \neg c$.

Definition 4. *A formula is under minimal positive conjunctive normal form (denoted $+mcnf$) if it is a conjunction of clauses whose all literals are positive and such that no clause subsumes another one. The set of $+mcnf$-formulas is denoted $+mcnf FORM$.*

For instance a is a $+mcnf$-formula; $a \wedge (a \vee b)$ and $a \wedge \neg c$ are not.

Proposition 2. *Mod^σ is not a bijection from $+mcnf FORM/r^\sigma_\leftrightarrow$ to $2^{mod(\sigma)}$*

As a counter example, I_1 in the previous example, is the truth set of no $+mcnf$-formulas.

3   Studying DSmT with Logic

3.1   Preliminaries: Dempster-Shafer Theory

Dempster-Shafer Theory considers *a finite discernment frame* $\Theta = \{\theta_1, ...\theta_n\}$ whose elements, called hypothesis, are *exhaustive et exclusive*.

An *basic belief assignment (or mass function)* is a function $m : 2^\Theta \to [0, 1]$ such that: $m(\emptyset) = 0$ et $\sum_{A \subseteq \Theta} m(A) = 1$.

Given a basic belief assignment m, one defines *a belief function* $Bel : 2^\Theta \to [0, 1]$ by: $Bel(A) = \sum_{B \subseteq A} m(B)$. One also defines *a plausibility function* $Pl : 2^\Theta \to [0, 1]$ such that $Pl(A) = 1 - Bel(\overline{A})$.

Let m_1 and m_2 two basic assigments on the frame Θ. *Dempster's combination rule* defines a basic belief assignment denoted $m_1 \oplus m_2$, from 2^Θ to $[0, 1]$ by:

$$m_1 \oplus m_2(\emptyset) = 0$$

$$m_1 \oplus m_2(C) = \frac{\sum_{A \cap B = C} m_1(A).m_2(B)}{N}$$

with

$$N = \sum_{A \cap B \neq \emptyset} m_1(A).m_2(B)$$

3.2   DSmT: Case of Free Models

Presentation of the Case of Free Models. Contrary to Demspter-Shafer Theory, DSmT supposes that frames of discernement are finite sets of hypothesis which are exhaustive but not exclusive.

Let $\Theta = \{\theta_1, ...\theta_n\}$ be a discernment frame.

According to [7], the *hyper-power-set* D^Θ is the set made of $\emptyset$, $\theta_1,...,$ θ_n and all the expressions composed from the hypothesis and operators $\cup$ and $\cap$ such that: $\forall A \in D^\Theta \; \forall B \in D^\Theta \; (A \cup B) \in D^\Theta$ and $(A \cap B) \in D^\Theta$.

Example 2. Si $\Theta = \{\theta_1, \theta_2\}$ alors $D^\Theta = \{\emptyset, \theta_1, \theta_2, \theta_1 \cap \theta_2, \theta_1 \cup \theta_1\}$.

However, it seems that the authors implicitly suppose that elements of D^Θ are under some form. Indeed, one can notice that some expressions are reduced by implicitly using properties of $\cap$ and $\cup$. For instance, the expression $\theta_1 \cap (\theta_1 \cup \theta_2)$ is in fact reduced to θ_1. In the same way, for n=3, $(\theta_2 \cup \theta_3) \cap \theta_1$ belongs to D^Θ but $(\theta_2 \cap \theta_1) \cup (\theta_3 \cap \theta_1)$ does not. However it characterizes the same element. In the same way, $(\theta_1 \cap \theta_2) \cup \theta_3$ appears but $(\theta_1 \cup \theta_3) \cap (\theta_2 \cup \theta_3)$ does not. However, theses two expressions which do not appear are well formed.

Definition 5. *Given a frame Θ, we call reduced conjonctive normal form (rcnf), any conjunction of disjunction of hypothesis of Θ such that no disjunction contains another one.*

For instance, $\theta_1 \cap (\theta_1 \cup \theta_2)$ is not rcnf; θ_1 is.

It has been confirmed, [8], that D^Θ *is the set of all the reduced conjunctive normal forms that can be buit.*

In DSmT, the basic assignements are defined not on elements of 2^Θ but on elements of D^Θ. Definitions of belief functions and plausibility functions are unchanged.

A combination rule has been defined in [7] Given two basic belief assigments m_1 and m_2, this rule builds the following assigment: $m_1 \ [\oplus] \ m_2 : D^\Theta \to [0, 1]$:

$$m_1[\oplus]m_2(C) = \sum_{A \cap B = C} m_1(A).m_2(B)$$

Furthermore, it is also mentionned in [7], that a normalization can be applied on the assigment which is obtained. This would lead to consider the rule:

$$m_1[\oplus]m_2(C) = \frac{\sum_{A \cap B = C} m_1(A).m_2(B)}{N}$$

N being the sum of all the masses $m_1[\oplus]m_2(C)$.

Reformulation of the Case of Free Models. In the following, we use propositional logic to show that DSmT can be reformulated in the Dempter-Shafer's classical framework.

Let us first recall that the notions of discernment frame and hypothesis can be given a logical interpretation [2].

Proposition 3. *Any discernment frame* $\Theta = \{\theta_1, ...\theta_n\}$ *can be associated with a logical propositional language we will denote* Θ *whose propositional letters will still be denoted* $\theta_1, ...\theta_n$.

Any expression E of D^Θ can be associated by a bijection with a +mcnf-formula of language Θ which will be still denoted E.

Furthermore, assuming exhaustivity of hypothesis comes to consider the formula $\sigma_0 = \theta_1 \vee ... \vee \theta_n$ *as true.*

Example 3. For instance, the frame $\Theta = \{\theta_1, \theta_2, \theta_3\}$ corresponds to the propositional language Θ whose letters are θ_1, θ_2 and θ_3.

The expression $\theta_1 \cap (\theta_2 \cup \theta_3)$ corresponds to the formula $\theta_1 \wedge (\theta_2 \vee \theta_3)$ And here, $\sigma_0 = \theta_1 \vee \theta_2 \vee \theta_3$.

Proposition 4. *Any discernment frame* Θ *of DSmT can be associated with a new discernment frame, denoted* $mod(\sigma_0)$, *whose hypothesis are exhaustive and exclusive.*

Any expression E of D^Θ can be associated with an expression of $2^{mod(\sigma_0)}$ which will be denoted $T(E)$.

The function T is defined by composition of the three following functions:

1. By proposition 3, any expression E of D^Θ is associated with a +mcnf-formula E of language Θ.

   Let us denote c_E, the equivalence class (given relation $r^{\sigma_0}_{\leftrightarrow}$) of this formula.

2. Let $mod(\sigma_0)$ the truth set of σ_0.

   By proposition 1, we know that c_E can be associated by function Mod with an element of $2^{mod(\sigma_0)}$, $Mod(c_E)$.

3. Finally, $2^{mod(\sigma_0)}$ can be associated with a discernment frame, denoted $2^{mod(\sigma_0)}$, whose hypothesis correspond to models of σ_0 and thus, are mutually exclusive. Thus, any element of $2^{mod(\sigma_0)}$ can be associated with a union of hypothesis of the discernment frame $2^{mod(\sigma_0)}$.

   In particular, if $Mod(c_E) = \{w_1, ..., w_n\}$ then we can associate it with the expression $w_1 \cup ... \cup w_n$ of frame $2^{mod(\sigma_0)}$.

   $T(E)$ is this very expression.

Figure 1 sums up this definition.

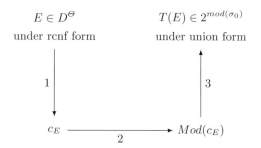

$$E \in D^{\Theta} \qquad\qquad T(E) \in 2^{mod(\sigma_0)}$$
$$\text{under rcnf form} \qquad\qquad \text{under union form}$$
$$1 \qquad\qquad\qquad 3$$
$$c_E \xrightarrow{\quad 2 \quad} Mod(c_E)$$

Fig. 1. Case of free models

Example 4. Consider $\Theta = \{\theta_1, \theta_2, \theta_3\}$. We can associate this frame with the propositional language Θ whose letters are $\theta_1, \theta_2, \theta_3$ (proposition 3).

We have: $\sigma_0 = \theta_1 \vee \theta_2 \vee \theta_3$. And $mod(\sigma_0) = \{w_1, w_2, ..., w_7\}$ with:

$w_1 = \{\theta_1, \theta_2, \theta_3\}$, $w_2 = \{\theta_1, \theta_2, \neg\theta_3\}$, $w_3 = \{\theta_1, \neg\theta_2, \theta_3\}$,
$w_4 = \{\theta_1, \neg\theta_2, \neg\theta_3\}$, $w_5 = \{\neg\theta_1, \theta_2, \theta_3\}$, $w_6 = \{\neg\theta_1, \theta_2, \neg\theta_3\}$,
$w_7 = \{\neg\theta_1, \neg\theta_2, \theta_3\}$,

- Consider the expression $E_1 = \theta_1 \cap \theta_2$ of D^{Θ}. E_1 is associated with the equivalence class of the formula $\theta_1 \wedge \theta_2$ (point 1). By function Mod this class is associated with $\{w_1, w_2\}$ (point 2). Finally (point 3), this set is associated with an expression of frame $mod(\sigma_0)$, which is $w_1 \cup w_2$. Thus, $T(E_1) = w_1 \cup w_2$.
- Consider the expression $E_2 = \theta_1 \cup \theta_2$. In the same way, we can show that $T(E_2) = w_1 \cup w_2 \cup w_3 \cup w_4 \cup w_5 \cup w_6$.
- As for the expression $E_3 = \theta_1 \cup (\theta_2 \cap \theta_3)$, we have: $T(E_3) = w_1 \cup w_2 \cup w_3 \cup w_4 \cup w_5$.

Proposition 5. *Function T is not a bijection.*

Indeed, some elements of $2^{mod(\sigma_0)}$ are the image by function Mod of some equivalence class of formulas which are not +mcnf-formulas. This is shown by the following example.

Example 4 (continued). Consider the expression $w_2 \cup w_4$. There is no expression of D^Θ whom it is the image by T. Indeed, the equivalence class which is its antecedent by Mod is the class of formula $\theta_1 \wedge \neg\theta_3$. This formula is not +mcnf. Thus, it is associated with no expression of D^Θ.

T is not a bijection because expressions of D^Θ are only built with $\cap$ and $\cup$ operators. They are not built with complementation operator (which would correspond to logical negation).

Definition 6. *Let Θ be a discernment frame. Any basic belief assignment m defined on D^Θ can be associated with a basic belief assignment denoted m', defined on $mod(\sigma_0)$ by: $m'(T(E)) = m(E)$ for any $E \in D^\Theta$.*

Example 4 (continued). Let m defined on D^Θ by:

$$m(\theta_1) = 0.80 \qquad m(\theta_2) = 0.15$$
$$m(\theta_1 \cup \theta_2) = 0 \qquad m(\theta_1 \cap \theta_2) = 0.05$$

Basic belief assignment m' is defined on $2^{mod(\sigma_0)}$ by:

$$m'(w_1 \cup w_2 \cup w_3 \cup w_4) = 0.80$$
$$m'(w_1 \cup w_2 \cup w_5 \cup w_6) = 0.15$$
$$m'(w_1 \cup w_2 \cup w_3 \cup w_4 \cup w_5 \cup w_6) = 0$$
$$m'(w_1 \cup w_2) = 0.05$$

Proposition 6. *Let Θ be a discernment frame and let m_1 and m_2 be two basic belief assigments defined on D^Θ. If $[\oplus]$ denotes the normalized version of DSmT combination rule, then for any expression E of D^Θ:*

$$m_1[\oplus]m_2(E) = m'_1 \oplus m'_2(T(E))$$

This proposition shows that the normalized combination rule defined in the framework of the free models of DSmT is equivalent to Dempster's combination rule (when they are respectively applied on expressions E of D^Θ and $T(E)$ of $2^{mod(\sigma_0)}$.

Example 5. (see example 11 of [7]). Let m_1 and m_2 be two basic belief assigments on D^Θ:

$$m_1(\theta_1) = 0.80 \qquad m_1(\theta_2) = 0.15$$
$$m_1(\theta_1 \cup \theta_2) = 0 \qquad m_1(\theta_1 \cap \theta_2) = 0.05$$
$$m_2(\theta_1) = 0.90 \qquad m_2(\theta_2) = 0.05$$
$$m_2(\theta_1 \cup \theta_2) = 0 \qquad m_2(\theta_1 \cap \theta_2) = 0.05$$

$$m_1[\oplus]m_2(\theta_1) = 0.72 \qquad m_1[\oplus]m_2(\theta_2) = 0.0075$$
$$m_1[\oplus]m_2(\theta_1 \cup \theta_2) = 0 \qquad m_1[\oplus]m_2(\theta_1 \cap \theta_2) = 0.2725$$

Consider now the new discernment frame: $mod(\sigma_0) = \{w_1, w_2, ..., w_7\}$ (see example 4).

Then m_1' and m_2' are the assigments:

$m_1'(w_1 \cup w_2 \cup w_3 \cup w_4) = 0.80$
$m_1'(w_1 \cup w_2 \cup w_5 \cup w_6) = 0.15$
$m_1'(w_1 \cup w_2 \cup w_3 \cup w_4 \cup w_5 \cup w_6) = 0$
$m_1'(w_1 \cup w_2) = 0.05$

$m_2'(w_1 \cup w_2 \cup w_3 \cup w_4) = 0.90$
$m_2'(w_1 \cup w_2 \cup w_5 \cup w_6) = 0.05$
$m_2'(w_1 \cup w_2 \cup w_3 \cup w_4 \cup w_5 \cup w_6) = 0$
$m_2'(w_1 \cup w_2) = 0.05$

If we apply Dempster's combination rule on these new assigments we get:

$m_1' \oplus m_2'(w_1 \cup w_2 \cup w_3 \cup w_4) = 0.72$
$m_1' \oplus m_2'(w_1 \cup w_2 \cup w_5 \cup w_6) = 0.0075$
$m_1' \oplus m_2'(w_1 \cup w_2 \cup w_3 \cup w_4 \cup w_5 \cup w_6) = 0$
$m_1' \oplus m_2'(w_1 \cup w_2) = 0.2725$

Obviously, we have $m_1[\oplus]m_2 = m_1' \oplus m_2'$.

3.3 DSmT: Case of Hybrid Models

Presentation of Hybrid Models. Hybrid models, [13], correspond to the case when contraints are expressed among hypothesis. These constraints, called integrity constraints, make some intersections empty.

In the following, IC denotes an expression of the form $E = \emptyset$, where $E \in D^\Theta$. Let us notice that even if there are several integrity constraints, $E_1 = \emptyset$ ... $E_n = \emptyset$, we can always get to the case of a unique integrity constraint by taking $E = E_1 \cap ... \cap E_n$.

For the authors of DSmT, taking integrity constraints into account comes to consider a restriction of the hyper power set D^Θ. This restricted set contains less elements than in the general case and we will denote it D^Θ_{IC} to signify that this new set depends on IC.

To the limit, if the constraints express that all the hypothesis are mutually exclusive we get $D^\Theta_{IC} = 2^\Theta$.

Thus, by considering constraints IC, any expression E of D^Θ (the hyper power set without contraints) can be reduced into an expression of D^Θ_{IC}.

Example 6. Let $\Theta = \{\theta_1, \theta_2, \theta_3\}$. Consider the constraint $IC = \theta_1 \cap \theta_2 = \emptyset$. Then the expression $\theta_1 \cap (\theta_2 \cup \theta_3)$ of D^Θ is reduced to $\theta_1 \cap \theta_3$.

In the case of hybrid models, the authors of DSmT have defined different combination rules. In what follows, we show that these rules can be associated with combination rules on assigments defined on a discernment frame whose hypothesis are exhaustive and exclusive.

Reformulation of Hybrid Models. As we did in the previous section, we build a function which associates any expression of D^Θ to an expression of a new discernment frame whose hypothesis are exhaustive and exclusive.

As previously, the frame $\Theta = \{\theta_1, ...\theta_n\}$ is associated with a propositional language Θ whose letters are $\theta_1, ..., \theta_n$.

As previously, we can associate the expression IC to a logical formula we still denote IC.

Example 6 (continued).   The constraint $\theta_1 \cap \theta_2 = \emptyset$ is associated with the formula $\neg(\theta_1 \wedge \theta_2)$.

Proposition 7. *The results of proposition 4 can be adapted in the context of hybrid models by considering the formula $\sigma_1 = \sigma_0 \wedge IC$ instead of the formula σ_0.*

Figure 2 sums up this result.

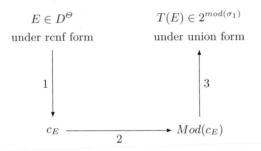

$$E \in D^{\Theta} \qquad\qquad\qquad T(E) \in 2^{mod(\sigma_1)}$$

under rcnf form   under union form

1   3

$$c_E \xrightarrow{\quad\quad 2 \quad\quad} Mod(c_E)$$

Fig. 2. Case of hybrid models

Example 7. Take example 4 and consider the constraint $IC : \theta_1 \cap \theta_2 = \emptyset$. It corresponds to the formula $\neg(\theta_1 \wedge \theta_2)$. Then $\sigma_1 = (\theta_1 \vee \theta_2 \vee \theta_3) \wedge \neg(\theta_1 \wedge \theta_2)$. And we get $mod(\sigma_1) = \{w_3, w_4, w_5, w_6, w_7\}$

Consider the expression $E_4 = \theta_1 \cap (\theta_2 \cup \theta_3)$. The formula which is associated with it is $E_4 = \theta_1 \wedge (\theta_2 \vee \theta_3)$ and we have $E_4 \wedge \sigma_1$ equivalent to $\theta_1 \wedge \neg\theta_2 \wedge \theta_3$. Thus $mod(E_4 \wedge \sigma_1) = \{w_3\}$. Finally $T(E_4) = w_3$.

The existence of this function implies the following corollary:

Proposition 8. *Any combination rule defined in the case of hybrid models of DSmT corresponds to a combinaison rule applied on assignments defined on the discernment frame $mod(\sigma_1)$.*

4   Discussion

The previous proposition shows the strong relation that exists between discernment frames with exhaustive but not exclusive hypothesis (DSmT) and the case of discernment frames with hypothesis which are exhaustive and exclusive (case of Dempster-Shafer theory). It ensures that any combination rule defined in the framework of DSmT can be reformulated in the framework of Dempster-Shafer Theory if we change the discernment frames.

This proposition is not constructive and it does not define what is the combination rule in Dempster-Shafer Theory which correponds to a given combination rule in DSmT. However, the correspondance has been established for some rules.

For instance, it has been proved [4], that the hybrid combination rule (Hybrid DSm rule) corresponds to Dubois and Prade rule.

From this correspondance, we can conclude that everything which is done in the DSmT framework with non exclusive hypothesis can be done in the framework of exclusive hypothesis[3]. This is consistant with Daniel's work [5], who gets to the same result by a different way.

It must noticed that the equivalence result we prove here was already known by DSmT authors themselves. Indeed, they have shown that any discernment frame Θ (for instance $\{\theta_1, \theta_2\}$) can be transformed in a discernment frame they call refined-Θ and denoted 2^Θ, with exclusive hypothesis (here $\{\theta_1 - \theta_2, \theta_1 \wedge \theta_2, \theta_2 - \theta_1\}$). But for the DSmT authors this new discernment frame is only a mathematical abstraction with no physical reality [8].

The main contributions of our work is to use logic mathematic not only to show that a correspondance between DSmT and Demspter-Shafer Theory can be done, but to show that the new discernment frame that must be considered makes sense. More precisely, we have shown that DSmT work with formulas in the syntax and that we can equivalently work in the semantics, by using Dempster-Shafer theory with the models of these formulas.

Thus we have shown that DSmT belongs to the proof theory side and we have shown its reformulation in Dempster-Shafer's framework belongs to the model theory side. Proving a strong relation between the two is not surprising. It is well known that knowledge or beliefs can be modelled by formulas (proof theory approach) or equivalently by their models (model theory approach).

Notice that a side effect of our contribution is to enlight the fact that expressions considered by DSmT are missing complementation. Taking complementation into account would lead to have all the richness of the logical language. In this case, function T would be a bijection.

However, let us add that, even not more powerful that Dempster-Shafer framework, the main advantage we see to DSmT is the compacity of the expressions on which basic assignments are. For instance (see example 4), it is more compact to consider expressions that can be buit with 3 non-exclusive hypothesis $\theta_1, \theta_2, \theta_3$ than to consider expressions that can be build with 7 exclusive hypothesis $w_1, ..., w_7$.

It must be noticed that this argument of compactness is exactly the same which is used in classical logic to argue that formulas can be used to model knowledge in a more compact form than sets of models. It is also the same argument that is used in preference representation [9] to argue that expressing preferences on formulas is more compact that expressing preferences on models (also called alternatives).

However, as for the complexity aspects (complexity of combination rules) comparing what is done in DSmT and what is done in the classical Dempster-Shafer framework remains to be done.

---

[3] Our work should be easily apply to to recent extensions of DSmT to qualitative beliefs.

Acknowledgments. I would like to thank Milan Daniel for the mails we have exchanged on the subject. I also thank the anonymous reviewers whose comments and questions helped me to improve this paper.

References

1. Appriou, A.: Multi-sensor Data Fusion in Situation Assesment Processes. In: Non-nengart, A., Kruse, R., Ohlbach, H.J., Gabbay, D.M. (eds.) FAPR 1997 and EC-SQARU 1997. LNCS, vol. 1244. Springer, Heidelberg (1997)
2. Cholvy, L.: Applying Theory of Evidence in multisensor data fusion: a logical interpretation. In: Proc. of the 3rd International Conference on Information Fusion (FUSION 2000) (2000)
3. Cholvy, L.: Data Merging: Theory of Evidence vs Knowledge-Bases Merging Operators. In: Benferhat, S., Besnard, P. (eds.) ECSQARU 2001. LNCS, vol. 2143, p. 478. Springer, Heidelberg (2001)
4. Daniel, M.: Generalization of the Classic Combination Rules to DSm Hyper-Power Sets. Information and Security, An International Journal 20, 50–64 (2006)
5. Daniel, M.: Contribution of DSm Approach to the belief Function Theory. In: Proc of 12th International Conference on Information Processing and Management of Uncertainty in Knowledge-Based Systems (IPMU 2008), Malaga (2008)
6. Denoeux, T., Smets, P.: Classification using Belief Functions: the Relationship between the Case-Based and Model-Based Approaches. IEEE Transactions on Systems, Man and Cybernetics B 36(6), 1395–1406 (2006)
7. Dezert, J.: Foundations for a new theory of plausible and paradoxical reasoning. Information and Security. An International Journal 9, 90–95 (2002)
8. Dezert, J.: Private Communication (2008)
9. Lang, J.: Logical preference representation and combinatorial vote. Annals of Mathematics and Artificial Intelligence (Special Issue on the Computational Properties of Multi-Agent Systems) 42 (2004)
10. Mercier, D., Cron, G., Denoeux, T., Masson, M.-H.: Decision Fusion for postal address recognition using belief functions. Expert Systems with Applications 36(3), 5643–5653 (2009)
11. Kholas, J., Monney, P.-A.: Representation of Evidence by Hints. In: Yager, et al. (eds.) Recent Advances in Dempster-Shafer Theory. Wiley, Chichester (1994)
12. Milisavljevic, N., Bloch, I.: Improving Mine Recognition through Processing and Dempster-Shafer Fusion of Multisensor Data. In: Sarfraz, M. (ed.) Computer-Aided Intelligent Recognition, Techniques and Applications, ch. 17. J. Wiley, Chichester (2005)
13. Smarandache, F., Dezert, J.: An introduction to the DSm Theory for the Combination of Paradoxical, Uncertain and Imprecise Sources of Information, http://www.gallup.unm.edu/~smarandache/DSmT-basic-short.pdf
14. Shafer, G.: A mathematical theory of evidence. Princeton University Press, Princeton (1976)
15. Smets, P.: The combination of Evidence in the Transferable Belief Model. IEEE Transactions PAMI 12 (1990)
16. Smets, P.: Probability of provability and belief functions. Logique et Analyse 133-134 (1991)
17. Smets, P.: The Transferable Belief Model. Artificial Intelligence 66 (1994)

Complexes of Outer Consonant Approximations

Fabio Cuzzolin

Oxford Brookes University, Oxford, UK
fabio.cuzzolin@brookes.ac.uk

Abstract. In this paper we discuss the problem of approximating a belief function (b.f.) with a necessity measure or "consonant belief function"
(co.b.f.) from a geometric point of view. We focus in particular on outer
consonant approximations, i.e. co.b.f.s less committed than the original
b.f. in terms of degrees of belief. We show that for each maximal chain
of focal elements the set of outer consonant approximation is a polytope.
We describe the vertices of such polytope, and characterize the geometry
of maximal outer approximations.

1 Introduction

The theory of evidence (ToE) [1] is a popular approach to uncertainty description. Probabilities are there replaced by *belief functions* (b.f.s), which assign
values between 0 and 1 to subsets of the sample space Θ instead of single elements. Possibility theory [2], on its side, is based on *possibility measures*, i.e.,
functions $Pos : 2^{\Theta} \rightarrow [0, 1]$ on Θ such that $Pos(\bigcup_i A_i) = \sup_i Pos(A_i)$ for any
family $\{A_i | A_i \in 2^{\Theta}, i \in I\}$ where I is an arbitrary set index. Given a possibility
measure Pos, the dual *necessity* measure is defined as $Nec(A) = 1 - Pos(A)$.

Necessity measures have as counterparts in the theory of evidence *consonant*
b.f.s, i.e. belief functions whose focal elements are nested [1]. The problem of
approximating a belief function with a necessity measure is then equivalent to
approximating a belief function with a consonant b.f. [3,4,5,6]. As possibilities
are completely determined by their values on the singletons $Pos(x)$, $x \in \Theta$, they
are less computationally expensive than b.f.s, making the approximation process
interesting for many applications. The points of contact between evidence (in the
transferable belief model implementation) and possibility theory have been for
instance investigated by Ph. Smets [7].

A geometric interpretation of uncertainty theory has been recently proposed
[8] in which several classes of uncertainty measures (among which belief functions
and possibilities) are represented as points of a Cartesian space.

In this paper we consider the problem of approximating a belief function
with a possibility/necessity [3] from such geometric point of view. We focus in
particular on the class of *outer consonant approximations* of belief functions.

More precisely, after reviewing the basic notions of evidence and possibility
theory we formally introduce the consonant approximation problem, and in particular the notion of outer consonant approximation. We then recall how the set

C. Sossai and G. Chemello (Eds.): ECSQARU 2009, LNAI 5590, pp. 275–286, 2009.
© Springer-Verlag Berlin Heidelberg 2009

of all consonant belief functions forms a *simplicial complex*, a structured collection of higher-dimensional triangles or "simplices". Each such maximal simplex is associated with a maximal chain of subsets of Θ. Starting from the simple binary case we prove that the set of outer consonant approximations of a b.f. forms, on each such maximal simplex, a polytope. We investigate the form of its vertices and prove that one of them corresponds to the minimal outer approximation, the one [3] generated by a permutation of the element of Θ. To improve the readability of the paper all major proofs are collected in an Appendix. Illustrative examples accompany all the presented results.

2   Outer Consonant Approximations of Belief Functions

Belief and possibility measures. A *basic probability assignment* (b.p.a.) over a finite set *(frame of discernment [1])* Θ is a function $m : 2^\Theta \rightarrow [0, 1]$ on its power set $2^\Theta = \{A \subseteq \Theta\}$ such that $m(\emptyset) = 0$, $\sum_{A \subseteq \Theta} m(A) = 1$, and $m(A) \geq 0$ $\forall A \subseteq \Theta$. Subsets of Θ associated with non-zero values of m, $\{E \subset \Theta : m(E) \neq 0\}$ are called *focal elements*. The *belief function* $b : 2^\Theta \rightarrow [0, 1]$ associated with a basic probability assignment m on Θ is defined as: $b(A) = \sum_{B \subseteq A} m(B)$. The *plausibility function* (pl.f.) $pl_b : 2^\Theta \rightarrow [0, 1]$, $A \mapsto pl_b(A)$ such that $pl_b(A) \doteq 1 - b(A^c) = \sum_{B \cap A \neq \emptyset} m_b(B)$ expresses the amount of evidence *not against* A.

A probability function is simply a peculiar belief function assigning non-zero masses to singletons only (*Bayesian* b.f.): $m_b(A) = 0\ |A| > 1$. A b.f. is said to be *consonant* if its focal elements $\{E_i, i = 1, ..., m\}$ are nested: $E_1 \subset E_2 \subset ... \subset E_m$. It can be proven that [2,9] the plausibility function pl_b associated with a belief function b on a domain Θ is a possibility measure iff b is consonant. Equivalently, a b.f. b is a necessity iff b is consonant.

Outer consonant approximations. Finding the "best" consonant approximation of a belief function is equivalent to approximating a belief measure with a necessity measure. B.f.s admit (among others) the following order relation

$$b \leq b' \equiv b(A) \leq b'(A)\ \forall A \subseteq \Theta \tag{1}$$

called *weak inclusion*. We can then define the *outer consonant approximations* [3] of a belief function b as those co.b.f.s such that $co(A) \leq b(A)\ \forall A \subseteq \Theta$ (or equivalently $pl_{co}(A) \geq pl_b(A)\ \forall A$). With the purpose of finding outer approximations which are *minimal* with respect to the weak inclusion relation (1)) Dubois and Prade [3] introduced a family of outer consonant approximations obtained by considering all permutations ρ of the elements $\{x_1, ..., x_n\}$ of the frame of discernment Θ: $\{x_{\rho(1)}, ..., x_{\rho(n)}\}$. A family of nested sets can be then built $\{S_1^\rho = \{x_{\rho(1)}\}, S_2^\rho = \{x_{\rho(1)}, x_{\rho(2)}\}, ..., S_n^\rho = \{x_{\rho(1)}, ..., x_{\rho(n)}\}\}$ so that a new consonant belief function co^ρ can be defined with b.p.a.

$$m_{co^\rho}(S_j^\rho) = \sum_{i:\min\{l: E_i \subseteq S_l^\rho\} = j} m_b(E_i). \tag{2}$$

S_j^ρ is assigned the mass of the focal elements of b included in S_j^ρ but not in S_{j-1}^ρ.

3   The Complex of Consonant Belief Functions

A useful tool to represent uncertainty measures and discuss issues like the approximation problem is provided by convex geometry. Given a frame of discernment Θ, a b.f. $b : 2^{\Theta} \to [0,1]$ is completely specified by its $N - 2$ belief values $\{b(A), A \subseteq \Theta, A \neq \emptyset, \Theta\}$, $N \doteq 2^{|\Theta|}$, and can then be represented as a point of $\mathbb{R}^{N-2}$. The *belief space* associated with Θ is the set of points $\mathcal{B}$ of $\mathbb{R}^{N-1}$ which correspond to b.f.s. Let us call

$$b_A \doteq b \in \mathcal{B} \ s.t. \ m_b(A) = 1, \ m_b(B) = 0 \ \forall B \neq A \tag{3}$$

the unique b.f. assigning all the mass to a single subset A of Θ (A-th *categorical belief function*). It can be proven that [8] the belief space $\mathcal{B}$ is the convex closure of all the categorical belief functions (3), $\mathcal{B} = Cl(b_A, \ \emptyset \subsetneq A \subseteq \Theta)$ where Cl denotes the convex closure operator: $Cl(b_1, ..., b_k) = \{b \in \mathcal{B} : b = \alpha_1 b_1 + \cdots + \alpha_k b_k, \sum_i \alpha_i = 1, \alpha_i \geq 0 \ \forall i\}$.

More precisely $\mathcal{B}$ is an $N - 2$-dimensional *simplex*, i.e. the convex closure of $N - 1$ (affinely independent[1]) points of the Euclidean space $\mathbb{R}^{N-1}$. The *faces* of a simplex are all the simplices generated by a subset of its vertices. Each belief function $b \in \mathcal{B}$ can be written as a convex sum as $b = \sum_{\emptyset \subsetneq A \subseteq \Theta} m_b(A) b_A$. Similarly the set of all Bayesian b.f.s is $\mathcal{P} = Cl(b_x, x \in \Theta)$.

Binary example. As an example consider a frame of discernment containing only two elements, $\Theta_2 = \{x, y\}$. Each b.f. $b : 2^{\Theta_2} \to [0,1]$ is determined by its belief values $b(x), b(y)$, as $b(\Theta) = 1$ and $b(\emptyset) = 0 \ \forall b$. We can then collect them in a vector of $\mathbb{R}^{N-2} = \mathbb{R}^2$:

$$[b(x) = m_b(x), b(y) = m_b(y)]' \in \mathbb{R}^2. \tag{4}$$

Since $m_b(x) \geq 0$, $m_b(y) \geq 0$, and $m_b(x) + m_b(y) \leq 1$ the set $\mathcal{B}_2$ of all the possible b.f.s on Θ_2 is the triangle of Figure 1, whose vertices are the points $b_\Theta = [0,0]'$, $b_x = [1,0]'$, and $b_y = [0,1]'$. The region $\mathcal{P}_2$ of all Bayesian b.f.s on Θ_2 is in this case the line segment $Cl(b_x, b_y)$. On the other side, consonant belief functions can have as chain of focal elements either $\{\{x\}, \Theta_2\}$ or $\{\{y\}, \Theta_2\}$. As a consequence the region $\mathcal{CO}_2$ of all co.b.f.s is the union of two segments: $\mathcal{CO}_2 = \mathcal{CO}_x \cup \mathcal{CO}_y = Cl(b_\Theta, b_x) \cup Cl(b_\Theta, b_y)$.

The consonant simplicial complex. The geometry of $\mathcal{CO}$ can be described in terms of a concept of convex geometry derived from that of simplex [10].

Definition 1. *A simplicial complex is a collection Σ of simplices such that*

1. *if a simplex belongs to Σ, then all its faces are in Σ;*
2. *the intersection of two simplices is a face of both.*

---

[1] An *affine combination* of k points $v_1, ..., v_k \in \mathbb{R}^m$ is a sum $\alpha_1 v_1 + \cdots + \alpha_k v_k$ whose coefficients sum to one: $\sum_i \alpha_i = 1$. The affine subspace generated by the points $v_1, ..., v_k \in \mathbb{R}^m$ is the set $\{v \in \mathbb{R}^m : v = \alpha_1 v_1 + \cdots + \alpha_k v_k, \sum_i \alpha_i = 1\}$. If $v_1, ..., v_k$ generate an affine space of dimension k they are said to be *affinely independent*.

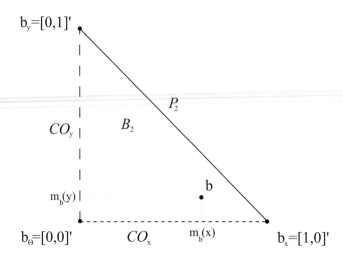

Fig. 1. The belief space $\mathcal{B}$ for a binary frame is a triangle in $\mathbb{R}^2$ whose vertices are the categorical belief functions b_x, b_y, b_Θ focused on $\{x\}$, $\{y\}$ and Θ, respectively. The probability region is the segment $Cl(b_x, b_y)$. Consonant belief functions are constrained to belong to the union of the two segments $\mathcal{CO}_x = Cl(b_\Theta, b_x)$ and $\mathcal{CO}_y = Cl(b_\Theta, b_y)$.

Fig. 2. Constraints on the intersection of simplices in a complex. Only the right-hand pair of triangles meets condition (2) of the definition of simplicial complex.

Let us consider for instance two triangles on the plane (2-dimensional simplices). Roughly speaking, the second condition says that the intersection of those triangles cannot contain points of their interiors (Figure 2 left). It cannot also be any subset of their boundaries (middle), but has to be a face (right, in this case a single vertex). It can be shown that [8]

Proposition 1. $\mathcal{CO}$ is a simplicial complex included in the belief space $\mathcal{B}$.

$\mathcal{CO}$ is the union of a collection of $\prod_{k=1}^{n} \binom{k}{1} = n!$ simplices, each associated with a maximal chain $\mathcal{C} = \{A_1 \subset \cdots \subset A_n = \Theta\}$ of 2^Θ:

$$\mathcal{CO} = \bigcup_{\mathcal{C}=\{A_1 \subset \cdots \subset A_n\}} Cl(b_{A_1}, \cdots, b_{A_n}).$$

4   Outer Approximations in the Binary Case

We can then study the geometry of the set $O[b]$ of all outer consonant approximations of a belief function b. In the binary case the latter is depicted in Figure 3 (dashed lines), as the intersection of the region of the points b' with $b'(A) \leq b(A)$ $\forall A \subset \Theta$, and the complex $\mathcal{CO} = \mathcal{CO}_x \cup \mathcal{CO}_y$ of consonant b.f.s. Among them, the co.b.f.s generated by the 2 possible permutations $\rho_1 = (x,y)$, $\rho_2 = (y,x)$ of elements of Θ_2 as in (2) correspond to the points co^{ρ_1}, co^{ρ_2} in Figure 3.

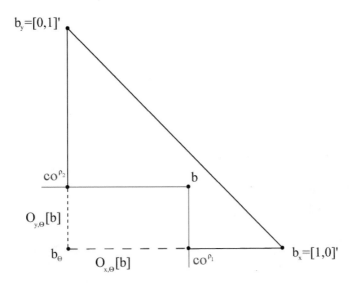

Fig. 3. Geometry of outer consonant approximations of a belief function $b \in \mathcal{B}_2$

Let us denote by $O_{\mathcal{C}}[b]$ the intersection of the set $O[b]$ of all outer consonant approximations with the component $\mathcal{CO}_{\mathcal{C}}$ of the consonant complex, with $\mathcal{C}$ a maximal chain of 2^{Θ}. We can notice a number of interesting facts. For each maximal chain $\mathcal{C}$:

1. $O_{\mathcal{C}}[b]$ is convex (in the binary case $\mathcal{C} = \{x, \Theta\}$ or $\{y, \Theta\}$);
2. $O_{\mathcal{C}}[b]$ is in fact a *polytope*, i.e. the convex closure of a number of vertices: in particular a segment in the binary case ($O_{x,\Theta}[b]$ or $O_{y,\Theta}[b]$);
3. the maximal (with respect to (1)) outer approximation of b is one of the vertices of this polytope $O_{\mathcal{C}}[b]$, the one (co^{ρ}, Equation (2)) associated with the permutation ρ of singletons which generates the chain.

In the binary case there are just two such permutations, $\rho_1 = \{x, y\}$ and $\rho_2 = \{y, x\}$, which generate respectively the chains $\{x, \Theta\}$ and $\{y, \Theta\}$. We will prove that all those properties indeed hold in the general case.

5   Polytopes of Outer Consonant Approximations

We first need a preliminary result on the basic probability assignment of consonant belief functions weakly included in b [11,12].

Weak inclusion and mass re-assignment

Lemma 1. *Consider a belief function b with basic probability assignment m_b. A consonant belief function co is weakly included in b, for all $A \subseteq \Theta$ $co(A) \le b(A)$, if and only if there is a choice of coefficients $\{\alpha_A^B, B \subseteq \Theta, A \supseteq B\}$ with*

$$0 \le \alpha_A^B \le 1 \ \forall B \subseteq \Theta, \forall A \supseteq B; \qquad \sum_{A \supseteq B} \alpha_A^B = 1 \ \forall B \subseteq \Theta \qquad (5)$$

such that co has basic probability assignment

$$m_{co}(A) = \sum_{B \subseteq A} \alpha_A^B m_b(B). \qquad (6)$$

Lemma 1 states that the b.p.a. of any outer consonant approximation of b is obtained by *re-assigning the mass of each f.e. A of b to some $B \supseteq A$*. We will extensively use this result in the following.

Vertices of the polytopes. Given a consonant belief function co weakly included in b, its focal elements will form a chain $\mathcal{C} = \{B_1, ..., B_n\}$ ($|B_i| = i$) associated with a specific maximal simplex of $\mathcal{CO}$. According to Lemma 1 the mass of each focal element A of b can be re-assigned to some of the events of the chain $B_1, ..., B_n$ which contain A in order to obtain co.

It is therefore natural to conjecture that, for each maximal simplex $\mathcal{CO}_\mathcal{C}$ of $\mathcal{CO}$ associated with a maximal chain $\mathcal{C}$, $O_\mathcal{C}[b]$ is the convex closure of the co.b.f.s $o^B[b]$ with b.p.a.

$$m_{o^B[b]}(B_i) = \sum_{A \subseteq \Theta : B(A) = B_i} m_b(A) \qquad (7)$$

each of them associated with an "assignment function"

$$\begin{aligned} B : 2^\Theta &\to \mathcal{C} \\ A &\mapsto B(A) \supseteq A \end{aligned} \qquad (8)$$

which maps each event A to one of the events of the chain $\mathcal{C} = \{B_1 \subset ... \subset B_n\}$ which contains A. As a matter of fact:

Theorem 1. *For each simplicial component $\mathcal{CO}_\mathcal{C}$ of the consonant space associated with any maximal chain of focal elements $\mathcal{C} = \{B_1, ..., B_n\}$ the set of outer consonant approximation of any b.f. b is the convex closure*

$$O_\mathcal{C}[b] = Cl(o^B[b], \forall B)$$

of the co.b.f.s (7) indexed by all admissible assignment functions (8).

In other words, $O_\mathcal{C}[b]$ is a *polytope*, the convex closure of a number of b.f.s whose number is equal to the number of assignment functions (8). Each B is characterized by assigning each event A to an element $B_i \supseteq A$ of the chain $\mathcal{C}$.

As we will see in the following ternary example the points (7) are not guaranteed to be all proper vertices of the polytope $O_\mathcal{C}[b]$. Some of them can be obtained as a convex combination of the others, i.e. they may lie on a side of the polytope.

Maximal outer consonant approximations. We can prove instead that the outer approximation (2) obtained by permuting the singletons of Θ as in Section 2 is not only a pseudo-vertex of $O_{\mathcal{C}}[b]$, but it is an actual vertex, i.e. it *cannot* be obtained as a convex combination of the others. More precisely, all possible permutations of elements of Θ generate exactly $n!$ different outer approximations of b, each of which lies on a single simplicial component of the consonant complex. Each such permutation ρ generates a maximal chain $\mathcal{C}_\rho = \{S_1^\rho, ..., S_n^\rho\}$ of focal elements so that the corresponding b.f. will lie on $\mathcal{CO}_{\mathcal{C}_\rho}$.

Theorem 2. *The outer consonant approximation co^ρ (2) generated by a permutation ρ of the singletons is a vertex of $O_{\mathcal{C}_\rho}[b]$.*

We can prove that the maximal outer approximation is indeed the vertex co^ρ associated with the corresponding permutation ρ of the singletons which generates the maximal chain $\mathcal{C} = \mathcal{C}_\rho$ (as in the binary case of Section 4).
By definition (2) co^ρ assigns the mass $m_b(A)$ of each focal element A to the smallest element of the chain containing A. By Lemma 1 each outer consonant approximation of b with chain $\mathcal{C}$, $co \in O_{\mathcal{C}_\rho}[b]$, is the result of re-distributing the mass of each focal element A to all its supersets in the chain $\{B_i \supseteq A, B_i \in \mathcal{C}\}$. But then each such co is weakly included in co^ρ for its b.p.a. can be obtained by re-distributing the mass of the minimal superset B_j, where $j = \min\{i : B_i \subseteq A\}$, to all supersets of A.

Corollary 1. *The maximal outer consonant approximation with maximal chain $\mathcal{C}$ of a belief function b is the vertex (2) of $O_{\mathcal{C}}[b]$ associated with the permutation ρ of the singletons which generates $\mathcal{C} = \mathcal{C}_\rho$.*

Example. For a better understanding of the above results, let us consider as an example a belief function b on a ternary frame $\Theta = \{x, y, z\}$ and study the polytope of outer consonant approximations with focal elements $\mathcal{C} = \{\{x\}, \{x, y\}, \{x, y, z\}\}$. According to Theorem 1, such polytope is the convex closure of all assignment functions $\boldsymbol{B} : 2^\Theta \to \mathcal{C}$: there are $\prod_{k=1}^{3} k^{2^{3-k}} = 1^4 \cdot 2^2 \cdot 3^1 = 12$ such functions. We list them here according to the notation

$$\boldsymbol{B} = \boldsymbol{B}(\{x\}), \boldsymbol{B}(\{y\}), \boldsymbol{B}(\{z\}), \boldsymbol{B}(\{x, y\}), \boldsymbol{B}(\{x, z\}), \boldsymbol{B}(\{y, z\}), \boldsymbol{B}(\{x, y, z\}),$$

i.e.,

$$
\begin{array}{ll}
\boldsymbol{B}_1 = \{x\}, \quad \{x,y\}, \Theta, \{x,y\}, \Theta, \Theta, \Theta; & \boldsymbol{B}_7 = \{x,y\}, \Theta, \quad\;\; \Theta, \{x,y\}, \Theta, \Theta, \Theta; \\
\boldsymbol{B}_2 = \{x\}, \quad \{x,y\}, \Theta, \Theta, \quad\;\; \Theta, \Theta, \Theta; & \boldsymbol{B}_8 = \{x,y\}, \Theta, \quad\;\; \Theta, \Theta, \quad\;\; \Theta, \Theta, \Theta; \\
\boldsymbol{B}_3 = \{x\}, \quad \Theta, \quad\;\; \Theta, \{x,y\}, \Theta, \Theta, \Theta; & \boldsymbol{B}_9 = \Theta, \quad\;\; \{x,y\}, \Theta, \{x,y\}, \Theta, \Theta, \Theta; \\
\boldsymbol{B}_4 = \{x\}, \quad \Theta, \quad\;\; \Theta, \Theta, \quad\;\; \Theta, \Theta, \Theta; & \boldsymbol{B}_{10} = \Theta, \quad\;\; \{x,y\}, \Theta, \Theta, \quad\;\; \Theta, \Theta, \Theta; \\
\boldsymbol{B}_5 = \{x,y\}, \{x,y\}, \Theta, \{x,y\}, \Theta, \Theta, \Theta; & \boldsymbol{B}_{11} = \Theta, \quad\;\; \Theta, \quad\;\; \Theta, \{x,y\}, \Theta, \Theta, \Theta; \\
\boldsymbol{B}_6 = \{x,y\}, \{x,y\}, \Theta, \Theta, \quad\;\; \Theta, \Theta, \Theta; & \boldsymbol{B}_{12} = \Theta, \quad\;\; \Theta, \quad\;\; \Theta, \Theta, \quad\;\; \Theta, \Theta, \Theta.
\end{array}
$$

They correspond to the following co.b.f.s with b.p.a. $[m(\{x\}), m(\{x, y\}), m(\Theta)]'$:

$$
\begin{aligned}
o^{B_1} &= [m_b(x), \; m_b(y) + m_b(x, y), \; 1 - b(x, y) &&]'; \\
o^{B_2} &= [m_b(x), \; m_b(y), && 1 - m_b(x) - m_b(y) &&]'; \\
o^{B_3} &= [m_b(x), \; m_b(x, y), && 1 - m_b(x) - m_b(x, y) &&]'; \\
o^{B_4} &= [m_b(x), \; 0, && 1 - m_b(x) &&]'; \\
o^{B_5} &= [0, \quad b(x, y), && 1 - b(x, y) &&]'; \\
o^{B_6} &= [0, \quad m_b(x) + m_b(y), && 1 - m_b(x) - m_b(y) &&]'; \\
o^{B_7} &= [0, \quad m_b(x) + m_b(x, y), \; 1 - m_b(x) - m_b(x, y) &&]'; \\
o^{B_8} &= [0, \quad m_b(x), && 1 - m_b(x) &&]'; \\
o^{B_9} &= [0, \quad m_b(y) + m_b(x, y), \; 1 - m_b(y) - m_b(x, y) &&]'; \\
o^{B_{10}} &= [0, \quad m_b(y), && 1 - m_b(y) &&]'; \\
o^{B_{11}} &= [0, \quad m_b(x, y), && 1 - m_b(x, y) &&]'; \\
o^{B_{12}} &= [0, \quad 0, && 1 &&]'.
\end{aligned}
\tag{9}
$$

Figure 4-left shows the resulting polytope $O_C[b]$ for a belief function $m_b(x) = 0.3$, $m_b(y) = 0.5$, $m_b(\{x, y\}) = 0.1$, $m_b(\Theta) = 0.1$, in the component $\mathcal{CO}_C = Cl(b_x, b_{\{x,y\}}, b_\Theta)$ of the consonant complex (black triangle in the figure). The polytope $O_C[b]$ is plotted in red, together with all the 12 points (9) (red squares). Many of them lie on some side of the polytope. However, the point obtained by

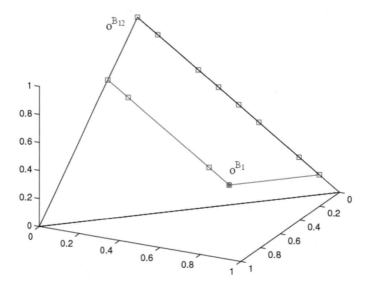

Fig. 4. Not all the points (7) associated with assignment functions are actual vertices of $O_C[b]$. Here the polytope $O_C[b]$ of outer consonant approximations for the belief function $m_b(x) = 0.3$, $m_b(y) = 0.5$, $m_b(\{x, y\}) = 0.1$, $m_b(\Theta) = 0.1$ defined on $\Theta = \{x, y, z\}$, with $\mathcal{C} = \{\{x\}, \{x, y\}, \Theta\}$ is plotted in red, together with all the 12 points (9) (red squares). Many of them lie on a side of the polytope. However, the point obtained by permutation of singletons (2) is an actual vertex (red star). The minimal and maximal outer approximations with respect to weak inclusion are $o^{B_{12}}$ and o^{B_1}, respectively.

permutation of singletons (2) is an actual vertex (red star): it is the first o^{B_1} of the list (9).

It is interesting to point out how the points (9) are ordered with respect to the weak inclusion relation (we just need to apply its definition, or the re-distribution property of Lemma 1). The result is summarized in the graph of Figure 5. We can appreciate that the vertex o^{B_1} generated by singleton permutation is indeed the maximal outer approximation of b, as stated by Corollary 1.

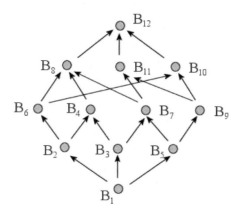

Fig. 5. Partial order of the points (9) with respect to the weak inclusion relation. For sake of simplicity we denote by B_i the co.b.f. o^{B_i} associated with the assignment function $\boldsymbol{B}_i$. An arrow from $\boldsymbol{B}_i$ to $\boldsymbol{B}_j$ stands for $o^{B_j} \leq o^{B_i}$.

6   Conclusions

In this paper we studied the convex geometry of the consonant approximation problem, focusing in particular on the properties of outer consonant approximations. We showed that such approximations form a polytope in each maximal simplex of the complex $\mathcal{CO}$ of all consonant belief functions. We proved that for a given chain the maximal outer approximation is a vertex of the corresponding polytope and is generated by a permutation of the elements of the frame.

As they also live on simplicial complexes, natural extensions of this study to guaranteed possibility measures and consistent belief functions are in sight.

References

1. Smets, P.: The Transferable Belief Model. Artificial Intelligence 66 (1994)
2. Dubois, D., Prade, H.: Possibility theory. Plenum Press, New York (1988)
3. Dubois, D., Prade, H.: Consonant approximations of belief functions. International Journal of Approximate Reasoning 4, 419–449 (1990)
4. Joslyn, C., Klir, G.: Minimal information loss possibilistic approximations of random sets. In: Bezdek, J. (ed.) Proc. 1992 FUZZ-IEEE Conference, pp. 1081–1088 (1992)

5. Joslyn, C.: Possibilistic normalization of inconsistent random intervals. Advances in Systems Science and Applications, 44–51 (1997)
6. Baroni, P.: Extending consonant approximations to capacities. In: Proceedings of IPMU, pp. 1127–1134 (2004)
7. Smets, P.: The transferable belief model and possibility theory. In: Kodratoff, Y. (ed.) Proceedings of NAFIPS 1990, pp. 215–218 (1990)
8. Cuzzolin, F.: Simplicial complexes of finite fuzzy sets. In: Proceedings of the 10th International Conference on Information Processing and Management of Uncertainty IPMU 2004, Perugia, Italy, pp. 1733–1740 (2004)
9. Joslyn, C.: Towards an empirical semantics of possibility through maximum uncertainty. In: Lowen, R., Roubens, M. (eds.) Proc. IFSA 1991, vol. A, pp. 86–89 (1991)
10. Dubrovin, B.A., Novikov, S.P., Fomenko, A.T.: Sovremennaja geometrija. Metody i prilozenija, Nauka, Moscow (1986)
11. Dubois, D., Prade, H.: A set-theoretic view of belief functions: Logical operations and approximation by fuzzy sets. Int. J. of General Systems 12(3), 193–226 (1986)
12. Cuzzolin, F.: The geometry of consonant belief functions: simplicial complexes of necessity measures (in preparation) (2009)

Appendix

Proof of Theorem 1. We need to prove that:

1. each co.b.f. $co \in \mathcal{CO}_{\mathcal{C}}$ such that $co(A) \leq b(A)$ for all $A \subseteq \Theta$ can be written as a convex combination of the points (7): $co = \sum_{B} \alpha_{B} o^{B}[b]$, $\sum_{B} \alpha_{B} = 1$, $\alpha_{B} \geq 0\ \forall B$;
2. vice-versa, each convex combination of the $o^{B}[b]$ satisfies $\sum_{B} \alpha_{B} o^{B}[b](A) \leq b(A)$ for all $A \subseteq \Theta$.

Let us consider (2) first. By definition of b.f. $o^{B}[b](A) = \sum_{B \subseteq A, B \in \mathcal{C}} m_{o^{B}[b]}(B)$ where $m_{o^{B}[b]}(B) = \sum_{X \subseteq B : B(X) = B} m_{b}(X)$ so that

$$o^{B}[b](A) = \sum_{B \subseteq A, B \in \mathcal{C}} \sum_{X \subseteq B : B(X) = B} m_{b}(X) = \sum_{X \subseteq B_{i} : B(X) = B_{j}, j \leq i} m_{b}(X) \qquad (10)$$

where B_{i} is the largest element of the chain $\mathcal{C}$ included in A. As $B_{i} \subseteq A$ (10) is obviously not larger than $\sum_{B \subseteq A} m_{b}(B) = b(A)$, so that $o^{B}[b](A) \leq b(A)$ for all A. Hence $\forall A \subseteq \Theta$

$$\sum_{B} \alpha_{B} o^{B}[b](A) \leq \sum_{B} \alpha_{B} b(A) = b(A) \sum_{B} \alpha_{B} = b(A).$$

Let us prove point (1). According to Lemma 1, if $\forall A \subseteq \Theta\ co(A) \leq b(A)$ then the mass $m_{co}(B_{i})$ of each event B_{i} of the chain is

$$m_{co}(B_{i}) = \sum_{A \subseteq B_{i}} m_{b}(A) \alpha_{B_{i}}^{A}. \qquad (11)$$

To prove (1) we then need to write (11) as a convex combination of the $m_{o^B[b]}(B_i)$, i.e.

$$\sum_{B} \alpha_B o^B[b](B_i) = \sum_{B} \alpha_B \sum_{X \subseteq B_i : B(X) = B_i} m_b(X) = \sum_{X \subseteq B_i} m_b(X) \sum_{B(X) = B_i} \alpha_B.$$

In other words we need to show that the system of equations

$$\left\{ \alpha_{B_i}^A = \sum_{B(A) = B_i} \alpha_B \qquad \forall i = 1, ..., n; \quad \forall A \subseteq B_i \right. \tag{12}$$

has at least one solution $\{\alpha_B\}$ such that $\sum_B \alpha_B = 1$ and $\forall B \ \alpha_B \geq 0$. The normalization constraint is in fact trivially satisfied as from (12) it follows that

$$\sum_{B_i \supseteq A} \alpha_{B_i}^A = 1 = \sum_{B_i \supseteq A} \sum_{B(A) = B_i} \alpha_B = \sum_{B} \alpha_B$$

i.e. $\sum_B \alpha_B = 1$. Using the normalization constraint the system of equations (12) reduces to

$$\left\{ \alpha_{B_i}^A = \sum_{B(A) = B_i} \alpha_B \qquad \forall i = 1, ..., n-1; \quad \forall A \subseteq B_i. \right. \tag{13}$$

We can show that each equation in the reduced system (13) involves at least one variable α_B which is not present in any other equation. Formally, the set of assignment functions which meet the constraint of equation A, B_i but not all others is not empty:

$$\left\{ B : (B(A) = B_i) \bigwedge_{\forall j = 1, ..., n-1; j \neq i} (B(A) \neq B_j) \bigwedge_{\forall A' \neq A; \forall j = 1, ..., n-1} (B(A') \neq B_j) \right\} \neq \emptyset. \tag{14}$$

But the assignment functions B such that $B(A) = B_i$ and $\forall A' \neq A \ B(A') = \Theta$ all meet condition (14). Indeed they obviously meet $B(A) \neq B_j$ for all $j \neq i$ while clearly for all $A' \subseteq \Theta \ B(A') = \Theta \neq B_j$, as $j < n$ so that $B_j \neq \Theta$.

A non-negative solution of (13) (and hence of (12)) can be obtained by setting for each equation one of such variables equal to the first member $\alpha_{B_i}^A$, and all the others to zero.

Proof of Theorem 2. The proof is divided in two parts.

1. We first need to find an assignment $B : 2^\Theta \to \mathcal{C}_\rho$ which generates co^ρ. Each singleton x_i is mapped by ρ to the position j: $i = \rho(j)$. Then, given any event $A = \{x_{i_1}, ..., x_{i_m}\}$ its elements are mapped to the new positions $x_{j_{i_1}}, ..., x_{j_{i_m}}$, where $i_1 = \rho(j_{i_1}), ..., i_m = \rho(j_{i_m})$. But then the map

$$B_\rho(A) = B_\rho(\{x_{i_1}, ..., x_{i_m}\}) = S_j^\rho \doteq \{x_{\rho(1)}, ..., x_{\rho(j)}\}$$

where

$$j \doteq \max\{j_{i_1}, ..., j_{i_m}\}$$

maps each event A to the smallest S_i^ρ in the chain which contains A: $j = \min\{i : A \subseteq S_i^\rho\}$. Therefore it generates a co.b.f. with b.p.a. (2), i.e. co^ρ.

2. In order for co^ρ to be an actual vertex, we need to ensure that it cannot be written as a convex combination of the other (pseudo) vertices $o^{\boldsymbol{B}}[b]$:

$$co^\rho = \sum_{\boldsymbol{B} \neq \boldsymbol{B}_\rho} \alpha_{\boldsymbol{B}} o^{\boldsymbol{B}}[b], \qquad \sum_{\boldsymbol{B} \neq \boldsymbol{B}_\rho} \alpha_{\boldsymbol{B}} = 1, \quad \forall \boldsymbol{B} \neq \boldsymbol{B}_\rho \; \alpha_{\boldsymbol{B}} \geq 0.$$

As $m_{o\boldsymbol{B}}(B_i) = \sum_{A:\boldsymbol{B}(A)=B_i} m_b(A)$ the above condition reads as

$$\left\{ \sum_{A \subseteq B_i} m_b(A) \left(\sum_{\boldsymbol{B}:\boldsymbol{B}(A)=B_i} \alpha_{\boldsymbol{B}} \right) = \sum_{A \subseteq B_i : \boldsymbol{B}_\rho(A)=B_i} m_b(A) \qquad \forall B_i \in \mathcal{C}.$$

Remembering that $\boldsymbol{B}_\rho(A) = B_i$ iff $A \subseteq B_i, \not\subseteq B_{i-1}$ we get

$$\left\{ \sum_{A \subseteq B_i} m_b(A) \left(\sum_{\boldsymbol{B}:\boldsymbol{B}(A)=B_i} \alpha_{\boldsymbol{B}} \right) = \sum_{A \subseteq B_i, \not\subseteq B_{i-1}} m_b(A) \qquad \forall B_i \in \mathcal{C}.$$

For $i = 1$ the condition is $m_b(B_1) \left(\sum_{\boldsymbol{B}:\boldsymbol{B}(B_1)=B_1} \alpha_{\boldsymbol{B}} \right) = m_b(B_1)$ i.e.

$$\sum_{\boldsymbol{B}:\boldsymbol{B}(B_1)=B_1} \alpha_{\boldsymbol{B}} = 1, \qquad \sum_{\boldsymbol{B}:\boldsymbol{B}(B_1)\neq B_1} \alpha_{\boldsymbol{B}} = 0.$$

Replacing this condition in the second constraint $i = 2$ yields

$$m_b(B_2 \setminus B_1) \left(\sum_{\substack{\boldsymbol{B} \, : \, \boldsymbol{B}(B_1) = B_1, \\ \boldsymbol{B}(B_2 \setminus B_1) = B_2}} \alpha_{\boldsymbol{B}} \right) + m_b(B_2) \left(\sum_{\substack{\boldsymbol{B} \, : \, \boldsymbol{B}(B_1) = B_1, \\ \boldsymbol{B}(B_2) = B_2}} \alpha_{\boldsymbol{B}} \right) =$$
$$= m_b(B_2 \setminus B_1) + m_b(B_2)$$

i.e.

$$m_b(B_2 \setminus B_1) \left(\sum_{\substack{\boldsymbol{B} \, : \, \boldsymbol{B}(B_1) = B_1, \\ \boldsymbol{B}(B_2 \setminus B_1) \neq B_2}} \alpha_{\boldsymbol{B}} \right) + m_b(B_2) \left(\sum_{\substack{\boldsymbol{B} \, : \, \boldsymbol{B}(B_1) = B_1, \\ \boldsymbol{B}(B_2) \neq B_2}} \alpha_{\boldsymbol{B}} \right) = 0$$

which implies $\alpha_{\boldsymbol{B}} = 0$ for all the assignment functions $\boldsymbol{B}$ such that $\boldsymbol{B}(B_2 \setminus B_1) \neq B_2$ or $\boldsymbol{B}(B_2) \neq B_2$. The only non-zero coefficients can then be the $\alpha_{\boldsymbol{B}}$ s.t. $\boldsymbol{B}(B_1) = B_1$, $\boldsymbol{B}(B_2 \setminus B_1) = B_2$, $\boldsymbol{B}(B_2) = B_2$.
By induction you get that $\forall \boldsymbol{B} \neq \boldsymbol{B}_\rho$ we have $\alpha_{\boldsymbol{B}} = 0$.

The Intersection Probability and Its Properties

Fabio Cuzzolin

Oxford Brookes University, Oxford, UK
fabio.cuzzolin@brookes.ac.uk

Abstract. In this paper we discuss the properties of the intersection probability, a recent Bayesian approximation of belief functions introduced by geometric means. We propose a rationale for this approximation valid for interval probabilities, study its geometry in the probability simplex with respect to the polytope of consistent probabilities, and discuss the way it relates to important operators acting on belief functions.

1 Introduction

In the *theory of evidence* [1,2] the best representation of chance is a *belief function* (b.f.) rather than a Bayesian probability distribution, assigning probability values to *sets* of possibilities rather than single events. The relationship between belief functions and probabilities is of course of great interest in the theory of evidence: the issue is often known as "probability transformation" [3,4,5,6]. The connection between belief functions and probabilities is as well the foundation of a popular approach to the theory of evidence, Smets' *Transferable Belief Model* [7]. Beliefs are there represented as convex sets of probabilities or "credal sets", while decisions are made after *pignistic transformation* [8]. On his side, in his 1989 paper [9] F. Voorbraak proposed to adopt the so-called *relative plausibility* function. This is the unique probability that, given a belief function b with plausibility pl_b, assigns to each singleton its normalized plausibility.

The transformation problem can be posed in a different setting too. Belief and probability functions on finite domains can be represented as points of a large enough Cartesian space [10]. For instance, a belief function $b : 2^\Theta \to [0, 1]$ is completely specified by its belief values $\{b(A), A \subset \Theta, A \neq \emptyset, \Theta\}$ and can be seen as a point of $\mathbb{R}^{N-2}$, $N = 2^{|\Theta|}$. We can then obtain different probability transformations by minimizing different distances between the original belief function and the set of all probabilities.

In particular, we introduced a new probability $p[b]$ related to a belief function b, which we called *intersection probability*, determined by the intersection of the line joining a b.f. b and the related pl.f. pl_b with the region of all Bayesian (pseudo) b.f. [11].

In this paper we show that the intersection probability can in fact be defined for any *interval probability* system, as the unique probability obtained by assigning to all the elements of the domain the same fraction of uncertainty (Section 2). As a belief function determines an interval probability system, $p[b]$

C. Sossai and G. Chemello (Eds.): ECSQARU 2009, LNAI 5590, pp. 287–298, 2009.
© Springer-Verlag Berlin Heidelberg 2009

exists for belief functions too (for which it was originally introduced). The intersection probability can then be compared with other classical transformations like pignistic function and relative plausibility. In particular, the pignistic function has a strong credal interpretation as the barycenter of the polytope of all probabilities consistent with b. We prove that the intersection probability also possesses a credal interpretation in the probability simplex, as the "focus" of the pair of simplices embodying the interval probability system (Section 3).

In Section 4 we compare $p[b]$ with probability transformations of both the "affine" and "epistemic" family. While pignistic transformation and orthogonal projection commute with affine combination of belief functions, this is true for the intersection probability if the considered interval probabilities attribute the same "weight" to the uncertainty of each element.

2   The Intersection Probability and Its Rationale

Belief functions and probability intervals are different but related mathematical representations of the bodies of evidence we possess on a given decision or estimation problem Q. We assume that the possible answers to Q form a finite set $\Theta = \{x_1, ..., x_n\}$, called "frame of discernment".

Given a certain amount of evidence we are allowed to describe our belief on the outcome of Q in several possible ways: the classical option is to assume a probability distribution on Θ. However, as we may need to incorporate imprecise measurements and people's opinions in our knowledge state, or cope with missing or scarce information, a more sensible approach is to assume we have no access to the "correct" probability distribution but the available evidence provides us with some sort of constraint on this true distribution. Both interval probabilities and belief functions are mathematical descriptions of such a constraint. They hence define different *credal sets* or sets of probability distributions on Θ.

An "interval probability system" is a system of constraints on the probability values of a probability distribution $p : \Theta \rightarrow [0, 1]$ on a finite domain Θ of the form:

$$(l, u) \doteq \{l(x) \leq p(x) \leq u(x), \forall x \in \Theta\}. \tag{1}$$

The system (1) determines an entire set of probability distributions whose values are constrained to belong to a closed interval.

There are clearly many ways of selecting a single measure in order to represent a probability interval. We can point out, however, that all the intervals $[l(x), u(x)]$, $x \in \Theta$ have the same importance in the definition of the interval probability. There is no reason for the different singletons x to be treated differently.

It is then reasonable to request that the desired probability, candidate to represent the interval (1), should behave homogeneously in each element x of the frame Θ. This translates into seeking a probability p such that

$$p(x) = l(x) + \alpha(u(x) - l(x))$$

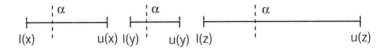

Fig. 1. An illustration of the notion of intersection probability for an upper/lower probability system

homogeneously for all elements x of Θ, for some value of $\alpha \in [0,1]$ (see Figure 1). Such value needs to be between 0 and 1 in order for the sought probability distribution p to belong to the interval. It is easy to see that there is indeed a *unique* solution to this problem. It suffices to enforce the normalization constraint $\sum_x p(x) = \sum_x [l(x) + \alpha(u(x) - l(x))] = 1$ to understand that the unique solution is given by:

$$\alpha = \beta[(l,u)] = \frac{1 - \sum_{x \in \Theta} l(x)}{\sum_{x \in \Theta} (u(x) - l(x))}. \tag{2}$$

We can define the *intersection probability* associated with the interval probability system (1) as the probability distribution with values

$$p[(l,u)](x) = \beta[(l,u)] \cdot u(x) + (1 - \beta[(l,u)]) \cdot l(x). \tag{3}$$

The most interesting interpretation of $p[(l,u)]$ comes from its alternative form

$$p[(l,u)](x) = l(x) + \left(1 - \sum_x l(x)\right) R[(l,u)](x) \tag{4}$$

where

$$R[(l,u)](x) \doteq \frac{u(x) - l(x)}{\sum_y (u(y) - l(y))} = \frac{\Delta(x)}{\sum_y \Delta(y)}, \tag{5}$$

the quantity $\Delta(x)$ measuring the size of the probability interval on x. $R(x)$ measures how much the uncertainty on the probability value on a singletons "weights" on the total uncertainty represented by the interval probability (1). It is natural to call it *relative uncertainty of singletons*.

Example. Consider a probability interval on a domain $\Theta = \{x, y, z\}$:

$$0.2 \leq p(x) \leq 0.8, \quad 0.4 \leq p(y) \leq 1, \quad 0 \leq p(x) \leq 0.4.$$

The widths of the corresponding intervals are $\Delta(x) = 0.6$, $\Delta(y) = 0.6$, $\Delta(z) = 0.4$ respectively. By Equation (2) the "fraction of uncertainty" to add to $l(x)$ to get an admissible probability is

$$\beta = \frac{1 - 0.2 - 0.4 - 0}{0.6 + 0.6 + 0.4} = \frac{0.4}{1.6} = \frac{1}{4}.$$

The intersection probability has then values (4):

$$p[(l,u)](x) = 0.2 + \tfrac{1}{4}0.6 = 0.35, \qquad p[(l,u)](y) = 0.4 + \tfrac{1}{4}0.6 = 0.55,$$
$$p[(l,u)](z) = 0 + \tfrac{1}{4}0.4 = 0.1.$$

Intersection probability for belief measures. As a belief measure [1] also determines a probability interval, the intersection probability can be defined for belief functions too.

A "basic probability assignment" (b.p.a.) over a finite set or "frame of discernment" Θ is a function $m : 2^\Theta \to [0, 1]$ on its power set $2^\Theta = \{A \subseteq \Theta\}$ such that 1. $m(\emptyset) = 0$; 2. $\sum_{A \subseteq \Theta} m(A) = 1$; 3. $m(A) \geq 0 \ \forall A \subseteq \Theta$. Subsets A of Θ associated with non-zero values $m(A) \neq 0$ of m are called "focal elements".

The *belief function* $b : 2^\Theta \to [0, 1]$ associated with a basic probability assignment m on Θ is defined as:

$$b(A) = \sum_{B \subseteq A} m(B). \tag{6}$$

A finite probability or *Bayesian* belief function is just a special b.f. assigning non-zero masses to singletons only: $m_b(A) = 0, |A| > 1$.

A dual mathematical representation of the evidence encoded by a belief function b is the "plausibility function" (pl.f.) $pl_b : 2^\Theta \to [0, 1]$, where $pl_b(A) \doteq 1 - b(A^c) = \sum_{B \cap A \neq \emptyset} m_b(B) \geq b(A)$ and A^c denotes the complement of A in Θ.

In the following we denote by b_A the unique "categorical" b.f. which assigns unitary mass to a single event A: $m_{b_A}(A) = 1$, $m_{b_A}(B) = 0 \ \forall B \neq A$. We can then decompose each belief function b with b.p.a. $m_b(A)$ as

$$b = \sum_{A \subseteq \Theta} m_b(A) b_A. \tag{7}$$

A pair belief-plausibility determines an interval probability system associated with a belief function, i.e.,

$$(b, pl_b) \doteq \{p \in \mathcal{P} : b(x) \leq p(x) \leq pl_b(x), \forall x \in \Theta\}. \tag{8}$$

In this case the intersection probability can be written as

$$p[b](x) = \beta[b] pl_b(x) + (1 - \beta[b]) m_b(x) \tag{9}$$

with

$$\beta[b] = \frac{1 - \sum_{x \in \Theta} m_b(x)}{\sum_{x \in \Theta} (pl_b(x) - m_b(x))} = \frac{1 - k_b}{k_{pl_b} - k_b} \tag{10}$$

where $k_{pl_b} \doteq \sum_{x \in \Theta} pl_b(x)$, $k_b \doteq \sum_{x \in \Theta} m_b(x)$ are the total plausibility and belief of singletons respectively.

3   Credal Geometry in the Probability Simplex

Credal interpretation of belief functions and pignistic function. It is well known that a belief function determines an entire set of probabilities *consistent* with it, i.e. such that $b(A) \leq p(A) \leq pl_b(A)$ *for all events* $A \subseteq \Theta$.

Notice that this set [12,13]:

$$\mathcal{P}[b] \doteq \{p \in \mathcal{P} : b(A) \le p(A) \le pl_b(A) \ \forall A \subseteq \Theta\} \tag{11}$$

is different from the set of probabilities (8) determined by the probability interval. A natural probabilistic approximation of b is then the center of mass of the set of consistent probabilities or "pignistic function" [8]:

$$BetP[b](x) = \sum_{A \supseteq \{x\}} \frac{m_b(A)}{|A|}. \tag{12}$$

$BetP[b]$ is the probability we obtain by re-assigning the mass of each focal element $A \subseteq \Theta$ of b *homogeneously* to each of its elements $x \in A$.

It is interesting to notice that the naive choice of choosing the barycenter of each interval $[l(x), u(x)]$ does not yield in general a valid probability function, for

$$\sum_x \left[l(x) + \frac{1}{2}(u(x) - l(x)) \right] \ne 1.$$

This marks the difference with the case of belief functions, in which the barycenter of the set of probabilities defined by a belief function has a valid interpretation in terms of degrees of belief.

Credal interpretation of interval probabilities. However, a similar credal interpretation can be given for the intersection probability too, once we determine the credal set associated with an interval probability (1). Here we develop our argument in particular for the interval (8) determined by a belief function b. The polytope $\mathcal{P}[b]$ can be naturally decomposed as the intersection

$$\mathcal{P}[b] = \bigcap_{i=1}^{n-1} T^i[b] \tag{13}$$

of the regions $T^i[b] \doteq \{p \in \mathcal{P} : p(A) \ge b(A) \ \forall A : |A| = i\}$ formed by all probability meeting the lower probability constraint *for size-i events*. Let us consider in particular the set of probabilities which meet the lower constraint *on singletons* $T^1[b]$,

$$T^1[b] \doteq \{p \in \mathcal{P} : p(x) \ge b(x) \ \forall x \in \Theta\}.$$

It is also easy to see that

$$\begin{aligned} T^{n-1}[b] &\doteq \{p \in \mathcal{P} : p(A) \ge b(A) \ \forall A : |A| = n-1\} \\ &= \{p \in \mathcal{P} : p(\{x\}^c) \ge b(\{x\}^c) \ \forall x \in \Theta\} \\ &= \{p \in \mathcal{P} : p(x) \le pl_b(x) \ \forall x \in \Theta\} \end{aligned}$$

expresses instead the *upper probability constraint on singletons*.

Clearly, then, the pair $(T^1[b], T^{n-1}[b])$ is the *geometric counterpart of an interval probability* in the probability simplex, exactly as the polytope of consistent probabilities $\mathcal{P}[b]$ represents there a belief function.

They form a higher dimensional triangle or *simplex*, i.e. the convex closure

$$Cl(\mathbf{v}_1, ..., \mathbf{v}_k) = \left\{ \mathbf{v} \in \mathbb{R}^d : \mathbf{v} = \alpha_1 \mathbf{v}_1 + \cdots + \alpha_k \mathbf{v}_k, \sum_i \alpha_i = 1, \alpha_i \geq 0 \,\forall i \right\} \quad (14)$$

of a collection $\mathbf{v}_1, ..., \mathbf{v}_k$ of *affinely independent* points. The points $\mathbf{v}_1, ..., \mathbf{v}_k$ are said to be affinely independent if none of them can be expressed as an affine combination of the others: $\nexists \mathbf{v}_i, \{\alpha_j, j \neq i : \sum_{j \neq i} \alpha_j = 1\}$ such that $\mathbf{v}_i = \sum_{j \neq i} \alpha_j \mathbf{v}_j$.

We then call $T^1[b]$ and $T^{n-1}[b]$ *lower* and *upper simplices* respectively. They have very simple expressions in terms of the basic probability assignment of b. Using the notation of Equation (7) it can be proven that [14]:

Proposition 1. *The set $T^1[b]$ of all probabilities meeting the lower probability constraint on singletons is the simplex $T^1[b] = Cl(t^1_x[b], x \in \Theta)$ with vertices*

$$t^1_x[b] = \sum_{y \neq x} m_b(y) b_y + \left(1 - \sum_{y \neq x} m_b(y)\right) b_x. \quad (15)$$

A dual proof can be provided for the set $T^{n-1}[b]$ of probabilities which meet the upper probability constraint on singletons [14]. We just need to replace belief with plausibility values on singletons.

Proposition 2. $T^{n-1}[b] = Cl(t^{n-1}_x[b], x \in \Theta)$ *is a simplex with vertices*

$$t^{n-1}_x[b] = \sum_{y \neq x} pl_b(y) b_y + \left(1 - \sum_{y \neq x} pl_b(y)\right) b_x. \quad (16)$$

Consider as an example the case of a belief function

$$\begin{array}{lll} m_b(x) = 0.2, & m_b(y) = 0.1, & m_b(z) = 0.3, \\ m_b(\{x, y\}) = 0.1, & m_b(\{y, z\}) = 0.2, & m_b(\Theta) = 0.1 \end{array} \quad (17)$$

defined on a ternary frame $\Theta = \{x, y, z\}$. Figure 2 illustrates the geometry of its consistent simplex $\mathcal{P}[b]$. We can notice that by Equation (13) $\mathcal{P}[b]$ (the polygon delimited by tiny squares) is in this case the intersection of two triangles (2-dimensional simplices) $T^1[b]$ and $T^2[b]$. The intersection probability

$$p[b](x) = m_b(x) + \beta[b](m_b(\{x, y\}) + m_b(\Theta)) = .2 + \tfrac{.4}{1.5 - 0.4} 0.2 = .27;$$
$$p[b](y) = .1 + \tfrac{.4}{1.1} 0.4 = .245; \quad p[b](z) = .485,$$

is the unique intersection of the lines joining the corresponding vertices of the upper $T^2[b]$ and lower $T^1[b]$ simplices.

Intersection probability as focus of upper and lower simplices. This fact, true in the general case, can be formalized by the notion of "focus" of a pair of simplices.

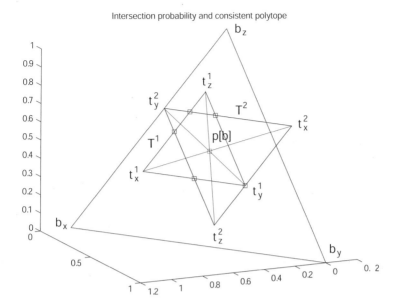

Fig. 2. The intersection probability is the focus of the two simplices $T^1[b]$ and $T^{n-1}[b]$. In the ternary case the latter reduce to the triangles $T^1[b]$ and $T^2[b]$. Their focus is geometrically the intersection of the lines joining their corresponding vertices.

Definition 1. *Consider a pair of simplices* $S = Cl(s_1, ..., s_n)$, $T = Cl(t_1, ..., t_n)$ *in* $\mathbb{R}^{n-1}$. *We call* focus *of the pair* (S, T) *the unique point* $f(S, T)$ *of* $\mathbb{R}^{n-1}$ *which has the same affine coordinates in both simplices:*

$$f(S, T) = \sum_{i=1}^{n} \alpha_i s_i = \sum_{j=1}^{n} \alpha_j t_j, \quad \sum_{i=1}^{n} \alpha_i = 1. \tag{18}$$

The focus of two simplices does not always fall in their intersection $S \cap T$ (i.e., α_i is not necessarily non-negative $\forall i$). However, if this is the case, the focus coincides with the unique intersection of the lines $a(s_i, t_i)$ joining corresponding vertices of S and T (see Figure 3): $f(S, T) = \bigcap_{i=1}^{n} a(s_i, t_i)$. Suppose indeed that a point p is such that $p = \alpha s_i + (1-\alpha) t_i \ \forall \ i = 1, ..., n$ (i.e. p lies on the line passing through s_i and $t_i \ \forall i$). Then necessarily $t_i = \frac{1}{1-\alpha}[p - \alpha s_i] \ \forall \ i = 1, ..., n$. If p has coordinates $\{\alpha_i, i = 1, ..., n\}$ in T, $p = \sum_{i=1}^{n} \alpha_i t_i$, then $p = \sum_{i=1}^{n} \alpha_i t_i = \frac{1}{1-\alpha}[p - \alpha \sum_i \alpha_i s_i]$ which implies $p = \sum_i \alpha_i s_i$, i.e. p is the focus of (S, T).

Theorem 1. *The intersection probability is the focus of the pair of upper and lower simplices:* $p[b] = f(T^{n-1}[b], T^1[b])$.

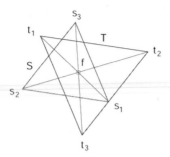

Fig. 3. If the focus of a pair of simplices belongs to their intersection, it is the unique intersection of the lines joining their corresponding vertices

Proof. We need to show that $p[b]$ has the same simplicial coordinates in $T^1[b]$ and $T^{n-1}[b]$. These coordinates turn out to be the values of the relative uncertainty function (5) for b:

$$R[b](x) = \frac{pl_b(x) - m_b(x)}{k_{pl_b} - k_b}. \tag{19}$$

Recalling the expression (15) of the vertices of $T^1[b]$, the point of the simplex $T^1[b]$ with coordinates (19) is

$$\sum_x R[b](x)t_x^1[b] = \sum_x R[b](x)\Big[\sum_{y \neq x} m_b(y)b_y + \big(1 - \sum_{y \neq x} m_b(y)\big)b_x\Big]$$
$$= \sum_x R[b](x)\Big[\sum_{y \in \Theta} m_b(y)b_y + (1 - k_b)b_x\Big]$$
$$= \sum_x b_x\Big[(1 - k_b)R[b](x) + m_b(x)\sum_y R[b](y)\Big] = \sum_x b_x\big[(1 - k_b)R[b](x) + m_b(x)\big]$$

as $R[b]$ is a probability ($\sum_y R[b](y) = 1$).

By Equation (4) the above quantity coincides with $p[b]$.

The point of $T^{n-1}[b]$ with the same coordinates $\{R[b](x), x \in \Theta\}$ is again

$$\sum_x R[b](x)t_x^{n-1}[b] = \sum_x R[b](x)\Big[\sum_{y \neq x} pl_b(y)b_y + \big(1 - \sum_{y \neq x} pl_b(y)\big)b_x\Big]$$
$$= \sum_x R[b](x)\Big[\sum_{y \in \Theta} pl_b(y)b_y + (1 - k_{pl_b})b_x\Big] =$$
$$= \sum_x b_x\Big[(1 - k_{pl_b})R[b](x) + pl_b(x)\sum_y R[b](y)\Big] =$$
$$= \sum_x b_x\big[(1 - k_{pl_b})R[b](x) + pl_b(x)\big] = \sum_x b_x\Big[pl_b(x)\frac{1 - k_b}{k_{pl_b} - k_b} - m_b(x)\frac{1 - k_b}{k_{pl_b} - k_b}\Big]$$

$= p[b]$ by Equation (19).   □

Pignistic function and intersection probability both adhere to rationality principles for belief functions and interval probabilities respectively. Geometrically, this translates into a similar behavior in the probability simplex, in which they are the center of mass of the consistent polytope and the focus of the pair of lower and upper probability simplices.

4   Intersection Probability and Other Transformations

The epistemic family. An approach to the problem of approximating a b.f. with a probability seeks transformations which enjoy commutativity properties with respect to some combination rule [9,15], in particular the original Dempster's sum [2]. Voorbraak's *relative plausibility of singletons* [9] (rel.plaus.) $\tilde{pl}_b$ is the unique probability that, given a belief function b with plausibility pl_b, assigns to each singleton its normalized plausibility

$$\tilde{pl}_b(x) = \frac{pl_b(x)}{\sum_{y \in \Theta} pl_b(y)} = \frac{pl_b(x)}{k_{pl_b}}, \tag{20}$$

and commutes with Dempster's orthogonal sum $\oplus$ [2,16].

Dually, a *relative belief of singletons* [15] (rel.bel.) can be defined which assigns to the elements of Θ their normalized belief values:

$$\tilde{b}(x) \doteq \frac{b(x)}{\sum_{y \in \Theta} b(y)}. \tag{21}$$

Clearly $\tilde{b}$ exists iff b assigns some mass to singletons: $k_b = \sum_{x \in \Theta} m_b(x) \neq 0$.

These two approximations form a strongly linked couple. They are sometimes called the *epistemic* family of transformations [15].

It is important to notice, though, that in the interpretation of a belief function as a probability interval (8), the probabilities we obtain by normalizing the lower bound $\tilde{l}(x) = l(x)/\sum_y l(y)$ or the upper bound $\tilde{u}(x) = u(x)/\sum_y u(y)$ of the interval are *not* consistent with the interval itself. If there exists an $x \in \Theta$ such that $b(x) = pl_b(x)$ (the interval has width zero for that element) we have that

$$\tilde{b}(x) = \frac{m_b(x)}{\sum_y m_b(y)} > pl_b(x), \qquad \tilde{pl}_b(x) = \frac{pl_b(x)}{\sum_y pl_b(y)} < b(x),$$

and both the relative belief and plausibility of singletons fall outside the interval (8). This holds for a general probability interval (1), and supports the argument in favor of the interval probability.

The affine family. Uncertainty measures can be represented as points of a Cartesian space [10]. In that context, affine combination is the geometric counterpart of the normalization constraint imposed on basic probability assignments. As a result, all significant entities form convex regions of the Cartesian space.

A different family of probability transformations commute indeed with affine combination of belief functions (as points of such a space). This is the case of pignistic function $BetP[b]$ and orthogonal projection $\pi[b]$ of a belief function b onto the probability simplex $\mathcal{P}$ [11]. Whenever $\alpha_1 + \alpha_2 = 1$ we have that:

$$\begin{aligned} BetP[\alpha_1 b_1 + \alpha_2 b_2] &= \alpha_1 BetP[b_1] + \alpha_2 BetP[b_2] \\ \pi[\alpha_1 b_1 + \alpha_2 b_2] &= \alpha_1 \pi[b_1] + \alpha_2 \pi[b_2]. \end{aligned}$$

The condition under which the intersection probability commutes with affine combination is indeed quite interesting.

Theorem 2. *Intersection probability $p[b]$ and affine combination commute, $p[\alpha_1 b_1 + \alpha_2 b_2] = \alpha_1 p[b_1] + \alpha_2 p[b_2]$ for $\alpha_1 + \alpha_2 = 1$, if the relative uncertainty of the singletons is the same for both intervals: $R[b_1] = R[b_2]$.*

Proof. By definition (9) $p[\alpha_1 b_1 + \alpha_2 b_2] =$

$$= \alpha_1 m_1(x) + \alpha_2 m_2(x) + (1 - k_{\alpha_1 b_1 + \alpha_2 b_2}) \frac{\alpha_1 \Delta_1(x) + \alpha_2 \Delta_2(x)}{\sum_{y \in \Theta}(\alpha_1 \Delta_1(y) + \alpha_2 \Delta_2(y))}$$

that after defining $R(x) \doteq \frac{\alpha_1 \Delta_1(x) + \alpha_2 \Delta_2(x)}{\sum_{y \in \Theta}(\alpha_1 \Delta_1(y) + \alpha_2 \Delta_2(y))}$ becomes

$$\alpha_1 m_1(x) + \alpha_2 m_2(x) + [1 - (\alpha_1 k_{b_1} + \alpha_2 k_{b_2})]R(x) =$$
$$= \alpha_1 \big(m_1(x) + (1 - k_{b_1})R(x)\big) + \alpha_2 \big(m_2(x) + (1 - k_{b_2})R(x)\big)$$

which is equal to $\alpha_1 p[b_1] + \alpha_2 p[b_2]$ iff

$$\alpha_1(1 - k_{b_1})(R(x) - R[b_1](x)) + \alpha_2(1 - k_{b_2})(R(x) - R[b_2](x)) = 0.$$

If $R(x) = R[b_1](x) = R[b_2](x)\ \forall x$ the thesis is trivially true.   □

The intersection probability does not possess the same nice relation with affine combination which characterizes pignistic function and orthogonal projection. However, Theorem 2 states that they commute exactly when each uncertainty interval $l(x) \le p(x) \le u(x)$ has the same "weight" in the two interval probabilities.

Comparison with the members of the affine family. It is natural to wonder what are the other differences between $p[b]$ and its "sister" functions $BetP[b]$ and $\pi[b]$. Some sufficient conditions [11] have been already worked out in the past. More stringent conditions can be formulated.

Theorem 3. *If a belief function b is such that its mass is equally distributed among focal elements of the same size*

$$m_b(A) = const\ \forall A : |A| = k, \forall k = 2, ..., n. \qquad (22)$$

then its pignistic and intersection probabilities coincide: $BetP[b] = p[b]$.

Proof. If b meets (22), then the expression for the probability values of the intersection probability gives, for each $x \in \Theta$,

$$p[b](x) = m_b(x) + \beta[b] \sum_{A \supsetneq x} m_b(A) = m_b(x) + \beta[b] \sum_{k=2}^{n} \sigma^k \frac{\binom{n-1}{k-1}}{\binom{n}{k}} =$$

(as there are $\binom{n-1}{k-1}$ events of size k containing x, and $\binom{n}{k}$ events of size k)

$$= m_b(x) + \beta[b] \sum_{k=2}^{n} \sigma^k \frac{k}{n} = m_b(x) + \frac{1}{n} \frac{\sigma^2 + ... + \sigma^n}{2\sigma^2 + ... + n\sigma^n}(2\sigma^2 + ... + n\sigma^n)$$
$$= m_b(x) + \frac{1}{n}(\sigma^2 + ... + \sigma^n)$$

after recalling the decomposition of $\beta[b]$:

$$\beta[b] = \frac{\sum_{|B|>1} m_b(B)}{\sum_{|B|>1} m_b(B)|B|} = \frac{\sum_{k=2}^{n} \sum_{|B|=k} m_b(B)}{\sum_{k=2}^{n} k \cdot \sum_{|B|=k} m_b(B)} = \frac{\sigma_2 + \cdots + \sigma_n}{2\sigma_2 + \cdots + n\sigma_n}. \quad (23)$$

On the other side, under the hypothesis, the pignistic function reads as

$$BetP[b](x) = m_b(x) + \sum_{k=2}^{n} \sum_{A \supseteq x, |A|=k} \frac{m_b(A)}{k} = m_b(x) + \sum_{k=2}^{n} \frac{\sigma^k}{k} \frac{\binom{n-1}{k-1}}{\binom{n}{k}}$$

$$= m_b(x) + \sum_{k=2}^{n} \frac{\sigma^k}{k} \frac{k}{k} = m_b(x) + \sum_{k=2}^{n} \frac{\sigma^k}{n} \quad (24)$$

and the two functions coincide. $\qquad\square$

Condition (22) is sufficient to guarantee the equality of intersection probability and orthogonal projection [11] too.

Theorem 4. *If a belief function b meets condition (22) (i.e., its mass is equally distributed among focal elements of the same size) then the related orthogonal projection and intersection probability coincide.*

Proof. The orthogonal projection of a belief function b on to the probability simplex $\mathcal{P}$ has the following expression [11]:

$$\pi[b](x) = \sum_{A \supseteq x} m_b(A)\left(\frac{1 + |A^c|2^{1-|A|}}{n}\right) + \sum_{A \not\supseteq x} m_b(A)\left(\frac{1 - |A|2^{1-|A|}}{n}\right). \quad (25)$$

Under condition (22) it becomes

$$\pi[b](x) = m_b(x) + \sum_{k=2}^{n}\left(\frac{1 + (n-k)2^{1-k}}{n}\right) \sum_{A \supseteq x, |A|=k} m_b(A)$$

$$+ \sum_{k=2}^{n}\left(\frac{1 - (n-k)2^{1-k}}{n}\right) \sum_{A \not\supseteq x, |A|=k} m_b(A) \quad (26)$$

where again $\sum_{A \supseteq x, |A|=k} m_b(A) = \sigma^k k/n$, while

$$\sum_{A \not\supseteq x, |A|=k} m_b(A) = \sigma^k \frac{\binom{n-1}{k}}{\binom{n}{k}} = \sigma^k \frac{(n-1)!}{k!(n-k-1)!} \frac{k!(n-k)!}{n!} = \sigma^k \frac{n-k}{n}.$$

Replacing those expressions in Equation (26) yields

$$m_b(x) + \sum_{k=2}^{n}\left(\frac{1 + (n-k)2^{1-k}}{n}\right)\sigma^k \frac{k}{n} + \sum_{k=2}^{n}\left(\frac{1 - (n-k)2^{1-k}}{n}\right)\sigma^k \frac{n-k}{n} =$$

$$= m_b(x) + \sum_{k=2}^{n}\left(\sigma^k \frac{k}{n^2} + \sigma^k \frac{n-k}{n^2}\right) = m_b(x) + \frac{1}{n}\sum_{k=2}^{n} \sigma^k$$

which is exactly the value (24) of the intersection probability under the same assumption. $\qquad\square$

5   Conclusions

In this paper we studied the intersection probability, a Bayesian transformation of belief functions originally derived from purely geometric arguments, from the more abstract point of view of interval probabilities, providing a rationality principle for it. We studied its behavior in the probability simplex, proving that it can be described as the focus of the upper and lower simplices which geometrically embody an interval probability. We compared it to transformations of both the affine and epistemic families, and studied the condition under which it commutes with convex combination.

References

1. Shafer, G.: A Mathematical Theory of Evidence. Princeton University Press, Princeton (1976)
2. Dempster, A.: Upper and lower probabilities generated by a random closed interval. Annals of Mathematical Statistics 39, 957–966 (1968)
3. Denoeux, T.: Inner and outer approximation of belief structures using a hierarchical clustering approach. Int. Journal of Uncertainty, Fuzziness and Knowledge-Based Systems 9(4), 437–460 (2001)
4. Denoeux, T., Yaghlane, A.B.: Approximating the combination of belief functions using the fast moebius transform in a coarsened frame. International Journal of Approximate Reasoning 31(1-2), 77–101 (2002)
5. Haenni, R., Lehmann, N.: Resource bounded and anytime approximation of belief function computations. IJAR 31(1-2), 103–154 (2002)
6. Bauer, M.: Approximation algorithms and decision making in the Dempster-Shafer theory of evidence–an empirical study. IJAR 17(2-3), 217–237 (1997)
7. Smets, P.: Belief functions versus probability functions. In: Bouchon, B., Saitta, L., Yager, R. (eds.) Uncertainty and Intelligent Systems, pp. 17–24. Springer, Berlin (1988)
8. Smets, P.: Decision making in the TBM: the necessity of the pignistic transformation. IJAR 38(2), 133–147 (2005)
9. Voorbraak, F.: A computationally efficient approximation of Dempster-Shafer theory. International Journal on Man-Machine Studies 30, 525–536 (1989)
10. Cuzzolin, F.: A geometric approach to the theory of evidence. IEEE Transactions on Systems, Man, and Cybernetics - Part C 38(4), 522–534 (2007)
11. Cuzzolin, F.: Two new Bayesian approximations of belief functions based on convex geometry. IEEE Trans. on Systems, Man, and Cybernetics B 37(4), 993–1008 (2007)
12. Chateauneuf, A., Jaffray, J.Y.: Some characterizations of lower probabilities and other monotone capacities through the use of Möbius inversion. Mathematical Social Sciences 17, 263–283 (1989)
13. Dubois, D., Prade, H., Smets, P.: New semantics for quantitative possibility theory. In: ISIPTA, pp. 152–161 (2001)
14. Cuzzolin, F.: Credal semantics of Bayesian approximations in terms of probability intervals. IEEE Trans. on Systems, Man, and Cybernetics B (to appear) (2009)
15. Cuzzolin, F.: Semantics of the relative belief of singletons. In: International Workshop on Uncertainty and Logic UNCLOG 2008, Kanazawa, Japan (2008)
16. Cobb, B., Shenoy, P.: On the plausibility transformation method for translating belief function models to probability models. IJAR 41(3), 314–330 (2006)

Can the Minimum Rule of Possibility Theory Be Extended to Belief Functions?

Sébastien Destercke[1] and Didier Dubois[2]

[1] Centre de coopération internationale en recherche agronomique pour le développement (CIRAD), UMR IATE, Campus Supagro, Montpellier, France
`sebastien.destercke@cirad.fr`
[2] Toulouse Institute of Research in Computer Science (IRIT), Toulouse, France
`Dubois@irit.fr`

Abstract. When merging belief functions, Dempster rule of combination is justified only when sources can be considered as independent. When dependencies are ill-known, it is usual to ask the merging operation to satisfy the property of idempotence, as this property ensures a cautious behaviour in the face of dependent sources. There are different strategies to find such rules for belief functions. One strategy is to rely on idempotent rules used in either more general or more specific frameworks and to respectively study their particularisation or extension to belief functions. In this paper, we try to extend the minimum rule of possibility theory to belief functions. We show that such an extension is not always possible, unless we accept the idea that the result of the fusion process can be a family of belief functions.

Keywords: Belief functions, idempotence, fusion, possibility.

1 Introduction

The main merging rule in the theory of evidence is Dempster's rule [4], even if other proposals exist [17]. Combining belief functions by Dempster's rule is justified only when sources can be considered as independent. In other cases, a specific dependence structure between sources can be assumed, and suitable alternative merging rules can be used (e.g., the commensuration method [13,7]). However, assuming that the (in)dependence structure between sources is well-known is often unrealistic. In those cases, an alternative is to adopt a conservative approach when merging belief functions, by applying the "least commitment principle", which states that one should never presuppose more beliefs than justified. This principle is basic in the frameworks of possibility theory, imprecise probability [19], and the Transferable Belief Model (TBM) [18]. This cautious approach can be interpreted and used in different ways [6,2,7]. However, all these approaches agree on the fact that a cautious conjunctive merging rule should satisfy the property of idempotence, as this property ensures that the same information supplied by two dependent sources will remain unchanged after merging.

There are mainly three strategies to construct idempotent rules that make sense in the belief function setting. The first one looks for idempotent rules

C. Sossai and G. Chemello (Eds.): ECSQARU 2009, LNAI 5590, pp. 299–310, 2009.
© Springer-Verlag Berlin Heidelberg 2009

that satisfy a certain number of desired properties and appear sensible in the framework of belief functions. This is the solution retained by Denoeux [6] and Cattaneo [2]. The second strategy relies on the natural idempotent rule consisting of intersecting sets of probabilities and tries to express it in the particular case of belief functions (Chateauneuf [3]). Finally, the third approach, explored in this paper, starts from the natural idempotent rule in a less general framework, possibility theory, trying to extend it to belief functions. Namely, we study the generalisation of the minimum rule, viewing possibility distributions as contour functions of consonant belief functions [16]. If we denote by $(m_1, \mathcal{F}_1), (m_2, \mathcal{F}_2)$ two belief functions, $\mathcal{P}_1, \mathcal{P}_2$ two sets of probabilities, and π_1, π_2 two possibility distributions, the three approaches are summarized in Figure 1 below.

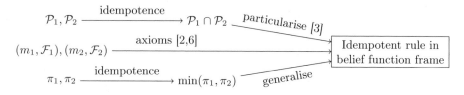

Fig. 1. Search of idempotent merging rules

Section 2 recalls basics of belief functions and defines conjunctive merging in this framework. Section 3 then studies to what extent the minimum rule of possibility theory can be extended to the framework of belief functions. The idea is to request that the contour function after merging be the minimum of the contour functions of the input belief functions, what we call the *strong contour function principle*. Note that a similar property holds for the unnormalized Dempster rule with respect to the product of contour functions. In the case of the minimum rule, we are led to propose a weak contour principle as the former condition turns out to be too strong.

2   Preliminaries

This section introduces the basics of belief functions needed in this paper. In the whole paper, we consider that information pertains to a variable V taking its values on a finite space $\mathcal{V}$, with generic element denoted v.

2.1   Belief Functions

Suppose that beliefs can be modelled by a belief function, or, equivalently, by a basic belief assignment (bba), that is, a function m from the power set $2^{|\mathcal{V}|}$ of $\mathcal{V}$ to $[0, 1]$ such that $\sum_{A \subseteq \mathcal{V}} m(A) = 1$. Let $\mathcal{M}_\mathcal{V}$ be the set of bba's on $2^{|\mathcal{V}|}$. A set A such that $m(A) > 0$ is called a focal set. We denote by $\mathcal{F}$ the set of focal sets corresponding to bba m, and $(m, \mathcal{F})$ a *belief structure*. $m(A)$ is the mass of A. Given a bba m, belief, plausibility and commonality functions of an event $E \subseteq \mathcal{V}$ are:

$$bel(E) = \sum_{A \subseteq E} m(A); \quad pl(E) = \sum_{A \cap E \neq \emptyset} m(A); \quad q(E) = \sum_{E \subseteq A} m(A)$$

A belief function measures to what extent an event is certainly supported by the available information. A plausibility function measures the amount of evidence that does not contradict a given event. A commonality function measures the quantity of mass that may be re-allocated to a set from its supersets. The commonality (resp. belief) function increases when bigger (resp. smaller) focal sets receive greater mass, hence the greater the commonality (resp. the belief) degrees, the less (resp. the more) informative is the bba. Note that the four representations contain the same amount of information [16].

In Shafer's seminal work [16], taken over by Smets in his Transferable Belief Model [18], there is no reference to any underlying probabilistic interpretation or framework. The bba and the associated belief function model beliefs *per se*. However, a belief structure $(m, \mathcal{F})$ can also be interpreted as a convex set $\mathcal{P}_{(m,\mathcal{F})}$ of probabilities [19] such that $Bel(A)$ and $Pl(A)$ are probability bounds: $\mathcal{P}_{(m,\mathcal{F})} = \{P | \forall A \subset X, \ Bel(A) \leq P(A) \leq Pl(A)\}$. Classical probability distributions are retrieved when only singletons receive positive masses. This interpretation is closer to random sets and to Dempster's view [4].

2.2 Possibility Distributions and Contour Functions

A possibility distribution [9] is a mapping $\pi : \mathcal{V} \to [0, 1]$ such that $\pi(v) = 1$ for at least one element $v \in \mathcal{V}$. The latter is the normalization condition. Two dual functions (resp. the possibility and necessity function) can be defined from π: $\Pi(A) = \sup_{v \in A} \pi(v)$ and $N(A) = 1 - \Pi(A^c)$. Their characteristic properties are: $\Pi(A \cup B) = \max(\Pi(A), \Pi(B)); \quad N(A \cap B) = \min(N(A), N(B)) \forall A, B \subseteq \mathcal{V}$.

Definition 1. *The contour function π_m of a belief structure $(m, \mathcal{F})$ is a mapping $\pi_m : \mathcal{V} \to [0, 1]$ such that, for any $v \in \mathcal{V}$, $\pi_m(v) = pl(\{v\}) = q(\{v\})$, with pl, q the plausibility and commonality functions of $(m, \mathcal{F})$.*

A belief structure $(m, \mathcal{F})$ is called consonant when its focal sets are completely ordered with respect to inclusion (for any $A, B \in \mathcal{F}$, $A \subset B$ or $B \subset A$). In this case, the plausibility (resp. belief) function has the characteristic properties of a possibility (resp. necessity) function, and the information contained in the consonant belief structure can be represented by the possibility distribution π equal to the contour function $\sum_{v \in E} m(E)$. Conversely, any possibility distribution and its associated possibility (resp. necessity) function defines a unique consonant belief structure and plausibility (resp. belief) function.

The contour function can be seen as a (possibly unnormalized) possibility distribution, and is a trace of the whole belief structure $(m, \mathcal{F})$ restricted to singletons. Except when $(m, \mathcal{F})$ is consonant, the contour function represents only part of the information contained in $(m, \mathcal{F})$. The contour function is however a summary, easier to manipulate than the whole random set.

2.3   Information Orderings between Belief Functions

There are many notions extending classical set inclusion to the framework of belief functions. The most classical notions are the pl-,q- and s-inclusions:

- A belief structure $(m_1, \mathcal{F}_1)$ is q- (resp. pl-) included in another belief structure $(m_2, \mathcal{F}_2)$, denoted $(m_1, \mathcal{F}_1) \sqsubseteq_q (m_2, \mathcal{F}_2)$ (resp. $(m_1, \mathcal{F}_1) \sqsubseteq_{pl} (m_2, \mathcal{F}_2)$), if and only if for all $A \subseteq \mathcal{V}$, $q_1(A) \leq q_2(A)$ (resp. $pl_1(A) \leq pl_2(A)$).
- A belief structure $(m_1, \mathcal{F}_1)$ with $\mathcal{F}_1 = \{E_1, \ldots, E_q\}$ is s-included in another belief structure $(m_2, \mathcal{F}_2)$ with $\mathcal{F}_2 = \{E'_1, \ldots, E'_p\}$, denoted $(m_1, \mathcal{F}_1) \sqsubseteq_s (m_2, \mathcal{F}_2)$, if and only if there exists a non-negative matrix G with generic term g_{ij} such that for $j = 1, \ldots, p, i = 1, \ldots, q$

$$\sum_{i=1}^{q} g_{ij} = 1, g_{ij} > 0 \Rightarrow E_i \subseteq E'_j, \quad \sum_{j=1}^{p} m_2(E'_j)g_{ij} = m_1(E_i).$$

The strict part $(m_1, \mathcal{F}_1) \sqsubset_s (m_2, \mathcal{F}_2)$ means $(m_1, \mathcal{F}_1) \sqsubseteq_s (m_2, \mathcal{F}_2)$ and $\exists i, j : E_i \subset E'_j$. Our objective being to investigate the generalisation of possibilistic idempotent rule to belief structures, we also use the notion of π-inclusion.

Definition 2 (π-inclusion). *A belief structure $(m_1, \mathcal{F}_1)$ is said to be π-included in another belief structure $(m_2, \mathcal{F}_2)$ if and only if, for all $v \in \mathcal{V}$, $\pi_{m_1}(v) \leq \pi_{m_2}(v)$ and this relation is denoted $(m_1, \mathcal{F}_1) \sqsubseteq_\pi (m_2, \mathcal{F}_2)$*

Since notions of inclusion allow to compare informative contents, we will also say, when $(m_1, \mathcal{F}_1) \sqsubseteq_x (m_2, \mathcal{F}_2)$ $((m_1, \mathcal{F}_1) \sqsubset_x (m_2, \mathcal{F}_2))$ with $\mathbf{x} \in \{pl, q, s, \pi\}$, that $(m_1, \mathcal{F}_1)$ is (strictly) $\mathbf{x}$-more committed than $(m_2, \mathcal{F}_2)$. The following implications hold between these notions of inclusion [8]:

$$(m_1, \mathcal{F}_1) \sqsubseteq_s (m_2, \mathcal{F}_2) \Rightarrow \left\{ \begin{array}{l} (m_1, \mathcal{F}_1) \sqsubseteq_{pl} (m_2, \mathcal{F}_2) \\ (m_1, \mathcal{F}_1) \sqsubseteq_q (m_2, \mathcal{F}_2) \end{array} \right\} \Rightarrow (m_1, \mathcal{F}_1) \sqsubseteq_\pi (m_2, \mathcal{F}_2).$$

Each of these notions induces a partial ordering between elements of $\mathcal{M}_\mathcal{V}$. The relation $\sqsubseteq_\pi$ only induces a partial pre-order (i.e., we can have $(m_1, \mathcal{F}_1) \sqsubseteq_\pi (m_2, \mathcal{F}_2)$ and $(m_2, \mathcal{F}_2) \sqsubseteq_\pi (m_1, \mathcal{F}_1)$ with $(m_1, \mathcal{F}_1) \neq (m_2, \mathcal{F}_2)$), while the others induce partial orders. The first implication holds for strict inclusions as well. If belief structures are consonant, then all the notions of pl, q, s and π-inclusion reduce to the same definition (the one of π-inclusion).

Example 1. Consider the two belief structures $(m_1, \mathcal{F}_1), (m_2, \mathcal{F}_2)$ on the domain $\mathcal{V} = \{v_1, v_2, v_3\}$

$\mathcal{F}_1$	m_1	$\mathcal{F}_2$	m_2
$E_{11} = \{v_2\}$	0.5	$E_{21} = \{v_2, v_3\}$	0.5
$E_{12} = \{v_1, v_2, v_3\}$	0.5	$E_{22} = \{v_1, v_2\}$	0.5

They have the same contour function, but $(m_1, \mathcal{F}_1) \sqsubset_{pl} (m_2, \mathcal{F}_2)$ and $(m_2, \mathcal{F}_2) \sqsubset_q (m_1, \mathcal{F}_1)$.

Another way to compute informative contents is to use scalar information evaluations, such as expected cardinality. The interest of ordering belief structures with such an evaluation is that it induces a complete (pre-)order on belief structures. The expected cardinality of a belief structure $(m, \mathcal{F})$, denoted $\mathbb{C}_{|(m,\mathcal{F})|}$, is $\mathbb{C}_{|(m,\mathcal{F})|} = \sum_{E \in \mathcal{F}} m(E)|E|$. It is equal to the area under the contour function π_m [11]:

$$\mathbb{C}_{|(m,\mathcal{F})|} = \sum_{v \in \mathcal{V}} \pi_m(v). \tag{1}$$

Then $(m_1, \mathcal{F}_1)$ is said to be more $\mathbb{C}$-*specific* than $(m_2, \mathcal{F}_2)$ if and only if $\mathbb{C}_{|(m_1,\mathcal{F}_1)|} \leq \mathbb{C}_{|(m_2,\mathcal{F}_2)|}$, denoted $(m_1, \mathcal{F}_1) \sqsubseteq_{\mathbb{C}} (m_2, \mathcal{F}_2)$. It can be shown that $\mathbb{C}$-specificity is in agreement with other inclusion notions, namely, the following implications hold:

1. $(m_1, \mathcal{F}_1) \sqsubseteq_\pi (m_2, \mathcal{F}_2) \rightarrow (m_1, \mathcal{F}_1) \sqsubseteq_{\mathbb{C}} (m_2, \mathcal{F}_2)$
2. $(m_1, \mathcal{F}_1) \sqsubseteq_s (m_2, \mathcal{F}_2) \rightarrow (m_1, \mathcal{F}_1) \sqsubseteq_{\mathbb{C}} (m_2, \mathcal{F}_2)$
3. $(m_1, \mathcal{F}_1) \sqsubseteq_{pl} (m_2, \mathcal{F}_2) \rightarrow (m_1, \mathcal{F}_1) \sqsubseteq_{\mathbb{C}} (m_2, \mathcal{F}_2)$
4. $(m_1, \mathcal{F}_1) \sqsubseteq_q (m_2, \mathcal{F}_2) \rightarrow (m_1, \mathcal{F}_1) \sqsubseteq_{\mathbb{C}} (m_2, \mathcal{F}_2)$

2.4   Conjunctive Merging and Least Commitment

Let $(m_1, \mathcal{F}_1), (m_2, \mathcal{F}_2)$ be belief structures supplied by two, not necessarily independent, sources. We define a belief structure $(m_\cap, \mathcal{F}_\cap)$ resulting from merging $(m_1, \mathcal{F}_1), (m_2, \mathcal{F}_2)$ as the result of the following procedure [7]:

1. A joint mass distribution m is built on $\mathcal{V}^2$, with focal sets of the form $A \times B$ with $A \in \mathcal{F}_1, B \in \mathcal{F}_2$ and preserving m_1, m_2 considered as marginals. It means that

$$\forall A \in \mathcal{F}_1, m_1(A) = \sum_{B \in \mathcal{F}_2} m(A, B), \quad \forall B \in \mathcal{F}_2, m_2(B) = \sum_{A \in \mathcal{F}_1} m(A, B). \tag{2}$$

2. Each joint mass $m(A, B)$ is allocated to the subset $A \cap B$ only.

We call such a rule *conjunctive*[1], and denote by $\mathcal{M}_{12}$ the set of conjunctively merged belief structures. The idea behind the conjunctive approach is to gain as much information as possible from the merging. Not every belief structure obtained by conjunctive merging is normalized (i.e. one may get $m(\emptyset) \neq 0$). In this paper, unless stated otherwise, we do not assume that a belief structure has to be normalised after conjunctive merging. We also do not renormalise such belief structures, because, after renormalisation, they no longer satisfy Eq. (2), and renormalisation is usually not required when working with possibility distributions. A belief structure obtained by a conjunctive merging rule is a specialisation of both $(m_1, \mathcal{F}_1)$ and $(m_2, \mathcal{F}_2)$. The set $\mathcal{M}_{12}$ is a subset of all belief structures that are specialisations of both $(m_1, \mathcal{F}_1)$ and $(m_2, \mathcal{F}_2)$, that is $\mathcal{M}_{12} \subseteq \{m \in \mathcal{M}_X | m \sqsubseteq_s m_i, i = 1, 2\}$, and the inclusion is usually strict[2]. Actually, three situations can occur [7]:

---

[1] A disjunctive merging rule can be defined likewise, changing $\cap$ into $\cup$.
[2] Take, for example, the empty belief structures $m_1(\mathcal{V}) = m_2(\mathcal{V}) = 1$ as marginals.

1. $\mathcal{M}_{12}$ contains only normalized belief functions. It means that $\forall A \in \mathcal{F}_1, B \in \mathcal{F}_2, A \cap B \neq \emptyset$. The two bbas are said to be *logically consistent*.
2. $\mathcal{M}_{12}$ contains both subnormalized and normalized bbas. It means that $\exists A, B, A \cap B = \emptyset$ and that the marginal-preservation equations (2) have solutions which allocate zero mass to such $A \times B$. The two bbas are said to be *non-conflicting*. Chateauneuf [3] has shown that being non-conflicting is a sufficient and necessary condition for $\mathcal{P}_{(m_1, \mathcal{F}_1)} \cap \mathcal{P}_{(m_2, \mathcal{F}_2)}$ to be non-empty.
3. $\mathcal{M}_{12}$ contains only subnormalized belief functions. This situation is equivalent to $\mathcal{P}_{(m_1, \mathcal{F}_{1,})} \cap \mathcal{P}_{(m_2, \mathcal{F}_2)} = \emptyset$. The two bbas are said to be *conflicting*.

When both sources can be considered independent, the TBM conjunctive rule yields the merged belief structure inside $\mathcal{M}_{12}$ such that $m(A, B) = m_1(A) \cdot m_2(B)$ in Equations (2). When the sources (in)dependence structure is ill-known, a common practice is to use the least-commitment principle (LCP) to merge belief structures. Denote by $\mathcal{M}_{12}^{\sqsubseteq \mathbf{x}}$ the set of all maximal elements inside $\mathcal{M}_{12}$ with respect to the order induced by $\mathbf{x}$-inclusion, with $\mathbf{x} \in \{s, pl, q, \pi, \mathbb{C}\}$. Note that $\{\mathcal{M}_{12}^{\sqsubseteq \mathbf{pl}} \cup \mathcal{M}_{12}^{\sqsubseteq \mathbf{q}} \cup \mathcal{M}_{12}^{\sqsubseteq \pi} \cup \mathcal{M}_{12}^{\sqsubseteq \mathbb{C}}\} \subseteq \mathcal{M}_{12}^{\sqsubseteq \mathbf{s}}$. LCP then often consists of choosing a given $\mathbf{x}$ and picking a particular element inside $\mathcal{M}_{12}^{\sqsubseteq \mathbf{x}}$ that also satisfies a number of desired properties.

In the rest of the paper, we study how to extend the minimum rule of possibility theory to belief structures, and under which conditions such an idempotent merging of belief structures exists.

3   Extending the Minimum Rule

If π_1, π_2 denote two possibility distributions, the natural conjunctive idempotent rule between these two distributions is their minimum $\min(\pi_1, \pi_2)$. It is also the most cautious, as it is the most conservative of all t-norms [14]. Now, let us consider two random sets $(m_1, \mathcal{F}_1), (m_2, \mathcal{F}_2)$ and their respective contour functions π_{m_1}, π_{m_2}. The first result is the following:

Proposition 1 (s-covering). *The following inequality holds for any $v \in \mathcal{V}$:*

$$\max_{(m, \mathcal{F}) \in \mathcal{M}_{12}} \pi_m(v) \leq \min(\pi_{m_1}(v), \pi_{m_2}(v)). \tag{3}$$

It is known [15] that the same inequality holds for sets of probabilities. Considering the idempotent rule of possibility theory and this property, it makes sense to ask for inequality (3) to become an equality. We now study two different ways to formulate this requirement on conjunctively merged belief structures.

3.1   Strong Contour Function Principle (SCFP)

Let us first start with the strongest requirement.

Definition 3 (SCFP). *An element $(m, \mathcal{F})$ in $\mathcal{M}_{12}$ is said to satisfy the strong contour function principle if, for any $v \in \mathcal{V}$,*

$$\pi_{(m, \mathcal{F})}(v) = \min(\pi_{m_1}(v), \pi_{m_2}(v)), \tag{4}$$

We require the selected merged belief structure to have a contour function equal to the minimum of the two original contour functions. We retrieve the minimum rule of possibility theory if both $(m_1, \mathcal{F}_1), (m_2, \mathcal{F}_2)$ are consonant. Also, a merged belief structure satisfying the SCFP has maximal expected cardinality among elements of $\mathcal{M}_{12}$, and hence is coherent with previous approaches [7] studying expected cardinality as a criteria to cautiously merge belief structures.

First assume that the two belief structures $(m_1, \mathcal{F}_1), (m_2, \mathcal{F}_2)$ can satisfy the SCFP. In this case, satisfying SCFP also implies satisfying the following property:

Proposition 2 (s-coherence). *Let $(m_1, \mathcal{F}_1)$ be s-included in the wide sense in $(m_2, \mathcal{F}_2)$, that is $(m_1, \mathcal{F}_1) \sqsubseteq_s (m_2, \mathcal{F}_2)$. Then, the unique element in $\mathcal{M}_{12}$ satisfying Equation (4) is $(m_1, \mathcal{F}_1)$.*

This proposition indicates that satisfying the SCFP ensures idempotence (simply take $(m_1, \mathcal{F}_1) = (m_2, \mathcal{F}_2)$ in the above property) and is coherent with the notion of specialisation, that is the notion of inclusion that we consider as the most sensible to extend cautious possibilistic merging to belief structures. To see that Proposition 2 is not valid for *pl-* and *q*-inclusions, consider Example 1, the fact that one of them is either strictly *pl-* or *q*-included in the other and that none of these two belief functions belong to $\mathcal{M}_{12}$.

The Case of Consonant Belief Structures. Let π_1, π_2 be two possibility distributions and $(m_1, \mathcal{F}_1), (m_2, \mathcal{F}_2)$ be the corresponding consonant belief structures. In this case, it is known [13] that the minimum of π_1, π_2 can be retrieved by a bba inside $\mathcal{M}_{12}$ where dependency between focal set precision is assumed.

This merged belief structure has a maximal cardinality, meaning that it is s-least committed inside $\mathcal{M}_{12}$ (i.e., it is inside $\mathcal{M}_{12}^{\sqsubseteq s}$). It is also the single least q-committed [12] element among $\mathcal{M}_{12}$ (i.e., $\mathcal{M}_{12}^{\sqsubseteq q}$ is reduced to a single element). The next example, which completes Example 1, indicates that $\mathcal{M}_{12}^{\sqsubseteq s}$ is not necessarily reduced to a single element.

Example 2. Consider the two following possibility distributions π_1, π_2, expressed as belief structures $(m_{\pi_1}, \mathcal{F}_{\pi_1}), (m_{\pi_2}, \mathcal{F}_{\pi_2})$

$\mathcal{F}_{\pi_1}$	m_{π_1}	$\mathcal{F}_{\pi_2}$	m_{π_2}
$\{v_0, v_1, v_2\}$	0.5	$\{v_2, v_3, v_4\}$	0.5
$\{v_0, v_1, v_2, v_3, v_4\}$	0.5	$\{v_1, v_2, v_3, v_4, v_5\}$	0.5

The two belief structures $(m_1, \mathcal{F}_1), (m_2, \mathcal{F}_2)$ of Example 1 can be obtained by conjunctively merging these two consonant belief structures. But none of the belief structures in Example 1 is s-included in the other, while $(m_2, \mathcal{F}_2) \sqsubseteq_q (m_1, \mathcal{F}_1)$. The merged belief structure satisfying the SCFP is usually unnormalized, except when $(m_1, \mathcal{F}_1), (m_2, \mathcal{F}_2)$ are logically consistent.

Satisfying the SCFP in General. Necessary and sufficient conditions under which the merged belief structure satisfies the SCFP are given by Dubois and

Prade [10], however they are difficult to check. Constraining two belief structures to be either logically consistent, non-conflicting or conflicting, as well as requiring the conjunctively merged belief structure to be normalised are conditions that are easier to check.

Let us first explore the most constraining case, that is, when belief structures to be merged are logically consistent. The next example indicates that the SCFP cannot always be satisfied, even in such a restricted case.

Example 3. Consider the merging of belief structures $(m_1, \mathcal{F}_1), (m_2, \mathcal{F}_2)$ of Example 1. They are logically consistent, have the same contour function, and if there were a belief structure $(m_{12}, \mathcal{F}_{12})$ in $\mathcal{M}_{12}$ that can satisfy SCFP, this belief structure should have the same contour function again:

$$pl_{12}(v_1) = 0.5 \quad pl_{12}(v_2) = 1 \quad pl_{12}(v_3) = 0.5$$

with expected cardinality 2. As expected cardinality is a linear function, as well as constraints (2), a conjunctively merged belief structure with maximal expected cardinality can be found by linear programming. Running such a program, the maximal value obtained with belief structures of Example 2 is 1.5, and is given, for example, by $m_{12}(\{v_2, v_3\}) = 0.5$, $m_{12}(\{v_2\}) = 0.5$.

This is less than what the expected cardinality a conjunctively merged belief structure satisfying the SCFP should reach. It indicates that, even when belief structures to be merged are logically consistent, the SCFP cannot be always satisfied. The next examples show that the SCFP cannot always be satisfied either for non-logically consistent input random sets, whether conflicting or not.

Example 4. Let us consider the space $V = \{v_1, v_2, v_3\}$ and the two non-conflicting random sets $(m_1, \mathcal{F}_1), (m_2, \mathcal{F}_2)$ summarized in the table below.

Set	$\{v_1\}$	$\{v_2\}$	$\{v_3\}$	$\{v_1, v_2\}$	$\{v_1, v_3\}$	$\{v_2, v_3\}$	X
m_1	0.3	0	0	0	0	0.4	0.3
m_2	0.2	0.1	0.1	0.2	0.2	0.1	0.1

The minimum $\pi_{\min} = \min(\pi_1, \pi_2)$ of their contour functions is such that

$$\pi_{\min}(v_1) = 0.6 \quad \pi_{\min}(v_2) = 0.5 \quad \pi_{\min}(v_3) = 0.5.$$

The expected cardinality of $\pi_{\min}$ is 1.6. However, the linear program computing the maximal expected cardinality reached by an element of $\mathcal{M}_{12}$ yields value 1.5 as solution. Therefore, no bba in $\mathcal{M}_{12}$ satisfies the SCFP for this example.

Example 5. Let us then consider the two conflicting random sets $(m_1, \mathcal{F}_1), (m_2, \mathcal{F}_2)$ summarised below.

$(m_1, \mathcal{F}_1)$		$(m_2, \mathcal{F}_2)$	
Focal sets	Mass	Focal sets	Mass
$E_{11} = \{v_2\}$	0.5	$E_{21} = \{v_1 v_2, v_3\}$	0.5
$E_{12} = \{v_3\}$	0.5	$E_{22} = \{v_1\}$	0.5

The minimum $\pi_{\min} = \min(\pi_1, \pi_2)$ of their contour functions is such that

$$\pi_{\min}(v_1) = 0 \quad \pi_{\min}(v_2) = 0.5 \quad \pi_{\min}(v_3) = 0.5.$$

Its expected cardinality is 1, while the maximum expected cardinality reachable by an element of $\mathcal{M}_{12}$ is 0.5 (by assigning $m_2(\{v_1v_2, v_3\})$ to v_2 or v_3).

All these examples indicate that the SCFP is too strong a requirement in general. An alternative is to search for *subsets* of conjunctively merged belief structures globally satisfying the contour function principle, admitting that the result of the fusion is a set of belief functions rather than a single one. This is in the spirit of proposals made by other authors dealing with situations where mass functions of belief structures are not precisely known [1,5]. Such an alternative is explored in the next section.

Table 1 summarizes cases when the SCFP can be satisfied. It shows that, except for specific situations, the SCFP is difficult to satisfy in general.

Table 1. Satisfiability of SCFP given $(m_1, \mathcal{F}_1), (m_2, \mathcal{F}_2)$. $\checkmark$: always satisfiable. $\times$: not always satisfiable. N.A.: Not Applicable

Situation \ Constraints	Consonant	$m_{1\cap 2}(\emptyset) = 0$	unconst.
Logically consistent	$\checkmark$	$\times$	$\times$
Non-conflicting	$\checkmark$	$\times$	$\times$
Conflicting	$\checkmark$	N.A.	$\times$

3.2 Weak Contour Function Principle (WCFP)

In this section, while we still require that the result of the conjunctive merging of $(m_1, \mathcal{F}_1)$ and $(m_2, \mathcal{F}_2)$ coincides on singletons with the minimum of the contour functions π_1, π_2, we no longer require the result of the merging to be a single belief structure.

Definition 4 (WCFP). *Consider two belief structures $(m_1, \mathcal{F}_1), (m_2, \mathcal{F}_2)$ and $\mathcal{M}_{12}$ the set of conjunctively merged belief structures. Then, a subset $\mathcal{M} \subseteq \mathcal{M}_{12}$ is said to satisfy the weak contour function principle if, for any $v \in \mathcal{V}$,*

$$\max_{(m, \mathcal{F}) \in \mathcal{M}} \pi_m(v) = \min(\pi_{m_1}(v), \pi_{m_2}(v)), \tag{5}$$

Any random set satisfying the SCFP after merging also satisfies the WCFP. However, we try to find subsets of $\mathcal{M}_{12}$ that always satisfy the WCFP.

Subsets of normalised merged belief functions. A first subset of interest is the one of normalised conjunctively merged belief structures (that is, all $(\mathcal{F}, m) \in \mathcal{M}_{12}$ such that $m(\emptyset) = 0$). As the lower measure induced by this subset is equal to the lower probability of the intersection of the sets of probabilities induced by the belief structures to be merged, we denote it by $\mathcal{M}_{\mathcal{P}_1 \cap \mathcal{P}_2}$. The next example shows that there are cases where the WCFP cannot be satisfied when we consider the subset $\mathcal{M}_{\mathcal{P}_1 \cap \mathcal{P}_2}$.

Example 6. Consider the two belief structures $(m_1, \mathcal{F}_1), (m_2, \mathcal{F}_2)$ on $\mathcal{V} = \{v_1, v_2, v_3\}$ such that

$$m_1(\{v_1\}) = 0.5, \quad m_1(\{v_1, v_2, v_3\}) = 0.5,$$
$$m_2(\{v_1, v_2\}) = 0.5, \quad m_2(\{v_3\}) = 0.5.$$

The minimum of contour functions $\pi_{\min} = \min(\pi_1, \pi_2)$ is given by $\pi_{\min}(v_i) = 0.5$ for $i = 1, 2, 3$. The only merged bba m_{12} to be in $\mathcal{M}_{\mathcal{P}_1 \cap \mathcal{P}_2}$ is

$$m_{12}(\{v_1\}) = 0.5; \quad m_{12}(\{v_3\}) = 0.5,$$

for which $\pi_{12}(v_2) = 0 < 0.5$.

The above example is not surprising: it recalls that requiring coherence (i.e., $m(\emptyset) = 0$) while conjunctively merging uncertain information, can be too strong a requirement. In particular, the element v_2 is considered as impossible by $\mathcal{M}_{\mathcal{P}_1 \cap \mathcal{P}_2}$, while both sources consider v_2 as possible.

Subsets of s-Least Committed Merged Belief Structures. Another possible solution is to consider a subset of minimally committed belief structures. That is, given two belief structures $(m_1, \mathcal{F}_1), (m_2, \mathcal{F}_2)$, we consider the subsets $\mathcal{M}_{12}^{\sqsubseteq^{\mathbf{x}}}$, of least $\mathbf{x}$-committed belief structures with $\mathbf{x} \in \{s, pl, q, \pi\}$. The following proposition is the main result of the paper. It implies that the subset of $\mathbf{x}$-least committed belief structures in $\mathcal{M}_{12}$ always satisfies the WCFP, with $\mathbf{x} \in \{s, pl, q, \pi\}$.

Proposition 3. *The subset $\mathcal{M}_{12}^{\sqsubseteq^s} \subseteq \mathcal{M}_{12}$ satisfies the WCFP, in the sense that* $\max_{(m, \mathcal{F}) \in \mathcal{M}_{12}^{\sqsubseteq^s}} \pi_m(v) = \min(\pi_1(v), \pi_2(v))$.

To prove this proposition, we show that $\forall v \in \mathcal{V}$ there is at least one merged belief structure $(m_v, \mathcal{F}_v)$ in $\mathcal{M}_{12}$ such that $\pi_{(m_v, \mathcal{F}_v)}(v) = \min(\pi_1(v), \pi_2(v))$. If $\pi_1(v) = \sum_{v \in E} m_1(E) \leq \sum_{v \in E'} m_2(E') = \pi_2(v)$, it is always possible to transfer part of the masses $m_2(E'), v \in E' \in \mathcal{F}_2$ to subsets $E \cap E'$ containing v, so as to ensure $\sum_{v \in E \cap E'} m_v(E, E') = \sum_{E \in \mathcal{F}_{v,1}} m_1(E)$, while respecting Eq. (2).

Given the implications between notions of inclusions of belief structures, it is clear that, any element in $\mathcal{M}_{12}^{\sqsubseteq^{\mathbf{x}}}$ with $x = \{pl, q, \pi\}$ is also in $\mathcal{M}_{12}^{\sqsubseteq^s}$. However, there may be some elements of $\mathcal{M}_{12}^{\sqsubseteq^s}$ that are not in $\mathcal{M}_{12}^{\sqsubseteq^{\mathbf{x}}}$. What we have to do is to show that, if one element is suppressed, then this element is of no use to satisfy Proposition 3. Let us consider two such elements m, m' that are in $\mathcal{M}_{12}^{\sqsubseteq^s}$ (i.e., they are s-incomparable) but are such that $m' \sqsubseteq_x m$, hence m' is not in $\mathcal{M}_{12}^{\sqsubseteq^{\mathbf{x}}}$. For any $\mathbf{x} \in \{pl, q, \pi\}$, we do have (see section 2.3)

$$m' \sqsubseteq_x m \Rightarrow \pi_{m'} \leq \pi_m,$$

$\pi_{m'} \leq \pi_m$ ensures that m' is useless when taking the maximum of all least s-committed contour functions that satisfy the WCFP.

Corollary 1. *The subsets $\mathcal{M}_{12}^{\sqsubseteq^{\mathbf{x}}}$ for $\mathbf{x} = \{pl, q, \pi\}$ satisfy the WCFP.*

Corollary 2. *If any of the subsets* $\mathcal{M}_{12}^{\sqsubseteq \mathbf{x}}$ *with* $\mathbf{x} = \{s, pl, q, \pi\}$ *is reduced to a singleton* $(m_{\mathbf{x}}, \mathcal{F}_{\mathbf{x}})$, *then this bba satisfies the SCFP.*

This is, for instance, the case with $\mathcal{M}_{12}^{\sqsubseteq \mathbf{q}}$ when both $(m_1, \mathcal{F}_1), (m_2, \mathcal{F}_2)$ are consonant. As for SCFP, Table 2 summarises for which subsets of merged belief structures the WCFP is always satisfiable. So, one can always satisfy the WCSP by selecting at most $|\mathcal{V}|$ merged belief structures, each of them obeying (2) for one element $v \in \mathcal{V}$. One can also restrict to bba's in the set $\mathcal{M}_{12}^{\sqsubseteq \pi}$. However, tools to compute them must be devised.

Table 2. Satisfiability of WCFP given $(m_1, \mathcal{F}_1), (m_2, \mathcal{F}_2)$. $\sqrt{}$: always satisfiable. $\times$: not satisfiable in general. N.A.: Not Applicable

Subset / Situation	$\mathcal{M}_{\mathcal{P}_1 \cap \mathcal{P}_2}$	$\mathcal{M}_{12}^{\sqsubseteq \mathbf{s}}$	$\mathcal{M}_{12}^{\sqsubseteq \mathbf{pl}}$	$\mathcal{M}_{12}^{\sqsubseteq \mathbf{q}}$	$\mathcal{M}_{12}^{\sqsubseteq \pi}$
Logically consistent	$\sqrt{}$	$\sqrt{}$	$\sqrt{}$	$\sqrt{}$	$\sqrt{}$
Non-conflicting	$\times$	$\sqrt{}$	$\sqrt{}$	$\sqrt{}$	$\sqrt{}$
Conflicting	N.A.	$\sqrt{}$	$\sqrt{}$	$\sqrt{}$	$\sqrt{}$

4   Conclusion

From a practical standpoint, our results are rather negative, as they show the lack of a universal cautious merging rule for belief functions, extending the idempotent possibilistic rule. However, they do have theoretical interest, as they tend to confirm the need to use of sets of belief structures rather than of a single one as the result of cautious merging, particularly when dependencies are ill-known. This indicates that single belief functions are perhaps not always sufficient to tackle some problems. Section 3.2 also indicates that restricting ourselves to normalised merged belief functions is too constraining if we want to comply with our principles. This is in agreement with the Transferable Belief Model [18] and the open world assumption, where unnormalized belief structures are authorised.

A pending question is whether the WCFP applies to the set of conjunctively merged belief structures with maximal expected cardinality? If the answer is affirmative, a subset of conjunctively merged belief functions satisfying the WCSP could be computed by means of linear programming techniques.

References

1. Augustin, Th.: Generalized basic probability assignments. I. J. of General Systems 34(4), 451–463 (2005)
2. Cattaneo, M.: Combining belief functions issued from dependent sources. In: Proc. Third International Symposium on Imprecise Probabilities and Their Application (ISIPTA 2003), Lugano, Switzerland, pp. 133–147 (2003)
3. Chateauneuf, A.: Combination of compatible belief functions and relation of specificity. In: Advances in the Dempster-Shafer theory of evidence, pp. 97–114. John Wiley & Sons, Inc., New York (1994)

4. Dempster, A.P.: Upper and lower probabilities induced by a multivalued mapping. Annals of Mathematical Statistics 38, 325–339 (1967)
5. Denoeux, T.: Reasoning with imprecise belief structures. I. J. of Approximate Reasoning 20, 79–111 (1999)
6. Denoeux, T.: Conjunctive and disjunctive combination of belief functions induced by non-distinct bodies of evidence. Artificial Intelligence 172, 234–264 (2008)
7. Destercke, S., Dubois, D., Chojnacki, E.: Cautious conjunctive merging of belief functions. In: Proc. Eur. Conf. on Symbolic and Quantitative Approaches to Reasoning with Uncertainty, pp. 332–343 (2007)
8. Dubois, D., Prade, H.: A set-theoretic view on belief functions: logical operations and approximations by fuzzy sets. I. J. of General Systems 12, 193–226 (1986)
9. Dubois, D., Prade, H.: Possibility Theory: An Approach to Computerized Processing of Uncertainty. Plenum Press, New York (1988)
10. Dubois, D., Prade, H.: Fuzzy sets, probability and measurement. European Journal of Operational Research 40, 135–154 (1989)
11. Dubois, D., Prade, H.: Consonant approximations of belief functions. I. J. of Approximate reasoning 4, 419–449 (1990)
12. Dubois, D., Prade, H., Smets, P.: A definition of subjective possibility. I. J. of Approximate Reasoning 48, 352–364 (2008)
13. Dubois, D., Yager, R.R.: Fuzzy set connectives as combination of belief structures. Information Sciences 66, 245–275 (1992)
14. Klement, E.P., Mesiar, R., Pap, E.: Triangular Norms. Kluwer Academic Publisher, Dordrecht (2000)
15. Moral, S., Sagrado, J.: Aggregation of imprecise probabilities. In: Bouchon-Meunier, B. (ed.) Aggregation and Fusion of Imperfect Information, pp. 162–188. Physica-Verlag, Heidelberg (1997)
16. Shafer, G.: A Mathematical Theory of Evidence. Princeton University Press, New Jersey (1976)
17. Smets, P.: Analyzing the combination of conflicting belief functions. Information Fusion 8, 387–412 (2006)
18. Smets, P., Kennes, R.: The transferable belief model. Artificial Intelligence 66, 191–234 (1994)
19. Walley, P.: Statistical reasoning with imprecise Probabilities. Chapman and Hall, New York (1991)

Capacity Refinements and Their Application to Qualitative Decision Evaluation

Didier Dubois and Hélène Fargier

IRIT, 118 route de Narbonne
31062 Toulouse Cedex, France
{dubois,fargier}@irit.fr

Abstract. This paper deals with the lack of discrimination of aggregation operations in decision-evaluation methods, typically in multi-factorial evaluation, and in decision under uncertainty. When the importance of groups of criteria is modeled by a monotonic but non-additive set-function, strict monotonicity of evaluations with respect to Pareto-dominance is no longer ensured. One way out of this problem is to refine this set-function. Two refinement techniques are presented, extending known refinements of possibility and necessity measures, respectively based on so-called discrimax and leximax orderings. Capacities then become representable by means of belief functions, plausibility functions or both. In particular it yields a natural technique for refining a Sugeno integral by means of a Choquet integral.

Keywords: Capacities, belief functions, Sugeno integral, decision evaluation.

1 Introduction

In decision evaluation problems [1], partial ratings are often aggregated into a global evaluation. A natural condition to be respected is then Pareto dominance, namely, if an alternative is rated as good as another alternative on all criteria, the first one having a better rating on one criterion, then the first alternative should be preferred. When numerical ratings make sense, the arithmetic mean yields a ranking of alternatives that obeys this condition. This is no longer true, generally, with ordinal or qualitative ratings for which only maximum and minimum aggregation operations sound natural: two alternatives, one of which Pareto-dominates the other, may have equal ranks under qualitative aggregations. This is the so-called drowning effect [6]. A similar defect appears when dependencies between criteria are accounted for, which leads to attaching importance weights to groups of criteria with a monotonic set-function κ called capacity or fuzzy measure [11]: Choquet and Sugeno integrals do not satisfy Pareto dominance.

Some proposals exist for overcoming the drowning effect due to the use of minimum and maximum operations in qualitative aggregation. For instance, when comparing vectors by their component of minimal value, one may delete components sharing the same value beforehand (this is the "discrimin" order [6]).

C. Sossai and G. Chemello (Eds.): ECSQARU 2009, LNAI 5590, pp. 311–322, 2009.

A further refinement is "leximin"[3]. Similar techniques exist for the maximum operator. More recently, Fargier and Sabbadin [9] generalized these lexicographic refinements to the weighted maximum $\max_{i=1:n} \min(\pi_i, \alpha_i)$, where α_i is a rating according to feature i and π_i is the weight of this feature, as well as the weighted minimum. The idea is to nest leximin inside the leximax or conversely. This study was pursued in [4] so as to refine a Sugeno integral of a tuple of ratings $S_\kappa(\alpha) = \max_{\lambda \in L} \min(\lambda, \kappa(A_\lambda))$, where A_λ is the set of features i for which $\alpha_i \geq \lambda$. This led to a refined ranking that can be represented by a Choquet integral preserving the set-function κ on which the integral relies. Hence, this approach handles the drowning effect due to maximum and minimum but the one due to the set-function κ itself still remains. Hence the idea of refining a capacity so as to fully overcome the drowning effect.

In this paper, we propose a method for refining the ranking of events induced by a capacity, so as to make it more discriminant. Existing results for possibility and necessity measures are recalled, for which a unique maximal refinement exists, that can be encoded by a big-stepped probability function. Then, based on the notion of qualitative Möbius transform [10] that generalize possibility distributions, capacities are refined by means of orderings on sets compatible with belief functions, plausibility functions or both. These results are then applied to Sugeno integrals that are refined by means of Choquet integrals, thus generalizing existing results for weighted minimum and maximum. Proofs are omitted due to the lack of space.

2   Pareto Dominance for Capacity Functions

A capacity κ over a finite set S is a set function $A \in 2^S \mapsto \kappa(A) \in L$, where L is a bounded totally ordered set with maximal and minimal elements $\top$ and $\bot$. Capacities are required to be monotonic (if $A \subseteq B$ then $\kappa(A) \leq \kappa(B)$) and such that $\kappa(\emptyset) = \bot$ and $\kappa(S) = \top$. The conjugate of κ is a capacity $\kappa^c(A) = \nu(\kappa(A^c))$, where $A^c = S \setminus A$ and ν is the order-reversing map on L (typically, for $L = [0, 1]$, $\kappa^c(A) = 1 - \kappa(A^c)$). Such set functions are widely used in multicriteria decision problems, where they measure the importance of coalitions of criteria, and also in decision making under uncertainty, where S is rather a set of possible states of the world : in this case, A is a event and $\kappa(A)$ measures its likelihood. Actually, the same formal expressions can be found in the two domains, that is, expected utility and weighted mean, their qualitative counterparts [8] and their extensions. In the following, we do not make any assumption about the domain of interpretation and simply consider S as a set of features, our goal being to define how a capacity κ over S can be refined so as to satisfy Pareto dominance as much as possible.

When considering sets as Boolean vectors, the monotonicity property of capacity functions can be viewed as a weak Pareto dominance property. However strict Pareto dominance cannot be written as a strict monotonicity property (if $A \subset B$ then $\kappa(A) < \kappa(B)$) because S may contain null subsets. A set B is said to be *null* for κ if an only if

$$\kappa(A \cup B) = \kappa(A), \forall A \subseteq S \tag{1}$$

In uncertainty representations, null sets may represent impossible events. Notice that A is null is not equivalent to $\kappa(A) = \perp$: for necessity measures, $N(B) = \perp$ does obviously not mean that B is null. However, A null always implies that $\kappa(A) = \perp$ (and thus $\kappa(A) > \perp$ implies that A is not null) but the converse (namely, $\kappa(A) = \perp \implies A$ null) does not hold in general.

We shall then write the strict Pareto dominance requirement as follows :

SPAR: $\forall A, B$ disjoint, if B is not null, then $\kappa(A \cup B) > \kappa(A)$.

Requiring this condition is quite demanding. Possibility measures Π for instance often violate **SPAR**. So do necessity measures and many belief and plausibility functions [16]. Axiom **SPAR** is clearly violated as soon as there are two *disjoints* sets A, B such that $\kappa(B) > \perp$ and $\kappa(A \cup B) = \kappa(A)$. Restricting **SPAR** to sets B such that $\kappa(B) > \perp$, yields a weak version of **SPAR**:

S (Strictness): if $\kappa(B) > \perp$ then $\kappa(A \cup B) > \kappa(A), \forall A, B$ disjoint,

It is called *converse null-additivity* by Z. Wang[18]. The converse of axiom **S** is :

NA: $\forall A, B$ disjoint: $\kappa(A \cup B) > \kappa(A) \implies \kappa(B) > \perp$.

This axiom also writes $\kappa(B) = \perp \implies \kappa(A \cup B) = \kappa(A)$ and is called *null-additivity* by Z. Wang [17] (see also Pap[14]). When both **NA** and **S** holds, the corresponding property is denoted **NAS**.

NAS: $\forall A, B$ disjoint: $\kappa(A \cup B) > \kappa(A) \iff \kappa(B) > \perp$.

Proposition 1. NAS *and* **SPAR** *are equivalent when there is no null set.*

The approach of the paper to overcome the drowning effect pertaining to some capacity k is to define a new capacity κ' that is in agreement with κ and refines it. In the following, $\succeq_\kappa$ will denote the order among sets, induced by κ:

$$A \succeq_\kappa B \iff \kappa(A) \geq \kappa(B) \tag{2}$$

Conversely, given any monotonic weak order $\succeq$ over 2^S, we say that $\succeq$ is *represented* by a capacity κ whenever $A \succeq B \iff \kappa(A) \geq \kappa(B)$ ($\succeq \equiv \succeq_\kappa$).

A reflexive relation $\succeq'$ is said to *refine* a reflexive relation $\succeq$ iff $\forall A, B : A \succ B \implies A \succ' B$ where $\succ$ denotes the strict part of $\succeq$ ($A \succ B \iff A \succeq B$ but not $B \succeq A$). By extension, $\succeq'$ *refines* κ whenever $\succeq'$ refines $\succeq_\kappa$ and *a capacity* κ' *refines a capacity* κ iff $\forall A, B : \kappa(A) > \kappa(B) \implies \kappa'(A) > \kappa'(B)$.

A straightforward way to construct an ordering on sets obeying **SPAR** is to refine the ranking induced by κ by means of the inclusion relation:

$$B \succ_\kappa^\subseteq A \iff \kappa(B) > \kappa(A) \text{ or } A \subsetneq B. \tag{3}$$

$\succ_\kappa^\subseteq$ is obviously a strict partial ordering. Each equivalence class $\mathcal{C}_\kappa$ of equally important sets in the sense of κ is partially ordered by the inclusion relation. The strict partial ordering $\succ_\kappa^\subseteq$ restricted to each $\mathcal{C}_\kappa$ can be embedded into a weak order, for instance considering cardinality (where $\sim$ stands for indifference):

$$B \succ_\kappa^{card} A \iff \kappa(B) > \kappa(A) \text{ or } (\kappa(B) = \kappa(A) \text{ and } |A| < |B|) \tag{4}$$
$$B \sim_\kappa^{card} A \iff \kappa(B) = \kappa(A) \text{ and } |A| = |B|.$$

$\succeq_\kappa^{card}$ is a weak order, namely, complete and transitive. It can thus be represented by a capacity κ^{card}, that refines κ (notice that this κ^{card} uses a larger scale $\Lambda \supseteq L$). κ^{card} is a refinement of κ and satisfies **SPAR**.

We could try to refine κ even more by means of a comparative probability relation, i.e. enforcing the following *preadditivity* axiom:

PRAD: $\forall A, D, C$ s. t. $C \cap (D \cup A) = \emptyset$: $\kappa(D) \succeq \kappa(A) \Leftrightarrow \kappa(D \cup C) \succeq \kappa(A \cup C)$.

This axiom can be viewed as an instance of the Sure Thing Principle [15], restricted to capacity functions. **PRAD** implies **SPAR**, since, under the former, $\kappa(A) > \bot$ is equivalent to A not null. However, a refinement of κ that satisfies **PRAD** may not exist, since it may be that $\kappa(B) > \kappa(A)$ and $\kappa(B \cup C) < \kappa(A \cup C)$ with $C \cap (A \cup B) = \emptyset$. This is frequent for instance with Shafer's belief and plausibility functions:

$$Bel(A) = \sum_{E, E \subseteq A} m(E), \forall A \subseteq S \qquad (5)$$

$$Pl(A) = 1 - Bel(A^c) = \sum_{E, E \cap A \neq \emptyset} m(E), \forall A \subseteq S \qquad (6)$$

where m is a mass function, i.e. a probability distribution over $2^S \setminus \emptyset$. It is nevertheless always possible to look for the following weak version of **PRAD**:

BELPL: $\forall A, B, C$ disjoints: $\kappa(C \cup B) > \kappa(C) \iff \kappa(C \cup B \cup A) > \kappa(A \cup C)$

This axiom puts together two axioms already encountered in the literature:

BEL: $\forall A, B, C$ disjoint, if $\kappa(A \cup B) > \kappa(A)$ then $\kappa(A \cup B \cup C) > \kappa(A \cup C)$.

PL: $\forall A, B, C$ disjoint, if $\kappa(A \cup B \cup C) > \kappa(A \cup C)$ then $\kappa(A \cup B) > \kappa(A)$.

Belief functions satisfy **BEL**. Conversely, likelihood relations that are monotonic under inclusion and obey property **BEL** can always (but not exclusively) be represented by belief functions Bel [19]. The converse axiom, **PL**, is satisfied by Shafer's plausibility functions. The same properties make sense for a strict partial order instead of a capacity.

Notice that **BEL** and **PL** are just slight reinforcements of the property $\forall A, B, C$ disjoint sets: $\kappa(A \cup B) > \kappa(A) \implies \kappa(A \cup B \cup C) \geq \kappa(A \cup C)$, which trivially holds for any capacity. So it is clear that relations among sets induced by capacities differ from relations induced by belief functions *only* by breaking some indifferences, so that it is always possible to refine a capacity by another one satisfying **BEL**, **PL** or both. Belief and plausibility functions representing a relation satisfying **BELPL** are not necessarily the conjugate of each other. Indeed, while a relation satisfying axiom **PRAD** is self-conjugate ($A \succ B \iff B^c \succ A^c$), not all relations satisfying **BELPL** are.

Setting $A = \emptyset$ in the definition of axiom **BEL** (resp. **PL**) we recover axiom **S** (resp. **NA**). More generally, it can be shown that:

Proposition 2. PRAD $\implies$ **BELPL** $\implies$ **NAS** $\implies$ **SPAR**

It is clear that the capacity κ^{card} satisfies both **BEL** and **PL** (and thus **NAS** and **SPAR**). It can thus be represented by a plausibility function, and also by a belief function. Unfortunately, this refinement of κ is not always natural. Consider for instance a possibility measure Π on $S = \{1, 2, 3, 4, 5\}$, with $\pi(1) = 1$, $\pi(2) = 1$, $\pi(3) = \pi(4) = \pi(5) = 0.1$. The behavior of $\succ_{\Pi}^{card}$ may be counterintuitive, e.g. when $A = \{1, 2\}$ and $B = \{1, 3, 4, 5\}$ since $B \succ_{\Pi}^{card} A$. The drawback of the cardinality-based refinement is that it does not pay any attention to the relative importance of subsets of A and B. Alternative refinements of κ that satisfy **BELPL** should be considered. Looking back to the literature on possibility and necessity measures, one can identify dedicated refinement principles that are much more attractive than $\succeq_{\Pi}^{card}$, as detailed in the next Section. The extension of these principles to general capacities, that is the core the present work, is detailed in Section 4. For the sake of simplicity, we will assume in the following that there is no null set.

3   Refining Possibility and Necessity Measures

Possibility and necessity measures are generally defined from a possibility distribution, i.e. a mapping π from S to L (generally, $L = [0, 1]$, $\bot = 0$ and $\top = 1$):

$$\Pi(A) = \max_{x \in A} \pi(x); \quad N(A) = \nu(\Pi(A^c)) \tag{7}$$

Possibility measures (resp. necessity measures) satisfy axiom **PL** (resp. **BEL**), but they clearly fail to satisfy **SPAR** : it may happen that $\Pi(A \cup B) = \Pi(B)$ even when $\Pi(A) > \bot$. More generally $\Pi(A) = \Pi(B)$ as soon as $\Pi(A \cap B) > \max(\Pi(A \cap B^c), \Pi(A^c \cap B))$ and $N(A) = N(B)$ as soon as $\Pi(A^c \cap B^c) > \max(\Pi(A \cap B^c), \Pi(A^c \cap B))$.

A known technique for refining the max-based possibilistic ranking is to compare the sets on the basic of their disjoint parts. This "discrimax" comparison (denoted $\succeq_{d\Pi}$) is defined as follows [2][12][7]. $\forall A, B \in 2^S$:

$$A \succeq_{d\Pi} B \iff \Pi(A \cap B^c) > \Pi(A^c \cap B) \text{ or } A = B \tag{8}$$

This relation is reflexive, transitive, and refines both rankings induced by possibility and necessity measures. It satisfies axiom **PRAD**, and thus **SPAR**, but it cannot be represented by a capacity, since it is not complete. Completeness can be recovered by using a further refinement, the "leximax" refinement, originally defined for the comparison of vectors. Practically, the leximax and leximin procedures consist in ranking both tuples in increasing order and then lexicographically comparing them in a suitable way[3]. Let $\alpha, \beta \in L^n$, and $\alpha_{(j)}$ be the jth least component of α:

$$\alpha \succ_{lmin} \beta \iff \exists i, \forall j < i, \alpha_{(j)} = \beta_{(j)} \text{ and } \alpha_{(i)} > \beta_{(i)} \tag{9}$$

$$\alpha \succ_{lmax} \beta \iff \exists i, \forall j > i, \alpha_{(j)} = \beta_{(j)} \text{ and } \alpha_{(i)} > \beta_{(i)} \tag{10}$$

$$\alpha \sim_{lmax} \beta \iff \alpha \sim_{lmin} \beta \iff \forall j, \alpha_{(j)} = \beta_{(j)}$$

The leximax refinement $\succeq_{l\Pi}$ of possibility measures comes down to representing each event A by a tuple $\boldsymbol{A} = (\pi_1^A, \ldots, \pi_n^A)$, with $\pi_i^A = \pi(s_i)$ if $s_i \in A$ and $\perp$ otherwise, and comparing tuples $\boldsymbol{A}$ and $\boldsymbol{B}$ using leximax [7]:

$$A \succeq_{l\Pi} B \iff \boldsymbol{A} \succeq_{lmax} \boldsymbol{B}. \tag{11}$$

The leximax comparison is very discriminant. Indeed, $A \sim_{l\Pi} B$ iff there exists a one-to-one mapping $\sigma : A \to B$ such that $\forall s \in A, \pi(s) = \pi(\sigma(s))$. Since it is complete and transitive, $\succeq_{l\Pi}$ can be represented by a capacity function, which is actually a probability measure P_l, based on a big-stepped probability distribution p, namely such that $\forall s \in S, p(s) > \sum_{s' \text{ s.t. } p(s')<p(s)} p(s')$, where $p(s) = \chi(\pi(s))$. Function χ is a *super-increasing* mapping from $L = \{\lambda_m = \top > \cdots > \lambda_0 = \perp\}$ to the unit interval, defined by $\chi(\lambda_i) > K \cdot \chi(\lambda_{i-1}), \forall i = 2, \ldots, m$, for some integer $K > 1$. Here, we set $\chi(\lambda_i) = \frac{K^i-1}{v}$, where $K = Card(S)$ and v is a normalisation factor allowing to recover probabilities whose sum equals 1. Notice that this probability-based representation is not unique, i.e. there may be several measures P_l such that $A \succeq_{l\Pi} B \iff P_l(A) \geq P_l(B)$. Since the refining capacity P_l is a probability, axioms **SPAR**, **BELPL** and even **PRAD** are satisfied. Moreover, it should be noticed that P refines both Π and its conjugate N.

4   Refining Capacities by Belief and Plausibility Functions

We investigate in the sequel of the paper whether these discri- and lexi-based refinements of possibility measures can be extended to any kind of capacity measure. First of all, notice that we cannot refine κ by means of the usual definition of the possibilistic likelihood based on comparing disjoint parts of A and B, as $A \succ_{dcap} B \iff \kappa(A \setminus B) > \kappa(B \setminus A)$ because κ may be strongly non-additive, i.e, it may be that $\kappa(A) > \kappa(B)$ and $\kappa(A \setminus B) < \kappa(B \setminus A)$ hold for κ. There is nevertheless a way to extend possibility refinements to capacities.

4.1   Refinements Based on the Inner Qualitative Moebius Transform

The key idea is that any capacity can be understood as the max aggregation of a vector of ordinal information. Indeed, Grabisch [10] has shown that any capacity κ can be characterized by another set-function $\kappa_\#$ defined by $\kappa_\#(\emptyset) = \perp$ and:

$$\kappa_\#(E) = \kappa(E) \text{ if } \kappa(E) > \max\{\kappa(B), B \subset E\} \text{ and } \kappa_\#(E) = \perp \text{ otherwise.} \tag{12}$$

It is clear that $\kappa_\#$ contains the minimal information to reconstruct κ as:

$$\kappa(A) = \max_{E \subseteq A} \kappa_\#(E). \tag{13}$$

Hence function $\kappa_\#$ plays the role of a "qualitative" mass function obtained via a kind of Moebius transform [10]. The subsets E that receive a positive support play the same role as the focal elements in Dempster-Shafer's theory: they are the primitive items of knowledge. The set-function $\kappa_\#$ can also be viewed as a possibilistic mass assignment, a possibility distribution over the power set 2^S, and (13) appears as the qualitative counterpart of the definition of a belief function or an inner measure. This expression is also a generalization of the definition of the degree of possibility of a set in terms of a possibility distribution on S. Indeed, the function $\Pi_\#(E) = \bot$ as soon as E is not a singleton, and $\Pi_\#(\{i\}) = \pi_i, \forall i \in S$.

Possibility measures can be refined by a discrimax procedure. Likewise, we shall define the discrimax refinement of a capacity κ, based on comparing values $\kappa_\#(E)$ for subsets E of A that are not subsets of B, and conversely [5]:

$$A \succeq^\kappa_{dcap} B \iff \max_{E, E \subseteq A, E \nsubseteq B} \kappa_\#(E) > \max_{E, E \subseteq B, E \nsubseteq A} \kappa_\#(E) \text{ or } A = B. \quad (14)$$

In this definition, all subsets common to A and B play the same role in the expressions of $\kappa(A)$ and $\kappa(B)$ and are canceled. It is easy to check that

Proposition 3. *Relation* $\succeq^\kappa_{dcap}$ *refines* $\succeq^\kappa$ *and satisfies* **BEL**.

The lexicographic refinement $\succeq^\kappa_{lcap}$ of $\succeq^\kappa_{dcap}$ is a ranking defined likewise:

$$A \succeq^\kappa_{lcap} B \iff \mathbf{A}_\# \succeq_{lmax} \mathbf{B}_\# \quad (15)$$

where $\mathbf{A}_\#$ (resp. $\mathbf{B}_\#$) is the tuple with size 2^S, containing all values $\kappa_\#(E)$, $\forall E \subseteq A$ (resp. $\forall E \subseteq B$), and $\bot$ if $E \nsubseteq A$ (resp. $E \nsubseteq B$).

Proposition 4. *Relation* $\succeq^\kappa_{lcap}$ *refines* $\succeq^\kappa_{dcap}$.

It is clear that if κ is a possibility measure, then $\succeq^\kappa_{lcap}$ boils down to the leximax possibility ranking encountered in the previous section. Clearly, $\succeq^\kappa_{lcap}$ is a weak order. Like in the possibilistic case, it is possible to represent it by a capacity κ_{lcap} on a refined ordinal scale Λ. Namely, we shall use a super-increasing transformation as in the previous section, thus defining a big-stepped mass function $m_\# : 2^S \mapsto [0, 1]$:

$$m_\#(E) = \chi(\kappa_\#(E)) \quad (16)$$

where χ is the super-increasing mapping $\chi(\lambda_i) = \frac{K^i - 1}{v}$ with, now, $K = Card(2^S)$; v is a normalisation factor so that the sum of masses is 1.

Proposition 5. *The belief function* κ_{lcap} *based on* $m_\#$ *represents* $\succeq^\kappa_{lcap}$.

Corollary 1. $\succeq^\kappa_{lcap}$ *and* κ_{lcap} *satisfy* **BEL**.

As already pointed out, $\succeq^\Pi_{lcap}$ and $\succeq^\Pi_{dcap}$ are self-conjugate: they refine the conjugate necessity measure as well. However, in general neither $\succeq^\kappa_{dcap}$ nor $\succeq^\kappa_{lcap}$ are self-conjugate, and of course generally not of the **PL** type as the existence of $E^* \subseteq A \cup B \cup C$ while $E^* \nsubseteq A \cup C$ does not ensure that $E^* \subseteq A \cup B$ and $E^* \nsubseteq A$ (for instance if $E^* = A \cup B \cup D$ with $D \subseteq C$ not empty). Worse, $\succeq^\kappa_{dcap}$ is ineffective on necessity measures and belief functions.

Proposition 6. *Relations* $\succ^N_{dcap}$ *and* $\succ^N_{lcap}$ *are equivalent to* $\succ_N$.

Proposition 7. *If* κ *is a belief function, then* $\succ^\kappa_{dcap}$ *is equivalent to* $\succ_\kappa$.

However Bel_{lcap} may strictly refine Bel (if it is not a necessity measure), since the former has more focal elements than the latter.

4.2 Refinements Based on the Outer Qualitative Moebius Transforms

In order to directly refine a necessity measure, another qualitative representation of a capacity κ can be used, namely a new set function, denoted $\kappa^{\#}$, the knowledge of which is again enough to reconstruct the capacity:

$$\kappa^{\#}(A) = \kappa(A) \text{ if } \kappa(A) < \min\{\kappa(F), A \subset F\} \text{ and } \top \text{ otherwise.} \quad (17)$$

and $\kappa^{\#}(S) = \top$. The original capacity is then retrieved as [5]:

$$\kappa(A) = \min_{A \subseteq F} \kappa^{\#}(F), \quad (18)$$

which reminds of outer measures. Function $\kappa^{\#}$ can be called *outer qualitative mass function* of κ, as $\kappa(A)$ is recovered from $\kappa^{\#}$ via weights assigned to supersets of set A, while $\kappa_{\#}$ stands for an *inner qualitative mass function*. So we can consider refining the κ ranking as follows:

$$A \succeq_{\kappa}^{dcap} B \iff \min_{E: A \subseteq E, B \not\subseteq E} \kappa^{\#}(E) > \min_{E: B \subseteq E, A \not\subseteq E} \kappa^{\#}(E) \text{ or } A = B; \quad (19)$$

$$A \succeq_{\kappa}^{lcap} B \iff A^{\#} \succeq_{lmin} B^{\#} \quad (20)$$

where $A^{\#}$ (resp. $B^{\#}$) is the tuple containing all values $\kappa^{\#}(E), \forall A \subseteq E$ (resp. $\forall B \subseteq E$), and $\top$ if $A \not\subseteq E$ (resp. $B \not\subseteq E$). These relations are generally not of the **BEL** type but are of the **PL** type.

Proposition 8. *Relations $\succeq_{\kappa}^{dcap}$ and $\succeq_{\kappa}^{lcap}$ satisfy **PL** and refine $\succeq_{\kappa}$.*

$\succeq_{\kappa}^{lcap}$ is a weak order that satisfies **PL**, so, it can be represented by a plausibility function. In order to build this plausibility function, we cannot use equation (18) directly. However, note that $Pl(A) = 1 - Bel(A^c)$ where $Bel(A^c) = \sum_{A \subseteq E} m(E^c)$, to be compared with the conjugate of κ, for which $\kappa^c(A^c) = \max_{A \subseteq E} \nu(\kappa^{\#}(E))$. The mass function $m^{\#}$ induced by $\kappa^{\#}$ can be defined considering a big-stepped mass function $m^{\#} : 2^S \mapsto [0, 1]$

$$m^{\#}(E) = \chi(\nu(\kappa^{\#}(E^c))), \quad (21)$$

where χ is the super-increasing mapping $\chi(\lambda_i) = \frac{K^{i} - 1}{v}$, v being the normalisation factor and K being equal to $Card(2^S)$.

Proposition 9. *The plausibility function $\kappa^{lcap}(A) = \sum_{A \cap E \neq \emptyset} m^{\#}(E)$ represents $\succeq_{\kappa}^{lcap}$.*

The question is whether there is or not a relationship between κ_{lcap} and κ^{lcap}. It is easy to see that orders $\succeq_{\kappa}^{dcap}$ and $\succeq_{dcap}^{\kappa}$ differ ($\succeq_{dcap}^{\Pi}$ strictly refines Π, while $\succeq_{\kappa}^{dcap}$ does not) and can strongly diverge. The reason is that the sets of values $\kappa^{\#}(E)$ deciding if $A \succ_{dcap}^{\kappa} B$ and $\kappa_{\#}(E)$ deciding if $A \succ_{\kappa}^{dcap} B$ are totally independent (the former are attached to supersets of A and B, the latter to their subsets). So the capacities functions κ^{lcap} (that refines $\succeq_\kappa^{dcap}$) and κ_{lcap} (that refines $\succeq_{dcap}^\kappa$) may disagree.

Altogether, the two refinements based on inner and outer qualitative Moebius transforms can be applied to a capacity and its conjugate, which makes two possibly distinct refinements of $\succeq_\kappa$: $\succeq_\kappa^{dcap}$, $\succeq_{dcap}^\kappa$ (and similarly for $lcap$ refinements). The same can be done for κ^c. Taking conjugates of these four refinements into account, the landscape includes eight relations. Not all these eight relations are distinct. Indeed, the inner qualitative mass function $\kappa_\#^c$ of κ^c is related to the outer qualitative mass function $\kappa^\#$:

Proposition 10. $\kappa^\#(E) = \nu(\kappa_\#^c(E^c))$.

since $\kappa(A) < \min\{\kappa(F), A \subset F\}$ also writes $\kappa^c(A^c) > \max\{\kappa^c(F^c), F^c \subset A^c\}$. For instance in the case of a necessity measure $N^\#(E) \neq \top$ only if $E = S \setminus \{i\}$ for some $s \in S$, and then $N^\#(S \setminus \{i\}) = \nu(\pi_i)$. As a consequence, it holds that $A \succ_\kappa^{dcap} B \iff \min_{E:A \subseteq E, B \not\subseteq E} \nu(\kappa_\#^c(E^c)) > \min_{E:B \subseteq E, A \not\subseteq E} \nu(\kappa_\#^c(E^c))$. But $B \subseteq E, A \not\subseteq E$ also writes $E^c \subseteq B^c, E^c \not\subseteq A^c$, so, $A \succ_\kappa^{dcap} B \iff \max_{E:E \subseteq A^c, E \not\subseteq B^c} \kappa_\#^c(E) < \max_{E:E \subseteq B^c E \not\subseteq A^c} \kappa_\#^c(E) \iff B^c \succ_{dcap}^{\kappa^c} A^c$. If κ is a necessity measure, then we get $A \succ_N^{dcap} B \iff B^c \succ_{dcap}^\Pi A^c$, which is equivalent to $A \succ_{dcap}^\Pi B$. Also, $\Pi(A) > \Pi(B)$ does imply $\Pi(B^c) \geq \Pi(A^c)$, which allows for such a conjoint refinement : $\succ_{dcap}^\kappa$ and $\succ_{\kappa^c}^{dcap}$ coincide when κ (resp. κ^c) is a possibility (resp. a necessity) measure. However in the general case, we may have $\kappa(A) > \kappa(B)$ and $\kappa^c(B) > \kappa^c(A)$ so that refinements of κ can be at odds with refinements of its conjugate. Basic results are summarized in Table 1.

Table 1. Comparison of refinements

Type	Relation	is the same as	Case $\kappa = \Pi$	refined by	Case $\kappa = \Pi$
BEL	$A \succeq_{dcap}^\kappa B$	$B^c \succeq_{\kappa^c}^{dcap} A^c$	$\succeq_{dcap}^\Pi = \succeq_{d\Pi}$	$\kappa_{lcap}(A) = 1 - (\kappa^c)^{lcap}(A^c)$	$\Pi_{lcap} = P_l$
BEL	$A \succeq_{dcap}^\kappa B$	$B^c \succeq_\kappa^{dcap} A^c$	$\succeq_{dcap}^N = \succeq_N$	$(\kappa^c)_{lcap}(A) = 1 - \kappa^{lcap}(A^c)$	$N_{lcap} = N$
PL	$B^c \succeq_{dcap}^\kappa A^c$	$A \succeq_{\kappa^c}^{dcap} B$	$\succeq_{\kappa^c}^{dcap} = \succeq_{d\Pi}$	$(\kappa^c)^{lcap}(A) = 1 - \kappa_{lcap}(A^c)$	$N^{lcap} = P_l$
PL	$B^c \succeq_{dcap}^{\kappa^c} A^c$	$A \succeq_\kappa^{dcap} B$	$\succeq_\Pi^{dcap} = \succeq_\Pi$	$\kappa^{lcap}(A) = 1 - (\kappa^c)_{lcap}(A^c)$	$\Pi^{lcap} = \Pi$

The results also hold for $lcap$. Indeed, due to Proposition 10, it is clear that $m^\#(E) = \chi(\kappa_\#^c(E)) = m_\#^c(E)$, where $m_\#^c$ is the mass function obtained from the inner qualitative Moebius transform of the conjugate of κ. Hence, $\kappa^{lcap}(A) = \sum_{A \cap E \neq \emptyset} m_\#^c(E)$. So, the following identity holds:

Proposition 11. $\kappa_{lcap}(A) = 1 - \kappa_c^{lcap}(A^c)$

This result is sufficient to complete Table 1 for $lcap$ refinements. In the case when κ is a possibility function, Π_{lcap}, N^{lcap} and their conjugates coincide with a big-stepped probability function P_l, but $N_{lcap} = N$ and $\Pi^{lcap} = \Pi$.

5   Refining Sugeno Integral

Let us now consider the use of a refined capacity in the context of Sugeno integral. S is a set of n features or criteria (denoted by integers i), and the aim is to order a set of vectors representing objects or items to be rated according to these features. For rating the merit of objects, the totally ordered value scale L is used and supposed to be common to all features. The set of objects will be identified with the set L^n of n-tuples α of values of L. Using a Sugeno integral (see [11]), the global evaluation of the merits of an object is based on the comparison of ratings of the object w.r.t. different features. The importance of groups of features is evaluated on the same scale and modeled by a capacity κ. Sugeno integral is often defined as follows:

$$S_\kappa(\alpha) = \max_{\lambda \in L} \min(\lambda, \kappa(A_\lambda)) \qquad (22)$$

where $A_\lambda = \{i : 1 \leq i \leq n, \alpha_i \geq \lambda\}$ is the set of features having best ratings for some object, down to utility threshold λ, and $\kappa(A)$ is the degree of importance of feature set A.

An equivalent expression is $S_\kappa(\alpha) = \max_{A \subseteq S} \min(\kappa(A), \min_{i \in A} \alpha_i)$ [13]. In this disjunctive form expression the set-function κ can be replaced without loss of information by the inner qualitative Moebius transform $\kappa_\#$ defined earlier.

$$S_\kappa(\alpha) = \max_{A \in \mathcal{P}_\#(S)} \min(\kappa_\#(A), \alpha_A) \qquad (23)$$

where $\alpha_A = \min_{i \in A} \alpha_i$ and $\mathcal{P}_\#(S) = \{A, \kappa_\#(A) > 0\}$. The above expression of Sugeno integral has the standard maxmin form, viewing $\kappa_\#$ as a possibility distribution over 2^S, since $\max_{A \subseteq S} \kappa_\#(A) = \top$. Applying the increasing transformation χ that changes a maxmin form into a sum of products [9] yields :

$$E_\#^{lsug}(\alpha) = \sum_{A \in 2^S} \chi(\alpha_A) \cdot \chi^*(\kappa_\#(A)) = \sum_{A \in 2^S} \chi(\alpha_A) \cdot m_\#(A), \qquad (24)$$

where $\chi(\lambda_m) = 1, \chi(\lambda_0) = 0, \chi(\lambda_j) = \frac{K}{K^{2^m - j}}, j = 1, m - 1$, and we set $K = 2^{|S|}$ [4]. Function χ^* is the normalization of χ in such a way that $\sum_{A \in 2^S} \chi^*(\kappa_\#(A)) = 1$. Ranking tuples by $E_\#^{lsug}(\alpha)$ comes down to a *Leximax($\geq_{lmin}$)* comparison of $(2^n \times 2)$ matrices with rows of the form $(\kappa_\#(A), \alpha_A)$. Now, since $\chi(\alpha_A) = \chi(\min_{i \in A} \alpha_i) = \min_{i \in A} \chi(\alpha_i)$, then $E_\#^{lsug}(\alpha) = \sum_{A \subseteq S} m_\#(A) \cdot \min_{s \in A} \chi(\alpha_i)$ is a Choquet integral w.r.t. the belief function κ_{lcap} refining κ. $E_\#^{lsug}(\alpha)$ refines the original Sugeno integral and is the expression of a Choquet integral in terms of the Moebius transform $m_\#$ of the belief function κ_{lcap}:

Proposition 12. $E_\#^{lsug}(\alpha) = \sum_{j=1}^m \kappa_{lcap}(A_{\lambda_j}) \cdot (\chi(\lambda_j) - \chi(\lambda_{j-1}))$.

This shows that any Sugeno integral can be refined by a Choquet integral *w.r.t a belief function*. Contrary to the refinements based on the first expression of Sugeno integral [4], the capacity κ is generally not preserved under the present transformation. The resulting Choquet integral is generally more discriminant than the original criterion. Two particular cases are interesting to consider:

- If κ is a possibility measure Π, then Sugeno integral is the prioritized maximum, $m_\#$ is a regular big-stepped probability function and Choquet integral reduces to a regular weighted average. We retrieve the maximal refinement $WA^+_{\chi(\pi)}$ of the prioritized maximum proposed in [9].
- If κ is a necessity measure N, then Sugeno integral is the prioritized minimum, but N_{lcap} does not induce any further refinement. In this case, the resulting Choquet integral is one with respect to a necessity measure. Only the "max-min" framing of Sugeno integral has been turned into a "sum-product" framing: the transformation preserves the nature of the original capacity. This refinement of the prioritized minimum is not a weighted average, contrary to the one obtained in [9].

In order to recover the refinement of the weighted minimum by a weighted average, one must use the conjunctive form $S_\kappa(\boldsymbol{\alpha}) = \min_{A \subseteq S} \max(\kappa(A^c), \max_{i \in A} \alpha_i)$ of Sugeno integral [13]. Clearly, in this expression, κ can be replaced by its outer qualitative Moebius transform $\kappa^\#$. Then compute

$$\nu(S_\kappa(\boldsymbol{\alpha})) = \max_{A \subseteq S} \min(\nu(\kappa^\#(A^c)), \min_{i \in A} \nu(\alpha_i)) = \max_{A \subseteq S} \min(\nu(\kappa^\#(A^c)), \nu(\alpha_A)).$$

Applying the increasing transformation χ that changes a maxmin form into a sum of products, encoding its maximax(leximin) refinement, it yields:

$$E^\#_{lsug}(\nu(\boldsymbol{\alpha})) = \sum_{A \in 2^S} \chi(\nu(\alpha_A)) \cdot \chi^*(\nu(\kappa^\#(A^c))) = \sum_{A \in 2^S} m^\#(A) \cdot \min_{i \in A} \chi(\nu(\alpha_i)) \quad (25)$$

Turning the expression upside down, and using the fact that, if f is a function that takes values in the unit interval $Ch_\kappa(f) = 1 - Ch_{\kappa^c}(1 - f)$, it yields

$$Ch(\boldsymbol{\alpha}) = 1 - E^\#_{lsug}(\nu(\boldsymbol{\alpha})) = \sum_{A \in 2^S} m^\#(A) \cdot \max_{i \in A}(1 - \chi(\nu(\alpha_i))). \quad (26)$$

Note that $\phi(\alpha_i) = 1 - \chi(\nu(\alpha_i))$ is a numerical utility function that is increasing with α_i, and $Ch(\boldsymbol{\alpha})$ is a Choquet integral with respect to the *plausibility* function with mass function $m^\#$.

6   Conclusion

This study lays bare two possible lines of refinements of a capacity κ and its conjugate, using the outer and inner Moebius transforms. There is no unique capacity refining both a prescribed capacity and its conjugate, except for special cases like possibility measures. So one may get up to four refinements, two obeying axiom **BEL**, and their conjugates obeying axiom **PL**. The next step to this study is to iterate this refinement process so as to get refined capacities that obey axiom **BELPL**.

References

1. Bouyssou, D., Marchant, T., Pirlot, M., Perny, P., Tsoukiàs, A., Vincke, P.: Evaluation and Decision Models: A critical perspective. Kluwer Academic Publishers, Dordrecht (2000)
2. Cohen, M., Jaffray, J.Y.: Rational behavior under complete ignorance. Econometrica 48(5), 1281–1299 (1980)
3. Deschamps, R., Gevers, L.: Leximin and utilitarian rules: A joint characterization. Journal of Economic Theory 17, 143–163 (1978)
4. Dubois, D., Fargier, H.: Lexicographic refinements of Sugeno integrals. In: Mellouli, K. (ed.) ECSQARU 2007. LNCS (LNAI), vol. 4724, pp. 611–622. Springer, Heidelberg (2007)
5. Dubois, D., Fargier, H.: Making Sugeno integrals more discriminant. Int. J. Approximate Reasoning (to appear, 2009)
6. Dubois, D., Fargier, H., Prade, H.: Refinements of the maximin approach to decision-making in a fuzzy environment. Fuzzy Sets Systems 81(1), 103–122 (1996)
7. Dubois, D., Fargier, H., Prade, H.: Possibilistic likelihood relations. In: Proceedings of IPMU 1998, pp. 1196–1203. EDK, Paris (1998)
8. Dubois, D., Prade, H., Roubens, M., Sabbadin, R., Marichal, J.-L.: The use of the discrete Sugeno integral in decision-making: a survey. Int. J. Uncertainty, Fuzziness and Knowledge-based Systems 9(5), 539–561 (2001)
9. Fargier, H., Sabbadin, R.: Qualitative decision under uncertainty: Back to expected utility. Artificial Intelligence 164, 245–280 (2005)
10. Grabisch, M.: The Moebius transform on symmetric ordered structures and its application to capacities on finite sets. Discrete Mathematics 287, 17–34 (2004)
11. Grabisch, M., Murofushi, T., Sugeno, M.: Fuzzy Measures and Integrals - Theory and Applications. Physica Verlag, Heidelberg (2000)
12. Halpern, J.Y.: Defining relative likelihood in partially-ordered preferential structures. Journal of A.I. Research 7, 1–24 (1997)
13. Marichal, J.-L.: On Sugeno integrals as an aggregation function. Fuzzy Sets and Systems 114(3), 347–365 (2000)
14. Pap, E.: Null-Additive Set-Functions. Kluwer Academic, Dordrecht (1995)
15. Savage, L.J.: The Foundations of Statistics. Wiley, New York (1954)
16. Shafer, G.: A Mathematical Theory of Evidence. Princeton University Press, Princeton (1976)
17. Wang, Z.: The autocontinuity of set function and the fuzzy integral. Journal of Mathematical Analysis and Applications 99, 195–218 (1984)
18. Wang, Z.: Asymptotic structural characteristics of fuzzy measures and their applications. Fuzzy Sets and Systems 16, 277–290 (1985)
19. Wong, S.K.M., Bollmann, P., Yao, Y.Y., Burger, H.C.: Axiomatization of qualitative belief structure. IEEE transactions on SMC 21(34), 726–734 (1991)

Belief Functions and Cluster Ensembles

Marie-Hélène Masson[1] and Thierry Denoeux[2]

[1] Université de Picardie Jules Verne, IUT de l'Oise
[2] Université de Technologie de Compiègne,
Laboratoire Heudiasyc, BP 20529,
60205 Compiègne, France

Abstract. In this paper, belief functions, defined on the lattice of partitions of a set of objects, are investigated as a suitable framework for combining multiple clusterings. We first show how to represent clustering results as masses of evidence allocated to partitions. Then a consensus belief function is obtained using a suitable combination rule. Tools for synthesizing the results are also proposed. The approach is illustrated using two data sets.

1 Introduction

Ensemble clustering methods aim at combining multiple clustering solutions into a single one, the *consensus*, to produce a more accurate clustering of the data. Several studies have been published on this subject for many years (see, for example, the special issue of the Journal of Classification devoted to the "Comparison and Consensus of Classifications" published in 1986 [4]). The recent interest of the machine learning and artificial intelligence communities for ensemble techniques in clustering can be explained by the success of such ensemble techniques in a supervised context. As recalled in [12,13], various ways of generating cluster ensembles have been proposed. We may use different clustering algorithms or the same algorithm while varying a characteristic of the method (starting values, number of clusters, hyperparameter) [9]. We may also resample the data set [8]. This approach is called *bagged* clustering. Another well-known application of cluster ensembles is called distributed clustering, which refers to the fact that clusterings are performed using different (overlapping or disjoint) subsets of features [21,22,1]. A member of the ensemble is called a *clusterer*. Once several partitions are available, they have to be aggregated into a single one, providing a better description of the data than individual partitions. A variety of strategies have been proposed to achieve this goal: voting schemes [7], hypergraph partitioning [21], pairwise or co-occurrence approach [9,11]. This last approach, which will be shown to have some connections with what is proposed in this paper, is perhaps the simplest approach. The collection of partitions can be be mapped into a squared co-association matrix where each cell (i, j) represents the fraction of times the pair of objects (x_i, x_j) has been assigned to the same cluster. This matrix is then considered as a similarity matrix which can be in turn clustered. A hierarchical clustering algorithm is the most common algorithm used for this purpose.

C. Sossai and G. Chemello (Eds.): ECSQARU 2009, LNAI 5590, pp. 323–334, 2009.
© Springer-Verlag Berlin Heidelberg 2009

In this paper, we propose a new approach based on belief functions theory. This theory has been already successfully applied to unsupervised learning problems [15,6,16,17]. In those methods, belief functions are defined on the set of possible clusters, the focal elements being subsets of this frame of discernment. The idea here is radically different. It consists in defining and manipulating belief functions on the set of all partitions of the data set. Each clustering algorithm is considered as a source providing an opinion about the unknown partition of the objects. The information of the different sources are converted into masses of evidence allocated to partitions. These masses can be combined and synthesized using some generalizations of classical tools of the belief functions theory.

The rest of the paper is organized as follows. Section 2 gives necessary backgrounds about partitions of a finite set and belief functions defined on the lattice of partitions of a finite set. Section 3 describes how to generate the belief functions, how to combine them and how to synthesize the results. The methodology is illustrated using a simple example. The results of some experiments are shown in Section 4. Finally, Section 5 concludes this paper.

2   Background

2.1   Partitions of a Finite Set

Let E denote a finite set of n objects. A partition p is a set of non empty subsets $E_1,...,E_k$ of E such that:

1) the union of all elements of p, called clusters, is equal to E;
2) the elements of p are pairwise disjoint.

Every partition can be associated to an equivalence relation (i.e., a reflexive, symmetric, and transitive binary relation), on E, denoted by R_p, and characterized, $\forall x, y \in E$, by:

$$R_p(x, y) = \begin{cases} 1 & \text{if } x \text{ and } y \text{ belong to the same cluster in } p \\ 0 & \text{otherwise.} \end{cases}$$

Example. Let $E = \{1, 2, 3, 4, 5\}$. A partition p of E, composed of two clusters, the clusters of which are $\{1, 2, 3\}$ and $\{4, 5\}$ will be denoted as $p = (123/45)$. The associated equivalence relation is:

$$R_p = \begin{pmatrix} 1 & 1 & 1 & 0 & 0 \\ 1 & 1 & 1 & 0 & 0 \\ 1 & 1 & 1 & 0 & 0 \\ 0 & 0 & 0 & 1 & 1 \\ 0 & 0 & 0 & 1 & 1 \end{pmatrix}.$$

The set of all partitions of E, denoted by $\mathcal{P}(E)$, can be partially ordered using the following ordering relation: a partition p is said to be *finer* than a partition p' on the same set E (or, equivalently p' is *coarser* than p) if the clusters of p

can be obtained by splitting those of p' (or equivalently, if each cluster of p' is the union of some clusters of p). In, this case, we write:

$$p \preceq p'.$$

Note that this ordering can be alternatively defined using the equivalence relations associated to p and p':

$$p \preceq p' \Leftrightarrow R_p(x,y) \leq R_{p'}(x,y) \quad \forall(x,y) \in E^2.$$

The *finest* partition in the order $(\mathcal{P}(E), \preceq)$, denoted $p_0 = (1/2/.../n)$, is the partition where each object is a cluster. The *coarsest* partition is $p_E = (123..n)$, where all objects are put in the same cluster. Each partition precedes in this order every partition derived from it by aggregating two of its clusters. Similarly, each partition succeeds (*covers*) all partitions derived by subdividing one of its clusters in two clusters. The *atoms* of $(\mathcal{P}(E), \preceq)$ are the partitions preceded by p_0. There are $n(n-1)/2$ such partitions, each one having $(n-1)$ clusters with one and only one cluster composed of two objects. Atoms are associated to matrices R_p with only one off-diagonal entry equal to 1.

2.2   Lattice of the Partitions of a Finite Set

The set $\mathcal{P}(E)$ endowed with the $\preceq$-order has a lattice structure [18]. *Meet* ($\wedge$) and *join* ($\vee$) operations can be defined as follows. The partition $p \wedge p'$, called the *infimum* of p and p', is defined as the coarsest partition among all partitions finer than p and p'. The clusters of $p \wedge p'$ are obtained by considering pairwise intersections between clusters of p and p'. The equivalence relation $R_{p \wedge p'}$ is simply obtained by taking the minimum of R_p and $R_{p'}$. The partition $p \vee p'$, called the *supremum* of p and p', is similarly defined as the finest partition among the ones that are coarser than p and p'. The equivalence relation $R_{p \vee p'}$ is given by the *transitive closure* of the maximum of R_p and $R_{p'}$.

2.3   Belief Functions on the Lattice of Partitions

Belief functions [19,20] are most of the time defined on the Boolean lattice of subsets of a finite set. However, following the first investigations of Barthélemy [2], Grabisch [10] has shown that it is possible to extend these notions to the case where the underlying structure is no more the Boolean lattice of subsets, but any lattice. In particular, considering the lattice of partitions, some of the classical constructions and definitions of belief functions (mass assignment, mass combination, commonalities,...) remain valid, up to some adaptations. Let $\mathcal{L} = (\mathcal{P}(E), \preceq)$ denote a lattice of partitions endowed with the meet and join operations defined in section 2.2. A basic belief assignment (bba) is defined as a mass function m from $\mathcal{L}$ to $[0;1]$ verifying:

$$\sum_{p \in \mathcal{L}} m(p) = 1. \tag{1}$$

A bba m is said to be normal if $m(p_0) = 0$. In the rest of this paper, only normal mass functions will be considered. Each partition p that receives a mass $m(p) > 0$ is called a *focal element* of m. A bba m is said to be *categorical* is there is a unique focal element p with $m(p) = 1$. A bba m is said to be of *simple support* if there exists $p \in \mathcal{L}$ and $w \in [0; 1]$ such that $m(p) = 1 - w$ and $m(p_E) = w$, all other masses being zero. The bba m can be equivalently represented by a credibility function bel, and a commonality function q defined, respectively, by:

$$\mathrm{bel}(p) \triangleq \sum_{p' \preceq p} m(p'), \tag{2}$$

$$\mathrm{q}(p) \triangleq \sum_{p \preceq p'} m(p'), \tag{3}$$

$\forall p \in \mathcal{L}$. When the reliability of a source (e.g., a clustering algorithm) is doubtful, the mass provided by this source can be discounted using the following operation (discounting process):

$$\begin{cases} m^\alpha(p) = (1 - \alpha)m(p) \quad \forall p \neq p_E \in \mathcal{L}, \\ m^\alpha(p_E) = (1 - \alpha)m(p_E) + \alpha, \end{cases} \tag{4}$$

where $0 \leq \alpha \leq 1$ is the discount rate. This discount rate is related to the confidence held by an external agent in the reliability of the source.

Two bbas m_1 and m_2 induced by distinct items of evidence on $\mathcal{L}$ can be combined using the normalized Dempster's rule of combination. The resulting mass function $m_1 \oplus m_2$ will be defined by:

$$(m_1 \oplus m_2)(p) \triangleq \frac{1}{1 - K} \sum_{p' \wedge p'' = p} m_1(p')m_2(p'') \quad \forall p \in \mathcal{L}, p \neq p_0 \tag{5}$$

with

$$K = \sum_{p' \wedge p'' = p_0} m_1(p')m_2(p''). \tag{6}$$

Alternatively, one may use the average of m_1 and m_2 defined by:

$$(m_{av})(p) \triangleq \frac{1}{2} (m_1(p) + m_2(p)) \quad \forall p \in \mathcal{L}. \tag{7}$$

3   Ensemble Clustering

3.1   General Approach

Belief functions, as defined in the previous section, offer a general framework for combining and synthesizing the results of several clustering algorithms. We propose to use the following strategy for ensemble clustering:

1) Mass generation: Given r clusterers, build a collection of r bbas $m_1, m_2, ..., m_r$;
2) Aggregation: Combine the r bbas into a single one using an appropriate combination rule;
3) Synthesis: Provide a summary of the results.

Mass generation. This step depends on the clustering algorithm used to build the ensemble. The simplest situation is encountered when a clusterer produces a single partition p of the data set. To account for the uncertainty of the clustering process, this categorical opinion can be transformed into a simple support mass function using the discounting operation (4). We propose to relate the discounting factor of the source to a cluster validity index, measuring the quality of the partition. Various cluster validity indices can be used for this purpose (see, for instance, [23] for a review of fuzzy cluster validity). In the experiments reported in Section 4, we have used the fuzzy c-means algorithm (converting the fuzzy partition into a hard one) and a partition entropy to define the discounting factor as follows. Let μ_{jk} denote the fuzzy membership degree assigned to the jth object and the kth cluster and c denote the number of clusters (note that c may vary from a clusterer to another). The normalized partition entropy is a value $0 \leq h \leq 1$ defined by:

$$h = \frac{1}{n \log(c)} \sum_{j=1}^{n} \sum_{k=1}^{c} \mu_{jk} \log(\mu_{jk}). \tag{8}$$

This quantity is maximal (equal to 1) when the quality of the partition is poor, i.e., when all membership values are equal to $1/c$. This value can be used as a discounting factor of the clusterer. This strategy leads to the generation of a bba m, with two focal elements, defined by:

$$\begin{cases} m(p) & = 1 - h, \\ m(p_E) = h. \end{cases} \tag{9}$$

Suppose now that a clusterer expresses its opinion by a hierarchical clustering [14]. This kind of algorithm produces a sequence of nested partitions, $p_0 \preceq ... \preceq p_E$. At each intermediate stage, the method joins together the two closest clusters. In the *single linkage* approach, the distance between clusters is defined as the distance between the closest pair of objects. The result of this algorithm is commonly displayed as a dendrogram (or classification tree) such as represented in Figure 1. The aggregation levels, used for the representation of the dendrogram, are usually equal to the distances computed when merging two clusters. Note that aggregation levels may be normalized so that the first level is 0 and the last level is 1. This normalization will be assumed in the sequel. Cutting

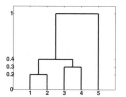

Fig. 1. Example of a small dendrogram (left) and associated nested partitions (right)

the tree at different levels of the hierarchy gives a sequence of nested partitions $\{p_0, p_1, p_2, ..., p_K\}$ (a chain in the lattice of the partitions) associated to levels $\{0, i_1, i_2, ..., i_K\}$ with $p_K = p_E$ and $i_K = 1$. The masses associated to the different partitions can be computed from the levels of the hierarchy as follows. The size of a step between two consecutive levels of the hierarchy in the dendrogram is often considered as an indication on the appropriate number of clusters and on the height at which the dendrogram should be cut. So, we propose the following mass allocation:

$$\begin{cases} m(p_E) = 0 \\ m(p_k) = i_{k+1} - i_k \quad k = 0, \ldots, K - 1. \end{cases} \quad (10)$$

For example, cutting the dendrogram of Fig. 1 at different levels produces the bba given in Table 1. It can be seen that the highest mass is allocated to the partition that seems the most natural with respect to the shape of the dendrogram.

Table 1. Bba derived from the dendrogram of Fig. 1.

k	p_k	i_k	$m(p_k)$
0	$p_0 = (1/2/3/4/5)$	0	0.2
1	$p_1 = (12/345)$	0.2	0.1
2	$p_2 = (12/34/5)$	0.3	0.1
3	$p_3 = (1234/5)$	0.4	0.6
4	$p_4 = (12345)$	1	0

Note that many other clustering methods can be described in the same framework. For instance, fuzzy equivalence relations, used for cluster analysis [3], are naturally represented by consonant belief functions on the lattice of partitions.

Combination and synthesis. Once r bbas are available, they can be aggregated into a single one using one of the combination rules recalled in Section 2.3. The interpretation of the results is a more difficult problem, since, depending on the number of clusterers in the ensemble, on their nature anf the conflict between them, and on the combination rule, a potentially high number of focal elements may be found. If the number of focal elements in the combined bba is too high to be explored, a first way to proceed is to select only the partitions associated with the highest masses or use a simplification algorithm such as described in [5]. We propose another approach which consists in building a matrix $Q = (q_{ij})$ whose elements are the commonalities associated to each atom of the lattice of partitions. This approach amounts computing, for each pair of object (i, j), a new similarity measure q_{ij} by accumulating the masses which support the association between i and j:

$$q_{ij} = \sum_p m(p) R_p(i, j). \quad (11)$$

Matrix Q can be in turn clustered using, for instance, a hierarchical clustering algorithm. If a partition is needed, the classification tree can be cut at a specified level or so as to insure a user-defined number of clusters. Note that the co-association method proposed in [9] is recovered as a special case of our approach if the consensus has been obtained by averaging the masses of the individual clusterers.

3.2   Toy Example

Let $E = \{1, 2, 3, 4, 5\}$ be a set composed of 5 objects. We assume that two clustering algorithms have produced partitions $p_1 = (123/45)$ and $p_2 = (12/345)$. As it can be seen, the partitions disagree on the third element which is clustered with $\{1, 2\}$ in p_1 and $\{4, 5\}$ in p_2. As proposed in Section 3.1, we construct two simple mass functions by discounting each clusterer i by a factor α_i. In a first situation, we consider that we have an equal confidence in the two clusterers, so we fix $\alpha_1 = \alpha_2 = 0.1$. We have:

$$m_1(p_1) = m_2(p_2) = 0.9 \quad m_1(p_E) = m_2(p_E) = 0.1,$$

with $p_E = (12345)$. Applying Dempster's rule of combination (5)-(6) leads to the following combined bba $m = m_1 \cap m_2$:

Focal elements	mass m	bel
$p_1 \wedge p_2 = (12/3/45)$	0.81	0.81
$p_1 = (123/45)$	0.09	0.90
$p_2 = (12/345)$	0.09	0.90
$p_E = (12345)$	0.01	1

Suppose now that the confidence is less in the second clusterer than in the first one. We fix $\alpha_1 = 0.1$ and $\alpha_2 = 0.2$. In that case, we obtain a bba m' characterized by:

Focal elements	mass m'	bel'
$p_1 \wedge p_2 = (12/3/45)$	0.72	0.72
$p_1 = (123/45)$	0.18	0.90
$p_2 = (12/345)$	0.08	0.80
$p_E = (12345)$	0.02	1

The commonalities of the atoms of the lattice are given for the two situations by the following matrices:

$$Q = \begin{pmatrix} 1 & 1 & 0.1 & 0.01 & 0.01 \\ 1 & 1 & 0.1 & 0.01 & 0.01 \\ 0.1 & 0.1 & 1 & 0.1 & 0.1 \\ 0.01 & 0.01 & 0.1 & 1 & 1 \\ 0.01 & 0.01 & 0.1 & 1 & 1 \end{pmatrix} \quad Q' = \begin{pmatrix} 1 & 1 & 0.2 & 0.02 & 0.02 \\ 1 & 1 & 0.2 & 0.02 & 0.02 \\ 0.2 & 0.2 & 1 & 0.1 & 0.1 \\ 0.02 & 0.02 & 0.1 & 1 & 1 \\ 0.02 & 0.02 & 0.1 & 1 & 1 \end{pmatrix}$$

Applying the single linkage algorithm to these two matrices gives the hierarchical clusterings represented in Figure 2. The dendrogram may be seen as a good synthesis of the information (consensual and conflicting) provided by the clusterers. On the left, we can see that no cut is able to recover a partition in which object 3 is associated to the other objects (except in the root of the tree). On the right, cutting the tree at a level greater than 0.8, allows us to recover the partition given by m_1, reflecting the fact that a greater confidence is allocated to this source.

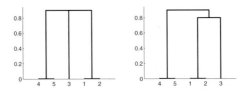

Fig. 2. Dendrograms computed from Q (left) and Q' (right) for the toy example

4   Two Examples

4.1   Distributed Clustering

In a distributed computing environment, the data set is spread into a number of different sites. In that case, each clusterer has access to a limited number of features and the distributed computing entities share only higher level information describing the structure of the data such as cluster labels. The problem is to find a clustering compatible with what could be found if the whole set of features was considered. To illustrate this point, we used a dataset named 8D5K found in [21]. This dataset is composed of five Gaussian clusters in dimension 8. Out of the 1000 points of the original data set, we retain only 200 points (40 points per cluster). We created five 2D views of the data by selecting five pairs of features. We applied the fuzzy c-means algorithm in each view (each one with $c = 5$) to obtain five hard partitions computed from the fuzzy partitions. These partitions are represented in Figure 3. The left row shows the partitions in the 2D views, and the right row shows the same partitions projected onto the first two principal components of the data. An ensemble of five mass functions was constructed using the approach proposed in Section 3.1: each clusterer, discounted according the entropy of partition (8), was represented by a mass function with two focal elements. A "consensus" clustering was obtained by applying Dempster's rule of combination, computing the matrix Q and the associated tree using single linkage, and cutting the tree to obtain five clusters. The consensus clustering is presented in Figure 3. It may be seen that a very good clustering is obtained, although some of the partitions provided by the clusterers were poor in the space described by the eight features.

4.2   Non Elliptical Clusters

This section is intended to show that the proposed approach is able to detect clusters with complex shapes. The half-ring data set is inspired from [9]. It

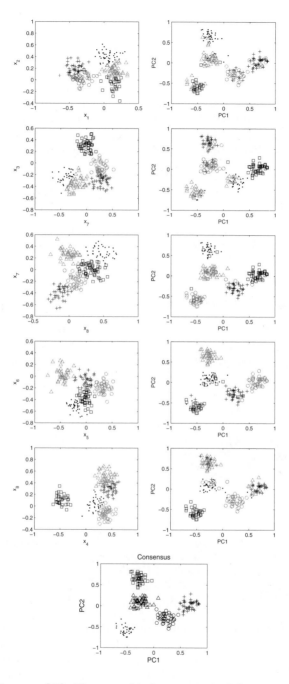

Fig. 3. 8D5K data set [21]. The ensemble is composed of five individual clustering solutions obtained from five 2D views of the data. The left row shows the partition obtained in each two-dimensional features space and the right row shows the corresponding partition in the plane spanned by the two first principal components.

consists of two clusters of 100 points each in a two-dimensional space. To build the ensemble, we use the fuzzy c-means algorithm with a varying number of clusters (3 to 7). The hard partitions are represented in Figure 4.

As in the previous example, each partition was discounted using the entropy of partition and five mass functions with two focal elements each were combined using Dempster's rule of combination. A tree was computed from the commonality matrix using the single linkage algorithm and a partition in two clusters was derived from the tree. This partition is also represented in Figure 4. We can see that the natural structure of the data is perfectly recovered.

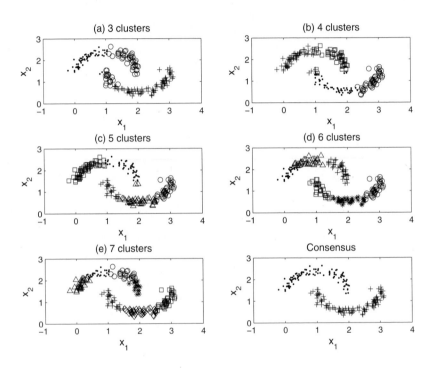

Fig. 4. Half-rings data set. Figures (a) to (e) show the individual clusterings in the 2D views; the last figure shows the consensual clustering.

5   Conclusion

We have proposed in this paper a new approach for aggregating multiple clusterings. This approach is based on the use of belief functions defined on the lattice of partitions of the set of objects to be clustered. In this framework, it possible to assign masses of evidence to partitions. We have shown that a wide variety of clusterers can be naturally represented in this framework and that combination tools can provide a "consensual" description of the data. The preliminary experiments on several data sets have shown the usefulness of the method. A drawback of the conjonctive combination, especially when the conflict between the clusterers is important, is to potentially generate a large number of focal elements. A similar problem was already encountered in [15]. Future work will investigate how to simplify the result by merging similar or unimportant focal elements using a procedure similar to the one proposed in [5].

References

1. Avogadri, R., Valentini, G.: Fuzzy ensemble clustering for DNA microarray data analysis. In: Masulli, F., Mitra, S., Pasi, G. (eds.) WILF 2007. LNCS, vol. 4578, pp. 537–543. Springer, Heidelberg (2007)
2. Barthélemy, J.-P.: Monotone functions on finite lattices: an ordinal approach to capacities, belief and necessity functions. In: Fodor, J., de Baets, B., Perny, P. (eds.) Preferences and decisions under incomplete knowledge, pp. 195–208. Physica-Verlag, Heidelberg (2000)
3. Bezdek, J.C., Keller, J., Krisnapuram, R., Pal, N.R.: Fuzzy Models and Algorithms for Pattern Recognition and Image Processing. Series: The Handbooks of Fuzzy Sets, vol. 4. Kluwer Academic Publishers, Nonwell (1999)
4. Day, W.: Foreword: Comparison and Consensus of Classifications. Journal of Classification 3, 183–185 (1986)
5. Denœux, T.: Inner and outer approximation of belief structures using a hierarchical clustering approach. Int. Journal of Uncertainty, Fuzziness and Knowledge-Based Systems 9(4), 437–460 (2001)
6. Denœux, T., Masson, M.-H.: EVCLUS: EVidential CLUStering of proximity data. IEEE Transactions on Systems, Man and Cybernetics Part B 34(1), 95–109 (2004)
7. Dimitriadou, E., Weingessel, A., Hornik, K.: Voting-Merging: an ensemble method for clustering. In: Dorffner, G., Bischof, H., Hornik, K. (eds.) ICANN 2001. LNCS, vol. 2130, p. 217. Springer, Heidelberg (2001)
8. Dudoit, S., Fridlyand, J.: Bagging to improve the accuracy of a clustering procedure. Bioinformatics 19(9), 1090–1099 (2003)
9. Fred, A., Jain, A.K.: Data clustering using evidence accumulation. In: Proc. of the 16th International Conference on Pattern Recognition, Quebec, Canada, pp. 276–280 (2002)
10. Grabisch, M.: Belief functions on lattices. International Journal of Intelligent Systems 24(1), 76–95 (2009)
11. Greene, D., Tsymbal, A., Bolshakova, N., Cunningham, P.: Ensemble clustering in medical diagnostics. In: Proc. of the 17th IEEE Symposium on Computer-Based Medical Systems, Bethesda, MD, USA, pp. 576–581 (2004)
12. Hadjitodorov, S.T., Kuncheva, L., Todorova, L.: Moderate diversity for better cluster ensemble. Information Fusion 7(3), 264–275 (2006)
13. Hornik, K., Leisch, F.: Ensemble methods for cluster analysis. In: Taudes, A. (ed.) Adaptive Information Systems and Modelling in Economics and Management Science. Interdisciplinary Studies in Economics and Management, pp. 261–268. Springer, Heidelberg (2005)
14. Jain, A.K., Dubes, R.C.: Algorithms for Clustering Data. Prentice Hall, Englewood Clis (1988)
15. Le Hégarat-Mascle, S., Bloch, I., Vidal-Madjar, D.: Application of Dempster-Shafer Evidence Theory to Unsupervised Classification in Multisource Remote Sensing. IEEE Transactions on Geoscience and Remote Sensing 35(4), 1018–1031 (1997)

16. Masson, M.-H., Denœux, T.: Clustering interval-valued data using belief functions. Pattern Recognition Letters 25(2), 163–171 (2004)
17. Masson, M.-H., Denœux, T.: ECM: An evidential version of the fuzzy *c*-means algorithm. Pattern Recognition 41, 1384–1397 (2008)
18. Montjardet, B.: The presence of lattice theory in discrete problems of mathematical social sciences. Why. Mathematical Social Sciences 46, 103–144 (2003)
19. Shafer, G.: A mathematical theory of evidence. Princeton University Press, Princeton (1976)
20. Smets, P., Kennes, R.: The Transferable Belief Model. Artificial Intelligence 66, 191–243 (1994)
21. Strehl, A., Ghosh, J.: Cluster ensemble - a knowledge reuse framework for combining multiple partitions. Journal of Machine Learning Research 3, 563–618 (2002)
22. Topchy, A., Jain, A.K., Punch, W.: Combining multiple weak clusterings. In: Proc. of the IEEE Int. Conference on Data Mining, Melbourne, Australia, pp. 331–338 (2003)
23. Wang, W., Zhang, Y.: On fuzzy cluster validity indices. Fuzzy Sets and Systems 158(19), 2095–2117 (2007)

Upper Probabilities Attainable by Distributions of Measurable Selections*

Enrique Miranda, Inés Couso, and Pedro Gil

Department of Statistics and Operations Research, University of Oviedo. C-Calvo
Sotelo, s/n, 33007 Oviedo, Spain
{mirandaenrique,couso,pedro}@uniovi.es

Abstract. A random set can be regarded as the result of the imprecise
observation of a random variable. Following this interpretation, we study
to which extent the upper and lower probabilities induced by the random
set keep all the information about the values of the probability distribution of the random variable. We link this problem to the existence of
selectors of a multi-valued mapping and with the inner approximations
of the upper probability, and prove that under fairly general conditions
(although not in all cases), the upper and lower probabilities are an adequate tool for modelling the available information. Finally, we study the
particular case of consonant random sets and we also derive a relationship between Aumann and Choquet integrals.

Keywords: Random sets, Dempster-Shafer upper and lower probabilities, measurable selections, reducible σ-fields, Aumann integral, Choquet
integral.

1 Introduction

Random sets, or multi-valued mappings, have been used by several authors in
the context of imprecise (or incomplete) information. In this paper, we follow
the interpretation given to them by Kruse and Meyer [21] and we regard them
as the imprecise observation of a measurable mapping.

Dempster [9] summarised the probabilistic information of the random set by
means of the upper and lower probabilities, which constitute a generalisation to
a context of imprecise information of the concept of probability induced by a
random variable. The upper and lower probabilities of a random set are plausibility and belief functions in the context of evidence theory [29], and capacities
of infinite order under Choquet's terminology [5]. This type of set functions have
been thoroughly studied in the literature [6,27], not only in the context of multi-
valued mappings [10,12,14,29], as a powerful alternative to probability measures
that is able to deal with uncertain, or vague, knowledge. Nevertheless, under
Kruse and Meyer's interpretation, the most precise piece of information that the
random set gives about the measurable mapping imprecisely observed is the set

---

* We acknowledge the financial support of the projects TIN2008-06796-C04-01,
MTM2007-61193 and TIN2007-67418-C03-03.

C. Sossai and G. Chemello (Eds.): ECSQARU 2009, LNAI 5590, pp. 335–346, 2009.

of the probability distributions of the measurable selections. The relationship
between this set and Dempster-Shafer upper and lower probabilities has already
been studied by some authors (see for instance [4,15,16]). In this paper we study
some additional aspects of this relationship. Let us introduce first some concepts
and notation.

Consider a probability space $(\Omega, \mathcal{A}, P)$, a measurable space $(X, \mathcal{A}')$, and a
measurable mapping $U_0 : \Omega \to X$. We will refer to U_0 as the **original random
variable**. There may be some imprecision in the observation of the values of
U_0. Following Kruse and Meyer ([21]), a possible model for this situation is
to consider a multi-valued mapping $\Gamma : \Omega \to \mathcal{P}(X)$, in the sense that for all
$\omega \in \Omega$ we are certain that $U_0(\omega)$ belongs to the set $\Gamma(\omega)$. As a consequence, we
shall assume throughout that $\Gamma(\omega)$ is non-empty for all ω. We will say that the
multi-valued mapping is **closed** (resp., compact, open, complete) when $\Gamma(\omega)$ is
a closed (resp., compact, open, complete) subset of X for all $\omega \in \Omega$.

Let us define

$$S(\Gamma) := \{U : \Omega \to X \mid U \text{ measurable}, U(\omega) \in \Gamma(\omega) \, \forall \omega\}. \tag{1}$$

This is the class of the **measurable selections** of Γ. Taking into account our
interpretation of Γ, all we know about U_0 is that it is one of the elements of
$S(\Gamma)$. Concerning the probability distribution of U_0, it will belong to

$$\mathcal{P}(\Gamma) = \{P_U \mid U \in S(\Gamma)\}, \tag{2}$$

the class of the probability distributions induced by the measurable selections
on $\mathcal{A}'$. In particular, the probability that the value of U_0 belongs to an event
$A \in \mathcal{A}'$, i.e. $P_{U_0}(A)$, is an element of $\mathcal{P}(\Gamma)(A) := \{P_U(A) \mid U \in S(\Gamma)\}$.

Hence, the interpretation of a multi-valued mapping as a model for the imprecise observation of a random variable provides us with a Bayesian sensitivity
analysis model for the probability distribution of this variable: the set of probability distributions $\mathcal{P}(\Gamma)$. There is, however, another set of probabilities that
shall also be interesting for our purposes. It is based on the notions of upper and
lower probabilities induced by multi-valued mapping.

Definition 1. *[9,27] Let $(\Omega, \mathcal{A}, P)$ be a probability space, $(X, \mathcal{A}')$ a measurable
space and $\Gamma : \Omega \to \mathcal{P}(X)$ a multi-valued mapping. Given $A \in \mathcal{A}'$, its **upper
inverse** by Γ is $\Gamma^*(A) := \{\omega \mid \Gamma(\omega) \cap A \neq \emptyset\}$, and its **lower inverse**, $\Gamma_*(A) :=
\{\omega \mid \Gamma(\omega) \subseteq A\}$.*

Following Nguyen [27], a multi-valued mapping is said to be **strongly mea-
surable** when $\Gamma^*(A) \in \mathcal{A}$ for all $A \in \mathcal{A}'$. In that case, we will refer to Γ as
random set. Taking into account the relationship $\Gamma_*(A) = (\Gamma^*(A^c))^c$, valid for
all $A \in \mathcal{A}'$, a random set satisfies $\Gamma_*(A) \in \mathcal{A}$ for all $A \in \mathcal{A}'$. We will use the
shorter notation $A^* = \Gamma^*(A)$ and $A_* = \Gamma_*(A)$ when no confusion arises. Al-
though there are other measurability conditions for multi-valued mappings (see
for instance [17]), we shall only consider in this paper the strong measurability;
this condition is necessary if we want to define the upper and lower probabilities
of the random set on $\mathcal{A}'$, as we see next.

Definition 2. *[9] Let $(\Omega, \mathcal{A}, P)$ be a probability space, $(X, \mathcal{A}')$ a measurable space and $\Gamma : \Omega \to \mathcal{P}(X)$ a non-empty random set. The **upper probability** induced by Γ on $\mathcal{A}'$ is defined as $P^*(A) = P(A^*) \; \forall A \in \mathcal{A}'$, and the **lower probability** is given by $P_*(A) = P(A_*) \; \forall A \in \mathcal{A}'$.*

The upper probability of a random set is ∞-alternating and lower continuous, and the lower probability is ∞-monotone and upper continuous [27]. They are moreover conjugate functions, meaning that $P^*(A) = 1 - P_*(A^c) \; \forall A \in \mathcal{A}'$. If the final space is finite, they are a plausibility and a belief function, respectively. We shall sometimes use the notation $P_\Gamma^* := P^*$ and $P_{*\Gamma} := P_*$, if there is ambiguity about the random set inducing the upper and lower probabilities. It is easy to see that $A_* \subseteq U^{-1}(A) \subseteq A^*$ for every $A \in \mathcal{A}'$ and every $U \in S(\Gamma)$. This implies that the class $\mathcal{P}(\Gamma)$ defined in Eq. (2) is included in

$$M(P^*) = \{Q : \mathcal{A}' \to [0, 1] \text{ probability s.t. } Q(A) \leq P^*(A) \; \forall A\}, \qquad (3)$$

which is called the **core** of P^*.

The upper probability of a random set generalises the concept of probability distribution of a random variable, and is sometimes used as a model of the probabilistic information of the random set. In this paper, we shall investigate if it is appropriate to do so when Γ has the interpretation considered by Kruse and Meyer. Specifically, we are going to study under which conditions we can use the upper probability to represent the information about the probability that our original random variable takes values in some arbitrary set $A \in \mathcal{A}'$. That is, we shall investigate under which conditions the equality $\mathcal{P}(\Gamma)(A) = [P_*(A), P^*(A)]$ holds. This is important because, as we shall show, when these two sets are not equal the use of the upper and the lower probability could carry some serious loss of information.

The study of the equality $\mathcal{P}(\Gamma)(A) = [P_*(A), P^*(A)]$ can be split into two different subproblems: on the one hand, we need to study the convexity of the set $\mathcal{P}(\Gamma)(A)$; and on the other, we need to determine whether the supremum and infimum values of this set coincide with the upper and lower probabilities of A, respectively. Because of the duality existing between P^* and P_*, it suffices to study one of the two equalities.

Although this problem has already been studied by some authors ([2,15,16]), this has always been done as a support for other mathematical considerations, and hence the sufficient conditions established for the equalities $P^*(A) = \sup \mathcal{P}(\Gamma)(A)$ and $P_*(A) = \inf \mathcal{P}(\Gamma)(A)$ assume some hypotheses on the random set that are not really necessary for the equalities to hold. We shall see nevertheless that the problem is not trivial, and we shall improve the established results. As far as we know, the most important result on this subject is the following:

Theorem 1. *[7, Prop. 3] Let $(\Omega, \mathcal{A}, P)$ be a probability space, (X, τ) a Polish space and let $\Gamma : \Omega \to \mathcal{P}(X)$ be a compact random set. Then, $P^*(A) = \sup \mathcal{P}(\Gamma)(A)$ for every A in β_X, the Borel σ-field associated to τ.*

Recall here that a *Polish* space is a separable and completely metrizable topological space. We shall also use later *Souslin* spaces, which are the images of Polish spaces by continuous mappings.

We shall study the equality $\mathcal{P}(\Gamma)(A) = [P_*(A), P^*(A)]$ in detail in our next section. First, we will investigate under which conditions $P^*(A)$ and $P_*(A)$ are, respectively, the supremum and infimum values of $\mathcal{P}(\Gamma)(A)$. For this, we shall use some results on the existence of measurable selections and the inner approximations of the upper probability. Secondly, we will study the convexity of $\mathcal{P}(\Gamma)(A)$. Finally, in Section 3 we shall show some of the consequences of our results. Due to limitations of space, all proofs have been omitted.

2 $[P_*(A), P^*(A)]$ as a Model of $P_{U_0}(A)$

2.1   Study of the Equality $P^*(A) = \sup \mathcal{P}(\Gamma)(A)$

Let us study first if the upper and lower probabilities of an event A are the most precise bounds of $P_{U_0}(A)$ that we can give, taking into account the information given by Γ. As we shall show in the following example, this is not always the case: in fact, it may happen that $[P_*(A), P^*(A)] = [0, 1]$ while $\mathcal{P}(\Gamma)(A) = \{0\}$. In such an extreme case, the set of the distributions of the measurable selections would provide precise information, while the upper and lower probabilities would give no information at all. The example we give is based on [18, Example 5]; the differences are that Himmelberg et al. consider a weaker notion of measurability and give an example of a multi-valued mapping which is measurable in that sense and for which $S(\Gamma) = \emptyset$.

Example 1. Let P be an absolutely continuous probability measure on $(\mathbb{R}, \beta_{\mathbb{R}})$, and let $P^{\mathbb{N}}$ denote the product of countably many copies of P. Consider $\Omega := \{F \subseteq \mathbb{R} \text{ countable}\}$ and $\mathcal{A} := \sigma(\{\mathcal{F}_A \mid A \in \beta_{\mathbb{R}}\})$, with $\mathcal{F}_A := \{F \in \Omega \mid F \cap A \neq \emptyset\}$. Let us define the mapping $g : \mathbb{R}^{\mathbb{N}} \to \Omega$ by $g(z) = \{x \in \mathbb{R} \mid x = z_n \text{ for some } n\}$. Then, $g^{-1}(\mathcal{F}_A) = \bigcup_n \left(\prod_{i=1}^{n-1} \mathbb{R} \times A \times \prod_{i>n} \mathbb{R} \right) \in \beta_{\mathbb{R}^{\mathbb{N}}}$, so g is a measurable mapping. Let us denote by Q the probability measure it induces on $\mathcal{A}$. Consider now the multi-valued mapping

$$\Gamma : \Omega \to \mathcal{P}(\mathbb{R})$$
$$F \hookrightarrow F \cup \{0\}$$

- Take $A \in \beta_{\mathbb{R}}$. Then, $\Gamma^*(A) = \Omega$ if $0 \in A$, and $\Gamma^*(A) = \mathcal{F}_A$ otherwise. Hence, Γ is strongly measurable.
- Given $B = \mathbb{R} \setminus \{0\}$, $\Gamma^*(B) = \mathcal{F}_B$, whence $P^*(B) = Q(\mathcal{F}_B) = P^{\mathbb{N}}(g^{-1}(\mathcal{F}_B)) = P^{\mathbb{N}}(\cup_n(\prod_{i=1}^{n-1} \mathbb{R} \times B \times \prod_{i>n} \mathbb{R})) = 1 - P^{\mathbb{N}}(\{0, 0, 0, \dots\}) = 1$, taking into account that $P^{\mathbb{N}}$ is the product of an infinite number of copies of a continuous probability measure.
- Now, if $U \in S(\Gamma)$ satisfies $P_U(B) > 0$, U is also a measurable selection of the multi-valued mapping $\Gamma_1 : \Omega \to \mathcal{P}(\mathbb{R})$ given by $\Gamma_1(F) = F$ if $F \in U^{-1}(B)$, $\Gamma_1(F) = F \cup \{0\}$ otherwise. However, reasoning as in [18, Example 5], it can

be checked that Γ_1 does not have measurable selections. As a consequence, $\mathcal{P}(\Gamma) = \{\delta_0\}$.

Hence, $\mathcal{P}(\Gamma)(B) = \{0\}$ and $[P_*(B), P^*(B)] = [0, 1]$. ◆

This example shows that the use of the upper and lower probabilities may carry some serious loss of information: for instance, if the continuous probability P considered in the example satisfies $P([0, 1]) = 1$, and we use P^*, P_* to model the expectation of the original random variable, we would obtain the interval $[0, 1]$ as the set of possible expectations; however, we know, because $\mathcal{P}(\Gamma) = \{\delta_0\}$, that the expectation of the original random variable is 0.

Hence, it is necessary to consider some additional hypotheses in the random set to guarantee that the upper probability of a set A, $P^*(A)$, is the supremum of the set $\mathcal{P}(\Gamma)(A)$ of its probabilities provided by the measurable selections. Our next result shows that the supremum of $\mathcal{P}(\Gamma)(A)$ is indeed a maximum:

Proposition 1. *Let $(\Omega, \mathcal{A}, P)$ be a probability space, $(X, \mathcal{A}')$ be a measurable space and let $\Gamma : \Omega \to \mathcal{P}(X)$ be a random set. Then, $\mathcal{P}(\Gamma)(A)$ has a maximum and a minimum value for every $A \in \mathcal{A}'$.*

Let us now define

$$\mathcal{H}_\Gamma := \{A \in \mathcal{A}' \mid P^*(A) = \max \mathcal{P}(\Gamma)(A)\}. \qquad (4)$$

We can then rephrase our goal in this section by stating that we are interested in providing conditions for the equality between $\mathcal{H}_\Gamma$ and $\mathcal{A}'$. To see that they do not coincide in general, check that $\mathcal{H}_\Gamma := \{B \in \beta_\mathbb{R} : P(B) = 0 \text{ or } 0 \in B\}$ in Example 1. It is also easy to modify the example in order to obtain a random set without measurable selections. In that case the class $\mathcal{H}_\Gamma$ would be empty.

In the following proposition, we state that a set $A \in \mathcal{A}'$ belongs to the class $\mathcal{H}_\Gamma$ given by Eq.(4) if and only if a random set that we can derive from Γ has measurable selections. The proof follows by taking into account that a measurable set A belongs to $\mathcal{H}_\Gamma$ if and only if there is a measurable selection of Γ taking values in A whenever possible (up to a set of zero probability).

Proposition 2. *Let $(\Omega, \mathcal{A}, P)$ be a probability space, $(X, \mathcal{A}')$ a measurable space and let $\Gamma : \Omega \to \mathcal{P}(X)$ be a random set.*

$$A \in \mathcal{H}_\Gamma \Leftrightarrow \exists H \in \mathcal{A}, P(H) = 0, \text{ s.t. } S(\Gamma_{A,H}) \neq \emptyset,$$

where $\Gamma_{A,H}(\omega) = \begin{cases} \Gamma(\omega) \cap A & \text{if } \omega \in A^* \setminus H \\ \Gamma(\omega) & \text{otherwise.} \end{cases}$

In the sequel, we shall denote $\Gamma_A := \Gamma_{A,\emptyset} = (\Gamma \cap A)I_{A^*} \oplus \Gamma I_{(A^*)^c}$.

Proposition 2 shall be useful later on when studying which sets belong to $\mathcal{H}_\Gamma$, because the existence of measurable selections of a random set is one of the most important problems in this framework, and there are therefore many results that may be applicable together with this proposition; see the survey on the existence of measurable selections by Wagner [30]. Together with some of the results in [30], we shall prove and use another sufficient condition for the existence of measurable selections. For this, we must introduce the notion of *reducible* σ-field.

Definition 3. *Consider a measurable space* $(X, \mathcal{A}')$. *Given* $x \in X$, *we define the* **minimal measurable set** *generated by* x *as* $[x] := \bigcap\{A \in \mathcal{A}' \mid x \in A\}$. *The* σ-*field* $\mathcal{A}'$ *is called* **reducible** *when* $[x] \in \mathcal{A}'$ *for all* $x \in X$, *and* $(X, \mathcal{A}')$ *is called then a* **reducible measurable space**.

We can easily see that most σ-fields in our context are reducible: for instance, the Borel σ-field generated by a T_1 topology (and hence also by a metric) is reducible, because in that case we have $[x] = \{x\} \; \forall x \in X$. To see that the notion is not trivial, we give next an example of a non-reducible σ-field:

Example 2. Let $\leq$ be a well-order on the set of real numbers (existing by Zermelo's theorem), and let $\mathcal{P}_x^< := \{y \in \mathbb{R} \mid y < x\}$ and $\mathcal{P}_x^\leq := \{y \in \mathbb{R} \mid y \leq x\}$ denote the sets of strict predecessors and predecessors of x under $\leq$, respectively. Let us also define the notation $\mathcal{P}_x^\geq := (\mathcal{P}_x^<)^c$.

There is $x_0 \in \mathbb{R}$ such that $\mathcal{P}_{x_0}^<$ uncountable and such that $\mathcal{P}_x^<$ is countable for every $x < x_0$: it suffices to take the set of points with an uncountable number of predecessors, and select its first element, existing because $\leq$ is a well-order. Consider $X := \mathcal{P}_{x_0}^\leq$, and let us define $\mathcal{B} := \{\emptyset, \; A \subseteq \mathcal{P}_{x_0}^< \text{ countable}\} \cup \{\mathcal{P}_x^\geq \cup A \mid A \subseteq \mathcal{P}_x^<, x \in \mathcal{P}_{x_0}^<\}$.

- Given a countable set $A \subseteq \mathcal{P}_{x_0}^\leq$, $\sup A$ always exists, because $\leq$ is a well-order, and it belongs to $\mathcal{P}_{x_0}^<$. Taking this into account, we can deduce that $\mathcal{B}$ is closed under complementation. Since it is immediate that it is closed under countable unions and that $\emptyset, X$ belong to $\mathcal{B}$, we deduce that $\mathcal{B}$ is a σ-field.

- Note now that x_0 does not have previous element under the order $\leq$: otherwise, we contradict the uncountability of $\mathcal{P}_{x_0}^<$. Hence, the minimal measurable set generated by x_0 is $[x_0] = \cap_{x < x_0} \mathcal{P}_x^\geq = \{x_0\}$. This set does not belong to $\mathcal{B}$ and as a consequence this σ-field is not reducible. ◆

We have already mentioned that a random set may not possess measurable selections, and that we need to make some requirements in order to guarantee that the set $S(\Gamma)$ given by Eq. (1) is non-empty. The existing results usually make some assumptions on the images of the random set and on the structure of the final σ-field. In our next result, we give a sufficient condition for the existence of measurable selections where the only thing we require in $\mathcal{A}'$ is its reducibility, and apply this condition to prove that countable sets belong to $\mathcal{H}_\Gamma$:

Proposition 3. *Let* $(\Omega, \mathcal{A}, P)$ *be a probability space,* $(X, \mathcal{A}')$ *a reducible measurable space, and let* $\Gamma : \Omega \to \mathcal{P}(X)$ *be a random set.*

1. *If there is some countable* $\{x_n\}_n \subseteq X$ *s.t.* $\cup_n \Gamma^*([x_n]) = \Omega$, *then* $S(\Gamma) \neq \emptyset$.
2. *If* $S(\Gamma) \neq \emptyset$, *then for any countable subset* $\{x_n\}_n$ *of* X, $\cup_n [x_n] \in \mathcal{H}_\Gamma$.

We turn now to another property of random sets that shall be useful in our quest for sufficient conditions for the equality between $\mathcal{P}(\Gamma)(A)$ and the interval $[P_*(A), P^*(A)]$: the existence of inner approximations of P^*. We shall investigate under which conditions there is some subclass $\mathcal{A}'_1$ of $\mathcal{A}'$ such that P^* is the inner set function of its restriction to $\mathcal{A}'_1$. The interest of this problem for our purposes lies in the following proposition:

Proposition 4. *Let $(\Omega, \mathcal{A}, P)$ be a probability space, $(X, \mathcal{A}')$ a measurable space and let $\Gamma : \Omega \to \mathcal{P}(X)$ be a random set. If $B \in \mathcal{A}'$ satisfies $P^*(B) = \sup_n P^*(A_n)$ for some increasing sequence $\{A_n\}_n \subseteq \mathcal{H}_\Gamma$ of subsets of B, then $B \in \mathcal{H}_\Gamma$.*

We deduce that if P^* satisfies

$$P^*(A) = \sup_{B \subseteq A, B \in \mathcal{H}_\Gamma} P^*(B) \ \forall A \in \mathcal{A}', \tag{5}$$

it also satisfies $P^*(A) = \max \mathcal{P}(\Gamma)(A)$ for every $A \in \mathcal{A}'$ (i.e., $\mathcal{H}_\Gamma$ is actually equal to $\mathcal{A}'$). This will be helpful for our purposes because in some cases it will be easier to prove the equality $P^*(A) = \max \mathcal{P}(\Gamma)(A)$ for some specific types of sets, such as closed or compact sets, and to show then that the upper probability can be approximated from below using these sets. In particular, Proposition 4 and the lower continuity of P^* implies that $\mathcal{H}_\Gamma$ is closed under countable unions.

In the language of measure theory, Eq. (5) means that P^* is the *inner set function* of its restriction to $\mathcal{H}_\Gamma$, or that it is *inner regular* with respect to $\mathcal{H}_\Gamma$. There are some results about the inner regularity of upper probabilities in the literature (see [4,22]). In this respect, we have proven the following:

Lemma 1. *Let $(\Omega, \mathcal{A}, P)$ be a probability space, (X, τ) a Polish space and consider a closed random set $\Gamma : \Omega \to \mathcal{P}(X)$. For every $A \in \beta_X$, $P^*(A) = \sup_{K \subseteq A \text{ compact}} P^*(K)$.*

This lemma generalises a result in [22, Section 2.1]. Let us establish now sufficient conditions for the equality between $\mathcal{H}_\Gamma$ and $\mathcal{A}'$.

Theorem 2. *Let $(\Omega, \mathcal{A}, P)$ be a probability space, $(X, \mathcal{A}')$ a measurable space and $\Gamma : \Omega \to \mathcal{P}(X)$, a random set. Under any of the following conditions:*

1. *Ω is complete, X is Souslin and $Gr(\Gamma) \in \mathcal{A} \otimes \beta_X$*
2. *X is a separable metric space and Γ is compact*
3. *X is a Polish space and Γ is closed*
4. *X is a σ-compact metric space and Γ is closed*
5. *X is a separable metric space and Γ is open*
6. *$\mathcal{A}'$ is reducible and $\mathcal{C}_\Gamma := \{\Gamma^*(B) : B \in \mathcal{A}'\}$ is countable*
7. *$\mathcal{A}'$ is reducible and Γ has a countable range,*

$P^*(A) = \max \mathcal{P}(\Gamma)(A)$ and $P_*(A) = \min \mathcal{P}(\Gamma)(A) \ \forall A \in \mathcal{A}'$.

The second and third points of Theorem 2 generalise Theorem 1. Moreover, this theorem also generalises the results mentioned in the proofs of [2, Proposition 2.7] and [15, Theorem 1].

This result shows that the upper and lower probabilities of a random set provide, under fairly general conditions, the tightest available bounds for the probabilities induced by the original random variable. They are hence an adequate tool under the interpretation of Kruse and Meyer. Note that in particular the bounds are attained for *finite* random sets, i.e., those where X is a finite space and $\mathcal{A}' = \mathcal{P}(X)$. These random sets have been studied in detail in [23].

In our last proposition in this section, we provide a sufficient condition for the equality $P^*(A) = \max \mathcal{P}(\Gamma)(A)$ to hold for every set A in a field that is included in the σ-field $\mathcal{A}'$. This property shall be useful in Section 3.3, when we relate the probability distributions in $\mathcal{P}(\Gamma)$ and $M(P^*)$. Recall that a complete random set is one whose images are complete subsets of the final space, i.e., subsets for which any Cauchy sequence has a limit within the set.

Theorem 3. *Let $(\Omega, \mathcal{A}, P)$ be a probability space, (X, d) a separable metric space, let $\Gamma : \Omega \rightarrow \mathcal{P}(X)$ be a complete random set. For every A in $\mathcal{Q}(\tau(d))$, the field generated by the open balls, $P^*(A) = \max \mathcal{P}(\Gamma)(A)$ and $P_*(A) = \min \mathcal{P}(\Gamma)(A)$.*

It is an open problem at this stage whether, for this type of random sets, $\mathcal{H}_\Gamma$ coincides with $\mathcal{A}'$. An affirmative answer to this question would generalise the second and third points from Theorem 2. One possible approach would be to study whether $\mathcal{H}_\Gamma$ is closed under countable intersections: in that case $\mathcal{H}_\Gamma$ would include the monotone class generated by the field $\mathcal{Q}(\tau(d))$, which is the Borel σ-field β_X.

2.2   Convexity of $\mathcal{P}(\Gamma)(A)$

As we mentioned in the introduction, the study of the equality between $\mathcal{P}(\Gamma)(A)$ and $[P_*(A), P^*(A)]$ can be split into two different subproblems: the equality between $P^*(A), P_*(A)$ and the maximum and minimum values of $\mathcal{P}(\Gamma)(A)$ and the convexity of this last set. We focus our attention now on this second problem. We introduce first the following definition:

Definition 4. *[3] Let $(\Omega, \mathcal{A}, P)$ be a probability space. We say that a set $B \in \mathcal{A}$ is not an **atom** when for every $\epsilon \in (0, 1)$ there is some measurable $B_\epsilon \subsetneq B$ such that $P(B_\epsilon) = \epsilon P(B)$.*

Proposition 5. *Let $(\Omega, \mathcal{A}, P)$ be a probability space, $(X, \mathcal{A}')$ be a measurable space and let $\Gamma : \Omega \rightarrow \mathcal{P}(X)$ be a random set. Let $U_1, U_2 \in S(\Gamma)$ satisfy $P_{U_1}(A) = \max \mathcal{P}(\Gamma)(A), P_{U_2}(A) = \min \mathcal{P}(\Gamma)(A)$. Then $\mathcal{P}(\Gamma)(A)$ is convex $\Leftrightarrow$ $U_1^{-1}(A) \setminus U_2^{-1}(A)$ is not an atom.*

We deduce that whenever the equalities $P^*(A) = \max \mathcal{P}(\Gamma)(A)$ and $P_*(A) = \min \mathcal{P}(\Gamma)(A)$ hold, $\mathcal{P}(\Gamma)(A) = [P_*(A), P^*(A)]$ if and only if $A^* \setminus A_*$ is not an atom of the initial probability space. This immediately implies the following:

Corollary 1. *Under any of the conditions listed in Theorem 2,*

$$[P_*(A), P^*(A)] = \mathcal{P}(\Gamma)(A) \ \forall A \in \mathcal{A}' \Leftrightarrow \forall A \in \mathcal{A}', A^* \setminus A_* \text{ is not an atom of } \mathcal{A}.$$

The right-hand side of this equivalence holds trivially whenever the initial probability space is non-atomic; however, as we show in [23, Remark 1], there are examples of random sets defined on a purely atomic probability space where $[P_*(A), P^*(A)] = \mathcal{P}(\Gamma)(A) \ \forall A \in \mathcal{A}'$.

3   Some Implications of the Previous Results

3.1   Consonant Random Sets

One particular type of random sets which is of interest in practice are the **consonant** random sets, which are those whose images are nested. They have been studied in connection with possibility and maxitive measures in a number of works ([8,11,13,24]). Since a possibility measure is usually defined on all subsets of its possibility space, we are going to assume in this section that the final σ-field is $\mathcal{P}(X)$, which is in particular reducible.

In this paper we are going to consider the following notion of consonant random sets. Other possibilities can be found in [24].

Definition 5. *A random set* $\Gamma : \Omega \to \mathcal{P}(X)$ *is called* **consonant** *when the following two conditions hold:*

- *For every* $\omega_1, \omega_2 \in \Omega$, *either* $\Gamma(\omega_1) \subseteq \Gamma(\omega_2)$ *or* $\Gamma(\omega_2) \subseteq \Gamma(\omega_1)$.
- *Every* $A \subseteq \Omega$ *has a countable subset* B *for which* $\cap_{\omega \in A} \Gamma(\omega) = \cap_{\omega \in B} \Gamma(\omega)$.

This definition is a generalisation of the so-called **antitone**[8] random sets, where the initial probability space is $([0,1], \beta_{[0,1]}, \lambda_{[0,1]})$ and where $x \leq y \Rightarrow \Gamma(x) \supseteq \Gamma(y)$.

Proposition 6. *Let* $(\Omega, \mathcal{A}, P)$ *be a probability space,* $(X, \mathcal{P}(X))$ *a measurable space and* $\Gamma : \Omega \to \mathcal{P}(X)$ *a consonant random set. Then* $P^*(A) = \max \mathcal{P}(\Gamma)(A)$ *for all* $A \subseteq X$.

The proof of this result follows by showing that if Γ is consonant the upper probability P^* is the inner approximation of its restriction to countable sets. We can deduce from this and [24, Propositions 2.4 and 5.2] that P^* is a possibility measure.

An open problem at this point is whether we can generalise Proposition 6 to weaker notions of consonancy for random sets, such as those considered in [24].

3.2   Relationship between the Aumann and the Choquet Integral

Our results allow us also to relate the Choquet [10] integral of a bounded function with respect to the upper and lower probabilities and the set of its integrals with respect to the measurable selections. This set is related to the Aumann integral of the random set, whose definition we recall:

Definition 6. *[1] Let* $(\Omega, \mathcal{A}, P)$ *be a probability space, and let* $\Gamma : \Omega \to \mathcal{P}(\mathbb{R}^n)$ *be a random set. Its* **Aumann integral** *is given by*

$$(A) \int \Gamma dP := \left\{ \int f dP : f \in L^1(P), f(\omega) \in \Gamma(\omega) \ a.s \right\}.$$

Note that is this definition we consider the set of the integrals with respect to the *almost-surely integrable* selections, which are those integrable mappings whose images are included in the random set with probability one. Given a random set $\Gamma : \Omega \to \mathcal{P}(X)$ and a measurable mapping $f : X \to \mathbb{R}$, it is not difficult to see that $f \circ \Gamma : \Omega \to \mathcal{P}(\mathbb{R})$ is also a random set.

Lemma 2. *Let $(\Omega, \mathcal{A}, P)$ be a probability space, $(X, \mathcal{A}')$ be a measurable space and $\Gamma : \Omega \to \mathcal{P}(X)$ a random set. If $P^*(A) = \max \mathcal{P}(\Gamma)(A)$ for all $A \in \mathcal{A}'$, then for any finite chain $A_1 \subseteq A_2 \subseteq \cdots \subseteq A_n$ there is some $U \in S(\Gamma)$ such that $P_U(A_i) = P^*(A_i)$ for every $i = 1, \ldots, n$.*

This lemma allows us to relate the Choquet integral of a simple mapping with respect to the upper probability and the set of its integrals with respect to the probability distributions of the measurable selections. As a consequence, we can establish the following:

Theorem 4. *Let $(\Omega, \mathcal{A}, P)$ be a probability space, $(X, \mathcal{A}')$ be a measurable space and $\Gamma : \Omega \to \mathcal{P}(X)$ a random set. If $P^*(A) = \max \mathcal{P}(\Gamma)(A)$ for all $A \in \mathcal{A}'$, then for any bounded random variable $f : X \to \mathbb{R}$,*

$$(C) \int f dP^* = \sup_{U \in S(\Gamma)} \int f dP_U = \sup(A) \int (f \circ \Gamma) dP,$$

and

$$(C) \int f dP_* = \inf_{U \in S(\Gamma)} \int f dP_U = \inf(A) \int (f \circ \Gamma) dP.$$

Using this result together with Theorem 2, we can generalise [4, Theorem 3.2].

3.3   Measurable Selections and the Core of P^*

As we said in the introduction, the set $\mathcal{P}(\Gamma)$ of distributions of the selections is included in the core $M(P^*)$ of the upper probability, given by Eq. (3). This set can be more imprecise than $\mathcal{P}(\Gamma)$; on the other hand, it has the advantage of being convex and it is uniquely determined by the function P^*. This makes $M(P^*)$ easier to handle for practical purposes than $\mathcal{P}(\Gamma)$.

We can use our results on the equality between $\mathcal{P}(\Gamma)(A)$ and $[P_*(A), P^*(A)]$ to derive conclusions on the relationship between $\mathcal{P}(\Gamma)$ and $M(P^*)$. In this respect, we have proven in [25] the following result:

Theorem 5. *[25, Theorem 4.4] Let $(\Omega, \mathcal{A}, P)$ be a probability space, (X, d) a separable metric space and $\Gamma : \Omega \to \mathcal{P}(X)$ a random set. Let $\{x_n\}_n$ be a countable dense subset of X and let $\mathcal{J} := \{B(x_i, q) : q \in \mathbb{Q}, i \in \mathbb{N}\}$. If $P^*(A) = \max \mathcal{P}(\Gamma)(A)$ for all A in the field $\mathcal{Q}(\mathcal{J})$ generated by $\mathcal{J}$:*

1. *$\overline{M(P^*)} = \overline{Conv(\mathcal{P}(\Gamma))}$, where the closures are taken in the weak topology.*
2. *$\overline{M(P^*)} = \overline{\mathcal{P}(\Gamma)} \Leftrightarrow \overline{\mathcal{P}(\Gamma)}$ is convex.*

Note that not only we can apply this result together with Theorem 2 and Proposition 6, but also with Theorem 3, because for any separable metric space the field generated by the open balls includes in particular the field generated by $\mathcal{J}$. Hence, under very general situations, we can relate the core of the upper probability with the distributions of the measurable selections. Moreover, $\overline{\mathcal{P}(\Gamma)}$ is a convex set as soon as the initial probability space is non-atomic [25, Theorem 4.7]; this allows us to derive conditions for applying the second point of Theorem 5. On the other hand, the equality $\overline{\mathcal{P}(\Gamma)} = M(P^*)$ does not imply in general that $\mathcal{P}(\Gamma)$ coincides with $M(P^*)$; an example and sufficient conditions for this equality can be found in [26].

As a side result, we also deduce that under any of the conditions listed in Theorem 2 and Proposition 6, P^* is the upper envelope of its core $M(P^*)$. This relates our work to the problem studied by Krätschmer in [20], and also to some results in [19,28].

4   Conclusions

The results we have established show that the upper and lower probabilities of the random set are informative enough in most (but not in all) cases about the values taken by the distribution of the original random variable. Indeed, the features of Example 1 and the sufficient conditions listed in Theorem 2 make us conclude that we can use the upper and lower probabilities in all cases of practical interest. Moreover, the problem we have studied allows us to derive relationships between the core of the upper probability and the set of distributions of the measurable selections, and between the Aumann and Choquet integrals.

We have already pointed out in a few places some of the open problems derived from our results. More generally, it would be interesting to investigate the suitability of the upper and the lower probabilities when we have some additional information on the distribution of the original random variable (for instance that it belongs to some parametric family). Another interesting possibility would be to consider the case where we model the imprecise observation of U_0 by means of a fuzzy random variable.

References

1. Aumann, J.: Integral of set-valued functions. Journal of Mathematical Analysis and Applications 12, 1–12 (1965)
2. Arstein, Z., Hart, S.: Law of large numbers for random sets and allocation processes. Mathematics of Operations Research 6(4), 485–492 (1981)
3. Billingsley, P.: Probability and measure. Wiley, New York (1986)
4. Castaldo, A., Maccheroni, F., Marinacci, M.: Random correspondences as bundles of random variables. Sankhya 66(3), 409–427 (2004)
5. Choquet, G.: Theory of capacities. Annales de l'Institut Fourier 5, 131–295 (1953)
6. Couso, I.: Teoría de la Probabilidad con datos imprecisos. Algunos aspectos. PhD Thesis, University of Oviedo (1999) (in Spanish)

7. Couso, I., Montes, S., Gil, P.: Second order possibility measure induced by a fuzzy random variable. In: Bertoluzza, C., Gil, M.A., Ralescu, D.A. (eds.) Statistical modeling, analysis and management of fuzzy data. Springer, Heidelberg (2002)
8. de Cooman, G., Aeyels, D.: A random set description of a possibility measure and its natural extension. IEEE Transactions on Systems, Man and Cybernetics 30, 124–130 (2000)
9. Dempster, A.P.: Upper and lower probabilities induced by a multivalued mapping. Annals of Mathematical Statistics 38, 325–339 (1967)
10. Denneberg, D.: Non-additive measure and integral. Kluwer, Dordrecht (1994)
11. Dubois, D., Prade, H.: The mean value of a fuzzy number. Fuzzy Sets and Systems 24, 279–300 (1987)
12. Dubois, D., Prade, H.: Focusing versus updating in belief function theory. In: Yager, R.R., Fedrizzi, M., Kacprzyk, J. (eds.) Advances in the Dempster-Shafer Theory of Evidence, pp. 71–95. Wiley, Chichester (1994)
13. Goodman, I.R.: Fuzzy sets as equivalence classes of possibility random sets. In: Yager, R.R. (ed.) Fuzzy Sets and Possibility Theory: Recent Developments, pp. 327–343. Pergamon, Oxford (1982)
14. Grabisch, M., Nguyen, H.T., Walker, E.A.: Fundamentals of uncertainty calculi with applications to fuzzy inference. Kluwer, Dordretch (1995)
15. Hart, S., Köhlberg, E.: Equally distributed correspondences. Journal of Mathematical Economics 1(2), 167–674 (1974)
16. Hess, C.: The distribution of unbounded random sets and the multivalued strong law of large numbers in nonreflexive Banach spaces. Journal of Convex Analysis 6(1), 163–182 (1999)
17. Himmelberg, C.J.: Measurable relations. Fund. Mathematicae 87, 53–72 (1975)
18. Himmelberg, C.J., Parthasarathy, T., Van Vleck, F.S.: On measurable relations. Fundamenta Mathematicae 111(2), 161–167 (1981)
19. Huber, P.J., Strassen, V.: Minimax tests and the Neyman-Pearson lemma for capacities. Annals of Statistics 1(2), 251–263 (1973)
20. Krätschmer, V.: When fuzzy measures are upper envelopes of probability measures. Fuzzy Sets and Systems 138(3), 455–468 (2003)
21. Kruse, R., Meyer, K.D.: Statistics with vague data. D. Reidel Publishing Company, Dordrecht (1987)
22. Mathéron, G.: Random sets and integral geometry. Wiley, New York (1975)
23. Miranda, E., Couso, I., Gil, P.: Upper probabilities and selectors of random sets. In: Grzegorzewski, P., Hryniewicz, O., Gil, M.A. (eds.) Soft methods in probability, statistics and data analysis, pp. 126–133. Physica-Verlag, Heidelberg (2002)
24. Miranda, E., Couso, I., Gil, P.: A random set characterisation of possibility measures. Information Sciences 168(1-4), 51–75 (2004)
25. Miranda, E., Couso, I., Gil, P.: Random sets as imprecise random variables. Journal of Mathematical Analysis and Applications 307(1), 32–47 (2005)
26. Miranda, E., Couso, I., Gil, P.: Random intervals as a model for imprecise information. Fuzzy Sets and Systems 154(3), 386–412 (2005)
27. Nguyen, H.T.: On random sets and belief functions. Journal of Mathematical Analysis and Applications 65(3), 531–542 (1978)
28. Philippe, F., Debs, G., Jaffray, J.-Y.: Decision making with monotone lower probabilities of infinite order. Mathematics of Operations Research 24(3), 767–784 (1999)
29. Shafer, G.: A mathematical theory of evidence. Princeton University Press, Princeton (1976)
30. Wagner, D.H.: Survey of measurable selection theorems. SIAM Journal Control and Optimization 15(5), 859–903 (1977)

Merging Qualitative Constraints Networks Using Propositional Logic

Jean-François Condotta, Souhila Kaci,
Pierre Marquis, and Nicolas Schwind

Université d'Artois
CRIL CNRS UMR 8188, F-62307 Lens
{condotta,kaci,marquis,schwind}@cril.univ-artois.fr

Abstract. In this paper we address the problem of merging qualitative constraints networks ($QCNs$). We propose a rational merging procedure for $QCNs$. It is based on translations of $QCNs$ into propositional formulas, and take advantage of propositional merging operators.

1 Introduction

Representing and reasoning about time and space is an important task in many domains such as natural language processing, geographic information systems, computer vision, robot navigation. Several qualitative approaches have been proposed so far to represent spatial or temporal entities and their relations [1,24,21,18,19]. The majority of these formalisms use qualitative constraints networks ($QCNs$ for short) as a representation language.

In some applications, especially multi-agent ones, spatial or temporal information comes from different sources, i.e. each source provides a spatial or temporal QCN representing relative positions of objects. The multiplicity of sources providing spatial or temporal information makes that the underlying $QCNs$ are generally conflicting. A way to address the conflict issue consists in defining a merging operator which takes as input a set of $QCNs$ $\mathcal{N} = \{N_1, \cdots, N_m\}$ modeling the information provided by the different sources and returns a consistent set of spatial or temporal information corresponding to the global information deduced from the information of the different sources.

Merging multiple sources information has attracted much attention in the framework of (weighted) propositional logic [22,23,12,13,14,11,3,2]. Inspired from these works, Condotta et al. [5] have proposed a first merging approach to $QCNs$. In this paper, we propose a new merging procedure for $QCNs$ by first translating each QCN into a propositional formula and then merging these formulas using propositional merging operators [12,13]. The new approach can benefit from recent advances on merging propositional formulas [9,10]. Different translations have been proposed in literature. Initially such translations have been defined to tackle the consistency problem for $QCNs$ in propositional logic. Note that such translations do not exist for all qualitative formalisms. For example, Nebel and Bürckert [19] represent constraints of interval algebra by a set of propositional

C. Sossai and G. Chemello (Eds.): ECSQARU 2009, LNAI 5590, pp. 347–358, 2009.

clauses. Other more generic translations [20,4] allow to represent $QCNs$ defined on a qualitative formalism for which the closure by weak composition is complete for the consistency problem.

The aim of this paper is to characterize such translations and study the behavior of propositional merging operators on a set of propositional formulas resulting from the translation of a set of $QCNs$. The rest of this paper is organized as follows. We present in Section 2 some necessary background on qualitative formalisms for representing space or time. In Section 3, we describe the problem and briefly recall the merging procedure given in [5]. Then we present in Section 4 a merging procedure based on the translation of $QCNs$ into propositional formulas. We show that this procedure is equivalent to the one proposed in [5]. In Section 5 we propose rationality postulates for merging $QCNs$ and show how one can define a rational $QCNs$ merging operator from propositional merging operators thanks to more generic translations. Lastly we conclude.

2   Background on Qualitative Formalisms

Let $\mathcal{B}$ be a finite set of binary relations (called basic relations) over a domain $\mathcal{D}$. Each of these basic relations represents a particular qualitative position between two elements of $\mathcal{D}$. We suppose that these basic relations are complete and mutually exclusive, namely two elements of $\mathcal{D}$ satisfy one and only one basic relation of $\mathcal{B}$. The weak composition $r_1 \diamond r_2$ between two basic relations $r_1, r_2 \in \mathcal{B}$ is defined by the set $\{r : \exists x, y, z \in \mathcal{D}, x\ r_1\ y, y\ r_2\ z, x\ r\ z\}$. $\mathcal{A}$ denotes the set $2^{\mathcal{B}}$, i. e. , the set of all subsets of $\mathcal{B}$. An element $R \in \mathcal{A}$ is a set of basic relations between two elements of $\mathcal{D}$. Thus we have $X\ R\ Y \Leftrightarrow \exists r \in R : X\ r\ Y$. For illustration, we consider the Point Algebra [24] which considers relations between two points of the rational line. Figure 1 details the three basic relations of the Point Algebra, forming the set $\mathcal{B}_{pt}$.

Relation	Symbol	Illustration
precedes	<	
follows	>	
same	=	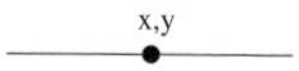

Fig. 1. The 3 basic relations of the Point Algebra

Pieces of knowledge about a set of spatial or temporal entities can be represented by means of qualitative constraints networks ($QCNs$ for short). A QCN N is a pair (V, C), where $V = \{v_0, \cdots, v_{n-1}\}$ is a finite set of variables representing the spatial or temporal entities and C is a mapping which associates to each pair of variables (v_i, v_j), with $i < j$, an element R of $\mathcal{A}$. R represents the set of all possible basic relations between v_i and v_j. We write C_{ij} instead of

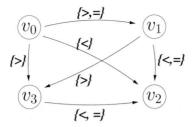

Fig. 2. N_1, a QCN of the Point Algebra

$C(v_i, v_j)$ for short. A QCN N_1 defined over 4 variables within the Point Algebra is depicted on Fig. 2.

Definition 1. *Let $N = (V, C)$ be a QCN.*

- *A consistent instantiation of N over $V' \subseteq V$ is a mapping α from V' to $\mathcal{D}$ such that $\alpha(v_i)\ C_{ij}\ \alpha(v_j)$, $\forall v_i, v_j \in V'$.*
- *N is consistent iff there exists a consistent instantiation of N over V.*
- *N is $\diamond$-closed iff $\forall v_i, v_j, v_k \in V$, $C_{ij} \subseteq C_{ik} \diamond C_{kj}$.*
- *A sub-network of N is a QCN $N' = (V, C')$, where $C'_{ij} \subseteq C_{ij}$, $\forall i, j \in \{0, \cdots, n-1\}$.*
- *A consistent scenario of N is a consistent sub-network of N, in which each constraint is composed of one and only one basic relation of $\mathcal{B}$.*

Figure 3.a depicts a consistent scenario σ of N_1 given in Fig. 2. Figure 3.b depicts a consistent instantiation of σ.

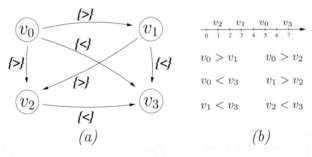

Fig. 3. A consistent scenario σ (a) and a consistent instantiation of σ (b)

$[N]$ denotes the set of consistent scenarios of a QCN N. N_{ALL}^V denotes the QCN defined on the set of variables V in which each constraint corresponds to the set $\mathcal{B}$. Thus the set of consistent scenarios defined on a set V corresponds to the set $[N_{ALL}^V]$.

3   Merging $QCNs$

Let $\mathcal{N} = \{N_1, \ldots, N_m\}$ be a set of $QCNs$ defined on the same set of variables $V = \{v_0, \ldots, v_{n-1}\}$ and on the same qualitative algebra having $\mathcal{B}$ as the set of basic relations. The problem we consider consists in merging the different information modeled by the QCN of $\mathcal{N}$. A natural way to solve this problem is to consider the set $\bigcap_{N_i \in \mathcal{N}} [N_i]$ as the result of merging, i.e., the set of consistent scenarios belonging to each QCN of $\mathcal{N}$. However this set may be empty due to the multiplicity of sources providing information. It is necessary to define a more parsimonious merging method in order to get a consistent result. The problem of merging $QCNs$ has been addressed in [5] where the authors propose a merging procedure inspired from propostional merging [22,12]. We first recall the merging process in propositional setting before we recall the merging procedure of $QCNs$ developped in [5].

3.1   Merging Propositional Bases

We consider a propositional language $PROP$ defined on a finite alphabet of variables V. An *interpretation* is a mapping from V to $\{0, 1\}$. We denote by $\mathcal{W}$ the finite set of all possible interpretations. An interpretation ω is a *model* of a formula ϕ (denoted $\omega \models \phi$) if and only if it makes the formula true. A *knowledge base* K is a finite set of propositional formulas $\{\phi_1, \cdots, \phi_m\}$. We consider K as logically equivalent to the conjunction of its formulas: $K = \phi_1 \wedge \cdots \wedge \phi_m$. K is *consistent* iff $\exists \omega \in \mathcal{W}$ such that $\omega \models K$. If K_1 and K_2 are two knowledge bases, we denote $K_1 \equiv K_2$ when two knowledge bases K_1, K_2 are logically equivalent. A multiset of knowledge bases $\{K_1, \cdots, K_m\}$ is called a *profile*. Two profiles $\mathcal{K}_1$ and $\mathcal{K}_2$ are *equivalent*, denoted $\mathcal{K}_1 \equiv \mathcal{K}_2$, if there exists a bijection f between $\mathcal{K}_1$ and $\mathcal{K}_2$ such that $\forall K \in \mathcal{K}_1, K \equiv f(K)$. $\sqcup$ is the union operator for multisets.

A merging operator Δ is a mapping which associates a propositional formula to a profile $\mathcal{K}$ and a propositional formula IC representing integrity constraints. A logical characterization of merging operators under integrity constraints has been proposed in [13], by means of a set of rationality postulates. The result of merging is denoted $\Delta_{IC}(\mathcal{K})$. For example, the first postulate (see Section 5) expresses that the propositional formula representing the result of merging should pick its models in the set of models of IC, namely $\Delta_{IC}(\mathcal{K}) \models IC$.

Merging operators in the propositional logic framework [17,12,13,14] are often based on a pseudo-distance d which is a mapping from $\mathcal{W} \times \mathcal{W}$ to $\mathbb{N}$ such that $\forall \omega, \omega'$ we have $d(\omega, \omega') = d(\omega', \omega)$ and $d(\omega, \omega) = 0$. Merging propositional knowledge bases is then a three step process. First, the "distance" between an interpretation ω and a knowledge base K is defined as follows: $d(\omega, K) = \min_{\omega' \models K} d(\omega, \omega')$. Then an aggregation operator denoted $\otimes$ [17,12,22,23] is used to compute the distance between an interpretation ω and a profile $\mathcal{K}$. This distance is defined by $d_\otimes(\omega, \mathcal{K}) = \otimes\{d(\omega, K) \mid K \in \mathcal{K}\}$. Lastly, the result of merging, denoted $\Delta_{IC}^{\otimes, d}(\mathcal{K})$, is the set of models of IC which are the closest to $\mathcal{K}$ w.r.t. d. Formally, we have $\Delta_{IC}^{\otimes, d}(\mathcal{K}) = \{\omega \models IC \mid \nexists \omega' \models IC, d(\omega', \mathcal{K}) < d(\omega, \mathcal{K})\}$.

3.2   Merging $QCNs$

Inspired from propositional operators described in the previous subsection, Condotta et al. [5] have defined an operator for merging a set of $QCNs$ $\mathcal{N}$ in a similar way. The merging process also follows three steps.

The first step consists in computing a local distance between each consistent scenario of $[N_{ALL}^V]$ and each QCN of $\mathcal{N}$. The distance between a scenario σ and a QCN N is the minimum distance between σ and all consistent scenarios of N.

$$d(\sigma, N) = \begin{cases} \min\{d^{QCN}(\sigma, \sigma') \mid \sigma' \in [N]\} & \text{if } N \text{ is consistent,} \\ 0 & \text{otherwise.} \end{cases}$$

Thus we need to define a distance between scenarios. Such a distance is a mapping from $[N_{ALL}^V] \times [N_{ALL}^V]$ to $\mathbb{N}$ such that $\forall \sigma, \sigma' \in [N_{ALL}^V]$,

$$\begin{cases} d^{QCN}(\sigma, \sigma') = d^{QCN}(\sigma', \sigma) \\ d^{QCN}(\sigma, \sigma) = 0. \end{cases}$$

Different distances between scenarios have been defined in [5]. Some of them are inspired from distances between interpretations as defined in the propositional logic framework (e.g. drastic distance, Hamming distance. See Section 4). Other more specific distances have also been defined in the context of $QCNs$ (e.g. the conceptual neighborhood distance [5]).

The second step consists in aggregating local distances computed in the previous step in order to compute a global distance between each consistent scenario of $[N_{ALL}^V]$ and $\mathcal{N}$. Different aggregation operators have been defined in literature. For example, the majority operator $\sum$ [17], which computes the sum of local distances, favors the point of view of the majority of sources. Arbitration operator $\mathcal{MAX}$ [23], which returns the greatest distance, has a more consensual behavior. The global distance between a scenario σ and a set $\mathcal{N}$ of $QCNs$ is defined by $d_\otimes(\sigma, \mathcal{N}) = \otimes\{d(\sigma, N) \mid N \in \mathcal{N}\}$, where $\otimes$ is an aggregation operator.

The result of merging, denoted $\Theta^{\otimes, d^{QCN}}(\mathcal{N})$, is the set of consistent scenarios of $[N_{ALL}^V]$ which are the "closest" to $\mathcal{N}$. These are consistent scenarios which have a minimal global distance to $\mathcal{N}$. Formally,

$$\Theta^{\otimes, d^{QCN}}(\mathcal{N}) = \{\sigma \in [N_{ALL}^V] \mid \nexists \sigma' \in [N_{ALL}^V], d(\sigma', \mathcal{N}) < d(\sigma, \mathcal{N})\}.$$

4   A Merging Procedure of $QCNs$ Based on a Propositional Translation

4.1   Characterization of a Translation

We consider fixed a set of variables V and a qualitative formalism defined on a set of basic relations $\mathcal{B}$. We denote $QCN_\mathcal{B}^V$ the set of $QCNs$ defined on the qualitative formalism given by $\mathcal{B}$ and V. We call translation a mapping from $QCN_\mathcal{B}^V$ to the set of propositional formulas $PROP$. The main advantage of existing translations proposed in literature [19,20,4] is to benefit from works made around the SAT problem in order to solve the consistency problem for $QCNs$. A translation τ has to satisfy at least the following property:

Property 1. $\forall N \in QCN_\mathcal{B}^V$, $\tau(N)$ is satisfiable if and only if N is consistent.

We mean that a QCN has to admit a consistent scenario if and only if its associated propositional formula admits a model. However this property is insufficient in our context when considered alone. Indeed the merging process of $QCNs$ described in the previous section is based on distances between consistent scenarios while merging propositional bases is based on distances between interpretations. Therefore we suppose that τ also satisfies the following property:

Property 2

(a.) $\forall N \in QCN_{\mathcal{B}}^V$, $\tau(N) \models \tau(N_{ALL}^V)$,

(b.) There is a bijection μ_τ from the set of models of $\tau(N_{ALL}^V)$ to $[N_{ALL}^V]$ such that $\forall N \in QCN_{\mathcal{B}}^V$, $\{\mu_\tau(\omega) \mid \omega \models \tau(N)\} = [N]$.

The last property makes it possible to identify (through a bijection μ_τ) a consistent scenario with an interpretation ω of $\mathcal{W}$ if $\omega \models \tau(N_{ALL}^V)$, and that $\forall N \in QCN_{\mathcal{B}}^V$, $\tau(N)$ represents through its models the set of consistent scenarios of N.

We now give an example of such a translation. Let $\tau_{Sup}(N)$ be the translation of a QCN $N = (V,C) \in QCN_{\mathcal{B}}^V$ using the support encoding [8,6,20]. The propositional formula $\tau_{Sup}(N)$ is built on the set of propositional variables $V_T = \{r_{ij} \mid r \in \mathcal{B}, 0 \le i < j \le n-1\}$. The propositional variable r_{ij} is valuated to *true* if and only if the basic relation r is satisfied for the constraint between the two variables v_i and v_j of V. We say that l is a literal of V_T if and only if l is a variable of V_T or its negation. $\forall N = (V,C) \in QCN_{\mathcal{B}}^V$, $\tau_{Sup}(N)$ is the conjunction of the following clauses:

- $\bigvee_{r \in C_{ij}} r_{ij}$, $\forall 0 \le i < j \le n-1$ (at least one),
- $\neg r_{ij} \vee \neg s_{ij}$, $\forall 0 \le i < j \le n-1$, $\forall r, s \in \mathcal{B}, r \ne s$ (at most one),
- $\neg r_{ik} \vee \neg s_{kj} \vee \bigvee_{t \in (r \diamond s) \cap C_{ij}} t$, $\forall 0 \le i < k < j \le n-1$, $\forall r \in C_{ik}$, $\forall s \in C_{kj}$ (supports).

Since $\mathcal{B}$ is a given set of a constant number of basic relations, the size of the translation $\tau_{Sup}(N)$ only depends on the number of variables n of the QCN N. Indeed the number of propositional variables of V_T is in $O(n^2)$, the number of clauses generated by $\tau_{Sup}(N)$ is in $O(n^3)$ and the number of literals in any clause of $\tau_{Sup}(N)$ is in $O(1)$.

If $\omega \models \tau_{Sup}(N)$, then ω represents the consistent scenario $\sigma = (V,C')$ of N such that $\forall 0 \le i < j \le n-1$, C'_{ij} is defined by the basic relation $r \in C_{ij}$ such that the value of the propositional variable r_{ij} in ω is *true*. At least one and at most one clauses certify that each constraint of σ is composed of one and only one basic relation of the associated constraint in N, namely σ is a scenario of N. The consistency of σ is given by the presence of *supports* clauses which guarantee the $\diamond$-closure of the scenario (we consider qualitative algebras in which $\diamond$-closed scenarios are consistent). Thus τ_{Sup} satisfies Properties 1 and 2. Since the formula $\tau_{Sup}(\sigma)$ admits exactly one model, we can represent it by a conjunction of literals of V_T.

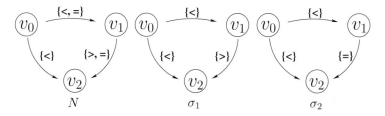

Fig. 4. A *QCN* N and its two consistent scenarios σ_1 and σ_2

Example. Figure 4 depicts a *QCN* N defined on Point Algebra where $\mathcal{B}_{pt} = \{<, =, >\}$ and $V = \{v_0, v_1, v_2\}$. N admits two consistent scenarios σ_1 and σ_2 depicted in the same figure.

$\tau_{Sup}(N)$ is built on $V_T = \{r_{ij} \mid r \in \{<, =, >\}, 0 \le i < j \le 2\}$. It is composed of the following clauses:

$$
\begin{cases}
(<_{01} \vee =_{01}), (<_{02}), (>_{12} \vee =_{12}) & \text{(at least one)} \\
(\neg <_{01} \vee \neg =_{01}), (\neg =_{01} \vee \neg >_{01}), (\neg <_{01} \vee \neg >_{01}), \\
(\neg <_{12} \vee \neg =_{12}), (\neg =_{12} \vee \neg >_{12}), (\neg <_{12} \vee \neg >_{12}), \\
(\neg <_{02} \vee \neg =_{02}), (\neg =_{02} \vee \neg >_{02}), (\neg <_{02} \vee \neg >_{02}) & \text{(at most one)} \\
(\neg <_{01} \vee \neg >_{12} \vee <_{02} \vee =_{02} \vee >_{02}), (\neg <_{01} \vee \neg =_{12} \vee <_{02}), \\
(\neg =_{01} \vee \neg >_{12} \vee >_{02}), (\neg =_{01} \vee \neg =_{12} \vee =_{02}) & \text{(supports)}
\end{cases}
$$

$\tau_{Sup}(N)$ admits exactly two models, which represents (by $\mu_{\tau_{Sup}}$) the consistent scenarios σ_1 and σ_2 of N, i.e., $\tau_{Sup}(\sigma_1)$ and $\tau_{Sup}(\sigma_2)$.

$\tau_{Sup}(\sigma_1)$ is equivalent to the following conjunction of literals of V_T:

$$(<_{\mathbf{01}} \wedge \neg =_{01} \wedge \neg >_{01} \wedge \neg <_{12} \wedge \neg =_{12} \wedge >_{\mathbf{12}} \wedge <_{\mathbf{02}} \wedge \neg =_{02} \wedge \neg >_{02}).$$

$\tau_{Sup}(\sigma_2)$ is equivalent to the following conjunction of literals of V_T:

$$(<_{\mathbf{01}} \wedge \neg =_{01} \wedge \neg >_{01} \wedge \neg <_{12} \wedge =_{\mathbf{12}} \wedge \neg >_{12} \wedge <_{\mathbf{02}} \wedge \neg =_{02} \wedge \neg >_{02}).$$

4.2   The Merging Process

We now consider a set $\mathcal{N} = \{N_1, \ldots, N_m\}$ of *QCNs* $\in QCN_{\mathcal{B}}^V$. Our merging procedure is a three step process. We first encode each *QCN* N_i ($i \in \{1, \ldots, m\}$) into a propositional formula $\tau(N_i)$. Then, we apply an *IC* merging operator $\Delta_{IC}^{\otimes, d}$ on the resulting set $\mathcal{K}$ of propositional formulas, with $IC = \tau(N_{ALL}^V)$. Lastly the set of interpretations resulting from this merging will represent the subset of consistent scenarios of $[N_{ALL}^V]$ resulting from the merging of $\mathcal{N}$.

Recall that the first requirement for defining a propositional merging operator is to define a local distance between interpretations. Given a distance d^{QCN} between scenarios, we define a distance d^{PROP} between interpretations as follows:

Definition 2. *Let τ be a translation satisfying Properties 1 and 2 and a d^{QCN} be a distance between scenarios of $[N_{ALL}^V]$. We define the distance d^{PROP} between models ω and ω' of $\tau(N_{ALL}^V)$ by $d^{PROP}(\omega, \omega') = d^{QCN}(\mu_\tau(\omega), \mu_\tau(\omega'))$.*

This definition is intuitively derived from Property 2.b.

Differents proposals can be made to define distance d^{QCN} between scenarios [5]. We recall here the Hamming distance between scenarios and the Hamming distance between interpretations.

Definition 3 (Hamming distance). *The Hamming distance between scenarios σ and σ', denoted $d_H^{QCN}(\sigma, \sigma')$, is the number of constraints that are different in the two scenarios. Formally,*

$$d_H^{QCN}(\sigma, \sigma') = |\{(v_i, v_j) \in V : \sigma(i,j) \neq \sigma'(i,j), i < j\}|,$$

where $|E|$ is the number of elements of the set E. The Hamming distance between two interpretations ω and ω', denoted $d_H(\omega, \omega')$, is the number of propositional variables of V_T which differ between the two interpretations. Formally,

$$d_H(\omega, \omega') = |\{x \in V_T : \omega(x) \neq \omega'(x)\}|,$$

where $\omega(x)$ is the truth value of the literal x in ω.

Given a translation τ, thanks to Definition 2 we can define a distance between the interpretations $\tau(\sigma)$ and $\tau(\sigma')$ equivalent to $d_H^{QCN}(\sigma, \sigma')$ for all scenarios σ, σ' of $[N_{ALL}^V]$. For the translation τ_{Sup} we have the following result.

Proposition 1. $\forall \sigma, \sigma' \in [N_{ALL}^V], \ 2 \cdot d_H^{QCN}(\sigma, \sigma') = d_H(\tau_{Sup}(\sigma), \tau_{Sup}(\sigma'))$.

Thus we can associate to the distance d_H^{QCN} between scenarios the distance d_H^{PROP} such that $\forall \omega, \omega'$ models of $\tau_{Sup}(N_{ALL}^V)$, $d_H^{PROP}(\omega, \omega') = (1/2) \cdot d_H(\omega, \omega')$.

A specific distance in the context of $QCNs$ has been defined in [5]. This distance, called the neighborhood distance, considers the notion of proximity between basic relations of a qualitative algebra [7]. A neighborhood between basic relations is often represented by a conceptual neighborhood graph. Some lattice structures allowing to determine these graphs have been defined in the literature [15,16]. The conceptual neighborhood is more precise and suitable than Hamming distance in the context of $QCNs$. Using the translation τ_{Sup}, we cannot directly define a corresponding distance between interpretations. However this will be possible if we add to τ_{Sup} some additional clauses encoding the conceptual neighborhood graph. We do not give further details on this issue due to the lack of space.

The next steps in the merging process of propositional bases consist in aggregating local distances computed in the previous step, using the aggregation operator used for merging the $QCNs$ under consideration. Therefore, we have the following result:

$$\forall \sigma \in [N_{ALL}^V] \ \sigma \in \Theta^{\otimes, d^{QCN}}(\mathcal{N}) \text{ iff } \tau(\sigma) \in \Delta_{\tau(N_{ALL}^V)}^{\otimes, d^{PROP}}(\mathcal{K}).$$

Figure 5 summarizes the merging procedure.

In the next section we describe a set of rationality postulates for $QCNs$ merging and properties of the merging operator using a more generic class of translations.

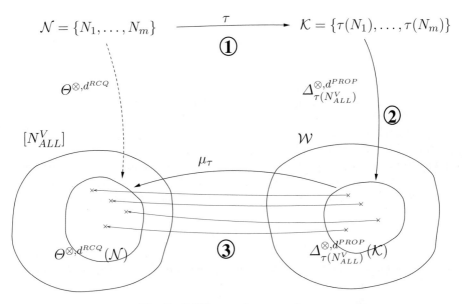

Fig. 5. *QCNs* merging procedure

5   Merging *QCNs*: Rationality Postulates and Rational Operators

In the propositional setting, a number of postulates characterizing the rational belief merging operators have been given [13]. The purpose of this section is to present similar postulates for merging *QCNs* and to show how one can define rational operators for merging *QCNs* from propositional merging operators thanks to translations. We first recall rationality postulates given in the propositional setting. We denote by $\bigwedge \mathcal{K}$ the conjunction of the knowledge bases of the profile $\mathcal{K}$.

Definition 4. *Let* $\mathcal{K}, \mathcal{K}_1$ *and* $\mathcal{K}_2$ *be three profiles,* K_1, K_2 *be consistent knowledge bases and* IC, IC_1, IC_2 *be propositional formulas.* Δ *is an IC merging operator iff it satisfies the following postulates.*

(IC0) $\Delta_{IC}(\mathcal{K}) \models IC$.
(IC1) *If IC is consistent, then* $\Delta_{IC}(\mathcal{K})$ *is consistent.*
(IC2) *If* $\bigwedge \mathcal{K} \wedge IC$ *is consistent, then* $\Delta_{IC}(\mathcal{K}) \equiv \bigwedge \mathcal{K} \wedge IC$.
(IC3) *If* $\mathcal{K}_1 \equiv \mathcal{K}_2$ *and* $IC_1 \equiv IC_2$, *then* $\Delta_{IC_1}(\mathcal{K}_1) \equiv \Delta_{IC_2}(\mathcal{K}_2)$.
(IC4) *If* $K_1 \models IC$ *and* $K_2 \models IC$, *then if* $\Delta_{IC}(\{K_1, K_2\}) \wedge K_1$ *is consistent then* $\Delta_{IC}(\{K_1, K_2\}) \wedge K_2$ *is consistent.*
(IC5) $\Delta_{IC}(\mathcal{K}_1) \wedge \Delta_{IC}(\mathcal{K}_2) \models \Delta_{IC}(\mathcal{K}_1 \sqcup \mathcal{K}_2)$.
(IC6) *If* $\Delta_{IC}(\mathcal{K}_1) \wedge \Delta_{IC}(\mathcal{K}_2)$ *is consistent, then* $\Delta_{IC}(\mathcal{K}_1 \sqcup \mathcal{K}_2) \models \Delta_{IC}(\mathcal{K}_1) \wedge \Delta_{IC}(\mathcal{K}_2)$.
(IC7) $\Delta_{IC_1}(\mathcal{K}) \wedge IC_2 \models \Delta_{IC_1 \wedge IC_2}(\mathcal{K})$.
(IC8) *If* $\Delta_{IC_1}(\mathcal{K}) \wedge IC_2$ *is consistent, then* $\Delta_{IC_1 \wedge IC_2}(\mathcal{K}) \models \Delta_{IC_1}(\mathcal{K}) \wedge IC_2$.

For the sake of generality, we consider that information conveyed by a QCN N is not reduced to the set $[N]$ of its consistent scenarios (as it is the case in [5]) but more generally to a subset $\langle N \rangle$ of $QCN_\mathcal{B}^V$ (this enables us for taking some context-dependent information into account). In the same vein, the result of the merging is defined as a subset of $QCN_\mathcal{B}^V$ (instead of a subset of $[N_{ALL}^V]$). Thus we now define a $QCNs$ merging operator Θ as a mapping which associates to a finite subset $\mathcal{N}$ of $QCN_\mathcal{B}^V$, a subset of $QCN_\mathcal{B}^V$.

The following rationality postulates are the direct counterparts in the $QCNs$ setting of the postulates *(IC0)* - *(IC8)* from [13] and the postulates *(A1)* - *(A6)* from [12], for propositional merging. Before presenting them, we first need to define a notion of equivalence between $QCNs$ and between sets of $QCNs$: Two $QCNs$ N and N' are said to be *equivalent*, denoted $N \equiv N'$, iff $\langle N \rangle = \langle N' \rangle$. Two subsets $\mathcal{N}_1$ and $\mathcal{N}_2$ of $QCNs$ are said to be *equivalent*, denoted $\mathcal{N}_1 \equiv \mathcal{N}_2$, iff there exists a bijection f from $\mathcal{N}_1$ to $\mathcal{N}_2$ such that $\forall N_1 \in \mathcal{N}_1$, $N_1 \equiv f(N_1)$.

Definition 5. *Let $\mathcal{N}, \mathcal{N}_1$ and $\mathcal{N}_2$ be finite sets of $QCNs$, and let N_1, N_2 be two consistent $QCNs$. Θ is a $QCNs$ merging operator iff it satisfies the following postulates:*

(N1) $\Theta(\mathcal{N}) \neq \emptyset$.
(N2) If $\bigcap_{N_i \in \mathcal{N}} \langle N_i \rangle \neq \emptyset$, then $\Theta(\mathcal{N}) = \bigcap_{N_i \in \mathcal{N}} \langle N_i \rangle$.
(N3) If $\mathcal{N}_1 \equiv \mathcal{N}_2$, then $\Theta(\mathcal{N}_1) = \Theta(\mathcal{N}_2)$.
(N4) If $\Theta(\{N_1, N_2\}) \cap \langle N_1 \rangle \neq \emptyset$, then $\Theta(\{N_1, N_2\}) \cap \langle N_2 \rangle \neq \emptyset$.
(N5) $\Theta(\mathcal{N}_1) \cap \Theta(\mathcal{N}_2) \subseteq \Theta(\mathcal{N}_1 \sqcup \mathcal{N}_2)$.
(N6) If $\Theta(\mathcal{N}_1) \cap \Theta(\mathcal{N}_2) \neq \emptyset$, then $\Theta(\mathcal{N}_1 \sqcup \mathcal{N}_2) \subseteq \Theta(\mathcal{N}_1) \cap \Theta(\mathcal{N}_2)$.

(N1) ensures that the result of the merging is non-trivial. *(N2)* requires $\Theta(\mathcal{N})$ to be the set of $QCNs$ shared by $\langle N_i \rangle \; \forall N_i \in \mathcal{N}$, when this set is non-empty. *(N3)* is a syntax-irrelevance principle. *(N4)* is an equity postulate: it asks that the merging operator does not exploit any hidden preferences between two $QCNs$ to be merged. *(N5)* and *(N6)* state that, if there exists a non-empty set E of $QCNs$ shared by the mergings of two groups $\mathcal{N}_1$ and $\mathcal{N}_2$, then the merging of the joint groups must be this set E.

We now show how one can define $QCNs$ merging operators Θ satisfying all those postulates *(N1)* - *(N6)* from propositional IC merging operators Δ thanks to translations τ. We first need to slightly modify the notion of translation presented in the previous section so that to ensure the existence of a bijection from the set of models of the propositional formula $\tau(N)$ to $\langle N \rangle$:

Definition 6. *A translation τ is a mapping from $QCN_\mathcal{B}^V$ to $PROP$ satisfying Property 1 and such that:*

(a.) $\exists \varphi_\tau \in PROP : \forall N \in QCN_\mathcal{B}^V, \tau(N) \models \varphi_\tau$.
(b.) *There exists a bijection μ_τ from the set of models of φ_τ to $QCN_\mathcal{B}^V$ such that $\forall N \in QCN_\mathcal{B}^V, \{\mu_\tau(\omega) \mid \omega \models \tau(N)\} = \langle N \rangle$.*

The conditions on τ ensure the existence of a propositional formula φ_τ associated to τ, such that every model of φ_τ corresponds (in a bijective way via μ_τ) to a QCN of $QCN_\mathcal{B}^V$. For instance, one can have $\varphi_\tau = \tau(N_{ALL}^V)$, provided that

$\forall N \in QCN_{\mathcal{B}}^V$, $\tau(N) \models \tau(N_{ALL}^V)$. In addition, $\forall N \in QCN_{\mathcal{B}}^V$, the set of models of $\tau(N)$ must be in bijection via μ_τ with $\langle N \rangle$. Observe that the notion of translation defined in the previous section satisfies the requirements of Definition 6 assuming that $\forall N$, $\langle N \rangle = [N]$.

Definition 7. *Let τ be a translation in the sense of Definition 6 and Δ_{IC} an IC merging operator (i.e., a propositional merging operator satisfying **(IC0)** - **(IC8)**) with $IC = \varphi_\tau$. The $QCNs$ merging operator Θ induced by τ and Δ_{IC} is defined by: let $\mathcal{N} = \{N_1, \dots, N_m\}$ be a set of $QCNs$, we have*

$$\Theta(\mathcal{N}) = \{\mu_\tau(\omega) \mid \omega \models \Delta_{\varphi_\tau}(\tau(N_1), \dots, \tau(N_m))\}.$$

(IC0) ensures that every model of $\Delta_{\varphi_\tau}(\tau(N_1), \dots, \tau(N_m))$ is associated to a QCN from $QCN_{\mathcal{B}}^V$ via μ_τ. We have:

Proposition 2. *Every $QCNs$ merging operator induced by a translation (in the sense of Definition 6) and an IC merging operator satisfies **(N1)** - **(N6)**.*

6   Conclusion

Using a particular class of propositional distance-based merging operators, we have shown that the $QCNs$ merging operator presented in [5] can be reduced to propositional merging. Thus we can retrieve some interesting results from the widely studied topic of merging propositional bases to our work. For example, we directly get a characterization of the complexity of the process [11]. Moreover an efficient implementation of propositional merging operators has recently been proposed [9] while the implementation of the merging method developed in [5] is hard in practice.

Our method is valid for qualitative formalisms in which the closure by weak composition is complete for its consistency problem, however it is not appropriate if no translation allows a propositional formula to capture the set of consistent scenarios of the translated QCN. In addition, we have proposed a set of rationality postulates for $QCNs$ merging operators. These postulates are satisfied if we use an appropriate translation from $QCNs$ to propositional formulas and a particular class of propositional merging operators.

This work can be extended in several directions. Given a context and a particular definition of the set $\langle N \rangle$ for all N, one can define and study some appropriate translations. Another perspective is to study properties about the $QCNs$ merging operator using other classes of propositional merging operators.

References

1. Allen, J.-F.: An interval-based representation of temporal knowledge. In: IJCAI, pp. 221–226 (1981)
2. Benferhat, S., Dubois, D., Prade, H., Williams, M.-A.: A practical approach to fusing prioritized knowledge bases. In: Barahona, P., Alferes, J.J. (eds.) EPIA 1999. LNCS, vol. 1695, pp. 222–236. Springer, Heidelberg (1999)

3. Cholvy, L.: Reasoning about merging information. In: Handbook of Defeasible Reasoning and Uncertainty Management Systems, vol. 3, pp. 233–263 (1998)
4. Condotta, J.-F., D'Almeida, D.: Qualitative constraints representation for the time and space in SAT. In: ICTAI, pp. 74–77 (2007)
5. Condotta, J.-F., Kaci, S., Schwind, N.: A framework for merging qualitative constraints networks. In: FLAIRS (2008)
6. Drake, L., Frisch, A.M., Gent, I.P., Walsh, T.: Automatically reformulating SAT-encoded CSPs. In: CP (2002)
7. Freksa, C.: Temporal reasoning based on semi-intervals. Artificial Intelligence 54(1), 199–227 (1992)
8. Gent, I.P.: Arc consistency in SAT. In: ECAI, pp. 121–125 (2002)
9. Gorogiannis, N., Hunter, A.: Implementing semantic merging operators using binary decision diagrams. International Journal of Approximate Reasoning 49(1), 234–251 (2008)
10. Hué, J., Papini, O., Würbel, E.: Removed sets fusion: Performing off the shelf. In: ECAI (July 2008)
11. Konieczny, S., Lang, J., Marquis, P.: Distance-based merging: A general framework and some complexity results. In: KR, pp. 97–108 (2002)
12. Konieczny, S., Pérez, R.P.: On the logic of merging. In: KR, pp. 488–498 (1998)
13. Konieczny, S., Pérez, R.P.: Merging with integrity constraints. In: Hunter, A., Parsons, S. (eds.) ECSQARU 1999. LNCS, vol. 1638, pp. 233–244. Springer, Heidelberg (1999)
14. Konieczny, S., Pérez, R.P.: Merging information under constraints: a logical framework. Journal of Logic and Computation 12(5), 773–808 (2002)
15. Ligozat, G.: On generalized interval calculi. In: AAAI, Anaheim, CA, pp. 234–240 (1991)
16. Ligozat, G.: Reasoning about cardinal directions. Journal of Visual Languages and Computing 9(1), 23–44 (1998)
17. Lin, J.: Integration of weighted knowledge bases. Artificial Intelligence 83, 363–378 (1996)
18. Mitra, D.: Modeling and reasoning with star calculus. In: Annals of Mathematics and Artificial Intelligence (2004)
19. Nebel, B., Bürckert, H.-J.: Reasoning about temporal relations: A maximal tractable subclass of allen's interval algebra. In: AAAI, Seattle, WA, pp. 356–361 (1994)
20. Pham, D.N., Thornton, J., Sattar, A.: Towards an efficient SAT encoding for temporal reasoning. In: Benhamou, F. (ed.) CP 2006. LNCS, vol. 4204, pp. 421–436. Springer, Heidelberg (2006)
21. Randell, D.-A., Cui, Z., Cohn, A.: A spatial logic based on regions and connection. In: KR, pp. 165–176 (1992)
22. Revesz, P.Z.: On the semantics of theory change: Arbitration between old and new information. In: PODS, Washington, DC, pp. 71–82 (1993)
23. Revesz, P.Z.: On the semantics of arbitration. Journal of Algebra and Computation 7(2), 133–160 (1997)
24. Vilain, M., Kautz, H., Van Beek, P.: Constraint propagation algorithms for temporal reasoning: A revised report. In: Readings in Qualitative Reasoning about Physical Systems, pp. 373–381. Kaufmann, San Mateo (1990)

Distance-Based Semantics for C-Structure Belief Revision

Omar Doukari, Eric Würbel, and Robert Jeansoulin

UMR CNRS 6168 LSIS, Domaine Universitaire de Saint-Jérôme
13397 Marseille France
{omar.doukari,eric.wurbel,robert.jeansoulin}@lsis.org
http://www.lsis.org

Abstract. In [1], the authors have extended Parikh's relevance-sensitive model for belief revision by defining a new model for belief representation and local belief revision called C-structure Model. This model allows to make local revision when Parikh's model fails to do it: the case of "fully overlapping belief sets". Using Grove's system of spheres construction, we consider additional constraints to define an ordering between interpretations, and show that these constraints allow to formalize perfectly the local revision by the mean of C-structure model, thus providing a well defined semantics for revision of C-structures.

Keywords: Belief revision, C-structure model, systems of spheres.

1 Introduction

Agents facing incomplete, uncertain, and inaccurate information must use a rational belief revision operation in order to manage belief changes. The agent's epistemic state represents its reasoning process with his beliefs and belief revision consists in modifying its initial epistemic state in order to maintain consistency, while keeping new information and modifying the least possible previous information.

Unfortunately, in the general case, the theoretical complexity of revision is high. More precisely, it belongs to the $\prod_2^p$ class in the framework of propositional logic [2,3,4,5]. Similarly for the few applications which have been developed for belief revision [6,7]. Hence reducing the amount of data to be processed during the revision operation seems to be an interesting approach, since formal complexity cannot be reduced.

Usually inconsistency is due to the accidental presence of a "few" pieces of contradictory information about a given subject. Hence, revision may be restricted to local portions of the belief corpus (those intersecting with the language of the new epistemic input).

To introduce relevance-sensitivity into belief revision, Parikh [8] defined the language splitting (LS) model which says that any set of beliefs may be represented as a family of letter-disjoint sets and that revision may be made locally

C. Sossai and G. Chemello (Eds.): ECSQARU 2009, LNAI 5590, pp. 359–370, 2009.

on one of these sets. In practice, since beliefs do have some overlap, the partition of the main set of beliefs cannot be actually strict. In view of this gap, Parikh's original model for belief revision [8] has been extended, by allowing for such overlap, in the B-structures model [9]. However, this model is not able to guarantee a global revision by only a local one, i.e., after revising our beliefs we are not sure of their global consistency. This fact interfers with the correction of belief revision (according to the rational belief revision principles called AGM postulates [10]). The B-structures model also lacks a semantic characterization at present.

In order to circumvent these problems, a new model called the C-structure model, has been defined in [1]. This model allows some overlap between the different belief subsets and preserves all the desirable properties of the language splitting model (in particular, it allows to prove global consistency by the mean of local consistency check). Furthermore, this model allows to perform local revision when the LS model fails to do that; for instance, in cases where the belief set cannot be split in more than one subtheory (the case of fully overlapping belief sets).

Using Grove's system of spheres construction [11], we provide semantics for local revision by the mean of the C-structure model, by defining additional constraints based on a distance measurement between interpretations. These constraints characterize local revision by the mean of the C-structure model in the case of fully overlapping theories.

The structure of the paper is as follows. In section 2, we provide some preliminaries and background material on the AGM paradigm. In the following section, we define the C-structure model. In section 4, we provide system of spheres semantics for local revision by the mean of the C-structure model in the case of fully overlapping belief sets.

2   Preliminaries

Throughout this paper, $\mathcal{L}$ is a propositional language defined on some finite set of propositional variables (atoms) $\mathcal{V}$ and the usual connectors ($\neg, \vee, \wedge, \rightarrow, \leftrightarrow$). If $\alpha \in \mathcal{L}$ is a sentence, then $\mathcal{V}(\alpha)$ represents the set of variables appearing in α, and similarly $\mathcal{V}(X)$ for a set of sentences X. If V is a subset of $\mathcal{V}$ then $\mathcal{L}(V)$ represents the propositional sublanguage defined over V, i.e., $\mathcal{L}(V) = \{\alpha \in \mathcal{L} : \mathcal{V}(\alpha) \subseteq V\}$. $\vdash$ represents the classical inference relation. A literal is a propositional variable or its negation. A clause is a disjunction of literals. A clause c is an implicate of a sentence α iff $\alpha \vdash c$. A clause c is a prime implicate of α iff for all implicates c' of α such that $c' \vdash c$, it is the case that $c \vdash c'$. We denote by $Cove_\alpha$ an arbitrary covering of α, which is a set of prime implicates of α such that for every clause c where $\alpha \vdash c$, there exists $c' \in Cove_\alpha$ such that $c' \vdash c$. $\mathcal{V}(Cove_\alpha)$ is the minimal set of atoms needed to express (a sentence logically equivalent to) α [12]. This set is unique [8].

If X is a set of sentences then $Cn(X)$ is the logical closure of X, i.e., $Cn(X) = \{\alpha \in \mathcal{L} : X \vdash \alpha\}$. In particular, X is a theory, i.e., a belief set iff $X = Cn(X)$.

If T is a theory of $\mathcal{L}$, it is said a fully overlapping theory if and only if for all partitions V_1, V_2 of $\mathcal{V}$ there does not exist sentences $\alpha_1 \in \mathcal{L}(V_1)$, and $\alpha_2 \in \mathcal{L}(V_2)$ such that $T = Cn(\{\alpha_1, \alpha_2\})$. For a theory T, we denote by B_T a belief base of T which is a finite set of sentences that generates T, i.e., $T = Cn(B_T)$. B_T is a minimal belief base of T iff (i) B_T is a belief base of T, (ii) T is axiomatized by B_T (i.e., $\forall \alpha \in B_T, (B_T \setminus \{\alpha\}) \nvdash \alpha$), and (iii) $B_T \subseteq Cove_{\bigwedge_{\alpha \in B_T}}$.

If B_T is an inconsistent belief base, $M \subseteq B_T$ is a minimal inconsistent subset (MIS) of B_T iff for all $M' \subset M$, M' is consistent. We denote the set of all consistent theories of $\mathcal{L}$ by $\mathcal{K}_{\mathcal{L}}$. We denote by $\mathcal{I}_{\mathcal{L}}$ the set of all interpretations of $\mathcal{L}$. For a set of sentences X of $\mathcal{L}$, $[X]$ represents the set of all interpretations of $\mathcal{L}$ that satisfy X (the set of models of X). Often we use the notation $[\alpha]$ for a sentence $\alpha \in \mathcal{L}$, as an abbreviation of $[\{\alpha\}]$. For a theory T and a set of sentences X of $\mathcal{L}$, $T + X$ represents the set $Cn(T \cup X)$.

Let $\mathcal{L}' \subseteq \mathcal{L}$ be defined over a subset V' of $\mathcal{V}$, $\overline{\mathcal{L}'}$ represents the sublanguage defined over the propositional variables in the complement of V', i.e., $\overline{\mathcal{L}'} = \mathcal{L}(\mathcal{V} \setminus V')$. $Cn_{\mathcal{L}'}(X)$ for a set of sentences $X \subset \mathcal{L}'$, represents the logical closure of X in $\mathcal{L}'$. When no subscript is present, it is understood that the operation is relevant to the original language $\mathcal{L}$. Finally, let U be a set of interpretations in $\mathcal{I}_{\mathcal{L}}$. By $U/\mathcal{L}'$ we denote the restriction of U to $\mathcal{L}'$; that is, $U/\mathcal{L}' = \{w \cap \mathcal{L}' : w \in U\}$.

In belief revision, much work takes as its starting point the AGM postulates [10], which appear to capture much of what characterizes rational belief revision. In this framework belief states are represented as theories of $\mathcal{L}$, and the process of belief revision is modelled by a revision function $*$ which is any function from $\mathcal{K}_{\mathcal{L}} \times \mathcal{L}$ to $\mathcal{K}_{\mathcal{L}}$, mapping $\langle T, \alpha \rangle$ to $T * \alpha$ that satisfies the AGM postulates [10]. This set of postulates describes a class of revision functions, however it does not provide a constructive way of defining such a function.

Grove introduced in [11] a construction of revision functions that generates precisely the class of functions satisfying the AGM postulates. It is based on a special structure on consistent theories, called a system of spheres. Let T be a theory of $\mathcal{L}$, and S_T a collection of sets of interpretations, i.e., $S_T \subseteq 2^{\mathcal{I}_{\mathcal{L}}}$. S_T is a system of spheres centered on $[T]$ iff the following conditions are satisfied:

(S1). S_T is totally ordered wrt set inclusion.

(S2). The smallest sphere in S_T is $[T]$.

(S3). $\mathcal{I}_{\mathcal{L}} \in S_T$.

(S4). $\forall \alpha \in \mathcal{L}$, if there is any sphere in S_T intersecting $[\alpha]$ then there is also a smallest sphere in S_T intersecting $[\alpha]$.

For a system of spheres S_T and a sentence $\alpha \in \mathcal{L}$, the smallest sphere in S_T intersecting α is denoted $C_T(\alpha)$[1]. With any system of spheres S_T, Grove associates a function $f_T : \mathcal{L} \mapsto 2^{\mathcal{I}_{\mathcal{L}}}$ defined as follows : $f_T(\alpha) = [\alpha] \cap C_T(\alpha)$. Consider now a theory T of $\mathcal{L}$ and let S_T be a system of spheres centered on $[T]$. Grove uses S_T to define constructively the process of revising T, by means of the following condition : $(S*) : T * \alpha = \bigcap f_T(\alpha)$.

---

[1] In the limiting case where α is inconsistent, Grove defines $C_T(\alpha)$ to be the set $\mathcal{I}_{\mathcal{L}}$. In this paper we only consider revision by consistent sentences.

3   Local Revision by the Mean of the C-Structure Model

The C-structure model [1] extends the language splitting [8] and the B-structure models [9]. It uses disjoint sublanguages to define a set of *cores* of a given language, each surrounded by a *covering* of atoms. The concept of covering allows some degree of overlap between the sublanguages defined over the coverings. We now recall the main definitions and results of this model.

Definition 1. $\{V_1, ..., V_n\}$ *is a set of cores of* $\mathcal{L}$ *iff it is a partition of* $\mathcal{V}$.

Example 1. Let the language $\mathcal{L}$ be built from the propositional variables a, b, c, d. Let T be a fully overlapping theory of $\mathcal{L}$, axiomatized by $B_T = \{\neg a \vee b, \neg b \vee c, \neg c \vee b, \neg c \vee d\}$. The set $\{\{a\}, \{b\}, \{c\}, \{d\}\}$ is a set of cores of $\mathcal{L}$.

To order the atoms of $\mathcal{L}$, we use the following relevance relation from [13].

Definition 2. *Let* T *be a theory of* $\mathcal{L}$. *We say that two atoms,* p *and* q, *are directly relevant wrt* B_T, *denoted by* $R(p, q, B_T)$ *(or by* $R_0(p, q, B_T)$*), iff* $\exists \alpha \in B_T$ *s.t.,* $p, q \in \mathcal{V}(\alpha)$. *Two atoms* p, q *are k-relevant wrt* B_T, *denoted by* $R_k(p, q, B_T)$, *if* $\exists p_0, p_1, ..., p_{k+1} \in \mathcal{V}$ *s.t.:* $p_0 = p$; $p_{k+1} = q$; *and* $\forall i \in \{0, ..., k\}, R(p_i, p_{i+1}, B_T)$.

In Example 1, we find : $R(a, b, B_T), R_1(a, c, B_T), R_2(a, d, B_T)$, etc.
   To define clearly the extent of overlapping between the various sublanguages, we define a *distance* between variables.

Definition 3. *Suppose two atoms* $p, q \in \mathcal{V}$, T *is a theory of* $\mathcal{L}$. *The distance between* p, q *wrt* B_T, *denoted by* $dist(p, q, B_T)$, *is defined as follows:*

$$dist(p, q, B_T) = \begin{cases} 0 & \text{if } p = q \\ min\{k : R_k(p, q, B_T)\} + 1 & \text{if such } k \text{ exists} \\ \infty & \text{otherwise.} \end{cases}$$

In Example 1, $dist(a, b, B_T) = 1, dist(a, c, B_T) = 2, dist(a, d, B_T) = 3$, etc.
   We now define the notion of a covering, parametrized by its *thickness*:

Definition 4. *Let* $\{V_1, ..., V_n\}$ *be a set of cores of* $\mathcal{L}$ *and* T *be a theory of* $\mathcal{L}$. $Cov_k(V_i, B_T)$ *is a covering of thickness* k *of* V_i *wrt* B_T *iff:* $Cov_k(V_i, B_T) \subseteq \mathcal{V}$; *and* $\forall p \in \mathcal{V}$, *if* $\exists q \in V_i$ *s.t.,* $dist(p, q, B_T) \le k$ *then* $p \in Cov_k(V_i, B_T)$.

For example, the set of coverings with thickness 1 corresponding to the set of cores $\{\{a\}, \{b\}, \{c\}, \{d\}\}$ wrt B_T (Example 1) is : $\{\{a, b\}, \{a, b, c\}, \{b, c, d\}, \{c, d\}\}$.
   In order to parametrize a C-structure by a particular thickness (as we will see later), we require a definition of the size of a MIS:

Definition 5. *Let* B_T *and* $B'_{T'}$ *be two belief bases such that* $\mathcal{V}(B'_{T'}) \subseteq \mathcal{V}(B_T)$ *and* $B'_{T'}$ *is inconsistent. The size of the MIS* M *of* $B'_{T'}$ *wrt* B_T, $Size(M, B_T) = max\{dist(a, b, B_T) : a, b \in \mathcal{V}(M)\}$.

In Example 1, let $M = \{a \wedge \neg b, \neg a \vee b\}$ be a MIS of $B'_{T'} = B_T \cup \{a \wedge \neg b\}$, so $Size(M, B_T) = 1$.
   The only assumption made by the C-structure model is that the maximal size of eventual existing MISs in a given belief base is known. Hence, when we

construct a C-structure C on a belief base B_T, we only require that the thickness of coverings of cores (value of k) should be (at least) equal to the maximal size of MISs which may exist in B_T.

Informally, a C-structure represents the knowledge of an agent with a good understanding of the interactions between subjects.

Definition 6. *Let T be a theory defined in $\mathcal{L}$ and B_T an arbitrary belief base of T. The set $C = \{(V_1, Cov_k(V_1, B_T), T_1), ..., (V_n, Cov_k(V_n, B_T), T_n)\}$ is a C-structure of T iff: (i) $\{V_1, ..., V_n\}$ is a set of cores of $\mathcal{L}$, (ii) $Cov(C) = \{Cov_k(V_1, B_T), ..., Cov_k(V_n, B_T)\}$ is a corresponding set of coverings wrt B_T s.t., $\forall i \in \{1, ..., n\} \forall \alpha \in \mathcal{L}(Cov_k(V_i, B_T))$, if $B_T \cup \{\alpha\}$ is inconsistent, then $\forall M$ a MIS of $B_T \cup \{\alpha\}$, $Size(M, B_T) \leq k$, and (iii) $\forall T_i, T_i = Cn_{\mathcal{L}(Cov_k(V_i, B_T))}(\mathcal{L}(Cov_k(V_i, B_T))) \cap T)$. C is called an atomic C-structure of T iff $\forall i \in \{1, ..., n\}$, $|V_i| = 1$ (in the case of redundancy of coverings, we merge the corresponding cores).*

We obtain the following C-structure corresponding to Example 2 by assuming that the maximal size of eventual exiting MISs in B_T is 1 (condition (ii) of Definition 6): $\{(\{a\}, \{a, b\}, Cn_{\mathcal{L}(\{a,b\})}(\{\neg a \vee b\})), (\{b\}, \{a, b, c\}, Cn_{\mathcal{L}(\{a,b,c\})}(\{\neg a \vee b, \neg b \vee c, \neg c \vee b\})), (\{c\}, \{b, c, d\}, Cn_{\mathcal{L}(\{b,c,d\})}(\{\neg b \vee c, \neg c \vee b, \neg c \vee d\})), (\{d\}, \{c, d\}, Cn_{\mathcal{L}(\{c,d\})}(\{\neg c \vee d\}))\}$.

Now, we can formulate local revision by the mean of the C-structure model as follows.

(Local Revision): Let T be a theory of $\mathcal{L}$, B_T an arbitrary belief base of T, and $C = \{(V_1, Cov_k(V_1, B_T), T_1), ..., (V_n, Cov_k(V_n, B_T), T_n)\}$ a C-structure of T. If $\alpha \in \mathcal{L}(Cov_k(V_i, B_T))$ for all $i \in \{1, .., m\}$ and $\mathcal{V}(Cove_\alpha) \cap V_i \neq \emptyset$, then: $T * \alpha = (Cn_{\mathcal{L}(\bigcap_{i=1}^{m} Cov_k(V_i, B_T))}(\bigcap_{i=1}^{m} T_i) \circ \alpha) + (B_T \setminus \bigcap_{i=1}^{m} T_i)$, where $\circ$ is a revision operator of the sublanguage $\mathcal{L}(\bigcap_{i=1}^{m} Cov_k(V_i, B_T))$.

Informally, local revision has to precise two points. Firstly, anything outside the related pat of T to α ($B_T \setminus \bigcap_{i=1}^{m} T_i$) will not be affected during the revision of the theory T by α. Secondly, the related part of the theory T to α ($\bigcap_{i=1}^{m} T_i$) should change into $Cn_{\mathcal{L}(\bigcap_{i=1}^{m} Cov_k(V_i, B_T))}(\bigcap_{i=1}^{m} T_i) \circ \alpha$, where $\circ$ is a revision function defined over the sublanguage $\mathcal{L}_1 = \mathcal{L}(\bigcap_{i=1}^{m} Cov_k(V_i, B_T))$. In the following, we consider the revision function $\circ$ that modifies the relevant part of T does not vary from theory to theory, even when the relevant part $\bigcap_{i=1}^{m} T_i$ stays the same (i.e., the case where the revision function $\circ$ is context-insensitive).

More formally, local revision, given above, can be defined by mean of the following conditions.

Let $C = \{(V_1, Cov_k(V_1, B_T), T_1), ..., (V_n, Cov_k(V_n, B_T), T_n)\}$ be a C-structure of T, $\alpha \in \mathcal{L}(Cov_k(V_i, B_T))$, and $\mathcal{V}(Cove_\alpha) \cap V_i \neq \emptyset$, for all $i \in \{1, .., m\}$. We denote by $\mathcal{L}_1$, the sublanguage $\mathcal{L}(\bigcap_{i=1}^{m} Cov_k(V_i, B_T))$.

(C1). $(T * \alpha) \cap \overline{\mathcal{L}_1} = (Cn_{\mathcal{L}_1}(\bigcap_{i=1}^{m} T_i) \circ \alpha) + (B_T \setminus \bigcap_{i=1}^{m} T_i)) \cap \overline{\mathcal{L}_1}$.

(C2). $(T * \alpha) \cap \mathcal{L}_1 = (Cn(\bigcap_{i=1}^{m} T_i) * \alpha) \cap \mathcal{L}_1$.

Condition (C1) is straightforward : when revising a theory T by a sentence α, the part of T that is not related to α is not affected by the revision; we do not remove any information from it since MISs generated by α are all in the related part to α. However, we can deduce more consequences because the existence of overlap between the parts related and unrelated to α. Condition (C2) is what imposes the context-insensitivity of local revision. To see this, consider a revision function $*$ (which defines a revision policy for all the theories of $\mathcal{L}$), and let $C = \{(V_1, Cov_k(V_1, B_T), T_1), (V_2, Cov_k(V_2, B_T),\ T_2), ..., (V_n, Cov_k(V_n, B_T), T_n)\}$ and $C' = \{(V_1', Cov_{k'}(V_1', B_{T'}), T_1'), (V_2', Cov_{k'}(V_2', B_{T'}), T_2'), ..., (V_{n'}', Cov_{k'}(V_{n'}', B_{T'}), T_{n'}')\}$, be two C-structures of the two theories T and T' on which are based the revisions of T and T', respectively, by a sentence $\alpha \in \mathcal{L}$. If the relevant parts to α of T and T' is in both cases the same, then according to (C2), the way that this relevant part is modified in both T and T' is also the same.

The following result shows that (C1) and (C2) are indeed equivalent to local revision for consistent fully overlapping theories.

Theorem 1. *Let $*$ be a revision function satisfying the AGM postulates $(T * 1)$– $(T * 8)$. Then $*$ satisfies local revision iff $*$ satisfies (C1) and (C2).*

Local revision makes associations between the revision policies of different fully overlapping theories. Thus, by the mean of (C2), it introduces dependencies between the revisions carried out on different (overlapping) C-structures. In [8], the authors introduced the same property but only between theories which can be split into at least two subtheories. However, the AGM postulates are too weak to induce such property, since they all refer to a single theory T.

In the next section, we formulate system-of-spheres semantics for local revision of consistent fully overlapping theories.

4 Semantics for Local Revision by the Mean of the C-Structure Model

Let T be a consistent fully overlapping theory, and let S_T be a system of spheres centered on $[T]$. The intended meaning of S_T is that it represents comparative plausibility between interpretations, i.e., the further away an interpretation is from the center of S_T, the less plausible it is to $[T]$ [14]. However, none of the conditions (S1)–(S4) indicate how plausibility between interpretations should be measured.

In order to formalize semantics for local revision in the realm of system of spheres, we need to define a specific criterion of plausibility *Plaus* to cover comparisons between an interpretation w and a fully overlapping theory T. This criterion allows to define an ordering between interpretations based on condition (C1).

Our definition of *Plaus* is based on the comparison between an interpretation w and a C-structure C of T which verifies some conditions. First, this C-structure should be an atomic one to get the smallest related part of T to the new information. Hence, during revision operation, we avoid throwing away non-tautological beliefs in T whenever it is possible. Second, it should contain at least one subtheory whose removing restaures consistency between T and w, i.e., $\exists T_i \in C$ such that, $w \in [B_T \setminus T_i]$. Third, C should be minimal in the sense that any atomic C-structure C' whose thickness of coverings is $k' < k$ (k is the thickness of coverings of C) has any subtheory T'_i such that, $w \in [B_T \setminus T'_i]$. These three conditions allow to locate, on B_T, the minimal part which is responsible of the inconsistency between T and w. Furthermore, since it may exist more than one subtheory in C whose removing restores consistency between T and w, so removing the intersection of these subtheories also restores consistency between T and w.

The criterion of plausibility $Plaus_T(w)$ gives the subset of variables on which is defined the intersection of subtheories of C (the C-structure of T verifying the last three conditions) whose removing restores consistency between T and w.

In particular, if $w \in [T]$, we have $Plaus_T(w) = \emptyset$, since k, the thickness of the C-structure by which the comparison with w should be done, is equal to 0. Hence, $[T]$ represents the most plausible subset of interpretations in $\mathcal{I}_\mathcal{L}$. Clearly, this fact is intuitively satisfactory, since it follows Grove's construction requirements.

Now, we define the criterion $Plaus$ formally as follows: Now, we define the criterion $Plaus$ formally as follows:

Definition 7. *Let T be a consistent fully overlapping theory of $\mathcal{L}$, w an interpretation, and C be an atomic C-structure of T constructed on a minimal belief base B_T, and k, the thickness of coverings of C, is such that:*

1. *$\exists X \in Cov(C)$ such that $w \in [B_T \setminus (T \cap X)]$, and*
2. *$\forall k' < k$, if C' is the atomic C-structure of thickness k' constructed on B_T, then $\nexists X \in Cov(C')$ such that $w \in [B_T \setminus (T \cap X)]$.*

Then $Plaus_T(w) = \bigcap \{ Cov_k(V_i, B_T) \in Cov(C) : w \in [B_T \setminus T_i] \}$.

From Example 1, Table 1 below illustrates the computation of $Plaus_T(w_i)$ for all $w_i \in \mathcal{I}_\mathcal{L} \setminus [T]^2$. $[T] = \{abcd, \bar{a}bcd, \bar{a}\bar{b}\bar{c}d, \bar{a}\bar{b}\bar{c}\bar{d}\}$.

Table 1. Computation of $Plaus_T(w_i)$ for Example 1

w_i	$Plaus_T(w_i)$	w_i	$Plaus_T(w_i)$
$w_1 = abc\bar{d}$	$\bigcap\{\{b,c,d\},\{c,d\}\} = \{c,d\}$	$w_7 = a\bar{b}\bar{c}\bar{d}$	$\bigcap\{\{a,b\},\{a,b,c\}\} = \{a,b\}$
$w_2 = ab\bar{c}d$	$\bigcap\{\{a,b,c\},\{b,c,d\}\} = \{b,c\}$	$w_8 = \bar{a}bc\bar{d}$	$\bigcap\{\{c,d\},\{b,c,d\}\} = \{c,d\}$
$w_3 = ab\bar{c}\bar{d}$	$\bigcap\{\{a,b,c\},\{b,c,d\}\} = \{b,c\}$	$w_9 = \bar{a}b\bar{c}d$	$\bigcap\{\{a,b,c\},\{b,c,d\}\} = \{b,c\}$
$w_4 = a\bar{b}cd$	$\bigcap\{\{a,b,c\}\} = \{a,b,c\}$	$w_{10} = \bar{a}b\bar{c}\bar{d}$	$\bigcap\{\{a,b,c\},\{b,c,d\}\} = \{b,c\}$
$w_5 = a\bar{b}c\bar{d}$	$\bigcap\{\{a,b,c,d\}\} = \{a,b,c,d\}$	$w_{11} = \bar{a}\bar{b}cd$	$\bigcap\{\{a,b,c\},\{b,c,d\}\} = \{b,c\}$
$w_6 = a\bar{b}\bar{c}d$	$\bigcap\{\{a,b\},\{a,b,c\}\} = \{a,b\}$	$w_{12} = \bar{a}\bar{b}c\bar{d}$	$\bigcap\{\{b,c,d\}\} = \{b,c,d\}$

4.1   Semantics for Condition (C1)

Saying that Condition (C1) requires keeping the part of T unrelated to α unaffected by the revision of T by α, means that the two following conditions should be satisfied:

---

[2] In this table we are representing interpretations as sequences of literals. Moreover the negation of a variable p is denoted $\bar{p}$.

(I1). during the revision operation by α, we do not remove any information from the part of T unrelated to α, and

(I2). we do not accept any non "necessary" information into the part of T unrelated to α; that means, T should be revised minimally, in the sense that no new formula should be added unless it can be deduced from the information received and the part of T unrelated to α.

For instance, consider the revision of the theory T given in Example 1 by the formula $\alpha = a \wedge \neg b$. The parts related and unrelated of T with respect to α and the C-structure C given above for T, are respectively $\{a \rightarrow b\}$ and $\{b \leftrightarrow c, c \rightarrow d\}$. For this example, (I1) indicates that both $b \leftrightarrow c$ and $c \rightarrow d$ should still be deductible from $T * \alpha$. However, (I2) indicates that necessary information as $\neg c$ should appear in $T * \alpha$, since $\neg c \in (\{\alpha\} + \{b \leftrightarrow c\})$, and non necessary information as d should not be in $T * \alpha$ because its presence is counter-intuitive for the revision operation.

Later in this section, the two general conditions (Q1) and (Q2) represent the semantic counterparts corresponding respectively to the intuitive conditions (I1) and (I2).

A simple example will help to formalize condition (Q1). Suppose that the language $\mathcal{L}$ is built from the propositional variables a, b, c, and T is the theory $T = Cn(\{a \rightarrow b, b \leftrightarrow c\})$, and the two systems of spheres S_T, S_T' centered on $[T]$ are as represented below in Figure 1:

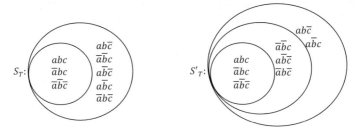

Fig. 1. Condition (Q1)

Clearly, the revision function $*$ induced from S_T violates condition (I1) at T; simply consider the revision of T by $\alpha = a \wedge \neg b$ which gives $T * \alpha = Cn(\{a \wedge \neg b\})$. So, $*$ removed $b \leftrightarrow c$ which contradicts condition (I1) requirement, since the two interpretations $w = a\bar{b}c$ and $w' = a\bar{b}\bar{c}$ are placed in the same sphere. On the other hand, it is not hard to verify that the revision function $*'$ induced by the system of spheres S_T', satisfies (I1). Hence, w and w' should not be placed in the same sphere, because of the following two reasons: First, w is more plausible than w', since $Plaus_T(w) = \{a, b\} \subset \{a, b, c\} = Plaus_T(w')$. Second, $w \cap \mathcal{L}(Plaus_T(w)) = w' \cap \mathcal{L}(Plaus_T(w)) = \{a, \bar{b}\}$. It should be noted that the first criterion alone (the fact that w is more plausible than w') is not sufficient to allow us to place w in a sphere before w'. That means, the plausibility criterion is stronger than condition (I1). In particular consider the interpretations w' and $w'' = ab\bar{c}$ in the system of spheres S_T'. While $Plaus_T(w'') = \{b, c\} \subset \{a, b, c\} =$

$Plaus_T(w')$, the two interpretations w' and w'' are placed in the same sphere without violating condition (I1). Thus we have the following condition:

(Q1). If $Plaus_T(w) \subset Plaus_T(w')$ and $w \cap \mathcal{L}(Plaus_T(w)) = w' \cap \mathcal{L}(Plaus_T(w))$
then there is a sphere $V \in S_T$ containing w but not w'.

Condition (Q1) formalizes the intuition mentioned earlier about preserving all information of the unrelated part by the revision operation.

Now, condition (Q1) alone does not suffice to guarantee the satisfaction of local revision (condition (C1)); we have to formalize the condition corresponding to condition (I2).

To do that, we consider the following example. The language $\mathcal{L}$ is built over three propositional variables a, b, c, the initial belief set T is $T = Cn(\{b \leftrightarrow c\})$, and the two systems of spheres S_T and S'_T centered on $[T]$ are as given below in Figure 2:

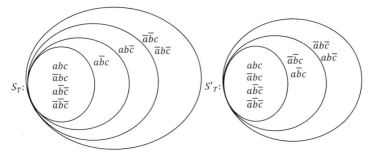

Fig. 2. Condition (Q2)

In this example all the interpretations outside $[T]$ (i.e. in $\mathcal{I}_\mathcal{L} \setminus [T]$) have the same plausibility with respect to T, namely $\forall w \in (\mathcal{I}_\mathcal{L} \setminus [T]), Plaus_T(w) = \{b, c\}$. While S_T satisfies (Q1) since its antecedent $Plaus_T(w) \subset Plaus_T(w')$ never holds for $w, w' \notin [T]$, the revision function $*$ induced from S_T violates the second requirement of (C1) (condition (I2)) at T. However, the revision function $*'$ induced by the system of spheres S'_T, satisfies condition (I2).

In particular consider the revision of T by $\neg b \wedge c$ using S_T. The resulting theory is equal to $Cn(\{a, \neg b, c\})$. Clearly, this contradicts condition (I2) since we have the non necessary information a into the result.

To block that, we should place $w = a\bar{b}c$ and $w' = \bar{a}\bar{b}c$ in the same sphere as in S'_T. The particularity of these two interpretations is that: $Plaus_T(w) = Plaus_T(w')$, and $w \cap \mathcal{L}(Plaus_T(w)) = w' \cap \mathcal{L}(Plaus_T(w))$.

We conclude that whenever w and w' are two interpretations such that the two previous conditions are verified, then they should be placed in the same sphere.

We now proceed with the presentation of condition (Q2), which together with (Q1), brings about the correspondence with (C1). In the following condition, T is a fully overlapping consistent theory of $\mathcal{L}$, S_T is a system of spheres centered on $[T]$, and w, w' are interpretations.

(Q2). If $Plaus_T(w) = Plaus_T(w')$, and $w \cap \mathcal{L}(Plaus_T(w)) = w' \cap \mathcal{L}(Plaus_T(w'))$, then w, w' belong to the same spheres in S_T, i.e., for any sphere $V \in S_T$, $w \in V$ iff $w' \in V$.

The promised correspondence between (C1) and the two conditions (Q1) and (Q2) is given by the theorem below:

Theorem 2. *Let $*$ be a revision function satisfying $(T * 1)$–$(T * 8)$. Let T be a consistent fully overlapping theory of $\mathcal{L}$, and S_T a system of spheres centered on $[T]$, that corresponds to $*$ by means of $(S*)$. Then $*$ satisfies (C1) at T iff S_T satisfies (Q1)–(Q2).*

Here we can show the advantage of our local revision approach with respect to the one defined by the mean of axiom (P) [14]. In the case of fully overlapping theories T, the approach based on axiom (P), is not capable to avoid the counter-intuitive effect of throwing away all non-tautological beliefs in T whenever the new information is inconsistent with T, regardless of whether these beliefs can be kept or not.

For example, the system of spheres S_T centred on $[T]$ (T is the theory of Example 1), and satisfying axiom (P)[3] is only composed of two spheres: the sphere $[T]$ and the sphere $\mathcal{I}_{\mathcal{L}}$, the set of all interpretations of $\mathcal{L}$. However, systems of spheres satisfying the two conditions (Q1) and (Q2) allows us to avoid such undesirable systems of spheres. To see this, consider a system of spheres S'_T corresponding to the theory T of Example 1, and satisfying the two conditions (Q1) and (Q2) as given below in Figure 3. Then, consider the revision of T by $a \wedge \neg b$ using S_T and S'_T. By S_T, the result is $Cn(a \wedge \neg b)$.

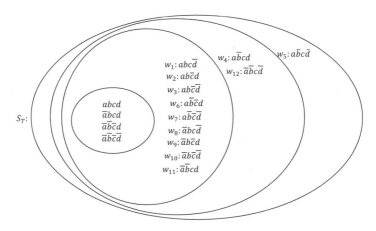

Fig. 3. System of spheres corresponding to Example 1 verifying Condition (C1)

4.2   Semantics for Condition (C2)

The context-insensitivity property of local revision is given by condition (C2). In [14], the authors have proved that such conditions introduce dependencies

---

[3] See [14] for all details on S_T.

between revision policies associated with different theories. That means, the condition corresponding to (C2) that we define below, makes associations between systems of spheres with different centers.

As usual, C and C' are C-structures of the fully overlapping theories T and T' respectively. S_T, $S_{T'}$ are systems of spheres centered on $[T]$ and $[T']$ respectively. We denote by $S_T/\mathcal{L}'$, such that $\mathcal{L}' \subseteq \mathcal{L}$, the restriction of S_T to $\mathcal{L}'$; that is, $S_T/\mathcal{L}' = \{U/\mathcal{L}' : U \in S_T\}$.

For any sublanguage $\mathcal{L}'$ of $\mathcal{L}$, $S_T/\mathcal{L}'$ is also a system of spheres [14]. The condition (Q3) below is the semantic counterpart of (C2).

(Q3). If $C = \{(V_1, Cov_k(V_1, B_T), T_1), (V_2, Cov_k(V_2, B_T), T_2), ..., (V_n, Cov_k(V_n, B_T), T_n)\}$, $C' = \{(V_1', Cov_{k'}(V_1', B_{T'}), T_1'), (V_2', Cov_{k'}(V_2', B_{T'}), T_2'), ..., (V_{n'}', Cov_{k'}(V_{n'}', B_{T'}), T_{n'}')\}$, and $\exists i \in \{1, .., m\}, j \in \{1, .., p\}$ such that $\bigcap_{i=1}^{m} T_i = \bigcap_{j=1}^{p} T_j'$, then $S_T/\mathcal{L}(\mathcal{V}(\bigcap_{i=1}^{m} T_i)) = S_{T'}/\mathcal{L}(\mathcal{V}(\bigcap_{i=1}^{m} T_i))$.

In other words, (Q3) says that if we have two C-structures C and C' of two theories T and T' respectively, which are intersecting on a subtheory T'' of T and T', then the restriction of S_T to the sublanguage over which T'' is defined is equal to the restriction of S_T' to the same sublanguage; i.e., $S_T/\mathcal{L}(\mathcal{V}(T'')) = S_{T'}/\mathcal{L}(\mathcal{V}(T''))$.

The following result shows that (Q3) is the semantic counterpart of (C2):

Theorem 3. *Let $*$ be a revision function satisfying the AGM postulates ($T *$ 1)–($T * 8$), and $\{S_T\}_{T \in \mathcal{K}_\mathcal{L}}$ a family of systems of spheres (one for each fully overlapping theory T in $\mathcal{K}_\mathcal{L}$), corresponding to $*$ by means of ($S*$). Then $*$ satisfies (C2) iff $\{S_T\}_{T \in \mathcal{K}_\mathcal{L}}$ satisfies (Q3).*

Theorems 1, 2 and 3 provide immediately the following theorem that provides semantics for local revision using Grove construction.

Theorem 4. *Let $*$ be a revision function satisfying the AGM postulates ($T * 1$)–($T * 8$), $\{S_T\}_{T \in \mathcal{K}_\mathcal{L}}$ a family of systems of spheres (one for each fully overlapping theory T in $\mathcal{K}_\mathcal{L}$), corresponding to $*$ by means of ($S*$). Then $*$ satisfies local revision iff $\{S_T\}_{T \in \mathcal{K}_\mathcal{L}}$ satisfies (Q1)–(Q3).*

It should be noted that these results can be extended to the general case of arbitrary consistent theories (where a theory can be seen as a set of fully overlapping theories) either by defining a C-structure on each fully overlapping theory and keeping the conditions (Q1)–(Q3), or by defining a unique C-structure for the theory and modifying slightly the conditions given above.

5   Conclusion

The contribution of this paper is that having Grove's system of spheres construction as a base, we provide a semantics for local revision using the C-structure model in the case of consistent fully overlapping theories. We do that by providing additional constraints based on a distance measurement between interpretations, then we prove these constraints characterize local revision by the mean of the C-structure model. What is particularly pleasing about our result is that these new constraints on systems of spheres extend those [14] given for local revision defined by the language splitting model. Thus, our local revision approach presents an extension of the one based on Parikh's model and preserves all its nice properties. Precisely, our approach allows to guarantee the satisfaction of minimal change principle of revision for fully overlapping belief sets, to whome Parikh's revision approach fails to do that.

In future, we intend to extend our approach by generalizing our results to arbitrary consistent belief sets.

References

1. Doukari, O., Würbel, E., Jeansoulin, R.: A new model for belief representation and belief revision based on inconsistencies locality. In: 19th IEEE International Conference on Tools with Artificial Intelligence, ICTAI 2007, Patras, Greece, pp. 262–269. IEEE Computer Society Press, Los Alamitos (2007)
2. Eiter, T., Gottlob, G.: On the complexity of propositional knowledge base revision, updates, and counterfactuals. Artificial Intelligence 57(2-3), 227–270 (1992)
3. Liberatore, P., Schaerf, M.: The complexity of model checking for belief revision and update. In: Proceedings of the Thirteenth National Conference on Artificial Intelligence (AAAI 1996), pp. 556–561. AAAI Press/The MIT Press (1996)
4. Nebel, B.: How hard is it to revise a belief base? In: Dubois, D., Prade, H. (eds.) Handbook of Defeasible Reasoning and Uncertainty Management Systems. Belief Change, vol. 3, pp. 77–145. Kluwer Academic Publishers, Dordrecht (1998)
5. Bennaim, J., Benferha, S., Papini, O., Würbel, E.: An answer set programming encoding of prioritized removed sets revision: application to GIS. In: Alferes, J.J., Leite, J. (eds.) JELIA 2004. LNCS (LNAI), vol. 3229, pp. 604–616. Springer, Heidelberg (2004)
6. Williams, M.A.: Applications of belief revision. In: ILPS 1997: International Seminar on Logic Databases and the Meaning of Change, Transactions and Change in Logic Databases, London, UK, pp. 287–316. Springer, Heidelberg (1998)
7. Würbel, E., Papini, O., Jeansoulin, R.: Revision: an application in the framework of GIS. In: 7th International Conference on Principles of Knowledge Representation and Reasoning, KR 2000, Breckenridge, Colorado, USA, pp. 505–516 (2000)
8. Parikh, R.: Beliefs, belief revision, and splitting languages. Logic, language and computation 2, 266–278 (1999)
9. Chopra, S., Parikh, R.: Relevance sensitive belief structures. Annals of Mathematics and Artificial Intelligence 28, 259–285 (2000)
10. Alchourrón, C.E., Gärdenfors, P., Makinson, D.: On the logic of theory change: Partial meet contraction and revision functions. J. Symb. Logic 50, 510–530 (1985)
11. Grove, A.: Two modelings for theory change. Philosophical Logic 17, 157–170 (1988)
12. Herzig, A., Rifi, O.: Propositional belief base update and minimal change. Artificial Intelligence 115, 107–138 (1999)
13. Chopra, S., Parikh, R., Wassermann, R.: Approximate belief revision. Logic J. IGPL 9, 755–768 (2001)
14. Peppas, P., Chopra, S., Foo, N.Y.: Distance semantics for relevance-sensitive belief revision. In: 9th International Conference on Principles of Knowledge Representation and Reasoning, KR 2004, Canada, pp. 319–328 (2004)

Merging Belief Bases Represented by Logic Programs

Julien Hué[1], Odile Papini[2], and Eric Würbel[1]

[1] LSIS-CNRS 6168, Université du Sud Toulon -Var. BP 132. 83957 La Garde Cedex France
{hue,wurbel}@univ-tln.fr
[2] LSIS-CNRS 6168, ESIL, Université de la Méditerranée. Av de Luminy. Case 13009 Marseille
Cedex 9 France
papini@esil.univmed.fr

Abstract. This paper presents a method which allows for merging beliefs expressed thanks to logic programming with stable model semantics. This method is based on the syntactic merging operators described in the framework of propositional logic. The study of these operators leads to a new definition of the consequence relation between logic programs which is based on the logic of Here-and-There brought by Turner. Moreover, the specificity of the non-monotonic framework given by logic programming with stable model semantics allows for describing a weakened version of the merging operation. Once the operators are defined, their behaviour with respect to the Konieczny and Pino-Perez postulates for merging are examined and discussed.

1 Introduction

Nowadays, the computer science field has to deal with distributed sources of knowledge, especially in the context of databases. These sources are rarely synchronized, they generally conflict. Therefore, the interrogation and sharing of those distributed sources are crucial questions for artificial intelligence.

This problem has been widely discussed within the framework of propositional logic [5,19,3]. These operators have been defined in a semantic way [10], a syntactic way [16,8] or based on morphologic properties of beliefs [4]. The last two methods has led to an implementation [9]. The main advantages of propositional logic is both its strong formal background and its simplicity. But, this simplicity can also be a drawback for representing real world situations. Hence, logic programming with stable model semantics [7] provides a belief representation formalism which is more interesting than classical logic for non-monotonic reasoning. Thus, the question of belief bases merging when beliefs are represented by logic programs deserves attention.

This paper presents a method for merging belief bases represented by logic programs. This method is based on the syntactic operators defined in [9]. Moreover, the non-monotonicity of the stable model semantics allows us to define a weakened version of the merging operation in order to save more beliefs than the strong merging operation. A study of properties will be conducted and an implementation of the merging operations will be provided.

The rest of the paper is organized as follows. Section 1 gives a refresher on belief bases merging and logic programming. Section 2 gives the definition of strong and

C. Sossai and G. Chemello (Eds.): ECSQARU 2009, LNAI 5590, pp. 371–382, 2009.
© Springer-Verlag Berlin Heidelberg 2009

weak version of merging operations. Section 3 presents an implementation of merging operations based on logic programming with stable model semantics before concluding.

2 Preliminaries and Notations

In this section, we give the definitions and notations with respect to logic programming with stable model semantics. We then remind the work described in [21] on the logic of Here-and-There which is used to provide an alternative definition of stable models. We also give a reminder on belief merging and on the Konieczny and Pino-Perez postulates for merging operations.

We consider a finite alphabet $\mathcal{P}$ consisted in propositional atoms. Atoms and formulas are denoted by lower case letters. Sets of atoms are denoted by capital letters. An interpretation is a function from $\mathcal{P}$ to $\{0, 1\}$ and the set of every interpretation is denoted by $\mathcal{W}$. For every interpretation I and every set of atoms A, we say that I implies A ($I \models A$) iff every atom of A is true for I. If A and B are two sets of formulas, then $A \models B$ iff $I \models A$ implies $I \models B$. $mod(A)$ represents the set of models of A.

2.1 Logic Programming with Stable Model Semantics

A normal logic program is a set of rules with the form $c \leftarrow a_1, \ldots, a_n, not\ b_1, \ldots, not\ b_m$ where $c, a_i (1 \leq i \leq n), b_j (1 \leq j \leq m)$ are propositional atoms and the symbol not stands for negation as failure. A basic program is a logic program without negation as failure. Let r be a rule, we introduce $head(r) = c$ and $body(r) = \{a_1, \ldots, a_n, b_1, \ldots, b_m\}$. Moreover, we define $body^+(r) = \{a_1, \cdots, a_n\}$ which represents the set of positive atoms in the body of this rule and $body^-(r) = \{b_1, \ldots, b_m\}$ which represents the set of negative atoms in the body of this rule, hence $body(r) = body^+(r) \cup body^-(r)$. r^+ represents the rule $head(r) \leftarrow body^+(r)$, obtained from r by withdrawing negative elements to the body of r.

A set X of atoms is closed under a logic program Π iff $\forall r \in \Pi$, $head(r) \in X$ when $body(r) \subseteq X$. The smallest set of atoms which is closed under a basic program Π is denoted $CN(\Pi)$. The reduction of Gelfond-Lifschitz [7] of a program Π with respect to a set X of atoms is defined by $\Pi^X = \{r^+ \mid r \in \Pi$ and $body^-(r) \cap X = \emptyset\}$. A set X of atoms is a stable model of Π iff $CN(\Pi^X) = X$. A logic program is said to be inconsistent if it does not have any stable model.

Extended Logic Programs. In order to represent more complete information, it is possible to consider classical negation $\neg$ in addition to the negation as failure. Therefore, an extended logic program is a set of rules in the form: $c \leftarrow a_1, \ldots, a_n, not\ b_1, \ldots, not\ b_m$ where $c, a_i (1 \leq i \leq n), b_j (1 \leq j \leq m)$ are literals (atoms or negation of atoms). The previous definition of a stable model remains valid for any set of atoms X which is consistent (does not contain an atom and its negation).

During the last years, logic programming with stable model semantics has been considered as a convenient tool to handle non-monotonic reasoning. It especially led to several efficient systems, called ASP solvers: smodels [17], DLV [6], NoMore [1], ASSAT [13], CLASP [2].

In [21], H.Turner gave an alternative definition of stable models of a logic program based on the logic of Here-and-There. This logic, which is monotonic, represents interpretations of a logic program in the form of pairs of sets of atoms. Intuitively, the collection of all HT-interpretations can be constructed as follows. Let Π be a logic program and X and Y be sets of atoms from Π. First, Y is a set of atoms which is consistent with the program Π. Then, for every set Y, X is a set of atoms such that $X \subseteq Y$ and which is a set of plausible consequences of Π knowing Y.

Definition 1 (HT-interpretations). *Let Π be a logic program and X and Y be consistent sets of atoms such that $X \subseteq Y$. A pair (X, Y) is a HT-interpretation of Π iff $Y \models \Pi$ and $X \models \Pi^Y$. We denote by $HT(\Pi)$ the set of all HT-interpretations of program Π.*

Example 1. Let us consider the following program $\Pi = \{a \leftarrow not\ b.\quad b \leftarrow not\ a.\}$. The sets of atoms consistent for every rule of Π are: $\{a\}, \{b\}$ and $\{a, b\}$. It is not possible to have an HT-interpretation with $\emptyset$ as a second element because the absence of a entails the deduction of b (by the rule $a \leftarrow not\ b$.) and vice versa. For $Y = \{a\}$, $\Pi^Y = \{a \leftarrow\}$ then $X = \{a\}$ is the only set of atoms which is consistent with Π^Y; Similarly for $Y = \{b\}$ then $X = \{b\}$; for $Y = \{a, b\}$, $\Pi^Y = \emptyset$ then every $X \subseteq Y$ is a set of plausible consequences of Π^Y.

Finally $HT(\Pi) = \{(\{a\}, \{a\}), (\{b\}, \{b\}), (\emptyset, \{a, b\}), (\{a\}, \{a, b\}), (\{b\}, \{a, b\}), (\{a, b\}, \{a, b\})\}$

If there is only one HT-interpretation $(\{Y\}, \{Y\})$ with a given Y as a second element then Y is consistent with Π and every of its atom is justified.

Definition 2 (Stable models). *Let Π be a logic program and Y be a set of atoms. Y is a stable model of Π iff (Y, Y) is the only element of $HT(\Pi)$ where the second element is Y.*

Lemma 1. *Let Π_1 and Π_2 be two logic programs then $HT(\Pi_1 \cup \Pi_2) = HT(\Pi_1) \cap HT(\Pi_2)$. This result is provided by [21].*

Example 2. In the example 1, the only HT-interpretation of Π with a as a second element is $(\{a\}, \{a\})$ then $\{a\}$ is a stable model of Π. $(\{b\}, \{b\})$ is the only HT-interpretation with $\{b\}$ as a second element then $\{b\}$ is a stable model of Π.

2.2 Belief Merging and Removed Sets Fusion

First works on belief merging came from the database area [20] and, later, Konieczny [12] focused on belief merging from a semantical point of view. Belief merging aims at associating a consistent interpretation or a consistent belief base to an inconsistent set of belief bases (called belief profile). The interpretation or belief base resulting of this operation has to be as close as possible to the original belief profile. Let $\Psi = \{\varphi_1, \ldots, \varphi_n\}$ be a belief profile. We denote $\Delta(\Psi)$ the result of the merging operation. There are two straightforward ways to define $\Delta(\Psi)$ depending on if the sources are conflicting or not, the classical conjunctive merging: $\Delta(\Psi) = \bigwedge_{\varphi_i \in \Psi} \varphi_i$ suitable when the sources are not conflicting and the classical disjunctive merging: $\Delta(\Psi) = \bigvee_{\varphi_i \in \Psi} \varphi_i$ appropriate in case of conflicting sources. These two opposite cases are not satisfactory, so several methods have been proposed for fusion depending on if the bases have the same importance or not.

When the solution provided by the merging operation is an interpretation, it is called semantic merging. When the solution is a belief base, it is called syntactic merging, like in [8]. In particular, the following classical fusion operators have been proposed according to various strategies. The *Sum operator*, denoted by Σ, [15,18] which follows the point of view of the majority of the belief bases of Ψ. The *Cardinality operator*, denoted by $Card$, [3] which is similar to Σ but without taking repetitions into account. The *Max-based* operator, denoted by Max [19], which tries to best satisfy all the belief bases of Ψ. The *Leximax-based* operator, denoted by $GMax$, [11] which is the lexicographic refinement of Max.

Some methods have been proposed within the context of semantic merging [5,14,3]. Among the approaches within the syntactic merging framework, Removed Sets Fusion [9] provides a method and an implementation with the central idea to determine maximal consistent subsets of formulas. As an heuristic, we consider the set of formulas to remove in order to restore consistency.

Definition 3 (Potential Removed Set). *Let $\Psi = \{\varphi_1, \ldots, \varphi_n\}$ be a belief profile constrained by the belief base μ such that $\varphi_1 \sqcup \ldots \sqcup \varphi_n \sqcup \mu$ is inconsistent. Let X be a subset of formulas of $\varphi_1 \sqcup \ldots \sqcup \varphi_n$. X is a potential Removed Set of $\varphi_1 \sqcup \ldots \sqcup \varphi_n$ with μ for constraints iff $((\varphi_1 \sqcup \ldots \sqcup \varphi_n) \setminus X) \sqcup \mu$ is consistent.*

The Removed Sets Fusion framework captures the classical merging strategies thanks to total pre-orders over Potential Removed Sets.

Definition 4 (Pre-order and strategies). *Let $\Psi = \{\varphi_1 \sqcup \ldots \sqcup \varphi_n\}$ be a belief profile constrained by the belief base μ and X and Y be potential Removed Sets of Ψ constrained by μ. For every strategy P, a pre-order $\leq_P$ over potential Removed Sets is defined. $X \leq_P Y$ means that X is preferred to Y according to the strategy P. We define $<_P$ as the strict pre-order associated with $\leq_P$ (i.e. $X <_P Y$ iff $X \leq_P Y$ and $Y \not\leq_P X$).*

Therefore, potential Removed Sets of $\varphi_1 \sqcup \ldots \sqcup \varphi_n$ with μ for constraints which are minimal according to the chosen strategy will be considered as the solutions of our merging operation. These potential Removed Sets which are minimal for the $<_P$ pre-order are called Removed Sets according to P.

Definition 5 (Removed Sets according to P). *Let $\Psi = \{\varphi_1, \ldots, \varphi_n\}$ be a belief profile constrained by the belief base μ such that $\varphi_1 \sqcup \ldots \sqcup \varphi_n \sqcup \mu$ is inconsistent. Let P be a merging strategy. $X \subseteq \varphi_1 \sqcup \ldots \sqcup \varphi_n$ is a Removed Set of $\varphi_1 \sqcup \ldots \sqcup \varphi_n$ with μ for constraints according to the strategy P iff (i) X is a potential Removed Set of $\varphi_1 \sqcup \ldots \sqcup \varphi_n$ with μ for constraints; (ii) There is no potential Removed Set Y of $\varphi_1 \sqcup \ldots \sqcup \varphi_n$ with μ for constraints such that $Y <_P X$.*

The collection of Removed Sets of Ψ with μ for constraints according to the strategy P is denoted by $\mathcal{F}_\mu^P \mathcal{R}(\Psi)$. The Removed Set Fusion operation is defined by:

Definition 6 (Removed Sets Fusion). *Let $\Psi = \{\varphi_1, \ldots, \varphi_n\}$ be a belief profile constrainted by the belief base μ, the Removed Sets Fusion operation $\Delta_\mu^P(\Psi)$ is defined by:*
$$\Delta_\mu^P(\Psi) = \bigvee\nolimits_{X \in \mathcal{F}_\mu^P \mathcal{R}(\Psi)} \{((\varphi_1 \sqcup \ldots \sqcup \varphi_n) \backslash X) \sqcup \mu\}$$

Konieczny and Pino-Perez Postulates for Merging Operation. In [12], Konieczny and Pino-Perez defined a set of postulates for belief bases merging. Let $\Psi = \{\varphi_1, \ldots, \varphi_n\}$ be a belief profile and μ be a belief base.

(KP0) $\Delta_\mu(\Psi) \vdash \mu$.
(KP1) If μ is consistent, then $\Delta_\mu(\Psi)$ is consistent.
(KP2) If Ψ is consistent with μ, then $\Delta_\mu(\Psi) = \bigwedge \Psi \wedge \mu$.
(KP3) If $\Psi_1 \equiv \Psi_2$ and $\mu_1 \equiv \mu_2$, then $\Delta_{\mu_1}(\Psi_1) \equiv \Delta_{\mu_2}(\Psi_2)$.
(KP4) If $\varphi_1 \vdash \mu$ and $\varphi_2 \vdash \mu$, then $\Delta_\mu(\varphi_1 \sqcup \varphi_2) \wedge \varphi_1 \not\vdash \bot \to \Delta_\mu(\varphi_1 \sqcup \varphi_2) \wedge \varphi_2 \not\vdash \bot$.
(KP5) $\Delta_\mu(\Psi_1) \wedge \Delta_\mu(\Psi_2) \vdash \Delta_\mu(\Psi_1 \sqcup \Psi_2)$.
(KP6) If $\Delta_\mu(\Psi_1) \wedge \Delta_\mu(\Psi_2)$ is consistent, then $\Delta_\mu(\Psi_1 \sqcup \Psi_2) \vdash \Delta_\mu(\Psi_1) \wedge \Delta_\mu(\Psi_2)$.
(KP7) $\Delta_{\mu_1}(\Psi) \wedge \mu_2 \vdash \Delta_{\mu_1 \wedge \mu_2}(\Psi)$.
(KP8) If $\Delta_{\mu_1}(\Psi) \wedge \mu_2$ is consistent, then $\Delta_{\mu_1 \wedge \mu_2}(\Psi) \vdash \Delta_{\mu_1}(\Psi) \wedge \mu_2$.

3 Syntactic Merging of Belief Bases Represented by Logic Programs

We here propose to study Removed Sets Fusion when beliefs are expressed in terms of logic programs with stable model semantics. The principle remains identical to the propositional case: the removal of some formulas in order to restore consistency. For the rest of this section, we consider that the belief bases are expressed in the form of logic progams. We now give the definition of Removed Sets Fusion in this context, this definition deals with constraints.

Definition 7 (Strong Potential Removed Set). *Let $\Psi = \{\varphi_1, \ldots, \varphi_n\}$ be a belief profile constrainted by the belief base μ such that $\varphi_1 \sqcup \ldots \sqcup \varphi_n \sqcup \mu$ is inconsistent. Let X be a subset of formulas of $\varphi_1 \sqcup \ldots \sqcup \varphi_n$. X is a strong potential Removed Set of $\varphi_1 \sqcup \ldots \sqcup \varphi_n$ constrainted by μ iff $((\varphi_1 \sqcup \ldots \sqcup \varphi_n) \backslash X) \sqcup \mu$ is consistent.*

Definition 8 (Strong Removed Sets according to P). *Let $\Psi = \{\varphi_1, \ldots, \varphi_n\}$ be a belief profile constrainted by the belief base μ such that $\varphi_1 \sqcup \ldots \sqcup \varphi_n \sqcup \mu$ is inconsistent. Let P be a merging strategy. $X \subseteq \varphi_1 \sqcup \ldots \sqcup \varphi_n$ is a Removed Set of $\varphi_1 \sqcup \ldots \sqcup \varphi_n$ constrainted by μ according to the strategy P iff (i) X is a strong potential Removed Set of $\varphi_1 \sqcup \ldots \sqcup \varphi_n$ constrainted by μ; (ii) There is no strong potential Removed Set Y of $\varphi_1 \sqcup \ldots \sqcup \varphi_n$ constrainted by μ such that $Y <_P X$.*

The collection of Strong Removed Sets of Ψ according to the strategy P is denoted by $\mathcal{F}_\mu^P \mathcal{R}(\Psi)$. The Removed Sets Fusion operation is defined by:

Definition 9 (Strong Removed Sets Fusion). *Let $\Psi = \{\varphi_1, \ldots, \varphi_n\}$ be a belief profile constrainted by the belief base μ, the Strong Removed Sets Fusion operation $\Delta_\mu^P(\Psi)$ is defined by: $\Delta_\mu^P(\Psi) = \bigvee\nolimits_{X \in \mathcal{F}_\mu^P \mathcal{R}(\Psi)} \{((\varphi_1 \sqcup \ldots \sqcup \varphi_n) \backslash X) \sqcup \mu\}$*

3.1   Consequence Relation between Logic Programs

In propositional logic, the consequence relation between two set of formulas is clearly defined ($A \models B$ iff $\forall I \in \mathcal{W}$, if $I \models A$ then $I \models B$). This definition can hardly be applied to logic programs.

Example 3. Let $\Pi = \{a\}$ be a belief profile constrainted by the belief base $\mu = \{c \leftarrow not\ a.\quad \neg b.\}$. Thus, the consequences of μ are $\{\neg b, c\}$ and the consequences of $\Pi \cup \mu$ are $\{\neg b, a\}$.

One can easily see that the consequences of $\Pi \cup \mu$ are completely different from the consequences of μ and that a definition of a consequence relation given in terms of stable models inclusion will not properly fit the logic programming framework. This problem can be overcome by chosing a definition of the consequence relation in terms of inclusions of HT-interpretations.

Definition 10 (Inference). *Let Π_1 and Π_2 be logic programs, we define that Π_1 implies Π_2, denoted by $\Pi_1 \models \Pi_2$ iff $HT(\Pi_1) \subseteq HT(\Pi_2)$.*

Example 4. Let us consider again the example 3 in order to illustrate the consequence relation between two sets of rules. In this example, we have:

$$HT(\mu) = \{((\{\neg b\}, \{\neg b, a\}), (\{\neg b, a\}, \{\neg b, a\}), (\{\neg b, c\}, \{\neg b, c\}),$$
$$(\{\neg b\}, \{\neg b, a, c\}), (\{\neg b, a\}, \{\neg b, a, c\}), (\{\neg b, c\}, \{\neg b, a, c\}), (\{\neg b, a, c\}, \{\neg b, a, c\})\}$$
$$HT(\Pi \cup \mu) = \{((\{\neg b, a\}, \{\neg b, a\}), (\{\neg b, a\}, \{\neg b, a, c\}), (\{\neg b, a, c\}, \{\neg b, a, c\})\}$$

We have $HT(\Pi \cup \mu) \models HT(\mu)$ and therefore according to the previous definition $(\Pi \cup \mu) \models \mu$.

This new definition allows us to study properly the KP postulate for the Strong Removed Sets Fusion operation.

3.2   KP Postulates with Respect to Removed Sets Fusion for Logic Programs

The Strong Removed Sets Fusion operation for logic programs verifies the following KP postulates:

Strategies	(KP0)	(KP1)	(KP2)	(KP3)	(KP4)	(KP5)	(KP6)	(KP7)
Σ	yes	yes	yes	no	no	no	no	no
Card	yes	yes	yes	no	no	no	no	no
Max	yes	yes	yes	no	no	no	no	no
Gmax	yes	yes	yes	no	no	no	no	no

Sketch of Proofs. The counter-examples are explained only for the Σ operator but also make sense for the other operators.

(KP0) Thanks to the theorem 1, we know that $HT(\Pi \cup \Pi') = HT(\Pi) \cap HT(\Pi')$. By contruction, we have that $HT(\Delta_\mu^\Sigma(\Psi)) \subseteq HT(\mu)$ and then $\Delta_\mu^\Sigma(\Psi) \models \mu$.

(KP1) and (KP2) True by construction

(KP3) Consider $\Psi_1 = \{p. \quad q.\}$, $\Psi_2 = \{p. \quad q \leftarrow p.\}$ and $\mu = \{\neg p.\}$.

$\Delta_\mu^\Sigma(\Psi_1) = \{\neg p, q\} \quad HT(\Delta_\mu^\Sigma(\Psi_1)) = \{(\{\neg p, q\}, \{\neg p, q\})\}$

$\Delta_\mu^\Sigma(\Psi_2) = \{\neg p, q \leftarrow p\} \quad HT(\Delta_\mu^\Sigma(\Psi_2)) = \{(\{\neg p\}, \{\neg p\}) \quad (\{\neg p\}, \{\neg p, \neg q\}$

$(\{\neg p, \neg q\}, \{\neg p, \neg q\})\}$

(KP4) and (KP5)

$$\Pi_1 = \left\{ \begin{array}{ll} a \leftarrow not\ \neg h. & c \leftarrow b. \\ c \leftarrow a. & b \leftarrow not\ \neg h. \\ \neg c. & \end{array} \right\} \qquad \Pi_2 = \left\{ \begin{array}{ll} d \leftarrow not\ \neg c. & h \leftarrow e. \\ h \leftarrow d. & e \leftarrow not\ \neg c. \\ \neg h. & \end{array} \right\}$$

$\Delta_\top^\Sigma(\Pi_1) = \{a \leftarrow not\ \neg h. \quad b \leftarrow not\ \neg h. \quad c \leftarrow a. \quad c \leftarrow b.\}$ and every HT-interpretation has c in the first set of atoms of the pair. $\Delta_\top^\Sigma(\Pi_2) = \{d \leftarrow not\ \neg c. \quad e \leftarrow not\ \neg c. \quad h \leftarrow d. \quad h \leftarrow e.\}$ and every HT-interpretation has h in the first set of atoms of the pair. Hence, $\Delta_\top^\Sigma(\Pi_1 \sqcup \Pi_2) = \Pi_1 \cup \Pi_2$ and $\neg c$ and $\neg h$ are in the first set of atoms of the pair.

(KP6) and (KP7)

$$\Pi = \left\{ \begin{array}{lll} a \leftarrow not\ c. & \neg d. & d \leftarrow a. \\ d \leftarrow b. & b \leftarrow not\ c. & \end{array} \right\}$$

with $\mu_2 = \{c.\}$. $\Delta_\top^\Sigma(\Pi) = \{a \leftarrow not\ c. \quad b \leftarrow not\ c. \quad d \leftarrow a. \quad d \leftarrow b.\}$. Each HT-interpretation of $\Delta_\top^\Sigma(\Pi)$ has d in the first set of atoms of the pair and each HT-interpretation of $\Delta_{\top \wedge \mu_2}^\Sigma(\Pi)$ has $\neg d$ in the first set of atoms of the pair.

Discussion. The KP postulates have been defined in the framework of monotonic propositional logic. It is normal that the operators described in this paper do not fully respect them. For instance, the postulates **(KP4)** and **(KP5)** mean that if two sets of rules agree on some consequences, their union should respect the consensus; which is clearly not the case in logic programming with stable model semantics.

3.3   Weak Removed Sets Fusion for Logic Programs

Some sets of rules do not have stable models because they imply inconsistent sets of atoms and some others because it is impossible to justify their consequences. For instance, the program $\Pi = \{\neg a. \quad a \leftarrow b. \quad b.\}$ has for immediate consequences the set of atoms $\{a, \neg a, b\}$ which is inconsistent. On the contrary, the program $\Pi' = \{a \leftarrow not\ b. \quad b \leftarrow not\ c. \quad c \leftarrow not\ a.\}$ which does not imply inconsistent sets of atoms but does not have any stable models because it is impossible to find any self-justifying set of atoms. In one hand, in the case of Π, there are no set of rules φ such that $\Pi \cup \varphi$ has stable models. Though, the only way to restore consistency in those beliefs is to remove some rule. On the other hand, the program Π' is not intrinsically inconsistent. It is possible to restore consistency without losing beliefs, for instance, the union $\Pi' \cup \{a.\}$ has a stable model $(\{a, b\})$.

It seems reasonable to consider an operation which would keep as much rules as possible as long as consistency can still be restored. We can call this operation a weak merging operation.

Formally, a set of formulas which has at least one HT-interpretation can still have its consistency restored. Generally speaking, consider a logic program which has several HT-interpretations where the second element is Y. If a set of facts Y is added, then this new program will have Y as stable model. Actually, a set of atoms Y is a stable model of Π iff the only HT-interpretation of Π where the second element is Y is (Y, Y). Let $\Psi = \{\varphi_1, \ldots, \varphi_n\}$ be a belief profile and μ be a belief base such that $\varphi_1 \sqcup \ldots \sqcup \varphi_n \sqcup \mu$ does not have any stable model, a set of rules X such that $((\varphi_1 \sqcup \ldots \sqcup \varphi_n) \backslash X) \sqcup \mu$ has at least one HT-interpretation, it is called weak potential Removed Set. A weak potential Removed Set of Ψ constrainted by μ which is minimal according to the strategy P, is called weak Removed Set of Ψ constrainted by μ according to P.

Definition 11 (Weak potential Removed Set). *Let $\Psi = \{\varphi_1, \ldots, \varphi_n\}$ be a belief profile constrainted by the belief base μ such that $\varphi_1 \sqcup \ldots \sqcup \varphi_n \sqcup \mu$ does not have any HT-interpretation. Let X be a subset of rules of $\varphi_1 \sqcup \ldots \sqcup \varphi_n$. X is a weak potential Removed Set of $\varphi_1 \sqcup \ldots \sqcup \varphi_n$ constrainted by μ iff $((\varphi_1 \sqcup \ldots \sqcup \varphi_n) \backslash X) \sqcup \mu$ has at least one HT-interpretation.*

Definition 12 (Weak Removed Set). *Let $\Psi = \{\varphi_1, \ldots, \varphi_n\}$ be a belief profile constrainted by the belief base μ such that $\varphi_1 \sqcup \ldots \sqcup \varphi_n \sqcup \mu$ does not have any HT-interpretation. Let P be a merging strategy. $X \subseteq \varphi_1 \sqcup \ldots \sqcup \varphi_n$ is a weak Removed Set of $\varphi_1 \sqcup \ldots \sqcup \varphi_n$ constrainted by μ iff (i) X is a weak potential Removed Set of $\varphi_1 \sqcup \ldots \sqcup \varphi_n$ constrainted by μ; (ii) There is no Y which is a weak potential Removed Set of $\varphi_1 \sqcup \ldots \sqcup \varphi_n$ constrainted by μ such that $Y <_P X$.*

The collection of Weak Removed Sets of Ψ according to the strategy P is denoted by $\mathcal{F}_\mu^{P,w}\mathcal{R}(\Psi)$. The Removed Sets Fusion operation is defined by:

Definition 13 (Weak Removed Sets Fusion). *Let $\Psi = \{\varphi_1, \ldots, \varphi_n\}$ be a belief profile constrainted by the belief base μ and P be a merging strategy, the Weak Removed Sets Fusion operation $\Delta_\mu^{P,w}(\Psi)$ is defined by: $\Delta_\mu^{P,w}(\Psi) = \bigvee_{X \in \mathcal{F}_\mu^{P,w}\mathcal{R}(\Psi)} \{((\varphi_1 \sqcup \ldots \sqcup \varphi_n) \backslash X) \sqcup \mu\}$*

4   Implementation of the Merging Problem

The implementation of Removed Sets Fusion for logic programs stems from an approach similar to the propositional cases described in [8]. Let Ψ be a belief profile and μ be a belief base representing constraints on Ψ. It constructs a logic program $\Pi_{\Psi,\mu}$, such that for any strategy P, the preferred stable models of $\Pi_{\Psi,\mu}$ according to P correspond to the Removed Sets of Ψ constrainted by μ according to P. In the same way, we construct a logic program $\Pi_{\Psi,\mu}^w$ to solve the weak merging operation.

The first part of the program gives the potential Removed Sets of Ψ constrainted by μ and the second part selects Removed Sets amongst them. It is done thanks to the enumeration of possible interpretations which will provide the maximal consistent subsets of logic program. There is however some differences with the propositional case:

- A model for a propositional logic base can contain either a or $\neg a$. A stable model can contain a, $\neg a$ or none of them.

- In Removed Sets Fusion, the subset of formulas generated by an interpretation has to be consistent (the interpretation which generates it being the model), which is not the case for logic programs because an interpretation can satisfy every rule without being a stable model of the program.

Let $\Psi = \{\varphi_1, \ldots, \varphi_n\}$ be a belief profile and μ be a belief base representing constraints. The set of all positive (resp. negative) literals of $\Pi_{\Psi,\mu}$ is denoted by V^+ (resp. V^-). The set of atoms representing rules is defined by $R^+ = \{r_f^i \mid f \in \varphi_i\}$ and $FO(r_f^i)$ denotes the rule of φ_i corresponding to r_f^i in $\Pi_{\Psi,\mu}$. Namely, $\forall r_f^i \in R^+, FO(r_f^i) = f$. To each answer set of $\Pi_{\Psi,\mu}$ we associate the potential Removed Set $FO(R^+ \cap S)$. Considering this, we will describe the logic program which will represent the merging problem. Our program will have four steps:

- The first step generates the set of interpretations of V which can be stable models of a subset of rules. (4.1)
- The second step assures that there exists a rule which allows the atom to be present in the current interpretation. (4.2)
- The third step allows to point out the rules that should be removed. (4.3)
- The last step, finally, is used to encode the strategy. (4.4)

Example 5. We illustrate each part of the translation with the following example. Consider $\Psi = \{\varphi_1, \varphi_2\}$ with $\varphi_1 = \{f_1 \quad : \quad a \leftarrow not\ b. \quad f_2 \quad : \quad b \leftarrow not\ c.\}$, $\varphi_2 = \{f_3 : c \leftarrow not\ a. \quad f_4 : d \leftarrow a.\}$ and $\mu = \{\leftarrow a.\}$.

4.1 First Step: Generating Interpretations

Generating all the interpretations of V for the set of atoms $\{a_1, \ldots, a_n\}$ is done through the rules $\{a_1, a_1', \ldots, a_n, a_n'\}$ where a_i' represents the negation of a_i. Finally, to avoid the presence of an atom and its negation in the same interpretation, we introduce, for every atom a_i, the contraint $\leftarrow a_i, a_i'$.

Case of Basic Program. When dealing with basic programs (which do not contain any negation), this part can be reduced to the instruction $\{a_1, \ldots, a_n\}$.

Example 6. Continuing the example 5. Their interpretations are generated thanks to the statement $\{a, b, c, d\}$.

4.2 Second Step: Rules to Remove

It is impossible that a set of atoms S is a stable model of a logic program $\Pi_{\Psi,\mu}$ if there exists a rule f such that S satisfies $body(f)$ and $head(f) \notin S$. Such a rule should therefore be removed in order to allow the interpretation to be a model.

Hence, for every rule $f \quad : \quad head(f) \leftarrow body(f)$, we introduce the rule $r_f \leftarrow not\ head(f), body(f)$. The presence of the atom r_f means that the rule f should not be considered in the stable model corresponding to S.

Example 7. Consider the example 5. The selection of rules to remove is done thanks to $r_1 \leftarrow not\ a, not\ b.\ r_2 \leftarrow not\ b, not\ c.\ r_3 \leftarrow not\ c, not\ a\ r_4 \leftarrow not\ d, a.$ [1]

---

[1] For the sake of readability, we will note r_i instead of r_{f_i}.

4.3   Third Step: Necessity of the Presence of an Atom

Generally speaking, a stable model represents the set of reasonable consequences of a logic program. It means that an atom only belongs to a stable model if it has been deduced thanks to a rule or a fact. It is necessary, for every set of atoms S, that there exists a very reason for any atom to be true.

For every atom a, we define an atom $auth(a)$ representing the fact that an atom a has been authorized to be deduced. Therefore, for every atom a, we introduce the rule $\leftarrow a, not\ auth(a)$ which implies the impossibility for an atom to be present if its presence is not justified.

Logically, $auth(a)$ is deduced if a rule has not been removed and if $body(f) \subseteq S$. For every rule f, we introduce the rule $auth(head(f)) \leftarrow not\ r_f, body(f)$.

Example 8. Continuing the example 5. The rules allowing to determine if an atom has a reason for being deduced are: $\leftarrow a, not\ auth(a). \leftarrow b, not\ auth(b). \leftarrow c, not\ auth(c).$ $\leftarrow d, not\ auth(d).\ auth(a) \leftarrow not\ b, not\ r_1.\ auth(b) \leftarrow not\ c, not\ r_2.\ auth(c) \leftarrow not\ a, not\ r_3.\ auth(d) \leftarrow a, not\ r_4.$

The whole $\Pi_{\Psi,\mu}$ program has the following stable models: $\{b, auth(b), r_3\}$ $\{r_1, r_2, r_3\}$ $\{c, auth(c), r_1\}$.

Proposition 1. *Let $\Psi = \{\varphi_1, \ldots, \varphi_n\}$ be an belief profile and μ be a belief base representing constraints. Let $S \subseteq V$ be a set of atoms. S is a stable model of $\Pi_{\Psi,\mu}$ iff I_S is an interpretation of V^+ which satisfies $((\varphi_1 \sqcup \ldots \sqcup \varphi_n) \backslash FO(R^+ \cap S)) \sqcup \mu$.*

4.4   Fourth Step: Optimization

The optimization statements are similar to the ones presented in [9].

Example 9. Continuing the example 5.

For the Σ strategy, the optimization statement will be:

$minimize\{r_1, r_2, r_3, r_4\}$.

For the Max strategy, the optimization statements will be:

$\#domain\ possible(U). \#domain\ base(V). \#domain\ possible(W).$
$possible(1..2).\ base(1..2).\ size(U) \leftarrow U\{r_f^V | F_0(f) \in \varphi_V\}U.$
$negmax(W) \leftarrow size(U), U > W.\ max(U) \leftarrow size(U), not\ negmax(U).$
$minimize[max(1) = 1, max(2) = 2]$

For both strategies, the preferred stable models of $\Pi_{\Psi,\mu}$ are: $\{b, auth(b), r_3\}$ and $\{c, auth(c), r_1\}$ which correspond to the Strong Removed Sets of Ψ constrainted by μ according to Σ and Max: $\{a \leftarrow not\ b.\}$ and $\{c \leftarrow not\ a.\}$.

The following proposition establishes the one-to-one correspondence between the preferred stable models of $\Pi_{\Psi,\mu}$ according to P and the Strong Removed Sets of $\Delta_\mu^P(\Psi)$

Proposition 2. *The set of Strong Removed Sets of Ψ constrainted by μ according to P is the set of preferred stable models of $\Pi_{\Psi,\mu}$ according to P. This proposition holds for $\Sigma, Max, Card$ and $Gmax$.*

4.5 Weak Merging Operation

The main difference between the strong and weak merging operations is that an atom does not need justification to belong to an interpretation. Therefore a program to solve the weak version of merging operator will have the same rules as a strong one except the rules described in 4.3.

Example 10. Consider again the example in 5. The program $\Pi^w_{\Psi,\mu}$ is:

$$\{a, b, c, d\}.\ r_1 \leftarrow not\ a, not\ b.\ r_2 \leftarrow not\ b, not\ c.$$
$$\leftarrow a.\quad r_3 \leftarrow not\ c, not\ a.\quad r_4 \leftarrow not\ d, a.$$
$$minimize\{r_1, r_2, r_3, r_4\}.$$

This program has 20 stable models and the minimal one for every strategy is $\{a, b, c, d\}$ which corresponds to the Weak Removed Set of Ψ constrainted by μ which is the empty set.

The following proposition establishes the one-to-one correspondence between the preferred stable models of $\Pi^w_{\Psi,\mu}$ and the Weak Removed Sets of $\Delta^{P,w}_{\mu}(\Psi)$

Proposition 3. *The set of Weak Removed Sets of Ψ constrainted by μ according to P is the set of preferred stable models of $\Pi^w_{\Psi,\mu}$ according to P. This proposition holds for Σ, Max, $Card$ and $Gmax$.*

5 Conclusions and Perspectives

We presented a first approach for merging logic programs based on Removed Sets Fusion. A study of the properties has been led thanks to the Konieczny and Pino-Perez postulates. This study showed that the Konieczny and Pino-Perez postulates are not suitable in the framework of belief bases merging when beliefs are expressed thanks to logic programs with stable model semantics. We proposed a definition of an inference relation between logic programs. We also defined a weakened version of the merging operation.

Removed Sets Fusion for belief bases represented by logic programs is translated into a logic program with stable model semantics and the one-to-one correspondence between removed sets (both Weak and Strong version) and preferred stable models is shown. Moreover, the paper shows how Removed Sets Fusion can be performed with any ASP solver.

Future works will study the properties of the weak Removed Sets Fusion. A more extensive experimentation of the Removed Sets Fusion for belief represented by logic programs has to be performed. It also can be relevant to study KP postulates in order to allow postulates for dealing with a broader range of frameworks for representing beliefs.

Acknowledgements

Work partially supported by the European Community under project VENUS (Contract IST-034924) of the "Information Society Technology (IST) program of the 6th FP of RTD". The authors are solely responsible for the contents of this paper. It does not represent the opinion of the European Community, and the European Community is not responsible for any use that might be made of data therein.

382 J. Hué, O. Papini, and E. Würbel

References

1. Anger, C., Konczak, K., Linke, T.: Nomore: Non-monotonic reasoning with logic programs. In: Flesca, S., Greco, S., Leone, N., Ianni, G. (eds.) JELIA 2002. LNCS, vol. 2424, pp. 521–524. Springer, Heidelberg (2002)
2. Baral, C., Brewka, G., Schlipf, J. (eds.): LPNMR 2007. LNCS (LNAI), vol. 4483. Springer, Heidelberg (2007)
3. Baral, C., Kraus, S., Minker, J., Subrahmanian, V.S.: Combining knowledge bases consisting of first order theories. In: Raś, Z.W., Zemankova, M. (eds.) ISMIS 1991. LNCS, vol. 542, pp. 92–101. Springer, Heidelberg (1991)
4. Bloch, I., Lang, J.: Towards mathematical morpho-logics. In: Technologies for constructing intelligent systems: tools, Heidelberg, Germany, pp. 367–380. Physica-Verlag GmbH (2002)
5. Cholvy, L.: Reasoning about merging information. In: Handbook of DRUMS, vol. 3, pp. 233–263 (1998)
6. Eiter, T., Leone, N., Mateis, C., Pfeifer, G., Scarcello, F.: The kr system dlv: progress report, comparison and benchmarks. In: Proc. of KR 1998, pp. 406–417 (1998)
7. Gelfond, M., Lifschitz, V.: The stable model semantics for logic programming. In: Proc. of the 5th Int. Conf. on Log. Prog, pp. 1070–1080 (1988)
8. Hue, J., Papini, O., Würbel, E.: Syntactic propositional belief bases fusion with removed sets. In: Mellouli, K. (ed.) ECSQARU 2007. LNCS, vol. 4724, pp. 66–77. Springer, Heidelberg (2007)
9. Hue, J., Papini, O., Würbel, E.: Removed sets fusion: Performing off the shelf. In: Proc. of ECAI 2008 (2008)
10. Konieczny, S.: On the difference between merging knowledge bases and combining them. In: Proc of KR 2000, pp. 135–144. Morgan Kaufmann, San Francisco (2000)
11. Konieczny, S., Pino Pérez, R.: On the logic of merging. In: Proc. of KR 1998, pp. 488–498 (1998)
12. Konieczny, S., Pino Pérez, R.: Merging with integrity constraints. In: Hunter, A., Parsons, S. (eds.) ECSQARU 1999. LNCS, vol. 1638, p. 233. Springer, Heidelberg (1999)
13. Lin, F., Zhao, Y.: Assat: computing answer sets of a logic program by sat solvers. Artif. Intell. 157(1-2), 115–137 (2004)
14. Lin, J.: Integration of weighted knowledge bases. AI 83, 363–378 (1996)
15. Lin, J., Mendelzon, A.O.: Merging databases under constraints. Int. Journal of Cooperative Information Systems 7(1), 55–76 (1998)
16. Meyer, T., Ghose, A., Chopra, S.: Syntactic representations of semantic merging operations. In: Proc. of IJCAI 2001 (2001)
17. Niemelä, I., Simons, P.: An implementation of stable model and well-founded semantics for normal logic programs. In: Fuhrbach, U., Dix, J., Nerode, A. (eds.) LPNMR 1997. LNCS, vol. 1265, pp. 420–429. Springer, Heidelberg (1997)
18. Revesz, P.Z.: On the semantics of theory change: arbitration between old and new information. In: 12th ACM SIGACT-SGMIT-SIGART symposium on Principes of Databases, pp. 71–92 (1993)
19. Revesz, P.Z.: On the semantics of arbitration. Journal of Algebra and Computation 7(2), 133–160 (1997)
20. Fagin, R., Kuper, G.M., Ullman, J.D., Vardi, M.Y.: Updating logical databases. In: Advances in computing research, pp. 1–18 (1986)
21. Turner, H.: Strong equivalence made easy: nested expressions and weight constraints. Theory Pract. Log. Program. 3(4), 609–622 (2003)

Knowledge Base Stratification and Merging Based on Degree of Support

Anthony Hunter[1] and Weiru Liu[2]

[1] Department of Computer Science, University College London,
Gower Street, London WC1E 6BT, UK
a.hunter@cs.ucl.ac.uk
[2] School of Electronics, Electrical Engineering and Computer Science,
Queen's University Belfast, Belfast BT7 1NN, UK
w.liu@qub.ac.uk

Abstract. Most operators for merging multiple knowledge bases (where each is a set of formulae) aim to produce a knowledge base as output that best reflects the information available in the input. Whilst these operators have some valuable properties, they do not provide explicit information on the degree to which each formula in the output has been, in some sense, supported by the different knowledge bases in the input. To address this, in this paper, we first define the degree of support that a formula receives from input knowledge bases. We then provide two ways of determining formulae which have the highest degree of support in the current collection of formulae in KBs, each of which gives a preference (or priority) over formulae that can be used to stratify the formulae in the output. We formulate these two preference criteria, and present an algorithm that given a set of knowledge bases as input, generates a stratified knowledge base as output. Following this, we define some merging operators based on the stratified base. Logical properties of these operators are investigated and a criterion for selecting merging operators is introduced.

1 Introduction

The notion of priority (preference) is important in inconsistency-tolerant reasoning (such that for potentially inconsistent knowledge-based systems [B+04], belief updating [Gad88], analyzing inconsistent regulations [BB04]; and analyzing social networks [KG06]). Priorities can be encoded in two different ways, one way is to prioritize these sources according to their reliability and another is to attach priorities to items of knowledge within each source [BDP96]. Both approaches usually would require that information on priorities is available explicitly, especially for the reliability of sources.

Merging operators for multiple knowledge bases therefore can be characterized for whether they are designed to merge flat (e.g., [BKMS92, KLM04]) or stratified knowledge/belief bases (e.g.,[B+98, DDL06]). Often, the result of merging, no matter whether for stratified or for flat knowledge bases, is simply a flat base[1]. However, even if the

---

[1] We only consider merging in the context of propositional logic. As for possibilistic logic, the merged result is a new possibilistic knowledge base which can be regarded as prioritized.

C. Sossai and G. Chemello (Eds.): ECSQARU 2009, LNAI 5590, pp. 383–395, 2009.

original knowledge bases are not prioritized, some items of knowledge (i.e., some formulae) receive more support (i.e. *more preferred*) then others. This information is often ignored when a merged result is obtained.

The importance of differentiating *more preferred* (w.r.t. support) from *less preferred* formulae can be seen in many applications such as in requirements engineering or in analyzing findings from clinical trials. In requirements engineering, common requirements from different sources are in general selected first before resolving conflicting requirements [Dav05]. Assume that there are four stakeholders involved in building a new computer system. Stakeholder A (sales person) prefers the new system to be *open* and *fashionable*, stakeholder B (system security) prefers the system to be *authorized* and *not open*, stakeholder C (programmer) prefers *open* and *fashionable*, and stakeholder D (investor) prefers *open*, and *fashionable and easy to use*. We can construct four knowledge bases as follows where $p = open$, $q = fashionable$, $s = authorized$ and $r = easy\ to\ use$, $K_A = \{p, q\}$, $K_B = \{\neg p, s\}$, $K_C = \{p, q\}$, and $K_D = \{p, q \wedge r\}$. A combination method in [BKMS92] generates two possible solutions $M_1 = \{p, q, q \wedge r, s\}$ and $M_2 = \{\neg p, q, q \wedge r, s\}$. With these two maximal consistent subsets, it is not possible to decide whether it should be p or $\neg p$ in the merged result, although p has support from three bases. The merging operators proposed in [Kon00] overcome this problem. For instance, one of the three operators in [Kon00] selects $M_1 = \{p, q, q \wedge r, s\}$ in the above example since this subset is consistent with three of the four original bases. However, current merging operators cannot tell which formulae are more preferred in the merged base. In this example, proposition p (the new system should be open) is directly required by three stakeholders, while proposition s is only supported by one stakeholder, nevertheless, they are all treated equally in the merged knowledge base. Furthermore, q is directly given in two bases and is *entailed* from another base. This extra information is not retained either after merging. We believe that the *degree of support* for p (no matter whether directly given or implicitly entailed) from these sources should be reflected in a merged base, so the output of merging also states that p is preferred to q and q is preferred to s.

In this paper, we investigate how the above two types of support to a formula can be identified when multiple knowledge bases are present. For this, we first define the *degree of support* that a formula receives from a profile, we then define both the notion of *most primed formulae* and the *most entailed formulae* and propose methods to select them. A formula with the highest degree of support in a set of knowledge bases (a profile) is either a most primed or a most entailed formula in the profile. We also propose an algorithm to stratify the union of formulae from multiple knowledge bases based on the degree of support a formula receives, with the result being a stratified base. A stratified base induces a total preorder relation which ranks a more preferred formula ahead of less preferred ones. The significance of such a base can be shown in several aspects. First, common beliefs (knowledge) from the majority of sources are given higher priorities than other beliefs in a profile. Second, it can be used to determine which merging result is better in terms of retaining more important formulae. Third, it can be taken as a prioritized observation base and merging operators tailored towards such a base can be applied [DDL06].

The paper is organized as follows. The preliminaries are introduced in Section 2. Two conceptualizations of *most preferred formulae* which have the highest degree of support in the current profile are presented, namely, the *most primed* and the *most entailed* formulae in Section 3. A stratification algorithm and its properties are studied in Section 4. In Section 5, some new merging operators are proposed and their logical properties are discussed. Comparisons with related work and conclusions are given in Section 6.

2   Preliminaries

2.1   Propositional Logic

We consider standard finite classical propositional logic here. We denote atoms as p, q, r, etc. and formulas as $\phi, \psi, \gamma,$, etc.. A literal is an atom p or its negation $\neg p$ and is denoted as l_i. The classical consequence relation is denoted as $\vdash$. An interpretation (possible world), ω, is a function from $\mathcal{A}$ (the set of atoms) to $\{0, 1\}$ and is denoted as the conjunction of literals $l_1 \wedge ... \wedge l_{|\mathcal{A}|}$, where $|\mathcal{A}|$ means the cardinality of set $\mathcal{A}$. ω is a model of a formula ϕ iff $\omega(\phi) = 1$. Two formulae ϕ and φ are said to be equivalent (or equal), denoted as $\phi \equiv \varphi$, iff they have the same set of models. In this paper, we say ϕ is in K_i iff there exists a $\varphi \in K_i$ such that $\phi \equiv \varphi$ regardless of their syntax. Also, $\wedge K_i \vdash \phi$ denotes $\wedge \varphi_j \vdash \phi$ where $\varphi_j \in K_i$.

A knowledge base, K, is a collection of propositional formulae and a knowledge profile, $\mathcal{E} = \{K_1, ..., K_n\}$ contains a set of knowledge bases which are not necessarily distinct. In the following, we let $\mathcal{E}_\cup = K_1 \cup ... \cup K_n$ and let $\mathcal{E}_\cap = K_1 \cap ... \cap K_n$ where $\cup$ and $\cap$ are usual set-based union and intersection operators. A subset E of knowledge profile $\mathcal{E}$ itself in turn is a knowledge profile and $E_\cup$ and $E_\cap$ are similarly defined. In subsequent sections, we always use $\mathcal{E}$ to denote an original knowledge profile, and use E (possibly with subscript) to denote a subset of $\mathcal{E}$.

2.2   Stratified Knowledge Bases

A stratified knowledge base, also called a ranked knowledge base [Bre04] or a prioritized knowledge base [B+93] models a set of formulae with explicit preferences (or priorities) among the formulae. Let K be a knowledge base containing a set of propositional formulae, $(K, \preceq)$ is a **stratified base** if there is a total preorder relation $\preceq$ on K. $\preceq$ is a **total preorder** on K iff for any $\phi, \varphi \in K$, either $\phi \preceq \varphi$ or $\varphi \preceq \phi$ holds. A preorder relation, $\preceq$, is transitive and reflexive and its associated strict preorder relation, $\prec$, is defined as $\phi \prec \varphi$ iff $\phi \preceq \varphi$ but $\varphi \not\preceq \phi$. $\phi \preceq \varphi$ is interpreted as ϕ is at least as preferred (or plausible) as φ and $\phi \prec \varphi$ as ϕ is more preferred than φ. Two preorder relations $\preceq_1$ and $\preceq_2$ are **equivalent**, denoted as $\preceq_1 \equiv \preceq_2$, if for any two formulae ϕ and φ, $\phi \preceq_1 \varphi$ implies $\phi \preceq_2 \varphi$ and vice versa.

For simplicity, in the following we use $\mathcal{S}_K$ to denote a stratified version of a knowledge base K without mentioning the total preorder relation on K and $\mathcal{S}_K$ can be equivalently represented as a tuple $\mathcal{S}_K = \langle S_1, ..., S_m \rangle$ such that $S_i \neq \emptyset, (i = 1, ..., m)$ and S_i contains all the most preferred elements in $K \setminus (\cup_{j=1}^{i-1} S_j)$ w.r.t $\preceq$, that is, $S_i = \{\phi \in K \setminus (\cup_{j=1}^{i-1} S_j), s.t., \forall \varphi \in K \setminus (\cup_{j=1}^{i-1} S_j), \phi \preceq \varphi\}$. Each S_i is called a stratum of K and index i is the priority level of formulae in S_i. Therefore, the lower the index is, the more preferred a formula is. A stratified knowledge base K can be inconsistent and the degree of inconsistency of K is defined as $\mathsf{Inc}(S_K) = i, i > 0 s.t., \cup_{j=1}^{i-1} S_j \not\vdash \bot, \cup_{j=1}^{i} S_j \vdash \bot$. Let $\mathcal{S}_{K_1} = \langle S_1, ..., S_m \rangle$ and $\mathcal{S}_{K_2} = \langle T_1, ..., T_n \rangle$ be two tuples, the concatenation of them is defined as $\mathcal{S}_{K_1} \oplus \mathcal{S}_{K_2} = \langle S_1, ..., S_m, T_1, ..., T_n \rangle$. In particular for any tuple $\mathcal{S}_K$, $\mathcal{S}_K \oplus \langle \rangle = \langle \rangle \oplus \mathcal{S}_K = \mathcal{S}_K$. Also, we write $\mathcal{S}_{K_1} \approx \mathcal{S}_{K_2}$ iff $m = n$ and for each i where $S_i = \{\beta_1, ..., \beta_p\}$ and $T_i = \{\beta_1', ..., \beta_p'\}$ it is the case that $\beta_j = \beta_j'$, for $j = 1, ..., p$.

3   Most Preferred Formulae

To characterize that some formulae are more preferred than others in a collection of knowledge bases, we first define the *degree of support* of a formula. The rational of this definition is that there could be many ways to define a function from formulae to [0,1] and some of them are not acceptable as formalizing the amount of support a formula gets from a knowledge base. So we only give some constraints about what a degree of support function shall obey, rather than specify a single function.

Definition 1. *Let $\mathcal{E}$ be a knowledge profile and let ϕ be a formula. A real function $S_{\mathcal{E}}$ is a **degree of support** function for $\mathcal{E}$ iff it satisfies the following two conditions.*

1. *If $|\mathcal{E}| = 1$ and $\phi \in \mathcal{E}_\cup$ then $S_{\mathcal{E}}(\phi) = 1$*
2. *If $\mathcal{E}_\cup \not\vdash \bot$ and $\phi \in \mathcal{E}_\cup$ then $S_{\mathcal{E}}(\phi) > 0$, where $\mathcal{E}_\cup \not\vdash \bot$ means $\mathcal{E}_\cup$ is consistent.*

ϕ is a **most preferred** formula if $\forall \varphi \in \mathcal{E}_\cup$, $abs(S_{\mathcal{E}}(\phi)) \geq abs(S_{\mathcal{E}}(\varphi))$ holds, where $abs(S_{\mathcal{E}}(\phi))$ returns the positive value (the absolute value) obtained from $S_{\mathcal{E}}(\phi)$.

This definition contains two constraints. The first says that when a knowledge profile contains a single knowledge base then every formula in the base should have the maximum support, and the second states that when a knowledge profile is consistent then every formula appearing in the profile should have a positive degree of support. We now provide some subsidiary definitions that we will use to define two possible definitions of a degree of support functions. Let $\mathsf{Atoms}(\mathcal{E}_\cup)$ be the set of atoms that appear in the formulae in $\mathcal{E}_\cup$. We can use the power set of $\mathsf{Atoms}(\mathcal{E}_\cup)$ to denote the set of interpretations of $\mathcal{E}_\cup$. For any $I \subseteq \mathsf{Atoms}(\mathcal{E}_\cup)$, if $\alpha \in I$ then α is true in I, otherwise α is false in I.

Definition 2. *For a set of formulae X, the set of models of X in the context of $\mathcal{E}$, denoted $\mathsf{M}_{\mathcal{E}}(X)$, is defined as $\mathsf{M}_{\mathcal{E}}(X) = \{I \subseteq \mathsf{Atoms}(\mathcal{E}_\cup) \mid I \models \wedge X\}$*

Definition 3. *Let K be a consistent set of formulae and let ϕ be a consistent formula. The **degree of entailment** of K for ϕ in the context of $\mathcal{E}$, denoted $E_{\mathcal{E}}(K, \phi)$, is defined as $E_{\mathcal{E}}(K, \phi) = \frac{|\mathsf{M}_{\mathcal{E}}(K \cup \{\phi\})|}{|\mathsf{M}_{\mathcal{E}}(K)|}$.*

For instance, let $K = \{p, q \wedge r\}$, then $E_{\mathcal{E}}(K, p) = 1$, $E_{\mathcal{E}}(K, p \wedge q) = 1$, $E_{\mathcal{E}}(K, q \wedge r \wedge s) = \frac{1}{2}$.

The Dalal distance (Hamming distance) between w_i and w_j, denoted $\mathsf{Dalal}(w_i, w_j)$, is the difference in the number of atoms assigned true (i.e. $\mathsf{Dalal}(w_i, w_j) = |w_i - w_j| + |w_j - w_i|$). To evaluate the conflict between two formulae, we take a pair of models, one for each formula, such that the Dalal distance is minimized. The degree of conflict is this distance divided by the maximum possible Dalal distance between a pair of models.

Definition 4. *Let X and Y be sets of formulae, each of which is consistent. The set of* **distances** *between X and Y, denoted* $\text{Dist}_{\mathcal{E}}(X, Y)$ *is defined as*

$$\text{Dist}_{\mathcal{E}}(X, Y) = \{\text{Dalal}(w_x, w_y) \mid w_x \in M_{\mathcal{E}}(X) \land w_y \in M_{\mathcal{E}}(Y)\}$$

Definition 5. *Let K be a consistent set of formulae and let ϕ be a consistent formula. The* **degree of conflict** *of K for ϕ in the context of $\mathcal{E}$, denoted $C_{\mathcal{E}}(K, \phi)$, is defined as follows:*

$$C_{\mathcal{E}}(K, \phi) = \frac{\text{Min}(\text{Dist}_{\mathcal{E}}(K, \{\phi\}))}{|\text{Atoms}(\mathcal{E}_{\cup})|}$$

Once again, let $K = \{p, q \land r\}$, then $C_{\mathcal{E}}(K, p) = C_{\mathcal{E}}(K, p \land q) = 0$, $C_{\mathcal{E}}(K, q \land \neg r \land s) = 1/4$.

We now define two instances of a degree of support.

Definition 6. *Let $\mathcal{E}$ be a knowledge profile and let ϕ be a formula. The* **drastic degree of support**, *denoted $S_{\mathcal{E}}^d$, is defined as*

$$S_{\mathcal{E}}^d(\phi) = \sum_{K \in \mathcal{E}, E_{\mathcal{E}}(K, \phi) = 1} E_{\mathcal{E}}(K, \phi)$$

Definition 7. *Let $\mathcal{E}$ be a knowledge profile and let ϕ be a formula. The* **balanced degree of support**, *denoted $S_{\mathcal{E}}^b$, is defined as*

$$S_{\mathcal{E}}^b(\phi) = \sum_{K \in \mathcal{E}} E_{\mathcal{E}}(K, \phi) - \sum_{K \in \mathcal{E}} C_{\mathcal{E}}(K, \phi).$$

Proposition 1. *Let $\mathcal{E}$ be a knowledge profile and let ϕ and ψ be formulae.*

1. *If $\phi \vdash \psi$ then $S_{\mathcal{E}}^d(\phi) \leq S_{\mathcal{E}}^d(\psi)$*
2. *If $\mathcal{E} \nvdash \perp$ and $\phi \vdash \psi$, then $S_{\mathcal{E}}^b(\phi) \leq S_{\mathcal{E}}^b(\psi)$*

Example 1. Let $\mathcal{E} = \{K_A, K_B, K_C, K_D\}$ where $K_A = \{p, q\}$, $K_B = \{\neg p, s\}$, $K_C = \{p, q\}$, and $K_D = \{p, q \land r\}$ (as defined in the Introduction). Then the drastic degrees of support and the balanced degree of support for formulae p, q, $\neg p$, s, and $q \land r$ respectively are

	p	q	$\neg p$	s	$q \land r$
$S_{\mathcal{E}}^d(\bullet)$	3	3	1	1	1
$S_{\mathcal{E}}^b(\bullet)$	$\frac{11}{4}$	$\frac{7}{2}$	$\frac{1}{4}$	$\frac{5}{2}$	$\frac{9}{4}$

Formulae p and q have the highest drastic degree of support from these bases while q is the only formula in $\mathcal{E}_{\cup}$ that has the highest balanced degree of support. The balanced degrees of support for p and $\neg p$ are both less than their drastic degrees of support because one of them contributed to the degree of conflict of the other. $S_{\mathcal{E}}^d(q \land r) = 1$ is increased to $S_{\mathcal{E}}^b(q \land r) = 9/4$ because $q \land r$ is partially entailed by K_A and K_C, and it is not in conflict with K_B. If we consider the drastic degree of support, p and q are among the most preferred formulae. It should be pointed out that p is directly given in three bases while q is given in two and is entailed by another, although they have the same degree of support. We want to differentiate these two types of formulae when their degrees of support are the same, since we believe that p (as an individual statement) holds more support than q (if we assume stakeholder D really prefers that both q and r be true, not just q).

In this paper, we will only consider the drastic degree of support and we conceptualize these two types of most preferred formulae which have the highest drastic degree of support in the current profile, namely the *most primed* and the *most entailed*.

Definition 8. *Let* $\mathcal{E} = \{K_1, ..., K_n\}$ *be a knowledge profile. We define a total preorder relation* $\preceq_p$ *on* $\mathcal{E}_\cup$ *as follows.*

$$\forall \phi, \varphi \in \mathcal{E}_\cup, \phi \preceq_p \varphi \text{ iff } S_\mathcal{E}^{dp}(\phi) \geq S_\mathcal{E}^{dp}(\varphi)$$

where $S_\mathcal{E}^{dp}(\phi) = |\sharp(\mathcal{E})|s.t., \sharp(\mathcal{E}) = \{K \in \mathcal{E}|E_\mathcal{E}(K, \phi) = 1 \text{ and } \phi \in K\}.$

$S_\mathcal{E}^{dp}(\phi)$ is a variant of $S_\mathcal{E}^d(\phi)$ in which we not only require that $E_\mathcal{E}(K, \phi) = 1$ but also $\phi \in K$, so $S_\mathcal{E}^{dp}(\phi)$ is more restricted than $S_\mathcal{E}^d(\phi)$. For instance, with Example 1, we have $S_\mathcal{E}^d(q) = 3$ whilst $S_\mathcal{E}^{dp}(q) = 2$, because $q \notin K_D$ although $E_\mathcal{E}(K_D, q) = 1$.

Definition 9. *Let* $\mathcal{E} = \{K_1, ..., K_n\}$ *be a knowledge profile.* $\phi \in \mathcal{E}_\cup$ *is called a* **most primed formula** *in* $\mathcal{E}_\cup$ *iff* $\phi \preceq_p \varphi, \forall \varphi \in \mathcal{E}_\cup.$

The most primed formulae are knowledge profile dependent, since a formula can be a most primed formula in one profile but not in another. For example, if we have a knowledge profile $\mathcal{E} = \{K_1, K_2, K_3\}$ such that $K_1 = \{p, q, r\}, K_2 = \{p, r, s\}$, and $K_3 = \{p, q, \neg r\}$, then the most primed formula in $\mathcal{E}_\cup$ is p. However, if we delete p from K_3, then the most primed formulae are $\{p, q, r\}$.

Definition 10. *Let* $\mathcal{E} = \{K_1, ..., K_n\}$ *be a knowledge profile. We define a total preorder relation* $\preceq_e$ *on* $\mathcal{E}_\cup$ *as follows*

$$\forall \phi, \varphi \in \mathcal{E}_\cup, \phi \preceq_e \varphi \text{ iff } S_\mathcal{E}^d(\phi) \geq S_\mathcal{E}^d(\varphi)$$

ϕ *is a* **most entailed formula** *in* $\mathcal{E}_\cup$ *iff* $\forall \psi \in \mathcal{E}_\cup, \phi \preceq_e \varphi.$

Like the most primed formulae in $\mathcal{E}_\cup$, the most entailed formulae are dependent on which knowledge profile is under consideration. A formula can be a most entailed formula in one knowledge profile but is not in another.

It is obvious that given $\mathcal{E} = \{K_1, ..., K_n\}, \forall \phi \in \mathcal{E}_\cup, S_\mathcal{E}^{dp}(\phi) \leq S_\mathcal{E}^d(\phi).$

Example 2. Let $\mathcal{E}_1 = \{K_1, K_2, K_3\}$ with $K_1 = \{p, q\}, K_2 = \{p \wedge q, r\}$ and $K_3 = \{p, s\}$, then, $S_\mathcal{E}^d(p) = 3$ and $S_\mathcal{E}^{dp}(p) = 2, S_\mathcal{E}^d(q) = 2$ and $S_\mathcal{E}^{dp}(q) = 1$, and $S_\mathcal{E}^d(r) = S_\mathcal{E}^{dp}(r) = 1, S_\mathcal{E}^d(s) = S_\mathcal{E}^{dp}(s) = 1.$

It should be noted that although the concept of most entailed formulae subsumes the concept of most primed formulae, we still prefer to have these two types of preferences defined separately. One advantage of this is that we would be able to distinguish a formula that is directly given by several knowledge bases (that is, this element of knowledge is *explicitly* believed and supported) from a formula which is inferred from the same number of knowledge bases (that is, this element of knowledge is *implicitly* believed and supported), if these two formulae have the same degree of support. For instance, for p and q in Example 1 we have $S_{\mathcal{E}}^{d}(p) = S_{\mathcal{E}}^{d}(q) = 3$, however, p holds more confidence in the four knowledge bases than q does. Therefore, we believe that p is more preferred than q in the profile and this rationale is used in the algorithm in the following section.

4 Stratification of a Knowledge Profile

The algorithm below ranks a more preferred formula (using the *most primed* and the *most entailed* formulae) ahead of a less preferred formula. $\mathsf{S_{MPE}}(\mathcal{E})$ stands for an algorithm for **S**tratification based on the **M**ost **P**rimed and/or **E**ntailed formulae in $\mathcal{E}_{\cup}$.

Algorithm: $\mathsf{S_{MPE}}(\mathcal{E})$
1 **Input**: a knowledge profile $\mathcal{E}$
2 **Output**: a stratified version of $\mathcal{E}_{\cup}$, denoted $\mathcal{S}_{\mathcal{E}}$
3 **begin**
4 Let $\mathcal{S}_{\mathcal{E}} = \langle\rangle$, $i = 1$.
5 **while** $\mathcal{E}_{\cup} \neq \emptyset$ **do**
6   $\mathsf{Sup}_1 = max\{S_{\mathcal{E}}^{dp}(\phi)|\phi \in \mathcal{E}_{\cup}\}$.
7   $\mathsf{Sup}_2 = max\{S_{\mathcal{E}}^{d}(\phi)|\phi \in \mathcal{E}_{\cup}\}$.
8   **if** $\mathsf{Sup}_1 = \mathsf{Sup}_2$
9   **then do**
10     $S_i = \{\phi|S_{\mathcal{E}}^{dp}(\phi) = \mathsf{Sup}_1, s.t., \phi \in \mathcal{E}_{\cup}\}$.
11     $\mathcal{S}_{\mathcal{E}} = \mathcal{S}_{\mathcal{E}} \oplus \langle S_i \rangle$.
12     $i = i + 1$.
13     $\mathcal{E}_{\cup} \setminus S_i$.
14   **end of then**
15   **else do**
16     $S_i = \{\phi|S_{\mathcal{E}}^{d}(\phi) = \mathsf{Sup}_2, s.t., \phi \in \mathcal{E}_{\cup}\}$.
17     $\mathcal{S}_{\mathcal{E}} = \mathcal{S}_{\mathcal{E}} \oplus \langle S_i \rangle$.
18     $i = i + 1$.
19     $\mathcal{E}_{\cup} \setminus S_i$.
20   **end of else**
21 **end of while**
22 **end**

Example 3. Let $\mathcal{E} = \{K_1, ..., K_7\}$ where $K_1 = K_2 = K_3 = \{p\}$, $K_4 = K_5 = K_6 = \{q \wedge r\}$ and $K_7 = \{p, s\}$. Then $\mathsf{S_{MPE}}(\mathcal{E})$ returns $\mathcal{S}_{\mathcal{E}} = \langle\{p\}, \{q \wedge r\}, \{s\}\rangle$.

The stratification makes it explicit that if a formula (statement) is more primed or more entailed, then it should be more preferred in a merged result than other formulae. Therefore, it has the obvious advantage over using a merging operator that gives a flat base. This is especially useful when knowledge bases are inconsistent.

Example 4. (Continue Example 1) Let $\mathcal{E} = \{K_A, K_B, K_C, K_D\}$, then the stratified base from the algorithm is $\mathcal{S}_{\mathcal{E}} = \langle\{p\}, \{q\}, \{\neg p, s, q \wedge r\}\rangle$ which clearly shows that p gathers more support from these sources.

Algorithm $\mathsf{S_{MPE}}(\mathcal{E})$ can be modified to stratify a knowledge profile based on the criteria of the most primed (resp. the most entailed formulae) only by using value Sup_1 (resp. Sup_1) alone. The following example reveals the subtle difference between the algorithm and its variants.

Example 5. Let $\mathcal{E} = \{K_1, K_2, K_3, K_4\}$ where $K_1 = \{p, q \wedge r\}$, $K_2 = \{p, q \wedge r, s\}$, $K_3 = \{p, q, s\}$, and $K_4 = \{p, q, s\}$. The stratification of $\mathcal{E}_\cup$ from the exact algorithm is $\mathcal{S}_\mathcal{E} = \langle \{p\}, \{q\}, \{s\}, \{q \wedge r\} \rangle$. The stratification of $\mathcal{E}_\cup$ when only the most primed formulae are considered (only Sup_1 is used) is $\mathcal{S}_\mathcal{E} = \langle \{p\}, \{s\}, \{q, q \wedge r\} \rangle$. On the other hand, when only the most entailed formulae are used (only Sup_2 is considered), the result is $\mathcal{S}_\mathcal{E} = \langle \{p, q\}, \{s\}, \{q \wedge r\} \rangle$.

Definition 11. *Let $\mathcal{E}$ be a knowledge profile and every knowledge base in $\mathcal{E}$ be consistent. We define an ordering relation $\preceq_{\mathsf{MPE}}$ on $\mathcal{E}_\cup$ induced by $\mathcal{S}_\mathcal{E}$ from $\mathsf{S}_{\mathsf{MPE}}(\mathcal{E})$ as $\phi \preceq_{\mathsf{MPE}} \psi$, iff $\phi \in S_i, \psi \in S_j$, and $i \leq j$.*

Proposition 2. *Let $\preceq_{\mathsf{MPE}}$ be defined as above, then $\phi \preceq_{\mathsf{MPE}} \varphi$ iff one of the following conditions is true*

$$S_\mathcal{E}^{dp}(\phi) = S_\mathcal{E}^{d}(\phi) = S_\mathcal{E}^{dp}(\varphi) = S_\mathcal{E}^{d}(\varphi);$$
$$S_\mathcal{E}^{d}(\phi) = S_\mathcal{E}^{d}(\varphi) > S_\mathcal{E}^{dp}(\varphi);$$
$$S_\mathcal{E}^{d}(\phi) > S_\mathcal{E}^{d}(\varphi).$$

Conditions 1 and 3 correspond exactly to the *if* and *else* statements of lines 8 and line 15 respectively. For Condition 2, when $S_\mathcal{E}^{dp}(\phi) = S_\mathcal{E}^{d}(\phi)$, it has the same effect as Condition 3, i.e., ϕ is one stratum lower than φ; however when $S_\mathcal{E}^{dp}(\phi) < S_\mathcal{E}^{d}(\phi)$, it has the same effect as Condition 1, i.e., ϕ and φ are in the same stratum. For instance, in Example 5, we have $S_\mathcal{E}^{d}(p) = S_\mathcal{E}^{d}(q) > S_\mathcal{E}^{dp}(q)$ (and $S_\mathcal{E}^{d}(p) = S_\mathcal{E}^{dp}(p)$), so Condition 2 is met and p is ranked ahead of q.

Definition 12. *Let $\preceq_{\mathsf{MPE}}^{p}$ be a variant of $\preceq_{\mathsf{MPE}}$ on $\mathcal{E}_\cup$ such that $\mathsf{S}_{\mathsf{MPE}}(\mathcal{E})$ considers only the most primed formulae in $\mathcal{E}_\cup$ for stratification.*

Proposition 3. *Let $\preceq_p$ be as defined in Definition 8, then $\preceq_{\mathsf{MPE}}^{p} \equiv \preceq_p$.*

Definition 13. *Let $\preceq_{\mathsf{MPE}}^{e}$ be a variant of $\preceq_{\mathsf{MPE}}$ on $\mathcal{E}_\cup$ such that $\mathsf{S}_{\mathsf{MPE}}(\mathcal{E})$ considers only the most entailed formulae in $\mathcal{E}_\cup$ for stratification.*

Proposition 4. *Let $\preceq_e$ be as defined in Definition 10, then $\preceq_{\mathsf{MPE}}^{e} \equiv \preceq_e$.*

Propositions 3 and 4 show that there is a stratification method corresponding to each of the two total preorder relations defined in Section 3.

5   Merging Operators Based on Stratification

The result of $\mathsf{S}_{\mathsf{MPE}}(\mathcal{E})$ is a stratified base which can be inconsistent. This base can also be viewed as a prioritized observation base [DDL06] where observations in S_i have higher priorities than observations in S_j for $j > i$. To obtain a consistent subset from $\mathsf{S}_{\mathsf{MPE}}(\mathcal{E})$, we need to have suitable operators applicable to a stratified base. We define two such operators here and call them *merging operators*. Let $\mathcal{E}$ be a knowledge profile, and $K = \mathcal{E}_\cup$, then $\mathcal{S}_\mathcal{E} = \mathcal{S}_K = \langle S_1, ..., S_n \rangle$ denotes its stratification. Let $A = \langle A_1, ..., A_n \rangle$ such that $A_i \subseteq S_i$, we define $A_\cup = A_1 \cup ... \cup A_n$.

5.1 New Merging Operators

Definition 14. *Let $\mathcal{E}$ be a knowledge profile and $\mathcal{S}_\mathcal{E} = \langle S_1, ..., S_n \rangle$ be its stratification. Let $A = \langle A_1, ..., A_n \rangle$ be a subset of $\mathcal{S}_\mathcal{E}$ such that $A_i \subseteq S_i$ and $A_\cup \not\vdash \bot$. A* **lexicographical maximal** *merging operator, denoted as $\Delta_{leximax}$, is defined as*

$$\Delta_{leximax}(\mathcal{S}_\mathcal{E}) = \bigvee \{A_\cup | \textit{if for any } A' = \langle A'_1, ..., A'_n \rangle, A'_\cup \not\vdash \bot, \textit{then}$$

$$\textit{either } \forall i \ |A_i| = |A'_i|,$$

$$\textit{or } \exists i, s.t., |A_i| > |A'_i| \textit{ and for } j < i, |A_j| = |A'_j|\}$$

When there is only one stratum in $\mathcal{S}_\mathcal{E}$, merging operator $\Delta_{leximax}(\mathcal{S}_\mathcal{E})$ is equivalent to operator $Comb_4(\mathcal{E}, \top)$ in [BKMS92] and operator $\Delta_\top^{C4}(\mathcal{E})$ in [Kon00] (when we let the integrity constraint IC be a tautology $\top$). However, when $\mathcal{S}_\mathcal{E}$ has more than one stratum, $\Delta_{leximax}(\mathcal{S}_\mathcal{E})$ preserves more information.

Example 6. Let $\mathcal{E} = \{K_1, ..., K_5\}$ where $K_1 = \{\neg p, q\}$, $K_2 = \{p, q\}$, $K_3 = \{p, q \rightarrow r\}$, $K_4 = \{\neg p, s\}$, and $K_5 = \{\neg p, \neg s\}$. Then $\mathcal{S}_\mathcal{E} = \langle \{\neg p\}, \{p, q\}, \{q \rightarrow r, s, \neg s\}\rangle$. Applying operator $\Delta_{leximax}$, we have $\Delta_{leximax}(\mathcal{S}_\mathcal{E}) = \vee\{\{\neg p, q, q \rightarrow r, s\}, \{\neg p, q, q \rightarrow r, \neg s\}\}$ which is equivalent to $\{\neg p, q, q \rightarrow r\}$.

Definition 15. *Let $\mathcal{E}$ be a knowledge profile and $\mathcal{S}_\mathcal{E} = \langle S_1, ..., S_n \rangle$ be its stratification. Let $A = \langle A_1, ..., A_n \rangle$ be a subset of $\mathcal{S}_\mathcal{E}$ such that $A_i \subseteq S_i$ and $A_\cup$ is consistent. Let* $\mathsf{Inc}(\mathcal{S}_\mathcal{E}) = i$. *A* **maximal-consistency** *based merging operator Δ_{conmax}, is defined as*

$$\Delta_{conmax}(\mathcal{S}_\mathcal{E}) = \bigvee \{A_\cup | \cup_{j=1}^{i-1} S_j \subseteq A_\cup, s.t., \forall A' = \langle A'_1, ..., A'_n \rangle, A'_\cup \not\vdash \bot,$$

$$\textit{if } \cup_{j=1}^{i-1} S_j \subseteq A'_\cup \textit{ then } |A_\cup| \geq |A'_\cup|\}$$

This operator guarantees that all the more preferred consistent formulae are selected first, before considering any further formulae. Δ_{conmax} and $\Delta_{leximax}$ are equivalent when $\mathcal{S}_\mathcal{E}$ has only one stratum, i.e., $\mathcal{S}_\mathcal{E} = \langle \mathcal{E}_\cup \rangle$. Furthermore, we define Δ_{conmax}^i as a variant of Δ_{conmax} such that $\mathsf{Inc}(\mathcal{S}_\mathcal{E}) = i$ and $\Delta_{conmax}^i = S_1 \cup ... \cup S_{i-1}$.

Example 7. Let a stratified knowledge profile be $\mathcal{S}_\mathcal{E} = \langle \{\neg p, q\}, \{q \rightarrow r, s, \neg r, \neg r \wedge q\}, \{r\} \rangle$. Then $\Delta_{leximax}(\mathcal{S}_\mathcal{E}) = \{\neg p, q, s, \neg r, \neg r \wedge q\}$ and $\Delta_{conmax}(\mathcal{S}_\mathcal{E}) = \vee\{\{\neg p, q, s, q \rightarrow r, r\}, \{\neg p, q, s, \neg r, \neg r \wedge q\}\}$ which is equivalent to $\{\neg p, q, s\}$. In this case, $\Delta_{leximax} \vdash \Delta_{conmax}$. However, these two operators are not comparable in general.

5.2 Properties

In [DDL06], three merging operators for a prioritized base[2] are defined. Among them operator **best-out**, $*_{bo}(\mathcal{S}_\mathcal{E})$ is defined as

$$*_{bo}(\mathcal{S}_\mathcal{E}) = \bigwedge(\bigwedge S_j | j < i, \mathsf{Inc}(\mathcal{S}_\mathcal{E}) = i)$$

for $\mathcal{S}_\mathcal{E} = \langle S_1, ..., S_n \rangle$ where $\bigwedge S_j = \bigwedge_{\phi \in S_j} \phi$.

---

[2] Note: in their original paper a prioritized base is represented as $\sigma = \langle \sigma(1), ..., \sigma(n) \rangle$ where $\sigma(i)$ denotes a set of formulae with rank k_i and $\sigma(n)$ contains the highest ranked formulae. In this paper, we let S_1 (not S_n) denote the set of highest ranked formulae and ignore the rank itself since it is not used in the merging.

That is, $\bigwedge S_j$ is the conjunction of all formulae in S_j and $\bigwedge(\bigwedge S_j)$ is the conjunction of all $\bigwedge S_j$ for $j = 1,..., i - 1$.

Let $\mathsf{Cons}(\mathcal{S}_\mathcal{E})$ be the set of all consistent subsets of $\mathcal{S}_\mathcal{E}$, that is, the set of all stratified subsets $A = \langle A_1, ..., A_n \rangle$, such that $A_i \subseteq S_i$ and $A_\cup$ is consistent. If $\succ$ is a strict order on set Y, then $\mathsf{Max}(\succ, Y)$ is defined as $\mathsf{Max}(\succ, Y) = \{y \in Y \,|\, \forall z \in Y, z \not\succ y\}$

Definition 16. *[DDL06] For $S, S' \in \mathsf{Cons}(\mathcal{S}_\mathcal{E})$, define $S' \succ_{discrimin} S$ iff $\exists k$ such that*

(a) $\langle S_1, ..., S_k \rangle \cap S' \supset \langle S_1, ..., S_k \rangle \cap S$, and

(b) $\forall i < k, \langle S_1, ..., S_i \rangle \cap S' = \langle S_1, ..., S_i \rangle \cap S$.

Then $*_{discrimin}(\mathcal{S}_\mathcal{E}) = \bigvee\{\bigwedge S, S \in \mathsf{Max}(\succ_{discrimin}, \mathsf{Cons}(\mathcal{S}_\mathcal{E}))\}$

Definition 17. *[DDL06] For $S, S' \in \mathsf{Cons}(\mathcal{E}_\cup)$, define $S' \succ_{leximin} S$ iff $\exists k$ such that*

(a) $|\langle S_1, ..., S_k \rangle \cap S'| > |\langle S_1, ..., S_k \rangle \cap S|$, and

(b) $\forall i < k, |\langle S_1, ..., S_i \rangle \cap S'| = |\langle S_1, ..., S_i \rangle \cap S|$.

Then $*_{leximin}(\mathcal{S}_\mathcal{E}) = \bigvee\{\bigwedge S, S \in \mathsf{Max}(\succ_{leximin}, \mathsf{Cons}(\mathcal{S}_\mathcal{E}))\}$

The following logical properties[3] are given in [DDL06] on merging operators $*$ for prioritized bases.

> **(PMon)** for $i < n$, $*(\langle S_1, ..., S_{i+1} \rangle) \vdash *(\langle S_1, ..., S_i \rangle)$
> **(Succ)** $*(\mathcal{S}_\mathcal{E}) \vdash *(S_1)$
> **(Cons)** $*(\mathcal{S}_\mathcal{E})$ is consistent
> **(Taut)** $*(\mathcal{S}_\mathcal{E}, \top) \equiv *(\mathcal{S}_\mathcal{E})$
> **(Opt)** if $\wedge \mathcal{S}_\mathcal{E}$ is consistent then $*(\mathcal{S}_\mathcal{E}) \equiv \wedge \mathcal{S}_\mathcal{E}$
> **(IS)** If $\mathcal{S}_{\mathcal{E}_1} \approx \mathcal{S}_{\mathcal{E}_2}$ then $*(\mathcal{S}_{\mathcal{E}_1}) = *(\mathcal{S}_{\mathcal{E}_2})$
> **(RA)**[4] $*(\langle S_1, ..., S_i \rangle) = *(*(\langle S_1, ..., S_{i-1} \rangle), S_i)$

Proposition 5. *$\Delta_{leximax}(\mathcal{S}_\mathcal{E})$ is equivalent to operator $*_{leximin}(\mathcal{S}_\mathcal{E})$ and variant Δ_{conmax}^i is equivalent to $*_{bo}(\mathcal{S}_\mathcal{E})$.*

Proposition 6. *Merging operators $\Delta_{leximax}(\mathcal{S}_\mathcal{E})$ and $\Delta_{conmax}^i(\mathcal{S}_\mathcal{E})$ satisfy all the seven properties given above. $\Delta_{conmax}(\mathcal{S}_\mathcal{E})$ satisfies (Cons), (Taut), (Opt), (IS) and (RA).*

Below we examine how a stratified profile can be used to compare different prioritized merging operators.

Definition 18. *Let $\mathcal{S}_\mathcal{E} = \langle S_1, ..., S_n \rangle$ where $\mathsf{Inc}(\mathcal{S}_\mathcal{E}) = i > 1$. Let Γ be the set of all formula-based merging operators for a prioritized base. Let Δ_1 and Δ_2 be two operators in Γ. We define a partial order relation $\preceq$ over Γ as: $\Delta_1 \preceq \Delta_2$ iff one of the following conditions holds.*

---

[3] We only have space in this extended abstract to discuss these properties. In the full paper, we consider further properties including the logical properties in [KP98] when we view the input of such an operator as a knowledge profile and the output as a consistent subset without considering the process of stratification in between. Our operators are also compared with the *Adjustment* and the *Maxi-Adjustment* algorithms in [B+04].

[4] Note: Given $\mathcal{S}_\mathcal{E} = \langle S_1, ..., S_n \rangle$, S_1 has the most reliable formulae and it is equivalent to σ_n in the original definition of a prioritized base in [DDL06], therefore, the Right Associativity (RA) property looks like a Left Associativity property.

- $S_1 \cup ... \cup S_{i-1} \subseteq \Delta_1(\mathcal{E})$ and $S_1 \cup ... \cup S_{i-1} \not\subseteq \Delta_2(\mathcal{E})$;
- $S_1 \cup ... \cup S_{i-1} \subseteq \Delta_1(\mathcal{E})$ and $S_1 \cup ... \cup S_{i-1} \subseteq \Delta_2(\mathcal{E})$, then $|\Delta_1(\mathcal{E})| > |\Delta_2(\mathcal{E})|$.

$\Delta_1 \preceq \Delta_2$ indicates that Δ_1 is at least as efficient as Δ_2 to merge a knowledge profile. Here $|\Delta(\mathcal{E})|$ denotes the cardinality of merging result $\Delta(\mathcal{E})$. The first condition says that if an operator can select all the more preferred and consistent formulae while another cannot, then the former is a better merging operator. The second conditions reveals that when all the more preferred and consistent formulae are included, the operator with more additional formulae is better than the other.

Proposition 7. *Based on Definition 18, we have* $\Delta_{leximax}(\mathcal{S}_\mathcal{E}) \preceq \Delta^i_{conmax}(\mathcal{S}_\mathcal{E})$ *and* $*_{leximin}(\mathcal{S}_\mathcal{E}) \preceq *_{discrimin}(\mathcal{S}_\mathcal{E}) \preceq *_{bo}(\mathcal{S}_\mathcal{E})$. $\Delta_{conmax}(\mathcal{S}_\mathcal{E})$ *is not comparable with* $\Delta_{leximax}(\mathcal{S}_\mathcal{E})$ *or* $*_{discrimin}(\mathcal{S}_\mathcal{E})$.

Operator $\Delta_{conmax}(\mathcal{S}_\mathcal{E})$ does not really take into account the priorities of formulae. Therefore, although it may contain more formulae than other operators, such as, $\Delta_{leximax}(\mathcal{S}_\mathcal{E})$, it is less desirable for a prioritized merging.

6 Related Work and Conclusion

Approaches to stratifying a knowledge base with default rules have been reported in several research proposals [Pea90, GP91, Bre89, Cho94], all of which are about stratifying a single knowledge base with defaults (rules and/or facts). Since we start with multiple original knowledge bases and aim to merge them into a single knowledge base with priorities automatically generated, these proposals cannot be applied. The priorities of formulae are calculated based on the degree of support they receive from the input knowledge bases. A knowledge profile is then stratified based on the priorities of formulae in the profile, with formulae having the highest priority in the current profile being the most preferred formulae. In this respect, our idea of stratification is in spirit similar to Pearl's method in [Pea90], that is, the more support a formula (rule) gets, the higher rank it is assigned.

Our method on stratification has some similarities with voting systems. In a voting system, many voting policies require that a voter simply votes for the chosen candidate(s) without requiring preferences over the chosen candidates. In *plurality voting*, a voter is allowed to vote for one candidate only, so such a knowledge base contains one formula (i.e., candidate). In *approval voting*, a voter can vote for multiple candidates without preferring one over the other, so such a knowledge base contains multiple formulae. For both cases, when our algorithm is applied to stratify a set of votes (knowledge bases), the algorithm produces the same result as either of the two voting policies. More specifically, in both voting policies, the candidates who receive the largest number of votes are the winners (at least for the current round, if a single winner has to be selected, more rounds of votes are required). These candidates are exactly the formulae selected in the first stratum in our algorithm. When this stratum contains a single formula, a single winner is selected. Therefore, let $\mathcal{E}$ be a knowledge profile and each knowledge base in $\mathcal{E}$ contains votes from a voter following the voting rules in plurality (resp. approval) voting. Then the first stratum from $\mathsf{S}_{\mathsf{MPE}}(\mathcal{E})$ is equivalent to the result of plurality (resp. approval) voting system.

In conclusion, in this paper, we focused on how to extract information provided by the original sources about which formulae gathered more support. This information is preserved in the form of a stratified base for formulae in the union of original bases. Stratifying a knowledge base in this way overcomes the problem of deciding which formula should be kept when a choice has to be made to resolve a conflict after merging. An obvious decision is that a higher ranked formula shall be kept. Also, such a merged base provides a basis for ranking merging operators such as a merging operator that preserve as many high ranked formulae as possible is certainly better than the one that cannot.

References

[BKMS92]   Baral, C., Kraus, S., Minker, J., Subrahmanian, V.: Combining knowledge bases consisting of first-order theories. Computational Intelligence 8(1), 45–71 (1992)

[BB04]   Benferhat, S., Baida, R.: A stratified first order logic approach for access control. Int. J. of Int. Sys. 19, 817–836 (2004)

[B+93]   Benferhat, S., Cayrol, C., Dubois, D., Lang, J., Prade, H.: Inconsistency management and prioritized syntax-based entailment. In: Proc. of IJCAI 1993, pp. 640–645 (1993)

[BDP96]   Benferhat, S., Dubois, D., Prade, H.: Reasoning in inconsistent stratified knowledge bases. In: Proc. of ICMVL 1996, pp. 184–189 (1996)

[B+98]   Benferhat, S., Dubois, D., Lang, J., Prade, H., Saffiotti, A., Smets, Ph.: A General approach for inconsistency handling and merging information in prioritized knowledge bases. In: Proc. of KR 1998, pp. 466–477 (1998)

[B+04]   Benferhat, S., Kaci, S., Berre, D., Williams, M.: Weakening conflicting information for iterated revision and knowledge integration. Art. Int. 153(1-2), 339–371 (2004)

[BLR07]   Benferhat, S., Lagrue, S., Rossit, J.: An Egalitarist Fusion of Incommensurable Ranked Belief Bases under Constraints. In: Proc. of AAAI 2007, pp. 367–372 (2007)

[Bre89]   Brewka, G.: Preferred subtheories - an extended logical framework for default reasoning. In: Proc. of IJCAI 1989, pp. 1043–1048 (1989)

[Bre04]   Brewka, G.: A rank-based description language for qualitative preferences. In: Proc. of ECAI 2004, pp. 303–307 (2004)

[CD97]   Cadoli, M., Donini, F.: A survey on knowledge compilation. AI Communication 10(3-4), 137–150 (1997)

[Cho94]   Cholewinski, P.: Stratified default logic. In: Pacholski, L., Tiuryn, J. (eds.) CSL 1994. LNCS, vol. 933, pp. 456–470. Springer, Heidelberg (1995)

[Dav05]   Davis, A.: Just Enough Requirements Management: Where Software Development Meets Marketing. Dorset House, New York (2005)

[DDL06]   Delgrande, J., Dubois, D., Lang, J.: Iterated revision as prioritized merging. In: Proc. of KR 2006, pp. 201–220 (2006)

[EKM05]   Everaere, P., Konieczny, S., Marquis, P.: Quota and Gmin merging operators. In: Proc. of IJCAI 2005, pp. 424–429 (2005)

[Gad88]   Gädenfors, P.: Knowledge in Flux-Modeling the Dynamic of Epistemic States. MIT Press, Cambridge (1988)

[GP91]   Goldszmidt, M., Pearl, J.: System Z^+: A formalism for reasoning with variable-strength defaults. In: Proc. of AAAI 1991, pp. 399–404 (1991)

[KG06]   Katz, Y., Golbeck, J.: Social Network-based Trust in Prioritized Default Logic. In: Proc. of AAAI 2006 (2006)

[KLM04]   Konieczny, S., Lang, J., Marquis, P.: DA^2 operators. Art. Int. 157(1-2), 49–79 (2004)

[Kon00]   Konieczny, S.: On the difference between merging knowledge bases and combining them. In: KR 2000, pp. 12–17 (2000)

[KP98]   Konieczny, S., Pino Pérez, R.: On the logic of merging. In: Proc. of KR 1998, pp. 488–498 (1998)

[Pea90]   Pearl, J.: System Z: A natural ordering of defaults with tractable applications to default reasoning. In: Proc. of the 3rd Int. Conf. on Theor. Aspects of Reas. about Know., pp. 121–135 (1990)

Using Transfinite Ordinal Conditional Functions

Sébastien Konieczny

CNRS
CRIL - Université d'Artois
Lens, France
konieczny@cril.fr

Abstract. Ordinal Conditional Functions (OCFs) are one of the predominant frameworks to define belief change operators. In his original paper Spohn defines OCFs as functions from the set of worlds to the set of ordinals. But in subsequent paper by Spohn and others, OCFs are just used as functions from the set of worlds to natural numbers (plus eventually $+\infty$). The use of transfinite ordinals in this framework has never been studied. This paper opens this way. We study generalisations of transmutations operators to transfinite ordinals. Using transfinite ordinals allows to represent different "levels of beliefs", that naturally appear in real applications. This can be viewed as a generalisation of the usual *"two levels of beliefs"* framework: *knowledge* versus *beliefs*; or *rules base* versus *facts base*, issued from expert systems works.

1 Introduction

Ordinal Conditional Functions (OCFs) [14] are one of the predominant frameworks to represent epistemic state and define belief change operators (see e.g.[14,16,3,10,13]). The intuitive appeal of the definition explains its success: an OCF is a function that maps worlds into ordinals. The smaller the ordinal, the more plausible the world for the agent. This representation of epistemic state is more expressive than the one using total pre-orders on worlds, that is one of the canonical ones for classical AGM belief revision [12,3]. The fundamental role of OCF for defining belief revision operators is shown by the fact that Spohn's conditionalization of OCF [14] is often used to illustrate works on iterated belief revision [3,10].

In his original paper Spohn defines OCFs as functions from the set of worlds to the set of ordinals. But in subsequent papers by Spohn [15] and others, OCFs are just functions from the set of worlds to natural numbers (and eventually $+\infty$). This restriction is natural, since it is enough to represent usual epistemic states and belief change operators.

But it is strange that in works using OCF it was never studied what the use of transfinite ordinals can bring to the representation of epistemic states, and its consequence on the definition of belief revision operators.

This paper aims at studying transfinite OCFs, i.e. OCFs using transfinite ordinals. Very roughly, transfinite ordinals allow to describe different "infinity levels". From a representational point of view, this allows to encode different "levels of beliefs", i.e. more or less strong beliefs, where the strong ones are considered as integrity constraints by weaker ones.

This allows to define generalisation of usual frameworks. First, when one use OCFs that are defined on the restriction ⟨natural numbers $\cup$ $\{+\infty\}$⟩, the worlds that are

C. Sossai and G. Chemello (Eds.): ECSQARU 2009, LNAI 5590, pp. 396–407, 2009.

mapped to $+\infty$ represent unquestionable beliefs[1], that are usually called knowledge in this case.

It is a difficult philosophical debate to determine if knowledge exist or not. Which agent can have unquestionable beliefs ? How can an agent be sure that what he "knows" (believes) is absolutely true. It seems to us that no human/artificial agent can be sure of that[2]. So speaking of knowledge for a human/artificial agent is just a convenient simplifying convention for designing beliefs much more entrenched than other ones. But, even in these really entrenched beliefs some can be even more entrenched than other ones. So having only one $+\infty$ level to represent deep entrenched beliefs is not enough. One would need to be able to represent differences in the entrenchment of these deep entrenched beliefs.

This distinction between knowledge and belief recalls the traditional view of agent representation in expert systems and in automation, that divides the epistemic state of the agent between a base of rules (that corresponds to knowledge) that is a set of entrenched beliefs (rules) on how the represented system evolves, and a base of facts (that corresponds to beliefs), that is a set of observations made by the agent (through captors for instance).

To illustrate this view let us give an example about a doctor's epistemic state. The doctor has a base of rules, that represents his medical expertise/beliefs, and has a base of facts, that represents the symptoms that he observes on a particular patient (this can be medical analysis, visual observations, etc.).

For most applications this representation is clearly sufficient. And it allows also to illustrate interesting discussions on the status of iterated belief revision.

In most papers iterated belief revision is presented as the process of incorporating successively incoming new evidences. So the main point seems to be that the successive inputs are just more and more recent observations. It is true that an autonomous agent has to be able to do this kind of change, but it is an error to use iterated belief revision operators [3] to do that. Iterated belief revision operators [3] do not allow to incorporate more and more recent observations, but more and more reliable observations. This subject is the starting point of the two interesting papers [9,4], where it is clearly explained that if one wants to incorporate more and more recent information, one has to use prioritised belief merging. Roughly, if the observations incorporation order depends only of recency, and that they can have different reliability, then just store the observations with their degree of reliability, and merge all those observations.

In [4] Dubois identifies three different kinds of revision. We will focus on the two first ones. The first one is the one we just discuss above, that is incorporation of more and more recent observations. This basically corresponds to cases where the base of rule does not change, and where the base of facts increases. So suppose that the doctor receives successively several different medical analysis (that have different reliability). The incorporation of these facts will change the beliefs of the doctor on the disease of the patient, but will not change his medical expertise. This case can be basically handled by classical AGM belief revision [1,7,12] if all the observations are jointly consistent

---

[1] Note that in this case OCFs can be viewed as a semantics for possibilistic logic [5].

[2] Under the hypothesis of his existence, the only agent that could hold real knowledge is a God agent.

(that is the case if the world does not evolve (usual hypothesis in belief revision - otherwise one has to use update operators [11,8]), and when the captors are reliable (for instance with direct visual observations)). If the observations are not jointly consistent (for instance if some captors/sources are not reliable), then one has to use the prioritised belief merging framework proposed in [9] to this aim.

The second kind of revision identified by Dubois is when the set of rules of the agent has to be changed. Suppose that our doctor goes to a scientific congress and learn new protocols about a specific disease, then he has to change his base of rules. This is the typical use of iterated belief revision à la Darwiche and Pearl [3]: a more reliable piece of information has to be incorporated in our current theory. So typical examples of DP iterated revision should be scientific theory change rather than every day life observations examples with birds.

Some years ago a very interesting paper from Friedman and Halpern already discuss the problems and dangers of developing new technical change operators without specifying their exact application cases (i.e. without giving them an "ontology") [6]. We think that the papers of Dubois [4] and Delgrande-Dubois-Lang [9] is an interesting reminder of this discussion for iterated belief change.

So to sum up Dubois' view in [4], consider that the agent epistemic state is represented by two bases: a base of rules and a base of facts, the base of rules being more important/reliable/entrenched than the base of facts. Then the two kinds of revision are defined by the base that has to be revised. The first one revise the base of facts, the second one the base of rules.

We think that one can go further than that. There is no objective reason to restrict this process to only two bases, one can need to use more levels of beliefs. So we want to define as many bases as needed, and each of this base can be revised differently.

Let us come back to our doctor example. We can not seriously restrict the beliefs of this agent to a base of medical expertise, and a base of facts on the patient. This agent can have other beliefs much more entrenched that his medical expertise, such as basic arithmetics for instance. So we have at least three "level of beliefs": basic arithmetics that is much more entrenched than medical expertise, that is much more entrenched than facts on the patient.

This is the kind of situation that Transfinite Ordinal Conditional Functions allow to represent and handle.

In next section we we give a short refresher on ordinals, and in Section 3 recall the basic definitions of OCF theory change. Then in Section 4 we define Transfinite Ordinal Conditional Functions, that allow to encode different levels of beliefs in an OCF. In Section 5 we show how to define a Transfinite OCF from a set of classical OCFs that represent the different levels of beliefs. In Section 6 we discuss the revision of Transfinite OCFs, and define relative transmutations, that allow to localize the change to the concerned level of beliefs. Finally we conclude in Section 7.

2   Naive Ordinal Arithmetics

In set theory, the natural numbers can be build from sets:

$\mathbf{0} = \{\}$ (the empty set)
$\mathbf{1} = \{ \{\} \} = \{0\}$

2 = { {}, { {} } } = {0, 1}
3 = {{}, { {} }, { {}, { {} } }} = {0, 1, 2}
4 = { {}, { {} }, { {}, { {} } }, {{}, { {} }, { {}, { {} } }} }= {0, 1, 2, 3}
etc.

So every natural number can be seen as a well ordered set, and the natural order on natural number is given by inclusion of the corresponding sets ($\alpha < \beta$ iff $\alpha \in \beta$).

A possible definition of ordinals is that a set S is an ordinal if and only if S is strictly well-ordered with respect to set membership and every element of S is also a subset of S.

So, starting from 0 ({}), and using a successor operation, noted $\alpha + 1 = \alpha \cup \{\alpha\}$, allows to build the ordinals.

The ordinals that correspond to natural numbers are finite ordinals. The existence of transfinite ordinals is ensured by the axiom of infinity. The first transfinite ordinal is denoted ω. It corresponds to the set of natural numbers $\{0, 1, 2, \ldots\}$. But we can define a successor to this ordinal ω. So we can define $\omega + 1$, $\omega + 2$, etc. until $\omega + \omega = \omega.2$.

If we describe ω as the set $\{a_0, a_1, a_2 \ldots\}$, where $a_0 < a_1 < a_2 < \ldots$, then $\omega + 1$ can be seen as the set $\{a_0, a_1, a_2, \ldots, b_0\}$, where $a_0 < a_1 < a_2 < \ldots < b_0$. See figure 1 for a graphical representation of ω^2.

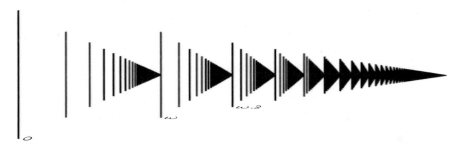

Fig. 1. A graphical "matchstick" representation of the ordinal ω^2. Each stick corresponds to an ordinal of the form $\omega.m + n$ where m and n are natural numbers. (Figure from Wikipedia).

Then one can similarly define $\omega.3$, $\omega.4$, etc. And the ordinal that is the set of all these ordinals is denoted ω^2, etc. We will not use ordinals greater than ω^2 in this work.

The ordinals ω, $\omega.2$, $\omega.3$, $\ldots$, ω^2, $\ldots$, that have no predecessor, are called limit ordinals. β is a limit ordinal if there is no ordinal α such that $\alpha + 1 = \beta$.

Let us now define addition on ordinals.

Definition 1. *The addition on ordinals $\alpha + \beta$ is defined inductively by:*

- $\alpha + 0 = \alpha$,
- $\alpha + (\beta + 1) = (\alpha + \beta) + 1$ [3],
- *if β is limit then $\alpha + \beta$ is the limit of the $\alpha + \gamma$ for all $\gamma < \beta$.*

---

[3] Recall that "+ 1" denotes the successor of an ordinal.

This definition coincides with natural addition when working with finite ordinals. But with transfinite ordinals the addition is not any more commutative. It is for instance easy to see that $3 + \omega = \omega$ and is different from $\omega + 3$ that is the successor of the successor of the successor of ω.

3   Classical OCF Theory

We consider a propositional language $\mathcal{L}$ defined from a finite set of propositional variables $\mathcal{P}$ and the standard connectives.

A world (interpretation) I is a total function from $\mathcal{P}$ to $\{0, 1\}$. The set of all worlds is noted $\mathcal{W}$. An interpretation I is a model of a formula $\varphi \in \mathcal{L}$ if and only if it makes it true in the usual truth functional way. $mod(\varphi)$ denotes the set of models of the formula φ, i.e., $mod(\varphi) = \{I \in \mathcal{W} \mid I \models \phi\}$. Let us denote $\mathcal{O}$ the class of ordinals.

Definition 2. *An Ordinal Conditional Function (OCF) κ is a function from the set of worlds $\mathcal{W}$ to the set of ordinals such that at least one world is assigned 0.*

The meaning of an OCF is simple. The ordinal associated to a world denotes the world plausibility. The higher the ordinal, the less plausible the world. So let us call this the *degree of disbelief* of the world. In particular world that are assigned 0 are the most plausible worlds, i.e. the currently believed worlds. This means that if one use OCFs as representation of epistemic states for iterated belief revision, the belief base φ associated to this epistemic state $\varphi = Bel(\kappa)$ is defined by those models: $mod(\varphi) = \{I \mid \kappa(I) = 0\}$. The set of OCFs will be denoted $\mathcal{K}$.

The degree of disbelief can be straightforwardly extended to formulae (set of worlds).

Definition 3. *The degree of disbelief of a formula φ is the minimum of the degree of disbelief of its models: $\kappa(\varphi) = \min_{I \models \varphi} \kappa(I)$.*

And one can also define the degree of acceptance of a formula.

Definition 4. *A formula φ is accepted (for an OCF κ) if $\kappa(\varphi) = 0$.*
    The degree of acceptance of an accepted formula φ is $d_\kappa(\varphi) = \kappa(\neg\varphi)$.

Now we can define change operators in this setting as functions that change the degree of acceptance of a formula:

Definition 5. *A Transmutation [16] is a function that, given an OCF κ, a formula φ and a degree of acceptance α, produces a new OCF $\kappa * (\varphi, \alpha)$ such that φ is accepted with degree $d_{\kappa*(\varphi,\alpha)}(\varphi) = \alpha$.*

Several different transmutation operators can be defined. The problem is to meet the condition of transmutation operators while keeping as much information as possible from the old OCF. As for works on AGM belief revision there are several ways of considering this minimality. The two most usual ones are conditionalization [14] and adjustment [16].

Definition 6. *The (φ, α)-conditionalization of κ is a new OCF $\kappa *_C (\varphi, \alpha)$ defined as*

$$(\kappa *_C (\varphi, \alpha))(I) = \begin{cases} -\kappa(\varphi) + \kappa(I) & \text{if } I \models \varphi \\ (-\kappa(\neg\varphi) + \kappa(I)) + \alpha & \text{if } I \models \neg\varphi \end{cases}$$

where $-\beta + \alpha$ represents the ordinal γ such that $\beta + \gamma = \alpha$.

Conditionalization moves all the models of φ. Adjustment moves only the most plausible models of φ (some models of φ are moved if necessary).

Definition 7. *The (φ, α)-adjustment of κ is a new OCF $\kappa *_A (\varphi, \alpha)$ defined as*

$$(\kappa *_A (\varphi, \alpha))(I) = \begin{cases} 0 & \text{if } \kappa(I) = \kappa(\varphi) \\ \alpha & \text{if } I \models \neg\varphi \text{ and } \kappa(I) < \alpha \\ \kappa(I) & \text{otherwise} \end{cases}$$

Adjustment can be seen as the counterpart of Boutilier's natural revision [2] for OCFs.

4 Transfinite OCF

So the aim of this work is to encode different "levels of beliefs" in a same OCF. These levels of beliefs have to be strictly hierarchized, in order to ensure that a belief in a higher level is considered as an integrity constraint by the beliefs in lower levels. Let us illustrate this need on a car-driving example.

Example 1. The most important beliefs of the agents are physical beliefs, that compose the highest level of beliefs:

- The road is slippery if and only if it is snowed or frozen ($sl \leftrightarrow sn \vee f$).

  Then the driving behaviour rules form the second level of beliefs:

- If the road is slippery, then adopt a moderate speed ($sl \rightarrow m$).
- If there are roadworks, then adopt a moderate speed ($w \rightarrow m$).

The first rule being more important/entrenched/believed than the second one (let us assign a weight of 2 to the first rule, and a weight of 1 to the second one).
    Finally the lowest level of beliefs is the one of facts that describe the agent's beliefs about the current situation:

- The road is snowed (sn).
- There are no roadworks ($\neg w$). The road is not frozen ($\neg f$).

The belief that the road is snowed is more important/entrenched/believed than the fact that the road is not frozen. Let us assign a weight of 5 to the first fact (the road is snowed), and a weight of 2 to the other ones. The numbers reflect in a sense the intensity of belief of these facts for the agent.

So we will use transfinite ordinals in order to encode the different levels of beliefs. The idea is to use a limit ordinal as boundary between two levels of beliefs.

Definition 8. *A Transfinite OCF is an OCF κ such that for any world I either $\kappa(I) = 0$ or $\omega.(m-1) < \kappa(I) < \omega.m$, and for at least one world I' $\kappa(I') > \omega$.*

When the last inequality hold m is called the level of belief of I, and is denoted $\lambda^\kappa(I)$. And if $\kappa(I) = 0$, then $\lambda^\kappa(I) = 1$.

The level of belief of a formula, noted $\lambda^\kappa(\varphi)$, is the minimum level of beliefs of its models: $\lambda^\kappa(\varphi) = \min_{I \models \varphi} \lambda^\kappa(I)$.

So a Transfinite OCF that corresponds to the car-driving example of example 1 is for instance:

Example 2. Let us introduce our representation of OCF. The ordinal at the left of the line is the one associated to the worlds at the right. The propositional symbols are considered in the order (f sn sl w m) for the interpretations. The notation $*$ represents all the worlds where $*$ can be replaced by 0 or 1, for instance $1 * 1$ is a shortcut for $\{101, 111\}$.

κ:

$\omega.2 + 1$	$\{001 * *, 010 * *, 100 * *, 110 * *\}$
$\omega + 2$	$\{10110, 01110, 01100, 10100, 11100, 11110\}$
$\omega + 1$	$\{00010\}$
5	$\{00000, 00001, 00011, 10111, 10101\}$
2	$\{11101, 11111, 01111\}$
0	$\{01101\}$

This OCF represents the information given by the rules of Example 1. It is obtained in the usual way [5]. So in this example the worlds that are associated to the ordinal 0 are the worlds that satisfy all the formulas of all the levels of beliefs. The worlds that are associated to 2 or 5 are the worlds that satisfy the two most important levels of beliefs, the worlds that are associated to 2 being more plausible than the ones that are associated to 5. The worlds that are associated to $\omega + 1$ and $\omega + 2$ satisfy the most important level of beliefs. The worlds that are associated to $\omega.2 + 1$ do not satisfy the most important level of beliefs.

5   Building a Transfinite OCF from a Set of Classical OCFs

It is possible to build a Transfinite OCF from a set of classical OCFs [4], each classical OCF representing one level of beliefs.

Definition 9. *Let $\kappa_1, \ldots, \kappa_n$ being the classical OCFs that represent respectively the first (least important), ..., last (most important) level of beliefs. Then κ is the Transfinite OCF defined inductively as $\kappa(I) = \kappa_{\kappa_1,\ldots,\kappa_n}(I)$:*

- $\kappa_\emptyset(I) = 0$
- $\kappa_{\kappa_1,\ldots,\kappa_n}(I) = \begin{cases} \omega.(n-1) + \kappa_n(I) & \text{if } \kappa_n(I) > 0 \\ \kappa_{\kappa_1,\ldots,\kappa_{n-1}}(I) & \text{otherwise} \end{cases}$

---

[4] Let us call *Classical OCFs* OCFs where all the ordinals associated to worlds are strictly smaller than ω. And let us call *Constrained OCFs* OCFs where all the ordinals associated to worlds are smaller or equal to ω.

Coming back to the car-driving example, this amounts to consider the three following classical OCFs representing the three levels of beliefs:

Example 3. The first level of belief, containing the facts, is encoded by κ_1. The second one, containing the driving behaviour rules, is encoded by κ_2. The last one, containing physical beliefs, is encoded by κ_3.

κ_1:
5 $\{101 * *, 000 * *, 001 * *, 100 * *\}$
2 $\{111 * *, 01111, 01110, 01010, 01011, 110 * *\}$
0 $\{01 * 0*\}$
κ_2:
2 $\{* * 1 * 0\}$
1 $\{* * 010\}$
0 $\{* * 000, * * 001, * * 011, * * 101, * * 111\}$
κ_3:
1 $\{001 * *, 010 * *, 100 * *, 110 * *\}$
0 $\{000 * *, 101 * *, 011 * *, 111 * *\}$

It is easy to check that starting from $\kappa_1, \kappa_2, \kappa_3$ and using the construction of definition 9, we obtain κ of example 2.

6   Revising Transfinite OCFs

Of course, since Transfinite OCFs are a subclass of OCFs, then one can use usual conditionalization (or adjustment, or any other transmutation) on Transfinite OCFs.

But this may cause some problems since conditionalization allows to change the degree of acceptance of an interpretation (so of any formula), to **any** new degree.

This freedom may cause problems for Transfinite OCFs, since this means that this allows to "merge" different levels of beliefs together, just as if there was only one such level. So in this case this means that the representation using levels of beliefs is useless. Since it would be possible for instance to define a classical OCF (using a mapping from the Transfinite OCF) with exactly the same behaviour for transmutations/conditionalization (up to the mapping).

So we would rather need a conditionalization (or more generally a transmutation) that allows only change inside each level of beliefs.

Let us see how to define this operation below. Let us call usual conditionalization (resp. transmutation) absolute conditionalization (resp. absolute transmutation). We will define now relative conditionalization (and relative transmutations).

First, let us illustrate what the Transfinite OCF means from each level of beliefs. This is done using projections.

Definition 10. *Let κ be a Transfinite OCF with n different levels of beliefs. The i-projection of κ (projection of κ on the i-th level of belief), denoted $\kappa_{\downarrow i}$ is defined as:*

$$\kappa_{\downarrow i}(I) = \begin{cases} \omega & \text{if} & \omega.i < \kappa(I) \\ \kappa(I) & \text{if } \omega.(i-1) < \kappa(I) < \omega.i \\ 0 & \text{if} & \kappa(I) < \omega.(i-1) \end{cases}$$

So let us see a projection of the Transfinite OCF of the example:

Example 4. From the most important level of beliefs point of view, the projection is just:

$\kappa_{\downarrow 3}$:
1 $\{001 * *, 010 * *, 100 * *, 110 * *\}$
0 $\{000 * *, 101 * *, 011 * *, 111 * *\}$

The worlds that are associated to 0 in the third level of beliefs, will be eventually discriminated by lower levels.

From the second level of beliefs point of view, the third level of beliefs appear as integrity constraints that can not be questioned, so all the worlds that are not associated to 0 in the (projection of the) third level of beliefs are just impossible worlds. So the projection is:

$\kappa_{\downarrow 2}$:
ω $\{001 * *, 010 * *, 100 * *, 110 * *\}$
2 $\{01100, 10100, 11100, 11110, 10110, 01110\}$
1 $\{00010\}$
0 $\{00000, 00001, 00011, 10111, 10101, 11101, 11111, 01111, 01101\}$

Now from the first level of beliefs, all the highest levels of beliefs appear as integrity constraints. So the projection is:

$\kappa_{\downarrow 1}$:
ω $\{001 * *, 010 * *, 100 * *, 110 * *, 00010, 10110, 01110, 01100, 10100, 111 * 0\}$
5 $\{00000, 00001, 00011, 10111, 10101\}$
2 $\{11101, 11111, 01111\}$
0 $\{01101\}$

These projection give an idea of how to revise Transfinite OCFs. A relative transmutation will only change the corresponding projection (i.e. level of beliefs). Since if the level of a formula is i, this means that a change of its degree of disbelief will change the information of the i-th level of beliefs.

Let us define relative transmutation formally.

Definition 11. *Let κ be a Transfinite OCF with n levels of beliefs. Let α be an ordinal $\alpha < \omega$. Given a (absolute) transmutation $*$. Then the corresponding relative transmutation $\boxplus$ is defined as:*

$$(\kappa \boxplus (\varphi, \alpha))(I) = \begin{cases} \kappa(I) & \text{if } \kappa^o(I) = \omega \\ \kappa(I) & \text{if } \kappa^o(I) = 0 \text{ and } \lambda^\kappa(I) < \lambda^\kappa(\varphi) \\ \mathbf{B}(\kappa, \lambda^\kappa(\varphi) - 1) + 1 & \text{if } \kappa^o(I) = 0 \text{ and } \lambda^\kappa(I) = \lambda^\kappa(\varphi) \\ \omega.\lambda^\kappa(\varphi) + \kappa^o(I) & \text{otherwise} \end{cases}$$

where

- $\kappa^o = \kappa_{\downarrow \lambda^\kappa(\varphi)} * (\varphi, \alpha)$
- $\mathbf{B}(\kappa, i) = \max_{\{I | \lambda^\kappa(I) = i\}} \kappa(I)$, *if $i > 0$; and $\mathbf{B}(\kappa, 0) = -1$.*

The main idea of this definition is to localize the change to the concerned level of beliefs. This is the aim of κ^o, that does the transmutation only on the projection on the concerned level. Then the result is incorporated in the full Transfinite OCF κ, with the four points of the main definition. The first point ensures that worlds in higher levels of beliefs are not moved during the change. The second point says similarly that worlds that are in lower levels of beliefs, and that are not involved in the change at the concerned level of belief, are not moved during the change. The fourth point just encode the changes on the concerned level of beliefs. The interesting part of the definition is given by the third point that says that if there are new worlds that are possible (i.e. such that $\kappa^o(I) = 0$) after the transmutation on the concerned level, then they are downgraded to the lower level of beliefs. The problem is then to know where to put them in the lower level. In order to ensure minimal change for this lower level we have to try to modify as little as possible the structure of that level. This can be done by including the downgraded worlds as the least plausible worlds of this level (this is the aim of the function **B** that allows to find the plausibility of the least plausible worlds in a given level of beliefs).

Let us see this on the example.

Example 5. Suppose that we just bought a new car with new driving assistance systems, that make us remove from our driving behaviour rules that $sl \rightarrow m$. So to make this contraction we make a relative 0-conditionalization

$\kappa \boxplus_C (sl \rightarrow m, 0)$:

$\omega.2 + 1$	$\{001 * *, 010 * *, 100 * *, 110 * *\}$
$\omega + 1$	$\{00010\}$
6	$\{10110, 01110, 01100, 10100, 11100, 11110\}$
5	$\{00000, 00001, 00011, 10111, 10101\}$
2	$\{11101, 11111, 01111\}$
0	$\{01101\}$

An interesting point to note is that after this relative conditionalization, the formula $sl \rightarrow m$ still holds in the Transfinite OCF $\kappa \boxplus_C (sl \rightarrow m, 0)$. But it is no longer a formula of the second level of belief (i.e. $[\kappa \boxplus_C (sl \rightarrow m, 0)]_{\downarrow 2}$). It is now a formula of the first level of belief. So now a change in the first level of belief can remove this rule from the beliefs of the agent, whereas it was not possible before since beliefs of the second level of beliefs can not be changed by revision of the first level.

Note in particular that, to remove completely this formula from the beliefs of the agent, one has to do one more contraction:

Example 6. $(\kappa \boxplus_C (sl \rightarrow m, 0)) \boxplus_C (sl \rightarrow m, 0)$:

$\omega.2 + 1$	$\{001 * *, 010 * *, 100 * *, 110 * *\}$
$\omega + 1$	$\{00010\}$
5	$\{00000, 00001, 00011, 10111, 10101\}$
2	$\{11101, 11111, 01111\}$
0	$\{01101, 10110, 01110, 01100, 10100, 11100, 11110\}$

Note that on this example we have $(\kappa \boxplus_C (sl \rightarrow m, 0)) \boxplus_C (sl \rightarrow m, 0) = \kappa *_C (sl \rightarrow m, 0)$. But this is generally not the case.

As explained in the introduction, it makes sense to use different revision operators for the different levels of beliefs. For instance the less important level, that usually contains factual information can use more drastic revision operators, since loss of information in this level is not that important (with respect to loss in higher levels).

So this means that we need some adaptative change operators. We define such operators as relative transmutations where the computation of κ^o depends of the level of the new piece of information:

Definition 12. *Let κ be a Transfinite OCF with n levels of beliefs. Let $\mathcal{A} = \{*_1, \ldots, *_n\}$ be a vector of n absolute transmutations. Let α be an ordinal $\alpha < \omega$. Then the corresponding adaptative relative transmutation $\boxplus^{\mathcal{A}}$ is defined as:*

$$(\kappa \boxplus^{\mathcal{A}} (\varphi, \alpha))(I) = \begin{cases} \kappa(I) & \text{if } \kappa^o(I) = \omega \\ \kappa(I) & \text{if } \kappa^o(I) = 0 \text{ and } \lambda^\kappa(I) < \lambda^\kappa(\varphi) \\ \mathbf{B}(\kappa, \lambda^\kappa(\varphi) - 1) + 1 & \text{if } \kappa^o(I) = 0 \text{ and } \lambda^\kappa(I) = \lambda^\kappa(\varphi) \\ \omega.\lambda^\kappa(\varphi) + \kappa^o(I) & \text{otherwise} \end{cases}$$

where

- $\kappa^o = \kappa_{\downarrow \lambda^\kappa(\varphi)} *_{\lambda^\kappa(\varphi)} (\varphi, \alpha)$
- $\mathbf{B}(\kappa, i) = \max_{\{I \mid \lambda^\kappa(I) = i\}} \kappa(I)$, *if $i > 0$; and $\mathbf{B}(\kappa, 0) = -1$.*

7 Conclusion

In this paper we have investigated how to represent and change beliefs of an agent that are hierarchized through several levels of beliefs, where each level appears as integrity constraint for less important levels. We have shown how to represent these levels by using Ordinal Conditional Functions. This is the first time, as far as we know, that the use of transfinite ordinals is investigated. Spohn in a footnote of [14] says:

> "It would be a natural idea to restrict the range of OCFs to the set of natural numbers. In fact, much of the following could thereby be simplified since usual arithmetic is simpler than the arithmetic on ordinals. For the sake of formal generality I do not impose this restriction. But larger ranges may be intuitively needed. For example, it is tempting to use OCFs with larger ranges to represent the stubbornness with which some beliefs are held in the face of seemingly arbitrarily augmentable counter-evidence."

So in this work we have proposed a representation of these stubbornly held beliefs by mean of levels of beliefs. And, more importantly, we have discussed the inadequacy of usual (absolute) transmutations to realize the change on these OCFs. So we have proposed the definition of relative transmutations, that limit the change to the concerned level of belief.

We are convinced that several other interesting change operators can be defined in the framework of Transfinite OCFs. We let this for future work.

Acknowledgements. The author would like to thank the anonymous reviewers for their useful comments.

References

1. Alchourrón, C.E., Gärdenfors, P., Makinson, D.: On the logic of theory change: Partial meet contraction and revision functions. Journal of Symbolic Logic 50, 510–530 (1985)
2. Boutilier, C.: Iterated revision and minimal change of conditional beliefs. Journal of Philosophical Logic 25(3), 262–305 (1996)
3. Darwiche, A., Pearl, J.: On the logic of iterated belief revision. Artificial Intelligence 89, 1–29 (1997)
4. Dubois, D.: Three scenarios for the revision of epistemic states. Journal of Logic and Computation 18(5), 721–738 (2008)
5. Dubois, D., Lang, J., Prade, H.: Possibilistic logic. In: Handbook of Logic in Artificial Intelligence and Logic Programming, vol. 3, pp. 439–513. Oxford University Press, Oxford (1994)
6. Friedman, N., Halpern, J.Y.: Belief revision: a critique. In: Proceedings of the Fifth International Conference on Principles of Knowledge Representation and Reasoning (KR 1996), pp. 421–431 (1996)
7. Gärdenfors, P.: Knowledge in flux. MIT Press, Cambridge (1988)
8. Herzig, A., Rifi, O.: Update operations: a review. In: Proceedings of the Thirteenth European Conference on Artificial Intelligence (ECAI 1998), pp. 13–17 (1998)
9. Lang, J., Delgrande, J.P., Dubois, D.: Iterated revision as prioritized merging. In: Proceedings of the tenth International Conference on Principles of Knowledge Representation and Reasoning (KR 2006), pp. 210–220 (2006)
10. Jin, Y., Thielscher, M.: Iterated belief revision, revised. Artificial Intelligence 171, 1–18 (2007)
11. Katsuno, H., Mendelzon, A.O.: On the difference between updating a knowledge base and revising it. In: Proceedings of the Second International Conference on Principles of Knowledge Representation and Reasoning (KR 1991), pp. 387–394 (1991)
12. Katsuno, H., Mendelzon, A.O.: Propositional knowledge base revision and minimal change. Artificial Intelligence 52, 263–294 (1991)
13. Meyer, T.: On the semantics of combination operations. Journal of Applied Non Classical Logics 11(1/2), 59–84 (2001)
14. Spohn, W.: Ordinal conditional functions: A dynamic theory of epistemic states. In: Harper, W.L., Skyrms, B. (eds.) Causation in Decision: Belief Change and Statistics, pp. 105–134. Kluwer, Dordrecht (1988)
15. Spohn, W.: Ranking Functions, AGM Style. Internet Festschrift for Peter Gärdenfors, Lund (1999)
16. Williams, M.A.: Transmutations of knowledge systems. In: Proceedings of the Fourth International Conference on the Principles of Knowledge Representation and Reasoning (KR 1994), pp. 619–629 (1994)

The Non-archimedean Polynomials and Merging of Stratified Knowledge Bases

Jianbing Ma[1], Weiru Liu[1], and Anthony Hunter[2]

[1] School of Electronics, Electrical Engineering and Computer Science
Queen's University Belfast Belfast BT7 1NN, UK
{jma03,w.liu}@qub.ac.uk

[2] Department of Computer Science University College London
Gower Street London WC1E 6BT, UK
a.hunter@cs.ucl.ac.uk

Abstract. In this paper, a new algebraic representation by the non-Archimedean fields is proposed to model stratified/ranked knowledge bases. The non-Archimedean representation is in the form of the non-Archimedean polynomials. With the non-Archimedean representation, the most widely used ordering strategies are easily induced and compared. Moreover, a framework of prioritized merging operators using the non-Archimedean representation is presented. It is shown that these merging operators satisfy the prioritized merging properties proposed by Delgrande, Dubois and Lang. In addition, several prioritized merging operators in the literature are proved to be special cases of the framework. Furthermore, the egalitarist fusion of incommensurable ranked bases by Benferhat, Lagrue and Rossit is also derived from the non-Archimedean representation.

1 Introduction

In many applications, there is a need to combine possibly conflicting information from different sources in order to get coherent knowledge. This is the origin of information/data fusion problem. As a very important part of the data fusion problem, in the last two decades, the merging of knowledge bases has attracted significant attention.

Knowledge bases (KBs) can be flat or stratified/ranked. In a flat KB, all the logical formulae are viewed as equally important. In stratified KBs (SKBs), however, formulae are assigned with different levels of importance (priority). A formula at a higher level is viewed as more important than those at a lower level, while in a ranked KB (RKB), each formula is attached to a rank (e.g., an ordinal number). A formula with a higher rank is more preferred than those with lower ranks.

A significant property of stratified/ranked KBs is that higher level/rank items are more important than lower ones. This property is exploited in all the prioritized merging operators in different forms. That is, each of such merging operators involves a step that captures prioritized information as well as a step that merges knowledge. In this paper, we want to investigate whether there is a unified framework to represent the prioritized information prior considering merging and hence use this unified framework to define prioritized merging operators. To achieve this, we introduce the non-Archimedean fields to represent stratified/ranked KBs. We demonstrate that this representation perfectly

C. Sossai and G. Chemello (Eds.): ECSQARU 2009, LNAI 5590, pp. 408–420, 2009.

captures this property and its format is intuitive. In this way, a merging operator only requires a simple definition since most of the work required for merging has already been encoded in the non-Archimedean representation.

It appears that simple vectors of integers can also be used to represent Stratified KBs, but they have some major drawbacks. First, it is difficult to represent ranked bases with simple vectors, especially in performing the scalings of ranked bases (Def. 15). Second, although simple vectors can be used to obtain orderings between possible worlds, it is hard to present a *unified* picture on *different* ordering strategies. In contrast, our non-Archimedean representation solves these problems easily.

The merging of stratified/ranked KBs has been studied in many papers such as, [B+93, Leh95, Bre04, DDL06, QLB06, BLR07]. The extra knowledge implied in a Stratified KB can be used to induce a total preorder relation on interpretations, and the three widely used ordering strategies are *best out*, *maxsat* and *leximin* [B+93, Bre04]. In [Bre04], the relationship between the three orderings was studied.

In this paper, we first provide the non-Archimedean polynomial (NAP) representation for Stratified KBs which gives us a clear and unified representation of the three preorder relations on interpretations, and therefore, makes the relationship between them immediately provable.

Second, we propose a family of merging operators for Stratified KBs in terms of NAPs. This family of merging operators captures a wide class of prioritized merging operators. It not only captures several existing prioritized merging operators in the literature, such as the linear and leximin operators, but also identifies new merging operators. Our family of prioritized merging operators is the counterpart of the DA^2 family of flat merging operators [KLM04].

When merging prioritized KBs, an issue to be considered is whether this set of bases is commensurable. In fact, most of the merging operators for Stratified KBs proposed so far require that the commensurability assumption is in place. For the incommensurable situation, a method called the *egalitarist fusion* of Ranked KBs was proposed [BLR07]. It is proved that the egalitarist fusion, obtained from a maximum based ordering that is unchanged in all compatible scalings, is equivalent to a *Pareto-like* operator. In this paper, we show that our non-Archimedean representation of the ranked bases are sufficient to simulate the egalitarist fusion.

In summary, the main contributions of this paper are as follows. First, this paper provides a uniform framework to represent prioritized information at a higher level than embedding it in concrete merging operators. Second, this paper shows that the NAPs provides a unified format to represent three commonly used ordering strategies so that relationship between them can be induced easily. Third, this paper proposes a new family of prioritized merging operators in terms of NAPs which covers a variety of prioritized merging operators in the literature. Fourth, this paper shows that the egalitarist fusion for ranked bases can also be represented and interpreted by NAPs.

The rest of the paper is organized as follows. In Section 2, we recall some basic concepts on propositional logic, Stratified KBs, and non-Archimedean fields. For convenience and subsequent representation, we also introduce some definitions from the DA^2 merging operators. In Section 3, we propose the NAPs and relate them to the three ordering strategies. In Section 4, we introduce the framework of prioritized merging operators using NAPs. In Section 5, we give the NAP representation for Ranked KBs and simulate the egalitarist fusion. Finally, we conclude the paper in Section 6.

2   Preliminaries

In this paper, we consider a propositional language $\mathcal{L}_{\mathcal{PS}}$ defined on a finite set $\mathcal{PS}$ of propositional atoms, denoted by p, q, r etc. A proposition ϕ is constructed by propositional atoms with logical connectives $\neg, \wedge, \vee, \rightarrow$ in the usual way. An interpretation w (or possible world) is a function that maps $\mathcal{PS}$ to $\{0, 1\}$. The set of all possible interpretations on $\mathcal{PS}$ is denoted as W. Function w can be extended to any propositional sentence in $\mathcal{L}_{\mathcal{PS}}$ in the usual way, $w : \mathcal{L}_{\mathcal{PS}} \rightarrow \{0, 1\}$. w is a model of (or satisfies) ϕ iff $w(\phi) = 1$, denoted as $w \models \phi$. We use $Mod(\phi)$ to denote the set of models for ϕ.

For any set A, a pre-order $\leq$ is a reflexive and transitive relation over $A \times A$. $\leq$ is total iff for all elements $a, b \in A$, either $a \leq b$ or $b \leq a$ holds. Conventionally, a strict order $<$ and an indifferent relation $=$ can be induced by $\leq$ such that $\forall a, b \in A$, $a < b$ iff $a \leq b$ but $b \not\leq a$, and $a = b$ iff $a \leq b$ and $b \leq a$. We use $max(A, \leq)$ to denote the set $\{a \in A | \nexists b \in A, b > a\}$ and $min(A, \leq)$ for $\{a \in A | \nexists b \in A, b < a\}$.

2.1   Flat/Stratified/Prioritized KBs

A *(flat) KB* K is a finite set of propositions. K is consistent iff there is at least one interpretation that satisfies all propositions in K.

A *SKB* K is a set of propositions with a pre-order $\leq$ on K, where $\phi \leq \varphi$ means ϕ is more important (plausible) than φ. Commonly, it is written as $K = (S_1, \ldots, S_n)$ where each S_i (called a stratum) is a set of propositions with all the most important (plausible) elements in $K \setminus \bigcup_{j=1}^{i-1} S_j$, i.e., $S_i = min(K \setminus \bigcup_{j=1}^{i-1} S_j, \leq)$, where $\phi < \varphi$ if $\phi \in S_i$ and $\varphi \in \phi_j$ s.t., $i < j$.

A knowledge profile E is a multi-set of KBs such that $E = \{K_1, \ldots, K_n\}$ where $K_i, 1 \leq i \leq n$, is a flat or stratified KB. $K_E = K_1 \bigsqcup \ldots \bigsqcup K_n$ denotes the set union of K_is.

In [DDL06], the concept of *prioritized observation base (POB)* is introduced. A POB K is in the form $K = \langle \sigma_1, \ldots, \sigma_n \rangle$ with $n \geq 1$, where each σ_i is a set of propositional formulae with reliability level i and formulae with higher reliability levels are more important than those with lower reliability levels (we require that each σ_i is not empty without losing generality). Obviously, a POB $K = \langle \sigma_1, \ldots, \sigma_n \rangle$ induces a SKB $K = (S_1, \ldots, S_n)$ such that $S_i = \sigma_{n+1-i}$.

Knowledge from a single source (e.g., an expert) can either be represented as a SKB or a POB. However a POB can also be used to represent a collection of knowledge from multiple sources/observations (e.g., a knowledge profile) as discussed in [DDL06]. In this case, a POB is equivalent to a knowledge profile, that is a POB contains all the formulae from a knowledge profile, and each formula is assigned with a reliability value if KBs in the profile are stratified.

For simplicity and consistency, in the rest of the paper, we use $K = \langle S_1, \ldots, S_n \rangle$ to stand for a POB without explicitly considering priority levels, because we do not need the values of these levels in the rest of the paper. Such a K can be taken as consisting of formulae from a knowledge profile of stratified bases where S_1 contains all the most reliability formulae. We still use $K = (S_1, ..., S_n)$ to denote a single SKB. We also follow the notations below for a prioritized base [DDL06].

1. $K_{i \to j} = \langle S_i, \ldots, S_j \rangle$, $1 \le i \le j \le n$, particularly $K_{1 \to n} = K$, $K_i = K_{i \to i} = S_i$.
2. $\bigwedge S_i = \bigwedge_{\phi \in S_i} \phi$, $\bigwedge K_{i \to j} = \bigwedge_{t=i}^{t=j} S_t$.
3. If $K = \langle S_1, \ldots, S_n \rangle$ and $K' = \langle S_1', \ldots, S_p' \rangle$, then (K, K') is the concatenation of K and K' such that $(K, K') = \langle S_1, \ldots, S_n, S_1', \ldots, S_p' \rangle$.
4. $Cons(K)$ is the set of consistent subsets of K, that is, the set of all POBs $K' = \langle S_1', \ldots, S_n' \rangle$ such that $\bigwedge K'$ is consistent and $S_i' \subseteq S_i$, $1 \le i \le n$.

2.2 Non-archimedean Field

Now we give a brief introduction to the non-Archimedean fields [Rob73].

Definition 1. *([Ham99]) An ordered field $\langle \mathbb{F}, +, \cdot, 0, 1, > \rangle$ is a set $\mathbb{F}$ together with:*

1. *the two algebraic operations $+$ (addition) and $\cdot$ (multiplication);*
2. *the two corresponding identity elements 0 and 1;*
3. *the transitive and irreflexive total order $>$ on $\mathbb{F}$ satisfying $1 > 0$.*

Moreover, the set $\mathbb{F}$ must be closed under $+$ and $\cdot$. Addition and multiplication both have to be commutative and associative; the distributive law must hold. And every element $x \in \mathbb{F}$ must have both an additive inverse $-x$ and a multiplicative inverse $1/x$, except that $x/0$ is undefined. The order $>$ must be such that $y > z \leftrightarrow y - z > 0$. Also, the set $\mathbb{F}+$ of positive elements in $\mathbb{F}$ must be closed under both addition and multiplication.

Both the real line $\mathbb{R}$ and the rationals $\mathbb{Q}$ are obvious examples of ordered fields. The name "non-Archimedean" stems from the following Archimedes' Axiom.

Axiom 1. *For any ordered field $\mathbb{F}$, and $0 < a < b$, $a, b \in \mathbb{F}$, $\exists n \in \mathbb{N}$, s.t., $na > b$.*

Thus a non-Archimedean field is a field dissatisfying the Archimedes' Axiom.

The non-Archimedean fields contain real numbers and also infinite numbers and *infinitesimals* (infinitely small). In this paper, we adopt the smallest non-Archimedean field generated by combining the real line $\mathbb{R}$ and a single infinitesimal ϵ, denoted as $\mathbb{R}(\epsilon)$ [Rob73]. An infinitesimal ϵ is positive but smaller than any positive real number. If ϵ is an infinitesimal, then ϵ^i is also an infinitesimal when $i > 0$. Moreover, for any positive real numbers a, b, we have $a\epsilon^{i+1} < b\epsilon^i$ as $\epsilon < b/a$ (as b/a is a positive real number). Note that if ϵ is an infinitesimal, then $1/\epsilon$ is an infinite number (larger than any positive real number), and vice versa. As $\mathbb{R}(\epsilon)$ is an ordered field, it is also closed under $+$ and $\cdot$. Moreover, the usual arithmetic properties also apply in $\mathbb{R}(\epsilon)$.

The non-Archimedean field, especially the non-standard probability (i.e., the probability values can involve infinitesimals), has already been introduced in the literature of uncertainty reasoning. For example, in [Spo88], Spohn demonstrated the relationship between his ordinal conditional function and the non-standard probability. In [Pea94], Pearl used the non-standard probability to model non-monotonic reasoning.

Definition 2. $(\leq^{\epsilon})$ *Let* $x = \sum_{i=1}^{s} a_i \epsilon^{b_i}$ *and* $y = \sum_{j=1}^{t} c_j \epsilon^{d_j}$ *be two polynomial representations of infinitesimals, where all* a_i, c_j *are positive real numbers,* b_i, d_j *are integers, and* $b_1 < \ldots < b_s$, $d_1 < \ldots < d_t$. *We write* $x \leq^{\epsilon} y$ *iff* $b_1 \geq d_1$. *For convenience, we also write* $0 <^{\epsilon} x$ *as* 0 *can be seen as* $0 = \epsilon^{+\infty}$.

The $\leq^{\epsilon}$ relation is not the usual mathematical $\leq$ relation, rather it aims to compare the order of the infinitesimal ϵ, namely, we view x as $O(\epsilon^{b_1})$ and y as $O(\epsilon^{d_1})$.

Example 1. *Let* $x = 2\epsilon^2 + 4\epsilon^4$, $y = \epsilon^2 + 3\epsilon^3$, *then we have* $x =^{\epsilon} y$. *That is,* x *and* y *both can be seen as* $O(\epsilon^2)$.

We have the following result.

Proposition 1. *Let* x, y *be two polynomial representations of infinitesimals, we have: if* $x <^{\epsilon} y$, *then* $x < y$, *if* $x \geq y$, *then* $x \geq^{\epsilon} y$.

2.3   The DA² Merging Operators

In [KLM04], a family of merging operators, called the DA^2 merging operators, was proposed to generalize both model-based and syntax-based merging operators. The DA^2 merging operators, consisting of a distance relation between interpretations and two aggregation functions, are defined below.

Definition 3. *([KLM04], distance) A distance relation between interpretations is a total function d from* $W \times W$ *to* N *s.t. for every* $w_1, w_2 \in W$
1. $d(w_1, w_2) = d(w_2, w_1)$,   2. $d(w_1, w_2) = 0$ *iff* $w_1 = w_2$.

The distance d between interpretations can be extended to be a distance between an interpretation and a formula as $d(w, \phi) = min_{w' \models \phi} d(w, w')$.

Definition 4. *([KLM04], aggregation function) An aggregation function is a total function* $\oplus$ *associating a nonnegative integer to every finite tuple of nonnegative integers and verifying (non-decreasingness), (minimality) and (identity).*

non-decreasingness *If* $x \leq y$, *then* $\oplus(x_1, \ldots, x, \ldots, x_n) \leq \oplus(x_1, \ldots, y, \ldots, x_n)$.
minimality $\oplus(x_1, \ldots, x_n) = 0$ *iff* $x_1 = \ldots = x_n = 0$.
identity *For every nonnegative integer* x, $\oplus(x) = x$.

Definition 5. *([KLM04], DA² merging operators) Let* d *be a distance between interpretations and* $\oplus$ *and* $\odot$ *be two aggregation functions. For every knowledge profile* $E = \{K_1, \ldots, K_n\}$ *and every integrity constraint IC, a* DA^2 *merging operator* $\triangle_{IC}^{d,\oplus,\odot}(E)$ *is defined in a model-theoretical way by:*

$$Mod(\triangle_{IC}^{d,\oplus,\odot}(E)) = min(IC, \leq_E^{d,\oplus,\odot}).$$

$\leq_E^{d,\oplus,\odot}$ *is defined as* $w \leq_E^{d,\oplus,\odot} w'$ *iff* $d(w, E) \leq d(w', E)$, *where*

$$d(w, E) = \odot(d(w, K_1), \ldots, d(w, K_n)),$$

and for every $K_i = \{\phi_{i,1}, \ldots, \phi_{i,n_i}\}$, $d(w, K_i) = \oplus(d(w, \phi_{i,1}), \ldots, d(w, \phi_{i,n_i}))$.

An example of distance function is the *drastic distance* defined as

$$d_D(w_1, w_2) = \begin{cases} 0 & if\ w_1 = w_2, \\ 1 & otherwise. \end{cases}$$

For the drastic distance d_D, we can easily get $d_D(w, \phi) = 0$ for $w \models \phi$ and $d_D(w, \phi) = 1$ otherwise. Thus it is the characterization function of $Mod(\phi)$. The commonly used aggregation functions are max and sum with usual meanings.

3 NAP Representation of Interpretations

In this section, we discuss how to obtain non-Archimedean polynomials (NAPs) from SKBs. With the non-Archimedean fields, we associate each stratum in a stratified base with an infinitesimal of degree i (the level of the stratum in the base) , so the prioritized information is represented. We then define a NAP for each interpretation w based on the given SKB making use of the representation of prioritized information. This way, the three ordering strategies can be easily simulated using NAPs of interpretations and their relationships can be easily established and proved.

Let $K = (S_1, \ldots, S_n)$ be a SKB, the three widely used ordering strategies are *best out*, *maxsat* and *leximin* and they are defined as follows:

best out ordering [B+93] Let $r_{BO}(w) = min_i\{w \not\models S_i\}$. Conventionally, $min_i \emptyset = +\infty$. $w \leq_{bo} w'$ iff $r_{BO}(w) \geq r_{BO}(w')$.

maxsat ordering [Bre04] Let $r_{MO}(w) = min_i\{w \models S_i\}$. $w \leq_{maxsat} w'$ iff $r_{MO}(w) \leq r_{MO}(w')$.

leximin ordering [B+93] Let $K_i(w) = \{\phi \in S_i | w \models \phi\}$. $w \leq_{leximin} w'$ iff
1. $|K_i(w)| = |K_i(w')|$ for all i, or
2. $\exists i$ such that $|K_i(w)| > |K_i(w')|$, and $|K_j(w)| = |K_j(w')|$ for all $j < i$.

Definition 6. *(NAP) Let $K = (S_1, \ldots, S_n)$ be a SKB, d be a distance and $\oplus$ be an aggregation function, then the NAP of an interpretation w is defined as*

$$NA_K^{d,\oplus}(w) = \sum_{i=1}^{n}(d(w, S_i)\epsilon^i) \tag{1}$$

where for $S_i = \{\phi_{i1}, \ldots, \phi_{in_i}\}$, $d(w, S_i) = \oplus(d(w, \phi_{i1}), \ldots, d(w, \phi_{in_i}))$.

Eq. 1 defines a family of NAPs of an interpretation w, e.g., $NA_K^{d_D,max}(w)$ is one specific polynomial where $d = d_D$ and $\oplus = max$. In the following, when there is no confusion, we simplify $NA_K^{d,\oplus}(w)$ as $NA_K(w)$.

If a SKB has $\{\top\}$ as its first stratum, then we have the following proposition.

Proposition 2. *Let K be a SKB, then $\forall w \in W$, $NA_{(\{\top\},K)}(w) = \epsilon NA_K(w)$.*

Now, we use NAPs to induce the above ordering strategies.

Definition 7. *(best out simulation) For a SKB K, the best out simulation polynomial $bo(w)$ is defined as $bo(w) = NA_K^{d_D,max}(w)$.*

When $d = d_D$ and $\oplus = max$ in Def. 6, we have $d(w, S_i) = max_{\phi \in S_i}(d(w, \phi)) = 0$ iff $\forall \phi \in S_i, d(w, \phi) = 0$ (i.e., $w \models S_i$). Therefore, $bo(w)$ is actually simplified as

$$
bo(w) = \begin{cases} \epsilon^t + \sum_{i=t+1}^n d(w, S_i)\epsilon^i, & \text{if } r_{BO}(w) = t < \infty; \\ 0, & \text{if } r_{BO}(w) = \infty. \end{cases}
$$

$bo(w)$ captures the best out strategy by making the most important stratum that w falsifies as the largest ϵ−term.

In Definition 7, we can also let $\oplus = +$, thus we get $bo^+(w) = NA_K^{d_D,+}(w)$. Based on the pre-order $\leq^\epsilon$ defined in Definition 2, we have

Proposition 3. $w \leq_{bo} w'$ iff $bo(w) \leq^\epsilon bo(w')$ iff $bo^+(w) \leq^\epsilon bo^+(w')$.

Definition 8. *(maxsat simulation) For a SKB K, the maxsat simulation polynomial $maxsat(w)$ is defined as: $maxsat(w) =\sim NA_K^{d_D,max}(w)$ where for a polynomial $x = \sum_{i=1}^n a_i\epsilon^i$, $\sim x = \sum_{i=1}^n \sim a_i\epsilon^i$ s.t. $\sim a_i = 1$ if $a_i = 0$ and $\sim a_i = 0$ otherwise.*

For any w, if $r_{MO}(w) = t$, we get $NA_K^{d_D,max}(w) = \sum_{i=1}^{t-1} \epsilon^i + \sum_{i=t+1}^n d(w, S_i)\epsilon^i$, then we have $maxsat(w) =\sim NA_K^{d_D,max}(w) = \epsilon^t + \sum_{i=t+1}^n \sim d(w, S_i)\epsilon^i$.

$maxsat(w)$ captures the maxsat strategy by making the most important stratum that w satisfies as the largest ϵ-term from the $\sim$ operation.

Proposition 4. $w \leq_{maxsat} w'$ iff $maxsat(w) \geq^\epsilon maxsat(w')$.

From Definitions 7 and 8, given any stratified base K, the following should hold

$$
bo(w) + maxsat(w) = \sum_{i=1}^n \epsilon^i. \tag{2}
$$

For the leximin ordering strategy, we also define the leximin simulation.

Definition 9. *(leximin simulation) For a SKB K, the leximin simulation polynomial $leximin(w)$ is defined as $leximin(w) = NA_K^{d_D,+}(w)$.*

Obviously, we have $bo^+(w) = leximin(w)$. As for leximin simulation, because of its lexicographic nature, we cannot use the $\leq^\epsilon$ relation to compare two leximin simulation polynomials, instead, the usual mathematical comparative relation $\leq$ should be used. Namely, we have the following result.

Proposition 5. $w \leq_{leximin} w'$ iff $leximin(w) \leq leximin(w')$.

With the help of NAPs, the three ordering strategies are represented in a very similar form as shown by Propositions 3, 4 and 5, and the results in the following proposition are immediate following Proposition 1 and Equation 2.

Proposition 6. *[Bre04] Let $w, w' \in W$, then the following relationships hold:*

1. $w <_{bo} w'$ *implies* $w <_{leximin} w'$.
2. $w <_{bo} w'$ *implies* $w \leq_{maxsat} w'$ *and* $w <_{maxsat} w'$ *implies* $w \leq_{bo} w'$.

In fact, as $bo^{+}(w) = leximin(w)$, from $w <_{bo} w'$, we get $leximin(w) <^{\epsilon} leximin(w')$, thus from Proposition 1, we immediately get $leximin(w) < leximin(w')$ which implies $w <_{leximin} w'$. From $w <_{bo} w'$ and Proposition 1, we get $bo(w) < bo(w')$, from Equation 2, we have $maxsat(w) > maxsat(w')$, thus Proposition 1 gives $maxsat(w) \geq^{\epsilon} maxsat(w')$ which implies $w \leq_{maxsat} w'$.

Example 2. *Let $K = \{\{p\}, \{q\}, \{\neg p \wedge \neg q\}, \{\neg p \wedge q\}\}$ be a SKB, and let $w = \neg p \wedge q$, $w' = \neg p \wedge \neg q$ be two possible worlds. Then we have*

$$bo(w) = leximin(w) = \epsilon + \epsilon^3, \quad maxsat(w) = \epsilon^2 + \epsilon^4,$$
$$bo(w') = leximin(w') = \epsilon + \epsilon^2 + \epsilon^4, \quad maxsat(w') = \epsilon^3.$$

Hence we get $w =_{bo} w'$, $w <_{maxsat} w'$ and $w <_{leximin} w'$.

4   The Non-archimedean Polynomial Merging Operators

Since a prioritized observation base (POB) is taken as containing formulae from a set of SKBs, the issue of merging a set of SKBs becomes manipulating formulae in a single POB to obtain a consistent formula (or a set of consistent formulae). To this end, we define the non-Archimedean polynomial (np for short) merging operators for a POB as follows (similar to Def. 5).

Definition 10. *Let d be a distance relation between interpretations and $\oplus$ be an aggregation function. For a POB $K = \langle S_1, \ldots, S_n \rangle$, a np merging operator $\triangle^{d,\oplus}(K)$ is defined in a model-theoretical way by:*

$$Mod\left(\triangle^{d,\oplus}(K)\right) = min\left(W, \leq_K^{d,\oplus}\right).$$

$\leq_K^{d,\oplus}$ *is defined as $w \leq_K^{d,\oplus} w'$ iff $NA_K(w) \leq NA_K(w')$, where NA_K is the NAP for POB K by Definition 6.*

Since the syntactic form of a POB K is the same as that of a SKB K', we can define the NAP $NA_K(w)$ from a POB K based on the definition $NA_{K'}(w)$ from a SKB K'.

When d and $\oplus$ are assigned with different distances and different aggregation functions respectively, we get a family of prioritized merging operators.

Other families of prioritized merging operators can be obtainable in terms of NAPs. For example, $\preceq_K^{d,\oplus}$ defined as $w \preceq_K^{d,\oplus} w'$ iff $NA_K(w) \leq^{\epsilon} NA_K(w')$ gives us a new family of prioritized merging operators. Due to space limitation, in this paper we only consider the np merging operator defined in Definition 10.

A number of desirable properties for prioritized merging were proposed in [DDL06]. Let K be a POB and $\triangle$ be a prioritized merging operator, these properties are

PMon	For every $i < n$, $\triangle(K_{1 \to i+1}) \vdash \triangle(K_{1 \to i})$.
Succ	$\triangle(K) \vdash \triangle(K_1)$.
Cons	$\triangle(K)$ is consistent.
Taut	$\triangle(\{\top\}, K) = \triangle(K)$.
Opt	If $\bigwedge K$ is consistent, then $\triangle(K) = \bigwedge(K)$.
IS	If $K \equiv K'$, then $\triangle(K) = \triangle(K')$.
RA	$\triangle(K_{1 \to i}) = \triangle(\triangle(K_{1 \to i-1}), K_i)$.

Note: because we represent a prioritized base as $K = \langle S_1, ..., S_n \rangle$ with S_1 having the most reliable formulae, and S_1 is equivalent to σ_n in the original definition of a prioritized base, the **RA** *(Right Associativity)* property looks like a *Left Associativity* property.

Proposition 7. *Let* $\triangle^{d,\oplus}$ *be any np merging operator defined in Definition 10, then it satisfies (PMon), (Succ), (Cons), (Taut), (Opt), (IS) and (RA).*

Some existing prioritized merging operators, e.g., the $\triangle_{linear}$ and the $\triangle_{leximin}$ merging operators defined below are special cases of $\triangle^{d,\oplus}$.

Definition 11. *(linear, [DP91, Neb94]) Let* $K = \langle S_1, \ldots, S_n \rangle$, *and* $\triangle_{linear}$ *be defined inductively by:* $\triangle_{linear}() = \top$, *and for* $j \geq 1$,
$$\triangle_{linear}(K_{1 \to j}) = \begin{cases} \bigwedge S_j \wedge \triangle_{linear}(K_{1 \to j-1}) & \text{if consistent,} \\ \triangle_{linear}(K_{1 \to j-1}) & \text{otherwise.} \end{cases}$$

Definition 12. *(leximin, [B+93, Leh95]) Let* $K = \langle S_1, \ldots, S_n \rangle$. *For* $K^1, K^2 \in Cons(K)$, *define* $K^2 >_{leximin} K^1$ *iff* $\exists j$ *such that*

1. $|K_{1 \to j} \cap K^2| > |K_{1 \to j} \cap K^1|$,
2. $\forall i < j, |K_{1 \to i} \cap K^2| = |K_{1 \to i} \cap K^1|$.

Then $\triangle_{leximin}(K) = \bigvee \{ \bigwedge K', s.t., K' \in Max(>_{leximin}, Cons(K)) \}$.

Proposition 8. *Let* $d = d_D$ *and* $\oplus = max$ *in Definition 10, then* $\triangle^{d,\oplus} = \triangle_{linear}$.

Proposition 9. *Let* $d = d_D$ *and* $\oplus = +$ *in Definition 10, then* $\triangle^{d,\oplus} = \triangle_{leximin}$.

However, not all the prioritized merging operators proposed so far in the literature can be induced from the family of np merging operators. For instance, the *discrimin* operator [B+93], as it makes use of set inclusion, cannot be represented by NAPs.

5   Non-archimedean Polynomial for Merging RKBs

In this section, we use NAPs to represent the merging of RKBs. That is, we aim to represent and further interpret the *Egalitarist Fusion* of incommensurable RKBs [BLR07].

A *RKB* K is a set of ranked propositions, i.e., $K = \{(\phi_1, r_1), \ldots, (\phi_n, r_n)\}$. r_i is the rank of ϕ_i, $r_i \in \mathbb{N} \cup \{+\infty\}$, $1 \leq i \leq n$. Here a proposition with a higher rank is more important (prioritized) than the one with a lower rank. The notion of RKB is a generalization of SKBs. Each RKB induces a SKB in which formulae with the same rank are in the same stratum and formulae with the highest rank are in the first stratum. First, we recall some results in [BLR07].

Definition 13. *(Ranking functions) A ranking function* κ_K *associated with a RKB K is a function:* $W \to \mathbb{N} \cup \{0\}$ *such that:*

$$\kappa_K(w) = \begin{cases} 0 & \text{if } \forall (\phi_i, r_i) \in K, w \models \phi_i, \\ max(r_i : w \not\models \phi_i) & \text{otherwise.} \end{cases}$$

With the help of $\kappa_K(w)$, a strict order $\lhd_{Max}^E$ can be defined between interpretations.

Definition 14. *[BLR07] Let $E = \{K_1, \ldots, K_n\}$ be a knowledge profile of RKBs and $w, w' \in W$ be two interpretations, then we have:*

$$w \vartriangleleft_{Max}^{E} w' \text{ iff } max(\kappa_{K_i}(w) : 1 \le i \le n) < max(\kappa_{K_i}(w') : 1 \le i \le n).$$

Example 3. *Let $E = \{K_1, K_2\}$ such that $K_1 = \{(p, 4), (\neg q, 2)\}$ and $K_2 = \{(q, 3), (p, 1)\}$, then we have the following*

	$p\ q$	$\kappa_{K_1}(w)$	$\kappa_{K_2}(w)$	max
w_0	0 0	4	3	4
w_1	0 1	4	1	4
w_2	1 0	0	3	3
w_3	1 1	2	0	2

Obviously, w_3 is the smallest w.r.t. $\vartriangleleft_{Max}^{E}$.

In [BLR07], it is explicitly stated that the scales used in different RKBs are not required to be commensurable. To merge incommensurable ranked bases, a scaling method is proposed as follows.

Definition 15. *([BLR07], Compatible scaling) Let $E = \{K_1, \ldots, K_n\}$ be a profile of ranked bases, a scaling $\mathcal{S}$ is defined as ($\bigsqcup$ represents the union of multi-sets):*

$$\mathcal{S} : K_1 \bigsqcup \ldots \bigsqcup K_n \to \mathbb{N} \qquad (\phi_{ij}, r_{ij}) \mapsto \mathcal{S}(\phi_{ij}).$$

$\mathcal{S}$ is said to be compatible with E iff $\forall K_i \in E$, we have $\forall (\phi, r), (\phi', r') \in K_i, r \le r'$ iff $\mathcal{S}(\phi) \le \mathcal{S}(\phi')$.

Given a compatible scaling $\mathcal{S}$, $K^{\mathcal{S}}$ (resp. $E^{\mathcal{S}}$) is used to denote the ranked base (resp. profile of ranked bases) obtained from K (resp. E) by replacing each pair (ϕ_i, r_i) with $(\phi_i, \mathcal{S}(\phi_i))$ (resp. replacing each $K_i \in E$ with $K_i^{\mathcal{S}}$).

Example 4. *(Exam. 3 cont.) Let $E = \{K_1, K_2\}$, then a scaling s_1 produces $K_1^{s_1} = \{(p, 2), (\neg q, 1)\}$ and $K_2^{s_1} = \{(q, 5), (p, 2)\}$ is a compatible scaling. However, a scaling s_2 with $K_1^{s_2} = \{(p, 2), (\neg q, 3)\}$ and $K_2^{s_2} = \{(q, 5), (p, 2)\}$ is not a compatible scaling as $s_2(p) < s_2(\neg q)$ for $K_1^{s_2}$.*

Definition 16. *([BLR07], Compatible scaling ordering) Let $E = \{K_1, \ldots, K_n\}$, $\mathbb{S}_E$ be the set of all compatible scalings with E, then a partial order $<_{\forall}^{E}$ is defined on W as*

$$\forall w, w' \in W, w <_{\forall}^{E} w' \text{ iff } \forall \mathcal{S} \in \mathbb{S}_E, w \vartriangleleft_{Max}^{E^{\mathcal{S}}} w'.$$

Example 5. *(Exam. 3 cont.) In Example 3, we have $w_2 \vartriangleleft_{Max}^{E} w_1$. But after using the compatible scaling s_1 in Example 4, we get $w_1 \vartriangleleft_{Max}^{E^{s_1}} w_2$, thus $w_2 \not<_{\forall}^{E} w_1$. It shows the difference between $\vartriangleleft_{Max}^{E}$ and $<_{\forall}^{E}$.*

Definition 17. *([BLR07], Pareto-like ordering) Let $E = \{K_1, \ldots, K_n\}$, we denote $w \vartriangleleft_{Pareto} w'$ iff the following conditions are satisfied:*

1. $\exists i \in \{1, \ldots, n\}, \kappa_{K_i}(w') \neq 0$.
2. $\forall i \in \{1, \ldots, n\}, \kappa_{K_i}(w) = \kappa_{K_i}(w') = 0, or \ \kappa_{K_i}(w) < \kappa_{K_i}(w')$.

The main result in [BLR07] is the proof of equivalence between Pareto-like ordering and the compatible scaling ordering as stated in the following proposition.

Proposition 10. *([BLR07], Prop. 16)* $\forall w, w' \in W$, *we have* $w <_{\forall}^{E} w'$ *iff* $w \lhd_{Pareto} w'$.

Here we show that strict orders $\lhd_{Max}^{E}$ and $<_{\forall}^{E}$ can be represented by NAPs.

Definition 18. *(NAPs from RKBs) Let d be a distance function, for every RKB $K = \{(\phi_1, r_1), \ldots, (\phi_n, r_n)\}$, we define the NAP of an interpretation w as $NA_K^d(w) = \sum_{i=1}^{n}(d(w, \phi_i)\epsilon^{-r_i})$.*

When a SKB $K = (S_1, \ldots, S_n)$ is viewed as a RKB

$$K^* = \{(\phi_{11}, -1), \ldots, (\phi_{1|S_1|}, -1), \ldots, (\phi_{n1}, -n), \ldots, (\phi_{n|S_n|}, -n)\}$$

where $\phi_{ij} \in S_i, 1 \leq j \leq |S_i|, 1 \leq i \leq n$, the NAPs from K is exactly the same as that from K^*, thus the above definition derives Definition 6 when the aggregation operation $\oplus$ is '+'. For simplicity, we write $NA_i(w)$ instead of $NA_{K_i}^d(w)$, and use $NA_E(w)$ for $\sum_{i=1}^{|E|} NA_i(w)$ in the rest of the section.

Definition 19. *(non-Archimedean pre-order relation) Let $E = \{K_1, \ldots, K_n\}$ and w, w' be two interpretations. We denote $w <_{NA}^{K_i} w'$ iff $NA_i(w) < NA_i(w')$. $w <_{NA}^{E} w'$ iff $NA_E(w) <^{\epsilon} NA_E(w')$.*

Note that $<_{NA}^{K_i}$ deploys $<$ while $<_{NA}^{E}$ deploys $<^{\epsilon}$.

Example 6. *(Exam. 3 Cont.) Let $E = \{K_1, K_2\}$, then we have the following*

	$p\ q$	$NA_1(w)$	$NA_2(w)$	$NA_E(w)$
w_0	$0\ 0$	ϵ^{-4}	$\epsilon^{-3} + \epsilon^{-1}$	$\epsilon^{-4} + \epsilon^{-3} + \epsilon^{-1}$
w_1	$0\ 1$	$\epsilon^{-4} + \epsilon^{-2}$	ϵ^{-1}	$\epsilon^{-4} + \epsilon^{-2} + \epsilon^{-1}$
w_2	$1\ 0$	0	ϵ^{-3}	ϵ^{-3}
w_3	$1\ 1$	ϵ^{-2}	0	ϵ^{-2}

We can see that $w_0 <_{NA}^{K_1} w_1$, $w_0 >_{NA}^{K_2} w_1$, etc., and w_3 is the smallest w.r.t. $<_{NA}^{E}$.

Definition 20. Let $E = \{K_1, \ldots, K_n\}$ and w, w' be two interpretations. We denote $w <_{NA}^{Com} w'$ iff $\sum_{i=1}^{n} a_i NA_i(w) <^{\epsilon} \sum_{i=1}^{n} a_i NA_i(w')$ where $a_i = \epsilon^{\kappa_{K_i}(w')}$ is a commensurable coefficient.

We call $<_{NA}^{Com}$ the *commensurable non-Archimedean pre-order relation* because for each i, $\epsilon^{\kappa_{K_i}(w')} NA_i(w')$ has a minimum degree 0 for ϵ. That is, for any $i, j \in [1, n]$, $\epsilon^{\kappa_{K_i}(w')} NA_i(w')$ and $\epsilon^{\kappa_{K_j}(w')} NA_j(w')$ are somehow commensurable. This is illustrated by the following simple example.

Example 7. Let $E = \{K_1, K_2\}$ s.t. $K_1 = \{(p, 4), (\neg q, 2)\}$ and $K_2 = \{(q, 3), (p, 1)\}$ and $w' = \neg p \wedge q$ be a possible world. We have $NA_1(w') = \epsilon^{-2} + \epsilon^{-4}$, and $NA_2(w') = \epsilon^{-1}$. Obviously, the minimum degrees for ϵ in $NA_1(w')$ and in $NA_2(w')$ are not the same. Now as $\kappa_{K_1}(w') = 4$ and $\kappa_{K_2}(w') = 1$, we get $\epsilon^{\kappa_{K_1}(w')} NA_1(w') = 1 + \epsilon^2$ and $\epsilon^{\kappa_{K_2}(w')} NA_2(w') = 1$, both having 0 as the minimum degree for ϵ (as $1 = \epsilon^0$).

The following theorems show that the egalitarist fusion can be characterized by the non-Archimedean pre-orders.

Theorem 1. *Let E be a profile of RKBs, $\forall w, w' \in W$, we have $w \lhd^E_{Max} w'$ iff $w <^E_{NA} w'$.*

Theorem 2. *Let E be a profile of RKBs, $\forall w, w' \in W$, we have $w <^E_\forall w'$ iff $w \lhd_{Pareto} w'$ and iff $w <^{Com}_{NA} w'$.*

These theorems show that the egalitarist fusion for the incommensurable RKBs can be described by our non-Archimedean approaches. Furthermore, from Theorem 2, it shows that the egalitarist fusion can in particular be described by our commensurable pre-order relation. Therefore, it is not surprising why the egalitarist fusion does not need the commensurable assumption to deal with the incommensurable RKBs.

6   Conclusion

In this paper, we have proposed a new method to model stratified/ranked KBs. Unlike the commonly used logical approaches, our method is largely numerical. We used the non-Archimedean representation for stratified/ranked KBs to represent the ordering strategies, to define new merging operators, and to simulate the *egalitarist fusion* for incommensurable RKBs. This wide range coverage shows that the non-Archimedean representation is very suitable for modeling stratified/ranked KBs.

In [Pap01], a polynomial representation for each possible world w was proposed which associates w and each epistemic state Φ (which assigns w an ordinal as its weight) with a polynomial $p_\Phi(w) = \sum_{i=0}^{n} p_i(w)x^i$. Here coefficients $p_i(w) \in \{0,1\}, 0 \leq i \leq n$, encodes the binary representation, read in reverse order of the weight assigned to w [B+02], e,g, if the weight of w is 6, then its binary form is 110, so $p_0(w) = 0$ and $p_1(w) = p_2(w) = 1$. The interpretation of such polynomials differs from NAPs in the following aspects. Papini's polynomials consider epistemic states and their coefficients are 0 or 1, representing the binary form of the weight of w provided by the epistemic states whilst NAPs are for SKBs with each coefficient standing for the aggregation result of distances of formulae in the same level to w, and is not limited to $\{0,1\}$. Therefore, Papini's polynomial and NAP are very different.

Our new framework of prioritized merging operators based on NAPs is the counterpart of the DA^2 [KLM04] framework of flat merging operators. For each stratum of stratified bases, NAP representation also uses the distance and aggregation function to obtain an aggregated effect which is also used in the DA^2 framework. Thus, our framework can be seen as an extension of the DA^2 framework for SKBs.

The non-Archimedean field is not an entirely new idea in artificial intelligence research. It appeared in the nonstandard probabilities [Spo88, Pea94], in decision making [Leh98] to model utilities to provide a unified theory for qualitative and quantitative decision theories, and in data envelopment analysis [TN09] to define merit functions. However, it has never been used to manipulate stratified/ranked KBs. Our work therefore is novel and significant. There are still many aspects that can be further developed, such as, the relationship between various prioritized merging operators in the literature.

References

[B+93]   Benferhat, S., Cayrol, C., Dubois, D., Lang, J., Prade, H.: Inconsistency management and prioritized syntax-based entailment. In: Proc. of IJCAI 1993, pp. 640–647 (1993)

[B+02]   Benferhat, S., Dubois, D., Lagrue, S., Papini, O.: Making revision reversible: an approach based on polynomials. Fundamenta Informaticae 53(3-4), 251–280 (2002)

[BLR07]   Benferhat, S., Lagrue, S., Rossit, J.: An Egalitarist Fusion of Incommensurable Ranked Belief Bases under Constraints. In: Proc. of AAAI 2007, pp. 367–372 (2007)

[Bre04]   Brewka, G.: A rank based description language for qualitative preferences. In: Proc. of ECAI 2004, pp. 303–307 (2004)

[DDL06]   Delgrande, J., Dubois, D., Lang, J.: Iterated revision as prioritized merging. In: Proc. of KR 2006, pp. 210–220 (2006)

[DP91]   Dubois, D., Prade, H.: Epistemic entrenchment and possibilistic logic. Artificial Intelligence 50, 223–239 (1991)

[Ham99]   Hammond, P.J.: Non-Archimedean Subjective Probabilities in Decision Theory and Games. Mathematical Social Sciences 38(2), 139–156 (1999)

[KLM04]   Konieczny, S., Lang, J., Marquis, P.: DA^2 merging operators. Artif. Intel. 157, 49–79 (2004)

[Leh95]   Lehmann, D.: Another perspective on default reasoning. Annals of Mathematics and Art. Int. 15(1), 61–82 (1995)

[Leh98]   Lehmann, D.: Nonstandard numbers for qualitative decision making. In: Proc. of TARK 1998, pp. 161–174 (1998)

[Neb94]   Nebel, B.: Base revision operations and schemes: Semantics, representation, and complexity. In: Proc. of ECAI 1994, pp. 341–345 (1994)

[Pap01]   Papini, O.: Iterated Revision Operations Stemming from the History of an Agent's Observations. In: Rott, H., Williams, M. (eds.) Frontiers of Belief Revision, pp. 279–301. Kluwer Academic, Dordrecht (2001)

[Pea94]   Pearl, J.: From Adams conditionals to default expressions, causal conditionals and coounterfactuals. In: Eells, E., Skyrms, B. (eds.) Probability and Conditionals: Belief Revision and Rational Decision (1994)

[QLB06]   Qi, G., Liu, W., Bell, D.A.: Merging stratified knowledge bases under constraints. In: Proc. of AAAI 2006, pp. 281–286 (2006)

[Rob73]   Robinson, A.: Function Theory on Some Nonarchimedean Fields. American Mathematical Monthly: Papers in the Foundations of Mathematics 80, 87–109 (1973)

[Spo88]   Spohn, W.: Ordinal Conditional Functions: A Dynamic Theory of Epistemic States. In: Harper, W., Skyrms, B. (eds.) Causation in Decision, Belief Change, and Statistics, vol. 2, pp. 105–134. Kluwer Academic Publishers, Dordrecht (1988)

[TN09]   Toloo, M., Nalchigar, S.: A new integrated DEA model for finding most BCC-efficient DMU. Applied Mathematical Modelling 33(1), 597–604 (2009)

Encoding the Revision of Partially Preordered Information in Answer Set Programming

Mariette Sérayet, Pierre Drap, and Odile Papini

LSIS-CNRS 6168, Université de la Méditerranée, ESIL - Case 925, Av de Luminy,
13288 Marseille Cedex 9 France
{serayet,drap,papini}@esil.univmed.fr

Abstract. Most of belief revision operations have been proposed for totally preordrered information. However, in case of partial ignorance, pieces of information are partially preordered and few effective approaches of revision have been proposed. The paper presents a new framework for revising partially preordered information, called Partially Preordered Removed Sets Revision (PPRSR). The notion of removed set, initially defined in the context of the revision of non ordered or totally preordered information is extended to partial preorders. The removed sets are efficiently computed thanks to a suitable encoding of the revision problem into logic programming with answer set semantics. This framework captures the possibilistic revision of partially preordered information and allows for implementing it with ASP. Finally, it shows how PPRSR can be applied to a real application of the VENUS european project before concluding.

1 Introduction

Belief revision has been extensively studied in the domain of knowledge representation for artificial intelligence, mainly for totally preordered information. A characterization of belief revision has been provided by Alchourron, Gärdenfors, Makinson (AGM) with a set of postulates that any revision operation should satisfy [6]. Katsuno and Mendelzon (KM) reformulated AGM's postulates and provided a representation theorem that characterizes revision operations based on total preorders [11]. Belief revision has been discussed within different frameworks (probabillity theory, Sphon's conditional functions, Grove's system of spheres, etc ···). Some approaches have been implemented, among them, Removed Sets Revision which has been initially proposed in [15] for revising a set of propositional formulae. This approach stems from removing a minimal number of formulae, called removed set, to restore consistency. The Removed Sets Revision (RSR) and then a prioritized form of Removed Sets Revision, called Prioritized Removed Sets Revision (PRSR) [1] have been encoded into answer set programming and allowed for solving a practical revision problem coming from a real application in the framework of geographical information system.

However in some applications, an agent has not always a total preorder between situations at his disposal, but is only able to define a partial preorder between situations, particularly in case of partial ignorance and incomplete information. In such cases, an epistemic state can be represented by either a partial preorder on interpretations or a partially preordered belief base.

The revision of partially preordered information has been less investigated in the literature, however Lagrue and al. [4] pointed out that the KM's postulates are not appropriate for partial preorders and proposed a suitable definition of faithful assignment, called P-faithful assignment, a new set of postulates and a representation theorem. Some revision operations initially defined for total preorders, such as revision with memory and possibilistic revision have been sucessfully extended to partial preorders [2]. This paper proposes a new framework for revising partially preordered information and provides an efficient implementation thanks to Answer Set Programming. The main contributions of this paper are the following:

- It extends the Removed Sets Revision to partially preordered information, called Partially Preordered Removed Sets Revision (PPRSR). The paper shows how the notion of removed set, roughly speaking, the subsets of formulae to remove to restore consistency, initially defined in the context of non ordered [15] or totally ordered [1] information is extended to the case of the revision of partially preordered information,
- It provides an implementation of PPRSR with ASP. The revision problem is translated into a logic program with answer set semantics and a one-to-one correspondence between removed sets and preferred answer sets is shown. The computation of answer sets is performed with any ASP solver supporting the minimize statement.
- It shows that the possibilistic revision of partially preordered information can be captured within the PPRSR framework allowing for an efficient implementation with ASP.

The rest of this paper is organized as follows. Section 2 fixes the notations and gives a refresher on RSR, on answer set programming and on partial preorders. Section 3 presents the Partially Preordered Removed Set Revision (PPRSR) and shows how it captures the possibilistic revision. Section 4 details the encoding of PPRSR into logic programming with answer set semantics and the computation of answer sets thanks to ASP solvers. It then shows the one-to-one correspondence between removed sets and preferred answer sets. Section 5 illustrates how PPRSR can be applied in the context of the VENUS project before concluding.

2   Background and Notations

2.1   Notations

In this paper we use propositional calculus, denoted by $\mathcal{L}_{\mathcal{PC}}$, as knowledge representation language with usual connectives $\neg, \wedge, \vee, \rightarrow, \leftrightarrow$. Let X be a set of propositional formulae, we denote by $Cons(X)$ the set of logical consequences of X. We denote by $\mathcal{W}$ the set of interpretations of $\mathcal{L}_{\mathcal{PC}}$ and by $Mod(\psi)$ the set of models of a formula ψ, that is $Mod(\psi) = \{\omega \in \mathcal{W}, \omega \models \psi\}$ where $\models$ denotes the inference relation used for drawing conclusions.

2.2   Removed Sets Revision

We briefly recall the Removed Sets Revision approach. Removed Sets Revision [15] deals with the revision of a set of propositional formulae by a set of propositional formulae[1]. Let K and A be finite sets of clauses. Removed Sets Revision (RSR) focuses on the minimal subsets of clauses to remove from K, called *removed sets*, in order to restore the consistency of $K \cup A$. More formally: let K and A be two consistent sets of clauses such that $K \cup A$ is inconsistent. R a subset of clauses of K, is a removed set of $K \cup A$ iff (i) $(K \backslash R) \cup A$ is consistent; (ii) $\forall R' \subseteq K$, if $(K \backslash R') \cup A$ is consistent then $\mid R \mid \leq \mid R' \mid$[2]. Let denote by $\mathcal{R}(K \cup A)$ the collection of removed sets of $K \cup A$, RSR is defined as follows: let K and A be two consistent sets of clauses, $K \circ_{RSR} A =_{def} \bigvee_{R \in \mathcal{R}(K \cup A)} Cons((K \backslash R) \cup A)$. According to a semantic point of view, $\mid \mathcal{NS}_K(\omega) \mid$ denotes the number of clauses of K falsfied by an interpretation ω and a total preorder on interpretations is defined by: $\omega_i \leq_K \omega_j$ iff $\mid \mathcal{NS}_K(\omega_i) \mid \leq \mid \mathcal{NS}_K(\omega_j) \mid$. Removed Sets Revision can be semantically defined by $Mod(K \circ_{RSR_{sem}} A) = min(Mod(A), \leq_K)$. It minimizes the number of clauses falsified by the models of A and $Mod(K \circ_{RSR} A) = Mod(K \circ_{RSR_{sem}} A)$. In case of prioritized belief bases, RSR has been extended to Prioritized Removed Sets Revision (PRSR) [1].

2.3   Partial Preorders

A partial preorder, denoted by $\preceq$ on a set A is a reflexive and transitive binary relation. Let x and y be two members of A, the equality is defined by $x = y$ iff $x \preceq y$ and $y \preceq x$. The corresponding strict partial preorder, denoted by $\prec$, is such that, $x \prec y$ iff $x \preceq y$ holds but $x = y$ does not hold. We denote by $\sim$ the incomparability relation $x \sim y$ iff $x \preceq y$ does not hold nor $y \preceq x$. The set of minimal elements of A with respect to $\prec$, denoted by $Min(A, \prec)$, is defined as: $Min(A, \prec) = \{x \in A, \nexists y \in A : y \prec x\}$.

Generally, epistemic states are represented by total preorders on interpretations, however, as mentionned in the introduction, in case of partial ignorance, the agent is unable to compare all situations between them and a partial preorder seems to be more suitable to represent epistemic states.

Let Ψ be an epistemic state and $Bel(\Psi)$ its corresponding belief set, Ψ is first represented by a partial preorder on interpretations, denoted by $\preceq_\Psi$. In [4], a suitable definition of faithful assignment is given: let $Bel(\Psi) = min(W, \prec_\Psi)$, $\preceq_\Psi$ is a P-faifhful assignment if (1) if $\omega, \omega' \models Bel(\Psi)$ then $\omega \prec_\Psi \omega'$ does not hold, (2) if $\omega' \not\models Bel(\Psi)$, then there exists ω such that $\omega \models Bel(\Psi)$ and $\omega \prec_\Psi \omega'$, (3) if $\Psi = \Phi$ then $\preceq_\Psi = \preceq_\Phi$. Moreover, [4] gives a set of postulates an operation $\circ$ has to satisfy and a representation theorem such that $Mod(Bel(\Psi \circ \mu)) = min(Mod(\mu), \preceq_\Psi)$. An alternative syntactic but equivalent representation of an epistemic state, Ψ is a partially preordered belief base, denoted by $(\Sigma, \preceq_\Sigma)$, where Σ is a set of

---

[1] We consider propositional formulae in their equivalent conjonctive normal form (CNF).
[2] $\mid R \mid$ denotes the number of clauses of R.

propositional formulae, and $\preceq_\Sigma$ is a partial preorder on the formulae of Σ. Several ways of defining a partial preorder on subsets of formulae belonging to Σ, called comparators, from a partial preorder on a set of formulae Σ have been proposed: inclusion-based [10], possibilistic [3], lexicographic [16] comparators. They are such that the preferred formulae are kept in the belief base. In our approach, according to the Removed Sets strategy, we adopt a dual point of view in the sense that we want to prefer the subsets of formulae to remove. For example, we rephrase the possibilistic comparator (or weak comparator) used in [3], already defined in [12] and reused by [8] as follows. Y is preferred to X if for each element of Y, there exists at least one element of X which is preferred to it, more formally: let $\preceq_\Sigma$ be a partial preorder on Σ, $Y \subseteq \Sigma$ and $X \subseteq \Sigma$. Y is preferred to X, denoted by $Y \trianglelefteq_w X$ iff $\forall y \in Y$, $\exists x \in X$ such that $x \preceq_\Sigma y$.

We now briefly recall the extension of the semantic possibilistic revision to partial preorders [2]. Let π be a possibility distribution [5] and let Ψ be an epistemic state, represented by $(\mathcal{W}, \preceq_\Psi)$, such that $\forall \omega, \omega' \in \mathcal{W}$, $\omega \preceq_\Psi \omega'$ iff $\pi(\omega') \preceq \pi(\omega)$. The possibilistic revision of Ψ by a propositional formula μ leads to the epistemic state $\Psi \circ_\pi \mu$, represented by $(\mathcal{W}, \preceq_{\Psi \circ_\pi \mu})$ which considers all the counter-models of μ as impossible and preserves the relative ordering between the models of μ. More formally, $\Psi \circ_\pi \mu$ corresponds to the following partial preorder: (i) if $\omega, \omega' \in Mod(\mu)$ then $\omega \preceq_{\Psi \circ_\pi \mu} \omega'$ iff $\omega \preceq_\Psi \omega'$, (ii) if $\omega, \omega' \notin Mod(\mu)$ then $\omega =_{\Psi \circ_\pi \mu} \omega'$, (iii) if $\omega \in Mod(\mu)$ and $\omega' \notin Mod(\mu)$ then $\omega \prec_{\Psi \circ_\pi \mu} \omega'$.

2.4   Answer Sets

A *normal logic program* is a set of rules of the form $c \leftarrow a_1, \ldots, a_n, not\ b_1, \ldots, not\ b_m$ where $c, a_i(1 \leq i \leq n), b_j(1 \leq j \leq m)$ are propositional atoms and the symbol *not* stands for *negation as failure*. For a rule r like above, we introduce $head(r) = c$ and $body(r) = \{a_1, \cdots, a_n, b_1, \cdots, b_m\}$. Furthermore, let $body^+(r) = \{a_1, \cdots, a_n\}$ denotes the set of positive body atoms and $body^-(r) = \{b_1, \cdots, b_m\}$ the set of negative body atoms, and $body(r) = body^+(r) \cup body^-(r)$. Let r be a rule, r^+ denotes the rule $head(r) \leftarrow body^+(r)$, obtained from r by deleting all negative body atoms in the body of r.

A set of atoms X is *closed under* a basic program P iff for any rule $r \in P$, $head(r) \in X$ whenever $body(r) \subseteq X$. The smallest set of atoms which is closed under a basic program P is denoted by $CN(P)$. The *reduct* or Gelfond-Lifschitz transformation [13], P^X of a program P relatively to a set X of atoms is defined by $P^X = \{r^+ \mid r \in P \text{ and } body^-(r) \cap X = \emptyset\}$. A set of atoms X is an *answer set* of P iff $CN(P^X) = X$.

3   Partially Preordered Removed Sets Revision (PPRSR)

Let Ψ be an epistemic state for partially preordered information. Ψ is syntactically represented by $(\Sigma, \preceq_\Sigma)$ where Σ is a set of formulae and $\preceq_\Sigma$ is a partial preorder on Σ. Ψ can be represented from a semantic point of view as $(\mathcal{W}, \preceq_\Psi)$ where $\mathcal{W}$ is the set of interpretations and $\preceq_\Psi$ is a partial preorder on $\mathcal{W}$ such that

$Mod(Bel(\Psi)) = min(\mathcal{W}, \prec_\Psi)$. We present the Partially Preordered Removed Sets Revision (PPRSR) of an epistemic state Ψ by a formula μ. According to the syntactic point of view, we focus on the preferred subsets of formulae to remove from Σ to restore consistency. We first define the potential removed sets as follows:

Definition 1. *Let $(\Sigma, \preceq_\Sigma)$ be a syntactic representation of Ψ. Let μ be a formula such that $\Sigma \cup \{\mu\}$ is inconsistent. R, a subset of formulae of Σ, is a potential removed set of $\Sigma \cup \{\mu\}$ iff $(\Sigma \backslash R) \cup \{\mu\}$ is consistent.*

Example 1. Let $\Sigma = \{a, b, a \vee \neg b, \neg a \vee b\}$ and $\preceq_\Sigma$ be a given partial preorder: ($b \leftarrow a$ means that $b \prec_\Sigma a$). We revise Σ by $\mu = \neg a \vee \neg b$.

$\Sigma \cup \{\mu\}$ is inconsistent. The potential removed sets are $R_0 = \{a, a \vee \neg b\}, R_1 = \{a, a \vee \neg b, \neg a \vee b\}, R_2 = \{a, b, a \vee \neg b\}, R_3 = \{a, b, a \vee \neg b, \neg a \vee b\}, R_4 = \{b, \neg a \vee b\}, R_5 = \{b, a \vee \neg b, \neg a \vee b\}, R_6 = \{a, b, \neg a \vee b\}, R_7 = \{a, b\}$.

(The diagram shows: a and $a \vee \neg b$ on the top, b and $\neg a \vee b$ on the bottom, with arrows pointing downward from a to b and from $a \vee \neg b$ to $\neg a \vee b$.)

Let $\mathcal{R}(\Sigma \cup \{\mu\})$ be the set of potential removed sets. Among them, we want to prefer the potential removed sets which allow us to remove the formulae that are not preferred according to $\preceq_\Sigma$. This leads to define a partial preorder on subsets of formulae of Σ, called comparator [3,16], denoted by $\trianglelefteq_C$. We now generalize the notion of Removed Sets to subsets of partially preordered formulae. We denote by $\mathcal{R}_C(\Sigma \cup \{\mu\})$ the set of removed sets of $\Sigma \cup \{\mu\}$.

Definition 2. *Let $(\Sigma, \preceq_\Sigma)$ be a syntactic representation of Ψ. Let μ be a formula such that $\Sigma \cup \{\mu\}$ is inconsistent. $R \subseteq \Sigma$ is a removed set of $\Sigma \cup \{\mu\}$ iff*

1. *R is a potential removed set.*
2. *$\nexists R' \in \mathcal{R}(\Sigma \cup \{\mu\})$ such that $R' \subseteq R$.*
3. *$\nexists R' \in \mathcal{R}(\Sigma \cup \{\mu\})$ such that $R' \triangleleft_C R$.*

Example 2. In the examples, we will use the weak comparator, denoted by $\trianglelefteq_w$ and defined in 2.3. We have $R_0 \trianglelefteq_w R_1$ because $a \preceq_\Sigma a$ and $\neg a \vee b \preceq_\Sigma a \vee \neg b$. The partial preorder on the potential removed sets is: $R_0 \trianglelefteq_w R_1$, $R_0 \trianglelefteq_w R_2$, $R_0 \triangleleft_w R_3$, $R_0 \trianglelefteq_w R_4$, $R_0 \sim_w R_7$, $R_1 \sim_w R_2$, $R_1 \trianglelefteq_w R_3$, $R_1 \sim_w R_7$, $R_2 \trianglelefteq_w R_3$, $R_7 \trianglelefteq_w R_2$, $R_7 \trianglelefteq_w R_3$, $R_3 =_w R_4 =_w R_5 =_w R_6$. We have $\nexists R' \in \mathcal{R}(\Sigma \cup \{\mu\})$ such that $R' \triangleleft_w R_0$ and $R' \triangleleft_w R_7$. Moreover, R_0 and R_7 are minimal according to the inclusion. So, $\mathcal{R}_w(\Sigma \cup \{\mu\}) = \{R_0, R_7\}$.

Remark: We could refine the notion of removed set with an extra preference according to a strategy P (cardinality or minimality). $\mathcal{R}_{C,P}(\Sigma \cup \{\mu\})$ denotes the set of removed sets of $\Sigma \cup \{\mu\}$ according to the strategy P. In this case, a preferred removed set according to a strategy P is a removed set R such that $\nexists R' \in \mathcal{R}_C(\Sigma \cup \{\mu\})$ such that $R' <_P R$. According to the cardinality, $R_Y \leq_{CARD} R_X$ iff $|R_Y| \leq |R_X|$ with $|X|$ the cardinality of the set X. According to the minimality, $R_Y \leq_{MIN} R_X$ iff $|R_Y \cap MIN| \leq |R_X \cap MIN|$ with $MIN = \{x | x \in \Sigma, \nexists y \in \Sigma, y \prec_\Sigma x\}$.

Example 3. We can apply strategies: $\mathcal{R}_{w,CARD}(\Sigma \cup \{\mu\}) = \{R_0, R_7\}$ and $\mathcal{R}_{w,MIN}(\Sigma \cup \{\mu\}) = \{R_0\}$.

The revision of an epistemic state represented by $(\Sigma, \preceq_\Sigma)$ by a formula μ is a new epistemic state represented by $(\Sigma \circ_{\trianglelefteq_C} \mu, \preceq_{\Sigma \circ_{\trianglelefteq_C} \mu})$ and is defined as follows:

Definition 3. *Let $(\Sigma, \preceq_\Sigma)$ be a syntactic representation of Ψ. Let μ be a formula such that $\Sigma \cup \{\mu\}$ is inconsistent. The Partially Preordered Removed Sets Revision (PPRSR) is defined by:*

$$- \quad \Sigma \circ_{\trianglelefteq_C} \mu = \bigvee_{R \in \mathcal{R}(\Sigma \cup \{\mu\})} Cons((\Sigma \backslash R) \cup \{\mu\})$$
$$- \quad \preceq_{\Sigma \circ_{\trianglelefteq_C} \mu}: (i) \; \forall \psi \in \Sigma, \; \mu \prec_{\Sigma \circ_{\trianglelefteq_C} \mu} \psi; (ii) \; \forall \psi, \phi \in \Sigma, \; \psi \preceq_{\Sigma \circ_{\trianglelefteq_C} \mu} \phi \; iff \; \psi \preceq_\Sigma \phi$$

Example 4. According to the example 1, Ψ is syntactically represented by $(\Sigma, \preceq_\Sigma)$ and revising by μ using the weak comparator gives $\Sigma \circ_{\trianglelefteq_w} \mu = Cons(\{b, \neg a \vee b, \neg a \vee \neg b\}) \vee Cons(\{a \vee \neg b, \neg a \vee b, \neg a \vee \neg b\})$ and $\preceq_{\Sigma \circ_{\trianglelefteq_w} \mu}$:

$$
\begin{array}{ll}
a & a \vee \neg b \\
\downarrow & \downarrow \\
b & \neg a \vee b \\
\searrow & \swarrow \\
& \neg a \vee \neg b
\end{array}
$$

In order to establish the equivalence between the syntactic and the semantic representations of Ψ, we use the following definition where $F_\Sigma(\omega)$ denotes the set of formulae of Σ falsified by an interpretation ω.

Definition 4. $\forall \omega, \omega' \in \mathcal{W}, \; \omega \preceq_\Psi^C \omega' \; iff \; F_\Sigma(\omega) \trianglelefteq_C F_\Sigma(\omega') \; and \; F_\Sigma(\omega') \not\trianglelefteq_C F_\Sigma(\omega).$

Using this definition, the semantic representation of Ψ is $(\mathcal{W}, \preceq_\Psi^C)$ and is such that $Mod(\Sigma) = min(\mathcal{W}, \prec_\Psi^C)$. Moreover the following proposition holds.

Proposition 1. *Let Ψ be an epistemic state and $\preceq_\Psi^C$ be a partial preorder on $\mathcal{W}$ associated to Ψ. Then, $\preceq_\Psi^C$ is a P-faithful assignment.*

We are now able to define the semantic counterpart of PPRSR as follows:

Definition 5. *Let Ψ be an epistemic state and μ be a formula. $Mod(\Psi \circ_{\trianglelefteq_C^{sem}} \mu) = min(Mod(\mu), \prec_\Psi^C).$*

The equivalence between the semantic and the syntactic PPRSR is given by the following proposition.

Proposition 2. *Let $(\Sigma, \preceq_\Sigma)$ be a syntactic representation of Ψ and μ be a formula. $Mod(\Sigma \circ_{\trianglelefteq_C} \mu) = Mod(\Psi \circ_{\trianglelefteq_C^{sem}} \mu).$*

The semantic representation of the revised epistemic state is $(\mathcal{W}, \preceq_{\Psi \circ_{\trianglelefteq_C^{sem}} \mu}^w)$ with $\preceq_{\Psi \circ_{\trianglelefteq_C^{sem}} \mu}^w$ defined by $\omega \preceq_{\Psi \circ_{\trianglelefteq_C^{sem}} \mu}^w \omega'$ iff $F_{\Sigma \circ_{\trianglelefteq_C} \mu}(\omega) \trianglelefteq_w F_{\Sigma \circ_{\trianglelefteq_C} \mu}(\omega')$ and $F_{\Sigma \circ_{\trianglelefteq_C} \mu}(\omega') \not\trianglelefteq_C F_{\Sigma \circ_{\trianglelefteq_C} \mu}(\omega)$. When we select the weak comparator defined in 2.3, the PPRSR framework can capture the possibilistic revision recalled in 2.3 and the following proposition holds.

Proposition 3. *Let $\circ_\pi$ be the possibilistic revision operator. $\forall \omega, \omega' \in \mathcal{W}$, $\omega \preceq_{\Psi \circ_{\trianglelefteq_w^{sem}} \mu}^w \omega' \; iff \; \omega \preceq_{\Psi \circ_\pi \mu}^w \omega'$*

Example 5. Let $(\Sigma, \preceq_\Sigma)$ be the syntactic representation of Ψ from the example 1 with $\Sigma = \{a, b, a \vee \neg b, \neg a \vee b\}$. The interpretations are: $\omega_0 = \{\neg a, \neg b\}$, $\omega_1 = \{\neg a, b\}$, $\omega_2 = \{a, \neg b\}$ and $\omega_3 = \{a, b\}$. Using the definition 4 with the weak comparator, we construct a partial preorder on the interpretations. The sets of formulae of Σ falsified by the interpretations are $F_\Sigma(\omega_0) = \{a, b\}$, $F_\Sigma(\omega_1) = \{a, a \vee \neg b\}$, $F_\Sigma(\omega_2) = \{b, \neg a \vee b\}$ and $F_\Sigma(\omega_3) = \emptyset$ and the partial preorder $\preceq_\Psi^w$ is given by the Fig. 1 (a). Therefore $(\mathcal{W}, \preceq_\Psi^w)$ is the semantic representation of Ψ and is such that $Mod(\Sigma) = min(\mathcal{W}, \prec_\Psi^w)$.

Let $(\Sigma \circ_{\lhd_w} \mu, \preceq_{\Sigma \circ_{\lhd_w} \mu})$ be the syntactic representation of the epistemic state Ψ revised by μ. Using the definition 4 with the weak comparator, we construct a new partial preorder on the interpretations. The sets of formulae of $\Sigma \circ_{\lhd_w} \mu$ falsified by the interpretations are $F_{\Sigma \circ_{\lhd_w} \mu}(\omega_0) = \{a, b\}$, $F_{\Sigma \circ_{\lhd_w} \mu}(\omega_1) = \{a, a \vee \neg b\}$, $F_{\Sigma \circ_{\lhd_w} \mu}(\omega_2) = \{b, \neg a \vee b\}$ and $F_{\Sigma \circ_{\lhd_w} \mu}(\omega_3) = \{\neg a \vee \neg b\}$ and the partial preorder $\preceq_{\Psi \circ_{\lhd_w^{sem}} \mu}$ is given by the Fig. 1 (b). Therefore $(\mathcal{W}, \preceq_{\Psi \circ_{\lhd_w^{sem}} \mu})$ is the semantic representation of Ψ revised by μ and with the proposition 2 is such that $Mod(\Sigma \circ_{\lhd_w} \mu) = min(Mod(\mu), \prec_\Psi^w)$.

If we apply, the semantic possibilistic revision of $(\mathcal{W}, \preceq_\Psi^w)$ by μ which preserves the relative ordering between the models of μ and considers all the counter-models of μ as impossible, we obtain the partial preorder $\preceq_{\Psi \circ_\pi \mu}$ illustrated in Fig. 1 (c). Therefore $(\mathcal{W}, \preceq_{\Psi \circ_{\lhd_w^{sem}} \mu}) = (\mathcal{W}, \preceq_{\Psi \circ_\pi \mu})$.

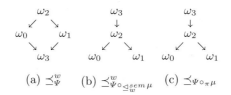

(a) $\preceq_\Psi^w$ (b) $\preceq_{\Psi \circ_{\lhd_w^{sem}} \mu}$ (c) $\preceq_{\Psi \circ_\pi \mu}$

Fig. 1. Partial preorders between interpretations

4 Encoding PPRSR in Answer Set Programming

In order to compute the removed sets, we extend the methods proposed by [9] and [1] to the revision of partially preordered information. We first translate our revision problem into a logic program with answer sets semantics, denoted by $\Pi_{\Sigma \cup \{\mu\}}$. The set of answer sets is denoted by $S(\Pi_{\Sigma \cup \{\mu\}})$. We then define a partial preorder between answer sets of $\Pi_{\Sigma \cup \{\mu\}}$ and we show a one-to-one correspondence between removed sets of $\Sigma \cup \{\mu\}$ and preferred answer sets of $\Pi_{\Sigma \cup \{\mu\}}$.

Let Σ be a set of partially preordered formulae and μ a formula such that $\Sigma \cup \{\mu\}$ is inconsistent. The set of all positive literals of $\Pi_{\Sigma \cup \{\mu\}}$ is denoted by V^+ and the set of all negative literals of $\Pi_{\Sigma \cup \{\mu\}}$ is denoted by V^-. The set of all rule atoms representing formulae is defined by $R^+ = \{r_f | f \in \Sigma\}$ and $F_O(r_f)$ represents the formula of Σ corresponding to r_f in $\Pi_{\Sigma \cup \{\mu\}}$, namely $\forall r_f \in R^+, F_O(r_f) = f$. This translation requires the introduction of intermediary atoms representing subformulae of f. We denote by ρ_f^j the intermediary atom representing f^j which is a subformula of $f \in \Sigma$. To each answer set S of $\Pi_{\Sigma \cup \{\mu\}}$, an interpretation of $\Sigma \cup \{\mu\}$ is associated. Each interpretation of $\Sigma \cup \{\mu\}$ corresponds to several potential removed sets denoted by $F_O(R^+ \cap S)$.

1. In the first step, we introduce rules in order to build a one-to-one correspondence between answer sets of $\Pi_{\Sigma \cup \{\mu\}}$ and interpretations of V^+. For each atom, $a \in V^+$ two rules are introduced: $a \leftarrow not\ a'$ and $a' \leftarrow not\ a$ where $a' \in V^-$ is the negative atom corresponding to a.
2. In the second step, we introduce rules in order to exclude the answer sets S corresponding to interpretations which are not models of $(\Sigma \backslash F) \cup \{\mu\}$ with $F = \{f | r_f \in S\}$. According to the syntax of f, the following rules are introduced:
   - If $f \equiv a$, the rule $r_f \leftarrow not\ a$ is introduced;
   - If $f \equiv \neg f^1$, the rule $r_f \leftarrow not\ \rho_{f^1}$ is introduced;
   - If $f \equiv f^1 \vee \ldots \vee f^m$, the rule $r_f \leftarrow \rho_{f^1}, \ldots, \rho_{f^m}$ is introduced;
   - If $f \equiv f^1 \wedge \ldots \wedge f^m$, it is though necessary to introduce several rules to the program. These rules are introduced: $\forall 1 \leq j \leq m,\ r_f \leftarrow \rho_{f^j}$.
3. The third step rules out answer sets of $\Pi_{\Sigma \cup \{\mu\}}$ which correspond to interpretations which are not models of μ. According to the syntax of μ, the following rules are introduced:
   - If $\mu \equiv a$, the rule $false \leftarrow not\ a$ is introduced;
   - If $\mu \equiv \neg f^1$, the rule $false \leftarrow not\ \rho_{f^1}$ is introduced;
   - If $\mu \equiv f^1 \vee \ldots \vee f^m$, the rule $false \leftarrow \rho_{f^1}, \ldots, \rho_{f^m}$ is introduced;
   - If $\mu \equiv f^1 \wedge \ldots \wedge f^m$, the rules $\forall 1 \leq j \leq m,\ false \leftarrow \rho_{f^j}$ are introduced.
   In order to rule out $false$ from the models of μ, the following rule is introduced: $contradiction \leftarrow false, not\ contradiction$.

Example 6. For the previous example, the logic program $\Pi_{\Sigma \cup \{\mu\}}$ is the following:

$a \leftarrow not\ a'$ $b \leftarrow not\ b'$ $r_a \leftarrow a'$ $r_{a \vee \neg b} \leftarrow a', b$
$a' \leftarrow not\ a$ $b' \leftarrow not\ b$ $r_b \leftarrow b'$ $r_{\neg a \vee b} \leftarrow a, b'$
$false \leftarrow not\ a', not\ b'$ $contradiction \leftarrow false, not\ contradiction$

If $f = \neg a \vee b$ belongs to a removed set, then $r_{\neg a \vee b}$ should belong to an answer set. f has to be falsified and so $\neg f$, i.e. $a \wedge \neg b$, has to be satisfied that is why the rule $r_{\neg a \vee b} \leftarrow a, b'$ is introduced to $\Pi_{\Sigma \cup \{\mu\}}$.

From the logic program, we show how we obtain a one-to-one correspondence between the preferred answer sets of $\Pi_{\Sigma \cup \{\mu\}}$ and the removed sets of $\Sigma \cup \{\mu\}$. Let S be a set of atoms, we define the interpretation over the atoms of $S \cap V^+$ as $I_S = \{a | a \in S\} \cup \{\neg a | a' \in S\}$ and the following result holds.

Proposition 4. *Let ρ a rule atom or an intermediary atom. $\rho \in CN(\Pi^S_{\Sigma \cup \{\mu\}})$ iff $I_S \not\models F_O(R^+ \cap S)$.*

The correspondence between answer sets of $\Pi_{\Sigma \cup \{\mu\}}$ and interpretations of $(\Sigma \backslash F_O(R^+ \cap S)) \cup \{\mu\}$ is given in the following proposition:

Proposition 5. *Let Σ be a set of partially preordered formulae. Let $S \subseteq V$ be a set of atoms. S is an answer set of $\Pi_{\Sigma \cup \{\mu\}}$ iff S corresponds to an interpretation I_S of V^+ which satisfies $(\Sigma \backslash F_O(R^+ \cap S)) \cup \{\mu\}$.*

The proof of the proposition 5 is based on the rules construction.

Example 7. The answer sets of $\Pi_{\Sigma \cup \{\mu\}}$ are: $S_0 = \{a', b, r_{a \vee \neg b}, r_a\}$, $S_1 = \{a, b',$ $r_{\neg a \vee b}, r_b\}$ and $S_2 = \{a', b', r_a, r_b\}$.

In order to compute the answer sets corresponding to the removed sets, we introduce new preference relations between answer sets according to a partial preorder. We define the notion of preferred answer sets of $\Pi_{\Sigma \cup \{\mu\}}$ according to the weak comparator denoted by $S_w(\Pi_{\Sigma \cup \{\mu\}})$.

Definition 6. *Let $\preceq_\Sigma$ be a partial preorder on Σ, μ be a formula such that $\Sigma \cup \{\mu\}$ is inconsistent, $S \in S(\Pi_{\Sigma \cup \{\mu\}})$. S is a preferred answer set of $\Pi_{\Sigma \cup \{\mu\}}$ iff $\nexists S' \in S(\Pi_{\Sigma \cup \{\mu\}})$ such that $F_O(S' \cap R^+) \lhd_w F_O(S \cap R^+)$.*

Example 8. We have $F_O(S_0 \cap R^+) \unlhd_w F_O(S_1 \cap R^+)$ and $F_O(S_2 \cap R^+) \unlhd_w F_O(S_1 \cap R^+)$. So, $S_w(\Pi_{\Sigma \cup \{\mu\}}) = \{S_0, S_2\}$.

Remark: As previously, it is possible to refine the notion of preferred answer set with an extra preference according to a strategy P. Let S_X, $S_Y \in S_w(\Pi_{\Sigma \cup \{\mu\}})$. S_Y is preferred to S_X according to $CARD$ (resp. MIN) iff $|F_O(S_Y \cap R^+)| \leq |F_O(S_X \cap R^+)|$ (resp. $|F_O(S_Y \cap R^+) \cap MIN| \leq |F_O(S_X \cap R^+) \cap MIN|$).

Example 9. We have S_0 is as preferred as S_2 according to $CARD$ and S_0 is preferred to S_2 according to MIN.

The one-to-one correspondence between preferred answer sets of $\Pi_{\Sigma \cup \{\mu\}}$ and the removed sets is given by the following proposition:

Proposition 6. *Let Σ be a finite set of partially preordered formulae and μ be a formula such that $\Sigma \cup \{\mu\}$ is inconsistent. X is a removed set of $\Sigma \cup \{\mu\}$ iff there exists a preferred answer set S of $\Pi_{\Sigma \cup \{\mu\}}$ such that $F_O(R^+ \cap S) = X$.*

Sketch of the proof: we show that the set of removed sets of $\Sigma \cup \{\mu\}$ equals the set of preferred answer sets of $\Pi_{\Sigma \cup \{\mu\}}$.

Example 10. We have $F_O(S_0 \cap R^+) = \{a, a \vee \neg b\}$ and $F_O(S_2 \cap R^+) = \{a, b\}$ which correspond to the removed sets R_0 and R_7 found in the previous section.

Performing PPRSR. Regarding the implementation, CLASP [7] gives us the answer sets of $\Pi_{\Sigma \cup \{\mu\}}$. But our method requires to partially preorder the answer sets with the comparator $\unlhd_w$ to obtain the preferred answer sets corresponding to removed sets. This step is not yet implemented in ASP. We used a java program to partially preorder the answer sets to obtain the preferred answer sets. We denote by N the number of answer sets given by CLASP. The computation of the partial preorder between them can be realized in less than $\frac{N(N-1)}{2}$ comparisons. Indeed, it is sufficient to compare the minimal formulae according to $\preceq_\Sigma$ of each answer set and so using the following proposition, we reduce the cost of the computation.

Proposition 7. *Let $\preceq_\Sigma$ be a partial preorder on Σ, μ be a formula such that $\Sigma \cup \{\mu\}$ is inconsistent and $S, S' \in S(\Pi_{\Sigma \cup \{\mu\}})$. $F_O(S \cap R^+) \unlhd_w F_O(S' \cap R^+)$ iff $\forall y \in Min(F_O(S \cap R^+), \prec_\Sigma), \exists x \in Min(F_O(S' \cap R^+), \prec_\Sigma)$ such that $x \preceq_\Sigma y$ where $Min(F_O(S \cap R^+), \prec_\Sigma) = \{x | x \in F_O(S \cap R^+), \nexists y \in F_O(S \cap R^+), y \prec_\Sigma x\}$.*

Moreover, the determination of the minimal answer sets according to this partial preorder does not increase the cost since the complexity of CLASP is similar to the complexity of the SAT problem.

5   VENUS Application

The european VENUS project (Virtual ExploratioN of Underwater Sites) no (IST-034924)[3] aims at providing scientific methodologies and technological tools for the virtual exploration of deep underwater archaeology sites. In this context, technologies like photogrammetry are used for data acquisition and the knowledge about the studied objects is provided by both archaeology and photogrammetry. We constructed an application ontology in [14] from a domain ontology which describes the vocabulary on the amphorae (the studied artefacts) and from a task ontology describing the data acquisition process. This ontology consists of a set of concepts, relations, attributes and constraints like "If the typology of the amphora is Dressel 20 then the total length of the amphora should be included between 0,368 and 0.552 m." Our knowledge base contains our ontology and observations. The ontology represents the generic knowlegde which is preferred to observations. The observations on the same amphora can be preordered according to the reliability of the experts who provide them. In this context, we revise the generic knowledge and the observations by new observations. We only consider a small part of the ontology (Fig. 2) and some observations in order to provide an example where the knowledge base is expressed in propositional logic.

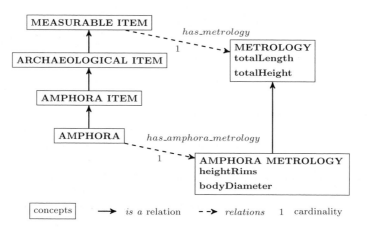

Fig. 2. Extract of the application ontology

We use the following propositional variables: a for the amphora, t for the typology, b for Beltran 2B, h for the total height, l for the total length, c_h (resp. c_l) for the constraint of compatibility between the height (resp. length) and the typology. The propositional translation of the extract of the ontology can be

---

[3] http://www.venus-project.eu

resumed by the set of formulae: $G = \{(\neg a \vee h) \wedge (\neg a \vee t), (\neg t \vee \neg b \vee h) \wedge (\neg t \vee \neg b \vee c_h), (\neg t \vee \neg b \vee l) \wedge (\neg t \vee \neg b \vee c_l)\}$. We then add the formulae provided by the observations of the first expert denoted by $O_1 = \{a, b, c_l, c_h, l, h, t\}$. We obtain $\Sigma = G \cup O_1$ and $\preceq_\Sigma$ is represented by the figure 3 (a). We revised by the observations given by the second expert who is more reliable than the first one, denoted by $O_2 = \{\neg c_l, \neg c_h\}$ and such that $\neg c_l \sim_{O_2} \neg c_h$, the revised preorder is represented by Fig. 3 (b). The revision presented in the section 3 is the first

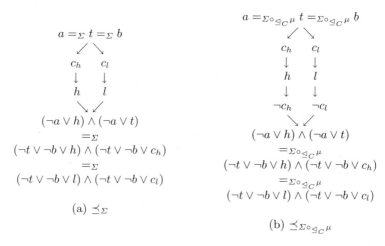

Fig. 3. $\preceq_\Sigma$ and $\preceq_{\Sigma \circ_{\lhd_C} \mu}$

step of the revision to apply in the VENUS context. Indeed, the revision could be defined as follows:

- $\Sigma \circ_{\lhd_C} O_2 = \bigvee_{R \in \mathcal{R}_C(\Sigma \cup O_2)} Cons((O_1 \backslash R) \cup G \cup O_2)$ with a modified definition of the potential removed sets of the definition 1. R is a potential removed set of $\Sigma \cup \{\mu\}$ iff $(O_1 \backslash R) \cup G \cup O_2$ is consistent.
- $\preceq_{\Sigma \circ_{\lhd_C} O_2}$: (i) $\forall \psi, \phi \in O_1$, $\psi \prec_{\Sigma \circ_{\lhd_C} O_2} \phi$ iff $\psi \preceq_\Sigma \phi$, (ii) $\forall \psi, \phi \in G$, $\psi \prec_{\Sigma \circ_{\lhd_C} O_2} \phi$ iff $\psi \preceq_\Sigma \phi$, (iii) $\forall \psi \in G$, $\mu \in O_2$, $\psi \prec_{\Sigma \circ_{\lhd_C} O_2} \mu$, (iv) $\forall \psi \in G$, $\phi \in O_1$, $\psi \prec_{\Sigma \circ_{\lhd_C} O_2} \phi$ and (v) $\forall \psi \in O_1$, $\phi \in O_2$ such that ψ and ϕ refers to the measures of the same attribute[4], $\phi \prec_{\Sigma \circ_{\lhd_C} O_2} \psi$.

6   Conclusion

This paper presents a new framework for revising partially preordered information called Partially Preordered Removed Sets Revision (PPRSR) which extends the Removed Sets approach to partial preorders. The paper shows that PPRSR can be successfully encoded into answer set programming and proposes an implementation stemming from ASP solvers. It shows that the extension of the possibilistic revision to partial preorders can be captured within the PPRSR

---

[4] It is obvious that measures of different attributes are incomparable.

framework allowing for an efficient implementation with ASP. It illustrates how PPRSR can be applied within the context of the VENUS european project dealing with archaeological information. An experimental study has now to be conducted in the context of the VENUS project in order to provide a more accurate evaluation of the performance of PPRSR. We have to deeper investigate the use of ASP solver statements in order to directly define a partial preorder between answer sets. A future work will investigate the use of the lexicogtraphic comparator for defining revision operations within the framework of PPRSR.

Acknowledgements

Work partially supported by the European Community under project VENUS (Contract IST-034924) of the "Information Society Technologies (IST) program of the 6th FP for RTD". The authors are solely responsible for the content of this paper. It does not represent the opinion of the European Community, and the European Community is not responsible for any use that might be made of data appearing therein.

References

1. Benferhat, S., Ben-Naim, J., Papini, O., Würbel, E.: Answer set programming encoding of prioritized removed sets revision: Application to gis. In: Applied Intelligence. Springer, Heidelberg
2. Benferhat, S., Lagrue, S., Papini, O.: Revising partially ordered belief. In: Proc. of NMR 2002, Toulouse, France (2002)
3. Benferhat, S., Lagrue, S., Papini, O.: Revision with partially ordered information in a possibilistic framework. Fuzzy Sets and Systems 144(1), 25–41 (2004)
4. Benferhat, S., Lagrue, S., Papini, O.: Revision of partially ordered information. In: Proc. of IJCAI 2005, Edinburgh, pp. 376–381 (2005)
5. Dubois, D., Prade, H.: Belief change and possibility theory. In: Gärdenfors, P. (ed.) Belief Revision, pp. 142–182. Cambridge University Press, U. K. (1992)
6. Gärdenfors, P.: Knowledge in Flux: Modeling the Dynamics of Epistemic States. Bradford Books/ MIT Press, Cambridge (1988)
7. Gebser, M., Kaufmann, B., Neumann, A., Schaub, T.: Clasp: A conflict-driven answer set solver. In: Baral, C., Brewka, G., Schlipf, J. (eds.) LPNMR 2007. LNCS, vol. 4483, pp. 260–265. Springer, Heidelberg (2007)
8. Halpern, J.Y.: Defining relative likelihood in partially-ordered structures. In: Proc. of UAI 1996, pp. 299–306 (1996)
9. Hué, J., Würbel, E., Papini, O.: Removed sets fusion: Performing off the shelf. In: Ghallab, M., Spyropoulos, C.D., Fakotakis, N., Avouris, N. (eds.) Proc. of ECAI 2008 (2008)
10. Junker, U., Brewka, G.: Handling partially ordered defaults in tms. In: Proceedings of IJCAI 1989, pp. 1043–1048 (1989)
11. Katsuno, H., Mendelzon, A.: Propositional Knowledge Base Revision and Minimal Change. Artificial Intelligence 52, 263–294 (1991)
12. Lewis, D.K.: Counterfactuals. Harvard University Press, Cambridge (1973)

13. Gelfond, M., Lifschitz, V.: The stable model semantics for logic programming. In: Kowalski, R.A., Bowen, K. (eds.) Proc. of ICLP 1988, pp. 1070–1080. MIT Press, Cambridge (1988)
14. Papini, O., Würbel, E., Jeansoulin, R., Curé, O., Drap, P., Sérayet, M., Hué, J., Seinturier, J., Long, L.: D3.4 representation of archaeological ontologies 1. Technical report, Projet VENUS (2008)
15. Würbel, E., Jeansoulin, R., Papini, O.: Revision: An application in the framework of gis. In: Cohn, A.G., Giunchiglia, F., Selman, B. (eds.) Proc. of KR 2000, pp. 505–516 (2000)
16. Yahi, S., Benferhat, S., Lagrue, S., Sérayet, M., Papini, O.: A lexicographic inference for partially preordered belief bases. In: Brewka, G., Lang, J. (eds.) Proc. of KR 2008, pp. 507–516 (2008)

A Distance-Based Operator to Revising Ontologies in DL $\mathcal{SHOQ}^{\star}$

Fangkai Yang[1], Guilin Qi[2], and Zhisheng Huang[3]

[1] Department of Computer Sciences, The University of Texas at Austin, USA
[2] AIFB-University of Karlsruhe, Karlsruhe, Germany
[3] Department of Computer Science, Vrije Universiteit Amsterdam,
Amsterdam, The Netherlands

Abstract. In this paper, we propose a distance-based operator to revise ontologies with acyclic generalized terminology as its TBOX in description logic
$\mathcal{SHOQ}$. Our operator resolves incoherence between the original ontology and
the newly received ontology. We first reformulate Dalal's operator to $\mathcal{SHOQ}$,
and propose a query-equivalent syntactical formulation based on a notion called
a *revision policy*. We then propose a tableau algorithm to generate such revision
policies and prove the correctness of the algorithm. We show that the complexity
of our algorithm stays at the same level as that of satisfiability check in $\mathcal{SHOQ}$.

1 Introduction

Ontology change is an important topic in the Semantic Web. When ontologies evolve,
one of the central problems is to deal with logical contradictions [1,2]. When the newly
received information is considered as more reliable or important than the original one,
we will change the original ontology to resolve contradiction. In this case, this problem
is similar to the problem of belief revision. Therefore, instead of proposing a solution
from scratch, it is reasonable to reuse existing methods for belief revision to solve the
problem of inconsistency handling in ontology change.

The most influential work on belief revision is done by Alchourrón, Gärdenfors and
Makinson (AGM for short) who develop the so-called AGM theory of belief change
[3]. However, it is not a trivial task to apply AGM theory to Description Logics (DLs)
because some AGM assumptions fail for DLs. For example, the negation of a *terminology axiom* cannot be defined in most of DLs. In [4], AGM postulates for *contraction*
are adapted to DLs, but they show that for some important DLs, such as $\mathcal{SHIQ}$ and
$\mathcal{SHOIN}$, we cannot define a revision operator that satisfies all of their postulates.
Furthermore, in [5], the authors propose a set of postulates for characterizing a revision operator for ontologies in DLs by introducing axiom negation in ontologies. They
differentiate two kinds of logical contradictions in DLs during ontology change: inconsistency and incoherence. A DL-based ontology is inconsistent if it has no model and it
is incoherent if there is a concept in the ontology which always denotes an empty set.

* Guilin Qi is partially supported by the EU in the IST project NeOn. Zhisheng Huang is partially
supported by EU-funded Projects OpenKnowledge and LarKC. We thank the reviewers for
very helpful comments to improve the quality of our work.

C. Sossai and G. Chemello (Eds.): ECSQARU 2009, LNAI 5590, pp. 434–445, 2009.

Indeed, incoherence is the logical error that occurs often in terminology part of a DL-based ontology, when new *terminology axioms* are added into the original ontology manually or automatically [6,7]. Although an incoherent ontology can have a model, querying over it may result in undesirable conclusions. The following scenario is adapted from [7], where new terminology axioms are added into an ontology PROTON which consists of a set of terminology axioms, through human annotation of the disjointness between two concepts. Suppose the following axioms are contained in PROTON: {HydrographicStructure $\sqsubseteq$ Facility $\sqcap$ Hydrographic, Reservoir $\sqsubseteq$ Lake $\sqcap$ HydrographicStructure, Lake $\sqsubseteq$ WaterRegion}. By learning that concept *WaterRegion* is disjoint with concept *Facility* and adding this disjointness axiom to the above ontology, we get an incoherence because concept *Reservoir* becomes unsatisfiable. Such a logical error, if not repaired, will result in trivial inference as any concept subsumed by concept *Reservoir* will become unsatisfiable. However, work on automatically resolving incoherence of ontologies in ontology change is rare.

In this paper, we propose a novel distance-based revision operator for ontologies in $\mathcal{SHOQ}$ [8] by adapting the well-known Dalal's revision operator [9], which is an intuitive, model-based revision operator satisfying all the AGM postulates. $\mathcal{SHOQ}$ is an expressive DL that underpins the Web ontology language OWL-DL [10], and our method is powerful in revising $\mathcal{SHOQ}$ ontology with *acyclic generalized terminology* as its TBOX [11]. As far as we know, it is the *first* revision operator that resolves incoherence by following the principle of minimal change and is not dependent on the syntactical forms of terminology axioms. Inspired by the work in [12], we propose a notion named *revision policy*, which first substitutes a concept name by a fresh concept name to resolve incoherence, and then asserts the cardinality difference between the new name and the original one to guarantee minimal change. Based on the revision policies, we can obtain an ontology which is query-equivalent to an ontology resulting from the revision operator. Finally, we propose an algorithm to generate such revision policies by extending the tableau algorithm for $\mathcal{SHOQ}$. We prove its correctness and show that the complexity our our algorithm stays at the same level as the complexity of satisfiability check in $\mathcal{SHOQ}$.

Proofs were omitted due to lack of space, but can be found in the technical report at http://www.cs.utexas.edu/~fkyang/rev.pdf.

2   Description Logic $\mathcal{SHOQ}$

We assume that the reader is familiar with Description Logics (DLs) and refer to DL Handbook [11] (Chapter 2) for more details. In this section, we give a brief review of DL $\mathcal{SHOQ}$. Let $\mathbf{C}$, $\mathbf{R}_A$ and $\mathbf{I}$ be disjoint sets of *concept names*, abstract *role names*, and *individual names*. For R and S roles, a *role axiom* is either a role inclusion, which is of the form $R \sqsubseteq S$ for $R, S \in \mathbf{R}_A$ or a transitivity axiom, which is of the form $Trans(R)$ for $R \in \mathbf{R}_A$. A *RBOX* $\mathcal{R}$ is a set of role axioms. A role R is *simple* if, for $\sqsubseteq^*$ the transitive reflexive closure of $\sqsubseteq$ on $\mathcal{R}$ and for each role S, $S \sqsubseteq^* R$ implies $Trans(S) \notin \mathcal{R}$. The set of $\mathcal{SHOQ}$-concepts (or concepts) is the smallest set such that each concept name $A \in \mathbf{C}$ is a concept, for each individual name $o \in \mathbf{I}$, $\{o\}$ is a concept, and for C and D concepts, R an abstract role, S a simple role, complex concepts can be built using conjunction $(C \sqcap D)$, disjunction $(C \sqcup D)$, concept negation $(\neg C)$, exists restriction $(\exists R.C)$, universal restriction $(\forall R.C)$, atleast restriction $(\geq n)S.C$, atmost restriction $(\leq n)S.C$. A *TBOX* is a finite set of *concept inclusion axioms* $C \sqsubseteq D$, where C and D are concepts. An *ABox* is a set of concept and role *assertions* $C(a)$, $R(a,b)$, and *(in)equality axioms* $a = b$ $(a \neq b)$, where C is a concept, R is a role and a and b are individuals. An ontology is a triple $O = \langle \mathcal{T}, \mathcal{R}, \mathcal{A} \rangle$, where $\mathcal{T}$ is a TBOX, $\mathcal{R}$ is an RBOX, and $\mathcal{A}$ is an ABOX. In our paper, O is considered as the set union of $\mathcal{T}$, $\mathcal{R}$ and $\mathcal{A}$.

The semantics of $\mathcal{SHOQ}$ ontology is given by an interpretation $\mathcal{I} = (\Delta^{\mathcal{I}}, \cdot^{\mathcal{I}})$ that consists of a non-empty set $\Delta^{\mathcal{I}}$ (the domain of $\mathcal{I}$) and the function $\cdot^{\mathcal{I}}$ which maps individuals, concepts and roles to elements of the domain, subsets of the domain and binary relations on the domain, respectively. For a complete definition of the semantics of $\mathcal{SHOQ}$ we refer the reader to [8]. An interpretation $\mathcal{I}$ is called a *model* of an ontology O, denoted as $\mathcal{I} \models O$, if it satisfies each axiom in the ontology. We use $\mathcal{M}(O)$ to denote the set of all models of ontology O. A concept C is unsatisfiable if for each model $\mathcal{I}$ of O, $C^{\mathcal{I}} = \emptyset$. An ontology O is incoherent if there exists an unsatisfiable concept in O and it is inconsistent if it has no model. Although incoherence is a notion different from classical view of inconsistency, it is firmly relevant with inconsistency [5]. That is, given an incoherent ontology O, and a set of concept assertions $A = \{C(i_C)|\text{for each concept name } C \text{ in} O\}$, then $O^{+} = O \cup A$, named as the enhanced ontology of O, is inconsistent. Furthermore, for nominal $\{o\}$, its unsatisfiability is defined to be a form of inconsistency, as we fail to find an interpretation for the individual o.

We are interested in the problem of incoherence handling in this paper. Since incoherence often occurs in terminologies, we assume all the ontologies consist of a TBOX and an RBOX, with empty ABOX, and all the logical contradictions take the form of incoherence. Furthermore, each TBOX has a restricted form, called *acyclic generalized terminology* (AGT) [11]. In a *generalized terminology*, each axiom is a GCI $C \sqsubseteq D$, where the left hand side is a concept name, which occurs at most once on the left hand side of axioms in the ontology. For each concept inclusion axiom $C \sqsubseteq D$, if a concept name C' occurs in D, we say C *uses* C', and the relation *uses* is transitive so that we can obtain a transitive closure. A generalized terminology is acyclic if any concept doesn't use itself. In the setting of ontology revision, given two ontologies O and O' which are represented as a set of DL axioms and assertions, we assume that the TBOX of $O \cup O'$ is an AGT. The concept names of O can be divided into two disjoint sets. $\mathcal{B}_{\mathcal{T}}$ are called *base symbols*, in which each concept, named a *primitive concept* is a concept name occurring only on the right hand side of the GCI, and $\mathcal{N}_{\mathcal{T}}$ are called *named symbols*, in which each concept, named a *defined concept*, is the symbol occurred on the left hand side of some axiom. Given the interpretations of symbols in $\mathcal{B}_{\mathcal{T}}$, we can build models of the terminology based on them.

Checking satisfiability of a $\mathcal{SHOQ}$ concept D is accomplished by a *tableau algorithm* [8], which tries to explicitly build up a model by the *completion forest* for the given concept and knowledge base by exhaustively applying a set of *tableau rules*. The algorithm initializes the completion forest F to contain $l + 1$ root nodes $x_0, x_{\{o_1\}}, \ldots, x_{\{o_l\}}$ with labels $\mathcal{L}(x_{o_i}) = \{\{o_i\}\}$, where o_i is nominal occurring in

D, and begin to expand the completion forest by applying two sets of rules: a set of *non-deterministic rules* $\mathbb{NR}$, i.e.⊔-rule, *choose*-rule, $\leq$-rule, and the set of *deterministic rules* $\mathbb{DR}$, i.e, ⊓-rule, ∃-rule, ∀-rule, $\forall_+$-rule, $\geq$-rule, **O**-rule, and terminates when the completion forest is *complete*: when either no more rules are applicable, where the concept is satisfiable, or all applications of the rules result in a *clash*, where inconsistency is met. Specifically, assuming there is no ABOX and no inconsistencies caused by nominals, a clash is of the following two forms: (C_1): for some concept names $A \in N_C$, $\{A, \neg A\} \subseteq \mathcal{L}(x)$, and (C_2):for some role names S, $(\leq n)S.C \in \mathcal{L}(x)$, and there are $n + 1$ S-successors $y_0, \ldots, y_n$ of x with $C \in \mathcal{L}(y_i)$, for each $1 \leq i < j \leq n$ and $y_i \neq y_j$.

3   A Semantic Revision Operator in $\mathcal{SHOQ}$

3.1   Definition

Our revision operator is based on the well-known Dalal's operator [9]. The idea of this revision operator is that the models of the revised knowledge base of the operator should be the models of the newly received knowledge base which have minimal distance with the original one. However, adapting such idea to DLs is not trivial, because DLs have first-order features. Following the idea of Dalal's operator, we first define the distance between two interpretations and use it to define the distance between an interpretation and an ontology.

Definition 1. *(Distance between interpretations) Let* $\mathcal{I} = \langle \Delta, \cdot^{\mathcal{I}} \rangle$ *and* $\mathcal{I}' = \langle \Delta, \cdot^{\mathcal{I}'} \rangle$ *be two interpretations over the same domain. Let* $d(M^{\mathcal{I}}, M^{\mathcal{I}'}) = |M^{\mathcal{I}} \ominus M^{\mathcal{I}'}|$ *where* M *is a concept name or a role name. The distance between* $\mathcal{I}$ *and* $\mathcal{I}'$, *denoted* $d(\mathcal{I}, \mathcal{I}')$, *is defined as follows:*

$$d(\mathcal{I}, \mathcal{I}') = \sum_{A \in L_C} |A^{\mathcal{I}} \ominus A^{\mathcal{I}'}| + \sum_{R \in L_R} |R^{\mathcal{I}} \ominus R^{\mathcal{I}'}|$$

where $S \ominus S'$ *denotes the symmetric difference between sets* S *and* S', *i.e.,* $S \ominus S' = (S \cup S') \setminus (S \cap S')$, L_C *and* L_R *are respectively the sets of all concept names and role names which are used to construct the DL ontology.*

Definition 2. *Let* $O = \langle \mathcal{T}, \mathcal{A} \rangle$ *and* $O' = \langle \mathcal{T}', \mathcal{A}' \rangle$ *be ontologies with empty ABOXes. Let* $\mathcal{I}$ *be a model of* O'. *The distance between* $\mathcal{I}$ *and* O, *denoted* $d(\mathcal{I}, O)$, *is defined as follows:* $d(\mathcal{I}, O) = min_{\mathcal{I}' \models O} d(\mathcal{I}, \mathcal{I}')$, *where* $\mathcal{I}$ *and* $\mathcal{I}'$ *are over the same domain.*

Based on the above definition, we can define a total pre-order on the models of O' as follows: $\mathcal{I} \preceq_O \mathcal{I}'$ iff $d(\mathcal{I}, O) \leq d(\mathcal{I}', O)$. We can also define the distance between these two ontologies as $d(O, O') = min_{\mathcal{I} \models O'} d(\mathcal{I}, O)$.

Incoherence doesn't lead to the classical sense of contradiction: we may have $d(O, O') = 0$ if $O \cup O'$ is incoherent but consistent. Therefore, unlike Dalal's operator, we cannot define the models of the revised ontology as the models of O' which are minimal w.r.t. the ordering $\preceq_O$. Instead, we append a fresh individual to each concept in the ontology to render inconsistency so that we can apply the idea of Dalal's operator to define a revision operator for resolving incoherence. Based on the notion of enhanced ontologies in [5], we define the *enhancement* of O relative to O' as $O_{O'}^+ = O \cup \{C(i_C)|$for all the primitive concepts C of $O \cup O', i_C$ is a fresh name for $C\}$.

Definition 3. *Let O and O' be two ontologies, and $O_{O'}^+$ and $(O')_O^+$ be their enhanced forms. The result of revision of O by O', denoted $O \circ_D O'$, is defined in a model-theoretical way as follows:*

$$\mathcal{M}(O \circ_D O') = Min(\mathcal{M}((O')_O^+), \preceq_{O_{O'}^+}).$$

That is, the models of the result of revision of O with O' are models of enhanced ontology of O' that are minimal w.r.t. the total pre-order $\preceq_{O_{O'}^+}$.

Example 1. For $O = \{C \sqsubseteq \forall R.D\}$ and $O' = \{C \sqsubseteq \exists R.\neg D\}$, $O_{O'}^+ = O \cup \{C(i_D), D(i_D)\}$ $(O')_O^+ = O' \cup \{C(i_D), D(i_D)\}$. Therefore, given $\Delta = \{a, b, c, d, o_C, o_D\}$, let $C^{\mathcal{I}} = \{a, b, o_C\}$, $D^{\mathcal{I}} = \{c, d, o_D\}$, and $R^{\mathcal{I}} = \{(a, c), (b, d), (o_C, o_D)\}$, $(i_C)^{\mathcal{I}} = o_C$ and $(i_D)^{\mathcal{I}} = o_D$, we have $\mathcal{I} \models O_{O'}^+$. Let $C^{\mathcal{I}'} = \{a, b, o_C\}$, $D^{\mathcal{I}'} = \{c, d\}$, and $R^{\mathcal{I}'} = \{(a, c), (b, d), (o_C, o_D)\}$, $(i_C)^{\mathcal{I}'} = o_C$ and $(i_D)^{\mathcal{I}'} = o_D$ $\mathcal{I}' \models (O')_O^+$. So $d(\mathcal{I}, \mathcal{I}') = 1$. Therefore, $\mathcal{I}' \in Min(\mathcal{M}((O')_O^+), \preceq_{O_{O'}^+})$ and thus $\mathcal{I}' \in \mathcal{M}(O \circ_D O')$

Our revision operator leads to a coherent ontology while preserving the consistency of the resulting ontology.

Theorem 1. *Let O and O' be two ontologies with empty ABOXes and O' be consistent and coherent, then $O \circ_D O'$ is coherent and consistent.*

3.2   Syntactic Formulation of Our Revision Operator

The syntactical formulation is inspired by the work in [12], in which a propositional knowledge base K revised relative to a propositional formula ϕ using Dalal's operator is query-equivalently formulated as a cardinality-circumscription theory, where each atomic proposition of K, say p, is substituted by a fresh name, say p', and a fresh proposition w defined as $p = p'$ is cardinality circumscribed. When dealing with first order semantics, we formalize it by a notion of *revision policy*.

Definition 4. *(Substitution) A substitution ϕ on ontology O is defined as $[C/C']$ where C is the concept occurring in O and C' is a fresh concept.*

Given $\phi = [C/C']$, we use $O\phi$ to denote an ontology obtained by substituting each occurrence of C in O by C'. For two substitution $\phi_1 = [C/C']$ and $\phi_2 = [D/D']$ where C, D are different concept names in O and C', D' are fresh concept names, the composition of $\phi_1 \circ \phi_2$ is defined as $[C, D/C', D']$ and we have $O(\phi_1 \circ \phi_2) \doteq (O\phi_1)\phi_2$.

We now define the revision policy which is critical to the computation of our revision operator.

Definition 5. *(Revision Policy) Given ontology O, a revision policy $\mathbb{P}$ is a pair $< \phi, n_\phi >$ where $\phi = [C/C']$ is a substitution on a primitive concept C in O, and n_ϕ is an integer, called the degree of $\mathbb{P}$. The ontology $O\mathbb{P}$ obtained by applying $\mathbb{P}$ to O is defined in a model-theoretical way as: $\mathcal{I}' \models O\mathbb{P}$ iff $\mathcal{I}' \models O\phi$ and there exists a model $\mathcal{I}$ of O such that (1) $C^{\mathcal{I}'} \ominus C^{\mathcal{I}} = n_\phi$, (2) $C'^{\mathcal{I}'} = C^{\mathcal{I}}$, and (3) $D^{\mathcal{I}'} = D^{\mathcal{I}}$), for any other concept name D in O which is different from C.*

Imposing an revision policy $\mathbb{P} =< [C/C'], n >$ to O will result in another ontology $O\mathbb{P}$ with fresh concept C' such that (1) interpretation of C' in $O\mathbb{P}$ is the same as that of C in O, and the interpretation of C in $O\mathbb{P}$ has a distance of n from that in O. It is intuitive to resolve incoherence by changing the interpretation of primitive concepts because unsatisfiability of defined concepts are usually caused by primitive concepts. The following theorem states the validity of the definition.

Theorem 2. *Given an ontology O and a revision policy $\mathbb{P} =< C/C', n_\phi >$, for every model $\mathcal{I}$ of O, there exists a model $\mathcal{I}'$ of $O\mathbb{P}$ such that $C^{\mathcal{I}'} \ominus C^{\mathcal{I}} = n_\phi$, $D^{\mathcal{I}'} = D^{\mathcal{I}}$, for other concept name D in $O[C/C']$, and $(C')^{\mathcal{I}'} = C^{\mathcal{I}}$.*

Given two revision policies $\mathbb{P} =< \phi, n_\phi >$ and $\mathbb{Q} =< \xi, n_\xi >$, where $\phi = [C/C']$, $\xi = [D/D']$ and C, D are not synonyms, the composition of $\mathbb{P}$ and $\mathbb{Q}$ is defined as $\mathbb{P} \circ \mathbb{Q} \doteq < \phi \circ \xi, \{n_\phi, n_\xi\} >$.

Definition 6. *Given an ontology O, the ontology $O(\mathbb{P} \circ \mathbb{Q})$ obtained by applying $\mathbb{P} \circ \mathbb{Q}$ to O is defined in a model-theoretical way as: $\mathcal{I}' \models O\mathbb{P} \circ \mathbb{Q}$ iff $\mathcal{I}' \models O\phi \circ \xi$ and there exists a model $\mathcal{I}$ of O such that (1) $C^{\mathcal{I}'} \ominus C^{\mathcal{I}} = n_\phi$ and $D^{\mathcal{I}'} \ominus D^{\mathcal{I}} = n_\xi$, (2) $C'^{\mathcal{I}'} = C^{\mathcal{I}}$ and $D'^{\mathcal{I}'} = D^{\mathcal{I}}$ for new concept names C', D', (3) $E^{\mathcal{I}} = E^{\mathcal{I}'}$, for any concept name E in O different from C and D. For two revision policies with same substitution, i.e., $\mathbb{P} =< \phi, n_\phi >$ and $\mathbb{Q} =< \phi, n_\xi >$, $\mathbb{P} \circ \mathbb{Q} \doteq < \phi, max\{n_\phi, n_\xi\} >$.*

It is easy to check that for any revision policies $\mathbb{P}$ and $\mathbb{Q}$ and ontology O, we have $\mathcal{M}(O(\mathbb{P} \circ \mathbb{Q})) = \mathcal{M}(O(\mathbb{P})\mathbb{Q}) = \mathcal{M}(O(\mathbb{Q} \circ \mathbb{P}))$.

We now consider the syntactical counterpart of revision policy. For ontology O on language $\mathcal{L}$, given a revision policy $\mathbb{P} =< [C/C'], n_C >$, we can obtain an axiom set $\mathcal{A}_\mathbb{P}$ consisting of the following axioms $\{o_1, \ldots, o_{|n_C|}\} \equiv ((C \sqcap \neg C') \sqcup (\neg C \sqcap C'))$, where $o_1, \ldots, o_{|n_C|}$ are fresh nominals not occurring in $\mathcal{L}$, and furthermore, we assume unique name assumption (UNA) on them.

The following theorem states that a revision policy $\mathbb{P}$ can be syntactically characterized by substitution and axiom set $\mathcal{A}_\mathbb{P}$.

Theorem 3. *Given ontology O, for a revision policy $\mathbb{P} =< \phi, n_\phi >$ where $\phi = [C/C']$, we have $\mathcal{M}(O\mathbb{P}) = \mathcal{M}(O\phi \cup \mathcal{A}_\mathbb{P})$.*

By Theorem 3, we have the following corollary.

Corollary 1. *Given ontology O, for two revision policy $\mathbb{P} =< \phi, n_\phi >$ and $\mathbb{Q} =< \xi, n_\xi >$, where ϕ and ξ only substitute the symbols occurring in $\mathcal{L}$. We have $O(\mathbb{P} \circ \mathbb{Q}) = O(\phi \circ \xi) \cup \mathcal{A}_\mathbb{P} \cup \mathcal{A}_\mathbb{Q}$.*

Now we characterize the query-equivalent syntactical counterpart of the operator by revision policies, beginning with the following definition and lemma.

Definition 7. *Given two ontologies O, O' and models $\mathcal{I}$ and $\mathcal{I}'$ of O and O' respectively such that $d(\mathcal{I}, \mathcal{I}') = d(O_{O'}^+, O_O'^+)$, for primitive concept C such that $C^{\mathcal{I}} \neq C^{\mathcal{I}'}$. We say $\mathbb{P}_i$ is generated from $\mathcal{I}$ and $\mathcal{I}'$ if $\mathbb{P}_i =< [C/C'], |C^{\mathcal{I}} \ominus C^{\mathcal{I}'}| >$.*

Lemma 1. *Let $O \circ_D O' \models C \sqsubseteq D$ for GCI $C \sqsubseteq D$. For all revision policies $\mathbb{P}_1, \ldots, \mathbb{P}_n$ generated from model $\mathcal{I}$ of O, and $\mathcal{I}'$ of O' such that $d(\mathcal{I}, \mathcal{I}') = d(O_{O'}^+, O_O'^+)$, they satisfies (1) $\Sigma_{i=1}^n deg(\mathbb{P}_i)$ is minimal; (2) $O\mathbb{P}_1 \ldots \mathbb{P}_n \cup O'$ is coherent and consistent, and (3) $O\mathbb{P}_1 \ldots \mathbb{P}_n \cup O' \models C \sqsubseteq D$.*

Theorem 4. *Given ontology O and O' and a query of the form $C \sqsubseteq D$, where C and D are concepts, then $O \circ_D O' \models C \sqsubseteq D$ if and only if for any sequence of revision policies $\mathbb{P}_1, \ldots, \mathbb{P}_n$ on $\mathcal{L}$ such that $\sum_{1 \leq k \leq n} n_{\mathbb{P}_k}$ is minimal and $(O\mathbb{P}_1 \ldots \mathbb{P}_n) \cup O'$ is coherent and consistent, $(O\mathbb{P}_1 \ldots \mathbb{P}_n) \cup O' \models C \sqsubseteq D$.*

4   A Tableau Algorithm for Policy-Based Revision

When an ontology is incoherent, its completion forest can only have clashes (C_1), (C_2) given in Section 2. Our method focuses on resolving the clashes by extending the tableau algorithm of $\mathcal{SHOQ}$ with several *repairing rules* so that no clash can be met. The strategy for such repair is to generate revision policies with minimal degree.

Specifically, given ontologies O that is to be revised by O', suppose that concept D is unsatisfiable in $O \cup O'$, the completion forest of D relative to $O \cup O'$ will contain clashes. Informally, we repair the above two kinds of clashes as follows.

- **Concept Clash Repair.** For clash (C_1), we rename A by A', and specify the difference between the interpretations of A and A' as 1. Therefore, we generate a revision policy $\mathbb{P} = < [A/A'], 1 >$ (see R_1 in Fig.1).
- **Role Clash Repair.** For clash (C_2), we rename C by C', and specify the difference between the interpretations of C and C' as 1. Therefore, we generate a revision policy $\mathbb{P} = < [C/C'], 1 >$. If there are more than $n + 1$ different S-successors of x, then we increase the degree of $\mathbb{P}$ (see R_2 in Fig.1).

We extend the tableau algorithm to repair the clashes in the completion forests. For concept C in an ontology, we arrange all of its completion forests into a *hyper-tree*, in which each node is a *weak complete* completion forest, and each leave is a complete completion forest in the sense of [8]. If C is unsatisfiable, then each leaf contains clash(es), from which the revision policies will be generated. First we define the notion of *weak-completeness* for the completion forest.

Definition 8. *Given a $\mathcal{SHOQ}$ concept D in Negation Normal Form (NNF), a completion forest is weak-complete iff all rules in $\mathbb{DR}$ have been applied till no more of these rules are applicable and none of the nondeterministic rules in $\mathbb{NR}$ has ever been applied.*

From a weak completion tree, we then use rules in $\mathbb{NR}$ to generate its *successors* by first duplicating all the items from a node into a new one and then using nondeterministic rules in $\mathbb{NR}$ to make it weak-complete.

Definition 9. *Given two completion forests F_1 and F_2 which are weak- complete, F_2 is the successor of F_1, denoted as $succ(F_1, F_2)$, if (1) F_1 is weak-complete; (2) F_2 is generated from F_1 by first copying all the nodes of F_1, the structure between the nodes,* *and the labels of the nodes, and then exhaustively applying rules in* $\mathbb{NR}$, *and (3)* F_2 *is weak completed by rules in* $\mathbb{DR}$. *A node* F_1 *is* successor complete *if all of its successors have been generated.*

Finally, we define the completion hyper-tree, which organizes a set of weak completion forests by the above successor relationship.

Definition 10. *Given a* $\mathcal{SHOQ}$ *concept* D, *the completion hyper-tree* $\mathbb{T}$ *of* D *is inductively defined as follows:*

1. *The root of* $\mathbb{T}$, *denoted as* $root(\mathbb{T})$ *is initialized in the same way as tableau algorithm of* $\mathcal{SHOQ}$ *does, and then completed to be weak-complete.*
2. *The subtrees of* $root(\mathbb{T})$ *are denoted as* $\mathbb{T}_1, \ldots, \mathbb{T}_n$, *and* $succ(root(\mathbb{T}), root(\mathbb{T}_i))$ *holds for* $1 \le i \le n$, *and* $\mathbb{T}_i$ *are all completion hyper-trees.*

A completion hyper-tree is *complete* if (1) all of its non-leaf nodes are weak complete and successor complete, and (2) all of the leaf nodes are complete in the sense of tableau algorithm of $\mathcal{SHOQ}$. When there are clashes in the set F of leaves of the hyper-tree, for each node x_i of F, R_1 and R_2 in Fig.1 will compute revision policies resolving each clash. However, this is not enough because the concept influenced by the revision policy may occur in O' rather than O, which has no effect to O, as the following example illustrates.

Example 2. Given $O = \{A = C_1 \sqcap C_2\}$, *and* $O' = \{C_1 \sqsubseteq D, C_2 \sqsubseteq \neg D\}$, *we have a clash in the completion forest of* A *relative to* $O \cup O'$ *as* $\{D, \neg D\}$, *and by revision policy* $\mathbb{P} =< [D/D'], 1 >$ *we can repair this clash. However, we find that* $\mathbb{P}$ *repairs concepts in* O' *rather than* O. *To deal with this case, we need to be aware that the clash is caused by* C_1 *and* C_2, *which occur in* O, *and that* $C_2 \sqsubseteq \neg C_1$ (*or* $C_2 \sqsubseteq \neg C_1$). *Instead, we can use* $< [C_1/C_1'], 1 >$ (*or* $< [C_2/C_2'], 1 >$) *to repair the defined concept which makes another concept involved in a clash.*

The above example shows that if those concepts in the clash of the completion forrest happen to be those in O' but not in O, we need to revise the concepts in O that are *dependent* on concepts in O' that are involved in the clash. Based on this observation, we define the notion of *concept dependency* in AGT.

Definition 11. *Given ontology* O, *we use* $\boldsymbol{C}$ *to denote a set of concept names appearing in* O. *Let* $\boldsymbol{C}^+ = \boldsymbol{C} \cup \{\neg C | C \in \boldsymbol{C}\} \cup \{\{o\} | for \ individual \ o \ in O\}$. *A dependency relation* $\mathcal{D}$ *is defined as a binary relation on* $\boldsymbol{C}^+ \times \boldsymbol{C}^+$ *such that given two concepts* C *and* C' *in* $\boldsymbol{C}^+$, *we say that* C *is dependent on* C' *if there exists a specification* $C \sqsubseteq D$ *such that* C' *appearing in* D. *We use* $\mathcal{D}$ *to denote the dependence relation.*

For each leaf of a completion hyper-tree which is complete, we can create a dependency graph by Algorithm 1 with $\bot$ on the top and $\top$ at the bottom. Based on the dependency graph, we propose two *tracing rules* (see Fig. 1). As we assume that $O \cup O'$ is an AGT and both O and O' are coherent, tracing rules can always find a concept occurring in O, otherwise O' itself is incoherent. Furthermore, the first concept it finds must be a primitive concept in O, otherwise $O \cup O'$ is not an AGT.

Algorithm 1. Generating Dependency Graph

1: **Procedure CreateDependencyGraph**($\mathbb{T}$)
2: $\mathbf{C}^+ := \{C | C \in \mathcal{L}(x)\}$, where x is the node of $\mathbb{T}$.
3: $\mathcal{D}_\mathbb{T} := \emptyset$
4: **for all** all $C \in \mathbf{C}^+$ such that there exists no $C' \in \mathbf{C}^+$ which is dependent on C **do**
5:     $\mathcal{D}_\mathcal{T} := \{(\bot, C)\} \cup \mathcal{D}_\mathbb{T}$
6:     CreateGraph($C, \mathbf{C}^+, \mathcal{D}_\mathbb{T}$)
7: **end for**
8: **Procedure CreateGraph**($C, \mathbf{C}^+, \mathcal{D}_\mathbb{T}$)
9: **for all** concept $C_i \in \mathbf{C}^+$ **do**
10:     **if** C is dependent on C_i and $(C, C_i) \notin \mathcal{D}_\mathbb{T}$ **then**
11:         $\mathcal{D}_\mathbb{T} := \mathcal{D}_\mathbb{T} \cup (C, C_i)$
12:         CreateGraph($C_i, \mathbf{C}^+, \mathcal{D}_\mathbb{T}$)
13:     **else**
14:         $\mathcal{D}_\mathbb{T} := \{(C, \top)\} \cup \mathcal{D}_\mathbb{T}$
15:     **end if**
16: **end for**

Based on the above algorithm, we can see that the dependency relation $\mathcal{D}$ for Example 2 includes $(A, C_1), (A, C_2), (C_1, D)$ and $(C_2, \neg D)$. The following theorem states that Algorithm 1 terminates in polynomial time.

Theorem 5. *Suppose $\mathbb{T}$ is a leaf of a complete completion hyper-tree, the algorithm $CreateDependencyGraph(\mathbb{T})$ terminates within polynomial time relative to $|\mathbf{C}^+|$, returning a directed acyclic dependency graph $\mathcal{D}$.*

Extending the tableau algorithm with R_1-rule, R_2-rule, tracing rule-1 and tracing rule-2 in Figure 1, we can obtain a set of revision policies $\mathbb{P}_1, \ldots, \mathbb{P}_n$ addressing each kind of clash. To repair all the clashes occurring in a completion forest, we need the composite revision policy:

$$\mathbb{P} = \mathbb{P}_1 \circ \ldots \circ \mathbb{P}_n \doteq \prod_{1 \leq i \leq n} \mathbb{P}_i \qquad (1)$$

where each revision policy contains different substitution. We will later prove that the above rules only make minimal change to the difference between the introduced concept and the original concept. For a completion forest F, and for all the nodes x_i of F with revision policy sets $\mathcal{S}^{x_i}$, we have

$$\mathbb{P}_F = \prod_{x_i \in F} (\prod_{\mathbb{P}^{x_i} \in \mathcal{S}^{x_i}} \mathbb{P}^{x_i}), \mathbb{D}_F = \sum_{x_i \in F} (\sum_{\mathbb{P}^{x_i} \in \mathcal{S}^{x_i}} deg(\mathbb{P}^{x_i})) \qquad (2)$$

For any concept D, it usually has more than one completion forest, due to the existence of nondeterministic rules in $\mathbb{DR}$. Given all completion forests with their revision policies, we will choose those whose degrees are the minimal:

$$\mathbb{P} = min_{\mathbb{D}_{F_i}} \{\mathbb{P}_{F_i} | 0 \leq i \leq n\} \qquad (3)$$

In Fig.1, composite rule composes all the revision policies for each node to obtain the revision policy of F. Finally, synthesis rule chooses the revision policy with smallest degree at each branch of the hyper-tree. See the following example.

R_1-rule	if $C, \neg C \in \mathcal{L}(x)$, then $\mathbb{P} =< [C/C'], 1 >$
R_2-rule	if $(\leq n)S.C \in \mathcal{L}(x)$, and there are $n + 1$ S-successors $y_0, \ldots, y_n$ of x, $C \in \mathcal{L}(y_i)$, for each $1 \leq i < j \leq n$ and $y_i \neq y_j$, $Rel^{\neq} \Leftarrow Rel^{\neq} \backslash \{y_i \neq y_j\}$, and if there is no revision policy for C, then $\mathbb{P} = [< C/C' >, 1]$; else $\mathbb{P} := increment(\mathbb{P})$. $Rel^{\neq}$ is the binary relation recording all inequalities between constants inherited from Tableau algorithm.
tracing-rule-1	For a revision policy $\mathbb{P} =< C/C', n >$ generated by R_1-rule, if C occurs in O', trace in the dependency graph of the completion tree to the nearest concept D such that D occurs in O and D is dependent on C or $\neg C$, and change $\mathbb{P}$ to be $< D/D', 1 >$.
tracing-rule-2	For a revision policy $\mathbb{P} =< C/C', n >$ generated by R_2-rule, if C occurs in O', trace in the dependency graph of the completion tree to the nearest concept D such that D occurs in O and D is dependent on C, and change $\mathbb{P}$ to be $< D/D', 1 >$.
composition-rule	For a node F, for all the node x_i of F with the revision policy sets $\mathcal{S}^{x_i}$ of, $\mathbb{P}_F = \prod_{x_i \in F}(\prod_{\mathbb{P}^{x_i} \in \mathcal{S}^{x_i}} \mathbb{P}^{x_i})$, and $\mathbb{D}_F = \sum_{x_i \in F}(\sum_{\mathbb{P}^{x_i} \in \mathcal{S}^{x_i}} deg(\mathbb{P}^{x_i}))$
synthesis-rule	For a node F of $\mathbb{T}_D$ and all its successors $F_1, \ldots, F_n \in \mathbb{T}_D$, $\mathbb{P}_F = min_{\mathbb{D}_{F_i}}\{\mathbb{P}_{F_i} | 0 \leq i \leq n\}$
Note: For $\mathbb{P} =< [C/C'], n >$, $increment(\mathbb{P}) =< [C/C', n+1] >$	

Fig. 1. The tableaux of revision policy generation

Example 3. We consider ontologies O and O' adapted from PROTON:

O	Reservoir$\sqsubseteq$ Lake$\sqcap$ HydrographicStructure, Lake$\sqsubseteq$ NaturalWaterRegion
	HydrographicStrure$\sqsubseteq$ (≥ 3) Owns.Harbor $\sqcap$ Hydrographic
O'	NaturalWaterRegion$\sqsubseteq$ (≤ 1) Own.Harbor

By applying R_2, we can obtain a revision policy $\mathbb{P} =< [Harbor/Harbor'], 1 >$. The revised ontology will be $O\mathbb{P} \cup O' = O[Harbor/Harbor'] \cup O' \cup \{\{o_1, o_2\} \equiv ((Harbor \sqcap \neg Harbor') \sqcup (\neg Harbor \sqcap Harbor'))\}$. In this case, *Reservoir* is satisfiable. As $\mathbb{P}$ is the only revision policy generated, for query-answering on $O\mathbb{P} \cup O'$, we have Reservoir$\sqsubseteq$ (≤ 1)Owns.Harbor and HydroGraphicStructure$\sqsubseteq$Facility now becomes unknown.

However, HydroGraphicStructure$\sqsubseteq$HydroGraphic still holds, illustrating the syntax irrelevance of our method. Furthermore, we can also obtain Reservoir$\sqsubseteq$ $(= 1)$ Owns.Harbor. Intuitively, the introduced name Harbor$'$ can be regarded as an *abnormal Harbor*, which is different from Harbor with one individuals. Such advantage can also benefit to build consistent ABOX afterwards: for each individual of Reservoir, it can only have one Harbor. If it is connected with more harbors, they are inferred to be synonyms.

We now discuss properties of the extended tableau for revising a $\mathcal{SHOQ}$ ontology.

Theorem 6. *A $\mathcal{SHOQ}$-concept C in NNF is satisfiable wrt a RBOX $\mathcal{R}$ if and only if the expansion rules can yield a complete completion hyper-tree, and at least one of its leaves does not contain revision policies.*

Theorem 7. *Given a $\mathcal{SHOQ}$ ontology O and a concept C, the extended tableau algorithm, when applied to C, will terminate in finite steps.*

The following theorem shows the correctness of our algorithm.

Theorem 8. *Given two $\mathcal{SHOQ}$ ontologies O and O', and an unsatisfiable concept C in O in $O \cup O'$. Let the complete completion hyper-tree of $O \cup O'$ be $\mathbb{T}_D$. Then the revision policy of $\mathbb{T}_D$ is $\mathbb{P}_D$ if and only if $\deg(\mathbb{P}_D)$ is minimal relative to all revision policies $\mathbb{P}$ such that D is satisfiable in $O_1\mathbb{P} \cup O_2$.*

The algorithm is applied multiple times to repair all unsatisfied concepts. As the revision policies can be generated based on the framework of the tableau algorithm of $\mathcal{SHOQ}$, we have:

Theorem 9. *Given a query Q, checking $O \circ_D O' \models Q$ is EXPTime-complete.*

5   Related Work

The work in [4,5] focus on the postulates of rational revision operators in DLs and no concrete revision operator is given. In [13], two revision operators are given to revise ontologies in DL $\mathcal{ALCO}$ but they only consider the inconsistencies due to objects being explicitly introduced in the ABOX and their operators are syntax-dependent. Unlike the AGM-oriented approaches, the revision operators presented in [1,14,15] delete some elements from the original ontology to accommodate the new ontology and so they are all syntax-dependent. In [5], the authors argue that incoherence is also important during revision. However, they do not give a revision operator to deal with this problem. The work on debugging and repairing (see, for example, [6,16]) may be applied to give a revision operator that can resolve incoherence. However, these approaches are also syntax-dependent. In contrast, our revision operator is syntax-independent. Our work is also related to the work on updating ABOX in DLs [17] where an ordering between interpretations w.r.t. an interpretation is given and this ordering is used to define an update operator. In contrast, we define an ordering on the interpretations based on a distance function which is not dependent on a specific interpretation and use this ordering to define our operator.

6   Conclusion and Future Work

In this paper, we proposed a novel revision operator for $\mathcal{SHOQ}$ ontologies by adapting Dalal's revision operator. Since a straightforward adaption does not work, we used the notion of enhancement of an ontology to define our revision operator. We then proposed a notion named revision policy to obtain an ontology which is query-equivalent to an ontology resulting the revision operator. We extended the tableau algorithm for DL $\mathcal{SHOQ}$ by proposing some novel rules to generate revision policies to resolve clashes in the original tableau. We showed that our algorithm is correct and that the complexity our our algorithm stays at the same level as the complexity of satisfiability check in $\mathcal{SHOQ}$. Our framework can be easily adapted to OWL-DL or even more expressive language such as $\mathcal{SHOIQ}$.

As a future work, we will consider other revision operators such as Satoh's operator [12]. However, this problem is very challenging because Satoh's operator is not based on cardinality-circumscription. We are also considering applying this work into real Semantic Web system.

References

1. Haase, P., Stojanovic, L.: Consistent evolution of owl ontologies. In: Gómez-Pérez, A., Euzenat, J. (eds.) ESWC 2005. LNCS, vol. 3532, pp. 182–197. Springer, Heidelberg (2005)
2. Haase, P., van Harmelen, F., Huang, Z., Stuckenschmidt, H., Sure, Y.: A framework for handling inconsistency in changing ontologies. In: Gil, Y., Motta, E., Benjamins, V.R., Musen, M.A. (eds.) ISWC 2005. LNCS, vol. 3729, pp. 353–367. Springer, Heidelberg (2005)
3. Gärdenfors, P.: Knowledge in Flux-Modeling the Dynamic of Epistemic States. MIT Press, Cambridge (1988)
4. Flouris, G., Plexousakis, D., Antoniou, G.: On applying the agm theory to dls and owl. In: Gil, Y., Motta, E., Benjamins, V.R., Musen, M.A. (eds.) ISWC 2005. LNCS, vol. 3729, pp. 216–231. Springer, Heidelberg (2005)
5. Flouris, G., Huang, Z., Pan, J.Z., Plexousakis, D., Wache, H.: Inconsistencies, negations and changes in ontologies. In: Proc. of AAAI 2006, pp. 1295–1300 (2006)
6. Schlobach, S., Huang, Z., Cornet, R., van Harmelen, F.: Debugging incoherent terminologies. J. Autom. Reasoning 39(3), 317–349 (2007)
7. Völker, J., Vrandecic, D., Sure, Y., Hotho, A.: Learning disjointness. In: Franconi, E., Kifer, M., May, W. (eds.) ESWC 2007. LNCS, vol. 4519, pp. 175–189. Springer, Heidelberg (2007)
8. Horrocks, I., Sattler, U.: Ontology reasoning in the shoq(d) description logic. In: Proc. of IJCAI 2001, pp. 199–204 (2001)
9. Dalal, M.: Investigations into a theory of knowledge base revision: Preliminary report. In: Proc. of AAAI 1988, pp. 475–479 (1988)
10. Patel-Schneider, P., Hayes, P., Horrocks, I.: Owl web ontology language semantics and abstract syntax. In: Technical report, W3C. W3C Recommendation (2004)
11. Baader, F., Calvanese, D., McGuinness, D., Nardi, D., Patel-Schneider, P.: The Description Logic Handbook: Theory, implementation and application. Cambridge University Press, Cambridge (2007)
12. Liberatore, P., Schaerf, M.: Reducing belief revision to circumscription (and vice versa). Artif. Intell. 93, 261–296 (1997)
13. Qi, G., Liu, W., Bell, D.A.: Knowledge base revision in description logics. In: Fisher, M., van der Hoek, W., Konev, B., Lisitsa, A. (eds.) JELIA 2006. LNCS, vol. 4160, pp. 386–398. Springer, Heidelberg (2006)
14. Halaschek-Wiener, C., Katz, Y., Parsia, B.: Belief base revision for expressive description logics. In: Proc. of OWL-ED 2006 (2006)
15. Ribeiro, M.M., Wassermann, R.: Base revision in description logics - preliminary results. In: Proc. of IWOD 2007 (2007)
16. Qi, G., Haase, P., Huang, Z., Ji, Q., Pan, J.Z., Völker, J.: A kernel revision operator for terminologies - algorithms and evaluation. In: Proc. of ISWC 2008, pp. 419–434 (2008)
17. de Giacomo, G., Lenzerini, M., Poggi, A., Rosati, R.: On the update of description logic ontologies at the instance level. In: Proc. of AAAI 2006 (2006)

An Experimental Study about Simple Decision Trees for Bagging Ensemble on Datasets with Classification Noise⋆

Joaquín Abellán and Andrés R. Masegosa

Department of Computer Science & Artificial Intelligence,
University of Granada, Spain
{jabellan,andrew}@decsai.ugr.es

Abstract. Decision trees are simple structures used in supervised classification learning. The results of the application of decision trees in classification can be notably improved using ensemble methods such as Bagging, Boosting or Randomization, largely used in the literature. Bagging outperforms Boosting and Randomization in situations with classification noise. In this paper, we present an experimental study of the use of different simple decision tree methods for bagging ensemble in supervised classification, proving that simple credal decision trees (based on imprecise probabilities and uncertainty measures) outperforms the use of classical decision tree methods for this type of procedure when they are applied on datasets with classification noise.

Keywords: Imprecise probabilities, credal sets, imprecise Dirichlet model, uncertainty measures, classification, ensemble decision trees.

1   Introduction

A decision tree (or classification tree) is a simple structure that can be used as a classifier. A classifier can be applied on a given dataset, containing several samples where each sample also contains a set of values belonging to an attribute or predictive variable set and a variable labeled *class variable*. In the field of Machine Learning, the classification subject is based on the use of several techniques that infer rules from a given data set in order to predict new values of the class variable (discrete or discretized) using a new set of values for the remaining variables (known as *attribute variables*). The dataset used to obtain these rules is labeled the training data set and the data set used to check the classifier is called the test dataset. The applications of classification are important and distinguished in fields such as medicine, bioinformatics, physics, pattern recognition, economics, etc., and are used for disease diagnosis, meteorological forecasts, insurance, text classification, to name but a few.

An important aspect of decision trees is their inherent instability which means that different training datasets from a given problem domain will produce very different trees. This characteristic is essential to consider them as suitable classifiers in a ensemble scheme as Bagging (Breiman [9]), Boosting (Freund and Schapire [14]) or Randomforest (Breiman [10]). It is proved that the techniques to combine multiple trees, or

---

⋆ This work has been jointly supported by the Spanish Ministry of Education and Science under project TIN2007-67418-C03-03 and by European Regional Development Fund (FEDER); and FPU scholarship programme (AP2004-4678).

C. Sossai and G. Chemello (Eds.): ECSQARU 2009, LNAI 5590, pp. 446–456, 2009.

committees of trees, allow us to obtain better results, in accuracy, than those that can be obtained from a single model. This approach is not just restricted to decision trees learning, it has been also applied to most other Machine Learning methods.

The performing of classifiers on datasets with classification noise is an important field of Machine Learning methods. Classification noise is named to those situations that appear when datasets have incorrect class labels in their training and/or test sets. It is proved in the literature [12] that bagging scheme outperforms boosting and randomization schemes when they are applied on datasets with classification noise.

In this paper, we want to check the use of different simple decision trees, i.e. decision trees built with different simple split criteria and without pre or post prune process, in a bagging scheme, with the aim to ascertain the best simple split criterion for this scheme when it is used on datasets with classification noise. In our case, we will experiment with datasets where we have introduced noise, in a random way, only in the training sets. In our experimentation, we have introduced different percentages of noise in 25 datasets, largely used in classification, to check the performance of different split criteria used in a decision tree growing procedure into a bagging scheme.

In Section 2 of this article, we shall present basic concepts about decision trees and the split criteria analyzed in this paper. In Section 3, we shall briefly describe bagging scheme for combining simple decision trees. In Section 4, we shall check bagging procedures obtained with decision trees with different split criteria on datasets which are widely used in classification. Section 5 is devoted to the conclusions.

2   Decision Trees and Split Criteria

A Decision tree is a structure that has its origin in Quinlan's ID3 algorithm [18]. As a basic reference, we should mention the book by Breiman et al. [8].

Within a decision tree, each node represents an attribute variable and each branch represents one of the states of this variable. A tree leaf specifies the expected value of the class variable depending on the information contained in the training data set. When we obtain a new sample or instance of the test data set, we can make a decision or prediction about the state of the class variable following the path to the tree from the root until a leaf using the sample values and the tree structure. Associated to each node is the most informative variable which has not already been selected in the path from the root to this node (as long as this variable provides more information than if it had not been included). In this last case, a leaf node is added with the most probable class value for the partition of the data set defined with the configuration given by the path until the tree root.

In order to measure the quantity of information, several criteria or metrics can be used, and these are called split criteria. In this article, we will analyze the following ones: Info-Gain, Info-Gain Ratio, Gini Index, and Imprecise Info-Gain.

Info-Gain [IG]. This metric was introduced by Quinlan as the basis for his ID3 model [18]. This model has the following main features: it was defined to obtain decision trees with discrete variables, it does not work with missing values, a pruning process is not carried out, and it is based on Shannon's entropy[21]. This split criterion can therefore be defined on an attribute variable X given the class variable C in the following way:

$$IG(X, C) = H(C) - H(C|X),$$

where $H(C)$ is the entropy of C: $H(C) = -\sum_j p(c_j) \log p(c_j)$, with $p(c_j) = p(C = c_j)$, the probability of each value of the class variable estimated in the training data set. In the same way,

$$H(C|X) = -\sum_t \sum_j p(c_j|x_t) \log p(c_j|x_t),$$

where $x_t, t = 1, .., |X|$, is each possible state of X and $c_j, j = 1, .., k$ each possible state of C. Finally, we can obtain the following reduced expression for the Info-Gain criterion:

$$IG(X, C) = -\sum_t \sum_j p(c_j, x_t) \log \frac{p(c_j, x_t)}{p(c_j)p(x_t)}.$$

This criterion is also known as the *Mutual Information Criterion* and is widely used for measuring the dependence degree between an attribute variable and the class variable. It tends to select attribute variables with many states and consequently results in excessive ramification.

Info-Gain Ratio [IGR]. In order to improve the ID3 model, Quinlan introduces the C4.5 model, where the Info-Gain split criterion is replaced by an Info-Gain Ratio criterion which penalizes variables with many states. A procedure can then be defined to work with continuous variables, it is possible to work with missing data, and a posterior pruning process is introduced.

The Info-Gain Ratio of an attribute variable X_i on a class variable C can be expressed as:

$$IGR(X_i, C) = \frac{IG(X_i, C)}{H(X_i)}.$$

Gini Index [GIx]. This criterion is widely used in statistics for measuring the impurity degree of a partition of a data set in relation to a given class variable (we can say that a partition is "pure" when it only has a single associated value of the class variable). The work by Breiman et al. [8] can be mentioned as a reference for the use of the Gini Index in decision trees.

In a given dataset, the Gini Index of a variable X_i can be defined as:

$$gini(X_i) = 1 - \sum_j p^2(x_j^i).$$

In this way, we can define the split criterion based on the Gini Index as:

$$GIx(X_i, C) = gini(C|X_i) - gini(C),$$

where

$$gini(C|X_i) = \sum_t p(x_t^i)gini(C|X_i = x_t^i).$$

We can see that the expression GIx is written in a different way to that used for the previous split criteria because now the variable with the highest $gini(C|X_i)$ value is selected (contrary to what happens with the entropy).

Imprecise Info-Gain [IIG]. The Imprecise Info-Gain criterion was first used for building simple decision trees in Abellán and Moral's method [3] and in a more complex procedure in Abellán and Moral [5]. In a similar way to ID3, this tree is only defined for discrete variables, it cannot work with missing values, and it does not carry out a posterior pruning process. It is based on the application of uncertainty measures on convex sets of probability distributions. More specifically, probability intervals are extracted from the dataset for each case of the class variable using Walley's imprecise Dirichlet model (IDM) [24], which represents a specific kind of convex sets of probability distributions, and on these the entropy maximum is estimated. This is a total measure which is well known for this type of set (see Abellán, Klir and Moral [6]).

The IDM depends on a hyperparameter s and it estimates that (in a given dataset) the probabilities for each value of the class variable are within the interval:

$$p(c_j) \in \left[\frac{n_{c_j}}{N+s}, \frac{n_{c_j}+s}{N+s} \right],$$

with n_{c_j} as the frequency of the set of values $(C = c_j)$ in the dataset. The value of parameter s determines the speed with which the upper and lower probability values converge when the sample size increases. Higher values of s give a more cautious inference. Walley [24] does not give a definitive recommendation for the value of this parameter but he suggests values between $s = 1$ and $s = 2$. In Bernard [7], we can find reasons in favor of values greater than 1 for s.

If we label $K(C)$ and $K(C|(X_i = x_t^i))$ for the following sets of probability distributions q on Ω_C:

$$K(C) = \left\{ q \mid q(c_j) \in \left[\frac{n_{c_j}}{N+s}, \frac{n_{c_j}+s}{N+s} \right] \right\},$$

$$K(C|(X_i = x_t^i)) = \left\{ q \mid q(c_j) \in \left[\frac{n_{\{c_j, x_t^i\}}}{N+s}, \frac{n_{\{c_j, x_t^i\}}+s}{N+s} \right] \right\},$$

with $n_{\{c_j, x_t^i\}}$ as the frequency of the set of values $\{C = c_j, X_i = x_t^i\}$ in the dataset, we can define the Imprecise Info-Gain for each variable X_i as:

$$\mathbf{IIG}(X_i, C) = S(K(C)) - \sum_t p(x_t^i) S(K(C|(X_i = x_t^i))),$$

where $S()$ is the maximum entropy function of a convex set.

For the previously defined intervals and for a value of s between 1 and 2, it is very easy to obtain the maximum entropy using procedures of Abellán and Moral [2,4] or the specific one for the IDM of Abellán [1].

3   Bagging Decision Trees

The technique of Bagging (Breiman [9]), or Boostrap-Aggregating, refers to the creation of an ensemble of models obtained using the same algorithm on a training dataset resampled with replacement. Finally, Bagging uses all the models with a majority voting criteria for classification.

This method needs of the instability of a learning algorithm, because it makes possible to obtain different classifiers using the same algorithm on different samples of the original dataset. As Breiman [9] said about Bagging: *The vital element is the instability of the prediction method. If perturbing the learning set can cause significant changes in the predictor constructed, then Bagging can improve accuracy.* The final majority vote procedure of the predictors obtained can improve the prediction of any single model. If the algorithm is stable, then the different samples yield similar models and their ensemble produces similar results than a single model. It is important to remark that to combine multiple models helps to reduce the instability and the variance of the method with respect to the single ones.

An important aspect of decision trees is their inherent instability which means that different training datasets from a given problem domain will produce quite different trees. This characteristic is essential to consider them as suitable classifiers in a ensemble scheme as Bagging, Boosting [14] and Randomforest [10]. It is proved that the techniques of combine multiple trees, or committees of trees, allow us to obtain better results than those that can be obtained from a single model.

4   Experimentation

In our experimentation, we have used a wide and different set of 25 known datasets, obtained from the *UCI repository of machine learning databases* which can be directly downloaded from ftp://ftp.ics.uci.edu/machine-learning-databases. A brief description of these can be found in Table 1, where column "N" is the number of instances in the datasets, column "Attrib" is the number of attribute variables, "Num" is the number of numerical variables, column "Nom" is the number of nominal variables, column "k" is the number of cases or states of the class variable (always a nominal variable) and column "Range" is the range of states of the nominal variables of each dataset.

For our experimentation, we have used *Weka* software [26] on Java 1.5, and we have added the necessary methods to build decision trees using the different split criteria used in this paper with the same procedure to built a decision tree. For the IIG criterion we have used the parameter of the IDM $s = 1$.

We have applied the following preprocessing: databases with missing values have been replaced with mean values (for continuous variables) and mode (for discrete variables) using *Weka's* own filters. In the same way, continuous variables have been discretized using the Fayyad and Irani's known discretization method [13]. We note that these two preprocessing steps were carried out considering only information from training datasets. Hence, test datasets were preprocessed with the values computed from these training datasets (information from test datasets were not used in this preprocessing). Using *Weka's* filters, we also added the following percentages of noise to the class variable: $0\%, 5\%, 10\%, 20\%$ and 30%, for training datasets. In this case, these noise levels were not introduced in test datasets (the classes of the test samples were not modified). For each database, we have repeated 10 times a k-10 folds cross validation procedure to estimate the accuracy of each classification model.

Table 1. Dataset Description

Dataset	N	Attrib	Num	Nom	k	Range
Anneal	898	38	6	32	6	2-10
Audiology	226	69	0	69	24	2-6
Autos	205	25	15	10	7	2-22
Breast-cancer	286	9	0	9	2	2-13
Cmc	1473	9	2	7	3	2-4
Colic	368	22	7	15	2	2-6
Credit-german	1000	20	7	13	2	2-11
Diabetes-pima	768	8	8	0	2	-
Glass-2	163	9	9	0	2	-
Hepatitis	155	19	4	15	2	2
Hypothyroid	3772	29	7	22	4	2-4
Ionosfere	351	35	35	0	2	-
Kr-vs-kp	3196	36	0	36	2	2-3
Labor	57	16	8	8	2	2-3
Lymph	146	18	3	15	4	2-8
Mushroom	8123	22	0	22	2	2-12
Segment	2310	19	16	0	7	-
Sick	3772	29	7	22	2	2
Solar-flare1	323	12	0	12	2	2-6
Sonar	208	60	60	0	2	-
Soybean	683	35	0	35	19	2-7
Sponge	76	44	0	44	3	2-9
Vote	435	16	0	16	2	2
Vowel	990	11	10	1	11	2
Zoo	101	16	1	16	7	2

For the sake of simplicity, we use the same names of the split criteria for the bagging methods using decision trees which each split criteria. For each method we have used 100 decision trees for the bagging scheme, as in Freund and Schapire [14].

For space reason we do not present all the results of the accuracy of each method on each dataset with different percentages of noise added. We present, in Table 2 and in Figure 1, the averages of the percentage of correct classifications (Accuracy in Figure 1) of each method on the 25 datasets with different percentages of noise added. The bagging scheme with decision trees using the IIG criterion has a clear better performing with respect the others when we increase the level of noise, as it can be directly appreciated in Figure 1. As we can see, the average in accuracy of the methods is very similar when we add no noise and, when we add noise, all the method decrease in their averages of accuracy, but this decrease is less reduced for the method with the IIG criterion.

To compare the methods, we have used different tests (see Demsar [11] and Witten and Frank [26]).

Table 2. Averages of the percentage of correct classifications of each method on the 25 datasets with 0%, 5%, 10%, 20% and 30% of classification noise

	0%	5%	10%	20%	30%
IG	85.31	83.16	80.92	75.19	68.18
IGR	85.80	83.87	81.99	76.85	70.18
GIx	85.33	82.28	81.00	75.49	68.73
IIG	86.08	85.58	84.64	81.40	75.08

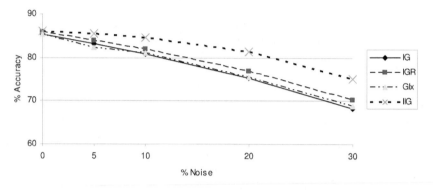

Fig. 1. Averages of the accuracy of each method on the 25 datasets with different percentages of noise added in the training sets

–To compare two classifier on a single data set:

Corrected Paired T-test: a corrected version of the Paired T-test implemented in *Weka* [26]. It is used to avoid some problems of the original test with cross validation schemes. This test checks whether one classifier is better or worse than another on average, across all training and test datasets obtained from a original dataset. We use this test on the training and test datasets obtained from a 10 times k-10 folds cross validation procedure on a original dataset. The level of significance used for this test is 0.05.

–To compare two classifier on multiple data sets:

Wilcoxon Signed-Ranks test (Wilcoxon [25]): a non-parametric test which ranks the differences in performance of two classifiers of each data set, ignoring the sings, and compares the ranks for the positive and the negative differences. We will use the level of significance of 0.05.

Counts of wins, losses and ties: Sign test (Sheshkin [22]; Salzberg [20]): a binomial test that counts the number of data sets on which an algorithm is the overall winner. We will use the level of significance of 0.05.

–To compare multiple classifiers on multiple data sets:

Friedman test (Friedman [15,16]): a non-parametric test that ranks the algorithms for each data set separately, the best performing algorithm getting the rank of 1, the second best rank 2,...The null hypothesis is that all the algorithms are equivalent. When the null-hypothesis is rejected, we can compare all the algorithms to each other using the **Nemenyi test** (Nemenyi [17]).

Table 3. Number of wins (W), ties (T) and defeats (D), noted as (W/T/D), obtained in the corrected Paired T-test carried out on the 25 datasets with 0%, 5% and 10% of classification noise, respectively

	0%			5%			10%		
	1	**2**	**3**	**1**	**2**	**3**	**1**	**2**	**3**
1 IG	.	.	.	.	.	.	.	.	.
2 IGR	(1/24/0)	.	.	(2/23/0)	.	.	(3/22/0)	.	.
3 GIx	(0/25/0)	(0/24/1)	.	(0/24/1)	(0/23/2)	.	(0/24/1)	(0/23/2)	.
4 IIG	(2/22/1)	(1/23/1)	(2/22/1)	(11/14/0)	(10/15/0)	(11/14/0)	(13/12/0)	(10/15/0)	(13/12/0)

Table 4. Number of wins (W), ties (T) and defeats (D), noted as (W/T/D), obtained in the corrected Paired T-test carried out on the 25 datasets with 20% and 30% of classification noise, respectively

	20%			30%		
	1	**2**	**3**	**1**	**2**	**3**
1 IG	.	.	.	.	.	.
2 IGR	(3/22/0)	.	.	(4/21/0)	.	.
3 GIx	(2/22/1)	(0/23/2)	.	(2/22/1)	(0/22/3)	.
4 IIG	(16/9/0)	(11/14/0)	(18/7/0)	(17/8/0)	(11/14/0)	(17/8/0)

Table 5. Results of the test carried out on the percentage of correct classifications on the 25 datasets with 0% of classification noise. Into the table are the numbers of the winners for each test ('−' indicates non statistical significant differences).

	Wil. test			Sing test				Nem. test		
	1	**2**	**3**	**1**	**2**	**3**	Friedman rank	**1**	**2**	**3**
1 IG	.	.	.	.	.	.	2.92	.	.	.
2 IGR	−	.	.	−	.	.	2.52	−	.	.
3 GIx	−	−	.	−	−	.	2.70	−	−	.
4 IIG	4	−	4	4	−	4	1.86	4	−	−

Table 6. Results of the test carried out on the percentage of correct classifications on the 25 datasets with 5% of classification noise. Into the table are the numbers of the winners for each test between two methods ('−' indicates non statistical significant differences).

	Wil. test			Sing test				Nem. test		
	1	**2**	**3**	**1**	**2**	**3**	Friedman rank	**1**	**2**	**3**
1 IG	.	.	.	.	.	.	3.18	.	.	.
2 IGR	2	.	.	−	.	.	2.54	−	.	.
3 GIx	−	2	.	−	2	.	3.12	−	−	.
4 IIG	4	4	4	4	4	4	1.18	4	−	−

In Tables 3 and 4, we can see the number of datasets where each method has obtained wins (W), ties (T) and defeats (D) (W/T/D) with respect the others in the corrected T-test carried out. For example (1/24/0) in the second row signifies that the bagging scheme with IGR split criterion wins to the one with IG criterion (column label as number **1**) in 1 dataset, ties in 24 datasets and defeats in 0 datasets. All the split criteria perform in a similar way when we add no noise, but the difference in favor of IIG criterion is more clear when we increase the percentage of noise.

Table 7. Results of the test carried out on the percentage of correct classifications on the 25 datasets with 10% of classification noise. Into the table are the numbers of the winners for each test between two methods ('−' indicates non statistical significant differences).

	Wil. test			Sing test				Nem. test		
	1	2	3	1	2	3	Friedman rank	1	2	3
1 IG	.	.	.	.	.	.	3.26	.	.	.
2 IGR	2	.	.	2	.	.	2.36	−	.	.
3 GIx	−	2	.	−	2	.	3.26	−	−	.
4 IIG	4	4	4	4	4	4	1.12	4	4	4

Table 8. Results of the test carried out on the percentage of correct classifications on the 25 datasets with 20% of classification noise. Into the table are the numbers of the winners for each test between two methods ('−' indicates non statistical significant differences).

	Wil. test			Sing test				Nem. test		
	1	2	3	1	2	3	Friedman rank	1	2	3
1 IG	.	.	.	.	.	.	3.20	.	.	.
2 IGR	2	.	.	2	.	.	2.16	2	.	.
3 GIx	−	2	.	−	2	.	3.52	−	2	.
4 IIG	4	4	4	4	4	4	1.12	4	4	4

Table 9. Results of the test carried out on the percentage of correct classifications on the 25 datasets with 30% of classification noise. Into the table are the numbers of the winners for each test between two methods ('−' indicates non statistical significant differences).

	Wil. test			Sing test				Nem. test		
	1	2	3	1	2	3	Friedman rank	1	2	3
1 IG	.	.	.	.	.	.	3.36	.	.	.
2 IGR	2	.	.	2	.	.	2.26	2	.	.
3 GIx	−	2	.	−	2	.	3.26	−	2	.
4 IIG	4	4	4	4	4	4	1.12	4	4	4

Tables 5, 6, 7, 8 and 9 present the winners of the Wilcoxon test (Wil.) and Sing test (Sing), the Friedman rank (in all cases the null hypothesis is rejected) and the winners of the Nemenyi test (Nem.), when 0%, 5%, 10%, 20% and 30% percentage of noise is added, respectively. The values also show that the method with IIG criterion is clearly better than the rest when we increase the level of noise. As we can see in the last row of each table (the row of IIG), the number of wins increases (method 4 is the winner) in all tests, when we increase the level of noise. The method with IIG criterion is the winner in all these tests when the percentage of noise is greater or equal than 10%. It must be remarked that the difference between the Friedman rank of the method with IIG criterion with respect the ones with IG and GIx criteria, increases when we increase the level of noise. Also, we can observe that the method with IGR outperforms the one with IG and GIx in this way, being the method with IGR the second best in this study. The methods with IG and GIx split criteria have a similar behavior in this experimentation.

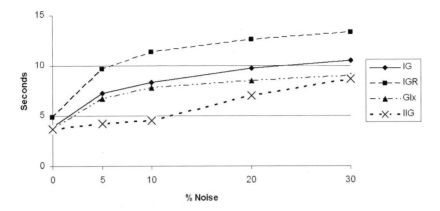

Fig. 2. Averages of the time, in seconds, of each method on the 25 datasets with different percentages of noise added in the training sets

The results of this experimental study can resume that IIG split criterion performs, in accuracy, in a more robust way than the rest when it is used to built decision trees for a bagging scheme on datasets with classification noise.

Other interesting aspect of the experimentation presented should be the time of processing of every method. The study of the time of processing has not been totally analyzed here, as the accuracy, due to space reasons, only, we present Figure 2, where again we can see that IIG has a little better behavior than the rest. In this figure, we present the average of the time for one partition between training/test sets[1], expressed in seconds.

5   Conclusions

We have presented an experimental study about the perform of different split criteria when they are used to build decision trees in a bagging scheme under classification noise. We have analyzed the behavior of three classic, and very used in the literature, simple split criteria as: Info-Gain (IG), Info-Gain Ratio (IGR) and Gini index (GIx) and the one of a new simple split criterion based on imprecise probabilities and uncertainty measures (IIG). We have proved that bagging scheme using decision trees built with IIG criterion has a strong behavior with respect similar scheme using the rest of criteria when we increase the percentage of noise in a dataset.

Other aspect of the experimentation presented here is the time of processing of every method. We have used simple decision trees because they have a reduced time of built, that can allow us to apply some ensemble methods, as bagging scheme, on very large datasets, such as the ones used in bioinformatics or text field classification. The study of the time of processing has not been totally analyzed in this paper due to space reasons, but we have checked that the method with IIG criterion is a little bit better than the one with GIx and these better than the ones with the rest of criteria.

---

[1] 10 times a 10-folds cross validation procedure represents 100 partitions between training/test sets.

References

1. Abellán, J.: Uncertainty measures on probability intervals from Imprecise Dirichlet model. Int. J. General Systems 35(5), 509–528 (2006)
2. Abellán, J., Moral, S.: Maximum entropy for credal sets. Int. J. of Uncertainty, Fuzziness and Knowledge-Based Systems 11(5), 587–597 (2003)
3. Abellán, J., Moral, S.: Building classification trees using the total uncertainty criterion. Int. J. of Intelligent Systems 18(12), 1215–1225 (2003)
4. Abellán, J., Moral, S.: An algorithm that computes the upper entropy for order-2 capacities. Int. J. of Uncertainty, Fuzziness and Knowledge-Based Systems 14(2), 141–154 (2006)
5. Abellán, J., Moral, S.: Upper entropy of credal sets. Applications to credal classification. Int. J. of Approximate Reasoning 39(2-3), 235–255 (2005)
6. Abellán, J., Klir, G.J., Moral, S.: Disaggregated total uncertainty measure for credal sets. Int. J. of General Systems 35(1), 29–44 (2006)
7. Bernard, J.M.: An introduction to the imprecise Dirichlet model for multinomial data. Int. J. of Approximate Reasoning 39, 123–150 (2005)
8. Breiman, L., Friedman, J.H., Olshen, R.A., Stone. C.J.: Classification and Regression Trees. Wadsworth Statistics, Probability Series, Belmont (1984)
9. Breiman, L.: Bagging predictors. Machine Learning 24(2), 123–140 (1996)
10. Breiman, L.: Random Forests. Machine Learning 45(1), 5–32 (2001)
11. Demsar, J.: Statistical Comparison of Classifiers over Multiple Data Sets. Journal of Machine Learning Research 7, 1–30 (2006)
12. Dietterich, T.G.: An Experimental Comparison of Three Methods for Constucting Ensembles of Decision Trees: Bagging, Boosting, and Randomization. Machine Learning 40, 139–157 (2000)
13. Fayyad, U.M., Irani, K.B.: Multi-valued interval discretization of continuous-valued attributes for classification learning. In: Proceedings of the 13th International Joint Conference on Artificial Intelligence, pp. 1022–1027. Morgan Kaufmann, San Mateo (1993)
14. Freund, Y., Schapire, R.E.: Experiments with a new boosting algorithm. In: Thirteenth International Conference on Machine Learning, San Francisco, pp. 148–156 (1996)
15. Friedman, M.: The use of rank to avoid the assumption of normality implicit in the analysis of variance. Journal of the American Statistical Association 32, 675–701 (1937)
16. Friedman, M.: A comparison of alternative tests of significance for the problem of m rankings. Annals of Mathematical Statistics 11, 86–92 (1940)
17. Nemenyi, P.B.: Distribution-free multiple comparison. PhD thesis, Princenton University (1963)
18. Quinlan, J.R.: Induction of decision trees. Machine Learning 1, 81–106 (1986)
19. Quinlan, J.R.: Programs for Machine Learning. Morgan Kaufmann series in Machine Learning (1993)
20. Salzberg, S.L.: On comparison classifiers: Pitfalls to avoid and a recommended approach. Data Mining and Knowledge Discovery 1, 317–328 (1997)
21. Shannon, C.E.: A mathematical theory of communication. The Bell System Technical Journal 27, 379–423, 623–656 (1948)
22. Sheskin, D.J.: Handbook of parametric and nonparametric statistical procedures. Chapman & Hall/CRC, Boca Raton (2000)
23. Walley, P.: Statistical Reasoning with Imprecise Probabilities. Chapman and Hall, London (1991)
24. Walley, P.: Inferences from multinomial data: learning about a bag of marbles. J. Roy. Statist. Soc. B 58, 3–57 (1996)
25. Wilcoxon, F.: Individual comparison by ranking methods. Biometrics 1, 80–83 (1945)
26. Witten, I.H., Frank, E.: Data Mining: Practical machine learning tools and techniques, 2nd edn. Morgan Kaufmann, San Francisco (2005)

Incremental Maintenance of Frequent Itemsets in Evidential Databases

Mohamed Anis Bach Tobji[1], Boutheina Ben Yaghlane[2], and Khaled Mellouli[2]

[1] LARODEC Laboratory, Institut Suprieur de Gestion de Tunis
[2] LARODEC Laboratory, Institut des Hautes Etudes Commerciales de Carthage

Abstract. In the last years, the problem of Frequent Itemset Mining (FIM) from imperfect databases has been sufficiently tackled to handle many kinds of data imperfection. However, frequent itemsets discovered from databases describe only the current state of the data. In other words, when data are updated, the frequent itemsets could no longer reflect the data, i.e., the data updates could invalidate some frequent itemsets and vice versa, some infrequent ones could become valid. In this paper, we try to resolve the problem of Incremental Maintenance of Frequent Itemsets (IMFI) in the context of evidential data. We introduce a new maintenance method whose experimentations show efficiency compared to classic methods.

1 Introduction

The field of Frequent Itemset Mining (FIM) is extensive in the context of perfect databases [1,8,17]. Nevertheless, many applications in the real world treat imperfect data. For example, medical systems that store physician diagnosis [12] or detection systems that are based on sensors [16] may generate imperfect data. That is why, recent years knew emergence of FIM techniques that process uncertain data. Mined data are *probabilistic* [13], *possibilistic* [7], *fuzzy* [4,6] and *evidential* [9,3].

Frequent itemsets are patterns that describe the current state of the data, at the instant t when FIM operation occurred producing a Frequent Itemset Base (FIB). When mined data are updated, the FIB could become invalid. Indeed, data updates may not only invalidate some already frequent itemsets, but also turn some infrequent itemsets into frequent ones. The two general approaches that solve the problem of frequent itemset maintenance after data updating are (1) performing again an operation of FIM on the whole of the updated data, and (2) applying an incremental maintenance on the initial FIB.

Like the FIM area, Incremental Maintenance of Frequent Itemsets (IMFI) one has attracted attention of several researches and literature is abundant in this way [5,10,2]. However, IMFI in uncertain databases is not sufficiently tackled in spite of its importance and applicability on interesting fields where data are frequently updated. In this paper we introduce a new method for maintaining incrementally frequent itemsets when increment of data is added to the initial one.

C. Sossai and G. Chemello (Eds.): ECSQARU 2009, LNAI 5590, pp. 457–468, 2009.
© Springer-Verlag Berlin Heidelberg 2009

Our solution processes evidential data, i.e., data whose imperfection is modelled via the Dempster-Shafer theory.

The remainder of the paper is organized as follows: in section 2, we introduce the basic concepts of the FIM from evidential data, in section 3, we introduce the problem of IMFI in evidential data followed by the solution we propose, section 4 contains the experimentations led on our algorithm and finally section 5 is the conclusion of the work accompanied with some perspectives.

2   The FIM Model in Evidential Databases

An *evidential database*, also called *D-S database* stores data that could be perfect or imperfect. It allows users to set null (missing) values and also uncertain values. Uncertainty in such database is expressed via the evidence theory [14]. An evidential database is defined as follows:

It is a database with n attributes and d lines. Each attribute i ($1 \leq i \leq n$) has a domain D_i of discrete values. Each attribute k among the n ones could store uncertain values. An instance of the attribute k in the line j is an *evidential value* V_{kj} which is a *bba* defined as follows:

$$m_{kj} : 2^{D_k} \to [0,1] \text{ with:}$$

$$m_{kj}(\emptyset) = 0 \text{ and } \sum_{x \subseteq D_k} m_{kj}(x) = 1$$

Table 1. Evidential database example

id	A	B	C
1	$A_1(0.6)$ $A_2(0.4)$	$B_1(0.4)$ $\{B_5, B_6, B_7\}(0.6)$	$C_1(0.5)$ $\{C_1, C_2\}(0.5)$
2	$A_1(0.2)$ $A_3(0.3)$ $\{A_2, A_3\}(0.5)$	$B_1(0.4)$ $\{B_2, B_3\}(0.6)$	C_2

FIM from evidential databases (table 1 is an example of evidential database) is based on new *item*, *itemset* and *support* definitions that are adapted to the uncertain context of the data. The basic concepts of this model [9,3] are the following:

Basic Concepts. An *evidential item* denoted iv_k is one focal element in a body of evidence V_{kj} corresponding to the evidential attribute k. Thus, it is defined as a subset of D_k ($iv_k \in 2^{D_k}$). For example, in table 1, C_1 is an item, $\{C_1, C_2\}$ too.

An *evidential itemset* is a set of evidential items that correspond to different attributes domains. For example, $A_1 B_1 \{C_1, C_2\}$ is an evidential itemset. Formally, an evidential itemset X is defined as: $X \in \prod_{1 \leq i \leq n} 2^{D_i}$

The *inclusion operator* for evidential itemsets is defined as follows: let X and Y be two evidential itemsets. The ith items of X and Y are respectively denoted by i_X and i_Y.

$$X \subseteq Y \text{ if and only if: } \forall i_X \in X, i_X \subseteq i_Y$$

For example, the itemset $A_1B_1\{C_1, C_2\}$ includes the itemset $A_1B_1C_1$. Now, the *line body of evidence* is defined thanks to the conjunctive rule of combination [15]. A line body of evidence is computed from the evidential values composing the line. The frame of discernment of a line BoE is the cross product of all attributes domains denoted by $\Theta = \prod_{1 \leq i \leq n} 2^{D_i}$. Focal elements are included in Θ, and thus are vectors of the form $X = \{x_1, x_2, \ldots, x_n\}$ where $x_i \subseteq D_i$. The mass of a vector X in a line j is computed via the conjunctive rule of combination of the *bba*'s evidential values [15].

$$m_j : \Theta \to [0, 1] \text{ with } m_j(\emptyset) = 0 \text{ and } m_j(X) = \bigcirc_{i \leq n} m_{ij}(X) = \prod_{iv_i \in X} m_{ij}(iv_i)$$

As example, we present here the first line body of evidence in our database example (Table 1). The frame of discernment is Θ which is the cross product of all attributes domains and the frame of discernment of the whole of the bodies of evidence of the database lines. Focal elements are combinations of all evidential items in the line, and thus the BoE of the first line contains eight focal elements, for short we cite only the itemsets $A_1B_1C_1$ with mass equal to 0.12 and $A_1\{B_5, B_6, B_7\}\{C_1, C_2\}$ with mass equal to 0.18.

Now, we introduce the notion of *evidential database body of evidence* which is induced from the line body of evidence notion since database is a set of lines. The body of evidence of evidential database EDB is defined on the frame of discernment Θ, the set of focal elements is composed of all possible evidential itemsets existing in the database and the mass function m_{EDB} is defined as follows: Let X be an evidential itemset and d be the size of EDB:

$$m_{DB} : \Theta \to [0, 1] \text{ with } m_{DB}(X) = \frac{1}{d} \sum_{j=1}^{d} m_j(X)$$

Belief function is naturally defined as follows: $Bel_{DB}(X) = \sum_{Y \subseteq X} m_{DB}(Y)$.

Example 1. In our database (Table 1) the mass of evidential itemset A_1B_1 $\{C_1, C_2\}$ is the sum of its line masses in the database divided by $d = 2$ so $m_{BD}(A_1B_1\{C_1, C_2\}) = 0.06$. Its belief in the database is the sum of all database masses of evidential itemsets that are included in, which are $A_1B_1C_1(0.06)$, $A_1B_1C_2(0.04)$ and $A_1B_1\{C_1, C_2\}(0.06)$ so $Bel_{BD}(A_1B_1\{C_1, C_2\}) = 0.16$.

According to [9,3], the support of an itemset X in the evidential database, is its belief measurement in the database BoE.

The FIM Problem. The problem of FIM in evidential databases is the same as in perfect databases. It consists in mining evidential itemsets whose supports (and thus believes in the database's BoE) exceed a user-defined threshold of support. Formally, let EDB be an evidential database, X be an evidential itemset and Θ be the cross product of all attribute domains. F is the set of frequent evidential itemsets in EDB mined under the user-defined support threshold denoted by min_{supp}; the problem is to extract the set $F = \{X \subseteq \Theta / support(X) \geq min_{supp}\}$.

3   Incremental Maintenance of Frequent Itemsets in Evidential Databases

3.1   Problem Definition

The problem of IMFI in perfect databases was introduced in [5]. We present now the same problem about evidential databases. It is formally defined as follows:

Let EDB be an evidential database and D be its size. Let F be the set of frequent itemsets in EDB and min_{sup} the support threshold under which F was mined. After some updates of EDB -consisting in inserting the increment edb^+ of size d^+-, we obtain $EDB\prime = EDB \cup edb^+$. The size of $EDB\prime$ is denoted by $D\prime$. The problem of IMFI consists in computing $F\prime$: the set of frequent evidential itemsets in $EDB\prime$ under the initial support threshold min_{sup}.

3.2   The Kinds of Itemsets

When EDB is updated by inserting the data increment edb^+, some itemsets that were infrequent in EDB will emerge to be frequent in $EDB\prime$, and vice versa, some itemsets that were frequent in EDB will become infrequent under the threshold min_{sup}. That is why, itemsets in $EDB\prime$ are classified as follows:

- **Winner Itemsets** are itemsets X that were infrequent in EDB ($X.support_{EDB} < min_{sup} \times D$) and become frequent in $EDB\prime$ thanks to the update.
- **Looser Itemsets** are itemsets X that were frequent in EDB ($X.support_{EDB} \geq min_{sup} \times D$) and become infrequent in $EDB\prime$ because of the update.
- **Persistent Itemsets** are itemsets X that were frequent in EDB ($X.support_{EDB} \geq min_{sup} \times D$) and remain frequent in $EDB\prime$ in spite of the update.
- **Invalid Itemsets** are itemsets X that were infrequent in EDB ($X.support_{EDB} < min_{sup} \times D$) and remain also infrequent in $EDB\prime$.
- **Hidden Itemsets** are itemsets X that are composed of non singleton items (such as $A_4\{B_5, B_6\}$) and occur in the data increment but not in the initial database. These itemsets could be frequent in EDB, but are not present in F because they didn't occur in.

The goal of the IMFI is to compute the set $F\prime$ that is composed of the sets W and P of respectively *winner* and *persistent* itemsets, but also of some *hidden* itemsets (this type of itemset is well presented in the next section).

3.3   The Method of IMFI

Our solution proceeds level-by-level in the itemset lattice and is based on candidate itemset generation at each level. It proceeds as follows:

First, we generate the set C_k (candidates of size k) from $F_{k-1}\prime$ (frequent itemsets of size $k-1$ in $EDB\prime$) [1] via the $Apriori\_Gen$ function [1]. C_k contains three types of itemsets; (1) itemsets of F_k whose supports in EDB are known. These itemsets compose the set PP_k of potentially persistent itemsets, (2) itemsets composed of singleton items (like $A_1 B_2 C_1$ or $A_2 B_2 C_1$ but not $A_2\{B_2, B_3\}$) that were not frequent in EDB and so future winner or invalid itemsets. They compose the set PW_k of potentially winner itemsets, and (3) itemsets composed of non singleton items, that didn't occur in EDB and thus we have no information about their frequency (or not) in the initial database. These latter compose the set FS and are handled in the set SS.

Thus, the set C_k is split into three complementary sets PP_k, PW_k and SS_k; the set $PP_k = C_k \cap F_k$ including candidate itemsets that are in F_k (potentially persistent itemsets) and the set $PW_k \cup SS_k = C_k \setminus F_k$ including the remainder candidate itemsets (potentially winner itemsets). Then the set PW_k contains itemsets that are composed of singleton items, and SS_k contains the rest. After preparing our three candidates itemsets (PW_k, PP_k and SS_k) , we scan the increment edb^+, we update supports of itemsets in PP_k (so we get their supports in the whole of the updated database $EDB\prime$) and also supports of itemsets in PW_k (we obtain their supports only in the increment edb^+). We can already distinguish between persistent itemsets ($X.support_{EDB\prime} \geq min_{sup}\prime \times D\prime$) and looser ones ($X.support_{EDB\prime} < min_{sup}\prime \times D\prime$). The set P_k of persistent itemset is already computed after a light scan of the data increment edb^+. After that, we compute the set W_k of winner itemsets starting from the set PW_k. A first pruning of this set, consists in eliminating all itemset whose supports in edb^+ do not exceed the *Candidate Pruning Threshold* denoted by *cpt* (see proposition 1). This optimization is very important because it makes the obligatory return to the initial database less heavy. Indeed, from the itemsets of PW_k, only those whose supports exceed the *cpt* will be updated when scanning the initial database EDB, to obtain their supports in the whole of $EDB\prime$. After this scan, we can filter the winner itemsets from the invalid ones by comparing their supports in $EDB\prime$ to the minimum support threshold $min_{sup} \times D\prime$. Finally, we are obliged to compute the supports of the itemsets of SS_k in the whole of $EDB\prime$, to get the set FS of frequent ones in the updated database. The computation of FS consists a costly operation.

Proposition 1 (The Candidate Pruning Threshold). *Let be X an itemset that was infrequent in EDB. X could not win if:* $X.support_{edb^+} < min_{supp} \times d^+$

Proof. X is frequent in $EDB\prime$ if and only if: $X.support_{EDB\prime} \geq min_{sup} \times D\prime \Leftrightarrow$ $X.support_{EDB} + X.support_{edb^+} \geq min_{sup} \times D\prime \Leftrightarrow X.support_{edb^+} \geq min_{sup} \times D\prime - X.support_{EDB}$ **(1)** Now, we know that X is infrequent in $EDB \Leftrightarrow$ $X.support_{EDB} < min_{sup} \times D$ **(2)**

---

[1] Assume that the set C_1 in the first level contains all possible items in $EDB\prime$.

(1) and **(2)** $\Rightarrow X.support_{edb+} \geq min_{sup} \times D\prime - min_{sup} \times D \Leftrightarrow X.support_{edb+} \geq min_{sup} \times d^+$.

In the next section, we present the data structure we use to accelerate the support computation. The data structure is adapted to the evidential character of the data, it allows also the optimization of the early pruning of looser and invalid itemsets.

3.4   Used Data Structure

The data structure we use to compute rapidly the supports of the itemsets is the *RidLists* one. This structure showed its performance in the case of evidential data [3] especially when data are sparse. It consists in storing for each evidential item the list of the couples (1) *record identifier* of the line that contains the item and (2) *the belief* of the item in the corresponding record. The table 2 presents the *RidLists* that corresponds to our database example.

Table 2. The *RidLists* corresponding to the database example

item	*rid* list
$\{A_2, A_3\}$	$(1, 0.4)(2, 0.8)$
A_1	$(1, 0.6)(2, 0.2)$
A_2	$(1, 0.4)$
A_3	$(2, 0.3)$
$\{B_5, B_6, B_7\}$	$(1, 0.6)$
$\{B_2, B_3\}$	$(2, 0.6)$
B_1	$(1, 0.4)(2, 0.4)$
$\{C_1, C_2\}$	$(1, 1)(2, 1)$
C_1	$(1, 0.5)$
C_2	$(2, 1)$

Once we have the *RidLists* representation of the evidential database, we can compute the support of any itemset via the intersection of the lists of its items. Its support is the sum of the product of the believes of the shared records identifiers of its items. An optimization we introduce in this paper consists in computing the supports of the supersets of items before computing their subsets. Indeed, the belief function being monotone, if a superset of items is infrequent, then all its subsets are also. For example, if the item $\{A_2, A_3\}$ is infrequent, then A_2 and A_3 are too.

The property of monotony of the support function relative to the inclusion itemset operator in one level helps us to prune again the set of candidate itemsets. In fact, in each level of the search space, we start counting the support of the largest evidential itemsets, i.e., evidential itemsets that contains the more the elementary items, instead of computing the itemsets supports in a random way. Then, we store each found infrequent itemset in the set OS (for Optimizing Set), and we delete all the subsets of OS's itemsets without computing their supports because we are sure they do not exceed our threshold. This optimization is applied in the filtering of the sets PP_k and PW_k and saves much computation time.

Proposition 2 (Monotony of the support function). *Let ϕ be an evidential database, X and Y two itemsets in ϕ with the same size k, X includes Y and let α a support threshold.*

If $X.support_\phi < \alpha$ then $Y.support_\phi < \alpha$

Proof. The support function in evidential databases corresponds to the belief function in the evidential database *BoE* (see section 2). Now, according to the *evidence theory*, the *belief function* is monotone relatively to the inclusion operator [14], thus the support function is too.

3.5   The Algorithm

The first iteration of our main algorithm computes the frequent evidential items. It is a particular iteration compared to the other ones, i.e., when $k \geq 2$. Indeed, when $k \geq 2$ the set C_k -from which starts our method- is computed from the set F_{k-1} via the *Apriori_Gen* function. Thus, this latter function could not generate the set C_1, hence the particularity of the first iteration.

Table 3. The data increment edb^+

id	A	B	C
3	$A_1(0.2)$ $A_4(0.2)$ $\{A_1, A_3\}(0.6)$	$B_3(0.1)$ $B_8(0.6)$ $\{B_2, B_3\}(0.1)$ $\{B_5, B_6\}(0.2)$	$C_1(0.4)$ $\{C_1, C_2\}(0.6)$

   The algorithm 1 presents the procedure that computes the frequent evidential items. It starts not from a candidate set, but from both the items of the data increment and the set of initial frequent items F_1. This method allows to compute the sets PP_1 and PW_1, but also the set SS of *super items*, i.e., non-singleton items that are not in F_1. This latter set is very special because it includes the only items whose we do not know any information about their frequency in *EDB*. In other words, we do not know if these items are frequent in *EDB* or not. We present here the procedure *ComputeFrequentItems* followed by a detailed example that explains more explicitly our method.

Example 2 (Maintaining the frequent evidential items). Let F be the set of frequent evidential itemsets mined under the minimum support $min_{supp} = 30\%$ and thus the absolute threshold $0.6(= 2 \times 30\%)$; $F = \{A_1(0.8), \{A_2, A_3\}(1.2),$ $B_1(0.8), \{B_2, B_3\}(0.6), \{B_5, B_6, B_7\}(0.6), C_2(1), \{C_1, C_2\}(2)\}$.

   Let edb^+ be a data increment (presented in table 3) that includes only one record, we try here to compute the set $F\prime$ that includes the frequent evidential items in $EDB\prime = EDB \cup edb^+$ following our procedure *ComputeFrequentItems*:

Algorithm 1. ComputeFrequentItems

Require: RL_{EDB} as RIDLIST, RL_{edb+} as RIDLIST, F_1 as Set of Items, min_{supp} as Real

Ensure: $F_1\prime$ as Set of Items

1 $PP_1 \leftarrow F_1$
2 **for all** item i in RL_{edb+} **do**
3   **if** $\forall j \in OS, i \nsubseteq j$ **then**
4   **if** $i \in PP_1$ **then**
5   Compute $i.support_{edb+}$
6   **if** $i.support_{EDB\prime} < min_{supp} \times D\prime$ **then**
7   Add i to OS if non-singleton
8   **else**
9   Add i to P_1
10   Delete i from PP_1
11   **end if**
12   **else if** i is a superset **then**
13   Add i to SS
14   **else**
15   Compute $i.support_{edb+}$
16   **if** $i.support_{edb+} \geq min_{supp} \times d^+$ **then**
17   Add i to PW_1
18   **end if**
19   **end if**
20   **end if**
21 **end for**
22 **for all** item i in PP_1 **do**
23   **if** $\forall j \in OS, i \nsubseteq j$ **then**
24   Compute $i.support_{edb+}$
25   **if** $i.support_{EDB\prime} < min_{supp} \times D\prime$ **then**
26   Add i to OS if non-singleton
27   **else**
28   Add i to P_1
29   **end if**
30   **end if**
31 **end for**
32 **for all** item i in PW_1 **do**
33   **if** $\forall j \in OS, i \nsubseteq j$ **then**
34   Compute $i.support_{EDB}$
35   **if** $i.support_{EDB\prime} < min_{supp} \times D\prime$ **then**
36   Add i to OS
37   **else**
38   Add i to W_1
39   **end if**
40   **end if**
41 **end for**
42 **for all** item i in SS **do**
43   **if** $\forall j \in OS, i \nsubseteq j$ **then**
44   Compute $i.support_{edb+}$
45   Compute $i.support_{EDB}$
46   **if** $i.support_{EDB\prime} < min_{supp} \times D\prime$ **then**
47   Add i to OS
48   **else**
49   Add i to FS
50   **end if**
51   **end if**
52 **end for**
53 $F_1\prime \leftarrow P_1 \cup W_1 \cup FS$

---

1. Lines 1 to 21 in the algorithm 1: The vertical database RL_{edb+} is scanned item by item . We obtain the set $SS = \{\{A_1, A_3\}, \{B_5, B_6\}\}$ including the non-singleton items that are not in F_1. Then, we obtain the set OS including the non-singleton looser items $\{B_2, B3\}$. Then, we obtain the set $PW_1 = \{B_8(0.6), C_1(0.4)\}$. Note that these items are singleton (else they had gone to the set SS), not in F_1, exceed the *cpt* which is equal to $0.3 = 30\% \times 1$ in our case (that's why A_4 is not in PW_1) and are not included in any item of the set OS (that is why B_3 is not in PW_1). After that, we obtain a part of the set P of persistent items $P = \{A_1, \{C_1, C_2\}\}$. Note that is only a part of persistent items since items that are in F_1 and not in RL_{edb+} are not processed yet.

2. Lines 22 to 31: In the second step, we complete the set P by handling the items that are in F_1 and not in RL_{edb+}. Each item that is a subset of any item in OS is a looser one and is eliminated without computing its support in edb^+, else we compute its support in the increment, and having its in EDB, we decide if it is a looser item (like B_1 and $\{B_5, B_6, B_7\}$) or a persistent one according to the minimum support in $EDB\prime$, i.e., $0.9 = 30\% \times 3$. Note that non-singleton looser items are stored in OS. The result of this step is $P = P \cup \{\{A_2, A_3\}, C_2\}$ and $OS = \{\{B_5, B_6, B_7\}\}$.

3. Lines 32 to 41: Now, we can handle the potentially winner items. Each item included in a one of OS is eliminated (like B_7) without computing its support in EDB. For the other ones, the return to EDB is necessary to update theirs in $EDB\prime$ and only those whose supports in $EDB\prime$ exceed 0.9 are inserted into W; the set of winners. The result of this step is $W = \{C_1\}$.

4. Lines 42 to 52: In the last step, we handle the non-singleton items that are not in F_1 and thus we have no information about their support in EDB; it is the set SS. We do not know if they are frequent in the initial database or not. For example, the item $\{A_1, A_3\}$ is not in F_1 even so it is frequent in EDB because it did not occur in anymore. More generally, for this kind of items we are also obliged to return to the initial database to compute their support in EDB and also in the increment edb^+ except they are subset of an item in OS (like $\{B_5, B_6\}$). After that, we distinguish frequent items from infrequent according to the support threshold 0.9. The result is $FS = \{\{A_1, A_3\}\}$.

The result of the procedure *ComputeFrequentItems* (line 53) in this example is the set $F_1\prime = P \cup W \cup FS = \{A_1, \{A_1, A_3\}, \{A_2, A_3\}, C_1, C_2, \{C_1, C_2\}\}$.

Algorithm 2. ComputeFrequentItemsets

Require: RL_{EDB} as RIDLIST, RL_{edb^+} as RIDLIST, F as Set of Itemsets, min_{supp} as Real
Ensure: $F\prime$ as Set of Itemsets
1 ComputeFrequentItems($RL_{EDB}, RL_{edb^+}, F_1, min_{supp}, F_1\prime$)
2 $k \leftarrow 1$
3 **while** $F_k\prime \neq \emptyset$ **do**
4      $k \leftarrow k + 1; C_k \leftarrow Apriori\_Gen(F_{k-1}\prime)$
5      $PP_k \leftarrow F_k \cap C_k; PW_k \leftarrow C_k \setminus F_k$
6   **for all** itemset $X \in PP_k$ **do**
7   **if** $\forall j \in OS, X \not\subseteq j$ **then**
8   Compute $X.support_{edb^+}$
9   **if** $X.support_{EDB\prime} < min_{supp} \times D\prime$ **then**
10   Add X to OS if non-singleton
11   **else**
12   Add X to P_k
13   **end if**
14   **end if**
15   **end for**
16   **for all** itemset $X \in PW_1$ **do**
17   **if** $\forall j \in OS, X \not\subseteq j$ **then**
18   Compute $i.support_{edb^+}$
19   **if** X is composed of singleton-items **then**
20   **if** $X.support_{edb^+} < min_{supp} \times d^+$ **then**
21   Delete X from PW_k
22   **end if**
23   **else**
24   Add X to SS
25   **end if**
26   **end if**
27   **end for**
28   **for all** itemset $X \in SS$ **do**
29   **if** $\forall j \in OS, X \not\subseteq j$ **then**
30   Compute $X.support_{edb^+}$
31   Compute $X.support_{EDB}$
32   **if** $X.support_{EDB\prime} < min_{supp} \times D\prime$ **then**
33   Add X to OS
34   **else**
35   Add X to FS
36   **end if**
37   **end if**
38   **end for**
39     $F_k\prime \leftarrow P_k \cup W_k \cup FS$
40 **end while**

We introduce above the process of our IMFI solution, detailed in section 3.3, and presented formally in the algorithm 2 that generates the set $F\prime$.

The next section presents the results of the experimentations we led on our solution performance compared to classic ones. We recall that experimentations concern only the performance side and not the quality one since all the solutions give the same set of frequent itemsets, but the answer time is not the same.

4   Experimentations

To evaluate the performance of our solution, we implemented the algorithms of [3],[9] and the ours. Then we generated several synthetic databases where each attribute in each record takes a random evidential value [3], for short, we present here experimentations done on the database $D5000I800C5\%U10$. The parameter D is the size of the whole of the database, I is the size of all attributes cardinal, C is the number of attributes and $\%U$ is the percent of records that includes evidential values in the database.

Then, we performed an initial mining operation on only 4000 records of the database under a threshold min_{sup} and we stored the set F. Then we add the last 1000 records, and we perform the maintenance operation to compute $F\prime$, the set of frequent itemset in the whole of the database ($D = 5000$), using the three algorithms [3,9] (classic maintenance solution) and our method (incremental solution). We repeated these operations for a range of support thresholds, the figure 1 shows how our solution (denoted IMFI in the figure) is more efficient compared to classic ones (HPSS05 for [9] and BBM08a for [3]).

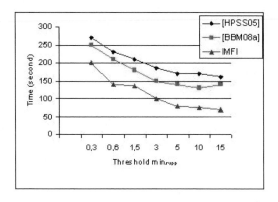

Fig. 1. Comparison performance between [3],[9] and IMFI

Another interesting experimentation (figure 2) presents the effect of the size of the increment on the performance of our algorithm. Indeed, incremental maintenance is efficient when the data increment is small relatively to the initial database. So, the more the data increment is large, the more efficiency of our algorithm decreases and approaches classic methods one. Indeed, IMFI takes advantage from the fast persistent itemsets computation. When the set P_k is relatively small, its complement in C_k, i.e., $PW_k \cup SS_k$, will be relatively large. In this case, the return to the initial database EDB will be heavy because we have to compute the supports of a great number of candidate itemsets. For this reason, and to optimize the use of our algorithm, a study on this question "When to maintain frequent evidential itemsets" would be interesting. It could help the data miner to choose the propice time for maintenance while fulfilling a balance between (1) performance and (2) novelty of mined itemsets. Indeed, someone could maintain the frequent itemsets base after each inserted record; maintenance would be very fast since we do it after the smallest data increment that possible, but this type of maintenance will not produce interesting results because the mined set $F\prime$ will not be really "novel" compared to the initial one F and that is logical since size of data increment is not large enough to change the behavior of the whole of the data and so the discovered knowledge.

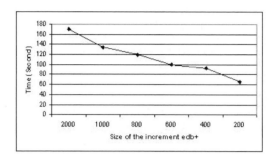

Fig. 2. Effect of the data increment size on the performance of IMFI

5   Conclusion

In this paper, we introduce a novel method for maintaining frequent evidential itemsets after data updates in an incremental manner. This solution is an alternative to the classic way, i.e., mining the whole of the updated data and ignore the previous results obtained in the initial FIM operation. The experimental results showed that our method outperform the straightforward one from a performance point of view.

This work opens several perspectives, like maintenance of frequent evidential itemsets after not only data insert but also data delete. We also think that IMFI methods are interesting in the field of inductive databases [11] where data and patterns are stored both. The IMFI methods reduce the answer time of the mining queries since previous results are stored in the database.

References

1. Agrawal, R., Srikant, R.: Fast Algorithms for Mining Association Rules. In: Proceedings of the 20th International Conference on Very Large Data Bases, pp. 487–499 (1994)
2. Bach Tobji, M.-A., Abrougui, A., Ben Yaghlane, B.: GUFI: A New Algorithm for General Updating of Frequent Itemsets. In: 11th IEEE International Conference on CSE Workshops, San Paulo, Brazil, pp. 45–52 (2008)
3. Bach Tobji, M.-A., Ben Yaghlane, B., Mellouli, K.: A New Algorithm for Mining Frequent Itemsets from Evidential Databases. In: Proceedings of the 12th International Conference on IPMU, Malaga, Spain, pp. 1535–1542 (2008)

4. Chen, G., Wei, Q.: Fuzzy association rules and the extended mining algorithms. Information Sciences-Informatics and Computer Science 147(1-4), 201–228 (2002)
5. Cheung, D.-W., Han, H., Ng, N.-T.: Maintenance of Discovered Association Rules in Large Databases: An Incremental Update technique. In: Proceedings of The Twelfth International Conference on Data Engineering, pp. 106–114 (1996)
6. Dubois, D., Hullermeier, E., Prade, H.: A systematic approach to the assessment of fuzzy association rules. Data Mining and Knowledge Discovery 13(2), 167–192 (2006)
7. Djouadi, Y., Redaoui, S., Amroun, K.: Mining Association Rules under Imprecision and Vagueness: towards a Possibilistic Approach. In: Intl. Fuzzy Systems Conference, vol. (23-26), pp. 1–6 (2007)
8. Han, J., Pei, J., Yin, Y.: Mining Frequent Patterns without Candidate Generation. In: Proceedings of the 2000 ACM-SIGMOD International Conference on Management of Data, pp. 1–12 (2000)
9. Hewawasam, K.K.R.G.K., Premaratne, K., Subasingha, S.P., Shyu, M.-L.: Rule Mining and Classification in Imperfect Databases. In: Proceedings of the Seventh International Conference on Information Fusion, pp. 661–668 (2005)
10. Hong, T.-P., Huang, T.-J.: Maintenance of Generalized Association Rules for Record Deletion Based on the Pre-Large Concept. In: International Conference on Artificial Intelligence, Knowledge Engineering and Data Bases, pp. 142–146 (2007)
11. Imielinski, T., Mannila, H.: A Database Perspective On Knowledge Discovery. Communications of the ACM 39(11), 58–64 (1996)
12. Konias, S., Chouvarda, I., Vlahavas, I., Maglaveras, N.: A Novel Approach for Incremental Uncertainty Rule Generation from Databases with Missing Values Handling: Application to Dynamic Medical Databases. Medical Informatics and The Internet in Medicine 30(3), 211–225 (2005)
13. Leung, C.K.-S., Carmichael, C.L., Hao, B.: Efficient Mining of Frequent Patterns from Uncertain Data. In: Proceedings of the Seventh IEEE International Conference on Data Mining, pp. 489–494 (2007)
14. Shafer, G.: A Mathematical Theory of Evidence. Princeton University Press, Princeton (1976)
15. Smets, P.: The Application of the Transferable Belief Model to Diagnostic Problems. Int. J. Intelligent Systems 13, 127–158 (1998)
16. Vaughn, R.B., Farrell, J., Henning, R., Knepper, M., Fox, K.: Sensor Fusion and Automatic Vulnerability Analysis. In: Proceedings of the 4th international symposium on Information and communication technologies, pp. 230–235 (2005)
17. Zhao, Q., Bhowmick, S.-S.: Association Rule Mining: A survey. Technical Report, CAIS, Nanyang Technological University, Singapore, 2003116 (2003)

A Bayesian Random Split to Build Ensembles of Classification Trees*

Andrés Cano, Andrés R. Masegosa, and Serafín Moral

Department of Computer Science and
Artificial Intelligence,
University of Granada, Spain
{acu,andrew,smc}@decsai.ugr.es

Abstract. Random forest models [1] consist of an ensemble of random-ized decision trees. It is one of the best performing classification models. With this idea in mind, in this section we introduced a random split op-erator based on a Bayesian approach for building a random forest. The convenience of this split method for constructing ensembles of classifica-tion trees is justified with an error bias-variance decomposition analysis. This new split operator does not clearly depend on a parameter K as its random forest's counterpart, and performs better with a lower number of trees.

1 Introduction

The idea of randomized decision trees was first proposed two decades ago by Minger [2], but it was since ensembles of classifiers were introduced that the combination of randomized decision trees arose as a very powerful approach for supervised classification models [3,1,4].

Bagging [3] was one of the first approaches that exploited this idea. A group of decision trees was built over a bootstrapped replicate of the former training dataset. Finally, the last prediction was made by a majority voting criterion over the set of predictions of each single decision tree. As each decision tree was built following the usual approach [5] from different bootstrapped training data, each tree comprised a different set of split nodes. Thus, the randomization was caused by the different random variations of the bootstrapped training sets.

Another trend appeared with the use of random split node selection. For example, Diettrich et al. [4] built ensembles of trees in which at each node, the split was randomly selected from among the K best splits attributes. Some years later, Breiman proposed the random forest model [1] as a combination of the bagging approach with a random split node selection. In this method, each decision tree was once again built over a bootstrapped replicate of the

---

* This work has been jointly supported by Spanish Ministry of Education and Science under project TIN2007-67418-C03-03, by European Regional Development Fund (FEDER), by the Spanish research programme Consolider Ingenio 2010: MIPRCV (CSD2007-00018), and by the FPU scholarship programme (AP2004-4678).

C. Sossai and G. Chemello (Eds.): ECSQARU 2009, LNAI 5590, pp. 469–480, 2009.

former training set. But, as opposed to the Diettrich et al. approach [4], first K nodes were randomly selected and the best one of these was chosen. The Random Forests outperformed the Bagging and Diettrich approaches [1]. One issue relating to Random Forests is their sensitivity to the selection of the K value [1,6], although Breiman suggested a K value around the logarithm of the number of variables as a default choice.

One the main questions relating to ensembles of trees was a theoretical justification of their excellent performance. The notion of bias-variance decomposition of the error [7,8] appeared to provide some insights. Bias represents the systematic component of the error resulting from the incapacity of the predictor to model the underlying distribution. However, variance represents the component of the error that stems from the particularities of the training sample. As both are added to the error, a bias-variance trade-off therefore takes place [8]. When we attempt to reduce bias by creating more complex models that fit better the underlying distribution of the data, we take the risk of increasing the variance component due to overfitting of the learning data. As decision trees can easily encode complex data distributions, their main disadvantage could lie in the high variance they are associated with. A special role is played, in this sense, by the selection of the K value in random forests. Higher K values usually imply low bias but higher variance and, on the other hand, lower K values present poorer bias but better variance [6].

In this work, we propose a new random split method derived from a Bayesian approach for building ensembles of trees. This random split is similar to the random forests one, but does not pose the problem of choosing an optimal K value and allows better performance to be obtained with a lower number of trees.

The rest of the paper is divided as follows. Firstly in Section 2, the Bayesian framework to inference classification trees is introduced. In Section 3, we present our Bayesian approach to random split. And, finally, in Section 4 we show the experimental results of the evaluation of this approach. We conclude giving the main conclusions and proposing future work in Section 5.

2   Bayesian Inference of Classification Trees

In this section we introduce the basic tools to face the problem of inferring class probabilities using trees and ensembles of tree models under a Bayesian inspired approaches. We start giving the basic framework for a single tree and, after that, we introduce the Bayesian notion of ensembles of trees.

2.1   Basic Framework

In order to introduce the basic framework for applying the Bayesian approach for inferring classification trees, the notation used by Buntine [9] is followed. Buntine was the first author to apply Bayesian techniques to this specific problem.

Classification trees partition the space of examples into disjoint subsets, each one represented by a leaf in the tree, and associates a conditional probability distribution for the class variable in relation to the configuration that defines the partition assigned to that leaf.

It is assumed that there are C mutually exclusive and exhaustive classes, $d_1, ..., d_C$. Assuming that example x falls to leaf l in the tree structure T, then the tree gives a vector of class probabilities $\phi_{j,l}$ for $j = 1, ..., C$, which are the probability of class c_j at leaf l. Thus, a classification tree has a discrete component determined by the structure of tree T and a continuous component that is given by the class probabilities of all the leaves of the tree $\Phi_T = \{\phi_{j,l} : j = 1, ..., C; l \in leaves(T)\}$. No more parameters are needed, as it is assumed that all variables are multinomial, although continuous variables could be also managed, including the cut-points in the branching nodes.

Thus, for the above mentioned example x falling into leaf l, its predicted probability class value is described as $P(C = c_j|x, T, \Phi_T) = \phi_{j,l}$. If a concrete class had to be predicted, it would be the one with the highest probability.

Under the Bayesian approach, the quality of the models is evaluated as their posterior probability, given the learning data. This learning data comprises a set of N i.i.d. samples, $(\bar{c}, \bar{x}) = \{(c_1, x_1), ..., (c_N, x_N)\}$. The probability of the model can be computed using the Bayes's theorem as:

$$P(T, \Phi_T|\bar{c}, \bar{x}) \propto P(T, \Phi_T|x) \prod_{i=1}^{N} P(c_i|x_i, T, \Phi_T) = P(T, \Phi_T|x) \prod_{l \in leaves(T)} \prod_{j=1}^{C} \phi_{j,l}^{n_{j,l}} \quad (1)$$

where $n_{j,l}$ is the number of samples of class d_j falling into leaf l. $P(T, \Phi_T|x)$ can be considered equal to $P(T, \Phi_T)$ as the prior over the models, as T and Φ_T are conditioned to unclassified samples.

The factor Φ_T can be removed from Equation 1 if a prior over the set of parameters is defined and integrated into them. This can be easily achieved if the conjugate of this prior has the same functional form, as it is the case of Dirichlet distributions.

Parameters Priors. It is assumed that the prior beliefs over parameter values are given by a Dirichlet distribution. It is also assumed that these distributions are independent from the parameters of the different leaves of the tree. That can be formulated as follows:

$$P(\Phi_T|T) = \prod_{l \in leaves(T)} \frac{1}{B_C(\alpha_1, ..., \alpha_C)} \prod_{j=1}^{C} \phi_{j,l}^{\alpha_{j,l}-1}$$

where B_C is the C-dimensional beta function and $\alpha_{i,l}$ are the parameters of the Dirichlet. B_C is computed in terms of product of gamma functions $\Gamma(x)$ $(\Gamma(x+1) = x\Gamma(x))$ [9].

Posterior Tree Probability. Therefore, using these priors, the posterior probability of a tree, T, can be computed as follows:

$$P(T|\bar{x}, \bar{c}) \propto P(\bar{c}|\bar{x}, T)P(\bar{x}|T)P(T) = P(\bar{x}|T)P(T) \int_{\Phi_T} P(\bar{c}|\bar{x}, T, \Phi_T)P(\Phi_T|T)d\Phi_T$$

This integral can be computed using the above formulation. At the same time, $P(x|T)$ is included in the proportional constant, as it is assumed to be the same for all the tree structures.

$$P(T|\bar{x}, \bar{c}) \propto P(T) \prod_{l \in leaves(T)} \frac{B_C(n_{1,l} + \alpha_1, ..., n_{C,l} + \alpha_C)}{B_C(\alpha_1, ..., \alpha_C)} \qquad (2)$$

Although Buntine tested several priors over the possible tree structures, $P(T)$, in an attempt to favour simpler trees, there was no definitive recommendation [9]. A uniform prior over the possible tree structures will therefore be assumed.

Posterior Class Probability Estimates. Finally, the estimations of the probabilities of the leaves of the tree T are also computed by averaging all possible parameter configurations, by means of expectation:

$$P(C = d_j|x, T, \bar{x}, \bar{c}) = \int_{\Phi_T} \Phi_{j,l} P(\Phi_T|T, \bar{c}, \bar{x}) d\Phi_T = \frac{n_{j,l} + \alpha_j}{n_l + \alpha_0} \,.$$

where l is the leaf where x falls and $n_l = \sum_{j=1}^{C} n_{j,l}$ and $\alpha_0 = \sum_{j=1}^{C} \alpha_j$.

2.2   Multiple Classification Trees

In the full Bayesian approach, inference considers all possible models with the corresponding posterior probability and not just the most probable one. In order to handle several models, the final prediction is performed by adding each particular prediction of each model weighted by its posterior probability:

$$P(C = d_j|x, \bar{c}, \bar{x}) = \sum_T \int_{\Phi_T} P(C = d_j|x, T, \Phi_T) P(T, \Phi_T|\bar{c}, \bar{x}) d\Phi_T \qquad (3)$$

here the summation covers all possible tree structures.

In Bayesian Model Averaging, Equation (3) is approximated by using importance sampling and Monte-Carlo estimation. Thus, tree structures will be generated in an approximate proportion to their posterior probabilities. But applying Monte-Carlo methods in this huge model space would lead to a very computationally expensive approach.

Buntine computed two approximations to this sum by reducing the set of tree structures [9]. One approximation, known as **Smoothing**, restricted the structures to the ones obtained by pruning a complete tree. It is a smoothing because probabilities at final leaves are computed by averaging them with some of the class probabilities from the interior nodes of the tree. The other approximation used by Buntine was called **Option Trees** [9]. This approximation was based on searching and storing many dominant terms of the sum, i.e., trees with high posterior probabilities. The multiple tree structures were compactly represented using AND-OR nodes. The final predictions were made by averaging the predictions of the different models encoded in these option trees.

3   Bayesian Ensembles of Classification Trees

In this section, we present the new approach for building an ensemble of classification trees. This approach is similar to Bayesian model averaging (Equation (3)) which attempts to collect trees with high posterior probabilities. But the predictions of these trees are not weighted. Rather, greater importance is given to the most probable trees which appear more frequently in the ensemble of trees. Thus, this approach should be viewed as a Monte-Carlo inspired one.

3.1   Justifying a Random Split with a Stop Criteria

The greedy approach has been usually employed to build decision trees, selecting at each level of the tree the most promising split nodes [10,5,9]. But when you are looking for a wide range of classification trees with higher posterior probabilities (Section 2.2), this greedy approach does not seem very suitable. Firstly, it is known to be very sensitive to the selection of the root node of the tree [11]. So, if there is a very high informative variable these greedy approaches such as Bagging will probably start most of the trees of the ensembles with the same root node. Therefore, greedy search schemes seems to discover most of the time a narrow set of local maxima of the global posterior probability distribution over the different decision trees.

With this in mind, we chose a random split criterion similar to the one used in random forests. As the random selection of the split nodes at the beginning of tree appears to be more suitable than a greedy scheme, the approach presented differs from the random split of random forests in the introduction of a random condition for stopping the branching.

Information-based scores used to grow random ensembles [1] such as information gain [10] or Gini index [5] predict better partitions whenever a new split node is added. Therefore, stop criteria usually include conditions such as a minimum threshold for the number of samples or a pure partition of the data. Excessive branching implies a higher risk of over-fitting, and post-pruning techniques were therefore applied as suitable stop criteria (they reduce the size of three defining shorter rules and, in consequence, establish better stop levels).

The use of a Bayesian approach enables us to tackle the stop branching problem in an elegant manner, because of the inherent penalty they impose upon more complex models. Buntine proposed a Bayesian method that combined the predictions of some of the pruned sub-trees of a given tree [9]. In a recent work [12], we also proposed a different Bayesian approach to tackling the stop branching problem, this combining different classification rules. In both cases significant performance improvements were noted. For these reasons, possibly stopping the branching, as an additional option to be considered, appear to be justified.

3.2   A Bayesian Approach to Random Splits

In this subsection, we depicted the details of our random split criterion.

As in random forests [1], at any node S_0 of the tree, K split attributes $(S_1, ..., S_K)$ are randomly selected from the set of all possible split candidates. Therefore a score, $Score(T_{S_i})$, is computed for each split node S_i as the logarithm of $P(T_{S_i}|\bar{x}, \bar{c})$ (Equation 2). Simultaneously, we also compute the score of the model without further splitting at this point, $Score(T_{S_0})$. Exponentiating and normalizing this vector of scores, we obtain a distribution $\Lambda_K = (\lambda_0, ..., \lambda_K)$ where each λ_i informs us of the degree of probability of the tree model with the split node S_i with respect to the rest of split candidate nodes and the tree without further splits, T_{S_0}. It must be remembered that Λ is a proper probability distribution, because each $Score(T_{S_i})$ comes from a probability itself. This would not be so evident if the scores were based on information theoretic criteria.

As $Score(T_{S_i})$ is computed with a logarithmic transformation in order to avoid overflows, normalization has to be performed as follows:

$$\lambda_i = \frac{\varphi(T_{S_i})}{\sum_{j=0}^{K} \varphi(T_{S_j})}, i \in \{0, ..., K\}$$

where $\varphi(T_{S_i})$ is a scaled value by the maximum logarithmic score of the candidate models, $Score(T_{S_{max}})$:

$$\varphi(T_{S_i}) = e^{(Score(T_{S_i})-Score(T_{S_{max}}))}, i \in \{0, ..., K\}$$

Finally, our approach randomly samples the split node among the K candidates according to Λ_K distribution. If the T_{S_0} tree is sampled (i.e., branching is stopped at this leaf), the current K split attributes are discarded and other different K split attributes are randomly selected. The whole process of computing the Λ_K is conducted again. Thus, branching stops when T_{S_0} is selected and there are no more split attributes to repeat the whole process again. It is important to remark that the discarded attributes in this process can be considered again in the selection of another split node.

We now provide the pseudo-code of our Bayesian approach to a random split criterion.

Algorithm 1. *Bayesian Random Split*

SelectSplit(S_0, $\boldsymbol{X} = \{X_1, ..., X_n\}$)
$\boldsymbol{Z} = AvailableAttributes(S_0, \boldsymbol{X})$;
end = false;
while (not end)
    $\{S_1, ..., S_K\} = Random\ Selection(\boldsymbol{Z})$;
    $\{S_0, S_1, ..., S_K\} = \{S_0\} \cup \{S_1, ..., S_K\}$;
    $\Lambda_K = (\lambda_0, ..., \lambda_K) = ComputeScores(\{S_0, ..., S_K\})$;
    $S^* = Sampling(\Lambda_K)$;
    $\boldsymbol{Z} = \boldsymbol{Z} \setminus \{S_1, ..., S_K\}$;
    if $S^* \neq S_0$ OR $\boldsymbol{Z} \neq \emptyset$
        end=true;
    else if $S^* = S_0$ AND $\boldsymbol{Z} \neq \emptyset$
        end=false;
return S^*;

The function $AvailableAttributes(S_0, \boldsymbol{X})$ returns the attributes not included as split nodes in the path from S_0 to the root node. That is, all possible attributes available to be used as split nodes.

Random forests perform the same steps but use an information-based score instead of a Bayesian one; they select the split node with the highest score among the K candidates rather than a random sampling of the split node; and they stop branching when this reaches a pure partition or there are few samples in the partition.

4   Bayesian Random Split Evaluation

In this section, we present the experimental results of the comparison of the Bayesian approach for random splits with the random forest one. In the first subsection we will detail the experimental approach employed and the evaluation methodology, and the second subsection presents results and conclusions. The approach presented will be denoted as Bayesian random split (BRS) as opposed to random forests (RF).

Experimental and Evaluation Setup

For these experiments we selected a set of 23 different datasets taken from the UCI repository. In Table 1, the datasets with their basic features are listed. In the last row, we present the range of each feature of the data sets in order to show the heterogeneity of this benchmark.

Table 1. Data Bases Description

Name	t	n	c	Name	t	n	c
anneal	898	39	6	....			
audiology	226	70	24	labor	57	17	2
autos	205	26	7	lymphography	148	19	4
breast-cancer	286	10	2	segment	2310	20	7
horse-colic	368	23	2	sick	3772	30	2
german-credit	1000	21	2	solar-flare	323	13	2
pima-diabetes	768	9	2	sonar	208	61	2
glass2	163	10	2	soybean	683	36	19
hepatitis	155	20	2	sponge	76	45	3
hypothyroid	3772	30	4	vote	435	17	2
ionosphere	351	35	2	vowel	990	12	11
kr-vs-kp	3196	37	2	zoo	101	17	7
....				Range	57-4k	9-70	2-24

t = number of samples, n = number of variables and c = number of classes.

The approach presented, an ensemble of classification trees induced with a Bayesian random split, was implemented in Elvira environment [13], whereas the experiments, along with the rest of the classifiers evaluated, were carried out in Weka platform [14]. We used non-informative Dirichlet priors over the parameters, setting the $\alpha_{i,l}$ parameters of this distribution at $1/C$.

The data were preprocessed with the Weka filters themselves: missing values were replaced (with the mean value for continuous attributes and with the mode for the discrete ones) and discretized with the Fayyad and Irani method [15].

We evaluated the performance of the classifiers with the error rate and with a bias-variance decomposition of this error. For that purpose, we used the Weka utility, following the bias-variance decomposition of the error proposed by Kohavi and Wolpert [8] and using the experimental methodology proposed in [16].

Comparison of those performance measures followed the methodology proposed by Demsar [17] for the comparison of several classifiers over several datasets. In this methodology, the non-parametric Friedman test was used to evaluate the rejection of the hypothesis that all the classifiers perform equally well for a given significant α level (5% in this case). When the Friedman test detected significant differences, a post-hoc test was also used to assess particular differences among these classifiers: the Bonferroni-Dunn test [17] with a 5% significance level establishing a given classifier (marked with $\star$ in the tables) as the reference model.

It is well known that non-parametric tests are hard for rejecting hypotheses. So, in this evaluation it was displayed the ranking score that Friedman test assess to each classifier (ranking scores close to 1 indicate better performance for those classifiers) with the idea of detecting some trends although they do not reach significance levels. Because of the lack of space and because this ranking scores give enough clues to know which is the relative performance of each classifier, Tables detailing the error, bias an variance exact values were removed.

Both Bayesian random split and Random Forests were evaluated with different K values and number of trees in the ensembles. Concretely, K was fixed to 1, 3, 5 and equal to the logarithm of the number of variables as Breiman recommended. Four different number of trees were evaluated: 10, 50, 100 and 200.

The Role of K and the Number of Trees in Bayesian Random Split

The aim of this initial analysis is to show that the Bayesian random split quickly reaches a competitive performance level with a lower number of trees and that this performance does not depend much on the K value as in the case of Random Forests.

Firstly, the following comparison was made. An ensemble with the Bayesian random split was built setting the number of trees at 10 and the K value at 1. This ensemble was then compared with random forest ensembles with different numbers of trees and different K values.

Table 2 shows the Friedman Test's results when it was applied over the different number of trees: it is depicted the ranking scores of each ensemble as well as the acceptation or the rejection of the null hypothesis (all classifiers performs equally well). As it can be seen in Table 2, BRS ensemble with 10 trees and $K = 1$ becomes difficult to beat using Random Forest ensembles with a higher number of trees and different K values, but what it is most important, there is any clear trend and Random Forests seems to beat BRS depending on the concrete K value and with a concrete number of trees.

Table 2. Evaluating BRS ensembles with 10 Trees - Ranking Scores

	$\star BRS$ (10 trees)	RF				
RF Trees	K=1	K=1	K=3	K=5	K=Log N	Friedman Test
10	2.0	$3.9^{\perp}$	$3.1^{\perp}$	2.9	3.1	Reject
50	3.4	3.5	2.6	2.9	2.7	Accept
100	3.7	3.2	$2.3^{\top}$	2.8	2.9	Reject
200	3.8	3.1	2.7	2.8	2.6	Accept

$\top, \perp$ statistically significant improvement or degradation respect to BRS.

In a second step, we exchanged the roles and tested a Random Forest ensemble with 10 trees and $K = LogN$ (the recommended value by [1]) against different BRS ensembles with different tree sizes and K values. The analogous results are presented for this new analysis in Table 3.

In this new analysis, the trend is much clearer than in the previous case. As can be seen in Table 3, the BRS ensembles now robustly outperform the Random Forest ensembles with different K values and different numbers of trees.

Table 3. Evaluating Random Forests with 10 Trees - Ranking Scores

	$\star RF$ (10 trees)	BRS				
BRS Trees	K=Log N	K=1	K=3	K=5	K=Log N	Friedman Test
10	4.3	3.2	$2.1^{\top}$	$2.6^{\top}$	$2.8^{\top}$	Reject
50	4.9	$2.7^{\top}$	$2.3^{\top}$	$2.5^{\top}$	$2.6^{\top}$	Reject
100	5.0	$2.3^{\top}$	$2.5^{\top}$	$2.7^{\top}$	$2.6^{\top}$	Reject
200	5.0	$2.4^{\top}$	$2.4^{\top}$	$2.7^{\top}$	$2.5^{\top}$	Reject

$\top, \perp$ statistically significant improvement or degradation respect to RF.

The first conclusion seems clear: BRS forests reach a high performance level with a low number of trees and this performance does not depend much upon the concrete K value, as in the case of Random Forests. Throughout the next subsection, we will show how this trend mainly results from a better trade-off between the bias and the variance obtained with the Bayesian random split operator.

Bias-Variance Analysis

Herein we conducted a bias-variance decomposition of the error for both the BRS and RF models. With the aim of simplifying the result analysis, we evaluated the BRS models with $K = 1$. Analyzing the results of the previous section devoted to the role of K in BRS models, we did not find any good reason to prefer a specific K value. The BRS with $K = 1$ appeared to stand out somewhat more than the others.

Following Demsar's methodology [17], Error (Table 4), Bias (Table 5) and Variance (Table 6) were compared between the Bayesian random split operator and Random Forests. It is given the ranking score of each approach as well as wether the Friedman test accepted or rejected the null-hypothesis (all classifier performs equally well). The tests were independently carried out for the different number of trees in the ensembles.

For the error, Table 4, only with 10 trees there are significant differences among the classifiers. In that case, Bonferroni-Dunn Test [17] says that Bayesian random split is significantly better than Random Forests with $K = 1$ (its ranking is marked with $\perp$). For higher number of trees although there are no significant differences, our approach always get the best ranking. For Random Forests, $K = LogN$ seems to be the best option.

Table 4. Error - Ranking Scores

	$\star BRS$	RF				
Trees	K=1	K=1	K=3	K=5	K=Log N	Friedman Test
10	2.0^1	$3.9^\perp$	3.1	2.9	3.1	Reject
50	2.4^1	3.7	2.9	3.1	2.9	Accept
100	2.5^1	3.6	2.7	3.1	3.2	Accept
200	2.4^1	3.5	3.1	3.1	2.9	Accept

$\perp$ indicates this classifier is statistically worst than the respective BRS model.

Table 5 shows the bias evaluation results. As was mentioned in Section 1, Random Forests $K = 1$ have the worst bias and it can be observed in this table. The Bonferroni-Dunn test reveals significant differences of RF $K = 1$ with respect to the BRS. There is no difference with respect to the rest, but the Bayesian random split model clearly shows a better ranking across the different numbers of trees. Although the BRS exhibits a K value fixed to 1, it achieves the best bias. This is a good indication, as the randomness introduction in the split criteria through a Bayesian approach indicates a promising method for further improvements.

Lastly, we evaluate the variance component (Table 6). In this case, the non-parametric test indicates non significant differences among the classifiers, although RF $(K = 1)$ appears to stand out somewhat, with 200 trees.

Table 5. Bias - Ranking Scores

	$\star BRS$	RF				
Trees	K=1	K=1	K=3	K=5	K=Log N	Friedman Test
10	2.5^1	3.8	2.9	3.0	2.9	Accept
50	2.2^1	$3.8^\perp$	3.0	3.1	2.9	Reject
100	2.1^1	$3.8^\perp$	2.8	3.0	3.2	Reject
200	2.3^1	$3.9^\perp$	3.0	3.0	2.7	Reject

$\perp$ indicates this classifier is statistically worst than the respective BRS model.

Table 6. Variance - Ranking Scores

Trees	$\star BRS$ K=1	RF K=1	RF K=3	RF K=5	RF K=Log N	Friedman Test
10	2.3^1	3.5	3.2	3.0	3.0	Accept
50	2.8^1	2.9	3.0	3.2	3.0	Accept
100	2.9	3.0	2.9	3.3	2.8^1	Accept
200	2.8	2.4^1	3.0	3.5	3.2	Accept

Experimental Conclusions

The value of K in Random Forests has been known to affect the performance of the ensembles [1]. In a bias-variance analysis, it was shown [6] that lower K values reduce variance, but increase bias and viceversa. $K = LogN$ seems to present the best trade-off between bias and variance and, in consequence, the best error rate. Our experiments confirm this trend.

This trend is broken with the introduction of more randomness in the split criteria. In BRS ensembles with $K = 1$, the low variance is maintained, while the bias shows a noteworthy decrease. Thus, we achieve the best trade-off between bias and variance.

5   Conclusions and Future Works

In this study, we have presented a new random split operator for building ensembles of classification trees based on Bayesian ideas. We also depicted the method for constructing ensembles of classification trees using this random split through a Bayesian approach.

In an experimental study, we showed that this new split operator does not clearly depend upon the K parameter, like its counterpart of the random forests models, and performs better with a lower number of trees. These advantages were justified with the use of a bias-variance decomposition of the error. In random forests, $K = LogN$ attempts to find a balance between bias and variance. With the Bayesian random split with $K = 1$ presented, the low variance is maintained while the bias is clearly improved.

Under our point of view, this study provides some insights into how to address the building of ensembles of classification trees through a Bayesian approach. Further experiments and, specially, theoretical developments are needed.

References

1. Breiman, L.: Random forests. Mach. Learn. 45(1), 5–32 (2001)
2. Mingers, J.: An empirical comparison of selection measures for decision-tree induction. Mach. Learn. 3(4), 319–342 (1989)
3. Breiman, L.: Bagging predictors. Mach. Learn. 24(2), 123–140 (1996)

4. Dietterich, T.G.: An experimental comparison of three methods for constructing ensembles of decision trees: Bagging, boosting, and randomization. Machine Learning 40(2), 139–157 (2000)

5. Breiman, L., Friedman, J.H., Olshen, R.A., Stone, C.J.: Classification and Regression Trees. Wadsworth International Group, Belmont (1984)

6. Geurts, P., Ernst, D., Wehenkel, L.: Extremely randomized trees. Machine Learning 63(1), 3–42 (2006)

7. Breiman, L.: Arcing classifiers. The Annals of Statistics 26(3), 801–824 (1998)

8. Kohavi, R., Wolpert, D.: Bias plus variance decomposition for zero-one loss functions. In: ICML, pp. 275–283 (1996)

9. Buntine, W.: Learning classification trees. Statistics and Computing (2), 63–73 (1992)

10. Quinlan, J.R.: Induction of decision trees. Mach. Learn. 1(1), 81–106 (1986)

11. Abellán, J., Masegosa, A.R.: Combining decision trees based on imprecise probabilities and uncertainty measures. In: Mellouli, K. (ed.) ECSQARU 2007. LNCS, vol. 4724, pp. 512–523. Springer, Heidelberg (2007)

12. Andrés Cano, A.M., Moral, S.: A bayesian approach to estimate probabilities in classification trees. In: Jaeger, M., Nielsen, T.D. (eds.) Proceedings of the Fourth European Workshop on Probabilistic Graphical Models, pp. 49–56 (2008)

13. Consortium, E.: Elvira: An environment for probabilistic graphical models. In: Gámez, J., Salmerón, A. (eds.) Proceedings of the 1st European Workshop on Probabilistic Graphical Models, pp. 222–230 (2002)

14. Witten, I.H., Frank, E.: Data mining: practical machine learning tools and techniques with Java implementations. Morgan Kaufmann Publishers Inc., San Francisco (2000)

15. Fayyad, U., Irani, K.: Multi-interval discretization of continuous-valued attributes for classification learning. In: Proc. of 13th Int. Joint Conf. on AI (1993)

16. Webb, G.I., Conilione, P.: Estimating bias and variance from data (2006)

17. Demsar, J.: Statistical comparisons of classifiers over multiple data sets. J. Mach. Learn. Res. 7, 1–30 (2006)

HODE: Hidden One-Dependence Estimator

M. Julia Flores, José A. Gámez, Ana M. Martínez, and José M. Puerta

Computing Systems Department, Intelligent Systems and Data Mining group, i3A,
University of Castilla-La Mancha, Albacete, Spain
{julia,jgamez,anamartinez,jpuerta}@dsi.uclm.es

Abstract. Among the several attempts to improve the Naive Bayes
(NB) classifier, the Aggregating One-Dependence Estimators (AODE)
has proved to be one of the most attractive, considering not only the
low error it provides but also its efficiency. AODE estimates the corresponding parameters for every SPODE (Superparent-One-Dependence
Estimators) using each attribute of the database as the superparent, and
uniformly averages them all. Nevertheless, AODE has properties that can
be improved. Firstly, the need to store all the models constructed leads
to a high demand on space and hence, to the impossibility of dealing
with problems of high dimensionality; secondly, even though it is fast,
the computational time required for the training and the classification
time is quadratic in the number of attributes. This is specially significant
in the classification time, as it is frequently carried out in real time. In
this paper, we propose the HODE classifier as an alternative approach
to AODE in order to alleviate its problems by estimating a new variable
(the hidden variable) as a superparent besides the class, whose main
objective is to gather all the dependences existing in the AODE models. The results obtained show that this new algorithm provides similar
results in terms of accuracy with a reduction in classification time and
space complexity.

Keywords: AODE, SPODE, ODE, Bayesian Networks, Bayesian Classifiers, Classification.

1 Introduction

The probabilistic paradigm, more precisely Bayesian classifiers [1], offers important advantages compared to other approaches for classification tasks. Bayesian
classifiers are able to naturally deal with uncertainty and estimate, not only the
label assigned to every object, but also, the probability distribution over the
different labels of the class variable.

NB [2] is the simplest Bayesian classifier and one of the most efficient and
effective inductive algorithms for machine learning. Despite the strong independence assumption between predictive attributes given the class value, it provides
a surprisingly high level of accuracy, even compared to other more sophisticated
models [3]. However, in many real applications it is not only necessary to be
accurate in the classification task, but also to produce a ranking as precise as
possible with the probabilities of the different class values.

C. Sossai and G. Chemello (Eds.): ECSQARU 2009, LNAI 5590, pp. 481–492, 2009.
© Springer-Verlag Berlin Heidelberg 2009

During the last few years, attention has focused on developing NB variants in order to alleviate the independence assumption between attributes, of which so far selective Bayesian classifiers (SBC) [4], tree-augmented naive Bayes (TAN) [5], NBTree [6], boosted naive Bayes [7] and AODE [8] achieve a considerable improvement over naive Bayes in terms of classification accuracy. Specially the group of one dependence estimators (ODEs) [9], such as TAN, provides a powerful alternative to NB. ODEs are very similar to NB but they also allow every attribute to depend on, at most, another attribute besides the class. SPODEs can be considered a subcategory of ODEs where all attributes depend on the same attribute. They have received much attention because of their efficiency in training time and their effectiveness and accuracy in classification [10]. Due to all of these advantages, SPODE classifiers are considered a potential substitute for NB in many real problems such as medical diagnosis, fraud detection, spam filtering, document classification and prefetch of web pages.

Among all these approaches, AODE has come out as the most interesting option due to its capability to improve NB's accuracy with only a slight increase in time complexity (from $\mathcal{O}(n)$ to $\mathcal{O}(n^2)$). An extensive study comparing different semi-naive Bayes techniques [11] proves that AODE is significantly better in terms of error reduction compared to the rest of semi-naive techniques, with the exception of Lazy Bayesian Rules (LBR) [12] and Super-Parent TAN (SP-TAN)[13], which obtain similar results but with a higher time complexity.

Even though AODE is a fast classifier, it is quadratic in training and classification time. This latter issue in particular can be a handicap in many real applications where the response time is critical. Furthermore, the memory required by AODE is quite large due to the necessity to store the n models, n being the number of attributes in the data set. This fact can become a real problem when the size of the database (mainly the number of attributes) is very large.

In this paper, we propose a new classifier which alleviates these two problems by estimating a new variable whose objective is to gather the dependences represented by every superparent in AODE into a single model. The number of states of the new variable as well as the distribution probabilities for the final model are estimated by means of the EM algorithm [14].

This paper is organized as follows: sections 2 and 3 present an overview of NB and AODE classifiers respectively, which are the bases for understanding the new classifier developed. Section 4 provides a detailed explanation of the HODE algorithm. In Section 5, we describe the experimental setup and results. And finally, Section 6 summarizes the main conclusions of our paper and outlines the future work related with the HODE paradigm for classification.

2   Naive Bayes

The problem of classification could be solved exactly using the Bayes theorem in Equation 1. Unfortunately, the direct application of this theorem entails an intractable computational cost in large databases, as it is necessary to estimate the joint probability distribution.

$$p(c|e) = \frac{p(c)p(e|c)}{p(e)} . \tag{1}$$

where c represents the class variable, $e = \{a_1, a_2, \cdots, a_n\}$ the instance example to classify with length n and a_i the value of the ith attribute A_i.

NB estimates the class conditional probability, assuming all the attributes are conditionally independent given the class: $p(e|c) = \prod_{i=1}^{n} p(a_i|c)$. This approach is more feasible, as a large training set is not required to obtain an acceptable probability estimation. Hence, the maximum a posteriori hypothesis (MAP hypothesis) is obtained by Equation 2.

$$c_{MAP} = argmax_{c \in \Omega_C} \ p(c|e) = argmax_{c \in \Omega_C} \left(p(c) \prod_{i=1}^{n} p(a_i|c) \right) . \tag{2}$$

At *training time*, NB has a time complexity $\mathcal{O}(tn)$, where t is the number of training examples. The space complexity is $\mathcal{O}(knv)$ where v is the average number of values per attribute and k the number of classes. The resulting time complexity at *classification time* is $\mathcal{O}(kn)$, while the space complexity is $\mathcal{O}(knv)$.

3   Averaged One-Dependence Estimators

AODE [8] classifiers are considered an improvement of NB and a good alternative to other attempts such as LBR and SP-TAN, as they offer similar accuracy ratios, but AODE is significantly more efficient at classification time compared to the first one and at training time compared to the second.

Back in 1996, Sahami [9] introduced the notion of k-dependence estimators, through which the probability of each attribute value is conditioned by the class and, at most, k other attributes. In order to maintain efficiency, AODE is restricted to exclusively use 1-dependence estimators. Specifically, AODE makes use of SPODEs, as every attribute depends on the class and another shared attribute, designated as superparent.

Considering the MAP hypothesis, the classification of a single example in a **SPODE classifier** (and hence, a 1-dependence ODE) is defined in Equation 3.

$$c_{MAP} = argmax_{c \in \Omega_C} \ p(c|e) = argmax_{c \in \Omega_C} \ p(c, a_j) \prod_{i=1, i \neq j}^{n} p(a_i|c, a_j). \tag{3}$$

The Bayesian network corresponding to an SPODE classifier is depicted in figure 1. In order to avoid selection between models or, in other words, trying to take advantage of all the created models, AODE averages the n SPODE classifiers, which have every different attribute as superparent, as is shown in Equation 4.

$$c_{MAP} = argmax_{c \in \Omega_C} \left(\sum_{j=1, N(a_j) > m}^{n} p(c, a_j) \prod_{i=1, i \neq j}^{n} p(a_i|c, a_j) \right) . \tag{4}$$

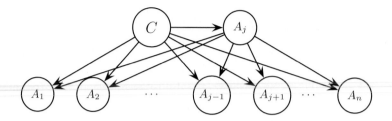

Fig. 1. Generalized structure of SPODE classifiers

The $N(a_j) > m$ condition is used as a threshold to avoid predictions from attributes with insufficient samples, where $N(a_j)$ is a count of the number of training examples having attribute-value a_j and m the limit we place on the support needed in order to accept a conditional probability estimate. If there is not any value which exceeds this threshold, the results are equivalent to NB.

At *training time*, AODE has a $\mathcal{O}(tn^2)$ time complexity, whereas the space complexity is $\mathcal{O}(k(nv)^2)$. The resulting time complexity at *classification time* is $\mathcal{O}(kn^2)$, while the space complexity is $\mathcal{O}(k(nv)^2)$.

4   Hidden One-Dependence Estimators

Just as discussed above, one of the main drawbacks of AODE is the high space cost it entails, as it is necessary to store all the SPODE models in main memory. In order to alleviate these large memory requirements, we suggest the estimation of a new variable, specifically, a **hidden variable** H, which gathers the suitable dependences among the different superparents and the rest of the attributes. In other words, instead of averaging the n SPODE classifiers, a new variable is estimated in order to represent the links existing in the n models. In [15], the authors try to improve NB by introducing hidden variables as well.

In this case, we have to estimate the probability of every attribute value conditioned by the class and the new variable which plays the superparent role. Figure 2(a) shows the structure of the Bayesian network to learn. The aim is not to search exactly the same AODE structure (Figure 2(b)), but the class values become the Cartesian product of the original class values and the estimated states for H ($\#H$).

Equation 5 shows the MAP hypothesis for the HODE algorithm. Each h_j represents the jth virtual value for H.

$$c_{MAP} = argmax_{c \in \Omega_C} \; p(c|e) = argmax_{c \in \Omega_C} \left(\sum_{j=1}^{\#H} p(c, h_j) \prod_{i=1}^{n} p(a_i|c, h_j) \right).$$

(5)

The following two sections explain how to adapt the EM algorithm to estimate the probability distributions of the model, and the technique used to find out the most suitable number of states for H.

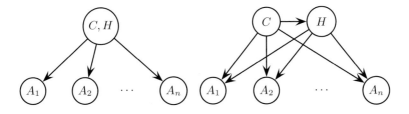

Fig. 2. HODE classifier possible structures

4.1   Application of the EM Algorithm

As the different values for H are not known, we make use of the EM algorithm to obtain the maximum likelihood estimation of the parameters, its use being quite common in this kind of approaches [16],[17]. Algorithm number 1 shows the detailed process followed by the straight adaptation of the EM algorithm to this problem. Until convergence is reached, in the Expectation step (E-step), the conditional probability tables (CPTs) are constructed using the weights estimated in the Maximization step (M-step). These weights are, in turn, estimated according to the attribute values, the class value and the corresponding label assigned to H.

Algorithm 1. EM algorithm adaptation to HODE

1: Random initialization of weights;
2: {**EM ALGORITHM**}
3: **while** (!$convergence()$) **do**
4:     {**E-STEP**}
5:     Update prob. according to weights
6:     {**M-STEP**}
7:     **for** ($j = 1$ to $j = numInstances$) **do**
8:         **for** ($s = 0$ to $s < \#H$) **do**
9:             $w_{\{c,h_s,a_i,\cdots,a_n\}_j} = P(c,h_s)P(a_1|c,h_s)\cdots P(a_n|c,h_s)$;
10:            Normalize $\boldsymbol{w}$;
11:        **end for**
12:    **end for**
13: **end while**

In EM, the database is virtually divided according to the following procedure: we divide every instance into $\#H$ virtual instances. Each one of the subinstances corresponds to a different value of H and a weight reflecting its likelihood $(w_{\{c,h_s,a_i,\cdots,a_n\}_j})$. At the beginning, these weights are randomly initialized ($\boldsymbol{w}$ vector), considering that the sum of weights from a common instance has to be equal to 1. Table 1 shows a virtual division example of a small database.

An example of how to adapt the said EM algorithm, is described below. For the database in Table 1, the probabilities shown in Figure 3 are obtained in every E-step.

Table 1. Virtual division example of the database with $H = \{h_1, h_2\}$

A	B	C	H	w
a	b	c	h_1	0.3
			h_2	0.7
a	$\bar{b}$	$\bar{c}$	h_1	0.5
			h_2	0.5
$\bar{a}$	$\bar{b}$	c	h_1	0.9
			h_2	0.1
a	b	c	h_1	0.6
			h_2	0.4
$\bar{a}$	b	$\bar{c}$	h_1	0.7
			h_2	0.3
a	$\bar{b}$	c	h_1	0.2
			h_2	0.8

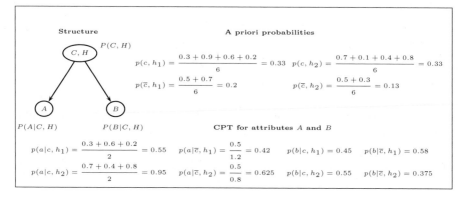

Fig. 3. Count of database weights to obtain the CPTs (E-step)

The adaptation of the M-step to our problem consists in the estimation of the corresponding weights of the virtual instances from the probabilities estimated in the previous step. Equations in Figure 4 show how the M-step is carried out for the first instance in our example. Once this M-step is finished for all the instances, the following generation of weights is depicted in the table on the right-hand side.

Finally, the following E-step would use the $\boldsymbol{w_2}$ vector weight and the cycle would continue until the algorithm converges, in other words, until the weight difference from adjacent iterations is lower than 5 thousandths.

4.2   Number of States for the H Variable

Even though the graphical structure is already fixed, we still have to perform certain structural learning in order to find the inner structure of H, in other words, its cardinality or number of states. To achieve this, we make use of the following greedy technique: firstly, $\#H$ is fixed to 1 (base case equivalent to naive Bayes), the EM algorithm is executed and the model built is evaluated; after that, the number of states for H is increased one by one in every iteration of the EM algorithm. If the result of the evaluation of one model is better than

		A B C	H	w_1	w_2
		a b c	h_1	0.3	0.32
			h_2	0.7	0.68
		a $\bar{b}$ $\bar{c}$	h_1	0.5	0.41
			h_2	0.5	0.59
		$\bar{a}$ b c	h_1	0.9	0.92
			h_2	0.1	0.08
		a b c	h_1	0.6	0.32
			h_2	0.4	0.68
		$\bar{a}$ b $\bar{c}$	h_1	0.7	0.79
			h_2	0.3	0.21
		a $\bar{b}$ c	h_1	0.2	0.41
			h_2	0.8	0.59

$$p(c, h_1 | a, b) = \frac{p(c, h_1)p(a|c, h_1)p(b|c, h_1)}{\sum_{i=1}^{H} (p(c, h_i)p(a|c, h_i)p(b|c, h_i))} \qquad = \frac{0.33 \cdot 0.55 \cdot 0.45}{0.254} = 0.32 \tag{6}$$

$$p(c, h_2 | a, b) = \frac{p(c, h_2)p(a|c, h_2)p(b|c, h_2)}{\sum_{i=1}^{H} (p(c, h_i)p(a|c, h_i)p(b|c, h_i))} \qquad = \frac{0.33 \cdot 0.95 \cdot 0.55}{0.254} = 0.68 \tag{7}$$

Weights count — *Weights modification after M-step*

Fig. 4. Count of database weights (M-step)

the one in the previous iteration, the process continues, otherwise, the previous model is restored and considered the final model.

How is the fitness of the model evaluated? We employ the log-likelihood (LL) meassure to find how the mathematical model estimated fits the training data. Equation 6 shows the formula used.

$$LL = \sum_{i=1}^{I} \log \left(\sum_{t=1}^{\#H} p(c^i, a_1^i, \cdots, a_n^i, h_t) \right) = \sum_{i=1}^{I} \log \left(\sum_{t=1}^{\#H} p(c^i, h_t) \prod_{r=1}^{n} p(a_r^i | c^i, h_t) \right). \tag{6}$$

where I is the number of instances and the superscript i indicates the class or the attribute value that correspond with the ith instance.

Nevertheless, when we use these measures, it is also necessary to add another quality measure to counteract the monotonous feature of the LL. In other words, it is necessary to somehow penalize, the increase in the number of states for H. To do this, we firstly carried out experiments with the Minimum Description Length (MDL) measure, for which the model complexity is computed in Equation 7.

$$C(M) = \sum_{i=1}^{n} ((\#H \cdot \#C)(\#A_i - 1)) + \#H \cdot \#C - 1. \tag{7}$$

where $\#C$ is the number of classes and $\#A_i$ the number of states of A_i.

Thus, the MDL measure could be defined as in Equation 8:

$$MDL = LL - \frac{1}{2} \log I \cdot C(M). \tag{8}$$

Later, we designed a different way to penalize the LL, which consisted in using information measures with the basic idea of selecting the model which best fits the data, penalizing according to the number of parameters needed to specify its corresponding probability distribution. Specifically, we tested the so-called

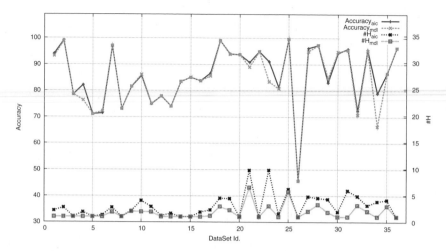

Fig. 5. Accuracy and $\#H$ obtained with AIC and MDL penalization in HODE

Akaike Information Criterion or **AIC** [18], which turns out to be equal to the previous one but removing the $\frac{1}{2} \log I$ factor.

$$AIC = LL - C(M). \tag{9}$$

From our experiments testing these two measures, AIC is the one that testing a smoother penalization over LL and hence, it achieved better results as it explores more states of H (this is in concordance with [16], where large cardinalities are used in order to achieve good modeling). The graph in Figure 5 shows the comparison between accuracy results using both penalty measures. The left-hand Y-axis represents accuracy results (upper pair of lines) whereas the right-hand Y-axis represents the average $\#H$ obtained in the evaluation of each dataset (lower pair of lines).

From now on in this paper, we will refer to HODE with the AIC measure as HODE.

5   Experimental Methodology and Results

This section presents the experimental results of HODE compared to the AODE classifier.

First, in Section 5.1 we study the accuracy results obtained for 36 UCI repository datasets [19], whose main characteristics are summarized in Table 2; whereas Section 5.2 is devoted to the study of the performance of HODE in terms of efficiency.

5.1   Evaluation in Terms of Accuracy

Table 3 shows the classification accuracy of both classifiers (AODE and HODE) on each dataset obtained via 10 runs of ten-fold cross validation. Each value

Table 2. Main characteristics of the datasets: number of different values of the class variable (k), number of predictive variables (n), and number of instances (I)

Id.	Dataset	k	n	I	Id.	Dataset	k	n	I
1	anneal.ORIG	6	38	898	19	ionosphere	2	34	351
2	anneal	6	38	898	20	iris	3	4	150
3	audiology	24	69	226	21	kr-vs-kp	2	36	3196
4	autos	7	25	205	22	labor	2	16	57
5	balance-scale	3	4	625	23	letter	26	16	20000
6	breast-cancer	2	9	286	24	lymph	4	18	148
7	breast-w	2	9	699	25	mushroom	2	22	8124
8	colic.ORIG	2	27	368	26	primary-tumor	21	17	339
9	colic	2	27	368	27	segment	7	19	2310
10	credit-a	2	15	690	28	sick	2	29	3772
11	credit-g	2	20	1000	29	sonar	2	60	208
12	diabetes	2	8	768	30	soybean	19	35	638
13	glass	6	10	214	31	splice	3	61	3190
14	heart-c	2	13	303	32	vehicle	4	18	846
15	heart-h	2	13	294	33	vote	2	16	435
16	heart-statlog	2	13	270	34	vowel	11	13	990
17	hepatitis	2	19	155	35	waveform-5000	3	40	5000
18	hypothyroid	4	29	3772	36	zoo	7	17	101

Table 3. Accuracy results obtained with AODE and HODE classifiers

Dataset	AODE	HODE	#H	Dataset	AODE	HODE	#H
anneal.ORIG	93,3185	●94,0646	2, 2	ionosphere	92,9915	●93,9886	4, 4
anneal	98,196	●99,1203	2, 8	iris	93,2	●93,7333	1
audiology	71,6372	●78,5841	1	kr-vs-kp	**91,0325**	90,8229	9, 7
autos	81,3658	●82,0975	1, 9	labor	**95,0877**	94,9123	1
balance-scale	69,344	●71,088	1	letter	88,902	●91,117	9, 8
breast-cancer	●72,7273	71,4336	1, 3	lymph	●87,5	81,1487	1, 5
breast-w	96,9671	96,9814	2, 8	mushroom	●99,9508	99,6824	6, 2
colic.ORIG	●75,9511	73,0707	1	primary-tumor	●47,8761	45,7227	1
colic	●82,5543	81,5489	2, 1	segment	95,7792	●96,1732	4, 8
credit-a	●86,5507	85,5942	4, 1	sick	●97,3966	97,3118	4, 6
credit-g	●76,33	74,94	2, 9	sonar	●86,5865	83,0769	4, 3
diabetes	●78,2292	77,8516	1, 2	soybean	93,3089	●94,3631	1, 9
glass	●76,2617	74,0187	1, 6	splice	●96,116	95,8872	3, 9
heart-c	83,2013	●83,4323	1	vehicle	72,3049	**72,3522**	4, 9
heart-h	84,4898	**85,0**	1	vote	94,5288	●95,5173	3, 1
heart-statlog	82,7037	●83,7037	1, 9	vowel	●80,8788	79,0101	3, 9
hepatitis	85,4839	●86,6452	2, 3	waveform-5000	86,454	**86,54**	4, 2
hypothyroid	98,7513	●99,0668	4, 5	zoo	94,6535	●96,2376	1

represents the arithmetical mean from the 10 executions. The bullet next to certain outputs means that the corresponding classifier on this particular dataset is significantly better than the other classifier. The results were compared using a two-tailed t-test with a 95% confidence level.

In 16 of the 36 databases, HODE is significantly better than AODE, whereas AODE outperforms HODE in 14 of them. They draw in 6 of them, hence 16/6/14, where the notation $w/t/l$ means that HODE wins in w datasets, ties in t datasets, and loses in l datasets, compared to AODE. The results undergo no variation when the confidence level is raised to 99%, obtaining 15/8/13.

On the other hand, although it is not shown in the tables, we also studied HODE with the MDL penalization, and observed that it was significantly better than AODE in 11 of the 36 datasets, drew in 7 of them, and lost in 18 of them (11/7/18).

5.2   Evaluation in Terms of Efficiency

As there is not a clear difference in terms of accuracy between the two classifiers, what could make us vote for one or the other? In fact, HODE's time complexity at *training time* is quadratic in the worst case ($1tn + 2tn + \cdots + ntn$, considering the different executions of the EM algorithm). However, AODE is usually faster than HODE in model construction, as HODE spends more time executing the EM algorithm to find the most suitable $\#H$, increasing this time as $\#H$ increases.

With respect to *classification time*, HODE's is linear, whereas AODE's is quadratic. Figure 6 shows the experimental classification times obtained, which corroborate this theoretical study. Note that in most real applications, it is essential that classification time is as short as possible, as model training can usually be performed offline. For example, consider spam detection in mail, the recommendation of a specific product according to previous purchases, interpretation of characters by an OCR tool, determining the folder for a certain e-mail, etc.

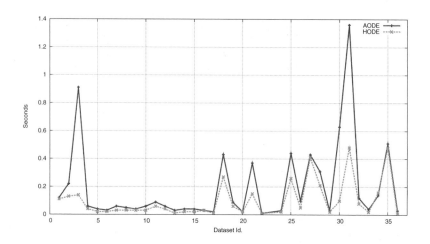

Fig. 6. Classification time comparison between AODE and HODE

Furthermore, space complexity for AODE is higher than HODE's, as the former needs to store more CPTs. In fact, HODE's is $\mathcal{O}(n\#Hvk)$, where $\#H$ is usually much lower than n. This requirement leads to a higher demand on RAM memory which could be a big problem in large databases with a high number of attributes, such as microarrays or DNA chips. To corroborate this fact, we have experimented with a group of 7 databases of this type (see left part of table 4)

Table 4. Main characteristics of the datasets (number of different values of the class variable (k), number of genes (n), and number of microarrays (I)); and accuracy results obtained with NB, AODE and HODE classifiers in these datasets

Dataset	k	n	I	NB	AODE	HODE
colon	2	2000	2	93, 5484	91, 9355	**96, 7742**
DLBCL-Stanford	2	4026	47	100	100	100
GCM	14	4026	47	60, 5263	OutOfMem	**70**
leukemia	2	7129	72	**100**	OutOfMem	98, 6111
lungCancerHarvard2	175	12533	181	98, 895	OutOfMem	**99, 4475**
lymphoma	9	4026	96	**96, 875**	OutOfMem	75
prostate_tumorVS	2	12600	136	80, 1471	OutOfMem	**95, 5882**

and AODE had problems of overflow with a maximum of 8 gigabytes of memory available, while HODE terminated its executions without problems, even with a lower need for memory.

6   Conclusions and Future Work

In this paper, we have proposed an alternative to the AODE classifier named HODE, which provides a linear order in classification time and a reduction in space complexity as well. This leads to a lower time response in many real applications and a lower RAM consumption. Basically, HODE estimates a new variable whose main objective is to model the dependences between each attribute and the rest of the attributes that AODE takes into account. In order to estimate the number of states of this new variable, we make use of the EM algorithm, evaluating the fitness for every model with a greedy technique (LL).

So far, we have proved the good performance of this basic but innovative idea. We have already tested the good performance of HODE in a parallel environment, as we are able to find a global optimum for $\#H$. An additional advantage of our proposal is the direct adaptation to work with missing values in the dataset, due to the use of EM in its main cycle. Furthermore, we believe that there can be lots of variations to this idea that could provide even better performance and a clear alternative to AODE in many real applications.

Acknowledgments. This work has been partially supported by the Consejería de Educación y Ciencia (JCCM) under Project PCI08-0048-8577574, the Spanish Ministerio de Educación y Tecnología under Project TIN2007-67418-C03-01 and the FPU grant with reference number AP2007-02736.

References

1. Langley, P., Iba, W., Thompson, K.: An Analysis of Bayesian Classifiers. In: 10th national conference on artificial intelligence, pp. 223–228. AAAI Press, Menlo Park (1992)
2. Duda, R.O., Hart, P.E.: Pattern Classification and Scene Analysis. John Wiley & Sons Inc., Chichester (1973)

3. Domingos, P., Pazzani, M.: Beyond independence: Conditions for the optimality of the simple Bayesian classifier. In: 13th International Conference on Machine Learning, pp. 105–112. Morgan Kaufmann, Italy (1996)
4. Langley, P., Sage, S.: Induction of selective Bayesian classifiers. In: 10th Conference on Uncertainty in Artificial Intelligence, pp. 399–406. Morgan Kaufmann, San Francisco (1994)
5. Friedman, N., Geiger, D., Goldszmidt, M.: Bayesian Network Classifiers. J. Mach. Learn. 29(2-3), 131–163 (1997)
6. Kohavi, R.: Scaling Up the Accuracy of Naive-Bayes Classifiers: a Decision-Tree Hybrid. In: 2nd International Conference on Knowledge Discovery and Data Mining, pp. 202–207 (1996)
7. Elkan, C.: Boosting and Naive Bayesian learning, Technical report, Dept. Computer Science and Eng., Univ. of California, San Diego (1997)
8. Webb, G.I., Boughton, J.R., Wang, Z.: Not So Naive Bayes: Aggregating One-Dependence Estimators. J. Mach. Learn. 58(1), 5–24 (2005)
9. Sahami, M.: Learning limited dependence Bayesian classifiers. In: 2nd International Conference on Knowledge Discovery in Databases, pp. 335–338. AAAI Press, Menlo Park (1996)
10. Yang, Y., Korb, K., Ting, K.-M., Webb, G.I.: Ensemble Selection for SuperParent-One-Dependence Estimators. In: Zhang, S., Jarvis, R. (eds.) AI 2005. LNCS, vol. 3809, pp. 102–112. Springer, Heidelberg (2005)
11. Zheng, F., Webb, G.I.: A Comparative Study of Semi-naive Bayes Methods in Classification Learning. In: Simoff, S.J., Williams, G.J., Galloway, J., Kolyshkina, I. (eds.) 4th Australasian Data Mining Conference (AusDM 2005), pp. 141–156. University of Technology, Sydney (2005)
12. Zheng, Z., Webb, G.I.: Lazy Learning of Bayesian Rules. J. Mach. Learn. 41(1), 53–84 (2000)
13. Keogh, E., Pazzani, M.: Learning Augmented Bayesian Classifiers: A Comparison of Distribution-based and Classification-based Approaches. In: 7th Int'l Workshop on AI and Statistics, Florida, pp. 225–230 (1999)
14. Dempster, A., Laird, N.M., Rubin, D.B.: Maximum Likelihood from Incomplete Data via the EM Algorithm. J. R. Stat. Soc. Ser. B-Stat. Methodol. 39, 1–38 (1977)
15. Langseth, H., Nielsen, T.D.: Classification using hierarchical Naïve Bayes models. J. Mach. Learn. 63(2), 135–159 (2006)
16. Lowd, D., Domingos, P.: Naive Bayes models for probability estimation. In: 22nd international conference on Machine learning, pp. 529–536. ACM, Bonn (2005)
17. Cheeseman, P., Stutz, J.: Bayesian classification (AutoClass): theory and results. In: Advances in knowledge discovery and data mining, pp. 153–180. AAAI Press, Menlo Park (1996)
18. Akaike, H.: A Bayesian analysis of the minimum AIC procedure. Ann. Inst. Statist. Math. 30A, 9–14 (1978)
19. Collection of Datasets available from the Weka Official HomePage, University of Waikato, http://www.cs.waikato.ac.nz/ml/weka/

On the Effectiveness of Diversity When Training Multiple Classifier Systems

David Gacquer, Véronique Delcroix, François Delmotte,
and Sylvain Piechowiak

Univ Lille Nord de France, F-59000 Lille, France
UVHC, LAMIH F-59313 Valenciennes, France
{david.gacquer,veronique.delcroix,francois.delmotte,
sylvain.piechowiak}@univ-valenciennes.fr
http://www.univ-valenciennes.fr/LAMIH

Abstract. Discussions about the trade-off between accuracy and diversity when designing Multiple Classifier Systems is an active topic in Machine Learning. One possible way of considering the design of Multiple Classifier Systems is to select the ensemble members from a large pool of classifiers focusing on predefined criteria, which is known as the Overproduce and Choose paradigm. In this paper, a genetic algorithm is proposed to design Multiple Classifier Systems under this paradigm while controlling the trade-off between accuracy and diversity of the ensemble members. The proposed algorithm is compared with several classifier selection methods from the literature on different UCI Repository datasets. This paper specifies several conditions for which it is worth using diversity during the design stage of Multiple Classifier Systems.

Keywords: Supervised Classification, Multiple Classifier Systems, Diversity, Genetic Algorithm, Classifier Selection.

1 Introduction

Many advances in Machine Learning suggest using a set of individual classifiers, or Multiple Classifier System, instead of a single predictor to address supervised classification problems [16, 7]. A large number of studies show that Multiple Classifier Systems generally achieve better results compared to a single classifier in terms of misclassification error [8, 20]. This improvement of performances relies on the concept of diversity [15, 5] which states that a good classifier ensemble is an ensemble in which the examples that are misclassified are different from one individual classifier to another. The comparison between Boosting [10] and Bagging [3] approaches is one of the most significant illustration of the concept of diversity and shows that a set of weak learners specialized on different hard examples often produces an accurate classifier ensemble [23].

However, it appears that using diversity explicitly during the design stage of Multiple Classifier Systems does not always give the expected performances in terms of misclassification error [14]. The objective of this paper is to clarify

C. Sossai and G. Chemello (Eds.): ECSQARU 2009, LNAI 5590, pp. 493–504, 2009.

the effectiveness of using diversity explicitly when training a classifier ensemble under the Overproduce and Choose approach. To do so, we propose a genetic algorithm to perform classifier selection from an initial set of decision trees while focusing on the trade-off between accuracy and diversity of the resulting classifier ensemble.

The remainder of this paper is organized as follows. The next section briefly introduces the context of supervised classification and discusses about how Multiple Classifier Systems address this problem in Machine Learning. The concept of diversity between classifiers and its elusive behaviour are also introduced. Next, we present a genetic algorithm to design Multiple Classifier Systems according to the Overproduce and Choose paradigm. This algorithm selects the ensemble members from an initial pool of decision trees obtained with the Adaboost algorithm while focusing on the trade-off between accuracy and diversity of the resulting classifier ensemble. The proposed algorithm is compared with several classifier selection methods proposed in the literature on different datasets from the UCI Repository. Finally, we present the different conclusions we draw from the obtained results and propose possible improvements of the research work presented in this paper.

2   Diversity in Multiple Classifier Systems

2.1   Multiple Classifier Systems: Overview

Given a set of N samples $Z = \{(X_1, Y_1), (X_2, Y_2), \ldots (X_N, Y_N)\}$ where $X_i \in \Re^p$ is an input vector and $Y_i \in \Omega = \{\omega_1, \omega_2, \ldots \omega_c\}$ is the class label of X_i, the supervised classification problem consists in learning a classifier $C : \Re^p \mapsto \Omega$ from the available examples to automatically assign the corresponding class label of new input vectors [19, 9]. Machine Learning provides numerous solutions to address this problem, from neural networks [12] or decision trees [22] to support vector machines [6].

Current research works suggest using simultaneously several classifiers and combining their individual decisions to reduce the number of misclassified samples. This particular framework called Multiple Classifier System (or classifier ensemble) is given in figure 1.

Unlike binary decomposition schemes that are necessary for multi-class support vector machines for instance [1], Multiple Classifier Systems are composed of different classifiers trained to solve the same problem. The differences between these individual classifiers, that we will refer to with the term *diversity* from now on, can be introduced at different levels of the design stage of the ensemble, as shown in figure 1. The *data* level is composed of various techniques that focus on the training data, like feature selection or resampling techniques of the original training set. The *classifier* level mainly concerns manipulations of architectures of the individual classifiers in the ensemble. Finally, the *supervisor* level corresponds to the different combination rules used for the fusion of label outputs. Brown *and al.* [5] define a possible taxonomy of techniques to generate

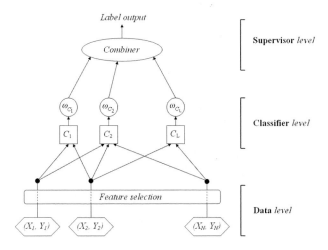

Fig. 1. General description of Multiple Classifier Systems

diversity between classifiers, on which rely the various ensemble learning methods proposed in the literature, like the well known Bagging [3] and Boosting [10] techniques, random forests of decision trees [4, 13] and more recently the Rotation Forest [24] or DECORATE [18] algorithms. Most of these methods enforce diversity in an implicit or explicit manner during the design stage of the ensemble to reduce the number of misclassified samples.

2.2 Ensembles of Diverse Classifiers

A large number of studies show that Multiple Classifier Systems generally achieve better performances in terms of misclassification error compared to a single classifier [8, 20]. The effectiveness of this framefork relies on the assumption that ensemble of classifiers which exhibit a certain diversity are generally more accurate than ensemble of classifiers having a similar behaviour. Two classifiers C_i and C_j are said *diverse* if they assign different class labels to the same examples. Various measures have been proposed to quantify the diversity between two classifiers from their respective outputs [14]. Margineantu and Dietterich [17] propose to use kappa error diagrams as a tool to observe the relation of ensemble learning algorithms with diversity. An example of such diagram is given figure 2.

Each dot in the diagram corresponds to a pair of classifiers $C_i C_j$ defined by its pairwise diversity $\kappa_{i,j}$ and its pairwise error $e_{i,j} = \frac{e_i + e_j}{2}$. An ensemble of L classifiers obtained with a given algorithm is represented by the scatterplot of the $\frac{L(L-1)}{2}$ possible pairs of classifiers within the ensemble. The scatterplot shape shows how ensemble algorithms interact with diversity. For instance, the Boosting technique, which focus the training of the individual classifiers on hard examples to enforce diversity, produces ensemble of classifiers generally less

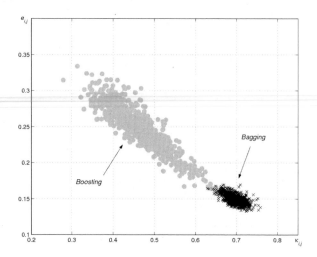

Fig. 2. Example of Kappa Error diagram used to compare ensemble of classifiers obtained with Bagging and Boosting

accurate but more diverse compared to the Bagging algorithm whose scatterplot is concentrated in the lower right part of the kappa error diagram. Moreover, figure 2 shows the existence of a trade-off between accuracy and diversity in classifier ensembles, the classifiers which exhibits important diversity being generally the ones that possess the lowest accuracy. Kappa error diagrams make a passive use of diversity. In the next section, we briefly describe methods that make an active use of diversity during the design stage of classifier ensembles.

2.3 Using Diversity to Train Classifier Ensembles

Comparisons between Bagging and Boosting show that, in most cases and given noise free datasets, Boosting always produces more accurate classifier ensembles [20, 8]. If we refer to the kappa error diagram given in figure 2, we can make the assumption that weakening the individual accuracy of the ensemble members to enforce diversity generally brings some benefits when the ensemble is used as a whole using voting methods for instance. Following this assumption, numerous research works were proposed to train diverse classifier ensembles.

The different methods that make an active use of diversity can be divided into two category: those that enforce diversity during training of the individual classifiers of the ensemble and those which use diversity to perform classifier selection from a large pool of classifiers. The Adaboost algorithm for instance belongs to the first category. A given number of classifiers are trained iteratively on different weights distributions over the training set and the ensemble is built in an ascending manner by adding iteratively those different classifiers. On the contrary, the methods that belong to the second category build the ensemble in a descending manner, by selecting the ensemble members that best satisfy certain criteria. This general method is called the Overproduce and Choose paradigm and regroups different classifier selection heuristics used to prune a large pool of classifiers [17, 25].

However, it appears that using diversity explicitely during the design stage of Multiple Classifier Systems does not always give the expected performances in terms of misclassification error and Kuncheva points out this elusive behaviour of diversity in classifier ensembles [14]. To answer some of the questions that remain about the use of diversity when training a classifier ensemble, we propose in the next section a genetic algorithm to study the trade-off between accuracy and diversity when training a classifier ensemble under the Overproduce and Choose paradigm.

3 The Proposed Contribution

3.1 Overview of the Overproduce and Choose Paradigm

The Overproduce and Choose paradigm, also called Ensemble Pruning, consists in selecting the ensemble members from a pool of individual classifiers of important size. There is no particular definition for the Overproduce step of this approach. Margineantu and Dietterich [17] use the Adaboost algorithm to train an ensemble of decision trees which is pruned with different selection heuristics. Roli *and al.* [25] propose to prune a pool of classifiers containing neural networks with different number of hidden nodes and nearest neighbor classifiers.

The selection step mainly relies on the use of various classifier selection methods based on different criteria. The *Choose Best* heuristic described in [21] consists in selecting the L^* classifiers which possess the highest individual accuracy. This heuristic does not take into account diversity. On the contrary, the *Kappa Pruning* heuristic [17] retrieves the L^* classifiers which belong to the pairs of classifiers of highest diversity. The previous selection methods do not consider the trade-off between accuracy and diversity. To solve this limitation, Margineantu and Dietterich [17] suggest pruning the initial pool of classifiers using the convex hull of the kappa error diagram to select simultaneously the most diverse (and less accurate) classifiers and the most accurate (and less diverse) classifiers. A similar pruning method is proposed by Kuncheva [14] to select the individual classifiers which belong to the Pareto front of the kappa error diagram. An illustration of these two pruning methods is given in figure 3.

Roli *and al.* [25] also propose several forward and backward search methods as well as clustering algorithms and tabu search to perform classifier selections. Most of the methods proposed in the literature are based on a single parameter to select the ensemble members. Search methods based on the convex hull or the Pareto front of the kappa error diagram consider both diversity and accuracy to train the ensemble but with a greedy strategy which does not allow to control the number of selected classifiers. To solve these limitations and study the trade-off between accuracy and diversity that is visible in kappa error diagrams, we propose a genetic algorithm which allows a direct control of the importance given to the diversity and accuracy components of classifier ensembles.

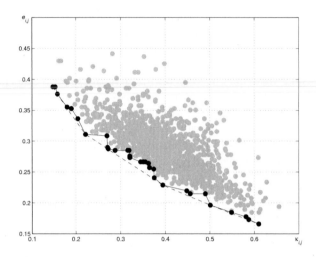

Fig. 3. Kappa error diagram obtained for the *glass* dataset from the UCI Repository. The dotted and solid lines correspond respectively to the classifier ensembles obtained with the kappa convex hull pruning and Pareto pruning heuristics.

3.2 The GenDiv Algorithm

We propose a genetic algorithm called GenDiv [11] to design Multiple Classifier Systems under the Overproduce and Choose paradigm. Given a pool of decision trees obtained with the Adaboost algorithm, a population of P classifier ensembles is obtained by randomly selecting L^* decision trees. Each solution is then defined by a chromosome consisting of the different individual classifiers that form the ensemble. To study the trade-off between accuracy and diversity, we define the following fitness function to evaluate each classifier ensemble:

$$Fitness(E) = \alpha \times Acc(E) + (1 - \alpha) \times Div(E) \qquad (1)$$

where $Acc(E)$ and $Div(E)$ correspond respectively to the accuracy and diversity of the classifier ensemble E computed on validation samples. In this paper, we use the kappa value between two classifiers to quantify their diversity. With this metric, two classifiers are diverse if their kappa value is low. The overall diversity $Div(E)$ of a given classifier ensemble E is defined by:

$$Div(E) = 1 - \frac{2}{L^*(L^* - 1)} \sum_{i=1}^{L^*-1} \sum_{j=i+1}^{L^*} \kappa_{E(i),E(j)} \qquad (2)$$

where $E(i)$ and $E(j)$ correspond respectively to the classifiers C_i and C_j in the ensemble E. The accuracy $Acc(E)$ of the ensemble is also computed respectively to a validation set using the usual weighted majority vote proposed for the Adaboost algorithm. In the fitness function given previously, a value greater than 0.5 for the α parameter will focus the classifier selection on the accuracy of the ensemble. On the contrary, a value of α lower than 0.5 will enforce diversity during the selection process. The α parameter is used to study the trade-off between accuracy and diversity when training classifier ensembles.

Ensembles that maximize the fitness function are selected to breed a new population of P classifier ensembles of size L^* using cross-over and mutation mechanisms. Two distinct ensembles exchange some of their individual classifiers to form new candidate solutions to be evaluated. This process is repeted during a given number of iterations and the ensemble which possesses the highest fitness in the final population is returned by the algorithm. In the next section, we compare the performances of the GenDiv algorithm with different classifier selection methods to investigate the effectiveness of diversity.

4 Experiments

4.1 Experimental Material

The proposed genetic algorithm is compared with several classifier selection methods described in the previous section of this paper on different UCI datasets [2]. The caracteristics of those datasets are reported in table 1. For each dataset, we give respectively the number of classes, the number of features and the number of available samples. The last columns correspond respectively to the sizes of the training set, validation set and test set used during our experiments.

Table 1. UCI datasets used to compare the different classifier selection methods

dataset	#classes	#samples	#features	#training	#validation	#test
letter	26	20000	16	2000	500	500
segment	7	2310	19	1310	500	500
german-credit	2	1000	20	600	200	200

The training set is used during the overproduce step of our experiments to train 50 decision trees using the Adaboost.M1 algorithm. This step is common for all the classifier selection methods that we have implemented. The validation set is used to perform classifier selection using different selection methods and the accuracy of the resulting classifier ensemble is computed on the test set consisting of previously unseen examples.

For each dataset, we compare the accuracy of the classifier ensembles obtained with the different classifier selection methods described in the previsous section: *Choose Best, Kappa Pruning, Convex Hull Pruning, Pareto Pruning* and *Early Stopping*. We also use a *Random Selection* method as a reference to determine how well perform these heuristics respectively to a classifier ensemble whose members are randomly chosen. The proposed GenDiv algorithm considers both accuracy and diversity when selecting the ensemble members and the trade-off between these two components is controlled by the α parameter in the fitness function. Since it is commonly assumed that optimizing diversity alone is a poor strategy when designing classifier ensembles, we set the value of α to 0.8 to increase the importance of the accuracy component for the experiments presented in this section.

We use the protocol proposed by Margineantu and Dietterich [17] to study the behaviour of the different classifier selection heuristics. Performances are computed in terms of gain respectively to a single classifier and the initial pool of decision trees obtained during the overproduce step of our implementation. A gain equal to 1 indicates that the resulting classifier ensemble is as accurate as the initial pool of classifiers. On the contrary, a value of 0 for the gain indicates that the accuracy of the pruned ensemble is similar to the one of a single decision tree. The curves presented below plot the evolution of the gain obtained with the different selection heuristics for different pruning levels. When it is possible to control the size of the pruned ensemble, we construct ensembles consisting of 40, 30, 20 and 10 decision trees. This corresponds to pruning levels respectively equal to 20%, 40%, 60% et 80%. Values given for a pruning level of 0% correspond to the accuracy of the initial pool of classifiers whereas a pruning level of 100% corresponds to the accuracy obtained with a single decision tree. The different plots presented below are obtained by averaging results over 30 different training-validation-test splits of the available datasets.

4.2   Results and Discussions

Results obtained for the *segment* dataset are given in figure 4.

The initial pool of classifiers can be pruned until 60% while maintaining a gain close to 1. The proposed GenDiv algorithm even exhibits a gain slightly superior to 1 for pruning levels of 20% and 40 %. However, for levels of pruning superior to 60%, the reduction of accuracy for the pruned ensemble is more significant if we

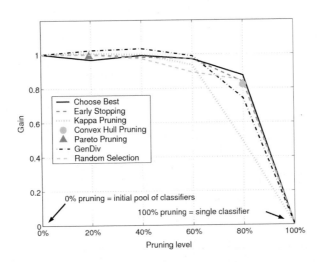

Fig. 4. Comparison of classifier selection methods for the *segment* dataset

focus on diversity to perform classifier selection. The random selection method can even produce ensembles that are more accurate than the one obtained with a selection algorithm which only rely on diversity. For small size ensembles, a selection based on the individual accuracy of the ensemble members remains the best available option, as shown by the performances achieved by the *Choose Best* heuristic.

Similar observations can be made from the results obtained for the *letter* dataset reported in figure 5.

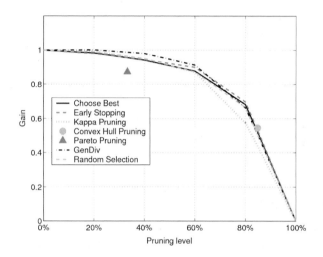

Fig. 5. Comparison of classifier selection methods for the *letter* dataset

Until a pruning level of 60%, the GenDiv algorithm allows the construction of classifier ensembles which are generally more accurate. However, when the desired size for the pruned ensemble decreases, a selection based on individual accuracy alone is more relevant than the proposed genetic algorithm.

We present in figure 6 the results obtained for the *german-credit* dataset.

Observations of these results show that diversity is not a relevant criterion to perform classifier selection for this particular problem. The kappa pruning heuristic, which only relies on diversity to select the ensemble members, even exhibits a negative gain. This means that a single decision tree achieves better performances than the pruned ensemble obtained by focusing on diversity only. For this problem, the *Choose Best* heuristic seems to be a more relevant selection strategy. We note that for this dataset, the overproduce step sometimes produces an initial ensemble of decisions trees whose accuracy is lower than the accuracy of a single decision tree train on a uniform distribution on the training set. In this case, the initial pool of decision trees is rejected and another overproduce step is performed. For the *segment* and *letter* datasets, we did not observe such rejects of the initial pool of predictors.

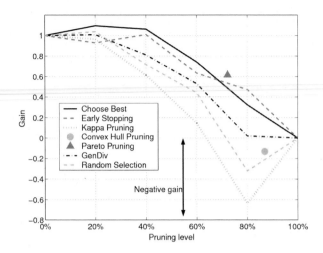

Fig. 6. Comparison of classifier selection methods for the *german-credit* dataset

Similar experiments were ran on additionnal UCI datasets and the same conclusions were drawn. It appears that accuracy is generally the component the more relevant for classifier selection. This conclusion comes from the general behaviour of the *Choose Best* heuristic presented in figures 4, 5 and 6. This observation is somehow a contradiction with the conclusions that we can draw from kappa error diagrams about the relevant aspect of diversity when designing Multiple Classifier Systems. To the question *Is it effective to use diversity explicitly to train classifier ensembles ?*, our answer would be *yes, but with a controlled importance and only under certain conditions*. Indeed, it appears that giving a strong importance to the diversity component of the trade-off often limits the benefits we obtain in terms of accuracy compared to a single classifier. This observation is illustrated by the *Kappa Pruning* heuristic which can even produce Multiple Classifier Systems whose accuracy is inferior to the one of a single classifier in case of a negative gain, which occurs for high pruning levels.

When the number of available samples is sufficient so that classifier selection is not subject to overfitting problems, it appears that giving a moderate importance to the diversity component leads to accurate classifier ensembles. For the *letter* and *segment* datasets, pruning the initial pool of decision trees slighty increases the accuracy for ensemble sizes ranging from 40 to 20 classifiers. However, for small ensembles, enforcing diversity during classifier selection often limits the accuracy of the resulting ensemble.

5   Conclusions and Future Works

In this paper, we have proposed a genetic algorithm to study the trade-off between accuracy and diversity when designing Multiple Classifier Systems under the Overproduce and Choose paradigm. We have compared this genetic algorithm with several classifier selection methods using the protocol proposed by Margineantu and Dietterich to prune a set of decision trees with different heuristics. When the number of validation samples provided is not sufficient, classifier selection is often subject to overfitting problems and discussions about the effectiveness of diversity are likely to lead to a dead end. However it appears that for certain large datasets, a cleaver use of diversity during the selection of classifiers can improve the accuracy when pruning a pool of predictors under the Overproduce and Choose paradigm. This observation comes from the results obtained with the GenDiv algorithm which tries to optimize the trade-off between accuracy and diversity.

So it seems that there are still many unanswered questions about the precise role of diversity to provide further research works, like a more detailed study of the trade-off between accuracy and diversity for instance. Running similar experiments with two overproduce phases based respectively on Bagging and Boosting would also be an interesting work to study the behaviour of different classifier selection methods when the initial pool of predictors is accurate but less diverse in general. Most of the diversity measures proposed in the literature are pairwise and only offer a local vision of the behaviour of diversity. Our intuition is that when those pairwise measures are averaged to compute the diversity of the ensemble as a whole, important interaction between classifiers are not visible anymore and our understanding of the diversity mechanisms applied to the ensemble as a whole is limited. If we consider the proposed genetic algorithm for instance, relevant pairs of classifier which exhibit an interesting accuracy-diversity trade-off might be broken at each cross-over. To address this limitation, a direct improvement of our work concerns the development of a diversity measure more convenient for the design of Multiple Classifier Systems.

Acknowledgments. The present research work has been supported by the International Campus on Safety and Intermodality in Transportation, the Région Nord Pas de Calais, the European Community, the Délégation Régionale à la Recherche et à la Technologie, the Ministère de l'Enseignement Supérieur et de la Recherche, and the Centre National de la Recherche Scientifique. The authors gratefully acknowledge the support of these institutions.

References

[1] Allwein, E.L., Schapire, R.E., Singer, Y.: Reducing multiclass to binary: a unifying approach for margin classifiers. Journal of Machine Learning Research 1, 113–141 (2001)
[2] Asuncion, A., Newman, D.J.: UCI Machine Learning Repository (2007), http://www.ics.uci.edu/~mlearn/MLRepository.html
[3] Breiman, L.: Bagging predictors. Machine Learning 24, 123–140 (1996)
[4] Breiman, L.: Random forests. Machine Learning 45, 5–32 (2001)
[5] Brown, G., Wyatt, J., Harris, R., Yao, X.: Diversity creation methods: a survey and categorisation. Information Fusion 6, 5–20 (2005)

[6] Burges, C.J.C.: A tutorial on support vector machines for pattern recognition. Data Mining and Knowledge Discovery 2, 121–167 (1998)

[7] Dietterich, T.G.: Ensemble methods in machine learning. In: Kittler, J., Roli, F. (eds.) MCS 2000. LNCS, vol. 1857, pp. 1–15. Springer, Heidelberg (2000)

[8] Dietterich, T.G.: An experimental comparison of three methods for constructing ensembles of decision trees: bagging, boosting and randomization. Machine Learning 40, 139–157 (2000)

[9] Duda, R.O., Hart, P.E., Stork, D.G.: Pattern Classification, 2nd edn. John Wiley & Sons, Chichester (2001)

[10] Freund, Y., Schapire, R.E.: Experiments with a new boosting algorithm. In: Proceedings of the Thirteenth International Conference on Machine Learning, pp. 148–156 (1996)

[11] Gacquer, D., Delmotte, F., Delcroix, V., Piechowiak, S.: A genetic approach for training diverse classifier ensembles. In: Proceedings of the 12th International Conference on Information Processing and Management of Uncertainty in Knowledge-Based Systems (IPMU 2008), Malaga, Spain, pp. 798–805 (2008)

[12] Hagan, M.T., Demuth, H.B., Beale, M.: Neural Network Design. PWS Publishing, Boston (1996)

[13] Ho, T.K.: The random subspace method for constructing decision forests. IEEE Transactions on Pattern Analysis and Machine Intelligence 20, 832–844 (1998)

[14] Kuncheva, L.I.: That elusive diversity in classifier ensembles. In: Proceedings of the 1st Iberian Conference on Pattern Recognition and Image Analysis, pp. 1126–1138 (2003)

[15] Kuncheva, L.I., Whitaker, C.: Measures of diversity in classifier ensembles and their relationship with the ensemble accuracy. Machine Learning 51, 181–207 (2003)

[16] Kuncheva, L.I.: Combining Pattern Classifiers: Methods and Algorithms. Wiley Interscience, Hoboken (2004)

[17] Margineantu, D.D., Dietterich, T.G.: Pruning adaptive boosting. In: Proceedings 14th International Conference on Machine Learning, pp. 211–218 (1997)

[18] Melville, P., Mooney, R.J.: Creating diversity in ensembles using artificial data. Information Fusion 6, 99–111 (2005)

[19] Mitchell, T.: Machine Learning. McGraw-Hill, New York (1997)

[20] Opitz, D., Maclin, R.: Popular ensemble methods: An empirical study. Journal of Artificial Intelligence Research 11, 169–198 (1999)

[21] Partridge, D., Yates, W.B.: Engineering multiversion neural-net systems. Neural Computation 8, 869–893 (1996)

[22] Quinlan, J.R.: Induction of decision trees. Machine Learning 1, 81–106 (1986)

[23] Quinlan, J.R.: Bagging, boosting, and c4.5. In: Proceedings of the Thirteenth National Conference on Artificial Intelligence, pp. 725–730 (1996)

[24] Rodriguez, J.J., Kuncheva, L.I., Alonso, C.J.: Rotation forest: A new classifier ensemble method. IEEE Transactions on Pattern Analysis and Machine Intelligence 28, 1619–1630 (2006)

[25] Roli, F., Giacinto, G., Vernazza, G.: Methods for designing multiple classifier systems. In: Proceedings of the 2nd International Workshop on Multiple Classifier Systems, Cambridge, UK, pp. 78–87 (2001)

On the Use of Clustering in Possibilistic Decision Tree Induction

Ilyes Jenhani[1,2], Salem Benferhat[2], and Zied Elouedi[1]

[1] LARODEC, Institut Supérieur de Gestion de Tunis, Tunisia
[2] CRIL, Université d'Artois, Lens, France
ilyes.j@lycos.com, benferhat@cril.univ-artois.fr, zied.elouedi@gmx.fr

Abstract. This paper presents an extension of a standard decision tree classifier, namely, the C4.5 algorithm. This extension allows the C4.5 algorithm to handle uncertain labeled training data where uncertainty is modeled within the possibility theory framework. The extension mainly concerns the attribute selection measure in which a clustering of possibility distributions of a partition is performed in order to assess the homogeneity of that partition. This paper also provides a comparison with previously proposed possibilistic decision tree approaches.

1 Introduction

Machine learning is gaining more and more popularity within the artificial intelligence research field and is attracting a great attention in industry, e.g., pharmaceutical industry, banking, stock exchange, marketing, etc. Classification is one among supervised learning tasks of machine learning. It consists in 1) inducing a classifier from a set of historical examples (training set) with known class labels and then 2) using the induced classifier to predict the class label (the category) of new objects on the basis of values of their attributes.

This task is ensured by a large number of techniques, e.g., artificial neural networks , k-nearest neighbors [5], Bayesian networks [16] and decision trees [4,21]. The latter, namely, decision tree classifier, is considered as one of the most popular classification techniques. They are able to represent knowledge in a flexible and easy form and they present high classification accuracy rates.

Over the last years, many variants of standard decision tree classifiers have been proposed depending on the hypotheses of the classification problem. For instance, *multi-label* decision trees [2] deal with problems where training data can be labeled by more than one class (not mutually exclusive). In *Multi-instance* decision tree learning [2], the objective is to build a classifier from a set of labeled bags rather than a set of labeled instances. A bag is a collection of instances that is labeled positive if all its instances are positive otherwise it is labeled negative. *Ambiguous-label* decision trees [8] deal with training instances that are labeled by a subset of *candidate* classes.

In this paper, we deal with a more general form of ambiguous-label classification, namely, *uncertain-label* classification. In spite of the great progress in

C. Sossai and G. Chemello (Eds.): ECSQARU 2009, LNAI 5590, pp. 505–517, 2009.

ordinary classification field in recent years, the problem of uncertain data has remained a great challenge for classification algorithms. This uncertainty is often inevitable in real-world applications due to reasons such as reliability of human observers, imprecise measurement, outdated sources, sampling errors, etc. For instance, in molecular biology, the functional category of a protein is often not exactly known. If we aim to perform an efficient classification, uncertainty should not be ignored but it should be well managed.

This paper addresses the problem of classification from uncertain data. More precisely, we deal with *possibilistic-labeled* decision trees [12]. In this setting, each training instance is labeled by a possibility distribution over the different possible classes. Hence, instead of just giving a subset of potential labels (the case of ambiguous-classification), an expert can express his opinion by giving an order on the possible labels. Thus, a possibility degree of a given label expresses the degree of confidence of the expert that the label is the true one.

Recently, a non-specificity based possibilistic decision tree approach has been proposed in [12]. It dealt with building decision trees from possibilistic labeled examples. This approach has good results for the general case of training instances that are imprecisely labeled. However, it generates less accurate trees for the particular case of full certainty (i.e. when all instances are exactly labeled). In this paper, we propose an approach that extends C4.5 algorithm to deal with possibilistic labeled examples. Two clustering strategies will be used in order to make the C4.5's gain ratio attribute selection measure applicable even on a set of possibility distributions. For the standard case of exactly labeled training set, our so-called Clust-PDT (for clustering-based possibilistic decision tree) approach recovers the C4.5 algorithm.

The rest of the paper is organized as follows: Section 2 gives the necessary background concerning possibility theory. Section 3 is devoted to decision tree classifiers. Section 4 summarizes all of the possibilistic decision tree approaches and explains the difference between these approaches and the one presented in this paper. Section 5 presents the different components of our proposed Clustering based possibilistic decision tree approach (Clust-PDT) with an illustrative example. Finally, Section 6 concludes the paper.

2   Possibility Theory

Possibility theory represents an uncertainty theory, first introduced by Zadeh [23] and then developed by several authors (e.g., Dubois and Prade [6]). In this section, we will give a brief recalling on possibility theory.

Given a universe of discourse Ω, one of the fundamental concepts of possibility theory is the *possibility distribution* denoted by π. π corresponds to a function which associates to each element ω_i from the universe of discourse Ω a value from a bounded and linearly ordered valuation set $(L,<)$. This value is called a *possibility degree*: it encodes our knowledge on the real world. Note that, in possibility theory, the scale can be numerical (e.g. L=[0,1]): in this case we have numerical possibility degrees from the interval [0,1] and hence we are dealing with the quantitative setting of the theory.

By convention, $\pi(\omega_i) = 1$ means that it is fully possible that ω_i is the real world, $\pi(\omega_i) = 0$ means that ω_i cannot be the real world (is impossible). Flexibility is modeled by allowing to give a possibility degree from $]0,1[$. In possibility theory, extreme cases of knowledge are given by:

- *Complete knowledge*: $\exists \omega_i$, $\pi(\omega_i) = 1$ and $\forall\, \omega_j \neq \omega_i$, $\pi(\omega_j) = 0$.
- *Total ignorance*: $\forall\, \omega_i \in \Omega$, $\pi(\omega_i) = 1$ (all values in Ω are possible).

A possibility distribution π is said to be *normalized* if there exists at least one state $\omega_i \in \Omega$ which is totally possible. In the case of sub-normalized π,

$$Inc(\pi) = 1 - \max_{\omega \in \Omega}\{\pi(\omega)\} \tag{1}$$

is called the *inconsistency degree* of π. It is clear that, for normalized π, $\max_{\omega \in \Omega}\{\pi(\omega)\} = 1$, hence $Inc(\pi){=}0$. The measure *Inc* is very useful in assessing the degree of conflict between two distributions π_1 and π_2 which is given by $Inc(\pi_1 \wedge \pi_2)$. We take the $\wedge$ as the minimum operator.

3   Decision Trees

A decision tree is a flow-chart-like hierarchical tree structure which is generally made up of two major procedures:

Building procedure: Given a training set, building a decision tree is usually done by starting with an empty tree and selecting for each decision node the 'appropriate' test attribute using an attribute selection measure. The principle is to select the attribute that maximally diminish the mixture of classes between each training subset created by the test, thus, making easier the determination of object's classes. The process continues for each sub decision tree until reaching leaves and fixing their corresponding classes.

Classification procedure: To classify a new instance, having only values of all its attributes, we start with the root of the constructed tree and follow the path corresponding to the observed value of the attribute in the interior node of the tree. This process is continued until a leaf is encountered. Finally, we use the associated label to obtain the predicted class value of the instance at hand.

Several algorithms for building decision trees have been developed. The most popular and applied ones are: **ID3** [21] and its successor **C4.5** [22]. The main component of these algorithm is the attribute selection measure.

Attribute selection measure generally based on information theory, serves as a criterion in choosing among a list of candidate attributes at each decision node, the attribute that generates partitions where objects are less randomly distributed, with the aim of constructing the smallest tree among those consistent with the data. The well-known measure used in the **C4.5** algorithm of Quinlan [22] is the gain ratio. The information gain relative to an attribute A_k is defined as follows:

$$Gain(T, A_k) \;=\; E(T) \;-\; E_{A_k}(T) \tag{2}$$

where

$$E(T) = - \sum_{i=1}^{nc} \frac{nb(C_i, T)}{|T|} \, log_2 \, \frac{nb(C_i, T)}{|T|} \tag{3}$$

and

$$E_{A_k}(T) = \sum_{v \in D(A_k)} \frac{|T_v^{A_k}|}{|T|} E(T_v^{A_k}) \tag{4}$$

nc in Equation (3) corresponds to the number of classes of the problem. $nb(C_i, T)$ denotes the number of objects in the training set T belonging to the class C_i, $D(A_k)$ in Equation (4) denotes the finite domain of the attribute A_k and $|T_v^{A_k}|$ denotes the cardinality of the set of objects for which the attribute A_k has the value v. Thus, $E(T)$ corresponds to the *entropy* of the set T. The gain ratio is given by:

$$Gr(T, A_k) = \frac{Gain(T, A_k)}{SplitInfo(T, A_k)} \tag{5}$$

$SplitInfo(T,A_k)$ measures the information in the attribute due to the partition-ing of the training set T into $|D(A_k)|$ training subsets. This quantity describes the information content of the attribute itself. It is given by:

$$SplitInfo(T, A_k) = - \sum_{v \in D(A_k)} \frac{|T_v^{A_k}|}{|T|} \, log_2 \, \frac{|T_v^{A_k}|}{|T|} \tag{6}$$

4   A Brief Overview of Possibilistic Decision Trees

Several approaches for building possibilistic decision trees were proposed. We can divide them into two sets: those acting under standard decision tree learning hypotheses (i.e. where training data are exactly labeled and therefore standard decision tree algorithms can be applied) and those approaches dealing with non-standard learning hypotheses (i.e. where training data are not exactly labeled and consequently, standard decision trees cannot be used).

For instance, in the work of Borgelt et al. [3], authors dealt with a standard decision tree learning problem. However, in order to assess the informational contribution of a given attribute at a given node, they took the probability distributions of the instances reaching each node as possibility distributions (an interpretation which is based on the context model of possibility theory [18]). Then, as the role of the non-specificity in possibility theory is similar to that of Shannon entropy in probability theory, they have used non-specificity to choose the attribute that gives more specific partitions.

The work developed by Hüllermeier [7] also dealt with standard decision tree learning hypotheses. In this work, possibility theory was introduced to define a possibilistic branching within the lazy decision tree technique [9] resulting on possibilistic decisions (i.e. each leaf is labeled by a unique class value character-ized by a possibility degree).

Another work by Ben Amor et al. [1] have dealt with uncertainty in the classification phase. More precisely, authors have proposed a method based on leximin-leximax criterion [19] in order to make standard decision tree algorithms (e.g. C4.5) able to classify instances having uncertain attribute values. In a previous work [10], we have proposed the *possibilistic option decision tree approach* which deals with the uncertainty related to the choice of an attribute at a given decision node. In this approach, each decision node can be split according to more than one attribute, using multiple attribute-value tests, or "options". These option nodes are quantified via possibility distributions.

The second set of possibilistic decision tree approaches deals with uncertain training data. Uncertainty can affect class labels and/or attribute values characterizing some or all instances of the training set. In this setting, uncertainty should not be ignored and consequently, standard decision tree algorithms should be adapted to such contexts since they can not be directly applied. This has been the focus of our recently published works [12] and [13,15].

The so-called Non-specificity based possibilistic decision tree (NS-PDT) [12] handles the problem of possibilistic labeled examples using the concept of non-specificity. The informational contribution of a given attribute (which is determined by the degree of homogeneity of partitions generated by that attribute) corresponds to the average non-specificity of all representative possibility distributions characterizing each partition. See also [3] for a similar work.

A similarity-based possibilistic decision tree approach (Sim-PDT) has been proposed in [13,15]. This approach has the advantage of recovering the standard C4.5 algorithm when dealing with a precise training set (all instances are labeled by possibility distributions corresponding to a situation of complete knowledge). The informational content of a training partition corresponds to the entropy of the distribution of Meta-classes forming that partition weighted by the average similarity of the original possibility distributions (labels of training data) pertaining to each Meta-class. A Meta-class is a label associated to each original possibility distribution which is in turn associated to the most similar wrapper possibility distribution (binary possibility distributions).

5   The Clust-PDT Approach Components

The above approaches either NS-PDT or Sim-PDT present some problems. For instance, the major problem for the NS-PDT approach is that it does not recover C4.5 in the certain case. The main problem with the Sim-PDT approach is related to the number of wrapper possibility distributions (and hence the number of Meta-classes) to be considered at the beginning of the building procedure. Suppose that we have a problem with nc classes (i.e., $\Omega = C = \{C_1, C_2, ..., C_{nc}\}$). We define a degree of imprecision $dimp \in \{1, ..., nc\}$ which will allow to determine the set of Wrapper possibility distributions WD and the set of Meta-classes MC to be used.

The number of meta-classes is exponential in the number of classes. This will badly affect the performance of the induced possibilistic decision trees in terms of classification accuracy and time. To avoid this problem, we propose two clustering strategies: the first (baseline) strategy will use wrapper possibility distributions[1] (with dimp=1) and the second strategy will perform a hierarchical clustering of the whole possibility distributions reaching each partition to automatically determine the meta-classes in each generated partition.

In this section, we will present the components of the Clust-PDT approach in its two variants. We will mainly concentrate on the way the possibility distributions are clustered and on the attribute selection measure component, namely, the *Clust-Gain ratio* measure. Before that, let us present a key element that will be used in both clustering strategies, namely, the similarity between two normalized possibility distributions.

Information Affinity: A possibilistic similarity measure

Comparing pieces of uncertain information given by several sources has attracted a lot of attention for a long time. This could be ensured by the use of similarity indexes. After an analysis of existing possibilistic similarity measures in [11], we have proposed in a recent work [11,14] a new similarity index satisfying interesting properties. The information affinity index, denoted by $GAff$ takes into account a classical informative distance, namely, the Manhattan distance along with the well known inconsistency measure. $GAff$ is applicable to any pair of normalized possibility distributions.

Definition 1. *Let π_1 and π_2 be two possibility distributions on the same universe of discourse Ω. We define a measure $GAff(\pi_1, \pi_2)$ as follows:*

$$GAff(\pi_1, \pi_2) = 1 - (0.5 * d(\pi_1, \pi_2) + 0.5 * Inc(\pi_1 \wedge \pi_2)) \qquad (7)$$

d represents a (Manhattan or Euclidean) normalized metric distance between π_1 and π_2. $Inc(\pi_1 \wedge \pi_2)$ is the inconsistency degree between the two distributions (see Equation (1)) where $\wedge$ is the min operator.

Two possibility distributions π_1 and π_2 are said to have a strong affinity (resp. weak affinity) if $GAff(\pi_1, \pi_2) = 1$ (resp. $GAff(\pi_1, \pi_2) = 0$).

5.1   Clustering of the Possibility Distributions

In possibilistic-labeled learning, instances labels are presented by means of possibility distributions on the different possible classes rather than a single class label. Given the set of all possible labels, the possibility degree on each class label expresses the confidence of an expert that this label is the true one.

Given a training set T containing n instances and given the set of attributes, let us denote by π_i the possibility distribution labeling the class of the instance i in T. The direct application of the gain ratio criterion [22] (Equation (5))

---

[1] Wrapper possibility distributions are binary possibility distributions (i.e., $\forall \omega \in \Omega, \pi(\omega) \in \{0, 1\}$) representing special cases of complete knowledge, partial ignorance and total ignorance. For instance, for a 3-class problem, if dimp=1 (i.e. we only consider cases of complete knowledge) then $WD = \{[1, 0, 0], [0, 1, 0], [0, 0, 1]\}$.

has no sense since π_i's are most of the time very different, so it has no sense to directly compute their frequencies in order to determine the entropy of T. Hence, we will try to regroup *similar* π_i's into clusters using our proposed possibilistic similarity measure [11,14]. Then, each cluster will be assigned a label ($Clust_j$) so that similar possibility distributions will no longer be considered as different but will be labeled by the same label (the label of the cluster).

The following presents two strategies for clustering possibility distributions.

Clustering with First-Level Wrapper Distributions

The first clustering strategy is very intuitive and uses the concept of first-level wrapper possibility distributions (i.e. wrapper possibility distributions corresponding to complete knowledge (the case of $dimp = 1$).

To each π_i, we will assign the most similar wrapper distribution, $Clust_j = WD_j$ such that $j = \arg\max_{j=1}^m \{GAff(\pi_i, WD_j)\}$ where m is the total number of wrapper distributions and $GAff$ corresponds to the information affinity index (Equation (7)). Note that, as in standard decision trees, ties are broken arbitrarily.

In this *baseline* strategy, the number of clusters is known a-priori: it is equal to the number of classes of the problem. Moreover, the clustering of π_i's should be done from the beginning of the decision tree building procedure as a pretreatment. Namely, the clustering is done once for the whole training set. We can easily check that the cluster into which a possibility distribution will fall will be the same when performing the clustering before and after splitting the training set. This can be explained by the fact that the cluster that will be assigned to a given possibility distribution π_i is independent from the other possibility distributions belonging to the same partition as π_i.

Hierarchical clustering of the possibility distributions

In this strategy, we have chosen to automatically cluster possibility distributions without fixing the number of clusters. Hence, we have opted for a hierarchical clustering method.

Given a set of n possibility distributions $\pi_{i=1..n}$ to be clustered, and an $n * n$ similarity matrix (the similarity between any two possibility distributions π_k, π_l is determined by $GAff(\pi_k, \pi_l)$), the process of the *agglomerative hierarchical clustering* [17] of a set of possibility distributions is performed as follows:

1. Start by assigning each π_i to a cluster, so that if you have n possibility distributions, you obtain n clusters, each containing just one possibility distribution. Let the similarities between the clusters be the same as the similarities between the possibility distributions they contain.
2. Find the most similar pair of clusters and merge them into a single cluster, so that now you have one cluster less.
3. Compute similarities between the new cluster and each of the old clusters.
4. Repeat steps 2 and 3 until reaching a predefined similarity threshold ST.

The similarity between two clusters of possibility distributions can be computed in several ways which makes the difference between what are called *single-linkage, complete-linkage* and *average-linkage* clustering. We have chosen the *complete-linkage* clustering in which the similarity between two clusters corresponds to the similarity between the two least similar possibility distributions (one from each cluster).

The similarity threshold ST plays the role of a stopping criterion in the clustering process. In fact, after computing the different similarities between the clusters formed in step $s - 1$, if the greatest similarity is strictly less than ST OR $ST < 0.5$, we should stop the clustering process and hence keep the clusters of step $s - 1$ as a final result. The similarity threshold is computed from the initial training set and is given by:

$$ST = \frac{\sum_{i=1}^{n} \sum_{j=i+1}^{n} GAff(\pi_i, \pi_j)}{\frac{n*(n-1)}{2}} \qquad (8)$$

In this strategy, performing the clustering before and after splitting will not give the same results because the cluster that will be assigned to a given possibility distribution will strongly depend on the other possibility distributions of the same partition. Hence, to obtain accurate results, we will integrate the above hierarchical clustering procedure in the attribute selection step. Namely, we will re-perform the clustering whenever a splitting of a training partition is done.

5.2   The Clust-Gain Ratio Attribute Selection Measure

Regardless of the chosen clustering strategy (baseline or hierarchical), we define the *Clust-Gain ratio* by:

$$ClustGr(T, A_k) = \frac{ClustGain(T, A_k)}{SplitInfo(T, A_k)} \qquad (9)$$

$SplitInfo(T, A_k)$ is given by Equation (6) and $ClustGain(T, A_k)$ is given by:

$$ClustGain(T, A_k) = ClustE(T) - ClustE_{A_k}(T) \qquad (10)$$

where

$$ClustE(T) = -\sum_{j=1}^{m}(\frac{|Clust_j|}{|T|} log_2 \frac{|Clust_j|}{|T|}) \qquad (11)$$

and

$$ClustE_{A_k}(T) = \sum_{v \in D(A_k)} \frac{|T_v^{A_k}|}{|T|} ClustE(T_v^{A_k}) \qquad (12)$$

$|Clust_j|$ in Equation (11) denotes the cardinality of objects in (sub)-partition T whose class labels (i.e. possibility distributions) were assigned to the cluster $Clust_j$. m denotes the number of obtained clusters. In the baseline strategy, m is equal to the number of classes of the problem nc. However, in the hierarchical strategy, m varies from one (sub)-partition T to another. Obviously, the attribute maximizing $ClustGr$ will be selected as the label of the current decision node.

5.3   The Partitioning Strategy

Once an attribute is selected at a given decision node and since we only deal with nominal attributes, the partitioning strategy will consist in partitioning the training (sub)-set according to all values of the selected attribute which leads to the generation of one partition for each possible value of the selected attribute. For the case of continuous attributes, a discretization step is needed.

5.4   The Stopping Criteria

Several stopping criteria could be defined for the Clust-PDT approach for its two variants. These criteria present cases for which we should stop the partitioning process for each generated training sub-partition T_p. Hence, we should stop growing the tree if:

1. There is no further attribute to test.
2. There is a partition T_p with only one cluster: all possibility distributions in T_p belong to the same cluster.
3. $ClustGr <= 0$: no information is gained.
4. $|T_p|=0$: the generated partition does not contain any instance.

5.5   Structure of Leaves

Stopping criteria are the same for both baseline and hierarchical variants of the Clust-PDT approach. However, the structure of leaves is different. This is due to the ways both variants are conceived. In fact, the baseline strategy is conceived to make precise decisions which justifies our choice for the first-level wrapper possibility distributions. Namely, leaves will be labeled by one among the pre-defined wrapper possibility distributions.

In the other hand, the hierarchical variant seeks to remain faithful to the original possibility distributions of the training set and hence will produce leaves labeled by one among these possibility distributions.

The baseline Clust-PDT variant:

- When stopping criterion 1 or 3 is satisfied, we declare a leaf labeled by the *majority* wrapper possibility distribution in T_p.
- When stopping criterion 2 is satisfied, we declare a leaf labeled by that unique wrapper possibility distribution.
- When stopping criterion 4 is satisfied, we declare a leaf labeled by a randomly chosen wrapper possibility distribution from WD.

The hierarchical Clust-PDT variant:

- When stopping criteria 1 or 2 or 3 is satisfied for a training partition T_p containing n possibility distributions: we declare a leaf labeled by the representative possibility distribution of that set (π_{Rep}), that is, the distribution which corresponds to the closest distribution to all the remaining distributions in T_p: $\pi_{Rep} = \arg\max_{i=1}^{n}\{\frac{\sum_{j\neq i} GAff(\pi_i,\pi_j)}{(n-1)}\}$

– When stopping criteria 4 is satisfied, we declare a leaf labeled by the representative possibility distribution (π_{Rep}) of the set of the previous level.

Property 1. If all training instances are precisely labeled then *Clust-PDT* (in its both variants) is equivalent to C4.5.

Property 2. In the baseline variant, clustering all the possibility distributions from the beginning of the building procedure gives the same result as clustering them in each generated sub-partition during the development of the tree.

Property 3. In the hierarchical variant, the result of the clustering changes from one (sub)-partition to another.

Property 4. In the hierarchical variant, if all training instances are precisely labeled, then labeling a leaf by the closest distribution to all the remaining distributions in that leaf is equivalent to choosing the majority class (the method adopted by C4.5).

Example 1. In order to illustrate the Clust-PDT approach (we only illustrate the hierarchical variant for reasons of space), we will consider an example in the intrusion detection field.

The training set T given in Table 1 is composed of an excerpt of different connections corresponding to a TCP/IP dump rows. Note that, for the sake of simplicity, each connection is described by only three attributes which are: protocole_type, service and flag. Domains of these attributes are: $D_{Protocole\_type} = \{tcp, udp\}$, $D_{Service} = \{http, domain\_u, private\}$ and $D_{Flag} = \{SF, REJ, RSTO\}$.

We asked a security administrator to give us his opinion about the class of each connection. Three classes are possible either, *Normal* (N) or *Probing* (P) or *DOS* (D). Normal corresponds to a normal connection while DOS and Probing are relative to two categories of attacks. Because of the lack of information about each connection (only three attributes are provided), the security administrator was unable to give us an exact response. He therefore provided, for each connection, a possibility degree for each possible class. This possibility degree expresses his confidence that a given connection belongs to a given class.

Now, we should find the most informative attribute that will be selected as the label of the root node. Namely, we should compute $ClustGr(T, Protocole\_type)$, $ClustGr(T, Service)$ and $ClustGr(T, Flag)$ and choose the attribute providing the greatest value. In this example, for reasons of space, we will only show steps for computing $ClustGr(T, Protocole\_type)$.

Step1: $Split\_info(T, Protocole\_type) = -\frac{4}{10} * log_2(\frac{4}{10}) - \frac{6}{10} * log_2(\frac{6}{10}) = 0.971$.

Step2: In order to compute $ClustGain(T, Protocole\_type)$, we should compute $ClustE(T)$ and $ClustE_{Protocole\_type}(T)$. Let us start by computing $ClustE(T)$. At this step, we should perform a hierarchical clustering of the possibility distributions in T as described in Section 5.1. The computation of the similarity threshold ST associated to T gave the value: 0.679. In this example, $GAff$ is used with: $d \equiv Manhattan\ distance$ and $\wedge \equiv min\ operator$.

Table 1. Training set with possibilistic class labels

	Protocole_type	Service	Flag	N	P	D
i_1	tcp	http	SF	1	0.8	0.2
i_2	tcp	private	SF	1	0.2	0
i_3	tcp	private	RSTO	0.3	0.6	1
i_4	tcp	http	RSTO	1	1	0.5
i_5	udp	private	REJ	0	1	0
i_6	udp	domain_u	SF	1	0.4	0.7
i_7	udp	domain_u	RSTO	0	1	0.5
i_8	udp	http	REJ	0	1	1
i_9	udp	http	SF	0.3	1	0.3
i_{10}	udp	private	SF	0.8	1	0.4

The application of the complete-linkage hierarchical clustering algorithm on T gives 3 clusters: $Clust_1 = \{i_4, i_5, i_7, i_9, i_{10}\}$, $Clust_2 = \{i_1, i_2, i_6\}$ and $Clust_3 = \{i_3, i_8\}$. Hence, $ClustE(T) = -\frac{5}{10}*log_2(\frac{5}{10}) - \frac{3}{10}*log_2(\frac{3}{10}) - \frac{2}{10}*log_2(\frac{2}{10}) = 1.485$.

Now, to compute $ClustE_{Protocole\_type}(T)$, we should compute both $ClustE$ $(T_{tcp}^{Protocole\_type})$ and $ClustE(T_{udp}^{Protocole\_type})$ as we did with $ClustE(T)$.

The application of the complete-linkage hierarchical clustering algorithm on $T_{tcp}^{Protocole\_type}$ with associated $ST = 0.652$ (resp. on $T_{udp}^{Protocole\_type}$ with associated $ST = 0.737$) has given 2 clusters $Clust_1 = \{i_1, i_2, i_4\}$ and $Clust_2 = \{i_3\}$ (resp. 2 clusters $Clust_1 = \{i_5, i_7, i_8, i_9, i_10\}$ and $Clust_2 = \{i_6\}$). $\Rightarrow ClustGr$ $(T, Protocole\_type) = \frac{0.771}{0.971} = 0.794$.

Similarly, we obtain $ClustGr(T, Service) = 0.187$ and $ClustGr(T, Flag) = 0.488$. Hence, the root node of our tree is labeled by $Protocole\_type$. The final Clust-PDT tree is given by Fig. 1.

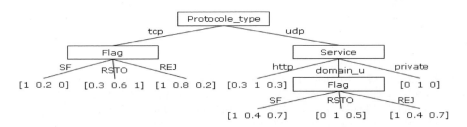

Fig. 1. Final Clust-PDT tree

6   Conclusion

In this paper, we have developed a new approach so-called Clustering-based possibilistic decision tree. This approach represents an extension of C4.5 to the uncertain setting. The proposed approach integrates a clustering routine in the attribute selection measure(the gain ratio criterion). Two clustering strategies were proposed. The baseline strategy is more adapted to problems where the decision maker wants to obtain precise classifications from the Clust-PDT tree (an exact label). However, the hierarchical strategy is more suited to problems where the decision maker suggests a decision (a label) that is as faithful as possible to original labels (a possibility distribution). A future work is to apply our learning approach for attack detection for incomplete connections.

Acknowledgment

This works is supported by the SETIN06 project PLACID (Probabilistic graphical models and Logics for Alarm Correlation in Intrusion Detection).

References

1. Ben Amor, N., Benferhat, S., Elouedi, Z.: Qualitative classification and evaluation in possibilistic decision trees. In: FUZZ-IEEE 2004, pp. 653–657 (2004)
2. Blockeel, H., Page, D., Srinivasan, A.: Multi-instance tree learning. In: Proceedings of the 22nd international conference on Machine learning, Bonn, Germany, pp. 57–64 (2005)
3. Borgelt, C., Gebhardt, J., Kruse, R.: Concepts for Probabilistic and Possibilistic Induction of Decision Trees on Real World Data. In: EUFIT 1996, pp. 1556–1560 (1996)
4. Breiman, L., Friedman, J.H., Olshen, R.A., Stone, C.J.: Classification and regression trees. Wadsworth & Brooks, Monterey (1984)
5. Dasarathy, B.V.: Nearest Neighbour(NN) norms: NN pattern classification techniques. IEEE Computer sociaty Press, Los Alamitos (1991)
6. Dubois, D., Prade, H.: Possibility theory. An approach to computerized processing of uncertainty. Plenum Press, New York (1988)
7. Hüllermeier, E.: Possibilistic Induction in decision tree learning. In: Elomaa, T., Mannila, H., Toivonen, H. (eds.) ECML 2002. LNCS, vol. 2430, pp. 173–184. Springer, Heidelberg (2002)
8. Hüllermeier, E., Beringer, J.: Learning from Ambiguously Labeled Examples. In: Intelligent Data Analysis, pp. 168–179 (2005)
9. Friedman, J.H., Kohavi, R., Yun, Y.: Lazy decision trees. In: Proceedings of the 13th National Conference on Artificial Intelligence, pp. 717–724 (1996)
10. Jenhani, I., Elouedi, Z., Ben Amor, N., Mellouli, K.: Qualitative inference in possibilistic option decision trees. In: Godo, L. (ed.) ECSQARU 2005. LNCS, vol. 3571, pp. 944–955. Springer, Heidelberg (2005)
11. Jenhani, I., Ben Amor, N., Elouedi, Z., Benferhat, S., Mellouli, K.: Information Affinity: a new similarity measure for possibilistic uncertain information. In: Mellouli, K. (ed.) ECSQARU 2007. LNCS, vol. 4724, pp. 840–852. Springer, Heidelberg (2007)
12. Jenhani, I., Ben Amor, N., Elouedi, Z.: Decision Trees as Possibilistic Classifiers. International Journal of Approximate Reasoning 48(3), 784–807 (2008)
13. Jenhani, I., Ben Amor, N., Benferhat, S., Elouedi, Z.: SIM-PDT: A Similarity Based Possibilistic Decision Tree Approach. In: Hartmann, S., Kern-Isberner, G. (eds.) FoIKS 2008. LNCS, vol. 4932, pp. 348–364. Springer, Heidelberg (2008)

14. Jenhani, I., Benferhat, S., Elouedi, Z.: Properties Analysis of Inconsistency-based Possibilistic Similarity Measures. In: Proceedings of IPMU 2008, pp. 173–180 (2008)
15. Jenhani, I., Benferhat, S., Elouedi, Z.: Learning and evaluating possibilistic decision trees using information affinity. International Journal of Computer Systems Science and Engineering 4(3), 206–112 (2008)
16. Jensen, F.V.: Introduction to Bayesian networks. UCL Press (1996)
17. Johnson, S.C.: Hierarchical Clustering Schemes. Psychometrika 2, 241–254 (1967)
18. Kruse, R., Gebhardt, J., Klawonn, F.: Foundations of Fuzzy Systems. John Wiley and Sons, Chichester (1994)
19. Moulin, H.: Axioms for cooperative decision-making. Cambridge Univ. Press, Cambridge (1988)
20. Murphy, P.M., Aha, D.W.: UCI repository of machine learning databases (1996)
21. Quinlan, J.R.: Induction of decision trees. Machine Learning 1, 81–106 (1986)
22. Quinlan, J.R.: C4.5: Programs for machine learning. Morgan Kaufmann, San Francisco (1993)
23. Zadeh, L.A.: Fuzzy sets as a basis for a theory of possibility. Fuzzy Sets ans Systems 1, 3–28 (1978)

When in Doubt ... Be Indecisive

Linda C. van der Gaag[1], Silja Renooij[1],
Wilma Steeneveld[2], and Henk Hogeveen[2]

[1] Department of Information and Computing Sciences, Utrecht University,
P.O. Box 80.089, 3508 TB Utrecht, The Netherlands
{linda,silja}@cs.uu.nl
[2] Department of Farm Animal Health, Utrecht University
{w.steeneveld,h.hogeveen}@uu.nl

Abstract. For a presented case, a Bayesian network classifier in essence computes a posterior probability distribution over its class variable. Based upon this distribution, the classifier's classification function returns a single, determinate class value and thereby hides the uncertainty involved. To provide reliable decision support, however, the classifier should be able to convey indecisiveness if the posterior distribution computed for the case does not clearly favour one class value over another. In this paper we present an approach for this purpose, and introduce new measures to capture the performance and practicability of such classifiers.

Keywords: Probabilistic classification, indecisiveness.

1 Introduction

Many real-life problems can be viewed as classification problems in which a case described in terms of a number of features is to be assigned to one of several distinct classes. In the management of animal health on dairy farms, for example, the problem of establishing an appropriate diagnosis for a combination of clinical signs can be viewed as a classification problem in which a cow has to be assigned to one of a number of diagnostic categories. Bayesian network classifiers have gained considerable popularity for solving such problems. These classifiers embed a Bayesian network composed of a single class variable, modelling the possible classes for the problem under study, and a set of feature variables, modelling the features that constitute the basis for distinguishing between the classes. For a presented case, this network serves to establish the posterior probability distribution over the class variable given the case's features. Based upon this distribution, the classifier assigns a single, determinate class to the case [3,4].

Bayesian network classifiers are being applied in a wide range of domains for a variety of problems; for some recent examples in the biomedical field we refer to [1,2,5,6,7]. In some applications, such as in automated spam filtering, the class value returned by the classifier conveys sufficient information to solve the problem at hand and does not require any further decisions from the user. We have noticed however, that in other applications the returned class value may not always provide a sufficient basis for reliable further decision making.

C. Sossai and G. Chemello (Eds.): ECSQARU 2009, LNAI 5590, pp. 518–529, 2009.

In our domain of animal health, for example, the actual problem is not just to establish the most likely diagnosis but, even more importantly, to control the disease patterns in a dairy herd by appropriate treatment. The diagnostic category returned by the classifier does not necessarily provide sufficient information for this purpose, as it hides the uncertainty involved in the classification result. A differential diagnosis in which two or more diagnostic categories have almost equal probabilities, for example, could call for a different treatment regime than a differential diagnosis in which one disease clearly stands out. From the classification result however, the decision maker cannot distinguish between clear-cut cases and cases which in essence are inconclusive.

In this paper, we enhance Bayesian network classifiers by allowing them to be indecisive. The basic idea is that the classifier returns a classification result for a case only if a single class value stands out convincingly in the posterior probability distribution computed over the class variable. If none of the possible class values receives sufficient support in the computed distribution, then the classifier does not return a determinate classification result but leaves the case unclassified instead. The case at hand then is left to the human decision maker, who evaluates the probabilistic information computed by the classifier in view of further decision making. For our new type of classifier we introduce measures to express its classification performance and its practicability. These measures closely resemble the well-known concept of classification accuracy, yet take into account the classifier's reduced practicability as a result of its occasional indecisiveness. We illustrate the usefulness of our new type of classifier for an example application in animal health management.

The paper is organised as follows. In Section 2 we review Bayesian network classifiers and introduce our domain of application. In Section 3, we discuss the well-known concept of classification accuracy and study its dependence on the probability thresholds commonly used by classification functions. In Section 4, we introduce the new concept of stratifying classifier and define associated measures of classification performance and practicability. We illustrate our concept of stratifying classifier and its associated measures for our domain of application in Section 5. The paper ends with our concluding observations in Section 6.

2 Preliminaries

In this section, we briefly review Bayesian network classifiers. In doing so, we restrict the discussion to naive Bayesian classifiers with binary variables only; the illustrated concepts, however, are readily extended to non-binary variables and to Bayesian network classifiers of more general topological structure. In addition, we introduce our application domain, which will serve as a running example.

2.1 Naive Bayesian Classifiers

A naive Bayesian classifier includes a designated class variable C and a set $\mathbf{F}$ of one or more feature variables F_i. If a variable V_j adopts the value *true*, we will write v_j; we use $\bar{v}_j$ to denote $V_j = \textit{false}$. A joint value assignment to all feature variables concerned is termed a case and will be denoted by **f**. The classifier's graphical structure includes arcs $C \rightarrow F_i$ which capture dependence of each feature variable on the class variable, yet mutual independence of any two feature variables given this class variable. The classifier further specifies a prior probability distribution $\Pr(C)$ over the class variable and a set of conditional distributions $\Pr(F_i \mid C)$ for each feature variable. Naive Bayesian classifiers are typically constructed by extracting the most discriminating feature variables, and their associated probability distributions, from a set of example cases.

A naive Bayesian classifier in essence allows the computation of any probability of interest over its variables. More specifically, it provides for establishing, for a presented case **f**, the posterior probability distribution $\Pr(C \mid \mathbf{f})$ over the class variable given the case's features. The classifier does not return this probability distribution, but instead establishes a single, determinate class value for its output, using a classification function. For the binary class variable C, this function takes the following form:

$$class(C, t; \mathbf{f}) = \begin{cases} c, & \text{if } \Pr(c \mid \mathbf{f}) \geq t \\ \bar{c}, & \text{otherwise} \end{cases}$$

where t is a pre-defined threshold value. In most applications, the winner-takes-all rule is used for the model's classification function, which takes $t = 0.50$. For applications with skewed prior distributions over the class variable, however, other values of t are preferred. In general, the choice of an appropriate threshold value is domain dependent. If the classification function of a classifier returns $class(C, t; \mathbf{f}) = c$ for a case **f**, then we say that **f** is classified as belonging to class c; analogous terminology is used for $class(C, t; \mathbf{f}) = \bar{c}$.

2.2   An Example Application in Dairy Science

Clinical mastitis is one of the most frequent and cost incurring diseases in a dairy herd. The disease affects the cow's udder, causing a reduction of the cow's milk production and an increased risk of the cow being culled. Clinical mastitis can be caused by a large variety of pathogens; diagnosis of the causing pathogen is done by bacteriological culturing. Bacteriological culturing takes at least three days. Yet, a timely administered treatment is important to eliminate the disease and to prevent recurrence as much as possible. Ideally, the disease is controlled with limited use of antibiotics, to reduce the risk of antibiotic contamination of the milk and to minimise the impact of treatment on antimicrobial resistance. The most appropriate treatment is highly dependent upon the specific pathogen causing the disease in the current instance, however. If a single specific pathogen is convincingly favoured over other possible causal pathogens, a narrow-spectrum antibiotic would be preferred; in case two or more pathogens are quite likely, broad-spectrum antibiotic treatment would be more appropriate. Unfortunately, a farmer will typically have to decide upon treatment in uncertainty, before the actual causal pathogen is known from bacteriological culturing.

To support a dairy farmer in his treatment decisions, we constructed a standard naive Bayesian classifier. Cases of clinical mastitis to be presented to the classifier are described by a number of features which range from the cow's mastitis history to such clinical signs as the appearance of the milk and the cow's demeanor. For a case, the classifier returns the most likely Gram-status of the pathogen causing the mastitis; this status is an important indicator for the type of antibiotics to be applied. For constructing the classifier, we had available a set of 3 833 clinical mastitis cases; in 2 706 (or 70.6%) of these cases the disease was caused by a Gram-positive pathogen, and in 1 127 cases the causal pathogen was Gram-negative. We used 2 631 cases (67%) for constructing the classifier and retained the remaining 1 202 cases for studying its performance. The constructed classifier was optimised for a classification threshold value of $t = 0.71$.

3   The Accuracy Measure and Its Threshold Dependence

The performance of a Bayesian network classifier is commonly summarised as the proportion of cases which are assigned to their true class value. In this section we review this measure of accuracy. We further argue that the accuracy of a Bayesian network classifier depends heavily on the threshold value used in its classification function. We investigate this dependence and study the effects of varying the threshold value on the classifier's accuracy.

3.1   The Measure of Accuracy

We consider a naive Bayesian classifier with the classification function *class*. We further consider a set $\mathcal{F}$ of cases for the classifier. The case set $\mathcal{F}$ is partitioned into the set $\mathcal{F}^+$ which includes p cases belonging to class c, and the set $\mathcal{F}^-$ which includes n cases belonging to $\bar{c}$, with $p + n = m$. A case belonging to class c will be termed a positive case; likewise, a case with class $\bar{c}$ is coined a negative case. The function *class* of the classifier now partitions the case set $\mathcal{F}$ into four mutually exclusive and collectively exhaustive subsets; the basic idea of this partitioning is shown in Table 1. The first subset includes all cases from $\mathcal{F}^+$ which are classified as belonging to class c by the classifier. This set is called the set of true positive cases, denoted by TP; the size of this set is denoted by tp. The cases from $\mathcal{F}^+$ which are incorrectly classified as belonging to $\bar{c}$ constitute the set of false negative cases, denoted by FN; the size of this set is fn. Note that TP ∩ FN = ∅ and TP ∪ FN = $\mathcal{F}^+$, and hence that tp + fn = p. Likewise, we define the set TN of true negative cases and the set FP of false positive cases; the sizes of these sets are tn and fp respectively, with tn + fp = n. The performance

Table 1. The sizes of the partition subsets resulting from the classification function

		classifier		*total*
		c	$\bar{c}$	
data	c	tp	fn	p
	$\bar{c}$	fp	tn	n
total				m

of a Bayesian network classifier now is commonly captured by its (*empirical*) *accuracy*, which is defined as the proportion of correctly classified cases for a given case set $\mathcal{F}$:

$$accuracy(\mathcal{F}) = \frac{\mathsf{tp} + \mathsf{tn}}{m}$$

The case set from which a classifier's accuracy is established, is usually omitted from the notation; we adopt this convention and from now on leave $\mathcal{F}$ implicit.

3.2   Dependency on the Classification Threshold

The measure of accuracy reviewed above pertains to a given Bayesian network classifier with a fixed classification function. The accuracy of a classifier in general depends on the threshold value t used in its classification function. More specifically, the sizes of the four partition subsets constructed from a case set $\mathcal{F}$ are threshold dependent. For example, a classification function with $t = 0$ would result in each and every case being assigned to class c, from which we would have that $\mathsf{tp} + \mathsf{fp} = m$ and $\mathsf{fn} = \mathsf{tn} = 0$; on the other hand, a threshold value $t > 1$ would distribute all cases over the two sets TN and FN, from which we would have that $\mathsf{fn} + \mathsf{tn} = m$ and $\mathsf{tp} = \mathsf{fp} = 0$. From now on, we make this dependency on the threshold value explicit in our notations, by writing $\mathsf{tp}(t)$, $\mathsf{tn}(t)$, $\mathsf{fp}(t)$, and $\mathsf{fn}(t)$ for the sizes of the four sets TP, TN, FP, and FN, respectively. The accuracy of a classifier as a function of the threshold value t then becomes

$$accuracy(t) = \frac{\mathsf{tp}(t) + \mathsf{tn}(t)}{m}$$

The above considerations show that a classifier's accuracy can be manipulated by choosing an appropriate threshold value t. We note that upon varying the value of t from zero to one, cases *migrate* from the sets TP and FP to the sets TN and FN. More specifically, upon varying t, cases from $\mathcal{F}^+$ can migrate, and migrate only, between the sets TP and FN, whereas cases from $\mathcal{F}^-$ can move only between the sets FP and TN. Despite the seeming mutual independence of the numbers of true positives and true negatives, these numbers are traded off through their dependence on the threshold value t: while $\mathsf{tp}(t)$ is non-increasing for increasing values of t, $\mathsf{tn}(t)$ is non-decreasing in t. The changes in size of the sets TP and TN upon varying the threshold value t are illustrated for our example application in Fig. 1.

4   Stratifying Classifiers

In this section we introduce the idea of stratified classification, by defining classification functions for Bayesian network classifiers with two separate threshold values. In addition, we define associated performance measures.

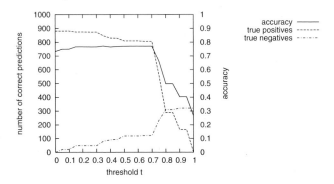

Fig. 1. The accuracy, the number of true positives, and the number of true negatives as functions of the threshold value t for our example classifier and set of 1 202 cases

4.1 Introducing Stratification

The main motivation underlying the introduction of stratifying classifiers is that Bayesian network classifiers should be able to convey indecisiveness, especially when inconclusive cases call for different further decision making than cases with a convincingly outstanding class value. In view of conveying indecisiveness, we now look upon a Bayesian network classifier as distributing a case set over different strata, based upon the computed posterior distribution over the class variable. After stratification, the classifier returns a determinate class value for the cases from some strata and leaves the cases from other strata unclassified. To distinguish between determinate and inconclusive cases, a stratifying classifier employs a partial classification function $class^*$ with two threshold values $t^- \leq t^+$:

$$class^*(C, t^-, t^+; \mathbf{f}) = \begin{cases} c, & \text{if } \Pr(c \mid \mathbf{f}) \geq t^+ \\ \bar{c}, & \text{if } \Pr(c \mid \mathbf{f}) < t^- \end{cases}$$

The threshold value t^- is termed the function's lower threshold value; t^+ is called its upper threshold value. The *-notation is used to denote a function adapted to stratification. Note that for $t^- < t^+$, the stratifying classification function $class^*$ serves to classify only those cases $\mathbf{f}$ for which either $\Pr(c \mid \mathbf{f}) \geq t^+$ or $\Pr(c \mid \mathbf{f}) < t^-$. All cases with $t^- \leq \Pr(c \mid \mathbf{f}) < t^+$ are left unclassified by the stratifying classification function. Further note that the function $class^*$ has the standard, single-threshold classification function as a special case, with $t^- = t^+$.

At first glance, the idea of stratifying classifiers shows similarities to multiway classification and to threshold decision making. In multiway classification, the purpose of the classifier is to distinguish between more than two distinct classes. Stratification in contrast does not increase the number of class values under consideration and thus differs conceptually from multiway classification. The idea of threshold decision making, which was introduced to support physicians during the diagnostic-testing phase in patient management [8], builds upon concepts of decision analysis to establish two patient-specific threshold values, p^- and p^+, on the probability of disease $\Pr(d)$ computed for a patient. These threshold values serve to decide between withholding treatment ($\Pr(d) < p^-$), further diagnostic testing ($p^- \leq \Pr(d) < p^+$), and immediate treatment ($\Pr(d) \geq p^+$). Taking a decision of further diagnostic testing as conveying indecisiveness concerning whether or not to treat the patient, the threshold decision making model can in fact be implemented with a stratifying classifier.

4.2   The Accuracy of a Stratifying Classifier

We consider again the case set $\mathcal{F} = \mathcal{F}^+ \cup \mathcal{F}^-$ with $m = p + n$ cases. The classification function of the stratifying classifier partitions this set in five mutually exclusive and collectively exhaustive subsets. Four of these subsets match the sets TP, FP, TN, and FN introduced above; the fifth set is the set of unclassified cases. The sizes of the five sets again depend upon the threshold values used by the classification function. To capture this dependence, we observe that the stratifying classifier displays the following behaviour:

$$\forall \mathbf{f} \in \mathcal{C}^+ = \{\mathbf{f} \mid \mathbf{f} \in \mathcal{F}, \; \Pr(c \mid \mathbf{f}) \geq t^+\} : \; class^*(C, t^-, t^+; \mathbf{f}) = c$$
$$\forall \mathbf{f} \in \mathcal{C}^- = \{\mathbf{f} \mid \mathbf{f} \in \mathcal{F}, \; \Pr(c \mid \mathbf{f}) < t^-\} : \; class^*(C, t^-, t^+; \mathbf{f}) = \bar{c}$$
$$\forall \mathbf{f} \in \mathcal{C}^u = \mathcal{F} \setminus (\mathcal{C}^+ \cup \mathcal{C}^-) : \; \text{unclassified}$$

This observation shows that all cases from the set $\mathcal{C}^+$ are classified as being positive; the set thus is distributed over the two sets TP and FP. Since the size of the set $\mathcal{C}^+$ depends on the upper threshold value t^+, the sizes tp and fp of the sets TP and FP depend on the value t^+ as well; we will write $\mathsf{tp}(t^+)$ and $\mathsf{fp}(t^+)$, respectively, to express this dependence. Similarly, the set $\mathcal{C}^-$ is distributed over TN and FN. The sizes tn and fn of these sets depend on the lower threshold value t^-; we will write $\mathsf{tn}(t^-)$ and $\mathsf{fn}(t^-)$, respectively, to express this dependence.

We recall that, for a standard Bayesian network classifier with a classification function based on a single threshold value t, accuracy is defined as the proportion of cases that are assigned to their true class value:

$$accuracy(t) = \frac{\mathsf{tp}(t) + \mathsf{tn}(t)}{m}$$

The proportion of cases that are correctly classified by a stratifying classifier now equals

$$accuracy(t^-, t^+) = \frac{\mathsf{tp}(t^+) + \mathsf{tn}(t^-)}{m}$$

Since a stratifying classifier may leave some cases unclassified, one or more of the sets TP, FP, TN, and FN may decrease in size compared to those with a standard classifier. More specifically, we find that

$$\mathsf{tp}(t^+) + \mathsf{fn}(t^-) = p^* \leq p \quad \text{and} \quad \mathsf{tn}(t^-) + \mathsf{fp}(t^+) = n^* \leq n$$

where p^* is the number of actually classified cases from $\mathcal{F}^+$ and n^* is the number of classified cases from $\mathcal{F}^-$; $m - p^* - n^*$ cases are left unclassified by the stratifying classifier. Now, if the stratification results in smaller sets TP and/or TN, then some of the cases considered inconclusive after stratification would have been classified correctly, yet not convincingly, without the stratification. The accuracy of the stratifying classifier then is smaller than that of the standard classifier. On the other hand, if the stratification affects neither TP nor TN, then the accuracy of the stratifying classifier remains unchanged compared to that of the standard classifier, which indicates that all cases considered inconclusive after stratification would have been classified incorrectly without the stratification. For the stratifying classifier, the standard measure of accuracy thus still captures the proportion of correctly classified cases from *all* cases presented to the classifier, including those considered inconclusive. To capture the proportion of correctly classified cases among the cases that were actually classified, we now introduce a measure of stratified accuracy, defined by

$$accuracy^*(t^-, t^+) = \frac{\mathsf{tp}(t^+) + \mathsf{tn}(t^-)}{m^*}$$

where m^* equals $p^* + n^* = \mathsf{tp}(t^+) + \mathsf{tn}(t^-) + \mathsf{fp}(t^+) + \mathsf{fn}(t^-)$.

The measure of stratified accuracy is in essence defined in terms of the sets $\mathcal{C}^+$ and $\mathcal{C}^-$. The measure can also be related to the set $\mathcal{C}^u$ of cases that are left unclassified by the stratifying classifier. To this end, we consider the distribution of these inconclusive cases over the sets TP, FP, TN, and FN with a standard classifier. For each $\mathbf{f} \in \mathcal{C}^u$, we have that

$$class(C, t^-; \mathbf{f}) = c \quad \text{and} \quad class(C, t^+; \mathbf{f}) = \bar{c}$$

With the threshold value t^-, therefore, a standard classifier would distribute all cases from $\mathcal{C}^u$ over the two sets TP and FP, with sizes $\mathsf{tp}(t^-)$ and $\mathsf{fp}(t^-)$, respectively. Similarly, with the threshold value t^+, all cases from $\mathcal{C}^u$ would be distributed over the sets TN and FN, with sizes $\mathsf{tn}(t^+)$ and $\mathsf{fn}(t^+)$. Upon varying the threshold value from t^- to t^+, therefore, the cases $\mathbf{f} \in \mathcal{C}^u$ would migrate from the set TP to the set FN, and from FP to TN. This observation underlies the following formula:

$$accuracy^*(t^-, t^+) = \frac{\mathsf{tp}(t^-) - \Delta\mathsf{tp} + \mathsf{tn}(t^+) - \Delta\mathsf{tn}}{m - \Delta\mathsf{tp} - \Delta\mathsf{tn}}$$

where $\Delta\mathsf{tp} = \mathsf{tp}(t^-) - \mathsf{tp}(t^+)$ is the number of cases from $\mathcal{C}^u \cap \mathcal{F}^+$ that would be incorrectly classified as negative if t^+ were to be taken as the single threshold value; $\Delta\mathsf{tn} = \mathsf{tn}(t^+) - \mathsf{tn}(t^-)$ has an analogous interpretation.

The effects of stratification on the accuracy of a classifier can be studied by comparing the resulting stratified accuracy to the standard accuracy. We consider to this end a standard Bayesian network classifier with the classification function $class(C, t; \mathbf{f})$. Introducing stratification into this classifier entails choosing two threshold values t^- and t^+, $t^- \leq t \leq t^+$, and replacing the function *class* by the stratifying classification function $class^*$. If neither the set TP nor the set TN is affected by the stratification, that is, if $\mathsf{tp}(t^+) = \mathsf{tp}(t)$ and $\mathsf{tn}(t^-) = \mathsf{tn}(t)$, we find that

$$accuracy(t^-, t^+) = accuracy(t) \quad \text{and} \quad accuracy^*(t^-, t^+) \geq accuracy(t)$$

where equality holds for the formula on the right whenever $\mathsf{tp}(t^-) = \mathsf{tp}(t)$ and $\mathsf{tn}(t^+) = \mathsf{tn}(t)$. If the stratification results in a decrease in size of the sets TP and/or TN, then the standard accuracy decreases with the stratification: $accuracy(t^-, t^+) < accuracy(t)$. The stratified accuracy $accuracy^*(t^-, t^+)$, however, can be smaller than, equal to, or larger than the standard accuracy, depending on the size of the set $\mathcal{C}^u$ of inconclusive cases and the standard classifier's performance on $\mathcal{C}^u$. If $accuracy^*(t^-, t^+) < accuracy(t)$, we say that stratification results in a deterioration in the performance of the classifier. Such a deterioration indicates that, among the unclassified cases, a relatively large number were classified correctly prior to the stratification. It further means that the cases which remain incorrectly classified after the stratification, are cases for which high posterior probabilities are established for the *incorrect* class value. Often, however, the introduction of stratification will result in an improvement of the performance of a classifier, that is, in $accuracy^*(t^-, t^+) > accuracy(t)$. An appropriate choice of threshold values can in fact result in extremely high stratified accuracies, possibly even equal to 1.

4.3   The Classification Percentage

In most applications, the introduction of stratification into a Bayesian network classifier will result in an increased stratified accuracy. The improvement in classification performance, however, typically comes at the price of a reduced practicability of the classifier for decision support. To capture the issue of practicability, we introduce the concept of *classification percentage*, which equals the proportion of cases that are classified:

$$classification\_percentage = \frac{m^*}{m} \cdot 100\%$$

Note that a standard classifier has a classification percentage of 100%. By introducing stratification, the classification percentage will typically decrease. When viewing a stratifying classifier as a tool for support to a decision maker in his daily practice, the classification percentage indicates, given the stratification under consideration, the percentage of cases for which the tool will actually advance the decision-making process. Alternatively, the classification percentage conveys information about the percentage of cases for which the tool will be indecisive, that is, for which the tool will leave the actual decision to the decision maker.

5   Stratification in the Example Application

In our application domain of animal health management, a dairy farmer typically has to decide upon treatment of a cow with clinical mastitis before knowing the pathogen that causes the disease. As a result, often broad-spectrum antibiotics are administered, where narrow-spectrum antibiotic treatment is preferred. The administration of narrow-spectrum antibiotics is possible, however, only if one specific pathogen is convincingly favoured over all others. Our naive Bayesian

Table 2. Predicted and actual numbers of positive and negative cases for our stratifying classifier, using threshold values $t^- = 0.30$ and $t^+ = 0.80$, with 1 202 cases

		classifier +	classifier −	total
data	+	289	8	297
	−	11	48	59
	total			356

classifier supports the choice of antibiotics by classifying mastitis cases according to the Gram-status of the causal pathogen. We recall that this Gram-status is an important indicator for the type of antibiotics to be used. The predicted Gram-status, however, may be quite uncertain for cases with a posterior probability close to the threshold value of the classification function. For such cases, in fact, a broad-spectrum treatment would still be preferred. In this section we use our example application to illustrate the concepts, measures and observations put forward in the previous section.

With our standard naive Bayesian classifier and with the case set of 1 202 mastitis cases, we find the following values tp and tn for the sizes of the sets TP and TN of correctly classified cases, for different threshold values t:

t	0.30	0.40	0.50	0.60	0.70	0.80	0.90	1.00
tp	873	830	809	809	806	289	165	0
tn	48	90	118	118	121	310	321	321

With a threshold value of $t = 0.50$ for the classification function, for example, the accuracy of our classifier equals $(809 + 118)/1\,202 = 0.77$. We now introduce stratification into our classifier, using a classification function with threshold values $t^- = 0.30$ and $t^+ = 0.80$. With the resulting stratifying classifier, a total of 356 cases from the case set are classified, giving a classification percentage of 29.6%. For the remaining 846 cases the classifier is indecisive, indicating that the dairy farmer should administer broad-spectrum antibiotics to the diseased cows. The distribution of the classified cases over the four sets TP, FP, TN, and FN is shown in Table 2. The stratifying classifier has a standard accuracy of 0.28 and a stratified accuracy of 0.95. Fig. 2 shows the effects of separately varying the two threshold values t^- and t^+, on the two accuracies. The figure clearly shows that an increasing distance between the two threshold values may result in a higher stratified accuracy, which then typically comes at the expense of a decrease in the classifier's classification percentage.

Our earlier observation that the introduction of stratification may both serve to improve and deteriorate classifier performance, is illustrated by the following example. We consider two different classification functions for our stratifying classifier: one function with the threshold values $t^- = 0.40$ and $t^+ = 0.80$ (I), and another one with the threshold values $t^- = 0.40$ and $t^+ = 1.00$ (II). The resulting classifiers both have a standard accuracy smaller than that of the standard classifier. While the standard classifier has an accuracy of $accuracy(0.50) = 0.77$, we find for the two stratifying classifiers:

$$\text{(I):}\quad accuracy(0.40, 0.80) = \frac{289 + 90}{1\,202} = 0.32, \quad\text{and}$$

$$\text{(II):}\quad accuracy(0.40, 1.00) = \frac{0 + 90}{1\,202} = 0.07$$

For classifier (I), we further find for its stratified accuracy:

$$accuracy^*(0.40, 0.80) = \frac{289 + 90}{1\,202 - (830 - 289) - (310 - 90)} = 0.86$$

which shows an increase in accuracy over the standard, single-threshold classifier. This increase is explained by the observation that the decrease in the number of correct classifications compared to the standard classifier, is smaller than the relative decrease in the total number of classified cases. For classifier (II), on the other hand, we find that

$$accuracy^*(0.40, 1.00) = \frac{90}{1\,202 - (321 - 90) - (830 - 0)} = 0.64$$

which reveals a decrease in accuracy compared to the standard classifier. This decrease is explained by the observation that the relative decrease in the number of correct classifications now is larger than that in the number of classified cases.

To conclude, by introducing stratification into our example naive Bayesian classifier with threshold values $t^- = 0.10$ and $t^+ = 0.90$, we find a stratified accuracy of 1.00 at a classification percentage of 15.5%. A much poorer choice of threshold values is $t^- = 0.20$ and $t^+ = 1.00$, which results in a stratifying classifier which is decisive on just 4.5% of the cases and has a stratified accuracy of 0.87. The same stratified accuracy is also obtained by setting the threshold value t^+ to the smaller value of $t^+ = 0.75$. We now find a classification percentage of 58.7%!

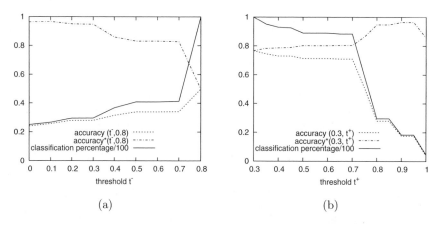

Fig. 2. The accuracy, stratified accuracy, and classification percentage for our example classifier and 1 202 cases, as functions of (a) threshold value t^-, with threshold value t^+ fixed at 0.80, and (b) threshold value t^+, with t^- fixed at 0.30

6   Conclusions

For some problems, the single class value returned by a classifier does not necessarily provide a sufficient basis for reliable further decision making. Building upon this observation, we introduced stratifying classifiers as classifiers with the ability to express indecisiveness by not classifying inconclusive cases. These stratifying classifiers are particularly appropriate for applications in which indecisiveness about the class value for a case is a usable result for the decision maker, as in our application domain. Associated with this new type of classifier, we introduced new measures of classification performance and practicability; these measures serve to give insight in the values of the two threshold values to be used with the classification function. In the future we want to extend our concept of stratification to multiway classification and to study the practicability of returning two or more class values for inconclusive cases.

References

1. Blanco, R., Inza, M., Merino, M., Quiroga, J., Larrañaga, P.: Feature Selection in Bayesian Classifiers for the Prognosis of Survival of Cirrhotic Patients Treated with TIPS. Journal of Biomedical Informatics 38, 376–388 (2005)
2. Chapman, W.W., Dowling, J.N., Wagner, M.M.: Classification of Emergency Department Chief Complaints into 7 Syndromes: A Retrospective Analysis of 527,228 Patients. Annals of Emergency Medicine 46, 445–455 (2005)
3. Friedman, N., Goldszmidt, M.: Building Classifiers Using Bayesian Networks. In: Proceedings of the National Conference on Artificial Intelligence, pp. 1277–1284. AAAI Press, Menlo Park (1996)
4. Friedman, N., Geiger, D., Goldszmidt, M.: Bayesian Network Classifiers. Machine Learning 29, 131–163 (1997)
5. Geenen, P.L., Van der Gaag, L.C., Loeffen, W.L.A., Elbers, A.R.W.: Naive Bayesian Classifiers for the Clinical Diagnosis of Classical Swine Fever. In: Proceedings of the 2005 Meeting of the Society for Veterinary Epidemiology and Preventive Medicine, Nairn, Scotland, pp. 169–176 (2005)
6. Kazmierska, J., Malicki, J.: Application of the Naive Bayesian Classifier to Optimize Treatment Decisions. Radiotherapy and Oncology 86, 211–216 (2008)
7. Kuncheva, L.I., Vilas, V.J.D., Rodriguez, J.J.: Diagnosing Scrapie in Sheep: A Classification Experiment. Computers in Biology and Medicine 37, 1194–1202 (2007)
8. Pauker, S.G., Kassirer, J.P.: The Threshold Approach to Clinical Decision Making. New England Journal of Medicine 302, 1109–1117 (1980)
9. Steeneveld, W., Van der Gaag, L.C., Barkema, H.W., Hogeveen, H.: Providing Probability Distributions for the Causal Pathogen of Clinical Mastitis Using Naive Bayesian Networks. Journal of Dairy Science (accepted for publication) (2009)

Acyclic Directed Graphs to Represent Conditional Independence Models

Marco Baioletti[1], Giuseppe Busanello[2], and Barbara Vantaggi[2]

[1] Dip. Matematica e Informatica, Università di Perugia, Italy
baioletti@dipmat.unipg.it
[2] Dip. Metodi e Modelli Matematici, Università "La Sapienza" Roma, Italy
{busanello,vantaggi}@dmmm.uniroma1.it

Abstract. In this paper we consider conditional independence models closed under graphoid properties. We investigate their representation by means of acyclic directed graphs (DAG). A new algorithm to build a DAG, given an ordering among random variables, is described and peculiarities and advantages of this approach are discussed. Finally, some properties ensuring the existence of perfect maps are provided. These conditions can be used to define a procedure able to find a perfect map for some classes of independence models.

Keywords: Conditional independence models, Graphoid properties, Inferential rules, Acyclic directed graphs, Perfect map.

1 Introduction

Graphical models [6,8,9,10,12,15] play a fundamental role in probability and statistics and they have been deeply developed as a tool for representing conditional independence models. It is well known (see for instance [6]) that, under the classical definition, the independence model $\mathcal{M}$ associated to any probability measure P is a semi–graphoid and, if P is strictly positive, $\mathcal{M}$ is a graphoid. On the other hand, an alternative definition of independence (a reinforcement of cs–independence [4,5]), which avoids the well known critical situations related to 0 and 1 evalutations, induces independence models closed under graphoid properties [13].

In this paper the attention is focusing on graphoid structures and we consider a set J of conditional independence statements, compatible with a (conditional) probability, and its closure $\bar{J}$ with respect to graphoid properties. Since the computation of $\bar{J}$ is infeasible (its size is exponentially larger than the size of J), then, as shown in [10,11,1], we will use a suitable set J_* of independence statements (obviously included in $\bar{J}$), that we call "fast closure", from which it is easy to verify whether a given relation is implied, i.e. whether a given relation belongs to $\bar{J}$. Some of the main properties of fast closure will be described.

The fast closure is also relevant for building the relevant acyclic directed graph (DAG), which is able to concisely represent the independence model. In fact we

C. Sossai and G. Chemello (Eds.): ECSQARU 2009, LNAI 5590, pp. 530–541, 2009.

will define the procedure BN-DRAW which builds, starting from this set and an ordering on the random variables, the corresponding independence map. The main difference between BN-DRAW and the classical procedures (see e.g. [7,9]) is that the relevant DAG is built without referring to the whole closure.

Finally, we give a condition assuring the existence of a perfect map, i.e. a DAG able to represent all the independence statements of a given independence model. By using this result it is possible to define a correct, but incomplete, method to find a perfect map. First, a suitable ordering, satisfying this condition, is searched by means of a backtracking procedure. If such an ordering exists, a perfect map for the independence model can be found by using the procedure BN-DRAW.

Since the above condition is not necessary, but only sufficient, as shown in Example 3, such condition can fail even if a perfect map exists. The provided result is a first step to look for a characterization of orderings giving rise to perfect maps.

2   Graphoid

Throughout the paper the symbol $\tilde{S} = \{Y_1, \ldots, Y_n\}$ denotes a finite not empty set of variables. Given a probability P, a conditional independence statement $Y_A \perp\!\!\!\perp Y_B | Y_C$ (compatible with P), where A, B, C are disjoint subsets of the set $S = \{1, \ldots, n\}$ of indices associated to $\tilde{S}$, is simply denoted by the ordered triple (A, B, C). Furthermore, $S^{(3)}$ is the set of all ordered triples (A, B, C) of disjoint subsets of S, such that A and B are not empty. A conditional independence model $\mathcal{I}$, related to a probability P, is a subset of $S^{(3)}$. As recalled in the introduction we refer to probabilistic independence models even if the results are valid for any graphoid structure.

We recall that a graphoid is a couple $(S, \mathcal{I})$, with $\mathcal{I}$ a ternary relation on the set $S^{(3)}$, satisfying the following properties:

G1: if $(A, B, C) \in \mathcal{I}$, then $(B, A, C) \in \mathcal{I}$ (Symmetry);

G2: if $(A, B, C) \in \mathcal{I}$, then $(A, B', C) \in \mathcal{I}$ for any nonempty subset B' of B (Decomposition);

G3: if $(A, B_1 \cup B_2, C) \in \mathcal{I}$ with B_1 and B_2 disjoint, then $(A, B_1, C \cup B_2) \in \mathcal{I}$ (Weak Union);

G4: if $(A, B, C \cup D) \in \mathcal{I}$ and $(A, C, D) \in \mathcal{I}$, then $(A, B \cup C, D) \in \mathcal{I}$ (Contraction);

G5: if $(A, B, C \cup D) \in \mathcal{I}$ and $(A, C, B \cup D) \in \mathcal{I}$, then $(A, B \cup C, D) \in \mathcal{I}$ (Intersection).

A semi–graphoid is a couple $(S, \mathcal{I})$ satisfying only the properties G1–G4. The symmetric version of rules G2 and G3 will be denoted by

G2s: if $(A, B, C) \in \mathcal{I}$, then $(A', B, C) \in \mathcal{I}$ for any nonempty subset A' of A;

G3s: if $(A_1 \cup A_2, B, C) \in \mathcal{I}$, then $(A_1, B, C \cup A_2) \in \mathcal{I}$.

3 Generalized Inference Rules

Given a set J of conditional independence statements compatible with a probability, a relevant problem about graphoids is to find, in an efficient way, the closure of J with respect to G1–G5

$$\bar{J} = \{\theta \in S^{(3)} : \theta \text{ is obtained from } J \text{ by } G1 - G5\}.$$

A related problem, called implication, concerns to establish whether a triple $\theta \in S^{(3)}$ can be derived from J, see [16].

It is clear that the implication problem can be easily solved once the closure has been computed. But, the computation of the closure is infeasible because its size is exponentially larger than the size of J. In [1,2,3] we describe how it is possible to compute a smaller set of triples having the same information as the closure. The same problem has been already faced successfully in [11], with particular attention to semi–graphoid structures.

In the following for a generic triple $\theta_i = (A_i, B_i, C_i)$, the set X_i stands for $(A_i \cup B_i \cup C_i)$.

We recall some definitions and properties introduced and studied in [1,2,3] useful to efficiently compute the closure of a set of conditional independence statements. Given a pair of triples $\theta_1, \theta_2 \in S^{(3)}$ we say that θ_1 is *generalized–included* in θ_2 (briefly g–included), in symbol $\theta_1 \sqsubseteq \theta_2$, if θ_1 can be obtained from θ_2 by a finite number of applications of G1, G2 and G3.

Proposition 1. *Given $\theta_1 = (A_1, B_1, C_1)$ and $\theta_2 = (A_2, B_2, C_2)$, then $\theta_1 \sqsubseteq \theta_2$ if and only if the following conditions hold*

(i) $C_2 \subseteq C_1 \subseteq X_2$;
(ii) either $A_1 \subseteq A_2$ and $B_1 \subseteq B_2$ or $A_1 \subseteq B_2$ and $B_1 \subseteq A_2$.

Generalized inclusion is strictly related to the concept of dominance $\sqsubseteq_a$ on $S^{(3)}$, already defined in [10,11]. We say $\theta_1 \sqsubseteq_a \theta_2$ if θ_1 can be obtained from θ_2 with a finite number of applications of G2, G3, G2s and G3s.

Therefore it is easy to see that $\theta' \sqsubseteq \theta$ if and only if either $\theta' \sqsubseteq_a \theta$ or $\theta' \sqsubseteq_a \theta^T$ where θ^T is the transpose of θ ($\theta^T = (B, A, C)$ if $\theta = (A, B, C)$).

The definition of g–inclusion between triples can be extended to sets of triples and its properties are showed in [2,3].

Definition 1. *Let H and J be subsets of $S^{(3)}$. J is a covering of H (in symbol $H \sqsubseteq J$) if and only if for any triple $\theta \in H$ there exists a triple $\theta' \in J$ such that $\theta \sqsubseteq \theta'$.*

3.1 Closure through One Generalized Rule

The target of [10,2,3] is to find a fast method to compute a reduced (with respect to g–inclusion $\sqsubseteq$) set J^* bearing the same information of $\bar{J}$, that is for any triple $\theta \in \bar{J}$ there exists a triple $\theta' \in J^*$ such that $\theta \sqsubseteq \theta'$.

Therefore, the computation of J^* provides a simple solution to the implication problem for J. The strategy to compute J^* is to use a generalized version of the remaining graphoid rules G4, G5 and their symmetric ones (see also [11]).

These two inference rules are called generalized contraction (G4*) and generalized intersection (G5*). The rule G4* allows to deduce from θ_1, θ_2 the greatest (with respect to $\sqsubseteq$) triple τ which derives from the application of G4 to all the possible pairs of triples θ'_1, θ'_2 such that $\theta'_1 \sqsubseteq \theta_1$ and $\theta'_2 \sqsubseteq \theta_2$. The rule G5* is analogously defined and it is based on G5 instead of G4.

It is possible to compute the closure of a set J of triples in $S^{(3)}$, with respect to G4* and G5*, that is

$$J^* = \{\tau : J \vdash^*_G \tau\} \tag{1}$$

where $J \vdash^*_G \tau$ means that τ is obtained from J by applying a finite number of times the rules G4* and G5*.

In [2,3] it is proved that J^*, even if it is smaller, is equivalent to $\bar{J}$ with respect to graphoids, in that $J^* \sqsubseteq \bar{J}$ and $\bar{J} \sqsubseteq J^*$.

A further reduction is to keep only the "maximal" (with respect to g–inclusion) triples of J^*

$$J^*_{/\sqsubseteq} = \{\tau \in J^* : \nexists \bar{\tau} \in J^* \text{ with } \bar{\tau} \neq \tau, \tau^T \text{ such that } \tau \sqsubseteq \bar{\tau}\}.$$

Obviously, $J^*_{/\sqsubseteq} \subseteq J^*$.

In [2,3] it is proved that $J^* \sqsubseteq J^*_{/\sqsubseteq}$, therefore there is no loss of information by using $J^*_{/\sqsubseteq}$ instead of J^*. Then, given a set J of triples in $S^{(3)}$, we compute the set $J^*_{/\sqsubseteq}$, which we call "fast closure" and denote with J_*.

In [2,3] it is proved that the fast closure set $\{\theta_1, \theta_2\}_*$ of two triples $\theta_1, \theta_2 \in S^{(3)}$ is formed with at most eleven triples. Furthermore, these triples have a particular structure strictly related to θ_1 and θ_2 and they can be easily computed. By using $\{\theta_1, \theta_2\}_*$, it is possible to define a new inference rule

C : from θ_1, θ_2 deduce any triple $\tau \in \{\theta_1, \theta_2\}_*$.

We denote with J^+ the set of triples obtained from J by applying a finite number of times the rule C. As proved in [2,3], J^+ is equivalent to $\bar{J}$ with respect to graphoids, that means $J^+ \sqsubseteq \bar{J}$ and $\bar{J} \sqsubseteq J^+$. Obviously, by transitivity J^+ is equivalent to J_*.

Therefore, it is possible to design an algorithm, called FC1, which starts from J and recursively applies the rule C and a procedure, called FINDMAXIMAL (which computes $H_{/\sqsubseteq}$ for a given set $H \subseteq S^{(3)}$), until it arrives to a set of triples closed with respect to C and maximal with respect to g–inclusion.

In [2,3] completeness and correctness of FC1 are proved.

Note that, as confirmed in [2,3] by some experimental results, FC1 is more efficient than the algorithms based on G4* and G5*.

4   Graphs

In the following, we refer to the usual graph definitions (see e.g. [9]). We denote by $G = (\mathcal{U}, E)$ a graph with set of nodes $\mathcal{U}$ and oriented arcs E formed by ordered pairs of nodes. In particular, we consider directed graphs having no cycles, i.e. acyclic directed graphs (DAG). We denote for any $u \in \mathcal{U}$, as usual, with $pa(u)$ the parents of u, $ch(u)$ the child of u, $ds(u)$ the sets of descendants and $an(u)$ the set of ancestors. We use the convention that each node u belongs to $an(u)$ and to $ds(u)$, but not to $pa(u)$ and $ch(u)$.

Definition 2. *If A, B and C are three disjoint subsets of nodes in a DAG G, then C is said to d–separate A from B, denoted $(A, B, C)_G$, if there is no path between a node in A and a node in B along which the following two conditions hold:*

1. *every node with converging arrows is in C or has a descendent in C;*
2. *every other node is outside C.*

In order to study the representation of a conditional independence model, we need to distinguish between dependence map and independence map, since there are conditional independence models that cannot be completely represented by a DAG (see e.g. [9,11]).

In the following we denote with J (analogously for $\bar{J}$, J_*) both a set of triples and a set of conditional independence relations, obviously, the triples are defined on the set S and the independence relations on $\tilde{S}$. Then a graph representing the conditional independence relations of J has S as node set.

Definition 3. *Let J be a set of conditional independence relations on a set $\tilde{S}$ of random variables. A DAG $G = (S, E)$ is a dependence map (briefly a D–map) if for all triple $(A, B, C) \in S^{(3)}$*

$$(A, B, C) \in \bar{J} \Rightarrow (A, B, C)_G.$$

Moreover, $G = (S, E)$ is an independence map (briefly an I–map) if for all triple $(A, B, C) \in S^{(3)}$

$$(A, B, C)_G \Rightarrow (A, B, C) \in \bar{J}.$$

G is a minimal I–map of J if deleting any arc, G is no more an I–map.

G is said to be a perfect map (briefly a p–map) if it is both a I–map and a D–map.

The next definition and theorem provide a tool to build a DAG given a independence model $\bar{J}$.

Definition 4. *Let $\bar{J}$ be an independence model defined on S and let $\pi = <\pi_1, \ldots, \pi_n>$ an ordering of the elements of S. The boundary strata of $\bar{J}$ relative to π is an ordered set of subsets $< B_1, B_2, \ldots, B_i, \ldots >$ of S, such that* each B_i is a minimal set satisfying $B_i \subseteq S_{(i)} = \{\pi_1, \ldots, \pi_{i-1}\}$ and $\gamma_i = (\{\pi_i\}, S_{(i)} \setminus B_i, B_i) \in \bar{J}$.

The DAG created by setting each B_i as parent set of the node π_i is called boundary DAG of J relative to π.

The triple γ_i is known as *basic triple*.

The next theorem is an extension of Verma's Theorem [14] stated for conditional independence relations (see [9]).

Theorem 1. *Let J be a independence model closed with respect to the semi–graphoid properties. If G is a boundary DAG of J relative to any ordering π, then G is a minimal I–map of J.*

The previous theorem helps to build a DAG for an independence model $\bar{J}_P$ induced by a probability assessment P on a set of random variables $\tilde{S}$ and a fixed ordering π on indices of S.

Now, we recall an interesting result [9].

Corollary 1. *An acyclic directed graph $G = (S, E)$ is a minimal I–map of an independence model J if and only if any index $i \in S$ is conditionally independent of all its non-descendants, given its parents $pa(i)$, and no proper subset of $pa(i)$ satisfies this condition.*

It is well known (see [9]) that the boundary DAG of J relative to π is a minimal I-map.

5   BN-Draw Function

The aim of this section is to define the procedure BN–DRAW, which builds a minimal I–map G (see Definition 3) given the fast closure J_* (introduced in Section 3) of a set J of independence relations. The procedure is described in the algorithm 1.

Given the fast closure set J_*, we cannot apply the standard procedure (see [7,9]) described in Definition 4 to draw an I–map because, in general, the basic triples related to an arbitrary ordering π could not be elements of J_*, but they could be just g–included to some triples of J_*, as shown in Example 1.

Example 1. Given $J = \{(\{1\}, \{2\}, \{3, 4\}), (\{1\}, \{3\}, \{4\})\}$, we want to find the corresponding basic triples and to draw the relevant DAG G related to the ordering $\pi = < 4, 2, 1, 3 >$.

By the closure with respect to graphoid properties we obtain

$\bar{J} = \{ (\{1\}, \{2\}, \{3, 4\}), (\{1\}, \{3\}, \{4\}), (\{1\}, \{2, 3\}, \{4\}), (\{1\}, \{2\}, \{4\}), (\{1\}, \{3\}, \{2, 4\}), (\{2\}, \{1\}, \{3, 4\}), (\{3\}, \{1\}, \{4\}), (\{2, 3\}, \{1\}, \{4\}), (\{2\}, \{1\}, \{4\}), (\{3\}, \{1\}, \{2, 4\}) \}$

and the set of basic triples is $\Gamma = \{(\{1\}, \{2\}, \{4\}), (\{3\}, \{1\}, \{2, 4\})\}$.

By FC1 we botain $J_* = \{(\{1\}, \{2, 3\}, \{4\})\}$ and it is simple to observe that $\Gamma \sqsubseteq J_*$.

Algorithm 1. DAG from J_* given an order π of S

1: **function** BN-DRAW(n, π, J_*) ▷ n is the cardinality of S
2:     **for** $i \leftarrow 2$ **to** n **do**
3:         $pa \leftarrow S_{(i)}$
4:         **for each** $(A, B, C) \in J_*$ **do**
5:             **if** $\pi_i \in A$ **then**
6:                 $p \leftarrow C \cup (A \cap S_{(i)})$
7:                 $r \leftarrow B \cap S_{(i)}$
8:                 **if** $(C \subseteq S_{(i)})$ and $(p \cup r = S_{(i)})$ and $(r \neq \phi)$ and $(|p| < |pa|)$ **then**
9:                     $pa \leftarrow p$
10:                 **end if**
11:             **end if**
12:             **if** $\pi_i \in B$ **then**
13:                 $p \leftarrow C \cup (B \cap S_{(i)})$
14:                 $r \leftarrow A \cap S_{(i)}$
15:                 **if** $(C \subseteq S_{(i)})$ and $(p \cup r = S_{(i)})$ and $(r \neq \phi)$ and $(|p| < |pa|)$ **then**
16:                     $pa \leftarrow p$
17:                 **end if**
18:             **end if**
19:         **end for**
20:         draw an arc from each index in pa to π_i
21:     **end for**
22: **end function**

The procedure BN-DRAW finds, for each π_i, with $i = 2, \ldots, n$, and possibly for each $\theta \in J_*$, a triple $(\{\pi_i\}, B, C) \sqsubseteq \theta$ such that $B \cup C = S_{(i)}$ and C has the minimum cardinality (analogously for the triples of the form $(A, \{\pi_i\}, C)$). It is easy to see that for each π_i, the triple with the smallest cardinality of C among all the selected triples, coincides with the basic triple γ_i, if γ_i exists. The formal justification of this statement is given by the following result:

Proposition 2. *Let $\bar{J}$ be an independence model on an index set S, J_* its fast closure and π an ordering on S. Then, the set*

$$\mathcal{B}_i = \{(\{\pi_i\}, B, C) \in S^{(3)} : B \cup C = S_{(i)}, \{(\{\pi_i\}, B, C)\} \sqsubseteq J_*\}$$

is not empty if and only if the basic triple $\gamma_i = (\{\pi_i\}, S_{(i)} \setminus B_i, B_i)$ related to π exists, for $i = 1, \ldots, |S|$.

Proof. Suppose that $\mathcal{B}_i$ is not empty. If there are two triples θ^1, θ^2 having the same cardinality of $pa(i)$ then, by definition of J_*, there is also the triple θ^3 obtained by applying the intersection rule between them. Since the third component of θ^3 has a smaller cardinality than those of θ^1 and θ^2, this means that the triple with the minimum cardinality of C is unique and it coincides with the basic triple γ_i. Vice versa, if the basic triple $\gamma_i = (\{\pi_i\}, S_{(i)} \setminus B_i, B_i)$ for π_i exists, then it is straightforward to see that $\gamma_i \in \mathcal{B}_i$.   □

Note that BN–DRAW allows to build the corresponding I–map, related to π, in linear time with respect to the cardinality of J_*; while the standard procedure requires a time proportional to the size of $\bar{J}$, which is usually much larger, as shown also in the empirical tests in [1]. Also the space needed in memory is almost exclusively used to contain the fast closure. Note that a theoretical comparison between the size of the whole closure and the size of the fast closure has not already been found and seems to be a difficult task.

The next example compares the standard procedure recalled in Definition 4 with BN-DRAW to build the I–map, given a subset J of $S^{(3)}$ and an ordering π among the elements of S.

Example 2. Consider the same independence set J of Example 1 and the ordering $\pi =< 4, 2, 1, 3 >$, we compute the basic triple by applying BN-DRAW to $J_* = \{\theta = (\{1\}, \{2, 3\}, \{4\})\}$. For $i = 2$ we have $2 \in B$, $p = \phi$, $r = \phi$, $C = \{4\} \subseteq \{4\}$, then there is no basic triple. For $i = 3$ we have $1 \in A$, $p = \phi$, $r = \{2\}$, $C = \{4\}$ then $(1, 2, 4)$ is a basic triple g–included to θ. For $i = 4$ we have $3 \in B$, $p = \{2\}$, $r = \{1\}$, $C = \{4\}$ then $(\{3\}, \{1\}, \{2, 4\})$ is a basic triple g–included to θ.

Therefore, we obtain the same set Γ computed in Example 1.

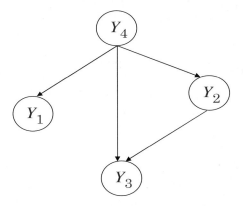

Fig. 1. $I - map$ related to $\pi =< 4, 2, 1, 3 >$

6   Perfect Map

In this section we introduce a condition ensuring the existence of a perfect map for an independence model $\bar{J}$, starting from the fast closure J_* of J (avoiding to build the whole $\bar{J}$, as recalled in Section 2). Given an ordering π on S, we denote with G_π the corresponding I–map of J_* with respect to π. Furthermore, we associate to any index $s \in S$ the set $S_{(s)}$ of indices appearing in π before s and the minimal subset B_s of $S_{(s)}$ such that $\gamma_s = (s, S_{(s)} \setminus B_s, B_s)$ is the basic triple, if any, as introduced in Definition 4.

Before stating the sufficient condition that ensures the existence of a perfect map G_π of the fast closure J_*, with respect to π, we want to underline some relations among the indices of a triple represented in G_π and g–included in J_*. In particular, we focus our attention on the relationship among the nodes associated to C and those related to $pa(A \cup B)$.

By Corollary 1, for any index $i \in S$ the triple $\{(\{i\}, S \setminus ds(i), pa(i))\}$ is represented in G_π and, since G_π is an I–map, $\{(\{i\}, S \setminus ds(i), pa(i))\} \sqsubseteq J_*$. Moreover, by d–separation, for any $K \subseteq ds(i) \setminus \{i\}$ such that $pa(i)$ d–separates $S \setminus ds(i)$ and K, also $\{(\{i\} \cup K, S \setminus ds(i), pa(i)), (\{i\}, S \setminus ds(i), pa(i) \cup K)\} \sqsubseteq J_*$.

Moreover, we have the following result considering I–maps and fast closure J_*.

Proposition 3. *Let J be a set of conditional independence relations and J_* its fast closure.*

If G_π is an I–map of J_*, with respect to π, then, for any triple $\theta = (A, B, C) \in J_*$ represented in G_π, it holds $pa(D_A^B) \cup pa(D_B^A) \subseteq C$, where $D_A^B = ds(A) \cap B$ and $D_B^A = ds(B) \cap A$.

Proof. Suppose by absurd that there exist $\alpha \in D_A^B$ and $j \in pa(\alpha)$ such that $j \notin C$. We want to show that $\theta' = (A, B \cup \{j\}, C)$ is represented in G_π. Let ρ a path from $a \in A$ to j. If ρ passes through α, then ρ is blocked by C (because $(A, B, C)_{G_\pi}$). Otherwise, if ρ does not pass through α, then the path ρ' from a to α obtained from ρ adding the edge (j, α) is blocked by C. Since j is not a converging node (and $j \notin C$), ρ is blocked by C.

Since G_π is an I–map, it follows that $\{\theta'\} \sqsubseteq J_*$, but $\theta \sqsubseteq \theta'$ and then θ would not be a maximal triple. □

By the previous observations and Proposition 3 it comes out the idea related to the relationship between the component C and $pa(A \cup B)$ of a triple (A, B, C), behind the condition introduced in the next proposition assuring the existence of a perfect map.

Proposition 4. *Let J be a set of conditional independence relations and J_* its fast closure.*

Given an ordering π on S, if for any triple $\theta = (A, B, C) \in J_*$ the ordering π satisfies the following conditions

1. *all indices of C appear before all indices belonging to one of the sets A or B;*
2. *all indices of $X = (A \cup B \cup C)$ appear in π before all indices belonging to $S \setminus X$;*

then the related I–map G_π is a perfect map.

Proof. Consider a triple $\theta = (A, B, C) \in J_*$. Under the hypotheses *1.* and *2.*, consider the restriction π_X of π to $X = (A \cup B \cup C)$. If we assume (without loss of generality) that all indices of C appear before of all those of B in π_X, then any index $b \in B$ has as parents the set of indices $B_b \subseteq C \cup (S_{(b)} \cap B)$. Therefore no index of A is a parent of any index of B. Moreover, no index of A can be a descendent of any index of B. In fact, the basic triple $\gamma_b = (b, S_{(b)} \setminus B_b, B_b)$ associated to b satisfies the condition $A \cap S_{(b)} \subseteq S_{(b)} \setminus B_b$, by construction, and for any index $a \in A$ appearing in π after at least a index $b' \in B$, the basic triple $\gamma_a = (a, S_{(a)} \setminus B_a, B_a)$ satisfies conditions: $B \cap S_{(a)} \subseteq S_{(a)} \setminus B_a$ and $B_a \subseteq C \cup (S_{(a)} \cap A)$.

We prove now that θ is represented in G_π. By the previous observations, in G_π no arc can join any element of A and any element of B. Let us consider a path between a node $a \in A$ and a node $b \in B$. If the path passes through a node y outside X, then y must be a collider (i.e. both edges end in y), because each index of X precedes each index outside X and therefore there are no arc from y to any index of X. Since neither y nor any of its descendent is in C, this path is blocked by C.

On the other hand, if the path passes only inside X, it must pass through a node c in C which cannot be a collider, since c precedes all the elements of B.   □

This result is a generalization to triples of J_* to that proved for basic triples in Pearl [9].

The next example shows that even if the conditions 1. and 2. of previous proposition are not satisfied, there could exist a perfect map.

Example 3. Let us consider the set $J = \{\theta_1 = (\{1,2,3\}, \{4,5,6,7\}, \{8,9\}), \theta_2 = (\{1,4\}, \{2,5,8\}, \{6,9\})\}$ of independence relations. Then, by applying $FC1$ we obtain

$J_* = \{\theta_1, \theta_2, \theta_3 = (\{1\}, \{2,4,5,6,7,8\}, \{9\}), \theta_4 = (\{2\}, \{1,4,5,6,7\}, \{8,9\}),$
$\theta_5 = (\{4\}, \{1,2,3,5,8\}, \{6,9\}), \theta_6 = (\{5\}, \{1,2,3,4\}, \{6,8,9\}),$
$\theta_7 = (\{2,4\}, \{1,5\}, \{6,8,9\})\}.$

The conditions 1. and 2. of Proposition 4 do not hold: in fact, by considering the triples θ_3 and θ_5 it is simple to observe that $7 \in X_3$, but $7 \notin X_5$ and $3 \notin X_3$, but $3 \in X_5$. Then, there is no ordering π satisfying conditions 1. and 2. of Proposition 4 for any $\theta \in J_*$.

However, by considering the ordering $\pi =< 1,9,2,8,3,5,6,4,7 >$, we can show that the related I-map G_π is perfect, i.e. it represents any triple of J_*.

By using Proposition 4 it is possible to define the following Algorithm 2, called SEARCHORDER

The algorithm SEARCHORDER searches for an ordering satisfying the conditions stated in Proposition 4. It firstly tries to find all the ordering constraints on the variables required by the condition 1. If an inconsistency is found, then no ordering exists. Then, it uses a backtracking procedure which decides, for each triple (A, B, C), if all the indices in C precede all the indices in A or all the indices in B. In this step, an inconsistency causes a backtracking phase. The algorithm can terminate with success by finding an ordering which satisfies Proposition4, therefore a p–map for J_* can be found by using BN–DRAW. On the other hand, the algorithm can report a failure, because no ordering respects the conditions. But as shown in the Example 3, a p–map can still exist, hence the procedure can only give a partial answer to the question if J is representable with a DAG.

The algorithm has been implemented by using a SAT solver to perform the second step (the function BACKTRACK) because it is possible to formulate the problem of finding an ordering which satisfies the condition 2 as a propositional satisfiability problem. The first empirical results show that this method is quite efficient.

Algorithm 2.

1: **function** SEARCHORDER(K)   ▷ $K \subseteq S^{(3)}$
2:     **for** $\theta = (A, B, C)$ from K **do**
3:         $X \leftarrow A \cup B \cup C$
4:         $R \leftarrow R \cup (X \preceq S \setminus X)$
5:         **if** R is inconsistent **then**
6:             **return** $\bot$
7:         **end if**
8:     **end for**
9:     **return** BACKTRACK(K, R)
10: **end function**
11: **function** BACKTRACK(K, R)   ▷ $K \subseteq S^{(3)}$ and R is a partial order
12:     **if** $K = \emptyset$ **then**
13:         **return** an order taken from R
14:     **else**
15:         $R' \leftarrow R$
16:         $R \leftarrow R \cup (C \preceq A)$
17:         **if** R is not inconsistent **then**
18:             $r \leftarrow$ BACKTRACK($K \setminus \theta, R$)
19:             **if** $r \neq \bot$ **then**
20:                 **return** r
21:             **end if**
22:         **end if**
23:         $R \leftarrow R' \cup (C \preceq B)$
24:         **if** R is not inconsistent **then**
25:             **return** BACKTRACK($K \setminus \theta, R$)
26:         **else**
27:             **return** $\bot$
28:         **end if**
29:     **end if**
30: **end function**

7   Conclusions

We have shown some properties of graphoid structures, which allow to compute efficiently the closure of a set J of conditional independence statements, compatible with a conditional probability ([1,2,3]). Moreover, from these properties it is possible to design an alternative method to build an I–map G_π, given an ordering π on the variable set S.

We have dealt with the problem of finding a perfect map given the fast closure J_*. In particular, we are looking for an ordering π giving rise to a perfect map, if there exists. Actually, we have made a first step in this direction by obtaining a partial goal. In fact, we have introduced a sufficient condition for the existence of a perfect map.

We are now working to relax this condition with the aim of finding a necessary and sufficient condition for the existence of an ordering generating a perfect map. Our idea is that such a condition will need to explore the relations among the triples in J_* and the components of each triple. We are also interested in translating this condition into an efficient algorithm.

Another strictly related, open problem is to find more efficient techniques to compute J_*, because it is clearly the first step needed by any algorithm which finds, if any, a DAG representing J.

References

1. Baioletti, M., Busanello, G., Vantaggi, B.: Algorithms for the closure of graphoid structures. In: Proc. of 12th Inter. Conf. IPMU 2008, Malaga, pp. 930–937 (2008)
2. Baioletti, M., Busanello, G., Vantaggi, B.: Conditional independence structure and its closure: inferential rules and algorithms. Accepted for Publication in International Journal of Approximate Reasoning (2008)
3. Baioletti, M., Busanello, G., Vantaggi, B.: Conditional independence structure and its closure: inferential rules and algorithms. In: Technical Report, 5/2009 of University of Perugia (2009)
4. Coletti, G., Scozzafava, R.: Zero probabilities in stochastical independence. In: Bouchon- Meunier, B., Yager, R.R., Zadeh, L.A. (eds.) Information, Uncertainty, Fusion, pp. 185–196. Kluwer Academic Publishers, Dordrecht (2000)
5. Coletti, G., Scozzafava, R.: Probabilistic logic in a coherent setting. Kluwer, Dordrecht (2002) (Trends in logic n.15)
6. Dawid, A.P.: Conditional independence in statistical theory. J. Roy. Stat. Soc. B 41, 15–31 (1979)
7. Jensen, F.V.: An Introduction to bayesian Networks. UCL Press/ Springer Verlag (1966)
8. Lauritzen, S.L.: Graphical models. Clarendon Press, Oxford (1996)
9. Pearl, J.: Probabilistic reasoning in intelligent systems: networks of plausible inference. Morgan Kaufmann, Los Altos (1988)
10. Studený, M.: Semigraphoids and structures of probabilistic conditional independence. Ann. Math. Artif. Intell. 21, 71–98 (1997)
11. Studený, M.: Complexity of structural models. In: Proc. Prague Stochastics 1998, Prague, pp. 521–528 (1998)
12. Studený, M., Bouckaert, R.R.: On chain graph models for description of conditional independence structures. Ann. Statist. 26(4), 1434–1495 (1998)
13. Vantaggi, B.: Conditional independence in a coherent setting. Ann. Math. Artif. Intell. 32, 287–313 (2001)
14. Verma, T.S.: Causal networks: semantics and expressiveness. Technical Report R–65, Cognitive Systems Laboratory, University of California, Los Angeles (1986)
15. Witthaker, J.J.: Graphical models in applied multivariate statistic. Wiley & Sons, New York (1990)
16. Wong, S.K.M., Butz, C.J., Wu, D.: On the Implication Problem for Probabilistic Conditional Independency. IEEE Transactions on Systems, Man, and Cybernetics, Part A: Systems and Humans 30(6), 785–805 (2000)

Towards a Conscious Choice of a Similarity Measure: A Qualitative Point of View

Bernadette Bouchon-Meunier[1], Giulianella Coletti[2], Marie-Jeanne Lesot[1],
and Maria Rifqi[1]

[1] Université Pierre et Marie Curie - Paris 6, CNRS UMR 7606, LIP6,
104 avenue du Président Kennedy, F-75016 Paris, France
{Bernadette.Bouchon-Meunier,Marie-Jeanne.Lesot,Maria.Rifqi}@lip6.fr
[2] Dipartimento Matematica e Informatica, Università di Perugia,
Via Vanvitelli 1, 06123 Perugia, Italy
coletti@dipmat.unipg.it

Abstract. In many applications, such as case based reasoning, data
mining or analogical reasoning, the choice of a particular measure of
similarity is crucial. In this paper, we propose to study similarity measures from the point of view of the ordering relation they induce on
object pairs. Using a classic method in measurement theory, introduced
by Tversky, we establish necessary and sufficient conditions for the existence of a specific numerical measure, or a class of measures, to represent
a given ordering relation, depending on the axioms this relation satisfies.
The interest is particularly focused on different conditions of independence.

Keywords: Similarity, comparison measure, ordering relation, representability, weak independence conditions.

1 Introduction

Similarity is a key concept in artificial intelligence [10] and similarity measures
have been extensively studied; Bouchon-Meunier et al. [3] or Lesot et al. [6] for
instance propose overviews of various approaches of similarity measures used
in data mining. Following the links to the concept of similarity in cognitive
science, especially in the process of categorization, the seminal work proposed
by Tversky [12] has often been considered as a reference for the description of
a general framework: it embeds numerous similarity measures and enables the
user to make an appropriate choice of a specific similarity measure when facing
a particular problem to solve. In [2,9] we have proposed a general form for
comparison measures, including similarity measures, compatible with Tversky's
model, i.e. such that the proposed classes of measures satisfy the basic axioms
introduced by Tversky. Nevertheless, there is still a need for a consensus about
the choice of similarity measures; we study the converse approach in this paper,
starting from Tversky's requirements and producing general classes of similarity
measures, that accept the classical measures as particular cases. We especially
focus on the independence axiom, introducing also relaxed variants.

C. Sossai and G. Chemello (Eds.): ECSQARU 2009, LNAI 5590, pp. 542–553, 2009.

We follow the idea of Tversky [12] of studying similarity in the environment of the theory of measurements [5,11]. This point of view has also been used more recently in [4] and [1] to study particular classes of dissimilarity and similarity indices used in Descriptive Statistics for the comparison of frequency distributions. In this framework, and considering that object ranking is a frequent reason to use similarity measures, we introduce a binary relation on a set of pairs of objects, expressing a *comparative degree of similarity*. We study the representability of this comparative similarity by means of different numerical similarity measures: we establish axioms stating necessary and sufficient conditions under which a given comparative similarity is represented by a specific class of similarity measures. In other words, given the set of properties possessed by a given comparative degree of similarity, we characterise the form of the similarity measures that can represent it.

Thus we obtain two kinds of equivalence classes of similarity measures: the first one is given by measures representing the same ordering relation, as introduced in [7,8]. The second, rougher, definition of equivalence is given by the measures representing orders that are not exactly identical but that possess the same properties and satisfy the same axioms. This definition permits to point out the actual rules we accept when we choose one particular measure of similarity and to make explicit underlying requirements on the induced order.

The paper is organized as follows. In Section 2, we consider as a starting point the numerical similarity measures: after recalling the classic notion of equivalence, we introduce basic axioms that are satisfied by comparative similarities induced from given classes of numerical similarity measures. We then turn to the reciprocal point of view, to relate given comparative similarities satisfying specific axioms to classes of numerical similarity measures. In Section 3, we introduce the independence axioms that are required to establish, in Section 4, these necessary and sufficient conditions for the existence of a class of measures to represent a given comparative degree of similarity.

2   From Numerical Similarity to Comparative Similarity

In this section, after introducing the notations used throughout the paper, we discuss the classic definition of equivalence between numerical similarity measures and establish basic axioms satisfied by comparative similarities induced from given classes of numerical similarities, following the ideas of Tversky to study similarity using the framework of the measurement theory [12].

For simplicity we consider the case of data described by a set of characteristics $\mathcal{A}$ that can be only present or absent in any object (so data are crisp and correspond to subsets of $\mathcal{A}$). We note that our approach can be easily extended to the case where objects are described by fuzzy subsets of $\mathcal{A}$.

2.1   Preliminaries

We consider that each object is described by p binary attributes, that is by the set of present characteristics from the predefined list $\mathcal{A}$. The data set is noted

$\mathcal{X} = \{0,1\}^p$: for every $X \in \mathcal{X}$, $X = \{x_1, ..., x_p\}$, $x_i \in \{0,1\}$. The particular object with $x_i = 0$ for every i is denoted $\underline{0}$. We note x_i^c the value $1 - x_i$, $X_k^c = \{x_1, ..., x_k^c, ..., x_p\}$ and $X^c = \{x_1^c, ..., x_p^c\}$. Finally, for any $X \in \mathcal{X}$, we note $I_X = \{i : x_i = 1\}$ and $|X|$ the cardinality of I_X.

Given a pair $(X, Y) \in \mathcal{X}$ we define $\mathbf{x} = |I_X \cap I_Y|$, i.e. the number of characteristics present in both objects, $\mathbf{y}^- = |I_X \setminus I_Y|$, i.e. the number of characteristics present in X but not in Y, $\mathbf{y}^+ = |I_Y \setminus I_X|$, i.e. the number of characteristics present in Y but not in X, and $\mathbf{y} = \mathbf{y}^- + \mathbf{y}^+ = |(I_X \setminus I_Y) \cup (I_Y \setminus I_X)|$. Finally we define $\mathbf{y}^* = |I_{X^c} \cap I_{Y^c}| = p - \mathbf{x} - \mathbf{y}$ that represents the number of characteristics absent of both objects.

Consider now a *comparative degree of similarity*, that is a binary relation $\preceq$ on $\mathcal{X}^2$, with the following meaning: for $X, Y, X', Y' \in \mathcal{X}$, $(X, Y) \preceq (X', Y')$ means that X is similar to Y no more than X' is similar to Y'.

The relations $\sim$ and $\prec$ are then induced by $\preceq$ as follows: $(X, Y) \sim (X', Y')$ if $(X, Y) \preceq (X', Y')$ and $(X', Y') \preceq (X, Y)$, meaning that X is similar to Y as X' is similar to Y'. Lastly $(X, Y) \prec (X', Y')$ if $(X, Y) \preceq (X', Y')$ but not $(X', Y') \preceq (X, Y)$, meaning that X is similar to Y less than X' is similar to Y'.

It is to be noticed that if $\preceq$ is complete, then $\sim$ and $\prec$ are the symmetrical and the asymmetrical parts of $\preceq$ respectively.

We now introduce the notion of representability of such a comparative degree of similarity by a numerical similarity measure:

Definition 1. *Given a comparative degree of similarity $\preceq$, a similarity measure $S : \mathcal{X}^2 \to \mathbb{R}$ represents $\preceq$ if and only if for any $(X, Y), (X', Y') \in \mathcal{X}^2$, both following conditions hold:*

$$(X, Y) \preceq (X', Y') \Rightarrow S(X, Y) \leq S(X', Y')$$
$$(X, Y) \prec (X', Y') \Rightarrow S(X, Y) < S(X', Y')$$

We recall that if the relation $\preceq$ is complete the above conditions are equivalent to the following one: $(X, Y) \preceq (X', Y') \Leftrightarrow S(X, Y) \leq S(X', Y')$.

2.2   Similarity Measure Equivalence

Any similarity measure on $\mathcal{X}^2$ induces a complete comparative degree of similarity $\preceq$, defined as follows: $(X, Y) \prec (X', Y')$ if $S(X, Y) < S(X', Y')$ and $(X, Y) \sim (X', Y')$ if $S(X, Y) = S(X', Y')$.

Now the same ordering relation is induced by any similarity measure that can be expressed as an increasing transformation of S: any similarity measure $S' = \varphi(S)$, with $\varphi : \mathbb{R} \to \mathbb{R}$ strictly increasing is also a representation of $\preceq$. Moreover, no other measure $S*$ represents $\preceq$.

Thus, from a comparative point of view, all functions $\varphi(S)$ are indistinguishable. Formally speaking, the relation r defined on the set of similarity measures as SrS' if and only if S and S' induce the same comparative degree of similarity on $\mathcal{X}$ is an equivalence relation. An equivalent formulation of this concept, expressed only in terms of numerical similarity functions, is given in [7,8].

For instance the similarity measures

$$S_\rho(X, Y) = \frac{\mathbf{x}}{\mathbf{x} + \rho \mathbf{y}} \tag{1}$$

with $\rho > 0$, are all equivalent, since each of them is an increasing transformation of any other. In particular, the Jaccard ($\rho = 1$), Dice ($\rho = 1/2$), Sorensen ($\rho = 1/4$), Anderberg ($\rho = 1/8$) and Sokal and Sneath ($\rho = 2$) measures are equivalent.

The same class also contains the function $S(X, Y) = \log(\mathbf{x}) - \log(\mathbf{y})$, which is of the kind proposed by Tversky [12] (a linear form of an increasing function): S is an increasing transformation of $S'(X, Y) = \mathbf{x}/\mathbf{y}$ which is an increasing transformation of S_1.

It is to be noted that the function $S(X, Y) = \alpha \log(\mathbf{x}) - \beta \log(\mathbf{y})$ for $\alpha, \beta > 0$ is not in the same class, but it is equivalent to all measures

$$S *_\rho (X, Y) = \frac{\mathbf{x}^\alpha}{\mathbf{x}^\alpha + \rho \mathbf{y}^\beta}.$$

2.3   Basic Axioms

We are now interested in a different classification of measures of similarity: instead of considering the measures that induce the same order, we consider the measures that induce orders satisfying the same class of axioms. In this section, we consider axioms that lead to preliminary results regarding relations between similarity measures and comparative degrees of similarity.

Basic Properties. The first two axioms we introduce describe basic properties a binary relation has to satisfy to define a comparative degree of similarity: the first one only states the relation must be a weak order.

Axiom S1 [weak order]
  $\preceq$ is a weak order, i.e it is complete, reflexive and transitive.

The second axiom expresses boundary conditions: it imposes that for any X, whatever Y, X cannot be more similar to Y than it is similar to itself, and it cannot be less similar to Y than it is to its complement. Lastly, it imposes that X is similar to itself as Y is to itself: all data are equally similar to themselves.

Axiom S2 [boundary conditions] $\forall X, Y \in \mathcal{X}$,
  $(X^c, X) \sim (Y^c, Y) \preceq (X, Y) \preceq (X, X) \sim (Y, Y)$ and $(X^c, X) \prec (X, X)$

The third axiom imposes a symmetry condition.

Axiom S3 [symmetry]
  $\forall X, Y \in \mathcal{X}$, $(X, Y) \sim (Y, X)$

These properties lead to the following two definitions:

Definition 2. *A binary relation $\preceq$ on $\mathcal{X}^2$ is a comparative similarity if and only if it satisfies axioms S1 and S2.*

Definition 3. *A comparative similarity is symmetric if and only if it satisfies axiom S3.*

The next axiom expresses the idea that all attributes have the same role with respect to the comparative similarity: a change in one attribute is equivalent to the modification of any attribute of the same category, i.e. attributes representing characteristics present in both objects (with indices in $I_X \cap I_Y$), only in one of them (with indices in $I_X \setminus I_Y$ or in $I_Y \setminus I_X$) or absent of both (indices not in $I_X \cup I_Y$):

Axiom S4 [attribute uniformity] $\forall h, k \in \{1, ..., p\}$,
   if $h, k \in I_X \cap I_Y$, or $h, k \in I_X \setminus I_Y$, or $h, k \in I_Y \setminus I_X$, or $h, k \notin I_X \cup I_Y$,
   then $(X, Y_k^c) \sim (X, Y_h^c)$ and $(X_k^c, Y) \sim (X_h^c, Y)$.

It must be underlined that any comparative similarity representable by a similarity measure depending only on $\mathbf{x}, \mathbf{y}^+, \mathbf{y}^-$ and $\mathbf{y}^*$ satisfies this axiom. Reciprocally, as $\mathcal{X}^2$ is finite, any comparative similarity satisfying axiom S4 can be represented by a function depending only on $\mathbf{x}, \mathbf{y}^+, \mathbf{y}^-$ and $\mathbf{y}^*$.

Monotonicity Axioms. The following three axioms of monotonicity govern the comparative similarity among pairs differing in the presence/absence of only one attribute: 4 pairs must then be compared, depending on whether the modification is applied to both data, one or the other, or none of them. The different axioms correspond to different choices regarding the semantics of the similarity measure, as commented below.

Axiom S5 [monotonicity] $\forall X, Y \in \mathcal{X}, X \neq Y$
   $\forall k \in I_X \cap I_Y,$ $(X, Y_k^c) \sim (X_k^c, Y) \prec (X_k^c, Y_k^c) \prec (X, Y)$
   if $I_X \cap I_Y = \emptyset, \forall k \in I_X$ $(X, Y) \sim (X_k^c, Y)$

The first condition means that if an attribute possessed by both objects is modified, the modified objects are less similar one to another than the initial object pairs were. This corresponds to a strong semantic choice: it implies that the common presence of an attribute is preferred to a common absence. Moreover, the axiom states that modifying both objects degrades the similarity to a lesser extent than changing only one of them. Lastly, if only one object is modified, there is no difference whether X or Y is concerned. Equivalently, the axiom can be written in the following three forms, describing the expected variations when other attribute types are considered:

$\forall k \in I_X \setminus I_Y, (X_k^c, Y_k^c) \sim (X, Y) \prec (X_k^c, Y) \prec (X, Y_k^c)$
$\forall k \in I_Y \setminus I_X, (X_k^c, Y_k^c) \sim (X, Y) \prec (X, Y_k^c) \prec (X_k^c, Y)$
$\forall k \notin I_X \cup I_Y, (X, Y_k^c) \sim (X_k^c, Y) \prec (X, Y) \prec (X_k^c, Y_k^c)$

The second condition considers the case where the intersection $I_X \cap I_Y$ is empty, i.e. when there is no common attributes: then it is indifferent whether the attributes are absent of both objects or present in one of them. In particular, it implies that whatever X and Y such that $I_X \cap I_Y = \emptyset$, $(X,Y) \sim (X,X^c)$.

It is easy to prove that any comparative similarity $\preceq$ representable by a similarity measure S defined as

$$S(X,Y) = \frac{f(\mathbf{x})}{f(\mathbf{x}) + \rho g(\mathbf{y})} \tag{2}$$

with $\rho > 0$ and f and g non negative increasing functions (or any strictly increasing transformation of this measure) satisfies Axiom S5. Thus in particular, it is verified by the measures belonging to the S_ρ class defined in Equation (1), in which f and g coincide with the identity function.

Axiom S6 relaxes the conditions required by S5, insofar as it does not impose conditions on the comparison of (X,Y_k^c) and (X_k^c,Y), whereas they are equivalent in Axiom S5:

Axiom S6 [weak monotonicity] $\forall X, Y \in \mathcal{X}$, $X \neq Y$, $\forall k \in I_X \cap I_Y$,
$\quad (X,Y_k^c) \prec (X_k^c,Y_k^c) \preceq (X,Y)$ and $(X_k^c,Y) \prec (X_k^c,Y_k^c) \preceq (X,Y)$
$\quad$ if $I_X \cap I_Y = \emptyset$, $\forall k \in I_X \quad (X,Y) \sim (X_k^c,Y)$

It is easy to prove that any comparative similarity $\preceq$ representable by a similarity measure S defined as

$$S(X,Y) = \frac{f(\mathbf{x})}{g(\mathbf{x}+\mathbf{y}^-)h(\mathbf{x}+\mathbf{y}^+)} \tag{3}$$

with f,g,h non negative increasing functions such that $\forall x$, $f(x) = g(x)h(x)$ (ensuring that $S(X,Y) = 1$ when $y^- = y^+ = 0$) satisfies Axiom S6 (as well as any strictly increasing transformation of this measure). A particular case is the Ochiai measure, where f is the identity function and $g(.) = h(.) = \sqrt{.}$.

Axiom S6 is also satisfied by any comparative similarity $\preceq$, representable by a similarity measure S defined as

$$S(X,Y) = \frac{f(\mathbf{x})}{g(\mathbf{x}+\mathbf{y}^-)} + \frac{f(\mathbf{x})}{h(\mathbf{x}+\mathbf{y}^+)} \tag{4}$$

with f,g,h non negative increasing functions such that $\forall x$, $f(x)(h(x)+g(x)) = g(x)h(x)$ (ensuring that $S(X,Y) = 1$ when $y^- = y^+ = 0$). In particular, it holds for the Kulczynski measure, where g and h are the identity function and $f(x) = x/2$.

Lastly Axiom S7 resembles Axiom S5 but considers the case where (X,Y) and (X_k^c,Y_k^c) are equivalent, i.e. characteristics present in both objects or absent of both objects play the same role. Besides, as Axiom S5, it requires that (X,Y_k^c) and (X_k^c,Y) are equivalent, i.e. the modification is symmetrical.

Axiom S7 [monotonicity 2] $\forall X, Y \in \mathcal{X}$, $X \neq Y$
$\quad \forall k \in I_X \cap I_Y, \quad (X,Y_k^c) \sim (X_k^c,Y) \prec (X_k^c,Y_k^c) \sim (X,Y)$
$\quad$ if $I_X \cap I_Y = \emptyset$, $\forall k \in I_X \quad (X,Y) \sim (X_k^c,Y)$

This is equivalent to saying that the same property holds for any k not in $I_X \cup I_Y$ or to saying that, $\forall k \in I_X \setminus I_Y$ and $\forall k \in I_Y \setminus I_X$, $(X_k^c, Y_k^c) \sim (X, Y) \prec (X, Y_k^c) \sim (X_k^c, Y)$.

It is easy to prove that any comparative similarity $\preceq$ representable by a similarity measure S defined as

$$S(X, Y) = \frac{f(\mathbf{x} + \mathbf{y}^*)}{f(\mathbf{x} + \mathbf{y}^*) + \rho g(\mathbf{y})} \tag{5}$$

with f, g increasing functions, $\rho > 0$ satisfies Axiom S7. This corresponds to so-called type II similarity measures [6]. In particular it holds for the Rogers and Tanimoto, Sokal and Michener and Sokal and Sneath measures, for which f and g are the identity function, α takes values 2, 1 and 1/2 respectively.

3   Independence Conditions

The objective is then to consider the reciprocal point of view: given the set of properties possessed by a given comparative degree of similarity, to characterise the form of the similarity measures that can represent it. To that aim, other conditions must be imposed to the comparative similarities: we take into account the basic properties considered by Tversky as fundamental for similarities in [12], focusing on the independence axiom he introduced. Starting from his classic definition, we extend it to weaker forms that will determine the class of measures a comparative similarity can be represented by, as will be shown in Section 4.

Axiom I is the independence axiom introduced by Tversky [12]:

Axiom I [independence] For any 4-tuple $(X_1, Y_1), (X_2, Y_2), (Z_1, W_1), (Z_2, W_2)$, if one of the following conditions holds
    (i) $\mathbf{x_i} = \mathbf{z_i}$ and $\mathbf{y_i^-} = \mathbf{w_i^-}$ $(i = 1, 2)$, and $\mathbf{y_1^+} = \mathbf{y_2^+}$, $\mathbf{w_1^+} = \mathbf{w_2^+}$
    (ii) $\mathbf{x_i} = \mathbf{z_i}$ and $\mathbf{y_i^+} = \mathbf{w_i^+}$ $(i = 1, 2)$, and $\mathbf{y_1^-} = \mathbf{y_2^-}$, $\mathbf{w_1^-} = \mathbf{w_2^-}$
    (iii) $\mathbf{y_i^+} = \mathbf{w_i^+}$ and $\mathbf{y_i^-} = \mathbf{w_i^-}$ $(i = 1, 2)$, and $\mathbf{x_1} = \mathbf{x_2}$, $\mathbf{z_1} = \mathbf{z_2}$
    then $(X_1, Y_1) \preceq (X_2, Y_2) \Leftrightarrow (Z_1, W_1) \preceq (Z_2, W_2)$.

where $z_i = |I_{Z_i} \cap I_{W_i}|$, $w_i^- = |I_{Z_i} \setminus I_{W_i}|$ and $w_i^+ = |I_{W_i} \setminus I_{Z_i}|$ (the same notations are used in the following).

Condition (i) for instance expresses that the joint effect of $\mathbf{x}$ and $\mathbf{y}^-$ is independent of the fixed component $\mathbf{y}^+$.

It must be underlined that comparative similarities representable by a similarity measure S defined by Equation (2), and in particular of the class S_h defined in Equation (1), do not satisfy this independence condition. This can be illustrated as follows in the case of the Jaccard measure, i.e. S_h with $h = 1$: considering hypothesis (i), by trivial computation, one has $(X_1, Y_1) \preceq (X_2, Y_2)$ if and only if $\mathbf{x_1}(\mathbf{y_2^-} + \mathbf{y_1^+}) \leq \mathbf{x_2}(\mathbf{y_1^-} + \mathbf{y_1^+})$ and $(Z_1, W_1) \preceq (Z_2, W_2)$ iff $\mathbf{x_1}(\mathbf{y_2^-} + \mathbf{w_1^+}) \leq \mathbf{x_2}(\mathbf{y_1^-} + \mathbf{w_1^+})$. Now the two inequalities can be independently satisfied, as can be shown using the following example: $X_1 = Z_1 = W_1 = (10000)$, $Y_1 = (11000)$, $X_2 = Z_2 = (11110)$, $Y_2 = (11101)$, and $W_2 = (11100)$.

We introduce now a weaker form of independence in which we only require that the common characteristics are independent of the totality of the characteristics present in only one element of the pair.

Axiom WI [weak independence] For any 4-tuple (X_1, Y_1), (X_2, Y_2), (Z_1, W_1), (Z_2, W_2), if one of the following conditions holds
  (i) $\mathbf{x_i} = \mathbf{z_i}$ ($i = 1, 2$), and $\mathbf{y_1} = \mathbf{y_2}$, $\mathbf{w_1} = \mathbf{w_2}$
  (ii) $\mathbf{y_i} = \mathbf{w_i}$ ($i = 1, 2$), and $\mathbf{x_1} = \mathbf{x_2}$, $\mathbf{z_1} = \mathbf{z_2}$
then $(X_1, Y_1) \preceq (X_2, Y_2) \Leftrightarrow (Z_1, W_1) \preceq (Z_2, W_2)$.

It must be underlined that the comparative similarities representable by a similarity measure S defined by Equation (2) satisfy this axiom, and thus in particular the elements of the class S_h. We prove this assertion for hypothesis (i): by trivial computation it holds that on one hand $(X_1, Y_1) \preceq (X_2, Y_2)$ iff $f(\mathbf{x_1}) \leq f(\mathbf{x_2})$, and on the other hand $(Z_1, W_1) \preceq (Z_2, W_2)$ iff $f(\mathbf{x_1}) \leq f(\mathbf{x_2})$, leading to the desired equivalence. The proof is similar for condition (ii).

Comparative similarities representable by a similarity measure S defined by Equation (3) do not satisfy the weak independence axiom WI. In particular the well known Ochiai measure does not satisfy WI, and, obviously, the independence axiom I. The same considerations hold for comparative similarities representable by a similarity measure S defined by Equation (4), and in particular for the Kulczynski measure.

We now introduce another weak kind of independence that considers as components the common characteristics and the sum of these common characteristics and the characteristics present in only one of the two objects.

Axiom CI [cumulative independence] For any 4-tuple (X_1, Y_1), (X_2, Y_2), (Z_1, W_1), (Z_2, W_2), if one of the following conditions holds
  (i) $\mathbf{x_i} = \mathbf{z_i}$ and $\mathbf{x_i} + \mathbf{y_i^-} = \mathbf{z_i} + \mathbf{w_i^-}$ ($i = 1, 2$), and $\mathbf{x_1} + \mathbf{y_1^+} = \mathbf{x_2} + \mathbf{y_2^+}$, $\mathbf{z_1} + \mathbf{w_1^+} = \mathbf{z_2} + \mathbf{w_2^+}$
  (ii) $\mathbf{x_i} = \mathbf{z_i}$ and $\mathbf{x_i} + \mathbf{y_i^+} = \mathbf{z_i} + \mathbf{w_i^+}$ ($i = 1, 2$), and $\mathbf{x_1} + \mathbf{y_1^-} = \mathbf{x_2} + \mathbf{y_2^-}$, $\mathbf{z_1} + \mathbf{w_1^-} = \mathbf{z_2} + \mathbf{w_2^-}$
  (iii) $\mathbf{x_i} + \mathbf{y_i^+} = \mathbf{z_i} + \mathbf{w_i^+}$ and $\mathbf{x_i} + \mathbf{y_i^-} = \mathbf{z_i} + \mathbf{w_i^-}$ ($i = 1, 2$), and $\mathbf{x_1} = \mathbf{x_2}$, $\mathbf{z_1} = \mathbf{z_2}$
then $(X_1, Y_1) \preceq (X_2, Y_2) \Leftrightarrow (Z_1, W_1) \preceq (Z_2, W_2)$.

It is easy to prove that a comparative similarity representable by a similarity measure S defined by Equation (3), in particular, the Ochiai measure, satisfies the cumulative independence condition CI.

Finally we introduce another weak definition of independence that considers as components the sum of characteristics which are common and those which are absent of both objects and the sum of those present in only one object of the pair.

Axiom TWI [totally weak independence] For any 4-tuple (X_1, Y_1), (X_2, Y_2), (Z_1, W_1), (Z_2, W_2), if one of the following conditions holds

(i) $\mathbf{x_i} + \mathbf{y_i^*} = \mathbf{z_i} + \mathbf{w_i^*}$ $(i = 1, 2)$, and $\mathbf{y_1} = \mathbf{y_2}$, $\mathbf{w_1} = \mathbf{w_2}$
(ii) $\mathbf{y_i} = \mathbf{w_i}$ $(i = 1, 2)$, and $\mathbf{x_1} + \mathbf{y_1^*} = \mathbf{x_2} + \mathbf{y_2^*}$ $\mathbf{z_1} + \mathbf{w_1^*} = \mathbf{z_2} + \mathbf{w_2^*}$
then $(X_1, Y_1) \preceq (X_2, Y_2) \Leftrightarrow (Z_1, W_1) \preceq (Z_2, W_2)$.

It is easy to prove that comparative similarities representable by a similarity measure S defined by Equation (5), in particular, the Rogers and Tanimoto, the Sokal and Michener and the Sokal and Sneath measures, satisfy the axiom TWI.

4   Representation Theorems

In this section we establish the theorems stating necessary and sufficient conditions for comparative similarities verifying the various independence axioms to be representable by classes of numerical measures.

Theorem 1. *Let $\preceq$ be a binary relation on $\mathcal{X}^2 \setminus \{(\underline{0}, \underline{0})\}$. The following conditions are equivalent:*

(i) $\preceq$ is a comparative similarity satisfying axioms S4 and S5 and possessing the weak independence property WI
(ii) there exist two non negative increasing functions f and g, with $f(0) = g(0) = 0$ such that the function $S : \mathcal{X}^2 \to [0,1]$ defined by Equation (2) represents $\preceq$.

Proof: we first prove the implication (ii) $\Rightarrow$ (i) and consider $\preceq$ the ordering relation induced by a similarity measure S satisfying the conditions (ii). Then $\preceq$, representable by a function with values in $\mathbb{R}$, is a weak order, i.e. satisfies Axiom S1. Moreover, as $\forall X \in \mathcal{X}$, $S(X, X) = 1 > S(X^c, X) = S(X, X^c) = 0$ and $S(X, Y) \in [0, 1]$, $\preceq$ also satisfies Axiom S2. Thus it is a comparative similarity.

Furthermore, it satisfies the Axioms S5 and WI as already underlined in the remarks following the introduction of these axioms (see pages 546 and 549): it satisfies S5, because S is increasing with respect to $\mathbf{x}$ and decreasing with respect to $\mathbf{y}$. Besides if $I_X \cap I_Y = \emptyset$, $\mathbf{x} = 0$, thus $S(X, Y) = 0 = S(X_k^c, Y)$ for all $k \in I_X$. It satisfies WI because of the independence properties of S.

We now prove the implication (i) $\Rightarrow$ (ii) and consider a comparative similarity $\preceq$ satisfying the conditions (i). Let us indicate by $\mathbb{R}^*$ the compactification of $\mathbb{R}$, that is $\mathbb{R}^* = \mathbb{R} \cup \{-\infty, +\infty\}$. Since $\mathbb{R}^*$ is a completely ordered set containing $\mathbb{R}$, all results related to the representability of a binary relation by a function with values in $\mathbb{R}$ remain valid for functions with values in $\mathbb{R}^*$ [5].

Due to S4, $\preceq$ is representable by a function $S : \mathcal{X}^2 \to \mathbb{R}^*$, depending only on $\mathbf{x}, \mathbf{y}^+, \mathbf{y}^-$ and $\mathbf{y}^*$. Due to S5, it is strictly increasing in $\mathbf{x}$ and strictly decreasing in $\mathbf{y}^+$ and $\mathbf{y}^-$. Due to condition WI, there exists a function $f_1 : \mathbb{R} \to \mathbb{R}^*$, that moreover is strictly increasing due to S5, and two real numbers $\alpha, \beta > 0$ so that S is a strictly increasing transformation $\varphi : \mathbb{R}^* \to \mathbb{R}^*$ of a linear form of the function f_1, i.e.

$$S(X, Y) = \varphi(\alpha f_1(\mathbf{x}) - \beta f_1(\mathbf{y})) \qquad (6)$$

Now from Axiom S2, necessarily $f_1(0) = -\infty$. Indeed from Axiom S2, it holds that $(X, X) \sim (Y, Y)$ and thus with $Y = X_k^c$, $(X, X) \sim (X_k^c, X_k^c)$, which implies

$S(X, X) = S(X_k^c, X_k^c)$. Applying Equation (6), $S(X, X) = \varphi(\alpha f_1(|I_X|) - \beta f_1(0))$ and $S(X_k^c, X_k^c) = \varphi(\alpha f_1(|I_X| - 1) - \beta f_1(0))$. As φ is strictly increasing, the equality implies $\alpha f_1(|I_X|) - \beta f_1(0) = \alpha f_1(|I_X| - 1) - \beta f_1(0)$. As f_1 is strictly increasing, $f_1(|I_X|) > f_1(|I_X| - 1)$. For the equality to hold, it is necessary that $f_1(0) = -\infty$: the unique possible elements of $[-\infty, +\infty]$ which summed to two different real numbers give the same results are $-\infty$ and $+\infty$, by monotonicity of f_1, we have $f_1(0) = -\infty$.

Letting $f_2 = \exp(f_1)$, that thus satisfies $f_2(0) = 0$, and $\psi = \varphi \circ \log$, $\preceq$ is thus representable by

$$S(X, Y) = \psi\left(\frac{f_2^\alpha(\mathbf{x})}{f_2^\beta(\mathbf{y})}\right) \tag{7}$$

considering the fraction takes value $+\infty$ when $y = 0$.

Choosing as ψ_2 the increasing function $\psi_2(z) = z/(z + \rho)$, with ρ positive real number, then $\preceq$ is representable by $\psi_2(S(X, Y))$. Denoting $f(x) = f_2^\alpha(x)$ and $g(y) = f_2^\beta(y)$, the latter can be written

$$\psi_2(S(X, Y)) = \frac{f(x)}{f(x) + \rho g(y)}$$

i.e. in the form of Equation (2). Furthermore, f and g satisfy the conditions required in (ii): they are strictly increasing, and $f(0) = g(0) = 0$.

The following theorem considers the case of cumulatively independent comparative similarities:

Theorem 2. *Let $\preceq$ be a binary relation on $\mathcal{X}^2 \setminus \{(\underline{0}, \underline{0})\}$. The following conditions are equivalent:*

(i) $\preceq$ is a comparative similarity satisfying axioms S4 and S6 and possessing the cumulative independence property CI

(ii) there exists a real-valued increasing function f, with $f(0) = 0$ and $\alpha, \beta, \gamma \geq 0$, $\alpha = \beta + \gamma$, such that the function $S : \mathcal{X}^2 \to [0, 1]$ defined by

$$S(X, Y) = \frac{f^\alpha(\mathbf{x})}{f^\beta(\mathbf{x} + \mathbf{y}^-) f^\gamma(\mathbf{x} + \mathbf{y}^+)} \tag{8}$$

represents $\preceq$.

Proof: The proof of implication (ii) $\Rightarrow$ (i) is direct and similar to that in Theorem 1. We prove (i) $\Rightarrow$ (ii). By the hypotheses, following the same considerations as in the previous theorem, there exists a class of functions S representing $\preceq$ that are increasing transformations of a function such as

$$S(X, Y) = \alpha f_1(\mathbf{x}) - \beta f_1(\mathbf{x} + \mathbf{y}^-) - \gamma f_1(\mathbf{x} + \mathbf{y}^+) \tag{9}$$

By the hypothesis of monotonicity S6, the function f_1 must be increasing and convex and $\alpha, \beta, \gamma > 0$. Moreover, by condition $(X, X) \sim (Y, Y)$ for every $X, Y \in \mathcal{X}$, we have necessarily $\alpha = \beta + \gamma$. From condition $(X, X^c) \sim (Y^c, Y)$ of Axiom

S2, with $Y = X_k^c$ and $k \in I_X$, it follows that $f_1(0) = -\infty$ using the same argument as in the proof of Theorem 1.

A result in the special case where $\preceq$ is symmetrical can be established: if, in condition (i), $\preceq$ is also required to satisfy Axiom S3 (symmetry), then the function representing $\preceq$ is such that $\beta = \gamma = \alpha/2$.

Lastly we consider the case of totally weak independent comparative similarities:

Theorem 3. *Let $\preceq$ be a binary relation on $\mathcal{X}^2$. The following conditions are equivalent:*

(i) *$\preceq$ is a comparative similarity satisfying axioms S4 and S7, and possessing the totally weak independence property TWI*

(ii) *there exist three real-valued increasing functions f, g, with $f(0) = 0$ and $\alpha, \beta \geq 0$, such that the function $S : \mathcal{X}^2 \to [0,1]$ defined by Equation (5) represents $\preceq$.*

The proof is very similar to that given in Theorem 1.

Again for this theorem a "symmetric version" can be proved: if, in condition (i), $\preceq$ is also required to satisfy S3, then the function representing $\preceq$ is such that $g = f$.

5   Conclusion

The approach of similarity we have presented is based on several basic hypotheses. The first one is the environment of measurement theory, stemming from Tversky's reference work, which has been considered since then by the community as a reasonable approach to model similarities managed by human beings. The second hypothesis is the importance of ranking in the management of similarities, which means that similarities are regarded as relative characteristics of families of objects, rather than intrinsic descriptions of these families. We are often interested in the comparison of similarity degrees attached to two pairs of objects, more than in the level of similarity attached to each of these pairs. We can remark that changing the measure of similarity in a model provides different values for these degrees, and it is therefore reasonable not to attach too much importance to the similarity degrees themselves, but to their relative values. The third hypothesis we make is the importance of the independence axiom among those proposed by Tversky.

We have therefore established a link between what we call comparative similarities in a qualitative approach on the one hand, and possible numerical representations of these similarities on the other hand, providing general forms of similarity measures compatible with the independence axiom and with weaker forms of this axiom. We show that such a framework embeds well-known similarity measures, and we point out classes of such measures with the same behavior with respect to independence.

It is to be hoped that this work will help users of similarities in all domains of artificial intelligence and image processing, in particular, to make an appropriate choice of a convenient measure when they have to manage resemblances. It will for instance avoid them to compare results based on the choice of several similarity measures, since results appear to be analogous when the measures belong to a same class. The choice of a similarity measure is then reduced to the choice of a class of measures.

Future works will take into account extensions of such similarity measures to graded values of attributes, in a fuzzy set based knowledge representation, replacing the binary attributes we have only considered in this paper. Such a work will meet a general framework for measures of similarity between fuzzy sets we have already proposed [2,9], providing a qualitative view of similarities associated with such numerical evaluations of similarities.

References

1. Bertoluzza, C., Di Bacco, M., Doldi, V.: An axiomatic characterization of the measures of similarity. Sankhya 66, 474–486 (2004)
2. Bouchon-Meunier, B., Rifqi, M., Bothorel, S.: Towards general measures of comparison of objects. Fuzzy Sets and Systems 84, 143–153 (1996)
3. Bouchon-Meunier, B., Rifqi, M., Lesot, M.J.: Similarities in fuzzy data mining: from a cognitive view to real-world applications. In: Zurada, J.M., Yen, G.G., Wang, J. (eds.) Computational Intelligence: Research Frontiers. LNCS, vol. 5050, pp. 349–367. Springer, Heidelberg (2008)
4. Coletti, G., Di Bacco, M.: Qualitative characterization of a dissimilarity and concentration index. Metron XLVII, 121–130 (1989)
5. Krantz, D., Luce, R., Suppes, P., Tversky, A.: Foundations of measurement, vol. I. Academic Press, London (1971)
6. Lesot, M.J., Rifqi, M., Benhadda, H.: Similarity measures for binary and numerical data: a survey. Intern. J. of Knowledge Engineering and Soft Data Paradigms (KESDP) 1, 63–84 (2009)
7. Omhover, J.F., Bouchon-Meunier, B.: Equivalence entre mesures de similarités floues: application à la recherche d'images par le contenu. In: 6eme Congrès Européen de Science des Systèmes (2005)
8. Omhover, J.F., Detyniecki, M., Rifqi, M., Bouchon-Meunier, B.: Image retrieval using fuzzy similarity: Measure of equivalence based on invariance in ranking. In: IEEE Int. Conf. on Fuzzy Systems (2004)
9. Rifqi, M.: Mesures de comparaison, typicalité, et classification d'objets flous: théorie et pratique. PhD thesis, University Paris 6 (1996)
10. Rissland, E.: Ai and similarity. IEEE Intelligent Systems 21, 33–49 (2006)
11. Suppes, P., Krantz, D., Luce, R., Tversky, A.: Foundations of measurement, vol. II. Academic Press, New York (1989)
12. Tversky, A.: Features of similarity. Psychological Review 84, 327–352 (1977)

Integrated Likelihood
in a Finitely Additive Setting

Giulianella Coletti[1], Romano Scozzafava[2], and Barbara Vantaggi[2]

[1] Università di Perugia, Italy
coletti@dipmat.unipg.it
[2] Università di Roma "La Sapienza", Italy
{romscozz,vantaggi}@dmmm.uniroma1.it

Abstract. Without a clear, precise and rigorous mathematical frame, is the likelihood "per se" a proper tool to deal with statistical inference and to manage partial and vague information? Since (as Basu puts it) "the likelihood function is after all a bunch of conditional probabilities", a proper discussion of the various extensions of a likelihood from a point function to a set function is carried out by looking at a conditional probability as a general non-additive "uncertainty" measure $P(E|\cdot)$ on the set of conditioning events.

Keywords: Conditional probability, likelihood function, statistical inference.

1 Introduction

Among statisticians there seems to be lack of consensus about the meaning of "statistical information" and how such an important notion should be meaningfully formalized: see, e.g., Basu [1]. On the contrary, if we agree to the stipulation that our search for the "whole of the relevant information in the data" should be limited within the framework of a given statistical model, then most statisticians could not find any cogent reason for not identifying the "information in the data" with the likelihood function generated by it.

An ensuing and long debated problem concerns the following question: is the likelihood just a *point function* or can it be also seen as a *measure*? Why can't we talk of the likelihood of a composite hypothesis in the same way as we talk about the probability of a composite event? Statisticians are usually inclined to accept the following "law of likelihood" [1]: of two (simple) hypotheses that are consistent with given data x, the better supported by the data is the one that has greater likelihood $L(x|\omega)$, where ω ranges in the parameter space Ω. An immediate consequence of this is the controversial inferential method based on the choice of the ω which corresponds to the "maximum likelihood" (also called "profile likelihood").

On the other hand, not all statisticians are willing to support also the so–called "strong law of likelihood", that can be expressed as follows: for any two

C. Sossai and G. Chemello (Eds.): ECSQARU 2009, LNAI 5590, pp. 554–565, 2009.
© Springer-Verlag Berlin Heidelberg 2009

subsets A and B of the parameter space Ω, the data x supports the hypothesis A better than the hypothesis B if

$$\sum_{\omega \in A} L(x|\omega) > \sum_{\omega \in B} L(x|\omega).$$

This amounts – essentially – to extend the domain of the likelihood function, since these sums can in fact be interpreted as if they were the "aggregated" likelihoods of the sets A and B.

In this paper we consider the extension of the likelihood $L(x|\omega)$ from a point function to a set function (a suitable measure) as a consequence of coherence, and we prove (in a finitely additive setting) that it is necessarily given by a class of "mixtures" of the relevant laws $L(x|\omega)$ (see Theorem 4). We deal also with the problem of managing "hypotheses" with zero probability. An interesting application of this "integration method" is to the well–known nuisance parameter elimination problem. As shown for example in [2], integrated likelihood has many advantages with respect to profile likelihood.

The connection between likelihoods and membership functions of fuzzy sets has been discussed by many authors, for instance [15], [10], [14]. In particular, in [10] the aforementioned connection is based on the general theory of coherent conditional probability. In this context, it is relevant the transition from a "pointwise uncertainty" of the membership function to a sort of "global" membership.

2   Preliminaries

What is usually emphasized in the literature – when a conditional probability $P(E|H)$ is taken into account – is only the fact that $P(\cdot|H)$ *is a probability for any given H*: this is a very restrictive (and misleading) view of conditional probability, corresponding trivially to just a modification of the "world" Ω. It is instead essential to regard the conditioning event H as a "variable", *i.e.* the "status" of H in $E|H$ is not just that of something representing a given *fact*, but that of an (uncertain) *event* (like E) for which the knowledge of its truth value is not required.

2.1   Conditional Probability

The classic *axioms for a conditional probability P*, as given by de Finetti [12] (see also [18], [13], [7]), in its most general sense related to the concept of *coherence*, are: given a set $\mathcal{C} = \mathcal{G} \times \mathcal{B}^o$ of conditional events $E|H$ such that $\mathcal{G}$ is a Boolean algebra and $\mathcal{B} \subseteq \mathcal{G}$ is closed with respect to (finite) logical sums, and putting $\mathcal{B}^o = \mathcal{B} \setminus \{\emptyset\}$, then

$$P : \mathcal{C} \to [0,1]$$

is such that

(i) $P(H|H) = 1$, for every $H \in \mathcal{B}^o$,
(ii) $P(\cdot|H)$ is a (finitely additive) probability on $\mathcal{G}$ for any given $H \in \mathcal{B}^o$,

(iii) $P((E \wedge A)|H) = P(E|H) \cdot P(A|(E \wedge H))$, for every $E, A \in \mathcal{G}$ and $E,$ $E \wedge H \in \mathcal{B}^o$.

In [20], condition *(ii)* is replaced by the stronger one of countable additivity.

A peculiarity – which entails a large flexibility in the management of any kind of uncertainty – of this approach to conditional probability is that, due to its *direct* assignment as a whole, the knowledge – or the assessment – of the "joint" and "marginal" unconditional probabilities $P(E \wedge H)$ and $P(H)$ is not required; moreover, the *conditioning* event H – which *must* be a *possible* one – may have *zero probability*.

2.2   Coherence

A conditional probability P is defined on $\mathcal{G} \times \mathcal{B}^o$: however it is possible, through the concept of *coherence*, to handle also those situations where we need to assess P on an *arbitrary* set $\mathcal{C} = \{E_i|H_i\}_{i \in J}$ of conditional events.

Definition 1. The assessment $P(\cdot|\cdot)$ on $\mathcal{C}$ is *coherent* if there exists a conditional probability $P'(\cdot|\cdot)$ which is an extension of P from $\mathcal{C}$ to $\mathcal{C}' \supset \mathcal{C}$, with $\mathcal{C}' = \mathcal{G} \times \mathcal{B}^o$ ($\mathcal{G}$ Boolean algebra and $\mathcal{B} \subseteq \mathcal{G}$ closed with respect to finite logical sums). We need also to recall the following

Definition 2. Given an *arbitrary* finite family $\{E_1, ..., E_n\}$, of events, all intersections

$$E_1^* \wedge E_2^* \ldots \wedge E_n^* ,$$

different from the impossible event $\emptyset$, obtained by putting – in all possible ways – in place of each E_i^*, for $i = 1, 2, \ldots, n$, the event E_i or its contrary E_i^c, are called *atoms* generated by the given events. The events $E_1, ..., E_n$ are called *logically independent* when the number of atoms equals 2^n.

Definition 3. *Let $\mathcal{F}$ be an algebra. A function $\mu : \mathcal{F} \to [-\infty, \infty]$ is said to be a charge on $\mathcal{F}$ if the following conditions are satisfied:*

$\mu(\emptyset) = 0$;

μ *is finitely additive. Moreover a charge is a* real charge *if* $-\infty < \mu(F) < \infty$ *for any $F \in \mathcal{F}$, it is* bounded *if* $\sup\{|\mu(F)| : F \in \mathcal{F}\} < \infty$, *and it is* positive *if* $\mu(F) \geq 0$ *for any $F \in \mathcal{F}$.*

A charge is a *probability* if it is positive and $\mu(\Omega) = 1$.

A characterization of coherence is given by the following theorem: see, e.g., [6,8].

Theorem 1. *Let $\mathcal{C}$ be an arbitrary family of conditional events. For a real function P on $\mathcal{C}$ the following statements are equivalent:*

(a) P is a coherent conditional probability on $\mathcal{C}$;

(b) there exists (at least) a class of function $\{m_\alpha\}$, with each m_α defined on suitable families $\mathcal{B}_\alpha \subseteq \mathcal{C}$; they are restriction of positive charges defined on the algebra generated by $\mathcal{B}_\alpha$, and for any conditional event $E|H \in \mathcal{C}$ there exists a unique m_α with

$$m_\alpha(H) > 0 \qquad and \qquad P(E|H) = \frac{m_\alpha(E \wedge H)}{m_\alpha(H)} \ ;$$

moreover $\mathcal{B}_\alpha \subset \mathcal{B}_\beta$ for $\alpha > \beta$ and $m_\beta(H) = 0$ iff $H \in \mathcal{B}_\alpha$;

(c) for any finite subset $\mathcal{F} = \{E_1|H_1, \ldots, E_n|H_n\}$ of $\mathcal{C}$, denoting by $\mathcal{A}_o$ the set of atoms A_r generated by the events $E_1, H_1, \ldots, E_n, H_n$, there exists (at least) a class of probabilities $\{P_0, P_1, \ldots P_k\}$, each probability P_α being defined on a suitable subset $\mathcal{A}_\alpha \subseteq \mathcal{A}_0$, such that for any $E_i|H_i \in \mathcal{C}$ there is a unique P_α with

$$\sum_{\substack{r \\ A_r \subseteq H_i}} P_\alpha(A_r) > 0 \qquad and \qquad P(E_i|H_i) = \frac{\displaystyle\sum_{\substack{r \\ A_r \subseteq E_i \wedge H_i}} P_\alpha(A_r)}{\displaystyle\sum_{\substack{r \\ A_r \subseteq H_i}} P_\alpha(A_r)} \ ;$$

moreover $\mathcal{A}_{\alpha'} \subset \mathcal{A}_{\alpha''}$ for $\alpha' > \alpha''$ and $P_{\alpha''}(A_r) = 0$ iff $A_r \in \mathcal{A}_{\alpha'}$.

Any class $\{P_\alpha\}$ singled-out by condition (c) is said to agree with the coherent conditional probability P restricted to the family $\mathcal{F}$.

Notice that coherence of an assessment $P(\cdot|\cdot)$ on an **infinite** set $\mathcal{C}$ of conditional events is equivalent to coherence on **any finite** subset $\mathcal{F}$ of $\mathcal{C}$.

Given a family $\mathcal{C}$ of conditional events $\{E_i|H_i\}_{i \in I}$, where $card(I)$ is arbitrary and the events H_i's are a partition of Ω, we recall the following corollary of the characterization Theorem 1.

Corollary 1. Any function $f : \mathcal{C} \to [0,1]$ such that $f(E_i|H_i) = 0$ if $E_i \wedge H_i = \emptyset$ and $f(E_i|H_i) = 1$ if $H_i \subseteq E_i$ is a coherent conditional probability.

Concerning coherence, another fundamental result is the following, essentially due – for unconditional events, and referring to an equivalent form of coherence in terms of betting scheme – to de Finetti [12] (see also [16,19,21]).

Theorem 2. Let $\mathcal{K}$ be any family of conditional events, and take an arbitrary family $\mathcal{C} \subseteq \mathcal{K}$. Let P be an assessment on $\mathcal{C}$; then there exists a (possibly not unique) coherent extension of P to $\mathcal{K}$ if and only if P is coherent on $\mathcal{C}$.

3   Likelihood and Its (Coherent) Extensions

From now on, given an arbitrary event E, let $\mathcal{C}$ be a family of conditional events $\{E|H_i\}_{i \in I}$, where $card(I)$ is arbitrary and events H_i's are a partition of Ω; $P(E|\cdot)$ is an arbitrary – coherent – conditional probability on $\mathcal{C}$; $\mathcal{H}$ is the algebra spanned by the H_i's, and $\mathcal{H}^o = \mathcal{H} \setminus \{\emptyset\}$.

Here we list some of the main relevant results, taken from [9].

By Theorem 2, P can be extended to a coherent conditional probability on $\mathcal{C}' = \{E|H : H \in \mathcal{H}^o\}$, and the latter in turn can be extended to a coherent conditional probability on $\mathcal{C}'' = \mathcal{C}' \cup \{H|K : H, K \in \mathcal{H}\}$. This satisfies, by axiom (iii) of a conditional probability,

$$P(E|H \vee K) = P(E|H)P(H|H \vee K) + P(E|K)P(K|H \vee K),$$

for every $H \wedge K = \emptyset$.

It follows that any coherent extension of P to $\mathcal{C}' = \{E|H : H \in \mathcal{H}^o\}$ is such that, for every $H, K \in \mathcal{H}$, with $H \wedge K = \emptyset$,

$$P(E|H \vee K) \leq P(E|H) + P(E|K). \tag{1}$$

Remark 1. *The previous inequality can be easily extended to a* partition *of* Ω. *Notice that, except in the trivial case that every partition has an event* H_j *with* $P(E|H_j) = 1$ *while for all others* $P(E|H_i) = 0$, *the function* $L(\cdot) = P(E|\cdot)$ *is* **not additive**. Moreover, since we have – by axioms *(i)* and *(ii)* of a conditional probability – $P(H|H \vee K) + P(K|H \vee K) = 1$ for $H \wedge K = \emptyset$, we get the following inequality

$$\min\{P(E|H), P(E|K)\} \leq P(E|H \vee K) \leq \max\{P(E|H), P(E|K)\}. \tag{2}$$

Then, the function $P(E|\cdot)$, with P a coherent conditional probability, in general *is not monotone* (with respect to implication). Moreover, among the coherent extension of a coherent conditional probability P on $\{E|H_i\}_{i \in I}$ to $\mathcal{C}'$ there is one such that

$$P(E|H \vee K) = \max\{P(E|H), P(E|K)\},$$

for any $H, K \in \mathcal{H}^o$ with $H \wedge K = \emptyset$. Then, this is the only extension of P which is monotone.

4   Likelihood and Statistical Inference

In Section 3 we have shown different ways of extending *coherently* a conditional probability $P(E|\cdot)$, and a very particular case of it – referring to discrete distributions – is the likelihood $L(\omega) = P(E|\omega)$. As already noticed in Remark 1, $L(\omega)$ is **not** additive, and this is true even when it is interpreted (in a Bayesian context) as the posterior corresponding to a uniform prior.

Example 1. *Consider a parameter space* $\{\omega_1, \omega_2 ..., \omega_n\}$, *with* $P(\omega_i) = \frac{1}{n}$; *we have, assuming* $P(E) > 0$,

$$P\big((\omega_1 \vee \omega_2)|E\big) = \frac{P(\omega_1 \vee \omega_2)P\big(E|(\omega_1 \vee \omega_2)\big)}{P(E)}, \tag{3}$$

$$P(\omega_i|E) = \frac{P(\omega_i)P(E|\omega_i)}{P(E)} = \frac{P(E|\omega_i)}{nP(E)} \quad , \quad i = 1, 2, \tag{4}$$

and so, by adding the two eqs. (4), we get

$$P(\omega_1|E) + P(\omega_2|E) = P\big((\omega_1 \vee \omega_2)|E\big) = \frac{P(E|\omega_1) + P(E|\omega_2)}{nP(E)}.$$

Then, since $P(\omega_1 \vee \omega_2) = \frac{2}{n}$, *from (3) it follows*

$$2P\big(E|(\omega_1 \vee \omega_2)\big) = P(E|\omega_1) + P(E|\omega_2),$$

i.e. $P\big(E|(\omega_1 \vee \omega_2)\big)$ *is a convex combination (with equal weights) of* $P(E|\omega_1)$ *and* $P(E|\omega_2)$. Going back to the results of Section 3, the question is whether the two extreme cases obtained extending $P(E|A_x)$ to the union of conditioning events by taking the *maximum* or by taking the *minimum* are the most natural ways to extend likelihood functions. We recall that, given a finite partition $\mathcal{H}_\circ = \{H_i\}$ of Ω, coherence implies

$$\min_i\{P(E|H_i)\} \le P\big(E\big|\bigvee_i H_i\big) \le \max_i\{P(E|H_i)\}$$

and the converse holds if we extend just to a single conditional event $E\big|\bigvee_i H_i$. But, considering the extension to more conditioning events the converse is not true. So, in general, the values between the two extremes are not necessarily coherent choices for the conditional probability $P(E|\cdot)$, which can be looked on as a sort of "aggregated" membership or of likelihood "measure".

Remark 2. *Coherent choices have been essentially characterized – for a finite family – in [5]: they are* **weighted means** *of the $P(E|H_i)$'s, where weights equal to zero or one are allowed.* Here is the relevant theorem (see also [11]), which can be seen as a "discrete" version of the main theorem given in the following Section.

Theorem 3. *Let E be an arbitrary event and $\mathcal{C}$ be a finite family of conditional events $\{E|H_i\}$ $(i = 1, 2, ..., n)$, where $\mathcal{H}_\circ = \{H_i\}$ is a partition of Ω. Let $\mathcal{A}$ be the algebra spanned by the H_i's, and put $\mathcal{A}^\circ = \mathcal{A}\backslash\{\emptyset\}$. If $p : \mathcal{C} \to [0,1]$ is a coherent conditional probability, i.e. any function such that*

$$p(E|H_i) = 0 \quad if \quad E \wedge H_i = \emptyset, \qquad p(E|H_i) = 1 \quad if \quad H_i \subseteq E,$$

the following two statements are equivalent:
   (i) P is a coherent conditional probability extending p to $\mathcal{K} = \{E\} \times \mathcal{A}^\circ$;
   (ii) there exist subfamilies $\mathcal{H}_\circ \supset \mathcal{H}_1 \supset \dots \mathcal{H}_\alpha \supset \dots \supset \mathcal{H}_k$ and relevant sets of coefficients $\lambda_i^\alpha \ge 0$ $(i = 1, ..., i_\alpha$, where i_α is the number of events $H_i \in \mathcal{H}_\alpha$, with $H_i \in \mathcal{H}_\alpha$ if and only if $\lambda_i^{\alpha-1} = 0$ and $\lambda_i^{-1} = 0$ for any i), with $\sum_i \lambda_i^\alpha = 1$, such that for every $H \in \mathcal{A}^\circ$ the value $x = P(E|H)$ is a solution of

$$x \sum_{H_i \subseteq H} \lambda_i^\alpha = \sum_{H_i \subseteq H} \lambda_i^\alpha p(E|H_i) \tag{5}$$

for all $\mathcal{H}_\alpha$, and if at least one H_i belongs to $\mathcal{H}_\alpha\backslash\mathcal{H}_{\alpha+1}$, then $P(E|H)$ is the only solution of (5).

We sketch the procedure to search for the λ_i^α's, given an extension of the assessment $\{p(E|H_i), i = 1, 2, ..., n\}$, and so to prove that the extension is coherent. Since all the H_i's belong to H_o, then (5) holds (with $\alpha = 0$) for all (possible) events belonging to the algebra $\mathcal{A}$. So the (first) set of coefficients λ_i^o, $(i = 1, 2, ..., n)$ satisfy all equations of the kind (5), with $x = P(E|H)$ for every $H \in \mathcal{A}^o$. Given now an $H \in \mathcal{A}^o$, if at least one λ_i^o (for i such that $H_i \subseteq H$) is positive, then $P(E|H)$ is the only solution of (5), and we have

$$P(E|H) = \frac{1}{\lambda(H)} \sum_{H_i \subseteq H} \lambda_i^o p(E|H_i), \tag{6}$$

where

$$\lambda(H) = \sum_{H_i \subseteq H} \lambda_i^o \, .$$

Note that $\lambda_i^\alpha = P(H_i | H^\alpha)$, where

$$H^\alpha = \bigvee_{H_i \in \mathcal{H}_\alpha} H_i \, .$$

Moreover, if on the contrary $\lambda_i^o = 0$ for all i such that $H_i \subseteq H$, then (5) is trivially satisfied for all value of x, and all $H_i \subseteq H$ belong to $\mathcal{H}_1$. In this case we must find coefficients λ_j^1 satisfying all the equations (5) related (only) to the events H obtained as unions of the H_i's such that $\lambda_i^o = 0$; and so on.

Notice that if we can assign to λ_i^o positive values for every i, then we obtain only one class of λ_i^α, and we can write (6) for every $H \in \mathcal{A}^o$. In particular, for $\lambda_i^o = \frac{1}{n}$ $(i = 1, 2, ..., n)$ we are in the situation of Example 1. The opposite situation corresponds to assign – for every α – value 1 to only one λ_{i*}^α and value 0 to all others. So in this case $P(E|H) = P(E|H_{i*})$ for all the events $H \supseteq H_{i*}$. A particular case is when at any step α the value 1 corresponds, for $E|H_i \in \mathcal{H}_\alpha$, to the *maximum* (or *minimum*) value of $p(E|H_i)$.

5   Main Result

First of all, we need to recall the following well–known definitions.

Definition 4. *Let $(\Omega, \mathcal{F}, \mu)$ be a charge space. A real valued function f on Ω is T_2-measurable if for every $\epsilon > 0$, there exists a partition $\{F_0, F_1, \ldots, F_n\}$ of Ω in $\mathcal{F}$ such that $\mu(F_0) < \epsilon$ and $|f(w) - f(w')| < \epsilon$ for every $w, w' \in F_i$ for every $i = 1, \ldots, n$. Concerning T_2-measurability, we show that the function in Corollary 1 has this property in the case of a charge defined on a suitable σ-field, but in general it is not T_2-measurable.*

Example 2. *Consider the partition $\mathcal{H} = \{H_i\}_{i \in \mathbb{N}}$, an event E logical independent from any H_i and the function $p(E|H_i) = \frac{1}{2}$ if $i = 2k$, while $p(E|H_i) = 1$ if $i = 2k + 1$. Let $\mathcal{A}$ be the minimal algebra generated by $\mathcal{H}$, then in $\mathcal{A}$ there are only finite and co-finite sets, so $p(E|\cdot)$ is not T_2-measurable with respect to any positive bounded charge space $(\Omega, \mathcal{A}, \mu)$, since no infinite set of the kind $\bigvee_{k \in \mathbb{N}} H_{2k}$ belongs to $\mathcal{A}$. On the other hand, by taking the positive charge space $(\Omega, \mathcal{F}, m)$, where $\mathcal{F}$ is the field generated by $\mathcal{A}$ and by the event $K = \bigvee_{k \in \mathbb{N}} H_{2k}$, then $p(E|\cdot)$ is T_2-measurable with respect to any charge space $(\Omega, \mathcal{F}, m)$.*

Lemma 1. *Let $\mathcal{E} = \{E|H_i\}_{i \in J}$ be an arbitrary set of conditional events such that the set of conditioning events $\mathcal{H}_0 = \{H_i\}_{i \in J}$ is a partition of Ω, and denote by $\mathcal{F}$ the σ-field spanned by $\mathcal{H}_0$ and $(\Omega, \mathcal{F}, \mu)$ a charge space with μ a positive bounded charge. Let $p : \mathcal{E} \to [0, 1]$ be any function such that*

$$p(E|H_i) = 0 \text{ if } E \wedge H_i = \emptyset, \ p(E|H_i) = 1 \text{ if } H_i \subseteq E \, ;$$

then the function p is T_2-measurable. Moreover, for any $H \in \mathcal{F}$, the function $f(\cdot) = p(E|\cdot)I_H$, where I_H is the indicator function of H, is T_2-measurable.

Proof: Given $\epsilon = \frac{1}{k}$, with $k \in \mathbb{N}$, consider $F_0 = \vee_1^n H_i$ with $\mu(F_0) < \frac{1}{k}$. The event $F_0 \in \mathcal{F}$ exists since no more than a finite number of H_j can have charge greater than $\frac{1}{k}$ (μ is a positive bounded charge). Now, let $F_1 = \vee\{H_j : H_j \wedge F_0 = \emptyset, \ p(E|H_j) < \frac{1}{k}\}$, and, for $2 \le i \le k-1$, $F_i = \vee\{H_j : H_j \wedge F_0 = \emptyset, \ (i-1)\frac{1}{k} \le p(E|H_j) < i\frac{1}{k}\}$, while $F_k = \vee\{H_j : H_j \wedge F_0 = \emptyset, \ (k-1)\epsilon \le p(E|H_j)\}$ (some $\mathcal{F}_j$ could be empty), hence $F_i \in \mathcal{F}$ for any $i = 1, ..., k$, and moreover, for any $w, v \in F_i$ one has $|p(E|w) - p(E|v)| \le \epsilon$. Therefore, it follows that $p(E|\cdot)$ is T_2-measurable. Moreover, taking $f(\cdot) = p(E|\cdot)I_H$, with $H \in \mathcal{F}$, taking as partition $F_0' = F_0 \wedge H^c$, $F_1' = F_1 \vee H$, $F_i' = F_i \wedge H^c$, for $i = 2, ..., k$, T_2-measurability of $p(E|\cdot)I_H$ follows. The hypothesis on $\mathcal{F}$ in the above Lemma is crucial, moreover there could be a smaller field such that $p(E|\cdot)$ is T_2-measurable (as shown in Example 2). The importance of T_2 measurability is related, for bounded functions, to equivalence with D-integrability, and in the case of positive bounded charges D-integrability and S-integrability coincide for real valued bounded functions (see [4]).

Concerning the extension of the function in Lemma 1, we can prove the following result:

Theorem 4. *Let $\mathcal{E} = \{E|H_i\}_{i \in J}$ be an arbitrary set of conditional events such that the set of conditioning events $\mathcal{H}_0 = \{H_i\}_{i \in J}$ is a partition of Ω. Denote by $\mathcal{F}_0$ the σ-field spanned by $\mathcal{H}_0$, $\mathcal{F}_0^0 = \mathcal{F}_0 \setminus \{\emptyset\}$ and $\mathcal{K} = \{E|H : H \in \mathcal{F}_0^0\}$. Let $p : \mathcal{E} \to [0,1]$ be any function such that*

$$p(E|H_i) = 0 \ \text{if} \ E \wedge H_i = \emptyset, \ p(E|H_i) = 1 \ \text{if} \ H_i \subseteq E.$$

The following statements are equivalent:

- *P is a coherent conditional probability extending p to $\mathcal{K}$;*
- *there exists a class of positive (not necessarily bounded) charges $\{m_\alpha\}$ on σ-fields $\{\mathcal{F}_\alpha\}$ defined by suitable families $\mathcal{B}_\alpha$ with $\mathcal{B}_\alpha \subset \mathcal{B}_\beta$ for $\alpha > \beta$ and $m_\beta(H) = 0$ iff $H \in \mathcal{B}_\alpha$, and for any conditional event $E|H \in \mathcal{K}$ there is a unique α such that $H \in \mathcal{B}_\alpha$, with $0 < m_\alpha(H) < \infty$, and $P(E|H) = x$ is solution of the equation*

$$x \int_H d(m_\alpha(y)) = \int_H p(E|y) d(m_\alpha(y)). \tag{7}$$

Proof: Since p is coherent (see Corollary 1) there is a coherent extension P of p on $\mathcal{K} = \{E|H : H \in \mathcal{F}_0^0\}$. Moreover, for any conditional probability P^* on $\mathcal{F}_0 \times \mathcal{F}_0^0$ there is a class $\{m_\alpha\}$ of positive charges agreeing with P^* (see Theorem 2, and also [17]). Take $m_0(A) = P^*(A|\Omega)$ for $A \in \mathcal{F}_0$; then, the positive charge m_0 is bounded and one has, from Lemma 1, that $f(\cdot) = p(E|\cdot)$ and $f_1(\cdot) = p(E|\cdot)I_H(\omega)$ (with $H \in \mathcal{F}_0$) are bounded and T_2-measurable with respect to the charge space $(\Omega, \mathcal{F}_0, m_0)$. Then, $f(\cdot)$ is Daniell integrable with respect to $(\Omega, \mathcal{F}_0, m_0)$ (and equivalently is Stieltjes integrable, see [4]) and for any $H \in \mathcal{F}_0^0$ such that $m_0(H) > 0$ it follows

$$S \int f_1 dm_0 = \int_H p(E|x)d(m_0(x)) \,.$$

Moreover, from construction of Stieltjes integral one has that

$$\int_H p(E|x)d(m_0(x)) = \lim_{\mathcal{P}} \sum_{i=0}^{n} (\sup_{w \subseteq F_i} p(E|w)I_H(\omega)m_0(F_i)) =$$

$$\lim_{\mathcal{P}} \sum_{i=0}^{n} (\inf_{w \subseteq F_i} p(E|w)I_H(\omega)m_0(F_i))$$

where $\mathcal{P}$ is the set of finite partitions of Ω in $\mathcal{F}_0$. Then, from disintegration property, for any finite partition $\{F_0, F_1, ..., F_n\}$ of Ω in $\mathcal{F}_0$,

$$P(E|H) = \sum_{i=0}^{n} P^*(E \wedge F_i|H) = \sum_{i=0}^{n} P^*(E|F_i \wedge H)P^*(F_i|H)$$

and

$$\sum_{i=0}^{n} P^*(E|F_i \wedge H)P^*(F_i|H) \geq \sum_{i=0}^{n} \inf_{w \subseteq F_i} p(E|w)I_H(\omega)\frac{m_0(F_i \wedge H)}{m_0(H)}$$

$$\sum_{i=0}^{n} P^*(E|F_i \wedge H)P^*(F_i|H) \leq \sum_{i=0}^{n} \sup_{w \subseteq F_i} p(E|w)I_H(\omega)\frac{m_0(F_i \wedge H)}{m_0(H)}$$

so from Stieltjes integrability of $p(E|\cdot)$ one has that for any $\epsilon > 0$ there exists a finite partition $\{F_0, F_1, ..., F_n\}$ of Ω in $\mathcal{F}_0$ such that

$$\sum_{i=0}^{n} \sup_{w \in F_i} p(E|w)I_H(\omega)m_0(F_i \wedge H) - \epsilon \leq P(E|H)m_0(H) \leq$$

$$\leq \sum_{i=0}^{n} \inf_{w \in F_i} p(E|w)I_H(\omega)m_0(F_i \wedge H) + \epsilon \,.$$

Then equation (7) follows.

Now, consider any ideal $\mathcal{B}_\alpha$ (formed by $H \in \mathcal{F}_0$ with $m_0(H) = 0$) in $\mathcal{F}_0$, that is an additive class, and let $\mathcal{F}_\alpha$ the corresponding σ-field, we have the following two situations: if in $\mathcal{B}_\alpha$ there is an event K such that for all $H \in \mathcal{B}_\alpha$ one has $H \subseteq K$, then by putting $m_\alpha(A) = P^*(A|K)$ with $A \in \mathcal{F}_\alpha$, since $m_\alpha(\Omega) = P^*(K|K) = 1$ it follows that m_α is a positive bounded charge on $\mathcal{F}_\alpha$, and so the proof goes along the same lines of the previous step to show that the above equation holds. Otherwise, if there is no event $K \in \mathcal{B}_\alpha$ containing all events $H \in \mathcal{B}_\alpha$, then, since $\mathcal{B}_\alpha$ is an additive set, the cardinality of $\mathcal{B}_\alpha$ is infinite. Then P^* agrees with a positive charge m_α on the minimal σ-field $\mathcal{F}_\alpha$ containing $\mathcal{B}_\alpha$, but it is not bounded on $\mathcal{F}_\alpha$. However, for any $H \in \mathcal{B}_\alpha$ one has $m_\alpha(H) < \infty$ (while it could happen, for some $K \in \mathcal{F}_\alpha \setminus \mathcal{B}_\alpha$, that $m_\alpha(K) = \infty$).

We show how to compute the extension for a chosen conditioning event H such that $0 < m_\alpha(H) < \infty$, and for simplicity we avoid to refer to H and we define $\lambda(A) = \frac{m_\alpha(A \wedge H)}{m_\alpha(H)}$ for any $A \in \mathcal{F}_\alpha$. Then $\lambda(\cdot)$ is a bounded positive charge in $\mathcal{F}_\alpha$ with $\lambda(H^c) = 0$ and $\lambda(\Omega) = \lambda(H) = 1$. Moreover $f_1(\cdot) = f(\cdot) I_{H \wedge H^\alpha}$ (with $H^\alpha = \vee_{H_i \in \mathcal{B}_\alpha} H_i$) is measurable with respect to $(\Omega, \mathcal{F}_\alpha, \lambda)$. Hence f_1 is Stieltjes integrable with respect to $(\Omega, \mathcal{F}_\alpha, \lambda)$ (see Theorem 4.5.7 in [4]) and for any $H \in \mathcal{B}_\alpha$ such that $\lambda(H) > 0$ (i.e., $0 < m_\alpha(H) < \infty$) it follows

$$S \int f_1 d\lambda = \int_H p(E|x) d(\lambda(x)).$$

Moreover, from construction of Stieltjes integral one has that

$$\int_H p(E|x) d(\lambda(x)) = \lim_{\mathcal{P}} \sum_{i=0}^{n} \left(\sup_{w \subseteq F_i} p(E|w) I_H(\omega) \lambda(F_i \wedge H) \right) =$$

$$\lim_{\mathcal{P}} \sum_{i=0}^{n} \left(\sup_{w \subseteq F_i} p(E|w) I_H(\omega) \frac{m_\alpha(F_i \wedge H)}{m_\alpha(H)} \right) =$$

$$\frac{1}{m_\alpha(H)} \lim_{\mathcal{P}} \sum_{i=0}^{n} \left(\sup_{w \subseteq F_i} p(E|w) I_H(\omega) m_\alpha(F_i \wedge H) \right)$$

and

$$\int_H p(E|x) d(\lambda(x)) = \lim_{\mathcal{P}} \sum_{i=0}^{n} \left(\inf_{w \subseteq F_i} p(E|w) I_H(\omega) \lambda(F_i \wedge H) \right) =$$

$$= \frac{1}{m_\alpha(H)} \lim_{\mathcal{P}} \sum_{i=0}^{n} \left(\inf_{w \subseteq F_i} p(E|w) I_H(\omega) m_\alpha(F_i \wedge H) \right)$$

where $\mathcal{P}$ is the set of finite partitions of Ω in $\mathcal{F}_\alpha$. Then, since for any finite partition $\{F_0, ..., F_n\}$ of Ω in $\mathcal{F}_\alpha$

$$P(E|H) = \sum_{i=0}^{n} P^*(E \wedge F_i | H) = \sum_{i=0}^{n} P^*(E|F_i \wedge H) P^*(F_i | H) =$$

$$\sum_{i=0}^{n} P^*(E|F_i \wedge H) \frac{m_\alpha(F_i \wedge H)}{m_\alpha(H)}$$

and from Stieltjes integrability one has that for any $\epsilon > 0$ there exists a finite partition $\{F_0, F_1, ..., F_n\}$ of Ω in $\mathcal{F}_\alpha$ such that

$$\sum_{i=0}^{n} \sup_{w \subseteq F_i} p(E|w) I_H(\omega) m_\alpha(F_i \wedge H) - \epsilon \leq P(E|H) m_\alpha(H) \leq$$

$$\leq \sum_{i=0}^{n} \inf_{w \subseteq F_i} p(E|w) I_H(\omega) m_\alpha(F_i \wedge H) + \epsilon,$$

then equation (7) follows.

Note that Theorem 3 is a particular case of Theorem 4.

A result similar to the above one can be given by using the characterization of conditional probabilities in a σ-additive setting, which corresponds to consider a class of σ additive positive charges (see [20] and also [17]).

Remark 3. *Theorem 4 characterizes coherent extensions in terms of class of mixtures, each one related to a charge m_α defined on the set of events H with $m_{\alpha-1}(H) = 0$. The first mixture is based on m_0, which is obviously defined on the whole set of conditioning events. This specific mixture has been studied, in a different context and terminology, in [3].*

An example in which the class of the charges in Theorem 4 is not finite follows:

Example 3. *Let $\{H_i\}_{i \in \mathbb{N}}$ be a partition of Ω and E an event logically independent from any H_i with $i \in \mathbb{N}$. Consider the coherent assessment $p(E|H_i) = \frac{1}{n}$ for $i \in \mathbb{N}$. A possible extension of the given assessment on the set $E \times \mathcal{H}^o$, with $\mathcal{H}$ the power set generated by $\{H_i\}_{i \in \mathbb{N}}$ and $\mathcal{H}^o = \mathcal{H} \setminus \{\emptyset\}$, is*

$$P(E|H) = \sup_{H_i \subseteq H} p(E|H_i)$$

for any $H \in \mathcal{H}^o$. This extension is obtained by considering the sets $\mathcal{H}_o = \{H_i\}_{i \in \mathbb{N}}$, $\mathcal{H}_1 = \{H_j \in \mathcal{H}_o : j > 1\}$, ... $\mathcal{H}_i = \{H_j \in \mathcal{H}_o : j > i\}$, ... and the charges $m_i(\cdot)$ with $i \in \mathbb{N}$ on $\mathcal{H}$ such that $m_i(H) = 1$ if $H_{i+1} \subseteq H$ and $m_i(H) = 0$ otherwise.

6 Conclusions

The main result (Theorem 4) provides a characterization of all possible extensions as set function of the (point function) likelihood. These extensions are suitable mixtures (that in the finite case reduce to weighted means, see Remark 2) of likelihoods. This result can be also the starting point for the elimination problem of nuisance parameters in a finitely additive setting, which is particularly fit for handling "improper" distributions. Another interesting connection is with possibility theory: in fact we obtain a possibility measure as a particular extension of a likelihood (see Example 3 and, for a finite case, [9]).

References

1. Basu, D.: Statistical Information and Likelihood. In: Ghosh, J.K. (ed.) A Collection of Critical Essays. Lecture Notes in Statistics, vol. 45. Springer, New York (1988)
2. Berger, J.O., Liseo, B., Wolpert, R.L.: Integrated likelihood methods for eliminating nuisance parameters. Statistical Science 14, 1–28 (1999)
3. Berti, P., Fattorini, L., Rigo, P.: Eliminating nuisance parameters: two characterizations. Test 9(1), 133–148 (2000)
4. Bhaskara Rao, K.P.S., Bhaskara Rao, M.: Theory of Charges. Academic Press, London (1983)

5. Ceccacci, S., Morici, C., Paneni, T.: Conditional probability as a function of the conditioning event: characterization of coherent enlargements. In: Proc. WUPES 2003, Hejnice, Czech Republic, pp. 35–45 (2003)
6. Coletti, G., Scozzafava, R.: Characterization of coherent conditional probabilities as a tool for their assessment and extension. Internat. J. Uncertainty Fuzziness Knowledge-Based Systems 4, 103–127 (1996)
7. Coletti, G., Scozzafava, R.: From conditional events to conditional measures: a new axiomatic approach. Annals of Mathematics and Artificial Intelligence 32, 373–392 (2001)
8. Coletti, G., Scozzafava, R.: Probabilistic Logic in a Coherent Setting. Trends in Logic, vol. 15. Kluwer, Dordrecht (2002)
9. Coletti, G., Scozzafava, R.: Coherent conditional probability as a measure of uncertainty of the relevant conditioning events. In: Nielsen, T.D., Zhang, N.L. (eds.) ECSQARU 2003. LNCS (LNAI), vol. 2711, pp. 407–418. Springer, Heidelberg (2003)
10. Coletti, G., Scozzafava, R.: Conditional Probability, Fuzzy Sets, and Possibility: a Unifying View. Fuzzy Sets and Systems 144, 227–249 (2004)
11. Coletti, G., Scozzafava, R.: Conditional Probability and Fuzzy Information. Computational Statistics & Data Analysis 51, 115–132 (2006)
12. de Finetti, B.: Sull'impostazione assiomatica del calcolo delle probabilità. Annali Univ. Trieste 19, 3–55 (1949); Engl. transl. in: Probability, Induction, Statistics, ch. 5. Wiley, London (1972)
13. Dubins, L.E.: Finitely additive conditional probability, conglomerability and disintegrations. Ann. Probab. 3, 89–99 (1975)
14. Dubois, D.: Possibility theory and statistical reasoning. Computational statistics & data analysis 51(1), 47–69 (2006)
15. Dubois, D., Moral, S., Prade, H.: A semantics for possibility theory based on likelihoods. J. of Mathematical Analysis and Applications 205, 359–380 (1997)
16. Holzer, S.: On coherence and conditional prevision. Bollettino Unione Mat. Ital. C(VI), 441–460 (1985)
17. Krauss, P.H.: Representation of Conditional Probability Measures on Boolean Algebras. Acta Math. Acad. Scient. Hungar. 19, 229–241 (1968)
18. Popper, K.R.: The Logic of Scientific Discovery. Routledge, London (1959)
19. Regazzini, E.: Finitely additive conditional probabilities. Rend. Sem. Mat. Fis. Milano 55, 69–89 (1985)
20. Rényi, A.: On conditional probability spaces generated by a dimensionally ordered set of measures. Theor. Probab. Appl. 1, 61–71 (1956)
21. Williams, P.M.: Notes on conditional previsions. School of Mathematical and Physical Sciences, University of Sussex, working paper (1975)

Triangulation Heuristics for BN2O Networks[*]

Petr Savicky[1] and Jiří Vomlel[2]

[1] Institute of Computer Science
Academy of Sciences of the Czech Republic
Pod vodárenskou věží 2,
182 07 Praha 8, Czech Republic
http://www.cs.cas.cz/savicky
[2] Institute of Information Theory and Automation of the AS CR,
Academy of Sciences of the Czech Republic
Pod vodárenskou věží 4,
182 08 Praha 8, Czech Republic
http://www.utia.cas.cz/vomlel

Abstract. A BN2O network is a Bayesian network having the structure
of a bipartite graph with all edges directed from one part (the top level)
toward the other (the bottom level) and where all conditional probability
tables are noisy-or gates. In order to perform efficient inference, graphical transformations of these networks are performed. The efficiency of
inference is proportional to the total table size of tables corresponding
to the cliques of the triangulated graph. Therefore in order to get efficient inference it is desirable to have small cliques in the triangulated
graph. We analyze existing heuristic triangulation methods applicable to
BN2O networks after transformations using parent divorcing and tensor
rank-one decomposition and suggest several modifications. Both theoretical and experimental results confirm that tensor rank-one decomposition
yields better results than parent divorcing in randomly generated BN2O
networks that we tested.

1 Introduction

A BN2O network is a Bayesian network having the structure of a directed bipartite graph with all edges directed from one part (the top level) toward the
other (the bottom level) and where all conditional probability tables are noisy-
or gates. Since the table size for a noisy-or gate is exponential in the number
of its parents, graphical transformations of these networks are performed in order to reduce the table size and allow efficient inference. This paper deals with
two transformations - parent divorcing (PD) [1], which is the most frequently
used transformation, and rank-one decomposition (ROD) [2,3,4]. Typically, in

---

[*] P. Savicky was supported by grants number 1M0545 (MŠMT ČR), 1ET100300517
(Information Society), and by Institutional Research Plan AV0Z10300504. J. Vomlel
was supported by grants number 1M0572 and 2C06019 (MŠMT ČR), ICC/08/E010
(Eurocores LogICCC), and 201/09/1891 (GA ČR).

C. Sossai and G. Chemello (Eds.): ECSQARU 2009, LNAI 5590, pp. 566–577, 2009.

order to get an inference structure, the graph obtained by parent divorcing is further transformed by the following two consecutive steps – moralization and triangulation that results in an undirected triangulated graph. The graph obtained by rank-one decomposition is transformed by triangulation only resulting in an undirected triangulated graph. The efficiency of inference is proportional to the total table size (tts) of tables corresponding to the cliques of the triangulated graph. The size of the largest clique minus one is often called the graph treewidth (tw). Since a BN2O network consists only of binary variables, the size of the largest probability table is 2^{tw+1}.

Both methods, parent divorcing and rank-one decomposition, were designed to minimize the size of probability tables before triangulation. In this paper, we consider the total table size after triangulation, which is the crucial parameter for efficiency of the inference. From this point of view, parent divorcing appears to be clearly inferior. In Section 2 we show that the treewidth tw of the optimally triangulated graph of a BN2O network after rank-one decomposition, which will be called a base ROD (BROD) graph, is not larger than the treewidth of the model preprocessed using parent divorcing, which will be called PD graph, and the same rule holds for the total table size tts. Hence, if we can use optimal elimination ordering (EO) for the transformed graphs, using ROD we never get results worse by more than a linear term compared to PD. Since the search for the optimal EO is NP-hard [5], we have to use heuristics. In this case, ROD is also not worse, since the upper bound on tw and tts for ROD holds efficiently. We propose an efficient procedure which transforms an EO for PD graph into an EO for a base ROD graph with the required upper bound on tw and tts. Similar conclusions concerning the comparison of ROD and PD transformations based on purely experimental results were obtained in [6].

Having the above-mentioned facts in mind, in Section 3 we concentrate on the search for a good EO to use in the BN2O graphs after the ROD transformation. We analyze existing heuristic triangulation methods applicable to BN2O networks and suggest several modifications. The experimental results in Section 4 confirm that these modifications further improve the quality of the obtained triangulation of the randomly generated BN2O networks we used.

2 Transformations of BN2O Networks

First, we introduce the necessary graph notions. For more detail see, e.g. [7].

Definition 1. An undirected graph G is triangulated if it does not contain an induced subgraph that is a cycle without a chord of a length of at least four.

Definition 2. A triangulation of G is a triangulated graph H that contains the same nodes as G and contains G as a subgraph.

Definition 3. A set of nodes $C \subseteq V$ of a graph $G = (V, E)$ is a clique if it induces a complete subgraph of G and it is not a subset of the set of nodes in any larger complete subgraph of G.

Definition 4. For any graph G, let $\mathcal{C}(G)$ be the set of all cliques of G.

Definition 5. The treewidth of a triangulation H of G is the maximum clique size in H minus one. The treewidth of G, denoted $tw(G)$, is the minimum treewidth over all triangulations H of G.

Definition 6. The table size of a clique C in an undirected graph is $\prod_{v \in C} |X_v|$, where $|X_v|$ is the number of states of a variable X_v corresponding to a node v.

In this paper all variables are binary, hence the table size of a clique C is $2^{|C|}$.

Definition 7. The total table size of a triangulation H of G is the sum of table sizes for all cliques of H. The total table size of a graph, denoted $tts(G)$, is the minimum total table size over all triangulations H of G.

Definition 8. The set of neighbors of node v in an undirected graph $G = (V, E)$ is the set $nb_G(v) = \{w \in V : \{v, w\} \in E\}$. The degree of v in G is $|nb_G(v)|$.

Definition 9. A node v is simplicial in G if $nb_G(v)$ induces a complete subgraph of G.

Definition 10. Elimination ordering of an undirected graph $G = (V, E)$ is any ordering of the nodes of G represented by a bijection $f : V \rightarrow \{1, 2, \ldots, n\}$.

The meaning of this representation is that, for every node u, the number $f(u)$ is the index of u in the represented ordering.

Definition 11. An elimination ordering $f : V \rightarrow \{1, 2, ..., n\}$ of an undirected graph $G = (V, E)$ is perfect if, for all $v \in V$, the set

$$B(v) = \{w \in nb_G(v) : f(w) > f(v)\}$$

induces a complete subgraph of G.

A graph possesses a perfect elimination ordering if and only if it is triangulated. If a graph $G = (V, E)$ is not triangulated, then we may triangulate it using any given elimination ordering f by considering the nodes in V in the order defined by f, and sequentially adding edges to E so that after considering node v, the set $B(v)$ induces a complete subgraph in the extended graph.

Now, we restrict our attention to the family of BN2O networks and define the corresponding graphs.

Definition 12. $G = (U \cup V, E)$ is a graph of a BN2O network (BN2O graph) if it is an acyclic directed bipartite graph, where U is the set of nodes of the top level, V is the set of nodes of the bottom level, and E is a subset of the set of all edges directed from U to V, $E \subseteq \{(u_i, v_j) : u_i \in U, v_j \in V\}$.

See Fig. 1 for an example of a BN2O graph.

Since the conditional probability tables in the BN2O networks take on a special form – they are noisy-or gates – we can transform the original BN2O graph

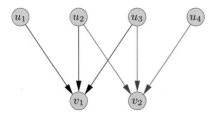

Fig. 1. A BN2O graph

and corresponding tables using methods exploiting their special form. Below we deal with two methods – parent divorcing and rank-one decomposition. Since we restrict ourselves to analyzing graph triangulation, we concentrate only on the graphical transformations performed when these methods are applied.

The first transformation is parent divorcing [1]. It avoids connecting all parents of each node of V (in the moralization step), which is achieved by introducing auxiliary nodes in between nodes from U and V. The next definition describes the graph obtained by a specific form of PD together with the moralization step.

Definition 13. *The parent divorcing (PD) graph of a BN2O graph $G = (U \cup V, E)$ is the undirected graph $G_{PD} = (U \cup V \cup W, H)$, where*

$$W = \cup_{v_i \in V} W_i \ and \ H = \cup_{v_i \in V} H_i$$

and for each node $v_i \in V$ with $pa(v_i) = \{u_j \in U : (u_j, v_i) \in E\}$ the set of auxiliary nodes

$$W_i = \{w_{i,j}, j = 1, \dots, k = |pa(v_i)| - 2\}$$

and the set of undirected edges

$$\begin{aligned}
H_i = \ & \{ \ \{w_{i,1}, u_{j_1}\}, \{w_{i,1}, u_{j_2}\}, \{u_{j_1}, u_{j_2}\}, \\
& \{w_{i,2}, w_{i,1}\}, \{w_{i,2}, u_{j_3}\}, \{w_{i,1}, u_{j_3}\}, \\
& \dots, \\
& \{w_{i,k}, w_{i,k-1}\}, \{w_{i,k}, u_{j_{k+1}}\}, \{w_{i,k-1}, u_{j_{k+1}}\}, \\
& \{v_i, w_{i,k}\}, \{v_i, u_{j_{k+2}}\}, \{w_{i,k}, u_{j_{k+2}}\} \ \} \ ,
\end{aligned}$$

where $\{u_{j_1}, \dots, u_{j_{k+2}}\} = pa(v_i)$.

See Fig. 2 for an example of a PD graph.

The second transformation – rank-one decomposition – was originally proposed by Díez and Galán [2] for noisy-max models and extended to other models by Savicky and Vomlel [3,4].

Definition 14. The rank-one decomposition (ROD) graph of a BN2O graph $G = (U \cup V, E)$ is the undirected graph $G_{ROD} = (U \cup V \cup W, F)$ constructed from G by adding an auxiliary node w_i for each $v_i \in V$, $W = \{w_i : v_i \in V\}$, and

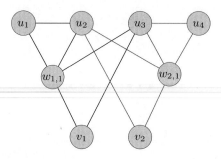

Fig. 2. The PD graph of BN2O graph from Fig. 1

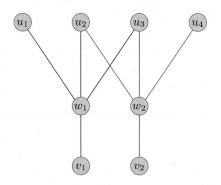

Fig. 3. The ROD graph of BN2O graph from Fig. 1

by replacing each directed edge $(u_j, v_i) \in E$ by undirected edge $\{u_j, w_i\}$ and adding an undirected edge $\{v_i, w_i\}$ for each $v_i \in V$:

$$F = \{\{u_j, w_i\} : (u_j, v_i) \in E\} \ \cup \ \{\{v_i, w_i\} : v_i \in V\}$$

See Fig. 3 for an example of an ROD graph.

Nodes $v_i \in V$ are simplicial in the ROD graph and have degree one; therefore we can perform optimal triangulation of the ROD graph by optimal triangulation of its subgraph induced by nodes $U \cup W$ [7]. This graph will be called the base ROD graph or shortly the BROD graph. For the treewidth it holds

$$tw(G_{ROD}) = \max\{1, tw\,(G_{BROD})\}$$

and for the total table size

$$tts(G_{ROD}) = tts\,(G_{BROD}) + 4\,|W| \ .$$

See Fig. 4 for the BROD graph of BN2O graph from Fig 1.

Definition 15. A graph H is a minor of a graph G if H can be obtained from G by any number of the following operations:

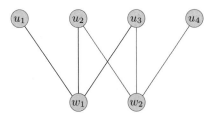

Fig. 4. The BROD graph of BN2O graph from Fig. 1

- node deletion,
- edge deletion, and
- edge contraction[1].

Lemma 1. *The BROD graph is a graph minor of the PD graph.*

Proof. For each set of edges H_i in the PD graph (see Definition 13) we delete the edge $\{u_{j_1}, u_{j_2}\}$ and contract edges

$$\{v_i, w_{i,k}\}, \{w_{i,k}, w_{i,k-1}\}, \ldots, \{w_{i,2}, w_{i,1}\}$$

and name the resulting node w_i. By these edge contractions the node w_i gets connected by undirected edges to all $u_j \in pa(v_i)$. Repeating this procedure for all $i, v_i \in V$ we get the BROD graph.   □

Theorem 1. *The treewidth of the BROD graph is not larger than the treewidth of the PD graph.*

Proof. Due to Lemma 1 the BROD graph is a graph minor of the PD graph. Therefore we can apply the well-known theorem (see, e.g. Lemma 16 in [8]) that the treewidth of a graph minor is not larger than the treewidth of the graph itself.   □

Lemma 2. *Let $G = (V, E)$ be a triangulated graph with a perfect elimination ordering f and $H = (U, F)$ be the graph constructed from G by contraction of the edge $\{u, v\}$ with the resulting node named w. Further, let $f(u) < f(v)$. Then H is triangulated and its elimination ordering g constructed from f by*

$$g(a) = \begin{cases} f(a) & \text{if } f(a) < f(u) \\ f(v) - 1 & \text{if } a = w \\ f(a) - 1 & \text{otherwise} \end{cases}$$

is perfect.

---

[1] Edge contraction is the operation that replaces two adjacent nodes u and v by a single node w that is connected to all neighbors of u and v.

Proof. By the definition of perfect elimination ordering (Definition 11) it is sufficient to show that for all nodes $a \in U$ the set

$$B_H(a) = \{b \in U : \{a, b\} \in F \text{ and } g(b) > g(a)\}$$

induces a complete subgraph of H. Since f is a perfect elimination ordering of G it holds for all nodes $a \in V$ that the set

$$B_G(a) = \{b \in V : \{a, b\} \in E \text{ and } f(b) > f(a)\}$$

induces a complete subgraph of G. Nodes $a \in U \setminus \{w\}$ have either $B_H(a) = B_G(a)$ or $B_H(a) = (B_G(a) \setminus \{u, v\}) \cup \{w\}$. In both cases these sets induce a complete subgraph of H. Every node $x \in B_G(u)$ is connected by an edge with v in G since $v \in B_G(u)$. Therefore $\{x \in B_G(u), f(x) > f(v)\} \subseteq B_G(v)$. Consequently, $B_H(w) = B_G(v)$ and induces a complete subgraph of H.   □

Lemma 3. *Let G be a triangulated undirected graph and H be the resulting graph after the contraction of an edge. Then $tts(H) \leq tts(G)$.*

Proof. Let H be the resulting graph after the contraction of an edge $\{u, v\}$ in G replaced by node w in H. Let ϕ be a mapping of nodes of G onto the nodes of H such that it is an identity mapping except for $\phi(u) = w$ and $\phi(v) = w$. Let us prove that for every clique D in H there exists a clique C in G such that $D = \phi(C)$. This assertion is obvious for cliques of H not containing node w. Let D be a clique in H containing node w. The assertion is also obvious for $|D| = 1$. For $|D| = 2$ it holds that $D = \phi(\{u, a\}) = \phi(\{v, a\}) = \phi(\{u, v, a\})$, where $a \neq w$ is a node from D. Furthermore, either $\{u, a\}$, $\{v, a\}$, or $\{u, v, a\}$ is a clique in G.

Now assume that $|D| \geq 3$. Denote

$$C_1 = (D \setminus \{w\}) \cup \{u\}$$
$$C_2 = (D \setminus \{w\}) \cup \{v\}$$
$$C_3 = (D \setminus \{w\}) \cup \{u, v\} \ .$$

It holds that $D = \phi(C_1) = \phi(C_2) = \phi(C_3)$. To show that either C_1 or C_2 is a complete subgraph of G, assume by contradiction that neither C_1 nor C_2 is a complete subgraph of G. Then nodes $a, b \in D \setminus \{w\}$ would exist, such that (a, u) and (b, v) are not edges in G. Since w is connected by an edge to all nodes from $D \setminus \{w\}$, $a \neq b$ and (a, b), (a, v), and (b, u) are edges of G. Consequently the cycle (a, v, u, b) does not have a chord in G, which is in contradiction with the assumption that G is triangulated.

Hence, for some $i = 1, 2$, C_i is complete in G. Therefore one of $C_i, i = 1, 2, 3$ must be a clique in G – none of the strict supersets of C_3 can be a clique in G, since this would contradict the assumption that D is a clique.

The properties of the mapping ϕ imply that there is an injective mapping from $\mathcal{C}(H)$ to $\mathcal{C}(G)$ non-decreasing the size of the cliques. Hence, we have $\sum_{A \in \mathcal{C}(H)} 2^{|A|}$ $\leq \sum_{A \in \mathcal{C}(G)} 2^{|A|}$, which implies $tts(H) \leq tts(G)$, since G and H (by Lemma 2) are triangulated.   □

The following two lemmas hold for general, not necessarily triangulated, undirected graphs. Their proofs are omitted and may be found in the extended version of this paper [9].

Lemma 4. *Let S be a set of some sets inducing complete subgraphs of a graph H. Then S is the set of all cliques of H iff S contains only incomparable pairs of sets and each set inducing a complete subgraph of H is a subset of an element of S.*

Lemma 5. *If a graph H is obtained from a graph G by removing an edge $\{u, v\}$, then*

$$\sum_{A \in \mathcal{C}(G)} 2^{|A|} \geq \sum_{A \in \mathcal{C}(H)} 2^{|A|} \ .$$

Theorem 2. *For any given elimination ordering f of a PD graph we can efficiently construct an elimination ordering g of the corresponding BROD graph such that the treewidth (and the total table size) of the BROD graph triangulated using g is not larger than the treewidth (and the total table size, respectively) of the PD graph triangulated using f.*

Proof. Let f be an elimination ordering for G_{PD}, which yields a triangulation G_{PD}^f. Let us construct a triangulation G' of the G_{BROD} from G_{PD}^f using the same sequence of edge contractions as in the proof of Lemma 1. Along these transformations we apply Lemma 2 to get an elimination ordering g for G', and by repeated application of Lemma 3 we obtain $tts(G') \leq tts(G_{PD}^f)$.

Graph G' has the same nodes as G_{BROD} and contains G_{BROD} as a subgraph. Let G_{BROD}^g be the triangulation of G_{BROD} obtained using the ordering g. In each step of the process of triangulation of G_{BROD} using g, we add only edges that belong to G'. Hence, the resulting graph G_{BROD}^g is a subgraph of G'. Consequently, by repeated use of Lemma 5 for all edges of G' which do not belong to G_{BROD}^g, we obtain $tts(G_{BROD}^g) \leq tts(G')$. This proves the statement concerning the total table size. The statement concerning the treewidth follows from the fact that G_{BROD}^g is a graph minor of G_{PD}^f and hence cannot have larger treewidth. □

Corollary 1. *The total table size of the BROD graph is not larger than the total table size of the PD graph.*

Proof. Use Theorem 2 for elimination ordering f, which yields a triangulation of PD graph with the smallest total table size. □

3   Triangulation Heuristics

In the previous section we have shown that using the PD graph for triangulation is inferior to using the BROD graph in the sense that we can always triangulate the BROD graph so that its treewidth (or total table size) is not greater than

574   P. Savicky and J. Vomlel

the treewidth (or total table size, respectively) of the PD graph. Therefore, in this section we pay attention to efficient triangulation of the BROD graph.

First, we applied several well-known triangulation heuristics to the BROD graph. We tested minfill [10], maximum cardinality search [11], minwidth [10], H1, and H6 [12]. The results of the comparisons can be found in the extended version of this paper [9]. Since minfill gave better results than the other heuristics, we selected it as a basis for further development of triangulation heuristics for the BROD graph. The minfill algorithm is described in Table 1. The output is an elimination ordering f of $G = (V, E)$

Table 1. The minfill heuristics

For $i = 1, \ldots, |V|$ do:

1. For $u \in V$ define set of edges $F(u) = \{\{u_1, u_2\} : \{u_1, u\} \in E, \{u_2, u\} \in E\}$ to be added for elimination of u.
2. Select a node $v \in V$ which adds the least number of edges when eliminated, i.e., $v \in \arg\min_{u \in V} |F(u) \setminus E|$, breaking ties arbitrarily.
3. Set $f(v) = i$.
4. Make v a simplicial node in G by adding edges to G, i.e., $G = (V, E \cup F(v))$.
5. Eliminate v from the graph G, i.e. replace G by its induced subgraph on $V \setminus \{v\}$.

Return f.

Minfill of the PD Graph Used for the BROD Graph

In our experiments, we have observed for some BN2O graphs that the minfill triangulation of a PD graph led to a graph with a smaller total table size than the triangulation of the BROD graph by minfill. This may seem to contradict the results from the previous section, but it does not, since the triangulation heuristics does not guarantee finding the optimal triangulation. In order to avoid this undesirable phenomenon, we can use the elimination ordering f found by minfill for the PD graph and construct an elimination ordering g for the BROD graph using the construction given in the proof of Theorem 2. This theorem guarantees that the total table size of the BROD graph triangulated using g is not larger than the total table size of the PD graph triangulated using f. We refer to this method as *PD-minfill* and use it as a base method for the comparisons in Section 4.

Minfill with n Steps Look-Ahead

Since the minfill algorithm is computationally fast for networks of moderate size, one can minimize the total number of edges added to the graph after more than one node is eliminated, i.e., one can look n steps ahead. Of course, this method scales exponentially, therefore it is computationally tractable only for small n. We refer to this method as *minfill-n-ahd*.

Minfill That Prefers Nodes from the Larger Level

The following proposition motivates another modification of the minfill algorithm. The proof is omitted and may be found in the extended version of this paper [9].

Proposition 1. *Let $G = (U \cup W, F)$ be a BROD graph. Then*

$$tw(G) \leq \min\{|U|, |W|\} \ .$$

This upper bound on the treewidth is guaranteed by any elimination ordering which starts with all nodes of the larger of the sets U and W.

This upper bound on the treewidth suggests a modification of the minfill heuristics. We can enforce edges to be filled in the smaller level only by taking nodes from the larger level into the elimination ordering first. Within the larger level we can use the minfill algorithm to choose the elimination ordering of nodes from this level. This gives a treewidth not larger than the number of nodes in the smaller level. The nodes from the smaller level are included in the elimination ordering after the nodes from the larger level. We will refer to this method as *minfill-pll*.

4   Experiments

In this section, we experimentally compare the proposed triangulation heuristics on 1300 randomly generated BN2O networks. The BN2O graphs were generated with varying values of the following parameters:

- x, the number of nodes on the top level,
- y, the number of nodes on the bottom level, and
- e, the average number of edges per node on the bottom level.

For each x-y-e type, $x, y = 10, 20, 30, 40, 50$ and $e = 3, 5, 7, 10, 14, 20$ (excluding those with $e \geq x$) we generated randomly ten BN2O graphs.

All triangulation heuristics were tested on the BROD graphs G_{BROD}. We used the total table size tts of the graph G_{BROD}^{h} triangulated by a triangulation heuristics h as the criterion for comparisons. We used the *PD-minfill* method as the base method against which we compared all other tested methods, since it is the closest to the current standard, which is to use the PD graph. Since randomness is used in the triangulation heuristics we run each heuristics ten times on each model and selected a triangulation with the minimum value of total table size tts.

For each tested model we computed the decadic logarithm ratio

$$r(pd, h) = \log_{10} tts \left(G_{BROD}^{PD\text{-}minfill} \right) - \log_{10} tts \left(G_{BROD}^{h} \right) \ ,$$

where h stands for the tested triangulation heuristics. In Table 2 we give frequencies of several intervals of log-ratio $r(pd, h)$ values for the tested heuristics in the test benchmark.

Table 2. Frequency of $r(pd, h)$ values for the heuristics tested on the test benchmark

Intervals of $r(pd,h)$	minfill	minfill-1ahd	minfill-2ahd	minfill-pll	minfill-comb
$(-3, -2]$	5	0	0	0	0
$(-2, -1]$	26	14	9	0	0
$(-1, -0.05]$	96	82	76	116	2
$(-0.05, 0.05]$	518	535	536	695	637
$(0.05, 1]$	328	339	350	177	334
$(1, 2]$	116	115	113	101	114
$(2, 3]$	101	103	104	99	101
$(3, 4]$	29	31	31	31	31
$(4, 5]$	27	27	27	27	27
$(5, 6]$	34	33	33	33	33
$(6, 7]$	9	10	10	10	10
$(7, 8]$	3	3	3	3	3
$(8, 9]$	8	8	8	8	8

From the table we can see that, on average, all tested heuristics perform significantly better than *PD-minfill*, since positive differences of the logarithms are more frequent and achieve larger absolute value. On the other hand, most of the heuristics are worse than *PD-minfill* for some of the models. Since triangulation heuristics *minfill*, *minfill-pll*, and *PD-minfill* are computationally fast on moderately large networks, the best solution seems to be to run all three of these algorithms and select the best solution. Already *minfill-comb*, which is the combination of *minfill* and *minfill-pll*, eliminates most of the cases where *minfill* is worse than *PD-minfill*.

5   Conclusions

In this paper we compare two transformations of BN2O networks that allow more efficient probabilistic inference: parent divorcing (PD) and rank-one decomposition (ROD). ROD appears to be superior to PD, since with ROD we can always get a total table size of the resulting triangulated graph not larger than using PD. The experiments confirm that in most cases, ROD leads directly to a better result. In the remaining cases, it is the best to calculate the elimination order for the PD graph and transform it to the elimination order for the ROD graph.

We also perform experiments with different triangulation heuristics and suggest few modifications of the minfill heuristics for BN2O networks, which lead to further improvements, although none of the heuristics is universally the best. In order to get the best result for all models, we suggest running several of the described heuristics, including minfill on the PD graph, and select the best solution. This process is efficient, since determining *tts* for a triangulation is fast and the actual inference is then performed with a well-chosen triangulation.

Acknowledgments

We would like to thank Mark Chavira for providing us with the code extracted from Ace [13], which we have used for the construction of PD graphs and for the computation of elimination orderings in these graphs.

References

1. Olesen, K.G., Kjærulff, U., Jensen, F., Jensen, F.V., Falck, B., Andreassen, S., Andersen, S.K.: A MUNIN network for the median nerve — a case study on loops. Applied Artificial Intelligence 3, 384–403 (1989); Special issue: Towards Causal AI Models in Practice
2. Díez, F.J., Galán, S.F.: An efficient factorization for the noisy MAX. International Journal of Intelligent Systems 18, 165–177 (2003)
3. Vomlel, J.: Exploiting functional dependence in Bayesian network inference. In: Proceedings of the 18th Conference on Uncertainty in AI (UAI), pp. 528–535. Morgan Kaufmann, San Francisco (2002)
4. Savicky, P., Vomlel, J.: Exploiting tensor rank-one decomposition in probabilistic inference. Kybernetika 43(5), 747–764 (2007)
5. Yannakakis, M.: Computing the minimum fill-in is NP-complete. SIAM J. Algebraic and Discrete Methods 2, 77–79 (1981)
6. Vomlel, J., Savicky, P.: Arithmetic circuits of the noisy-or models. In: Proceedings of the Fourth European Workshop on Probabilistic Graphical Models (PGM 2008), Hirtshals, Denmark, pp. 297–304 (2008)
7. Bodlaender, H.L., Koster, A.M.C.A., Eijkhof, F.V.D.: Preprocessing rules for triangulation of probabilistic networks. Computational Intelligence 21(3), 286–305 (2005)
8. Bodlaender, H.L.: A partial k-arboretum of graphs with bounded treewidth. Theoretical Computer Science 209(1-2), 1–45 (1998)
9. Savicky, P., Vomlel, J.: Triangulation heuristics for BN2O networks. Technical report, Institute of Information Theory and Automation of the AS CR (2009), http://www.utia.cas.cz/vomlel/ecsqaru2009-full-version.pdf
10. Rose, D.J.: A graph-theoretic study of the numerical solution of sparse positive definite systems of linear equations. Graph Theory and Computing, 183–217 (1972)
11. Tarjan, R.E., Yannakakis, M.: Simple linear-time algorithms to test chordality of graphs, test acyclicity of hypergraphs, and selectively reduce acyclic hypergraphs. SIAM J. Comput. 13, 566–579 (1984)
12. Cano, A., Moral, S.: Heuristic algorithms for the triangulation of graphs. In: Bouchon-Meunier, B., Yager, R.R., Zadeh, L.A. (eds.) IPMU 1994. LNCS, vol. 945, pp. 98–107. Springer, Heidelberg (1995)
13. Ace: A Bayesian network compiler (2008), http://reasoning.cs.ucla.edu/ace/

A Default Logic Patch for Default Logic

Philippe Besnard[1], Éric Grégoire[2], and Sébastien Ramon[2]

[1] IRIT, F-31062 Toulouse
CNRS UMR 5505, F-31062
118 route de Narbonne, F-31062 Toulouse France
besnard@irit.fr
[2] Université Lille - Nord de France, Artois, F-62307 Lens
CRIL, F-62307 Lens
CNRS UMR 8188, F-62307
rue Jean Souvraz SP18, F-62307 Lens France
{gregoire,ramon}@cril.fr

Abstract. This paper is about the fusion of multiple information sources represented using default logic. More precisely, the focus is on solving the problem that occurs when the standard-logic knowledge parts of the sources are contradictory, as default theories trivialize in this case. To overcome this problem, it is shown that replacing each formula belonging to Minimally Unsatisfiable Subformulas by a corresponding supernormal default allows appealing features. Moreover, it is investigated how these additional defaults interact with the initial defaults of the theory. Interestingly, this approach allows us to handle the problem of default theories containing inconsistent standard-logic knowledge, using the default logic framework itself.

Keywords: Default logic, logic-based fusion, inconsistency tolerance, MUS, Minimally Unsatisfiable Subformulas.

1 Introduction

In the Artificial Intelligence (A.I.) research community, one of the most popular tools to handle forms of defeasible reasoning remain Reiter's default logic [1] and its major variants (e.g. [2], [3], [4] and [5] just to name a few other seminal papers). Default logic has been defined to allow forms of reasoning by default to be modelled. It permits an inference system to jump to default conclusions and to retract them when new information shows that these conclusions now lead to inconsistency.

For example, default logic is a very convenient framework to encode patterns of reasoning like "Given an employee x, by default we should allow x to access the database unless this would contradict security rules. If some further additional information makes such contradictions occur then the permission must be retracted".

C. Sossai and G. Chemello (Eds.): ECSQARU 2009, LNAI 5590, pp. 578–589, 2009.

A *default-logic theory* is made of two parts: a set of first-order logic formulas representing knowledge and a set of default rules, i.e. a sort of inference rules capturing patterns of defeasible reasoning as in the above example.

In this paper, we investigate how several[1]default theories in Reiter's default logic should be fused when it is assumed that each default theory represents the knowledge of an agent or of a community of agents. More precisely, it is shown that merging these theories is not an issue that is to be taken as granted when the set-theoretical union of the standard-logic formulas to be fused is inconsistent. Indeed, keeping all such formulas would make the whole language to be the set of acceptable inferences because when the standard-logic knowledge part of a default theory is inconsistent, the default theory itself trivializes.

Quite surprisingly, to the best of our knowledge, this trivialization property of default logic has not been addressed so far in the literature. In this respect, the goal of this paper is to revisit default logic in such a way that trivialization is avoided in the presence of inconsistent premises, sharing the concerns of the large research effort from the A.I. research community to study how to reason in the presence of inconsistent knowledge and to develop inconsistency tolerence techniques (see e.g. [6]). In particular, when several information sources are to be aggregated, a single, possibly minor contradiction between two sources should not cause the whole system to collapse.

In the paper, a family of approaches in that direction are discussed. Mainly, they rely on the study of MUSes (Minimally Unsatisfiable Subformulas) in the standard-logic formulas. Accordingly, a series of reasoning paradigms are investigated. Specifically, it is shown that replacing each formula in the set of MUSes by a corresponding default rule is an appealing solution. As a special case, it offers a powerful way to recover from the inconsistencies that might occur in sets of standard-logic formulas. Interestingly, this latter technique can easily be exported to the main variants of default logic, like e.g. constrained [2], rational [3], justified [4] and cumulative default logic [5], of which some ensure that general default theories have at least one extension.

The paper is organized as follows. In the next section, MUSes and the way according to which they can be computed are presented. In Sections 3 and 4, an approach to replace MUSes by additional default rules is introduced and studied in the context of recovering from inconsistency in standard Boolean logic. Section 5 is devoted to how these additional rules interact with the default ones of the initial theories. In Section 6, a complexity analysis of this technique is provided, together with possible approximation techniques.

Throughout the paper, we use the following standard notations: $\neg$, $\vee$, $\wedge$ and $\supset$ represent the standard negation, disjunction, conjunction and material implication connectives, respectively. When Ω is a set of first-order formulas, $Cn(\Omega)$ denotes the deductive closure of Ω. Also, let us recall that in the Boolean case a *CNF* is a finite conjunction of clauses, where a clause is a disjunction of signed Boolean variables.

---

[1] On the other hand, the following applies to a single default theory, too.

In the following, we assume that the reader is familiar with default logic [1]. A brief reminder about default logic is provided in Appendix A.

2   MUSes

Assume that Σ is a set of Boolean formulas. A *Minimally Unsatisfiable Subformulas (MUS)* Φ of Σ is defined as follows:

– $\Phi \subseteq \Sigma$,
– Φ is unsatisfiable,
– $\forall \Psi \subset \Phi$, Ψ is satisfiable.

Accordingly, a MUS of Σ is a subset of Σ that is contradictory and that becomes satisfiable whenever any of its formulas is removed. Thus, a MUS of Σ describes a contradiction within Σ using a set of formulas of Σ that cannot be made smaller.

Note 1. The set of all MUSes of a set of formulas Σ is denoted $MUS(\Sigma)$. The set of all formulas occurring in the MUSes of Σ is denoted $\cup MUS(\Sigma)$.

Example 1. Let $\Sigma = \{a, a \supset b, \neg b, a \supset (c \vee d), \neg d, c \supset b, d \supset e, (c \wedge e) \supset a\}$. Clearly, Σ is unsatisfiable and contains two MUSes, namely $\Phi_1 = \{a, a \supset b, \neg b\}$ and $\Phi_2 = \{a \supset (c \vee d), a, \neg d, c \supset b, \neg b\}$.

This example also illustrates that MUSes can share non-empty intersections.

Many techniques to handle contradictions in logic-based systems have been discussed in the literature (see e.g [6] and [7] for surveys in that matter). One family of approaches amount to recovering satisfiability by dropping MUSes or parts of MUSes. Indeed, removing one formula in each MUS allows consistency to be recovered. Two extreme approaches can thus be proposed in that direction. On the one hand, we might drop the set-theoretical union of all MUSes, thus removing every minimal (w.r.t. the number of involved formulas) cause of inconsistency. On the other hand, we might prefer a minimal change policy, which requires us to drop at most one formula per MUS.

3   How to Handle Default Theories Containing Contradictory Standard-Logic Knowledge

In the following, we assume that Σ is a set of Boolean formulas and we are mostly interested in default theories $\Gamma = (\Delta, \Sigma)$ where Σ is inconsistent. In such a case, Γ has a unique extension, which is the whole logical language.

We distinguish between skeptical and credulous reasonings from a default theory Γ: a formula f can be *skeptically* (resp. *credulously*) inferred from a default theory Γ iff f belongs to all (resp. some) extensions of Γ.

Now, since a default theory consists of two parts, namely a set of defaults and a set of facts, the fusion of default theories amounts to merging sets of facts and merging sets of defaults. In the following, we assume that facts (resp. defaults) are unioned by this fusion process.

Assume that we are given n default theories $\Gamma_i = (\Delta_i, \Sigma_i)$ ($i \in [1..n]$) to be fused, that are such that the set-theoretical union of their standard logic parts, namely $\cup_{i=1}^{n} \Sigma_i$, is inconsistent. One direct way to address the trivialization of the resulting aggregated default theory consists of removing enough formulas from $\cup_{i=1}^{n} \Sigma_i$ so that the resulting subset becomes consistent. However, dropping formulas is unnecessarily destructive.

Indeed, a credulous reasoner might be interested in exploring the various extensions that could be obtained if regarding as being acceptable the various maximal consistent subsets of the various MUSes of $\cup_{i=1}^{n} \Sigma_i$. Also, a skeptical reasoner might want to explore what would belong to all those extensions. In this respect, if we replace each formula f in the MUSes of $\cup_{i=1}^{n} \Sigma_i$ by a corresponding supernormal default $\frac{:f}{f}$, we get a new default theory where the reasoner is considering that *each formula f in the MUSes could be inferred if f could be consistently assumed*. However, since the set-theoretical union of the consequents of these new defaults is inconsistent, default logic forbids the acceptance of all such f within the same extension. Let us stress that this policy does not enforce by itself any priority between the replaced formulas since all of those are treated in a uniform way. Interestingly, this approach allows us to handle the problem of default theories containing inconsistent standard-logic knowledge, using the default logic framework itself.

Definition 1 (fused default theory). *Let us consider a non-empty set of n default theories of the form $\Gamma_i = (\Delta_i, \Sigma_i)$ to be fused. The resulting fused default theory is given by $\Gamma = (\Delta, \Sigma)$ where:*

- $\Sigma = \cup_{i=1}^{n} \Sigma_i \setminus \cup MUS(\cup_{i=1}^{n} \Sigma_i)$,
- $\Delta = \cup_{i=1}^{n} \Delta_i \cup \{ \frac{:f}{f} \mid f \in \cup MUS(\cup_{i=1}^{n} \Sigma_i) \}$.

This definition thus corresponds to a policy that requires a uniform treatment of formulas inside MUSes. On the contrary, alternative definitions could make use of selection operators *select* to deliver a subset of $\cup MUS(\cup_{i=1}^{n} \Sigma_i)$ such that $\cup_{i=1}^{n} \Sigma_i \setminus select(\cup MUS(\cup_{i=1}^{n} \Sigma_i))$ is consistent, and such that each formula from $select(\cup MUS (\cup_{i=1}^{n} \Sigma_i))$ is to be replaced by a corresponding supernormal default in the fused theories.

4   Addressing the Trivialization Issue in the Standard Boolean Case

First, let us consider the basic situations where the set of defaults $\cup_{i=1}^{n} \Delta_i$ is empty. Obviously enough, this coincides with the problem of fusing sets of Boolean formulas that are such that their set-theoretical union is inconsistent. The above definition thus provides an original approach to address this issue.

Example 2. Let $\Gamma_1 = (\emptyset, \{ \neg a \vee b, \neg b \})$ and $\Gamma_2 = (\emptyset, \{a\})$ two default theories to be fused. Clearly, $\cup_{i=1}^{2} \Sigma_i = \{\neg a \vee b, a, \neg b\}$ is inconsistent. The fused default theory $\Gamma = (\{\frac{:a}{a}, \frac{:\neg a \vee b}{\neg a \vee b}, \frac{:\neg b}{\neg b}\}, \emptyset)$ exhibits the extensions $E_1 = Cn(\{a, \neg a \vee b\})$, $E_2 = Cn(\{\neg b, \neg a \vee b\})$ and $E_3 = Cn(\{a, \neg b\})$; each of them containing two of the consequents of the three defaults.

Interestingly, it is possible to characterize the set-theoretic intersection of all extensions of the fused default theory, when the initial theories do not contain any default.

To do that, we resort to (usual) choice functions θ for a finite family of non-empty sets $\Xi = \{\Omega_1, \ldots, \Omega_n\}$, which "pick" an element in every Ω_i of the family. In the limiting case that Ξ is empty, $\theta(\Xi)$ is empty.

Note 2. Let Θ denote the subclass of choice functions θ for $\Xi = \{\Omega_1, \ldots, \Omega_n\}$ such that for $i \neq j$, $\theta(\Omega_i) \in \Omega_j \Rightarrow \theta(\Omega_j) = \theta(\Omega_k)$ for some $k \neq j$ (but k may, or may not, be i). This subclass is reduce to choices functions θ whose image is minimal s.t. if $\theta \in \Theta$ then $\nexists \theta' \in \Theta$ s.t. $\theta'(\Xi) \subset \theta(\Xi)$.

Proposition 1. *Let $n > 1$. Consider n finite default theories of the form $\Gamma_i = (\Delta_i, \Sigma_i)$ to be fused. If Δ_i is empty for $i = 1..n$, then the set-theoretic intersection of all extensions of the resulting fused default theory $\Gamma = (\Delta, \Sigma)$ is $Cn(\{\psi\})$ where:*

$$\psi = \bigvee_{\theta \in \Theta} \bigwedge \left((\cup_{i=1}^{n} \Sigma_i) \setminus \theta(MUS(\cup_{i=1}^{n} \Sigma_i)) \right).$$

Remark 1. It is essential that the default theories to be fused are *finite* for Proposition 1 to hold. Otherwise, $MUS(\cup_{i=1}^{n} \Sigma_i)$ can be infinite. Then, not only would the axiom of countable choice be needed, but even worse, an infinite disjunction would be needed (which is outside classical logic). For example, assume that the default theories to be fused are $\Gamma_1 = (\emptyset, \Sigma_1)$, $\Gamma_2 = (\emptyset, \Sigma_2)$, and $\Gamma_3 = (\emptyset, \Sigma_3)$ where:

$$\Sigma_1 = \{p_1, q_1, r_1, \ldots\},$$
$$\Sigma_2 = \{p_2, q_2, r_2, \ldots\},$$
$$\Sigma_3 = \{\neg p_1, \neg p_2, \neg q_1, \neg q_2, \neg r_1, \neg r_2, \ldots\}.$$

Clearly, $MUS(\cup_{i=1}^{3} \Sigma_i) = \{\{p_1, \neg p_1\}, \{p_2, \neg p_2\}, \{q_1, \neg q_1\}, \{q_2, \neg q_2\}, \ldots\}$ is infinite. Therefore, ψ would have infinitely many conjuncts and disjuncts. For instance, taking θ_1 to pick only negative literals yields the infinite conjunction $\bigwedge \{p_1, p_2, q_1, q_2, r_1, r_2, \ldots\}$. The disjunction would also be infinite because infinitely many choice functions must be taken into account.

In this example, Σ is empty but it is easy to alter it to make Σ infinite:

$$\Sigma_1 = \{p_1, q_1, r_1, s_1, \ldots\},$$
$$\Sigma_2 = \{p_2, q_2, r_2, s_2, \ldots\},$$
$$\Sigma_3 = \{\neg p_1, p_2, q_1, \neg q_2, \neg r_1, r_2, s_1, \neg s_2, \ldots\}.$$

Interestingly, the following proposition shows us that any formula in the set $\cup MUS(\cup_{i=1}^{n} \Sigma_i)$ belongs to at least one extension of the fused theory. Accordingly, no formula is lost in the fusion process in the sense that each non-contradictory formula that is replaced by a default – and that would be dropped in standard fusion approaches – can be found in at least one extension of the fused theory.

Proposition 2. *Let $n > 1$. Consider n finite default theories $\Gamma_i = (\Delta_i, \Sigma_i)$ such that $\cup_{i=1}^{n} \Delta_i$ is empty and $\cup_{i=1}^{n} \Sigma_i$ is inconsistent. Let Γ denote the resulting fused default theory. There exists no extension of Γ that contains $\cup MUS(\cup_{i=1}^{n} \Sigma_i)$ but for any satisfiable formula f in $\cup MUS(\cup_{i=1}^{n} \Sigma_i)$, there exists an extension of Γ containing f.*

Based on the above proposition, it could be imagined that the intersection of all extensions will merely coincide with the extensions of the default theory $\Gamma' = (\emptyset, \cup_{i=1}^{n} \Sigma_i \setminus \cup MUS(\cup_{i=1}^{n} \Sigma_i))$. As the following example shows, this is not the case since the computation of the multiple extensions mimics a case analysis process that allows inferences to be entailed that would simply be dropped if $\cup MUS(\cup_{i=1}^{n} \Sigma_i)$ were simply removed from $\cup_{i=1}^{n} \Sigma_i$.

Example 3. Let us consider $\Gamma_1 = (\emptyset, \{a \wedge b, c \supset d\})$ and $\Gamma_2 = (\emptyset, \{\neg a \wedge c, b \supset d\})$. Clearly, $\{a \wedge b, \neg a \wedge c\}$ is a MUS. If we simply drop the MUS, we get $\Gamma' = (\emptyset, \{c \supset d, b \supset d\})$. Clearly, Γ' has a unique extension $Cn(\{c \supset d, b \supset d\})$ that does not contain d. This is quite inadequate since the contradiction is explained by the co-existence of a and $\neg a$. Assume that a is actually *true*. Then, from $a \wedge b$ and $b \supset d$ we should be able to deduce d. Similarly, if a is actually *false* then we should also be able to deduce d. Now, $\Gamma = (\{\frac{:a \wedge b}{a \wedge b}, \frac{:\neg a \wedge c}{\neg a \wedge c}\}, \{c \supset d, b \supset d\})$ exhibits two extensions, each of them containing d. Accordingly, d can be inferred using a skeptical approach to default reasoning.

This last example also shows us that this treatment of inconsistency permits more (legitimate) conclusions to be inferred than would be by removing MUSes or parts of MUSes. This is not surprising since by weakening formulas into default rules, we are dropping less information than if we were merely removing them. An alternative approach to allow such a form of case-analysis from inconsistent premises can be found in [8].

Applying Proposition 1 to Example 3 shows what the consequences of the resulting fused default theory $\Gamma = (\Delta, \Sigma)$ are:

Example 4 (con'd). $MUS(\cup_{i=1}^{2} \Sigma_i) = \{\{a \wedge b, \neg a \wedge c\}\}$ because $\Sigma_1 \cup \Sigma_2$ has only one MUS, that is, $\{a \wedge b, \neg a \wedge c\}$. Hence, there are only two choice functions over $\{a \wedge b, \neg a \wedge c\}$. One picks $a \wedge b$ and the other picks $\neg a \wedge c$. Let us denote them θ_1 and θ_2 respectively. Then, the formula ψ in Proposition 1 becomes:

$$\psi = \bigvee_{\theta \in \{\theta_1, \theta_2\}} \bigwedge \left((\cup_{i=1}^{2} \Sigma_i) \setminus \theta(MUS(\cup_{i=1}^{2} \Sigma_i)) \right).$$

That is:

$$\psi = \bigvee_{\theta \in \{\theta_1, \theta_2\}} \bigwedge \left((\Sigma_1 \cup \Sigma_2) \setminus \theta(\{\{a \wedge b, \neg a \wedge c\}\}) \right).$$

So:

$$\psi = \left(\bigwedge \{\neg a \wedge c, c \supset d, b \supset d\} \right) \bigvee \left(\bigwedge \{a \wedge b, c \supset d, b \supset d\} \right).$$

Applying various logical laws, ψ becomes the conjunction of the following four formulas:

$$(\neg a \wedge c) \vee (a \wedge b),$$
$$((c \supset d) \wedge (b \supset d)) \vee (\neg a \wedge c),$$
$$((c \supset d) \wedge (b \supset d)) \vee (a \wedge b),$$
$$((c \supset d) \wedge (b \supset d)) \vee ((c \supset d) \wedge (b \supset d)).$$

Of course, the latter disjunction is equivalent with $(c \supset d) \wedge (b \supset d)$ and subsumes the preceding two formulas. As a consequence:

$$\psi = ((\neg a \wedge c) \vee (a \wedge b)) \wedge (c \supset d) \wedge (b \supset d).$$

Finally, the set-theoretic intersection of all extensions of the resulting fused default theory $\Gamma = (\Delta, \Sigma)$ is $Cn(\{b \supset d, c \supset d, b \vee c, \neg a \vee b, a \vee c\})$.

Observe that $Cn(\{b \supset d, c \supset d, b \vee c, \neg a \vee b, a \vee c\})$ can be simplified as $Cn(\{b \supset d, c \supset d, \neg a \vee b, a \vee c\})$. In any case, both $b \vee c$ and d are in $Cn(\{b \supset d, c \supset d, \neg a \vee b, a \vee c\})$.

Now, another interesting feature of the fusion process given by Definition. 1 is that a skeptical reasoner will be able to infer (at least) all formulas that it would be able to infer if the MUSes of $\cup_{i=1}^{n} \Sigma_i$ were simply dropped.

Proposition 3. *Let $n > 1$. Consider n finite default theories $\Gamma_i = (\Delta_i, \Sigma_i)$ to be fused. Let $\cap_j E_j$ denote the set-theoretic intersection of all extensions of the resulting fused default theory $\Gamma = (\Delta, \Sigma)$. Let E denote the unique extension of $\Gamma' = (\emptyset, \Sigma)$. If Δ_i is empty for $i = 1..n$, then $E \subseteq \cap_j E_j$.*

Example 5. Assume $\Gamma_i = (\Sigma_i, \Delta_i)$ where $\cup_{i=1}^{n} \Sigma_i = \{a, \neg a \vee \neg b, b, c\}$ and $\cup_{i=1}^{n} \Delta_i = \emptyset$. $\cup MUS(\cup_{i=1}^{n} \Sigma_i) = \{a, \neg a \vee \neg b, b\}$. The resulting fused default theory is $\Gamma = (\Delta, \Sigma)$ where $\Sigma = \{c\}$ and $\Delta = \{\frac{:a}{a}, \frac{:\neg a \vee \neg b}{\neg a \vee \neg b}, \frac{:b}{b}\}$. The extensions of Γ are $E_1 = Cn(\{c, a, \neg a \vee \neg b\})$, $E_2 = Cn(\{c, \neg a \vee \neg b, b\})$ and $E_3 = Cn(\{c, a, b\})$. The unique extension of the default theory where all formulas from $\cup MUS(\cup_{i=1}^{n} \Sigma_i)$ are dropped is $E = Cn(\{c\})$. We have $\cap_{j=1}^{n} E_j = Cn(\{c, a \vee b\})$ and $E \subseteq \cap_{j=1}^{n} E_j$.

Let us now consider the general case where theories contain defaults, and study how new defaults interact with defaults of the initial theories.

5   How the New Defaults Interact with the Defaults of the Theories to be Fused

First, it is well-known that normal default theories enjoy interesting properties, like semi-monotonicity [1]. This property ensures that whenever we augment a normal default theory Γ with an additional normal default, every extension of Γ is included in an extension of the new theory. Accordingly, we can insure that the extension of Proposition 2 to normal default theories holds since we only add supernormal defaults to the set-theorical union of initial theories.

On the other hand, the extension of Proposition 3 to normal default theories does not hold: as the following example shows, the unique extension of $\Gamma' = (\cup_{i=1}^{n} \Delta_i, \Sigma)$ is not necessarily contained in the set-theorical intersection of all the extensions of the resulting fused theory.

Example 6. Let us assume that $\cup_{i=1}^{n} \Sigma_i = \{a, \neg a \vee \neg b, b, c\}$ and $\cup_{i=1}^{n} \Delta_i = \{\frac{a:d}{d},$ $\frac{c:\neg d}{\neg d}\}$. The resulting fused default theory is $\Gamma = (\Delta, \Sigma)$ where $\Sigma = \{c\}$ and $\Delta = \{\frac{:a}{a}, \frac{:\neg a \vee \neg b}{\neg a \vee \neg b}, \frac{:b}{b}, \frac{a:d}{d}, \frac{c:\neg d}{\neg d}\}$. The extensions of Γ are $E_1 = Cn(\{c, a, \neg a \vee \neg b, d\})$, $E_2 = Cn(\{c, \neg a \vee \neg b, b, \neg d\})$ and $E_3 = Cn(\{c, a, b, d\})$. The unique extension of the default theory $\Gamma' = (\cup_{i=1}^{n} \Delta_i, \Sigma)$ is $E = Cn(\{c, \neg d\})$ while $\cap_{j=1}^{n} E_j = Cn(\{c\})$. Thus $E \not\subseteq \cup_{j=1}^{n} E_j$.

Indeed, removing MUSes prevents the application of initial (normal) defaults whose prerequisite belongs to MUSes. Accordingly, we derive the following proposition.

Proposition 4. *Let $n > 1$. Consider n finite normal default theories $\Gamma_i = (\Delta_i, \Sigma_i)$ to be fused and $\Gamma' = (\cup_{i=1}^{n} \Delta_i, \cup_{i=1}^{n} \Sigma_i \setminus \cup MUS(\cup_{i=1}^{n} \Sigma_i))$. For any extension E of Γ', there exists an extension of the resulting fused default theory that contains E.*

Interestingly, this proposition ensures that whenever we iterate the fusion process of normal default theories, we are always ensured that any extension can only be extended in the process.

Now, in the general case, replacing MUSes or subparts of MUSes by corresponding defaults does not ensure that we shall obtain supersets of the extensions that would be obtained if those MUSes or some of their subparts were removed: the semi-monotonicity does not hold.

Example 7. Let us consider $\Gamma = (\Delta, \{a, \neg a\})$ where $\Delta = \{\frac{:b}{b}, \frac{a:c}{\neg b}, \frac{\neg a:c}{\neg b}\}$. The default theory $\Gamma' = (\Delta, \emptyset)$ exhibits one extension, which is $Cn(\{b\})$. On the contrary, $(\Delta \cup \{\frac{:a}{a}, \frac{:\neg a}{\neg a}\}, \emptyset)$ does not contain any extension containing b.

Generalizing Proposition 2 to default theories with non-empty sets of defaults does not hold either: as shown by the following example, it may happen that consistent formulas of $\cup MUS(\cup_{i=1}^{n} \Sigma_i)$ are in no extension of the resulting fused default theory.

Example 8. Let us consider the default theory $\Gamma_1 = (\emptyset, \{a, c\})$ and $\Gamma_2 = (\{\frac{c:b}{\neg a}\}, \{\neg a\})$ to be fused. $\cup MUS(\cup_{i=1}^{2} \Sigma_i) = \{a, \neg a\}$. The resulting fused default theory is $\Gamma = (\{\frac{:a}{a}, \frac{:\neg a}{\neg a}, \frac{c:b}{\neg a}\}, \{c\})$. The unique extension of Γ is $E = Cn(\{c, \neg a\})$, which does not contain a.

6   Complexity Issues and Approximation Techniques

In the Boolean case, computing MUSes is computationally heavy in the worst case since checking whether a clause belongs or not to the set of MUSes of a CNF is Σ_2^p-*complete* [9]. Accordingly, the whole process of finding and replacing contradictory formulas by corresponding defaults, and then achieving Boolean credulous default reasoning is not computationally harder than credulous default reasoning itself since, in the general case, the latter is also Σ_2^p-*complete* (whereas it is Π_2^p-*complete* in the skeptical case) [10].

Interestingly, recent algorithmic techniques make it possible to compute one MUS for many real-life problems [11]. However, the number of MUSes in a set of n clauses can be intractable too, since it is $C_n^{n/2}$ in the worst case. Fortunately, efficient techniques have also been defined recently to compute all MUSes for many benchmarks, modulo a possible exponential blow-up limitation [12].

However, in some situations we cannot afford to compute the set-theoretical union of all MUSes. In this context, several techniques can then be applied.

First, it should be noted that it is not required to replace all formulas in all MUSes by corresponding defaults to recover consistency. Indeed, one could first detect one MUS, replace all its formulas by defaults, then iterate this process until consistency is recovered. Such an approach may avoid us computing all MUSes; it has been studied in the clausal Boolean framework in the context of the detection of *strict inconsistent covers* [13].

Definition 2 (strict inconsistent cover). *Let Σ be a set of Boolean formulas. $\Sigma' \subseteq \Sigma$ is a strict inconsistent cover of Σ iff $\Sigma \setminus \Sigma'$ is satisfiable and $\Sigma' = \cup \mathcal{A}$ for some $\mathcal{A} \subseteq MUS(\Sigma)$ such that, if $|\mathcal{A}| > 1$, any two members of $\mathcal{A}$ are disjoint.*

Lemma 1. *A strict inconsistent cover of Σ is empty iff Σ is satisfiable.*

Lemma 2. *A strict inconsistent cover of Σ always exists.*

Lemma 3. *For all $M \in MUS(\Sigma)$, there exists a strict inconsistent cover of Σ that contains M.*

Strict inconsistent covers $IC(\Sigma)$ are thus minimal sets of formulas in Σ that can capture enough sources of contradiction in Σ to recover consistency if they were fixed. In [13] a technique to compute strict inconsistent covers in the Boolean clausal case has been introduced and proved efficient for many difficult benchmarks. Clearly, strict inconsistent covers is an approximation of the set-theoretical union of MUSes in the sense that all formulas of the cover always belong to this union but not conversely, and that dropping the cover causes consistency to be restored. The price to be paid for this approximation is that several different inconsistent covers can co-exist for a given set of MUSes.

Now, most of the time, it is possible to extract a super-set Ω of all MUSes of $\cup_{i=1}^{n} \Sigma_i$ very quickly, in such a way that retracting Ω would restore the consistency of $\cup_{i=1}^{n} \Sigma_i$. At the extreme, this super-set can be $\cup_{i=1}^{n} \Sigma_i$ itself. Accordingly, we could replace all formulas in Ω by corresponding supernormal defaults. Clearly, such a process would restore consistency. The price to be paid is that both uncontroversial and problematic information, i.e. both formulas belonging and not belonging to MUSes would be downgraded and treated in the same manner.

An alternative approach consists in replacing at most one formula per MUS. Clearly, from a practical point of view, detecting one such formula does not require us to compute one MUS exactly but simply a superset of a MUS, such that dropping the formula would make this superset consistent. Since MUSes can share non-empty intersections, let us note that it is however difficult to guarantee that a minimal number of formulas are replaced without computing all MUSes explicitly.

7 Conclusions and Future Works

In this paper, a "patch" to default logic, one of the most popular logics for representing defeasible reasoning, has been proposed. It allows a reasoner to handle theories involving contradictory standard-logic bases whereas standard default logic trivializes in this case. Interestingly, the new framework offers a powerful way to treat inconsistent standard logic theories as well. Such a default logic variant is of special interest when the fusion of several sources of knowledge is considered: indeed, without the patch, a default logic reasoner would be able to infer any conclusion (and its logical contrary) whenever two pieces of (standard logic) information appear to be contradicting one another in the sources.

In the basic approaches described in the previous section, no distinction is made between the defaults from the initial theories and the defaults that are introduced to replace MUSes or subparts of MUSes, as if all defaults were of the same epistemological nature. Indeed, the new defaults are introduced to *correct* and *weaken* some pieces of knowledge that exhibit some deficiencies. Our way to correct MUSes amounts to considering that formulas participating in MUSes should be accepted *by default*. In this respect, it can be argued that the role of the additional defaults is similar to the role of defaults of initial theories, which are normally intended to represent pieces of default reasoning.

On the contrary, it can be argued that new defaults should be given a higher (resp. lower) priority than defaults of initial theories. In these cases, we must resort to a form of prioritized default logic (see eg. [14]). We plan to investigate this issue in the future.

References

1. Reiter, R.: A logic for default reasoning. Artificial Intelligence 13, 81–132 (1980)
2. Schaub, T.: On constrained default theories. In: Neumann, B. (ed.) European conference on Artificial Intelligence (ECAI 1992), New York, NY, USA, pp. 304–308. John Wiley & Sons, Inc., Chichester (1992)
3. Mikitiuk, A., Truszczyński, M.: Rational default logic and disjunctive logic programming. In: Workshop on Logic Programming and Non-Monotonic Reasoning (LPNMR 1993), pp. 283–299. MIT Press, Cambridge (1993)
4. Lukaszewicz, W.: Considerations on default logic: an alternative approach. Computational intelligence 4(1), 1–16 (1988)
5. Brewka, G.: Cumulative default logic: in defense of nonmonotonic inference rules. Artificial Intelligence 50(2), 183–205 (1991)
6. Bertossi, L., Hunter, A., Schaub, T. (eds.): Inconsistency Tolerance. LNCS, vol. 3300. Springer, Heidelberg (2005)
7. Konieczny, S., Grégoire, E.: Logic-based information fusion in artificial intelligence. Information Fusion 7(1), 4–18 (2006)
8. Besnard, P., Hunter, A.: Quasi-classical logic: Non-trivializable classical reasoning from incosistent information. In: Froidevaux, C., Kohlas, J. (eds.) ECSQARU 1995. LNCS, vol. 946, pp. 44–51. Springer, Heidelberg (1995)
9. Eiter, T., Gottlob, G.: On the complexity of propositional knowledge base revision, updates, and counterfactuals. Artificial Intelligence 57(2-3), 227–270 (1992)

10. Gottlob, G.: Complexity results for nonmonotonic logics. Journal of Logic Computation 2(3), 397–425 (1992)
11. Grégoire, E., Mazure, B., Piette, C.: Extracting muses. In: European Conference on Artificial Intelligence (ECAI 2006), pp. 387–391. IOS Press, Amsterdam (2006)
12. Grégoire, E., Mazure, B., Piette, C.: Boosting a complete technique to find mss and mus thanks to a local search oracle. In: International Joint Conference on Artificial Intelligence (IJCAI 2007), pp. 2300–2305 (2007)
13. Grégoire, E., Mazure, B., Piette, C.: Tracking muses and strict inconsistent covers. In: ACM/IEEE Conference on Formal Methods in Computer Aided Design (FMCAD 2006), pp. 39–46 (2006)
14. Brewka, G., Eiter, T.: Prioritizing default logic. In: Intellectics and Computational Logic (to Wolfgang Bibel on the occasion of his 60th birthday), Deventer, The Netherlands, pp. 27–45. Kluwer Academic Publishers, Dordrecht (2000)

Appendix A: Default Logic

The basic ingredients of Reiter's default logic [1] are *default rules* (in short, *defaults*). A default d is of the form:

$$\frac{\alpha(\boldsymbol{x}) : \beta_1(\boldsymbol{x}), \ldots, \beta_m(\boldsymbol{x})}{\gamma(\boldsymbol{x})},$$

where $\alpha(\boldsymbol{x})$, $\beta_1(\boldsymbol{x})$, $\ldots$, $\beta_m(\boldsymbol{x})$, $\gamma(\boldsymbol{x})$ are first-order formulas with free variables belonging to $\boldsymbol{x} = \{x_1, \ldots, x_n\}$, and are called the *prerequisite*, the *justifications* and the *consequent* of d, respectively.

Intuitively, d is intended to allow the reasoning *"Provided that the prerequisite can be established and provided that each justification can be separately consistently assumed w.r.t. what is derived, infer the consequent"*.

Accordingly, the example in the introduction could be encoded by:

$$\frac{employee(x) : permit\_access\_DB(x)}{permit\_access\_DB(x)}.$$

Such a default where the justification and consequent are identical is called a *normal default*. A normal default with an empty prerequisite is called a *supernormal default*. For a default d, we use $pred(d)$, $just(d)$, and $cons(d)$ to denote the prerequisite, the set of justifications and the consequent of d, respectively.

A *default theory* Γ is a pair (Δ, Σ) where Σ is a set of first-order formulas and Δ is a set of defaults. It is usually assumed that Δ and Σ are in skolemized form and that open defaults, i.e. defaults with free variables, represent the set of their closed instances over the Herbrand universe. A default theory with open defaults is *closed* by replacing open defaults with their closed instances. In the following, we assume that defaults theories are closed.

Defining and computing what should be inferred from a default theory is not a straightforward issue. First, there is a kind of circularity in the definition and computation of what can be inferred. To decide whether a consequent of a default should be inferred, we need to check the consistency of its justifications. However, this consistency check amounts to proving that the opposites of the justifications cannot be inferred in turn. Actually, in the general case, fixpoints approaches are used to characterize what can be inferred from a default theory. Secondly, zero, one or several maximal sets of inferred formulas, called *extensions*, can be expected from a same default theory. One way to characterize extensions is as follows [1].

Let us define a series of sets of formulas E_i where $E_0 = Cn(\Sigma)$ and $E_{i+1} =$

$$Cn(E_i \cup \{\gamma \text{ s.t. } \frac{\alpha : \beta_1, \ldots, \beta_m}{\gamma} \in \Delta \text{ where } \alpha \in E_i \text{ and } \neg\beta_1, \ldots, \neg\beta_m \notin E\}),$$

for $i = 0, 1, 2, etc.$ Then, E is an extension of Γ iff $E = \bigcup_{i=0}^{\infty} E_i$.

A default d is called *generating* in a set of formulas Π, if $pred(d) \in \Pi$ and $\{\neg a \text{ s.t. } a \in just(d)\} \cap \Pi = \emptyset$. We note $GD(\Delta, E)$ the set of all defaults from Δ that are generating in E. It is also well-known that every extension of a default theory $\Gamma = (\Delta, \Sigma)$ is characterized through $GD(\Delta, E)$, i.e. $E = Cn(\Sigma \cup cons(GD(\Delta, E)))$, where $cons(\Delta') = \{cons(d) \text{ s.t. } d \in \Delta'\}$ for any set Δ'.

A Note on Cumulative Stereotypical Reasoning

Giovanni Casini[1] and Hykel Hosni[2]

[1] University of Pisa
giovanni.casini@gmail.com
[2] Scuola Normale Superiore, Pisa
hykel.hosni@sns.it

Abstract. We address the problem of providing a logical characterization of reasoning based on stereotypes. Following [7] we take a semantic perspective and we base our model on a notion of semantic distance. While still leading to cumulative reasoning, our notion of distance does, unlike Lehmann's, allow reasoning under inconsistent information.

Keywords: Stereotypes, prototypes, cumulative reasoning, nonmonotonic logic, default-assumption logic.

1 Introduction

One important feature of intelligent reasoning consists in the capability of associating specific situations to general patterns and by doing so, extending one's initial knowledge. Reasoning based on stereotypes is a case in point. Loosely speaking, a stereotype can be thought of as an individual whose characteristics are such that it represents a typical (i.e. generic) individual of the class it belongs to. For this reason a stereotypical individual can be expected to satisfy the key properties which are typically true of the class to which the individual belongs (see Section 2 below for an example). Of course exceptions might be waiting just around the corner and an intelligent agent must be ready to face a situation in which the properties projected on a specific individual by using stereotypical information turn out not to apply. Stereotypical reasoning is therefore *defeasible*.

The purpose of this paper is to provide a logical insight on the problem of modelling rational stereotypical reasoning. Our central idea consists in representing the latter as a two-stage inference process along the following lines. Given a piece of specific information, an agent *selects* among some background information available to it, those stereotypes which better fit the factual information at hand. We expect this to normally expand the initial information available to the agent. The second step is properly inferential: using the new (possibly expanded) information set the agent draws defeasible conclusions about the situation at hand. The key ingredient in the formalization of the first stage is a function which ranks the fitness of a set of stereotypes with respect to some factual information. Following [7] we interpret fitness in terms of a semantic distance function. Due to the defeasible nature of reasoning based on stereotypes, the inferential stage will have to be formalized by a non monotonic consequence

C. Sossai and G. Chemello (Eds.): ECSQARU 2009, LNAI 5590, pp. 590–601, 2009.

relation. Since we are interested in representing *rational* reasoning, we shall be asking for this consequence relation to be particularly well-behaved. In our model this amount to requiring that stereotypical reasoning should be *cumulative*.

The paper is organized as follows. Section 2 sets the stage for our discussion on stereotypes and provides a general characterization of semantic distance. Section 3 is devoted to recall some basic facts about non monotonic reasoning in general and default-assumption consequence in particular. We then review in Section 4 Lehmann's original proposal for distance-based stereotypical reasoning and we highlight a basic shortcoming of such model. We attempt to fix this in Section 5 where we propose a semantic distance for information which is potentially inconsistent with an agent's defaults. While overcoming the limitation of Lehmann's model, the distance function introduced there fails to lead to full cumulative reasoning. We then combine the intuition behind both distances in Section 6 where we define a lexicographic distance function which at the same time admits inconsistency and leads to cumulative consequence relations.

Before getting into the main topic of this paper, let us fix some notation. We denote by ℓ the set of sentences built-up from the finite set propositional letters $P = \{p_1, \ldots, p_n\}$ using the classical propositional connectives $\{\neg, \wedge, \vee, \rightarrow\}$ in the usual, recursive way. We denote by lowercase Greek letters $\alpha, \beta, \gamma, \ldots$, the sentences of ℓ while sets of such sentences will be denoted by capital Roman letters $A, B, C, \ldots$. As usual we denote consequence relations by $\vDash$ and $\vdash$. In particular, $\vDash$ denotes the classical (Tarskian) consequence relation while we use $\vdash$ (with various decorations) for non-monotonic consequence relations. Since it is sometimes handier to work with inference operations rather than consequence operations, we shall use Cl for the classical inference operation, that is $Cl(A) = \{\beta | A \vDash \beta\}$ and C (with decorations) for the non monotonic ones, that is $C(A) = \{\beta | A \vdash \beta\}$.

Semantically, we take sets of classical (binary) propositional valuations on the language $W = \{w, v, \ldots\}$ interpreted as possible *states* of the world. Then we also use $\vDash$ for the satisfaction relation between valuations and formulae where $w \vDash \alpha$ reads as 'The valuation w satisfies the formula α'. Given $A \subseteq \ell$ and a set W, we shall write $[A]_W$ to indicate the set of the valuations in W which satisfy all the sentences in A ($[A]_W = \{w \in W \mid w \vDash \phi \text{ for every } \phi \in A\}$). We shall drop the subscript and write simply $[A]$ whenever the reference to the particular set of valuations is irrelevant.

2 Stereotypes

Stereotypes have been vastly investigated in a number of areas, from the philosophy of mind to the cognitive sciences, for their key role in the development of theories of concept-formation and commonsense reasoning (for an overview, see e.g. [6]). Stereotypes feature prominently in Putnam's social characterization of meaning (see, e.g. [11]) as well as in most of the current approaches to conceptualization while Lackoff [5] points out their fundamental importance in commonsense and uncertain reasoning.

To fix a little our ideas on stereotypes, let us take a class of individuals, 'birds', for example; a *stereotype bird* can be thought as a set of properties defining an individual bird that we consider to be particularly representative of the very concept of a bird. In this case, then, those properties could be identifying a robin or some other little tree-bird. Hence, if we take a logical perspective on the problem, we can think of stereotypes as a set of states that typically, but not necessarily, are true of some particularly representative members of a class (a stereotypical bird will be a flying winged animal, of little dimensions, covered with feathers, with a beak, laying eggs, singing, nesting on trees, etc.). This idea suggests identifying a stereotype with a finite set of sentences $\Delta = \{\alpha_1, \ldots, \alpha_m\}$ which are true exaclty at those states which characterize the stereotype. We denote by $\mathfrak{S}, \mathfrak{T}, \ldots$ finite sets of stereotypes ($\mathfrak{S} = \{\Delta_1, \ldots, \Delta_n\}$).

In our interpretation, a set of stereotypes represents the *stereotypical* or *default* information available to an agent. This interpretation is justified by recalling that in defeasible reasoning, *defaults* refer to those pieces of information that an agent considers to be typically, normally, usually, etc. true. So, by taking stereotypical properties as defaults, we capture the idea that an agent considers stereotypical information as defeasible, and hence possibly revisable in the event of evidence to the contrary.

The close connection between defeasible and stereotypical reasoning has been brought to the logician's attention by D. Lehmann ([7]) who proposes a model for stereotypical reasoning along the following lines. An agent starts with a set of n stereotypes, $\mathfrak{S} = \{\Delta_1, \ldots, \Delta_n\}$, and is then given information about some particular individual, represented by a *factual* formula α, that we assume is consistent. This fixes what the agent considers true of the state of the world at hand. The idea then is that an agent's reasoning depends on "how good" α is as a stereotype in $\mathfrak{S}$. In order to capture this formally, Lehmann introduces a notion of semantic distance $d(\alpha, \Delta)$ between the factual and the stereotypical information available to the agent. The smaller the distance d, the better factual information "fits" the stereotype Δ. To take good advantage of stereotypical reasoning, then, the agent should associate to the factual information at hand the *nearest* stereotype. More precisely, given α and every stereotype Δ_i in $\mathfrak{S}$, the agent *selects* a subset of $\mathfrak{S}$, $\mathfrak{S}_d^\alpha$, of maximally close (i.e. nearest) elements of $\mathfrak{S}$ to α with respect to d. This is interpreted as the set of stereotypes which is natural for the agent to associate to α. Formally:

$$\mathfrak{S}_d^\alpha = \{\Delta_i \in \mathfrak{S} \mid d(\alpha, \Delta_i) \leq d(\alpha, \Delta_j) \text{ for every } \Delta_j \in \mathfrak{S})\} \qquad (\sharp)$$

The selection of the nearest stereotypes to a formula α leads naturally to defeasible reasoning which is captured by the consequence relation $\vdash_{\mathfrak{S},d}$. For obvious reasons we refer to this latter as the *consequence relation generated by $\mathfrak{S}$ and d*. To recap, the model goes as follows. An agent is equipped with a finite set of stereotypes $\mathfrak{S}$ and a semantic distance function d. Given a factual formula α, the agent selects the set $\mathfrak{S}_d^\alpha$ of stereotypes which d ranks nearest to α. Now this set $\mathfrak{S}_d^\alpha$ is used to *generate* a consequence relation $\vdash_{\mathfrak{S},d}$, which, as we shall shortly see, provides an adequate tool to produce defeasible conclusions from α and the default information contained in $\mathfrak{S}_d^\alpha$.

Thus, before recalling the constraints imposed by Lehmann on the distance function d and the properties of the generated consequence relation, we need to recall some basic facts about non monotonic consequence relations.

3   Cumulative and Default-Assumption Consequence

Among the many proposals to characterize defeasible reasoning (see [2] for an overview) some core structural properties emerge as particularly compelling (see [4], [9], and [10]). In particular, the class of *cumulative consequence relations* has gained quite a consensus in the community as the industry standard.

Definition 1 (Cumulative Consequence Relations). *A consequence relation $\mid\!\sim$ is* cumulative *if and only if it satisfies the following properties:*

$$REF \quad \alpha \mid\!\sim \alpha \qquad\qquad\qquad \textit{Reflexivity}$$

$$LLE \quad \frac{\alpha \mid\!\sim \gamma \ \models \alpha \leftrightarrow \beta}{\beta \mid\!\sim \gamma} \qquad \textit{Left Logical Equivalence}$$

$$RW \quad \frac{\alpha \mid\!\sim \beta \ \ \beta \models \gamma}{\alpha \mid\!\sim \gamma} \qquad \textit{Right Weakening}$$

$$CT \quad \frac{\alpha \mid\!\sim \beta \ \ \alpha \wedge \beta \mid\!\sim \gamma}{\alpha \mid\!\sim \gamma} \quad \textit{Cut (Cumulative Transitivity)}$$

$$CM \quad \frac{\alpha \mid\!\sim \beta \ \ \alpha \mid\!\sim \gamma}{\alpha \wedge \beta \mid\!\sim \gamma} \qquad \textit{Cautious Monotony}$$

where $\models$ denotes as usual the tarskian consequence relation of classical logic.

Combining the flexibility of nonmonotonic (i.e. default, revisable, defeasible, etc.) reasoning with many desirable metalogical properties, such as *idempotence*, *supraclassicality*, and *full-absorption* (see [9]), cumulative consequence relations constitute a tool of choice in the formalization of commonsense inference.

Among the class of cumulative consequence relations are the so-called *default-assumption consequence relations*, which will play a key role in our model and which we therefore turn to recall. The idea behind default-assumption reasoning is that an agent's information can be viewed as being two-fold. On the one hand agents have *defeasible information*, a set Δ of defaults that an agent *presumes* to be typically true. On the other hand agents might acquire *factual information*, that is, information that the agent *takes as true* of the particular situation at hand, and which is represented, in our setting, by a single formula α. Intuitively, then, default-assumption reasoning takes place when an agent extends its factual information α with those defaults which are compatible with α and takes the result as premises for its inferences.

In order to formalize this we need to define the set of *maximally α-consistent subsets* of Δ, or, equivalently, the notion of *remainder set* (see [3], p.12).

Definition 2 (Remainder Sets). *For B a set of formulae and α a formula, the* remainder set $B\perp\alpha$ *('B less α') is the set of sets of formulae such that $A \in B\perp\alpha$ if and only if:*

1. $A \subseteq B$
2. $\alpha \notin Cl(A)$
3. There is no set A' such that $A \subseteq A' \subseteq B$, and $\alpha \notin Cl(A')$

Thus, for a set of defaults Δ, $\Delta \perp \neg \alpha$ is the set of every maximal subsets of Δ consistent with α. Default-assumption consequence relation can then be defined as follows:

Definition 3 (Default-assumption consequence relation). *β is a default-assumption consequence of α given a set of default-assumptions Δ, (written $\alpha \mathrel{\vert\!\sim}_\Delta \beta$) if and only if β is a classical consequence of the union of α and every set in $\Delta \perp \neg \alpha$:*

$$\alpha \mathrel{\vert\!\sim}_\Delta \beta \text{ iff } \alpha \cup \Delta' \vDash \beta \text{ for every } \Delta' \in \Delta \perp \neg \alpha$$

It is well-known that default-assumption consequence relations are cumulative (see e.g. [1] and [10]).

4   Lehmann's Model

In [7] Lehmann proposes a set of intuitive constraints that the semantic distance d should satisfy in order to generate a well-behaved consequence relation $\mathrel{\vert\!\sim}_{\mathfrak{S},d}$. We denote by δ Lehmann's distance function and by $\mathfrak{S}^\alpha_\delta$ the set of stereotypes in $\mathfrak{S}$ selected by δ with respect to a factual formula α as in (♯) above. Finally we denote by $\mathrel{\vert\!\sim}_{\mathfrak{S},\delta}$ the consequence relation generated by $\mathfrak{S}$ and δ.

Recall that the stereotypes in $\mathfrak{S}^\alpha_\delta$, are meant be those which fit better the factual information represented by α i.e. those with minimal semantic distance from α. Thus, it is natural to capture this by looking at the overlap between the states of the world which make α and a set of stereotypes Δ true. But such states are precisely the models of α and the models of Δ ([α] and [Δ], respectively). The idea is obviously that greatest overlap means maximal closeness. So, given a set of stereotypes $\mathfrak{S}$ and a factual formula α, Lehmann requires that:

○ For every $\Delta \in \mathfrak{S}$, $\delta(\alpha, \Delta)$ should be anti-monotonic with respect to $[\Delta] \cap [\alpha]$ (the larger the overlap, the smaller the distance).
○ For every $\Delta \in \mathfrak{S}$, $\delta(\alpha, \Delta)$ should be monotonic with respect to $[\Delta] - [\alpha]$ (the larger the set of states which satisfy the defaults but not the factual information, the larger the distance).

The following simplifying assumption is also made:

○ $|\mathfrak{S}^\alpha_\delta| = 1$ (i.e. for every α and $\mathfrak{S}$, the agent selects exactly one element in $\mathfrak{S}$).

The above constraints are formalized by:

$$[\Delta'] \cap [\alpha'] \subseteq [\Delta] \cap [\alpha] \text{ and } [\Delta] - [\alpha] \subseteq [\Delta'] - [\alpha'] \Rightarrow \delta(\alpha, \Delta) \leq \delta(\alpha', \Delta') \quad (L1)$$

The generated consequence relation $\mathrel{\vert\!\sim}_{\mathfrak{S},\delta}$ is then defined by adding the information of the only default set Δ^α_δ in $\mathfrak{S}^\alpha_\delta$ to the premise set α:

$$\alpha \mathrel{\vert\!\sim}_{\mathfrak{S},\delta} \beta \text{ iff } \{\alpha\} \cup \Delta^\alpha_\delta \vDash \beta. \quad (L\!\mathrel{\vert\!\sim})$$

For any distance function δ satisfying (L1), Lehmann proves the following result:

Theorem 1 ([7], Theorem 5.5). *If* $([\alpha] \cap [\Delta_\delta^\alpha]) \subseteq [\alpha'] \subseteq [\alpha]$, *then* $\mathfrak{S}_\delta^{\alpha'} = \mathfrak{S}_\delta^\alpha$.

That is, if the agent becomes aware of new factual information α' that is not inconsistent with the stereotype previously selected $(([\alpha] \cap [\Delta_\delta^\alpha]) \subseteq [\alpha'])$, then the agent should not abandon the selected stereotype (Δ_δ^α) to extend its factual information (that is, $\mathfrak{S}_\delta^{\alpha'} = \mathfrak{S}_\delta^\alpha$). From this theorem Lehmann proves his main result: given a set of stereotypes $\mathfrak{S}$ and a distance function δ, if the distance function δ satisfies the constraint (L1), then the generated consequence relation $\vdash_{\mathfrak{S},\delta}$ is cumulative ([7], Corollary 5.6).

To see the importance of the result, let us observe one of its consequences through namely the fact that $\vdash_{\mathfrak{S},\delta}$ satisfies Cautious Monotonicity. Suppose that α stands for the fact that Sherkan is a big feline with a black-striped, tawny coat. Then it is natural to associate Sherkan to the stereotype of the tiger and then using this information to conclude that Sherkan has also long teeth $(\alpha \vdash_{\mathfrak{S},\delta} \beta)$ and is a predator $(\alpha \vdash_{\mathfrak{S},\delta} \gamma)$. Reasonably then, if we add to our premises the information that Sherkan has long teeth, we should continue to consider it to be a tiger and, consequently, a predator $(\alpha \wedge \beta \vdash_{\mathfrak{S},\delta} \gamma)$.

Intuitive as (L1) may be, Lehmann's model has a significant shortcoming. The problem lies in the requirement that in order for stereotypical reasoning to take place, there should be a nonempty intersection between the factual information at hand and the agent's set of stereotypes. In other words, Lehmann does not take into account the possibility that every stereotype in $\mathfrak{S}$ is inconsistent with the premise α. In such a case, then by the definition of $\vdash_{\mathfrak{S},\delta}$, we could set the distance between the premise and a stereotype to ∞ (i.e. the largest the distance according to δ):

$$\delta(\alpha, \Delta) = \infty \text{ iff } [\alpha] \cap [\Delta] = \emptyset.$$

So, any choice of stereotypes here is admissible, making stereotypical reasoning basically vacuous (that is, $\mathfrak{S}_\delta^\alpha = \mathfrak{S}$). This shortcoming reduces significantly the scope of Lehmann's model for one key feature of stereotypical reasoning is precisely the fact that an individual can be related to a stereotype even if its properties do not match all the properties of the stereotype so that we can derive defeasible conclusions on the basis of the pieces of stereotypical information compatible with the premises. For example, knowing that Tweety is a penguin, we can reason about it using the information contained in the stereotype of a bird excluding the information that is known to be inconsistent with being a penguin (flying, nesting in trees, etc.).

5   A Semantic Distance for Inconsistent Information.

So we now focus on the situation in which every stereotype available to an agent turns out to be inconsistent with its factual information. More precisely we define a notion of semantic distance, ε, with the idea of capturing the distance between a formula α and a set of α-inconsistent default sets. If there are no α-consistent stereotypes, we allow the choice of the 'nearest' α-inconsistent default sets. This new notion of distance has clearly an effect on the associated consequence relation $\mathrel{\vdash}_{\mathfrak{S},\varepsilon}$: as $\alpha \cup \Delta$ might be inconsistent, we need to move from the classical relation $\vDash$, used by Lehmann in (L$\mathrel{\vdash}$), to a default-assumption consequence relation $\mathrel{\vdash}_\Delta$. Note that by definition 3 above, if the set $\{\alpha\} \cup \Delta$ is consistent, we have $\alpha \mathrel{\vdash}_\Delta \beta$ if and only if $\{\alpha\} \cup \Delta \vDash \beta$, as in Lehmann's definition.

We begin by recalling the notion of semantic distance proposed by Lehmann, Magidor and Schlechta [8] in the context of belief revision. We claim that this is appropriate as a measure of 'consistency distance' between formulae. For an arbitrary set Z we say that ε is *a semantic pseudo-distance function*

$$\varepsilon : W \times W \to Z$$

if it satisfies the following:

(ε1) The set Z is totally ordered by a strict order $<$
(ε2) Z has a $<$-smallest element 0, and $\varepsilon(w, v) = 0$ if and only if $w = v$

Note that ε is not required to be symmetric (i.e. $\varepsilon(w, v) = \varepsilon(v, w)$ for every $w, v \in U$). This is matches our intuitive interpretation of distance. Indeed, as we shall shortly see, an agent should have different attitudes towards the information represented by the fist argument of the distance function, that refers to what the agent takes to be certainly true, and the second argument, which concerns default information.

Again, the distance between two given sets of formulae A and B is semantic as it is defined with respect to their models $[A]$ and $[B]$, and the distance between two sets of valuations U and U' $(U, U' \subseteq W)$ is set to be the minimal distance between the valuations in U and U':

$$\varepsilon(U, U') = min\{\varepsilon(w, v) | w \in U, v \in U'\}.$$

In analogy with equation ($\sharp$) above, given a finite set $\mathfrak{S}$ of default sets $\{\Delta_1, \ldots, \Delta_n\}$ and a formula α, $\mathfrak{S}_\varepsilon^\alpha$ is identified with the set of ε-'nearest' default sets in $\mathfrak{S}$ to α.

From now on we relax Lehmann's assumption that $\mathfrak{S}^\alpha$ must contain a single default set, thus allowing the possibility that, under the uncertainty connected to the presence of inconsistencies, a set of premises is taken to be equally distant from distinct default sets.

A few observations are in order. Note that since ε is a total function, $\mathfrak{S} \neq \emptyset$ implies $\mathfrak{S}_\varepsilon^A \neq \emptyset$. Note also that it makes no difference if we use as arguments ε sets of formulae A or sets of valuations $[A]$. That is, we $\varepsilon(\alpha, \Delta) = \varepsilon([\alpha], [\Delta])$ for every α, Δ. Finally, since ε satisfies (ε2), if a factual formula and a default set are mutually consistent, then the distance between them is 0, as they share at least a valuation. Hence, the default sets which turn out to be consistent with our set of premises have, intuitively, priority over those which are inconsistent.

We can now define $\mathrel{\vdash}_{\mathfrak{S},\varepsilon}$ using default-assumption consequence relations:

$$\alpha \mathrel{\vdash}_{\mathfrak{S},\varepsilon} \beta \text{ iff } \alpha \mathrel{\vdash}_\Delta \beta \text{ for every } \Delta \in \mathfrak{S}_\varepsilon^\alpha. \tag{*}$$

Note that the corresponding inference operation is

$$C_{\mathfrak{S},\varepsilon}(\alpha) = \bigcap \{C_\Delta(\alpha) | \Delta \in \mathfrak{S}_\varepsilon^\alpha\}$$

where C_Δ is the inference operation corresponding to the default-assumption consequence relation $\hspace{-0.1em}\sim_\Delta$. $\hspace{-0.1em}\sim_{\mathfrak{S},\varepsilon}$ so defined satisfies some properties of cumulative consequence relations.

Lemma 1. *Assume a pseudo-distance ε and a set of stereotypes $\mathfrak{S}$. The consequence relation $\hspace{-0.1em}\sim_{\mathfrak{S},\varepsilon}$ satisfies REF, LLE, RW.*

Proof. Assume a formula α and a set of stereotypes $\mathfrak{S}$. By means of our distance function ε, we can identify the set $\mathfrak{S}_\varepsilon^\alpha$. Since default-assumption consequence relations are cumulative (see section 3), we have that, for every default set Δ, the default-assumption relation $\hspace{-0.1em}\sim_\Delta$ satisfies REF, LLE, and RW. The inference operation $C_{\mathfrak{S},\varepsilon}(\alpha)$ is defined as the intersection of every default-assumption inference operation $C_\Delta(\alpha)$, s.t. $\Delta \in \mathfrak{S}_\varepsilon^\alpha$ $(C_{\mathfrak{S},\varepsilon}(\alpha) = \bigcap\{C_\Delta(\alpha)|\Delta \in \mathfrak{S}_\varepsilon^\alpha\})$ and it is straightforward to prove that REF, LLE, and RW are preserved under intersection.

To see that $\hspace{-0.1em}\sim_{\mathfrak{S},\varepsilon}$ is not cumulative, suppose that we have a set of stereotypes $\mathfrak{S} = \{\Delta, \Delta'\}$, where $\Delta = \{\neg p, p \to r, p \to t\}$ and $\Delta' = \{\neg p, p \wedge r \to \neg t\}$. Since we have $[\Delta] \cap [p] = [\Delta'] \cap [p] = \emptyset$, we have that $\varepsilon(p, \Delta) \neq 0$ and $\varepsilon(p, \Delta') \neq 0$. Without loss of generality let $\varepsilon(p, \Delta) < \varepsilon(p, \Delta')$ and $\varepsilon(p \wedge r, \Delta') < \varepsilon(p \wedge r, \Delta)$. Note that this satisfies both (d1) and (d2).

Now, from these assumptions we get $p \hspace{-0.1em}\sim_{\mathfrak{S},\varepsilon} r$, $p \hspace{-0.1em}\sim_{\mathfrak{S},\varepsilon} t$, since $C_{\mathfrak{S},\varepsilon}(p) = C_\Delta(p)$, but, since $C_{\mathfrak{S},\varepsilon}(p \wedge r) = C_{\Delta'}(p \wedge r)$, we also get $p \wedge r \hspace{-0.1em}\sim_{\mathfrak{S},\varepsilon} \neg t$, violating cautious monotony.

To get cumulativity, we need ε to satisfy a further constrain which intuitively ensures that given a premise α and a default set Δ, there is some valuation satisfying both α and a maximal α-consistent subset of Δ that is at least as near to Δ as any other valuation in $[\alpha]$. To formalize this we first define $[\Delta \bot \alpha]$ as the set of valuations satisfying at least one element of the remainder set $\Delta \bot \alpha$ (see definition 2):

$$[\Delta \bot \alpha] = \bigcup\{[B]|B \in \Delta \bot \alpha\}$$

We can now define the required new constraint

(ε3) For every α and Δ, there is a $w \in [\alpha] \cap [\Delta \bot \neg \alpha]$ s.t. $\varepsilon(w, [\Delta]) \leq \varepsilon(v, [\Delta])$ for every $v \in [\alpha]$.

In order to guarantee that $[\Delta \bot \neg \alpha] \neq \emptyset$ (and hence that $[\alpha] \cap [\Delta \bot \neg \alpha] \neq \emptyset$), we can simply assume that every default set Δ contains a tautology ($\top \in \Delta$, for every Δ). We now prove a series of lemmas leading to the result that if ε satisfies $(\varepsilon 1) - (\varepsilon 3)$, then the generated consequence relation $\hspace{-0.1em}\sim_{\mathfrak{S},\varepsilon}$ is cumulative.

Lemma 2. *If $[\alpha] \subseteq [\alpha']$, then $\varepsilon(\alpha', \Delta) \leq \varepsilon(\alpha, \Delta)$ for every Δ.*

Proof. $\varepsilon(\alpha, \Delta) = min\{\varepsilon(w, v)|w \in [\alpha], v \in [\Delta]\}$. Since $w \in [\alpha]$ implies $w \in [\alpha']$, we have that $min\{\varepsilon(w, v)|w \in [\alpha'], v \in [\Delta]\} \leq min\{\varepsilon(w, v)|w \in [\alpha], v \in [\Delta]\}$, i.e. $\varepsilon(\alpha', \Delta) \leq d(\alpha, \Delta)$.

We now want to prove that if we add to the factual information information which is itself derivable by means of $\hspace{-0.1em}\sim_{\mathfrak{S},\varepsilon}$, then we continue to associate the same stereotypes to our premise (see the example about the tiger Sherkan in section 4).

Lemma 3. *If $\alpha \mathrel{\vdash}_{\mathfrak{S},\varepsilon} \beta$ and $\Delta \in \mathfrak{S}^{\alpha}_{\varepsilon}$, then $\varepsilon(\alpha \wedge \beta, \Delta) = \varepsilon(\alpha, \Delta)$.*

Proof. Recall that $\mathfrak{S}^{\alpha}_{\varepsilon} = \{\Delta \in \mathfrak{S} \mid \varepsilon(\alpha, \Delta) \leq \varepsilon(\alpha, \Delta')$ for every $\Delta' \in \mathfrak{S}\}$.
By $(\varepsilon 3)$, we have that $\varepsilon([\alpha], [\Delta]) = \varepsilon(w, [\Delta])$ for some $w \in [\alpha] \cap [\Delta \perp \neg \alpha]$.
$\alpha \mathrel{\vdash}_{\mathfrak{S},\varepsilon} \beta$ implies that $\alpha \mathrel{\vdash}_{\Delta} \beta$ for every $\Delta \in \mathfrak{S}^{\alpha}_{\varepsilon}$, which implies that if $w \in [\alpha] \cap [\Delta \perp \neg \alpha]$, then $w \in [\alpha \wedge \beta]$.
   Given that $\varepsilon([\alpha], [\Delta]) = \varepsilon(w, [\Delta])$, we have that $\varepsilon(w, [\Delta]) \leq \varepsilon(v, [\Delta])$ for every $v \in [\alpha]$. Since $[\alpha \wedge \beta] \subseteq [\alpha]$, we have that $\varepsilon(w, [\Delta]) \leq \varepsilon(v, [\Delta])$ for every $v \in [\alpha \wedge \beta]$, that is, $\varepsilon(\alpha \wedge \beta, \Delta) = \varepsilon(w, [\Delta]) = \varepsilon(\alpha, \Delta)$.

Lemma 4. *If $\alpha \mathrel{\vdash}_{\mathfrak{S},\varepsilon} \beta$, then $\mathfrak{S}^{\alpha}_{\varepsilon} = \mathfrak{S}^{\alpha \wedge \beta}_{\varepsilon}$.*

Proof. Assume $\alpha \mathrel{\vdash}_{\mathfrak{S},\varepsilon} \beta$. We show that $\mathfrak{S}^{\alpha}_{\varepsilon} \subseteq \mathfrak{S}^{\alpha \wedge \beta}_{\varepsilon}$. If $\Delta \in \mathfrak{S}^{\alpha}_{\varepsilon}$, then $\varepsilon(\alpha, \Delta) \leq \varepsilon(\alpha, \Delta')$ for every $\Delta' \in \mathfrak{S}$. By Lemma 2, since $[\alpha \wedge \beta] \subseteq [\alpha]$, we have that $\varepsilon(\alpha, \Delta) \leq \varepsilon(\alpha \wedge \beta, \Delta')$ for every $\Delta' \in \mathfrak{S}$. Since $\Delta \in \mathfrak{S}^{\alpha}_{\varepsilon}$, by Lemma 3, we obtain $\varepsilon(\alpha \wedge \beta, \Delta) \leq \varepsilon(\alpha \wedge \beta, \Delta')$ for every $\Delta' \in \mathfrak{S}$, i.e. $\Delta \in \mathfrak{S}^{\alpha \wedge \beta}_{\varepsilon}$.
   For $\mathfrak{S}^{\alpha \wedge \beta}_{\varepsilon} \subseteq \mathfrak{S}^{\alpha}_{\varepsilon}$, if $\Delta \notin \mathfrak{S}^{\alpha}_{\varepsilon}$, then $\varepsilon(\alpha, \Delta') < \varepsilon(\alpha, \Delta)$ for some $\Delta' \in \mathfrak{S}^{\alpha}_{\varepsilon}$. By Lemma 2, we have that $\varepsilon(\alpha, \Delta') < \varepsilon(\alpha \wedge \beta, \Delta)$. Since $\Delta' \in \mathfrak{S}^{\alpha}_{\varepsilon}$, by Lemma 3, we obtain $\varepsilon(\alpha \wedge \beta, \Delta') < \varepsilon(\alpha \wedge \beta, \Delta)$, i.e. $\Delta \notin \mathfrak{S}^{\alpha \wedge \beta}_{\varepsilon}$.

We are now ready to prove the key result about our notion of distance ε.

Theorem 2. *Given a set of stereotypes $\mathfrak{S}$ and a distance function ε satisfying $(\varepsilon 1)$-$(\varepsilon 3)$, the generated consequence relation $\mathrel{\vdash}_{\mathfrak{S},\varepsilon}$ is cumulative.*

Proof. We have to show that $\mathrel{\vdash}_{\mathfrak{S},\varepsilon}$ satisfies CM and CT.
CM: assume $\alpha \mathrel{\vdash}_{\mathfrak{S},\varepsilon} \beta$ and $\alpha \mathrel{\vdash}_{\mathfrak{S},\varepsilon} \gamma$, which correspond to saying that $\alpha \mathrel{\vdash}_{\Delta} \beta$ and $\alpha \mathrel{\vdash}_{\Delta} \gamma$ for every $\Delta \in \mathfrak{S}^{\alpha}_{\varepsilon}$. Since every default-assumption consequence relation $\mathrel{\vdash}_{\Delta}$, being cumulative (see section 3), satisfies CM, we have $\alpha \wedge \beta \mathrel{\vdash}_{\Delta} \gamma$ for every $\Delta \in \mathfrak{S}^{\alpha}_{\varepsilon}$. Given $\alpha \mathrel{\vdash}_{\mathfrak{S},\varepsilon} \beta$, we have, by Lemma 4, that $\mathfrak{S}^{\alpha}_{\varepsilon} = \mathfrak{S}^{\alpha \wedge \beta}_{\varepsilon}$, which implies that $\alpha \wedge \beta \mathrel{\vdash}_{\Delta} \gamma$ for every $\Delta \in \mathfrak{S}^{\alpha \wedge \beta}_{\varepsilon}$, i.e. $\alpha \wedge \beta \mathrel{\vdash}_{\mathfrak{S},\varepsilon} \gamma$.

CT: assume $\alpha \wedge \beta \mathrel{\vdash}_{\mathfrak{S},\varepsilon} \gamma$ and $\alpha \mathrel{\vdash}_{\mathfrak{S},\varepsilon} \beta$. Note again that $\alpha \wedge \beta \mathrel{\vdash}_{\mathfrak{S},\varepsilon} \gamma$ means that $\alpha \wedge \beta \mathrel{\vdash}_{\Delta} \gamma$ for every $\Delta \in \mathfrak{S}^{\alpha \wedge \beta}_{\varepsilon}$. $\alpha \mathrel{\vdash}_{\mathfrak{S},\varepsilon} \beta$ implies, again by Lemma 4, that $\mathfrak{S}^{\alpha}_{\varepsilon} = \mathfrak{S}^{\alpha \wedge \beta}_{\varepsilon}$. Hence, we have that $\alpha \wedge \beta \mathrel{\vdash}_{\Delta} \gamma$ and $\alpha \mathrel{\vdash}_{\Delta} \beta$ for every $\Delta \in \mathfrak{S}^{\alpha}_{\varepsilon}$. Since every such $\mathrel{\vdash}_{\Delta}$, being cumulative, satisfies CT, we have $\alpha \mathrel{\vdash}_{\Delta} \gamma$ for every $\Delta \in \mathfrak{S}^{\alpha}_{\varepsilon}$, i.e. $\alpha \mathrel{\vdash}_{\mathfrak{S},\varepsilon} \gamma$.

Thus our notion of distance captures the stereotypical reasoning underlying Lehmann's approach while preserving the cumulativity of the generated consequence relation in the general case in which an agent's factual information comes out to be inconsistent with its stereotypical information. However, this revised distance function looses its appeal if more than one stereotype is consistent with an agent's factual information. In such a case, by $(\varepsilon 2)$, an agent cannot distinguish between the stereotypes in $\mathfrak{S}$ that are consistent with α, since their mutual distance is always 0.

6 A Lexicographic Combination of δ and ε

Summing up what has been done so far, we started by reviewing Lehmann's notion of distance δ and noted that while logically well-behaved (generates cumulative consequence relations) it suffers from the drawback of not handling inconsistency between factual and default information. In order to overcome this limitation we considered a semantic pseudo-distance ε, which again leads to cumulative reasoning, but which, at the same time, allows an agent to face the situation in which its factual information turns out to be inconsistent with its default information. The purpose of the remainder of this paper is to study a combination of the two approaches which enables us to refine pseudo-distance ε in order to let this latter distinguish between the stereotypes consistent with α. To do this we define a lexicographic ordering of a distance d_{lex}, with the idea that the precedence should be given whenever possible to ε over δ. More precisely:

$$d_{lex}(\alpha, \Delta) \leq d_{lex}(\alpha', \Delta') \Leftrightarrow \begin{cases} \varepsilon(\alpha, \Delta) < \varepsilon(\alpha', \Delta') \\ or \\ \varepsilon(\alpha, \Delta) = \varepsilon(\alpha', \Delta') \text{ and } \delta(\alpha, \Delta) \leq \delta(\alpha', \Delta') \end{cases}$$

Given a set of stereotypes $\mathfrak{S}$, a semantic distance d_{lex} and a formula α, we define, again in analogy with equation ($\sharp$), the set $\mathfrak{S}_\varepsilon^\alpha$ of the stereotypes in $\mathfrak{S}$ which are nearest to α. $\vdash_{\mathfrak{S},d_{lex}}$ is defined analogously to $\vdash_{\mathfrak{S},\varepsilon}$, using default-assumption consequence relations as in equation (*). We devote the rest of this paper to show that d_{lex} does indeed combine the best of ε and δ since it eventually leads to cumulative reasoning.

Recall that, given a set of stereotypes $\mathfrak{S}$, a semantic distance d_{lex}, defined lexicographically over two distances ε and δ, and a formula α, we indicate by $\mathfrak{S}_{d_{lex}}^\alpha$ the set of the nearest stereotypes to α with respect to d_{lex}, by $\mathfrak{S}_\varepsilon^\alpha$ the nearest stereotypes with respect to ε, and by $\mathfrak{S}_\delta^\alpha$ the nearest stereotypes with respect to δ. As we have seen, they define, respectively, three consequence relations: $\vdash_{\mathfrak{S},d_{lex}}$, $\vdash_{\mathfrak{S},\varepsilon}$, and $\vdash_{\mathfrak{S},\delta}$ (and the correspondent inference operations $C_{\mathfrak{S},d_{lex}}$, $C_{\mathfrak{S},\varepsilon}$, and $C_{\mathfrak{S},\delta}$). Note that by the lexicographic definition of the d_{lex}-ordering, we have:

$$\mathfrak{S}_{d_{lex}}^\alpha = \begin{cases} \mathfrak{S}_\varepsilon^\alpha \text{ if } |\mathfrak{S}_\varepsilon^\alpha| = 1 \\ (\mathfrak{S}_\varepsilon^\alpha)_\delta^\alpha \text{ if } |\mathfrak{S}_\varepsilon^\alpha| > 1 \end{cases}$$

where $(\mathfrak{S}_\varepsilon^\alpha)_\delta^\alpha$ is the composition of the selection functions of the stereotypes in $\mathfrak{S}$ that ε and δ associate to α: we first select the subset of $\mathfrak{S}$ nearest to α with respect to ε (that is, $\mathfrak{S}_\varepsilon^\alpha$), and, if using ε we have not been able to distinguish between distinct stereotypes ($|\mathfrak{S}_\varepsilon^\alpha| > 1$), we refine our procedure by selecting the δ-nearest stereotypes to α between those in $\mathfrak{S}_\varepsilon^\alpha$ (that is, $(\mathfrak{S}_\varepsilon^\alpha)_\delta^\alpha$).

Lemma 5. *If* $\alpha \vdash_{\mathfrak{S},d_{lex}} \beta$, *then* $\mathfrak{S}_{d_{lex}}^\alpha = \mathfrak{S}_{d_{lex}}^{\alpha \wedge \beta}$.

Proof. We have three possibilities.

(1) $|\mathfrak{S}_\varepsilon^\alpha| = 1$.
(2) $|\mathfrak{S}_\varepsilon^\alpha| > 1$ and $\varepsilon(\alpha, \Delta) > 0$ for every $\Delta \in \mathfrak{S}_\varepsilon^\alpha$.
(3) $|\mathfrak{S}_\varepsilon^\alpha| > 1$ and $\varepsilon(\alpha, \Delta) = 0$ for every $\Delta \in \mathfrak{S}_\varepsilon^\alpha$.

(1): $|\mathfrak{S}_\varepsilon^\alpha| = 1$ implies that $\mathfrak{S}_{d_{lex}}^\alpha = \mathfrak{S}_\varepsilon^\alpha$ and $C_{\mathfrak{S},d_{lex}}(\alpha) = C_{\mathfrak{S},\varepsilon}(\alpha)$, that is, $\alpha \mathrel{\vert\!\sim}_{\mathfrak{S},d_{lex}} \beta$ iff $\alpha \mathrel{\vert\!\sim}_{\mathfrak{S},\varepsilon} \beta$. By Lemma 4, from $\alpha \mathrel{\vert\!\sim}_{\mathfrak{S},\varepsilon} \beta$ we obtain $\mathfrak{S}_\varepsilon^\alpha = \mathfrak{S}_\varepsilon^{\alpha \wedge \beta}$, that implies $|\mathfrak{S}_\varepsilon^{\alpha \wedge \beta}| = 1$ and $\mathfrak{S}_{d_{lex}}^{\alpha \wedge \beta} = \mathfrak{S}_\varepsilon^{\alpha \wedge \beta}$, that is, $\mathfrak{S}_{d_{lex}}^{\alpha \wedge \beta} = \mathfrak{S}_{d_{lex}}^\alpha$.

(2): $\varepsilon(\alpha, \Delta) > 0$ for every $\Delta \in \mathfrak{S}_\varepsilon^\alpha$ implies that the default sets in $\mathfrak{S}_\varepsilon^\alpha$ are not consistent with the premise α. Therefore, δ cannot distinguish them out and we have $(\mathfrak{S}_\varepsilon^\alpha)_\delta^\alpha = \mathfrak{S}_\varepsilon^\alpha$. Again we have that $\mathfrak{S}_{d_{lex}}^\alpha = \mathfrak{S}_\varepsilon^\alpha$ and $C_{\mathfrak{S},d_{lex}}(\alpha) = C_{\mathfrak{S},\varepsilon}(\alpha)$, and the case is already covered by (1).

(3): $\varepsilon(\alpha, \Delta) = 0$ for every $\Delta \in \mathfrak{S}_\varepsilon^\alpha$ implies that the stereotypes in $\mathfrak{S}_\varepsilon^\alpha$ are consistent with α, and we can refine the choice by means of δ.
Since $\alpha \mathrel{\vert\!\sim}_{\mathfrak{S},d_{lex}} \beta$, we have that some default sets in $\mathfrak{S}_\varepsilon^\alpha$ are consistent with $\alpha \wedge \beta$ (surely the one in $(\mathfrak{S}_\varepsilon^\alpha)_\delta^\alpha$). Since $\mathfrak{S}_\varepsilon^\alpha$ is composed by every set in $\mathfrak{S}$ consistent with α, every default set consistent with $\alpha \wedge \beta$ is in $\mathfrak{S}_\varepsilon^\alpha$. Hence, we have $(\mathfrak{S}_\varepsilon^\alpha)_\delta^\alpha \subseteq \mathfrak{S}_\varepsilon^{\alpha \wedge \beta} \subseteq \mathfrak{S}_\varepsilon^\alpha$, that is,

$$\mathfrak{S}_{d_{lex}}^\alpha \subseteq \mathfrak{S}_\varepsilon^{\alpha \wedge \beta} \subseteq \mathfrak{S}_\varepsilon^\alpha$$

Since every element in $(\mathfrak{S}_\varepsilon^\alpha)_\delta^\alpha$ is in $\mathfrak{S}_\varepsilon^{\alpha \wedge \beta}$, we have that $(\mathfrak{S}_\varepsilon^{\alpha \wedge \beta})_\delta^\alpha = (\mathfrak{S}_\varepsilon^\alpha)_\delta^\alpha = \mathfrak{S}_{d_{lex}}^\alpha$.

Now take theorem 1, that is,

if $([\alpha] \cap [\Delta]) \subseteq [\alpha'] \subseteq [\alpha]$, then $\mathfrak{S}_\delta^{\alpha'} = \mathfrak{S}_\delta^\alpha$, where $\mathfrak{S}_\delta^\alpha = \{\Delta\}$.

Let α' be $\alpha \wedge \beta$ and $\mathfrak{S}$ be $\mathfrak{S}_\varepsilon^{\alpha \wedge \beta}$, and, consequently, let $(\mathfrak{S}_\varepsilon^{\alpha \wedge \beta})_\delta^\alpha = \{\Delta\}$. Given that $\alpha \mathrel{\vert\!\sim}_{\mathfrak{S},d_{lex}} \beta$ and $\mathfrak{S}_{d_{lex}}^\alpha = (\mathfrak{S}_\varepsilon^{\alpha \wedge \beta})_\delta^\alpha$, we have that $([\alpha] \cap [\Delta]) \subseteq [\alpha \wedge \beta] \subseteq [\alpha]$, and this, by theorem 1, implies $(\mathfrak{S}_\varepsilon^{\alpha \wedge \beta})_\delta^{\alpha \wedge \beta} = (\mathfrak{S}_\varepsilon^{\alpha \wedge \beta})_\delta^\alpha$.
Combining the equations, we have $(\mathfrak{S}_\varepsilon^\alpha)_\delta^\alpha = (\mathfrak{S}_\varepsilon^{\alpha \wedge \beta})_\delta^{\alpha \wedge \beta}$, that is

$$\mathfrak{S}_{d_{lex}}^\alpha = \mathfrak{S}_{d_{lex}}^{\alpha \wedge \beta}$$

as desired.

We now have all the ingredients to prove our central result.

Theorem 3. *Given a set of stereotypes $\mathfrak{S}$ and a distance function d_{lex}, the consequence relation $\mathrel{\vert\!\sim}_{\mathfrak{S},d_{lex}}$ is cumulative.*

Proof. Since $C_{\mathfrak{S},d_{lex}}(\alpha)$ is obtained by the intersection of default-assumption inference operations, it satisfies REF, LLE, RW (see Lemma 1).
Cumulativity then follows by Lemma 5 with exactly the same procedure argument used in the proof of Theorem 2.

7   Conclusions

We have addressed the problem of providing a logical characterization of reasoning based on stereotypes and we presented a model which combines two basic intuitions. On the one hand, stereotypical reasoning requires an agent to choose, given a piece of factual information, how this can be extended by relying on some background information about its class. This puts the agent in a new epistemic state (usually richer than the original one) which can be used to reason non-monotonically. Our central result shows that if we put appropriate constraints on the selection of stereotypes – in our case by using appropriate distance functions – we can generate a cumulative non monotonic consequence relation which is widely regarded in the field as capturing some fundamental aspects of commonsensical reasoning.

References

1. Freund, M.: Preferential Reasoning in the Perspective of Poole Default Logic. Artificial Intelligence 98(1-2), 209–235 (1998)
2. Gabbay, D., Woods, J. (eds.): Handbook of the History of Logic, vol. 8. North Holland, Amsterdam (2007)
3. Hansson, S.O.: A Textbook of Belief Dynamics. Kluwer, Dordrecht (1999)
4. Kraus, S., Lehmann, D., Magidor, M.: Nonmonotonic reasoning, preferential models and cumulative logic. Artificial Intelligence 44, 167–207 (1990)
5. Lakoff, G.: Cognitive Models and Prototype Theory. In: Laurence, S., Margolis, E. (eds.) Concepts, pp. 391–431. MIT Press, Cambridge (1999)
6. Laurence, S., Margolis, E.: Concepts and Cognitive Science. In: Laurence, S., Margolis, E. (eds.) Concepts, pp. 3–81. MIT Press, Cambridge (1999)
7. Lehmann, D.: Stereotypical Reasoning: Logical Properties. L. J. of the IGPL 6(1), 49–58 (1998)
8. Lehmann, D., Magidor, M., Schlechta, K.: Distance Semantics for Belief Revision. The Journal of Symbolic Logic 66(1), 295–317 (2001)
9. Makinson, D.: General Patterns in Nonmonotonic Reasoning. In: Gabbay, D., Hogger, C., Robinson, J. (eds.) Handbook of Logic in Artificial Intelligence and Logic Programming, vol. 3, pp. 35–110. Clarendon Press, Oxford (1994)
10. Makinson, D.: Bridges from Classical to Nonmonotonic Logic. King's College Publications, London (2005)
11. Putnam, H.: Representation and Reality. MIT Press, Cambridge (1988)

Realizing Default Logic over Description Logic Knowledge Bases*

Minh Dao-Tran, Thomas Eiter, and Thomas Krennwallner

Institut für Informationssysteme, Technische Universität Wien
Favoritenstraße 9-11, A-1040 Vienna, Austria
{dao,eiter,tkren}@kr.tuwien.ac.at

Abstract. We consider a realization of Reiter-style default logic on top of description logic knowledge bases (DL-KBs). To this end, we present elegant transformations from default theories to conjunctive query (cq-)programs that combine rules and ontologies, based on different methods to find extensions of default theories. The transformations, which are implemented in a front-end to a DL-reasoner, exploit additional constraints to prune the search space via relations between default conclusions and justifications. The front-end is a flexible tool for customizing the realization, allowing to develop alternative or refined default semantics. To our knowledge, no comparable implementation is available.

1 Introduction

Ontologies are very important for representing terminological knowledge. In particular, description logics (DLs) have proved to be versatile formalisms with far-reaching applications like expressing knowledge on the Semantic Web; the *Ontology Web Language* (OWL), which builds on DLs, has fostered this development.

However, well-known results from the literature show that DLs have limitations: they do not allow for expressing default knowledge due to their inherent monotonic semantics. One needs nontrivial extensions to the first-order semantics of description logics to express exceptional knowledge.

Example 1. Take, as an example, a bird ontology expressed in the DL-KB $L = \{ Flier \sqsubseteq \neg NonFlier, Penguin \sqsubseteq Bird, Penguin \sqsubseteq NonFlier, Bird(tweety) \}$. Intuitively, L distinguishes between flying and non-flying objects. We know that penguins, which are birds, do not fly. Nevertheless, we cannot simply add the axiom $Bird \sqsubseteq Flier$ to L to specify the common view that *"birds normally fly,"* as this update will make L inconsistent. From our bird ontology, we would like to conclude that Tweety flies; and if we learn that Tweety is a penguin, the opposite conclusion would be expected.

Hence, the simple ontology L from above cannot express exceptional knowledge. Default logic, a prominent formalism for expressing nonmonotonic knowledge in first-order logic, was introduced in the seminal work by Reiter [1]. To allow for nonmonotonicity in L, an extension of the semantics of terminological knowledge was given in [2], which is an early attempt to support default logic in the domain of description logics.

---

* This research has been supported by the Austrian Science Fund (FWF) project P20841 and P20840, and the EU research project IST-2009-231875 (ONTORULE).

C. Sossai and G. Chemello (Eds.): ECSQARU 2009, LNAI 5590, pp. 602–613, 2009.

Several other attempts to extend DLs with nonmonotonic features have been made, based on default logics [3,4], epistemic operators [5,6], circumscription [7,8], or argumentative approach [9]. They all showed that gaining both sufficient expressivity and good computational properties in a nonmonotonic DL is non-trivial.

However, reasoning engines for expressive nonmonotonic DLs are still not available (cf. Section 7). This forces users needing nonmonotonic features over DL-KBs to craft ad hoc implementations; systems for combining rules and ontologies (see [10]) might be helpful in that, but bear the danger that a non-standard semantics emerges not compliant with the user's intention. In fact, the less versatile a user is in KR formalisms, the higher is the likelihood that this will happen, even if just simple default rules of the form: *"If A is true and B can be consistently assumed, then conclude C,"* should be captured.

With the aim to offer a user-friendly reasoner over ontologies, we consider default reasoning on top of ontologies based on cq-programs [11], which integrate rules and ontologies. They slightly extend dl-programs [12] and allow for bidirectional communication between logic programs and DL-KBs by means of updates and powerful (unions of) conjunctive queries. Our main contributions are as follows.

- We consider a realization of Reiter-style default logic on top of DL-KBs, which amounts to a decidable fragment of Baader and Hollunder's terminological default logic [2], using cq-programs. A realization using dl-programs is discussed in [12], but is complex and more of theoretical interest than practical.
- We present two novel transformations of default theories into cq-programs, which are based on different principles and significantly improve over a similar transformation of default theories into dl-programs [12], both at the conceptual and the computational level. The former is apparent from the elegant formulation of the new transformations, while the latter is evidenced by experimental results. Furthermore, we present optimization methods by pruning rules, which are tailored specifically for the new translations.
- We describe a front-end as a new part of the dl-plugin [11] for the dlvhex engine implementing cq-programs. In this front-end, users can specify input comprising default rules in a text file and an ontology in an OWL file, run a command and get (descriptions of) the extensions of the default theory. Importantly, expert users in logic programming (LP) can exploit the front-end also at lower levels and customize the transformations, by adding further rules and constraints. In this way, alternative or refined semantics (e.g., preferences) may be realized, or the search space pruned more effectively.

Our front-end approach provides a simple and intuitive way to encode default knowledge on top of terminological KBs, relieving users from developing ad hoc implementations (which is rather difficult and error-prone). Furthermore, besides the benefit that special constructs of logic programs like weak constraints or aggregates can be utilized in customizations, the dlvhex implementation also offers the possibility to combine DL-KBs with other knowledge sources like, e.g., RDF KBs on a solid theoretical basis.

2   Preliminaries

Description Logics. We assume familiarity with DLs (cf. [13]), in particular any extension of $\mathcal{ALC}$ which can be recast to first-order logic w.r.t. CQ-answering (conceptually, we can apply Reiter-style default logic on top of any such DLs).[1] A *DL-KB* L is a finite set of axioms (TBox) and factual assertions (ABox) α in the respective DL, which are formed using disjoint alphabets of atomic concepts, roles, and individuals, respectively. By $L \models \alpha$ we denote logical consequence of an axiom resp. assertion α from L.

Default Logic. In this paper, we restrict Reiter's default logic [1] and consider only conjunctions of literals in default rules. We adjust default theories in a way such that the background theory, given by a DL-KB L, represents an ontology.

A *default* δ (*over* L) has the form $\frac{\alpha(X):\beta_1(Y_1),\ldots,\beta_m(Y_m)}{\gamma(Z)}$, where $\alpha(X) = \alpha_1(X_1) \wedge \cdots \wedge \alpha_k(X_k), \beta_i(Y_i) = \beta_{i,1}(Y_{i,1}) \wedge \cdots \wedge \beta_{i,\ell_i}(Y_{i,\ell_i}), \gamma(Z) = \gamma_1(Z_1) \wedge \cdots \wedge \gamma_n(Z_n)$, and X, Y_i, and Z are all variables occurring in the respective conjuncts. We allow c to be c^* or $\neg c^*$ for every $c \in \{\alpha_i, \beta_{i,j}, \gamma_i\}$, where each $\alpha_i^*, \beta_{i,j}^*, \gamma_i^*$ is either an atomic concept or role name in L. A *default theory over* L is a pair $\Delta = \langle L, D \rangle$, where L is a DL-KB and D is a finite set of defaults over L.

Example 2. Consider the DL-KB L in Ex. 1 and $D = \left\{ \frac{Bird(X):Flier(X)}{Flier(X)} \right\}$, then $\Delta = \langle L, D \rangle$ is a default theory over L.

Given that L is convertible into an equivalent first-order formula $\pi(L)$, we can view Δ as a *Reiter-default theory* $T = \langle W, D \rangle$ over a first-order language $\mathcal{L}$, where $W = \{ \pi(L) \}$, and apply concepts from [1] to Δ; we recall some of them in the sequel.

The semantics of $T = \langle W, D \rangle$ is given in terms of its *extensions*, which we recall next; intuitively, they are built by applying defaults in D as much as possible to augment the definite knowledge in W with plausible conclusions.

Suppose $T = \langle W, D \rangle$ is *closed* (i.e., defaults have no free variables). Then for any set of sentences $S \subseteq \mathcal{L}$, let $\Gamma_T(S)$ be the smallest set of sentences from $\mathcal{L}$ such that (i) $W \subseteq \Gamma_T(S)$; (ii) $\Gamma_T(S)$ is deductively closed, i.e., $Cn(\Gamma_T(S)) = \Gamma_T(S)$; and (iii) if $\frac{\alpha:\beta_1,\ldots,\beta_m}{\gamma} \in D$, $\alpha \in \Gamma_T(S)$, and $\neg\beta_1, \ldots, \neg\beta_m \notin S$ then $\gamma \in \Gamma_T(S)$. Here, $\vdash$ denotes classical derivability and $Cn(\mathcal{F}) = \{\phi \mid \mathcal{F} \vdash \phi$ and ϕ is closed$\}$, for every set $\mathcal{F}$ of closed formulas. For an arbitrary $T = \langle W, D \rangle$, Γ_T is applied to its closed version $cl(T)$, i.e., each default in D is replaced by all its grounded instances w.r.t. $\mathcal{L}$. Then, a set of sentences $E \subseteq \mathcal{L}$ is an extension of T, iff $E = \Gamma_T(E)$ [1].

cq-Programs. Informally, a cq-program consists of a DL-KB L and a disjunctive program P that may involve queries to L. Roughly, such a query may ask whether a specific conjunction of atoms or union of such conjunctions is entailed by L or not.

Syntax. A *conjunctive query* (CQ) $q(X)$ is an expression $\{X \mid Q_1(X_1), \ldots, Q_n(X_n)\}$, where each Q_i is a concept or role expression and each X_i is a list of variables and individuals of matching arity; $X \subseteq Vars(X_1, \ldots, X_n)$ are its *distinguished* (or *output*) variables, where $Vars(X_1, \ldots, X_n)$ is the set of variables appearing in $X_1, \ldots, X_n$. Intuitively, $q(X)$ is a conjunction $Q_1(X_1) \wedge \cdots \wedge Q_n(X_n)$ of concept and role expressions, which is true if all conjuncts are satisfied, and then projected on X.

A *union of conjunctive queries* (UCQ) $q(X)$ is a disjunction $\bigvee_{i=1}^m q_i(X)$ of CQs $q_i(X)$. Intuitively, $q(X)$ is satisfied, whenever some $q_i(X)$ is satisfied.

---

[1] This includes inverse roles ($\mathcal{I}$), qualified number restrictions ($\mathcal{Q}$), nominals ($\mathcal{O}$), and role hierarchy ($\mathcal{H}$), where a DL-KB L is convertible into an equivalent first-order formula $\pi(L)$ [13]; role transitivity ($\mathcal{S}$) may occur in L as well, but is disallowed in defaults.

A *cq-atom* α is of form $\mathrm{DL}[\lambda; q(\boldsymbol{X})](\boldsymbol{X})$, where $\lambda = S_1\ op_1\ p_1, \ldots, S_m\ op_m\ p_m$ ($m \geq 0$) is an (input) list of expressions $S_i\ op_i\ p_i$, each S_i is either a concept or a role name, $op_i \in \{\uplus, \cup\!\!\!\!-\,\}$, p_i is an (input) predicate symbol matching the arity of S_i, and $q(\boldsymbol{X})$ is a (U)CQ. Intuitively, $op_i = \uplus$ increases S_i by the extension of p_i, while $op_i = \cup\!\!\!\!-$ increases $\neg S_i$. If $m = 1$, α amounts to a *dl-atom* $\mathrm{DL}[\lambda; Q](\boldsymbol{t})$ as in [12] where $\boldsymbol{X} = Vars(\boldsymbol{t})$.

A *literal* l is an atom p or a negated atom $\neg p$. A *cq-rule* r is an expression of the form $a_1 \vee \cdots \vee a_k \leftarrow b_1, \ldots, b_m, \text{not } b_{m+1}, \ldots, \text{not } b_n$, where every a_i is a literal and every b_j is either a literal or a cq-atom. We define $H(r) = \{a_1, \ldots, a_k\}$ and $B(r) = B^+(r) \cup B^-(r)$, where $B^+(r) = \{b_1, \ldots, b_m\}$ and $B^-(r) = \{b_{m+1}, \ldots, b_n\}$. If $H(r) = \emptyset$ and $B(r) \neq \emptyset$, then r is a *constraint*.

A *cq-program* $KB = (L, P)$ consists of a DL-KB L and a finite set of cq-rules P.

Example 3. Let L be from Ex. 1 and $P = \{flies(tweety) \vee nflies(tweety); bird(X) \leftarrow \mathrm{DL}[Flier \uplus flies; Flier(X) \vee NonFlier(X)](X)\}$. Then, $KB = (L, P)$ is a cq-program. The body of the rule defining $bird$ is a cq-atom with the UCQ $q(X) = Flier(X) \vee NonFlier(X)$ and an input list $\lambda = Flier \uplus flies$. In this cq-atom we update the concept $Flier$ in L with the extension of *flies* before asking for the answers of $q(X)$.

Semantics. Given a cq-program $KB = (L, P)$, the *Herbrand base* of P, denoted HB_P, is the set of all ground literals with predicate symbols in P and constant symbols in a (predefined) set $\mathcal{C}$. An *interpretation* I relative to P is a consistent subset of HB_P. We say I is a *model* of $l \in HB_P$ under L, or I *satisfies* l under L, denoted $I \models_L l$, if $l \in I$.

For any CQ $q(\boldsymbol{X}) = \{\boldsymbol{X} \mid Q_1(\boldsymbol{X_1}), \ldots, Q_n(\boldsymbol{X_n})\}$, let $\phi_q(\boldsymbol{X}) = \exists \boldsymbol{Y} \bigwedge_{i=1}^n Q_i(\boldsymbol{X_i})$, where $\boldsymbol{Y}$ are the variables not in $\boldsymbol{X}$, and for any UCQ $q(\boldsymbol{X}) = \bigvee_{i=1}^m q_i(\boldsymbol{X})$, let $\phi_q(\boldsymbol{X}) = \bigvee_{i=1}^m \phi_{q_i}(\boldsymbol{X})$. Then, for any (U)CQ $q(\boldsymbol{X})$, the set of *answers of* $q(\boldsymbol{X})$ *on* L is the set of tuples $ans(q(\boldsymbol{X}), L) = \{\boldsymbol{c} \in \mathcal{C}^{|\boldsymbol{X}|} \mid L \models \phi_q(\boldsymbol{c})\}$.

An interpretation I *satisfies* a ground instance $a(\boldsymbol{c})$ of $a(\boldsymbol{X}) = \mathrm{DL}[\lambda; q(\boldsymbol{X})](\boldsymbol{X})$ (i.e., all variables in $q(\boldsymbol{X})$ are replaced by constant symbols from $\mathcal{C}$), denoted $I \models_L a(\boldsymbol{c})$, if $\boldsymbol{c} \in ans(q(\boldsymbol{X}), L \cup \lambda(I))$, where $\lambda(I) = \bigcup_{i=1}^m A_i(I)$ and (i) $A_i(I) = \{S_i(\boldsymbol{e}) \mid p_i(\boldsymbol{e}) \in I\}$, for $op_i = \uplus$, and (ii) $A_i(I) = \{\neg S_i(\boldsymbol{e}) \mid p_i(\boldsymbol{e}) \in I\}$, for $op_i = \cup\!\!\!\!-\,$.

I satisfies a ground cq-rule r, denoted $I \models_L r$, if $I \models_L H(r)$ whenever $I \models_L B(r)$, where $I \models_L H(r)$ if $I \models_L a$ for some $a \in H(r)$, and $I \models_L B(r)$ if $I \models_L a$ for all $a \in B^+(r)$ and $I \not\models_L a$ for all $a \in B^-(r)$.

I is a *model* of (or *satisfies*) a cq-program $KB = (L, P)$, denoted $I \models KB$, if $I \models_L r$ for all $r \in ground(P)$. The (strong) answer sets of KB, which amount to particular models of KB, are then defined like answer sets of an ordinary disjunctive logic program using the Gelfond-Lifschitz reduct P^I of P w.r.t. I, where cq-atoms are treated like ordinary atoms; I is then a *(strong) answer set* of KB, if I is a minimal model (w.r.t. set inclusion) of (L, P^I) (cf. also [12]).

Example 4 (cont'd). The strong answer sets of KB in Ex. 3 are $M_1 = \{flies(tweety), bird(tweety)\}$ and $M_2 = \{nflies(tweety)\}$. The answer set M_1 updates L in such a way that we can infer $q(tweety)$ from $L \cup \lambda(M_1)$, thus $bird(tweety) \in M_1$, whereas in M_2, $L \cup \lambda(M_2) \not\models q(tweety)$, and so $bird(tweety) \notin M_2$.

3   Transformations from Default Theories to cq-Programs

In the sequel, assume that we have a default theory $\Delta = \langle L, D \rangle$ over L.

We first revisit the transformation in [12], which we call Π. For each default of form $\frac{\alpha(\boldsymbol{X}):\beta_1(\boldsymbol{Y_1}),\ldots,\beta_m(\boldsymbol{Y_m})}{\gamma(\boldsymbol{Z})}$ (where the β_i and γ are just literals), it uses the following rules:

$$in\_\gamma(\boldsymbol{Z}) \leftarrow \text{not } out\_\gamma(\boldsymbol{Z}); \qquad out\_\gamma(\boldsymbol{Z}) \leftarrow \text{not } in\_\gamma(\boldsymbol{Z}) \tag{1}$$

$$g(\boldsymbol{Z}) \leftarrow \text{DL}[\lambda; \alpha_1](\boldsymbol{X_1}), \ldots, \text{DL}[\lambda; \alpha_k](\boldsymbol{X_k}), \tag{2}$$
$$\text{not } \text{DL}[\lambda'; \neg\beta_1](\boldsymbol{Y_1}), \ldots, \text{not } \text{DL}[\lambda'; \neg\beta_m](\boldsymbol{Y_m})$$

$$fail \leftarrow \text{DL}[\lambda'; \gamma](\boldsymbol{Z}), out\_\gamma(\boldsymbol{Z}), \text{not } fail \tag{3}$$

$$fail \leftarrow \text{not } \text{DL}[\lambda; \gamma](\boldsymbol{Z}), in\_\gamma(\boldsymbol{Z}), \text{not } fail \tag{4}$$

$$fail \leftarrow \text{DL}[\lambda; \gamma](\boldsymbol{Z}), out\_\gamma(\boldsymbol{Z}), \text{not } fail \tag{5}$$

where λ' contains for each default δ an update $\gamma^* \uplus in\_\gamma$ if $\gamma(\boldsymbol{Z})$ is positive, and an update $\gamma^* \cup in\_\gamma$ if $\gamma(\boldsymbol{Z})$ is negative; λ is similar with g in place of in_γ.

Π is based on a guess-and-check approach: the rules (1) guess whether the conclusion $\gamma(\boldsymbol{Z})$ belongs to an extension E or not. If yes, a ground atom with auxiliary predicate in_γ, which is used in the input list λ' to update L, is inferred. Intuitively, $L \cup \lambda'(I)$ represents E. Next, the rule (2) imitates the Γ_Δ operator to compute $\Gamma_\Delta(E)$. The outcome is stored in an auxiliary predicate g, which is used in a second input list λ to update L (independent from λ'); intuitively, $L \cup \lambda(I)$ represents $\Gamma_\Delta(E)$, Finally, the rule (3) checks whether the guess for E is compliant with L, and the rules (4) and (5) check whether E and $\Gamma_\Delta(E)$ coincide. If this is the case, then E is an extension of Δ.

A natural question is whether we can have a simpler transformation; in particular, with fewer and more homogeneous cq-atoms, in the sense that the update lists are similar; this would help to reduce communication between the rules and L, such that the evaluation of the transformation is more effective.

We give a positive answer to this question and present two novel transformations, called Ω and Υ, which are based on different ways of computing extensions, inspired by algorithms *select-default-and-check* and *select-justification-and-check* that were earlier mentioned in [14]. In fact, in both of them a *single* input list λ is sufficient for *all* cq-atoms. Furthermore, by the use of UCQs, we can easily handle also defaults with conjunctive justifications and conclusions.

The transformations Ω and Υ are compactly presented in Table 1, where we use the following notation. Given a default $\frac{\alpha(\boldsymbol{X}):\beta_1(\boldsymbol{Y_1}),\ldots,\beta_m(\boldsymbol{Y_m})}{\gamma(\boldsymbol{Z})}$, for $\Psi \in \{in, cons, \overline{cons}\}$ and $e(\boldsymbol{W}) \in \{\gamma(\boldsymbol{Z}), \gamma_i^*(\boldsymbol{Z_i}), \beta_i(\boldsymbol{Y_i}), \beta_{i,j}^*(\boldsymbol{Y_{i,j}})\}$, we use $\Psi(e(\boldsymbol{W}))$ to denote $\Psi_e(\boldsymbol{W})$ (where Ψ_e is a predicate name).

Transformation Ω. The main idea of Ω is to use only one update λ instead of both λ and λ' in Π, hence only one type of auxiliary predicates is needed, namely $in(\gamma(\boldsymbol{X}))$.

This transformation is quite intuitive and follows exactly the usual way of evaluating extensions in default theories: *"If the prerequisites can be derived, and the justifications can be consistently assumed, then the conclusion can be concluded."*

Intuitively, in the rule with head $in(\gamma(\boldsymbol{X}))$, we apply the Γ_Δ operator to find out whether the whole consequent $\gamma(\boldsymbol{Z})$ is in the extension E or not. If this is the case, then each $\gamma_i(\boldsymbol{Z_i})$ in $\gamma(\boldsymbol{Z})$ will also be concluded to be in E by rules in $\mathcal{R}$. In order to check

Table 1. Transformations Ω/Υ of default theory Δ to cq-program $KB_\Omega(\Delta)/KB_\Upsilon(\Delta)$

For $\Delta = \langle L, D \rangle$ and $X \in \{\Omega, \Upsilon\}$, let $KB_X(\Delta) = (L, P_X)$, where $P_X = \bigcup_{\delta \in D} X(\delta)$ and

---

$\Omega(\delta) = \mathcal{R} \cup \{ in(\gamma(\boldsymbol{Z})) \leftarrow \mathrm{DL}[\lambda; \alpha(\boldsymbol{X})](\boldsymbol{X}),$

$\qquad\qquad\qquad\qquad \mathrm{not}\ \mathrm{DL}[\lambda; d(\beta_1(\boldsymbol{Y_1}))](\boldsymbol{Y_1}), \ldots, \mathrm{not}\ \mathrm{DL}[\lambda; d(\beta_m(\boldsymbol{Y_m}))](\boldsymbol{Y_m}) \}$

$\Upsilon(\delta) = \mathcal{R} \cup \{ in(\gamma(\boldsymbol{Z})) \leftarrow \mathrm{DL}[\lambda; \alpha(\boldsymbol{X})](\boldsymbol{X}), cons(\beta_1(\boldsymbol{Y_1})), \ldots, cons(\beta_m(\boldsymbol{Y_m})) \} \cup$

$$\left\{\begin{array}{ll} \qquad fail \leftarrow cons(\beta_i(\boldsymbol{Y_i})), \mathrm{DL}[\lambda; d(\beta_i(\boldsymbol{Y_i}))](\boldsymbol{Y_i}), \mathrm{not}\ fail; & \\ \qquad fail \leftarrow \overline{cons}(\beta_i(\boldsymbol{Y_i})), \mathrm{not}\ \mathrm{DL}[\lambda; d(\beta_i(\boldsymbol{Y_i}))](\boldsymbol{Y_i}), \mathrm{not}\ fail; & \\ cons(\beta_i(\boldsymbol{Y_i})) \leftarrow \mathrm{not}\ \overline{cons}(\beta_i(\boldsymbol{Y_i})); & \\ \overline{cons}(\beta_i(\boldsymbol{Y_i})) \leftarrow \mathrm{not}\ cons(\beta_i(\boldsymbol{Y_i})) & | 1 \le i \le m \end{array}\right\}$$

---

where $\lambda = (\gamma_i^* \uplus in_{\gamma_i}, \gamma_i^* \uplus in_{\neg\gamma_i} \mid \delta \in D)$, $d(\beta_i(\boldsymbol{Y_i})) = \neg\beta_{i,1}(\boldsymbol{Y_{i,1}}) \vee \cdots \vee \neg\beta_{i,\ell_i}(\boldsymbol{Y_{i,\ell_i}})$,

and $\mathcal{R} = \{ in(\gamma_i(\boldsymbol{Z_i})) \leftarrow in(\gamma(\boldsymbol{Z})) \mid 1 \le i \le n \}$.

---

the satisfaction of the prerequisite, we use a CQ, while consistency of a justification is checked by a UCQ. In case the prerequisite or a justification is just a literal, the query amounts to instance checking (which is more efficient).

Example 5. Consider default theory Δ in Ex. 2. Since the prerequisite, justification and conclusion of the default in D are just literals, $\mathcal{R}$ can be simplified to $\emptyset$ and the cq-atoms to instance checks. Therefore, P_Ω consists only of the rule

$$in_{Flier}(X) \leftarrow \mathrm{DL}[\lambda; Bird](X), \mathrm{not}\ \mathrm{DL}[\lambda; \neg Flier](X) ,$$

where $\lambda = Flier \uplus in_{Flier}, Flier \uplus in_{\neg Flier}$. The single answer set of $KB_\Omega(L, D)$ is $I_\Omega = \{ in_{Flier}(tweety) \}$ which corresponds to the single extension.

Transformation Υ. In this transformation, we make use of the *Select-justifications-and-check* algorithm. The definition of $\Upsilon(\delta)$ in Table 1 is explained as follows. The first rule emulates the Γ_Δ operator to find the set of consequences under a consistency assumption for the default justifications $\beta_i(\boldsymbol{Y_i})$ with the extension E; like above, with $\gamma(\boldsymbol{Z})$ also each $\gamma_i(\boldsymbol{Z_i})$ is concluded by the rules in $\mathcal{R}$.

The assumptions for all justifications $\beta_i(\boldsymbol{Y_i})$ are guessed with the last two rules, and they are checked with two constraints: the first prevents cases in which we guess that $\beta_i(\boldsymbol{Y_i})$ is consistent with E but we can in fact derive $\neg\beta_i(\boldsymbol{Y_i})$. Similarly, the second constraint eliminates all models in which $\beta_i(\boldsymbol{Y_i})$ is guessed to be inconsistent with E but we cannot derive its negation.

We can see that transformation Υ involves less communication with the DL-KB than Ω; instead, it has explicit guessing on the logic program side. If the number of justifications is small, we may expect better performance.

Example 6. For the default theory Δ in Ex. 2, P_Υ consists of the following rules:

$$cons_{Flier}(X) \leftarrow \mathrm{not}\ \overline{cons}_{Flier}(X); \qquad \overline{cons}_{Flier}(X) \leftarrow \mathrm{not}\ cons_{Flier}(X)$$

$$in_{Flier}(X) \leftarrow \mathrm{DL}[\lambda; Bird](X), cons_{Flier}(X)$$

$$fail \leftarrow cons_{Flier}(X), \mathrm{DL}[\lambda; \neg Flier](X), \mathrm{not}\ fail$$

$$fail \leftarrow \overline{cons}_{Flier}(X), \mathrm{not}\ \mathrm{DL}[\lambda; \neg Flier](X), \mathrm{not}\ fail$$

where $\lambda = Flier \uplus in_{Flier}, Flier \cup in_{\neg Flier}$. The single answer set of $KB_\Upsilon(L, D)$ is $I_\Upsilon = \{in_{Flier}(tweety), cons_{Flier}(tweety)\}$ which corresponds to the single extension.

The following theorem shows the correctness of our transformations.

Theorem 1. *Let $\Delta = \langle L, D \rangle$ be a default theory over L, and $X \in \{\Omega, \Upsilon\}$. Then:*

(i) For each extension E of Δ, there exists a (unique) strong answer set M of KB_X (Δ), such that $E = Cn(L \cup \lambda(M))$.

(ii) For each strong answer set M of $KB_X(\Delta)$, the set $E = Cn(L \cup \lambda(M))$ is an extension of Δ.

Note that in general, answering UCQs over expressive DL-KBs may be undecidable; in our case variables range effectively only over known individuals in the KB (i.e., constants in $\mathcal{C}$). We also mention that the further transformation of $KB_X(\Delta)$ into HEX-programs [15] for execution in dlvhex requires rules to be domain-expansion safe; this is easily accomplished by introducing a domain predicate dom, adding to the body of each rule for each variable Y the atom $dom(Y)$, and appending a fact $dom(a)$ to P_X for each individual a in the KB (see [15] for details).

4   Optimization

This section introduces pruning rules to reduce the search space in model computation. In what follows, we consider defaults δ_1 and δ_2, where

$$\delta_i = \frac{\alpha_{i,1}(\boldsymbol{X_{i,1}}) \wedge \cdots \wedge \alpha_{i,k_i}(\boldsymbol{X_{i,k_i}}) : \beta_{i,1}(\boldsymbol{Y_{i,1}}), \ldots, \beta_{i,m_i}(\boldsymbol{Y_{i,m_i}})}{\gamma_{i,1}(\boldsymbol{Z_{i,1}}) \wedge \cdots \wedge \gamma_{i,n_i}(\boldsymbol{Z_{i,n_i}})} .$$

Based on the interaction of δ_1 and δ_2, we can add the following rules. Let $\gamma_i(\boldsymbol{Z_i})$ be short for $\gamma_{i,1}(\boldsymbol{Z_{i,1}}) \wedge \cdots \wedge \gamma_{i,n_i}(\boldsymbol{Z_{i,n_i}})$, where $\boldsymbol{Z_i} = \bigcup_{1 \leq j_i \leq n_i} \boldsymbol{Z_{i,j_i}}$, for $i = 1, 2$.

Forcing other defaults to be out. The well-known Nixon Diamond example motivates a shortcut in dealing with defaults whose conclusions are opposite. In this example, the conclusion of one default blocks the other. To prune such cases, we can add

$$fail \leftarrow in(\gamma_1(\boldsymbol{Z_1})), in(\gamma_2(\boldsymbol{Z_2})), not\ fail \qquad (6)$$

to P_X, where $X \in \{\Omega, \Upsilon\}$, whenever there exist $1 \leq j_1 \leq n_1, 1 \leq j_2 \leq n_2$ s.t. $\gamma_{1,j_1}(\boldsymbol{Z_{1,j_1}})$ and $\neg \gamma_{2,j_2}(\boldsymbol{Z_{2,j_2}})$ are unifiable.

Furthermore, also the relations between conclusions and justifications can be exploited for pruning purpose. If there exist $j \in \{1, \ldots, n_1\}$ and $j' \in \{1, \ldots, m_2\}$ such that $\gamma_{1,j}(\boldsymbol{Z_{1,j}})$ is unifiable with a disjunct in $\neg \beta_{2,j'}(\boldsymbol{Y_{2,j'}})$, then the conclusion $\gamma_1(\boldsymbol{Z_1})$ of δ_1 will block the application of δ_2 and the constraint (6) can also be added to P_X.

Forcing other defaults to be in. If $\gamma_1(\boldsymbol{Z_1})$ is part of $\gamma_2(\boldsymbol{Z_2})$, then adding an instance of $\gamma_2(\boldsymbol{Z_2})$ to an extension E requires also to add the respective instance of $\gamma_1(\boldsymbol{Z_1})$ to E. Thus, if for every $j_1 \in \{1, \ldots, n_1\}$, $\gamma_{1,j_1}(\boldsymbol{Z_{1,j_1}})$ is unifiable with $\gamma_{2,j_2}(\boldsymbol{Z_{i,j_2}})$ for some $j_2 \in \{1, \ldots, n_2\}$, then we add the following rule to P_X, where $X \in \{\Omega, \Upsilon\}$:

$$in(\gamma_1(\boldsymbol{Z_1})) \leftarrow in(\gamma_2(\boldsymbol{Z_2})) .$$

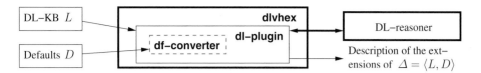

Fig. 1. Architecture of the front-end

Defaults whose conclusions are already in the background theory. Each extension contains all consequences of the background theory W of the default theory ($Cn(W) \subseteq E$). Hence, it is worth testing whether for a default $\frac{\alpha(\boldsymbol{X}):\beta_1(\boldsymbol{Y_1}),...,\beta_m(\boldsymbol{Y_m})}{\gamma(\boldsymbol{Z})}$ a conjunct $\gamma_i(\boldsymbol{X_i}) \in \gamma(\boldsymbol{X})$ can be concluded from the DL-KB before the guessing phase or the application of the Γ_Δ operator. To this end, we can add to $P_X(X \in \{\Omega, \Upsilon\})$ the rule

$$in(\gamma_i(\boldsymbol{X_i})) \leftarrow \mathrm{DL}[\gamma_i](\boldsymbol{X_i}) \ .$$

5   Implementation

We have implemented the transformations from Section 3 in a framework that provides the user with a front-end to cq-programs for transparent default reasoning over ontologies. An architectural overview of this front-end is shown in Fig. 1.

The implementation makes use of the dlvhex environment,[2] a framework for solving HEX-programs. It has a plugin facility that allows to define external atoms, and dl-plugin is one of the plugins deployed in this environment. The dl-plugin provides a mechanism for converting cq-programs to HEX-programs. It receives a cq-program or a HEX-program together with an OWL ontology, communicates with a DL-reasoner to evaluate queries in the program, and dlvhex processes the query answers and generates models of the program. Based on this framework, we implemented a converter for default rules on top of description logics, df-converter, as a pre-processor in the dl-plugin which takes a set of defaults and an OWL ontology as input, converts this input into cq-rules according to a transformation, and transfers the result to the dl-plugin; dlvhex then does the rest. Hence, all the complications including cq-programs, HEX-programs, and the transformations are transparent to the users. They just need to specify defaults in a simple format and get (descriptions of) the extensions of the input default theory, which were modified from the models of the transformed HEX-programs.

The grammar for the syntax of input defaults is as follows:

```
lit ::= atom | −atom
conjunction ::= lit ( & lit )*
default ::= [ conjunction ; conjunction ( , conjunction )* ] / [ conjunction ]
```
                 prerequisite   justifications   conclusion

Here, '−' is classical negation, '&' is conjunction, and 'atom' is an atomic formula with a concept or role.

As for the output, the interesting information about an extension E is the default conclusions that are in E. To this end, we filter all ground literals from the answer sets

---

[2] http://www.kr.tuwien.ac.at/research/systems/dlvhex/

belonging to the default conclusions derived in the program for the user (reasoning tasks can be easily realized on top by customization).

The following example illustrates this elegant interface.

Example 7. For the Bird example, the input includes an OWL file for the ontology L in Ex. 1 and a text file for D in Ex. 2, whose content simply is

```
[ Bird(X); Flier(X) ] / [ Flier(X) ]
```

We can now invoke dlvhex to ask for the extensions. We get only one extension in this particular case, and the only fact returned to users is `Flier(tweety)`.

However, users are not confined to simple defaults. Expert users in LP can provide more sophisticated pruning rules, or rules that select specific extensions (e.g., under preference) in terms of cq- or HEX-rules. They are directly sent to the dl-plugin and added to cq-rules supplied by the df-converter as an input for dlvhex. Our front-end therefore is flexible and can meet requirements of different classes of users.

Typing predicates. The front end supports explicit *typing predicates* as a means to control the instantiation of defaults and to limit the search space in advance. For example, in a predicate $hasScholarship(P, S)$ the first argument should be a *Student* while the second should be a *Scholarship*.

Users can attach to each default δ a *type guard* $\theta(W)$ whose variables W appear in δ, and list all facts for the predicate θ, or even write a program which computes them. Semantically, $\theta(W)$ is added to the precondition of δ and all facts $\theta(c)$ are added to the background knowledge of the default theory. If a rule r in a transformation of δ satisfies $W \subseteq Vars(r)$, then each atom $dom(X)$ in it with $X \in W$ is replaced by $\theta(W)$.

Example 8 (cont'd). Suppose we have many instances of the concept $Bird$ in L, but just want to know whether `tweety` and `joe` fly or not. We can modify the input to

```
[ Bird(X); Flier(X) ] / [ Flier(X) ]<mb(X)>
```

and add facts `mb(tweety), mb(joe)` to specify these two birds.

We remark that in general, adding typing predicates makes the transformations incomplete w.r.t. to the original theory. However, for so called *semi-monotonic* default theories (where the extensions increase with an increasing set of defaults; e.g., the important *normal* default theories [1] have this property) credulous conclusions are sound, as well as skeptical conclusions if a unique extension is guaranteed.

6   Experimental Results

We have tested the transformations Π, Ω, and Υ using the prototype implementation of the front-end as a new component in the dl-plugin for dlvhex, which uses RacerPro 1.9.2Beta [16] as DL-reasoner, to explore different aspects which can influence the overall system performance, namely (i) the number of cq-atoms in each transformation, (ii) the number of individuals (size of the ABox), (iii) size of the TBox, and (iv) the query caching to the DL-reasoner. The tests were done on a P4 1.8GHz PC with 1GB RAM under Linux. We report here only the outcome of the Bird benchmark, which consists of a set of ontologies in spirit of Ex. 1 with increasing size of $Bird$ instances.

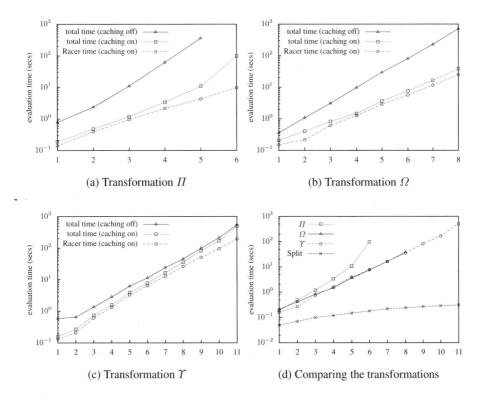

(a) Transformation Π

(b) Transformation Ω

(c) Transformation Υ

(d) Comparing the transformations

Fig. 2. Bird example – running time (x-axis: number of individuals)

Fig. 2 shows experimental results of this test, including running time of each transformation Π, Ω, Υ and the comparison between them when query caching to RacerPro is applied. Missing entries mean time exhaustion during evaluation. For each transformation, we are interested in the total running time and time that RacerPro consumes when caching is on. When caching is off, the difference between the total running time and time used by RacerPro is insignificant, hence only the total running time is displayed.

The experimental results reveal that Ω and Υ are much faster than Π, since Ω has fewer guessing rules and Υ has fewer cq-atoms, but Υ has a trade-off between consistency guessing rules and cq-atoms in rules that compute extensions. Hence, the performance of Ω and Υ depends very much on a particular set of defaults.

Regarding (i), the number of cq-atoms is important as they make up the search space which increases exponentially. Regarding (ii), also the number of individuals is clearly important (as it directly influences (i)), while for (iii), the size of the TBox was of minor concern. When increasing the TBoxes (by enlarging the taxonomies systematically), the performance was not much affected. This is not much surprising, as DL engines like RacerPro are geared towards efficient reasoning with (large) TBoxes (of course, one could have used "hard" TBoxes, but this seems less relevant from a practical perspective). Regarding (iv), it appeared that query caching plays an important role, as the system spends most of the time querying RacerPro for all ground cq-atoms.

To undercut the impact of (ii) on (i), a new version of dlvhex has been developed in which independence information about different cq-atoms, which is based on the individuals occurring in them, can be exploited to factorize a HEX-program into components that can be evaluated in parallel (see [17]). In particular, in the Bird benchmark, for each individual a separate component will be created; sequentially processed, this yields linear scalability with respect to (ii) (see Fig. 2d). Currently, the new dlvhex version is beneficial only to transformation Ω, but further improvements are expected for Π and Υ.

7   Related Work and Conclusions

As we already mentioned, this work is not the first considering default reasoning with description logics. Earlier ones [2,3,4] posed varying restrictions on the terminological representation to keep decidability, and provided no implementations or only for limited forms of defaults (on concepts). As our approach is theoretically based on a strict separation between rules and ontologies, it guarantees decidability as long as the DL-KB is convertible into a decidable first-order theory w.r.t. CQ-answering. Moreover, on the practical side, we provide a concrete implementation for Reiter-style default logic via a front-end hosted by a framework to combine rules and ontologies.

Hybrid Default Inheritance Theory (HDIT) [3] allows to specify defaults of form $\frac{A(X):C(X)}{C(X)}$ and $\frac{A(X) \wedge R(X,Y):C(Y)}{C(Y)}$. To retain decidability, concepts in HDIT must be conjunctions of (negated) atomic concepts; [2] allows only DLs from $\mathcal{ALC}$ to $\mathcal{ALCF}$. An implementation of the $\mathcal{DTL}$ language is reported in [4], but roles cannot be defined.

The DL $\mathcal{ALCK}$ [6] adds an epistemic operator $\mathbf{K}$ to $\mathcal{ALC}$ which allows to specify closed-world reasoning in queries. The later $\mathcal{DLK}_{\mathcal{NF}}$ [5] can be regarded as an extension of $\mathcal{ALCK}$ with epistemic operator $\mathbf{A}$ expressing "default assumption." Defaults can only be specified over concept expressions and are translated to TBox axioms using $\mathbf{K}$ for prerequisites and conclusions, and $\neg\mathbf{A}$ for negated justifications.

In [7], circumscriptive (minimal) models have been used to define the *Extended Closed World Assumption (ECWA) over hybrid systems* with a proof theory based on a translation to propositional theories. However, hybrid systems are actually a fragment of $\mathcal{ALE}$ and not very expressive. A recent paper [8] proposed extensions of expressive DLs $\mathcal{ALCIO}$ and $\mathcal{ALCQO}$ to form *circumscribed knowledge bases (cKBs)*. They can avoid the restriction of nonmonotonic reasoning to named individuals in the domain, but still allow only that concept names can be circumscribed.

Recently, [9] uses an argumentative approach for reasoning with inconsistent ontologies by translating DL ontologies into defeasible logic programs. However, only inconsistent concept definitions were considered. Concerning semantics, this approach is different from ours since it uses the notion of defeasible derivation which corresponds to SLD-derivation in logic programming.

Concerning further work, the experimental comparison of the transformations Π, Ω and Υ revealed several tasks that can help to improve performance. One is to investigate more refined pruning rules that depend on the structure of the default theory. Another issue is to look into particular kinds of default theories, such as normal or semi-normal default theories, for which more efficient transformations may be found. At the bottom level, we note that caching for cq-atoms (which is currently only available for plain dl-atoms) would give additional benefit. Furthermore, dlvhex is currently using RacerPro as its (only) DL-reasoner; it would be interesting to have support for other DL-reasoners such as KAON2 or Pellet, and compare the results. Finally, improvement of dlvhex evaluation at the general level (e.g., by refined dependency handling) would be beneficial.

References

1. Reiter, R.: A logic for default reasoning. Artif. Intell. 13(1-2), 81–132 (1980)
2. Baader, F., Hollunder, B.: Embedding Defaults into Terminological Knowledge Representation Formalisms. J. Autom. Reasoning 14(1), 149–180 (1995)
3. Straccia, U.: Default inheritance reasoning in hybrid KL-ONE-style logics. In: IJCAI 1993, pp. 676–681. Morgan Kaufmann, San Francisco (1993)
4. Padgham, L., Zhang, T.: A terminological logic with defaults: A definition and an application. In: IJCAI 1993, pp. 662–668. Morgan Kaufmann, San Francisco (1993)
5. Donini, F.M., Nardi, D., Rosati, R.: Description logics of minimal knowledge and negation as failure. ACM Trans. Comput. Logic 3(2), 177–225 (2002)
6. Donini, F.M., Lenzerini, M., Nardi, D., Nutt, W., Schaerf, A.: An epistemic operator for description logics. Artif. Intell. 100(1-2), 225–274 (1998)
7. Cadoli, M., Donini, F.M., Schaerf, M.: Closed world reasoning in hybrid systems. In: Methodologies for Intelligent Systems (ISMIS 1990), pp. 474–481. North-Holland, Amsterdam (1990)
8. Bonatti, P.A., Lutz, C., Wolter, F.: Expressive non-monotonic description logics based on circumscription. In: KR 2006, pp. 400–410. AAAI Press, Menlo Park (2006)
9. Gómez, S., Chesñevar, C., Simari, G.: An argumentative approach to reasoning with inconsistent ontologies. In: KROW 2008. CRPIT, vol. 90, pp. 11–20. ACS (2008)
10. Eiter, T., Ianni, G., Krennwallner, T., Polleres, A.: Rules and Ontologies for the Semantic Web. In: Baroglio, C., Bonatti, P.A., Małuszyński, J., Marchiori, M., Polleres, A., Schaffert, S. (eds.) Reasoning Web. LNCS, vol. 5224, pp. 1–53. Springer, Heidelberg (2008)
11. Eiter, T., Ianni, G., Krennwallner, T., Schindlauer, R.: Exploiting conjunctive queries in description logic programs. Ann. Math. Artif. Intell (2009); Published online January 27 (2009)
12. Eiter, T., Ianni, G., Lukasiewicz, T., Schindlauer, R., Tompits, H.: Combining answer set programming with description logics for the semantic web. Artif. Intell. 172(12-13) (2008)
13. Baader, F., Calvanese, D., McGuinness, D.L., Nardi, D., Patel-Schneider, P.F.: The Description Logic Handbook: Theory, Implementation, and Applications. Cambridge (2003)
14. Cholewinski, P., Truszczynski, M.: Minimal number of permutations sufficient to compute all extensions a finite default theory (unpublished note)
15. Eiter, T., Ianni, G., Schindlauer, R., Tompits, H.: Effective integration of declarative rules with external evaluations for semantic web reasoning. In: Sure, Y., Domingue, J. (eds.) ESWC 2006. LNCS, vol. 4011, pp. 273–287. Springer, Heidelberg (2006)
16. Haarslev, V., Möller, R.: Racer system description. In: Goré, R.P., Leitsch, A., Nipkow, T. (eds.) IJCAR 2001. LNCS, vol. 2083, pp. 701–706. Springer, Heidelberg (2001)
17. Eiter, T., Fink, M., Krennwallner, T.: Decomposition of Declarative Knowledge Bases with External Functions. In: IJCAI 2009 (July 2009) (to appear)
18. Horrocks, I., Patel-Schneider, P.F.: Reducing OWL entailment to description logic satisfiability. J. Web Semant. 1(4), 345–357 (2004)
19. Horrocks, I., Patel-Schneider, P.F., van Harmelen, F.: From $\mathcal{SHIQ}$ and RDF to OWL: The making of a Web ontology language. J. Web Semant. 1(1), 7–26 (2003)
20. Horrocks, I., Sattler, U., Tobies, S.: Practical reasoning for expressive description logics. In: Ganzinger, H., McAllester, D., Voronkov, A. (eds.) LPAR 1999. LNCS, vol. 1705, pp. 161–180. Springer, Heidelberg (1999)

Dealing Automatically with Exceptions by Introducing Specificity in ASP

Laurent Garcia, Stéphane Ngoma, and Pascal Nicolas

LERIA, UFR Sciences, University of Angers
2 bd Lavoisier, F-49045 ANGERS cedex 01
{garcia,ngoma,pn}@info.univ-angers.fr

Abstract. Answer Set Programming (ASP), via normal logic programs, is known as a suitable framework for default reasoning since it offers both a valid formal model and operational systems. However, in front of a real world knowledge representation problem, it is not easy to represent information in this framework. That is why the present article proposed to deal with this issue by generating in an automatic way the suitable normal logic program from a compact representation of the information. This is done by using a method, based on specificity, that has been developed for default logic and which is adapted here to ASP both in theoretical and practical points of view.

1 Introduction

Delgrande and Schaub [5] presented a general and automatic approach to introduce specificity in non-monotonic theories. This approach was illustrated, among others, for default logic [20]; however, they did not envisage any operational system. In this present article, we pursue the same general goal but we also finish the work by implementing the process of generation of default rules with exceptions. For that, we need a non monotonic framework where both the formal model is correct and for which there exist performing operational systems. To satisfy theses constraints, we chose to use *Answer Set Programming* (ASP).

In ASP, information is coded in the form of rules and an inference process allows to make reasoning; some rules can be sometimes blocked (which captures the non-monotonic property) and are able to express exceptions. ASP is a tested paradigm and has different roots in knowledge representation, particularly non-monotonic reasoning (with the default logic semantics) and logic programming. For an interesting overview, the reader should refer to the articles in [8] celebrating the 20 Years of Stable Models Semantics [10] that is considered as the first work on ASP. From a practical point of view, several ASP solvers are available and the most powerful among them, like Clasp [9], Dlv [15] and Smodels [21] are able to deal with highly combinatoric problems.

In a formal way, we use the normal logic programs which are suitable for representing default rules with exceptions. Unfortunately, in a real world application it may be difficult to write these complex rules directly and correctly. It seems more appropriate to consider that knowledge is given in a compact way

C. Sossai and G. Chemello (Eds.): ECSQARU 2009, LNAI 5590, pp. 614–625, 2009.

(that is without exceptions) and to generate automatically the suitable normal logic program (using specificity), then to use the machinery of ASP to reason.

By the sequel, an initial compact common-sense default rule ("A's are B's") which is not linked to a particular formalism will be denoted by a $\rightarrow (A \rightarrow B)$. On its side a final rule linked to the formalism of normal logic programs will be denoted by a $\leftarrow (B \leftarrow A.)$. For instance, starting from a set of compact and easily writable default rules like $B \rightarrow F$, $P \rightarrow \neg F$,... representing the naive informations "generally, birds fly", "generally, penguins do not fly", ... our goal is to automatically build the normal logic program $\{F \leftarrow B, \; not \; P., \; \neg F \leftarrow P., \; ...\}$ that correctly encodes the specificity hidden in the initial knowledge.

The article is given as follows. Section 2 is dedicated to a formal presentation of ASP, section 3 presents the approach proposed by Delgrande and Schaub on which ours is based. Then, in section 4, the theoretical aspect of our work is developed. Some definitions were proposed by Delgrande and Schaub in a way that is independent of the formalism. However, to be completed, the method must be linked to a particular representation. That's why we redefine all the definitions in the framework of ASP to be able to associate the theoretical work with an implementation. We end in section 5 by giving some perspectives to continue this work.

2   Answer Set Programming and Specificity

2.1   Background on ASP

A *definite logic program* is a finite set of rules like

$$b \leftarrow a_1, \; ..., \; a_n. \; (n \geq 0)$$

For such a rule r, $head(r) = b$ is an atom called the *head* and $body(r) = \{a_1, ..., a_n\}$ is an atom set called the *body*. The intuitive meaning of rule $b \leftarrow a.$ is "if we can prove a, then we can conclude b". Given a rule r and an atom set A, we say that r is *applicable* in A if $body(r) \subseteq A$. An atom set A is *closed* under a program P if and only if for all rule r in P, if $body(r) \subseteq A$ then $head(r) \in A$. We call $Cn(P)$, or Herbrand model, the minimal atom set closed under P. For a program P and an atom set A, the operator T_P defined by

$$T_P(A) = \{head(r) \mid r \in P, body(r) \subseteq A\}$$

computes the set of atoms deducible from A by means of P. It allows to define the sequence $T_P^0 = T_P(\emptyset)$, $T_P^{k+1} = T_P(T_P^k), \forall k \geq 0$. $Cn(P)$ is the least fix-point of T_P and $Cn(P) = \bigcup_{k \geq 0} T_P^k$ contains all the consequences of the program P.

A *normal logic program* is a finite set of rules like

$$c \leftarrow a_1, \; ..., \; a_n, \; not \; b_1, \; ..., \; not \; b_m. \; (n \geq 0, m \geq 0)$$

As previously, $c, a_1, ..., a_n, b_1, ..., b_m$ are atoms. For such a rule r, we denote $body^+(r) = \{a_1, ..., a_n\}$ its positive body, $body^-(r) = \{b_1, ..., b_m\}$ its negative body, $body(r) = body^+(r) \cup body^-(r)$ and $r^+ = c \leftarrow a_1, ..., a_n$. The intuitive meaning of a rule with default negation like $c \leftarrow a, \; not \; b.$ is "if we can prove a, and nothing proves b, then we can conclude c". Such a non monotonic rule r is *applicable* in an atom set A if $body^+(r) \subseteq A$ and $body^-(r) \cap A = \emptyset$.

The Gelfond-Lifschitz reduct of a program P by an atom set A is the program:
$$P^A = \{head(r) \leftarrow body^+(r). \mid body^-(r) \cap A = \emptyset\}$$
Since it has no default negation, such a program is definite and then it has a minimal Herbrand model $Cn(P)$. By definition, an *answer set* (originally called a *stable model* [10]) of P is an atom set S such that $S = Cn(P^S)$. For instance the program $\{a \leftarrow not\ b., \ b \leftarrow not\ a.\}$ has two answer sets $\{a\}$ and $\{b\}$.

For purposes of knowledge representation, one may have to use conjointly strong negation (like $\neg a$) and default negation (like $not\ a$) inside a same program. This is possible in ASP by means of an *extended logic program* [11] in which rules are built with literals instead of atoms only. But, if we are not interested in inconsistent models, and that it is explicitly the case in this work, the semantics associated to an extended logic program is reducible to stable model semantics for a normal logic program by taking into account the following conventions :

- every literal $\neg x$ is encoded by the atom nx,
- for every atom x, the rule $\perp \leftarrow x, \ nx.$ is added,
- a stable model should not contain the symbol $\perp$.

Rules with a head equal to $\perp$, sometimes noted without head, are called *constraints*. Given a program P, we denote by P_K the set of all constraints of P. The use of a constraint like $\perp \leftarrow x, \ nx.$ forbids to x and nx (ie $\neg x$) to appear in the same stable model. By this way, only consistent stable models are kept.

2.2   Basic Principles of Computing Specificity in ASP

The goal of our work is to benefit from the power of ASP to reason while using a compact representation of knowledge easy to express.

Example 1. The so-called example of birds (where B stands for "birds", W "to have wings", P "penguins" and F "to fly") can be encoded by the following program: $\{F \leftarrow B, \ not\ P., \ W \leftarrow B., \ B \leftarrow P., \ nF \leftarrow P., \ \perp \leftarrow F, \ nF.\}$. When we consider a penguin (represented by $P \leftarrow .$), there is one stable model: $\{W, P, B, nF\}$, which corresponds to the intuition.

It is currently admitted that normal logic programs are adapted for representing default rules with exceptions. However, they are not easy to write since it is necessary to express every exception or at least to describe the abnormalities. But it is more convenient to write a simple representation of information, without expressing exceptions in an explicit way. The idea is then first to express rules in a compact form by writing a normal logic program without exceptions (that is without a negative body). Unfortunately, this representation leads to inconsistency.

That's why our work consists in solving this by determining in an automatic way the classes and sub-classes of information. The notion of *specificity* [19] is then crucial. This aspect has yet been done formally for default logic [5] but no implementation was given and it remains to apply it to ASP to define an automatic process dealing correctly with exceptions in this framework (both in theoretical and practical points of view).

Hence, from the following example $\{B \rightarrow F, B \rightarrow W, P \rightarrow B, P \rightarrow \neg F\}$ which is directly encoded in $\{F \leftarrow B., W \leftarrow B., B \leftarrow P., nF \leftarrow P., \bot \leftarrow F, nF.\}$, we want to obtain the suitable normal logic program: $\{F \leftarrow B, not\ nF, not\ P., W \leftarrow B., B \leftarrow P., nF \leftarrow P, not\ F., \bot \leftarrow F, nF.\}$. The machinery of normal logic programs will give that birds fly and have wings while penguins do not fly and have wings which is conform to the intuition. Let us note that it is easy to generate the constraint in an automatic way. It is done indeed by the implementation that we have developed.

3   Default Logic and Specificity

In this section, we present the approach proposed by Delgrande and Schaub [5], applied particularly in the framework of Default Logic [20]. It consists in two general steps. First, from a default theory, the conflicting rules which have not the same specificity have to be localized; this is done by using a part of the machinery of system Z [19]. For instance, in our example, it is clear that the rules $B \rightarrow F$ and $P \rightarrow \neg F$ are conflicting (a contradiction is deduced, in fact F and $\neg F$, due to $P \rightarrow B$) and the second one is more specific than the first one (P is a sub-class of B). Secondly, some rules must be modified so that if two contradicting rules can be applied simultaneously then only the most specific one is applied.

In System Z, a set of rules R is partitioned (stratified) in subsets $R_0, \ldots, R_n$ where the rules of a lower stratum are less specific than the ones of an upper stratum. The resulting partitioning is called a *Z-ordering*; this ordering gives specificity information. Rather to compute the entire Z-ordering, Delgrande and Schaub first determine minimal conflicting sets, that are separately stratified; thus, information is classified according to its specificity, in relation with conflict(s) in which it occurs. In our example, $C = \{B \rightarrow F, P \rightarrow B, P \rightarrow \neg F\}$ is a conflict (in presence of P). Delgrande and Schaub showed that Z-ordering of a such set C is a binary partitioning (C_0, C_1) of the rules; the rules of C_0 are less specific than the ones of C_1. For instance, for C, the partitioning is $C_0 = \{B \rightarrow F\}$ and $C_1 = \{P \rightarrow B, P \rightarrow \neg F\}$. Then, if the rules of C_1 can be applied, we must make sure that some rules of C_0 are blocked.

After the conflicting rules being localized, it is necessary to determine the ones that are candidates to be blocked and the way to block their application. We want the most specific rules to be applied rather than the less specific ones, in an independent way of the other rules of the set. This is done by localizing the rules whose joint application leads to an inconsistency. In our example, the rules that are concerned are $B \rightarrow F$ and $P \rightarrow \neg F$ ($B \rightarrow F \in C_0$ and $P \rightarrow \neg F \in C_1$). The rules that are selected by this way in C_0 constitute the ones which are candidates to be blocked.

This selection criterion has the important property to be *context-independent*. For default theories R and R' such that $R \subseteq R'$, if $r \in R$ is chosen then it should also be chosen in R'. Moreover, if a rule should be blocked in the default theory R then it should also be blocked in any superset R'.

The second issue (*"How to block the application of a rule?"*) is dependent of the used framework. For example, in default logic, the default theory corresponding to our set of rules is composed of normal defaults, excepted for the ones selected in the previous step that become semi-normal defaults; for these last ones, the justification is composed by the consequent plus an assertion making sure that the rules of C_1 that are selected as in the previous way are not applicable (these rules, in the form of material implications, are added to the justification then a simplification is made if possible). Let us consider the set C. The rules $P \rightarrow B$ and $P \rightarrow \neg F$ are transformed respectively in: $\frac{P:B}{B}$ and $\frac{P:\neg F}{\neg F}$. The rule $B \rightarrow F$ is transformed in: $\frac{B:F \wedge (P \rightarrow \neg F)}{F}$, that can be simplified in: $\frac{B:F \wedge \neg P}{F}$.

To summarize, the approach, proposed in [5] and described above, can be decomposed in the following way, for a set of rules R:

Algorithm 1. Procedure `set_transformation`

Input: A set R of default rules
Output: The set R modified to handle specificity
1 $conf \leftarrow$ minimal_conflicts(R)
2 **for all** $C \in conf$ **do**
3   stratification(C)
4   rule_selection(C)
5 **end for**
6 rule_transformation$(R, conf)$

Let us note that the same basical semantics are shared by default logic and ASP. In [1,11], by using the mapping Tr such that each rule $r = c \leftarrow a_1, \ldots, a_n, not\ b_1, \ldots, not\ b_m$. of a normal logic program Π is associated with the default rule $Tr(r) = \frac{a_1 \wedge \ldots \wedge a_n : \neg b_1, \ldots, \neg b_m}{c}$, it is shown that if S is a stable model of Π then $Th(S)$ is an extension of $(\emptyset,\ Tr(\Pi))$ and each extension of $(\emptyset,\ Tr(\Pi))$ is the deductive closure of one stable model of Π. These results show that ASP can be seen as a simplification of default logic and allow us to use the formal results of [5] to ensure the validity of the work presented here. Hence, in the framework of ASP, we proceed in the same way as algorithm 1: we start by finding the conflicts, we stratify them, we compute their partitioning to select rules and, last, we modify the rules that are candidates.

4   Specificity in ASP

We have adapted to ASP all the notions and the associated algorithms that have been defined in the works [5,19] on which we have based ours.

Definition 1. *For a rule* $r = b \leftarrow a_1, \ldots, a_n. (n \geq 0)$ *of a definite logic program* P, *an atom set* A

- *satisfies* r *if* $\{a_1, \ldots, a_n\} \not\subseteq A$ *or* $b \in A$;
- *verifies* r *if* $\{a_1, \ldots, a_n\} \subseteq A$ *and* $b \in A$;
- *falsifies* r *if* $\{a_1, \ldots, a_n\} \subseteq A$ *and* $b \notin A$.

Definition 2. *A rule r of a program P is said to be* tolerated *by P if and only if there is an atom set A, closed under P and consistent, which verifies r.*

The tolerance of the rule characterizes the fact that its application does not generate any contradiction. From this notion of tolerance, it is possible to obtain a stratification of the program called *Z-ordering*.

In the following, the four steps of the algorithm 1 concerning the conflicts are developed (that is lines 1, 3, 4 and 6 of the procedure *set_transformation*). A conflict being a minimal set, the rules that compose it can introduce only one inconsistency; then, a conflict has one and only one constraint (see section 2 for the definition of a constraint). For a conflict C, we define:

- $constraint(C)$ the constraint of C;
- $rules(C) = C \setminus \{constraint(C)\}$;
- $constr(C) = body(constraint(C))$ (atoms in conflict).

4.1 Conflict Computation *(line 1 of the algorithm 1)*

If the determination of the conflicts for few rules seems easy, it is not true in general. The definition of a conflict (adapted to our framework) given in [5] is the following, where a trivial Z-ordering has only one stratum:

Definition 3. *Let P be a logic program. A set of rules $C \subseteq P$ is a conflict in P if and only if C has a non-trivial Z-ordering and $\forall C', C' \subset C$, C' has a trivial Z-ordering.*

In the worst case, for a program P of n rules, it is necessary to isolate and stratify the 2^n subsets of P. Let us note that some similar works have been yet done in other settings. In classical logic, particularly for the SAT-problem, a MUS (*Minimally Unsatisfiable Subformula*) is an unsatisfiable set of clauses such that all its subsets are satisfiable. Such a set gives then an "explanation" of the inconsistency which can not be smaller in terms of involved clauses, this corresponding to our notion of conflicts. Some works have shown that the MUS calculus is not possible to make in practice; indeed, to decide if a set of clauses is a MUS is DP-complete [18], and to test if a formula belongs to the sets of MUS is Σ_2^p-hard [7]. But Bruni [4] has shown that for some classes of clauses (like Horn clauses), the extraction of a MUS could be realized in polynomial time.

Let us note that every rule of the initial program is of the form $b \leftarrow a_1, \ldots, a_n.$ and can be translated into a clause $b \vee \neg a_1 \vee \ldots \vee \neg a_n$. Moreover, every constraint like $\perp \leftarrow x, nx.$ can be translated into the clause $\neg x \vee \neg nx$. Following this consideration, we have decided to use the algorithm HYCAM[1] [12] which computes all the MUS of an instance of SAT in a reasonable time. The function *minimal_conflicts* in algorithm 2 determines the conflicts using calls to HYCAM.

---

[1] It is an improvement of an existing algorithm, CAMUS (*Computing All MUS*) [16].

Algorithm 2. Function `minimal_conflicts`

Input: A normal logic program P expressed in a compact form
Output: The set of conflicts of P
1 $conflicts \leftarrow \emptyset$
2 $treated \leftarrow \emptyset$
3 **for all** $r \in P \setminus P_K$ **do**
4      $A \leftarrow \text{body}(r)$
5   **if** $A \notin treated$ **then**
6          $treated \leftarrow treated \cup \{A\}$
7          $conf \leftarrow \text{HYCAM}(P \cup A)$
8   **for all** $C \in conf$ **do**
9   **if** $\forall C' \in conflicts,\ C' \not\subseteq C$ **then**
10                 $conflicts \leftarrow (conflicts \cup \{C\}) \setminus \{C' \in conflicts \mid C \subset C'\}$
11   **end if**
12   **end for**
13   **end if**
14 **end for**
15 **return** $conflicts$

To determine the rules which are *potentially* responsible of a problem when they are applicable simultaneously, it is necessary to add facts to the program before using HYCAM (line 7). It is important to consider the interesting facts without testing every possible combinations (some of them being useless). Like System Z which uses the interpretations verifying a rule, we only use the atom sets allowing this verification (it is impossible to generate the others via the program). That is why we choose successively the atoms of the body of each rule (lines 3 to 7) to ensure that each rule will be used at least once; the algorithm then finds all the MUS where the rule occurs. We are computing the minimal sets so it is necessary to ensure that no generated set is a super-set of a conflict (lines 8 to 12).

Example 2. The program $P = \{Sh \leftarrow Mo.(r_1),\ Mo \leftarrow Ce.(r_2),\ nSh \leftarrow Ce.(r_3),\ Ce \leftarrow Na.(r_4),\ Sh \leftarrow Na.(r_5),\ \bot \leftarrow Sh,\ nSh.(c_1)\}$ means "generally, the molluscs (Mo) are shell-bearers (Sh), the cephalopods (Ce) are molluscs, the cephalopods are not shell-bearers, the nautiluses (Na) are cephalopods and the nautiluses are shell-bearers". The sets of facts that have to be added are $\{Mo\}$, $\{Ce\}$ and $\{Na\}$. For $\{Mo\}$, no MUS (conflict) is detected. For $\{Ce\}$, the conflict $C = \{r_1,\ r_2,\ r_3,\ c_1\}$ is obtained. For $\{Na\}$, the conflicts $C' = \{r_1,\ r_2,\ r_3,\ r_4,\ c_1\}$ and $C'' = \{r_3,\ r_4,\ r_5,\ c_1\}$ are obtained. $C \subset C'$, so C' is not minimal. The two conflicts of P are then C and C''.

4.2   Conflict Stratification *(line 3 of the algorithm 1)*

As shown in [5], the Z-ordering of a conflict contains exactly two strata. So, either a rule is tolerated in the conflict (line 4) and it belongs to the most general stratum (line 5), either it is not tolerated (line 6) and it belongs to the most specific stratum (line 7).

Algorithm 3. Function `stratification`

Input: A conflict C
Output: The stratification (C_0, C_1) of C
  1 $C_0 \leftarrow \emptyset$
  2 $C_1 \leftarrow \emptyset$
  3 **for all** $r \in \text{rules}(C)$ **do**
  4   **if** r is tolerated by $C \setminus \{r\}$ **then**
  5          $C_0 \leftarrow C_0 \cup \{r\}$
  6   **else**
  7          $C_1 \leftarrow C_1 \cup \{r\}$
  8   **end if**
  9 **end for**
10 **return** (C_0, C_1)

Example 3. Let us consider a conflict of the previous example: $C = \{Sh \leftarrow Mo.(r_1), Mo \leftarrow Ce.(r_2), nSh \leftarrow Ce.(r_3), \bot \leftarrow Sh, nSh.(c_1)\}$. $\{Mo, Sh\}$ verifies r_1, is closed under C and is consistent. So r_1 is tolerated in C, and $r_1 \in C_0$. The only closed set which verifies r_2 and doe not falsify neither r_1 nor r_3 is $A = \{Ce, Mo, Sh, nSh\}$; but A is inconsistent. So $r_2 \in C_1$. In the same way, we obtain $r_3 \in C_1$. Finally, the Z-ordering associated to C is the following: $C_0 = \{Sh \leftarrow Mo.\}$ and $C_1 = \{Mo \leftarrow Ce., nSh \leftarrow Ce.\}$.

4.3 Rule Selection *(line 4 of the algorithm 1)*

Three subsets of a stratified conflict can be expressed (the notations given in [5] are taken here): the rules that are candidates to be modified (general rules with exceptions), $min(C)$; the rules indicating how to modify the previous rules (expressing the exceptions), $max(C)$; and the remaining rules which are present in the conflict to link the previous sets, $inf(C)$. They are based on the smallest set of rules whose simultaneous application leads to an inconsistency.

Algorithm 4. Procedure `rule_selection`

Input: A stratified conflict C
Output: The rules to modify $(min(C))$, the exceptions $(max(C))$ and the other rules $(inf(C))$
  1 $core \leftarrow \{r \in \text{rules}(C) \mid \text{head}(r) \in \text{constr}(C)\}$
  2 $min(C) \leftarrow core \cap C_0$
  3 $max(C) \leftarrow core \cap C_1$
  4 $inf(C) \leftarrow \text{rules}(C) \setminus core$

Definition 4. *Let (C_0, C_1) be the stratification of the rules in the conflict C. A core of C is a pair of minimal sets $(min(C), max(C))$ such that $min(C) \subseteq C_0$, $max(C) \subseteq C_1$ and $\{a \leftarrow .|a \in body(r) \cup \{head(r)\}, \forall r \in min(C) \cup max(C)\} \cup \{constraint(C)\}$ leads to a contradiction.*

The core is composed of the rules concluding on conflicting atoms (line 1); so, it ensures that such a core is unique since the head of the rules contains only one atom. It is then easy to partition the conflict (lines 2 to 4).

Example 4. Let us take again the previously stratified conflict where $constr(C)$ = $\{Sh, nSh\}$: the core is $(\{Sh \leftarrow Mo.\}, \{nSh \leftarrow Ce.\})$ and $min(C) = \{Sh \leftarrow Mo.\}$, $max(C) = \{nSh \leftarrow Ce.\}$ and $inf(C) = \{Mo \leftarrow Ce.\}$.

4.4   Rule Modification *(line 6 of the algorithm 1)*

Once the core of each conflict is determined, the rules are transformed such that the most general ones are blocked if at least one of their exceptions is applicable but remain to be useful otherwise. For this aim, the most general rules are transformed by putting their exceptions in their negative body. The negative body of the other rules, which do not suffer any exception, remains empty. The algorithm 5 does this task by transforming a logic program P from its set of conflicts $conflicts$.

Algorithm 5. Procedure `rule_transformation`

Input: A logic program without exceptions denoted P and a set of conflicts denoted $conflicts$

Output: The logic program P transformed in a normal logic program with exceptions taking into account the specificity of information

```
 1 for all r ∈ P do
 2     if r is a constraint then
 3         for all r′ ∈ P such that head(r′) ∈ body⁺(r) do
 4             body⁻(r′) ← body⁻(r′) ∪ (body⁺(r) \ head(r′))
 5         end for
 6     else
 7         conf ← {C ∈ conflicts | r ∈ min(C)}
 8         if conf ≠ ∅ then
 9             for all C ∈ conf do
10                 for all r′ ∈ max(C) do
11                     if |body⁺(r′)| = 1 then
12                         body⁻(r) ← body⁻(r) ∪ body⁺(r′)
13                     else
14                         P ← P ∪ {σᵣ′}
15                         body⁻(r) ← body⁻(r) ∪ {head(σᵣ′)}
16                     end if
17                 end for
18             end for
19         end if
20     end if
21 end for
```

If r is a constraint, all the rules that can use the constraint (i.e. that can be implied in a contradiction) are transformed (lines 3 to 5). If r is not a constraint, the conflicts for which $r \in min(C)$ (i.e. that can be modified) are found (line 7) then the rule is modified by expressing its exceptions (lines 9 to 18), with a different treatment with respect to the number of atoms in the positive body of the rule $r′$ defining the exception (line 12 or lines 14 and 15). For a rule $r′$

containing several atoms in its positive body (lines 14 and 15), a new rule $\sigma_{r'}$ is defined such that:

- $head(\sigma_{r'})$ is a new atom (which does not yet exist in the program);
- $body^+(\sigma_{r'}) = body^+(r')$;
- $body^-(\sigma_{r'}) = \emptyset$.

By construction, $\sigma_{r'}$ is a rule that will be used each time r will be applicable.

So, for each one of its exceptions r', r is blocked if every atom of the positive body of r' are proven (i.e. r' is applicable) or if the head of r' has been obtained (or if the "opposite" of the head of r has been yet concluded).

Example 5. For our example, the complete transformation process leads to the normal logic program: $P = \{Sh \leftarrow Mo, \ not \ nSh, \ not \ Ce., \ Mo \leftarrow Ce., \ nSh \leftarrow Ce, \ not \ Sh, \ not \ Na., \ Ce \leftarrow Na., \ Sh \leftarrow Na, \ not \ nSh., \ \bot \leftarrow Sh, \ nSh.\}$.

5   Conclusion and Perspectives

From a compact representation of information, we can now use ASP to express in an automatic way default information, then to reason via stable models semantics. Due to lack of place, we do not present the implementation but it is important to notice that we have develop an operational system implementing the algorithms described in this article. The programs are available at http://www.info.univ-angers.fr/pub/pn/Softwares/specifASP.html. In particular, the conflict computation was not proposed in [5] where they consider the conflicts as asset.

We have to notice that several works dealing with default reasoning, but not linked to default logic, exist and it would be interesting to compare our proposal to these other ones. Particularly, we are interested in studying the properties of non-monotonic systems such as system P and rational monotony [14]. Moreover, in other frameworks, for instance machine learning, some works focused on the way to generate the exceptions but by observing some instances [13] while we are interested in determining the classes and subclasses of information.

Dealing with specificity and logic programming has been yet developped in several works. For example, in [22], an inheritance network is transformed in a normal logic program. However this graphical model based on path computing is less complex and does not allow to take into account rules with multiple atoms in the body or constraints which are not binary. The closer work to ours can be founded in [2]. Based on extended well founded semantics (WFSX) and computing a global stratification over the program, it suffers from cautiousness: on one hand, it does not allow to infer floating information (for example, with Nixon diamond); on the other hand, it faces the blocking inheritance problem (penguins do not have wings because of being abnormal birds).

Today, one of the domain of application of ASP is the semantic web [3]. In this context, a realistic scenario of reasoning may be the following. Let us suppose that we have different nodes of knowledge encoded by common-sense default rules and distributed over a network. On a particular node the knowledge could be very general as "birds fly". On another node, more specialized in some pieces of knowledge, we can find informations like "penguins are birds" and "penguins do not fly" but no information about the ability to fly of the birds. So, locally, the knowledge is well represented, consistent and an automatic reasoning is possible. But, if a system tries to exploit the entire knowledge of the network, then it has to gather all the rules. And, then, as we have shown before some conflicts arise and the reasoning is impossible if the system does not take into account the different levels of specificity of every piece of knowledge. It is an evidence that the methodology and the tools that we have described and developped in our work could be used to build a coherent knowledge base of rules with exceptions (automatically detected), allowing to infer conclusions based on the global knowledge distributed over the network.

Last, let us note that, in a previous work, we have proposed to merge the handling of default and uncertain information in ASP (for the default aspect) using possibility theory (for the uncertainty aspect) [17]. The present work should improve this proposal by allowing a compact representation of default information. This work should then be linked to the one proposed in [6] which has also been developped to deal with default and uncertain information but in the possibilistic logic setting.

References

1. Bidoit, N., Froidevaux, C.: General logical databases and programs: Default logic semantics and stratification. Information and Computation 91(1), 15–54 (1991)
2. Borges Garcia, B., Pereira Lopes, J.G., Varejão, F.: Compiling default theory int extended logic programming. In: Monard, M.C., Sichman, J.S. (eds.) SBIA 2000 and IBERAMIA 2000. LNCS, vol. 1952, pp. 207–216. Springer, Heidelberg (2000)
3. De Bruijn, J., Heymans, S., Pearce, D., Polleres, A., Ruckhaus, E. (eds.): Proc. of ICLP 2008 Workshop on Applications of Logic Programming to the (Semantic) Web and Web Services, ALPSWS 2008 (2008)
4. Bruni, R.: On Exact Selection of Minimally Unsatisfiable Subformulae. Ann. Math. Artif. Intell. 43(1-4), 35–50 (2005)
5. Delgrande, J.P., Schaub, T.: Compiling specificity into approaches to nonmonotonic reasoning. Artif. Intell. 90(1-2), 301–348 (1997)
6. Dupin de Saint-Cyr, F., Prade, H.: Handling uncertainty and defeasibility in a possibilistic logic setting. Int. J. Approx. Reasoning 49(1), 67–82 (2008)
7. Eiter, T., Gottlob, G.: On the Complexity of Propositional Knowledge Base Revision, Updates and Counterfactuals. Artif. Intell. 57(2-3), 227–270 (1992)
8. Garcia de la Banda, M., Pontelli, E. (eds.): ICLP 2008. LNCS, vol. 5366. Springer, Heidelberg (2008)
9. Gebser, M., Kaufmann, B., Neumann, A., Schaub, T.: Conflict-driven answer set solving. In: Proc. of IJCAI 2007, pp. 386–392 (2007)
10. Gelfond, M., Lifschitz, V.: The stable model semantics for logic programming. In: Proc. of ICLP 1988, pp. 1070–1080. MIT Press, Cambridge (1988)
11. Gelfond, M., Lifschitz, V.: Classical negation in logic programs and disjunctive databases. New Generation Computing 9(3-4), 363–385 (1991)

12. Grégoire, É., Mazure, B., Piette, C.: Boosting a Complete Technique to Find MSS and MUS thanks to a Local Search Oracle. In: Proc. of IJCAI 2007, pp. 2300–2305 (2007)
13. Inoue, K., Kudoh, Y.: Learning extended logic programs. In: Proc. of IJCAI 1997, pp. 176–181 (1997)
14. Kraus, S., Lehmann, D.J., Magidor, M.: Nonmonotonic reasoning, preferential models and cumulative logics. Artif. Intell. 44(1-2), 167–207 (1990)
15. Leone, N., Pfeifer, G., Faber, W., Eiter, T., Gottlob, G., Perri, S., Scarcello, F.: The dlv system for knowledge representation and reasoning. ACM Transactions on Computational Logic 7(3), 499–562 (2006)
16. Liffiton, M.H., Sakallah, K.A.: Algorithms for Computing Minimal Unsatisfiable Subsets of Constraints. J. Autom. Reasoning 40(1), 1–33 (2008)
17. Nicolas, P., Garcia, L., Stéphan, I., Lefèvre, C.: Possibilistic uncertainty handling for answer set programming. Ann. Math. Artif. Intell. 47(1-2), 139–181 (2006)
18. Papadimitriou, C.H., Wolfe, D.: The Complexity of Facets Resolved. J. Computer and System Sciences 37(1), 2–13 (1988)
19. Pearl, J.: System Z: A natural ordering of defaults with tractable applications to nonmonotonic reasoning. In: Proc. of Theoritical Aspects of Reasoning about Knowledge (TARK 1990), pp. 121–135 (1990)
20. Reiter, R.: A logic for default reasoning. Artif. Intell. 13, 81–132 (1980)
21. Simons, P., Niemelä, I., Soininen, T.: Extending and implementing the stable model semantics. Artif. Intell. 138(1-2), 181–234 (2002)
22. You, J.H., Wang, X., Yuan, L.Y.: Compiling defeasible inheritance networks to general logic programs. Artif. Intell. 113(1-2), 247–268 (1999)

Generalised Label Semantics as a Model of Epistemic Vagueness

Jonathan Lawry[1] and Inés González-Rodríguez[2]

[1] Department of Engineering Mathematics,
University of Bristol,
Bristol, UK
j.lawry@bris.ac.uk

[2] Dept. Mathematics, Statistics and Computer Science
University of Cantabria,
Santander, Spain
ines.gonzalez@unican.es

Abstract. A generalised version of the label semantics framework is proposed as an epistemic model of the uncertainty associated with vague description labels. In this framework communicating agents make explicit decisions both about which *labels* are appropriate to describe an element $x \in \Omega$ (the underlying universe), and also about which *negated labels* are appropriate to describe x. It is shown that such a framework can capture a number of different calculi for reasoning with vague concepts as special cases. In particular, different uncertainty assumptions are shown to result in the truth-functional max-min calculus and the standard label semantics calculus.

1 Introduction

Label semantics [5]-[7] is a random set framework for modelling the uncertainty associated with vague description labels based on the epistemic view of vagueness as proposed by Williamson and others [10]. The latter suggests that vague concepts have precise but uncertain boundaries. For instance, according to the epistemic view there is a precise but uncertain threshold above which a height is described as *tall* and below which it is described as *not tall*. In fact label semantics requires communicating agents to adopt an *epistemic stance* which is rather weaker that Williamson's epistemic theory. It is also assumed that the use of vague descriptions in language is governed by linguistic conventions adopted by a population of communicating agents and furthermore, in accordance with Parikh [8] and Kyburg [4], the focus of label semantics is on identifying which labels are assertible or appropriate based on these conventions.

This paper proposes an extension of label semantics in which an agent makes explicit decisions not only concerning which labels are appropriate to describe a given object or value, but also on which negated labels are appropriate. This more general framework is then shown to incorporate, as special cases, a number of different calculi for combining labels using logical connectives. In particular,

C. Sossai and G. Chemello (Eds.): ECSQARU 2009, LNAI 5590, pp. 626–637, 2009.

different uncertainty assumptions are shown to result in the truth-functional max-min calculus and the original standard label semantics calculus. An outline of the paper is as follows: Section 2 discusses the epistemic theory of vagueness underpinning label semantics; section 3 introduces the new general label semantics framework and gives a number of key results; section 4 discusses the laws of non-contradiction and excluded middle within this new framework; section 5 identifies three uncertainty assumptions which result in a truth-functional max-min calculus; section 6 shows that standard label semantics is a special case of the new theory and finally section 7 gives a summary and some conclusions.

2   The Epistemic Stance

Label semantics concerns the decision making process an intelligent agent must go through in order to identify which labels or expressions can actually be used to describe an object or value. In other words, in order to make an assertion describing an object in terms of some set of linguistic labels, an agent must first identify which of these labels are appropriate or assertible in this context. Given the way that individuals learn language through an ongoing process of interaction with the other communicating agents and with the environment, then we can expect there to be considerable uncertainty associated with any decisions of this kind. Furthermore, there is a subtle assumption central to the label semantic model, that such decisions regarding appropriateness or assertibility are meaningful. For instance, the fuzzy logic view is that vague descriptions like 'John is *tall*' are generally only partially true and hence it is not meaningful to consider which of a set of given labels can truthfully be used to described John's height. However, we contest that the efficacy of natural language as a means of conveying information between members of a population lies in shared conventions governing the appropriate use of words which are, at least loosely, adhere to by individuals within the population.

In our everyday use of language we are continually faced with decisions about the best way to describe objects and instances in order to convey the information we intend. For example, suppose you are witness to a robbery, how should you describe the robber so that police on patrol in the streets will have the best chance of spotting him? You will have certain labels that can be applied, for example *tall, short, medium, fat, thin, blonde, etc*, some of which you may view as inappropriate for the robber, others perhaps you think are definitely appropriate while for some labels you are uncertain whether they are appropriate or not. On the other hand, perhaps you have some ordered preferences between labels so that *tall* is more appropriate than *medium* which is in turn more appropriate than *short*. Your choice of words to describe the robber should surely then be based on these judgments about the appropriateness of labels. Yet where does this knowledge come from and more fundamentally what does it actually mean to say that a label is or is not appropriate? Label semantics proposes an interpretation of vague description labels based on a particular notion of appropriateness and suggests a measure of subjective uncertainty resulting from an agent's partial knowledge about what labels are appropriate to assert. Furthermore, it is suggested that the vagueness of these description labels lies fundamentally in the uncertainty about if and when they are appropriate as governed by the rules and conventions of language use.

The above argument brings us very close to the epistemic view of vagueness as expounded by Williamson [10]. However, while there are marked similarities between the epistemic theory and the label semantics view, there are also some subtle differences. For instance, the epistemic view would seem to assume the existence of some objectively correct, but unknown, definition of a vague concept. Instead of this we argue that individuals when faced with decision problems regarding assertions find it useful as part of a decision making strategy to assume that there is a clear dividing line between those labels which are and those which are not appropriate to describe a given instance. We refer to this strategic assumption across a population of communicating agents as the *epistemic stance* [7], a concise statement of which is as follows:

> *Each individual agent in the population assumes the existence of a set of labeling conventions, valid across the whole population, governing what linguistic labels and expressions can be appropriately used to describe particular instances.*

In practice these rules and conventions underlying the appropriate use of labels would not be imposed by some outside authority. In fact, they may not exist at all in a formal sense. Rather they are represented as a distributed body of knowledge concerning the assertability of predicates in various cases, shared across a population of agents, and emerging as the result of interactions and communications between individual agents all adopting the epistemic stance. The idea is that the learning processes of individual agents, all sharing the fundamental aim of understanding how words can be appropriately used to communicate information, will eventually converge to some degree on a set of shared conventions. The very process of convergence then to some extent vindicates the epistemic stance from the perspective of individual agents. Of course, this is not to suggest complete or even extensive agreement between individuals as to these appropriateness conventions. However, the overlap between agents should be sufficient to ensure the effective transfer of useful information.

3   Generalised Label Semantics

We assume that agents describe elements of an underlying universe Ω in terms of a finite set of labels LA from which a set of compound expressions LE can be generated through recursive applications of logical connectives. The labels $L_i \in LA$ are intended to represent words such as adjectives and nouns which can be used to describe elements from the underlying universe Ω. In other words, L_i correspond to description labels for which the expression 'x is L_i' is meaningful for any $x \in \Omega$. For example, if Ω is the set of all possible *rgb* values[1] then LA

---

[1] rgb is an additive colour model in which red, green and blue are combined to reproduce a broad range of colours.

could consist of the basic colour labels such as *red, yellow, green, orange* etc. In this case LE then contains those compound expressions such as *red & yellow, not blue nor orange* etc.

In the original label semantics framework [5]-[7], when describing an example $x \in \Omega$, an agent attempts to identify the set of labels $\mathcal{D}_x \subseteq LA$ corresponding to those labels which they judge to be appropriate to describe x based on their knowledge of labeling conventions. In this extended model we also assume that the agent attempts to explicitly define a set $\mathcal{C}_x$ corresponding to those labels L_i for which the negation $\neg L_i$ is appropriate to describe x^2. This explicitly separate treatement of labels and negated labels is consistent with a bi-polar approach to reasoning and representation [1]. In the face of their uncertainty regarding labeling conventions the agent will be uncertain as to the composition of both $\mathcal{D}_x$ and $\mathcal{C}_x$, and this uncertainty is quantified by a joint mass function $m_x : 2^{LA} \times 2^{LA} \to [0,1]$. For $F, G \subseteq LA$ $m_x(F, G)$ then quantifies the agent's subjective belief that $(\mathcal{D}_x, \mathcal{C}_x) = (F, G)$.

Definition 1. *Label Expressions*
We now define the set of label expressions LE generated recursively from the connectives $\wedge, \vee$ and $\neg$ as follows: $LA \subseteq LE$; $\forall \theta, \varphi \in LE$ $\theta \wedge \varphi$, $\theta \vee \varphi \in LE$; $\forall \theta \in LE$ $\neg \theta \in LE$

Definition 2. *Mass Function*
For $x \in \Omega$ let $m_x : 2^{LA} \times 2^{LA} \to [0,1]$ such that $\sum_{F \subseteq LA} \sum_{G \subseteq LA} m_x(F, G) = 1$

For an expression $\theta \in LE$, the assertion 'x is θ' naturally provides direct constraints on $\mathcal{D}_x$ and $\mathcal{C}_x$. For example, given $L_i \in LA$, asserting 'x is L_i' conveys the information that $L_i \in \mathcal{D}_x$ (i.e. L_i is appropriate to describe x), while asserting 'x is $\neg L_i$' conveys the information that $L_i \in \mathcal{C}_x$ (i.e. $\neg L_i$ is appropriate to describe x). Also for example, asserting the compound expression 'x is $L_i \wedge \neg L_j$' implies that $L_i \in \mathcal{D}_x$ and $L_j \in \mathcal{C}_x$. In general we can recursively define a mapping $\lambda : LE \to 2^{2^{LA}} \times 2^{2^{LA}}$ from expressions to sets of pairs of subsets of labels, such that the assertion 'x is θ' directly implies the constraint $(\mathcal{D}_x, \mathcal{C}_x) \in \lambda(\theta)$ and where $\lambda(\theta)$ is dependent on the logical structure of θ.

Definition 3. λ *mapping*
Let $\lambda : LE \to 2^{2^{LA}} \times 2^{2^{LA}}$ defined recursively by: $\forall \theta, \varphi \in LE$; $\forall L_i \in LA$
$\lambda(L_i) = \{(F, G) : L_i \in F\}$; $\lambda(\theta \wedge \varphi) = \lambda(\theta) \cap \lambda(\varphi)$; $\lambda(\theta \vee \varphi) = \lambda(\theta) \cup \lambda(\varphi)$;
$\lambda(\neg \theta) = \{(G^c, F^c) : (F, G) \in \lambda(\theta)^c\}$

The behaviour of the λ-mapping given in definition 3 is relatively intuitive except perhaps in the case of negation, a possible justification of which is given as follows: If $(\mathcal{D}_x, \mathcal{C}_x) = (F, G)$ then G^c corresponds those labels L_j for which $\neg L_j$ is not appropriate to describe x and F^c corresponds to those labels L_i for which L_i is not appropriate to describe x. Now an agent might take the view that if $L_i \notin \mathcal{D}_x$ (i.e. L_i is not appropriate) then this could provide some evidence

---

² In standard label semantics it is implicitly assumed that $\mathcal{C}_x = (\mathcal{D}_x)^c$.

that $L_i \in C_x$ (i.e. $\neg L_i$ is appropriate) and similarly that if $L_j \notin C_x$ (i.e. $\neg L_j$ is not appropriate) then this provides some evidence that $L_j \in D_x$ (i.e. that L_j is appropriate). Hence, given an agent for whom $(D_x, C_x) = (F, G)$, then $(D_x, C_x) = (G^c, F^c)$ would correspond to a dual state of knowledge obtained by positively asserting, firstly those labels L_j for which $\neg L_j$ has been ruled out as inappropriate, and secondly those negated labels $\neg L_i$ for which L_i has been ruled out as inappropriate. For an expression $\theta \in LE$, $\lambda(\neg\theta)$ then consists of those dual states (G^c, F^c) for which the state (F, G) is not consistent with θ.

Example 1. For $L_i, L_j \in LA$, $\lambda(L_i) = \{(F, G) : L_i \in F\}$, $\lambda(\neg L_j) = \{(F, G) : L_j \in G\}$, $\lambda(L_i \wedge \neg L_j) = \{(F, G) : L_i \in F, L_j \in G\}$

Given definitions 2 and 3 we can define an appropriateness measure $\mu_\theta(x)$ quantifying an agent's subjective belief that expression $\theta \in LE$ is appropriate to describe $x \in \Omega$. Here $\mu_\theta(x)$ is taken to be the sum of m_x across those pairs (F, G) consistent with θ (i.e. $(F, G) \in \lambda(\theta)$).

Definition 4. *Appropriateness Measures*
$\mu : LE \times \Omega \to [0, 1]$ *such that* $\forall \theta \in LE$, $\forall x \in \Omega$ $\mu_\theta(x) = \sum_{(F,G) \in \lambda(\theta)} m_x(F, G)$ *where* $\mu_\theta(x)$ *is shorthand for* $\mu(\theta, x)$.

Notice that both D_x and C_x are random sets and hence definition 4 links appropriateness measures to the work of Goodman and Nguyen on random set interpretations of fuzzy membership functions [2], [3]. Here, however, the model is two dimensional (i.e. based on two related random sets) and also the random sets take sets of labels as values rather than sets of elements from the underlying domain Ω as in Goodman and Nguyen's work.

The following theorems show that appropriateness measures as defined above satisfy De Morgan's laws and Double Negation.

Theorem 1. *De Morgan's Laws*
$\forall \theta, \varphi \in LE$

  – $\lambda(\neg(\theta \wedge \varphi)) = \lambda(\neg\theta \vee \neg\varphi)$
  – $\lambda(\neg(\theta \vee \varphi)) = \lambda(\neg\theta \wedge \neg\varphi)$

Proof.

$\lambda(\neg(\theta \wedge \varphi) = \{(G^c, F^c) \in \lambda(\theta \wedge \varphi)^c\} = \{(G^c, F^c) : (F, G) \in \lambda(\theta)^c \cup \lambda(\varphi)^c\}$
$= \{(G^c, F^c) : (F, G) \in \lambda(\theta)^c\} \cup \{(G^c, F^c) : (F, G) \in \lambda(\varphi)^c\} = \lambda(\neg\theta) \cup \lambda(\neg\varphi)$
$= \lambda(\neg\theta \vee \neg\varphi)$

$\lambda(\neg(\theta \vee \varphi)) = \{(G^c, F^c) : (F, G) \in \lambda(\theta \vee \varphi)^c\}$
$= \{(G^c, F^c) : (F, G) \in \lambda(\theta)^c \cap \lambda(\varphi)^c\}$
$= \{(G^c, F^c) : (F, G) \in \lambda(\theta)^c\} \cap \{(G^c, F^c) : (F, G) \in \lambda(\varphi)^c\} = \lambda(\neg\theta) \cap \lambda(\neg\varphi)$
$= \lambda(\neg\theta \wedge \neg\varphi)$

Corollary 1

$\forall \theta, \varphi \in LE, \ \forall x \in \Omega \ \mu_{\neg(\theta \wedge \varphi)}(x) = \mu_{\neg \theta \vee \neg \varphi}(x) \ and \ \mu_{\neg(\theta \vee \varphi)}(x) = \mu_{\neg \theta \wedge \neg \varphi}(x)$

Proof. Follows trivially from theorem 1 and definitions 3 and 4.

Theorem 2. *Double Negation*
$\forall \theta \in LE \ \lambda(\neg(\neg \theta)) = \lambda(\theta)$

Proof. Suppose $(F, G) \notin \lambda(\theta) \Rightarrow (G^c, F^c) \in \lambda(\neg \theta) \Rightarrow (G^c, F^c) \notin \lambda(\neg \theta)^c \Rightarrow$
$(F, G) \notin \lambda(\neg(\neg \theta))$
Suppose $(F, G) \notin \lambda(\neg(\neg \theta)) \Rightarrow (G^c, F^c) \in \lambda(\neg \theta) \Rightarrow (F, G) \in \lambda(\theta)^c \Rightarrow (F, G) \notin$
$\lambda(\theta)$

Corollary 2

$\forall \theta \in LE, \ \forall x \in \Omega \ \mu_{\neg(\neg \theta)}(x) = \mu_\theta(x).$

Proof. Follows trivially from theorem 2 and definitions 3 and 4.

The following theorem shows that for $\Psi \in LE$, $\lambda(\Psi)$ contains embedded nested sequences of pairs (F, G).

Theorem 3. *Nestedness*
$\forall \Psi \in LE \ if \ (F, G) \in \lambda(\Psi) \ and \ F' \supseteq F \ and \ G' \supseteq G \ then \ (F', G') \in \lambda(\Psi)$

Proof. Let $LE^0 = LA$ and $LE^k = LE^{k-1} \cup \{\theta \wedge \varphi, \theta \vee \varphi, \neg \theta : \theta, \varphi \in LE^{k-1}\}$ then by induction on k; For $L_i \in LA$ the result holds trivially. Now suppose true for k and prove for $k + 1$. For $\Psi \in LE^{k+1}$ either $\Psi \in LE^k$ in which case the result holds trivially, otherwise one of the following cases hold:

- $\Psi = \theta \wedge \varphi$ where $\theta, \varphi \in LE^k$. In this case $\lambda(\Psi) = \lambda(\theta) \cap \lambda(\varphi)$. Now suppose that $(F, G) \in \lambda(\Psi)$ then $(F, G) \in \lambda(\theta)$ and $(F, G) \in \lambda(\varphi)$. Hence by induction, if $F' \supseteq F$ and $G' \supseteq G$ then $(F', G') \in \lambda(\theta)$ and $(F', G') \in \lambda(\varphi)$. Therefore, $(F', G') \in \lambda(\theta) \cap \lambda(\varphi) = \lambda(\Psi)$ as required.
- $\Psi = \theta \vee \varphi$ where $\theta, \varphi \in LE^k$. In this case $\lambda(\Psi) = \lambda(\theta) \cup \lambda(\varphi)$. Now suppose that $(F, G) \in \lambda(\Psi)$ then $(F, G) \in \lambda(\theta)$ or $(F, G) \in \lambda(\varphi)$. Hence by induction, if $F' \supseteq F$ and $G' \supseteq G$ then $(F', G') \in \lambda(\theta)$ or $(F', G') \in \lambda(\varphi)$. Therefore, $(F', G') \in \lambda(\theta) \cup \lambda(\varphi) = \lambda(\Psi)$ as required.
- $\Psi = \neg \theta$. In this case $\lambda(\Psi) = \{(G^c, F^c) : (F, G) \in \lambda(\theta)^c\}$. Now suppose $(F, G) \in \lambda(\Psi)$ then $(G^c, F^c) \in \lambda(\theta)^c$. Now if $F' \supseteq F$ and $G' \supseteq G$ then $(F')^c \subseteq F^c$ and $(G')^c \subseteq G^c$ which implies by induction that $((G')^c, (F')^c) \in \lambda(\theta)^c$ [3] and hence $(F', G') \in \lambda(\neg \theta) = \lambda(\Psi)$ as required.

---

[3] Otherwise suppose $((G')^c, (F')^c) \notin \lambda(\theta)^c$ then $((G')^c, (F')^c) \in \lambda(\theta)$ and hence by induction $(G^c, F^c) \in \lambda(\theta)$. Therefore $(G^c, F^c) \notin \lambda(\theta)^c$ and hence $(F, G) \notin \lambda(\neg \theta)$ which is a contradiction.

Corollary 3. *For any* $\theta \in \lambda(\theta)$

- *If* $\lambda(\theta) \neq \emptyset$ *then* $(LA, LA) \in \lambda(\theta)$
- *If* $(\emptyset, \emptyset) \in \lambda(\theta)$ *then* $\lambda(\theta) = 2^{LA} \times 2^{LA}$

The next result shows that there is no expression $\Psi \in LE$ for which $\lambda(\Psi)$ is either empty or contains all possible pairs (F, G). Consequently there is no expression Ψ such that for all possible mass functions m_x either $\mu_\Psi(x) = 0$ or $\mu_\Psi(x) = 1$.

Theorem 4. *Non-Triviality*
$\forall \Psi \in LE \; \lambda(\Psi) \neq \emptyset \text{ and } \lambda(\Psi) \neq 2^{LA} \times 2^{LA}$

Proof. Let $LE^0 = LA$ and $LE^k = LE^{k-1} \cup \{\theta \wedge \varphi, \theta \vee \varphi, \neg \theta : \theta, \varphi \in LE^{k-1}\}$ then by induction on k; For $L_i \in LA \; \lambda(L_i) = \{(F, G) : L_i \in F\}$ so clearly $\lambda(L_i) \neq \emptyset$. Also $(\emptyset, \emptyset) \notin \lambda(L_i)$ and hence $\lambda(L_i) \neq 2^{LA} \times 2^{LA}$. Now suppose true for k and prove for $k + 1$. For $\Psi \in LE^{k+1}$ either $\Psi \in LE^k$ in which case the result holds trivially, otherwise one of the following cases hold:

- $\Psi = \theta \wedge \varphi$ where $\theta, \varphi \in LE^k$. In this case $\lambda(\Psi) = \lambda(\theta) \cap \lambda(\varphi)$. Now by induction $\lambda(\theta) \neq \emptyset$ and $\lambda(\varphi) \neq \emptyset$. Then by Corollary 3 $(LA, LA) \in \lambda(\theta)$ and $(LA, LA) \in \lambda(\varphi)$. Therefore $(LA, LA) \in \lambda(\theta) \cap \lambda(\varphi) = \lambda(\Psi)$ and hence $\lambda(\Psi) \neq \emptyset$. Also, by induction $\lambda(\theta) \neq 2^{LA} \times 2^{LA}$ and $\lambda(\varphi) \neq 2^{LA} \times 2^{LA}$ then trivially $\lambda(\Psi) = \lambda(\theta) \cap \lambda(\varphi) \neq 2^{LA} \times 2^{LA}$.
- $\Psi = \theta \vee \varphi$ where $\theta, \varphi \in LE^k$. In this case $\lambda(\Psi) = \lambda(\theta) \cup \lambda(\varphi)$. Now by induction $\lambda(\theta) \neq \emptyset$ and $\lambda(\varphi) \neq \emptyset$. Hence trivially $\lambda(\Psi) \neq \emptyset$. Also, by induction $\lambda(\theta) \neq 2^{LA} \times 2^{LA}$ and $\lambda(\varphi) \neq 2^{LA} \times 2^{LA}$. Therefore by corollary 3 $(\emptyset, \emptyset) \notin \lambda(\theta)$ and $(\emptyset, \emptyset) \notin \lambda(\varphi)$. Hence $(\emptyset, \emptyset) \notin \lambda(\theta) \cup \lambda(\varphi)$ and therefore $\lambda(\Psi) \neq 2^{LA} \times 2^{LA}$.
- $\Psi = \neg \theta$. In this case $\lambda(\Psi) = \{(G^c, F^c) : (F, G) \in \lambda(\theta)^c\}$. Now by induction $\lambda(\theta) \neq 2^{LA} \times 2^{LA}$ which implies that $\lambda(\theta)^c \neq \emptyset$ and hence $\lambda(\neg \theta) \neq \emptyset$. Also by induction $\lambda(\theta) \neq \emptyset$ and hence by corollary 3 $(LA, LA) \in \lambda(\theta)$. Therefore $(LA, LA) \notin \lambda(\theta)^c$ and hence $(\emptyset, \emptyset) \notin \lambda(\neg \theta)$. Hence $\lambda(\Psi) \neq 2^{LA} \times 2^{LA}$.

Corollary 4. $\forall \Psi \in LE \; (\emptyset, \emptyset) \notin \lambda(\Psi)$

The following theorem shows that for an expression $\Psi \wedge \neg \Psi$, if $(F, G) \in \lambda(\Psi \wedge \neg \Psi)$ then F and G must have labels in common.

Theorem 5. *Weak Non-Contradiction*
Let $\mathcal{A} = \{(F, G) : F \cap G = \emptyset\}$ then $\forall \Psi \in LE \; \lambda(\Psi \wedge \neg \Psi) \cap \mathcal{A} = \emptyset$

Proof. Let $LE^0 = LA$ and $LE^k = LE^{k-1} \cup \{\theta \wedge \varphi, \theta \vee \varphi, \neg \theta : \theta, \varphi \in LE^{k-1}\}$ then by induction on k; For $\Psi = L_i \in LA$ then $\lambda(L_i \wedge \neg L_i) = \{(F, G) : L_i \in F, L_i \in G\}$. Hence, $\forall (F, G) \in \lambda(L_i \wedge \neg L_i) \; F \cap G \supseteq \{L_i\} \neq \emptyset$. Now suppose true for k and prove for $k + 1$. For $\Psi \in LE^{k+1}$ either $\Psi \in LE^k$ in which case the result holds trivially, otherwise one of the following cases hold: For $\theta, \varphi \in LE^k$

- $\Psi = \theta \wedge \varphi$ then $\lambda(\Psi \wedge \neg \Psi) = \lambda(\Psi) \cap \lambda(\neg \Psi) = \lambda(\theta \wedge \varphi) \cap \lambda(\neg(\theta \wedge \varphi))$ by theorem $1 = [\lambda(\theta) \cap \lambda(\varphi)] \cap [\lambda(\neg \theta) \cup \lambda(\neg \varphi)] = [\lambda(\theta \wedge \neg \theta) \cap \lambda(\varphi)] \cup [\lambda(\varphi \wedge \neg \varphi) \cap \lambda(\theta)]$. Hence, $\lambda(\Psi \wedge \neg \Psi) \cap \mathcal{A} = [\lambda(\theta \wedge \neg \theta) \cap \mathcal{A} \cap \lambda(\varphi)] \cup [\lambda(\varphi \wedge \neg \varphi) \cap \mathcal{A} \cap \lambda(\theta)] = \emptyset$ by induction.

- $\Psi = \theta \vee \varphi$ then $\lambda(\Psi \wedge \neg \Psi) = \lambda(\Psi) \cap \lambda(\neg \Psi) = \lambda(\theta \vee \varphi) \cap \lambda(\neg(\theta \vee \varphi))$ by theorem
1$= [\lambda(\theta) \cup \lambda(\varphi)] \cap [\lambda(\neg \theta) \cap \lambda(\neg \varphi)] = [\lambda(\theta \wedge \neg \theta) \cap \lambda(\neg \varphi)] \cup [\lambda(\varphi \wedge \neg \varphi) \cap \lambda(\neg \theta)]$.
Hence $\lambda(\Psi \wedge \neg \Psi) \cap \mathcal{A} = [\lambda(\theta \wedge \neg \theta) \cap \mathcal{A} \cap \lambda(\neg \varphi)] \cup [\lambda(\varphi \wedge \neg \varphi) \cap \mathcal{A} \cap \lambda(\neg \theta)] = \emptyset$
by induction.
- $\Psi = \neg \theta$ then $\lambda(\Psi \wedge \neg \Psi) = \lambda(\Psi) \cap \lambda(\neg \Psi) = \lambda(\neg \theta) \cap \lambda(\neg(\neg \theta))$ by theorem 2
$= \lambda(\neg \theta) \cap \lambda(\theta) = \lambda(\theta \wedge \neg \theta)$. Hence, $\lambda(\Psi \wedge \neg \Psi) \cap \mathcal{A} = \emptyset$ by induction.

The following theorem shows that any pair of labels (F, G) where $F \cup G = LA$ is contained in $\lambda(\Psi \vee \neg \Psi)$ for any label expression Ψ.

Theorem 6. *Weak Excluded Middle*
Let $\mathcal{B} = \{(F, G) : F \cup G = LA\}$ then $\forall \Psi \in LE$ $\lambda(\Psi \vee \neg \Psi) \supseteq \mathcal{B}$

Proof. Note that $\mathcal{B} = \{(G^c, F^c) : G \cap F = \emptyset\} = \{(G^c, F^c) : (F, G) \in \mathcal{A}\}$. Then by theorem 1

$$\lambda(\Psi \vee \neg \Psi) = \lambda(\neg(\Psi \wedge \neg \Psi)) = \{(G^c, F^c) : (F, G) \in \lambda(\Psi \wedge \neg \Psi)^c\}$$
$$\supseteq \{(G^c, F^c) : (F, G) \in \mathcal{A}\} (\text{ since } \mathcal{A} \subseteq \lambda(\Psi \wedge \neg \Psi) \text{ by theorem 5}) = \mathcal{B}$$

4 Laws of Non-contradiction and Excluded Middle

In this section we show that two additional assumptions about mass function m_x ensures that the resulting appropriateness measures satisfy Non-Contradiction and Excluded Middle.

Definition 5. *Absolute Complementation*
m_x satisfies absolute complementation if $\sum_F \sum_{G:G \subseteq F^c} m_x(F, G) = 1$

Definition 5 is motivated by the assumption that if a label L_i is appropriate to describe x, then it's negation $\neg L_i$ cannot also be appropriate i.e. $\mathcal{C}_x \subseteq (\mathcal{D}_x)^c$

Definition 6. *Total Coverage*
The mass function m_x satisfies total coverage if $\sum_F \sum_{G:G \supseteq F^c} m_x(F, G) = 1$

Definition 6 is motivated by the assumption that for every label $L_i \in LA$ either L_i is appropriate to describe x or its negation is appropriate i.e. $\mathcal{C}_x \supseteq (\mathcal{D}_x)^c$

Theorem 7. *Non-Contradiction*
If m_x satisfies absolute complementation then $\forall \Psi \in LE$ $\mu_{\Psi \wedge \neg \Psi}(x) = 0$.

Proof. By theorem 5 $\mu_{\Psi \wedge \neg \Psi}(x) = \sum_{(F,G) \in \lambda(\Psi \wedge \neg \Psi)} m_x(F, G) \leq \sum_{(F,G) \in \mathcal{A}^c} m_x(F, G) = \sum_{(F,G):F \cap G \neq \emptyset} m_x(F, G) = 0$ by absolute complementation.

Theorem 8. *Excluded Middle*
If m_x satisfies total coverage then $\forall \Psi \in LE$ $\mu_{\Psi \vee \neg \Psi}(x) = 1$

Proof.

$$\mu_{\Psi \vee \neg \Psi}(x) = \sum_{(F,G) \in \lambda(\Psi \vee \neg \Psi)} m_x(F, G) \geq \sum_{(F,G) \in \mathcal{B}} m_x(F, G)$$
$$= \sum_F \sum_{G:G \supseteq F^c} m_x(F, G) = 1$$

by total coverage and theorem 6.

5   Max-Min Truth-Functionality

This section introduces two constraints on m_x which result in a truth-functional min-max calculus for appropriateness measures of the form proposed by Zadeh for fuzzy set membership function in his original 1965 paper [11].

Definition 7. *Complement Symmetric*
A mass function m_x is complement symmetric if $\forall F, G \subseteq LA \; m_x(F, G) = m_x(G^c, F^c)$.

Underlying the above property is the assumption that an agent allocates equal belief to the state of knowledge (F, G) and its dual state (G^c, F^c) as discussed in section 3.

Definition 8. *Pairwise Consonance*
Let $\mathcal{FG}_x = \{(F, G) : m_x(F, G) > 0\}$. Then a mass function m_x is pairwise consonant iff $\forall (F_1, G_1), (F_2, G_2) \in \mathcal{FG}_x$ either $F_1 \subseteq F_2$ and $G_1 \subseteq G_2$ or $F_2 \subseteq F_1$ and $G_2 \subseteq G_1$.

Pairwise consonance might be justified if agents adopted the following approach when estimating the composition of $\mathcal{D}_x$ and $\mathcal{C}_x$: Given $x \in \Omega$ an agent would generate two distinct rankings on labels; one based on their appropriateness to describe x and a second on the appropriateness of their negations to describe x. Given these rankings the agent's judgments would then be based on what they perceive to be the correct degree of 'open-mindedness' in the current context. In particular, a high degree of 'open-mindedness' would result in labels lower down the appropriateness ranking for labels being included in $\mathcal{D}_x$, and labels lower down the appropriateness ranking for negated labels being included in $\mathcal{C}_x$.

Theorem 9. *If m_x is complement symmetric then $\forall \theta \in LE \; \mu_{\neg\theta}(x) = 1 - \mu_\theta(x)$*

Proof.

$$\mu_{\neg\theta}(x) = \sum_{(F,G)\in\lambda(\neg\theta)} m_x(F, G) = \sum_{(F,G)\notin\lambda(\theta)} m_x(G^c, F^c)$$

$$= 1 - \sum_{(F,G)\in\lambda(\theta)} m_x(G^c, F^c) = 1 - \sum_{(F,G)\in\lambda(\theta)} m_x(F, G) = 1 - \mu_\theta(x)$$

by definitions 3 and 7.

Theorem 10. *If m_x is pairwise consonant then $\forall \theta, \varphi \in LE$*
$\mu_{\theta\wedge\varphi}(x) = \min(\mu_\theta(x), \mu_\varphi(x))$ and $\mu_{\theta\vee\varphi}(x) = \max(\mu_\theta(x), \mu_\varphi(x))$.

Proof. If m_x is pairwise consonant then w.l.o.g we can assume $\mathcal{FG}_x = \{(F_1, G_1), \ldots, (F_N, G_N)\}$ where $F_i \subseteq F_j$ and $G_i \subseteq G_j$ for $j \geq i$. Now by theorem 3 there exists $k, k' \in \{1, \ldots, N\}$ such that:

$$\lambda(\theta) \cap \mathcal{FG}_x = \{(F_k, G_k), \ldots, (F_N, G_N)\} \text{ and}$$
$$\lambda(\varphi) \cap \mathcal{FG}_x = \{(F_{k'}, G_{k'}), \ldots, (F_N, G_N)\}$$

Hence

$$\lambda(\theta \wedge \varphi) \cap \mathcal{FG}_x = \{(F_{\max(k,k')}, G_{\max(k,k')}), \ldots, (F_N, G_N)\} \text{ and}$$
$$\lambda(\theta \vee \varphi) \cap \mathcal{FG}_x = \{(F_{\min(k,k')}, G_{\min(k,k')}), \ldots, (F_N, G_N)\}$$

Therefore,

$$\mu_{\theta \wedge \varphi}(x) = \sum_{i=\max(k,k')}^{N} m_x(F_i, G_i) = \min(\sum_{i=k}^{N} m_x(F_i, G_i), \sum_{i=k'}^{N} m_x(F_i, G_i))$$

$$= \min(\mu_\theta(x), \mu_\varphi(x))$$

$$\mu_{\theta \vee \varphi}(x) = \sum_{i=\min(k,k')}^{N} m_x(F_i, G_i) = \max(\sum_{i=k}^{N} m_x(F_i, G_i), \sum_{i=k'}^{N} m_x(F_i, G_i))$$

$$= \max(\mu_\theta(x), \mu_\varphi(x))$$

Corollary 5. *If m_x is both complement symmetric and pairwise consonant then the resulting appropriateness measures are truth functional where:* $\forall \theta, \varphi \in LE$
$\mu_\theta(x) = 1 - \mu_\theta(x)$, $\mu_{\theta \wedge \varphi}(x) = \min(\mu_\theta(x), \mu_\varphi(x))$ *and* $\mu_{\theta \vee \varphi}(x) = \max(\mu_\theta(x), \mu_\varphi(x))$

Example 2. Let $LA = \{L_1, L_2, L_3, L_4\}$ then the following mass function satisfies both complement symmetry and pairwise consonance:

$$m_x := (\emptyset, \{L_2\}) : 0.3, \; (\{L_1\}, \{L_2, L_3\}) : 0.2, \; (\{L_1, L_4\}, \{L_2, L_3, L_4\}) : 0.2,$$
$$(\{L_1, L_4, L_3\}, \{L_1, L_2, L_3, L_4\}) : 0.3.$$

6   Standard Label Semantics as a Special Case

In this section we consider the standard label semantics model [5]-[7] as a special case of generalised label semantics. As mentioned in section 3, in standard label semantics agents only explicitly define $\mathcal{D}_x$ (the complete set of appropriate label for x) and then implicitly assume that $\mathcal{C}_x = (\mathcal{D}_x)^c$. This latter relationship clearly identifies the following restricted class of mass functions.

Definition 9. *Classical Mass Function*
m_x *is a classical mass function if* $\mathcal{FG}_x \subseteq \{(F, F^c) : F \subseteq LA\}^4$.

In standard label semantics the assertion 'x is θ' for $\theta \in LE$ is assumed to generate the constraint $\mathcal{D}_x \in \lambda'(\theta)$ where the mapping λ' is defined as follows:

Definition 10. *One-Dimensional λ-mapping*
The one-dimensional λ-mapping is a function $\lambda' : LE \to 2^{2^{LA}}$ *defined recursively as follows:*

---

[4] See definition 8.

- $\forall L \in LA \; \lambda'(L) = \{F \subseteq LA : L \in F\}$
- $\forall \theta, \varphi \in LE; \; \lambda'(\theta \wedge \varphi) = \lambda'(\theta) \cap \lambda'(\varphi), \; \lambda'(\theta \vee \varphi) = \lambda'(\theta) \cup \lambda'(\varphi)$ and $\lambda'(\neg \theta) = (\lambda'(\theta))^c$

If an agent then defines a one-dimensional mass function $m'_x : 2^{LA} \to [0,1]$ where $m'_x(F)$ is the agent's subjective belief that $\mathcal{D}_x = F$ then for expression $\theta \in LE$ in standard label semantics the appropriateness measure is given by $\mu_\theta(x) = \sum_{F \in \lambda'(\theta)} m'_x(F)$. Now clearly a classical mass function (definition 9) naturally generates such a one-dimensional mass function where $m'_x(F) = m_x(F, F^c)$. The following theorem shows that in such a case the standard and generalised label semantics definitions of appropriateness measure coincide.

Theorem 11. *Restriction*
$\forall \Psi \in LE \; \forall F \subseteq LA \; (F, F^c) \in \lambda(\Psi)$ *iff* $F \in \lambda'(\Psi)$

Proof. Let $LE^0 = LA$ and $LE^k = LE^{k-1} \cup \{\theta \wedge \varphi, \theta \vee \varphi, \neg \theta : \theta, \varphi \in LE^{k-1}\}$ then by induction on k; For $\Psi = L_i \in LA$ then $(F, F^c) \in \lambda(L_i)$ iff $L_i \in F$ (by definition 3) iff $F \in \lambda'(L_i)$ (by definition 10). Now suppose true for k and prove for $k+1$. For $\Psi \in LE^{k+1}$ either $\Psi \in LE^k$ in which case the result holds trivially, otherwise one of the following cases hold: For $\theta, \varphi \in LE^k$;

- $\Psi = \theta \wedge \varphi$. If $(F, F^c) \in \lambda(\Psi) \Rightarrow (F, F^c) \in \lambda(\theta) \cap \lambda(\varphi)$ (by definition 3) $\Rightarrow (F, F^c) \in \lambda(\theta)$ and $(F, F^c) \in \lambda(\varphi) \Rightarrow F \in \lambda'(\theta)$ and $F \in \lambda'(\varphi)$ (by inductive hypothesis) $\Rightarrow F \in \lambda'(\theta) \cap \lambda'(\varphi) = \lambda'(\theta \wedge \varphi) = \lambda'(\Psi)$ (by definition 10). If $F \in \lambda'(\Psi) \Rightarrow F \in \lambda'(\theta) \cap \lambda'(\varphi)$ (by definition 10) $\Rightarrow F \in \lambda'(\theta)$ and $F \in \lambda'(\varphi) \Rightarrow (F, F^c) \in \lambda(\theta)$ and $(F, F^c) \in \lambda(\varphi)$ (by inductive hypothesis) $\Rightarrow (F, F^c) \in \lambda(\theta) \cap \lambda(\varphi) = \lambda(\theta \wedge \varphi) = \lambda(\Psi)$ (by definition 3).
- $\Psi = \theta \vee \varphi$. If $(F, F^c) \in \lambda(\Psi) \Rightarrow (F, F^c) \in \lambda(\theta) \cup \lambda(\varphi)$ (by definition 3) $\Rightarrow (F, F^c) \in \lambda(\theta)$ or $(F, F^c) \in \lambda(\varphi) \Rightarrow F \in \lambda'(\theta)$ or $F \in \lambda'(\varphi)$ (by inductive hypothesis) $\Rightarrow F \in \lambda'(\theta) \cup \lambda'(\varphi) = \lambda'(\theta \vee \varphi) = \lambda'(\Psi)$ (by definition 10). If $F \in \lambda'(\Psi) \Rightarrow F \in \lambda'(\theta) \cup \lambda'(\varphi)$ (by definition 10) $\Rightarrow F \in \lambda'(\theta)$ or $F \in \lambda'(\varphi) \Rightarrow (F, F^c) \in \lambda(\theta)$ or $(F, F^c) \in \lambda(\varphi)$ (by inductive hypothesis) $\Rightarrow (F, F^c) \in \lambda(\theta) \cup \lambda(\varphi) = \lambda(\theta \vee \varphi) = \lambda(\Psi)$ (by definition 3).
- $\Psi = \neg \theta$. If $(F, F^c) \in \lambda(\Psi) \Rightarrow (F, F^c) \in \lambda(\theta)^c$ (by definition 3) $\Rightarrow F \in \lambda'(\theta)^c$ (by inductive hypothesis) $\Rightarrow F \in \lambda'(\neg \theta) = \lambda'(\Psi)$ (by definition 10). If $F \in \lambda'(\Psi) \Rightarrow F \in \lambda'(\theta)^c$ (by definition 10) $\Rightarrow (F, F^c) \in \lambda(\theta)^c$ (by inductive hypothesis) $\Rightarrow (F, F^c) \in \lambda(\neg \theta) = \lambda(\Psi)$ (by definition 3).

Corollary 6. *If m_x is a classical mass function then*

$$\forall \theta \in LE \; \mu_\theta(x) = \sum_{F \in \lambda'(\theta)} m_x(F, F^c)$$

Notice that a classical mass function trivially satisfies *absolute complementation*, *total coverage* and *complement symmetry*. In fact it is straightforward to see that classical mass functions are the only mass functions which satisfy both *absolute complementation* and *total coverage*. However, classical mass functions do not satisfy *pairwise consonance* since for $(F_1, F_1^c), (F_2, F_2^c) \in \mathcal{FG}_x \; F_1 \subseteq F_2$ implies that $F_2^c \subseteq F_1^c$. For classical mass functions we can instead define a weaker version of consonance in the standard way as follows:

Definition 11. *One-Dimensional Consonance*
A classical mass function m_x is (one-dimensionally) consonant if (F_1, F_1^c), $(F_2, F_2^c) \in \mathcal{FG}_x$ implies that either $F_1 \subseteq F_2$ or $F_2 \subseteq F_1$.

The following theorem [5], [9] shows that for classical mass functions satisfying one dimensional consonance the resulting appropriateness measures satisfy the max-min combination rules on a restricted class of label expressions.

Theorem 12. *Let $LE^{\wedge,\vee}$ denote the label expressions in LE which only involve connectives $\wedge$ and $\vee$. If m_x is a (one-dimensionally) consonant classical mass function then $\forall \theta, \varphi \in LE^{\wedge,\vee}$ $\mu_{\theta \wedge \varphi}(x) = \min(\mu_\theta(x), \mu_\varphi(x))$ and $\mu_{\theta \vee \varphi}(x) = \max(\mu_\theta(x), \mu_\varphi(x))$.*

7 Summary and Conclusions

In this paper we have proposed a generalised version of the label semantics framework in which communicating agents make explicit decisions both about which labels are appropriate to describe an element $x \in \Omega$, and also about which negated labels are appropriate. We have shown that such a framework can capture a number of different calculi for reasoning with vague concepts as special cases.

Acknowledgements

Inés González-Rodríguez is supported by MEC-Programa José Castillejo Grant JC2007-00152 and MEC-FEDER Grant TIN2007-67466-C02-01.

References

1. Dubois, D.: Prade. H.: An Introduction to Bipolar Representations of Information and Preference. International Journal of Intelligent Systems 23(8), 866–877 (2008)
2. Goodman, I.R.: Fuzzy Sets as Equivalence Classes of Random. In: Yager, R. (ed.) Sets in Fuzzy Set and Possibility Theory, pp. 327–342 (1982)
3. Goodman, I.R., Nguyen, H.T.: Uncertainty Models for Knowledge Based Systems. North-Holland, Amsterdam (1985)
4. Kyburg, A.: When Vague Sentences Inform: A Model of Assertability. Synthese 124, 175–192 (2000)
5. Lawry, J.: A Framework for Linguistic Modelling. Artificial Intelligence 155, 1–39 (2004)
6. Lawry, J.: Modelling and Reasoning with Vague Concepts. Springer, Heidelberg (2006)
7. Lawry, J.: Appropriateness measures: An uncertainty model for vague concepts. Synthese 161(2), 255–269 (2008)
8. Parikh, R.: Vagueness and Utility: The Semantics of Common Nouns. Linguistics and Philosophy 17, 521–535 (1994)
9. Tang, Y., Zheng, J.: Linguistic Modelling based on Semantic Similarity Relation amongst Linguistic Labels. Fuzzy Sets and Systems 157, 1662–1673 (2006)
10. Williamson, T.: Vagueness. Routledge, London (1994)
11. Zadeh, L.A.: Fuzzy Sets. Information and Control 8(3), 338–353 (1965)

Handling Analogical Proportions in Classical Logic and Fuzzy Logics Settings

Laurent Miclet[1] and Henri Prade[2]

[1] ENSSAT/IRISA, Lannion, France
miclet@enssat.fr
[2] IRIT, Toulouse, France
prade@irit.fr

Abstract. Analogical proportions are statements of the form "A is to B as C is to D" which play a key role in analogical reasoning. We propose a logical encoding of analogical proportions in a propositional setting, which is then extended to different fuzzy logics. Being in an analogical proportion is viewed as a quaternary connective relating four propositional variables. Interestingly enough, the fuzzy formalizations that are thus obtained parallel numerical models of analogical proportions. Potential applications to case-based reasoning and learning are outlined.

1 Introduction

Although analogical reasoning is largely used by humans in creative thinking (e.g. [17]) or for assessing day life situations, its place w. r. t. the other forms of reasoning has remained singular. Indeed while deductive reasoning uses sound and correct inferences, the conclusions obtained by analogical reasoning are provisional in nature and are plausible at the best. Deduction, but also abduction, or induction, have received rigorous logical formalizations, while it does not seem that it is really the case for analogical reasoning. Deduction and analogy are two very different forms of reasoning: Deductive entailment is based on the inclusion of classes, while analogy parallels particular situations. The latter form of reasoning applies when the former does not, and jumps to conclusions that may be more creative since they are not implicitly contained in the premises as in deduction. Although analogical reasoning has remained much less formalized, it has been considered early in artificial intelligence, e. g. ([6], [12], [22]), and case-based reasoning [1], a special form of it, has become a subfield in itself, while more general forms of analogical reasoning continue to be investigated (e.g. in conceptual graphs structures, [18], or in logic programming settings [8],[21]).

Analogical reasoning equates the way two pairs of situations differ, by stating analogical proportions of the form "*A* is to *B* as *C* is to *D*". It expresses that *A* and *B* are similar, and differ, in the same way as *C* and *D* are similar, and differ. The name "analogical proportion" comes from a quantitative view of this correspondence as an equality of ratios between numerical features. In this paper, we are interested in looking for a logical modeling that provides a symbolic

C. Sossai and G. Chemello (Eds.): ECSQARU 2009, LNAI 5590, pp. 638–650, 2009.

and qualitative representation of analogical proportion, and may be extended to fuzzy logic in order to obtain a logical graded counterpart of numerical models (defined in terms of ratios, or of differences). Then situations A, B, C and D are supposed to be described in terms of a family of properties, and to be represented by vectors of degrees of truth. Each vector component is the degree to which the corresponding property is true for the situation. In case of binary properties, the vector components belong to $\{0, 1\}$, and is equal to 1 if and only if the property holds in the considered situation.

The paper is organized in the following way. Section 2, after introducing the notations, discusses existing postulates for analogical proportions in relation with a set theoretic point of view. Section 3 proposes a classical logic representation of analogical proportions, and then studies its properties. Section 4 presents some fuzzy logic extensions of the logical modeling of analogical proportions, and compare them to numerical models. Section 5 and Section 6 point out potential applications in case-based reasoning and learning respectively.

2   Towards a Formalization of the Analogical Proportion

An analogical proportion is a statement of the form "A is to B as C is to D". This will be denoted by $(A : B :: C : D)$. In this particular form of analogy, the objects A, B, C and D usually correspond to descriptions of items under the form of objects such as sets, multisets, vectors, strings, or trees (see [20]). In the following, we are mainly interested in the basic cases where A, B, C and D may be binary values in $\{0, 1\}$, or "fuzzy values" in the unit interval $[0, 1]$, and more generally, vectors of such values (which may be used for the logical encoding of compound cases). These values can be thought in practice as degrees of truth of statements pertaining respectively to A, B, C and D. In the following if the objects A, B, C, and D are vectors having n components, i.e., $A = (a_1, \ldots, a_n)$, $\ldots$, $D = (d_1, \ldots, d_n)$, we shall say that A, B, C, and D are in analogical proportion if and only if for each component i an analogical proportion "a_i is to b_i as c_i is to d_i" holds. If there is no need to specify one particular component, we shall simply write $(a : b :: c : d)$ for stating that the 4-tuple (a, b, c, d) satisfies a relation of analogical proportion.

2.1   Postulates

We have now to specify what kind of relation an analogical proportion may mean. Intuitively speaking, we have to understand how to interpret "is to" and "as" in "A is to B as C is to D". A may be similar (or identical) to B in some respects, and differ in other respects. The way C differs from D should be the same as A differs from B, while C and D may be similar in some other respects, if we want the analogical proportion to hold. More formally, let us denote by U the features that both A and B have, by V the features possessed by A and not by B, and by W the features possessed by B and not by A, which can be symbolically written by $A = (U, V)$, and $B = (U, W)$. If C and D differ in the same way as A and B, this forces to have $C = (Z, V)$, and $D = (Z, W)$, where Z denotes the features that both C and D have. Note that U and Z may be different. This view is enough for justifying the following three postulates that date back to Aristotle's time. See, e.g. [13].

Definition 1. *An* analogical proportion *is a quaternary relation on a set X that verifies, for all A, B, C and D in X the three postulates of analogy:*

(ID) $(A : B :: A : B)$; (S) $(A : B :: C : D) \Leftrightarrow (C : D :: A : B)$
(CP) $(A : B :: C : D) \Leftrightarrow (A : C :: B : D)$

(ID) and (S) express reflexivity and symmetry for the comparison "as", while (CP) allows for a central permutation. These postulates are natural requirements, if we keep in mind that $A = (U, V)$, $B = (U, W)$, $C = (Z, V)$, and $D = (Z, W)$. Indeed the first two are particularly obvious; concerning the third, let us notice that V (resp. W) is the common part of A and C (resp. B and D) and when going from A to C, we leave U and get Z, as when going from B to D. The third postulate is peculiar to analogical proportions (and is reminiscent of numerical proportions). Immediate consequences of the postulates are:

(I): $(A : B :: C : D) \Leftrightarrow (B : A :: D : C)$,
(EP): $(A : B :: C : D) \Leftrightarrow (D : B :: C : A)$,
$(SR1)$: $(A : B :: C : D) \Leftrightarrow (D : C :: B : A)$,
$(SR2)$: $(A : B :: C : D) \Leftrightarrow (B : D :: A : C)$,
$(SR3)$: $(A : B :: C : D) \Leftrightarrow (C : A :: D : B)$,

where (I) allows for the inversion of the relations (obtained by applying (CP), (S) and (CP)), (EP) allows for external permutation (obtained by applying (S), (CP) and (S)), $(SR1)$, $(SR2)$ and $(SR3)$ expressing symmetries for the reading (and can be respectively obtained by (I) and (S), (CP) and (S), and (S) and (CP)). Note also that $(A : A :: B : B)$ is obtained from (ID) and (CP).

It has been noticed that starting with $(A : B :: C : D)$, the repeated application of (S) and (CP) generate only 8 of the 24 possible permutations of the 4-element set $\{A, B, C, D\}$. Indeed, if $(A : B :: C : D)$ is an analogical proportion, it is not expected that $(B : A :: C : D)$, or $(C : B :: A : D)$ be analogical proportions also. For instance, the statement "a calf is to a bull what a kitten is to a tomcat" does not mean that "a bull is to a calf what a kitten is to a tomcat", or that "a kitten is to a bull what a calf is to a tomcat", while the statement "a calf is to a kitten what a bull is to a tomcat", obtained by (CP), sounds more acceptable.

2.2   The Set Theoretic Point of View

When the objects are finite sets, they can be seen as subsets of some universal set $\mathcal{P}$. The relation "as" is simply chosen as the equality between sets. In this framework, Lepage has given in [13] the following informal definition: four subsets of $\mathcal{P}$ are in analogical proportion $(A : B :: C : D)$ if A is transformed into B and C is transformed into D by adding and deleting the same elements. For example, the four sets $A = \{t_1, t_2, t_3, t_4, \}$, $B = \{t_1, t_2, t_3, t_5\}$ and $C = \{t_1, t_4, t_6, t_7\}$, $D = \{t_1, t_5, t_6, t_7, \}$ are in analogical proportion: t_4 is deleted from A (resp. C) and t_5 is added to A (resp. C) in order to obtain B (resp. D).

More formally, a definition has been first proposed by Lepage in [13] in computational linguistics a few years ago, and further developed in [19]. We restate it in a different way here. We denote $\overline{A}$ the complementary set of A in $\mathcal{P}$ and $A - B = A \cap \overline{B}$. We notice that "$A : B$" stands for the set operation that transforms A into B by deleting the elements of $A - B$ and adding the elements of $B - A$. The analogical proportion states the identity of the operations that transform A into B and C into D. This leads to the following definition:

Definition 2. *Let A, B, C, D be subsets of a referential $\mathcal{P}$.*
$(A : B :: C : D) \Leftrightarrow (A - B = C - D)$ *and* $(B - A = D - C)$.

Definition 2 clearly satisfies the three postulates. Stroppa and Yvon [19] have given an equivalent set-theoretic characterization of the analogical proportion:

Definition 3. *Let A, B, C, D be subsets of $\mathcal{P}$. $(A : B :: C : D)$ holds if and only if there exist four subsets U, V, W and Z of $\mathcal{P}$, such that $A = U \cup V$, $B = U \cup W$, $C = Z \cup V$, $D = Z \cup W$.*

This decomposition is not unique, and the sets U, V, W, Z do not need to be disjoint. When they are, this provides a constructive description of the analogical process: X (resp. Z) is the elements that are untouched when going from A and B (resp. from C and D), while the elements in V go out, and those in W go in.

3   Proposal for a Classical Logic Embedding

Attempts at providing a logical embedding of analogical proportion at least dates back to the proposal made by a computer scientist, Klein, working in anthropology, more than twenty-five years ago in [10]. Klein used an operator (called by him ATO for "Appositional Transformation Operator") on binary truth-like tables, which is nothing but the logical equivalence connective: $a \equiv b = 1$ if $(a = b)$ and $a \equiv b = 0$ otherwise. His view amounts to define an analogical proportion semantically as a logical connective having the truth table of the logical expression $(a \equiv b) \equiv (c \equiv d)$. It partially agrees with the idea that A differs from B as C w. r. t. D, since it can be also written $(a \Delta b) \equiv (c \Delta d)$ where Δ denotes XOR. It can still be rewritten as well as $(a \wedge \neg b) \vee (\neg a \wedge b) \equiv (c \wedge \neg d) \vee (\neg c \wedge d)$.

However, this latter expression remains symmetrical, since it makes no difference between the way A differs from B and the way B differs from A. It is clearly weaker than stating the two equivalences $(a \wedge \neg b) \equiv (c \wedge \neg d)$ and $(b \wedge \neg a) \equiv (d \wedge \neg c)$ separately. This is the logical counterpart of Definition 2 given from the set theoretic point of view. Klein's view of analogy was indeed too permissive, since $(a \equiv b) \equiv (c \equiv d)$ is still equivalent to $(b \equiv a) \equiv (c \equiv d)$, thus making no difference between "A is to B" and "B is to A".

The 8 cases where Klein's expression, $(a \equiv b) \equiv (c \equiv d)$, takes truth value 1 are listed in Table 1. For the 8 other possible combinations of values of a, b, c,

Table 1. Contrasting Klein's definition of analogy with ours

$a\ b\ c\ d$	$(a \equiv b) \equiv (c \equiv d)$	$(a : b :: c : d)$
1 1 1 1 1	1	1
2 1 1 0 0	1	1
3 1 0 1 0	1	1
4 1 0 0 1	1	0
5 0 1 1 0	1	0
6 0 1 0 1	1	1
7 0 0 1 1	1	1
8 0 0 0 0	1	1

and d that are not in Table 1, $(a \equiv b) \equiv (c \equiv d)$ has truth value 0. Cases 1, 2, 7, and 8 correspond to situations where a and b are identical as well as c and d. Cases 3 and 6 correspond to changes from a to b, and from c to d, that go in the same sense. All this fits the semantics of the analogical proportion. The two other cases, namely 4 and 5, do not fit the idea that a is to b as c is to d, since the changes from a to b and from c to d are not in the same sense. They in fact correspond to cases of maximal analogical dissimilarity, where "d is not at all to c what b is to a", but rather "c is to d what b is to a". It emphasizes the non symmetry of the relations between b and a, and between d and c.

Our definition is equivalent to stating the two equivalences $(a \wedge \neg b) \equiv (c \wedge \neg d)$ and $(b \wedge \neg a) \equiv (d \wedge \neg c)$ separately, which is coherent with the definition of analogical proportion between finite sets given at section 2.2.

3.1 Logical Expressions for the Analogical Proportion

We are now looking for logical expressions corresponding to our definition, i.e. only covering cases 1, 2, 3, 6, 7, 8 in Table 1. Viewing $(a : b :: c : d)$ as a logical connective that reflects the analogical process, it can be checked that, taking

$$(a : b :: c : d) = ((a \equiv b) \equiv (c \equiv d)) \wedge ((a \Delta b) \rightarrow (a \equiv c)) \tag{1}$$

this expression is true only for the 6 cases required in Table 1. Indeed, it expresses in its second component that a and c should be identical where a and b differs, which is a natural constraint for making sure that the change from c to d will be in the same sense as the one from a to b. There exist equivalent expressions whose structures well reflect the meaning of analogical proportion:

$$(a : b :: c : d) = ((a \equiv b) \wedge (c \equiv d)) \vee ((a \equiv c) \wedge (b \equiv d)) \tag{2}$$

$$(a : b :: c : d) = ((a \rightarrow b) \equiv (c \rightarrow d)) \wedge ((b \rightarrow a) \equiv (d \rightarrow c)) \tag{3}$$

$$(a : b :: c : d) = ((a \wedge d) \equiv (b \wedge c)) \wedge ((a \vee d) \equiv (b \vee c)) \tag{4}$$

$$(a : b :: c : d) = ((a \wedge d) \equiv (b \wedge c)) \wedge ((a \vee d) \equiv (b \vee c))^1 \tag{5}$$

---

[1] Pointed out by Didier Dubois to the authors.

These expressions help to understand the structure of analogical proportion. For instance, expression (3) at the logical level parallels the difference-based view of the analogical proportion, expressed by the condition $(a - b) = (c - d)$. When a and b are equal to 0 or 1, $a - b \in \{-1, 0, 1\}$. It is why expression (3), which works in $\{0, 1\}$, states the equivalences in each sense (remember that $a \to b = 1$ if $a \leq b$ and $a \to b = 0$ if $a > b$, and observe that the condition $a \leq b$ covers two situations: $a = b$ (no change) or $a < b$ (change)).

Clearly, since $a - b \in \{-1, 0, 1\}$, $(a - b)$ is not a connective, but keeps track of the sense of the change if any. It is the basis of the notion of *analogical dissimilarity* $AD(a, b, c, d)$ [15] that measures how far objects a, b, c and d are from being in an analogical proportion. $AD(a, b, c, d)$ must be equal to 0 if they are in such a relation, and positive otherwise. Required properties are: i) *Coherence with analogy*: $AD(a, b, c, d) = 0 \Leftrightarrow (a : b :: c : d)$ ii) *Symmetry of "as"*: $AD(a, b, c, d) = AD(c, d, a, b)$ iii) *Triangle inequality*: $AD(a, b, c, d) \leq AD(a, b, e, f) + AD(e, f, c, d)$ iv) *Central permutation*: $AD(a, b, c, d) = AD(a, c, b, d)$ v) *Dissymmetry of "is to"*: $AD(a, b, c, d) \neq AD(b, a, c, d)$, in general. Defining $AD(b, a, c, d) = |(a - b) - (c - d)|$ agrees with the required properties.

3.2   Some Properties

We now state some results (symbolic propositional expressions and their semantical counterparts in terms of truth degrees are denoted in the same way):

Proposition 1. $(a : b :: \neg b : \neg a) = 1$.

This looks similar to the logical equivalence between $a \to b$ and $\neg b \to \neg a$.

Proposition 2. *If* $(a \to b) = 1$ *and* $(a : b :: c : d) = 1$ *then* $(c \to d) = 1$. *Similar results hold if* $\to$ *is replaced by* $\leftarrow$, $\equiv$ *or* Δ. *It does not hold for* $\vee$ *nor* $\wedge$.

It ensures that if "*A* is to *B* as *C* is to *D*", and *B* is more general than *A*, then *D* should be more general than *C*, as expected. More generally, it expresses a form of agreement with connectives related to entailment.

Proposition 3. $(a : b :: c : d) = 1, (c : d :: e : f) = 1 \Longrightarrow (a : b :: e : f) = 1$

It expresses that transitivity holds for analogical proportion. A less obvious result is about the behavior of analogical proportion w.r.t. conjunction and disjunction.

Proposition 4. $(a \wedge b : a \wedge c :: d \wedge b : d \wedge c) = 1$ *if and only if*
   $(a \vee b : a \vee c :: d \vee b : d \vee c) = 1$.

It can be seen as resulting from the combination of the two universal analogical proportions $(a : a :: d : d) = 1$ and $(b : c :: b : c) = 1$.

3.3   Solving an Analogical Proportion Equation

In its basic form, analogical reasoning amounts to solving an analogical proportion equation that is supposed to hold for some characteristics of objects, the assumption being generally based on the observation that the analogical proportion already holds for other known characteristics. Solving an analogical proportion equation consists in finding x s.t. $(a : b :: c : x) = 1$. When it exists, the value of x is unique in the classical logic setting, but does not always exist.

Proposition 5. *A triple $(a\ b\ c)$ can be completed by d in such a way that $(a : b :: c : d) = 1$ if and only if $((a \equiv b) \vee (a \equiv c)) = 1$.*

Proposition 6. *When it exists, the unique solution of the equation $(a : b :: c : x) = 1$ is logically expressed by $x = (a \equiv (b \equiv c))$.*

Proposition 5 points out that a should be equivalent to b or to c, in order to get rid of the two triples $(a\ b\ c) = (1\ 0\ 0)$ and $(a\ b\ c) = (0\ 1\ 1)$ that cannot be completed analogically (see Table 1). In the set-theoretic view, a, b, c, and d stand for the membership degrees of an element in a referential $\mathcal{P}$ to A, B, C, and D respectively. Then the impossibility of $(a\ b\ c) = (1\ 0\ 0)$ and $(a\ b\ c) = (0\ 1\ 1)$ translates respectively into $A \cap (\overline{B} \cap \overline{C}) = \emptyset$ and $\overline{A} \cap (B \cap C) = \emptyset$, i.e. the logical condition for analogical completion $(a \equiv b) \vee (a \equiv c) = 1$ can be written in set terms as $(B \cap C) \subset A \subset (B \cap C)$, a condition already given in Section 3.

Proposition 6 provides a compact writing of the solution of an analogical proportion. This is the solution first suggested by Klein [10] who noticed that the repeated use of what he called the ATO operator enables him to compute the solution of analogical proportions, according to the equality $d = (c \equiv (a \equiv b))$.

Other expressions of x under the requirement of Proposition 5 exist, e.g.:

Property 1. When it exists, the unique solution of the analogical equation $(a : b :: c : x) = 1$ is logically expressed by

$$x = ((b \vee c) \wedge \neg a) \vee (b \wedge c) = (b \wedge \neg a) \vee (c \wedge \neg a) \vee (a \wedge b \wedge c)$$

$$x = (a \to (b \wedge c)) \wedge (b \vee c) = (a \to b) \wedge (a \to c) \wedge (b \vee c)$$

Both can be easily checked on a truth table. The first one is nothing but the logical counterpart of expressions recently proposed in [15] in the "set element-interpretation". Both Proposition 6 and Proposition 1 could be applied when $(a\ b\ c)$ cannot be analogically completed, i.e. when $(a \equiv b) \vee (a \equiv c) = 0$. Mind that while Proposition 6 applied to the two "undesirable cases" $(a\ b\ c) = (1\ 0\ 0)$ and $(a\ b\ c) = (0\ 1\ 1)$ yields $x = 1$ and $x = 0$ respectively, the two expressions of Proposition 1 give the converse, namely $x = 0$ and $x = 1$ respectively in these two cases. This means that the expression $x = (a \equiv (b \equiv c))$ is logically equivalent to the two expressions in Proposition 1, only under the condition $(a \equiv b) \vee (a \equiv c) = 1$ (which is equivalent to conditions $((b \wedge c) \to a) = 1$ and $(a \to (b \vee c)) = 1$). It can be seen from Proposition 1 that x is also such that $((b \wedge c) \to x = 1)$ and $(x \to (b \vee c)) = 1$.

4   Extensions to Fuzzy Logic

When moving from the binary case to the graded (or fuzzy) case where truth values now belong to the continuous interval $[0, 1]$, many choices are possible for defining the connectives, and it should be clear that some of the equivalences previously found may now fail to hold since, whatever the choices, we shall be no longer in a Boolean algebra. Here we only consider choices that seem to be especially worth of interest, due to their resemblance with numerical models.

4.1   Construction of a Fuzzy Analogical Proportion

Let us recall that in fuzzy logic [11] there are three main choices for the conjunction, namely $a \wedge b = min(a, b)$, $a \wedge b = a \cdot b$, or $a \wedge b = max(0, a + b - 1)$, associated with the three disjunctions $a \vee b = max(a, b)$, $a \vee b = a + b - a \cdot b$, or $a \vee b = min(1, a + b)$ respectively. Then there are two main ways for defining implications, either as $a \rightarrow b = \neg a \vee b$, or by residuation: $a \rightarrow b = sup(x | a \wedge x \leq b)$.

It leads to distinct connectives for the first two pairs of conjunction/disjunction: $a \rightarrow b = max(1 - a, b)$ (Dienes implication) and $a \rightarrow b = 1$ if $a \leq b$ and $a \rightarrow b = b$ if $a > b$ (Gödel implication) for min/max, $a \rightarrow b = 1 - a + a \cdot b$ (Reinchenbach implication) and $a \rightarrow b = min(1, b/a)$ if $a > 0$, and $a \rightarrow b = 1$ if $a = 0$ (Goguen implication) with the second pair (using $\neg a = 1 - a$). For the last pair of conjunction/disjunction, Lukasiewicz implication $a \rightarrow b = min(1, 1 - a + b)$ is obtained in both cases.

The equivalence connective associated to Dienes implication is $(a \equiv b) = min(max(1 - a, b), max(1 - b, a)) = max(min(a, b), min(1 - a, 1 - b))$, to Gödel implication is $(a \equiv b) = 1$ if $a = b$, and $(a \equiv b) = min(a, b)$ otherwise (in a crisp version, one may take $(a \equiv b) = 0$ if $a \neq b$). Using min conjunction and Lukasiewicz implication, one gets $(a \equiv b) = min(min(1, 1 - a + b), min(1 - b + a)) = 1 - |a - b|$. Using min or product conjunction and Goguen implication, one gets $(a \equiv b) = min(1, b/a, a/b) = min(b/a, a/b) = min(1, b/a) \cdot min(1, a/b)$ for $a \neq 0, b \neq 0$ (if $a = 0$ and $b \neq 0$, $(a \equiv b) = 0$; if $a = 0$ and $b = 0$, $(a \equiv b) = 1$).

In the following, we only discuss the fuzzification of equation 3 that clearly states the identity of the differences, using successively the two following choices:

1. $a \wedge b = min(a, b)$; $a \rightarrow b = min(1, 1 - a + b)$; $a \equiv b = 1 - |a - b|$
2. $a \wedge b = a \cdot b$; $a \rightarrow b = max(1, b/a)$; $a \equiv b = min(b/a, a/b)$

It leads to the two formulas below for the value of the fuzzy analogical proportion

$$min \begin{cases} 1 - |min(1, 1 - a + b) - min(1, 1 - c + d)| \\ 1 - |min(1, 1 + a - b) - min(1, 1 + c - d)| \end{cases}$$

$$min \left(\frac{max(1, \frac{b}{a})}{max(1, \frac{d}{c})}, \frac{max(1, \frac{d}{c})}{max(1, \frac{b}{a})} \right) \cdot min \left(\frac{max(1, \frac{a}{b})}{max(1, \frac{c}{d})}, \frac{max(1, \frac{c}{d})}{max(1, \frac{a}{b})} \right).$$

The first formula yields 1 iff $a - b :: c - d$, the values of a, b c and d being in the unit interval. The second formula yields 1 iff $a/b :: c/d$. Both are also consistent with the logical analogy defined above.

4.2   Coherence with Numerical Analogy

When defining an analogical relation between four fuzzy values, first one has to make sure that it remains consistent with the classical logical definition when fuzzy values reach the bounds of the interval $[0, 1]$, as discussed in subsection 4.1. But it is also important to maintain a link with the definition of analogy between real numbers. Several definitions have been proposed for analogical proportions between real numbers, in particular: i) the additive analogy: $(a : b :: c : d) \Leftrightarrow (a + d = b + c)$, ii) the multiplicative analogy: $(a : b :: c : d) \Leftrightarrow (ad = bc)$.

The first fuzzification is rather coherent with the numerical additive case. We can analyse their difference in taking s and t as small positive numbers. We get in the fuzzy case $(a : a + s :: c : c - t) = 1 - min(s, t)$. In the numerical case, we would have obtained $(a : a + s :: c : c - t) = 1 - AD(a, a + s, c, c - t) = 1 - (s + t)$, still using the same definition of analogical dissimilarity, now applied to fuzzy values. However, note that in the fuzzy case we deal with truth values, while in the numerical case we deal with attribute values!

It is important to note here that the fuzzy counterpart of Proposition 1 cannot be straightforwardly applied for finding the solution of an analogical proportion in the graded case. Proper equivalent expressions have to be found. For instance, if we use $((b \rightarrow a) \rightarrow c)$ if $a \rightarrow b = 1$, and $\neg(c \rightarrow \neg(a \rightarrow b))$ if $b \rightarrow a = 1$, which is indeed equivalent to $(a \equiv (b \equiv c))$, we shall obtain with Lukasiewicz implication, $min(1, c + (b - a))$ if $a \leq b$, and $max(0, c - (a - b))$ if $a \geq b$, which are normalized versions of the solution of the numerical equation $a - b = c - x$.

5   Analogical Proportion-Based Reasoning

We have already noticed that, for any pair of propositions p and q it holds that $(p : q :: p : q) = 1$ (consistently with the 1st postulate), and $(p : q :: \neg q : \neg p) = 1$. This shows a form of agreement between the analogical proportion and the *modus ponens*, and (which is less expected), with the *modus tollens*. Indeed from a non specified, hypothetical relation between p and q, denoted by $p : q$ and from p, one concludes q, since it can be checked that $x \equiv q$ is the only solution of the equation $(p : q :: p : x) = 1$. Similarly, from $(p : q :: \neg q : x) = 1$, one gets $x \equiv \neg p$. But $(p : q :: \neg p : \neg q) = 1$ does not hold (while it holds with Klein's definition).

An object, a situation, a problem may be described in terms of sets of features (resp. properties) that are present (resp. true) or absent (resp. false) in the binary case, or more generally that are present (resp. true) to some extent in the fuzzy case. In that respect, the classical logic equivalence $(p : q :: r : s) \equiv (p : q :: \neg r : \neg s)$ insures the neutrality of the encoding whatever the convention used for stating what is present and what is absent ($\neg p$ present is the same as p absent). It holds in the fuzzy case under the form $(a : b :: c : d) = (1 - a : 1 - b :: 1 - c : 1 - d)$ for degrees of truth, using Lukasiewicz implication, since $(1 - a) \rightarrow (1 - b) = min(1, 1 - (1 - a) + 1 - b) = min(1, 1 - b + a) = b \rightarrow a$. It could be also preserved using a symmetrized form of Goguen implication: $min(1, b/a, (1 - a)/(1 - b))$.

The featured view can be applied to logical formulas themselves, e.g. "$\neg p \wedge q$ is to $p \wedge q$ as $\neg p \wedge \neg q$ is to $p \wedge \neg q$" can be encoded as the 2-component analogical proportion $((01) : (11) :: (00) : (10))$ where $0/1$ stands for the negative/positive presence of p and q in the clauses. This agrees with the use of Hamming distance for computing the amount of change from a formula to another. It suggests that analogical reasoning could be related to revision operations based on the idea of minimal change, as recently proposed in case-based reasoning [14].

Towards case-based reasoning. Let us illustrate a more direct and practical use of the featured view. Assume a base of cases describing houses to let. In the example, we consider four features: nature (villa (1) or apartment (0)), air conditioning (equipped (1) or not (0)), price (cheap (1) or expensive (0)), tax to pay (yes (1) or no (0)). Assume we know the three cases:

$A = $ (villa, equip., expen., tax) $= (1, 1, 0, 1)$
$B = $ (villa, not-eq., cheap, tax) $= (1, 0, 1, 1)$
$C = $ (apart., equip., expen., tax) $= (0, 1, 0, 1)$

Assume now a fourth house described by $D = $ (apart., not-eq., x, y) $= (0, 0, x, y)$ for which one has to guess a price and if there is a tax to pay. After checking that for the first two components we have an analogical proportion between A, B, C, and D (indeed in terms of truth values we have $(1 : 1 :: 0 : 0) = 1$, and $(1 : 0 :: 1 : 0) = 1$), one may assume that it also holds for the two other components and the unique solution of the equations $(0 : 1 :: 0 : x) = 1$ and $(1 : 1 :: 1 : y) = 1$ is $x = 1$ and $y = 1$, which means "cheap" and "tax to pay".

A refined version of the example can be described in a graded manner as e.g., $A = (1, 1, .2, .9)$, $B = (1, 0, .8, .8)$, $C = (0, 1, .3, .6)$, where the degrees respectively stand for the extent to which the price is cheap and the tax is high. Applying a difference-based approach, using Lukasiewicz implication, one gets $D = (0, 0, .9, .5)$, i.e. D should be quite cheap (.9) with a not too high tax (.5).

This example suggests that one may apply such an approach to case-based reasoning. Then A, B and C are three problems with their respective solutions that are identical according to some features and that differ with respect to other features. Both problems and solutions (e.g., a disease and a medical treatment for it) are described in terms of feature values, D is a new problem for which one looks for a tentative solution. Then the solution for D will be computed as an adapted version of those for cases A, B and C, from the differences and similarities between them, as outlined in the above example where the role of the "problem" was played by the nature and the air conditioning availability of the house, and the problem was to guess the price of house D and if there is a tax to pay. It is clear that in general there may exist in a repertory of cases several triples A_i, B_i, C_i, from which the solution for D can be computed analogically as just explained. Then it would lead to aggregate the different solutions *disjunctively* into an imprecise solution, as already done in the simplest type of case-based decision [4] and in case-based reasoning [5]. Indeed, in these approaches, a solution is proposed for a new case on the basis of a formal principle that states that "the more similar two cases are in some respects, the more guaranteed the possibility that they are similar in the other respect for which a solution is looked for for the new case". Then, since the new case may resemble several cases, in the repertory of cases, which are different with respect to the feature to predict (or the solution to choose) in the new case, a disjunctive combination should take place for aggregating the solutions found. We are here in a similar situation, since several triples A_i, B_i, C_i may be found in analogical proportion with D. The details of the procedure are left for further research.

6   Analogical Proportion-Based Learning

Let $\mathcal{S} = \{(x, \omega(x))\}$ be a finite set of training examples, where x is the description of an example as a binary vector and $\omega(x)$ its label in a finite set. Given the binary vector y of a new pattern, we want to assign a label $\omega(y)$ to y, based only on knowledge in $\mathcal{S}$. Finding $\omega(y)$ amounts to the *inductive learning* of a classification rule from examples (e.g. [16]). The nearest neighbor (*1-nn*) method, the most simple lazy learning technique, merely finds in $\mathcal{S}$ *one* description $x^\star$ which minimizes some distance to y and hypothesizes $\omega(x^\star)$, the label of $x^\star$, for the label of y. Moving one step further, learning from *one* analogical proportion would consist in searching in $\mathcal{S}$ for one triple $(x^\star, z^\star, t^\star)$ such that $x^\star : z^\star :: t^\star : y$ or $AD(x^\star, z^\star, t^\star, y)$ is minimal and would predict for y the label $\hat{\omega}(y)$ solution of equation $\omega(x^\star) : \omega(z^\star) :: \omega(t^\star) : \hat{\omega}(y)$.

The *1-nn* method is easily extended to examine a larger neighbourhood, resulting in the *k-nn* method: in $\mathcal{S}$, find the k descriptions which minimize the distance to y and let vote the k corresponding labels to choose the winner as the label of y. Extending the learning from one analogical proportion to k ones can be designed similarly. First, we define an analogical dissimilarity AD between two binary vectors. It can straightforwardly be defined as the sum of the AD of their components (see section 3.1). Secondly, we follow the following procedure:

- Consider only trivial equations on classes, e.g. $(\omega_1 : \omega_2 : \omega_1 : \hat{\omega}(y))$, which produces the solution $\hat{\omega}(y) = \omega_2$.
- Use a weighting of the binary attributes (the weights are learned from $\mathcal{S}$).
- For an integer k, use all the triples in $\mathcal{S}$ that make AD less than k for y.

To experiment the efficiency of this technique (*learning from k analogical proportions*), we have made experiments on several classical datasets with binary or nominal attributes, the latter being straightforwardly binarised. The results, given in [2], show that it gives excellent results on all the data bases, including those with missing data or composed with more than two classes. This method can be seen as an extension of the *k-nn* method. It also suggests possible links with fuzzy instance-based learning [9], which also extends the *k-nn* method.

7   Conclusion

This paper is a first attempt towards a logical formalization of analogical reasoning based on analogical proportions. It offers a unified treatment of symbolic and numerical analogical proportions, thanks to the extension of the proposed classical logic formulation to different fuzzy logics keeping the additive or the multiplicative flavors of numerical modeling. Beyond the theoretical interest of such logical encodings, the paper has indicated how reasoning from several cases and learning can benefit from the resolution of analogical proportions. Another direction for further research would be to discuss the relation between fuzzy set-based approximate reasoning and analogical reasoning, already studied in [3] with another approach. In the long range, it would be also of interest to develop cognitive validation tests in order to study if the predictions that can be obtained with the approach are in agreement with human reasoning (as in e.g.[7]).

References

1. Aamodt, A., Plaza, E.: Case-based reasoning: Foundational issues, methodological variations, and system approaches. Artificial Intelligence Com., 39–59 (1994)
2. Bayoudh, S., Miclet, L., Delhay, A.: Learning by analogy: a classification rule for binary and nominal data. In: Veloso, M. (ed.) Proc. IJCAI 2007, pp. 678–683. AAAI Press, Menlo Park (2007)
3. Bouchon-Meunier, B., Valverde, L.: A fuzzy approach to analogical reasoning. Soft Computing 3, 141–147 (1999)
4. Dubois, D., Esteva, F., Garcia, P., Godo, L., Lopez de Mantaras, R., Prade, H.: Fuzzy modelling of cased-based reasoning and decision. In: Leake, D.B., Plaza, E. (eds.) ICCBR 1997. LNCS, vol. 1266, pp. 599–610. Springer, Heidelberg (1997)
5. Dubois, D., Hüllermeier, E., Prade, H.: Fuzzy set-based methods in instance-based reasoning. IEEE Trans. on Fuzzy Systems 10, 322–332 (2002)
6. Evans, T.: A heuristic program to solve geometry analogy problem. In: Minsky, M. (ed.) Semantic Information Processing. MIT Press, Cambridge (1968)
7. Gentner, D., Kurtz, K.J.: Relations, objects, and the composition of analogies. Cognitive Science 30, 609–642 (2006)
8. Hirowatari, E., Arikawa, S.: Incorporating explanation-based generalization with analogical reasoning. Bulletin of Informatics and Cybernetics 26, 13–33 (1994)
9. Hüllermeier, E., Dubois, D., Prade, H.: Model adaptation in possibilistic instance-based reasoning. IEEE Trans. on Fuzzy Systems 10, 333–339 (2002)
10. Klein, S.: Culture, mysticism and social structure and the calculation of behavior. In: Proc. Europ. Conf. on Artificial Intelligence (ECAI 1982), pp. 141–146 (1982)
11. Klement, E.P., Mesiar, R., Pap, E.: Triangular Norms. Kluwer Acad. Publ., Dordrecht (2000)
12. Kling, R.E.: A paradigm for reasoning by analogy. Artif. Intellig. 2, 147–178 (1971)
13. Lepage, Y.: De l'analogie rendant compte de la commutation en linguistique. Habilitation (2003), http://www.slt.atr.jp/~lepage/pdf/dhdryl.pdf
14. Lieber, J.: Application of the revision theory to adaptation in case-based reasoning: The conservative adaptation. In: Weber, R.O., Richter, M.M. (eds.) ICCBR 2007. LNCS, vol. 4626, pp. 239–253. Springer, Heidelberg (2007)
15. Miclet, L., Delhay, A.: Analogical dissimilarity: definition, algorithms and first experiments in machine learning. Technical Report 5694, INRIA (September 2005)
16. Mitchell, T.: Machine Learning. McGraw-Hill, New York (1997)

17. Polya, G.: Mathematics and Plausible Reasoning. Patterns of Plausible Inference, vol. II. Princeton University Press, Princeton (1954)
18. Sowa, J.F., Majumdar, A.K.: Analogical reasoning. In: Ganter, B., de Moor, A., Lex, W. (eds.) ICCS 2003. LNCS (LNAI), vol. 2746, pp. 16–36. Springer, Heidelberg (2003)
19. Stroppa, N., Yvon, F.: Analogical learning and formal proportions: Definitions and methodological issues. Technical Report ENST-2005-D004 (June 2005), http://www.tsi.enst.fr/publications/enst/techreport-2007-6830.pdf
20. Stroppa, N., Yvon, F.: Formal models of analogical proportions. Technical report 2006D008, Ecole Nat. Sup. des Telecommunications, Paris (2006)
21. Tausend, B., Bell, S.: Analogical reasoning for logic programming. In: Kodratoff, Y. (ed.) EWSL 1991. LNCS, vol. 482, pp. 391–397. Springer, Heidelberg (1991)
22. Winston, P.H.: Learning and reasoning by analogy. Com. of ACM, pp. 689–703 (1980)

Qualitative Possibilities and Necessities

Aleksandar Perović[1], Zoran Ognjanović[2], Miodrag Rašković,[2]
and Zoran Marković[2]

[1] Facutly of transportation and traffic engineering,
Vojvode Stepe 305, 11000 Belgrade, Serbia
`pera@sf.bg.ac.yu`
[2] Mathematical Institute of Serbian Academy of Sciences and Arts,
Kneza Mihaila 36, 11000 Belgrade, Serbia
{`zorano,miodragr,zoranm`}`@mi.sanu.ac.yu`

Abstract. Qualitative possibilities and necessities are well known types of confidence relations. They have been extensively studied semantically, as relations on Boolean algebras (or equivalently, relations on algebras of sets). The aim of this paper is to give a syntactical flavor to the subject providing a sound and complete axiomatization of qualitative possibility relations.

1 Introduction

The notion of a confidence relation arose from the need for modelling beliefs. If $\leqslant_C$ is a confidence relation on some Boolean algebra, then $\alpha \leqslant_C \beta$ reads "the agent has at least as much confidence in β as in α". For our purpose, the underlying algebra will be a propositional Lindenbaum algebra LA over some countable set of propositional letters.

Qualitative possibilities and necessities are well known types of confidence relations. They have been extensively studied as relations on finite Boolean algebras (or equivalently, relations on finite algebras of sets). The aim of this paper is to give a syntactical flavor to the subject, through the form of a sound and complete axiomatization of qualitative possibility relations.

There are many reasons for the purely syntactical approach, such as verification of proposed inference mechanisms and initial postulates, finding the exact relationship between syntax and semantics (completeness theorems) and so on. Though intended as a support for formal studies of probabilistic logics, arguments of Nilsson published in [14] may be applied to the reasoning about possibility as well.

Qualitative possibility and necessity relations are studied by many authors, including Lewis, Dubois, Prade, Fariñas del Cerro, Herzig, Boutilier and many others. Arguably, the starting point are papers [2, 9]. A survey on qualitative possibility functions and integrals, which greatly influenced work presented here, was given by Didier Dubois and Henri Prade in [4]. We emphasize that [4] also contains an extensive list of references relevant for the study of qualitative possibilities and necessities. A modal approach to possibility theory was given

C. Sossai and G. Chemello (Eds.): ECSQARU 2009, LNAI 5590, pp. 651–662, 2009.

in [1,6]. Concerning qualitative probability and probabilistic reasoning in general, which is related to the problem of formal axiomatization of qualitative possibility relations in many different ways, we refer the reader to [5, 7, 8, 10, 11, 12, 13, 15, 16, 17, 18, 19, 21, 22, 23, 24].

The rest of the paper is organized as follows: Preliminaries are discussed in Section 2. The next three sections give a gradual introduction of the Hilbert-style formal system L_{Π}, in which the notion of qualitative possibility relation is completely axiomatized. The main results are completeness and compactness theorems. Concluding remarks are in the last section.

2   Preliminaries

In this paper we will use α, β and γ to denote classical propositional formulas or the corresponding equivalence classes in LA, the Lindenbaum algebra of equivalence classes of propositional formulas, letting the context to determine the corresponding meaning.

A binary relation $\leqslant_C$ on LA is a confidence relation if it has the following properties:

- $\leqslant_C$ is a weak order, i.e. $\leqslant_C$ is total ($\alpha \leqslant_C \beta$ of $\beta \leqslant_C \alpha$ for any α and β) and transitive;
- $\leqslant_C$ is monotone, i.e. $\alpha \leqslant_{LA} \beta$ implies $\alpha \leqslant_C \beta$. Here $\leqslant_{LA}$ is the usual partial order of LA: $\alpha \leqslant_{LA} \beta$ if $\alpha \wedge \beta$ is equivalent with α (or equivalently, if $\alpha \vee \beta$ is equivalent with β);
- $\leqslant_C$ is nontrivial, i.e. $\bot <_C \top$;
- (weak stability) If $\alpha \wedge (\beta \vee \gamma)$ is a contradiction, then

$$\gamma <_C \beta \quad \text{implies} \quad \alpha \vee \gamma \leqslant_C \alpha \vee \beta.$$

Here $\gamma <_C \beta$ means that $\gamma \leqslant_C \beta$ and not $\beta \leqslant_C \gamma$.

Example 1. The first example of a confidence relation is the comparative probability, introduced by Bruno de Finetti in 1937. A binary relation $\leqslant_{CP}$ on propositional Lindenbaum algebra LA is a comparative probability if it has the following properties:

- $\leqslant_{CP}$ is a non-trivial weak order;
- (consistency) $\bot \leqslant_{CP} \alpha$ for any α;
- (pre–additivity) If $\alpha \wedge (\beta \vee \gamma)$ is a contradiction, then

$$\gamma \leqslant_{CP} \beta \quad \text{iff} \quad \alpha \vee \gamma \leqslant_{CP} \alpha \vee \beta.$$

Clearly, pre-additivity is stronger than weak stability. To see that monotonicity hold for $\leqslant_{CP}$, we will combine consistency and pre–additivity in the following way: suppose that $\alpha \leqslant_{LA} \beta$. By consistency, $\bot \leqslant_{CP} \beta$. Since $\bot \wedge (\alpha \vee \beta)$ is a contradiction, pre–additivity implies $\bot \vee \alpha \leqslant_{CP} \beta \vee \alpha$. Finally, $\bot \vee \alpha$ is equivalent with α and $\alpha \vee \beta$ is equivalent with β ($\alpha \leqslant_{LA} \beta$), so $\alpha \leqslant_{CP} \beta$.   □

Definition 2. *A binary relation $\leqslant_\Pi$ on LA is a qualitative possibility if it has the following properties:*

- *$\leqslant_\Pi$ is a non-trivial weak order;*
- *$\perp \leqslant_\Pi \alpha$ for all α;*
- *(disjunctive stability) If $\gamma \leqslant_\Pi \beta$, then*

$$\alpha \vee \gamma \leqslant_\Pi \alpha \vee \beta$$

for all α.   □

Clearly, disjunctive stability is stronger than weak stability. Monotonicity of $\leqslant_\Pi$ can be shown in the similar way as in the case of comparative probabilities. Thus, qualitative possibilities are confidence relations.

Another important example of confidence relations are qualitative necessities, introduced by Didier Dubois in 1986 (see [2]).

Definition 3. *A binary relation $\leqslant_N$ on LA is a qualitative necessity if it has the following properties:*

- *$\leqslant_N$ is a non-trivial weak order;*
- *$\alpha \leqslant_N \top$ for all α;*
- *(conjunctive stability) If $\gamma \leqslant_N \beta$, then*

$$\alpha \wedge \gamma \leqslant_N \alpha \wedge \beta$$

for all α.   □

Notice that qualitative possibilities and necessities are dual relations in the following sense:

- If $\leqslant_\Pi$ is a qualitative possibility relation on LA, then, the relation $\leqslant_N$ defined by
$$\alpha \leqslant_N \beta \quad \text{iff} \quad \neg\beta \leqslant_\Pi \neg\alpha$$
is a qualitative necessity relation on LA;
- If $\leqslant_N$ is a qualitative necessity relation on LA, then, the relation $\leqslant_N$ defined by
$$\alpha \leqslant_\Pi \beta \quad \text{iff} \quad \neg\beta \leqslant_N \neg\alpha$$
is a qualitative possibility relation on LA.

Therefore, we will focus only on qualitative possibilities.

Definition 4. *Suppose that $\leqslant_\Pi$ is a qualitative possibility relation on LA. A function $\pi : LA \longrightarrow [0,1]$ will be called a distribution of $\leqslant_\Pi$ if it has the following properties:*

- *$\alpha \leqslant_\Pi \beta$ implies $\pi(\alpha) \leqslant \pi(\beta)$;*
- *$\alpha <_\Pi \beta$ implies $\pi(\alpha) < \pi(\beta)$;*
- *$\pi(\perp) = 0$ and $\pi(\top) = 1$.*   □

Using the density of the order on $[0, 1]$ and the fact that LA is countable, one can easily show that any qualitative possibility has a distribution. Notice that there are infinitely many distributions that correspond to some fixed qualitative possibility relation.

The key property of qualitative possibility relations is the following *maxitivity principle*, see [2, 4]. Namely, if $\leqslant_\Pi$ is a qualitative possibility on LA and $\pi : LA \longrightarrow [0, 1]$ is any distribution of $\leqslant_\Pi$, then, for all α and β,

$$\pi(\alpha \vee \beta) = \max(\pi(\alpha), \pi(\beta)).$$

Maxitivity principle is a consequence of monotonicity, linearity and disjunctive stability. Indeed, monotonicity implies that $\alpha \leqslant_\Pi \alpha \vee \beta$ and $\beta \leqslant_\Pi \alpha \vee \beta$. By linearity, $\alpha \leqslant_\Pi \beta$ or $\beta \leqslant_\Pi \alpha$. In the first case ($\alpha \leqslant_\Pi \beta$), disjunctive stability gives us

$$\alpha \vee \beta \leqslant_\Pi \beta \vee \beta \sim_\Pi \beta,$$

where $\sim_\Pi$ is an equivalence relation on LA defined by $\alpha \sim_\Pi \beta$ iff $\alpha \leqslant_\Pi \beta$ and $\beta \leqslant_\Pi \alpha$. In the second case ($\beta \leqslant_\Pi \alpha$), disjunctive stability gives us

$$\alpha \vee \beta \leqslant_\Pi \alpha \vee \alpha \sim_\Pi \alpha.$$

Since, by monotonicity, we have $\alpha \leqslant_\Pi \alpha \vee \beta$ and $\beta \leqslant_\Pi \alpha \vee \beta$, it follows that, for any α and β,

$$\alpha \vee \beta \sim_\Pi \alpha \quad \text{or} \quad \alpha \vee \beta \sim_\Pi \beta,$$

or, in terms of distribution π,

$$\pi(\alpha \vee \beta) = \max(\pi(\alpha), \pi(\beta)).$$

Maxitivity fully characterizes qualitative possibilities. Namely, a confidence relation $\leqslant_C$ is a qualitative possibility iff it has a maxitive distribution (consequently, all distributions of $\leqslant_C$ will be maxitive). In purely qualitative terms, maxitivity of $\leqslant_\Pi$ can be reformulated as follows: For any $\alpha_1, \alpha_2, \beta_1, \beta_2$, the following are equivalent:

(i) $\alpha_1 \vee \alpha_2 \leqslant_\Pi \beta_1 \vee \beta_2$;
(ii) $\alpha_1 \leqslant_\Pi \beta_1$ and $\alpha_2 \leqslant_\Pi \beta_1$, or $\alpha_1 \leqslant_\Pi \beta_2$ and $\alpha_2 \leqslant_\Pi \beta_2$.

This reformulation of maxitivity actually provides a finitary strongly complete axiomatization of the notion of qualitative possibility relation.

3   Syntax and Semantics

The primitive syntactical notions are:

- A countably many propositional letters. Propositional letters will be denoted by p and q, indexed if necessary. The set of all propositional letters will be denoted by Var;

- Classical propositional formulas, built over Var. Classical formulas will be denoted by α, β and γ, indexed if necessary. The set of all classical formulas will be denoted by For_C;
- Basic possibility formulas, i.e. formulas of the form $\alpha \preceq \beta$, where α and β are classical formulas;
- Possibility formulas, which are Boolean combinations of basic possibility formulas. Possibility formulas will be denoted by ϕ, ψ and θ, indexed if necessary. The set of all possibility formulas will be denoted by For.

In order to simplify notation, we introduce the following abbreviations:

- $\alpha \prec \beta$ is $\alpha \preceq \beta \wedge \neg(\beta \preceq \alpha)$;
- $\alpha \sim \beta$ is $\alpha \preceq \beta \wedge \beta \preceq \alpha$.

By $Var(\alpha)$ we will denote the set of all propositional letters appearing in $\alpha \in For_C$. Similarly, $Var(\phi)$ denotes the set of all propositional letters appearing in $\phi \in For$.

A possibility model is a pair $M = \langle LA, \leqslant_\Pi \rangle$, where LA is the Lindenbaum algebra of Var and $\leqslant_\Pi$ is a qualitative possibility relation on LA. Satisfiability relation $\models$ between models and possibility formulas is defined as follows:

- $M \models \alpha \preceq \beta$ if $\alpha \leqslant_\Pi \beta$;
- $M \models \neg\phi$ if $M \not\models \phi$;
- $M \models \phi \wedge \psi$ if $M \models \phi$ and $M \models \psi$.

A possibility formula ϕ is satisfiable if there is a possibility model M that satisfies it, i.e. $M \models \phi$. A possibility formula ϕ is valid if it is satisfied in all possibility models. The fact that ϕ is valid will be denoted by $\models \phi$.

Suppose that T is a theory ($T \subseteq For$). We say that T is satisfiable if there is a possibility model M such that $M \models \phi$ for all $\phi \in T$. Finally, $T \models \phi$ means that every model of T is also a model of ϕ.

The next theorem is an immediate consequence of the definition of satisfiability relation $\models$. Therefore, its proof will be omitted.

Theorem 5. Let $T \subseteq For$ and $\phi, \psi, \theta \in For$. Then:

1. $T, \phi \models \phi$;
2. $T, \phi, \neg\phi \models \psi$;
3. $T \models \neg\neg\phi$ iff $T \models \phi$;
4. $T \models \phi \rightarrow \psi$ iff $T, \phi \models \psi$;
5. $T \models \neg(\phi \rightarrow \psi)$ iff $T \models \phi$ and $T \models \neg\psi$;
6. $T, \neg\neg\phi \models \psi$ iff $T, \phi \models \psi$;
7. $T, \neg(\phi \rightarrow \psi) \models \theta$ iff $T, \phi, \neg\psi \models \theta$;
8. $T, \phi \rightarrow \psi \models \theta$ iff $T, \neg\theta \models \phi$ and $T, \neg\theta \models \neg\psi$.   □

4   Axiomatization

Formal system L_Π (possibility logic) is a Hilbert-style system with following axioms and inference rules:

656   A. Perović et al.

Propositional axioms:

A1: Substitutional instances of classical tautologies.

Possibility axioms:

A2: (linearity) $\alpha \preceq \beta \vee \beta \preceq \alpha$;
A3: (transitivity) $(\alpha \preceq \beta \wedge \beta \preceq \gamma) \rightarrow \alpha \preceq \gamma$;
A4: (monotonicity) $\alpha \preceq \beta$, whenever $\alpha \rightarrow \beta$ is a tautology;
A5: (non-triviality) $\bot \prec \top$, where $\bot$ is any contradiction and $\top$ is any tautology;
A6: (disjunctive stability) $\alpha \preceq \beta \rightarrow (\alpha \vee \gamma) \preceq (\beta \vee \gamma)$.

Inference rules:

MP. (modus ponens) From ϕ and $\phi \rightarrow \psi$ infer ψ.

Notions of theorem, proof (formal inference), consistency etc. are defined as usual. The next theorem can be proved in the exactly the same way as in the case of classical propositional logic.

Theorem 6. Let $T \subseteq For$ and $\phi, \psi, \theta \in For$. Then:

1. $T, \phi \vdash \phi$;
2. $T, \phi, \neg\phi \vdash \psi$;
3. $T \vdash \neg\neg\phi$ iff $T \vdash \phi$;
4. $T \vdash \phi \rightarrow \psi$ iff $T, \phi \vdash \psi$;
5. $T \vdash \neg(\phi \rightarrow \psi)$ iff $T \vdash \phi$ and $T \vdash \neg\psi$;
6. $T, \neg\neg\phi \vdash \psi$ iff $T, \phi \vdash \psi$;
7. $T, \neg(\phi \rightarrow \psi) \vdash \theta$ iff $T, \phi, \neg\psi \vdash \theta$;
8. $T, \phi \rightarrow \psi \vdash \theta$ iff $T, \neg\theta \vdash \phi$ and $T, \neg\theta \vdash \neg\psi$.   □

The next technical lemma will allow us to prove the maxitivity principle.

Lemma 7. Let $\alpha, \beta \in For_C$. Then,

$$\vdash ((\alpha \vee \beta) \sim \alpha) \vee ((\alpha \vee \beta) \sim \beta).$$

Proof. By monotonicity (Axiom A4), $\vdash \alpha \preceq (\alpha \vee \beta)$ and $\vdash \beta \preceq (\alpha \vee \beta)$. On the other hand, by monotonicity and disjunctive stability,

$$\alpha \preceq \beta \vdash (\alpha \vee \beta) \preceq \beta$$

and

$$\beta \prec \alpha \vdash (\alpha \vee \beta) \preceq \alpha.$$

Hence,

$$\alpha \preceq \beta \vee \beta \preceq \alpha \vdash ((\alpha \vee \beta) \sim \alpha) \vee ((\alpha \vee \beta) \sim \beta).$$

Finally, by linearity, $\vdash \alpha \preceq \beta \vee \beta \preceq \alpha$, so we have our claim.   □

Theorem 8 (Maxitivity principle). Let $\alpha_1, \alpha_2, \beta_1, \beta_2 \in For_C$. Then,

$$\vdash (\alpha_1 \vee \alpha_2) \preceq (\beta_1 \vee \beta_2) \leftrightarrow ((\alpha_1 \preceq \beta_1 \wedge \alpha_2 \preceq \beta_1) \vee (\alpha_1 \preceq \beta_2 \wedge \alpha_2 \preceq \beta_2)).$$

Proof. By Lemma 7,

$$\vdash ((\alpha_1 \vee \alpha_2) \sim \alpha_1) \vee ((\alpha_1 \vee \alpha_2) \sim \alpha_2)$$

and

$$\vdash ((\beta_1 \vee \beta_2) \sim \beta_1) \vee ((\beta_1 \vee \beta_2) \sim \beta_2).$$

It is sufficient to prove that

$$(\alpha_1 \vee \alpha_2) \sim \alpha_i, (\beta_1 \vee \beta_2) \sim \beta_j \vdash \phi$$

for all i, j, where ϕ is the formula

$$(\alpha_1 \vee \alpha_2) \preceq (\beta_1 \vee \beta_2) \leftrightarrow ((\alpha_1 \preceq \beta_1 \wedge \alpha_2 \preceq \beta_1) \vee (\alpha_1 \preceq \beta_2 \wedge \alpha_2 \preceq \beta_2)).$$

Since the proofs are identical, we will consider only the case $i = j = 1$. Let

$$T = \{(\alpha_1 \vee \alpha_2) \sim \alpha_1, (\beta_1 \vee \beta_2) \sim \beta_1, (\alpha_1 \vee \alpha_2) \preceq (\beta_1 \vee \beta_2)\}.$$

It is easy to see that $T \vdash \alpha_1 \preceq \beta_1$ and $T \vdash \alpha_2 \preceq \beta_1$, so

$$T \vdash (\alpha_1 \preceq \beta_1 \wedge \alpha_2 \preceq \beta_1) \vee (\alpha_1 \preceq \beta_2 \wedge \alpha_2 \preceq \beta_2).$$

By Deduction theorem,

$$(\alpha_1 \vee \alpha_2) \preceq (\beta_1 \vee \beta_2) \rightarrow ((\alpha_1 \preceq \beta_1 \wedge \alpha_2 \preceq \beta_1) \vee (\alpha_1 \preceq \beta_2 \wedge \alpha_2 \preceq \beta_2))$$

is a deductive consequence of $\{(\alpha_1 \vee \alpha_2) \sim \alpha_1, (\beta_1 \vee \beta_2) \sim \beta_1\}$.

For the converse implication, it is sufficient to prove that

$$(\alpha_1 \preceq \beta_1 \wedge \alpha_2 \preceq \beta_1) \vee (\alpha_1 \preceq \beta_2 \wedge \alpha_2 \preceq \beta_2) \vdash (\alpha_1 \vee \alpha_2) \preceq (\beta_1 \vee \beta_2).$$

To obtain this, we will prove that

$$\alpha_1 \preceq \beta_i \wedge \alpha_2 \preceq \beta_i \vdash (\alpha_1 \vee \alpha_2) \preceq (\beta_1 \vee \beta_2)$$

for all i. Since proofs are identical, we will consider only the case $i = 1$. By conjunctive stability, $\alpha_1 \preceq \beta_1 \vdash (\alpha_1 \vee \alpha_2) \preceq (\beta_1 \vee \alpha_2)$. The same argument yields $\alpha_2 \preceq \beta_1 \vdash (\beta_1 \vee \alpha_2) \preceq (\beta_1 \vee \beta_1)$. By monotonicity and transitivity, we obtain that $\alpha_2 \preceq \beta_1 \vdash (\beta_1 \vee \alpha_2) \preceq \beta_1$. Now we have that

$$\alpha_1 \preceq \beta_1 \wedge \alpha_2 \preceq \beta_1 \vdash (\alpha_1 \vee \alpha_2) \preceq \beta_1.$$

Finally, $\vdash \beta_1 \preceq (\beta_1 \vee \beta_2)$, which concludes the proof. $\qquad\square$

5   Completeness and Compactness

As it is usual in logic, the soundness part of the completeness theorem is a straightforward but tedious induction on the length of inference. Therefore, we will omit the proof of the following soundness theorem.

Theorem 9 (Soundness). Let $T \subseteq For$ and $\phi \in For$. Then, $T \vdash \phi$ implies $T \models \phi$. □

In order to prove the simple completeness theorem ($\vdash \phi$ iff $\models \phi$), we will need the following lema.

Lemma 10. Suppose that $\alpha_1, \ldots, \alpha_{2^n}$ are all formulas of the form

$$\pm p_1 \wedge \cdots \wedge \pm p_n,$$

where $p_1, \ldots, p_n$ are pairwise distinct propositional letters, $+p_i$ is p_i and $-p_i$ is $\neg p_i$. Then, any weak partial order $\leqslant_L$ on $\{\alpha_1, \ldots, \alpha_{2^n}\}$ can be extended to a qualitative possibility relation $\leqslant_L^\Pi$ on LA.

Proof. Without loss of generality, we may assume that $\leqslant_L$ is a weak linear order, since any partial weak order can be extended to a weak linear order. Let

$$\alpha_i = \min_{\leqslant_L}(\alpha_1, \ldots, \alpha_{2^n}) \quad \text{and} \quad \alpha_j = \max_{\leqslant_L}(\alpha_1, \ldots, \alpha_{2^n}).$$

If $\alpha_j \leqslant_L \alpha_i$, then $\leqslant_L^\Pi$ can be defined as follows:

- $\alpha \sim_L^\Pi \beta$, if either both α and β are contradictions, or neither α nor β is a contradiction;
- $\bot <_L^\Pi \top$.

Let $\alpha_i <_L \alpha_j$. Recall that $\alpha, \beta \in LA$ are compatible if there is $\gamma \in LA$ such that $\bot <_{LA} \gamma$, $\gamma \leqslant_{LA} \alpha$ and $\gamma \leqslant_{LA} \beta$. Now we can define $\leqslant_L^\Pi$ as follows:

- $\alpha_r \leqslant_L^\Pi \alpha_s$ iff $\alpha_r \leqslant_L \alpha_s$;
- $\alpha \sim_L^\Pi \alpha_k$ iff $\alpha_k = \max_{\leqslant_L}\{\alpha_r \mid \alpha \text{ and } \alpha_r \text{ are compatible}\}$;
- $\bot <_L^\Pi \alpha$, for any α that is not a contradiction.

Clearly, $\leqslant_L^\Pi$ extends $\leqslant_L$. It is easy to see that $\leqslant_L^\Pi$ is a qualitative possibility. □

Theorem 11. Let $\alpha_1, \ldots, \alpha_{2^n}$ be all formulas of the form

$$\pm p_1, \wedge \cdots \wedge \pm p_n,$$

where $p_1, \ldots, p_n$ are pairwise distinct propositional letters. Then,

$$\pm (\alpha_{i_1} \preceq \alpha_{j_1}), \ldots, \pm(\alpha_{i_m} \preceq \alpha_{j_m}) \models \pm(\alpha_i \preceq \alpha_j) \tag{1}$$

iff exactly of the following two cases hold:

(i) $\{\pm(\alpha_{i_1} \preceq \alpha_{j_1}), \ldots, \pm(\alpha_{i_m} \preceq \alpha_{j_m})\}$ is inconsistent theory;

(ii) $\pm(\alpha_i \leqslant_L \alpha_j)$, where $\leqslant_L$ is a weak partial order on $\{\alpha_1, \ldots, \alpha_{2^n}\}$ induced by $\{\pm(\alpha_{i_1} \preceq \alpha_{j_1}), \ldots, \pm(\alpha_{i_m} \preceq \alpha_{j_m})\}$, i.e. $\alpha_r \leqslant_L \alpha_s$ iff

$$\pm(\alpha_{i_1} \preceq \alpha_{j_1}), \ldots, \pm(\alpha_{i_m} \preceq \alpha_{j_m}) \vdash \alpha_r \preceq \alpha_s.$$

Proof. If either (i) or (ii) holds, then obviously (1) holds as well. Conversely, suppose that both (i) and (ii) fail. Then, $\leqslant_L$ can be extended to a weak linear order $\leqslant_{L'}$ on $\{\alpha_1, \ldots, \alpha_{2^n}\}$ so that $\leqslant_{L'}$ negates the order of α_i and α_j induced by $\pm(\alpha_i \preceq \alpha_j)$. For instance, if $\pm(\alpha_i \preceq \alpha_j)$ is $\neg(\alpha_i \preceq \alpha_j)$, then $\alpha_i \leqslant_{L'} \alpha_j$. By Lemma 10, $\leqslant_{L'}$ can be extended to a qualitative possibility relation $\leqslant_\Pi$. It follows that $M = \langle LA, \leqslant_\Pi \rangle$ is a model of $\pm(\alpha_{i_1} \preceq \alpha_{j_1}), \ldots, \pm(\alpha_{i_m} \preceq \alpha_{j_m})$, but it is not a model of the formula $\pm(\alpha_i \preceq \alpha_j)$. Therefore, (1) also fails. $\square$

The immediate consequence of theorems 9 and 11 is the following corollary:

Corollary 12. Let $\alpha_1, \ldots, \alpha_{2^n}$ be all formulas of the form

$$\pm p_1, \wedge \cdots \wedge \pm p_n,$$

where $p_1, \ldots, p_n$ are pairwise distinct propositional letters. Then,

$$\pm(\alpha_{i_1} \preceq \alpha_{j_1}), \ldots, \pm(\alpha_{i_m} \preceq \alpha_{j_m}) \vdash \pm(\alpha_i \preceq \alpha_j)$$

iff exactly of the following two cases hold:

(i) $\{\pm(\alpha_{i_1} \preceq \alpha_{j_1}), \ldots, \pm(\alpha_{i_m} \preceq \alpha_{j_m})\}$ is inconsistent theory;

(ii) $\pm(\alpha_i \leqslant_L \alpha_j)$, where $\leqslant_L$ is a weak partial order on $\{\alpha_1, \ldots, \alpha_{2^n}\}$ induced by $\{\pm(\alpha_{i_1} \preceq \alpha_{j_1}), \ldots, \pm(\alpha_{i_m} \preceq \alpha_{j_m})\}$, i.e. $\alpha_r \leqslant_L \alpha_s$ iff

$$\pm(\alpha_{i_1} \preceq \alpha_{j_1}), \ldots, \pm(\alpha_{i_m} \preceq \alpha_{j_m}) \vdash \alpha_r \preceq \alpha_s. \qquad \square$$

Now we are ready to prove the simple completeness theorem for L_Π.

Theorem 13 (Simple completeness). Let ϕ be a possibility formula. Then, $\vdash \phi$ iff $\models \phi$.

Proof. If $\vdash \phi$, then $\models \phi$ by Theorem 9. Conversely, suppose that $\models \phi$. Let $Var(\phi) = \{p_1, \ldots, p_n\}$ and let $\alpha_1, \ldots, \alpha_{2^n}$ be all formulas of the form

$$\pm p_1 \wedge \cdots \wedge \pm p_n.$$

By Theorem 8 and monotonicity, we can equivalently transform ϕ into possibility formula ψ such that all basic possibility subformulas of ψ have a form

$$\alpha_r \preceq \alpha_s.$$

Clearly, ψ is also valid. Now we apply Theorem 5 and equivalently reduce $\models \psi$ to finite (meta)conjunction of sequents of the form (1). By Theorem 11 and Corollary 12, $\models$ in each sequent of the form (1) can be equivalently replaced by $\vdash$. Now we start with the reverse reduction, but instead of Theorem 5, we use Theorem 6. As result, we obtain $\vdash \psi$. Finally, using Theorem 8 and monotonicity, we obtain $\vdash \phi$. $\square$

As a consequence of the simple completeness theorem, every consistent theory T is finitely satisfiable. Thus, in order to prove completeness theorem ($T \vdash \phi$ iff $T \models \phi$), it is sufficient to prove compactness theorem.

Theorem 14. Every finitely satisfiable L_Π-theory T is satisfiable.

Proof. We will use compactness theorem for the classical propositional logic. Let T be a finitely satisfiable L_Π-theory and let

$$\mathbb{P} = \{P(\alpha, \beta) \mid \alpha, \beta \in For_C\}$$

be a new set of propositional letters. We define a $\mathbb{P}$-theory T^* as a union of the following $\mathbb{P}$-theories:

- $T_1 = \{\phi^* \mid \phi \in T\}$, where each ϕ^* is formed from ϕ substitutions of the form $\alpha \preceq \beta \mapsto P(\alpha, \beta)$. For instance, if ϕ is the formula $(\alpha \preceq \beta) \vee (\alpha \preceq \gamma)$, then ϕ^* is the formula $P(\alpha, \beta) \vee P(\alpha, \gamma)$;
- $T_2 = \{P(\alpha, \beta) \vee P(\beta, \alpha) \mid \alpha, \beta \in For_C\}$;
- $T_3 = \{(P(\alpha, \beta) \wedge P(\beta, \gamma)) \rightarrow P(\alpha, \gamma) \mid \alpha, \beta, \gamma \in For_C\}$;
- $T_4 = \{P(\bot, \top) \wedge \neg P(\top, \bot)\}$;
- $T_5 = \{P(\alpha, \beta) \mid \alpha \leqslant_{LA} \beta\}$;
- $T_6 = \{P(\alpha, \beta) \rightarrow P(\alpha \vee \gamma, \beta \vee \gamma) \mid \alpha, \beta, \gamma \in For_C\}$.

Suppose that Γ is a finite subset of T. Since T is finitely satisfiable, there is a possibility model $M = \langle LA, \leqslant_\Pi \rangle$ of Γ. It is easy to see that evaluation $e : \mathbb{P} \longrightarrow \{0, 1\}$ defined by

$$e(P(\alpha, \beta)) = 1 \quad \text{iff} \quad \alpha \leqslant_\Pi \beta$$

is a model of $\{\phi^* \mid \phi \in \Gamma\} \cup T_2 \cup \cdots \cup T_6$. Thus, T^* is finitely satisfiable. By compactness theorem for classical logic, T^* is satisfiable. Let evaluation $e^* : \mathbb{P} \longrightarrow \{0, 1\}$ be a model of T^*. Then, we can define a possibility model of T in the following way:

$$\alpha \leqslant_\Pi \beta \quad \text{iff} \quad e^*(P(\alpha, \beta)) = 1. \qquad \square$$

6   Conclusion

The present paper offers a sound and strongly complete finitary axiomatization of the notion of qualitative possibility relation. We believe that the proposed formalism is of interest, since it naturally embeds the qualitative possibilities (and their dual relations - qualitative necessities) into propositional logic framework.

To emphasis the expressivity of L_Π, we will show that possibilistic likelihood relation (see [3]), is definable in L_Π. Indeed, let

$$\alpha \prec_{\Pi L} \beta$$

be an abbreviation of the formula

$$(\neg \alpha \wedge \beta) \prec (\alpha \wedge \neg \beta)$$

and let $\alpha \preceq_{\Pi L} \beta$ be the formula $\neg(\beta \prec_{\Pi L} \alpha)$. If $M = \langle LA, \leqslant_\Pi \rangle$ is any possibility model and π is any distribution of $\leqslant_\Pi$, then

$$M \models \alpha \prec_{\Pi L} \beta \text{ iff } M \models \neg(\beta \prec_{\Pi L} \alpha)$$
$$\text{iff } (\alpha \wedge \neg\beta) \not<_\Pi (\neg\alpha \wedge \beta)$$
$$\text{iff } \pi(\neg\alpha \wedge \beta) \leqslant \pi(\alpha \wedge \neg\beta).$$

In other words, $\alpha \preceq_{\Pi L} \beta$ indeed behaves like possibilistic likelihood relation. Using completeness theorem for L_Π, we can formally show well known properties of possibilistic likelihood relations (coherence with deduction, pre-additivity, self-duality etc.).

Another possible approach to the problem of formalization of qualitative possibilities is to introduce a probabilistic-like operator logic, with the intention to formally capture distributions of possibility relations. More precisely, basic formulas would have a form $\pi(\alpha) \geqslant s$ and $\pi(\alpha) \leqslant s$, where $s \in [0,1] \cap \mathbb{Q}$. The intended meaning is rather obvious: $\pi(\alpha) \geqslant s$ reads "the possibility measure of α is at least s". Notice that there is no finitary complete axiomatization of such logics, since the set

$$\{\pi(p) > 0\} \cup \{\pi(p) \leqslant 2^{-n} \mid n \in \omega\}$$

is finitely satisfiable, but there is no real valued maxitive function that satisfies it. Infinitary complete axiomatization can be obtained by modification of argument presented in [19]. Similar modification can be done using [7], with addition of some infinitary inference rules.

References

1. Boutilier, C.: Modal logics for qualitative possibility theory. Intrenat. J. Approx. Reason. 10, 173–201 (1994)
2. Dubois, D.: Belief structures, possibility theory and decomposable confidence relations on finite sets. Comput. Artificial Intelligence 5(5), 403–416 (1986)
3. Dubois, D., Fragier, H., Prade, H.: Possibilistic likelihood relations. In: IPMU 1998, Paris, pp. 1196–1202 (1998)
4. Dubois, D., Prade, H.: Qualitative possibility functions and integrals. In: Pap, E. (ed.) Handbook of measure theory, pp. 1499–1522. North-Holland, Amsterdam (2002)
5. Fagin, R., Halpern, J., Megiddo, N.: A logic for reasoning about probabilities. Information and Computation 87(1–2), 78–128 (1990)
6. Fariñas del Cerro, L., Herzig, A.: A modal analysis of possibility theory. In: Jorrand, P., Kelemen, J. (eds.) FAIR 1991. LNCS, vol. 535, pp. 11–18. Springer, Heidelberg (1991)
7. Godo, L., Marchioni, E.: Coherent conditional probability in a fuzzy logic setting. Logic journal of the IGPL 14(3) (2006)
8. Lehmann, D.: Generalized qualitative probability: Savage revisited. In: Horvitz, E., Jensen, F. (eds.) Procs. of 12 th Conference on Uncertainty in Artificial Intelligence (UAI 1996), pp. 381–388 (1996)

9. Lewis, D.: Counterfactuals. Basil Blackwell, London (1973)
10. Lukasiewicz, T.: Probabilistic Default Reasoning with Conditional Constraints. Annals of Mathematics and Artificial Intelligence 34, 35–88 (2002)
11. Lukasiewicz, T.: Nonmonotonic probabilistic logics under variable-strength inheritance with overriding: Complexity, algorithms, and implementation. International Journal of Approximate Reasoning 44(3), 301–321 (2007)
12. Marchioni, E., Godo, L.: A Logic for Reasoning about Coherent Conditional Probability: A Modal Fuzzy Logic Approach. In: Alferes, J.J., Leite, J. (eds.) JELIA 2004. LNCS (LNAI), vol. 3229, pp. 213–225. Springer, Heidelberg (2004)
13. Narens, L.: On qualitative axiomatizations for probability theory. Journal of Philosophical Logic 9(2), 143–151 (1980)
14. Nilsson, N.: Probabilistic logic. Artificial intelligence 28, 71–87 (1986)
15. Ognjanović, Z., Rašković, M.: A logic with higher order probabilities. Publications de l'institut mathematique, Nouvelle série, tome 60(74), 1–4 (1996)
16. Ognjanović, Z., Rašković, M.: Some probability logics with new types of probability operators. J. Logic Computat. 9(2), 181–195 (1999)
17. Ognjanović, Z., Rašković, M.: Some first-order probability logics. Theoretical Computer Science 247(1–2), 191–212 (2000)
18. Ognjanović, Z., Marković, Z., Rašković, M.: Completeness Theorem for a Logic with imprecise and conditional probabilities. Publications de L'Institute Matematique (Beograd) 78(92), 35–49 (2005)
19. Ognjanović, Z., Perović, A., Rašković, M.: Logic with the qualitative probability operator. Logic journal of the IGPL 16(2), 105–120 (2008)
20. Rašković, M.: Classical logic with some probability operators. Publications de l'institut mathematique, Nouvelle série, tome 53(67), 1–3 (1993)
21. Rašković, M., Ognjanović, Z.: A first order probability logic LP_Q. Publications de l'institut mathematique, Nouvelle série, tome 65(79), 1–7 (1999)
22. Rašković, M., Ognjanović, Z., Marković, Z.: A logic with Conditional Probabilities. In: Alferes, J.J., Leite, J. (eds.) JELIA 2004. LNCS (LNAI), vol. 3229, pp. 226–238. Springer, Heidelberg (2004)
23. van der Hoek, W.: Some considerations on the logic $P_F D$: a logic combining modality and probability. Journal of Applied Non-Classical Logics 7(3), 287–307 (1997)
24. Wellman, M.P.: Some varieties of qualitative probability. In: Proceedings of the 5th International Conference on Information Processing and the Management of Uncertainty, Paris (1994)

Probabilistic Reasoning by SAT Solvers

Emad Saad

Department of Computer Science
Gulf University for Science and Technology
West Mishref, Kuwait
saad.e@gust.edu.kw

Abstract. In a series of papers we have shown that fundamental probabilistic reasoning problems can be encoded as hybrid probabilistic logic programs with probabilistic answer set semantics described in [24]. These probabilistic reasoning problems include, but not limited to, probabilistic planning [28], probabilistic planning with imperfect sensing actions [29], reinforcement learning [30], and Bayes reasoning [25]. Moreover, in [31] we also proved that stochastic satisfiability (SSAT) can be modularly encoded as hybrid probabilistic logic program with probabilistic answer set semantics, therefore, the applicability of SSAT to variety of fundamental probabilistic reasoning problems also carry over to hybrid probabilistic logic programs with probabilistic answer set semantics. The hybrid probabilistic logic programs encoding of these probabilistic reasoning problems is related to and can be translated into SAT, hence, state-of-the-art SAT solver can be used to solve these problems. This paper establishes the foundation of using SAT solvers for reasoning about variety of fundamental probabilistic reasoning problems. In this paper, we show that fundamental probabilistic reasoning problems that include probabilistic planning, probabilistic contingent planning, reinforcement learning, and Bayesian reasoning can be directly encoded as SAT formulae, hence state-of-the-art SAT solver can be used to solve these problems efficiently. We emphasize on SAT encoding for probabilistic planning and probabilistic contingent planning, as similar encoding carry over to reinforcement learning and Bayesian reasoning.

1 Introduction

In a series of papers we have shown that fundamental probabilistic reasoning problems can be encoded as hybrid probabilistic logic programs with probabilistic answer set semantics that is described in [24]. These probabilistic reasoning problems include, but not limited to, probabilistic planning [28], probabilistic planning with imperfect sensing actions [29], reinforcement learning [30], and Bayes reasoning [25]. The hybrid probabilistic logic programs encoding of these probabilistic reasoning problems is related to and can be translated into SAT, hence, state-of-the-art SAT solver can be applied to solve these problems. The relationship between these probabilistic reasoning problems encodings in hybrid probabilistic logic programs and SAT stems from the fact that the hybrid probabilistic logic programs with probabilistic answer set semantics encodings of these

C. Sossai and G. Chemello (Eds.): ECSQARU 2009, LNAI 5590, pp. 663–675, 2009.

problems are reduced to classical normal logic programs with classical answer set semantics [8]. However, classical normal logic programs with classical answer set semantics is related to SAT. This is due to the presence of a one-to-one correspondence between the classical answer sets of a classical normal logic program and the models of a SAT translation of that classical normal logic program [19]. This means that SAT encoding of a variety of probabilistic reasoning problems — including probabilistic planning, probabilistic contingent planning, reinforcement learning, and Bayes reasoning — can be achieved from the encoding of these probabilistic reasoning problems in classical normal logic programs — by translating the classical normal logic programs encoding of these probabilistic reasoning problems into equivalent SAT formulae as shown in [28,25,30,29]. This implies that variety of probabilistic reasoning problems can be represented and solved by SAT solvers. However, the SAT encoding — as well as the classical normal logic program encoding — of probabilistic reasoning problems lacks the proper representation and reasoning about probabilities inherit in these kinds of problems. Therefore, as shown in [28,25,30,29], probabilistic reasoning problems are solved by SAT solvers in three steps. These steps are:

1. Acquire the classical normal logic program with classical answer set semantics encoding of a probabilistic reasoning problem, whose classical answer sets correspond to solutions to the probabilistic reasoning problem.
2. Translate the classical normal logic program encoding of the probabilistic reasoning problem into an equivalent SAT formula, whose models correspond to solutions to the probabilistic reasoning problem. As classical answer sets of a classical normal logic program are equivalent to the models of the SAT translation of that classical normal logic program [19].
3. Calculate the probability of the probabilistic reasoning task represented in the probabilistic reasoning problem, using the models of the SAT translation of the classical normal logic program encoding of the probabilistic reasoning problem and the probability distributions associated to the probabilistic reasoning problem.

We have shown the applicability of these three steps in a number of fundamental probabilistic reasoning problems including probabilistic planning [28], probabilistic contingent planning [29], reinforcement learning [30], and reasoning with causal Bayes nets [25]. However, achieving the SAT encoding of probabilistic reasoning problems through classical normal logic programs encoding adds extra complexity that can be avoided by encoding the probabilistic reasoning problems directly into SAT which improves the efficiency of the probabilistic reasoning system.

This paper lays the foundation for using SAT to represent and reason about a variety of fundamental probabilistic reasoning problems. In this paper, we show that fundamental probabilistic reasoning problems that include, probabilistic planning, probabilistic contingent planning, reinforcement learning, and Bayesian reasoning can be directly encoded as SAT formulae, hence state-of-the-art SAT solver can be used to solve these problems efficiently. This is accomplished in two steps.

1. The first step is to translate a probabilistic reasoning problem into a SAT formula where the models of the SAT translation correspond to solutions to the probabilistic reasoning problem.
2. The second step is to calculate the probability of the probabilistic reasoning task associated to the probabilistic reasoning problem, using the models of the SAT translation of the probabilistic reasoning problem and the probability distributions associated to the probabilistic reasoning problem.

We emphasize on SAT encoding for probabilistic planning and probabilistic contingent planning, as similar encoding carry over to reinforcement learning and Bayesian reasoning. In so doing, we present a direct SAT encoding of probabilistic planning and probabilistic contingent planning domains represented in the probabilistic action language $\mathcal{P}$ [29]. $\mathcal{P}$ is a probabilistic action language that allows [29] representing and reasoning about actions with probabilistic effects, sensing actions with probabilistic outcomes, executability conditions of actions, indirect effects of actions, and the initial probability distribution over states.

2   Probabilistic Action Language $\mathcal{P}$

In this section we review the syntax and semantics of the probabilistic action language $\mathcal{P}$ [29] that is capable of representing and reasoning about the initial probability distribution over states, indirect effects of actions, actions with probabilistic effects, sensing actions with probabilistic outcomes, and executability conditions of actions. The action language $\mathcal{P}$ overcomes the shortcomings in the representation of sensing actions with probabilistic outcomes described in [5], such as the sensing actions outcomes are represented by arbitrary strings called observation labels that do not relate to the fluents that describe the world and sensing action has preconditions and effects as well as outcomes resulting from observing the environment. As [32] proved that sensing actions affect only knowledge fluents and has no effect on the other fluents.

2.1   Syntax of $\mathcal{P}$

A proposition that describes a property of the world is called a fluent. A fluent literal is either a fluent or the negation of a fluent. A conjunction of fluent literals is a conjunctive fluent formula. Sometimes we abuse the notation and refer to a conjunctive fluent formula as a set of fluent literals ($\emptyset$ denotes *true*). A *probabilistic action theory* in $\mathcal{P}$ is a set of probabilistic propositions of the from:

$$\textbf{initially } \{\psi_i \ : \ p_i\} \tag{1}$$

$$l \textbf{ if } \psi \tag{2}$$

$$\textbf{executable } a \textbf{ if } \psi \tag{3}$$

$$a \textbf{ causes } \{\phi_i \ : \ p_i \textbf{ if } \psi_i\} \tag{4}$$

$$a \textbf{ determines } \{\phi_i \ : \ p_i \textbf{ sensing } \psi_i\} \tag{5}$$

where l is a fluent literal and ψ, ψ_i, ϕ_i are conjunctive fluent formulae, $p_i \in [0, 1]$, and a is an action. The set of all ψ_i must be exhaustive and mutually exclusive, where $\forall i \sum_s p_i Pr(\psi_i|s) = 1$ and $\forall i, j, s, \ \psi_i \neq \psi_j \Rightarrow Pr(\psi_i \wedge \psi_j|s) = 0$, where s is a state.

The initial probability distribution is presented by probabilistic proposition (1). It says that a possible initial state ψ_i holds with probability p_i. Probabilistic proposition (2) describes *Indirect effect of action*. It states that l holds in every state in which ψ also holds. *Executability condition* is represented by (3), which mentions that an action a is executable in any state in which ψ holds. *The probabilistic effects of a non-sensing action a* is described by (4). It says that for all $1 \leq i \leq n$, a causes ϕ_i to hold with probability p_i in a successor state to a state in which a is executed and ψ_i holds. Sensing action with probabilistic outcomes is presented by (5). It states that executing a sensing action a in a state causes any of ϕ_i to be known true with probability p_i whenever a correlated ψ_i is known to be true in a successor state to a state in which a is executed. The literals in ψ_i determine what the sensor is observing and the literals in ϕ_i determine what the sensor reports on. p_i is the probability that ϕ_i holds whenever ψ_i holds after executing a. Given that $\tilde{s}_I$ is a probabilistic proposition of the form (1) and $\mathcal{AD}$ is a set of probabilistic propositions from (2)-(5), we say that $\mathcal{D} = \langle \tilde{s}_I, \mathcal{AD} \rangle$ a probabilistic action theory. For convenience, we present an action a by the set $a = \{a_1, \ldots, a_n\}$, where for each $1 \leq i \leq n$, a_i corresponds to ϕ_i and ψ_i.

2.2   Semantics of $\mathcal{P}$

A set of literals ϕ is consistent if it does not contain a pair of complementary literals. An indirect effect of action proposition (2) is satisfied by a set of literals ϕ if l belongs to ϕ whenever ψ is contained in ϕ or ψ is not contained in ϕ. $\mathcal{C}_\mathcal{D}(\phi)$ is the smallest set of literals that contains ϕ and satisfies all the indirect effects of actions propositions in the probabilistic action theory $\mathcal{D}$. A state s is a complete and consistent set of literals that satisfies all the indirect effects of actions propositions. Let s be a state and $\mathcal{G}$ be conjunctive fluent formula. The probability that $\mathcal{G}$ holds in s is given by $Pr(\mathcal{G}|s) = 1$ if $\mathcal{G} \subseteq s$, otherwise, $Pr(\mathcal{G}|s) = 0$. Let $\mathcal{D}$ be a probabilistic action theory, s be a state, a **causes**$\{\phi_i : p_i$ **if** $\psi_i\}$ and a' **determines**$\{\phi_i' : p_i'$ **sensing** $\psi_i'\}$, $1 \leq i \leq n$, be probabilistic propositions, and $a = \{a_i \mid (1 \leq i \leq n)\}$, $a' = \{a_i' \mid (1 \leq i \leq n)\}$, where each $a_i(a_i')$ corresponds to ϕ_i (ϕ_i') and ψ_i (ψ_i'). Then, $\mathcal{C}_\mathcal{D}(\Phi(a_i, s))$ $(\mathcal{C}_\mathcal{D}(\Phi(a_i', s)))$ is the state resulting from executing a (a') in s, where $\Phi(a_i, s)$ is defined as follows:

1. $l \in \Phi(a_i, s)$ and $\neg l \notin \Phi(a_i, s)$ if $l \in \phi_i$ and $\psi_i \subseteq s$.
2. $\neg l \in \Phi(a_i, s)$ and $l \notin \Phi(a_i, s)$ if $\neg l \in \phi_i$ and $\psi_i \subseteq s$.
3. Otherwise, $l \in \Phi(a_i, s)$ iff $l \in s$ and $\neg l \in \Phi(a_i, s)$ iff $\neg l \in s$.

In addition, $\Phi(a_i', s)$ is defined as:

1. $l \in \Phi(a_i', s)$ and $\neg l \notin \Phi(a_i', s)$ iff $l \in \phi_i'$ and $\psi_i' \subseteq s$.
2. $\neg l \in \Phi(a_i', s)$ and $l \notin \Phi(a_i', s)$ iff $\neg l \in \phi_i'$ and $\psi_i' \subseteq s$.
3. Otherwise, $l \in \Phi(a_i', s)$ iff $l \in s$ and $\neg l \in \Phi(a_i', s)$ iff $\neg l \in s$.

The probability of a state s' resulting from executing $a(a')$ in a state s is given by $Pr'(s'|s,a) = p_i$ if a **causes**$\{\phi_i : p_i$ **if** $\psi_i\}$ and $s' = \mathcal{C}_\mathcal{D}(\Phi(a_i,s))$. Otherwise, $Pr'(s'|s,a) = 0$. $Pr'(s'|s,a') = p_i'$ if a' **determines**$\{\phi_i' : p_i'$ **sensing** $\psi_i'\}$ and $s' = \mathcal{C}_\mathcal{D}(\Phi(a_i',s))$. Otherwise, $Pr'(s'|s,a') = 0$.

Probabilistic (contingent) plan is a sequence of actions $\langle a, c \rangle$, where a is a non-sensing action with probabilistic effects and c is a probabilistic (contingent) plan or it is a sequence of actions $\langle a, case \ \{\phi_i \to c_i\}_{i=1}^n\rangle$, where a is a sensing action with probabilistic outcomes and c_i is a probabilistic (contingent) plan. The probability a conjunctive fluent formula $\mathcal{G}$ holds after executing a non-empty probabilistic (contingent) plan is given by:

- The probability of a state s' holds after executing a plan in a state s is:
  - $Pr(s'|s,\langle a,c\rangle) = \sum_{s''} Pr'(s''|s,a) \, Pr(s'|s'',c)$.
  - $Pr(s'|s,\langle a, case \ \{\phi_i \to c_i\}_{i=1}^n\rangle) = \sum_{s''\models\phi_i} Pr'(s''|s,a) Pr(s'|s'',c_i)$.
- The probability that $\mathcal{G}$ is true after executing a plan in a state s is given by:
  - $Pr(\mathcal{G}|s,\langle a,c\rangle) = \sum_{s'} Pr(s'|s,\langle a,c\rangle) \, Pr(\mathcal{G}|s')$.
  - $Pr(\mathcal{G}|s,\langle a, case \ \{\phi_i \to c_i\}_{i=1}^n\rangle) = \sum_{s'} Pr(s'|s,\langle a, case \ \{\phi_i \to c_i\}_{i=1}^n\rangle) Pr(\mathcal{G}|s')$.
- The probability $\mathcal{G}$ is true after executing a plan in the initial states $\tilde{s}_I$ is:
  - $Pr(\mathcal{G}|\tilde{s}_I,\langle a,c\rangle) = \sum_s Pr(\mathcal{G}|s,\langle a,c\rangle) \, Pr(\tilde{s}_I = s)$.
  - $Pr(\mathcal{G}|\tilde{s}_I,\langle a, case\{\phi_i \to c_i\}_{i=1}^n\rangle) = \sum_s Pr(\mathcal{G}|s,\langle a, case\{\phi_i \to c_i\}_{i=1}^n\rangle) Pr(\tilde{s}_I = s)$.

Example 1 ([5]). A robot is processing a widget with the goal to get a widget painted (pa) and processed (pr) without errors $(\neg er)$(by determining if it is flawed (fl) or not flawed $(\neg fl)$), then deciding to reject or ship the widget. Since flawed (fl) property is not directly observable, the robot determines whether the widget is flawed by performing *inspect* action that senses whether the widget is blemished (bl). If the widget is blemished (bl) the robot reports widget is flawed (fl) with 0.9 probability, however, it erroneously reports widget is not flawed $(\neg fl)$ with 0.1 probability due to imperfection in the robot's sensor. The *paint* action causes widget painted (pa) and all blemishes removed $(\neg bl)$ with 0.95 probability in the state of the world in which the widget is not processed $(\neg pr)$, and causes no change in the same state of the world with 0.05 probability. But, *paint* causes an error (er) in the state of the world in which widget is being processed (pr). The effects of *ship* and *reject* are certain. Let initially the widget be blemished and flawed with probability 0.3 and it is not blemished and not flawed with probability 0.7. Consider also that the target is to find a probabilistic contingent plan that achieves its goal with probability at least 0.95. This can be represented by the probabilistic action theory $\mathcal{D} = \langle \tilde{s}_I, \mathcal{AD}\rangle$ where

$$\tilde{s}_I = \textbf{initially} \left\{ \begin{array}{ll} \{bl, fl, \neg pa, \neg pr, \neg er\} & : 0.3 \\ \{\neg bl, \neg fl, \neg pa, \neg pr, \neg er\} & : 0.7 \end{array} \right\}$$

and $\mathcal{AD}$ consists of: **executable** AC **if** $\emptyset$, where $AC \in \{paint, inspect, ship, reject\}$.

$$paint \ \textbf{causes} \left\{ \begin{array}{ll} \{pa, \neg bl\} & : 0.95 \ \textbf{if} \ \{\neg pr\} \\ \emptyset & : 0.05 \ \textbf{if} \ \{\neg pr\} \\ \{er\} & : 1 \quad\ \ \textbf{if} \ \{pr\} \end{array} \right\} \qquad ship \ \textbf{causes} \left\{ \begin{array}{ll} \{pr\} & : 1 \ \textbf{if} \ \{\neg pr, \neg fl\} \\ \{pr, er\} & : 1 \ \textbf{if} \ \{\neg pr, fl\} \\ \{er\} & : 1 \ \textbf{if} \ \{pr\} \end{array} \right\}$$

$$reject \ \textbf{causes} \left\{ \begin{array}{ll} \{pr, er\} & : 1 \ \textbf{if} \ \{\neg pr, \neg fl\} \\ \{pr\} & : 1 \ \textbf{if} \ \{\neg pr, fl\} \\ \{er\} & : 1 \ \textbf{if} \ \{pr\} \end{array} \right\} \qquad inspect \ \textbf{determines} \left\{ \begin{array}{ll} \{fl\} & : 0.9 \ \textbf{sensing} \ \{bl\} \\ \{\neg fl\} & : 0.1 \ \textbf{sensing} \ \{bl\} \\ \{\neg fl\} & : 1 \quad\ \textbf{sensing} \ \{\neg bl\} \end{array} \right\}$$

A 4-tuple $\mathcal{PP} = \langle \tilde{s}_I, \mathcal{AD}, \mathcal{G}, \mathcal{T} \rangle$ is a probabilistic (contingent) planning problem, where $\langle \tilde{s}_I, \mathcal{AD} \rangle$ is a probabilistic action theory, $\mathcal{G}$ is conjunctive fluent formula represents the goal to be satisfied, and $0 \leq \mathcal{T} \leq 1$ is the probability threshold for the goal $\mathcal{G}$ to be achieved. A probabilistic (contingent) planning problem $\mathcal{PP}$ has a probabilistic (contingent) plan q iff each a in q appears in $\mathcal{AD}$ and $Pr(\mathcal{G}|\tilde{s}_I, q) \geq \mathcal{T}$.

3   SAT Probabilistic Planning

Satisfiability planning (SAT-planning) has been proven efficient and effective approach to classical planning. The appeal of SAT-planning arises from the fact that [16] SAT is a central widely studied problem in computer science, therefore, many techniques have been developed to solve the problem, in addition to the presence of many efficient SAT solvers. Building on the success of SAT-planning, a probabilistic extension to SAT-planning for probabilistic planning and probabilistic contingent planning has been developed using stochastic satisfiability (SSAT) [21,22], which is a probabilistic extension to SAT [31]. Probabilistic (contingent) planning using SSAT is NP^{PP}-complete [21,22]. In this section we describe a SAT-based probabilistic (continent) planning approach called *SAT probabilistic (contingent) planning*. Therefore, state-of-the-art SAT solvers can be used to efficiently solve probabilistic (contingent) planning problems. And hence, reduces the complexity of finding a probabilistic (contingent) plan to NP-complete. SAT probabilistic (contingent) planning is developed by providing a translation from a probabilistic (contingent) planning problem in the probabilistic action language $\mathcal{P}$ into a SAT formula where the models of the SAT formula correspond to solutions to the probabilistic (contingent) planning problem. Although completely different, the following translation is inspired by a SAT translation to classical planning presented in [12].

Let $\mathcal{PP} = \langle \tilde{s}_I, \mathcal{AD}, \mathcal{G}, \mathcal{T} \rangle$ be a probabilistic planning problem defined on the probabilistic action theory $\mathcal{D} = \langle \tilde{s}_I, \mathcal{AD} \rangle$. Let Lit be the set of all literals that appear in $\mathcal{D}$. Then, $\mathcal{PP}$ is encoded in SAT as follows.

- The possible initial states of the initial probability distribution over the initial states, $\tilde{s}_I =$ **initially** $\{\psi_i : p_i\}$, for $1 \leq i \leq n$, are encoded as the SAT formula,

$$I^0 = \bigvee_{i=1}^{n} \psi_i^0 \qquad (6)$$

- $\mathcal{F}_{\mathcal{D}}^t$ is the conjunction of the following formulae:
  - For each fluent literal $l \in Lit$, we have the formula

$$l^{t+1} \equiv \bigvee_{a \text{ causes } \{\phi_i \,:\, p_i \text{ if } \psi_i\} \in \mathcal{D} \text{ and } l \in \phi_i} (a^t \wedge \psi_i^t) \vee \bigvee_{l \text{ if } \psi \in \mathcal{D}} \psi^{t+1} \vee (l^{t+1} \wedge l^t)$$

$$(7)$$

where t represents a time step, a is an action, a **causes** $\{\phi_i \; : \; p_i \; \textbf{if} \; \psi_i\}$ is a non-sensing action with probabilistic effects proposition, and $l \; \textbf{if} \; \psi$ is an indirect effect of action proposition. In the presence of sensing actions with probabilistic outcomes, for each fluent literal $l \in Lit$, the above formula becomes,

$$l^{t+1} \equiv \bigvee_{a \; \textbf{determines} \; \{\phi_i \; : \; p_i \; \textbf{sensing} \; \psi_i\} \in \mathcal{D} \; and \; l \in \phi_i} (a^t \wedge \psi_i^t) \vee F^{t+1} \quad (8)$$

where

$$F^{t+1} = \bigvee_{a \; \textbf{causes} \; \{\phi_i \; : \; p_i \; \textbf{if} \; \psi_i\} \in \mathcal{D} \; and \; l \in \phi_i} (a^t \wedge \psi_i^t) \vee \bigvee_{l \; \textbf{if} \; \psi \in \mathcal{D}} \psi^{t+1} \vee (l^{t+1} \wedge l^t)$$

- The following formula generates one action at a time step t.

$$\bigvee_{a \in \mathcal{A}} (a^t \bigwedge_{b \; \in \mathcal{A} \setminus \{a\}} \neg b^t) \quad (9)$$

where $\mathcal{A}$ is a set of actions names appearing in $\mathcal{D}$.
- Each executability condition of an action a at time step t of the form (3) is encoded as the formula

$$a^t \equiv \psi^t \quad (10)$$

- The goal $\mathcal{G}$ at time step t is encoded as propositional formula $\mathcal{G}^t$

Moreover, the use of propositional formulae to represent the domains of probabilistic (contingent) planning problems allows easily to encode domain-dependant constraints which helps further prune the search space and increase the efficiency of finding probabilistic (contingent) plans.

4   Correctness

In this section we prove the correctness of the SAT probabilistic (contingent) planning presented in the previous section. The correctness of the SAT translation of probabilistic (contingent) planning is given as follows. Let t represents time steps, where the domain of t is $\{0, \ldots, n - 1\}$. Let $\mathcal{PP} = \langle \tilde{s}_I, \mathcal{AD}, \mathcal{G}, \mathcal{T} \rangle$ be a probabilistic (contingent) planning problem, Φ be a probabilistic transition function associated with $\mathcal{PP}$, s^0 be a possible initial state (a state at time step 0), and $a_0, \ldots, a_{n-1}$ be a collection of (sensing and non-sensing) actions in $\mathcal{A}$. We say that $s^0 \; a_{j_0}^0 \; s^1 \ldots a_{j_{n-1}}^{n-1} \; s^n$ is a trajectory in $\mathcal{PP}$ if $s^{i+1} = \mathcal{C}_\mathcal{D}(\Phi(a_{j_i}^i, s^i))$, where $\forall (0 \leq i \leq n - 1)$, s^i is a state, a^i is an action occurs at time step i and $a_{j_i}^i \in a^i = \{a_1^i, \ldots, a_m^i\}$, where $1 \leq j_i \leq m$. We say that a trajectory $s^0 \; a_{j_0}^0 \; s^1 \ldots a_{j_{n-1}}^{n-1} \; s^n$ in $\mathcal{PP}$ achieves a conjunctive fluent formula $\mathcal{G}$ if $\mathcal{G} \subseteq s^n$. Moreover, let $\mathcal{R}_\mathcal{G}$ be the set of all trajectories $s^0 \; a_{j_0}^0 \; s^1 \ldots a_{j_{n-1}}^{n-1} \; s^n$ in $\mathcal{PP}$ that achieve $\mathcal{G}$. A probabilistic (contingent) plan q is said to achieve a goal $\mathcal{G}$ if the execution of q in the initial states will yield a non-empty set of trajectories $\mathcal{R}_\mathcal{G}^q$ each of which achieves $\mathcal{G}$.

Theorem 1. *Let* $\mathcal{PP} = \langle \tilde{s}_I, \mathcal{AD}, \mathcal{G}, \mathcal{T} \rangle$ *be a probabilistic (contingent) planning problem. Then,* $s^0 \, a^0_{j_0} \, s^1 \ldots a^{n-1}_{j_{n-1}} \, s^n$ *is a trajectory in* $\mathcal{PP}$ *that achieves* $\mathcal{G}$ *iff* $s^0 \wedge a^0 \wedge s^1 \wedge \ldots \wedge a^{n-1} \wedge s^n$ *is true in a model of*

$$I^0 \wedge \bigwedge_{t=0}^{n-1} \mathcal{F}_{\mathcal{D}}^t \wedge \bigvee_{t=0}^{n} \mathcal{G}^t$$

Intuitively, the models of $\mathcal{F}_{\mathcal{PP}} = I^0 \wedge \bigwedge_{t=0}^{n-1} \mathcal{F}_{\mathcal{D}}^t \wedge \bigvee_{t=0}^{n} \mathcal{G}^t$, the SAT encoding of the probabilistic (contingent) planning problem $\mathcal{PP}$, are equivalent to the trajectories of $\mathcal{PP}$.

Lemma 1. *Let* $\mathcal{PP}$ *be a probabilistic (contingent) planning problem,* S *be a model for* $\mathcal{F}_{\mathcal{PP}} = I^0 \wedge \bigwedge_{t=0}^{n-1} \mathcal{F}_{\mathcal{D}}^t \wedge \bigvee_{t=0}^{n} \mathcal{G}^t$, *and* $q = \langle a, c \rangle$ $(q = \langle a, case \, \{\phi_i \rightarrow c_i\}_{i=1}^m \rangle)$ *be a probabilistic (contingent) plan for* $\mathcal{PP}$. *Let* $\mathcal{R}_{\mathcal{G}}^q$ *be the set of all trajectories* $s^0 \, a^0_{j_0} \, s^1 \ldots a^{n-1}_{j_{n-1}} \, s^n$ *of* q *that achieve* $\mathcal{G}$. *Then,*

$$\sum_{s^0 \, a^0_{j_0} \, s^1 \ldots a^{n-1}_{j_{n-1}} \, s^n \in \mathcal{R}_{\mathcal{G}}^q} Pr(s_0) \prod_{i=0}^{n-1} p^i_{j_i} = \sum_{S \models s^0 \wedge a^0 \wedge s^1 \wedge \ldots \wedge a^{n-1} \wedge s^n} Pr(s_0) \prod_{i=0}^{n-1} p^i_{j_i} = Pr(\mathcal{G}|\tilde{s}_I, q)$$

The probability that a goal $\mathcal{G}$ is true, after executing a probabilistic (contingent) plan q of $\mathcal{PP}$ in the possible initial states $\tilde{s}_I$, is equivalent to the summation (over all trajectories of q) of the product of transition probabilities, $p^i_{j_i}$ in each trajectory $s^0 \, a^0_{j_0} \, s^1 \ldots a^{n-1}_{j_{n-1}} \, s^n$ in q. This is also equivalent to the summation (over all models of the SAT formula $\mathcal{F}_{\mathcal{PP}}$ that satisfies $s^0 \wedge a^0 \wedge s^1 \wedge \ldots \wedge a^{n-1} \wedge s^n$) of the product of transition probabilities $p^i_{j_i}$, where $(0 \leq i \leq n-1)$ represents the time steps, $(1 \leq j_i \leq m)$, and a^i **causes** $\{\phi^i_{j_i} : p^i_{j_i} \text{ if } \psi^i_{j_i}\}$. The following theorem follows directly from Lemma 1.

Theorem 2. *Let* S *be a model for the SAT formula* $\mathcal{F}_{\mathcal{PP}}$ *and* $q = \langle a, c \rangle$ $(q = \langle a, case \, \{\phi_i \rightarrow c_i\}_{i=1}^m \rangle)$ *be a probabilistic (contingent) plan for* $\mathcal{PP}$. *Let* $\mathcal{R}_{\mathcal{G}}^q$ *be the set of all trajectories* $s^0 \, a^0_{j_0} \, s^1 \ldots a^{n-1}_{j_{n-1}} \, s^n$ *of* q *that achieve* $\mathcal{G}$. *Then,* $Pr(\mathcal{G}|\tilde{s}_I, q) \geq \mathcal{T}$ *iff*

$$\sum_{s^0 \, a^0_{j_0} \, s^1 \ldots a^{n-1}_{j_{n-1}} \, s^n \in \mathcal{R}_{\mathcal{G}}^q} Pr(s_0) \prod_{i=0}^{n-1} p^i_{j_i} = \sum_{S \models s^0 \wedge a^0 \wedge s^1 \wedge \ldots \wedge a^{n-1} \wedge s^n} Pr(s_0) \prod_{i=0}^{n-1} p^i_{j_i} \geq \mathcal{T}$$

Theorem 2 shows that propositional satisfiability can be used to solve probabilistic (contingent) planning problems in two steps. The first step is to translate a probabilistic (contingent) planning problem, $\mathcal{PP}$, into a SAT formula whose models correspond to valid trajectories in $\mathcal{PP}$. From the models of SAT translation of the probabilistic (contingent) planning problem $\mathcal{PP}$, we can determine the trajectories $\mathcal{R}_{\mathcal{G}}^q$ of a plan q in $\mathcal{PP}$ that achieve the goal $\mathcal{G}$. The second step is to calculate the probability that the goal is satisfied by q by:

$$\sum_{s^0 \, a^0_{j_0} \, s^1 \dots a^{n-1}_{j_{n-1}} \, s^n \in \mathcal{R}^q_{\mathcal{G}}} Pr(s_0) \prod_{i=0}^{n-1} p^i_{j_i}$$

Hence, state-of-the-art SAT solvers can be used to solve probabilistic (contingent) planning problems. This two-steps SAT solution to probabilistic (contingent) planning carry over to reinforcement learning and Bayesian reasoning.

5 Examples

In this section, we present examples of SAT translation of probabilistic planning and probabilistic contingent planning problems.

Example 2. The SAT encoding of the probabilistic contingent planning problem described in Example 1 is the conjunction of the following formulae, where for a plan of length n, we have for $0 \leq t \leq n-1$:

$$I^0 = (bl^0 \wedge fl^0 \wedge \neg pa^0 \wedge \neg pr^0 \wedge \neg er^0) \bigvee (\neg bl^0 \wedge \neg fl^0 \wedge \neg pa^0 \wedge \neg pr^0 \wedge \neg er^0)$$

$$(paint^t \wedge \neg inspect^t \wedge \neg ship^t \wedge \neg reject^t) \quad \vee$$

$$(\neg paint^t \wedge inspect^t \wedge \neg ship^t \wedge \neg reject^t) \quad \vee$$

$$(\neg paint^t \wedge \neg inspect^t \wedge ship^t \wedge \neg reject^t) \quad \vee$$

$$(\neg paint^t \wedge \neg inspect^t \wedge \neg ship^t \wedge reject^t)$$

$$pa^{t+1} \equiv (paint^t \wedge \neg pr^t) \vee (pa^{t+1} \wedge pa^t)$$

$$\neg bl^{t+1} \equiv (paint^t \wedge \neg pr^t) \vee (\neg bl^{t+1} \wedge \neg bl^t)$$

$$True \equiv (paint^t \wedge \neg pr^t)$$

$$er^{t+1} \equiv (paint^t \wedge pr^t) \vee (ship^t \wedge \neg pr^t \wedge fl^t) \vee (ship^t \wedge pr^t) \vee (reject^t \wedge \neg pr^t \wedge \neg fl^t) \vee$$

$$(reject^t \wedge pr^t) \vee (er^{t+1} \wedge er^t)$$

$$pr^{t+1} \equiv (ship^t \wedge \neg pr^t \wedge \neg fl^t) \vee (ship^t \wedge \neg pr^t \wedge fl^t) \vee (reject^t \wedge \neg pr^t \wedge \neg fl^t) \vee$$

$$(reject^t \wedge \neg pr^t \wedge fl^t) \vee (pr^{t+1} \wedge pr^t)$$

$$fl^{t+1} \equiv (inspect^t \wedge bl^t) \vee (fl^{t+1} \wedge fl^t)$$

$$\neg fl^{t+1} \equiv (inspect^t \wedge bl^t) \vee (inspect^t \wedge \neg bl^t) \vee (\neg fl^{t+1} \wedge \neg fl^t)$$

$$bl^{t+1} \equiv bl^{t+1} \wedge bl^t$$

$$\neg pa^{t+1} \equiv \neg pa^{t+1} \wedge \neg pa^t$$

$$\neg pr^{t+1} \equiv \neg pr^{t+1} \wedge \neg pr^t$$

$$\neg er^{t+1} \equiv \neg er^{t+1} \wedge \neg er^t$$

Example 3. Consider the following probabilistic planning problem from [18]. A robot arm is trying to grasp a block; the grasping operation is not always successful, especially when the robot's gripper is wet. The robot is able to hold a block (hb) with a probability 0.95 after executing the *pickup* action in the state of the world in which the gripper is dry (gd), and is unable to hold the block ($\neg hb$), after executing the *pickup* action in the same state of the world, with 0.05 probability. On the other hand, executing the *pickup* action in the state of the world in which the gripper is wet ($\neg gd$) yields hb with 0.5 probability and $\neg hb$ with 0.5 probability. Let us assume that, initially, the robot is not holding the block ($\neg hb$), and the gripper is dry (gd) with probability 0.7. Therefore, there are two possible initial states $s_1 = \{gd, \neg hb\}$ and $s_2 = \{\neg gd, \neg hb\}$, with the probability distribution $Pr(s_1) = 0.7$ and $Pr(s_2) = 0.3$. Consider also that the target is to find a probabilistic contingent plan that achieves its goal with probability at least 0.7. This can be represented by the probabilistic action theory $\mathcal{D} = \langle \tilde{s}_I, \mathcal{AD} \rangle$ where **executable** *pickup* **if** $\emptyset$ and

$$\tilde{s}_I = \textbf{initially} \left\{ \begin{array}{ll} \{gd, \neg hb\} & : 0.7 \\ \{\neg gd, \neg hb\} & : 0.3 \end{array} \right\} \quad \mathcal{AD} = pickup \textbf{ causes} \left\{ \begin{array}{ll} \{hb\} & : 0.95 \textbf{ if } \{gd\} \\ \{\neg hb\} & : 0.05 \textbf{ if } \{gd\} \\ \{hb\} & : 0.5 \textbf{ if } \{\neg gd\} \\ \{\neg hb\} & : 0.5 \textbf{ if } \{\neg gd\} \end{array} \right\}$$

The SAT translation of the above probabilistic planning problem is the conjunction of the following formulae, where for a plan of length n, we have for $0 \le t \le n - 1$:

$$I^0 = (gd^0 \wedge \neg hb^0) \vee (\neg gd^0 \wedge \neg hb^0)$$

$$pickup^t$$

$$hb^{t+1} \equiv (pickup^t \wedge gd^t) \vee (pickup^t \wedge \neg gd^t) \vee (hb^{t+1} \wedge hb^t)$$

$$\neg hb^{t+1} \equiv (pickup^t \wedge gd^t) \vee (pickup^t \wedge \neg gd^t) \vee (\neg hb^{t+1} \wedge \neg hb^t)$$

$$gd^{t+1} \equiv gd^{t+1} \wedge gd^t \qquad \neg gd^{t+1} \equiv \neg gd^{t+1} \wedge \neg gd^t$$

6   Conclusions and Related Work

We presented a framework for representing and reasoning about probabilistic reasoning problems using SAT along with the application of the framework to probabilistic (contingent) planning. The framework is introduced by providing a translation from a probabilistic reasoning problem directly into a SAT formula where the models of SAT formula correspond to solutions to the probabilistic reasoning problem. We described a translation from a probabilistic (contingent) planning problem presented in the action language $\mathcal{P}$ into a SAT formula, where the models of the SAT translation correspond to solutions to the probabilistic (contingent) planning problem. The action language $\mathcal{P}$ is a high level probabilistic action language $\mathcal{P}$ [29] that overcomes the shortcomings in the representation of sensing actions with probabilistic outcomes in [5]. Contrary to [5], in the probabilistic action $\mathcal{P}$, the sensing actions outcomes are represented by fluents that describe the world and has no preconditions and effects. A detailed survey on probabilistic planning and probabilistic contingent planning can be found in different venues including [2,28,29]. However, closely related work is described below.

In [21] and [22], probabilistic planning and probabilistic contingent planning approaches have been presented building on the success of SAT-planning [16]. The approaches of [21] and [22] for probabilistic planning and probabilistic contingent planning respectively are developed by converting a probabilistic (contingent) planning problem into a stochastic satisfiability problem and solving the stochastic satisfiability problem instead to produce probabilistic (contingent) plans. However, solving a probabilistic (contingent) planning problem as a stochastic satisfiability problem is NP^{PP}−complete [21,22]. However, SAT probabilistic (contingent) planning framework presented in this paper encodes probabilistic (contingent) planning problems directly into SAT formulae, where the models of the SAT encoding correspond to probabilistic (contingent) plans, and hence, reduces the complexity of finding a probabilistic (contingent) plan to NP-complete.

In [28] and [29], probabilistic planning and probabilistic contingent planning approaches have been described based on the success on another successful logic based approach to classical planning, using normal logic programs with answer set semantics (answer set planning) [34]. Three approaches are developed in [28] and [29] to solve probabilistic planning and probabilistic contingent planning problems respectively. The first approach is developed by translating a probabilistic (contingent) planning problem into a normal hybrid probabilistic logic program with probabilistic answer set semantics [24] whose probabilistic answer sets correspond to solutions to the probabilistic (contingent) planning problem. The second approach is to translate a probabilistic (contingent) planning problem into a classical normal logic program with classical answer set semantics whose classical answer sets correspond to solutions to the probabilistic (contingent) planning problem. The third approach is to convert the classical normal logic program encoding of a probabilistic (contingent) planning problem into an equivalent SAT formula where the models of the SAT formula correspond to solutions to the probabilistic (contingent) planning problem. This indirect conversion of a probabilistic (contingent) planning problem into SAT through classical normal logic program encoding of the probabilistic (contingent) planning problem is avoided in this paper by converting a probabilistic (contingent) planning problem directly into SAT formula.

References

1. Baral, C., Tran, N., Tuan, L.C.: Reasoning about actions in a probabilistic setting. In: AAAI (2002)
2. Boutilier, C., Dean, T., Hanks, S.: Decision-theoretic planning: structural assumptions and computational leverage. Journal of AI Research 11, 1–94 (1999)
3. Blum, A., Furst, M.: Fast planning through planning graph analysis. Artificial Intelligence 90(1-2), 297–298 (1997)

4. Blum, A., Langford, J.: Probabilistic planning in the Graphplan framework. In: 5th European Conference on Planning (1999)
5. Draper, D., Hanks, S., Weld, D.: Probabilistic planning with information gathering and contingent execution. In: Proc. of the 2nd International Conference on Artificial Intelligence Planning Systems, pp. 31–37 (1994)
6. Eiter, T., Lukasiewicz, T.: Probabilistic reasoning about actions in nonmonotonic causal theories. In: 19th Conference on Uncertainty in Artificial Intelligence (UAI 2003) (2003)
7. Ernst, M., Millstein, T., Weld, D.: Automatic SAT-compilation of planning problems. In: Proceedings of the International Joint Conference on Artificial Intelligence (1997)
8. Gelfond, M., Lifschitz, V.: The stable model semantics for logic programming. In: ICSLP. MIT Press, Cambridge (1988)
9. Gelfond, M., Lifschitz, V.: Classical negation in logic programs and disjunctive databases. New Generation Computing 9(3-4), 363–385 (1991)
10. Gelfond, M., Lifschitz, V.: Representing action and change by logic programs. Journal of Logic Programming 17, 301–321 (1993)
11. Giunchiglia, E., Lierler, Y., Maratea, M.: Answer set programming based on propositional satisfiability. Journal of Automated Reasoning 36(4), 345–377 (2006)
12. Castellini, C., Giunchiglia, E., Tacchella, A.: SAT-based planning in complex domains: Concurrency, constraints and nondeterminism. Artificial Intelligence 147(1-2), 85–117 (2003)
13. Iocchi, L., Lukasiewicz, T., Nardi, D., Rosati, R.: Reasoning about actions with sensing under qualitative and probabilistic uncertainty. In: 16th European Conference on Artificial Intelligence (2004)
14. Kaelbling, L., Littman, M., Cassandra, A.: Planning and acting in partially observable stochastic domains. Artificial Intelligence 101, 99–134 (1998)
15. Kautz, H., Selman, B.: Planning as satisfiability. In: Proceedings of the 10th European Conference on Artificial (1992)
16. Kautz, H., Selman, B.: Pushing the envelope: planning, propositional logic, and stochastic search. In: Proc. of 13th National Conference on Artificial Intelligence (1996)
17. Kautz, H., McAllester, D., Selman, B.: Encoding plans in propositional logic. In: Proceedings of Principles of Knowledge Representation and Reasoning (1996)
18. Kushmerick, N., Hanks, S., Weld, D.: An algorithm for probabilistic planning. Artificial Intelligence 76(1-2), 239–286 (1995)
19. Lin, F., Zhao, Y.: ASSAT: Computing answer sets of a logic program by SAT solvers. Artificial Intelligence 157(1-2), 115–137 (2004)
20. Littman, M., Goldsmith, J., Mundhenk, M.: The computational complexity of probabilistic planning. Journal of Artificial Intelligence Research 9, 1–36 (1998)
21. Majercik, S., Littman, M.: MAXPLAN: A new approach to probabilistic planning. In: Proc. of the 4th International Conference on Artificial Intelligence Planning, pp. 86–93 (1998)
22. Majercik, S., Littman, M.: Contingent planning under uncertainty via stochastic satisfiability. Artificial Intelligence 147(1–2), 119–162 (2003)
23. Saad, E., Pontelli, E.: Towards a more practical hybrid probabilistic logic programming framework. In: Practical Aspects of Declarative Languages (2005)
24. Saad, E., Pontelli, E.: A new approach to hybrid probabilistic logic programs. Annals of Mathematics and Artificial Intelligence Journal 48(3-4), 187–243 (2006)

25. Saad, E.: Incomplete knowlege in hybrid probabilistic logic programs. In: Fisher, M., van der Hoek, W., Konev, B., Lisitsa, A. (eds.) JELIA 2006. LNCS, vol. 4160, pp. 399–412. Springer, Heidelberg (2006)

26. Saad, E.: Towards the computation of the stable probabilistic model semantics. In: Freksa, C., Kohlhase, M., Schill, K. (eds.) KI 2006. LNCS, vol. 4314, pp. 143–158. Springer, Heidelberg (2007)

27. Saad, E.: A logical approach to qualitative and quantitative reasoning. In: Mellouli, K. (ed.) ECSQARU 2007. LNCS, vol. 4724, pp. 173–186. Springer, Heidelberg (2007)

28. Saad, E.: Probabilistic planning in hybrid probabilistic logic programs. In: 1st Scalable Uncertainty Management (2007)

29. Saad, E.: Probabilistic planning with imperfect sensing actions using hybrid probabilistic logic programs. In: Proceedings of the 9th International Workshop on Computational Logic in Multi-Agent Systems (CLIMA 2008) (2008)

30. Saad, E.: A logical framework to reinforcement learning using hybrid probabilistic logic programs. In: Greco, S., Lukasiewicz, T. (eds.) SUM 2008. LNCS, vol. 5291. Springer, Heidelberg (2008)

31. Saad, E.: On the relationship between hybrid probabilistic logic programs and stochastic satisfiability. In: Greco, S., Lukasiewicz, T. (eds.) SUM 2008. LNCS, vol. 5291. Springer, Heidelberg (2008)

32. Scherl, R., Levesque, H.: The frame problem and knowledge producing actions. In: AAAI (1993)

33. Son, T., Baral, C.: Formalizing sensing actions - a transition function based approach. Artificial Intelligence 125(1-2), 19–91 (2001)

34. Subrahmanian, V.S., Zaniolo, C.: Relating stable models and AI planning domains. In: International Conference of Logic Programming, pp. 233–247 (1995)

35. Tu, P., Son, T., Baral, C.: Reasoning and planning with sensing actions, incomplete information, and static causal laws using logic programming. TPLP 7(4), 377–450

Supporting Fuzzy Rough Sets in Fuzzy Description Logics

Fernando Bobillo[1] and Umberto Straccia[2]

[1] Dpt. of Computer Science & Systems Engineering, University of Zaragoza, Spain
[2] Istituto di Scienza e Tecnologie dell'Informazione (ISTI - CNR), Pisa, Italy
fbobillo@unizar.es, straccia@isti.cnr.it

Abstract. Classical Description Logics (DLs) are not suitable to represent vague pieces of information. The attempts to achieve a solution have lead to the birth of fuzzy DLs and rough DLs. In this work, we provide a simple solution to join these two formalisms and define a fuzzy rough DL. We also show how to extend two reasoning algorithms for fuzzy DLs, which are implemented in the fuzzy DL reasoners FUZZYDL and DeLorean.

1 Introduction

In the last years the interest in ontologies has significantly grown. An ontology is defined as an explicit and formal specification of a shared conceptualization [13], which means that ontologies represent the vocabulary of some domain. They have gained widespread popularity due to their success in several applications such as expert and multiagent systems or the Semantic Web. Description Logics (DLs) are a family of logics for representing structured knowledge [1]. They are the basis of most of the ontology languages, such as the current standard language OWL [16]. For instance, the logic behind the recent language OWL 2 is $\mathcal{SROIQ}(\mathbf{D})$ [8].

However, it is widely agreed that "classical" ontology languages are not appropriate to deal with *fuzzy/vague/imprecise knowledge*, which is inherent to several real world domains. With the aim of managing vagueness in ontologies, several extension of DLs have been proposed, being possible to group them in two categories. On the one hand, the combination with fuzzy logic [30] produced *fuzzy DLs*. Some notable works are [15,25,26,28], for a survey we refer to [21]. Under this approach, vagueness is quantified and expressed using a degree of membership to a vague concept. On the other hand, the combination with rough set theory [22] produced *rough DLs* [10,12,18,19,20,24]. These logics offer a qualitative approach to model vagueness. Instead of providing a degree of a membership, vague concepts are approximated by means of a couple of classical sets: an upper and a lower approximation. This approach is very useful when it is not possible to quantify the membership function of a vague concept.

Fuzzy logic and rough logic are complementary formalism to manage vagueness and hence it is natural to combine them by means of *fuzzy rough sets* [11,23].

C. Sossai and G. Chemello (Eds.): ECSQARU 2009, LNAI 5590, pp. 676–687, 2009.
© Springer-Verlag Berlin Heidelberg 2009

This is useful in several domains of application. For instance, in e-commerce, it is possible to combine rough concepts such as "potential buyer" (an individual which is possibly interested in some product) with fuzzy concepts such as "cheap price" (which can be modeled with a trapezoidal membership function). Another example is medicine, which combines rough concepts such as "possible patient" (an individual affected by some of the symptoms of some disease, and hence suspected of being patient) with fuzzy concepts such as "high blood pressure".

In this paper we follow this approach and extend a fuzzy DL with fuzzy rough sets. As we will see, the integration is seamless, as already pointed out by [24] for the classical semantics case, as the rough set component can mapped into the fuzzy DL component, with the non-negligible advantage that current fuzzy DLs reasoners can be used with minimal adaption.

Related works are [9], which presents a rough fuzzy ontology but without entering into the formal details of the subjacent logic, and [17], which considers a less expressive logic than ours and not dealing with implementation issues.

We proceed as follows. The next section provides some background on mathematical fuzzy logics and (fuzzy) rough set theory. Section 3 presents the definition of a extension of the DL $\mathcal{SROIQ}(\mathbf{D})$, the logic behind OWL 2, with fuzzy and fuzzy rough semantics. Section 4 describes two reasoning algorithms under two fragments of our logic. Finally, Section 5 sets out some conclusions.

2   Preliminaries

Mathematical Fuzzy Logic. In fuzzy logics, the convention prescribing that a statement is either true or false is changed. Changing the usual true/false convention leads to a new concept of statement, whose compatibility with a given state of facts is a matter of degree and can be measured on an ordered scale $\mathcal{S}$ that is no longer $\{0, 1\}$, but, e.g., the unit interval $[0, 1]$. This degree of fit is called *degree of truth* of the statement ϕ in the interpretation $\mathcal{I}$.

Fuzzy logics logics provide compositional calculi of degrees of truth, including degrees between "true" and "false". A statement is now not true or false only, but may have a truth degree taken from a *truth space* $\mathcal{S}$, usually $[0, 1]$ (in that case we speak about *Mathematical Fuzzy Logic* [14]). In this paper, *fuzzy statements* will have the form $\phi \geqslant l$ or $\phi \leqslant u$, where $l, u \in [0, 1]$ and ϕ is a statement, which encode that the degree of truth of ϕ is *at least l* resp. *at most u*.

Semantically, a *fuzzy interpretation* $\mathcal{I}$ maps each basic statement p_i into $[0, 1]$ and is then extended inductively to all statements as follows:

$$\mathcal{I}(\phi \wedge \psi) = \mathcal{I}(\phi) \otimes \mathcal{I}(\psi), \mathcal{I}(\phi \vee \psi) = \mathcal{I}(\phi) \oplus \mathcal{I}(\psi),$$
$$\mathcal{I}(\phi \rightarrow \psi) = \mathcal{I}(\phi) \Rightarrow \mathcal{I}(\psi), \mathcal{I}(\neg\phi) = \ominus \mathcal{I}(\phi),$$

where $\otimes$, $\oplus$, $\Rightarrow$, and $\ominus$ are so-called *combination functions*, namely, *triangular norms* (or *t-norms*), *triangular conorms* (or *t-conorms*), *implication functions*, and *negation functions*, respectively, which extend the classical Boolean conjunction, disjunction, implication, and negation, respectively, to the fuzzy case (see [14] for a formal definition of these functions and their properties). An important type of implication functions are *R-implications*, defined as $a \Rightarrow b = \sup \{c \mid a \otimes c \leq b\}$.

Table 1. Combination functions of various fuzzy logics

	Lukasiewicz Logic	Gödel Logic	Product Logic	Zadeh Logic
$a \otimes b$	$\max(a + b - 1, 0)$	$\min(a, b)$	$a \cdot b$	$\min(a, b)$
$a \oplus b$	$\min(a + b, 1)$	$\max(a, b)$	$a + b - a \cdot b$	$\max(a, b)$
$a \Rightarrow b$	$\min(1 - a + b, 1)$	$\begin{cases} 1 & \text{if } a \leq b \\ b & \text{otherwise} \end{cases}$	$\min(1, b/a)$	$\max(1 - a, b)$
$\ominus a$	$1 - a$	$\begin{cases} 1 & \text{if } a = 0 \\ 0 & \text{otherwise} \end{cases}$	$\begin{cases} 1 & \text{if } a = 0 \\ 0 & \text{otherwise} \end{cases}$	$1 - a$

Several t-norms, t-conorms, implication functions, and negation functions have been proposed, giving raise to different fuzzy logics with different logical properties. In fuzzy logic, one usually distinguishes three different logics, namely Łukasiewicz, Gödel, and Product logic [14]. Zadeh logic (the fuzzy operators originally considered by Zadeh [30]) is a sublogic of Łukasiewicz logic. Łukasiewicz, Gödel, and Product logics have an R-implication, while Zadeh logic does not.

A *fuzzy set* R over a countable crisp set X is a function $R \colon X \to [0, 1]$. A fuzzy set A is included in B (denoted $A \subseteq B$) iff $\forall x \in X, A(x) \leq B(x)$. The *degree of subsumption* between two fuzzy sets A and B is defined as $\inf_{x \in X} A(x) \Rightarrow B(x)$.

A (binary) *fuzzy relation* R over two countable crisp sets X and Y is a function $R \colon X \times Y \to [0, 1]$. The *inverse* of R is the function $R^{-1} \colon Y \times X \to [0, 1]$ with membership function $R^{-1}(y, x) = R(x, y)$, for every $x \in X$ and $y \in Y$. The *composition* of two fuzzy relations $R_1 \colon X \times Y \to [0, 1]$ and $R_2 \colon Y \times Z \to [0, 1]$ is defined as $(R_1 \circ R_2)(x, z) = \sup_{y \in Y} R_1(x, y) \otimes R_2(y, z)$. A fuzzy relation R is *reflexive* iff $\forall x \in X, R(x, x) = 1$. R is *symmetric* iff $\forall x \in X, y \in Y, R(x, y) = R(y, x)$. R is *transitive* iff $R(x, z) \geqslant (R \circ R)(x, z)$. A fuzzy *similarity relation* is a reflexive, symmetric and transitive relation.

A fuzzy interpretation $\mathcal{I}$ *satisfies* a fuzzy statement $\phi \geqslant l$ (resp., $\phi \leqslant u$) or $\mathcal{I}$ is a *model* of $\phi \geqslant l$ (resp., $\phi \leqslant u$), denoted $\mathcal{I} \models \phi \geqslant l$ (resp., $\mathcal{I} \models \phi \leqslant u$), iff $\mathcal{I}(\phi) \geqslant l$ (resp., $\mathcal{I}(\phi) \leqslant u$). The notions of satisfiability and logical consequence are defined in the standard way. $\phi \geqslant l$ is a *tight logical consequence* of a set of fuzzy statements $\mathcal{K}$ iff l is the infimum of $\mathcal{I}(\phi)$ subject to all models $\mathcal{I}$ of $\mathcal{K}$. Notice that the latter is equivalent to $l = \sup \{r \mid \mathcal{K} \models \phi \geqslant r\}$.

Rough Set and Fuzzy Rough Set Theories. The key idea in rough set theory [22] is the approximation of a vague concept by means of a pair a concepts: a sub-concept or *lower approximation* and a super-concept or *upper approximation*, describing the sets of elements which definitely and possibly belong to the vague set, respectively, as Figure 1 illustrates. The approximation is based on an indiscernibility equivalence relation (reflexive, symmetric and transitive) between elements of the domain. Given an indiscernibility relation R, the upper approximation of a set S is defined as: $\overline{S} = \{x \mid \exists y : (x, y) \in R \land y \in S\}$. Similarly, the lower approximation is defined as: $\underline{S} = \{x \mid \forall y : (x, y) \in R \to y \in S\}$.

A very natural extension is to consider a fuzzy similarity relation instead of an indiscernibility relation, which gives raise to *fuzzy rough sets* [11,23]. Given a fuzzy similarity relation R, a t-norm $\otimes$ and an implication function $\Rightarrow$, the upper

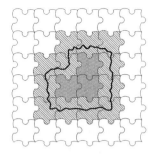

Fig. 1. Vague concept (bold line), upper approximation (striped line) and lower approximation (dotted line)

approximation of a fuzzy set S is given by the following membership function: $\forall x \in X, \overline{S}(x) = \sup_{y \in \Delta^{\mathcal{I}}}\{R(x,y) \otimes S(y)\}$. Similarly, the lower approximation is defined as: $\forall x \in X, \underline{S}(x) = \inf_{y \in \Delta^{\mathcal{I}}}\{R(x,y) \Rightarrow S(y)\}$.

3   The Fuzzy Rough DL $\mathcal{SROIQ}(\mathbf{D})$

In this section we describe a fuzzy rough extension of the fuzzy DL $\mathcal{SROIQ}(\mathbf{D})$, which is based on the fuzzy DLs presented in [5,7,28], and extended with upper and lower approximations of concepts. In the following, we assume $\bowtie \in \{\geq, >, \leq, <\}$, $\triangleright \in \{\geq, >\}$, $\triangleleft \in \{\leq, <\}$, $\alpha \in (0,1]$, $\beta \in [0,1)$, $\gamma \in [0,1]$.

Syntax. A *fuzzy concrete domain* [27] $\mathbf{D}$ is a pair $\langle \Delta_{\mathbf{D}}, \Phi_{\mathbf{D}} \rangle$, where $\Delta_{\mathbf{D}}$ is a concrete interpretation domain, and $\Phi_{\mathbf{D}}$ is a set of fuzzy concrete predicates $\mathbf{d}$ with an arity n and an interpretation $\mathbf{d_D} : \Delta_{\mathbf{D}}^n \to [0,1]$, which is an n-ary fuzzy relation over $\Delta_{\mathbf{D}}$. Usual functions for specifying fuzzy set membership degrees are the trapezoidal, the triangular, the L-function (left-shoulder function), and the R-function (right-shoulder function). For backwards compatibility, we also allow crisp intervals. These functions are defined over the set of non-negative rationals $\mathbb{Q}^+ \cup \{0\}$ For instance, we may define $\mathsf{Young} : \mathbb{N} \to [0,1]$ to be a fuzzy concrete predicate over the natural numbers denoting the degree of a person being young, as $\mathsf{Young}(x) = L(10,30)$.

We further allow fuzzy modifiers, such as very, moreOrLess and slightly, which apply to fuzzy sets to change their membership function. Formally, a *modifier* is a function $f_m : [0,1] \to [0,1]$. We will allow modifiers defined in terms of linear hedges and triangular functions. For instance, $\mathsf{very}(x) = linear(0.8)$.

Similarly as for its crisp counterpart, fuzzy $\mathcal{SROIQ}(\mathbf{D})$ assumes three alphabets of symbols, for concepts, roles and individuals.

The *abstract roles* (denoted R) of the language can be built inductively as:

$$\begin{aligned} R \to\ & R_A \mid \text{(atomic role)} \\ & R^- \mid \text{(inverse role)} \\ & U \mid \text{(universal role)} \end{aligned}$$

Concrete roles are denoted T and cannot be complex.

Now, let n, m be natural numbers $(n \geqslant 0, m > 0)$. The *concepts* (denoted C or D) of the language can be built inductively from atomic concepts (A), top concept $\top$, bottom concept $\bot$, named individuals (o_i), abstract roles (R), concrete roles (T), simple roles (S) [1] and fuzzy concrete predicates $(\mathbf{d})$ as:

$$
\begin{aligned}
C, D \rightarrow \quad & A \mid \text{(atomic concept)} \\
& \top \mid \text{(top concept)} \\
& \bot \mid \text{(bottom concept)} \\
& C \sqcap D \mid \text{(concept conjunction)} \\
& C \sqcup D \mid \text{(concept disjunction)} \\
& \neg C \mid \text{(concept negation)} \\
& \forall R.C \mid \text{(universal quantification)} \\
& \exists R.C \mid \text{(existential quantification)} \\
& \forall T.\mathbf{d} \mid \text{(concrete universal quantification)} \\
& \exists T.\mathbf{d} \mid \text{(concrete existential quantification)} \\
& \{o_1, \dots, o_m\} \mid \text{(nominals)} \\
& (\geq m\ S.C) \mid \text{(at-least qualified number restriction)} \\
& (\leq n\ S.C) \mid \text{(at-most qualified number restriction)} \\
& (\geq m\ T.\mathbf{d}) \mid \text{(concrete at-least qualified number restriction)} \\
& (\leq n\ T.\mathbf{d}) \mid \text{(concrete at-most qualified number restriction)} \\
& \exists S.Self \quad \text{(local reflexivity)}
\end{aligned}
$$

Assume m fuzzy similarity relations s_i $(i = 1, \dots, m)$. The above syntax is extended to include salient features of fuzzy DLs [3,7] as follows:

$$
\begin{aligned}
C, D \rightarrow \quad & \{\alpha_1/o_1, \dots, \alpha_m/o_m\} \mid \text{(fuzzy nominals)} \\
& C \rightarrow D \mid \text{(fuzzy implication concept)} \\
& \alpha_1 C_1 + \dots + \alpha_k C_k \mid \text{(fuzzy weighted sum)} \\
& mod(C) \mid \text{(modified concept)} \\
& [C \geq \alpha] \mid \text{(cut concept)} \\
& [C \leq \beta] \mid \text{(cut concept)} \\
& \overline{C}^i \mid \text{(upper approximation)} \\
& \underline{C}_i \quad \text{(lower approximation)}
\end{aligned}
$$

$$
\begin{aligned}
\mathbf{d} \rightarrow \quad & crisp(a, b) \mid \text{(fuzzy crisp set)} \\
& L(a, b) \mid \text{(fuzzy left-shoulder function)} \\
& R(a, b) \mid \text{(fuzzy right-shoulder function)} \\
& triangular(a, b, c) \mid \text{(fuzzy triangular function)} \\
& trapezoidal(a, b, c, d) \quad \text{(fuzzy trapezoidal function)}
\end{aligned}
$$

$$
\begin{aligned}
mod \rightarrow \quad & linear(c) \mid \text{(fuzzy linear modifier)} \\
& triangular(a, b, c) \quad \text{(fuzzy triangular modifier)}
\end{aligned}
$$

$$
\begin{aligned}
R \rightarrow \quad & mod(R) \mid \text{(modified role)} \\
& [R \geq \alpha] \quad \text{(cut role)}
\end{aligned}
$$

In the case of linear modifiers, we assume that $a = c/(c + 1), b = 1/(c + 1)$. Furthermore, for each of the connectives $\sqcap, \sqcup, \rightarrow$, we have indexed connectives

---

[1] Simple roles are needed to guarantee the decidability of the logic. Intuitively, simple roles cannot take part in cyclic role inclusion axioms (see [6] for a formal definition).

$\sqcap_X, \sqcup_X, \to_X$, where $X \in \{$Gödel, Łukasiewicz, Product$\}$, which are interpreted according to the semantics of the subscript.

Example 1. Concept Human $\sqcap$ $\exists$hasAge.$L(10, 30)$ denotes the set of young humans, with an age given by $L(10, 30)$. If $linear(4)$ represents the modifier very, $Human \sqcap linear(4)(\exists$hasAge.$L(10, 30))$ denotes the set of *very* young humans.

Furthermore, *abstract individuals* are denoted $a, b \in \Delta^{\mathcal{I}}$, while *concrete individuals* are denoted $v \in \Delta_{\mathbf{D}}$.

A *Fuzzy Knowledge Base* (KB) contains axioms organized in a fuzzy ABox $\mathcal{A}$, a fuzzy TBox $\mathcal{T}$ and a fuzzy RBox $\mathcal{R}$.

A *fuzzy ABox* consists of a finite set of *fuzzy assertions* of one of these types:

- a fuzzy concept assertion of the form $\langle a : C \bowtie \alpha \rangle$;
- a fuzzy role assertion, or constraint on the truth value of a role assertion, $\langle \Psi \bowtie \alpha \rangle$, where Ψ is of the form $(a, b) : R$, $(a, b) : \neg R$, $(a, v) : T$ or $(a, v) : \neg T$;
- an inequality assertion $\langle a \neq b \rangle$;
- an equality assertion $\langle a = b \rangle$.

A *fuzzy TBox* consists of a finite set of fuzzy General Concept Inclusions or *fuzzy GCIs*, which are expressions of the form $\langle C \sqsubseteq D \geq \alpha \rangle$ or $\langle C \sqsubseteq D > \beta \rangle$.

A *fuzzy RBox* consists of a finite set of *role axioms* of one these types:

- Fuzzy Role Inclusion Axioms or *fuzzy RIAs* $\langle w \sqsubseteq R \geq \alpha \rangle$, $\langle w \sqsubseteq R > \beta \rangle$, where $w = R_1 R_2 \ldots R_m$ is a role chain, $\langle T_1 \sqsubseteq T_2 \geq \alpha \rangle$, or $\langle T_1 \sqsubseteq T_2 > \beta \rangle$;
- *transitive* role axioms $\mathtt{trans}(R)$;
- *disjoint* role axioms $\mathtt{dis}(S_1, S_2)$, $\mathtt{dis}(T_1, T_2)$;
- *reflexive* role axioms $\mathtt{ref}(R)$;
- *irreflexive* role axioms $\mathtt{irr}(S)$;
- *symmetric* role axiom $\mathtt{sym}(R)$;
- *asymmetric* role axioms $\mathtt{asy}(S)$.

Example 2. $\langle$paul: Tall $\geq 0.5\rangle$ states that Paul is tall with at least degree 0.5. The fuzzy RIA $\langle$isFriendOf isFriendOf $\sqsubseteq$ isFriendOf $\geq 0.75\rangle$ states that the friends of my friends can also be considered my friends with degree 0.75.   □

A *fuzzy axiom* has a truth degree in [0,1]. A fuzzy axiom is *positive* (denoted $\langle \tau \rhd \alpha \rangle$) if it is of the form $\langle \tau \geq \alpha \rangle$ or $\langle \tau > \beta \rangle$. A fuzzy axiom is *negative* (denoted $\langle \tau \lhd \alpha \rangle$) if it is of the form $\langle \tau \leq \beta \rangle$ or $\langle \tau < \alpha \rangle$.

Semantics. A fuzzy interpretation $\mathcal{I}$ with respect to a fuzzy concrete domain $\mathbf{D}$ is a pair $(\Delta^{\mathcal{I}}, \cdot^{\mathcal{I}})$ consisting of a non empty set $\Delta^{\mathcal{I}}$ (the interpretation domain) disjoint with $\Delta_{\mathbf{D}}$ and a fuzzy interpretation function $\cdot^{\mathcal{I}}$ mapping:

- an *abstract individual* a onto an element $a^{\mathcal{I}}$ of $\Delta^{\mathcal{I}}$;
- a *concrete individual* v onto an element $v_{\mathbf{D}}$ of $\Delta_{\mathbf{D}}$;
- a *concept* C onto a function $C^{\mathcal{I}} : \Delta^{\mathcal{I}} \to [0, 1]$;
- an *abstract role* R onto a function $R^{\mathcal{I}} : \Delta^{\mathcal{I}} \times \Delta^{\mathcal{I}} \to [0, 1]$;

- a *concrete role* T onto a function $T^{\mathcal{I}} : \Delta^{\mathcal{I}} \times \Delta_{\mathbf{D}} \to [0,1]$;
- an *n-ary concrete fuzzy predicate* $\mathbf{d}$ onto the fuzzy relation $\mathbf{d_D} : \Delta_{\mathbf{D}}^n \to [0,1]$;
- a *modifier mod* onto a function $f_{mod} : [0,1] \to [0,1]$.

Given arbitraries t-norm $\otimes$, t-conorm $\oplus$, negation function $\ominus$ and implication function $\Rightarrow$, the fuzzy interpretation function is extended to *complex concepts and roles* as shown in Table 2, and to *fuzzy axioms* as shown in Table 3.

Table 2. Semantics of the fuzzy concepts and roles in fuzzy $\mathcal{SROIQ}(\mathbf{D})$

Constructor	Semantics
$(\top)^{\mathcal{I}}(x) = 1$	
$(\bot)^{\mathcal{I}}(x) = 0$	
$(A)^{\mathcal{I}}(x) = A^{\mathcal{I}}(x)$	
$(C \sqcap D)^{\mathcal{I}}(x) = C^{\mathcal{I}}(x) \otimes D^{\mathcal{I}}(x)$	
$(C \sqcup D)^{\mathcal{I}}(x) = C^{\mathcal{I}}(x) \oplus D^{\mathcal{I}}(x)$	
$(\neg C)^{\mathcal{I}}(x) = \ominus C^{\mathcal{I}}(x)$	
$(\forall R.C)^{\mathcal{I}}(x) = \inf_{y \in \Delta^{\mathcal{I}}} \{R^{\mathcal{I}}(x,y) \Rightarrow C^{\mathcal{I}}(y)\}$	
$(\exists R.C)^{\mathcal{I}}(x) = \sup_{y \in \Delta^{\mathcal{I}}} \{R^{\mathcal{I}}(x,y) \otimes C^{\mathcal{I}}(y)\}$	
$(\forall T.\mathbf{d})^{\mathcal{I}}(x) = \inf_{v \in \Delta_{\mathbf{D}}} \{T^{\mathcal{I}}(x,v) \Rightarrow \mathbf{d_D}(v)\}$	
$(\exists T.\mathbf{d})^{\mathcal{I}}(x) = \sup_{v \in \Delta_{\mathbf{D}}} \{T^{\mathcal{I}}(x,v) \otimes \mathbf{d_D}(v)\}$	
$(\{\alpha_1/o_1, \ldots, \alpha_m/o_m\})^{\mathcal{I}}(x) = \sup_{i \mid x = o_i^{\mathcal{I}}} \alpha_i$	
$(\geq m\, S.C)^{\mathcal{I}}(x) = \sup_{y_1,\ldots,y_m \in \Delta^{\mathcal{I}}} [(\min_{i=1}^{m} \{S^{\mathcal{I}}(x,y_i) \otimes C^{\mathcal{I}}(y_i)\}) \otimes (\otimes_{j<k} \{y_j \neq y_k\})]$	
$(\leq n\, S.C)^{\mathcal{I}}(x) = \inf_{y_1,\ldots,y_{n+1} \in \Delta^{\mathcal{I}}} [(\min_{i=1}^{n+1} \{S^{\mathcal{I}}(x,y_i) \otimes C^{\mathcal{I}}(y_i)\}) \Rightarrow (\oplus_{j<k} \{y_j = y_k\})]$	
$(\geq m\, T.\mathbf{d})^{\mathcal{I}}(x) = \sup_{v_1,\ldots,v_m \in \Delta_{\mathbf{D}}} [(\min_{i=1}^{m} \{T^{\mathcal{I}}(x,v_i) \otimes \mathbf{d_D}(v_i)\}) \otimes (\otimes_{j<k} \{v_j \neq v_k\})]$	
$(\leq n\, T.\mathbf{d})^{\mathcal{I}}(x) = \inf_{v_1,\ldots,v_{n+1} \in \Delta_{\mathbf{D}}} [(\min_{i=1}^{n+1} \{T^{\mathcal{I}}(x,v_i) \otimes \mathbf{d_D}(v_i)\}) \Rightarrow (\oplus_{j<k} \{v_j = v_k\})]$	
$(\exists S.Self)^{\mathcal{I}}(x) = S^{\mathcal{I}}(x,x)$	
$(mod(C))^{\mathcal{I}}(x) = f_{mod}(C^{\mathcal{I}}(x))$	
$([C \geq \alpha])^{\mathcal{I}}(x) = 1$ if $C^{\mathcal{I}}(x) \geq \alpha$, 0 otherwise	
$([C \leq \beta])^{\mathcal{I}}(x) = 1$ if $C^{\mathcal{I}}(x) \leq \beta$, 0 otherwise	
$(\alpha_1 C_1 + \cdots + \alpha_k C_k)^{\mathcal{I}}(x) = \alpha_1 C_1^{\mathcal{I}}(x) + \cdots + \alpha_k C_k^{\mathcal{I}}(x)$	
$(C \to D)^{\mathcal{I}}(x) = C^{\mathcal{I}}(x) \Rightarrow D^{\mathcal{I}}(x)$	
$(\overline{C^i})^{\mathcal{I}}(x) = \sup_{y \in \Delta^{\mathcal{I}}} s_i^{\mathcal{I}}(x,y) \otimes C^{\mathcal{I}}(y)$	
$(\underline{C_i})^{\mathcal{I}}(x) = \inf_{y \in \Delta^{\mathcal{I}}} s_i^{\mathcal{I}}(x,y) \Rightarrow C^{\mathcal{I}}(y)$	
$(R_A)^{\mathcal{I}}(x,y) = R_A^{\mathcal{I}}(x,y)$	
$(R^-)^{\mathcal{I}}(x,y) = R^{\mathcal{I}}(y,x)$	
$(U)^{\mathcal{I}}(x,y) = 1$	
$(mod(R))^{\mathcal{I}}(x,y) = f_{mod}(R^{\mathcal{I}}(x,y))$	
$([R \geq \alpha])^{\mathcal{I}}(x,y) = 1$ if $R^{\mathcal{I}}(x,y) \geq \alpha$, 0 otherwise	
$(T)^{\mathcal{I}}(x,v) = T^{\mathcal{I}}(x,v)$	

$C^{\mathcal{I}}$ denotes the membership function of the fuzzy concept C with respect to the fuzzy interpretation $\mathcal{I}$. $C^{\mathcal{I}}(x)$ gives us the degree of being the individual x an element of the fuzzy concept C under $\mathcal{I}$.

Similarly, $R^{\mathcal{I}}$ denotes the membership function of the fuzzy role R with respect to $\mathcal{I}$. $R^{\mathcal{I}}(x,y)$ gives us the degree of being (x,y) an element of the fuzzy role R under $\mathcal{I}$.

A fuzzy interpretation $\mathcal{I}$ *satisfies* (is a *model* of):

- $\langle a : C \bowtie \gamma \rangle$ iff $(a : C)^{\mathcal{I}} \bowtie \gamma$,
- $\langle (a,b) : R \bowtie \gamma \rangle$ iff $((a,b) : R)^{\mathcal{I}} \bowtie \gamma$,
- $\langle (a,b) : \neg R \bowtie \gamma \rangle$ iff $((a,b) : \neg R)^{\mathcal{I}} \bowtie \gamma$,
- $\langle (a,v) : T \bowtie \gamma \rangle$ iff $((a,v) : T)^{\mathcal{I}} \bowtie \gamma$,

Table 3. Semantics of the fuzzy axioms in fuzzy $\mathcal{SROIQ}(\mathbf{D})$

Axiom	Semantics
$(a\!:\!C)^{\mathcal{I}}$	$= C^{\mathcal{I}}(a^{\mathcal{I}})$
$((a,b)\!:\!R)^{\mathcal{I}}$	$= R^{\mathcal{I}}(a^{\mathcal{I}}, b^{\mathcal{I}})$
$((a,b)\!:\!\neg R)^{\mathcal{I}}$	$= \ominus R^{\mathcal{I}}(a^{\mathcal{I}}, b^{\mathcal{I}})$
$((a,v)\!:\!T)^{\mathcal{I}}$	$= T^{\mathcal{I}}(a^{\mathcal{I}}, v_{\mathbf{D}})$
$((a,v)\!:\!\neg T)^{\mathcal{I}}$	$= \ominus T^{\mathcal{I}}(a^{\mathcal{I}}, v_{\mathbf{D}})$
$(C \sqsubseteq D)^{\mathcal{I}}$	$= \inf_{x \in \Delta^{\mathcal{I}}} C^{\mathcal{I}}(x) \Rightarrow D^{\mathcal{I}}(x)$
$(R_1 \ldots R_m \sqsubseteq R)^{\mathcal{I}}$	$= \inf_{x_1, x_{n+1} \in \Delta^{\mathcal{I}}} \sup_{x_2 \ldots x_n \in \Delta^{\mathcal{I}}}(R_1^{\mathcal{I}}(x_1, x_2) \otimes \cdots \otimes R_n^{\mathcal{I}}(x_n, x_{n+1})) \Rightarrow R^{\mathcal{I}}(x_1, x_{n+1})$
$(T_1 \sqsubseteq T_2)^{\mathcal{I}}$	$= \inf_{x \in \Delta^{\mathcal{I}}, v \in \Delta_{\mathbf{D}}} T_1^{\mathcal{I}}(x, v) \Rightarrow T_2^{\mathcal{I}}(x, v)$

- $\langle (a,v)\!:\!\neg T \bowtie \gamma \rangle$ iff $((a,v)\!:\!\neg T)^{\mathcal{I}} \bowtie \gamma$,
- $\langle a \neq b \rangle$ iff $a^{\mathcal{I}} \neq b^{\mathcal{I}}$,
- $\langle a = b \rangle$ iff $a^{\mathcal{I}} = b^{\mathcal{I}}$,
- $\langle C \sqsubseteq D \rhd \gamma \rangle$ iff $(C \sqsubseteq D)^{\mathcal{I}} \rhd \gamma$,
- $\langle R_1 \ldots R_m \sqsubseteq R \rhd \gamma \rangle$ iff $(R_1 \ldots R_m \sqsubseteq R)^{\mathcal{I}} \rhd \gamma$,
- $\langle T_1 \sqsubseteq T_2 \rhd \gamma \rangle$ iff $(T_1 \sqsubseteq T_2)^{\mathcal{I}} \rhd \gamma$,
- $\mathrm{trans}(R)$ iff $\forall x,y \in \Delta^{\mathcal{I}}, R^{\mathcal{I}}(x,y) \geq \sup_{z \in \Delta^{\mathcal{I}}} R^{\mathcal{I}}(x,z) \otimes R^{\mathcal{I}}(z,y)$,
- $\mathrm{dis}(S_1, S_2)$ iff $\forall x,y \in \Delta^{\mathcal{I}}, S_1^{\mathcal{I}}(x,y) = 0$ or $S_2^{\mathcal{I}}(x,y) = 0$,
- $\mathrm{dis}(T_1, T_2)$ iff $\forall x \in \Delta^{\mathcal{I}}, v \in \Delta_{\mathbf{D}}, T_1^{\mathcal{I}}(x,v) = 0$ or $T_2^{\mathcal{I}}(x,v) = 0$,
- $\mathrm{ref}(R)$ iff $\forall x \in \Delta^{\mathcal{I}}, R^{\mathcal{I}}(x,x) = 1$,
- $\mathrm{irr}(S)$ iff $\forall x \in \Delta^{\mathcal{I}}, S^{\mathcal{I}}(x,x) = 0$,
- $\mathrm{sym}(R)$ iff $\forall x,y \in \Delta^{\mathcal{I}}, R^{\mathcal{I}}(x,y) = R^{\mathcal{I}}(y,x)$,
- $\mathrm{asy}(S)$ iff $\forall x,y \in \Delta^{\mathcal{I}}$, if $S^{\mathcal{I}}(x,y) > 0$ then $S^{\mathcal{I}}(y,x) = 0$,
- a fuzzy KB iff it satisfies each element in $\mathcal{A}$, $\mathcal{T}$ and $\mathcal{R}$.

Reasoning. The notions of logical consequence and tight logical consequence are defined as in Sect. 2. Additionally, the *maximal satisfiability degree* [7] of a concept C w.r.t. a fuzzy KB $\mathcal{K}$ is defined as $glb(\mathcal{K}, C) = \sup_{\mathcal{I}} \sup_{x \in \Delta^{\mathcal{I}}} C^{\mathcal{I}}(x)$.

Some logical properties. Due to the properties of fuzzy rough sets [23], in Zadeh, Gödel, Łukasiewicz and Product logics we have that:

- $\underline{\bot} \equiv \overline{\bot} \equiv \bot$, $\underline{\top} \equiv \overline{\top} \equiv \top$, $\underline{\underline{C}} \equiv \underline{C}$, $\overline{\overline{C}} \equiv \overline{C}$.
- $\neg\underline{C} \equiv \overline{\neg C}$, in Zadeh and Łukasiewicz logics.
- $\neg\overline{C} \equiv \underline{\neg C}$, in Zadeh and Łukasiewicz logics.
- $\overline{C \sqcap D} \subseteq \overline{C} \sqcap \overline{D}$, $\underline{C \sqcap D} \equiv \underline{C} \sqcap \underline{D}$,
- $\overline{C \sqcup D} \equiv \overline{C} \sqcup \overline{D}$, in Zadeh and Gödel logics.
- $\underline{C \sqcup D} \supseteq \underline{C} \sqcup \underline{D}$.

Note that fuzzy rough intersection and union are not truth-functional in general.

4   Reasoning and Implementation

In this section we will show how to extend two existing reasoning algorithms for fuzzy DLs so they can support fuzzy rough DLs, and how we have implemented them in the FUZZYDL system [7] and in the DELOREAN system [4]. To this end, we recall that indeed we can map lower and upper approximation concepts into fuzzy DL concepts. This is not surprising as already pointed out by [24] for the crisp case. In fact, it is not difficult to see from the semantics of upper $(\overline{C}^i)$ and lower $(\underline{C}_i)$ approximation concepts, that these can be represented as fuzzy DL concepts $\exists s_i.C$ and $\forall s_i.C$, respectively. That is, we consider the transformation:

$$\overline{C}^i \mapsto \exists s_i.C \tag{1}$$
$$\underline{C}_i \mapsto \forall s_i.C \tag{2}$$

and, thus, we may replace upper and lover approximation concepts with ordinary fuzzy DL concepts. This is exactly the same transformation as provided in [24]. In the following, we show how two currently highly expressive fuzzy DL reasoners have been adapted to support our logic.

4.1   Tableau Rules and an Optimization Problem in fuzzyDL

FUZZYDL is a reasoner for fuzzy $\mathcal{SHIF}(\mathbf{D})$ extended with a lot of salient features of fuzzy DLs, under Zadeh, Łukasiewicz and Gödel logics [7]. It is available from http://www.straccia.info, and supports the logic defined in Sect. 3 without the additional elements of $\mathcal{SROIQ}$, i.e., fuzzy nominals, qualified cardinality restrictions, role assertions with a negated role, disjoint role axioms, complex fuzzy RIAs (with $w \neq R$), irreflexive role axioms and asymmetric role axioms.

Its reasoning algorithm combines a tableaux algorithm and a mixed integer linear optimization problem. The basic idea is to build a tableaux using a set of satisfiability preserving rules which generate new simpler fuzzy assertion axioms together with some inequations over $[0, 1]$-valued variables. Finally, an optimization problem through the set of inequations is solved. A detailed description of the reasoning algorithm cannot fit into this paper, but it can be found in [29].

To support upper and lower approximation concepts in FUZZYDL, essentially we need to support reflexive roles (symmetric and transitive roles are already supported). In particular, we firstly extend FUZZYDL with a couple of fuzzy role axioms. Reflexive and symmetric role axioms are of the form (reflexive R) and (symmetric R), respectively, where R is a fuzzy role. Symmetric role axioms can already be simulated with FUZZYDL, and this axioms is just syntactic sugar. Indeed, axiom $R \sqsubseteq R^-$ implies that R is symmetric.

Then, we allow three additional concept constructors: upper approximations, lower approximations and local reflexivity concepts, which are of the form (ua s$_i$ C), (la s$_i$ C) and (self S), respectively, where s$_i$ is a fuzzy similarity relation, S is a simple fuzzy role and C is a fuzzy concept. Local reflexivity concepts are not necessary for the rough extension, but adding them is easy (reasoning is similar to the case of reflexive roles).

Similarity relations must be previously defined using the following syntax: (define-fuzzy-similarity s$_i$).

The reasoning algorithm is extended as follows:

- For every fuzzy similarity relation (`define-fuzzy-similarity` s_i) we assert s_i to be reflexive, symmetric and transitive by adding the following axioms: (`reflexive R`), (`symmetric R`), (`transitive R`).
- Every symmetric role axiom (`symmetric R`) is replaced with an inverse role axiom (`inverse R invR`) and a role inclusion axiom (`implies-role R invR`). Under an R-implication, it is well known that $\text{sym}(R)$ is equivalent to $R \sqsubseteq R^-$.
- Every upper approximation concept (`ua` s_i `C`) is replaced with an existential restriction concept (`some` s_i `C`).
- Every lower approximation concept (`la` s_i `C`) is replaced with a universal restriction concept (`all` s_i `C`).
- The rule for a local reflexivity concept (`self S`) asserts that an individual is related to itself. Formally, in the calculus if $\langle \exists\mathsf{S}.\mathtt{Self}, l \rangle \in \mathcal{L}(v)$ (that is, if v is an instance of $\exists\mathsf{S}.\mathtt{Self}$ to degree not smaller than l) then append $\langle \mathsf{S}, l \rangle$ to $\mathcal{L}(\langle v, v \rangle)$ (that is, the pair $\langle v, v \rangle$ is an instance of S at least to degree l).
- The rule for reflexive roles (`reflexive R`) asserts that every individual is related to itself. Formally, if $\langle \mathtt{ref}(\mathsf{R}) \rangle \in \mathcal{R}$ and v is a node to which this rule has not yet been applied then append $\langle \mathsf{R}, l \rangle$ to $\mathcal{L}(\langle v, v \rangle)$ (that is, the pair $\langle v, v \rangle$ is an instance of R to degree not smaller than l).

4.2 Reduction to Classical Description Logic in DeLorean

DELOREAN is a reasoner for basic fuzzy $\mathcal{SROIQ}(\mathbf{D})$ [4] (not supporting the additional features of fuzzy DLs defined in Sect. 3) under Zadeh and Gödel (with an involutive negation) logics. The syntax of the supported language is in [2].

Its reasoning algorithm is based on a reduction to a classical DL, so current DL reasoners can be reused. A full description may be found in [3,5,6].

DELOREAN already supported local reflexivity concepts, as well as reflexive and symmetric roles. Hence, it only remained to extend it with upper and lower approximations of the form (`upper` s_i `C`) and (`lower` s_i `C`), where s_i is a fuzzy similarity relation and C is a fuzzy concept.

Now, the reasoning algorithm is extended as follows:

- Every concept (`upper` s_i `C`) is replaced with an existential restriction concept (`some` s_i `C`). Furthermore, we add the following axioms if they do not exist in the fuzzy RBox: (`reflexive R`), (`symmetric R`), (`transitive R`).
- Every concept (`lower` s_i `C`) is replaced with a universal restriction concept (`all` s_i `C`). Once again, we add the following axioms in case they do not exist in the fuzzy RBox: (`reflexive R`), (`symmetric R`), (`transitive R`).

5   Conclusions

In this paper we have studied a DL managing vagueness in two different but complementary ways, combining a fuzzy DL with fuzzy rough sets. In particular, we have presented a very expressive fuzzy rough extension of the DL $\mathcal{SROIQ}(\mathbf{D})$, the logic behind the language OWL 2. The rough extension is general (independent of the family of fuzzy operators) and uses m possible fuzzy similarity relations.

Reasoning under our general fuzzy rough DL is not currently possible, but we have extended and implemented two well-known reasoning algorithms for fuzzy DLs in order to deal with two important fragments of the logic. On the one hand, FuzzyDL implements a combination of a tableaux algorithm and a mixed integer linear optimization problem, and already supports fuzzy rough $\mathcal{SHIF}(\mathbf{D})$ (extended with salient features of fuzzy DLs) under Zadeh, Łukasiewicz and Gödel logics. On the other hand, DeLorean implements a translation to a crisp DL and supports fuzzy rough $\mathcal{SROIQ}(\mathbf{D})$ under Zadeh and Gödel (with an involutive negation) logics. Extending the of reasoning algorithms and the expressivity reasoners remains an open research problem.

References

1. Baader, F., Calvanese, D., McGuinness, D., Nardi, D., Patel-Schneider, P.F. (eds.): The Description Logic Handbook: Theory, Implementation, and Applications. Cambridge University Press, Cambridge (2003)
2. Bobillo, F.: Managing Vagueness in Ontologies. PhD thesis, University of Granada, Spain (2008)
3. Bobillo, F., Delgado, M., Gómez-Romero, J.: A crisp representation for fuzzy $\mathcal{SHOIN}$ with fuzzy nominals and general concept inclusions. In: Uncertainty Reasoning for the Semantic Web I. LNCS, vol. 5327, pp. 174–188. Springer, Heidelberg (2008)
4. Bobillo, F., Delgado, M., Gómez-Romero, J.: DeLorean: A reasoner for fuzzy OWL 1.1. In: Proc. of the 4th International Workshop on Uncertainty Reasoning for the Semantic Web (URSW 2008). CEUR Workshop Proceedings, vol. 423 (2008)
5. Bobillo, F., Delgado, M., Gómez-Romero, J.: Optimizing the crisp representation of the fuzzy description logic $\mathcal{SROIQ}$. In: Uncertainty Reasoning for the Semantic Web I. LNCS, vol. 5327, pp. 189–206. Springer, Heidelberg (2008)
6. Bobillo, F., Delgado, M., Gómez-Romero, J., Straccia, U.: Fuzzy description logics under Gödel semantics. International Journal of Approximate Reasoning 50(3), 494–514 (2009)
7. Bobillo, F., Straccia, U.: Fuzzydl: An expressive fuzzy description logic reasoner. In: Proceedings of the 17th IEEE International Conference on Fuzzy Systems (FUZZ-IEEE 2008), pp. 923–930. IEEE Computer Society, Los Alamitos (2008)
8. Cuenca-Grau, B., Horrocks, I., Motik, B., Parsia, B., Patel-Schneider, P., Sattler, U.: OWL 2: The next step for OWL. Journal of Web Semantics 6(4), 309–322 (2008)
9. Dey, L., Abulaish, M., Goyal, R., Shubham, K.: A rough-fuzzy ontology generation framework and its application to bio-medical text processing. In: Proceedings of the 5th Atlantic Web Intelligence Conference (AWIC 2007), pp. 74–79 (2007)
10. Doherty, P., Grabowski, M., Łukaszewicz, W., Szałas, A.: Towards a framework for approximate ontologies. Fundamenta Informaticae 57(2-4), 147–165 (2003)
11. Dubois, D., Prade, H.: Rough fuzzy sets and fuzzy rough sets. International Journal of General Systems 17(2–3), 191–209 (1990)

12. Fanizzi, N., D'Amato, C., Esposito, F., Lukasiewicz, T.: Representing uncertain concepts in rough description logics via contextual indiscernibility relations. In: Proceedings of the 4th International Workshop on Uncertainty Reasoning for the Semantic Web (URSW 2008). CEUR Workshop Proceedings, vol. 423 (2008)
13. Gruber, T.R.: A translation approach to portable ontologies. Knowledge Acquisition 5(2), 199–220 (1993)
14. Hájek, P.: Metamathematics of Fuzzy Logic. Kluwer Academic Publishers, Dordrecht (1998)
15. Hájek, P.: Making fuzzy description logics more general. Fuzzy Sets and Systems 154(1), 1–15 (2005)
16. Horrocks, I., Patel-Schneider, P.: Reducing OWL entailment to description logic satisfiability. Journal of Web Semantics 1(4), 345–357 (2004)
17. Jiang, Y., Wang, J., Peng, D., Tang, S.: Reasoning within expressive fuzzy rough description logics. Fuzzy sets and systems (to appear)
18. Jiang, Y., Wang, J., Tang, S., Xiao, B.: Reasoning with rough description logics: An approximate concepts approach. Information Sciences 179(5), 600–612 (2009)
19. Klein, M.C.A., Mika, P., Schlobach, S.: Rough description logics for modeling uncertainty in instance unification. In: Proceedings of URSW 2007. CEUR Workshop Proceedings, vol. 327 (2007)
20. Liau, C.-J.: On rough terminological logics. In: Proc. of the 4th Int. Workshop on Rough Sets, Fuzzy Sets and Machine Discovery (RSFD 1996), pp. 47–54 (1996)
21. Lukasiewicz, T., Straccia, U.: Managing uncertainty and vagueness in description logics for the semantic web. Journal of Web Semantics 6(4), 291–308 (2008)
22. Pawlak, Z.: Rough sets. International Journal of Computer and Information Sciences 11, 341–356 (1982)
23. Radzikowska, A.M., Kerre, E.E.: A comparative study of fuzzy rough sets. Fuzzy Sets and Systems 126(2), 137–155 (2002)
24. Schlobach, S., Klein, M.C.A., Peelen, L.: Description logics with approximate definitions - Precise modeling of vague concepts. In: Proceedings of the 20th International Joint Conference on Artificial Intelligence (IJCAI 2007), pp. 557–562 (2007)
25. Stoilos, G., Stamou, G., Pan, J.Z., Tzouvaras, V., Horrocks, I.: Reasoning with very expressive fuzzy description logics. Journal of Artificial Intelligence Research 30(8), 273–320 (2007)
26. Straccia, U.: Reasoning within fuzzy description logics. Journal of Artificial Intelligence Research 14, 137–166 (2001)
27. Straccia, U.: Description logics with fuzzy concrete domains. In: Proceedings of the 21st Conference on Uncertainty in Artificial Intelligence (UAI 2005) (2005)
28. Straccia, U.: A fuzzy description logic for the semantic web. In: Sanchez, E. (ed.) Fuzzy Logic and the Semantic Web. Capturing Intelligence, pp. 73–90. Elsevier, Amsterdam (2006)
29. Straccia, U.: Reasoning in L-$\mathcal{SHIF}$: An expressive fuzzy description logic under Łukasiewicz semantics. Technical Report TR-2007-10-18, Istituto di Scienza e Tecnologie dell'Informazione, Consiglio Nazionale delle Ricerche, Pisa, Italy (2007)
30. Zadeh, L.A.: Fuzzy sets. Information and Control 8(3), 338–353 (1965)

Possibilistic Measures Taking Their Values in Spaces Given by Inclusion-Closed Fragments of Power-Sets

Ivan Kramosil

Institute of Computer Science AS CR, v.v.i., Pod Vodárenskou věží 2,
182 07 Prague, Czech Republic
kramosil@cs.cas.cz

Abstract. Fuzzy sets with non-numerical membership degrees, as well as the related possibilistic distributions and measures, have been developed mostly under the simplifying assumption that their membership or possibility degrees are taken from a complete lattice, so that all the supremum and infimum values to be processed are defined. In this paper the conditions imposed on the space in which possibilistic measures take their values are weakened in such a way that this space is defined by an inclusion-closed system of subsets of a space X, so that all subsets of sets from this system are in this system also incorporated. Let us note that the system of all finite subsets of an infinite space X (an incomplete lattice) or the system of all subsets of X the cardinality of which does not exceed a fixed positive integer are particular and intuitive examples of inclusion-closed systems of subsets of X. Some simple properties of such possibilistic measures are analyzed and compared with the properties of their standard versions taking values in complete lattices.

1 Introduction and Motivation

Let us begin the way of explanation leading to the notion of possibility (or possibilistic) measure with the already well-known idea of fuzzy set introduced by L. A. Zadeh in his famous pioneering paper [7]. Purposedly leaving aside the philosophical and methodological aspects of fuzzy sets and fuzziness as such, broadly discussed in numerous sources (let us mention [3] just as an example), we will take real-valued normalized fuzzy subset π of a crisp set (space) Ω as mapping which takes Ω into the unit interval of real numbers in such a way that the condition $\sup\{\pi(\omega) : \omega \in \Omega\} = 1$ is met.

With the aim to quantify the portion of fuzziness defined by the fuzzy subset π of Ω and contained in a crisp subset A of Ω let us define the value $\Pi(A) = \sup_{\omega \in A} \pi(\omega) \in [0, 1]$, so defining the set function $\Pi : \mathcal{P}(\Omega) \rightarrow [0, 1]$, here $\mathcal{P}(\Omega)$ denotes the power set of all subsets of Ω. In order to emphasize the formal and syntactical analogies between probability measures defined as set functions in $\mathcal{P}(\Omega)$ and induced by a probability distribution, and the set function Π, taken as quantification of uncertainty, in the sense of fuzziness and vagueness,

C. Sossai and G. Chemello (Eds.): ECSQARU 2009, LNAI 5590, pp. 688–699, 2009.

L. A. Zadeh introduces, in [8], the terms *possibility distribution* on Ω for π, and *possibility measure* induced by π on $\mathcal{P}(\Omega)$ for Π. These terms, with preferences given to the adjective "possibilistic", will be used also below.

The key operations applied when processing probability values and measures are those of addition (series taking, integration), which can be hardly extended to definition domains with non-numerical elements. On the other side, the operations of infimum and supremum, playing the key role when processing possibilistic measures, can be defined also in non-numerical domains and structures like, e.g., partially ordered (p.o.) sets, in particular, in Boolean algebras and lattices, e.g., represented by various systems of subsets of spaces of elementary possibility or fuzziness degrees. It is why the idea of fuzzy sets with non-numerical degrees of fuzziness emerged (cf. [5]) as soon as four years after the Zadeh's originating work and the related non-numerical possibilistic distributions and measures also followed (cf. the excellent work [2] by G. De Cooman). In order to simplify the work, the fuzziness and possibility degrees are supposed to be taken from a complete lattice, so that the existence of all supremum and infimum values is ensured.

In this paper we will investigate possibilistic distributions defined on a fixed space Ω and taking their values in the power-set of all subsets of another fixed set X, hence, our attention will be focused to mappings $\pi : \Omega \to \mathcal{P}(X)$ such that the relation $\bigcup_{\omega \in \Omega} \pi(\omega) = X$ holds. However, when processing these distributions we will be faced with the following problem: not all elements of $\mathcal{P}(X)$ will be completely identifiable and distinguishable from each other but only those belonging to certain fixed subsystem $\mathcal{S} \subset \mathcal{P}(X)$. So, given $\omega \in \Omega$ and asking (an oracle, say) for the value $\pi(\omega)$, we obtain the demanded value only when $\pi(\omega) \in \mathcal{S}$ is the case, otherwise the answer reads just that the demanded values is outside the scope defined by $\mathcal{S}$ without any more specification of this value.

An easy illustration of such a situation may be as follows: the technical devices processing the elements of $\mathcal{P}(X)$, i.e., subsets of X, are able to process only subsets of limited cardinality, say, only those $A \subset X$ for which the inequality $\|A\| \le R < \|X\|$ holds for fixed R (in case when the subsets of X are processed through their binary or decadic enumerations a similar restriction is imposed on the number of digits allowed). It may easily happen that the system $\mathcal{S}$ of observable subsets of X is not closed with respect to set union, i.e., supremum in $\mathcal{S}$ induced by set inclusion, as for some $A, B \subset X$ for which $\|A\|, \|B\| \le R$ holds, $\|A \cup B\| \le R$ need not be the case. The problem will be analyzed, below, from the two close points of view. Either, instead of the original possibilistic distribution $\pi : \Omega \to \mathcal{P}(X)$ we will consider its restriction to $\mathcal{S}$, i.e., the *partial mapping* $\pi^{\mathcal{S}} : \Omega \to \mathcal{P}(X)$ defined by $\pi^{\mathcal{S}}(\omega) = \pi(\omega)$, if $\pi(\omega) \in \mathcal{S}$, and leaving $\pi^{\mathcal{S}}(\omega)$ undefined (or defined by an auxiliary conventional symbol, say, $\pi^{\mathcal{S}}(\omega) = *$, if $\pi(\omega) \in \mathcal{S}$ does not hold. The problem then reads, whether, and in which sense and degree, the mapping $\pi^{\mathcal{S}}$ may serve as an approximation of the original mapping (possibilistic distribution) π on Ω. Or, the restriction of values from $\mathcal{P}(X)$ to $\mathcal{S}$ may be taken as an a priori condition and we will try to define an appropriate lattice or lattice-like structure on $\mathcal{S}$ and a mapping $\pi^0 : \Omega \to \mathcal{S}$ in such a way that π^0 appropriately simulates the nature of fuzziness which may be defined within the restricted framework of set-valued fuzziness (or possibility) degrees.

For a more detailed and systematic information dealing with p.o.sets, lattices, Boolean algebras and structures in general the reader is recommended to consult either the already classical monographs [1], [4], [6] or some more recent monographs and textbooks.

2   Inclusion-Closed Fragments of Power-Sets and Related Complete Lattices

Let Y be a nonempty set. A binary relation $\leq$ on Y is called *partial pre-ordering*, if it is reflexive and transitive. If the relation $\leq$ is also antisymmetric, then it is called *partial ordering* on Y and the pair $\mathbf{Y} = \langle Y, \leq \rangle$ is called *partially ordered set (p.o.set or poset)*. Given $B \subset Y$, we define the *supremum* $\bigvee_{y \in B} y$ or simply $\bigvee B$ as the least upper bound (l.u.b.) and the *infimum* $\bigwedge_{y \in B} y$ or simply $\bigwedge B$ as the greatest lower bound (g.l.b.) of elements from B. In general, neither $\bigvee B$ nor $\bigwedge B$ need be defined for every $B \subset Y$, but if $\bigvee B$ and/or $\bigwedge B$ are defined, they are defined uniquely. P.o.set $\mathbf{Y} = \langle Y, \leq \rangle$ is called *lattice*, if for *each finite* $B \subset Y$ both the values $\bigvee B$ and $\bigwedge B$ are defined, and $\mathbf{Y}$ is called *complete lattice*, if $\bigvee B$ and $\bigwedge B$ are defined for *each* $B \subset Y$.

Definition 1. Let X be a nonempty set. A fragment $\mathcal{S} \subset \mathcal{P}(X)$ of the power-set $\mathcal{P}(X)$ is called *inclusion-closed* (or shortly *closed*, if no misunderstanding menaces), if $\mathcal{S}$ does not contain X and if, for each $A \in \mathcal{S}$ and each $B \subset A, B \in \mathcal{S}$ holds as well.

Let us introduce some examples. The largest closed fragment of $\mathcal{P}(X)$ is obviously the system of all proper subsets of X, which may be easily written as $\mathcal{P}(X) - \{X\} = \bigcup \{\mathcal{P}(X - \{x\}) : x \in X\}$. Given a proper subset A of X, the power-set $\mathcal{P}(A)$ also defines a closed fragment of $\mathcal{P}(X)$ and the same is the case when $\mathcal{S} = \bigcup \{\mathcal{P}(A_0) : A_0 \in \mathcal{S}_0\}$ for some $\mathcal{S}_0 \subset \mathcal{P}(X) - \{X\}$. Indeed, if $A \in \mathcal{S}$ holds, then $A \in \mathcal{P}(A_0)$ for some $A_0 \in \mathcal{S}_0$ follows, hence, for each $B \subset A, B \in \mathcal{P}(A_0)$ and $B \in \mathcal{S}$ are also valid. Let us note that closed fragments of $\mathcal{P}(X)$ are closed with respect to (w.r.to) set intersection in the sense that, for each $A \in \mathcal{S}$ and no matter which $A \subset X, A \cap B \in \mathcal{S}$ holds. However, closed fragments are not closed, in general, w.r.to set union. Indeed, for proper subsets A_1, A_2 of X such that $A_1 - A_2 \neq \emptyset$ and $A_2 - A_1 \neq \emptyset$ hold, and for $\mathcal{S} = \mathcal{P}(A_1) \cup \mathcal{P}(A_2)$, the set $A_1 \cup A_2$ does not belong to $\mathcal{S}$.

Other examples of closed fragments of $\mathcal{P}(X)$ may read as follows. If the X is infinite, then the system $\mathcal{P}_f(X)$ of all *finite* subsets of X evidently defines a closed fragment of $\mathcal{P}(X)$. Similarly, if R is a positive integer such that $\|X\| > R$ holds, then the system $\mathcal{P}_R(X)$ of all subsets of X the cardinality of which does not exceed R, i.e., the system $\mathcal{P}_R(X) = \{A \subset X : \|A\| \leq R\}$, also defines a closed fragment of $\mathcal{P}(X)$.

Definition 2. Let X be a nonempty set, let $\mathcal{S}$ be a closed fragment of $\mathcal{P}(X)$. $\mathcal{S}$-*related inclusion* on $\mathcal{P}(X)$ is a binary relation $\subset_{\mathcal{S}}$ defined in this way: if $A, B \in \mathcal{S}$ is the case, then $A \subset_{\mathcal{S}} B$ holds if and only if $A \subset B$ holds, and if $B \in \mathcal{P}(X) - \mathcal{S}$ is valid, then $A \subset_{\mathcal{S}} B$ holds for each $A \subset X$.

Consequently, if both the sets A, B are out of $\mathcal{S}$, then $A \subset_{\mathcal{S}} B$ and $B \subset_{\mathcal{S}} A$ holds together, so that the sets A and B are indistinguishable by $\subset_{\mathcal{S}}$ contrary to the case of standard set inclusion $\subset$ applied to sets from $\mathcal{S}$, where $A \subset B$ and $B \subset A$ holding together yields that $A = B$ (axiom of extensionality in standard set theory). According to the intuition behind taking the sets beyond $\mathcal{S}$ as "too large", the contemporary validity of both the inclusions $A \subset B$ and $B \subset A$ for sets out of $\mathcal{S}$ can be taken as formalized description of the fact that sets out of $\mathcal{S}$ are not distinguishable from each other by the limited tools being at our disposal. Let us emphasize explicitly, that for *no* $A \in \mathcal{P}(X) - \mathcal{S}$ and $B \in \mathcal{S}$ the relation $A \subset_{\mathcal{S}} B$ holds.

Lemma 1. *Let X and $\mathcal{S}$ be as in Definition 2, then the $\mathcal{S}$-related inclusion $\subset_{\mathcal{S}}$ defines a partial pre-ordering relation on $\mathcal{P}(X)$.*

Proof. For both $A \in \mathcal{S}$ and $A \in \mathcal{P}(X) - \mathcal{S}$ the relation $A \subset_{\mathcal{S}} A$ holds trivially. Let $A \subset_{\mathcal{S}} B$ and $B \subset_{\mathcal{S}} C$ holds for $A, B, C \subset X$. If $C \in \mathcal{P}(X) - \mathcal{S}$ is the case, then $A \subset_{\mathcal{S}} C$ follows immediately from the definition of $\subset_{\mathcal{S}}$. According to this definition, $B \subset_{\mathcal{S}} C$ does not hold no matter which $B \in \mathcal{P}(X) - \mathcal{S}$ and $C \in \mathcal{S}$ may be. Hence, the validity of $B \subset_{\mathcal{S}} C$ for $C \in \mathcal{S}$ yields that $B \in \mathcal{S}$ holds and, applying the same reasoning to valid relation $A \subset_{\mathcal{S}} B$, we obtain that A, B, C are in $\mathcal{S}$, so that $A \subset C$ and $A \subset_{\mathcal{S}} C$ holds. The assertion is proved.

Let us define the binary relation $\equiv$ on $\mathcal{P}(X)$ as follows: given $A, B \subset X, A \equiv B$ holds, if $A \subset_{\mathcal{S}} B$ and $B \subset_{\mathcal{S}} A$ holds together. Obviously, $\equiv$ defines an equivalence relation on $\mathcal{P}(X)$, so that we may define the factor-space $\mathcal{P}(X)|_{\equiv}$, its elements are equivalence classes $[A] = \{B \subset X : B \equiv A\}$ for each $A \subset X$. Namely, if $A \in \mathcal{S}$, then $[A] = \{A\}$, if $A \in \mathcal{P}(X) - \mathcal{S}$, then $[A] = \mathcal{P}(X) - \mathcal{S} = [X]$, as $X \in \mathcal{P}(X) - \mathcal{S}$ holds. Hence,

$$\mathcal{P}(X)|_{\equiv} = \{[A] : A \in \mathcal{S}\} \cup [X] = \{\{A\} : A \in \mathcal{S}\} \cup \{(\mathcal{P}(X) - \mathcal{S})\} \qquad (2.1)$$

holds. Given $A, B \subset X$, set $[A] \subset_{\mathcal{S}}^* [B]$, if $A_0 \subset_{\mathcal{S}} B_0$ holds for some $A_0 \in [A]$, $B_0 \in [B]$, the choice of the representants A_0 and B_0 being obviously irrelevant. As $\langle \mathcal{P}(X), \subset_{\mathcal{S}} \rangle$ defines a partial pre-ordering on $\mathcal{P}(X)$, the pair $\langle \mathcal{P}(X)|_{\equiv}, \subset_{\mathcal{S}}^* \rangle$ defines a p.o.set and this p.o.set is isomorphic to the p.o.set $\langle \mathcal{S} \cup \{X\}, \subseteq \rangle$, where $\subseteq$ is the restriction of the standard set inclusion from $\mathcal{P}(X)$ to $\mathcal{S} \cup \{X\}$. Indeed, the isomorphism in question is settled by the simple one-to-one mapping $\sigma : \mathcal{P}(X)|_{\equiv} \longleftrightarrow \mathcal{S} \cup \{X\}$ defined by $\sigma([A]) = \sigma(\{A\}) = A$ for $A \in \mathcal{S}$, and $\sigma([x]) = X$ for $[X]$. Hence, without any loss of generality we may (and will) replace the p.o.set $\langle \mathcal{P}(X)|_{\equiv}, \subset_{\mathcal{S}}^* \rangle$ by the p.o.set $\langle \mathcal{S} \cup \{X\}, \subseteq \rangle$.

Lemma 2. *The p.o.set $\langle \mathcal{S} \cup \{X\}, \subseteq \rangle$ defines a complete lattice.*

Proof. Take $\emptyset \neq \mathcal{A} \subset \mathcal{S} \cup \{X\}$. If $X \in \mathcal{A}$, then X is the only upper bound for $\mathcal{A}$, so that the supremum value $\bigvee^{\mathcal{S}} \mathcal{A} = \bigvee_{A \in \mathcal{A}}^{\mathcal{S}} A$ is defined and identical with X.

Let $\emptyset \neq \mathcal{A} \subset \mathcal{S}$ be the case, let there exist $A_0 \in \mathcal{S}$ such that $A \subset A_0$ holds for each $A \in \mathcal{A}$. Then $\bigcup_{A \in \mathcal{A}} A = \bigcup \mathcal{A} \subseteq A_0$ holds, hence, $\bigcup \mathcal{A} \in \mathcal{S}$ follows (let us recall that $\mathcal{S}$ is an inclusion-closed fragment of $\mathcal{P}(X)$). So, $\bigvee^{\mathcal{S}} \mathcal{A}$ is defined and identical with $\bigcup \mathcal{A}$. Let $\emptyset \neq \mathcal{A} \subset \mathcal{S}$ hold, but there is no $A_0 \in \mathcal{S}$ dominating all $A \in \mathcal{A}$. In this case X is the only element in $\mathcal{S} \cup \{X\}$ dominating every $A \in \mathcal{A}$, so the $\bigvee^{\mathcal{S}} \mathcal{A}$ is defined and identical with X.

As far as the infimum $\bigwedge^{\mathcal{S}} \mathcal{A}$ for $\mathcal{A} \subset \mathcal{S} \cup \{X\}$ is concerned, if there is $A_0 \in \mathcal{A}, A_0 \neq X$, then $A_0 \in \mathcal{S}$ follows, hence, as $\bigcap \mathcal{A} = \bigcap_{A \in \mathcal{A}} A \subset A_0$ holds, $\bigcap \mathcal{A} \in \mathcal{S}$ follows ($\mathcal{S}$ is a closed fragment of $\mathcal{P}(X)$), so that the infimum $\bigwedge^{\mathcal{S}} \mathcal{A} = \bigcap \mathcal{A} \in \mathcal{S}$ is defined. If $\mathcal{A} = \{X\}$, then $X \subset A \subset X$ holds for each $A \in \mathcal{A}$, so that $\bigcap^{\mathcal{S}} \mathcal{A} = X = \bigvee^{\mathcal{S}} \mathcal{A}$. The assertion is proved.

Let us describe the following reason for the condition that $\mathcal{S}$ should be an inclusion-closed fragment of $\mathcal{P}(X)$. Given $A, B \subset X$, we analyze these four cases.

  (i) $A, B \in \mathcal{S}$ holds: then $A \subset B$ in $\mathcal{P}(X)$ implies that $A \subset_{\mathcal{S}} B$ holds as well.
 (ii) $A \in \mathcal{S}, B \in \mathcal{P}(X) - \mathcal{S}$: then $A \subset_{\mathcal{S}} B$ holds in general, in particular also when $A \subset B$ holds in $\mathcal{P}(X)$.
(iii) $A \in \mathcal{P}(X) - \mathcal{S}, B \in \mathcal{S}$: then $A \subset_{\mathcal{S}} B$ does not hold, but also $A \subset B$ cannot hold in $\mathcal{P}(X)$ supposing that $\mathcal{S}$ is closed fragment of $\mathcal{P}(X)$.
(iv) $A, B \in \mathcal{P}(X) - \mathcal{S}$: the same situation as in (ii).

Hence, if $\mathcal{S}$ is closed fragment of $\mathcal{P}(X)$, then in no case $A \subset B$ and *not* $(A \subset_{\mathcal{S}} B)$ holds together, in other terms, the $\mathcal{S}$-inclusion $\subset_{\mathcal{S}}$ does not contradict the standard set inclusion $\subset$ in $\mathcal{P}(X)$. However, if $\mathcal{S}$ does not meet the condition of closedness, this need not be the case. Indeed, let $A \in \mathcal{S}$ be such that, for some $A_0 \subset A, A_0$ is not in $\mathcal{S}$. Then we obtain that $A \subset_{\mathcal{S}} A_0$ but *not* $(A_0 \subset_{\mathcal{S}} A)$ holds – a counter-intuitive result.

3   Possibilistic Distributions with Values in Spaces Given by Closed Fragments of Power-Sets

Definition 3. Let X be a nonempty set, let $\mathcal{S} \subset \mathcal{P}(X)$ be an inclusion-closed fragment of $\mathcal{P}(X)$, let $\mathcal{S}^* = \langle \mathcal{S} \cup \{X\}, \subset \rangle$ be the related complete lattice. Let Ω be a nonempty set. A mapping $\pi : \Omega \to \mathcal{S} \cup \{X\}$ is called an $\mathcal{S}^*$-*(valued)* *possibilistic distribution on* Ω, if the relation $\bigvee^{\mathcal{S}}_{\omega \in \Omega} \pi(\omega) = X$ holds. For such π, let $\Pi : \mathcal{P}(\Omega) \to \mathcal{S} \cup \{X\}$ be defined, for each $\emptyset \neq A \subset \Omega$, by $\Pi(A) = \bigvee^{\mathcal{S}}_{\omega \in A} \pi(\omega), \Pi(\emptyset) = \emptyset = \emptyset_{\mathcal{S}^*}$ for the empty subset of Ω. Then Π is called the $\mathcal{S}^*$-*(valued) possibilistic measure on* $\mathcal{P}(\Omega)$ induced by π.

Lemma 3. *Let* $X, \mathcal{S}$ *and* $\mathcal{S}^*$ *be as in Definition 3, let* $\varphi_{\mathcal{S}} : \mathcal{P}(X) \to \mathcal{S} \cup \{X\}$ *be defined by* $\varphi_{\mathcal{S}}(A) = A$, *if* $A \in \mathcal{S}$ *holds,* $\varphi_{\mathcal{S}}(A) = X$ *otherwise, let* $\pi : \Omega \to \mathcal{P}(X)$ *be a mapping such that* $\Pi(\Omega) = X$, *where* $\Pi(A) = \bigcup_{\omega \in A} \pi(\omega)$ *for each* $\emptyset \neq A \subset \Omega, \Pi(\emptyset) = \emptyset$ *by convention. Then the mapping* $\pi_{\mathcal{S}} : \Omega \to \mathcal{S} \cup \{X\}$ *defined by* $\varphi_{\mathcal{S}}(\pi(\omega))$ *for each* $\omega \in \Omega$ *is an* $\mathcal{S}^*$-*possibilistic distribution on* Ω *and the relation* $\Pi_{\mathcal{S}}(A) = \varphi_{\mathcal{S}}(\Pi(A))$ *holds for each* $A \subset \Omega$.

Proof. Given $A \subset \Omega, \Pi(A) \subset X$ holds, so that $\varphi_S(\Pi(A)) = \Pi(A)$, if $\Pi(A) \in \mathcal{S}$ is the case, so that $\pi(\omega) \in \mathcal{S}$ holds for each $\omega \in A$. Hence, $\pi(\omega) = \pi_S(\omega) = \varphi_S(\pi(\omega))$ for each $\omega \in A$, consequently, the relation

$$\Pi_S(A) = \bigvee_{\omega \in A}^{\mathcal{S}} \pi_S(\omega) = \bigvee_{\omega \in A}^{\mathcal{S}} \pi(\omega) = \bigcup_{\omega \in A} \pi(\omega) = \Pi(A) = \varphi_S(\Pi(A)) \qquad (3.1)$$

is obvious. If $\Pi(A) \in \mathcal{S}$ does not hold, then $\varphi_S(\Pi(A)) = X$ follows, but in this case also $\bigcup_{\omega \in A} \pi_S(\omega) \in \mathcal{S}$ does not hold, so that $\Pi_S(A) = X = \varphi_S(\Pi(A))$ follows. The assertion is proved.

Lemma 4. *In each complete lattice $\mathcal{T} = \langle T, \le \rangle$ the following elementary relations are valid.*

(i) for each $T_1 \subset T_2 \subset T, \bigvee^{\mathcal{T}} T_1 \le \bigvee^{\mathcal{T}} T_2$ and $\bigwedge^{\mathcal{T}} T_1 \ge \bigwedge^{\mathcal{T}} T_2$ holds
(ii) for each $\mathcal{A} \subset \mathcal{P}(T)$ and for $T_0 = \bigcup \mathcal{A} = \bigcup_{S \in \mathcal{A}} S$ the identities

$$\bigvee^{\mathcal{T}} T_0 = \bigvee_{S \in \mathcal{A}}^{\mathcal{T}} \left(\bigvee_{t \in S}^{\mathcal{T}} t \right), \bigwedge^{\mathcal{T}} T_0 = \bigwedge_{S \in \mathcal{A}}^{\mathcal{T}} \left(\bigwedge_{t \in S}^{\mathcal{T}} t \right) \qquad (3.2)$$

are valid.

Proof. Indeed, (i) is obvious. For each $t \in T_0$ there exists $S_* \in \mathcal{A}$ such that $t \in S_*$ is the case, hence, $\bigvee^{\mathcal{T}} S_* = \bigvee_{t \in S_*}^{\mathcal{T}} t \ge t$ and, consequently, $\bigvee_{S \in \mathcal{A}}^{\mathcal{T}} (\bigvee^{\mathcal{T}} S) \ge t$ follows. On the other side, let $t^* \ge t$ for each $t \in T_0 = \bigcup \mathcal{A}$ hold, then $t^* \ge \bigvee^{\mathcal{T}} S_*$ follows, where $S_* \in \mathcal{A}, t \in S_*$ is valid. For each $\emptyset \ne S \in \mathcal{A}$ there exists $t \in T_0$ such that S may be taken as S_*, hence, $t^* \ge \bigvee_{S \in \mathcal{A}}^{\mathcal{T}} (\bigvee^{\mathcal{T}} S)$ holds and the value $\bigvee_{S \in \mathcal{A}}^{\mathcal{T}} (\bigvee_{t \in S}^{\mathcal{T}} t)$ meets the demands imposed on $\bigvee^{\mathcal{T}} T_0 = \bigvee^{\mathcal{T}} (\bigcup \mathcal{A})$, so that the first relation in (3.2) is proved. For $\bigwedge^{\mathcal{T}} T_0$, the proof is dual.

Theorem 1. *Let $X, \mathcal{S}$ and $\mathcal{S}^*$ be as in Definition 3, let Ω be a non-empty set, let $\pi : \Omega \to \mathcal{S} \cup \{X\}$ be an $\mathcal{S}^*$-possibilistic distribution on Ω, let $\Pi : \mathcal{P}(\Omega) \to \mathcal{S} \cup \{X\}$ be the $\mathcal{S}^*$-possibilistic measure on $\mathcal{P}(\Omega)$ induced by π. Then Π is completely maxitive in the sense that for each $\emptyset \ne \mathcal{E} \subset \mathcal{P}(\Omega)$ the relation*

$$\Pi \left(\bigcup \mathcal{E} \right) = \Pi \left(\bigcup_{E \in \mathcal{E}} E \right) = \bigvee_{E \in \mathcal{E}}^{\mathcal{S}} \Pi(E) \qquad (3.3)$$

is valid.

Proof. Applying to the complete lattice $\mathcal{S}^*$ the first part of (3.2) (ii), we obtain that

$$\Pi \left(\bigcup \mathcal{E} \right) = \Pi \left(\bigcup_{E \in \mathcal{E}} E \right) = \bigvee_{\omega \in \bigcup_{E \in \mathcal{E}} E}^{\mathcal{S}} \pi(\omega) = \bigvee_{E \in \mathcal{E}}^{\mathcal{S}} \left(\bigvee_{\omega \in E}^{\mathcal{S}} \pi(\omega) \right) =$$

$$= \bigvee_{E \in \mathcal{E}}^{\mathcal{S}} \Pi(E). \qquad (3.4)$$

The assertion is proved.

4   Convergence and Continuity of $\mathcal{S}^*$-Valued Possibilistic Measures

Definition 4. Let Y be a nonempty set, let $\approx$ denote an equivalence relation on the power-set $\mathcal{P}(Y)$ so that, for each $E_1, E_2, E_3 \subset Y$, (i) $E_1 \approx E_1$, (ii) if $E_1 \approx E_2$, then $E_2 \approx E_1$, and (iii) if $E_1 \approx E_2$ and $E_2 \approx E_3$, then $E_1 \approx E_3$ holds. A sequence $\{E_i\}_{i=1}^{\infty}$ of subsets of Y *tends (converges) to $E_0 \subset Y$ with respect to equivalence relation* $\approx$ ($\{E_i\} \to_{\approx} E_0$, in symbols), if there exists, for each $n = 0, 1, 2, \ldots$, a subset $E_n^* \subset Y$ such that $E_n^* \approx E_n$ holds and $\{E_n^*\}_{n=1}^{\infty}$ tends to E_0^* in the standard sense, i.e., the relation

$$\liminf\{E_n^*\} = \bigcup_{n=1}^{\infty} \bigcap_{j=n}^{\infty} E_j^* = E_0^* = \bigcap_{n=1}^{\infty} \bigcup_{j=n}^{\infty} E_j^* = \limsup\{E_n^*\} \qquad (4.1)$$

holds.

Obviously, if the identity relation on $\mathcal{P}(Y)$ is taken as the equivalence relation on $\mathcal{P}(Y)$, then convergence relation w.r.to $\approx$ reduces to the standard convergence of $\{E_n\}_{n=1}^{\infty}$ to E_0, so that Definition 4 indeed generalizes and weakens the standard notion of convergence of sequence of sets.

Theorem 2. *Let $X, \mathcal{S}, \mathcal{S}^*, \pi$ and Π be as in Theorem 1, let $\equiv (\mathcal{S})$ be the equivalence relation on $\mathcal{P}(X)$ such that, denoting by $[\cdot]$ the corresponding equivalence classes, $[x]_{\equiv(\mathcal{S})} = \{x\}$, if $x \in \mathcal{S}$, and $[x]_{\equiv(\mathcal{S})} = [X]_{\equiv(\mathcal{S})}$ for each other $x \subset X$. Let $E_1 \subset E_2 \subset \ldots$ be a monotone sequence of subsets of Ω, let $E_0 = \bigcup_{i=1}^{\infty} E_i$. Then $\Pi(E_n) \to_{\equiv(\mathcal{S})} \Pi(E_0)$ holds, hence, $\Pi(\bigcup_{j=1}^{n} E_j)$ tends to $\Pi(\bigcup_{j=1}^{\infty} E_j) = \Pi(E_0)$ with respect to the equivalence relation $\equiv (\mathcal{S})$ on $\mathcal{P}(X)$.*

Proof. Consider a sequence $E_1 \subset E_2 \subset \cdots \subset E_0 = \bigcup_{i=1}^{\infty} E_i$ of subsets of Ω. Let us analyze, first of all, the case when $\Pi(E_0) \in \mathcal{S}$ holds, so that $\Pi(E_0) \neq X$ is the case. Then we obtain that

$$\Pi(E_0) = \bigcup_{w \in E_0} \pi(w) = \bigcup_{i=1}^{\infty} \left(\bigcup_{w \in E_i} \pi(w) \right) = \bigcup_{i=1}^{\infty} \Pi(E_i) \qquad (4.2)$$

and all the values $\Pi(E_i)$ are in $\mathcal{S}$ as subsets of $\Pi(E_0)$. Consequently, $\Pi(E_i)$ tends to $\Pi(E_0)$ in the standard sense of (4.1), hence, also with respect to any equivalence relation on $\mathcal{P}(X)$ including the relation $\equiv (\mathcal{S})$. Consider, again, the sequence $E_1 \subset E_2 \subset \cdots \subset E_0 = \bigcup_{i=1}^{\infty} E_i$ as above, but this time with $\Pi(E_0)$ outside $\mathcal{S}$, hence, with $\Pi(E_0) = X$. Let $\Pi(E_i) = X$ for some $i = 1, 2, \ldots$ As $E_i \subset E_j$ holds for each $j \geq i$ and as Π is monotone w.r.to set inclusion on $\mathcal{P}(\Omega)$, then $\Pi(E_j) = X$ for each $j \geq i$ and the assertion $\Pi(E_i) \to_{\equiv(\mathcal{S})} \Pi(E_0)$ is trivially valid.

What remains to be solved is the case when $\Pi(E_0) = X$, but $\Pi(E_i) \neq X$ for each $i = 1, 2, \ldots$, hence, $\Pi(E_i) = \bigcup_{w \in E_i} \pi(w) \in \mathcal{S}$ holds. As $\bigcup_{i=1}^{\infty} E_i = E_0$ holds, for each $w_0 \in \Omega$ there exists $n_0(w_0) \in \mathcal{N} = \{1, 2, \ldots\}$ such that $w_0 \in E_j$

is the case for each $j \geq n_0(\omega_0)$. Consequently, the subset $\pi(\omega_0)$ of X is contained in $\bigcup_{\omega \in E_j} \pi(\omega)$ for each $j \geq n_0(\omega_0)$. So, the relation

$$\bigcup_{\omega \in E_0} \pi(\omega) = \bigcup_{j=1}^{\infty} \left(\bigcup_{\omega \in E_j} \pi(\omega) \right) = \lim_{j \to \infty} \Pi(E_j) \qquad (4.3)$$

is valid. In other notation, $\Pi(E_j) \to \bigcup_{\omega \in E_0} \pi(\omega)$ holds in the standard sense of convergence of sequences of sets. However, as $\Pi(E_0) = X$, the relation $\bigcup_{\omega \in E_0} \pi(\omega) \in \mathcal{S}$ does not hold, so that the equivalence $X \equiv (\mathcal{S}) \bigcup_{\omega \in E_0} \pi(\omega)$ is valid. As an immediate conclusion the relation $\Pi(E_i) \to_{\equiv(\mathcal{S})} X = \Pi(E_0)$ follows and the assertion is proved.

It is perhaps worth being noted explicitly that the set $\bigcup_{\omega \in E_0} \pi(\omega)$, even when it is not in $\mathcal{S}$, need not be identical with $\Pi(E_0)$, so that the assertion of Theorem 2 is not valid when replacing the equivalence relation $\equiv (\mathcal{S})$ on $\mathcal{P}(X)$ by the identity on $\mathcal{P}(X)$. Indeed, let $E_0 = \Omega = \{\omega_1, \omega_2, \dots\}$ and $X = \{x_1, x_2, \dots\}$ be infinite countable sets, let $R \in \mathcal{N}$ or $R = f$ be given, let $\mathcal{S} = \mathcal{P}_R(X) = \{A \subset X : \|A\| \leq R\}$ ($\mathcal{S} = \mathcal{P}_f(X) = \{A \subset X : \|A\| < \infty\}$, resp.). Let $\pi_{\mathcal{S}}(\omega_i) = \{x_{2i}\}$ for each $i = 1, 2, \dots$, let $E_n = \{\omega_1, \omega_2, \dots, \omega_n\} \subset \Omega$ for each $n = 1, 2, \dots$ Then $\{E_n\}_{n=1}^{\infty}$ tends to $\bigcup_{n=1}^{\infty} E_n = \Omega$ in the standard sense and $\Pi_{\mathcal{S}}(E_n) = \bigcup_{\omega \in E_n} \pi_{\mathcal{S}}(\omega) = \{x_2, x_4, \dots, x_{2n}\} \subset X$ for each $n = 1, 2, \dots$ So, $\bigcup_{\omega \in \Omega} \pi_{\mathcal{S}}(\omega) = \bigcup_{n=1}^{\infty} \bigcup_{\omega \in E_n} \pi_{\mathcal{S}}(\omega) = \{x_2, x_4, \dots\}$ is an infinite proper subset of X even when $\bigvee_{\omega \in \Omega}^{\mathcal{S}} \pi_{\mathcal{S}}(\omega) = X$ holds for each $\mathcal{S} = \mathcal{P}_R(X)$ with $R \in \{1, 2, \dots\} \cup \{f\}$.

Theorem 2 deals with convergence and continuity of $\mathcal{S}^*$-valued possibilistic measures from below, i.e., for increasing nested sequences of subsets of the space Ω. Let us consider also the convergence and continuity of $\mathcal{S}^*$-valued possibilistic measures from above, i.e., for decreasing nested sequences of sets, keeping in mind that the roles of the supremum and infimum operations when processing possibilistic measures are not dual. The following assertion is almost trivial.

Lemma 5. *Let $X, \mathcal{S}, \mathcal{S}^*, \pi$ and Π be as in Theorems 1 and 2, let $E_1 \supset E_2 \supset \cdots \supset E_0 = \bigcap_{i=1}^{\infty} E_i$ be a monotone sequence of subsets of Ω. If $\Pi(E_0) = X$, then $\Pi(E_i) = X$ for each $i = 1, 2, \dots$, if $\Pi(E_i) \in \mathcal{S}$ holds for some $i = 1, 2, \dots$, then $\Pi(E_0) = \Pi(\bigcap_{j=i_0}^{\infty} E_j)$, for some $i_0 \in \mathcal{N}$, so that, in both the cases, $\{\Pi(E_i)\}_{i=1}^{\infty}$ trivially tends to $\Pi(E_0)$ in the standard sense of convergence of sequences of sets.*

Proof. If $\Pi(E_0) = X$, i.e, if E_0 is not in $\mathcal{S}$, then, for each $i = 1, 2, \dots, E_i$ does not belong to $\mathcal{S}$, so that $\Pi(E_i) = X$ follows. If $E_{i_0} \in \mathcal{S}$ holds for some $i_0 = 1, 2, \dots$, then $E_j \in \mathcal{S}$ holds for each $j \geq i_0$, as $E_j \subseteq E_{i_0}$ is the case, so that $\Pi(E_j) \subset \Pi(E_{i_0})$ for each $j \geq i_0$ and for $j = 0$. Consequently, $\Pi(E_0) = \Pi(\bigcap_{i=1}^{\infty} E_i) = \Pi(\bigcap_{j=i_0}^{\infty} E_j) = \Pi(E_0)$ holds and the assertion is proved.

As a simple example shows, the property of continuity from above for $\mathcal{S}^*$-valued possibilistic measures does not hold in general. Let $X = \{x_1, x_2, \dots\}$ and $\Omega = \{\omega_1, \omega_2, \dots\}$ be infinite countable sets, let $\pi : \Omega \to \mathcal{P}(X)$ be such that $\pi(\omega_i) =$

$\{x_i\}$ for each $i = 1, 2, \ldots$ let $\mathcal{S} = \mathcal{P}_f(X) = \{A \subset X : \|A\| < \infty\}$, so that $\Pi(A) = \{x_i : \omega_i \in A\}$ for each finite $A \subset \Omega$ and $\Pi(A) = X$ for infinite $A \subset \Omega$. Consider the sequence $\{E_i\}_{i=1}^{\infty}$ of subsets of Ω such that $E_i = \{\omega_i, \omega_{i+1}, \omega_{i+2, \ldots}}$ for each $i = 1, 2, \ldots$ Consequently, $\{E_i\}_{i=1}^{\infty}$ is a decreasing nested sequence of subsets of Ω such that $E_0 = \lim_{i \to \infty} E_i = \bigcap_{i=1}^{\infty} E_i = \emptyset$ (the empty subset of Ω), so that $\Pi(E_0) = \emptyset$ (the empty subset of X). However, $\Pi(E_i) = \bigvee_{\omega \in E_i}^{\mathcal{S}} \pi(\omega) = X$ for each $i = 1, 2, \ldots$, as $\Pi(E_i) = \{x_i, x_{i+1}, \ldots\}$ is an infinite subset of X for each $i = 1, 2, \ldots$ So, we obtain that

$$\lim_{i \to \infty} \Pi(E_i) = \bigwedge_{i=1}^{\mathcal{S}\infty} \Pi(E_i) = X \neq \Pi(\lim_{i \to \infty} E_i) = \Pi\left(\bigcap_{i=1}^{\infty} E_i\right) = \Pi(E_0) = \emptyset \quad (4.4)$$

Consequently, $(\lim_{i \to \infty} \Pi(E_i)) \equiv (\mathcal{S})(\Pi(\bigcap_{i=1}^{\infty} E_i))$ does not hold as well.

5   Approximations of Possibilistic Distributions over Boolean Complete Lattices by $\mathcal{S}^*$-Valued Distributions

Possibilistic distributions taking their values in inclusion-closed fragments of power-sets may be applied as tools to obtain approximations of possibilistic distributions defined over richer structures, in particular over Boolean complete lattices. Given a nonempty set X, the most sophisticated structure over the power-set $\mathcal{P}(X)$ is the set-valued complete Boolean algebra $\mathcal{B}_X = \langle \mathcal{P}(X), \cup, \cap, (\cdot)^C, \oslash_X \mathbf{1}_X \rangle$, where $\cup$ and $\cap$ denote the union and intersection of sets from $\mathcal{P}(X)$, $A^C = X - A$ for each $A \subset X$, and $\oslash_X = \emptyset$ (the minimum or zero element of $\mathcal{B}_X$), $\mathbf{1}_X = X$ (the maximum or unit element of $\mathcal{B}_X$). Taking $\mathcal{B}_X$ as complete lattice and given a $\mathcal{B}_X$-possibilistic distribution π on Ω, set for each $\emptyset \neq E \subset \Omega, \Pi(E) = \bigcup_{\omega \in E} \pi(\omega)$; setting $\Pi(\emptyset) = \emptyset$ by convention, the obtained mapping $\Pi : \mathcal{P}(\Omega) \to \mathcal{P}(X)$ is called the $\mathcal{B}_X$-(valued) possibilistic measure on $\mathcal{P}(\Omega)$ induced by π. As may be easily seen, Π defines a completely maxitive (consequently, also monotonne w.r.to set inclusion) $\mathcal{B}$-possibilistic measure on $\mathcal{P}(\Omega)$ and the identity $\Pi(E) \cup \Pi(E^C) = X = \mathbf{1}_X$ holds for each $E \subset X$.

Let $\mathcal{S}$ be an inclusion-closed fragment of $\mathcal{P}(X)$, let $\mathcal{S}^* = \langle \mathcal{S} \cup \{X\}, \subset \rangle$ be the related complete lattice as defined above. Given $\pi : \Omega \to \mathcal{P}(X)$ such that $\Pi(\Omega) = X$ holds (in particular, given a $\mathcal{B}$-valued possibilistic distribution π on Ω), set $\pi^{\mathcal{S}}(\omega) = \pi(\omega)$, if $\pi(\omega) \in \mathcal{S}$, and set $\pi^{\mathcal{S}}(\omega) = X$ otherwise.

Lemma 6. *The mapping $\pi^{\mathcal{S}}$ defines an $\mathcal{S}^*$-valued possibilistic distribution on Ω.*

Proof. By definition, $\pi^{\mathcal{S}}(\omega) \in \mathcal{S} \cup \{X\}$ holds for each $\omega \in \Omega$. Moreover, for each $\omega \in \Omega$, $\pi^{\mathcal{S}}(\omega) \subset \pi(\omega)$ is valid and, for each $\mathcal{E} \subset \mathcal{P}(X), \bigvee^{\mathcal{S}} \mathcal{E} = \bigvee_{A \in \mathcal{E}}^{\mathcal{S}} A \supset \bigcup_{A \in \mathcal{E}} = \bigcup \mathcal{E}$ holds. Hence, the identity

$$\Pi^{\mathcal{S}}(\Omega) = \bigvee_{\omega \in \Omega}^{\mathcal{S}} \pi^{\mathcal{S}}(\omega) \supseteq \bigcup_{\omega \in \Omega} \pi(\omega) = X(= \mathbf{1}_X = \mathbf{1}_{\mathcal{S}^*}) \quad (5.1)$$

follows and the assertion is proved.

As a matter of fact, not every $\mathcal{S}^*$-possibilistic distribution on Ω can be defined as $\pi^{\mathcal{S}}$ for some $\mathcal{B}$-possibilsitic distribution π on Ω. Indeed, let X be an infinite set, let Y be a proper subset of X which is not in $\mathcal{S}$, and let $\pi : \Omega \to \mathcal{P}(X)$ be defined in this way: $\pi(\omega_0) = Y$ for just one $\omega_0 \in \Omega, \pi(\omega) = \emptyset$ (the empty subset of X) for every $\omega \in \Omega, \omega \neq \omega_0$. This mapping obviously does not define a $\mathcal{B}$-valued possibilistic distribution on Ω, as $\bigcup_{\omega \in \Omega} \pi(\omega) = Y \neq X$, but π defines an $\mathcal{S}^*$-valued possibilistic distribution on Ω. Indeed, the relation $\bigcup_{\omega \in \Omega} \pi(\omega) = Y \in \mathcal{S}$ does not hold, so that $\Pi_{\mathcal{S}}(\Omega) = \bigvee_{\omega \in \Omega}^{\mathcal{S}} \pi(\omega) = X$ holds. Moreover, the relation $\pi(\omega) = \pi_0^{\mathcal{S}}$ cannot hold (uniformly for each $\omega \in \Omega$) for no matter which $\mathcal{B}$-possibilistic distribution π_0 on Ω, as for each inclusion-closed fragment $\mathcal{S}$ of $\mathcal{P}(X)$ and each $\omega \in \Omega$ the inclusion $\pi_0^{\mathcal{S}}(\omega) \supset \pi_0(\omega)$ holds, hence, $\bigcup_{\omega \in \Omega} \pi_0^{\mathcal{S}}(\omega) = \bigcup_{\omega \in \Omega} \pi_0(\omega) = X$, but this is not the case for π, as $\bigcup_{\omega \in \Omega} \pi(\omega) = Y \neq X$ holds.

The following assertions are very simple and almost obvious, so that they are introduced just as "facts" without explicit proofs.

Fact 1

(i) For each $X \neq \emptyset$, each mapping $\pi : \Omega \to \mathcal{P}(X)$, each inclusion-closed fragments $\mathcal{S}_1, \mathcal{S}_2$ of $\mathcal{P}(X)$ such that $\mathcal{S}_1 \subset \mathcal{S}_2$ holds and for each $\omega \in \Omega$ the inclusions $\pi^{\mathcal{S}_1}(\omega) \supset \pi^{\mathcal{S}_2}(\omega) \supset \pi(\omega)$ hold. Consequently, for the induced mappings $\Pi^{\mathcal{S}_1}, \Pi^{\mathcal{S}_2}$ and Π on $\mathcal{P}(\Omega)$ and for each $E \subset \Omega$ similar inclusions $\Pi^{\mathcal{S}_1}(E) \supset \Pi^{\mathcal{S}_2}(E) \supset \Pi(E)$ are valid.

(ii) Let π define an $\mathcal{S}_1^*$-valued possibilistic distribution on Ω. Then π defines also a $\mathcal{B}$-valued possibilistic distribution on Ω if and only if the relation $\bigcup_{\omega \in \Omega} \pi(\omega) = X$ holds. If this last identity is the case, then $\pi(\omega) = \pi^{\mathcal{S}_2}(\omega)$ holds for each $\mathcal{S}_2 \supset \mathcal{S}_1$ and each $\omega \in \Omega$.

As can be easily seen, for each nonempty proper subset Y of X the power-set $\mathcal{P}(Y)$ defines an inclusion-closed fragment of $\mathcal{P}(X)$, so that $\mathcal{S}^*(Y) = \langle \mathcal{P}(Y) \cup \{X\}, \subseteq \rangle$ is the related complete lattice as introduced above. The intuition behind $\mathcal{S}^*(Y)$-valued possibilistic distribution $\pi_{\mathcal{S}(Y)}$ and measure $\Pi_{\mathcal{S}(Y)}$ is quite simple: only the values of possibility degrees from Y can be fully identified and processed, all the values beyond Y must be replaced by X as no better approximation is accessible. By $\mathcal{S}_f$ we denote the inclusion-closed fragment of $\mathcal{P}(X)$ consisting of all finite (proper, if X is finite) subsets of X, so that $\mathcal{S}_f^*(X) = \langle \{A \subset X, A \neq X : \|A\| < \infty\} \cup \{X\}, \subseteq \rangle$.

Theorem 3. *Let X and Ω be nonempty spaces (infinite in the case of X), let $\pi : \Omega \to \mathcal{S}_f \cup \{X\}$ be an $\mathcal{S}_f^*$-possibilistic distribution on Ω, so that, for each $\omega \in \Omega, \pi(\omega)$ is a finite subset of X or $\pi(\omega) = X$. Let $\{Y_i\}_{i=1}^{\infty}$ be a sequence of subsets of X which converges to X in the standard sense, so that $\liminf Y_i = X$ (for $\limsup Y_i$ this identity trivially follows). For each $i = 1, 2, \ldots$ and each $\omega \in \Omega$, let $\pi^i(\omega) = \pi^{\mathcal{S}(Y_i)}(\omega) = \pi(\omega)$, if $\pi(\omega) \subset Y_i$, let $\pi^i(\omega) = X$ otherwise. Then, for each $\omega \in \Omega, \bigcap_{i=1}^{\infty} \pi^i(\omega) = \pi(\omega)$ holds.*

Proof. As $\{Y_i\}_{i=1}^{\infty}$ tends to X, there exists, for each $x \in X$, an index $n_0(x) \in \mathcal{N} = \{1, 2, \ldots\}$ such that $x \in Y_j$ holds for each $j \geq n_0(x)$. Consequently, for each finite $A \subset X, A \subset Y_j$ holds for each $j \geq \max\{n_0(x) : x \in A\}$. Applying this result to finite sets $\pi(\omega)$, $\omega \in \Omega$, we obtain that for each $\omega \in \Omega$ there exists $n_1(\omega)$ such that $\pi(\omega) \subset Y_j$ holds for each $j \geq n_1(\omega)$. So, the identity $\pi^j(\omega) = \pi(\omega)$ for each $j \geq n_1(\omega)$ follows. As $\pi^i(\omega)$ takes only the values $\pi(\omega)$ or X, the relation $\bigcap_{i=1}^{\infty} \pi^i(\omega) = \pi(\omega)$ follows as well. In other terms, $\pi^i(\omega)$ tends to $\pi(\omega)$ with i increasing in the strong sense according to which $\pi^i(\omega)$ is identical with the limit value $\pi(\omega)$ for whole the sequence $\{\pi^i(\omega)\}_{i=1}^{\infty}$ up to a finite initial segment $\{\pi^i(\omega)\}_{i=1}^{n_1(\omega)}$. The assertion is proved.

It is perhaps worth being noted explicitly that the assertion just proved does not hold in general when admitting also infinite proper subsets of X as possible values taken by the possibilistic distribution π on Ω. Indeed, take $\Omega = \{\omega_1, \omega_2\}$, take $X = \{x_1, x_2, \ldots\}$, take $Y_n = \{x_1, x_2, \ldots, x_n\} \subset X$, and take $\pi : \Omega \to \mathcal{P}(X)$ in such a way that $\pi(\omega_1) = \{x_1, x_3, x_5, \ldots\}$ and $\pi(\omega_2) = \{x_2, x_4, x_6, \ldots\}$. As $\pi(\omega_1) \cup \pi(\omega_2) = X$, π defines a Boolean-valued possibilistic distribution on Ω. Then we obtain that $\pi^{\mathcal{S}(Y_i)}(\omega_1) = \pi(\omega_1) = \{x_1, x_3, x_5, \ldots\}$, if $\pi(\omega_1) \subset Y_i = \{x_1, x_2, \ldots, x_i\}$ holds, but this is not the case for no matter which Y_i, so that the relation $\pi^{Y_i}(\omega_1) = X$ follows for each $i = 1, 2, \ldots$ For ω_2 the situation is quite analogous, so that also $\pi^{Y_i}(\omega_2) = X$ holds for each $i = 1, 2, \ldots$ Hence, for both $j = 1, 2, \lim_{i \to \infty} \pi^{Y_i}(\omega_j) \neq \pi(\omega_j)$ is the case (writing π^{Y_i} instead of $\pi^{\mathcal{S}(Y_i)}$ for simplification).

6   Conclusions

Among the ways of possible development of the results achieved here as well as the results on which this paper is based but which could be developed also in other directions than we did it here, let us mention the following ones.

It would be perhaps worth of being analyzed which of the ideas applied and results obtained here could be shifted back to the case of real-valued possibilistic distributions and measures, or at least to the case of possibility degrees taken in the space of linear ordering (i.e., a fixed binary relation $\leq$ by which any two values from its support set are comparable).

Other interesting extensions of the results dealing with approximations of Boolean-valued possibilistic distributions and measures by their reductions to inclusion-closed fragments of the complete lattice $\langle \mathcal{P}(X), \subseteq \rangle$ can be obtained when applying them to two- or more -dimensional possibilistic distributions and measures.

It would be perhaps not so difficult present a number of other open problems and possible generalizations and extensions dealing with approximations of Boolean-valued possibilistic distributions and measures, but as the long experience of scientific research proved, the most important problems to be solved are not those presented ad hoc, but rather those emerging when solving the problems already picked up as worth being solved in the domain of research in question. The author hopes to be able to contribute, at least in fragments, to this research effort in his future work.

Acknowledgements

This work was partially supported by grant No. IAA100300503 of GA AS CR and by the Institutional Research Plan AV0Z10300504 "Computer Science for the Information Society: Models, Algorithms, Applications".

References

1. Birkhoff, G.: Lattice Theory, 3rd edn. American Mathematical Society, Providence (1967)
2. De Cooman, G.: Possibility theory I, II, III. International Journal of General Systems 25, 291–323, 325-351, 353-371 (1997)
3. Dubois, D., Nguyen, H., Prade, H.: Possibility theory, probability theory and fuzzy sets: misunderstandings, bridges and gaps. In: Dubois, D., Prade, H. (eds.) The Handbook of Fuzzy Sets Series, pp. 343–438. Kluwer Academic Publishers, Boston (2000)
4. Faure, R., Heurgon, E.: Structures Ordonnées et Algèbres de Boole. Gauthier-Villars, Paris (1971)
5. Goguen, J.A.: $\mathcal{L}$-fuzzy sets. Journal of Mathematical Analysis and Applications 18, 145–174 (1967)
6. Sikorski, R.: Boolean Algebras, 2nd edn. Springer, Heidelberg (1964)
7. Zadeh, L.A.: Fuzzy sets. Information and Control 8, 338–353 (1965)
8. Zadeh, L.A.: Fuzzy sets as a basis for a theory of possibility. Fuzzy Sets and Systems 1, 3–28 (1978)

Different Representations of Fuzzy Vectors

Jiuzhen Liang[1], Mirko Navara[2], and Thomas Vetterlein[3,⋆]

[1] School of Information Technology, Jiangnan University
1800 Lihu Road, Wuxi, Jiangsu Province, China 214122
jz.liang@yahoo.com.cn
[2] Center for Machine Perception, Department of Cybernetics
Faculty of Electrical Engineering, Czech Technical University in Prague
Technická 2, CZ-166 27, Praha, Czech Republic
navara@cmp.felk.cvut.cz
http://cmp.felk.cvut.cz/~navara
[3] Section on Medical Expert and Knowledge-Based Systems
Medical University of Vienna, Spitalgasse 23, A-1090 Wien, Austria
thomas.vetterlein@meduniwien.ac.at

Dedicated to the memory of Dan Butnariu

Abstract. Fuzzy vectors were introduced as a description of imprecise quantities whose uncertainty originates from vagueness, not from a probabilistic model. Support functions are a classical tool for representation and computation with compact convex sets. The combination of these two techniques—support functions of fuzzy vectors—has been proposed by Puri and Ralescu. Independently, Bobylev proposed another type of support functions which allows a more economical representation. However, the form of the functions is not very intuitive. We suggest a new type of support functions which combines the advantages of both preceding approaches. We characterize the functions which are support functions of fuzzy vectors in the new sense.

1 Introduction

Fuzzy sets were suggested as a tool for computing with imprecise quantities. At each point of the n-dimensional real vector space $\mathbb{R}^n$, the membership function of a fuzzy set attains a real value from $[0, 1]$ describing to which extent this point is a satisfactory approximation of the desired value. In contrast to probability models, here we do not assume the existence of a random experiment deciding whether the point belongs (totally) to the set or not; belongness is a matter of degree expressed by the values from the unit interval.

In order to represent imprecise quantities, only some fuzzy subsets of $\mathbb{R}^n$ are adequate. A natural requirement is that they are *normal*, i.e., at least one point

⋆ The first author is supported by the Agreement between the Czech Ministry of Education and the Chinese Ministry of Education and partly by Jiangsu Science Fund (BK20080544). The second author acknowledges the support by the Czech Ministry of Education under project MSM 6840770038.

C. Sossai and G. Chemello (Eds.): ECSQARU 2009, LNAI 5590, pp. 700–711, 2009.

has the degree of membership 1 and can serve as a representative crisp value. Further, the membership function is expected to decrease with distance from this point. A slightly stronger requirement is that the level sets (cuts) are compact and convex. Under these assumptions, operations with fuzzy sets can be made by the Zadeh's extension principle, as well as by the Minkowski operations on level sets; both give the same results. Nevertheless, pointwise operations with sets are difficult to compute (even approximately). As an alternative, support functions were suggested as a tool representing fuzzy vectors. Operations with fuzzy vectors correspond to pointwise operations with support functions.

In this paper, we present and compare three types of support functions of fuzzy vectors, the first by Puri and Ralescu, the second by Bobylev, the third was introduced by Butnariu, Navara, and Vetterlein, but without any details on its properties. We fill this gap now. We give formulas for conversions between various types of support functions. We characterize those functions which can occur as support functions (of all the three types). The new type of support functions is often convex. However, we show that this is not a rule.

The paper is organized as follows: Section 2 summarizes known facts about support functions of crisp sets, Section 3 recalls the vertical and horizontal representations of fuzzy sets, with emphasis on fuzzy vectors. Section 4 describes particular types of support functions of fuzzy vectors, their characterizations and other properties. The final conclusions suggest possible applications of support functions in computing with fuzzy vectors.

2   Support Functions of Crisp Sets

We denote by $\mathcal{K}^n$ the set of all non-empty compact convex subsets of $\mathbb{R}^n$. The set $\mathcal{K}^n$ is endowed with a linear structure (by $\mathbb{R}_+$ we denote the set of all non-negative reals):

$$\forall A, B \in \mathcal{K}^n : A + B = \{\boldsymbol{x} + \boldsymbol{y} \mid \boldsymbol{x} \in A, \ \boldsymbol{y} \in B\} ,$$
$$\forall A \in \mathcal{K}^n \ \forall \lambda \in \mathbb{R}_+ : \lambda A = \{\lambda \boldsymbol{x} \mid \boldsymbol{x} \in A\} .$$

For $r > 0$, we denote by $S_r^n \subset \mathbb{R}^n$ the sphere with diameter r, centered in the origin. If $r = 1$, we write $S^n = S_1^n$, and we denote by B^n the closed unit ball in $\mathbb{R}^n$. The open unit ball will be denoted by $B^n \setminus S^n$.

For any $A \in \mathcal{K}^n$, its *support function*, $h_A \colon \mathbb{R}^n \to \mathbb{R}$, is defined by

$$h_A(\boldsymbol{x}) = \max\{\langle \boldsymbol{p}, \boldsymbol{x} \rangle \mid \boldsymbol{p} \in A\} , \tag{1}$$

where $\langle \cdot, \cdot \rangle$ is the usual inner product in $\mathbb{R}^n$ (see e.g. [10]). The mapping $A \mapsto h_A$ (defined on $\mathcal{K}^n$) is injective, i.e., each compact convex set is uniquely represented by its support function. Moreover, this mapping preserves the linear operations on $\mathcal{K}^n$ and ordering by inclusion:

$$\forall A, B \in \mathcal{K}^n : h_{A+B} = h_A + h_B , \tag{2}$$
$$\forall A \in \mathcal{K}^n \ \forall \lambda \in \mathbb{R}_+ : h_{\lambda A} = \lambda h_A , \tag{3}$$
$$\forall A, B \in \mathcal{K}^n, \ A \subseteq B : h_A \leq h_B . \tag{4}$$

The functions which are support functions of compact convex sets can be characterized using the *epigraph* [8] (or *supergraph* [6]) of a function $h\colon \mathbb{R}^n \to \mathbb{R}$, i.e., the set

$$\mathrm{epi}_h = \left\{ (x_0, x_1, \ldots, x_n) \in \mathbb{R}^{n+1} \mid x_0 \geq h(x_1, \ldots, x_n) \right\}$$

of points which are above the graph of h.

Proposition 1. *A function $h\colon \mathbb{R}^n \to \mathbb{R}$ is a support function of some $A \in \mathcal{K}^n$ if and only if epi_h is a proper convex cone in $\mathbb{R}^{n+1}$, i.e., $\mathrm{epi}_h \neq \mathbb{R}^{n+1}$ and*

$$\forall \boldsymbol{x}, \boldsymbol{y} \in \mathrm{epi}_h \ \forall \lambda, \mu \in \mathbb{R}_+ : \lambda \boldsymbol{x} + \mu \boldsymbol{y} \in \mathrm{epi}_h \ .$$

In this case, $h = h_A$ for

$$A = \left\{ \boldsymbol{p} \in \mathbb{R}^n \mid \forall \boldsymbol{x} \in \mathbb{R}^n : \langle \boldsymbol{p}, \boldsymbol{x} \rangle \leq h(\boldsymbol{x}) \right\} \in \mathcal{K}^n . \tag{5}$$

For each $\boldsymbol{x} \in \mathbb{R}^n$, the set $\left\{ \boldsymbol{p} \in \mathbb{R}^n \mid \langle \boldsymbol{p}, \boldsymbol{x} \rangle \leq h(\boldsymbol{x}) \right\}$ is a closed halfspace; A is an intersection of such halfspaces, thus a closed convex set. Formula (1) (with sup instead of max) can be applied also in the more general case when A is an arbitrary bounded subset of $\mathbb{R}^n$; however, the result is the same as for the closed convex hull of A, thus the mapping is not injective when generalized to such sets. As the set epi_h is above a graph of a function which does not attain infinite values, $\mathrm{epi}_h - \mathrm{epi}_h = \mathbb{R}^n$. The support function is continuous.

Proposition 2. *The necessary and sufficient condition from Proposition 1 is equivalent to the conjunction of the following conditions:*

1. *Positive homogeneity:*

$$\forall \boldsymbol{x} \in \mathbb{R}^n \ \forall \lambda \in \mathbb{R}_+ : h(\lambda \boldsymbol{x}) = \lambda h(\boldsymbol{x}) , \tag{6}$$

2. *Subadditivity:*

$$\forall \boldsymbol{x}, \boldsymbol{y} \in \mathbb{R}^n : h(\boldsymbol{x} + \boldsymbol{y}) \leq h(\boldsymbol{x}) + h(\boldsymbol{y}) , \tag{7}$$

3. *Continuity.*

As a consequence (for $\lambda = 0$ and $\boldsymbol{y} = -\boldsymbol{x}$), we obtain

$$\forall \boldsymbol{x} \in \mathbb{R}^n : h(\boldsymbol{x}) + h(-\boldsymbol{x}) \geq 0 . \tag{8}$$

Due to (6), it is sufficient to know the values of h on a sphere S_λ^n for some $\lambda > 0$, i.e., the restricted support function $\eta_\lambda = h \restriction S_\lambda^n$. The reverse transformation is given by the formula

$$h(\boldsymbol{x}) = \frac{\|\boldsymbol{x}\|}{\lambda} \, \eta_\lambda \left(\frac{\lambda}{\|\boldsymbol{x}\|} \boldsymbol{x} \right) .$$

For a continuous function η_λ, condition (6) is useless and (7) attains the form

$$\forall \boldsymbol{x}, \boldsymbol{y} \in \mathbb{R}^n \setminus \{\boldsymbol{0}\}, \ \boldsymbol{x} + \boldsymbol{y} \neq \boldsymbol{0} \ \forall \lambda \in \mathbb{R}_+ :$$

$$\|\boldsymbol{x} + \boldsymbol{y}\| \, \eta_\lambda \left(\lambda \frac{\boldsymbol{x} + \boldsymbol{y}}{\|\boldsymbol{x} + \boldsymbol{y}\|} \right) \leq \|\boldsymbol{x}\| \, \eta_\lambda \left(\lambda \frac{\boldsymbol{x}}{\|\boldsymbol{x}\|} \right) + \|\boldsymbol{y}\| \, \eta_\lambda \left(\lambda \frac{\boldsymbol{y}}{\|\boldsymbol{y}\|} \right) . \tag{9}$$

Following Bobylev [1], we call this property *quasiadditivity*.

Remark 1. In the sequel, we shall generalize the notion of support function to fuzzy vectors. In order to distinguish different terms, types of support functions will be specified by prefixes (e.g., PR-support function). When we want to emphasize that we speak of a support function of a *crisp* set, we speak of a *classical* support function.

3   Two Representations of Fuzzy Sets and Fuzzy Vectors

Before extending the notion of support function to fuzzy sets, let us recall the properties of two representations of fuzzy sets (see, e.g., [4]) and their consequences for fuzzy vectors. In the sequel, they will be applied to descriptions of fuzzy vectors by support functions.

Let X be a non-empty crisp set (the *universe*) and V a fuzzy subset of X. The *vertical representation* of V is given by the *membership function* $m_V \colon X \to [0,1]$; $m_V(\boldsymbol{x})$ denotes the degree to which a point $\boldsymbol{x}$ belongs to V. The *horizontal representation* of V is given by the function $\ell_V \colon [0,1] \to 2^X$ which assigns to each $\alpha \in [0,1]$ the corresponding α-level set (α-cut)

$$\ell_V(\alpha) = [V]_\alpha = \{\boldsymbol{x} \in X \mid m_V(\boldsymbol{x}) \geq \alpha\} \subseteq X .$$

For $\alpha = 0$, the latter formula gives the whole space; instead of this, we make an exception (see [5,7]) and define $\ell_V(0) = [V]_0$ as the closure of the set supp $\boldsymbol{x} = \{\boldsymbol{x} \in \mathbb{R}^n \mid m_V(\boldsymbol{x}) > 0\} = \bigcup_{\alpha>0}[V]_\alpha$ (supp $\boldsymbol{x}$ is called the *support* of $\boldsymbol{x}$).

The membership function can be any function $X \to [0,1]$. However, when the fuzzy set expresses an "imprecise quantity", we often restrict attention to "meaningful" fuzzy subsets of $\mathbb{R}^n$; usual requirements are the following [4]:

1. *Normality:* $\exists \boldsymbol{x} \in \mathbb{R}^n : m_V(\boldsymbol{x}) = 1$.
2. *Fuzzy convexity:*

$$\forall \boldsymbol{x}, \boldsymbol{y} \in \mathbb{R}^n \; \forall \lambda \in [0,1] : m_V(\lambda \, \boldsymbol{x} + (1-\lambda) \, \boldsymbol{y}) \geq \min \{m_V(\boldsymbol{x}), m_V(\boldsymbol{y})\} .$$

3. *Boundedness:* supp $\boldsymbol{x}$ is bounded.

A fuzzy set with these properties is called an (n-dimensional) *fuzzy vector* [5,7]. We denote by $\mathcal{F}^n$ the set of all n-dimensional fuzzy vectors.

Remark 2. Normality says that there is at least one value $\boldsymbol{x}$ which belongs to V "completely" and can be considered a crisp representative value of the imprecise quantity V. Our definition (following [5,7]) admits more such points. (In particular, crisp sets which are fuzzy vectors are not only singletons, but all non-empty compact convex sets.)

We have the following characterization of functions which appear in the horizontal representation [4]:

Proposition 3. *Let $\tau \colon]0,1] \to 2^X$ be a set-valued function. A necessary and sufficient condition for the existence of a (unique) fuzzy set V such that τ coincides with ℓ_V on $]0,1]$ is the conjunction of the following properties:*

1. *Monotonicity:* τ *is non-increasing, i.e.,*

$$\forall \alpha, \beta \in \,]0,1], \ \alpha \leq \beta : \tau(\alpha) \supseteq \tau(\beta) \, .$$

2. *Continuity:* τ *is left continuous, i.e.,* $\forall \beta \in \,]0,1] : \tau(\beta) = \bigcap_{\alpha < \beta} \tau(\alpha) \, .$

In this case, $m_V(\boldsymbol{x}) = \sup\{\alpha \in \,]0,1] \mid \boldsymbol{x} \in \tau(\alpha)\}$ *(where we put* $\sup \emptyset = 0$*).*

Remark 3. If we want τ to be defined also at 0, then $\tau \colon [0,1] \to 2^X$ must be also right continuous at 0, i.e., $\tau(0) = \bigcup_{\alpha > 0} \tau(\alpha)$. Due to left continuity, the value of τ at 1 is also unnecessary; the values of τ on $]0,1[$ suffice to determine the fuzzy set.

Fuzzy vectors in the horizontal representation can be characterized as follows:

Proposition 4. *Let* $\tau \colon [0,1] \to 2^X$ *be a set-valued function satisfying the conditions of Proposition 3 and Remark 3 and V be the corresponding fuzzy set V. Then V is a fuzzy vector if and only if all values of* τ *are non-empty compact convex sets, i.e.,* $\forall \alpha \in [0,1] : \tau(\alpha) \in \mathcal{K}^n$.

4 Different Types of Support Functions of Fuzzy Vectors

4.1 Approach by Puri and Ralescu

Following Puri and Ralescu [7] (see also the book by Diamond and Kloeden [5]), we define the *PR-support function* of a fuzzy vector $V \in \mathcal{F}^n$ as the function $H_V \colon [0,1] \times \mathbb{R}^n \to \mathbb{R}$ such that

$$H_V(\alpha, \boldsymbol{x}) = \sup\{\langle \boldsymbol{p}, \boldsymbol{x} \rangle \mid \boldsymbol{p} \in [V]_\alpha\}. \tag{10}$$

This means that, for each $\alpha \in [0,1]$, $H_V(\alpha, \cdot)$ is the (classical) support function of the crisp α-level set $[V]_\alpha$,

$$H_V(\alpha, \boldsymbol{x}) = h_{[V]_\alpha}(\boldsymbol{x})$$

for all $\boldsymbol{x} \in \mathbb{R}^n$. As each fuzzy vector is uniquely determined by its collection of level sets and these are described by their support functions, the PR-support function characterizes the fuzzy vector completely. (The 0-level set is not needed, because it does not carry any additional information.)

PR-support functions appeared useful in the computation of the Steiner point of a fuzzy vector [11] (as a useful reference point describing the position of the fuzzy vector).

Functions which can be PR-support functions of fuzzy vectors were characterized in [3]:

Theorem 1. *A function $\varphi\colon [0,1] \times \mathbb{R}^n \to \mathbb{R}$ is the PR-support function of a (unique) fuzzy vector $V \in \mathcal{F}^n$ if and only if it satisfies (the conjunction of) the following conditions:*

$$\forall \boldsymbol{z} \in \mathbb{R}^n \ \forall \lambda \in \mathbb{R}_+ : \varphi(\alpha, \lambda \boldsymbol{z}) = \lambda \varphi(\alpha, \boldsymbol{z}) , \tag{11}$$

$$\forall \boldsymbol{x}, \boldsymbol{y} \in \mathbb{R}^n : \varphi(\alpha, \boldsymbol{x} + \boldsymbol{y}) \leq \varphi(\alpha, \boldsymbol{x}) + \varphi(\alpha, \boldsymbol{y}) , \tag{12}$$

$$\varphi(\cdot, \boldsymbol{z}) \ \textit{is non-increasing, left continuous on }]0, 1], \ \textit{and right continuous at } 0 . \tag{13}$$

The proof follows directly from Propositions 2 and 3.

Remark 4. As mentioned in Remark 3, the values for the first argument 0 or 1 are unnecessary for the representation. If they are omitted (only the restriction $\varphi \upharpoonright]0, 1[\times \mathbb{R}^n$ is used), Theorem 1 works if the requirement that $\varphi(\cdot, \boldsymbol{z})$ is right continuous at 0 is replaced by boundedness of φ.

As in the case of classical support functions, the representation of a fuzzy vector by the PR-support function $H_V\colon [0,1] \times \mathbb{R}^n \to \mathbb{R}$ is redundant; the domain can be restricted to $[0,1] \times B^n$, $[0,1] \times S^n$, or another appropriate set. The restriction $\varphi = H_V \upharpoonright [0,1] \times B^n$ was used in [5,3]. The only difference in Theorem 1 is that (11), (12) are applied only if all arguments fall in the restricted domain. In particular, (11) can be formulated as

$$\forall \boldsymbol{z} \in B^n \ \forall \lambda \in [0,1] : \varphi(\alpha, \lambda \boldsymbol{z}) = \lambda \varphi(\alpha, \boldsymbol{z}) . \tag{14}$$

For the restriction $\varphi = H_V \upharpoonright]0, 1] \times S^n$, (11) is irrelevant and (12) has to be modified as in (9):

$$\forall \boldsymbol{x}, \boldsymbol{y} \in S^n : \|\boldsymbol{x} + \boldsymbol{y}\| \, \varphi \left(\frac{\boldsymbol{x} + \boldsymbol{y}}{\|\boldsymbol{x} + \boldsymbol{y}\|} \right) \leq \|\boldsymbol{x}\| \, \varphi \left(\frac{\boldsymbol{x}}{\|\boldsymbol{x}\|} \right) + \|\boldsymbol{y}\| \, \varphi \left(\frac{\boldsymbol{y}}{\|\boldsymbol{y}\|} \right) . \tag{15}$$

4.2   Approach by Bobylev

The redundancy of the PR-support function inspired Bobylev to a more economical representation [1]. Each sphere $\{\alpha\} \times S^n$, $\alpha \in]0, 1]$, is mapped onto S_α^n by the mapping $(\alpha, \boldsymbol{x}) \mapsto \alpha \boldsymbol{x}$. All these spheres fit into the unit ball B^n. This reduces the dimensionality and facilitates computations with fuzzy vectors via their support functions. It also saves space in (approximate) computer representations of fuzzy vectors.

The *B-support function* of a fuzzy vector $V \in \mathcal{F}^n$ is defined as the function $\bar{H}_V\colon B^n \to \mathbb{R}$,

$$\bar{H}_V(\boldsymbol{x}) = \max\{ \langle \boldsymbol{p}, \boldsymbol{x} \rangle \mid \boldsymbol{p} \in [V]_{\|\boldsymbol{x}\|} \}. \tag{16}$$

Apparently, it determines a fuzzy vector V uniquely. The conversions between the two representations can be made by the following formulas:

$$\forall \boldsymbol{x} \in B^n : \bar{H}_V(\boldsymbol{x}) = H_V(\|\boldsymbol{x}\|, \boldsymbol{x}) ,$$

$$\forall \boldsymbol{y} \in \mathbb{R}^n \setminus \{\boldsymbol{0}\} \ \forall \alpha \in]0, 1] : H_V(\alpha, \boldsymbol{y}) = \frac{\|\boldsymbol{y}\|}{\alpha} \bar{H}_V \left(\frac{\alpha}{\|\boldsymbol{y}\|} \boldsymbol{y} \right) .$$

For each $\alpha \in {]0,1]}$, the mapping $\boldsymbol{x} \mapsto H_V(\alpha, \boldsymbol{x})$ coincides with $\bar{H}_V$ on the sphere S_α^n.

Functions which can be B-support functions of fuzzy vectors were characterized in [1] (we keep the original names of properties, although some of them might be debatable):

Theorem 2. *A function $\varphi \colon B^n \to \mathbb{R}$ is the B-support function of a (unique) fuzzy vector $V \in \mathcal{F}^n$ if and only if it satisfies the following conditions:*

1) Upper semicontinuity:

$$\forall \boldsymbol{x} \in B^n : \varphi(\boldsymbol{x}) = \limsup_{\boldsymbol{y} \to \boldsymbol{x}} \varphi(\boldsymbol{y}) . \tag{B1}$$

2) Positive semihomogeneity:

$$\forall \boldsymbol{x} \in B^n \ \forall \lambda \in [0,1] : \lambda \varphi(\boldsymbol{x}) \leq \varphi(\lambda \boldsymbol{x}) . \tag{B2}$$

3) Quasiadditivity:

$$\forall \boldsymbol{x}, \boldsymbol{y} \in \mathbb{R}^n, \boldsymbol{x}, \boldsymbol{y}, \boldsymbol{x} + \boldsymbol{y} \neq 0 \ \forall \lambda \in [0,1] :$$

$$\|\boldsymbol{x} + \boldsymbol{y}\| \, \varphi \left(\lambda \frac{\boldsymbol{x} + \boldsymbol{y}}{\|\boldsymbol{x} + \boldsymbol{y}\|} \right) \leq \|\boldsymbol{x}\| \, \varphi \left(\lambda \frac{\boldsymbol{x}}{\|\boldsymbol{x}\|} \right) + \|\boldsymbol{y}\| \, \varphi \left(\lambda \frac{\boldsymbol{y}}{\|\boldsymbol{y}\|} \right) . \tag{B3}$$

4) "Normality":

$$\forall \boldsymbol{x} \in B^n : \varphi(\boldsymbol{x}) + \varphi(-\boldsymbol{x}) \geq 0 . \tag{B4}$$

5) Boundedness:

$$\sup \left\{ \frac{|\varphi(\boldsymbol{x})|}{\|\boldsymbol{x}\|} \ \middle| \ \boldsymbol{x} \in B^n, \boldsymbol{x} \neq 0 \right\} < \infty . \tag{B5}$$

6)

$$\varphi(\mathbf{0}) = 0 . \tag{B6}$$

In this case, the fuzzy vector V such that $\bar{H}_V = \varphi$ is given by

$$[V]_\alpha = \left\{ \boldsymbol{p} \in \mathbb{R}^n \mid \forall \boldsymbol{x} \in S_\alpha^n : \langle \boldsymbol{p}, \boldsymbol{x} \rangle \leq \bar{H}_V(\boldsymbol{x}) \right\}$$

for all $\alpha \in {]0,1]}$ and its membership function is

$$m_V(\boldsymbol{p}) = \max \left\{ \alpha \in [0,1] \mid \forall \boldsymbol{x} \in S_\alpha^n : \langle \boldsymbol{p}, \boldsymbol{x} \rangle \leq \bar{H}_V(\boldsymbol{x}) \right\}$$

Remark 5. Following [2], (B2) could be called more precisely $[0,1]$-*superhomogeneity*; (6) is $[0,1]$-homogeneity or $\mathbb{R}_+$-homogeneity.

Remark 6. [1] The values of the B-support function at $\boldsymbol{x}$ such that $\|\boldsymbol{x}\| \in \{0,1\}$ are unnecessary for the representation (cf. Remark 3). Condition (B6) is used only for more consistent results, it is not needed for the description of the fuzzy vector.

Remark 7. In this context, upper semicontinuity implies that

$$\forall \boldsymbol{x} \in B^n : \varphi(\boldsymbol{x}) = \lim_{\gamma \to 1-} \varphi(\gamma \boldsymbol{x}) .$$

Together with quasiadditivity, it also ensures that the restriction of φ to the sphere S_α^n, $\alpha \in {]0,1]}$, is continuous.

4.3   New Approach to Support Functions

One disadvantage of the Bobylev representation is that it is not very intuitive. In particular, the B-support functions are usually neither convex nor concave. Besides, the conditions characterizing B-support functions are not very elegant.

We propose an alternative. It was first announced in [3], without any details on its properties. Here we give a characterization of functions obtained as support functions in this sense (the conditions are simpler than for the Bobylev representation). Like the Bobylev approach, the support functions are again defined on the unit ball, but the mapping is different. Each sphere $\{\alpha\} \times S^n$, $\alpha \in]0, 1[$, is mapped onto $S^n_{1-\alpha}$ by the mapping $(\alpha, \boldsymbol{x}) \mapsto (1 - \alpha)\, \boldsymbol{x}$. As a result, we obtain usually convex functions. Moreover, convexity with (B3) and (B6) appears to be a sufficient condition for a function to be a support function of a fuzzy vector in this sense.

The *BNV-support function* of a fuzzy vector $V \in \mathcal{F}^n$ is defined as the function $\widehat{H}_V \colon B^n \to \mathbb{R}$,

$$\widehat{H}_V(\boldsymbol{x}) = \sup\{\langle \boldsymbol{p}, \boldsymbol{x}\rangle \mid \boldsymbol{p} \in [V]_{1-\|\boldsymbol{x}\|}\}. \tag{17}$$

Apparently, it determines a fuzzy vector V uniquely. For each $\alpha \in]0, 1]$, the mapping $\boldsymbol{x} \mapsto H_V(1-\alpha, \boldsymbol{x})$ coincides with $\widehat{H}_V$ on the sphere S^n_α. The conversions between the representations can be made by the following formulas:

$$\forall \boldsymbol{x} \in B^n : \widehat{H}_V(\boldsymbol{x}) = H_V(1 - \|\boldsymbol{x}\|, \boldsymbol{x}),$$

$$\forall \boldsymbol{y} \in \mathbb{R}^n \setminus \{\boldsymbol{0}\}\ \forall \alpha \in]0, 1[: H_V(\alpha, \boldsymbol{y}) = \frac{\|\boldsymbol{y}\|}{1-\alpha} \widehat{H}_V\left(\frac{1-\alpha}{\|\boldsymbol{y}\|} \boldsymbol{y}\right),$$

$$\forall \boldsymbol{x} \in B^n \setminus (S^n \cup \{\boldsymbol{0}\}) : \widehat{H}_V(\boldsymbol{x}) = \frac{\|\boldsymbol{x}\|}{1-\|\boldsymbol{x}\|} \bar{H}_V\left(\frac{1-\|\boldsymbol{x}\|}{\|\boldsymbol{x}\|} \boldsymbol{x}\right),$$

$$\forall \boldsymbol{x} \in B^n \setminus (S^n \cup \{\boldsymbol{0}\}) : \bar{H}_V(\boldsymbol{x}) = \frac{1-\|\boldsymbol{x}\|}{\|\boldsymbol{x}\|} \widehat{H}_V\left(\frac{\|\boldsymbol{x}\|}{1-\|\boldsymbol{x}\|} \boldsymbol{x}\right).$$

Example 1. Let $V \in \mathcal{F}^1$ be the triangular fuzzy number given by its membership function

$$m_V(x) = \begin{cases} -1 + 2x & \text{if } x \in [-1/2, 0], \\ 1 - x & \text{if } x \in]0, 1], \\ 0 & \text{otherwise.} \end{cases}$$

From this we derive the horizontal representation:

$$\ell_V(\alpha) = \left[\frac{\alpha - 1}{2}, 1 - \alpha\right],$$

and its support functions:

$$H_V(\alpha, x) = \begin{cases} \frac{\alpha-1}{2} x & \text{if } x \leq 0, \\ (1 - \alpha)\, x & \text{if } x > 0, \end{cases}$$

$$\bar{H}_V(x) = \begin{cases} \frac{-x-1}{2} x & \text{if } x \in [-1, 0], \\ (1 - x)\, x & \text{if } x \in]0, 1], \end{cases}$$

$$\widehat{H}_V(x) = \begin{cases} \frac{x^2}{2} & \text{if } x \in [-1, 0], \\ x^2 & \text{if } x \in]0, 1]. \end{cases}$$

Notice that the BNV-support function is convex.

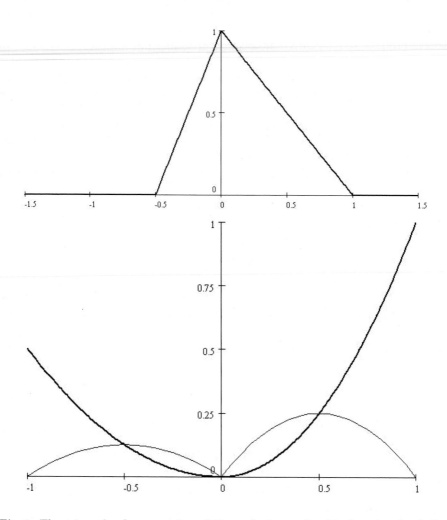

Fig. 1. The triangular fuzzy number of Example 1: membership function (top), B-support function (bottom thin), and BNV-support function (bottom thick)

BNV-support functions can be characterized as follows:

Theorem 3. *A function $\varphi\colon B^n \to \mathbb{R}$ is the BNV-support function of some fuzzy vector $V \in \mathcal{F}^n$ if and only if it satisfies conditions (B1), (B3), (B4) of Theorem 2 and*

$$\forall \boldsymbol{x} \in B^n \ \forall \lambda \in [0,1] : \lambda\varphi(\boldsymbol{x}) \geq \varphi(\lambda \boldsymbol{x}),\tag{B2'}$$

$$\varphi \text{ is continuous at each point of } S^n.\tag{B5'}$$

In this case, the fuzzy vector V such that $\widehat{H}_V = \varphi$ is given by

$$[V]_\alpha = \{\boldsymbol{p} \in \mathbb{R}^n \mid \forall \boldsymbol{x} \in S^n_{1-\alpha} : \langle \boldsymbol{p}, \boldsymbol{x} \rangle \leq \varphi(\boldsymbol{x})\} \tag{18}$$

for all $\alpha \in\,]0,1]$, and its membership function is

$$m_V(\boldsymbol{p}) = \max\{\alpha \in [0,1] \mid \forall \boldsymbol{x} \in S^n_{1-\alpha} : \langle \boldsymbol{p}, \boldsymbol{x} \rangle \leq \varphi(\boldsymbol{x})\} \tag{19}$$

Remark 8. Following [2], (B2') could be called $[0,1]$-*subhomogeneity*.

Proof. We prove the sufficiency of the conditions.

In this context, upper semicontinuity implies that

$$\forall \boldsymbol{x} \in B^n \setminus S^n : \varphi(\boldsymbol{x}) = \lim_{\gamma \to 1+} \varphi(\gamma \boldsymbol{x}). \tag{20}$$

Together with quasiadditivity, it also ensures that the restriction of φ to a sphere S^n_β, $\beta \in\,]0,1]$, is continuous (cf. Remark 7). For each $\alpha \in\,]0,1[$, upper semicontinuity and quasiadditivity imply that $\varphi \upharpoonright S^n_{1-\alpha}$ is continuous; it can be extended to a positively homogeneous subadditive continuous function $\eta_\alpha : \mathbb{R}^n \to \mathbb{R}$ by the formula

$$\eta_\alpha(\boldsymbol{y}) = \begin{cases} \dfrac{\|\boldsymbol{y}\|}{1-\alpha}\, \varphi\left(\dfrac{1-\alpha}{\|\boldsymbol{y}\|}\, \boldsymbol{y}\right) & \text{if } \boldsymbol{y} \neq \boldsymbol{0}, \\ \boldsymbol{0} & \text{if } \boldsymbol{y} = \boldsymbol{0}. \end{cases}$$

It is a (classical) support function of some compact convex set, say $A_\alpha \subset \mathbb{R}^n$,

$$A_\alpha = \{\boldsymbol{p} \in \mathbb{R}^n \mid \forall \boldsymbol{x} \in S^n_{1-\alpha} : \langle \boldsymbol{p}, \boldsymbol{x} \rangle \leq \varphi(\boldsymbol{x})\} \in \mathcal{K}^n .$$

Let $0 < \alpha < \beta < 1$ and let A_α, A_β be the corresponding sets. For each $\boldsymbol{y} \in \mathbb{R}^n \setminus \{\boldsymbol{0}\}$, we apply (B2') to

$$\lambda = \frac{1-\beta}{1-\alpha} < 1, \qquad \boldsymbol{x} = \frac{1-\alpha}{\|\boldsymbol{y}\|}\, \boldsymbol{y} \in S^n_{1-\alpha}$$

and obtain

$$\eta_\alpha(\boldsymbol{y}) = \frac{\|\boldsymbol{y}\|}{1-\alpha}\, \varphi\left(\frac{1-\alpha}{\|\boldsymbol{y}\|}\, \boldsymbol{y}\right) = \frac{\|\boldsymbol{y}\|}{1-\alpha}\, \varphi(\boldsymbol{x})$$

$$\geq \frac{\|\boldsymbol{y}\|}{1-\alpha}\, \frac{1}{\lambda}\, \varphi(\lambda \boldsymbol{x}) = \frac{\|\boldsymbol{y}\|}{1-\beta}\, \varphi\left(\frac{1-\beta}{\|\boldsymbol{y}\|}\, \boldsymbol{y}\right) = \eta_\beta(\boldsymbol{y}) .$$

Thus $\eta_\alpha \geq \eta_\beta$ and, by (4), $A_\alpha \supseteq A_\beta$. From (20) we infer that

$$\eta_\alpha = \lim_{\beta \to \alpha-} \eta_\beta$$

and hence

$$A_\alpha = \bigcap_{\beta < \alpha} A_\beta .$$

We proved that the mapping $\ell_V : \alpha \mapsto \ell_V(\alpha) = A_\alpha$ is a horizontal representation of a fuzzy set, in fact a fuzzy vector V whose level sets are of the form (18). The form (19) is obtained by the standard conversion to the vertical representation.

We omit the proof of the necessity of the conditions. It is easier and it follows the ideas used above.

Remark 9. The values of the BNV-support function at $\boldsymbol{x}$ such that $\|\boldsymbol{x}\| \in \{0,1\}$ are unnecessary for the representation (cf. Remark 3).

Remark 10. For B-support functions, (B5) guaranteed the boundedness of support of the corresponding fuzzy vector. BNV-support functions satisfy (B5) as a consequence of (B2'). On the other hand, boundedness is expressed by (B5'). By Remark 9, the values on S^n are unnecessary; as limits of values on smaller spheres, they do not bring any new information. Thus it is enough to define the BNV-support functions on the open unit ball $B^n \setminus S^n$. However, even then a boundedness condition is necessary. One possible formulation is that the BNV-support function is bounded on the open unit ball.

BNV-support functions satisfy also (B6). (This condition refers to the value at $\boldsymbol{0}$ which is unnecessary due to Remark 9.) Indeed, (B2') for $\lambda = 0$ implies $0 = 0\,\varphi(\boldsymbol{x}) \geq \varphi(0\,\boldsymbol{x}) = \varphi(\boldsymbol{0})$ and (B4) for $\boldsymbol{x} = \boldsymbol{0}$ gives $\varphi(\boldsymbol{0}) \geq 0$.

Proposition 5. *A convex function* $\varphi \colon B^n \to \mathbb{R}$ *satisfying (B3) and (B6) is the BNV-support function of some fuzzy vector.*

Proof. Convexity immediately implies (B2') and (B4). With quasiadditivity, we obtain boundedness, thus the function is also continuous. Theorem 3 applies and gives the desired fuzzy vector.

BNV-support functions are often convex (cf. Example 1). However, convexity is not a necessary condition in Proposition 5:

Example 2. In a 1-dimensional space $\mathbb{R}$, consider the fuzzy vector V with

$$m_V(p) = \sqrt{\max\{0, 1 - |p|\}}.$$

Its horizontal representation is $\ell_V(\alpha) = \left[-\left(1 - \alpha^2\right), 1 - \alpha^2\right]$ and BNV-support function $\widehat{H}_V(x) = \left(1 - (1 - |x|)^2\right)|x|$. Its second derivative, $\widehat{H}_V''(x) = 4 - 6|x|$, is negative for $|x| > 2/3$.

5   Conclusions – Computations with Support Functions

For all types of support functions introduced in this paper, the linear operations on $\mathcal{F}^n$ correspond to the pointwise operations on the support functions as in (2), (3). For the computation of Steiner points of fuzzy vectors (see [11]), all types of support functions are equally useful, but B- and BNV-support functions require one dimension less than PR-support functions.

Hausdorff metric on $\mathcal{K}^n$ gives rise to a metric on $\mathcal{F}^n$ which corresponds to the L_∞ (sup) norm on the space of PR-support functions on $[0,1] \times B^n$ (see [5] for details on this isometry). PR-support functions on $[0,1] \times S^n$ can be considered as well. Bobylev [1] introduced a metric on B-support functions on $[0,1] \times B^n$ and the isometry to the former metric was proved in [9]. Similar results can be obtained for BNV-support functions.

The advantage of BNV-support functions is that they form an irredundant representation in an n-dimensional domain (as well as B-support functions) while having a (usually convex) shape that reflects the properties of the fuzzy vector in a more intuitive way.

References

1. Bobylev, V.N.: Support function of a fuzzy set and its characteristic properties. Matematicheskie Zametki 37(4), 507–513 (1985); English translation: Math. Notes (USSR) 37(4), 281–285 (1985)
2. Burai, P., Száz, Á.: Homogeneity properties of subadditive functions. Ann. Math. Inform. 32, 189–201 (2005)
3. Butnariu, D., Navara, M., Vetterlein, T.: Linear space of fuzzy vectors. In: Gottwald, S., Hájek, P., Höhle, U., Klement, E.P. (eds.) Fuzzy Logics and Related Structures, Linz, Austria, pp. 23–26 (2005)
4. Klir, G.J., Yuan, B.: Fuzzy Sets and Fuzzy Logic. Theory and Applications. Prentice-Hall, Englewood Cliffs (1995)
5. Diamond, P., Kloeden, P.: Metric Spaces of Fuzzy Sets: Theory and Applications. World Scientific, Singapore (1994)
6. Hazewinkel, M.: Encyclopaedia of Mathematics. Kluwer Academic Publishers, Dordrecht (1995)
7. Puri, M.L., Ralescu, D.A.: Fuzzy random variables. J. Math. Anal. Appl. 114, 409–422 (1986)
8. Rockafellar, R.T.: Convex Analysis. Princeton University Press, Princeton (1970)
9. Rodriguez-Muñiz, L.J.: Revisiting Bobylev's definitions. In: 5th conference of the European Society for Fuzzy Logic and Technology, Ostrava, Czech Republic, vol. II, pp. 381–383 (2007)
10. Schneider, R.: Convex Bodies: The Brunn-Minkowski Theory. Cambridge Univ. Press, Cambridge (1993)
11. Vetterlein, T., Navara, M.: Defuzzification using Steiner points. Fuzzy Sets Syst. 157, 1455–1462 (2006)

Elicitating Sugeno Integrals: Methodology and a Case Study

Henri Prade, Agnès Rico, Mathieu Serrurier, and Eric Raufaste

IRIT - Université Paul Sabatier
118 route de Narbonne 31062 Toulouse France
LIRIS - Université Claude Bernard Lyon 1
43 bld du 11 novembre 69100 Villeurbanne France
`agnes.rico@univ-lyon1.fr`, `henri.prade@irit.fr`, `serrurier@irit.fr`,
`raufaste@univ-tlse2.fr`

Abstract. Sugeno integrals are aggregation functions that return a global evaluation that is between the minimum and the maximum of the combined evaluations. The paper addresses the problem of the elicitation of (families of) Sugeno integrals agreeing with a set of data, made of tuples gathering the partial evaluations according to the different evaluation criteria together with the corresponding global evaluation. The situation where there is no Sugeno integral that is compatible with a whole set of data is especially studied. The representation of mental workload data is used as an illustrative example, where several distinct families of Sugeno integrals are necessary for covering the set of data (since the way mental workload depends on its evaluation criteria may vary with contexts). Apart this case study illustration, the contributions of the paper are an analytical characterization of the set of Sugeno integrals compatible with a set of data, the expression of conditions ensuring that pieces of data are compatible with a representation by a common Sugeno integral, and a simulated annealing optimization algorithm for computing a minimal number of families of Sugeno integrals sufficient for covering a set of data.

1 Introduction

Sugeno integrals are used in multiple criteria decision making and in decision under uncertainty [7,2,3,15]. They are qualitative aggregation functions in the sense that they can be defined on any completely ordered scale, and that they return the median of a set of values made of the partial evaluations to be combined on the one hand, and of weights associated with subsets of the partial evaluation functions (or criteria) on the other hand. They can be seen as a qualitative counterpart of Choquet integrals that apply to quantitative settings. Discrete Choquet integrals and Sugeno integrals respectively include weighted averages, and weighted mimimum or weighted maximum as extreme particular cases. Their merits as general aggregation functions largely rely on the possibility of weighting each evaluation criteria, but also any group of criteria, thus offering the possibility of capturing some synergy between them.

C. Sossai and G. Chemello (Eds.): ECSQARU 2009, LNAI 5590, pp. 712–723, 2009.

The problem of eliciting Choquet integrals agreeing with a set of data has received some attention [4,9,6]. If we except results in [14], the elicitation of Sugeno integrals has been much less studied. This is the topic of the present paper, both from a theoretical and a practical point of view. Moreover, the perspective here is not exactly the same as in the previous works. Indeed, the works just mentioned were mainly motivated by the elicitation of a unique preference profile from a set of data that as expected to be consistent. Even in [8], where fuzzy integrals are computed for a set of classes, the classes are given and the problem is to find one integral for each class. In other words, inconsistency was viewed either as a kind of defect of the set of data (which then reflects preferences that are not representable by the considered integral), or as the indication of a limited number of outliers.

In this paper, it is not necessarily expected that all the data are representable (up to outliers) by a unique Sugeno integral, for two kinds of reasons. First, even if the data can be consistently represented by an integral, since the numerical data reflect subjective evaluations (as it is in our case study) it is natural to be interested in determining the bounds of the family of solutions (thus acknowledging some potential imprecision). Second, it may be the case that the data set refers to different classes of profiles that are not identified as such in the data set. Note that our problem differs from a supervised learning one. Indeed, in learning, it is implicitly assumed that their exists an hypothesis to be found that is at least approximately consistent with the data (possibly up to some outliers). In our case we are interested in identifying possibly distinct subsets in the data, obeying different aggregation policies, each of them being thus consistent with different hypotheses.

The paper first presents a brief reminder on Sugeno integrals in Section 2. Section 3 identifies the constraints induced by a piece of data on a family of Sugeno integral. Next, in section 4, we show how to check the compatibility of two pieces of data with respect to a Sugeno integral representation. This allows us to propose an efficient method for checking the compatibility of a set of data. In section 5 we propose an algorithm for building the smallest partition of a data set into subsets (each subset gathering compatible data with respect to a Sugeno integral representation). In the last section, we apply the algorithm to the evaluation of the subjective mental workload of flying personals in a NASA-TLX setting.

2   Sugeno Integrals: A Brief Reminder

Let $C = \{C_1, \ldots, C_n\}$ be a set of n evaluation criteria. A n-tuple of evaluations of some item on the basis of the n criteria is denoted $a = (a_1, \ldots, a_n)$ where $a_i \in [0, 1]$ $\forall i \in \{1, \ldots, n\}$. Thus a n-tuple a is a function from C to the real interval $[0, 1]$. In the following the set $\{1, \ldots, n\}$ is denoted N.

Discrete Sugeno integrals are particular aggregation functions [16,17], which are defined through the specification of a fuzzy measure, or capacity v. This capacity is a mapping from 2^C to $[0, 1]$, such that:

- $v(\emptyset) = 0; v(C) = 1;$
- if $G \subset G' \subset C$ then $v(G) \leq v(G')$.

Given two capacities v_1 and v_2 such that $v_1 \leq v_2$ [1], the set of the capacities v satisfying $v_1 \leq v \leq v_2$ is a lattice. More precisely, the considered set is a partially ordered set according to $\leq$ in which any two elements have a supremum and an infimum. A Sugeno integral of a function a from C to $[0,1]$ with respect to a capacity v is defined by:

$$S_v(a) = \bigvee_{i=1}^{n} a_{\sigma(i)} \wedge v(C_{(i)})$$
(1)

where σ is a permutation on N such that $a_{\sigma(1)} \leq \ldots \leq a_{\sigma(n)}$. $\bigvee$ and $\wedge$ denote max and min respectively. Moreover $C_{(i)} = \{C_{\sigma(i)}, \ldots, C_{\sigma(n)}\}$.

As first pointed out in [11] $S_v(a)$ is the median for $2n - 1$ terms. Namely

$$S_v(a) = median(\{a_{\sigma(1)}, \ldots, a_{\sigma(n)}\} \cup \{v(C_{(i)}), i = 2, \ldots, n\})$$

A noticeable property of Sugeno integral is:

$$\bigwedge_{i=1}^{n} a_i \leq S_v(a_1, \ldots, a_n) \leq \bigvee_{i=1}^{n} a_i.$$
(2)

An obvious consequence is: $\forall c \in [0,1]$, $S_v(c, \ldots, c) = c$ for any capacity v.

3   Constraints Induced on a Sugeno Integral Family by a Set of Data

The problem considered in this paper is the elicitation of a family of Sugeno integrals that are compatible with a set of data. Here, a set of data is a collection of n-tuples $a = (a_1, \ldots, a_n)$ associated with a global rating α. It is assumed that $\forall i \ a_i \in [0,1]$ and $\alpha \in [0,1]$. More formally, a pair (a, α) is compatible with a Sugeno integral S_v if and only if $S_v(a) = \alpha$. In this section, we study the constraints induced by a pair (a, α) on the Sugeno integrals compatible with it and we fully characterize this family.

For convenience, we assume that the a_i's are already increasingly ordered i.e. $a_1 \leq \ldots \leq a_n$. According to equation (2), there exists a Sugeno integral that satisfies $S_v(a) = \alpha$ if and only if $a_1 \leq \alpha \leq a_n$. *In the following we assume that this condition holds for the pairs (a, α) considered.* For discussing the equation $S_v(a) = \alpha$, it is useful to distinguish two cases.

Definition 1. *A data (a, α) is*

- *a DIF type piece of data if $\forall i \in \{1, \ldots, n\}$ $a_i \neq \alpha$;*
- *a EQU type piece of data if $\exists i \in \{1, \ldots, n\}$ $a_i = \alpha$.*

---

[1] I.e., $v_1(G) \leq v_2(G)$ for all set of criteria G.

3.1   DIF Case: $\forall i \in \{1, \ldots, n\}$ $a_i \neq \alpha$

In this section we only work with DIF type data. Let i be the index such that $a_1 \leq \ldots \leq a_{i-1} < \alpha < a_i \leq \ldots \leq a_n$. We can then define two particular capacities $\check{v}_{a,\alpha,DIF}$ and $\hat{v}_{a,\alpha,DIF}$:

Definition 2

$$\forall X \in 2^C, X \neq \emptyset, C \quad \check{v}_{a,\alpha,DIF}(X) = \begin{cases} \alpha & if \{C_i, \ldots, C_n\} \subseteq X \\ 0 & otherwise \end{cases}$$

$$\forall X \in 2^C, X \neq \emptyset, C \quad \hat{v}_{a,\alpha,DIF}(X) = \begin{cases} \alpha & if X \subseteq \{C_i, \ldots, C_n\} \\ 1 & otherwise \end{cases}.$$

It can be shown that

Proposition 1

$$\{v | S_v(a_1, \ldots, a_n) \geq \alpha\} = \begin{cases} \emptyset & if \ a_n < \alpha \\ \{v | v \geq \check{v}_{a,\alpha,DIF}\} & otherwise. \end{cases}$$

$$\{v | S_v(a_1, \ldots, a_n) \leq \alpha\} = \begin{cases} \emptyset & if \ \alpha < a_1 \\ \{v | v \leq \hat{v}_{a,\alpha,DIF}\} & otherwise. \end{cases}$$

As a corollary we have:

$$\forall v \ s.t. \ S_v(a) = \alpha \ \text{we have} \ \check{v}_{a,\alpha,DIF} \leq v \leq \hat{v}_{a,\alpha,DIF}.$$

Thus $\check{v}_{a,\alpha,DIF}$ and $\hat{v}_{a,\alpha,DIF}$ are the lower and upper bounds of the lattice of capacities which defines the family of Sugeno integrals compatible with the pair (a, α) in the DIF case. Moreover, to each pair (a, α) of the DIF type a unique subset of criteria is naturally associated, namely $G_a = \{C_i, \ldots, C_n\}$. Note that $v(G_a) = \alpha$.

3.2   EQU Case: $\exists i \in \{1, \ldots, 6\}$ $a_i = \alpha$

In this section we only work with EQU type data.

Let i and j be the indexes such that $a_1 \leq \ldots \leq a_{j-1} < a_j = \ldots = a_{i-1} = \alpha < a_i \leq \ldots \leq a_n$. We can then define two particular capacities $\check{v}_{a,\alpha,EQU}$ and $\hat{v}_{a,\alpha,EQU}$:

Definition 3

$$\forall X \in 2^C, X \neq \emptyset, C \quad \check{v}_{a,\alpha,EQU}(X) = \begin{cases} \alpha & if \{C_j, \ldots, C_{i-1}, \ldots C_n\} \subseteq X \\ 0 & otherwise \end{cases}$$

$$\forall X \in 2^C, X \neq \emptyset, C \quad \hat{v}_{a,\alpha,EQU}(X) = \begin{cases} \alpha & if X \subseteq \{C_i, \ldots, C_n\} \\ 1 & otherwise \end{cases}.$$

It can be shown that

Proposition 2

$$\{v|S_v(a_1,\ldots,a_n) \geq \alpha\} = \begin{cases} \emptyset & \text{if } a_n < \alpha \\ \{v|v \geq \check{v}_{a,\alpha,EQU}\} & \text{otherwise.} \end{cases}$$

$$\{v|S_v(a_1,\ldots,a_n) \leq \alpha\} = \begin{cases} \emptyset & \text{if } \alpha < a_1 \\ \{v|v \leq \hat{v}_{a,\alpha,EQU}\} & \text{otherwise.} \end{cases}$$

As a corollary we have:

$$\forall v \text{ s.t. } S_v(a) = \alpha \text{ we have } \check{v}_{a,\alpha,EQU} \leq v \leq \hat{v}_{a,\alpha,EQU}.$$

Thus $\check{v}_{a,\alpha,EQU}$ and $\hat{v}_{a,\alpha,EQU}$ are the lower and the upper bounds of the lattice of capacities which defines the family of Sugeno integrals compatible with the pair (a,α) in the EQU case. Moreover, to each pair (a,α) of the EQU type, two nested subsets of criteria are naturally associated, namely:

- $G'_a = \{C_j,\ldots,C_i,\ldots C_n\}$
- $G_a = \{C_i,\ldots,C_n\}$.

Note that $v(G'_a) \geq \alpha$ and $v(G_a) \leq \alpha$.

4   Consistency of a Set of Data with Respect to a Sugeno Integral Representation

A general issue when we look for a family of Sugeno integrals compatible with a set of data, is the question of how to compute it. In the following, we explain how this family can be obtained from the families of integrals associated with each piece of data in the dataset. Let us first consider two data pairs (a,α) and (b,β). The problem is then to determine if it exists a capacity v compatible with the two pairs i.e. such that $S_v(a) = \alpha$ and $S_v(b) = \beta$ and how to compute it. It turns out that the family of Sugeno integrals compatible with two pairs, when it is non empty, can be obtained from the family associated with each of the two pairs. Moreover it can be shown that it is enough in a set of data to be pairwise compatible for having the whole set compatible (in the sense that it exists a non empty family of Sugeno integrals compatible with each pair (a,α)).

4.1   Compatibility of Two Pairs

We have to examine three cases depending on the types (DIF or EQU) of the two pairs.

Case 1: (a,α) and (b,β) Are Both of the DIF Type
A capacity v that is compatible with the two pairs is in the intersection of the two following families :

- the lattice with the lower and upper bounds $\check{v}_{a,\alpha,DIF}$ and $\hat{v}_{a,\alpha,DIF}$.
- the lattice with the lower and upper bounds $\check{v}_{b,\beta,DIF}$ and $\hat{v}_{b,\beta,DIF}$.

There exists a capacity compatible with the two data if and only if the intersection of these two lattices is not empty. This intersection is a family with the lower bound $\check{v}_{a,\alpha,DIF} \vee \check{v}_{b,\beta,DIF}$ and the upper bound $\hat{v}_{a,\alpha,DIF} \wedge \hat{v}_{b,\beta,DIF}$. This family is non empty if and only if the following inequality holds:

$$\check{v}_{a,\alpha,DIF} \vee \check{v}_{b,\beta,DIF} \leq \hat{v}_{a,\alpha,DIF} \wedge \hat{v}_{b,\beta,DIF}. \tag{3}$$

To compare the previous capacities we do not have to do it on each criterion. We just need to compare the subset of criteria associated with each data pair.

Proposition 3

- *If $\alpha < \beta$ then (a, α) and (b, β) are compatible if and only if $G_b \nsubseteq G_a$.*
- *If $\beta < \alpha$ then (a, α) and (b, β) are compatible if and only if $G_a \nsubseteq G_b$.*
- *If $\beta = \alpha$ then (a, α) and (b, β) are always compatible.*

We omit the proof for a sake of brevity. Anyway it is only a matter of careful checking. For instance, when $\beta = \alpha$ the result is a direct consequence of equation (3). In this case, a new piece of data can only increase the number of subsets that receive the value α.

Case 2: (a, α) and (b, β) Are Both of the EQU Type

This case is quite similar to the previous one, using the appropriate bounds. Namely, the intersection of the two involved lattices is not empty if and only if

$$\check{v}_{a,\alpha,EQU} \vee \check{v}_{b,\beta,EQU} \leq \hat{v}_{a,\alpha,EQU} \wedge \hat{v}_{b,\beta,EQU}.$$

Similarly to the previous case we just need to compare two subsets of criteria.

Proposition 4

- *If $\alpha < \beta$ then (a, α) and (b, β) are compatible if and only if $G'_b \nsubseteq G_a$.*
- *If $\beta < \alpha$ then (a, α) and (b, β) are compatible if and only ifi $G'_a \nsubseteq G_b$.*
- *If $\beta = \alpha$ then (a, α) and (b, β) are always compatible.*

Case 3: (a, α) is of the DIF Type and (b, β) is of the EQU Type

This case is quite similar to the previous one, using the appropriate bounds. Namely, the intersection of the two involved lattices is not empty if and only if

$$\check{v}_{a,\alpha,DIF} \vee \check{v}_{b,\beta,EQU} \leq \hat{v}_{a,\alpha,DIF} \wedge \hat{v}_{b,\beta,EQU}.$$

Again, we just need to compare two subsets.

Proposition 5

- *If $\alpha < \beta$ then (a, α) and (b, β) are compatible if and only if $G'_b \nsubseteq G_a$.*
- *If $\beta < \alpha$ then (a, α) and (b, β) are compatible if and only if $G_a \nsubseteq G_b$.*
- *If $\beta = \alpha$ then (a, α) and (b, β) are always compatible.*

5   Global Compatibility

As pointed out in the beginning of the previous section, a set of data are compatible if there exists a non empty family of Sugeno integrals that are compatible with each pair (a, α) in the dataset. Otherwise, it means that there is not a representation of the dataset by a unique family of integrals and that several families are necessary, each covering a distinct subpart of the dataset.

Proposition 6. *Given* $(a_i, \alpha_i)_{i \in \{1,\dots,P\}}$ *a dataset set that contains* P *pairs. The set of data is compatible if and only if for any* i *and* j *in* $\{1, \dots, P\}$, (a_i, α_i) *is compatible with* (a_j, α_j).

Sketch of the proof
If the dataset is compatible, the result is straightforward. We suppose that for any i and j in $\{1, \dots, P\}$, (a_i, α_i) is compatible with (a_j, α_j). In order to simplify notations we note $\check{v}_i$ the lower bound (a_i, α_i) and $\hat{v}_i$ the upper bound associated with (a_i, α_i). Since (a_i, α_i) and (a_j, α_j) are compatible we have $\check{v}_i \vee \check{v}_j \leq \hat{v}_i \wedge \hat{v}_j$. We have abbreviated the notations here since it is no necessary to enforce the distinction between DIF and EQU types here. For all criteria C, we have $\check{v}_i \vee \check{v}_j(C) \leq \hat{v}_i \wedge \hat{v}_j(C)$; which is equivalent to $[\check{v}_i(C), \hat{v}_i(C)] \cap [\check{v}_j(C), \hat{v}_j(C)] \neq \emptyset$. Thus we have a family of intervals for which *all pairwise* intersections are not empty. From which we can conclude that the intersection of all the intervals is not empty; see [1] for a proof of this state of fact. This intersection is : $[\bigvee_{i=1}^{P} \check{v}_i(C), \bigwedge_{i=1}^{P} \hat{v}_i(C)]$.

This result is true for all criteria thus we have $\bigvee_{i=1}^{P} \check{v}_i \leq \bigwedge_{i=1}^{P} \hat{v}_i$, which is equivalent to say that the dataset is compatible.   ∎

The previous proposition is especially of interest when the dataset is inconsistent (which is often the case in practice). Indeed, as we shall see in the next section, it facilitates the buiding of a partition of the dataset into compatible subsets.

6   Looking for Consistent Subsets of Data

According to the previous results, we are now able to check the existence of a family of Sugeno integrals compatible with a set of data. If this family exists, it can easily be constructed by considering the intersection of all the families of Sugeno integrals associated with each pair (a, α) in the data set. However, in practice, the data set may be inconsistent, in the sense that there is no family of integrals compatible with all the data. In this case, an issue can be to build the smallest partition (in terms of number of subsets) of the dataset where there exists a family of integrals compatible for each subset of the partition.

This problem is an hard problem. Indeed, the number of possible partitions of an set of p elements is given by the Bell's number $B_p = \frac{1}{e} \sum_{k=0}^{\infty} \frac{k^P}{k!}$. In order to find the smallest partition, in the worst case, each partition have to be considered. B_n being greater than e^n the problem is intractable. Moreover, in order to check the compatibility of a set of data, the simplest algorithm consists in constructing the family of Sugeno integrals by considering the intersection of the lattices associated to each pair. The size of this lattice being 2^{n-1}, where n is the number of criteria, checking the compatibility of the subset becomes costly when n increases. However, this complexity can be reduced by considering the results given by the propositions 3, 4, 5, and 6. The propositions 3, 4, 5 state that the compatibility of two pairs (a, α) (b, β) can be checked with at most n operations where n is the number of criteria. The proposition 6 states that p^2 pairwise compatibility tests are needed for checking the compatibility of a data set of p elements.

Finding the smallest partition being intractable with a deterministic algorithm, we propose to use the simulated annealing [12] meta-heuristics in order to build a partition as small as possible. Simulated annealing is a meta-heuristic method developed for optimization problems. This method is inspired from a well-known physical phenomena, coming from metallurgy. Let us consider a function $F : S \mapsto \Re$ to be minimized, and representing the energy of a statistical mechanical system in a given state $s \in S$. The probability for the system to go from the state s to the state s' at the temperature T is given by the Boltzman-Gibbs distribution $P(s) = min(1, e^{\frac{-(F(s')-F(s))}{kT}})$ where k is the Boltzmann constant. For high values of T, all states have a high probability to be accepted. On the opposite side, when T is close to zero, only states improving the current minimization of the function will be accepted.

In order to apply simulated annealing for solving our problem, we have to define the neighborhood of a partition, and a criterion measure that evaluates the quality of our solution. The state space considered here is the space of the partitions of the data set into compatible subsets. A neighbor of such a partition is defined by simply moving a piece a data from one subset to another subset and checking the compatibility of this last subset. The measure we want to maximize is the following:

$$F(p_1, \ldots, p_k) = \frac{1}{k} - c * \sum_{i=1}^{k} \frac{p_i}{p} log_2(\frac{p_i}{p})$$

where k is the number of subsets, p the size of the dataset, p_i the size of the ith subset and c is a constant. The first part of the equation tends to minimize the number of subsets in the partition. The second, entropy based, term is a regulation term that favors the most contrasted partition (in terms of the cardinality of its subsets), for a given number of subsets. Note that a proper tuning of c should never block the possibility of getting a smaller partition. With this latter term we favor partition with some large subsets (as far as possible). Indeed, there exists data that are compatible with several subsets, and we try to put them together by this means. This regulation term helps the algorithm converging to the smallest partition by favoring the emergence of small subsets that may be easily fused in larger ones. The remaining small subsets account for a form of noise in the data. The starting state of the algorithm is the partition of the data set into singletons.

The constraints induced by a pair (a, α) may be too precise in practice. In order to relax these constraints we propose to consider that a Sugeno integral is compatible with a data pair if it respects the position of α with respect to $a_1, \ldots, a_n$. More formally, let us consider a pair (a, α) of the DIF type (we apply the procedure only to this type). We assume that $a_1 \leq \ldots \leq a_n$. Let i the index such that $a_1 \leq \ldots \leq a_{i-1} < \alpha < a_i \leq \ldots \leq a_n$. The bounds of the family of Sugeno integrals are now $\check{v}_{a,a_{i-1},EQU}$ and $\hat{v}_{a,a_i,EQU}$. This kind of compatibility constraint is more adapted to the subjective nature of the values.

7   Illustration: Mental Workload Evaluation

Measuring the mental workload associated with various situations is a very important topic in cognitive ergonomics and human factors. However, measuring mental workload is not a trivial task since subjective workload is generally defined as a multidimensional construct. Among the most widely used methods is the National Aeronautics and Space Administration-Task Load Index (NASA-TLX [10]).

The NASA-TLX rating procedure provides an overall workload score based on a weighted average of the ratings on six sub-scales: Mental Demands, Physical Demands, Temporal Demands, Performance, effort, and Frustration. Depending on situations, the various sources may differently contribute to the operators' subjective workload. Taking into account the relative weights of the sources first requires obtaining a measure of their relative importance. For example, during the standard NASA-TLX procedure participants provide the 15 possible pairwise comparisons of the six sub-scales. In each comparison, subjects select the source that contributed to the workload more than the other. Each source receives one point for each comparison where it was deemed to contribute more. The relative weight of a source is then given by the sum of those points, divided by 15 for normalization purposes. In order to avoid confusion, in this paper we will call "rating" the value provided for each workload source, and "weight" the relative importance of that source. After information about ratings and weights is collected, the question is to choose the aggregation method. The NASA-TLX makes use of a classical weighted mean, which simply sums the products of ratings by their normalized weights ($\sum w_i = 1$). Thus, noting a_i the rating about the ith source and w_i the relative importance of the same source, the subjective workload SW in the NASA-TLX method is provided by

$$SW = \sum_{i=1}^{6} w_i a_i.$$

where w_i and a_i respectively denote the weight and rating associated with the ith workload source. The weighted average is easy to compute and familiar to most users. On the other hand, despite its apparent simplicity, it is built upon several strong mathematical assumptions that are not necessarily verified in workload assessment. For example, it requires that weights do not depend on ratings. This condition could be attained, for example, by having operators providing the ratings and external experts providing the weights independently. Unfortunately, in the standard NASA-TLX procedure, each operator provides both the ratings and weights. Second, weighted average does not allow taking into account interactions between sources. Is it a reasonable choice to neglect dependencies and interactions between workload sources? By neglecting such interaction effects, a weighted average model might induce measurement biases.

This raises the question of the possiblity of finding an aggregation scale that does not require the elicitation of both ratings and weights by the same subjects, nor demand external expertise and still differentiates between aggregation policies. Moreover, one could want to process subjective data in a qualitative fashion since subjective rating scales are not proved to possess full numerical properties. Due to the qualitative nature of the data, least square error minimization regression does not sound much appropriate. This has motivated a recent study [13] where two of the authors experimented Sugeno integrals as a tool for representing a set of mental workload data. However no formal characterization of the family of the Sugeno compatible with a set of data where provided and the algorithm proposed for identifying subsets of compatible data, together with their Sugeno integral representation, was fairly adhoc. As in the previous study [13], the data used here were collected during a series of five rotations of planes pertaining to a big European airline company. Overall, the rotations covered 48 flights. Three types of planes were used (Airbus A319, A320, and A321). Each rotation covered three days. Twenty-two participants of flying personnel, on board of the planes, participated to the study. They were either stewards / stewardesses or cabin chiefs. All participants responded to a subjective mental workload assessment questionnaire once in each phase of the flight: preparation, taking off, cruise, and landing. Overall, the data set contains 840 pairs (a, α) where a is a 6-component vector corresponding to the evaluation of the six NASA-TLX criteria and α is a global corresponding evaluation of their mental workload. All the values are in $[0, 1]$. In practice, the participants where invited to estimate the criteria and their global mental workload by crossing a linear bounded segment. Such a qualitative assessment procedure makes differences between estimates not really meaningful, which leads us to use a Sugeno rather than a Choquet integral.

We applied the simulated annealing algorithm defined in the previous to the data. From the 840 data, 811 satisfy the representability condition (2). The 29 outliers were not further taken into consideration. We first consider the algorithm without relaxing the constraints $S_v(a) = \alpha$ as described in section 3. We obtained a partition with 30 subsets. The larger subset contains 168 pieces of data, 9 subsets have more than 30 elements and 6 subsets have less than 2 elements. Using the previous algorithm [13] on the same dataset yields to 37 subsets. When relaxing the constraints of the DIF type (as explained at the end of section 6), the number of subsets found decreases to 4 (having respectively 497, 179, 76, 59 elements). A relaxed version of the heuristic algorithm [13] still found 12 subsets, many of them being very small since the 4 largest subset gather 97% of the 811 compatible data. As it can be observed, the algorithm proposed in this paper performs better than the previous one. Still more importantly the proposed algorithm is based on the rigorous analysis of the compatibility between pieces of data and then use a well-founded meta-heuristics. The algorithm in [13] was not offering such guarantees.

8   Conclusion

In this paper, we have provided a general approach to the elicitation of Sugeno integrals where several partial coverages of the data set can be found in case the data are not altogether compatible. Although we take into account the subjective nature of the data by using Sugeno integral rather than Choquet integral, and by allowing some relaxation in the algorithm, it would be also interesting at the theoritical level to study constraints of the form $S_v(a) \leq \alpha$ or $S_v(a) \geq \alpha'$ and the sensitivity of the algorithm to slight variations of the a_i's. On this latter point, note however that at least for the DIF type of data, the capacity is not changed as long the position of the global rate α is not modified among the a_i's. This raises the question of how to use the families of Sugeno integrals for prediction tasks. A preliminary analysis of the results shows that, as it was already the case with the previous heuristic algorithm, the larger subsets found are correlated with some meaningful external features of the data. This has to be more deeply investigated. Lastly, Choquet integrals have been also considered as a possible substitute to the NASA-TLX weighted average [5]. This calls for a comparison of the relative merits of Choquet and Sugeno integrals in the NASA-TLX perspective. Besides, the problem that we have been dealing with is close to a learning problem. In case of situations where the data are assumed to be somewhat consistent with a unique hypotheses, we may introduce a compatibility measure, e.g. Spearman correlation, to be maximized.

Acknowledgements

CLEE-LTC and IRIT have been partially supported by an INRS (Inst. Nat. de la Recherche sur la Sante) grant on the evaluation of subjective mental workload.

References

1. Dubois, D., Fargier, H., Prade, H.: Multiple-sources informations fusion - a pratical inconsistency-tolerant approach. In: Proc. 8th Inter. Conf. on Information Processing and Management of Uncertainty in Knowledge-based Systems (IPMU 2000), pp. 1047–1054 (2000)
2. Dubois, D., Marichal, J.-L., Prade, H., Roubens, M., Sabbadin, R.: The use of the discrete Sugeno integral in decision making: A survey. International Journal of Uncertainty, Fuzziness and Knowledge-Based Systems 9, 539–561 (2001)
3. Dubois, D., Prade, H., Sabbadin, R.: Decision-theoretic foundations of qualitative possibility theory. Eur. J. of Operational Research 128, 459–478 (2001)

4. Grabisch, M., Baret, J.M., Larnicol, M.: Analysis of interaction between criteria by fuzzy measure and its application to cosmetics. In: Int. Conf on Methods and Applications of Multicriteria Decision Making, Belgium (1997)
5. Grabisch, M., Raufaste, E.: An experimental study of statistical properties of Choquet and Sugeno integrals. IEEE Trans. on Fuzzy Systems 16, 839–850 (2008)
6. Grabisch, M., Kojadinovic, I., Meyer, P.: A review of capacity identification methods for Choquet integral based multi-attribute utility theory — applications of the kappalab R package. Eur. J. of Operational Research 186, 766–785 (2008)
7. Grabisch, M., Murofushi, T., Sugeno, M.: Fuzzy Measures and Integrals. Theory and Applications. Physica-Verlag, Berlin (2000)
8. Grabisch, M., Nicolas, J.M.: Classification by fuzzy integral — performance and tests. Fuzzy Sets and Systems 65, 255–271 (1994)
9. Grabisch, M., Roubens, M.: Application of the Choquet integral in multicriteria decision making. In: Grabisch, M., Murofushi, T., Sugeno, M. (eds.) Fuzzy Measures and Integrals—Theory and Appl., pp. 415–434. Physica Verlag (2000)
10. Hart, S.G., Staveland, L.E.: Development of nasa-tlx (task load index): results of empirical and theoretical research. In: Hancock, P., Meshkati, N. (eds.) Human Mental Workload, pp. 139–183. North Holland B.V., Amsterdam (1988)
11. Kandel, A., Byatt, W.J.: Fuzzy sets, fuzzy algebra, and fuzzy statistics. Proceedings of IEEE 66, 1619–1639 (1978)
12. Kirkpatrick, S., Gelatt Jr., C.D., Vecchi, M.P.: Optimization by simulated annealing. Science 220, 671–680 (1983)
13. Raufaste, E., Prade, H.: Sugeno integrals in subjective mental workload evaluation: Application to flying personnel data. In: Proc. of the 11th Inter. Conf. on Information Processing and Management of Uncertainty in Knowledge-Based Systems (IPMU 2006), pp. 564–570 (2006)
14. Rico, A.: Modélisation des préférences pour l'aide à la décision par l'intégrale de Sugeno. PhD thesis, Université Paris I-Panthon-Sorbonne (2002)
15. Rico, A., Labreuche, C., Grabisch, M., Chateauneuf, A.: Preference modeling on totally ordered sets by the Sugeno integral. Discrete Applied Math. 147 (2005)
16. Sugeno, M.: Theory of fuzzy integrals and its applications. PhD thesis, Tokyo Institute of technology (1974)
17. Sugeno, M.: Fuzzy measures and fuzzy integrals: A survey. In: Gupta, M.M., Saridis, G.N., Gaines, B.R. (eds.) Fuzzy Automa and Decision Processes, pp. 89–102. North Holland B.V., Amsterdam (1977)

Robust Gene Selection from Microarray Data with a Novel Markov Boundary Learning Method: Application to Diabetes Analysis

Alex Aussem[1], Sergio Rodrigues de Morais[1],
Florence Perraud[1], and Sophie Rome[2]

[1] University of Lyon,
LIESP, Université de Lyon 1, F-69622 Villeurbanne France
`aaussem@univ-lyon1.fr`
[2] University of Lyon,
RMND INSERM U870/INRA 1235, F-69622 Villeurbanne France
`srome@univ-lyon1.fr`

Abstract. This paper discusses the application of a novel feature subset selection method in high-dimensional genomic microarray data on type 2 diabetes based on recent Bayesian network learning techniques. We report experiments on a database that consists of 22,283 genes and only 143 patients. The method searches the genes that are conjunctly the most associated to the diabetes status. This is achieved in the context of learning the Markov boundary of the class variable. Since the selected genes are subsequently analyzed further by biologists, requiring much time and effort, not only model performance but also robustness of the gene selection process is crucial. Therefore, we assess the variability of our results and propose an ensemble technique to yield more robust results. Our findings are compared with the genes that were associated with an increased risk of diabetes in the recent medical literature. The main outcomes of the present research are an improved understanding of the pathophysiology of obesity, and a clear appreciation of the applicability and limitations of Markov boundary learning techniques to human gene expression data.

1 Introduction

The identification of biologically relevant subset of genes (features) among tens of thousands of genes with no more than one hundred samples that are not captured by traditional statistical testing is a topic of considerable interest within the bioinformatics community. It is also a very challenging topic of pattern recognition research that has attracted much attention in recent years [1,2].

Type 2 diabetes mellitus (T2DM) affects over 140 million people worldwide and is a principal contributor to vascular disease, blindness, amputation and kidney failure. It is characterized by high blood sugar level (or hyperglycemia) arising from a deteriorated tissue response to the biological effect of insulin (i.e., insulin resistance). It is often associated with obesity. The primary cellular cause

C. Sossai and G. Chemello (Eds.): ECSQARU 2009, LNAI 5590, pp. 724–735, 2009.

of insulin resistance remains uncertain. Genomic analysis techniques offer powerful tools to decipher the pathology of T2DM at a molecular level. In this study we have used the results obtained from the Affymetrix microarray technology that can examine the level of expression of thousand genes simultaneously. As impaired insulin action in skeletal muscle is a hallmark feature of T2DM we have focused this study on this tissue and pooled different microarray data from different published studies in order to have a microarray dataset from individuals showing different stage of the disease (i.e. ; healthy subjects, insulin-sensitive obese patients, insulin-resistant morbid obese patients, glucose-intolerant patients and type 2 diabetic patients). Our data set consists of 22,283 genes and only 143 samples. It was obtained in collaboration with INSERM U870/INRA 1235 laboratory and represents a compilation of different microarray data published during the last five years on the skeletal muscle from patients suffering from type 2 diabetes or obesity or from healthy subjects.

As noted in [3], two problems arise when dealing with such high-dimensional databases. First, very few methods can scale up to ten of thousands of variables in reasonable time. Second, given the small sample size, the validity of existing Feature Subset Selection (FSS) methods is questionable, as they are typically tested on larger sample sizes and much lower dimensions. Therefore, when using FSS in these domains, not only model performance but also robustness of the feature selection process should be taken into account, as domain experts would prefer a stable FSS algorithm over an unstable one. Clearly, biologists need to feel confident in the selected features, as in most cases these features are subsequently analyzed further, requiring much time and effort.

FSS techniques can be divided into three categories, depending on how they interact with the classifier (i.e. filter, wrapper and embedded methods) [4]. In this study, the FSS is achieved in the context of determining the Markov boundary (MB for short) of the class variable that we want to predict. The MB of a variable T, denoted by $\mathbf{MB}_T$, is the minimal subset of $\mathbf{U}$ (the full set) that renders the rest of $\mathbf{U}$ independent of T. Inducing the MB automatically from data can be achieved by constraint-based (CB) algorithms [5,6,7]. CB methods yield compact MB by heeding independencies in the data using conditional independence tests. They systematically check the data for independence relations and use those relationships to infer necessary features in the MB.

A powerful and highly scalable CB algorithm called MBOR has been proposed recently in [8,9]. MBOR has been shown to outperform the latest MB learning proposal discussed in detail in [6], in terms of accuracy on large Markov boundaries on data sets scaling up to tens of thousand variables with small sample sizes. In this study, we show experimentally that it is also slightly more robust. Multiple runs of MBOR on resamples of the microarray data are combined, using ensemble techniques, to yield more robust results. Genes are aggregated into a consensus gene ranking and the top ranked features are analyzed by our domain expert. We show that our findings are in nice agreement with the genes that were associated with an increased risk of diabetes in the recent medical literature.

The paper is organized as follows. In Sections 2, we briefly introduce the principles of Bayesian networks and Markov boundaries. We discuss constraint-based search methods in Section 3. The algorithm MBOR is detailed in section 4. Extensive experiments are then conducted in Section 5 to evaluate a robust set of genes that are associated with diabetes.

2   Preliminaries

For the paper to be accessible to those outside the domain, we recall first the principles of Bayesian networks. In this paper, we only deal with discrete random variables. Formally, a BN is a tuple $< \mathcal{G}, P >$, where $\mathcal{G} =< \mathcal{V}, \mathcal{E} >$ is a directed acyclic graph (DAG) with nodes representing the random variables $\mathcal{V}$ and P a joint probability distribution on $\mathcal{V}$. A BN structure $\mathcal{G}$ entails a set of conditional independence assumptions. They can all be identified by the *d-separation criterion* [10]. We use $X \perp_{\mathcal{G}} Y|\mathbf{Z}$ to denote the assertion that X is d-separated from Y given $\mathbf{Z}$ in $\mathcal{G}$. Formally, $X \perp_{\mathcal{G}} Y|\mathbf{Z}$ is true when for every undirected path in $\mathcal{G}$ between X and Y, there exists a node W in the path such that either (1) W does not have two parents in the path and $W \in \mathbf{Z}$, or (2) W have two parents in the path and neither W nor its descendants is in $\mathbf{Z}$. $X \perp_{\mathcal{G}} Y|\mathbf{Z}$. If $< \mathcal{G}, P >$ is a BN, $X \perp_{P} Y|\mathbf{Z}$ if $X \perp_{\mathcal{G}} Y|\mathbf{Z}$. The converse does not necessarily hold. We say that $< \mathcal{G}, P >$ satisfies the *faithfulness condition* if the d-separations in $\mathcal{G}$ identify *all and only* the conditional independencies in P, i.e., $X \perp_{P} Y|\mathbf{Z}$ iff $X \perp_{\mathcal{G}} Y|\mathbf{Z}$.

A Markov blanket $\mathbf{M}_T$ of the T is any set of variables such that T is conditionally independent of all the remaining variables given $\mathbf{M}_T$. A Markov boundary, $\mathbf{MB}_T$, of T is any Markov blanket such that none of its proper subsets is a Markov blanket of T.

Theorem 1. *Suppose $< \mathcal{G}, P >$ satisfies the faithfulness condition. Then X and Y are not adjacent in $\mathcal{G}$ iff $\exists \mathbf{Z} \in \mathbf{U} \setminus \{X \cup Y\}$ such that $X \perp Y|\mathbf{Z}$. Moreover, for all X, the set of parents, children of X, and parents of children of X is the unique Markov boundary of X.*

A proof can be found for instance in [11]. We denote by $\mathbf{PC}_T^{\mathcal{G}}$, the set of parents and children of T in $\mathcal{G}$, and by $\mathbf{SP}_T^{\mathcal{G}}$, the set of *spouses* of T in $\mathcal{G}$. The *spouses* of T are the parents of the children of T. These sets are unique for all $\mathcal{G}$, such that $< \mathcal{G}, P >$ is faithful and so we will drop the superscript $\mathcal{G}$. We denote by $\mathbf{dSep}(X)$, the set that d-separates X from the (implicit) target T.

A structure learning algorithm from data is said to be correct (or sound) if it returns the correct DAG pattern (or a DAG in the correct equivalence class) under the assumptions that the independence test are reliable and that the learning database is a sample from a distribution P faithful to a DAG $\mathcal{G}$, The (ideal) assumption that the independence tests are reliable means that they decide (in)dependence iff the (in)dependence holds in P. The problem of learning the most probable *a posteriori* Bayesian network (BN) from data is worst-case NP-hard [12]. This challenging topic of pattern recognition has attracted much attention over the last few years.

3   Constraint-Based Methods

Having to learn a Bayesian network in order to learn a MB of a target can be very time consuming for high-dimensional databases. Fortunately, there exists several independence-test based (also called constraint-based) algorithms that search the MB of a variable without having to construct the whole BN first (see [2] for a review). Hence their ability to scale up to thousands of variables. In recent years, there have been a growing interest in inducing the MB automatically from data. Very powerful correct, scalable and data-efficient constraint-based (CB) algorithms have been proposed recently, e.g., PCMB [6], MBOR [8,9], IAMB [13] or its variants: Fast-IAMB [14] and Inter-IAMB [15]. CB methods systematically check the data for independence relationships to infer the structure. Typically, the algorithms run a χ^2 independence test in order to decide upon the acceptance or rejection of the null hypothesis of conditional independence. In our implementation we prefer a statistically oriented conditional independence test based on the G-statistic:

$$G = 2 \sum_{i=1}^{m} \sum_{j=1}^{p} \sum_{k=1}^{q} n(i,j,k) \ln \frac{n(i,j,k)n(\cdot,\cdot,k)}{n(i,\cdot,k)n(\cdot,j,k)}. \tag{1}$$

where $n(i,j,k)$ is the number of times simultaneously $X = x_i$, $Y = y_j$ and $\mathbf{Z} = \mathbf{z_k}$ in the samples, that is, the value of the cell (i,j,k) in the contingency table. The statistic is compared against a critical value to decide upon of the acceptance or rejection of the null hypothesis of conditional independence.

4   Markov Boundary Discovery

A novel powerful and highly scalable CB algorithm called MBOR has been proposed recently in [8,9]. MBOR was designed in order to endow the search procedure with the ability to: 1) handle efficiently data sets with thousands of variables but comparably few instances, 2) deal with datasets which present some deterministic relationships among the variables, 3) be correct under the faithfulness condition, and most important, 4) be able to learn large Markov boundaries. MBOR should be viewed as a meta-procedure that applies an ensemble technique to combine the advantages of both divide-and-conquer and incremental methods to improve accuracy, especially on densely connected networks. It is based on a subroutine, called Interleaved Incremental Association Parents and Children, Inter-IAPC, that takes a variables X as input and outputs an estimate of the set of parents and children of X, $\mathbf{PC}_X$.

4.1   Inter-IAPC Algorithm

Inter-IAPC is a fast incremental method that receives a target node T as input and promptly returns a rough estimate of $\mathbf{PC}_T$. It is a variant of an algorithm called Inter-IAMB that was proposed in [15]. The algorithm starts with a two-phase approach. A growing phase (lines 3-8) attempts to add the most dependent variables to T, followed by a shrinking phase (lines 9-13) that attempts to remove as many irrelevant variables as possible. The shrinking phase is interleaved with the growing phase. Interleaving the two phases allows to eliminate some of the false positives in the current blanket as the algorithm progresses during the growing phase. Once the MB is obtained, the spouses of the target (parents of children) are removed (lines 15-21). While Inter-IAPC is very fast, it is considered as data inefficient in [2] because $\mathbf{PC}_T$ can be identified by conditioning on sets much smaller than those used by Inter-IAPC(T). We will see next how MBOR combines several runs of Inter-IAPC using ensemble techniques to alleviate its data inefficiency.

4.2   MBOR Algorithm

MBOR receives a target node T as input and returns an estimate of $\mathbf{MB}_T$. MBOR works in five phases and uses Inter-IAPC as a subroutine. In phase I and II, MBOR constructs a superset of the parents and children to reduce as much as possible the number of variables before proceeding further. The size of the conditioning set $\mathbf{Z}$ is severely restricted to increase the reliability of the conditional independence tests (data efficiency) : $card(\mathbf{Z}) \leq 1$ (at lines 3 and 10). We denote by $\mathbf{dSep}(X)$, the set that d-separates X from the (implicit) target T (lines 5, 12 and 20). In phase III, a superset of the spouses of T is built with $card(\mathbf{Z}) \leq 2$ (at lines 20 and 26). Phase IV finds the parents and children in the superset of $\mathbf{PC}_T$. The rule for X to be considered as adjacent to T (and vice-versa) is as follows: $X \in \mathbf{PC}_T$ if $[X \in$ Inter-IAPC(T)$]$ OR $[T \in$ Inter-IAPC(X)$]$. Therefore, all variables that have T in their vicinity are included in $\mathbf{PC}_T$. This procedure of phase IV not only improves accuracy in practice, but it also handles some deterministic relationships (not discussed here for conciseness). Phase V identifies the spouses of the target among the variables in the Markov boundary superset ($\mathbf{MBS}$) in exactly the same way PCMB does [2].

We would like to stress that the OR operator is one of the key advantage of MBOR, compared to the state-of-the-art algorithms that use the AND operator instead (for instance, MMMB [13], PCMB [2] and IAMB [13]). By loosening the criteria by which two nodes are said adjacent, the effective restrictions on the size of the neighborhood are far less severe. This simple "trick" has significant impact on the accuracy of MBOR as we will see. It enables the algorithm to handle large neighborhoods while still being correct under faithfulness condition. MBOR's correctness under faithfulness condition is established by the following theorem (see [8] for the proof):

Theorem 2. *Under the assumptions that the independence tests are reliable and that the database is an independent and identically distributed sample from a probability distribution P faithful to a DAG $\mathcal{G}$, MBOR(T) returns $\mathbf{MB}_T$.*

Algorithm 1. *MBOR*

Require: T: target; **U**: variables
Ensure: **MB**: Markov boundary of T

    Phase I: *Remove X if $T \perp X$*
1. **PCS** = **U** \ T
2. **for all** $X \in$ **PCS do**
3.   **if** $(T \perp X)$ **then**
4.   **PCS** = **PCS** \ X
5.   **dSep**$(X) = \emptyset$
6.   **end if**
7. **end for**
    Phase II: *Remove X if $T \perp X|Y$*
8. **for all** $X \in$ **PCS do**
9.   **for all** $Y \in$ **PCS** \ X **do**
10.   **if** $(T \perp X \mid Y)$ **then**
11.   **PCS** = **PCS** \ X
12.   **dSep**$(X) = Y$; **go to 15**
13.   **end if**
14.   **end for**
15. **end for**
    Phase III: *Find super set for SP*
16. **SPS** = $\emptyset$
17. **for all** $X \in$ **PCS do**
18.   **SPS**$_X = \emptyset$
19.   **for all** $Y \in$ **U** \ $\{T \cup$ **PCS**$\}$ **do**
20.   **if** $(T \not\perp Y|$**dSep**$(Y) \cup X)$ **then**
21.   **SPS**$_X = $ **SPS**$_X \cup Y$
22.   **end if**
23.   **end for**
24.   **for all** $Y \in$ **SPS**$_X$ **do**
25.   **for all** $Z \in$ **SPS**$_X$ \ Y **do**
26.   **if** $(T \perp Y|X \cup Z)$ **then**
27.   **SPS**$_X = $ **SPS**$_X$ \ Y; **go to 30**
28.   **end if**
29.   **end for**
30.   **end for**
31.   **SPS** = **SPS** $\cup$ **SPS**$_X$
32. **end for**
    Phase IV: *Find PC of T*
33. **PC** = *Inter-IAPC*$(T, \mathcal{D}($**PCS** $\cup$ **SPS**$)$
34. **for all** $X \in$ **PCS** \ **PC do**
35.   **if** $T \in$ *Inter-IAPC*$(X, \mathcal{D})$ **then**
36.   **PC** = **PC** $\cup X$
37.   **end if**
38. **end for**
    Phase V: *Find spouses of T*
39. **SP** = $\emptyset$
40. **for all** $X \in$ **PC do**
41.   **for all** $Y \in$ *Inter-IAPC*$(X, D) \backslash \{$**PC** $\cup T\}$ **do**
42.   Find minimal **Z** $\subset$ **PCS** $\cup$ **SPS** $\backslash \{T \cup Y\}$ such that $T \perp Y|$**Z**
43.   **if** $(T \not\perp Y|$**Z** $\cup X)$ **then**
44.   **SP** = **SP** $\cup Y$
45.   **end if**
46.   **end for**
47. **end for**

Algorithm 2. *Inter-IAPC*

Require: T: target; D: data set; **V** set of variables
Ensure: **PC**: Parents and children of T;
1. **MB** = $\emptyset$
2. **repeat**
3.   *Add true positives to* **MB**
4. $Y = $ **argmax**$_{X \in (\mathbf{V} \backslash \mathbf{MB} \backslash \{T\})}$
5.   *AssocMeasure*$(T, X|$**MB**$)$
6.   **if** $T \not\perp Y|$**MB then**
7.   **MB** = **MB** $\cup Y$
8.   **end if**

    Remove false positives from **MB**
9.   **for all** $X \in$ **MB do**
10.   **if** $T \perp X|($**MB** \ $X)$ **then**
11.   **MB** = **MB** \ X
12.   **end if**
13.   **end for**
14. **until MB** has not changed

    Remove parents of children from **MB**
15. **PC** = **MB**
16. **for all** $X \in$ **MB do**
17.   **if** $\exists$**Z** $\subset ($**MB** \ $X)$
18.   such that $T \perp X \mid$ **Z then**
19.   **PC** = **PC** \ X
20.   **end if**
21. **end for**

4.3   FSS Robustness

When using FSS on data sets with large number of features, but a very small number of samples, not only model performance but also robustness of the FSS process is important. In microarray analysis, domain experts clearly prefer a stable gene selection as in most cases these genes are subsequently analyzed further, requiring much time and effort. Surprisingly, the robustness of FSS techniques has received relatively little attention so far in the literature. As noted in [3], modification of the dataset can be considered at different levels: perturbation at the instance level (e.g. by removing or adding samples), at the feature level (e.g. by adding noise to features), or variation of the parameter of the FSS algorithm (here the critical p-value of the independence test), or a combination of them. In the current work, we focus on perturbations at the instance level because our microarray sample size is very small. In addition, critical p-value and the variable order is chosen at random before each run. We define the robustness of FSS selector as the variation of the output due to small changes in the data set and variations in the critical value of the independence test.

Here, following [16], we take a similarity based approach where feature stability is measured by comparing the outputs of the feature selectors on the k subsamples. We use the Jaccard index as the similarity measure between two subsets S_1 and S_2. The more similar the outputs, the higher the stability measure. The overall stability can be defined as the average over all pairwise similarity comparisons between the different feature selectors:

$$I_{tot} = \frac{\sum_{i=1}^{n} \sum_{j=i+1}^{n} I(S_i, S_j)}{n(n-1)} \quad \text{with} \quad I(S_i, S_j) = \frac{|S_i \bigcap S_j|}{|S_i \bigcup S_j|}.$$

4.4   Ensemble FSS by Consensus Ranking

In this section, we discuss a simple ensemble technique that works by aggregating the feature rankings provided by the FSS selector into a final consensus ranking. We adopt a subsampling based strategy discussed in [3]. Consider a data set $X = \{x_1, \ldots, x_M\}$ with M instances and N features. Then k subsamples of size xM $(0 < x < 1)$ are drawn randomly from X, where the parameters k and x can be varied. Note that the proportion between classes in the original data set is maintained in each resampled data set, which is known as stratified bootstrap. Subsequently, the FSS algorithm is run p times on each of the k subsamples. Our implementation breaks ties at random: a random permutation of the variables is carried out and a random critical value α for the independence test is chosen before each algorithm is run. The more similar the features subsets, the higher the stability measure.

On the basis of all FSS, the confidence of each feature is computed as the relative frequency of its presence in the outputs. The consensus among all the induced models produces an ordered list of variables. The confidence level of a given feature is the number of times that feature appears in the outputs. By changing the confidence level, the biologists can build a hierarchy of feature subsets.

5   Experiments

In this section, we first jointly compare the robustness and the accuracy of MBOR against several other scalable FSS learning algorithms on synthetic data with as few samples as in our gene expression data set. More experiments on several UCI data bases are provided in [8,9] with larger sample sizes. We then use the ensemble method discusses in Section 4.4 on the expression data to provide the biologist with a list - as robust as possible - of genes statistically associated to diabetes, ranked in decreasing order of relevancy.

5.1   Robustness vs. Classification Accuracy

We first conducted several supervised experiments to evaluate the stability index I_{tot} and the predictive performance on synthetic data of MBOR and various correct and scalable MB learning algorithms, i.e. PCMB [6], Fast-IAMB [14] and Inter-IAMB [15], as a function of the size of the MB. The networks used in the evaluation are PIGS, and GENE (See [5] for references). The algorithms are run on each node and the robustness and performance indexes are average over all nodes. The aim was twofold: 1) to analyze the effect of the true MB size on the variability of the FSS output by the algorithms, 2) to assess the performance of the selected features in terms of accuracy. To evaluate the accuracy, we combine precision (i.e., the number of true positives in the output divided by the number of nodes in the output) and recall (i.e., the number of true positives divided by the true size of the Markov Boundary) as $\sqrt{(1 - precision)^2 + (1 - recall)^2}$, to measure the Euclidean distance from perfect precision and recall, as proposed in [2].

We evaluate the effect of the size of MB on the robustness of the feature selector. Results are shown in Figures 1 and 2. I_{tot} was averaged over 50 runs on each node with distinct subsets of 143 instances that were independent and identically sampled from these networks. The robustness seems to slow down with the number of features in the Markov boundary as we expected initially. This is more visible on the Pigs data where the maximum MB size is as much as 68, compared to 15 for Genes. MBOR seems more robust in most cases. Interestingly, MBOR's variability is far less dependent upon the MB size.

FSS needs to be combined with a classification model to estimate the practical relevancy of the features. We used the above BN-based FSS methods and, for sake of completeness, we also assessed the accuracy of two filter and one embedded FSS method. Note that the robustness of these methods was evaluated [3]. We discarded wrapper approaches because there are computationally not feasible for the large feature sizes we are dealing with. For the filter methods, we selected one univariate and one multivariate method. Univariate methods consider each feature separately, while multivariate methods take into account feature dependencies, which might yield better results. The univariate method we choose is the Gain Ratio Attribute Evaluation method. It evaluates the worth of an attribute by measuring the gain ratio with respect to the class, i.e. $GainR(C, X) = (H(C) - H(C|X))/H(X)$ where X and C represent a feature and the class respectively, and the function H calculates the entropy. As a multivariate method, we choose the RELIEF algorithm [17], which evaluates the worth of an attribute by repeatedly sampling an instance and considering the value of the given attribute for the nearest instance of the same and different class. In our experiments, ten neighboring instances were chosen. As embedded method, we evaluate the worth of an attribute by using support vector machines (SVM) classifier as discussed in [18]. Attributes are ranked by the square of the weight assigned by the SVM.

Figure 3 and 4 (upper plots) summarize the distribution of the Euclidean distance over 25 data sets for CB-based FSS methods and (lower plots) for non bayesian FSS methods. They summarize the variability of the Euclidean distance in the form of boxplots on the largest MB only using the CB-based algorithms. As may be seen, a clear picture emerges from the experiments. MBOR consistently outperforms the other algorithms in terms of Euclidean distance in both experiments.

5.2   Ensemble FSS Technique on Diabetes Data

In view of the good trade-off between performance and robustness on small sample/large feature size data, MBOR was chosen as our FSS selector. The ensemble strategy is used with $k = 50$ subsamples of size $0.9M$ (i.e. each subsample contains 90% of the data). Then, MBOR was run with $p = 25$ times on each subsample with a random permutation of the genes and α drawn at random in the interval $[0.001, 0.05]$. To increase the reliability of the independence tests, the expression levels of the genes in each training fold were transformed into binary ones using the class-attribute interdependence maximization algorithm (CAIM) [19]. CAIM maximizes the class-attribute interdependence and generate a list of thresholds. The association was measured by the G-statistic. The thresholds were then applied on both the training and test fold instances. Microarray data were separated into 8 different subsets in order to answer to 4 different relevant biological questions (i.e., case studies). These four case studies are:

1. Case study 1: the diabetes status, i.e. diabetic patients vs. obese + morbid obese + glucose-intolerant + healthy patients. 143 patients in total.
2. Case study 2: the diabetes status only, i.e. diabetic patients vs. obese + morbid obese patients. 45 instances in total.
3. Case study 3: the obesity status, i.e. obese + morbid obese + glucose-intolerant patients vs healthy patients. 125 patients in total.
4. Case study 4: the insulin-resistance status, i.e. morbid obese + diabetic patients vs healthy patients. 132 patients in total.

The top 15 ranked genes are shown in Figures 5 to 8 in decreasing order of frequency. The common genes between these case studies are shown in Table 1. We considered only the genes that appeared with a frequency of at least 1%. According to our biologist, the results obtained are biological relevant as some

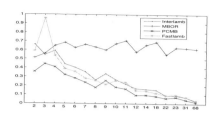

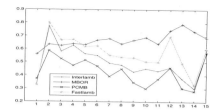

Fig. 1. Gene: Robustness vs MB size

Fig. 2. Pigs: Robustness vs MB size

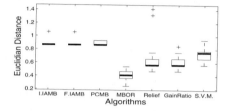

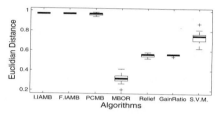

Fig. 3. Gene: Comparative accuracy

Fig. 4. Pigs: Comparative accuracy

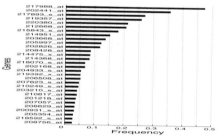

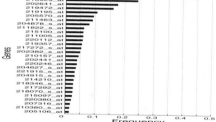

Fig. 5. Caste study 1 (diabetes status): **Fig. 6.** Case study 2 (diabetes status the 15 most frequent genes   only): the 15 most frequent genes

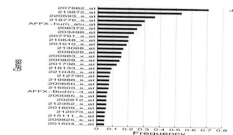

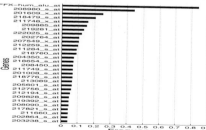

Fig. 7. Case study 3 (obese status): the 15 most frequent genes

Fig. 8. Case study 4 (insulin-resistant status): the 15 most frequent genes

Table 1. Common genes that appeared with a frequency of at least 1% in the output of MBOR

Case studies	Common genes
Case 1 vs Case 2	'202441_at', '219357_at', '220380_at', '218070_s_at', '220246_at'
Cas 1 vs Cas 3	none
Case 1 vs Case 4	'205997_at', '219392_x_at', '207057_at', '201447_at', '208090_s_at', '212194_s_at', '220137_at'
Case 2 vs Case 3	none
Cas 2 vs Cas 4	none
Case 3 vs Case 4	'218776_s_at', 'AFFX-hum_alu_at', '201609_x_at', '202074_s_at', '204144_s_at', '211284_s_at'

genes which can discriminate between each case studies have been already identified as abnormally expressed in the skeletal muscle of type 2 diabetic patients, like PPARGC1A (219195_at in Fig.4) [20]. Moreover when we compared the two lists of genes obtained between case 1 and case 2, we found CAMK1D (220246_at, in Table 1) which has been identified as a candidate gene for type 2 diabetes in genetic studies [21]. These results are thus highly promising and give new tracks to the biologists to identify new genes that would be related to the development of the disease.

5.3   Conclusion

This paper discusses the application of a novel feature subset selection method on a genomic database that consists of 22,283 genes and only 143 patients. We assessed the variability of our results and proposed an ensemble techniques to yield more robust results. Preliminary results are promising according to our domain expert. Future substantiation through more experiments are currently being undertaken and comparisons with other FSS techniques will be reported in due course.

Acknowledgment

This work is supported by "Institut des Sciences Complexes" (IXXI), Lyon, France.

References

1. Nilsson, R., Peña, J.M., Bjrkegren, J., Tegnr, J.: Consistent feature selection for pattern recognition in polynomial time. Journal of Machine Learning Research 8, 589–612 (2007)
2. Peña, J.M., Nilsson, R., Bjrkegren, J., Tegnr, J.: Towards scalable and data eficient learning of Markov boundaries. International Journal of Approximate Reasoning 45(2), 211–232 (2007)

3. Saeys, Y., Abeel, T., Van de Peer, Y.: Robust feature selection using ensemble feature selection techniques. In: Daelemans, W., Goethals, B., Morik, K. (eds.) ECML PKDD 2008, Part I. LNCS, vol. 5211, pp. 313–325. Springer, Heidelberg (2008)
4. Guyon, I., Elisseeff, A.: An introduction to variable and feature selection. Journal of Machine Learning Research 3, 1157–1182 (2003)
5. Tsamardinos, I., Brown, L.E., Aliferis, C.F.: The max-min hill-climbing Bayesian network structure learning algorithm. Machine Learning 65(1), 31–78 (2006)
6. Peña, J.M., Björkegren, J., Tegnér, J.: Scalable, efficient and correct learning of Markov boundaries under the faithfulness assumption. In: Godo, L. (ed.) EC-SQARU 2005. LNCS, vol. 3571, pp. 136–147. Springer, Heidelberg (2005)
7. Koller, D., Sahami, M.: Toward optimal feature selection. In: ICML, pp. 284–292 (1996)
8. Rodrigues de Morais, S., Aussem, A.: A novel scalable and data efficient feature subset selection algorithm. In: European Conference on Machine Learning and Principles and Practice of Knowledge Discovery in Databases ECML-PKDD 2008, Antwerp, Belgium, pp. 298–312 (2008)
9. Rodrigues de Morais, S., Aussem, A.: A novel scalable and correct Markov boundary learning algorithms under faithfulness condition. In: 4th European Workshop on Probabilistic Graphical Models PGM 2008, Hirtshals, Denmark, pp. 81–88 (2008)
10. Pearl, J.: Probabilistic Reasoning in Intelligent Systems: Networks of Plausible Inference. Morgan Kaufmann, San Francisco (1988)
11. Neapolitan, R.E.: Learning Bayesian Networks. Prentice-Hall, Englewood Cliffs (2004)
12. Chickering, D.M., Heckerman, D., Meek, C.: Large-sample learning of Bayesian networks is NP-hard. Journal of Machine Learning Research 5, 1287–1330 (2004)
13. Tsamardinos, I., Aliferis, C.F., Statnikov, A.R.: Algorithms for large scale Markov blanket discovery. In: Florida Artificial Intelligence Research Society Conference FLAIRS 2003, pp. 376–381 (2003)
14. Yaramakala, S.: Fast Markov blanket discovery. In: MS-Thesis, Iowa State University (2004)
15. Yaramakala, S., Margaritis, D.: Speculative Markov blanket discovery for optimal feature selection. In: IEEE International Conference on Data Mining, pp. 809–812 (2005)
16. Kalousis, A., Prados, J., Hilario, M.: Stability of feature selection algorithms: a study on high-dimensional spaces. Knowl. Inf. Syst. 12, 95–116 (2007)
17. Kononenko, I.: Estimating attributes: Analysis and extensions of relief. In: European Conference on Machine Learning, pp. 171–182 (1984)
18. Guyon, I., Weston, J., Barnhill, S., Vapnik, V.: Gene selection for cancer classification using support vector machines. Machine Learning 46, 389–422 (2002)
19. Kurgan, L.A., Cios, K.J.: Caim discretization algorithm. IEEE Trans. on Knowl. and Data Eng. 16(2), 145–153 (2004)
20. Lai, C.Q., et al.: PPARGC1A variation associated with DNA damage, diabetes, and cardiovascular diseases: the Boston Puerto Rican health study. diabetes. Diabetes 57, 809–816 (2008)
21. Zeggini, E., et al.: Meta-analysis of genome-wide association data and large-scale replication identifies additional susceptibility loci for type 2 diabetes. Nat. Genet. 40, 638–645 (2008)

Brain Tumor Segmentation Using Support Vector Machines

Raouia Ayachi and Nahla Ben Amor

LARODEC, Institut Supérieur de Gestion Tunis, 41 Avenue de la Liberté,
2000 Le Bardo, Tunisie
raouia.ayachi@gmail.com, nahla.benamor@gmx.fr

Abstract. One of the challenging tasks in the medical area is brain tumor segmentation which consists on the extraction process of tumor regions from images. Generally, this task is done manually by medical experts which is not always obvious due to the similarity between tumor and normal tissues and the high diversity in tumors appearance. Thus, automating medical image segmentation remains a real challenge which has attracted the attention of several researchers in last years. In this paper, we will focus on segmentation of Magnetic Resonance brain Images (MRI). Our idea is to consider this problem as a classification problem where the aim is to distinguish between normal and abnormal pixels on the basis of several features, namely intensities and texture. More precisely, we propose to use Support Vector Machine (SVM) which is within popular and well motivating classification methods. The experimental study will be carried on Gliomas dataset representing different tumor shapes, locations, sizes and image intensities.

1 Introduction

Magnetic Resonance Imaging (MRI) provides detailed images of brain studies. MRI data are used in brain pathology studies, where *regions of interest* (ROI's) are explored in detail. This assistant diagnostic device is very helpful for doctors during disease diagnosis and treatment. The plenty of acquired images show the inside to doctors, but, doctors seek to know more details about images, such as emphasizing the tumor area, quantifying its size, and so on. If these tasks are made by doctors themselves, it is possibly inaccurate or even impossible. Therefore, image processing by computers is relevant in radiology. There are already several *computer-aided diagnosis (CAD)* systems which are used in disease monitoring, operation guiding tasks, etc. [1]. In fact, CAD seeks to assist doctors by providing computer outputs, which are considered as a second opinion during abnormalities detection, disease progress survey, etc.

Among all medical image processing, image segmentation remains a crucial task consisting on extracting regions of interest from images. But it is still not well solved because of the complexity of medical images [1].

This task is too time consuming, tedious, and even impossible in some cases due to the large amounts of information provided by each image [3]. In fact,

C. Sossai and G. Chemello (Eds.): ECSQARU 2009, LNAI 5590, pp. 736–747, 2009.

automatic segmentation is not preferable for doctors since their knowledge and experience is more important than computers simulations. So CAD systems should find a good compromise between manual and automatic intervention in the segmentation procedure. To reach this goal, we will propose a supervised method for brain image segmentation based on a classification method, namely Support Vector Machines (SVM).

This paper will be structured as follows; In Section 2 we will briefly describe some image segmentation methods. Section 3 will detail the classification method SVM. Section 4 will present the environment and the input data in our system. In Section 5, 6, we will describe main processing steps which are image pre-processing (registration and noise reduction) and features computation. Section 7 will be dedicated to describe the brain tumor segmentation using SVM. Finally, Section 8 presents our experimental study.

2   Image Segmentation

Image analysis usually refers to computer image processing with the objective of finding image objects. Image segmentation is one of the most critical tasks in automatic image analysis. It plays a crucial role in many imaging applications and consists on subdividing an image into its constituent parts and extracting the regions of interest that should be homogeneous with respect to some characteristics such as intensity or texture.

A great variety of segmentation algorithms have been developed in the last few decades and this number continually increases each year. These methods vary widely dependly on the specific application and other factors. A classification on these methods was proposed in [2] on the basis of five criteria namely: *region, contour, shape, structural approaches* and *graph theory.*

Since we are interested in brain tumor segmentation, then tumor regions can be scattered all over the image. This explains the fact that determining the regions of interest is also called *pixel classification* and the sets are called *classes.* In fact, pixel classification rather than classical segmentation methods are often preferable especially when disconnected regions of interest belonging to the same class should be extracted.

Classifiers can be considered as pixel classification methods. For this reason, we will adopt the *region* criteria and more precisely the Support Vector Machines (SVM) as a segmentation method. Our choice is especially motivated by its fastness, robustness in generalization preperties and its capacity to handle voluminous data [5]. Also, many studies have indicated that SVM outperforms other binary classifiers in most cases [21]. Basics of this approach are given in the following Section.

3   Support Vector Machines

Support vector machines (SVM) represent a class of state-of-the-art classifiers that have been successfully used for classification [16]. It performs binary classification tasks.

Given a set of n labelled data points $\{(x_1, y_1), ..., (x_n, y_n)\}$ where $y_i = \pm 1$, an SVM learns a decision function by searching a separating hyperplane $\prec w,x \succ + b = 0$, where $x_i \in R^n$, $w \in R^n$ and $b \in R$. In the linear case, SVM looks for an hyperplane that maximizes the margin by minimizing $\frac{1}{2} \cdot ||\vec{w}||^2$, subject to $y_i(\prec w, x_i \succ +b) >= 1$, $i = 1, 2, ...n$. In the linear non-separable case, the optimal hyperplane is computed by adding *slack variables* $\varepsilon_i = 1, 2, \ldots, n$ and a penality parameter C and the optimization problem is expressed as follows:

$$\min \frac{1}{2} \cdot ||\vec{w}||^2 + C \sum_{i=1}^{n} \varepsilon_i \tag{1}$$

$$subject\ to\ y_i(\prec w, x_i \succ +b) >= 1 - \varepsilon_i, \ i = 1, 2, ...n$$

Using the *Lagrangian formulation*, these optimization problems are solved by introducing a new unknown scalar variable, α_i called the *Lagrange multiplier* which is introduced for each constraint and forms a linear combination involving the multipliers as coefficients. The problem will be expressed as follows:

$$L_d = \sum_{i=1}^{n} \alpha_i - \frac{1}{2} \sum_{i=1}^{n} \sum_{j=1}^{n} \alpha_i \alpha_j y_i y_j x_i x_j \tag{2}$$

After resolution of the optimization problem, the data points associated with the non zero α_i correspond to specific data (x_i, y_i) called *support vectors*. These data are used for computing the decision function, while the remaining data are discarded.

For the non linear case, the surface separating the two classes is not linear, so the idea is to transform data points to another high dimensional feature space where the problem is linearly separable. If the transformation to the high dimensional space is φ then the Lagrangian function can be expressed as:

$$Ln = \sum_{i=1}^{n} \alpha_i - \frac{1}{2} \sum_{i=1}^{n} \sum_{j=1}^{n} \alpha_i \alpha_j y_i y_j \varphi(x_i)\varphi(x_j) \tag{3}$$

The dot product $\varphi(x_i)\ \varphi(x_j)$ in that high dimensional space defines a *kernel function* $k(x_i, x_j)$. Within the most common kernel functions, we cite:

- Linear: $x_i.x_j$,
- Polynomial of degree d: $(x_i.x_j + 1)^d$,
- Radial Basis Function (RBF): $exp(\frac{-||x_i - x_j||^2}{2\sigma^2})$.

Once the support vectors have been determined, the SVM decision function has the form:

$$f(x) = \sum_{j=1}^{Support vectors} \alpha_j y_j K(x_j, x) \tag{4}$$

4   Medical Image Description

MRI is a powerful visualization technique that produces images of internal anatomy in a safe and non-invasive way. It generates a set of sagittal, coronal and axial images for each patient moving from the top of the head to the bottom. These images follow the Digital Imaging and Communications in Medicine (DICOM) standard which is an international standard for communication of biomedical diagnostic using digital images [3].

MRI offers different visualizations of images due to a combination of the device parameters (i.e., repetition time (TR), echo time (TE)). In our study, we handle T1 and T2-weighted images such as a T1-weighted image is produced by a relatively short TR/short TE sequence, and a long TR/long TE sequence produces a T2-weighted image. For the sake of simplicity, T1-weighted and T2-weighted will be referred to as T1 and T2, respectively.

In our work, we will focus on tumor images. In fact, brain tumor segmentation is a very complex problem since many indicators should be investigated, namely: the localization of tumor, evaluation of its shape, its volume, its homogeneity, its nature and its interactions with nearby brain structures, the presence of swelling, etc. Thus a unique segmentation method can not generalize all tumor types [4], that is why we only focus on a subset of tumor types, namely, *Gliomas* which are developped from glial cells.

5   Image Pre-processing

Image pre-processing is important for real-life data which are often noisy and inconsistent. The purpose of this step is to improve the quality of the image by transforming it into another image that is better suited for machine analysis. In our work, we will adopt *image registration* and *noise reduction* techniques as explained below.

5.1   Image Registration

Image registration represents the process of spatially aligning two images by computing a transformation applied to an input image in order to match it to a template image that is assumed stationary. The major challenge associated to this step is defining a quantitative measure that assesses spatial alignment and given this measure, the task is reduced to a search of a set of transformations parameters that optimize it. These transformations can be *affine*, *rigid* or *curved*. The rigid transformation is a special case of the affine transformation. It is composed of *translations* and *rotations* operations, while the affine transformation adds *scaling* and *shearing* operations.

In what follows, we will adopt two kinds of registration methods dependly on two criteria namely: *modality* and *subject* criteria.

- The first one is referred to by *co-registration* and consists on aligning different modalities of the same patient (e.g., T1 and T2). This step is essential if the used modalities are not in perfect alignment which is often the case with real data [5].

- The second corresponds to *template registration* and aligns the modalities with a template image in a standard coordinate system in order to average signals from brain images of different subjects. This is often done by mapping all the images into a recognized brain coordinate system. The coordinate system used in our work is the MNI [19] since it handles templates in different modalities i.e., T1 and T2.

5.2   Noise Reduction

The noise that corrupts the signal recorded at each pixel could be totally or partially removed. Within famous and successful methods, we cite the *anisotropic diffusion filtering* which is a technique introduced by Perona and Malik [6] commonly used to reduce the effects of local noise. It represents a simple method to reduce the effects of local noise without requiring a tissue model. This filter strengthens the difference between regions and eliminates the noise by increasing regions homogeneity and preserving the edges.

6   Image-Based Features

The calculation of image-based features is a primordial step in our work due to the fact that each pixel should have characteristics used for the differentiation between the tumor and the normal pixels. A great variety of features can be computed for each image such as intensities, textures, distances to labels, spatial tissue prior probabilities and existing works combine several of them depending on the nature of problem [7,8,9].

In fact, the main consideration when selecting features is that they should reflect properties that can help us to discriminate between normal and tumor pixels.

In our work, we will adopt *intensities* and *texture* features since there are the most commonly used features in brain tumor segmentation [5]. More precisely, the first pixel-level feature is pixel intensities from each modality (T1 and T2) and the second set is relative to calculations that can characterize patterns in region intensities. There is a large variety of methods that compute features characterizing image textures. Recent surveys can be found in [13] and [14].

We will explore first-order and second-order texture parameters. The first-order parameters which are called (*statistical moments*) ignore spatial information and are essentially features that characterize properties of the local histogram. We calculate the parameters from [11], which are *mean, variance, skewness, kurtosis, energy* and *entropy*. While the second-order ones, the most commonly used features that characterize textures in the medical imaging area are the *Haralick* features which represent a set of 14 statistics including measures such as *angular second momentum, contrast, dissimilarity, entropy, cluster shade, cluster prominence, local homogeneity* and *inertia*.

These features are computed from a gray-level spatial *coocurrence matrix* [15] which is an estimation of the likelihood that two pixels of intensities i and j will occur at a distance d and an angle of θ within a neighborhood. In general, d and θ take the following values : $d = 1, 2, etc.$ and $\theta = 0°, 45°, 90°, 135°$.

To obtain these features, cooccurrence matrix should be computed. This task involves defining the value of d and θ and computing the number of frequencies of each couple of intensities values in the image matrix with respect to d and θ.

7   Brain Tumor Segmentation Using Support Vector Machines

Our aim in this work is the extraction of tumor regions from brain images. To ensure this task, images should first be pre-processed in order to improve their quality. Moreover, classification methods need an attributes set characterizing each instance which corresponds in our case to a features set relative to each pixel. This will be done via an extraction step which consists on computing image-based features. Therefore, our segmentation method is based on four main phases, namely *pre-processing, features extraction, SVM training and testing* and *segmentation*. The whole process is illustrated by the diagram of Fig. 1 which can be detailed as follows:

7.1   Pre-processing Phase

This first phase ensures the pre-processing step. It is composed of six steps:

- *Acquisition*: this step is ensured via MRI and represents a primordial step since the remainder of the process will depend on the quality of acquired images. Note that the MR images can be axial, coronal or sagittal and that their number and type depends on the tumor localization. The output of this step is a set of 2D slices.
- *Reconstruction*: this step maps the 2D sequential slices of each patient to an entirely volume (i.e., 3D image). This step is needed in our system in order to register images using the SPM[1] software which handles 3D images. This step will be done using the XMedCon[2] software.
- *Co-registration*: this step spatially aligns volumes of each patient having different modalities (i.e., T1 and T2). The common tool to ensure this task is SPM software.
- *Template registration*: having as input co-registered volumes, this step consists on aligning the modalities (T1 and T2) with a template image in the MNI standard coordinate system in order to average signals from brain images of different subjects. This can also be ensured by SPM software.

---

[1] Statistical Parametric Mapping, available at http://www.fil.ion.ucl.ac.uk/spm/
[2] X Medical Conversion, available at http://xmedcon.sourceforge.net/

- *Extraction of slices of interest*: in this step, we will choose the 'interesting' slices from the registered volumes, i.e., the slices where we can see the tumor regions. The output of this step is a set of 2D pathological images. The tool that will be used in this step is MIPAV[3]. In fact, any medical imaging software ensuring volumes visualization can be used in this step.
- *Noise reduction*: this step improves the quality of data through the application of methods of denoising (see (Sub)Section 5.2).

7.2 Features Extraction Phase

Given denoised images, this phase refers to various quantitative measurements used for making decisions regarding the pathology of a tissue. More precisely, for each pixel, we should compute a feature vector composed of features already described in Section 6. Thus, the output of this phase will be unlabelled data set. In this phase, we have used the MIPAV tool.

7.3 Training and Testing Phase

In this phase, the Gliomas regions for each image in the training set will be selected manually and validated by the radiologist. In fact, axial images are displayed on the computer screen, one slice at a time. Next, representative regions of interest (i.e., tumor regions) are selected interactively on the computer screen using a mouse-driven interface aided by *active contour* models [18]. These models are used for definition and tracking of tumor regions contours due to their ability to approximate accurately the random shapes of boundaries [17]. After that, pixels within any of the defined regions of interest are labeled as tumor, while pixels outside of tumor contours are labeled as normal.

Once the labeled training set containing normal and tumourous pixels is constructed, SVM can be trained in order to fix support vectors as already explaind in Section 3, so that the output of this step will be the classification model allowing the differentiation between normal and tumor pixels. Note that the choice of appropriate parameters for the SVM model can be done through a testing phase by an analysis of the PCC (Percentage of correct classification). In the present study, the implementation of SVM is based on the strategy of the package LibSVM[4].

7.4 Segmentation Phase

To ensure this phase, we propose two steps:

- Classification step: in this step, the classification model generated from the previous phase will be used to affect the tumor or the normal class for each pixel in the new images which are pre-processed.

---

[3] Medical image processing, analysis and visualization, available at
http://mipav.cit.nih.gov/
[4] Library for Support Vector Machines, available at
http://www.csie.ntu.edu.tw/ cjlin/libsvm/

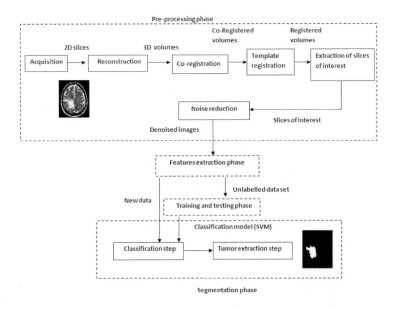

Fig. 1. Diagram illustrating our segmentation system

- Tumor extraction step: after that, the set of tumor pixels is grouped to form the tumor region. To clarify our output segmentation, we will propose for each segmented image, a binary image where the black region is the image background and the white region represents the tumor one.

8   Experimental Study

In order to test the efficiency of our segmentation method, we apply it on a real database relative to 80 images of 4 patients collected from August 1, 2008, to January 31, 2009 such that patients' ages are ranged from 18 to 64 years. For each patient, we have 20 tumor images (10 axial, 5 coronal and 5 sagittal) corresponding to Gliomas in different grades, for 3 different modalities (i.e., T1, T1 after injection of contrast agent and T2). For training, we have chosen 9 axial slices for each patient and the remaining slices are used for testing.

Once our data set is constructed, we follow different steps of our system described in the previous Section.

The implementation of the whole system needs several tools and consecutively several parameterizations. For the experimental study, our choices can be summarized as follows:

- for the reconstruction step, 2D images for each patient should be stacked into one volume. XMedCon, the software used in this step requires a sequence of images sorted alphabetically which is used to construct 3D images. Note that a specialized doctor brings a great help if he validates the image sequence.

- for the co-registration, the rigid-body model that can be parameterized by only translations and rotations is used and the objective function that is maximized in this case will be the *normalized mutual information* (NMI) measure since it is one of the most popular measures of co-registration in medical imaging [5].
- the measure used in template registration is *the sum of squared difference.* It needs the computation of a linear 12-parameters affine transformation that has twelve parameters due to the fact that each image has 3 dimensions and this consists of one parameter for each of the three dimensions with respect to translation, rotation, scaling, and shearing.
- images are denoised using the *anisotropic diffusion filtering* that is implemented using Matlab.
- for each pixel, the selected feature vector is restricted to intensities values and the following texture values: (*angular second momentum, contrast, dissimilarity, entropy, cluster shade, cluster prominence, local homogeneity* and *inertia*).
- After these steps, SVM will be applied. First of all, a kernel function (polynomial, linear, etc.) should be chosen. Then, kernel parameters have to be selected (σ for the RBF function, the degree d of a polynomial function, etc.). After that, data training can be done using these parameters. Finally, testing images are segmented using the SVM model. In fact, pixels classified as tumor ones represent the tumor region.

In our experimentation we apply the RBF kernel due to the fact that many studies have demonstrated that the preferable choice is RBF [20], and the technique used to fix its optimal parameters is a grid search using a cross-validation. In fact, a grid search with 10-fold cross-validation searches the best parameters among an interval of values which achieve a high accuracy during training and testing. For our data, we have obtained the following values: $\sigma = 0.5$, $C = 8$ with a PCC value of 79.89 %. We note that PCC is computed by Eq. 5 where TP, TN, FP and FN denote respectively the number of true positives, true negatives, false positives and false negatives pixels.

$$PCC = \frac{TP + TN}{TP + TN + FN + FN} \tag{5}$$

In order to quantitatively assess the quality of our segmentation in comparison to the ground truth GT obtained from the boundary drawings of a radiologist, we choose to use the commonly used criteria the *Match Percent* measure MP [12]. It is calculated as the direct ratio of the true positives to the number of ground truth tumor pixels. It is expressed by Eq. 6.

$$MP = \frac{\#TP}{\#GT} \tag{6}$$

This measure is equal to 1 if the segmentations are identical, while it will approach to 0 for completely dissimilar segmentations. In our case, we have obtained 81.97 % of similar segmentations.

Fig. 2, Fig. 3 and Fig. 4 show the output of our segmentation method for axial, coronal and sagittal orientations using the RBF kernel.

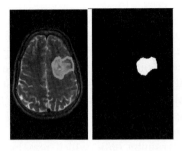

Fig. 2. Axial image and its segmentation output

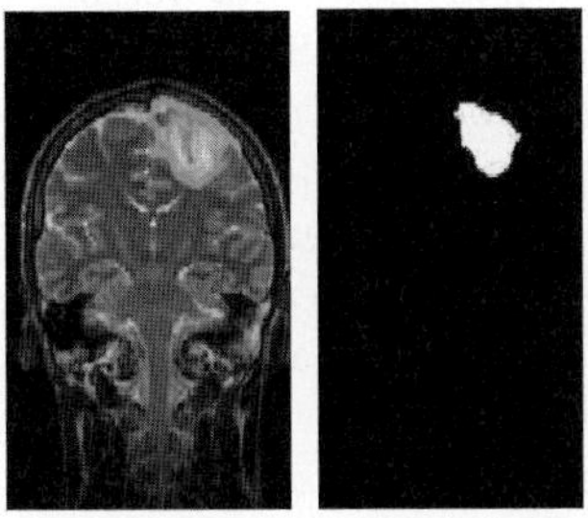

Fig. 3. Coronal image and its segmentation output

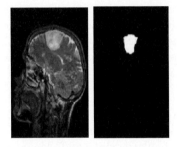

Fig. 4. Sagittal image and its segmentation output

9   Conclusion and Future Work

Medical image segmentation tools have already proved their efficiency in research applications and are now used for computer aided diagnosis and radiotherapy planning. They will be valuable in areas such as computer integrated surgery, where the visualization of the anatomy is fundamental.

This paper proposes a medical image processing system which is a user interactive tool for image segmentation. More precisely, we focus on brain tumor images issued from MRI device.

The proposed system considers the segmentation as a classification problem. More precisely, the SVM classification method is applied to ensure a segmentation task since it is significantly faster than other classification methods, also due to its robustness in generalization preperties and its capacity to handle voluminous data.

Our system is based on four main phases, namely pre-processing, features extraction, training and testing phase and segmentation phase. The pre-processing phase consists on registering images of different modalities with each other and registering them using a template image, i.e., MNI before filtering them to reduce the effect of eventual noise. While the features extraction phase computes for each pixel a feature vector that is characterized by intensities and texture values. After that, SVM constructs a classification model allowing the discrimination between normal and tumor pixels. Finally, this model is used to classify new pixels in order to extract tumor regions (i.e., the set of tumor pixels). Within the challenges of our system, we cite the choice of parameters which ensure a good accuracy. Effectively, as shown by Fig. 1 our system needs several steps with several tools, and at each step we should fix multiple parameters. To this end, the experimental study presents an appropriate way to fix them and our results from this study are motivating since the PCC is around 80 %. This rate can obviously be improved by enriching the training data, which remains the other challenge in this work due to the lack of real data in brain tumor imaging area.

This point presents our first line of research, since it will allow us to optimize choices at each step of the whole proposed segmentation method. In particular, the features selection phase can be improved by a better selection of features. For instance, we can include tumor characteristics and tumor localization and even some personal characteristics of patients which will enable us to extend our approach to a whole decision support system.

Acknowledgment

The authors wish to thank Doctor Kais Chelaifa for his helpful suggestion on tumor recognition and also thank the clinic of El Manar for their help in collecting data.

References

1. Atkins, M.S., Mackiewich, B.T.: Fully automatic segmentation of the brain in MRI. IEEE Transactions on Medical Imaging 17(1), 98–107 (1998)
2. Lecoeur, J., Barillot, C.: Segmentation d'images cérébrales: Etat de l'art. Rapport de Recherche INRIA 6306 (September 2007)
3. Jiang, C., Zhang, X., Huang, W., Meinel, C.: Segmentation and Quantification of Brain Tumor. In: IEEE International Conference on Virtual Environments, Human-Computer Interfaces and Measurement Systems, pp. 12–14 (2004)
4. Prastawa, M., Bullitt, E., Moon, N.: Automatic brain tumor segmentation by subject specific modification of atlas priors. Acad. Radiol. 10(12), 1341–1348 (2003)
5. Schmidt, M.: Automatic brain tumor segmentation. University of Alberta, Department of computing science (2005)
6. Perona, P., Malik, J.: Scale-space and edge detection using anisotropic diffusion. IEEE Transactions on Pattern Analysis and Machine Intelligence 12(7), 629–639 (1990)
7. Dickson, S., Thomas, B.: Using neural networks to automatically detect brain tumours in MR images. International Journal of Neural Systems 4(1), 91–99 (1997)

8. Kaus, M., Warfield, S., Nabavi, A., Black, P., Jolesz, F., Kikinis, R.: Automated segmentation of MR images of brain tumors. Radiology 218(2), 586–591 (2001)
9. Prastawa, M., Bullitt, E., Ho, S., Gerig, G.: Brain tumor segmentation framework based on outlier detection. Medical Image Analysis 8(3), 275–283 (2004)
10. Gunn, S.R.: Support vector machine for classification and regression. Technical report Faculty of Engineering, Science and Mathematics School of Electronics and Computer Science (1998)
11. Materka, A., Strzelecki, M.: Texture analysis methods: a review. Technical report COST B11 Technical University of Lodz Poland (1998)
12. Clark, M.C., Hall, L.O., Goldgof, D.B., Velthuizen, R., Murtagh, F.R., Silbiger, M.S.: Automatic tumor segmentation using knowledge based techniques. IEEE Trans. on Medical Imaging 17(2), 238–251 (1998)
13. Forsyth, D., Ponce, J.: Computer Vision: A Modern Approach. Prentice-Hall, Englewood Cliffs (2002)
14. Hayman, E., Caputo, E., Fritz, M., Eklundh, J.: On the significance of real-world conditions for material classification. In: Pajdla, T., Matas, J(G.) (eds.) ECCV 2004. LNCS, vol. 3024, pp. 253–266. Springer, Heidelberg (2004)
15. Haralick, R., Shanmugam, K., Dinstein, I.: Textural features for image classification. IEEE Trans. on Systems Man and Cybern 3(6), 610–621 (1973)
16. Boser, B.E., Guyon, I., Vapnik, V.: A training algorithm for optimal margin classifiers. In: Proceedings of the Fifth Annual Workshop on Computational Learning Theory, pp. 144–152. ACM Press, New York (1992)
17. Stoitsis, J., et al.: Computer aided diagnosis based on medical image processing and artificial intelligence methods. Nuclear Instruments and Methods in Physics Research 569(2), 591–595 (2006)
18. Kass, M., et al.: Snakes: Active Contour Models. International Journal of Computer Vision 1(4), 321–331 (1988)
19. Evans, A., et al.: An mri-based stereotactic atlas from 250 young normal subjects. Society for Neuroscience Abstracts 18, 408 (1992)
20. Scholkopf, B., Smola, A.J.: Learning with Kernels Support Vector Machines, Regularization, Optimization and Beyond. MIT Press, Cambridge (2001)
21. Chan, K., Lee, T.W., Sample, P.A., Goldbaum, M., Weinreb, R.N.: Comparison of machine learning and traditional classifers in glaucoma diagnosis. IEEE Trans. on Biomedical Engineering 49(9), 963–974 (2002)

Ensemble Learning for Multi-source Information Fusion

Jörg Beyer[1,2], Kai Heesche[1], Werner Hauptmann[1], Clemens Otte[1], and Rudolf Kruse[2]

[1] Siemens AG - Corporate Technology, Information and Communications, Learning Systems, Otto-Hahn-Ring 6, 80200 Munich, Germany
[2] Otto-von-Guericke-University Magdeburg - School of Computer Science, Universitätsplatz 2, 39106 Magdeburg, Germany

Abstract. In this paper, a new ensemble learning method is proposed. The main objective of this approach is to jointly use knowledge-based and data-driven submodels in the modeling process. The integration of knowledge-based submodels is of particular interest, since they are able to provide information not contained in the data. On the other hand, data-driven models can complement the knowledge-based models with respect to input space coverage. For the task of appropriately integrating the different models, a method for partitioning the input space for the given models is introduced. The benefits of this approach are demonstrated for a real-world application.

1 Introduction

Real-world applications are characterized by an increasing complexity. To generate adequate models the consideration of all available information sources is necessary. For this purpose, more and more sophisticated combinations of knowledge-based and data-driven models are required which are representing these sources. While data-driven models are learned from available training data the integration of knowledge-based models is of particular interest since they are able to provide information not contained in the training data. The knowledge-based models are designed for particular regions of the input space. In order to ensure that the models are only active in regions they are designed for, their specific validity ranges have to be included in the modeling process.

The use of multiple submodels is motivated by the paradigm that different submodels can complement each other avoiding the weakness of a single model. The combination of models constitutes an ensemble as depicted in Fig. 1. According to the divide-and-conquer principle a complex task is solved by dividing it into a number of simpler tasks and then combining the solutions of those tasks. The ensemble fuses information y_j acquired by model j, $j = 1, \ldots M$, to produce an overall solution y that is supposedly superior to that attainable by any one of them acting alone. Literature describes many approaches that address the problem of learning local models. Examples of such methods are boosting [1], mixture of experts [2], or ensemble averaging [3]. The algorithms for learning

C. Sossai and G. Chemello (Eds.): ECSQARU 2009, LNAI 5590, pp. 748–756, 2009.

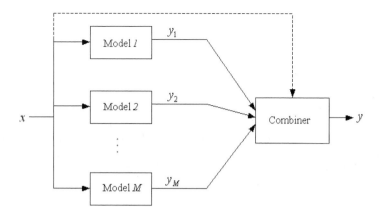

Fig. 1. A common ensemble model. The dashed line indicates that the Combiner can involve the current input in its decision dependent on the combining method.

local models can be discriminated with respect to several aspects: in the way they divide the training data into subsets, the type of submodels they use, or how they combine the outputs of the submodels. However, none of the existing methods are able to integrate predefined models that are designed for particular regions of the input space.

The paper is organized as follows: In Sect. 2, an introduction of multi-source fusion is given and Sect. 3 describes an ensemble learning model for combining data-driven and knowledge-based models. In Sect. 4, some experiments on a real-world application are outlined. Sect. 5 concludes the paper.

2   Multi-source Information Fusion

The term information fusion (IF) encompasses the process of merging and integrating heterogeneous information components from multiple sources, for instance, in the form of sensors, human experts, symbolic knowledge, or physical process models (according to Dasarathy [4]). IF is an important technique in different application domains, such as sensor fusion [5], identity verification [6], or signal and image processing [7].

Fusion implies the combination of information from more than one source. There are different reasons for fusion of multiple sources:

- The combined solution is able to attain more accurate, transparent, and robust results since the different information sources can complement each other with respect to their strengths and weaknesses.
- A model that depends on a single source is not robust with respect to error-proneness, i.e. if the single source is erroneous the whole model is affected. Models based on fused information sources are more robust since other sources are able to compensate for incorrect information.

– Fusion of information sources will provide extended coverage of information of the process to be modeled.

We consider two kinds of fusion approaches: complementary and cooperative fusion. They are discriminated with respect to the relationship among the information sources. In complementary fusion each source provides information from a different region of the input space, i.e. their responsibilities do not overlap. These sources provide locally a high performance. However, outside their regions the results are not valid. Cooperative fusion means that the information is shared among several information sources in the same region of the input space and has to be fused for a more complete modeling of the underlying process.

The next section describes an ensemble learning approach for IF. The information sources will be represented by predefined models. The process of partitioning the input space and the fusion of the models is performed by a separate data-driven model.

3   Combining Knowledge-Based and Data-Driven Models

The proposed ensemble model, referred to as heterogeneous mixture of experts (HME) model, is based on the mixture of experts (ME) approach [2], [8]. This model consists of a set of submodels that perform a local function approximation. The decomposition of the problem is learned by a gate function which partitions the input space and assigns submodels to these regions. In contrast to the ME model, the proposed ensemble learning method starts with some knowledge-based submodels, representing different information sources. Fig. 2 illustrates a general HME model. It consists of different models and a gate. To ensure that these submodels are assigned to those domains of the input space they are designed for, information about the specific validity ranges of the predefined knowledge-based submodels is used for the partitioning of the input space. It is assumed that the knowledge-based models will only cover a part of the input space while data-driven models learn the remainder.

From the probabilistic perspective the output of the HME model can be interpreted as the probability of generating output $y^{(n)}$ given input vector $\boldsymbol{x}^{(n)}$:

$$P\left(y^{(n)} \,\middle|\, \boldsymbol{x}^{(n)}, \Theta\right) = \sum_{j=1}^{M} P\left(z_j^{(n)} \,\middle|\, \boldsymbol{x}^{(n)}, \theta_g\right) P\left(y^{(n)} \,\middle|\, \boldsymbol{x}^{(n)}, \theta_j\right) , \qquad (1)$$

where M is the number of submodels, Θ is the set of parameters $\left\{\theta_g, \{\theta_j\}_{j=1}^{M}\right\}$ of the gate and of the submodels, respectively. The input vector $\boldsymbol{x}^{(n)} \in \Re^k$ and the output $y^{(n)} \in \Re$, where $n = 1, \ldots N$. The probability $P\left(z_j^{(n)} \,\middle|\, \boldsymbol{x}^{(n)}, \theta_g\right)$ represents the mixture coefficient of model j. The latent variable $z_j^{(n)}$ indicates which input vector $\boldsymbol{x}^{(n)}$ was generated by model j. Its introduction simplifies the training algorithm and allows the HME to be trained with the Expectation-Maximization (EM) algorithm [9]. The probability $P\left(y^{(n)} \,\middle|\, \boldsymbol{x}^{(n)}, \theta_j\right)$ represents the conditional densities of target $y^{(n)}$ for model j.

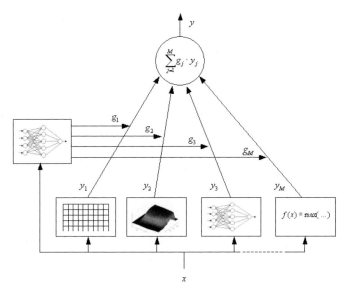

Fig. 2. Architecture of the proposed ensemble model. We include the case that gate and submodels may depend on different feature subsets of the input vector.

To compute the validity of each knowledge-based submodel j for an input vector a mapping $v_j : \Re^k \to [0, 1]$, $\forall j = 1, \ldots, M$ is defined. The specific validity function of a knowledge-based submodel j for the i-th dimension is

$$v_j\left(x_i^{(n)}\right) = \left(\frac{1}{1 + \exp\left(s_j\left(x_i^{(n)} - u_{ji}\right)\right)} - \frac{1}{1 + \exp\left(s_j\left(x_i^{(n)} - l_{ji}\right)\right)}\right) \ , \quad (2)$$

where l_{ji} and u_{ji} are the lower and the upper bound of the validity range of submodel j in dimension i. The parameter s_j determines the slope of the border of the validity range. Its influence on v_j is illustrated in Fig. 3.

For small s_j the slope of the border is more flat. The higher s_j gets, the steeper is the slope of the border. In this way, the transition between the submodels can be controlled. If there are smoothness assumptions about the target function one can choose a lower value for s_j.

To update the model parameter the EM algorithm is used. In the expectation step, the validity values are integrated into the computation of the posterior probability $h_j^{(n)}$ of selecting submodel j for input vector $\boldsymbol{x}^{(n)}$:

$$h_j^{(n)} = \frac{v_j\left(\boldsymbol{x}^{(n)}\right) P\left(z_j^{(n)} \big| \boldsymbol{x}^{(n)}, \theta_g\right) P\left(y^{(n)} \big| \boldsymbol{x}^{(n)}, \theta_j\right)}{\sum_{k=1}^{M} v_k\left(\boldsymbol{x}^{(n)}\right) P\left(z_k^{(n)} \big| \boldsymbol{x}^{(n)}, \theta_g\right) P\left(y^{(n)} \big| \boldsymbol{x}^{(n)}, \theta_k\right)} \ . \quad (3)$$

This enforces the gate to reduce the weights of submodel outputs if the input vectors are located outside their domains. The particular amount of weight decrease depends on the value of v_j.

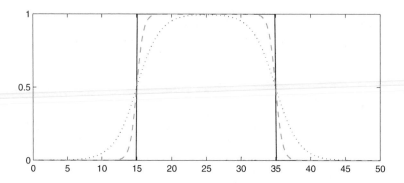

Fig. 3. The figure shows several validity ranges with different values of s: $s = 0.8$ (dottet line), $s = 2$ (dashed line), and $s = 100$ (solid line)

In the maximization step, the log likelihood function

$$L = \sum_{n=1}^{N} \sum_{j=1}^{M} h_j^{(n)} \log \left(P\left(z_j^{(n)} \middle| \boldsymbol{x}^{(n)}, \theta_g \right) P\left(y^{(n)} \middle| \boldsymbol{x}^{(n)}, \theta_j \right) \right) \tag{4}$$

is to be maximized with respect to the parameters of the gate and of the data-driven submodels.

4   Real-World Application

The application addresses the simulation of the electrical energy flow in the powertrain of a hybrid electric vehicle. Four distinct driving modes can be defined by the available expert knowledge: pure electric drive mode, hybrid drive mode, brake mode, and drag mode. Dependent on the current drive mode electrical energy is used in several different ways. In pure electric drive mode and hybrid drive mode energy is provided by the battery to drive the electric motor. In brake mode and drag mode the electric motor is operating as a generator to recuperate the kinetic energy to be used for charging the battery. Domain experts designed specific models for each mode. These models represent complementary information sources since they are defined for different regions of the input space with each model providing information for different mutually exclusive driving modes. Furthermore, the battery must maintain certain chemical limits. These limits determine the maximum charge and discharge capabilities of the battery dependent on its state of charge and temperature.

The data set is randomly divided into a training data set (80% of the data) and a test data set (20% of the data). The overall experiment is performed ten times and the results are averaged. The following models were compared: an HME, an ME, a multi-layer perceptron (MLP), and an ensemble of MLPs. The HME model uses four expert models. Two characteristic maps and a mathematical model represent the pure electric drive mode, brake, and drag mode. However, since there is no model provided for the hybrid drive mode a two-layer MLP with 5 input units, 6 hidden units and one output unit was learned. Each mode has different input features. As gate, an MLP with 4 hidden units was applied. For each knowledge-based model j a validity function v_j is defined by the the domain experts. For the data-driven model no validity function is given. Instead, it is assumed to be valid in the entire input space.

The ME consists of 4 MLPs with 6 hidden units and as gate an MLP with 5 hidden units was used. The single MLP comprises 14 hidden units. In the ensemble 10 members were combined. All members have the same architecture, i.e. MLPs with a single hidden layer of 8 hidden units. The ensemble is generated using K-fold cross-validation, where K is the number of ensemble members. The output of the ensemble is computed as follows:

$$y_{Ens}\left(x^{(n)}\right) = \frac{1}{K}\sum_{j=1}^{K} y_j\left(x^{(n)}\right) , \tag{5}$$

where $y_j\left(x^{(n)}\right)$ is the output of the j ensemble member. We used the mean absolute error to compare the perfomance of the models:

$$e = \frac{1}{N}\sum_{n=1}^{N}\left|y^{(n)} - f\left(\boldsymbol{x}^{(n)}\right)\right| . \tag{6}$$

Table 1 summarizes the results. The HME achieves superior performance due to the incorporation of available information sources. Fig. 4 shows the outputs of the gate model (the activation of the submodels) of the HME. In most cases, the gate selects only one submodel for each input vector. This behaviour is consistent with the knowledge of the domain expert that the submodels were defined for different mutually exclusive modes. The ME model was not able to identify the driving modes and dividing the input space in a technically non-plausible way. This is illustrated in Fig. 5. The overall output is composed of the outputs of the submodels.

The chemical battery limits are violated by all models, except the HME, since they predict energy flows that cannot be provided by the battery. Some violations of the limits are shown in Fig. 6 of (a) the MLP, (b) the ME, and (c) the ensemble. The necessary information about these limits is not contained

Table 1. Mean absolute error for the hybrid electric vehicle data set

Model	Mean absolute error	
	training	testing
HME	1.82	1.84
ME	2.57	2.71
MLP	2.05	2.11
Ensemble	1.97	2.03

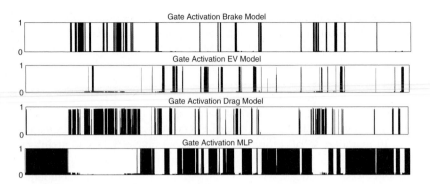

Fig. 4. The figure shows the activations of the different submodels by the gate of the HME model

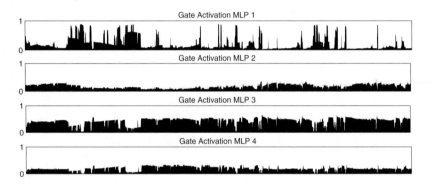

Fig. 5. The figure shows the activations of the different submodels by the gate of the ME model for the same data as shown in Fig 4

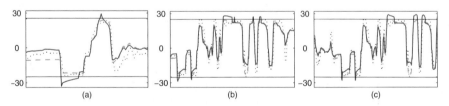

Fig. 6. The figures (a)-(c) show examples of violations of the chemical battery limits (depicted as horizontal lines) of (a) the MLP (solid line), (b) the ME (solid line), and (c) the ensemble (solid line). The target values for the energy flow and the outputs of the HME are depicted as dotted and dashed lines.

Table 2. Responsibilities of the mode models for data of the corresponding driving mode

HME Model	Driving mode (in %)			
	brake	pure electric drive	drag	hybrid
HME	97	94	92	96

Table 3. Mean absolute error for different sizes of the training data set T for the hybrid electric vehicle data set

Model	Mean absolute error				
	T	$T/2$	$T/4$	$T/8$	$T/16$
HME	1.82	1.81	1.83	1.86	1.90
ME	2.57	2.61	2.68	2.82	3.10
MLP	2.05	2.11	2.24	2.39	2.63
Ensemble	1.97	2.03	2.10	2.19	2.34

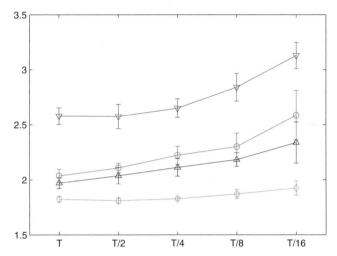

Fig. 7. The figure shows the predictive error of the models for different sizes of the training data set T. The HME model (square) has a slight increasing error for small training data set sizes. If the size of the training data set gets smaller the error of the ME model (downward-pointing triangle), the MLP (circle), and the ensemble (upward-pointing triangle) increases fast.

in the training data, but it is implicitly contained in the given knowledge-based models.

For the HME model Table 2 shows the distribution of the responsibilities of the mode models for data of the corresponding driving modes. The values indicate that the mode models are correctly assigned to the partitions of the driving modes.

An additional advantage of incorporating available knowledge is that fewer training data are required. In Table 3 and Fig. 7 the results for different sizes of the training data sets are shown. The smaller the training data set size the less robust are the results of the purely data-driven models. The results indicate that the HME model requires fewer training data compared to other regression methods in order to achieve a good predictive performance. This is useful if few training data are available.

5   Conclusions

By applying the proposed ensemble learning model it is possible to fuse information from multiple sources represented by knowledge-based models. Data-driven submodels are used to complement these models with respect to the coverage of the input space. To be able to integrate given knowledge-based models into the process of simultaneously training the data-driven submodels and a gate model it is crucial to incorporate the validity ranges of the knowledge-based models. The integration of knowledge-based models does not only lead to a superior performance but also results in an improved plausibility and reliability of the proposed model compared to the other models. Furthermore, the HME benefits from the additional information provided by the knowledge-based models as shown in the application example.

References

1. Freund, Y., Schapire, R.E.: Decision-theoretic generalization of on-line learning and an application to boosting. Journal of Computer and System Sciences 55(1), 119–139 (1997)
2. Jacobs, R.A., Jordan, M.I., Nowlan, S.J., Hinton, G.E.: Adaptive mixtures of local experts. Neural Computation 3, 79–87 (1991)
3. Perrone, M.P.: Improving Regression Estimation: Averaging Methods for Variance Reduction with Extensions to General Convex Measure Optmization. PhD thesis, Brown University (1993)
4. Dasarathy, B.V.: Information fusion - what, where, why, when, and how? Information Fusion 2(2), 75–76 (2001)
5. Durrant-Whyte, H.F.: Sensor models and multisensor integration. International Journal of Robotics Research 7(6), 97–113 (1988)
6. Bengio, S., Marcel, C., Marcel, S., Mariéthoz, J.: Confidence measures for multi-modal identity verification. Information Fusion 3(4), 267–276 (2002)
7. Bloch, I.: Information Fusion in Signal and Image Processing: Major Probabilistic and Non-Probabilistic Numerical Approaches. John Wiley & Sons Inc., Chichester (2008)
8. Jordan, M.I., Jacobs, R.A.: Hierarchical mixtures of experts and the EM algorithm. Neural Computation 6(2), 181–214 (1994)
9. Dempster, A.P., Laird, N.M., Rubin, D.B.: Maximum likelihood from incomplete data via the EM algorithm. Journal of the Royal Statistical Society 39(1), 1–38 (1977)

Bayesian Belief Network for Tsunami Warning Decision Support

Lilian Blaser, Matthias Ohrnberger, Carsten Riggelsen, and Frank Scherbaum

University of Potsdam, Institute of Geosciences
Karl-Liebknecht-Str. 24/25, 14476 Golm, Potsdam, Germany
{lilian,mao,riggelsen,fs}@geo.uni-potsdam.de

Abstract. Early warning systems help to mitigate the impact of disastrous natural catastrophes on society by providing short notice of an imminent threat to geographical regions. For early tsunami warning, real-time observations from a seismic monitoring network can be used to estimate the severity of a potential tsunami wave at a specific site. The ability of deriving accurate estimates of tsunami impact is limited due to the complexity of the phenomena and the uncertainties in seismic source parameter estimates. Here we describe the use of a Bayesian belief network (BBN), capable of handling uncertain and even missing data, to support emergency managers in extreme time critical situations. The BBN comes about via model selection from an artifically generated database. The data is generated by ancestral sampling of a generative model defined to convey formal expert knowledge and physical/mathematical laws known to hold in the realm of tsunami generation. Hence, the database implicitly holds the information for learning a BBN capturing the required domain knowledge.

Keywords: Bayesian belief network, learning, tsunami warning system, decision support, seismic source parameters.

1 Introduction

The tsunami disaster of December 2004 where over 230'000 people lost their lives has shocked the international community and awakened the awareness of the tsunami hazard world wide. The tragedy has exposed the vulnerability and the lack of a warning system in a cruel way.

During the last four years many studies have been launched aiming at establishing technically advanced concepts and modern equipments for robust tsunami early warning systems. Much effort has been undertaken in installing a variety of monitoring instrumentation. Focus has been set to the fast and exact analysis and evaluation of near-real time seismological, geodetical, and other geophysical data. All of these measures are important contributions to the mitigation of future tsunami hazard. Still scientific personnel at tsunami warning centers are in urgent need of operational tools that will provide an accurate answer to the

C. Sossai and G. Chemello (Eds.): ECSQARU 2009, LNAI 5590, pp. 757–768, 2009.

question whether a tsunami has been generated co-seismically (once an earth-quake occurred) in order to guide immediate decisions for evacuation and rescue operations.

For Indonesia or countries surrounding the Mediterranean Sea where the distance from an earthquake epicenter to the coast is small due to the geological situation, a tsunami wave may reach the populated coast in 20 min or even less. Consequently we need any evidence about a potential tsunami triggering as fast as possible. The first indirect measurements of tsunami generation come from the analysis of seismic waves, mostly available after about 5 min. Unfortunately, the seismic source parameter estimates are prone to large uncertainties at such early stages. Tsunami confirmation by tide gauges may arrive too late for timely evacuation measures. Therefore tsunami warning center's personnel face a difficult challenge: to issue tsunami warning based on incomplete and ambiguous data.

Currently, the automated evaluation of seismic source parameter estimates with respect to tsunami generation is achieved by rule-based systems. Only three variables are evaluated: First there is a boolean variable indicating whether the epicenter was located offshore or not, second an earthquake size estimate through magnitude is compared to a threshold and finally the depth of the hypocenter is categorized as being shallow or not. Current rule-based systems do not take into account the uncertainties of the earthquake parameter estimates and are not able to generate any decision support as long as one of the three variables is missing.

Within this study we apply a new approach based on Bayesian belief network (BBN) being capable of handling uncertain or even missing data. We show how to integrate more than the commonly used three variables by taking into account all estimated source parameters (size, orientation, rupture characteristics) independently. Generally they become available at irregular time instances and are updated continuously. Using a BBN for the automatized evaluation of the incoming evidences allows for integrating each information piece instantaneously and estimate the probability of the imminent tsunami risk supporting the decision maker from the very beginning. In this paper we will describe the construction of a first BBN draft.

Due to the fact that large earthquakes capable to trigger tsunamis are infrequent and tsunamis are even less frequent the set of historical data is very sparce. During the last 30 years about 230 tsunamis were observed world wide. Focusing on one of our sites of interest, offshore Sumatra (Indonesia), out of approximately 1800 earthquakes with magnitude larger than 5 a dozen tsunamis with run-ups between 0.1 m and 50 m since 1976 have been observed (see Figure 1). For most of these events the available earthquake parameters are epicentral location, magnitude and depth only. For a small part additional information about focal mechanism is available. Other interesting information as e.g., length or rupture velocity are discussed in particular cases only, or lack entirely. All the aforementioned variables are evaluated with post-processing methods. To our knowledge, there is no database available representing realistic real-time

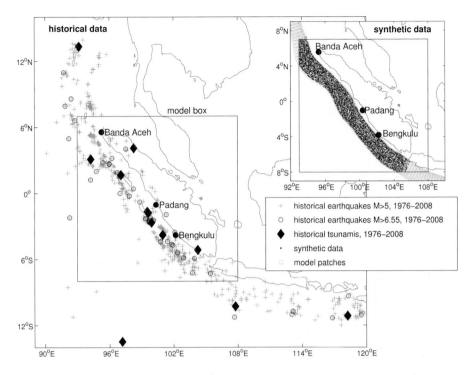

Fig. 1. Historical and synthetic earthquake locations and model constraints for target region Sumatra

evaluated earthquake parameter estimates for historical tsunami events. Thus, it is not feasible to learn a BBN from the existing database. But fortunately many experts have expressed their knowledge about the physical process of tsunami generation through different theoretical or empirical formulas. We will extract this "formula knowledge" and transform it into a BBN. The following steps have to be considered and will be discussed within this paper:

1. The set of formulas is assembled covering the whole process of a tsunami event from the triggering earthquake to the run-up approximation of the wave at the coast. The various formulas are outlined in Section 2 to give an idea of the physical background and to show the spectrum of the complexity of the formulas which have to be transformed into a BBN.
2. In a second step the variables for the BBN have to be defined and a database has to be generated. To generate records we employ ancestral sampling from our generative model derived from the formulas given in Section 2. This is described in Section 3.
3. Learning the structure as well as the parameters of the BBN is based on the synthetic database as described in Section 4.
4. Finally in Section 5 a first draft of a BBN is discussed which has to be tested and refined over several iterations in future research.

The elicitation-scheme for "formula knowledge" presented in this paper certainly is of interest in other domains where data is difficult to obtain or does not exist but a basic understanding of the fundamental physical processes is available.

BBNs are getting more and more popular in natural risk assessment in different subjects as earthquakes [1], volcanos [2], avalanches [3], rock fall [4] or desertification due to burned forests [5]. The field of tsunami hazard has been probabilistically analyzed in recent years by several authors [6], [7], [8], [9], [10]. These approaches differ fundamentally from the approach presented here.

2   Physical Background

Most earthquake triggered tsunamis originate at subduction zones where one plate is sliding underneath an other plate, at rates typically measured in centimeters per year (see Figure 2). The accumulated stress is released from time to time in an earthquake when the brittle material breaks. If the focal depth is shallow enough, the energy release is able to deform the Earth's crust up to the surface and in case of an offshore epicenter to displace a large water volume resulting in a gravity wave which eventually reaches the coast as destructive tsunami.

There is no direct way to measure tsunami generation. However we can infer from earthquake source parameter estimates specifying the location, size, orientation and rupture characteristic. In Section 2.1 we will sketch the different source parameters, estimated generally for earthquakes and outline on which assumptions they were determined. To estimate surface deformation (at the sea floor) given the earthquake rupture we use a model described in Section 2.2 to derive the displaced water volume. Knowledge about the displaced water volume allows then to calculate the wave height near the shore using differential equations of wave propagation for long waves in the ocean. This is described in Section 2.3. Finally, the approximation of the tsunami run-up from the near shore peak amplitude estimates is described in Section 2.4.

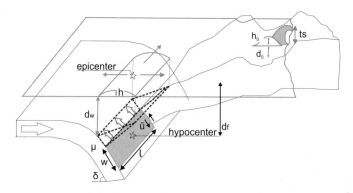

Fig. 2. Schematic view on tsunami triggering mechanism at a subduction zone

2.1   Earthquake Source Parameter Estimates

An earthquake can be observed indirectly by seismic waves. By analyzing their arrival times, amplitudes and waveform characteristics the location, size and focal depth (d_f, compare with Figure 2) of the earthquake is determined shortly after event detection. Whereas the localization of rupture initiation is rather accurate, uncertainties of the size of an earthquake might still be large at this early stage of observations. The uncertainties come about mainly from incomplete information about the Earth structure on local scales, which is a problem especially if the number of observations is small. Thus, further investigations on an increasing amount of available seismic records lead generally to a diminishment of the epistemic uncertainties, although sometimes contradicting information may be determined due to the analysis of local and distant observations.

Furthermore additional source parameters can be determined specifying rupture orientation, area (assumed to be rectangular with length l and width w) and the average displacement within the fault called average slip $\bar{u}$. Their (real-time) estimation techniques are more complex and subject to actual research.

As long as no independent estimates of these measures are available, scaling relations from magnitude can give a rough estimate. The magnitude evaluated by seismic wave inversion scales logarithmically with the seismic moment M_0 [11], one of the most adequate measure representing the size of an earthquake as it is proportional to the released energy. M_0 is related by scaling laws to fault geometry by

$$M_0 = \mu\,\bar{u}\,l\,w, \tag{1}$$

μ being the rigidity of the material [12]. Many studies have been conducted to obtain observational evidence of the direct link between magnitude and length and width of the fault area or the slip along the fault. In one widely cited analysis of Wells & Coppersmith (1994) [13], regression formulas are calculated from 244 detailed studied crustal earthquakes yielding a set of linear relations of the form

$$E[\log(l)] = a + bM_w,$$

for different fault geometries. As this study does not include subduction earthquakes data of additional studies [14], [15], [16], [17] has been added and new regression parameters were calculated.

To estimate the potential tsunami wave height not only earthquake parameters but also the water depths at the epicenter (d_w) and offshore near the specific target site (d_0) are of interest. The final tsunami run-up approximation used in the BBN is estimated as shown in the following sections. As intermediate variables the wave height at the epicenter h and in shallow water near the site h_0 are needed.

2.2   From Material Rupture to Sea Floor Deformation

A standard equation for surface deformation caused by an extended earthquake fault buried at depth has been developed by Okada (1985) [18]. His compact analytical set of formulas describes surface displacements due to inclined shear and tensile faults in a half-space for finite rectangular fault geometries. Based on theory of elastic dislocations, we can write

$$u_i = \frac{1}{F} \int \int_{\Sigma} \Delta u_j \left[\lambda \delta_{ij} \frac{\partial u_i^n}{\partial \xi_n} + \mu \left(\frac{\partial u_i^j}{\partial \xi_k} + \frac{\partial u_i^k}{\partial \xi_j} \right) \right] \nu_k d\Sigma.$$

Here, u_i is the ith component of the displacement field at (x_1, x_2, x_3) due to dislocation $\Delta u_j(\xi_1, \xi_2, \xi_3)$ across a surface Σ in an isotropic medium indicated by a point force of magnitude F at (ξ_1, ξ_2, ξ_3). δ_{jk} is the Kronecker delta, λ and μ are Lamé's material constants, ν_k is the direction cosine of the normal to the surface element $d\Sigma$.

This intimidating formula just points out the complexity of the theory we are dealing with and it becomes obvious that the conversion of the "formula knowledge" to a BBN is non-trivial. For numerical computation of surface deformation we use the code of Wang et al. 2003 [19].

2.3   Tsunami Propagation

A tsunami is a gravity wave generated by fast water displacement. The propagation is well understood in fluid dynamics and can be described by the shallow-water wave equation assuming a much smaller water depth (d_w) than the wave length. Including the bottom friction and the Coriolis force, the equation of motion for long waves can be written for a tree-dimensional case as

$$\frac{\partial U}{\partial t} + U \frac{\partial U}{\partial x} + V \frac{\partial U}{\partial y} = -fV - g \frac{\partial h}{\partial x} - C_f \frac{U \sqrt{U^2 + V^2}}{d+h},$$
$$\frac{\partial V}{\partial t} + U \frac{\partial V}{\partial x} + V \frac{\partial V}{\partial y} = fU - g \frac{\partial h}{\partial y} - C_f \frac{V \sqrt{U^2 + V^2}}{d+h},$$

with the corresponding equation of continuity

$$\frac{\partial U}{\partial t} + \frac{\partial}{\partial x}[V(h+d)] = 0,$$

where h is the wave amplitude, f is the Coriolis parameter, C_f is a non-dimensional frictional coefficient, and U and V are the average velocities in the x and y directions, respectively. Our generative model will use a tool, which calculates the tsunami propagation in a linearized form.

2.4   Run-Up Approximation

As numerical tsunami simulations work with limited bathymetry resolution which is not fine enough to calculate run-up height at the coast but just a near shore peak amplitude offshore at sites with water depth d_0 of about 20 m. Several run-up approximation theories of different complexities exists. We have chosen Green's law [20]

$$ts = h_0 \cdot \left(\frac{d_0}{d_r} \right)^{0.25}, \tag{2}$$

where h_0 is the sea surface height offshore calculated by the tsunami propagation model, and subscript r refers to the value at the coast ($d_r = 1$ m).

3   Method

In a next step we transform the knowledge encoded by the physical formulas into a generative model, and from there to a database via ancestral sampling. First off we have selected those variables which provide the most information about tsunami generation and which can be realistically obtained from near real-time measurements. The tsunami warning BBN system consists of the nine variables:

- Epicentral location categorized in regions (reg) and evaluated using bathymetry information to water depth (d_w),
- Magnitude (m),
- Rupture geometry: length (l) and width (w),
- Mean slip ($\overline{u}$),
- Focal depth (d_f),
- Focal mechanism (fm) is mainly defined by the rake angle λ describing the direction of the displacement,
- Tsunami wave height at the coast, called run-up (ts).

The generative model is constructed by coupling the various (sub)systems described in Section 2 in a hierarchical fashion as a directed graphical model. By exploiting the directed local Markov property, we employ simple ancestral sampling allowing us to generate an arbitrary number of cases/records. Metaphorically speaking "input" is passed down the system from the root variables down to the leaf-variables. The topmost random variables are identified as the three variables, epicentral location, loc, magnitude, m, and rake, λ (assumed to be marginally independent). For the ith record we therefore have

$$(m, loc, \lambda)^i \sim P(M, Loc, \Lambda) = P(M)P(Loc)P(\Lambda).$$

Without going into details, the probability distributions are defined and constrained by:

- The chosen study area Sumatra ("model box" in Figure 1) and the subduction zone [21] for location. Within these boundaries location is assumed to be distributed uniformly.
- Magnitude is restricted between $6.5 \leq M \leq 9.3$ and distributed according to Gutenberg Richter (1954) [22] so that the small events are more likely than the larger ones.
- We allow all possible rake values $-180° \leq \Lambda \leq 180°$. In order to account for the tectonic situation (subduction zone regime), we choose a multinomial distribution for drawing rake values.

The remaining variables are derived using the set of formulas given in Section 2. In the following, deterministic operations/mappings on/of (random) variables are denoted by $\mathcal{F}.(\cdot)$. The two variables reg and d_w can simply be derived from the location, loc

$$reg^i = \mathcal{F}_{reg}(loc^i),$$
$$d_w^i = \mathcal{F}_{d_w}(loc^i; Bathymetry),$$

where *Bathymetry* denotes the constant bathymetry [23]. In analogy we will write *Earthmodel* for the use of the Earth model assumptions [24].

The core or the generative model consists of a numerical simulation of sea floor deformation and the open water tsunami propagation for earthquakes offshore Sumatra ($\mathcal{F}_{ts}$), calculating tsunami wave height at a specific site. The requested input parameters are slip u_p and rake λ_p ($= \lambda$ for all patches) defined at every grid patch of the modeled subduction geometry. Again, for the ith record we have

$$ts^i = \mathcal{F}_{ts}(u_p^i, \lambda^i; Bathymetry, Earthmodel).$$

To specify u_p two intermediate steps have to be done:

1. The rupture area is determined by l, w and epicentral location, *loc*. l and w are inferred from magnitude by scaling laws ($\mathcal{F}_{sl}$) dependent on rake. To reflect the uncertainties, the values are disturbed by $\pm 30\%$ providing that the rupture plane fits totally into the grid mapping the subduction zone

$$l^i \sim P(L \mid f_{sl}(m^i, \lambda^i)),$$
$$w^i \sim P(W \mid f_{sl}(m^i, \lambda^i)).$$

2. Mean slip follows from Equation 1 given m, l, w and the depth dependent rigidity

$$\bar{u}^i = \mathcal{F}_{\bar{u}}(l^i, w^i, m^i; Earthmodel).$$

The spatial slip distribution is a sinusoidal shape approximation over the rupture area with the maximum slip in the middle and decreasing values towards the boundaries, $u_p^i = \mathcal{F}_{u_p}(loc^i, l^i, w^i, \bar{u}^i)$.

Focal depth is fixed by the projection of the epicenter on the subduction geometry. For the database the shallowest value is chosen, because for tsunami generation it is of interest how close to the surface a rupture is extended

$$d_f^i = \mathcal{F}_{d_f}(loc^i, w^i; Earthmodel).$$

Pre-calculated Green's functions allow to calculate sea floor deformation as well as tsunami wave propagation with linear superposition at given locations. The underling model uses a finite-difference scheme on a structured grid [23] [personal communication with Andrey Babeyko].

The output of the tsunami simulation consists of amplitudes at given locations near shore ($d_0 \approx 30$ m) every 25 sec. The final maximal wave height at the coast, ts, is determined for specific sites (e.g., the city Padang) by Equation 2.

In summary, a complete case i of the synthetic tsunami database becomes

$$\mathbf{d}^i = \{m^i, \lambda^i, l^i, w^i\} \cup \{\mathcal{F}_{reg}(loc^i), \mathcal{F}_{ts}(u_p^i, \lambda^i; Bathymetry, Earthmodel),$$
$$\mathcal{F}_{\bar{u}}(l^i, w^i, m^i; Earthmodel), \mathcal{F}_{d_w}(loc^i; Bathymetry),$$
$$\mathcal{F}_{d_f}(loc^i, w^i; Earthmodel)\}.$$

The first part denotes the set of random variables and the 2nd part denotes the set of deterministic variables (note that although *loc* is a random variable, it is not used in the BBN - only the derivatives thereof are used). In the remainder, all variables are treated as random variables.

4   Structure and Parameter Learning

We take a Bayesian approach to learning the "best" BBN (i.e., model selection), and consider the structure G and parameter Θ of the BBN as random variables, and define a joint prior $P(G, \Theta) = P(\Theta|G)P(G)$, where $P(\Theta|G)$ is a product Dirichlet distribution, and $P(G)$ is uniform. In particular, we want the *maximum a posteriori* BBN pair (MAP) given the database, d, i.e.,

$$\widehat{(g, \theta)} = \arg\max_{(g, \theta)} P(g, \theta|d).$$

We note this is different from merely learning the MAP model structure via the BD [25] scoring criterion and in a 2nd step estimate the BBN parameter (this is the classical approach to learning BBNs). Riggelsen [26] showed that optimizing the joint pair is beneficial in several regards. Moreover, no extra assumptions or constraints are imposed compared to the BD criterion, and computationally there is no disadvantage of using MAP BBN scoring metric instead of the BD scoring metric. The MAP BBN scoring criterion allows us to use the same traversal strategies as for any other BBN scoring criterion. We use a hill-climber to traverse the search space of essential graphs, simulated via the repeated covered arc reversal operator [27].

An important but difficult task is the discretization of the data. A sensitivity study showed strong influence even on the learned structure depending on number of discrete bins per variable as well as chosen thresholds. However, as we are able to generate an almost unlimited amount of data we discretize the parameter range in fine steps to combine bins with similar effects afterwards together again.

5   A First BBN Draft

Figure 3 shows a very first draft of the tsunami decision support BBN learned on a synthetic database of 50'000 records for the city of Padang.[1] Notice that although some variables are sampled/generated independently of each other in the (generating) hierarchical model they may be connected in the BBN structure (such as magnitude m and focal mechanism fm). However, this can be explained by the fact that learning BBNs from data yields minimal *I-maps* only (i.e., the structure only gives rise to the conditional independencies, not the dependencies).

Given any evidence the resulting marginal distributions of the parameters behave in a coherent and expected way. Assuming an earthquake was estimated around magnitude $m \approx 8.0$ the probability for a dangerous tsunami augments instantaneously. The tsunami probability distribution changes significantly by additional information about the region where the earthquake was detected as

---

[1] The tsunami BBN can be downloaded from `http://www.geo.uni-potsdam.de/mitarbeiter/blaser/blaser.html` and interactively tested with the freely available software GeNIe (`http://genie.sis.pitt.edu/`).

it is expected. Further additional evidences for other variables would not change the estimated tsunami risk significantly in this hypothetic case. This confirms the variable selection of the existing rule-based tsunami warning systems. Magnitude and location decide mainly over the warning status, given a shallow focal depth. The latter is implied in our model by the restrictions of the subduction geometry model which reaches maximal depth of 60 km, still classified as "shallow".

An advantage in comparison to the usual rule-based warning algorithm is the fact that a BBN provides probability distributions for all other variables, giving a visual overview and an intuitive better understanding for the existing situation. Furthermore the BBN is able to give an assessment at the time when the first evidence is coming in, where the rule-based system has to wait until all variables are known. The first draft of a BBN as shown in Figure 3 is a first step in building a tsunami early warning BBN which will be followed by a number of iterations of testing and refining to get an optimal solution. A single "best" BBN will not be able to deal with the uncertainties of the incoming earthquake parameter estimates or the uncertainties of the model structure. Hence, in a next step we will enable the tsunami warning BBN model to include "evidence uncertainty" as well as "model uncertainty". The former part will allow to insert an evidence distribution (e.g. $m \in [7.5 < m \leq 8.3]$: 80%, $m \in [7.0 < m \leq 7.5]$: 20%). The "model uncertainty" covers not only different possible net structures but can also be used to balance the sensitivity of the discretization problem by adding BBNs learned with variable discretization bins. In further steps expert knowledge will be elicited and incorporated to the tsunami warning BBN solution ensemble, too. We expect not only different net structures and parameter settings but also selections of additional variables (e.g., rupture velocity).

We will validate our different solutions on recent tsunami events where (near) real-time data is available (e.g., 2007-09-07, Bengkulu) as well as on cases not triggering a tsunami although it would have been expected due to offshore location and large magnitude. Considering those events is important to reduce the rate of false alarm.

Figure 4 depicts the iterative process of building, testing, refining and enlarging the BBN.

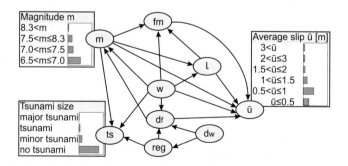

Fig. 3. A first BBN draft, structure and parameters learned from synthetic data

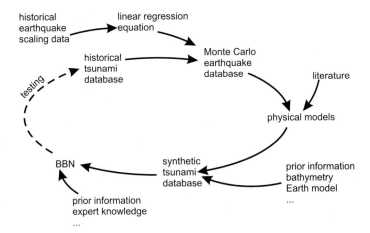

Fig. 4. BBN construction, testing and refinement iterations

6   Conclusion

Our first draft of a tsunami early warning BBN learned on synthetic data behaves in a coherent way and confirms the variable selection of the common rule-based systems. However, the resulting discrete probability distribution for the tsunami risk may help the emergency manager at a tsunami warning center to analyze the situation and decide on releasing a tsunami warning under extreme time pressure.

Moreover, this paper illustrates that a transform from rigorous mathematical and physical knowledge to a BBN via a generative hierarchical model and a synthetic database, provides a feasible addition or alternative to more traditional BBN knowledge acquisition techniques.

Acknowledgments. This work is funded by the German Federal Ministry of Education and Research, Geotechnologien "Frühwarnsysteme", No. 03G0648D. Special thanks to Andrey Babeyko providing his surface deformation and tsunami propagation model WaveGen.

References

1. Bayraktarli, Y.Y., Yazgan, U., Dazio, A., Faber, M.H.: Capabilities of the Bayesian probabilistic networks approach for earthquake risk management. In: Proceedings 1st European Conf. Earthqu. Eng. & Seism., Geneva, Switzerland (2006)
2. Hincks, T.: Probabilistic Volcanic Hazard and Risk Assessment. Phd thesis, University of Bristol (2006)
3. Gret-Regamey, A., Straub, D.: Spatially explicit avalanche risk assessment linking Bayesian networks to a GIS. NHESS 6, 911–926 (2006)
4. Straub, D.: Natural hazards risk assessment using Bayesian networks. In: Safety and Reliability of Engineering Systems and Structures, Proc. ICOSSAR 2005, Rome (2005)

5. Stassopoulou, A., Petrou, M., Kittler, J.: Application of a Bayesian network in a GIS based decision making system. IJGIS 12, 23–45 (1998)
6. Annaka, T., Satake, K., Sakakiyama, T., Yanagisawa, K., Shut, N.: Logic-tree approach for probabilistic tsunami hazard analysis and its applications to the Japanese coasts. PAGEOPH 164, 577–592 (2007)
7. Thio, H.K., Somerville, P., Ichinose, G.: Probabilistic analysis of strong ground motion and tsunami hazards in Southeast Asia. JET 1, 119–137 (2007)
8. Geist, E.L., Parsons, T.: Probabilistic analysis of tsunami hazards. Natural Hazards 37, 277–314 (2006)
9. Power, W., Downes, G., Stirling, M.: Estimation of tsunami hazard in New Zealand due to South American earthquakes. PAGEOPH 164, 547–564 (2007)
10. Maretzki, S., Grilli, S.T., Baxter, D.P.: Probabilistic SMF Tsunami Hazard Assessment for the upper East Coast of the United States. In: Proc. 3rd Intl. Symp. on Submarine Mass Movements and their Consequences, pp. 377–386. Springer, Heidelberg (2007)
11. Kanamori, H.: Energy-Release In Great Earthquakes. JGR 82, 2981–2987 (1977)
12. Aki, K., Richards, P.G.: Quantitative seismology, 2nd edn. University Science Books (2002)
13. Wells, D.L., Coppersmith, K.J.: New Empirical Relationships among Magnitude, Rupture Length, Rupture Width, Rupture Area, and Surface Displacement. BSSA 84, 974–1002 (1994)
14. Geller, R.J.: Scaling relations for earthquake source parameters and magnitudes. BSSA 66(5), 1501–1523 (1976)
15. Mai, P.M., Beroza, G.C.: Source scaling properties from finite-fault-rupture models. BSSA 90, 604–615 (2000)
16. Nuttli, O.W.: Empirical Magnitude And Spectral Scaling Relations For Mid-Plate and Plate-Margin Earthquakes. Tectonophysics 93, 207–223 (1983)
17. Scholz, C.H.: Scaling Laws For Large Earthquakes - Consequences For Physical Models. BSSA 72, 1–14 (1982)
18. Okada, Y.: Surface deformation due to shear and tensile faults in a half-space. BSSA 75, 1135–1154 (1985)
19. Wang, R.J., Martin, F.L., Roth, F.: Computation of deformation induced by earthquakes in a multi-layered elastic crust - FORTRAN programs EDGRN/EDCMP. Computers & Geosciences 29, 195–207 (2003)
20. Tadepalli, S., Synolakis, C.E.: Model for the Leading Waves of Tsunamis. PRL 77, 2141–2144 (1996)
21. Gudmundsson, O., Sambridge, M.: A regionalized upper mantle (RUM) seismic model. JGR 103, 7121–7136 (1998)
22. Gutenberg, B., Richter, C.F.: Seismicity of the Earth and Associated Phenomena, pp. 17–19. Princeton Univ. Press, Princeton (1954)
23. National Geophysical Data Center (NGDC), 2-Minute Gridded Global Relief Data (ETOPO2), http://www.ngdc.noaa.gov/mgg/global/etopo2.html
24. Kennett, B.L.N., Engdahl, E.R.: Traveltimes for global earthquake location and phase identification. GJI 195, 429–465 (1991)
25. Heckerman, D., Geiger, D., Chickering, D.M.: Learning Bayesian Networks: The Combination of Knowledge and Statistical Data. Machine Learning 20, 197–243 (1995)
26. Riggelsen, C.: Learning Bayesian Networks: A MAP Criterion for Joint Selection of Model Structure and Parameter. In: IEEE Int. Conf. on Data Mining (2008)
27. Castelo, R., Kocka, T.: On inclusion-driven learning of Bayesian networks. JMLR 4, 527–574 (2003)

Anti-division Queries
with Ordinal Layered Preferences

Patrick Bosc, Olivier Pivert, and Olivier Soufflet

Irisa – Enssat, University of Rennes 1
Technopole Anticipa 22305 Lannion Cedex France
bosc@enssat.fr, pivert@enssat.fr, soufflet@enssat.fr

Abstract. In this paper, we are interested in taking preferences into
account for a family of queries inspired by the anti-division. An anti-
division query aims at retrieving the elements associated with none of
the elements of a specified set of values. We suggest the introduction of
preferences inside such queries with the following specificities: i) the user
gives his/her preferences in an ordinal way and ii) the preferences apply
to the divisor which is defined as a hierarchy of sets. Different uses of
the hierarchy are investigated, which leads to queries conveying different
semantics and the property of the result delivered is characterized.

1 Introduction

Queries including preferences have received a growing interest during the last
decade [1,3,4,5,6,8,9]. One of their main advantages is to allow for some discrimination among the elements of their result thanks to the compliance with
the specified preferences. However, up to now, most of the research works have
focused on fairly simple queries where preferences apply only to selections. The
objective of this paper is to enlarge the scope of preference queries by considering more complex ones, founded on the association of an element with a given
set of values, in the spirit of the division operation. Moreover, a purely ordinal
framework is chosen and the user has only to deal with an ordinal scale, which
we think to be not too demanding. Lastly, taking preferences into account will
allow for keeping only the best k answers, in the spirit of top-k queries [4].

In the following, anti-division queries are considered. Let r be a relation of
schema $R(X, A)$ and s a relation of schema $S(B, Y)$, with A and B compatible
(sets of) attributes. The anti-division query $r[A \div B]s$ retrieves the X-values
present in relation r which are associated in r with none of the B-values present in
s. By analogy with a division, relation r may be called the dividend and relation
s the divisor. Knowing that an anti-division delivers a non-discriminated set of
elements, the idea is here to introduce preferences in this operator. Several lines
for assigning preferences may be thought of, depending on whether preferences
concern the divisor, the dividend or both, tuples individually (see e.g., [2,3], or
(sub)sets of tuples. In this paper, we investigate the case where: i) preferences
are purely ordinal and ii) they apply to the divisor only, which is structured as

C. Sossai and G. Chemello (Eds.): ECSQARU 2009, LNAI 5590, pp. 769–780, 2009.

a hierarchy (a set of layers). An element x of the dividend will be all the more acceptable as it is not connected with a certain number of the subsets (S_i's) defined over the divisor. Three different roles allotted to the divisor (described as a hierarchical set) are envisaged in the remainder of this paper. They differ in the way the layers of the divisor are taken into account for discrimination.

The rest of the paper is organized as follows. Section 2 is dedicated to some reminders on the division and anti-division operators. Three types of layered anti-division queries are studied and modeled in Section 3. In Section 4, it is shown that the result returned by these queries can be characterized as an "anti-quotient", i.e., a largest relation according to a given inclusion constraint. Section 5 deals with implementation aspects and presents some experimental results as to the performances of different algorithms implementing stratified anti-division queries. The conclusion summarizes the contribution of the paper and draws some lines for future research in particular as to implementation issues.

2   Some Reminders about the Anti-division

In the rest of the paper, the dividend relation r has the schema (A, X), while that of the divisor relation s is (B) where A and B are compatible sets of attributes. The division of relation r by relation s is defined as:

$$r[A \div B]s = \{x \mid x \in r[X] \land s \subseteq \Omega_r(x)\} \tag{1}$$
$$= \{x \mid x \in r[X] \land \forall a, a \in s \Rightarrow (a, x) \in r\} \tag{2}$$

where $r[X]$ denotes the projection of r over X and $\Omega_r(x) = \{a \mid (a, x) \in r\}$. In other words, an element x belongs to the result of the division of r by s iff it is associated in r with at least all the values a appearing in s. The justification of the term "division" assigned to this operation relies on the fact that a property similar to that of the quotient of integers holds. Indeed, the resulting relation d-res obtained with expression (1) has the double characteristic:

$$\forall t \in d\text{-}res, s \times \{t\} \subseteq r \qquad (3a) \qquad\qquad \forall t \notin d\text{-}res, s \times \{t\} \not\subseteq r \qquad (3b)$$

$\times$ denoting the Cartesian product of relations. Expressions (3a) and (3b) express the fact that relation d-res is a quotient, i.e., the largest relation whose Cartesian product with the divisor returns a result included in the dividend. In a similar way, we call anti-division the operator $*$ defined the following way:

$$r[A * B]s = \{x \mid x \in r[X] \land s \subseteq cp(\Omega_r(x))\} \tag{4}$$
$$= \{x \mid x \in r[X] \land \forall a, a \in s \Rightarrow (a, x) \notin r\}. \tag{5}$$

The result ad-res of the anti-division may be called an "anti-quotient", i.e., the largest relation whose Cartesian product with the divisor is included in the complement of the dividend. Thus, the following two properties hold:

$$\forall t \in \textit{ad-res}, \, s \times \{t\} \subseteq cp(r) \quad (6a) \qquad \forall t \in (r[X] - \textit{ad-res}), \, s \times \{t\} \not\subseteq cp(r) \quad (6b)$$

where $E - F$ denotes the difference between E and F and $cp(r)$ is the complement of r. In an SQL-like language, the division of r by s may be expressed:

select X **from** r [**where** condition] **group by** X
having set(A) **contains** $\{v_1, \, ..., \, v_n\}$

and the anti-division similarly as:

select X **from** r [**where** condition] **group by** X
having set(A) **contains-none** $\{v_1, \, , \, v_n\}$ (7)

where the operator "contains-none" states that the two operand sets do not overlap. An alternative expression of the latter can be based on a difference:

(**select** X **from** r) **differ** (**select** X **from** r **where** A **in** (**select** B **from** s)). (8)

Example 1. Let us consider the following relations P(product, component, proportion), which describes the composition of some chemical products and N(component) which gathers the identifications of noxious components:

P = $\{(p_1, c_1, 3), (p_1, c_2, 4), (p_1, c_3, 54), (p_2, c_2, 30), (p_3, c_2, 8), (p_3, c_6, 22)\}$,
N = $\{c_1, c_2, c_5\}$.

The query "retrieve any product which does not contain any noxious component in a proportion higher than 5%" can be expressed as the anti-division of the relation Prod' derived from Prod made of $\{(p_1, c_3), (p_2, c_2), (p_3, c_2), (p_3, c_6)\}$ by Nox, whose result according to (4) or (5) is $\{p_1\}$ and it is easy to check that formulas (6a-6b) both hold. ◇

3 Three Types of Layered Anti-division Queries

3.1 Anti-division and Preferences

What has been said until now concerns what we could call traditional anti-division queries inasmuch as no preferences come into play. We now move to more advanced queries mixing anti-division and the expression/handling of preferences. The three types of queries investigated here are the following:

- CJ queries: a direct extension of the anti-division in a conjunctive way, where the connection with the first layer of the divisor is forbidden and the non-association with the following ones is considered only desirable: find the elements x not connected with S_1 and if possible ... and if possible S_n (which is somehow related to bipolarity [7] since the non-connection with S_1 is a constraint while the non-connection with other levels is a wish),

- DJ queries: a disjunctive view where x is all the more satisfactory as it is connected with none of the values of a highly preferred sub(set) of the divisor: find the elements x not connected with S_1 or else ... or else S_n,
- FD queries: an intermediate approach where x is all the more highly ranked as it is not connected with numerous and preferred (sub)sets of the divisor: find the elements x not connected with S_1 and-or ... and-or S_n.

Knowing that the dividend may be any intermediate relation and the divisor is explicitly given by the user along with his/her preferences, the expression of these three types of anti-division queries is inspired from (7):

select top k X **from** r [**where** condition] **group by** X
having set(A) **contains-none** $\{v_{1,1}, ..., v_{1,j_1}\}$ connector ...
$$\text{connector } \{v_{n,1}, ..., v_{n,j_n}\}$$

where "connector" is either "and if possible", or "or else", or "and-or", and not from (8) inside which the integration of the layers of the divisor would not be easy. Such a statement induces an order over the divisor, namely $(S_1 = \{v_{1,1}, ..., v_{1,j_1}\}) \succ ... \succ (S_n = \{v_{n,1}, ..., v_{n,j_n}\})$ where $a \succ b$ denotes the preference of a over b. Actually, this order is about dislikes, i.e., S_1 contains the values the most highly undesired (sometimes excluded) and S_n those which are the most weakly unwanted. Associated with this preference relation is an ordinal scale L with labels l_i's (such that $l_1 > ... > l_n > l_{n+1}$) which will be used to assign levels of satisfaction to elements pertaining to the result of stratified anti-divisions (l_1 and l_{n+1} are extreme elements similar to 0 and 1 in the unit interval).

Example 2. Let us consider the case of a consumer who wants food products (e.g., noodles or vegetal oil) without certain additive substances. In the presence of the relation Products(p-name, add-s) describing which additives (add-s) are involved in products, a possible query is:

select top 6 p-name **from** Products **group by** p-name
having set(add-s) **contains-none** {AS27, BT12, C3}
and if possible {AS5, D2} **and if possible** {D8}

which induces the scale $L = l_1 > l_2 > l_3 > l_4$.   ◇

3.2   Conjunctive Queries (CJ)

As mentioned before, CJ queries are basically seen as an extension of the regular anti-division. To be more or less satisfactory, an element x must be connected with none of the elements having the maximal importance (S_1). In addition, as soon as it is connected with at least one of the elements of a set S_k, its association with values of any set S_{k+p} does not intervene for its final ranking. An element x is all the more preferred as it is not associated with any of the values of the succession of sets S_1 to S_i where i is large (if possible n for "perfection"). In other words, x is preferred to y if x is associated with none of the values of the sets S_1 to S_p and y is not associated with a shorter list of sets. This behavior is formalized using two approaches.

First, we consider the formal framework of relations assorted with preferences where every tuple t of a relation r is assigned a symbolic level of preference denoted by $pref_r(t)$. Tuples of the divisor are graded according to the ordering given by the user and since no preference applies to the dividend, its tuples have the maximal grade l_1 (conversely, any tuple absent from it is considered as having the grade l_{n+1}). The usual implication $(p \Rightarrow q = (\text{not } p) \text{ or } q)$, is extended to this context, which requires an adequate definition for both the negation and the disjunction. This latter is expressed thanks to the maximum (denoted by max), which satisfies most of the usual properties of the regular disjunction (associative, commutative, increasingly monotone with respect to each arguments, admittance of l_{n+1} as the neutral element). As to the negation, it corresponds to order reversal (denoted by $rev(-)$) defined as: $\forall i \in [1, n+1]$, $rev(l_i) = l_{n+2-i}$ which is involutive, i.e., $rev(rev(l_i)) = l_i$.

Example 3. Let us consider the following scale related to the importance of a phenomenon: $complete > high > medium > low > no$. The inverse scale is:

$$rev(complete) \quad < \quad rev(high) \quad < \quad rev(medium) \quad < \quad rev(low) \quad < \quad rev(no)$$
$$= \qquad\qquad = \qquad\qquad\qquad = \qquad\qquad\qquad = \qquad\qquad =$$
$$no \qquad\qquad low \qquad\qquad medium \qquad\qquad high \qquad\qquad complete$$

$\diamond$

This leads to the symbolic (or ordinal) version of Kleene-Dienes' implication defined as: $p \rightarrow q = max(rev(p), q)$. This implication coincides with the regular one when p and q take only the values l_1 and l_{n+1} (corresponding to true and false) and it obeys most of its properties, in particular contraposition and monotony with respect to the arguments.

Adapting the anti-division (formula (5)) to ordinal relations leads to assign each x of the dividend r the level of satisfaction $sat(x)$ defined as:

$$sat(x) = min_{v \in s}\, pref_s(v) \rightarrow rev(pref_r(v, x)))$$
$$= min_{v \in s}\, max(rev(pref_s(v)), rev(pref_r(v, x))). \tag{9}$$

Knowing that $pref_r(v, x)$ takes only the values l_1 and l_{n+1} depending on the presence or absence of the tuple (v, x) in relation r, each term $max(rev(pref_s(v)), rev(pref_r(v, x)))$ equals l_1 if x is not connected with v in r $((v, x) \notin r)$ and $rev(pref_s(v))$ otherwise. In particular, if x is connected with none of the values of the divisor, the maximal level l_1 is obtained and as soon as an association is encountered, the level of satisfaction decreases all the more as the undesired element is highly rejected.

Another description of CJ queries may also be provided. Its interest lies in its closeness to those given later for DJ and FD queries, which cannot be modeled by means of (logical) expressions in the spirit of (9). Let us denote: $I(x) = \{i \mid S_i \nsubseteq cp(\Omega_r(x))\}$ and $imin(x) = min(I(x))$ ($n + 1$ if $I(x) = \emptyset$).

The grade of satisfaction obtained by an element x ($sat(x)$) is expressed thanks to the scale L (implicitly) provided by the user as follows:

$$sat(x) = l_{n+2-imin(x)}. \tag{10}$$

So doing, the satisfaction is seen as a composition of the results of the anti-division of the dividend with each of the layers of the divisor. It is easy to prove that formulas (9) and (10) deliver the same result.

3.3 Disjunctive Queries (DJ)

While CJ queries have a conjunctive behavior, DJ queries are meant disjunctive instead, and S_1 is no longer a completely forbidden subset. Here, the order of the subsets according to user's preferences is used so that an element x is all the more preferred as it is connected with none of the values of S_k and k is small (ideally 1 for "perfection"). In this case again, the associations with the subsets of higher index ($> k$), and then lower importance, do not play any role in the discrimination strategy. In other words, x is preferred to y if x is associated with at least one of the values of each set S_1 to S_{k-1} and with none of the values of S_k and y is associated with at least one of the values of each set S_1 to S_{p-1} and with none of the values of S_p and $k < p$. Let us denote:

$$I'(x) = \{i \mid S_i \subseteq cp(\Omega_r(x))\} \text{ and } imin'(x) = min(I'(x)) \ (n+1 \text{ if } I'(x) = \emptyset).$$

Here again, the grade of satisfaction obtained by an element x is expressed using the ordinal scale L and:

$$sat(x) = l_{imin'(x)}. \tag{11}$$

The satisfaction is still a combination of the results of the anti-division of the dividend with each of the layers of the divisor. The grade l_1 is obtained if x is associated with none of the values of S_1, while l_{n+1} expresses rejection when the connection with at least one element of each of the S_i's holds.

3.4 Full Discrimination Queries (FD)

Queries of type FD are designed so as to counter the common disability of CJ and DJ queries in distinguishing between elements which are equally ranked because additional associations are not taken into account. So, the principle for interpreting FD queries is to consider all the layers for which no association occurs. An element is all the more preferred as it is connected with none of the elements of a set S_i highly excluded and this same point of view applies to break ties. In this case, the grade of satisfaction for x may be expressed thanks to a vector $V(x)$ of dimension n where $V_i(x) = 1$ if x is associated with none of the values of S_i, 0 otherwise. Ordering the elements is then a matter of comparison between such vectors according to the lexicographical order ($\succ_{lex}$):

$$x \succ_{lex} y \Leftrightarrow \exists k \in [1, n] \text{ s.t. } \forall j < k, V_j(x) = V_j(y) \text{ and } V_k(x) > V_k(y). \tag{12}$$

Here, scale L is not used directly even if the order of the elements of the vectors reflects it in the sense that, if $i < j$, $V_i(x)$ is more important than $V_j(x)$ as $l_i > l_j$.

Example 4. Let us take the divisor $\{\{a, b\} \succ c \succ \{d, e\}\}$ and the dividend: $r = \{(c, x_1), (g, x_1), (f, x_2), (c, x_3), (e, x_3), (a, x_4), (k, x_4)\}$. Here $n = 3$ and according to formula (9) or (10):

$$sat(x_1) = l_3, \ sat(x_2) = l_1, \ sat(x_3) = l_3, \ sat(x_4) = l_4 \text{ and } x_2 \succ \{x_1, x_3\} \succ x_4.$$

With formula (11), one gets:

$$sat(x_1) = sat(x_2) = sat(x_3) = l_1, \ sat(x_4) = l_2 \text{ and } \{x_1, x_2, x_3\} \succ x_4.$$

Last, using formula (12), we get a refinement of the previous two orderings, namely: $x_2 \succ x_1 \succ x_3 \succ x_4$ where the tie between x_1, x_2 and x_3 (resp. x_1 and x_3) in the result obtained with (11) (resp. (10)) is broken.   ◇

4   Characterizing the Result of Anti-division Queries

In this section, we provide a characterization (in terms of an "anti-quotient") of the result delivered by the three previous types of queries. In other words, the result returned by each of these queries is a maximal relation and it obeys formulas similar to (6a-6b). Due to space limitation, the characterization formulas are given, but not proved. Since relations are graded, (symbolic or ordinal) levels of satisfaction (l_i's) in both the result and the dividend have to be to considered for the characterization.

For CJ queries, if tuple x of the result is assigned the grade l_i ($i \in [1, n+1]$), the following properties hold:

if $i \in [1, n]$, $\forall k \in [1, n-i+1]$, $S_k \times \{x\} \subseteq cp(r)$
if $i = n+1$ $S_1 \times \{x\} \nsubseteq cp(r)$ \hfill (13a)

$$\forall i \in [1, n], \ S_{n-i+2} \times \{x\} \nsubseteq cp(r). \tag{13b}$$

In a similar way, for DJ queries, if x has received the grade of satisfaction l_i (letting S_{n+1} be empty) one has the double property:

$$S_i \times \{x\} \subseteq cp(r) \qquad (14a) \qquad\qquad \forall k \in [1, i-1], S_k \times \{x\} \nsubseteq cp(r). \qquad (14b)$$

As to FD queries, let us recall that the grade of satisfaction of x is basically expressed as a function of the values of the vector V stating whether x is connected ($V_i(x) = 0$) or not ($V_i(x) = 1$) with at least one of the values of layer i of the divisor according to formula (12). So, the following properties hold:

$$\forall i \in [1, n] \text{ such that } V_i(x) = 1, \ S_i \times \{x\} \subseteq cp(r), \tag{15a}$$
$$\forall i \in [1, n] \text{ such that } V_i(x) = 0, \ S_i \times \{x\} \nsubseteq cp(r). \tag{15b}$$

which means that if $sat(x)$ is increased, some constraint(s) of type (15a) will be violated. The validity of (13a-15b) can easily be checked over Example 4.

5 Implementation Aspects and Experimental Results

In this section, we outline some evaluation strategies and algorithms suited to anti-division queries of type CJ. First, we describe three algorithms implementing formula 9, then we present the experimentation and the results obtained.

5.1 Sequential Scan of the Dividend (SSD)

In this first algorithm, the idea is to access the tuples from the dividend relation (r) "in gusts", i.e., by series of tuples which share the same X-attribute value (in the spirit of what is performed by a *group by* clause). Moreover, we order the tuples (x, a) inside a group in increasing order of their A-attribute value. All this is performed by the query:

select * **from** r **order by** X, A.

Thanks to a table which gives, for each value (val-A) of the divisor, the layer to which it belongs (str-A), one can update the number of values from each layer which are associated with the current element x, while scanning the result of the query above. At the end of a group of tuples from the dividend, one checks the layers in decreasing order of their importance. This step stops as soon as the current element x is associated with at least one of the values from a layer V_i. Three cases can appear:

1. element x is associated with none of the values from any layer of the divisor and it gets the preference level l_1,
2. the stop occurs while checking layer V_i whose importance is not maximal ($i > 1$) and x gets the preference level $rv(l_i) = l_{n+2-i}$,
3. the stop occurs while checking layer V_1; element x gets the level l_{n+1} and is thus rejected.

5.2 Access Guided by the Divisor (AGD)

In this second algorithm, instead of scanning the dividend exhaustively and then checking the layers satisfied by a given x by means of the aforementioned table, one first retrieves the X-values from the dividend, and for each such x, one checks the associations with the different layers by means of an SQL query involving the aggregate *count*. Again, a layer is checked only if the layers of higher importance had none of their values associated with x.

The first step is to retrieve the distinct values of attribute X present in r by means of the query: **select distinct** X **from** r. Then, for each value x returned, one counts the A-values from V_1 which are associated with x in r:

select count(*) **from** r **where** $X = :x$ **and** A **in**
$\qquad$ (**select** A **from** s **where** pref $= l_1$);

If the value returned equals zero, one checks layer V_2 by means of a similar query and so on; otherwise the loop stops. The preference level assigned to x is computed according to the same principle as described above.

5.3   Series of Regular Anti-division Queries (SRA)

This third strategy consists of two steps: i) process as many regular anti-division queries as there are layers in the divisor, ii) merge the different results and compute the final preference degrees. The algorithm has the following general shape:

Step 1: For each layer V_i of the divisor, one processes an anti-division query which retrieves the x's which are associated in r with none of the values from V_i. The layers are examined in decreasing order of their importance and an element x is checked only if it belongs to the result associated with the previous layer.

```
create view T₁ as select distinct X from r where X not in
            (select X from r, s where r.A = s.B and s.pref = l₁);
for i := 2 to n do
begin
   create view Tᵢ as select X from Tᵢ₋₁ where X not in
            (select X from r, s where r.A = s.B and s.pref = lᵢ)
end;
```

Step 2: The results of the previous anti-division queries are merged by taking them in decreasing order of the corresponding layers. An element x which belongs to the result of layer V_i but not to that of layer V_{i+1} gets the preference level l_{n-i+1}. It is assumed hereafter that there exists a table T_{n+1} which is empty.

```
for i := 1 to n do
begin
   declare cursor cᵢ as
      select X from Tᵢ where X not in (select X from Tᵢ₊₁);
   open cᵢ; fetch cᵢ into :x;
   while not end (active set) do
   begin
      result := result + {lₙ₋ᵢ₊₁/x};
      fetch cᵢ into :x;
   end;
end;
```

We also tested different variants of this algorithm where set differences in both steps are performed either by means of the operator *minus* (instead of *not in*) or are expressed by an outer join. It appears that the most efficient expression is that where the set differences are based on: i) the operator *minus* in Step 1, ii) an outer join in Step 2.

5.4   Experimental Measures

The objectives of the experimentation are: i) to assess the additional processing cost related to the handling of preferences and ii) to compare the performances of the three algorithms presented above. The experimentation was performed with the DBMS Oracle^{TM} Enterprise Edition Release 8.0.4.0.0 running on an Alpha server 4000 bi-processor with 1.5 Gb memory. Even though the scope of the experiment presented here is still limited and should be extended in the future, it gives an interesting trend as to the cost of such queries.

A generic stratified anti-division query has been run on dividend relations of 300, 3000 and 30000 tuples, and a divisor including five layers made of respectively 3, 2, 1, 2 and 2 values.

The query taken as a reference is the analogous anti-division query without preferences, where the divisor is made of the sole first layer (which corresponds to a "hard constraint" as mentioned before). The reference query has been evaluated using two methods:

- algorithm AGD without preferences, denoted by REF2,
- first step of algorithm SRA with one layer only, denoted by REF3.

Notice that algorithm SSD without preferences would have the same complexity as SSD itself since it would also involve an exhaustive scan of the dividend (this is why there is no reference method "REF1"). Moreover:

- we used synthetic data generated in such a way that the selectivity of each value v_i from the divisor relatively to any x from the dividend is equal to 25% (for a given value v_i from the divisor and a given x from the dividend, tuple (x, v_i) has one chance out of four to be present in the dividend),
- each algorithm was run 8 times, so as to avoid any bias induced by the load of the machine,
- the time unit equals 1/60 second.

The results reported in Table 1 show that:

- among the reference methods for non-stratified anti-divisions, REF3 is much more efficient than REF2;
- the performances of REF2, AGD, and SSD vary linearly w.r.t. the size of the dividend. As to REF3 and SRA, their complexity is less than linear;
- the best algorithm for stratified anti-divisions is SRA, which is significantly better than AGD, itself much more efficient than SSD.
- the extra cost of SRA w.r.t. the most efficient reference algorithm, namely REF3, is still rather important (multiplicative factor between 4.6 and 8).

What all these measures show was somewhat predictable: the best way to process an anti-division query (stratified or not) is to express it by means of a single

Table 1. Experimental results

Size of the dividend	300	3000	30000
REF2	41.4	400.7	4055
REF3	13.2	81.4	760.2
SSD	108.6	960.5	10418
AGD	54.2	645.2	6315
SRA	106	375.1	4353
Number of answers (top layer)	37	427	4365

query that can be efficiently handled by the optimizor of the system, and not by external programs which induce a more or less important overhead. The extra cost attached to SRA w.r.t. REF3, is also explainable by the fact that SRA processes five regular anti-division queries — one for each layer — instead of one for REF3, and then has to merge the results of these queries. If the stratified anti-division functionality were to be integrated into a commercial DBMS, it is quite clear that it would have to be handled by the optimizor at an internal level, and processed as *one* query involving a new type of "having" clause, as in expression 7.

6   Conclusion

In this paper, preferences for a family of queries stemming from the relational anti-division have been considered. The key idea is to make use of a divisor made of a hierarchy of subsets of elements. So doing, the result is no longer a flat set but a list of items provided with a level of satisfaction. Three uses of the hierarchy have been investigated, which leads to three distinct semantics of the corresponding queries. Moreover, a characterization of the result produced in all cases has been suggested: it is an "anti-quotient", i.e., the largest relation whose product with the divisor remains included in the complement of the dividend. Besides, some experimental measures have been carried out in order to assess the feasibility of such anti-division queries. Even though these measures still need to be completed — only one of the three semantics has been considered so far —, they show that the additional cost induced by the stratified nature of the divisor is quite high (factor 4-8 w.r.t. a classical anti-division) but that the overall processing time is still acceptable for medium-sized dividend relations. To reach better performances, it would be of course necessary to integrate the new operator into the processing engine of the system, so as to benefit from a real internal optimization, instead of processing stratified anti-division queries externally, as we did here.

We are now planning to: i) design algorithms for implementing anti-division queries of types DJ and FD, ii) make experiments in order to evaluate these algorithms, as we did for CJ queries, and iii) check whether the results obtained are confirmed when another DBMS (e.g. PostgresQL or MySQL) is used.

References

1. Bőrzsőnyi, S., Kossmann, D., Stocker, K.: The skyline operator. In: Proc. of the 17th IEEE Inter. Conf. on Data Engineering, pp. 421–430 (2001)
2. Bosc, P., Pivert, O.: On a parameterized antidivision operator for database flexible querying. In: Proc. of the 19th Conference on Database and Expert Systems Applications, pp. 652–659 (2008)
3. Bosc, P., Pivert, O., Rocacher, D.: About quotient and division of crisp and fuzzy relations. Journal of Intelligent Information Systems 29(2), 185–210 (2007)
4. Bruno, N., Chaudhuri, S., Gravano, L.: Top-k selection queries over relational databases: mapping strategies and performance evaluation. ACM Transactions on Database Systems 27(2), 153–187 (2002)
5. Chomicki, J.: Preference formulas in relational queries. ACM Transactions on Database Systems 28(4), 427–466 (2003)
6. Dubois, D., Prade, H.: Using fuzzy sets in flexible querying: why and how. In: Proc. of the Workshop on Flexible Query-Answering Systems (FQAS 1996), pp. 89–103 (1996)
7. Dubois, D., Prade, H.: Handling bipolar queries in fuzzy information processing. In: Handbook of Research on Fuzzy Information Processing in Databases, pp. 97–114. IGI Global Publication (2008)
8. Hadjali, A., Kaci, S., Prade, H.: Database preference queries — a possibilistic logic approach with symbolic priorities. In: Hartmann, S., Kern-Isberner, G. (eds.) FoIKS 2008. LNCS, vol. 4932, pp. 291–310. Springer, Heidelberg (2008)
9. Kießling, W., Köstler, G.: Preference SQL — design, implementation, experiences. In: Proc. of the 28th Conference on Very Large Data Bases (VLDB 2002), pp. 990–1001 (2002)

Predicting Stock and Portfolio Returns Using Mixtures of Truncated Exponentials

Barry R. Cobb[1], Rafael Rumí[2], and Antonio Salmerón[2]

[1] Department of Economics and Business, Virginia Military Institute,
Lexington, Virginia, USA
cobbbr@vmi.edu

[2] Department of Statistics and Applied Mathematics University of Almería,
Almería, Spain
{rrumi,Antonio.Salmeron}@ual.es

Abstract. This paper presents mixtures of truncated exponentials (MTE) potentials in two applications of Bayesian networks to finance problems. First, naive Bayes and TAN models where continuous probability densities are approximated by MTE potentials are used to provide a distribution of stock returns. Second, a Bayesian network is used to determine a return distribution for a portfolio of stocks. Using MTE potentials to approximate the distributions for the continuous variables in the network allows use of the Shenoy-Shafer architecture to obtain a solution for the marginal distributions. We also illustrate the problem that arises in these models where deterministic relationships between variables appear, which is related to the partitioning of the domain of the MTE distributions. We propose a solution based on simulation.

1 Introduction

Finance models for predicting asset and portfolio returns focus on modeling relationships between historical, economic data. A multi-factor model of the rate of return on an asset [1] is

$$R_i = a_i + b_{i1}F_1 + b_{i2}F_2 + \cdots + b_{ik}F_k + e_i \ , \tag{1}$$

where the independent factors are denoted by $F_1, \ldots, F_k$ and $b_{i1}, \ldots, b_{ik}$ are constants. Portfolio return (R_p) is defined as the weighted average of the n individual assets in the portfolio, $R_p = \sum_{i=1}^{n} w_i R_i$, where w_i denotes the proportional amount invested in asset i. For the remainder of this paper, we will consider only equally weighted portfolios of assets with portfolio return calculated as $R_p = \left(\sum_{i=1}^{n} R_i\right)/n$.

This paper presents two applications of Bayesian networks to asset and portfolio valuation problems involving the gold mining stocks BGO (Bema Gold Corp.), ABX (Barrick Gold Corp.), and AEM (Agnico Eagle Mines). Shenoy and Shenoy [2] use Bayesian network models to predict the returns on a portfolio of these three stocks, with each of the stock returns being conditioned on

C. Sossai and G. Chemello (Eds.): ECSQARU 2009, LNAI 5590, pp. 781–792, 2009.

market returns (represented by returns on the S&P 500 index) and gold returns (represented by the London PM Gold closing price), as well as a stock-specific effect for each stock. We use weekly returns for the period of January 1996 through February 1998[1].

In our first application we compare a least-squares regression parameterization of the multi-factor model (see Equation 1) for the return on each stock to a Naive Bayes model that gives a distribution of stock returns. In the latter model, the distributions of the variables in the network are represented by mixtures of truncated exponentials (MTE) potentials [3]. Next, we use MTE potentials to parameterize a Bayesian network model to determine a distribution for the returns on an equally-weighted portfolio of the stocks BGO, ABX, and AEM.

The remainder of the paper is organized as follows. Section 2 gives some background on Bayesian networks and their relationship to regression models. Section 3 defines mixtures of truncated exponentials (MTE) potentials and Section 4 describes a Naive Bayes predictor model that uses MTE potentials. Section 5 describes the example of using the Naive Bayes model with MTE potentials for predicting stock returns and Section 6 contains the example of using a Bayesian network with MTE potentials to predict portfolio returns. Section 7 describes a Monte Carlo simulation model for determining marginal return distributions. Section 8 summarizes the paper.

2   Bayesian Networks and Regression

Bayesian networks [4,5] are efficient tools for handling problems that can be defined by a multivariate probability distribution over a set of variables $\mathbf{X} = \{X_1, \ldots, X_n\}$. A Bayesian network is a directed acyclic graph (DAG) where each vertex represents a random variable $X_i \in \mathbf{X}$ and there is a conditional distribution for each variable X_i given its parents $pa(X_i)$, so that the joint distribution can be expressed as $p(x_1, \ldots, x_n) = \prod_{i=1}^{n} p(x_i | pa(x_i))$.

A Bayesian network can be used for classification purposes if it consists of a *class* variable, C, and a set of *feature* variables $X_1, \ldots, X_n$, so that an individual with observed features $x_1, \ldots, x_n$ will be classified as a member of class c^* obtained as $c^* = \arg\max_{c \in \Omega_C} p(c|x_1, \ldots, x_n)$, where Ω_C denotes the support of variable C. Similarly, a Bayesian network can be used for *regression*, i.e., when C is continuous. However, in this case the goal is to compute the posterior distribution of the class variable given the observed features $x_1, \ldots, x_n$. Once this distribution is computed, a numerical prediction can be given using the mean, the median or the mode.

Note that $p(c|x_1, \ldots, x_n)$ is proportional to $p(c) \times p(x_1, \ldots, x_n|c)$, and therefore solving the regression problem would require specification of an n dimensional distribution for $X_1, \ldots, X_n$ given the class. Using the factorization determined by the Bayesian network, this problem is simplified. The extreme case is the so-called

---

[1] Shenoy and Shenoy [2] use a similar time period, but have 115 observations, whereas our data has 113 observations.

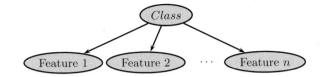

Fig. 1. Structure of a Naive Bayes classifier/predictor

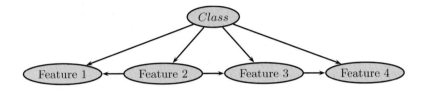

Fig. 2. Example of a TAN structure with 4 features

Naive Bayes structure [6,7], where all the feature variables are considered independent given the class. An example of Naive Bayes structure can be seen in Figure 1.

The restrictive independence assumption behind Naive Bayes models is compensated by the reduction on the number of parameters to be estimated from data, since in this case, it holds that $p(c|x_1, \ldots, x_n) = p(c) \prod_{i=1}^{n} p(x_i|c)$, which means that we operate with smaller distributions.

The Naive Bayes model assumes that the feature variables are independent given the class. The Tree Augmented Network (TAN) model [6] allows some dependencies in the feature variables, so that each feature variable can have at most one parent apart from the class variable. This parent is chosen according to the conditional mutual information of the variables, given the class. An example of such structure can be seen in Figure 2.

3   The MTE Model

Random variables are denoted by capital letters, and their values by lowercase letters. In the multi-dimensional case, boldfaced characters are used. The domain of the variable $\mathbf{X}$ is denoted by $\Omega_{\mathbf{X}}$. The MTE model is defined as follows [3]:

Definition 1. (MTE potential) *Let* $\mathbf{X}$ *be a mixed n-dimensional random vector. Let* $\mathbf{Y} = (Y_1, \ldots, Y_d)$ *and* $\mathbf{Z} = (Z_1, \ldots, Z_c)$ *be the discrete and continuous parts of* $\mathbf{X}$, *respectively, with* $c + d = n$. *A function* $f : \Omega_{\mathbf{X}} \mapsto \mathbb{R}_0^+$ *is a Mixture of Truncated Exponentials potential (MTE potential) if one of the next conditions holds:*

i. $\mathbf{Y} = \emptyset$ *and* f *can be written as*

$$f(\mathbf{x}) = f(\mathbf{z}) = a_0 + \sum_{i=1}^{m} a_i \exp \left\{ \sum_{j=1}^{c} b_i^{(j)} z_j \right\} \tag{2}$$

for all $\mathbf{z} \in \Omega_{\mathbf{Z}}$, where a_i, $i = 0, \ldots, m$ and $b_i^{(j)}$, $i = 1, \ldots, m$, $j = 1, \ldots, c$ are real numbers.

ii. $\mathbf{Y} = \emptyset$ and there is a partition $D_1, \ldots, D_k$ of $\Omega_{\mathbf{Z}}$ into hypercubes such that f is defined as

$$f(\mathbf{x}) = f(\mathbf{z}) = f_i(\mathbf{z}) \quad \text{if} \quad \mathbf{z} \in D_i \ ,$$

where each f_i, $i = 1, \ldots, k$ can be written in the form of equation (2).

iii. $\mathbf{Y} \neq \emptyset$ and for each fixed value $\mathbf{y} \in \Omega_{\mathbf{Y}}$, $f_{\mathbf{y}}(\mathbf{z}) = f(\mathbf{y}, \mathbf{z})$ is defined as in ii.

Example 1. The function ϕ defined as

$$\phi(z_1, z_2) = \begin{cases} 2 + e^{3z_1 + z_2} + e^{z_1 + z_2}, & 0 < z_1 \leq 1, \quad 0 < z_2 < 2, \\ 1 + e^{z_1 + z_2} & 0 < z_1 \leq 1, \quad 2 \leq z_2 < 3, \\ \dfrac{1}{4} + e^{2z_1 + z_2} & 1 < z_1 < 2, \quad 0 < z_2 < 2, \\ \dfrac{1}{2} + 5e^{z_1 + 2z_2} & 1 < z_1 < 2, \quad 2 \leq z_2 < 3. \end{cases}$$

is an MTE potential since all of its parts are MTE potentials.

Definition 2. (MTE density) *An MTE potential f is an MTE density if*

$$\sum_{\mathbf{y} \in \Omega_{\mathbf{Y}}} \int_{\Omega_{\mathbf{Z}}} f(\mathbf{y}, \mathbf{z}) d\mathbf{z} = 1 \ .$$

A *conditional MTE density* can be specified by dividing the domain of the conditioning variables and specifying an MTE density for the conditioned variable for each configuration of splits of the conditioning variables.

4   Bayesian Network Predictors Based on MTE Potentials

In [8], a regression model based on Bayesian networks is proposed. The networks have a Naive Bayes structure and the conditional distributions are MTE potentials. The advantage of using MTE potentials is that the independent variables can be discrete or continuous. In the MTE framework, the domain of the variables is split into pieces and in each resulting interval an MTE potential is fitted to the data. In this work we will use the so-called *five-parameter* MTE, which means that in each split there are five parameters to be estimated from data:

$$f(x) = a_0 + a_1 e^{a_2 x} + a_3 e^{a_4 x} \ . \tag{3}$$

The choice of the five-parameter MTE is motivated by its low complexity and high fitting power [9]. We follow the estimation procedure developed in [10], which has these main steps: (i) a Gaussian kernel density is fitted to the data, (ii) The domain of the variable is split according to changes in concavity/convexity or increase/decrease in the kernel density, (iii) in each split, a five-parameter MTE is fitted to the kernel by least squares estimation.

In [11,12] the TAN classifier is adapted to the MTE model. The conditional mutual information in the MTE model cannot be computed exactly, but in [11] it is shown how to approximate it by means of Monte Carlo simulation. The distributions are learned afterwards using the same procedure as with Naive Bayes.

Once the models are constructed, they can be used to predict the value of the class variable given observed values of the feature variables. The prediction is carried out by obtaining a numerical prediction from the posterior distribution of the class given the observed values for the features. In the experiments reported in [8], the best results are obtained using the expected value and the median. The posterior distribution for the class variable is computed using the Shenoy-Shafer algorithm [13] for probability updating in Bayesian networks, adapted to the MTE case as in [14].

5   Predicting Stock Returns Using a Bayesian Network Regression Model

This section describes the results of using a Naive Bayes and a TAN regression model as defined in Section 4 to predict the BGO, ABX and AEM returns (see Section 1). Since the size of the database is small (113 total observations), we have carried out a five-fold cross-validation procedure to check the performance of the models.

The accuracy of each model is measured as follows. Let $c_1, \ldots, c_m$ represent the values of the class for the observations in the corresponding *test* database and $\hat{c}_1, \ldots, \hat{c}_m$ represent the corresponding estimates provided by the model. The RMSE (root mean squared error) is obtained as

$$RMSE = \sqrt{\frac{1}{m} \sum_{i=1}^{m} (c_i - \hat{c}_i)^2} \ . \tag{4}$$

The final RMSE is the sum of the RMSE of the five *test* databases for each model. We have constructed nine different regression models, three for each variable (BGO, ABX and AEM): one linear model (LM), a Naive Bayes predictor (NB), and a TAN predictor (TAN). The accuracy of the fitted models is reported in Table 1.

Table 1. Root mean squared error for the nine fitted models

	BGO	ABX	AEM
LM	0.1219	0.0369	0.0524
NB	0.1059	0.03829	0.05443
TAN	0.1055	0.03821	0.055558

The differences in RMSE between the LM, NB and TAN models are very small. In order to test whether the differences in the estimates produced by the LM, NB and TAN can be considered statistically significant we have carried out a Friedman test; the p-value obtained is 0.7156.

These results support the use of Bayesian network predictors as competitive with the LM, because the differences in the model predictions are small and the Bayesian network models have an important added value with respect to the LM. These models not only give a numerical prediction of the class variable, but they also give its posterior distribution, so that other kinds of inferences can be performed. For instance, probabilities that the return is in a given interval, e.g., the return is positive, can be calculated from the posterior distribution.

6   Predicting Portfolio Returns Using a Bayesian Network

Figure 3 shows a Bayesian network model for determining returns for a portfolio of BGO, ABX, and AEM stocks. Again, each stock return, X_i, $i = 1, 2, 3$, is affected by market returns (M) and gold returns (G).

The node for portfolio return (P) is shown as a conditionally deterministic node, i.e. the values of portfolio return are completely determined by the values of the individual stock returns. The model depicted is similar to the simplest model described by Shenoy and Shenoy [2]. In their model, the distributions for market and gold returns are assumed to be normal and are parameterized by calculating the mean and variance of the observed returns. The residual stock effects are captured by separate nodes and are assumed to be normal with means of zero and variances equal to the residuals of the regressions. The stock return nodes have a functional relationship determined by least-squares regression. The model is solved by using Monte Carlo simulation to generate values for the independent variables and using the functional relationships to calculate marginal stock return and portfolio return distributions.

The MTE model requires no assumptions about the normality of market return, gold return, or stock-specific effects. The residual stock specific effects are considered in the procedure for determining the splits of the domain of the variables in the resulting MTE functions, as this ultimately determines the variance of the conditional distributions. The marginal stock return distributions

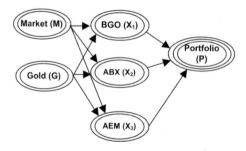

Fig. 3. A Bayesian network model for a portfolio

in the MTE model are determined by using the Shenoy-Shafer architecture [13] as adapted for MTE potentials in [14]. MTE potentials are closed under this propagation scheme because (closed-form) integration of MTE potentials and multiplication of MTE potentials results in a function in the MTE class [3]. To calculate the portfolio return distribution, a convolution operation defined by Cobb and Shenoy [15] is utilized. Assuming a linear relationship among variables, this operation also results in an MTE potential. Since the result of the solution for the marginal portfolio return distribution is an MTE potential, we can easily calculate probabilities of interest for this distribution. For instance, we can calculate the probability that the portfolio has a positive return.

In this example, marginal MTE potentials for the variables G and M and conditional MTE potentials for X_1, X_2, and X_3 are determined using the procedure in [10]. The domains of the variables are divided into four regions (or splits) in each MTE distribution. Although using additional regions would increase the accuracy of the marginal distributions, the computational complexity required to obtain an exact solution increases. Rumí and Salmerón [14,16] discuss methods for approximate inference that can lower the computational cost of inference when using MTE potentials.

As examples of the individual stock return distributions, the marginal distribution of BGO returns (X_1) and AEM returns (X_3) are shown in Figures 4 (a) and (b), respectively. The return on the portfolio is determined by the linear relationship $P = X_1/3 + X_2/3 + X_3/3$, i.e. the portfolio is equally weighted. The marginal MTE potentials for BGO, ABX, and AEM returns are denoted by ϕ_1, ϕ_2, and ϕ_3 respectively. The joint distribution for $\{X_1, X_2, X_3\}$ is calculatead as $\phi_4 = \phi_1 \otimes \phi_2 \otimes \phi_3$. To calculate the marginal distribution for the portfolio (P), we use the operation defined by Cobb and Shenoy [15] to remove X_1 as follows:

$$\phi_5(p, x_2, x_3) = 3 \cdot \phi_4\left(3 \cdot \left(p - \frac{1}{3}x_2 - \frac{1}{3}x_3\right), x_2, x_3\right)$$

This operation (as opposed to the usual marginalization operation involving integration) is required because the joint distribution of the variables $\{P, X_1, X_2, X_3\}$ is three-dimensional, as P is a conditionally deterministic variable.

The variables X_2 and X_3 are subsequently eliminated by integration as follows:

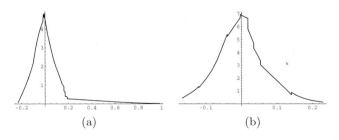

(a)   (b)

Fig. 4. The marginal distribution for (a) BGO return (X_1) and (b) AEM return (X_3) in the MTE model and (b) t

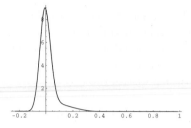

Fig. 5. The marginal distribution for portfolio return (P) in the MTE model

$$\phi_6(p) = \int_{\Omega_{X_3}} \left(\int_{\Omega_{X_2}} \phi_5\left(p, x_2, x_3\right)\, dx_2 \right)\, dx_3 \ .$$

The marginal distribution of portfolio return (P) is shown graphically in Figure 5.

The probability that the portfolio earns a positive return is calculated as

$$P(p > 0) = \int_0^\infty \phi_6\left(p\right)\, dp = 0.511 \ .$$

One weakness of the operations used to determine the marginal distribution for portfolio returns is that it results in a potential that does not strictly adhere to Definition 1. Specifically, the resulting potential contains the variable X_1 in the domain of the pieces of the function. This makes the limits of the integration above very difficult to determine. The next section addresses this computational issue.

7   Monte Carlo Simulation of Marginal Distributions

The convolution operation in the previous section involves a change in variable X_1 in both the functional form of the density and the domain of the function. The following example will illustrate this concept. Let ϕ_X be a potential for X and ϕ_Y a potential for Y, with

$$\phi_X(x) = \begin{cases} f_1(x) & 0 \leq x < 0.5 \\ f_2(x) & 0.5 \leq x \leq 1 \end{cases} \qquad \phi_Y(y) = \begin{cases} g_1(y) & 0 \leq y < 0.5 \\ g_2(y) & 0.5 \leq y \leq 1 \end{cases}$$

Then, potential ϕ defined for (X, Y) is

$$\phi(x, y) = (\phi_X \otimes \phi_Y)(x, y) = \begin{cases} f_1(x)g_1(y) & 0 \leq x < 0.5, \ 0 \leq y < 0.5 \\ f_1(x)g_2(y) & 0 \leq x < 0.5, \ 0.5 \leq y \leq 1 \\ f_2(x)g_1(y) & 0.5 \leq x \leq 1, \ 0 \leq y < 0.5 \\ f_2(x)g_2(y) & 0.5 \leq x \leq 1, \ 0.5 \leq y \leq 1 \end{cases}$$

If we want to obtain the marginal of a new variable $Z = (X + Y)/2$, then a new potential $\phi_2 = 2 \cdot \phi \left(2 \left(z - \frac{1}{2}y\right), y\right)$ is defined as

$$\phi_2(z,y) = \begin{cases} f_1(2(z - \frac{1}{2}y))g_1(y) & 0 \le 2(z - \frac{1}{2}y) < 0.5, \quad 0 \le y < 0.5 \\ f_1(2(z - \frac{1}{2}y))g_2(y) & 0 \le 2(z - \frac{1}{2}y) < 0.5, \quad 0.5 \le y \le 1 \\ f_2(2(z - \frac{1}{2}y))g_1(y) & 0.5 \le 2(z - \frac{1}{2}y) \le 1, \quad 0 \le y < 0.5 \\ f_2(2(z - \frac{1}{2}y))g_2(y) & 0.5 \le 2(z - \frac{1}{2}y) \le 1, \quad 0.5 \le y \le 1 \end{cases}$$

where the limits of variable Z are

$$\frac{1}{2}y \le z < 0.25 + \frac{1}{2}y \tag{5}$$

$$0.25 + \frac{1}{2}y \le z \le 0.5 + \frac{1}{2}y \tag{6}$$

That is, the limits of variable Z are not independent of Y, which is a problem for the posterior integral. An MTE continuous probability tree representing ϕ can be seen in Figure 6, while the tree representing $\phi_2(z,y)$ is depicted in Figure 7. The limits for variable Z in boldface are computed according to Equations (5) and (6) restricted to the corresponding values for variable Y. These intervals are not a partition of the domain of Z, since the intersection is not empty. The

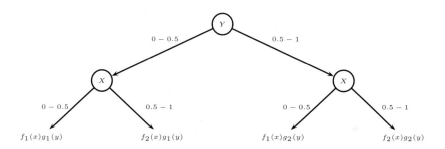

Fig. 6. Tree representing $\phi(x,y)$

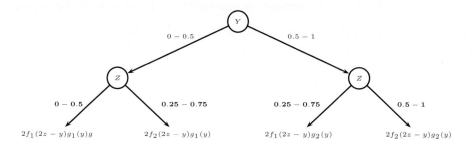

Fig. 7. Tree representing $\phi_2(z,y)$

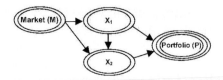

Fig. 8. A simpler model with only 2 stocks

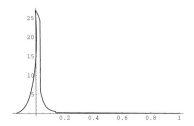

Fig. 9. A new marginal distribution for portfolio return (P) in the MTE model

actual problem is that, after performing the convolution, the resulting potential does not fulfill point ii) in Definition 1, since the joint domain of the variables is not split into hypercubes, and so the potential is not actually an MTE potential, and so the marginalization operation cannot be performed properly.

However, the integral over $\phi_2(z, y)$ can be performed approximately. For example, the marginal for P depicted in Figure 5 was obtained using the software Mathematica 5.2. However, in a different setting, even considering only 2 correlated stocks (see Figure 8), the Mathematica software is not able to obtain an approximation to the solution.

To solve this problem, we have followed a similar approach, that is, approximate the actual distribution of P, but using a Monte Carlo algorithm:

1. Consider N_1 a sub-network by removing from the original one the deterministic node (P in our example).
2. Learn this sub-network from the database.
3. Simulate a large database from this network.
4. Generate the deterministic variable P in the database from the stocks.
5. Learn an MTE distribution for the new variable P in the generated database.

Using this scheme, a new distribution for P can be seen in Figure 9, and we have again computed the probability of a positive return

$$P(p > 0) = \int_0^\infty \phi_6(p) \; dp = 0.513 \;.$$

8   Summary

This paper has demonstrated two applications of models using MTE potentials to stock and portfolio valuation. First, a Naive Bayes and a TAN regression model were compared to a linear regression model as a means of predicting gold mining stock returns based on market and gold returns. Second, a Bayesian network where continuous distributions were approximated by MTE potentials was used to value a portfolio of gold mining stocks. This Bayesian network was similar to the simplest model used by Shenoy and Shenoy [2].

In the first application, the Bayesian network models and the linear model yield similar predictions. An advantage of the Naive Bayes and TAN models with MTE potentials is the ability to calculate probabilities from the marginal asset return distribution. In the second application, use of MTE potentials allows the Bayesian network model to be solved approximately without assuming any standard probability distribution for the independent factors affecting stock returns.

We have also pointed out the problem that arises when handling deterministic relationships between MTE variables, which results in a bad partitioning of the domain of the variables. The problem has been solved by simulation.

A comparison of Bayesian network models with MTE potentials to the more complex models in [2] which consider correlations among residual assets was not performed, but may be the subject of future research in this area. Other future research may compare the performance of the valuation methods presented in this paper to distributions of actual returns using historical data to further determine the viability of these methods as decision-making tools in finance.

Acknowledgments

This work has been supported by the Spanish Ministry of Science and Innovation under project TIN2007-67418-C03-02 and by FEDER funds.

References

1. Ross, S.: The arbitrage theory of capital asset pricing. Journal of Economic Theory 13, 341–360 (1976)
2. Shenoy, C., Shenoy, P.: Bayesian network models of portfolio risk and return. In: Abu-Mostafa, Y.S., LeBaron, B., Lo, W., Weigend, A.S. (eds.) Computational Finance, pp. 85–104. MIT Press, Cambridge (1999)
3. Moral, S., Rumí, R., Salmerón, A.: Mixtures of truncated exponentials in hybrid Bayesian networks. In: Benferhat, S., Besnard, P. (eds.) ECSQARU 2001. LNCS, vol. 2143, pp. 135–143. Springer, Heidelberg (2001)
4. Jensen, F.V.: Bayesian networks and decision graphs. Springer, Heidelberg (2001)
5. Pearl, J.: Probabilistic reasoning in intelligent systems. Morgan Kaufmann, San Mateo (1988)
6. Friedman, N., Geiger, D., Goldszmidt, M.: Bayesian network classifiers. Machine Learning 29, 131–163 (1997)
7. Duda, R., Hart, P., Stork, D.: Pattern classification. Wiley Interscience, Hoboken (2001)
8. Morales, M., Rodríguez, C., Salmerón, A.: Selective naive Bayes for regression using mixtures of truncated exponentials. International Journal of Uncertainty, Fuzziness and Knowledge Based Systems 42, 54–68 (2006)

9. Cobb, B., Shenoy, P., Rumí, R.: Approximating probability density functions with mixtures of truncated exponentials. Statistics and Computing 16, 293–308 (2006)
10. Romero, V., Rumí, R., Salmerón, A.: Learning hybrid Bayesian networks using mixtures of truncated exponentials. International Journal of Approximate Reasoning 42, 54–68 (2006)
11. Fernández, A., Morales, M., Salmerón, A.: Tree augmented naive Bayes for regression using mixtures of truncated exponentials: Application to higher education management. In: Berthold, M., Shawe-Taylor, J., Lavrač, N. (eds.) IDA 2007. LNCS, vol. 4723, pp. 59–69. Springer, Heidelberg (2007)
12. Fernández, A., Salmerón, A.: Extension of Bayesian network classifiers to regression problems. In: Geffner, H., Prada, R., Machado Alexandre, I., David, N. (eds.) IBERAMIA 2008. LNCS (LNAI), vol. 5290, pp. 83–92. Springer, Heidelberg (2008)
13. Shenoy, P., Shafer, G.: Axioms for probability and belief function propagation. In: Shachter, R.D., Levitt, T.S., Kanal, L. (eds.) Uncertainty in Artificial Intelligence, vol. 4, pp. 169–198. North-Holland, Amsterdam (1990)
14. Rumí, R., Salmerón, A.: Approximate probability propagation with mixtures of truncated exponentials. International Journal of Approximate Reasoning 45, 191–210 (2007)
15. Cobb, B., Shenoy, P.: Operations for inference in continuous Bayesian networks with linear deterministic variables. International Journal of Approximate Reasoning 42, 21–36 (2006)
16. Rumí, R., Salmerón, A.: Penniless propagation with mixtures of truncated exponentials. In: Godo, L. (ed.) ECSQARU 2005. LNCS, vol. 3571, pp. 39–50. Springer, Heidelberg (2005)

Non-deterministic Distance Semantics for Handling Incomplete and Inconsistent Data

Ofer Arieli[1] and Anna Zamansky[2]

[1] School of Computer Science, The Academic College of Tel-Aviv, Israel
oarieli@mta.ac.il
[2] School of Computer Science, Tel-Aviv University, Israel
annaz@post.tau.ac.il

Abstract. We introduce a modular framework for formalizing reasoning with incomplete and inconsistent information. This framework is composed of non-deterministic semantic structures and distance-based considerations. The combination of these two principles leads to a variety of entailment relations that can be used for reasoning about non-deterministic phenomena and are inconsistency-tolerant. We investigate the basic properties of these entailments and demonstrate their usefulness in the context of model-based diagnostic systems.

1 Introduction

In this paper, we propose a general framework for representing and reasoning with uncertain information and demonstrate this in the context of model-based diagnostic systems. Our framework consists of two main ingredients:

• *Semantic structures for describing incompleteness:* The principle of truth functionality, according to which the truth-value of a complex formula is uniquely determined by the truth-values of its subformulas, is in an obvious conflict with non-deterministic phenomena and other unpredictable situations in everyday life. To handle this, Avron and Lev [6] introduced *non-deterministic matrices* (Nmatrices), where the value of a complex formula can be chosen *non-deterministically* out of a certain nonempty set of options. This idea turns out to be very useful for providing semantics to logics that handle uncertainty (see [4]). In this paper, we incorporate this idea and consider some additional types of (non-determinisitic) semantic structures for describing incompleteness.

• *Distance-based considerations for handling inconsistency:* Logics induced by Nmatrices are inconsistency-intolerant: whenever a theory has no models in a structure, everything follows from it, and so it becomes useless. To cope with this, we incorporate *distance-based reasoning*, a common technique for reflecting the principle of minimal change in different scenarios where information is dynamically evolving, such as belief revision, data-source mediators, and decision making in the context of social choice theory. Unlike 'standard' semantics, in which conclusions are drawn according to the models of the premises, reasoning

C. Sossai and G. Chemello (Eds.): ECSQARU 2009, LNAI 5590, pp. 793–804, 2009.

in distance-based semantics is based on the valuations that are *'as close as pos-sible'* to the premises, according to a pre-defined metric. As this set of valuations is never empty, reasoning with inconsistent set of premises is not trivialized.

Example 1. Consider the circuit that is represented in Figure 1.

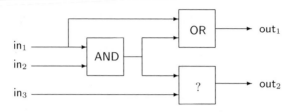

Fig. 1. The circuit of Example 1

Here, partial information (e.g., when it is unknown whether the ?-gate is an AND or an OR gate) may handled by non-deterministic semantics (see Example 7), and conflicting evidences (e.g., that the input line in_1 and the output line out_1 always have opposite values) can be handled by the incorporation of distance-based considerations (see Example 12).

In [2] Nmatrices were first combined with distance considerations and some prop-erties of the resulting framework were investigated. This paper generalizes these results in two aspects: First, we incorporate new types of structures into the framework and study the relations among them. Secondly, we define new meth-ods of constructing distance functions, tailored specifically for non-deterministic semantics, some of them are a conservative extension of well-known distances used in the classical case. The robustness of what is obtained for reasoning with uncertainty is demonstrated in the context of model-based diagnosis.

2   Semantic Structures for Incomplete Data

2.1   Preliminaries

Below, $\mathcal{L}$ denotes a propositional language with a set $\mathcal{W}_{\mathcal{L}} = \{\psi, \phi, \ldots\}$ of well-formed formulas. Atoms $= \{p, q, r \ldots\}$ are the atomic formulas in $\mathcal{W}_{\mathcal{L}}$. A theory Γ is a finite set of formulas in $\mathcal{W}_{\mathcal{L}}$. Atoms$(\Gamma)$ and SF(Γ) denote, respectively, the atoms appearing in the formulas of Γ, and the subformulas of Γ.

Given a propositional language $\mathcal{L}$, a *propositional logic* is a pair $\langle \mathcal{L}, \vdash \rangle$, where $\vdash$ is a consequence relation for $\mathcal{L}$, as defined below:

Definition 1. A (Tarskian) *consequence relation* for $\mathcal{L}$ is a binary relation $\vdash$ between sets of formulas in $\mathcal{W}_{\mathcal{L}}$ and formulas in $\mathcal{W}_{\mathcal{L}}$, satisfying:

Reflexivity: if $\psi \in \Gamma$ then $\Gamma \vdash \psi$.
Monotonicity: if $\Gamma \vdash \psi$ and $\Gamma \subseteq \Gamma'$, then $\Gamma' \vdash \psi$.
Transitivity: if $\Gamma \vdash \psi$ and $\Gamma', \psi \vdash \varphi$ then $\Gamma, \Gamma' \vdash \varphi$.

2.2   Matrices, Nmatrices and Their Families

We start with the simplest semantic structures used for defining logics: *many-valued (deterministic) matrices* (see, e.g., [12] and [14]).

Definition 2. A (deterministic) *matrix* for $\mathcal{L}$ is a tuple $\mathcal{M} = \langle \mathcal{V}, \mathcal{D}, \mathcal{O} \rangle$, where $\mathcal{V}$ is a non-empty set of truth values, $\mathcal{D}$ is a non-empty proper subset of $\mathcal{V}$, called the *designated* elements of $\mathcal{V}$, and for every n-ary connective $\diamond$ of $\mathcal{L}$, $\mathcal{O}$ includes an n-ary function $\widetilde{\diamond}_{\mathcal{M}} : \mathcal{V}^n \to \mathcal{V}$.

A matrix $\mathcal{M}$ induces the usual semantic notions: An $\mathcal{M}$-*valuation* for $\mathcal{L}$ is a function $\nu : \mathcal{W}_{\mathcal{L}} \to \mathcal{V}$ such that for each n-ary connective $\diamond$ of $\mathcal{L}$ and every $\psi_1, \ldots, \psi_n \in \mathcal{W}_{\mathcal{L}}$, $\nu(\diamond(\psi_1, \ldots, \psi_n)) = \widetilde{\diamond}(\nu(\psi_1), \ldots, \nu(\psi_n))$. We denote by $\Lambda^{\mathrm{s}}_{\mathcal{M}}$ the set of all the $\mathcal{M}$-valuations of $\mathcal{L}$.[1] A valuation $\nu \in \Lambda^{\mathrm{s}}_{\mathcal{M}}$ is an $\mathcal{M}$-*model* of ψ (or $\mathcal{M}$-*satisfies* ψ), if it belongs to $mod^{\mathrm{s}}_{\mathcal{M}}(\psi) = \{\nu \in \Lambda^{\mathrm{s}}_{\mathcal{M}} \mid \nu(\psi) \in \mathcal{D}\}$. A formula ψ is $\mathcal{M}$-*satisfiable* if $mod^{\mathrm{s}}_{\mathcal{M}}(\psi) \neq \emptyset$ and it is an $\mathcal{M}$-*tautology* if $mod^{\mathrm{s}}_{\mathcal{M}}(\psi) = \Lambda^{\mathrm{s}}_{\mathcal{M}}$. The $\mathcal{M}$-models of a theory Γ are the elements of the set $mod^{\mathrm{s}}_{\mathcal{M}}(\Gamma) = \cap_{\psi \in \Gamma} mod^{\mathrm{s}}_{\mathcal{M}}(\psi)$.

Definition 3. The relation $\vdash^{\mathrm{s}}_{\mathcal{M}}$ that is induced by a matrix $\mathcal{M}$ is defined for every theory Γ and formula $\psi \in \mathcal{W}_{\mathcal{L}}$ by $\Gamma \vdash^{\mathrm{s}}_{\mathcal{M}} \psi$ if $mod^{\mathrm{s}}_{\mathcal{M}}(\Gamma) \subseteq mod^{\mathrm{s}}_{\mathcal{M}}(\psi)$.

It is well-known that $\vdash^{\mathrm{s}}_{\mathcal{M}}$ is a consequence relation in the sense of Definition 1.

Deterministic matrices do not always faithfully represent incompleteness. This brings us to the second type of structures, called *non-deterministic matrices*, where the truth-value of a complex formula is chosen non-deterministically out of a set of options.

Definition 4. [6] A *non-deterministic matrix* (Nmatrix) for $\mathcal{L}$ is a tuple $\mathcal{N} = \langle \mathcal{V}, \mathcal{D}, \mathcal{O} \rangle$, where $\mathcal{V}$ is a non-empty set of truth values, $\mathcal{D}$ is a non-empty proper subset of $\mathcal{V}$, and for every n-ary connective $\diamond$ of $\mathcal{L}$, $\mathcal{O}$ includes an n-ary function $\widetilde{\diamond}_{\mathcal{N}} : \mathcal{V}^n \to 2^{\mathcal{V}} \setminus \{\emptyset\}$.

Example 2. Consider an AND-gate, $\diamond_1$, that operates correctly when its inputs have the same value and is unpredictable otherwise, and another gate, $\diamond_2$, that operates correctly, but it is not known whether its is an OR or a XOR gate. These gates may described by the following non-deterministic truth-tables:

$\widetilde{\diamond}_1$	t	f
t	{t}	{t, f}
f	{t, f}	{f}

$\widetilde{\diamond}_2$	t	f
t	{t, f}	{t}
f	{t}	{f}

Non-determinism can be incorporated into the truth-tables of the connectives by either a *dynamic* [6] or a *static* [5] approach, as defined below.

Definition 5. Let $\mathcal{N}$ be an Nmatrix for $\mathcal{L}$.

– A *dynamic* $\mathcal{N}$-*valuation* is a function $\nu : \mathcal{W}_{\mathcal{L}} \to \mathcal{V}$ that satisfies the following condition for every n-ary connective $\diamond$ of $\mathcal{L}$ and every $\psi_1, \ldots, \psi_n \in \mathcal{W}_{\mathcal{L}}$:

$$\nu(\diamond(\psi_1, \ldots, \psi_n)) \in \widetilde{\diamond}_{\mathcal{N}}(\nu(\psi_1), \ldots, \nu(\psi_n)). \tag{1}$$

---

[1] The 's', standing for 'static' semantics, is for uniformity with later notations.

- A *static $\mathcal{N}$-valuation* is a function $\nu : \mathcal{W}_\mathcal{L} \to \mathcal{V}$ that satisfies condition (1) and the following compositionality principle: for every n-ary connective $\diamond$ of $\mathcal{L}$ and every $\psi_1, \ldots, \psi_n, \phi_1, \ldots, \phi_n \in \mathcal{W}_\mathcal{L}$,

$$\text{if } \forall\, 1 \le i \le n\ \nu(\psi_i) = \nu(\phi_i), \text{ then } \nu(\diamond(\psi_1, \ldots, \psi_n)) = \nu(\diamond(\phi_1, \ldots \phi_n)). \quad (2)$$

We denote by $\Lambda_\mathcal{N}^d$ the space of the dynamic $\mathcal{N}$-valuations and by $\Lambda_\mathcal{N}^s$ the static $\mathcal{N}$-valuations. Clearly, $\Lambda_\mathcal{N}^s \subseteq \Lambda_\mathcal{N}^d$.

In both of the semantics considered above, the truth-value $\nu(\diamond(\psi_1, \ldots, \psi_n))$ assigned to the formula $\diamond(\psi_1, \ldots, \psi_n)$ is selected non-deterministically from a set of possible truth-values $\widetilde{\diamond}(\nu(\psi_1), \ldots, \nu(\psi_n))$. In the dynamic approach this selection is made separately, independently for each tuple $\langle \psi_1, \ldots, \psi_n \rangle$, and $\nu(\psi_1), \ldots, \nu(\psi_n)$ do not uniquely determine $\nu(\diamond(\psi_1, \ldots, \psi_n))$. In the static semantics this choice is made globally, and so the interpretation of $\diamond$ is a function.

Note 1. In ordinary (deterministic) matrices each $\widetilde{\diamond}$ is a function having singleton values only (thus it can be treated as a function $\widetilde{\diamond} : \mathcal{V}^n \to \mathcal{V}$). In this case the sets of static and dynamic valuations coincide, as we have full determinism.

Example 3. Consider the circuit of Figure 2.

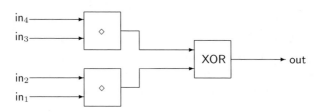

Fig. 2. A circuit of Example 3

If both of the $\diamond$ components implement the same Boolean function, which is unknown to the reasoner, the static approach would be more appropriate. In this case, for instance, whenever the inputs of these components are the same (that is, $\mathsf{in}_1 = \mathsf{in}_3$ and $\mathsf{in}_2 = \mathsf{in}_4$), the outputs will be the same as well, and so the output line (out) of the circuit will be turned off.

If, in addition, each one of these components has its own unpredictable behaviour, the dynamic semantics would be more appropriate. In this case, for instance, the outputs of the $\diamond$-components need not be the same for the same inputs, and so the value of the circuit's output line cannot be predicted either.[2]

---

[2] Also, in Example 2, the situation represented by $\widetilde{\diamond}_1$ is more suitable for dynamic semantics, while the one represented by $\widetilde{\diamond}_2$ is more adequate for the static semantics.

Definition 6. Let $\mathcal{N}$ be an Nmatrix for $\mathcal{L}$.

- The *dynamic models* of ψ and Γ are defined, respectively, by:
  $mod_{\mathcal{N}}^{\mathsf{d}}(\psi) = \{\nu \in \Lambda_{\mathcal{N}}^{\mathsf{d}} \mid \nu(\psi) \in \mathcal{D}\}$ and $mod_{\mathcal{N}}^{\mathsf{d}}(\Gamma) = \cap_{\psi \in \Gamma} mod_{\mathcal{N}}^{\mathsf{d}}(\psi)$.
- The consequence relation induced by the dynamic semantics of $\mathcal{N}$ is
  $\Gamma \vdash_{\mathcal{N}}^{\mathsf{d}} \psi$ if $mod_{\mathcal{N}}^{\mathsf{d}}(\Gamma) \subseteq mod_{\mathcal{N}}^{\mathsf{d}}(\psi)$.
- The corresponding definitions for the static semantics are defined similarly, replacing d in the previous items by s.

Again, it is easily verified that $\vdash_{\mathcal{N}}^{\mathsf{d}}$ and $\vdash_{\mathcal{N}}^{\mathsf{s}}$ are consequence relations for $\mathcal{L}$.

Note 2. It is important to observe that by Note 1, if $\mathcal{N}$ is a deterministic Nmatrix and $\mathcal{M}$ is its corresponding (standard) matrix, it holds that $\vdash_{\mathcal{N}}^{\mathsf{d}} = \vdash_{\mathcal{N}}^{\mathsf{s}} = \vdash_{\mathcal{M}}^{\mathsf{s}}$.

Example 4. Consider again the circuit of Figure 2. The theory below represents this circuit and the assumption that both of the $\diamond$-gates have the same input:

$$\Gamma = \left\{ \mathsf{out} \leftrightarrow (\mathsf{in}_1 \diamond \mathsf{in}_2) \oplus (\mathsf{in}_3 \diamond \mathsf{in}_4), \; \mathsf{in}_1 \leftrightarrow \mathsf{in}_3, \; \mathsf{in}_2 \leftrightarrow \mathsf{in}_4 \right\}.$$

Suppose now that $\mathcal{N}$ is a two-valued non-deterministic matrix in which $\leftrightarrow$ and $\oplus$ have the standard interpretations for double-arrow and xor, and $\diamond$ has the truth-table of $\diamond_2$ in Example 2. Denote by t and f the propositional constants that are always assigned the truth-values t and f, respectively. Then $\Gamma \vdash_{\mathcal{N}}^{\mathsf{s}} \mathsf{out} \leftrightarrow \mathsf{f}$, while $\Gamma \nvdash_{\mathcal{N}}^{\mathsf{d}} \mathsf{out} \leftrightarrow \mathsf{f}$ (consider a valuation $\nu \in \Lambda_{\mathcal{N}}^{\mathsf{d}}$ such that $\nu(\mathsf{out}) = \nu(\mathsf{in}_i) = \mathsf{t}$ for $1 \leq i \leq 4$, and $\nu(\mathsf{in}_1 \diamond \mathsf{in}_2) = \mathsf{t}$ but $\nu(\mathsf{in}_3 \diamond \mathsf{in}_4) = \mathsf{f}$; see also Example 3).

A natural question to ask at this stage is whether logics induced by non-deterministic matrices are representable by (finite) deterministic matrices. The answer is negative for dynamic semantics (Proposition 1) and is positive for static semantics (Proposition 2). To show this, we use yet another type of semantic structures, which is a simplification of the notion of a *family of matrices* of [14].

Definition 7. A *family of matrices* is a finite set of deterministic matrices $\mathcal{F} = \{\mathcal{M}_1, \ldots, \mathcal{M}_k\}$, where $\mathcal{M}_i = \langle \mathcal{V}, \mathcal{D}, \mathcal{O}_i \rangle$ for all $1 \leq i \leq k$. An $\mathcal{F}$-valuation is any $\mathcal{M}_i$-valuation for $i \in \{1, \ldots, k\}$. We denote $\Lambda_{\mathcal{F}}^{\mathsf{s}} = \cup_{1 \leq i \leq k} \Lambda_{\mathcal{M}_i}^{\mathsf{s}}$. The relation $\vdash_{\mathcal{F}}^{\mathsf{s}}$ that is induced by $\mathcal{F}$ is defined by: $\Gamma \vdash_{\mathcal{F}}^{\mathsf{s}} \psi$ if $\Gamma \vdash_{\mathcal{M}}^{\mathsf{s}} \psi$ for every $\mathcal{M} \in \mathcal{F}$.

Example 5. The circuit of Figure 1 may be represented as follows:

$$\Gamma = \{\mathsf{out}_1 \leftrightarrow (\mathsf{in}_1 \wedge \mathsf{in}_2) \vee \mathsf{in}_1 \;, \; \mathsf{out}_2 \leftrightarrow (\mathsf{in}_1 \wedge \mathsf{in}_2) \diamond \mathsf{in}_3\}.$$

Suppose that the connectives in Γ are interpreted by a family $\mathcal{F}$ of matrices with the standard meanings of $\wedge$, $\vee$, and $\leftrightarrow$, and the following interpretations for $\diamond$:

$\tilde{\diamond}_1$	t	f		$\tilde{\diamond}_2$	t	f		$\tilde{\diamond}_3$	t	f		$\tilde{\diamond}_4$	t	f
t	t	t		t	t	f		t	t	t		t	t	f
f	t	f		f	f	f		f	f	f		f	t	f

In this case we have, for instance, that $\Gamma \vdash_{\mathcal{F}}^{\mathsf{s}} \mathsf{out}_1 \leftrightarrow \mathsf{in}_1$, but $\Gamma \nvdash_{\mathcal{F}}^{\mathsf{s}} \mathsf{out}_2 \leftrightarrow \mathsf{in}_2$ (a counter-model assigns f to in_2, t to in_3, t to out_2, and interprets $\diamond$ by $\tilde{\diamond}_1$).

Lemma 1. *For a family $\mathcal{F}$ of matrices, denote $mod_{\mathcal{F}}^{s}(\psi) = \{\nu \in \Lambda_{\mathcal{F}}^{s} \mid \nu(\psi) \in \mathcal{D}\}$ and $mod_{\mathcal{F}}^{s}(\Gamma) = \cap_{\psi \in \Gamma} mod_{\mathcal{F}}^{s}(\psi)$. Then $\Gamma \vdash_{\mathcal{F}}^{s} \psi$ iff $mod_{\mathcal{F}}^{s}(\Gamma) \subseteq mod_{\mathcal{F}}^{s}(\psi)$.* [3]

Corollary 1. *For a family $\mathcal{F}$ of matrices, $\vdash_{\mathcal{F}}^{s}$ is a consequence relation for $\mathcal{L}$.*

The next proposition, generalizing [6, Theorem 3.4], shows that dynamic Nmatrices characterize logics that are not characterizable by ordinary matrices.

Proposition 1. *Let $\mathcal{N}$ be a two-valued Nmatrix with at least one proper non-deterministic operation. Then there is no family of matrices $\mathcal{F}$ such that $\vdash_{\mathcal{N}}^{d} = \vdash_{\mathcal{F}}^{s}$.*

In static semantics the situation is different, as reasoning with $\vdash_{\mathcal{N}}^{s}$ can be simulated by a family of ordinary matrices. To show this, we need the following:

Definition 8. [4] *Let $\mathcal{N}_1 = \langle \mathcal{V}_1, \mathcal{D}_1, \mathcal{O}_1 \rangle$ and $\mathcal{N}_2 = \langle \mathcal{V}_2, \mathcal{D}_2, \mathcal{O}_2 \rangle$ be Nmatrices for $\mathcal{L}$. $\mathcal{N}_1$ is called a simple refinement of $\mathcal{N}_2$ if $\mathcal{V}_1 = \mathcal{V}_2$, $\mathcal{D}_1 = \mathcal{D}_2$, and $\tilde{\diamond}_{\mathcal{N}_1}(\overline{x}) \subseteq \tilde{\diamond}_{\mathcal{N}_2}(\overline{x})$ for every n-ary connective $\diamond$ of $\mathcal{L}$ and every tuple $\overline{x} \in \mathcal{V}^n$.*

Intuitively, an Nmatrix refines another Nmatrix if the former is more restricted than the latter in the non-deterministic choices of its operators.

Definition 9. *For an Nmatrix $\mathcal{N}$, the family of matrices $\looparrowright\mathcal{N}$ is the set of all the deterministic matrices that are simple refinements of $\mathcal{N}$. A family of matrices $\mathcal{F}$ for $\mathcal{L}$ is called Cartesian, if there is some Nmatrix $\mathcal{N}$ for $\mathcal{L}$, such that $\mathcal{F} = \looparrowright\mathcal{N}$.*

Example 6. Consider the Nmatrix $\mathcal{N}$ for describing $\diamond_1$ in Example 2. Then $\looparrowright\mathcal{N}$ is the (Cartesian) family of the four deterministic matrices in Example 5.

Proposition 2. *For every Nmatrix $\mathcal{N}$ it holds that $\vdash_{\mathcal{N}}^{s} = \vdash_{\looparrowright\mathcal{N}}^{s}$.*

Proposition 2 shows that Nmatrices are representable by Cartesian families of deterministic matrices. Yet, there are useful families that are not Cartesian:

Example 7. Suppose that a gate $\diamond$ is either an AND or an OR gate, but it is not known which one. This situation *cannot* be represented by truth table of $\tilde{\diamond}_1$ in Example 2, as in both static and dynamic semantics the two choices for $\tilde{\diamond}_1(t, f)$ are completely independent of the choices for $\tilde{\diamond}_1(f, t)$. What we need is a more precise representation that makes choices between two *deterministic matrices*, each one of which represents a possible behaviour of the unknown gate. Thus, among the four matrices of Example 5, only the first *two* faithfully describe $\diamond$:

$$\mathcal{F} = \left\{ \begin{array}{c|cc} \tilde{\diamond}_1 & t & f \\ \hline t & t & t \\ f & t & f \end{array} \quad , \quad \begin{array}{c|cc} \tilde{\diamond}_2 & t & f \\ \hline t & t & f \\ f & f & f \end{array} \right\}$$

We now combine the concepts of Nmatrices and of their families.

---

[3] Due to a lack of space proofs are omitted. For full proofs see the longer version of the paper in http://www2.mta.ac.il/~oarieli/, or ask the first author.

Definition 10. A *family of Nmatrices* is a finite set $\mathcal{G} = \{\mathcal{N}_1, \ldots, \mathcal{N}_k\}$ of Nmatrices, where $\mathcal{N}_i = \langle \mathcal{V}, \mathcal{D}, \mathcal{O}_i \rangle$ for all $1 \leq i \leq k$.[4] A $\mathcal{G}$-valuation is any $\mathcal{N}_i$-valuation for $i \in \{1, \ldots, k\}$. For $\mathrm{x} \in \{\mathrm{d}, \mathrm{s}\}$, we denote $\Lambda_{\mathcal{G}}^{\mathrm{x}} = \cup_{1 \leq i \leq n} \Lambda_{\mathcal{N}_i}^{\mathrm{x}}$, and define: $\Gamma \vdash_{\mathcal{G}}^{\mathrm{x}} \psi$ if $\Gamma \vdash_{\mathcal{N}}^{\mathrm{x}} \psi$ for every $\mathcal{N} \in \mathcal{G}$.

Lemma 2. *Let* $\mathcal{G} = \{\mathcal{N}_1, \ldots, \mathcal{N}_k\}$ *be a family of Nmatrices. For* $\mathrm{x} \in \{\mathrm{d}, \mathrm{s}\}$, *denote* $mod_{\mathcal{G}}^{\mathrm{x}}(\psi) = \{\nu \in \Lambda_{\mathcal{G}}^{\mathrm{x}} \mid \nu(\psi) \in \mathcal{D}\}$ *and* $mod_{\mathcal{G}}^{\mathrm{x}}(\Gamma) = \cap_{\psi \in \Gamma} mod_{\mathcal{G}}^{\mathrm{x}}(\psi)$. *Then* $\Gamma \vdash_{\mathcal{G}}^{\mathrm{x}} \psi$ *iff* $mod_{\mathcal{G}}^{\mathrm{x}}(\Gamma) \subseteq mod_{\mathcal{G}}^{\mathrm{x}}(\psi)$.

Corollary 2. *Both of* $\vdash_{\mathcal{G}}^{\mathrm{d}}$ *and* $\vdash_{\mathcal{G}}^{\mathrm{s}}$ *are consequence relations for* $\mathcal{L}$.

Concerning the simulation of $\vdash_{\mathcal{G}}^{\mathrm{x}}$ by other consequence relations, note that:

(a) In the dynamic case we have already seen that even logics induced by a single Nmatrix cannot be simulated by a family of ordinary matrices.

(b) In the static case, logics induced by a family of Nmatrices can be simulated using a family of ordinary matrices:

Proposition 3. *For every family of Nmatrices* $\mathcal{G}$ *there is a family of matrices* $\mathcal{F}$ *such that* $\vdash_{\mathcal{G}}^{\mathrm{s}} = \vdash_{\mathcal{F}}^{\mathrm{s}}$.

2.3 Hierarchy of the Two-Valued Semantic Structures

In the rest of the paper we focus on the two-valued case, using a language $\mathcal{L}$ that includes the propositional constants $\mathbf{t}$ and $\mathbf{f}$. We shall also use a meta-variable $\mathfrak{M}$ that ranges over the two-valued structures defined above, and the metavariable x that ranges over $\{\mathrm{s}, \mathrm{d}\}$, denoting the restriction on valuations. Accordingly, $\Lambda_{\mathfrak{M}}^{\mathrm{x}}$ and $mod_{\mathfrak{M}}^{\mathrm{x}}(\psi)$ denote, respectively, the relevant space of valuations and the models of ψ. The following conventions will be useful in what follows:

- An M-logic is a logic that is induced by a (standard) two-valued matrix. The class of M-logics is denoted by $\mathbf{M}$.
- An SN-logic (resp., a DN-logic) is a logic based on a static (resp., a dynamic) two-valued Nmatrix. The class of SN-logics (DN-logics) is denoted $\mathbf{SN}$ ($\mathbf{DN}$).
- An F-logic is a logic that is induced by a family of two-valued matrices. The corresponding class of F-logics is denoted by $\mathbf{F}$.
- An SG-logic (DG-logic) is a logic based on a family of static (dynamic) two-valued Nmatrices. The class of SG-logics (DG-logics) is denoted $\mathbf{SG}$ ($\mathbf{DG}$).

For relating the classes of logics above, we need the following proposition.

Proposition 4. *Let* $\mathcal{F}$ *be a family of matrices for* $\mathcal{L}$ *with standard negation and conjunction. Then* $\mathbf{L} = \langle \mathcal{L}, \vdash_{\mathcal{F}}^{\mathrm{s}} \rangle$ *is an SN-logic iff* $\mathcal{F}$ *is Cartesian.*

Example 8. The family of matrices $\mathcal{F}$ in Example 7 (enriched with classical negation and conjunction) is not Cartesian and so, by Propositions 1 and 4, it is not representable by a (finite) non-deterministic matrix.

Theorem 1. *In the notations above, we have that:* (a) $\mathbf{M} = \mathbf{DN} \cap \mathbf{SN}$, (b) $\mathbf{SN} \subsetneq \mathbf{F}$, (c) $\mathbf{F} \not\subset \mathbf{DN}$, (d) $\mathbf{SG} = \mathbf{F}$, *and* (e) $\mathbf{DN} \subsetneq \mathbf{DG}$

A graphic representation of Theorem 1 is given in Figure 3.

---

[4] To the best of our knowledge, these structures have not been considered yet.

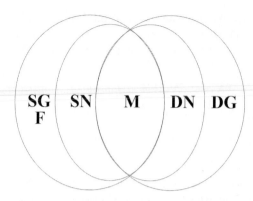

Fig. 3. Relations among the different classes of logics

3   Distance Semantics for Inconsistent Data

A major drawback of the logics considered above is that they do not tolerate inconsistency properly. Indeed, if Γ is not $\mathfrak{M}$-consistent, then $\Gamma \vdash^{\times}_{\mathfrak{M}} \psi$ for every ψ. To overcome this, we incorporate distance-based considerations. The idea is simply to define a distance-like measurement between valuations and theories, and for drawing conclusions, to consider the valuations that are 'closest' to the premises. This intuition is formalized in [2] for deterministic matrices and for Nmatrices under two-valued dynamic semantics only. It can also be viewed as a kind of a *preferential semantics* [13]. Below, we extend this method to all the semantic structures of Section 2. We also introduce a new method for constructing distances, which allows us to define a wide range of distance-based entailments.

3.1   Distances between Valuations

Definition 11. A *pseudo-distance* on a set S is a total function $d : S \times S \to \mathbb{R}^+$ that is symmetric ($\forall \nu, \mu \in S\ d(\nu, \mu) = d(\mu, \nu)$) and preserves identity ($\forall \nu, \mu \in S\ d(\nu, \mu) = 0$ iff $\nu = \mu$). A pseudo-distance d is a *distance (metric)* on S if it also satisfies the has the triangular inequality ($\forall \nu, \mu, \sigma \in S\ d(\nu, \sigma) \le d(\nu, \mu) + d(\mu, \sigma)$).

Example 9. The following functions are two common distances on the space of the two-valued valuations.

– *The drastic distance:* $d_U(\nu, \mu) = 0$ if $\nu = \mu$, otherwise $d_U(\nu, \mu) = 1$.
– *The Hamming distance:* $d_H(\nu, \mu) = |\{p \in \text{Atoms} \mid \nu(p) \ne \mu(p)\}|.$[5]

These distances can be applied on *any* space of static valuations (see also Note 3 below).

In the context of non-deterministic semantics, one needs to be more cautious in defining distances, as two dynamic valuations can agree on all the atoms of a complex formula, but still assign two different values to that formula. Therefore, complex formulas should also be taken into account in the distance definitions,

---

[5] Note that this definition assumes a finite number of atomic formulas in the language.

but there are infinitely many of them to consider. To handle this, we restrict the distance computations to some *context*, i.e., to a certain set of relevant formulas.[6]

Definition 12. A *context* C is a finite set of formulas closed under subformulas. The *restriction to* C of $\nu \in \Lambda_{\mathfrak{M}}^{\mathsf{x}}$ is a valuation $\nu^{\downarrow C}$ on C, such that $\nu^{\downarrow C}(\psi) = \nu(\psi)$ for every ψ in C. The restriction to C of $\Lambda_{\mathfrak{M}}^{\mathsf{x}}$ is the set $\Lambda_{\mathfrak{M}}^{\mathsf{x} \downarrow C} = \{\nu^{\downarrow C} \mid \nu \in \Lambda_{\mathfrak{M}}^{\mathsf{x}}\}$.

Distances between valuations are now defined as follows:

Definition 13. Let $\mathfrak{M}$ be a semantic structure, $\mathsf{x} \in \{\mathsf{d}, \mathsf{s}\}$, and d a function on $\bigcup_{\{C = \mathsf{SF}(\Gamma) \mid \Gamma \subseteq \mathcal{W}_{\mathcal{L}}\}} \Lambda_{\mathfrak{M}}^{\mathsf{x} \downarrow C} \times \Lambda_{\mathfrak{M}}^{\mathsf{x} \downarrow C}$.

- The *restriction* of d to C is a function $d^{\downarrow C}$ s.t. $\forall \nu, \mu \in \Lambda_{\mathfrak{M}}^{\mathsf{x} \downarrow C}$, $d^{\downarrow C}(\nu, \mu) = d(\nu, \mu)$.
- d is a *generic (pseudo) distance* on $\Lambda_{\mathfrak{M}}^{\mathsf{x}}$ if for every context C, $d^{\downarrow C}$ is a (pseudo) distance on $\Lambda_{\mathfrak{M}}^{\mathsf{x} \downarrow C}$.

General Constructions of Generic Distances

We now introduce a general method of constructing generic distances. These constructions include the functions of Example 9 as particular cases of generic distances, restricted to the context $C = \mathsf{Atoms}$ (see Note 3 and Proposition 6).

Definition 14. A *numeric aggregation function* is a complete mapping f from multisets of real numbers to real numbers, such that: (a) f is non-decreasing in the values of the elements of its argument, (b) $f(\{x_1, \ldots, x_n\}) = 0$ iff $x_1 = x_2 = \ldots x_n = 0$, and (c) $f(\{x\}) = x$ for every $x \in \mathbb{R}$.

As we aggregate non-negative (distance) values, functions that meet the conditions in Definition 14 are, e.g., summation, average, and the maximum.

Definition 15. Let $\mathfrak{M}$ be a (two-valued) structure, C a context, and $\mathsf{x} \in \{\mathsf{d}, \mathsf{s}\}$. For every $\psi \in C$, define the function $\bowtie^{\psi} \colon \Lambda_{\mathfrak{M}}^{\mathsf{x} \downarrow C} \times \Lambda_{\mathfrak{M}}^{\mathsf{x} \downarrow C} \to \{0, 1\}$ as follows:

- for $v_1, v_2 \in \{\mathsf{t}, \mathsf{f}\}$, let $\nabla(v_1, v_2) = 0$ if $v_1 = v_2$, otherwise $\nabla(v_1, v_2) = 1$.
- for an atomic formula p, let $\bowtie^{p}(\nu, \mu) = \nabla(\nu(p), \mu(p))$
- for a formula $\psi = \diamond(\psi_1, \ldots, \psi_n)$, define

$$\bowtie^{\psi}(\nu, \mu) = \begin{cases} 1 & \text{if } \nu(\psi) \neq \mu(\psi) \text{ but } \forall i \ \nu(\psi_i) = \mu(\psi_i), \\ 0 & \text{otherwise.} \end{cases}$$

For an aggregation g, define the following functions from $\Lambda_{\mathfrak{M}}^{\mathsf{x} \downarrow C} \times \Lambda_{\mathfrak{M}}^{\mathsf{x} \downarrow C}$ to $\mathbb{R}^{+}$:

- $d_{\nabla, g}^{\downarrow C}(\nu, \mu) = g(\{\nabla(\nu(\psi), \mu(\psi)) \mid \psi \in C\})$,
- $d_{\bowtie, g}^{\downarrow C}(\nu, \mu) = g(\{\bowtie^{\psi}(\nu, \mu) \mid \psi \in C\})$.

Proposition 5. *Both of $d_{\nabla, g}^{\downarrow C}$ and $d_{\bowtie, g}^{\downarrow C}$ are pseudo-distances on $\Lambda_{\mathfrak{M}}^{\mathsf{x} \downarrow C}$.*

The difference between $d_{\nabla, g}^{\downarrow C}$ and $d_{\bowtie, g}^{\downarrow C}$ is that while $d_{\nabla, g}$ compares *truth assignments*, $d_{\bowtie, g}$ compares (non-deterministic) *choices* (see also Example 10).

---

[6] Thus, unlike [1,8] and other frameworks that use distances as those of Example 9, we will not need the rather restricting assumption that the number of atoms in the language is finite.

Note 3. The pseudo distances defined above generalize those of Example 9:
- Both $d_{\nabla,\max}$ and $d_{\bowtie,\max}$ are natural generalizations of d_U. Moreover, for any $\nu,\mu \in \Lambda^s_{\mathfrak{M}}$ and finite set Atoms, $d_U(\nu,\mu) = d_{\nabla,\max}^{\downarrow\text{Atoms}}(\nu,\mu) = d_{\bowtie,\max}^{\downarrow\text{Atoms}}(\nu,\mu)$.
- Both $d_{\nabla,\Sigma}$ and $d_{\bowtie,\Sigma}$ are natural generalizations of d_H. Moreover, for any $\nu,\mu \in \Lambda^s_{\mathfrak{M}}$ and finite set Atoms, $d_H(\nu,\mu) = d_{\nabla,\Sigma}^{\downarrow\text{Atoms}}(\nu,\mu) = d_{\bowtie,\Sigma}^{\downarrow\text{Atoms}}(\nu,\mu)$.

Proposition 6. *If* $g(\{x_1,\ldots,x_n,0\}) = g(\{x_1,\ldots,x_n\})$ *for all* $x_1,\ldots,x_n \in \{0,1\}$, *then* $d_{\bowtie,g}^{\downarrow C}(\nu,\mu) = d_{\bowtie,g}^{\downarrow\text{Atoms}(C)}(\nu,\mu)$.

By Proposition 5, generic pseudo distances may be constructed as follows:

Proposition 7. *For an aggregation function* g, *define the following functions:*

$$d_{\nabla,g}(\nu,\mu) = g\big(\{\nabla(\nu(\psi),\mu(\psi)) \mid \psi \in C\}\big), \tag{3}$$

$$d_{\bowtie,g}(\nu,\mu) = g\big(\{\bowtie^\psi(\nu,\mu) \mid \psi \in C\}\big). \tag{4}$$

Then $d_{\nabla,g}$ *and* $d_{\bowtie,g}$ *are generic pseudo distances on* $\Lambda^x_{\mathfrak{M}}$.

3.2 Distance-Based Entailments

We now use the distances between valuations for defining entailments relations.

Definition 16. A (semantical) *setting* for $\mathcal{L}$ is a tuple $\mathcal{S} = \langle \mathfrak{M}, (d,\mathsf{x}), f \rangle$, where $\mathfrak{M}$ is a structure, d is a generic pseudo distance on $\Lambda^x_{\mathfrak{M}}$ for some $\mathsf{x} \in \{\mathsf{d},\mathsf{s}\}$, and f is an aggregation function.

A setting identifies the underlying semantics, and can be used for measuring the correspondence between valuations and theories.

Definition 17. Given a setting $\mathcal{S} = \langle \mathfrak{M}, (d,\mathsf{x}), f \rangle$ define

$$- \ d^{\downarrow C}(\nu,\psi_i) = \begin{cases} \min\{d^{\downarrow C}(\nu^{\downarrow C}, \mu^{\downarrow C}) \mid \mu \in mod^x_{\mathfrak{M}}(\psi_i)\} & \text{if } mod^x_{\mathfrak{M}}(\psi_i) \neq \emptyset, \\ 1 + \max\{d^{\downarrow C}(\mu_1^{\downarrow C}, \mu_2^{\downarrow C}) \mid \mu_1,\mu_2 \in \Lambda^x_{\mathfrak{M}}\} & \text{otherwise.} \end{cases}$$

$$- \ \delta_{d,f}^{\downarrow C}(\nu,\Gamma) = f(\{d^{\downarrow C}(\nu,\psi_1),\ldots,d^{\downarrow C}(\nu,\psi_n)\}).$$

The intuition here is to measure how 'close' a valuation is to satisfying a formula and a theory. To be faithful to this intuition, we are interested only in contexts where the distance between a formula and its model is zero, and is strictly positive otherwise.

Proposition 8. *Let* $\mathfrak{M}$ *be a semantic structure,* C *a context, and* $\mathsf{x} \in \{\mathsf{d},\mathsf{s}\}$.
- *If* $\text{Atoms}(\psi) \subseteq C$, *then* $d^{\downarrow C}(\nu,\psi) = 0$ *iff* $\nu \in mod^s_{\mathfrak{M}}(\psi)$.
- *If* $\text{SF}(\psi) \subseteq C$, *then* $d^{\downarrow C}(\nu,\psi) = 0$ *iff* $\nu \in mod^d_{\mathfrak{M}}(\psi)$.

It follows that the most appropriate contexts to use are the following:

Definition 18. Given a setting $\mathcal{S} = \langle \mathfrak{M}, (d,\mathsf{x}), f \rangle$, denote:

$$\mathsf{C}_\mathsf{x}(\Gamma) = \begin{cases} \text{Atoms}(\Gamma) & \text{if } \mathsf{x} = \mathsf{s}, \\ \text{SF}(\Gamma) & \text{if } \mathsf{x} = \mathsf{d}. \end{cases}$$

Definition 19. The *most plausible valuations* of Γ with respect to a semantic setting $\mathcal{S} = \langle \mathfrak{M}, (d, \mathsf{x}), f \rangle$ are the elements of the following set:

$$\Delta_{\mathcal{S}}(\Gamma) = \begin{cases} \{ \nu \in \Lambda^{\mathsf{x}}_{\mathfrak{M}} \mid \forall \mu \in \Lambda^{\mathsf{x}}_{\mathfrak{M}} \; \delta^{\downarrow C_{\mathsf{x}}(\Gamma)}_{d,f}(\nu, \Gamma) \leq \delta^{\downarrow C_{\mathsf{x}}(\Gamma)}_{d,f}(\mu, \Gamma) \} & \text{if } \Gamma \neq \emptyset, \\ \Lambda^{\mathsf{x}}_{\mathfrak{M}} & \text{otherwise.} \end{cases}$$

Example 10. Let $\mathcal{N}$ be an Nmatrix that interprets negation in the standard way, and $\diamond$ according to $\diamond_1$ in Example 2. Then $\Gamma = \{p, q, \neg(p \diamond q)\}$ is not $\mathcal{N}$-satisfiable, and $mod^{\mathsf{d}}_{\mathcal{N}}(\Gamma) = \emptyset$. Consider now the settings $\mathcal{S}_1 = \langle \mathcal{N}, (d_{\nabla, \Sigma}, \mathsf{d}), \Sigma \rangle$ and $\mathcal{S}_2 = \langle \mathcal{N}, (d_{\bowtie, \Sigma}, \mathsf{d}), \Sigma \rangle$, where $d_{\nabla, \Sigma}$ and $d_{\bowtie, \Sigma}$ are, respectively, the generic distances defined in (3) and (4). Then:

	p	q	$p \diamond q$	$\neg(p \diamond q)$	$\delta^{\downarrow C_{\mathsf{d}}}_{\mathcal{S}_1}(\nu_i, \Gamma)$	$\delta^{\downarrow C_{\mathsf{d}}}_{\mathcal{S}_2}(\nu_i, \Gamma)$
ν_1	t	t	t	f	3	1
ν_2	t	f	t	f	3	2
ν_3	t	f	f	t	1	1
ν_4	f	t	t	f	3	2
ν_5	f	t	f	t	1	1
ν_6	f	f	f	t	2	2

and so $\Delta_{\mathcal{S}_1}(\Gamma) = \{\nu_3, \nu_5\}$ and $\Delta_{\mathcal{S}_2}(\Gamma) = \{\nu_1, \nu_3, \nu_5\}$.

Proposition 9. *For every* $\mathcal{S} = \langle \mathfrak{M}, (d, \mathsf{x}), f \rangle$ *and* Γ, $\Delta_{\mathcal{S}}(\Gamma)$ *is nonempty. If* Γ *is* $\mathfrak{M}$-*satisfiable*, $\Delta_{\mathcal{S}}(\Gamma) = mod^{\mathsf{x}}_{\mathfrak{M}}(\Gamma)$.

Next, we formalize the idea that, according distance-based entailments, conclusions should follow from all of the most plausible valuations of the premises.

Definition 20. For $\mathcal{S} = \langle \mathfrak{M}, (d, \mathsf{x}), f \rangle$, denote: $\Gamma \hspace{1pt}\vdash\hspace{-9pt}\sim_{\mathcal{S}} \psi$ if $\Delta_{\mathcal{S}}(\Gamma) \subseteq mod^{\mathsf{x}}_{\mathfrak{M}}(\psi)$ or $\Gamma = \{\psi\}$.

Example 11. In Example 10, under the standard interpretation of disjunction, $\Gamma \hspace{1pt}\vdash\hspace{-9pt}\sim_{\mathcal{S}_1} \neg p \vee \neg q$ while $\Gamma \hspace{1pt}\not\vdash\hspace{-9pt}\sim_{\mathcal{S}_2} \neg p \vee \neg q$.

Example 12. Consider the $\mathcal{F}$-consistent theory Γ of Example 5 that represents the circuit of Figure 1. Learning that lines in_1 and out_1 always have opposite values, the revised theory, $\Gamma' = \Gamma \cup \{\mathsf{out}_1 \leftrightarrow \neg \mathsf{in}_1\}$, is not $\mathcal{F}$-satisfiable anymore, so $\vdash_{\mathcal{F}}$ is useless for making plausible conclusions from Γ'. However, using the setting $\mathcal{S} = \langle \mathcal{F}, (d_{\nabla, \Sigma}, \mathsf{s}), \Sigma \rangle$, or $\mathcal{S} = \langle \mathcal{F}, (d_{\bowtie, \Sigma}, \mathsf{s}), \Sigma \rangle$, it can be verified that:

- The assertion $\mathsf{out}_1 \leftrightarrow (\mathsf{in}_1 \wedge \mathsf{in}_2) \vee \mathsf{in}_1$ is falsified by some most plausible valuations of Γ', and so, while $\Gamma \vdash_{\mathcal{F}} \mathsf{out}_1 \leftrightarrow \mathsf{in}_1$, we have $\Gamma' \hspace{1pt}\not\vdash\hspace{-9pt}\sim_{\mathcal{S}} \mathsf{out}_1 \leftrightarrow \mathsf{in}_1$.
- The assertion $\mathsf{out}_2 \leftrightarrow (\mathsf{in}_1 \wedge \mathsf{in}_2) \diamond \mathsf{in}_3$ is validated by all the most plausible valuations of Γ', and so, despite the $\mathcal{F}$-inconsistency of Γ', the information about the relation between out_2 and in_1, in_2 may be retained.

The distance-based entailments defined above generalize the usual methods for distance-based reasoning in the context of deterministic matrices. This includes, among others, the operators in [8,10,11] and the distance-based entailments for deterministic matrices in [1,3]. The entailment $\hspace{0.3em}\sim_S$ for Nmatrices and dynamic valuations is studied in [2]. To the best of our knowledge, distance entailments for Nmatrices and static valuations, and entailments based on families of matrices and (static or dynamic) Nmatrices, have not been considered before.

Theorem 2. *Let $S = \langle \mathfrak{M}, (d, x), f \rangle$. For every $\mathfrak{M}$-consistent theory Γ, it holds that $\Gamma \hspace{0.3em}\sim_S \psi$ iff $\Gamma \vdash_{\mathfrak{M}}^{x} \psi$.*

Theorem 3. *Let $S = \langle \mathfrak{M}, (d, x), f \rangle$ be a setting in which f is hereditary.[7] Then $\hspace{0.3em}\sim_S$ is a* cautious consequence relation, *i.e., it has the following properties:*

Cautious Reflexivity: $\psi \hspace{0.3em}\sim_S \psi$

Cautious Monotonicity [9]:   *if $\Gamma \hspace{0.3em}\sim_S \psi$ and $\Gamma \hspace{0.3em}\sim_S \phi$, then $\Gamma, \psi \hspace{0.3em}\sim_S \phi$.*

Cautious Transitivity [7]:   *if $\Gamma \hspace{0.3em}\sim_S \psi$ and $\Gamma, \psi \hspace{0.3em}\sim_S \phi$, then $\Gamma \hspace{0.3em}\sim_S \phi$.*

References

1. Arieli, O.: Distance-based paraconsistent logics. International Journal of Approximate Reasoning 48(3), 766–783 (2008)
2. Arieli, O., Zamansky, A.: Reasoning with uncertainty by Nmatrix–Metric semantics. In: Hodges, W., de Queiroz, R. (eds.) Logic, Language, Information and Computation. LNCS, vol. 5110, pp. 69–82. Springer, Heidelberg (2008)
3. Arieli, O., Zamansky, A.: Some simplified forms of reasoning with distance-based entailments. In: Bergler, S. (ed.) Canadian AI. LNCS (LNAI), vol. 5032, pp. 36–47. Springer, Heidelberg (2008)
4. Avron, A.: Non-deterministic semantics for familes of paraconsistent logics. In: Beziau, J.-Y., Carnielli, W., Gabbay, D.M. (eds.) Handbook of Paraconsistency. Studies in Logic, vol. 9, pp. 285–320. College Publications (2007)
5. Avron, A., Konikowska, B.: Multi-valued calculi for logics based on non-determinism. Logic Journal of the IGPL 13(4), 365–387 (2005)
6. Avron, A., Lev, I.: Non-deterministic multi-valued structures. Journal of Logic and Computation 15, 241–261 (2005)
7. Gabbay, D.: Theoretical foundation for non-monotonic reasoning, Part II: Structured non-monotonic theories. In: Proc. SCAI 1991. IOS Press, Amsterdam (1991)
8. Konieczny, S., Pino Pérez, R.: Merging information under constraints: a logical framework. Journal of Logic and Computation 12(5), 773–808 (2002)
9. Kraus, S., Lehmann, D., Magidor, M.: Nonmonotonic reasoning, preferential models and cumulative logics. Artificial Intelligence 44(1-2), 167–207 (1990)
10. Lin, J., Mendelzon, A.O.: Knowledge base merging by majority. In: Dynamic Worlds: From the Frame Problem to Knowledge Management. Kluwer, Dordrecht (1999)
11. Ravesz, P.: On the semantics of theory change: arbitration between old and new information. In: Proc. PODS 1993, pp. 71–92. ACM Press, New York (1993)
12. Rosser, J., Turquette, A.R.: Many-Valued Logics. North-Holland, Amsterdam (1952)
13. Shoham, Y.: Reasoning about Change. MIT Press, Cambridge (1988)
14. Urquhart, A.: Many-valued logic. In: Gabbay, D., Guenthner, F. (eds.) Handbook of Philosophical Logic, vol. II, pp. 249–295. Kluwer, Dordrecht (2001)

---

[7] An aggregation function f is *hereditary*, if $f(\{x_1, \ldots, x_n\}) < f(\{y_1, \ldots, y_n\})$ implies that $f(\{x_1, \ldots, x_n, z\}) < f(\{y_1, \ldots, y_n, z\})$.

A Simple Modal Logic for Reasoning about Revealed Beliefs

Mohua Banerjee[1,*] and Didier Dubois[2]

[1] Department of Mathematics and Statistics,
Indian Institute of Technology, Kanpur 208 016, India
[2] Institut de Recherche en Informatique de Toulouse, CNRS
Université de Toulouse, France
mohua@iitk.ac.in, dubois@irit.fr

Abstract. Even though in Artificial Intelligence, a set of classical logical formulae is often called a belief base, reasoning about beliefs requires more than the language of classical logic. This paper proposes a simple logic whose atoms are beliefs and formulae are conjunctions, disjunctions and negations of beliefs. It enables an agent to reason about some beliefs of another agent as revealed by the latter. This logic, called MEL, borrows its axioms from the modal logic KD, but it is an encapsulation of propositional logic rather than an extension thereof. Its semantics is given in terms of subsets of interpretations, and the models of a formula in MEL is a family of such non-empty subsets. It captures the idea that while the consistent epistemic state of an agent about the world is represented by a non-empty subset of possible worlds, the meta-epistemic state of another agent about the former's epistemic state is a family of such subsets. We prove that any family of non-empty subsets of interpretations can be expressed as a single formula in MEL. This formula is a symbolic counterpart of the Möbius transform in the theory of belief functions.

1 Motivation

Formal models of interaction between agents are the subject of current significant research effort. One important issue is to represent how an agent can reason about another agent's knowledge and beliefs. Consider two agents $\mathcal{E}$ (for emitter) and $\mathcal{R}$ (for receiver). Agent $\mathcal{E}$ supplies pieces of information to agent $\mathcal{R}$, explaining what (s)he believes and what (s)he thinks is only plausible or conceivable. For instance, $\mathcal{E}$ is a witness and $\mathcal{R}$ collects his or her testimony. How can agent $\mathcal{R}$ reason about what $\mathcal{E}$ accepts to tell the former, that is, $\mathcal{E}$'s revealed beliefs? On this basis, how can $\mathcal{R}$ decide that $\mathcal{E}$ believes or not a prescribed statement? It is supposed that $\mathcal{E}$ provides some pieces of information of the form *I believe* α, *I am not sure about* β, to $\mathcal{R}$. The question is: how can $\mathcal{R}$ reconstruct the epistemic state of $\mathcal{E}$ from this information?

---

* The author acknowledges the support of Université Paul Sabatier, Toulouse during a visit at IRIT where the work was initiated.

C. Sossai and G. Chemello (Eds.): ECSQARU 2009, LNAI 5590, pp. 805–816, 2009.

In this paper, the information provided by agent $\mathcal{E}$ will be represented in a minimal modal logic, sufficient to let agent $\mathcal{R}$ reason about it. A formula α in a propositional belief base, understood as a belief, is encoded by $\Box\alpha$ in our logic. This is in contrast to, e.g. belief revision literature [8], where beliefs are represented by formulas in Propositional Logic (PL), keeping the modality implicit. A set of formulae in this language is called a *meta-belief base*, because it represents what $\mathcal{R}$ knows about $\mathcal{E}$'s beliefs. However, the nesting of modalities is not allowed because we are not concerned with introspective reasoning of $\mathcal{R}$ about his or her own beliefs. Some minimal axioms are proposed in such a way that the fragment of this modal logic restricted to propositions of the form $\Box\alpha$ is isomorphic to propositional logic, if the $\Box$ operator is dropped. We call the resulting logic a *Meta-Epistemic Logic* (MEL) so as to emphasize the fact that we deal with how an agent reasons about what (s)he knows about the beliefs of another agent.

At the semantic level, agent $\mathcal{E}$ has incomplete knowledge about the real world, which can be represented by a subset E of interpretations of $\mathcal{E}$'s language, one and only one of which is true. This subset is not empty as long as the epistemic state of agent $\mathcal{E}$ is consistent, which is assumed here. All agent $\mathcal{R}$ knows about $\mathcal{E}$'s epistemic state stems from what $\mathcal{E}$ told him or her. So $\mathcal{R}$ has incomplete knowledge about $\mathcal{E}$'s epistemic state E. The epistemic state of an agent regarding another agent's beliefs is what we call a *meta-epistemic state*. The meta-epistemic state of $\mathcal{R}$ (about $\mathcal{E}$'s beliefs) built from $\mathcal{E}$'s statements can be represented by a family $\mathcal{F}$ of non-empty subsets of the set $\mathcal{V}$ of all propositional valuations (models), one and only one of which is the actual epistemic state of $\mathcal{E}$. Moreover, any such family $\mathcal{F}$ can stand for a meta-epistemic state. In order not to confuse models of propositional formulae with models of MEL formulae, we call the latter *meta-models* since they are non-empty subsets of interpretations. Indeed, models of α in PL and $\Box\alpha$ in MEL have a different nature, the use of meta-models enabling more expressiveness, such as making the difference between $\Box(\alpha \vee \beta)$ and $\Box\alpha \vee \Box\beta$ (the last one being impossible to encode in a belief base). The encoding of a belief as α instead of $\Box\alpha$, also leads to a confusion between $\neg\Box\alpha(\equiv \Diamond\neg\alpha)$ and $\Box\neg\alpha$. In MEL their sets of meta-models are again different.

The paper demonstrates that the semantics of MEL exactly corresponds to meta-epistemic states modelled by families of non-empty subsets of propositional valuations. So, MEL can account for any consistent meta-epistemic state of an agent about another agent. Related works are discussed further and perspectives are outlined. In particular, an important connection is made between MEL and belief functions. Indeed, a meta-epistemic state can be viewed as the set of focal sets of a belief function. It is shown that the formula in MEL that exactly accounts for a *complete* meta-epistemic state (when the epistemic state of the emitter is precisely known by the receiver) can be retrieved by means of the Möbius transform of the belief function. This result looks promising for extending MEL to uncertainty theories. Most proofs are omitted due to the lack of space.

2   The Logic MEL

Let us consider classical propositional logic PL, with (say) k propositional variables, $p_1, \ldots, p_k$, and propositional constant $\top$. A propositional valuation, as usual, is a map $w : PV \to \{0, 1\}$, where $PV := \{p_1, \ldots, p_k\}$. $\mathcal{V}$, as mentioned in Section 1, denotes the set of all propositional valuations. For a PL-formula α, $w \models \alpha$ indicates that w satisfies α or w is a *model* of α, i.e. $w(\alpha) = 1$ (true). If $w \models \alpha$ for every α in a set Γ of PL-formulae, we write $w \models \Gamma$. $[\alpha] := \{w : w \models \alpha\}$, is the set of models of α. Let E denote the epistemic state of an agent $\mathcal{E}$. We assume that an epistemic state is represented by a subset of propositional valuations, understood as a disjunction thereof. Each valuation represents a 'possible world' consistent with the epistemic state of $\mathcal{E}$. So, $E \subseteq \mathcal{V}$, and it is further assumed that E is non-empty. Note that, for any E, $|E| \leq 2^k$.

2.1   The Language for MEL

The base is PL, and $\alpha, \beta \ldots$ denote PL-formulae. We add the unary connective $\Box$ to the PL-alphabet. *Atomic formulae* of MEL are of the form $\Box\alpha$, $\alpha \in PL$. The set of MEL-formulae, denoted $\phi, \psi \ldots$, is then generated from the set At of atomic formulae, with the help of the Boolean connectives $\neg, \wedge$:

$$MEL := \Box\alpha \mid \neg\phi \mid \phi \wedge \psi.$$

One defines the connective $\vee$ and the modality $\Diamond$ in MEL in the usual way. Namely $\phi \vee \psi := \neg(\neg\phi \wedge \neg\psi)$ and $\Diamond\alpha := \neg\Box\neg\alpha$, where $\alpha \in PL$. Like $\Box$, modality $\Diamond$ applies only on PL-formulae. It should be noticed that PL-formulae are not MEL-formulae, and that iteration of the modal operators $\Box, \Diamond$ is not allowed in MEL (as explained in Section 1).

An agent $\mathcal{E}$ provides some information about his or her beliefs about the outside world to another agent $\mathcal{R}$ by means of the above language. Any set Γ of formulae in this language is interpreted as what an agent $\mathcal{E}$ declares to another agent $\mathcal{R}$. It forms the meta-belief base possessed by $\mathcal{R}$; on this basis, agent $\mathcal{R}$ tries to reconstruct the epistemic state of the other agent. Some of the basic statements that agent $\mathcal{E}$ can express in this language are as follows.

- For any propositional formula α, $\Box\alpha \in \Gamma$ means agent $\mathcal{E}$ declares that (s)he believes α is true.
- $\Diamond\alpha \in \Gamma$ means agent $\mathcal{E}$ declares that, to him or her, α is *possibly* true, that is (s)he has no argument as to the falsity of α. Note that this is equivalent to $\neg\Box\neg\alpha \in \Gamma$, that is, all that $\mathcal{R}$ can conclude is that either $\mathcal{E}$ believes α is true, or ignores whether α is true or not. So, $\Diamond\alpha$ cannot be interpreted as a belief, but rather as an expression of partial ignorance.
- $\Diamond\alpha \wedge \Diamond\neg\alpha \in \Gamma$ means agent $\mathcal{E}$ declares to ignore whether α is true or not.
- $\Box\alpha \vee \Box\neg\alpha \in \Gamma$ means agent $\mathcal{E}$ says (s)he knows whether α is true or not, but prefers not to reveal it.

2.2   The Semantics

For a given agent $\mathcal{E}$, we define satisfaction of MEL-formulae recursively, as follows. $\Box\alpha \in At$, ϕ, ψ are MEL-formulae, and E is the epistemic state of an agent $\mathcal{E}$. Note that $\emptyset \neq E \subseteq \mathcal{V}$, the set of all propositional valuations.

- $E \models \Box\alpha$, if and only if $E \subseteq [\alpha]$.
- $E \models \neg\phi$, if and only if $E \not\models \phi$.
- $E \models \phi \wedge \psi$, if and only if $E \models \phi$ and $E \models \psi$.

It is clear that in the logic MEL, the meta-models, i.e. non-empty sets of valuations, play the same role as propositional valuations in classical logic. $E \models \Box\alpha$ means that in the epistemic state E, agent $\mathcal{E}$ believes α. Viewed from agent $\mathcal{R}$, if agent $\mathcal{E}$ declares (s)he believes α (i.e. $\Box\alpha \in \Gamma$), any E such that $E \models \Box\alpha$, is a possible epistemic state of $\mathcal{E}$. It is then clear that $E \models \Diamond\alpha$, if and only if $E \cap [\alpha] \neq \emptyset$, i.e. there is at least one possible world for agent $\mathcal{E}$, where α holds. If $\Diamond\alpha \in \Gamma$, it means that agent $\mathcal{E}$ declares that α is plausible (or conceivable) in the sense that there is no reason to disbelieve α. As a consequence, the epistemic state of $\mathcal{E}$ is known by agent $\mathcal{R}$ to be consistent with $[\alpha]$. Note that $\Box\alpha \vee \Box\neg\alpha \in \Gamma$ is not tautological. Generally, in the case of a disjunction $\Box\alpha \vee \Box\beta$, the only corresponding possible epistemic states form the set $\{E \subseteq [\alpha]\} \cup \{E \subseteq [\beta]\}$. It is clearly more informative than $\Box(\alpha \vee \beta)$, since the latter allows epistemic states where none of α or β can be asserted. As usual, we have the notion of semantic equivalence of formulae:

Definition 1. ϕ *is semantically equivalent to* ψ, *written* $\phi \equiv \psi$, *if for any epistemic state* E, $E \models \phi$, *if and only if* $E \models \psi$.

If Γ is a set of MEL-formulae, $E \models \Gamma$ means $E \models \phi$, for each $\phi \in \Gamma$. So the set of meta-models of Γ, which may be denoted $\mathcal{F}_\Gamma$, is precisely $\{E : E \models \Gamma\}$. Now $\mathcal{R}$ can reason about what is known from agent $\mathcal{E}$'s assertions:

Definition 2. *For any set* $\Gamma \cup \{\phi\}$ *of* MEL-*formulae,* ϕ *is a semantic consequence of* Γ, *written* $\Gamma \models_{MEL} \phi$, *provided for every epistemic state* E, $E \models \Gamma$ *implies* $E \models \phi$.

For any family $\mathcal{F}$ of sets of propositional valuations, $\mathcal{F} \models \phi$ means that for each $E \in \mathcal{F}$, $E \models \phi$. A natural extension gives the notation $\mathcal{F} \models \Gamma$, for any set Γ of MEL-formulae. So, for instance, $\mathcal{F}_\Gamma \models \Gamma$.

3   Axiomatization

For any set Γ of PL-formulae, $\Gamma \vdash \alpha$ denotes that α is a syntactic PL-consequence of Γ. And $\vdash \alpha$ indicates that α is a PL-theorem. For $\alpha, \beta \in PL, \phi, \psi, \mu \in MEL$, we consider the following KD-style axioms and rule of inference.

Axioms:

$(PL):$ $\phi \rightarrow (\psi \rightarrow \phi)$; $(\phi \rightarrow (\psi \rightarrow \mu)) \rightarrow ((\phi \rightarrow \psi) \rightarrow (\phi \rightarrow \mu))$;
    $(\neg\phi \rightarrow \neg\psi) \rightarrow (\psi \rightarrow \phi)$.

(RM) : $\Box\alpha \to \Box\beta$, whenever $\vdash \alpha \to \beta$.
(M) : $\Box(\alpha \wedge \beta) \to (\Box\alpha \wedge \Box\beta)$.
(C) : $(\Box\alpha \wedge \Box\beta) \to \Box(\alpha \wedge \beta)$.
(N) : $\Box\top$.
(D) : $\Box\alpha \to \Diamond\alpha$.

Rule: (MP) If $\phi, \phi \to \psi$ then ψ.

Observing valid formulae and rules in MEL indeed suggests that the modal system KD may provide an axiomatization for it – we establish this formally. Axioms $(RM), (M), (C), (N)$ mean that agent $\mathcal{E}$ is logically sophisticated, in the classical sense, i.e. the agent $\mathcal{R}$ assumes that $\mathcal{E}$ is a propositional logic reasoner. In particular, it means that $\mathcal{E}$ believes tautologies of the propositional calculus. Moreover, if $\mathcal{E}$ claims to believe α and to believe β, this is equivalent to believing their conjunction. It is thus that $\mathcal{E}$ follows (RM) as well: if it is true that $\alpha \to \beta$ and $\mathcal{E}$ believes α, (s)he must believe β. This is also the symbolic counterpart of the monotonicity of numerical belief measures for events, in the sense of set-inclusion. Axiom (D) comes down to considering that asserting the certainty of α is stronger than asserting its plausibility (it requires non-empty metamodels E). It is also a counterpart of numerical inequality between belief and plausibility functions [16], necessity and possibility measures [6] etc. in uncertainty theories. Finally, (PL) and (MP) enable agent $\mathcal{R}$ to infer from agent $\mathcal{E}$'s publicly declared beliefs, so as to reconstruct a picture of the latter agent's epistemic state. Syntactically, MEL's axioms can be viewed as a Boolean version of those of the fuzzy logic of necessities briefly suggested by Hájek [10].

Taking any set of MEL-formulae, one defines a compact syntactic consequence in MEL (written $\vdash_{MEL}$), in the standard way. Soundness of MEL w.r.t the semantics of Section 2.2, is then easy to obtain. Using soundness we get the following result, which demonstrates that deriving a $\Box$-formula, say $\Box\alpha$, in MEL is equivalent to deriving α in PL. It may be noted that the result was proved in [5] for the modal system $KD45$ having the standard Kripke semantics. The proof is immediately carried over to MEL. In fact, it holds for the MEL-fragment containing $\Box$-formulae and only their conjunctions.

For any set Γ of PL-formulae, let $\Box\Gamma := \{\Box\beta : \beta \in \Gamma\}$.

Theorem 1. $\Box\Gamma \vdash_{MEL} \Box\alpha$, *if and only if* $\Gamma \vdash \alpha$.

From the point of view of application, this result means that agent $\mathcal{R}$ can reason about $\mathcal{E}$'s beliefs (leaving statements of ignorance aside) as if they were $\mathcal{R}$'s own beliefs. In case $\Box\Gamma \vdash_{MEL} \Box\alpha$, if agent $\mathcal{R}$ were asked whether $\mathcal{E}$ believes α from what $\mathcal{E}$ previously declared to believe ($\Box\Gamma$), the former's answer would be yes because $\mathcal{E}$ would reason likewise about α. By virtue of Theorem 1, one may say that propositional logic PL is *encapsulated* in MEL; MEL is not a modal extension of PL: it is a two-tiered logic.

We recall that a Kripke model [13] for the system KD, is a triple $M := (U, R, V)$, where the accessibility relation R is *serial*. $M, u \models \phi$ denotes that the KD-formula ϕ is satisfied at $u(\in U)$ by V, i.e. $V(\phi, u) = 1$. The possibility of considering simplified models of modal systems like $S5$ and $KD45$, omitting the accessibility relation in Kripke structures (assuming all possible worlds are related), is pointed out in [12] p. 62. It is interesting to see that an analogous result may be obtained for the MEL-fragment of KD.

Proposition 1. *Let $M := (U, R, V)$ be a KD-Kripke model and $u \in U$. Then there is a structure $M_0 := (U_0, R_0, V_0)$ with $R_0 := U_0 \times U_0$, and a state $u_0 \in U_0$ such that for any MEL-formula ϕ, $M, u \models \phi$, if and only if $M_0, u_0 \models \phi$.*

So we may omit the accessibility relation R_0 and obtain a simpler structure (U_0, V_0) that suffices for consideration of satisfiability of MEL-formulae in terms of Kripke models. In fact, the MEL-semantics achieves this in an even simpler manner, as we do not have to deal with the valuation V_0 either. This is because, the following two key results establishing a passage to and from the MEL semantics and Kripke semantics, yield Proposition 1: (i) For any KD-Kripke model M and $u \in U$, there is an epistemic state E_u such that for any MEL-formula ϕ, $M, u \models \phi$, if and only if $E_u \models_{MEL} \phi$; (ii) every epistemic state E gives a KD-Kripke model M_E such that for any MEL-formula ϕ, $E \models_{MEL} \phi$, if and only if for every $w \in E$, $M_E, w \models \phi$.

These two results also give the completeness theorem for MEL.

Theorem 2. *(Completeness) If $\Gamma \models_{MEL} \phi$ then $\Gamma \vdash_{MEL} \phi$.*

4   The Logical Characterization of Meta-epistemic States

Let $\mathcal{F}$ be any collection of non-empty sets of propositional valuations, representing the meta-epistemic state of an agent regarding another agent's beliefs. It is shown here that a MEL-formula $\delta_{\mathcal{F}}$ may be defined such that : (i) $\mathcal{F}$ satisfies $\delta_{\mathcal{F}}$; (ii) furthermore, if $\mathcal{F}$ satisfies any set Γ' of MEL-formulae, the syntactic consequences of Γ' must already be consequences of $\delta_{\mathcal{F}}$. So the MEL-formula $\delta_{\mathcal{F}}$ completely characterizes the meta-epistemic state $\mathcal{F}$. For this purpose, we follow the line of characterization of Kripke frames by Jankov-Fine formulae (cf. [1]). Here, a Jankov-Fine kind of formula for any non-empty epistemic state is considered, keeping in mind the correspondence with the simpler Kripke frame (with universal accessibility relation), outlined at the end of Section 3. The formula is then extended naturally to a collection $\mathcal{F}$ of non-empty epistemic states.

4.1   Syntactic Representation of Meta-epistemic States

Let $E \subseteq \mathcal{V}, E \neq \emptyset$. Further, let $\alpha_E := \bigvee_{w \in E} \alpha_w$, where α_w is the PL-formula characterizing w, i.e. $\alpha_w := \bigwedge_{w(p)=1} p \ \wedge \ \bigwedge_{w(p)=0} \neg p$, where p ranges over PV. Observe that $E \models \Diamond \alpha_w$ if and only if $w \in E$, since $[\alpha_w] = \{w\}$. On the other hand, $E \models \Box \alpha_w$, if and only if $E = \{w\}$, since $E \neq \emptyset$. Consider now a meta-epistemic state, say the collection $\mathcal{F} := \{E_1, \ldots, E_n\}$, where the E_i's are non-empty sets of propositional valuations. Note that $|\mathcal{F}| \leq 2^{2^k - 1}$.

In order to *exactly* describe $\mathcal{F}$, we need a MEL-formula such that it is satisfied by all members of $\mathcal{F}$ *only*. In particular, it must *not* be satisfied by

(a) sets having elements from outside $\bigcup \mathcal{F}$,

(b) sets of valuations lying within $\bigcup \mathcal{F}$, but not equal to any of the E_i's,

(c) especially, subsets of members of $\mathcal{F}$.

Such a (non-unique) MEL-formula is denoted $\delta_\mathcal{F}$ and if $\mathcal{F} := \{E\}$, $\delta_\mathcal{F}$ is denoted δ_E. Now for any epistemic state $E := \{w_1, \ldots, w_m\}$, consider δ_E to be the conjunction of (i) $\Box(\alpha_{w_1} \vee \ldots \vee \alpha_{w_m})$ and (ii) $\Diamond \alpha_{w_i}$, $i = 1, \ldots, m$, i.e.,

$$\delta_E := \Box \alpha_E \wedge \bigwedge_{w \in E} \Diamond \alpha_w.$$

Then $E \models_{MEL} \delta_E$, and it is easy to check that for any epistemic state E',

Observation 1. $E' \models_{MEL} \delta_E$, *if and only if* $E' = E$.

A natural extension to the general case, where $\mathcal{F} := \{E_1, \ldots, E_n\}$ is a collection of mutually exclusive epistemic states, gives

Definition 3. $\delta_\mathcal{F} := \bigvee_{1 \leq i \leq n} \delta_{E_i}$.

Thus we see that the set of meta-models of $\delta_\mathcal{F}$ is precisely $\mathcal{F}$, and any consequence of sets of formulae satisfied by all epistemic states of $\mathcal{F}$, is also a consequence of $\delta_\mathcal{F}$.

Theorem 3

  (a) $\mathcal{F} \models_{MEL} \delta_\mathcal{F}$, i.e. for each $E_i \in \mathcal{F}$, $E_i \models_{MEL} \delta_\mathcal{F}$.

  (b) If $\mathcal{F}'$ is any other meta-epistemic state such that $\mathcal{F}' \models_{MEL} \delta_\mathcal{F}$, $\mathcal{F}' \subseteq \mathcal{F}$.

  (c) If Γ' is a set of MEL-formulae such that $\mathcal{F} \models_{MEL} \Gamma'$, $\Gamma' \vdash_{MEL} \phi$ would imply $\{\delta_\mathcal{F}\} \vdash_{MEL} \phi$, for any MEL-formula ϕ.

Proof. (c) Suppose $\Gamma' \vdash_{MEL} \phi$, and let $E \models_{MEL} \delta_\mathcal{F}$. By part (b) of this theorem, $E \in \mathcal{F}$. Then $E \models_{MEL} \Gamma'$, by assumption. Soundness of MEL gives $\Gamma' \models_{MEL} \phi$, and so $E \models_{MEL} \phi$. Thus $\{\delta_\mathcal{F}\} \models_{MEL} \phi$, and by completeness of MEL, we get the result. ∎

4.2 The Meta-models of Meta-belief Bases

Conversely, let Γ be any set of MEL-formulae representing a meta-belief base. We consider the family $\mathcal{F}_\Gamma$ of all meta-models (sets of propositional valuations) of Γ (cf. Section 2.2), $\mathcal{F}_\Gamma := \{E \subseteq \mathcal{V} : \emptyset \neq E \models \Gamma\}$. If $\Gamma := \{\phi\}$, we write $\mathcal{F}_\phi$.

The following theorem extends the classical properties of semantic entailment over to meta-models. It is the companion of Theorem 3. We see that $\mathcal{F}_\Gamma$ is the maximal set of meta-models of Γ that satisfies *precisely* the consequences of Γ.

Theorem 4

  (a) If Γ' is any set of MEL-formulae such that $\mathcal{F}_\Gamma \models_{MEL} \Gamma'$, $\Gamma' \vdash_{MEL} \phi$ would imply $\Gamma \vdash_{MEL} \phi$, for any MEL-formula ϕ.

  (b) Let $Con(\Gamma) := \{\phi : \Gamma \vdash_{MEL} \phi\}$ and $Th(\mathcal{F}_\Gamma) := \{\phi : \mathcal{F}_\Gamma \models \phi\}$. Then $Con(\Gamma) = Th(\mathcal{F}_\Gamma)$.

Definition 3 proposes an encoding of a meta-epistemic state into a MEL formula. We can also obtain the set of meta-models of any meta-belief base. We can now iterate the construction. It shows the bijection between classes of semantically equivalent formulae in MEL and sets of non-empty subsets of valuations.

Theorem 5

> (a)   If $\Gamma \cup \{\phi\}$ is any set of MEL-formulae, $\Gamma \vdash_{MEL} \phi$, if and only if $\{\delta_{\mathcal{F}_\Gamma}\} \vdash_{MEL} \phi$. In other words, the MEL-consequence sets of Γ and $\delta_{\mathcal{F}_\Gamma}$ are identical: $Con(\Gamma) = Con(\delta_{\mathcal{F}_\Gamma})$.
>
> (b)   If $\mathcal{F}$ is any collection of non-empty sets of propositional valuations, $\mathcal{F} = \mathcal{F}_{\delta_{\mathcal{F}}}$.

This result shows that MEL can precisely account for families of non-empty subsets of valuations. Moreover, the following bijections can be established.

Corollary 1

> (a)   The Boolean algebra on the set of MEL-formulae quotiented by semantical equivalence $\equiv$, is isomorphic to the power set Boolean algebra with domain $2^{2^{\mathcal{V}} \setminus \{\emptyset\}}$. The correspondence, for any MEL-formula ϕ, is given by: $[\phi]_{\equiv} \mapsto \mathcal{F}_\phi$.
>
> (b)   There is a bijection between the set of all meta-epistemic states and the set of all belief sets of MEL, i.e. Γ such that $Con(\Gamma) = \Gamma$. For any family $\mathcal{F}$, the correspondence is given by: $\mathcal{F} \mapsto Con(\delta_{\mathcal{F}})$.

5   From Meta-epistemic States to Belief Functions

A connection between MEL and belief functions was pointed out in the Introduction. A belief function [16] Bel is a non-additive monotonic set-function (a capacity) with domain $2^{\mathcal{V}}$ and range in the unit interval, that is super-additive at any order (also called ∞-monotone), that is, it verifies a relaxed version of the additivity axiom of probability measures. The degree of belief $Bel(A)$ in a proposition A evaluates to what extent this proposition is logically implied by the available evidence. The plausibility function $Pl(A) := 1 - Bel(A^c)$ evaluates to what extent events are consistent with the available evidence. The pair (Bel, Pl) can be viewed as quantitative randomized versions of KD modalities $(\Box, \Diamond)$ [17]. Interestingly, elementary forms of belief functions arose first, in the works of Bernoulli, for the modeling of unreliable testimonies [16], while MEL encodes the testimony of an agent. Function Bel can be mathematically defined from a (generally finite) random set on $\mathcal{V}$, that has a very specific interpretation. A so-called basic assignment $m(E)$ is assigned to each subset E of $\mathcal{V}$, and is such that $m(E) \geq 0$, for all $E \subseteq \mathcal{V}$ and $\sum_{E \subseteq \mathcal{V}} m(E) = 1$.

The degree $m(E)$ is understood as the weight given to the fact that all an agent knows is that the value of the variable of interest lies somewhere in set E, and nothing else. In other words, the probability allocation $m(E)$ could eventually be shared between elements of E, but remains suspended for lack of knowledge.

For instance, agent $\mathcal{R}$ receives a testimony in the form of a statement $\Box\alpha$ such that $E = [\alpha]$; $m(E)$ reflects the probability that E correctly represents the available knowledge. A set E such that $m(E) > 0$ is called a focal set. In the absence of conflicting information it is generally assumed that $m(\emptyset) = 0$. It is then clear that a collection of focal sets is a meta-epistemic state in our terminology. Interestingly, a belief function Bel can be expressed in terms of the basic assignment m [16]:

$$Bel(A) = \sum_{E \subseteq A} m(E).$$

This formula is clearly related with the meta-models $\mathcal{F}_{\Box\alpha} = \{E \subseteq \mathcal{V} : E \subseteq [\alpha]\}$ of atomic belief $\Box\alpha$ (cf. Section 2.2). The converse problem, namely, reconstructing the basic assignment from the belief function, has a unique solution via the so-called Möbius transform

$$m(E) = \sum_{A \subseteq E} (-1)^{|E \setminus A|} Bel(A).$$

It is clear that the assertion of a MEL formula $\Box\alpha$ is faithfully expressed by $Bel([\alpha]) = 1$. The fact that the calculus of belief functions is a graded extension of the KD45 modal logic was already briefly pointed out by Smets [17]; especially, $Bel([\alpha])$ can be interpreted as the probability of $\Box\alpha$. Moreover, there is a similarity between the problem of reconstructing a mass assignment from the knowledge of a belief function and the problem of singling out an epistemic state in the language of MEL as in Section 4.1. Namely, consider the MEL-formula $\Box\alpha_E \wedge \neg \bigvee_{w \in E} \Box\neg\alpha_w \equiv \delta_E$, whose set of meta-models is $\{E\}$. We shall show that this expression can be written as an exact symbolic counterpart of the Möbius transform. To see it, in fact, rewrite the Möbius transform as

$$m(E) = \sum_{A \subseteq E : |E \setminus A| \text{ even}} Bel(A) - \sum_{A \subseteq E : |E \setminus A| \text{ odd}} Bel(A).$$

Now translate $\sum$ into $\bigvee$, $Bel(A)$ into $\Box\alpha$, "$-$" into $\wedge\neg$, and get the following:

Proposition 2. $\delta_E \equiv \bigvee_{\alpha \models \alpha_E : |E \setminus [\alpha]| \text{ even}} \Box\alpha \wedge \neg \bigvee_{\alpha \models \alpha_E : |E \setminus [\alpha]| \text{ odd}} \Box\alpha.$

Proof. If $\beta \models \alpha$, $\Box\alpha \vee \Box\beta \equiv \Box\alpha$ in MEL, so, $\bigvee_{\alpha \models \alpha_E : |E \setminus [\alpha]| \text{ even}} \Box\alpha \equiv \Box\alpha_E$. Now the set of meta-models of the formula $\Box\alpha_E \wedge \bigvee_{w \in E} \Box\neg\alpha_w$ is

$$\{A : A \subseteq E\} \cap \cup_{w \in E} \{A \subseteq \mathcal{V} : w \notin A\} = \cup_{w \in E} \{A \subseteq E : w \notin A\}.$$

It is not difficult to see that the above is also the set of meta-models of the formula $\bigvee_{w \in E} \Box\alpha_{E \setminus \{w\}}$, and of the more redundant formula $\bigvee_{\alpha \models \alpha_E : |E \setminus [\alpha]| \text{ odd}} \Box\alpha$ equivalently. So the Möbius-like MEL-formula is semantically equivalent to $\Box\alpha_E \wedge \neg(\Box\alpha_E \wedge \bigvee_{w \in E} \Box\neg\alpha_w) \equiv \delta_E.$ ∎

So one may consider belief (resp. plausibility) functions as numerical generalisations of atomic (boxed) formulae of MEL (resp. diamonded formulae), and formulae describing single epistemic states (totally informed meta-epistemic states) can be obtained via a symbolic counterpart to Möbius transform.

6 Related Work

The standard modal logic approach to the representation of knowledge viewed as true belief relies on the $S5$ modal logic, while beliefs are captured by $KD45$ [12]. At the semantic level it uses Kripke semantics based on an accessibility relation R among possible worlds. Our approach does not require axioms **4** and **5** (positive and negative introspection), since we are not concerned with an agent reasoning about his or her own beliefs. The fact that we rule out nested modalities and do not consider introspection does not make this kind of semantics very natural. Nevertheless, our setting is clearly similar to the one proposed by Halpern and colleagues [12] reinterpreting knowledge bases as being fed by a "Teller" that makes statements supposed to be true in the real world. The knowledge base is what we call receiver and the teller what we call emitter. Important differences are that we are mainly concerned with beliefs held by the Teller (hence making no assumptions as to the truth of such beliefs), that these beliefs are incomplete, and that the Teller is allowed to explicitly declare partial ignorance about specific statements. Finally, even if not concerned with nonmonotonic reasoning, MEL may be felt as akin to early nonmonotonic modal logics such as Moore's autoepistemic logic (AEL) [14]. Expansions of an AEL theory can be viewed as meta-models expressing epistemic states. However, there are a couple of important differences between MEL and AEL. In autoepistemic logic an agent is reasoning about his or her own beliefs, or lack thereof, not about another agent's beliefs. So AEL naturally allows for the nesting of modalities, contrary to MEL. Moreover, sentences of the form $\Box\alpha \vee \neg\alpha$ (meaning that if α is not believed, then it is false) involving boxed and non-boxed formulae are allowed in AEL (and are the motivation for this logic), thus mixing propositional and modal formulae, which precisely MEL forbids.

The closest work to MEL is Pauly's logic for *consensus voting* [15] that has a language and axiomatization identical to those of MEL. However, the semantics is set in a different context altogether. A consensus model for n individuals is a collection of n propositional valuations that need not be distinct. So instead of epistemic states that are sets of valuations, Pauly uses multisets thereof. The subpart of consensus logic restricting models to subsets of distinct valuations coincides with MEL. However, the general completeness result obtained for MEL (cf. Theorem 2) will not find an analogue in the setting of consensus logic.

At first glance the semantics of MEL also seems to bring us close to neighborhood semantics of modal logics proposed by D. Scott and R. Montague [3]. However, neighborhood semantics replaces Kripke structures by collections of subsets of valuations in the definition of satisfiability (which enables logics weaker than K to be encompassed) while in MEL a model is a non-empty collection of valuations. Partial logic Par [2], like MEL, uses special sets of valuations in place of valuations, under the form of partial models. A partial model σ assigns truth-values to a subset of propositional variables. The corresponding meta-model is formed of all completions of σ. Unfortunately, Par adopts a truth-functional view, and assumes the equivalence $\sigma \models \alpha \vee \beta$ if and only if $\sigma \models \alpha$ or $\sigma \models \beta$. So it loses classical tautologies, which sounds paradoxical when propositional variables are Boolean [4]. Actually, the basic Par keeps the syntax of classical logic, which forbids to make a difference between the fact of believing $\alpha \vee \beta$ and that of believing α or believing β.

However a more promising connection is between MEL and possibilistic logic. Possibilistic logic has been essentially developed as a formalism for handling qualitative uncertainty with an inference mechanism that remains close to the one of classical logic [6]. A standard possibilistic logic expression is a pair (α, a), where α is a propositional formula and a a level of certainty in $[0, 1]$. Actually, the fragment of MEL restricted to boxed propositional formulae and conjunctions thereof is isomorphic to special cases of possibilistic logic bases where weights attached to formulae express full certainty. It suggests an extension of MEL to multimodalities (like the FN system suggested by Hájek [10] p. 232), using formulae such as $\Box_a \alpha$ expressing that the agent believes α at level at least a, and changing epistemic states into possibility distributions. Such an extension of MEL might also extend possibilistic logic by naturally allowing for other connectives between possibilistic formulae, such as disjunction and negation, with natural semantics already outlined in [7] in the scope of multiagent systems.

7   Conclusion

This paper lays the foundations for a belief logic that is in close agreement with more sophisticated uncertainty theories. It is a modal logic because it uses the standard modal symbols $\Box$ and $\Diamond$ for expressing ideas of certainty understood as validity in an epistemic state and possibility understood as consistency with an epistemic state. It differs from usual modal logics (even if borrowing much of their machinery) by a deliberate stand on not nesting modalities, and not mixing modal and non-modal formulae, thus yielding a two-tiered logic. At the semantic level we have proved that the MEL language is capable of accounting for any meta-epistemic state, viewed as a family of non-empty subsets of classical valuations, just as propositional logic language is capable of accounting for any epistemic state, viewed as a family of classical valuations. In this sense, MEL is a higher-order logic with respect to classical logic. It prevents direct access to the actual state of the world: in the belief environment of this logic, an agent is not allowed to claim that a proposition is true in the real world. We do not consider our modal formalism to be an extension of the classical logic language, but an encapsulation thereof, within an epistemic framework; hence combinations of objective and epistemic statements like $\alpha \wedge \Box\beta$ are considered meaningless in this perspective. This higher-order flavor is typical of uncertainty theories. The subjectivist stand in MEL does not lead us to object to the study of languages where meta-statements relating belief and actual knowledge, observations and objective truths could be expressed. We only warn that epistemic statements expressing beliefs and doubts on the one hand and other pieces of information trying to bridge the gap between the real world and such beliefs (like deriving the latter from objective observations) should be handled separately.

This study is a first step. Some aspects of MEL require more scrutiny, like devising proof methods and assessing their computational complexity. One of the merits of MEL is to potentially offer a logical grounding to uncertainty theories of incomplete information. An obvious extension to be studied is towards possibilistic logics, using (graded) multimodalities and generalizing epistemic states to possibility distributions. In fact, modal logics capturing possibility and necessity measures have been around since the early nineties [11], but they were devised with a classical view of modal logic and Kripke semantics. One important contribution of the paper is to show that MEL is the Boolean version of Shafer's theory of evidence, whereby a mass function is the probabilistic counterpart to a meta-epistemic state. It suggests that beyond possibilistic logic, MEL could be extended to belief functions in a natural way, and it would be useful to compare MEL with the logic of belief functions devised by Godo and colleagues [9].

References

1. Blackburn, P., de Rijke, M., Venema, Y.: Modal Logic. Cambridge U. P., Cambridge (2001)
2. Blamey, S.: Partial logic. In: Gabbay, D.M., Guenthner, F. (eds.) Handbook of Philosophical Logic, vol. 3, pp. 1–70. D. Reidel Publishing Company, Dordrecht (1985)
3. Chellas, B.F.: Modal Logic: an Introduction. Cambridge University Press, Cambridge (1980)
4. Dubois, D.: On ignorance and contradiction considered as truth-values. Logic Journal of the IGPL 16(2), 195–216 (2008)
5. Dubois, D., Hájek, P., Prade, H.: Knowledge-driven versus data-driven logics. J. Logic, Language and Information 9, 65–89 (2000)
6. Dubois, D., Prade, H.: Possibilistic logic: a retrospective and prospective view. Fuzzy Sets and Systems 144, 3–23 (2004)
7. Dubois, D., Prade, H.: Toward multiple-agent extensions of possibilistic logic. In: Proc. IEEE Int. Conf. on Fuzzy Systems, pp. 187–192 (2007)
8. Gärdenfors, P.: Knowledge in Flux. MIT Press, Cambridge (1988)
9. Godo, L., Hájek, P., Esteva, F.: A fuzzy modal logic for belief functions. Fundam. Inform. 57(2-4), 127–146 (2003)
10. Hájek, P.: The Metamathematics of Fuzzy Logics. Kluwer Academic Publishers, Dordrecht (1998)
11. Hájek, P., Harmancova, D., Esteva, F., Garcia, P., Godo, L.: On modal logics for qualitative possibility in a fuzzy setting. In: Lopez de Mantaras, R., Poole, D. (eds.) UAI, pp. 278–285. Morgan Kaufmann, San Francisco (1994)
12. Halpern, J.Y., Fagin, R., Moses, Y., Vardi, M.Y.: Reasoning About Knowledge. MIT Press, Cambridge (2003) (Revised paperback edition)
13. Hughes, G.E., Cresswell, M.J.: A New Introduction to Modal Logic. Routledge (1996)
14. Moore, R.C.: Semantical considerations on nonmonotonic logic. Artificial Intelligence 25, 75–94 (1985)
15. Pauly, M.: Axiomatizing collective judgment sets in a minimal logical language. Synthese 158(2), 233–250 (2007)
16. Shafer, G.: A Mathematical Theory of Evidence. Princeton University Press, Princeton (1976)
17. Smets, P.: Comments on R. C. Moore's autoepistemic logic. In: Smets, P., Mamdani, E.H., Dubois, D., Prade, H. (eds.) Non-standard Logics for Automated Reasoning, pp. 130–131. Academic Press, London (1988)

Complexity and Cautiousness Results for Reasoning from Partially Preordered Belief Bases

Salem Benferhat and Safa Yahi

Université Lille-Nord de France,
Artois, F-62307 Lens, CRIL, F-62307 Lens,
CNRS UMR 8188, F-62307 Lens
{benferhat,yahi}@cril.univ-artois.fr

Abstract. Partially preordered belief bases are very convenient for an efficient representation of incomplete knowledge. They offer flexibility and avoid to compare unrelated pieces of information. A number of inference relations for reasoning from partially preordered belief bases have been proposed. This paper sheds light on the following approaches: the partial binary lexicographic inference, the compatible-based lexicographic inference, the democratic inference, the compatible-based inclusion inference, the strong possibilistic inference and the weak possibilistic inference. In particular, we propose to analyse these inference relations according to two key dimensions: the computational complexity and the cautiousness. It turns out that almost all the corresponding decision problems are located at most at the second level of the polynomial hierarchy. As for the cautiousness results, they genereally extend those obtained in the particular case of totally preordered belief bases.

1 Introduction

Handling inconsistency is a fundamental problem in commonsense reasoning. This problem arises in several situations like belief revision, exceptions tolerant reasoning, information fusion, etc. For instance, in a cooperative intrusion detection framework, several intrusion detection systems (IDSs) need to be dispatched throughout the network in order to enhance the detection process. However, such a cooperation may easily lead to conflicting situations according to the topological and functional visibility of each IDS.

A number of approaches have been proposed to reason under inconsistency without trivialization. While some of them consist in weakening the inference relation such as paraconsistent logics [7], others weaken the available beliefs like the so-called coherence-based approaches [15] which are quite popular.

Coherence-based approaches can be considered as a two step process consisting first in generating some preferred consistent subbases and then using classical inference from some of them. Among these approaches, we can distinguish those that are dedicated for totally preordered (or stratified) belief bases and those which are more general and which deal with partially preordered belief bases.

C. Sossai and G. Chemello (Eds.): ECSQARU 2009, LNAI 5590, pp. 817–828, 2009.

The most frequently encountered coherence-based approaches dedicated for totally preordered belief bases are the possibilistic inference [8], the inclusion inference [4] and the lexicographic inference [1,11]. All these inferences have been deeply analyzed from both the computational complexity and the cautiousness sides [2,12,13,5,1].

However, no such necessary study has been devoted, at the best of our knowledge, to the several inference relations from partially preordered belief bases despite the flexibility they offer.

In this paper, we are interested in such approaches. In particular, we shed light on the partial binary lexicographic inference [16], the compatible-based lexicographic inference [16], the democratic inference [6], the compatible-based inclusion inference [10], the strong possibilistic inference and the weak possibilistic inference [3]. The common denominator of all these inferences is that each one of them is an extension of some popular approach for totally preordered belief bases.

This paper analyzes the previous inference relations by studying the corresponding computational complexity and comparing them in terms of cautiousness or equivalently in terms of productivity. We do believe that such an analysis is worth the effort in order to enable one to choose the most suitable inference that fits with the cautiousness required by the application at hand with the lowest computational cost.

The remainder of the paper is structured as follows. In Section 2, we give some formal preliminaries. In Section 3, we briefly review the inference relations from partially preordered belief bases that make the object of our study. In Section 4, we give the complexity results and in Section 5 we present the cautiousness-based comparison. Section 6 concludes the paper and gives some perspectives.

2   Preliminaries

We consider a finite set of propositional variables which are denoted by lower case Roman letters. Formulae are denoted by upper case Roman letters. Let Σ be a finite set of formulae, $Cons(\Sigma)$ denotes the set of all the consistent subbases of Σ while $MaxCons(\Sigma)$ denotes the set of all its maximal (with respect to set inclusion) consistent subbases.

A partial preorder $\preceq$ on a finite set A is a reflexive and transitive binary relation. In this paper, $a \preceq b$ expresses that a is at least as preferred as b. A strict order $\prec$ on A is an irreflexive and transitive binary relation. $a \prec b$ means that a is strictly preferred to b. A strict order is defined from a preorder as $a \prec b$ if and only if $a \preceq b$ holds but $b \preceq a$ does not hold. The equality, denoted by $\approx$, is defined as $a \approx b$ if and only if $a \preceq b$ and $b \preceq a$. Moreover, we define the incomparability, denoted by $\sim$, as $a \sim b$ if and only if $a \not\preceq b$ and $b \not\preceq a$. The set of minimal elements of A with respect to $\prec$, denoted by $Min(A, \prec)$, is defined as: $Min(A, \prec) = \{a \in A, \nexists b \in A : b \prec a\}$. A total preorder $\leq$ on a finite set A is a reflexive and transitive binary relation such that $\forall a, b \in A$, either $a \leq b$ or $b \leq a$.

We assume that the reader is familiar with some basic notions about complexity theory, like the classes P, NP and co-NP. Now, we will sketch the classes of the polynomial hierarchy (PH) (see [9,14] for more details). Let X be a class of decision problems. Then P^X denotes the class of decision problems that can be solved using a polynomial algorithm that uses an oracle for X (informally, a subroutine for solving a problem in X at unit cost). Similarly, NP^X denotes the class of decision problems that can be solved using a nondeterministic polynomial algorithm that uses an oracle for X. Based on these notions, the classes Δ_k^p, Σ_k^p and Π_k^p ($k \geq 0$) are defined as follows: $\Delta_0^p = \Sigma_0^p = \Pi_0^p = P$, $\Delta_{k+1}^p = P^{\Sigma_k^p}$, $\Sigma_{k+1}^p = NP^{\Sigma_k^p}$ and $\Pi_{k+1}^p = $ co-Σ_{k+1}^p.

Hence, $\Sigma_1^p = NP$ and $\Pi_1^p = $ co-NP. The class $\Delta_2^p[O(log\ n)]$ contains the problems in Δ_2^p that can be solved with $O(log\ n)$ many calls to an NP oracle.

3   A Refresher on the Inference from Partially Preordered Belief Bases

3.1   Inference from Totally Preordered Belief Bases

We first recall some popular inference relations from totally preordered belief bases, namely the lexicographic inference [1,11], the inclusion inference [4] and the possibilistic inference [8].

Let $(\Sigma, \leq)$ be a totally preordered belief base where Σ is a set of formulae and $\leq$ is a total preorder reflecting the priority relation that exists between these formulae. $(\Sigma, \leq)$ can be viewed as a stratified belief base $\Sigma = S_1 \cup \cdots \cup S_m$ such that the formulae in S_i have the same level of priority and have a higher priority than those in S_j with $j > i$.

Definition 1. *Let* $A, B \in Cons(\Sigma)$.

- *A is **lexicographically** preferred to B, denoted by* $A <_{lex} B$, *iff* $\exists i, 1 \leq i \leq m$ *such that* $|S_i \cap A| > |S_i \cap B|$ [1] *and* $\forall j, j < i, |S_j \cap B| = |S_j \cap A|$.
- *A is preferred to B with respect to the **inclusion preference**, denoted by* $A <_{incl} B$, *iff* $\exists i, 1 \leq i \leq m$ *such that* $(S_i \cap B) \subset (S_i \cap A)$ *and* $\forall j, j < i$, $(S_j \cap B) = (S_j \cap A)$.

Let $Lex(\Sigma, \leq)$ (resp. $Incl(\Sigma, \leq)$) denote the set of all the preferred consistent subbases of Σ with respect to $<_{lex}$ (resp. $<_{incl}$), namely $Lex(\Sigma, \leq) = Min(Cons(\Sigma), <_{lex})$ and $Incl(\Sigma, \leq) = Min(Cons(\Sigma), <_{incl})$. Then,

Definition 2. *Let* ψ *be a formula.*

- ψ *is said to be a **lexicographic** consequence of* Σ, *denoted by* $\Sigma \vdash_{lex} \psi$, *iff* $\forall B \in Lex(\Sigma, \leq) : B \models \psi$.
- ψ *is said to be an **inclusion** consequence of* Σ, *denoted by* $\Sigma \vdash_{incl} \psi$, *iff* $\forall B \in Incl(\Sigma, \leq) : B \models \psi$.

---

[1] $|A|$ denotes the number of formulae of A.

As to the possibilistic inference, it is defined by

Definition 3. *A formula ψ is a **possibilistic** consequence of $(\Sigma, \leq)$, denoted by $(\Sigma, \leq) \models_{pos} \psi$, iff $(\bigcup_{i=1}^{s-1} S_i) \models \psi$, where s is the smallest index such that $\bigcup_{i=1}^{s} S_i$ is inconsistent. If $\bigcup_{i=1}^{m} S_i$ is consistent then $(\Sigma, \leq) \models_{pos} \psi$ iff $(\bigcup_{i=1}^{m} S_i) \models \psi$.*

Let LEX, INCL and POS denote the decision problems respectively associated with $\vdash_{lex}$, $\vdash_{incl}$ and $\vdash_{pos}$. Then, it has been shown that LEX is Δ_2^p-complete [5], INCL is Π_2^p-complete [12] and POS is $\Delta_2^p[O(log\ n)]$-complete [13]. Moreover, the possibilistic inference is more cautious than the inclusion inference which is itself more cautious than the lexicographic inference [1].

3.2   Inference Relations from Partially Preordered Belief Bases

A number of inference relations from partially preordered belief bases have been defined by extending the inference relations from totally preordered belief bases recalled in the previous section. Then, the compatible-based lexicographic inference [16] and the partial binary lexicographic inference [16] extend the lexicographic inference. Both the democratic inference [6] and the compatible-based inclusion inference [4] generalise the inclusion inference. As to the possibilistic inference, it is extended by the strong possibilistic inference [3] and also by the weak possibilistic inference [3].

Before sketching these inference relations, let us recall the notion of totally preordered belief bases compatible with a given partially preordered belief base $(\Sigma, \preceq)$ [3]. Intuitively, a totally preordered belief base $(\Sigma, \leq)$ is said to be compatible with a $(\Sigma, \preceq)$ iff the total preorder $\leq$ extends or completes the partial preorder $\preceq$. More formally: 1) $\forall \varphi, \phi \in \Sigma$: if $\varphi \preceq \phi$ then $\varphi \leq \phi$ and 2) $\forall \varphi, \phi \in \Sigma$: if $\varphi \prec \phi$ then $\varphi < \phi$.

We denote by $Comp(\Sigma, \preceq)$ the set of all the totally preordered belief bases compatible with $(\Sigma, \preceq)$.

1. Compatible-based Lexicographic Inference: This inference, denoted here by Cmp-lexicographic inference, is based on the idea of totally preordered compatible belief bases [16].

Definition 4. *Let $B \in Cons(\Sigma)$. B is said to be Cmp-lexicographically preferred iff there exists a totally preordered base $(\Sigma, \leq)$ compatible with $(\Sigma, \preceq)$ such that B is lexicographically preferred in $(\Sigma, \leq)$.*

Let $CmpLex(\Sigma, \preceq)$ denote the set of all the *Cmp*- lexicographically preferred consistent subbases: $CmpLex(\Sigma, \preceq) = \bigcup_{(\Sigma, \leq) \in Comp(\Sigma, \preceq)} Lex(\Sigma, \leq)$. Then, a formula ψ is said to be a *Cmp*-lexicographic conclusion of $(\Sigma, \preceq)$, denoted by $(\Sigma, \preceq) \Vdash_{lex}^{cmp} \psi$, iff

$$\forall B \in CmpLex(\Sigma, \preceq), B \models \psi.$$

2. Partial Binary Lexicographic Inference: The idea of this inference which will be denoted by P-lexicographic inference is to compare directly two consistent subbases [16]. First, Σ is partitioned as follows $\Sigma = E_1 \cup \ldots \cup E_n$ $(n \geq 1)$ such that:

- $\forall i,\ 1 \leq i \leq n$, we have $\forall \varphi, \varphi' \in E_i\colon \varphi \approx \varphi'$,
- $\forall i,\ 1 \leq i \leq n,\ \forall j,\ 1 \leq j \leq n$ with $i \neq j$, we have $\forall \varphi \in E_i,\ \forall \varphi' \in E_j\colon \varphi \not\approx \varphi'$.

So, each subset E_i represents an equivalence class of Σ with respect to $\approx$. Then, a preference relation between two equivalence classes E_i and E_j, denoted by $\prec_s$, is defined by: $E_i \prec_s E_j$ iff $\exists \varphi \in E_i,\ \exists \varphi' \in E_j$ such that $\varphi \prec \varphi'$. One can easily see that this partition is a generalization of the idea of stratification associated with totally preordered belief bases. Now, the P-lexicographic preference between two consistent subbases of a partially preordered belief base $(\Sigma, \preceq)$, denoted by $\preceq^p_{lex}$, is defined as follows:

Definition 5. Let $A, B \in Cons(\Sigma)$. Then, A is said to be P-lexicographically preferred to B, denoted by $A \preceq^p_{lex} B$, iff $\forall i,\ 1 \leq i \leq n :$ if $|E_i \cap B| > |E_i \cap A|$ then $\exists j,\ 1 \leq j \leq n$ such that $|E_j \cap A| > |E_j \cap B|$ and $E_j \prec_s E_i$.

Let $PLex(\Sigma, \preceq) = Min((\Sigma, \preceq), \preceq^p_{lex})$. Then, a formula ψ is a P-lexicographic conclusion of $(\Sigma, \preceq)$, denoted by $(\Sigma, \preceq) \Vdash^p_{lex} \psi$, iff

$$\forall B \in PLex(\Sigma, \preceq) : B \models \psi.$$

3. Democratic Inference: The democratic inference [6] is based on the following preference:

Definition 6. Let $A, B \in Cons(\Sigma)$. Then, A is said to be democratically preferred to B, denoted by $A \prec_{demo} B$, iff $\forall b \in B/A, \exists a \in A/B$ such that $a \prec b$.

Let $Demo(\Sigma, \preceq) = Min(Cons(\Sigma, \preceq), \prec_{demo})$ denote the set of all the democratically preferred consistent subbases of $(\Sigma, \preceq)$. Then, a formula ψ is said to be a democratic conclusion of $(\Sigma, \preceq)$, denoted by $(\Sigma, \preceq) \Vdash_{demo} \psi$, iff

$$\forall B \in Demo(\Sigma, \preceq), B \models \psi.$$

4. Compatible-based Inclusion Inference: This inference, denoted here by Cmp-inclusion inference, is also based on the notion of compatible totally preordered belief bases [10].

Definition 7. $A \in Cons(\Sigma)$ is said to be a Cmp-inclusion preferred subbase iff there exists a compatible $(\Sigma, \leq)$ such that $A \in Incl(\Sigma, \leq)$.

Let $CmpIncl$ denote the set of all the Cmp-inclusion preferred subbases. Then,

$$(\Sigma, \preceq) \Vdash^{cmp}_{incl} \psi \text{ iff } \forall B \in CmpIncl(\Sigma, \leq),\ B \models \psi.$$

5. Strong and Weak Possibilistic Inferences: The corresponding preference relations are defined as follows [3].

Definition 8. Let $A, B \in Cons(\Sigma)$. Then,

- A is preferred to B with respect to the strong possibilistic preference, denoted by $A \prec^s_{pos} B$, iff $\exists b \notin B$ such that $\forall a \notin A, b \prec a$.

– *A is preferred to B with respect to the weak possibilistic preference, denoted by $A \prec_{pos}^{w} B$ iff $\forall a \notin A, \exists b \notin B$ such that $b \prec a$.*

Let $Pos^{s}(\Sigma, \preceq)$ and $Pos^{w}(\Sigma, \preceq)$ denote respectively $Min((\Sigma, \preceq), \prec_{pos}^{s})$ and $Min((\Sigma, \preceq), \prec_{pos}^{w})$. Then,

– $(\Sigma, \preceq) \Vdash_{pos}^{s} \psi$ iff $\forall B \in Pos^{s}(\Sigma, \preceq), B \models \psi$
– $(\Sigma, \preceq) \Vdash_{pos}^{w} \psi$ iff $\forall B \in Pos^{w}(\Sigma, \preceq), B \models \psi$.

4   Computational Complexity Results

In this section, we present complexity results for reasoning from partially pre-ordered belief bases using the inference relations recalled in the previous section. Let us consider the following inference relation:

$$\Sigma \Vdash_{mc} \psi \text{ iff } \forall B \in MaxCons(\Sigma) : B \models \psi$$

and let MaxCons denote the corresponding decision problem which is known to be Π_{2}^{p}-complete [12].

Let Pos-S, Pos-W, Demo, CmpIncl, PLex and CmpLex denote the decision problems respectively associated with: $(\Sigma, \preceq) \Vdash_{pos}^{s} \psi$, $(\Sigma, \preceq) \Vdash_{pos}^{w} \psi$, $(\Sigma, \preceq) \Vdash_{demo} \psi$, $(\Sigma, \preceq) \Vdash_{incl}^{cmp} \psi$, $(\Sigma, \preceq) \Vdash_{lex}^{p} \psi$ and $(\Sigma, \preceq) \Vdash_{lex}^{cmp} \psi$.

Let us first give the complexity of PLex:

Proposition 1. *PLex is Π_{2}^{p}-complete.*

Proof Sketch

1. **Membership to Π_{2}^{p}**

   Let us show that the complementary problem co-PLex $((\Sigma, \preceq) \not\Vdash_{lex}^{p} \psi)$ belongs to Σ_{2}^{p}. Membership in Σ_{2}^{p} follows from Algorithm 1.1.

   Algorithm 1.1: co-PLex$((\Sigma, \preceq), \psi)$
   begin
       1. Guess a subbase A of Σ
       2. Check that A is consistent
       3. Check that $A \in PLex(\Sigma, \preceq)$
       4. Check that $A \not\models \psi$
   end

Clearly, points 2 and 4 can be solved using an NP oracle. As for point 3, it can be solved by checking that there is no a consistent subbase B such that $B \prec_{lex}^{p} A$. The problem which consists in checking whether such a base exists will be denoted by NotLexPref and it can be solved via Algorithm 1.2 which is nondeterministic polynomial. So, NotLexPref $\in$ NP. Moreover, we can reduce the well known GSat problem (the satisfiability problem of

Algorithm 1.2: NotLexPref$((\Sigma, \preceq), A)$
begin

> Guess an interpretation ω
> $B \leftarrow \emptyset$
> **for** *each* $\phi \in \Sigma$ **do**
> > **if** ω *satisfies* ϕ **then**
> > > $B \leftarrow B \cup \{\phi\}$
>
> Check that $B \prec_{lex} A$

end

a propositional formula) which is NP-complete to NotLexPref using a polynomial transformation. Hence, NotLexPref is NP-complete.

Thus, Algorithm 1.1 is non deterministic (point 1) polynomial that uses an NP oracle. So, co-PLex $\in NP^{NP} = \Sigma_2^p$ and hence PLex $\in$ co-$\Sigma_2^p = \Pi_2^p$.

2. **Completeness**
   Given a belief base Σ, let us consider a new partially preordered belief base $(\Sigma, \preceq)$ such that: $\forall \alpha, \beta \in \Sigma : \alpha \sim \beta$. Now, let us show that $MaxCons(\Sigma) = PLex(\Sigma, \preceq)$.
   - It has been shown in [16] that each P-lexicographically preferred consistent subbase is maximal consistent, i.e., $PLex(\Sigma, \preceq) \subseteq MaxCons(\Sigma)$.
   - Now, let us show the other inclusion, namely $\forall A, B \in MaxCons(\Sigma)$: $A \sim_{lex}^p B$.
     Let $A, B \in MaxCons(\Sigma)$. $(\Sigma, \preceq) = \bigcup_{i=1}^{m=|\Sigma|} E_i$ such that:
     • $\forall i, 1 \leq i \leq m$: E_i contains only one formula $\phi \in \Sigma$.
     • $\forall i, 1 \leq i \leq m, \forall j, 1 \leq j \leq m$ such that $i \neq j$, we have $E_i \sim_s E_j$.
     On the other hand, neither $A \not\subset B$ nor $B \not\subset A$. Then, given $\alpha \in A - B$ and $\beta \in B - A$, let $E_a = \{\alpha\}$ and $E_b = \{\beta\}$. So, $\exists b, 1 \leq b \leq m$ with $|E_b \cap B| = 1 > |E_b \cap A| = 0$ such that $\nexists j, 1 \leq j \leq m$ with $E_j \prec_s E_b$ and $|E_j \cap A| > |E_j \cap B|$. This means that $A \not\prec_{lex}^p B$. In the same way, we prove that $B \not\prec_{lex}^p A$. Hence, $\forall A \in MaxCons(\Sigma), A \in PLex(\Sigma, \preceq)$.
   Thus, $MaxCons(\Sigma) = PLex(\Sigma, \preceq)$, i.e., $\Sigma \Vdash_{mc} \psi$ iff $(\Sigma, \preceq) \Vdash_{lex}^p \psi$. So, MaxCons $\propto$ PLex and since MaxCons is Π_2^p-complete and PLex $\in \Pi_2^p$, we deduce that PLex is Π_2^p-complete. ∎

Now, the complexity of CmpLex is as follows:

Proposition 2. CmpLex *is* Π_2^p-*complete.*

Proof Sketch

1. **Membership in Π_2^p**
   Let us show that the complementary problem co-CmpLex $\in \Sigma_2^p$ via Algorithm 1.3. This algorithm is nondeterministic given point 1. Then, clearly point 2 can be achieved in polynomial time. Point 3 can be achieved using a polynomial number of an NP oracle since it is known that Lex is Δ_2^p-complete. So, co-CmpLex $\in NP^{NP} = \Sigma_2^p$. Thus, CmpLex $\in \Pi_2^p$.

Algorithm 1.3: co-CMPLEX$((\Sigma, \preceq), \psi)$
begin
    1. Guess a totally preordered belief base $(\Sigma, \leq)$
    2. Check that $(\Sigma, \leq)$ is compatible with $(\Sigma, \preceq)$
    3. Check that $(\Sigma, \leq) \nvdash_{lex} \psi$
end

2. Completeness

Using the same polynomial transformation given for PLEX, we show that MAXCONS $\propto$ CMPLEX and hence CMPLEX is Π_2^p-complete [2]. ∎

Then, it turns out that the democratic inference belongs to the same class as the P-lexicographic and the Cmp-lexicographic inferences as shown by the following proposition:

Proposition 3. DEMO *is* Π_2^p-*complete.*

Indeed, membership in Π_2^p can be shown by proving that co-DEMO $\in \Sigma_2^p$ using similar ideas as those given for PLEX. Completeness derives from the fact that INCL which is Π_2^p-complete is a particular case of DEMO.

As for the compatible-based inclusion inference, we give the following upper and lower bounds:

Proposition 4. CMPINCL $\in \Pi_3^p$ *and is* Π_2^p-*hard.*

In fact, Algorithm 1.4 shows that co-CMPINCL $\in \Sigma_3^p$.

Algorithm 1.4: co-CMPINCL$((\Sigma, \preceq), \psi)$
begin
    1. Guess a totally preordered belief base $(\Sigma, \leq)$
    2. Check that $(\Sigma, \leq)$ is compatible with $(\Sigma, \preceq)$
    3. Check that $(\Sigma, \leq) \nvdash_{Incl} \psi$
end

Clearly, point 2 can be achieved in polynomial time while point 3 can be achieved using a Σ_2^p-complete oracle since co-INCL is Σ_2^p-complete. So, this is a nondeterministic (point 1) polynomial algorithm that uses a Σ_2^p oracle. Hence, co-CMPINCL $\in NP^{\Sigma_2^p} = \Sigma_3^p$. Consequently, CMPINCL $\in \Pi_3^p$. Hardness for Π_2^p holds since INCL which is Π_2^p-complete is a particular case of CMPINCL.

Finally, concerning the strong and weak possibilistic inferences, the following always holds:

Proposition 5. *We show that*

1. POS-S $\in \Pi_2^p$ *and is* $\Delta_2^p[O(log\ n)]$-*hard.*
2. POS-W $\in \Pi_2^p$ *and is* $\Delta_2^p[O(log\ n)]$-*hard.*

---

[2] For lake of space we only present proofs of some propositions.

Once again, membership in Π_2^p can be achieved using similar ideas as those given for PLEX. As to the hardness, it follows from the fact that POS that is known to be $\Delta_2^p[O(log\ n)]$-complete is a particular case of both POS-S and POS-W for totally preordered belief bases.

All these results are summarized by Table 1. It turns out that almost all these decision problems are at most at the second level of the polynomial hierarchy PH.

Table 1. Complexity results

Inference problem	Complexity
PLEX	Π_2^p-complete
CMPLEX	Π_2^p-complete
DEMO	Π_2^p-complete
CMPINCL	in Π_3^p, Π_2^p-hard
POS-S	in Π_2^p, $\Delta_2^p[O(logn)]$-hard
POS-W	in Π_2^p, $\Delta_2^p[O(logn)]$-hard

5   Cautiousness Analysis

The purpose of this section is to compare the previous inference relations in terms of cautiousness.

First, the democratic inference is more cautious than the P-lexicographic inference, namely:

Proposition 6. *Let* $(\Sigma, \preceq)$ *be a partially preordered belief base and* ψ *be a propositional formula. Then, we have only*

$$(\Sigma,\ \preceq)\ \Vdash_{demo} \psi \Rightarrow (\Sigma, \preceq)\ \Vdash_{lex}^p \psi.$$

On the other hand, the compatible-based inclusion inference and the P-lexicographic inference are incomparable: neither the former is more cautious than the later nor the converse.

Indeed, if $(\Sigma, \preceq)$ is totally preordered then $CmpIncl(\Sigma, \preceq) = Incl(\Sigma, \preceq)$ and $PLex(\Sigma, \preceq) = Lex(\Sigma, \preceq)$. Moreover, it is known that $Lex(\Sigma, \preceq) \subset Incl(\Sigma, \preceq)$. So, $CmpIncl(\Sigma, \preceq) \not\subset PLex(\Sigma, \preceq)$.

Now, the following example shows that $PLex(\Sigma, \preceq) \not\subset CmpIncl(\Sigma, \preceq)$:

Example 1. *Let* $(\Sigma, \preceq)$ *be such that* $\Sigma = \{a\wedge\neg d, \neg a, a\wedge f, d\}$ *with* $a\wedge\neg d \prec a\wedge f$ *and* $\neg a \prec d$.

Clearly, $MaxCons = \{A, B, C\}$ *such that* $A = \{a \wedge \neg d, a \wedge f\}$, $B = \{\neg a, d\}$ *and* $C = \{a \wedge f, d\}$. *Then, we can show that* $CmpIncl(\Sigma, \preceq) = \{A, B\}$. *Indeed, there is no a totally preordered belief base* $(\Sigma, \leq)$ *compatible with* $(\Sigma, \preceq)$ *such* $C \in Incl(\Sigma, \leq)$. *Moreover,* $A \sim_{lex}^p B$, $A \sim_{lex}^p C$ *and* $B \sim_{lex}^p C$ *which means that* $PLex(\Sigma, \preceq) = \{A, B, C\}$. *Therefore,* $PLex(\Sigma, \preceq) \not\subset CmpIncl(\Sigma, \preceq)$.

In addition, the weak possibilistic inference is more cautious than the democratic one:

Proposition 7. *Let* $(\Sigma, \preceq)$ *be a partially preordered belief base and* ψ *be a propositional formula. Then,* $(\Sigma, \preceq) \Vdash^w_{pos} \psi \Rightarrow (\Sigma, \preceq) \Vdash_{Demo} \psi$

Proof Sketch

Given $A, B \in Cons(\Sigma)$, we show that $A \prec^w_{pos} B \Rightarrow A \prec_{demo} B$.

Let us suppose that $A \prec^w_{pos} B$ and $A \not\prec_{demo} B$. Thus,

- $\forall \delta \notin A, \exists \lambda \notin B$ such that $\lambda \prec \delta \ldots$ **(h1)**
- $\exists \beta \in B - A, \forall \alpha \in A - B$ we have $\alpha \not\prec \beta \ldots$ **(h2)**

Let $\beta \in B - A$. So, $\beta \notin A$ and according to (h1), there must exist $\lambda \notin B$ such that $\lambda \prec \beta$. $\lambda \notin B$ so $\lambda \notin A \cap B$.

Moreover, $\lambda \notin A - B$ since $\forall \alpha \in A - B$ we have $\alpha \not\prec \beta$ according to (h2). Thus, $\lambda \notin A$, i.e., $\lambda \in \Sigma \cap \overline{(A \cup B)}$.

Now, let $\xi \in \Sigma \cap \overline{(A \cup B)}$ be such that $\xi \preceq \lambda$ and $\forall \theta \in \Sigma \cap \overline{(A \cup B)}$ we have $\theta \not\prec \xi$.

Since $\xi \notin A$, then according to (h1), there must exist $\chi \notin B$ such that $\chi \prec \xi$. $\chi \notin A \cap B$. In addition, since $\chi \prec \xi \preceq \lambda \prec \beta$, i.e., $\chi \prec \beta$, we deduce from (h1) that $\chi \notin A - B$. Hence, $\chi \notin A \cup B$, i.e., $\chi \in \Sigma \cap \overline{(A \cup B)}$ and $\chi \prec \xi$ which contradicts the definition of ξ. Then, $A \prec^w_{pos} B \Rightarrow A \prec_{demo} B$.

As for the converse, we can prove that it does not hold using the monotony property. Indeed, on the one hand, we know that $A \subset B \Rightarrow A \prec_{demo} B$. On the other hand, it has been shown in [3] that $A \subset B \not\Rightarrow A \prec^w_{pos} B$. Hence, $A \prec_{demo} B \not\Rightarrow A \prec^w_{pos} B$. ∎

Finally, the compatible-based lexicographic inference is less cautious than both the compatible-based inclusion inference and the P-lexicographic inference.

Proposition 8. *Let* $(\Sigma, \preceq)$ *be a partially preordered belief base and* ψ *be a propositional formula. Then,* 1) $(\Sigma, \preceq) \Vdash^{cmp}_{incl} \psi \Rightarrow (\Sigma, \preceq) \Vdash^{cmp}_{lex} \psi$ *and the converse does not hold,* 2) $(\Sigma, \preceq) \Vdash^p_{lex} \psi \Rightarrow (\Sigma, \preceq) \Vdash^{cmp}_{lex} \psi$ *and the converse does not hold.*

Clearly, $\forall (\Sigma, \leq) \in Comp(\Sigma, \preceq)$, we have $Lex(\Sigma, \leq) \subset Incl(\Sigma, \leq)$.

Then, $\bigcup_{(\Sigma, \leq) \in Comp(\Sigma, \preceq)} Lex(\Sigma, \preceq) \subset \bigcup_{(\Sigma, \leq) \in Comp(\Sigma, \preceq)} Incl(\Sigma, \preceq)$.

So, $CmpLex(\Sigma, \preceq) \subset CmpIncl(\Sigma, \preceq)$. Moreover, it has been shown in [16] that $CmpLex(\Sigma, \preceq) \subseteq PLex(\Sigma, \preceq)$. An example that shows that this inclusion is strict can be the following:

Example 2. *Let* $(\Sigma, \preceq)$ *be such that:*

$\Sigma = \{\alpha_1, \alpha_2, \beta_1, \beta_2, \beta_3, \gamma_1, \gamma_2\}$ *with* $\alpha_1 = a \wedge \neg b \wedge c$, $\beta_1 = \neg a \wedge \neg b \wedge c$, $\gamma_1 = b \wedge d$, $\alpha_2 = a \wedge \neg b \wedge d$, $\beta_2 = \neg a \wedge \neg b \wedge d$, $\gamma_2 = b \wedge e$, $\beta_3 = \neg a \wedge \neg b \wedge e$.

In addition, $\alpha_1 \approx \gamma_1 \approx \gamma_2$ *and* $\alpha_2 \approx \beta_1 \approx \beta_2 \approx \beta_3$.

Clearly, $MaxCons = \{A, B, C\}$ *such that* $A = \{\alpha_1, \alpha_2\}$, $B = \{\beta_1, \beta_2, \beta_3\}$ *and* $C = \{\gamma_1, \gamma_2\}$.

One can easily see that $PLex = \{A, B, C\}$ *while* $CmpLex = \{B, C\}$. *Hence, we deduce that* $CmpLex(\Sigma, \preceq) \subset PLex(\Sigma, \preceq)$.

All these results are summarized by Figure 1 where $A \rightarrow B$ means that A is more cautious than B. Note that the relation between the democratic inference and the compatible-based inclusion inference has been given in [6].

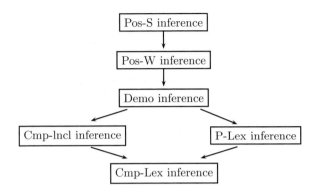

Fig. 1. Cautiousness results

These results preserve those known in the case of totally preordered belief bases except the relation between the P-lexicographic inference and the compatible-based inclusion inference. Surprisingly, they are incomparable while the later is more cautious than the former in the totally preordered case.

6   Conclusion and Perspectives

In this paper, we have analysed a number of inference relations from partially preordered belief bases regarding the computational complexity point of view. It turns out that almost all the corresponding decision problems are located at most at the second level of the polynomial hierarchy PH. On the other hand, it is known that the decision problems associated with the inference relations from totally preordered belief bases typically reside at the first level of PH. This seems the price to be paid to win in flexibility.

Moreover, we have compared them according to another key dimension namely the cautiousness one. All these results generalise those obtained in the particular case of totally preordered belief bases except the relation between the P-lexicographic inference and the compatible-based inclusion inference.

Now, this work calls for several perspectives. A first one consists in investigating the extent to which knowledge compilation can be used to circumvent these complexity results. Another perspective is to extend this work to the case of description logics in order to manage incoherence in cooperative intrusion detection.

Acknowledgements

This work is supported by the ANR SETIN 2006 project PLACID.

References

1. Benferhat, S., Dubois, D., Cayrol, C., Lang, J., Prade, H.: Inconsistency management and prioritized syntax-based entailment. In: IJCAI 1993, pp. 640–645 (1993)
2. Benferhat, S., Dubois, D., Prade, H.: Some syntactic approaches to the handling of inconsistent knowledge bases: A comparative study. Part 2: the prioritized case, vol. 24, pp. 473–511. Physica-Verlag, Heidelberg (1998)
3. Benferhat, S., Lagrue, S., Papini, O.: Reasoning with partially ordered information in a possibilistic framework. Fuzzy Sets and Systems 144, 25–41 (2004)
4. Brewka, G.: Preferred sutheories: an extende logical framework for default reasoning. In: IJCAI 1989, pp. 1043–1048 (1989)
5. Cayrol, C., Lagasquie-Schiex, M.-C., Schiex, T.: Nonmonotonic reasoning: From complexity to algorithms. Ann. Math. Artif. Intell 22(3-4), 207–236 (1998)
6. Cayrol, C., Royer, V., Saurel, C.: Management of preferences in assumption-based reasoning. In: Valverde, L., Bouchon-Meunier, B., Yager, R.R. (eds.) IPMU 1992. LNCS, vol. 682, pp. 13–22. Springer, Heidelberg (1993)
7. da Costa, N.C.A.: Theory of inconsistent formal systems. Notre Dame Journal of Formal Logic 15, 497–510 (1974)
8. Dubois, D., Lang, J., Prade, H.: Possibilistic logic. In: Handbook of Logic in Articial Intelligence and Logic Programming, vol. 3, pp. 439–513 (1994)
9. Garey, M.R., Johnson, D.S.: Computers and Intractability: A Guide to the Theory of NP-completeness. W.H. Freeman, New York (1979)
10. Junker, U., Brewka, G.: Handling partially ordered defaults in TMS. In: IJCAI 1989, pp. 1043–1048 (1989)
11. Lehmann, D.J.: Another perspective on default reasoning. Ann. Math. Artif. Intell 15(1), 61–82 (1995)
12. Nebel, B.: Belief revision and default reasoning: Syntax-based approaches. In: KR 1991, pp. 417–428 (1991)
13. Nebel, B.: How hard is it to revise a belief base? In: Handbook of Defeasible Reasoning and Uncertainty Management Systems, pp. 77–145 (1998)
14. Papadimitriou, C.H.: Computational Complexity. Addison-Wesley, Reading (1994)
15. Resher, N., Manor, R.: On inference from inconsistent premises. Theory and Decision 1, 179–219 (1970)
16. Yahi, S., Benferhat, S., Lagrue, S., Sérayet, M., Papini, O.: A lexicographic inference for partially preordered belief bases. In: KR 2008, pp. 507–517 (2008)

A Logic for Complete Information Systems

Md. Aquil Khan and Mohua Banerjee[*]

Department of Mathematics and Statistics,
Indian Institute of Technology,
Kanpur 208 016, India
{mohua,mdaquil}@iitk.ac.in

Abstract. A logic LIS for complete information systems is proposed. The language of LIS contains constants corresponding to attribute and attribute-values. A sound and complete deductive system for the logic is presented. Decidability is also proved.

1 Introduction

Rough set theory introduced by Pawlak is based on the concept of *approximation space* [11] which is defined as a tuple (W, R), where R is an equivalence relation on the set W. Any concept represented as a subset (say) X of the partitioned domain W, is then approximated from 'within' and 'outside', by its *lower* and *upper approximations* given as $\{[x] : [x] \subseteq X\}$ and $\{[x] : [x] \cap X \neq \emptyset\}$ respectively. $[x]$ denotes the equivalence class of $x \in W$.

A practical source of a Pawlak approximation space is a *complete information system* [11], formally defined as follows. There is also the notion of an incomplete/non-deterministic information system, but in this paper we deal only with complete information systems. So henceforth, we drop the word 'complete'.

Definition 1. *An* information system $\mathcal{S} := (W, A, \bigcup_{a \in A} Val_a, f)$, *comprises a non-empty set W of objects, a non-empty set A of attributes, a non-empty set Val_a of attribute values for each $a \in A$, and $f : W \times A \to \bigcup_{a \in A} Val_a$ such that $f(x, a) \in Val_a$.*

Any information system $\mathcal{S} := (W, A, \bigcup_{a \in A} Val_a, f)$ and $B \subseteq A$ would induce an 'indiscernibility' relation $Ind_{\mathcal{S}}(B)$ on W:

$$x \ Ind_{\mathcal{S}}(B) \ y \text{ if and only if } f(x, a) = f(y, a) \text{ for all } a \in B.$$

$x \ Ind_{\mathcal{S}}(B) \ y$ signifies that the objects x and y cannot be distinguished using only the information provided by the attributes of the set B. As B differs, we get different $Ind_{\mathcal{S}}(B)$, and therefore, different lower and upper approximations of any $X(\subseteq W)$ as well. So any information system $\mathcal{S}$ and a set of attributes B yields an approximation space $(W, Ind_{\mathcal{S}}(B))$. It is not difficult to prove that, on the other hand, for a given approximation space (W, R), there exists an information system $\mathcal{S}$ and a set of attributes B such that $Ind_{\mathcal{S}}(B) = R$.

---

[*] The authors would like to thank the referees for their valuable comments.

C. Sossai and G. Chemello (Eds.): ECSQARU 2009, LNAI 5590, pp. 829–840, 2009.
© Springer-Verlag Berlin Heidelberg 2009

A logic that can express properties of information systems would be expected to have attribute and attribute-value constants in its language. In this paper, our aim is to define a logic for information systems which not only has these entities in its language, but can also express properties of lower and upper approximations of sets with respect to different subsets of attributes. The logic (denoted LIS) is introduced in Section 4, and it is shown that LIS can express various concepts related to dependencies in data and data reduction [11]. In Section 5, a sound and complete deductive system for LIS is presented, and decidability of LIS is proved in Section 6. The next section gives the preliminaries, Section 3 surveys related existing logics, and Section 7 concludes the article.

2   Preliminaries

We refer to [11] for the definitions presented in this section.

Let $\mathcal{S} := (W, A, Val := \bigcup_{a \in A} Val_a, f)$ be an information system.

Given $X \subseteq W$ and $B \subseteq A$, the lower and upper approximations (cf. Section 1) of X with respect to the indiscernibility relation $Ind_{\mathcal{S}}(B)$ are denoted as $\underline{Ind_{\mathcal{S}}(B)}(X)$ and $\overline{Ind_{\mathcal{S}}(B)}(X)$ respectively. $\underline{Ind_{\mathcal{S}}(B)}(X)$, $(\overline{Ind_{\mathcal{S}}(B)}(X))^c$ and $\overline{Ind_{\mathcal{S}}(B)}(X) \setminus \underline{Ind_{\mathcal{S}}(B)}(X)$ respectively consist of the *positive, negative* and *boundary elements* of X. If there are no boundary elements, X is said to be *definable*, i.e. in this case, $\underline{Ind_{\mathcal{S}}(B)}(X) = \overline{Ind_{\mathcal{S}}(B)}(X)$.

The notion of dependency of sets of attributes is given as follows.

Definition 2. *Let* $P, Q \subseteq A$.

    (a) Q *is said to* depend on P *(denoted* $P \Rightarrow Q$*), if* $Ind_{\mathcal{S}}(P) \subseteq Ind_{\mathcal{S}}(Q)$.
    (b) P *and* Q *are called* equivalent *(denoted* $P \equiv Q$*), if* $Ind_{\mathcal{S}}(P) = Ind_{\mathcal{S}}(Q)$.
    (c) P *and* Q *are* independent *(*$P \not\equiv Q$*), if neither* $P \Rightarrow Q$ *nor* $Q \Rightarrow P$ *hold.*

Given an information system, one may be interested in removing all 'superfluous' attributes, i.e. those which do not affect the partition of the domain, and consequently, set approximations. This is the main idea of *reduction of knowledge*. Formally, we have the definitions as below.

Definition 3. *Let* $P, Q \subseteq A$.

1. $POS_P(Q) := \bigcup\limits_{X \in W/Ind_{\mathcal{S}}(Q)} \underline{Ind_{\mathcal{S}}(P)}(X)$ *is the* P-positive region of Q, *where*
   $W/Ind_{\mathcal{S}}(Q)$ *denotes the quotient set for the equivalence relation* $Ind_{\mathcal{S}}(Q)$.
2. $b \in P$ *is said to be* Q−dispensable *in* P *if* $POS_P(Q) = POS_{(P \setminus \{b\})}(Q)$; *otherwise* b *is* Q−indispensable *in* P.
3. *If every* $b \in P$ *is* Q−indispensable, P *is* Q−independent; *otherwise* P *is* Q−dependent.
4. $S \subseteq P$ *will be called a* Q−reduct *of* P *if* S *is* Q−independent *and* $POS_S(Q) = POS_P(Q)$.

We note that $P \subseteq A$ may have multiple $Q-$reducts. Moreover, if P is infinite, then it may not have any $Q-$reduct at all. In the special case that $P = Q$, we drop the prefix '$Q-$' in the above. In this case, observe that $POS_P(Q) = U$, and the condition under which b is dispensable in P, reduces to $Ind_S(P) = Ind_S(P\backslash\{b\})$.

3   Related Work

Several logics are proposed in which the language includes attribute and attribute value constants (e.g. [8,9,7], or cf. [3,5]). In the logic NIL of Orłowska, the structures defining the models do not accommodate attributes, and the wffs (which are built using attribute constants) just point to collections of objects of the domain. On the other hand the logic DIL presented in [7] does not have modal operators for indiscernibility or any other relations induced by information systems. So it can only talk about the changes in attribute values of the objects with time, and not about (changes in) set approximations. A class of multimodal logics with attribute expressions are also defined in [8,9]. Models are based on structures of the form $(W, A, \{ind(P)\}_{P \subseteq A})$, where the indiscernibility relation $ind(P)$ for each subset P of the attribute set A, has to satisfy certain conditions. The language of the logics has a set of variables each representing a set of attributes, and constants to represent all singleton sets of attributes. The language can also express the result of operations on sets of attributes. However, as remarked in [9], a complete axiomatization for such logics is not known.

As mentioned in the Introduction, every information system and a set of attributes gives rise to an approximation space, and conversely, one obtains an information system and a set of attributes from any approximation space, such that the induced indiscernibility is just the equivalence relation of the approximation space. So it would appear that a semantics with models based on information systems would be 'equivalent' to one based on approximation spaces. However, as observed in [10], there is a difference. When we say that two objects are indistinguishable in an information system, we actually mean that these are indistinguishable not absolutely, but with respect to certain properties/attributes. So Orłowska proposed a structure with *relative accessibility* relations for the study of indiscernibility relations. These are of the form $(W, \{R_B\}_{B \subseteq A})$, called *information structure*, where W is a non-empty set, A is a non-empty set of parameters or attributes and for each $B \subseteq A$, R_B is an equivalence relation satisfying

$$R_\emptyset = W \times W \tag{1}$$
$$R_{B \cup C} = R_B \cap R_C. \tag{2}$$

We note that given an information system $S := (W, A, \bigcup_{a \in A} Val_a, f)$, the structure $(W, \{Ind_S(B)\}_{B \subseteq A})$ is an information structure. For every information structure $(W, \{R_B\}_{B \subseteq A})$, can we determine an information system $S := (W, A, \bigcup_{a \in A} Val_a, f)$ such that $Ind_S(B) = R_B$ for all $B \subseteq A$? The answer is yes, *provided A is finite.* This is due to the fact that an information structure may not have the property $R_B = \bigcap_{b \in B} R_b$, $B \subseteq \mathcal{A}$ as shown in Example 1 below, but we always have $Ind_{\mathcal{S}}(B) = \bigcap_{b \in B} Ind_{\mathcal{S}}(\{b\})$.

Example 1. Consider $\mathfrak{F} := (W := \{x, y\}, \{R_B\}_{B \subseteq \mathcal{A}})$ where $R_B := W \times W$ for any finite subset B of $\mathcal{A}$, while $R_B := Id_W$ for infinite B. Note that for any infinite B, we have $R_B \neq \bigcap_{b \in B} R_b$.

In the next section, we present LIS, the models of which are based on information structures $(W, \{R_B\}_{B \subseteq \mathcal{A}})$ with $\mathcal{A}$ finite. So LIS serves as a logic for information systems as well.

It should be mentioned that Orłowska [10] cited the axiomatization of a logic with semantics based on information structures as an open problem. Later, Balbiani gave a complete axiomatization of the set of wffs valid in every information structure. In fact, in [2], complete axiomatizations of logics with semantics based on various types of structures with relative accessibility relations is presented. One of these is a logic for information structures (cf. [1]). This, as required, is a multi-modal logic with a modal operator $[P]$ for each $P \subseteq \mathcal{A}$. Apart from the $S5$−axioms for each modal operator, the axiom $[P]\alpha \vee [Q]\alpha \to [P \cup Q]\alpha$ is considered. The canonical model obtained for this system only satisfies the condition $R_{B \cup C} \subseteq R_B \cap R_C$. Such a model is called *decreasing*. Using the method of *copying*, one obtains from a decreasing model, a model that satisfies condition (2) (viz. $R_{B \cup C} = R_B \cap R_C$) and preserves satisfiability as well. Note that condition (1) is not proved but one can obtain it using generated sub-models, as we have done for LIS in Section 5.1.

However, the language of Balbiani's logic does not contain attribute or attribute-value constants – a limitation that LIS overcomes.

4   The Logic LIS for Information Systems

The language of LIS contains (i) a non-empty finite set $\mathcal{AC}$ of attribute constants, (ii) for each $a \in \mathcal{AC}$, a non-empty finite set $\mathcal{VC}_a$ of attribute value constants and (iii) a non-empty countable set PV of propositional variables. Atomic formulae are the propositional variables p from PV, and *descriptors* [11], i.e. (a, v), for each $a \in \mathcal{AC}$, $v \in \mathcal{VC}_a$. The set of all descriptors is denoted as $\mathcal{D}$.

Using the Boolean logical connectives $\neg$ (negation) and $\wedge$ (conjunction) and unary modal connectives $[B]$ for each $B \subseteq \mathcal{AC}$, well-formed formulae (wffs) of LIS are then defined recursively as: $(a, v)|p|\neg\alpha|\alpha \wedge \beta|[B]\alpha$.

Let $\mathfrak{L}$ denote the set of all LIS-wffs.

4.1   Semantics

A LIS-model $\mathfrak{M}$ is a tuple $(W, \{R_B\}_{B \subseteq \mathcal{AC}})$ equipped with meaning functions for the descriptors and the propositional variables. Formally,

Definition 4. $\mathfrak{M} := (W, \{R_B\}_{B \subseteq \mathcal{AC}}, m, V)$ *where* W *is a non-empty set,* $R_B \subseteq W \times W$, $m : \mathcal{D} \to 2^W$, *and* $V : PV \to 2^W$.

We now proceed to define *satisfiability* of a wff α in a model $\mathfrak{M}$ at an object w of the domain W, denoted as $\mathfrak{M}, w \models \alpha$. The Boolean cases are omitted.

Definition 5

$\mathfrak{M}, w \models (a, v)$ *if and only if* $w \in m(a, v)$, *for* $(a, v) \in \mathcal{D}$.

$\mathfrak{M}, w \models p$, *if and only if* $w \in V(p)$, *for* $p \in PV$.

$\mathfrak{M}, w \models [B]\alpha$, *if and only if for all* w' *in* W *with* $(w, w') \in R_B$, $\mathfrak{M}, w' \models \alpha$.

For any wff α in $\mathcal{L}$ and LIS-model $\mathfrak{M}$, let $\mathfrak{M}(\alpha) := \{w \in W : \mathfrak{M}, w \models \alpha\}$.
α is valid in $\mathfrak{M}$, denoted $\mathfrak{M} \models \alpha$, if and only if $\mathfrak{M}(\alpha) = W$.
α is said to be valid, if $\mathfrak{M} \models \alpha$ for every model $\mathfrak{M}$. It will be denoted by $\models \alpha$.

Since we wish to define a logic for information systems, some properties must be imposed on the structure defined above. Thus we have the following.

Definition 6. *By an IS-structure, we mean a tuple $\mathfrak{F} := (W, \{R_B\}_{B \subseteq AC}, m)$, where W, R_B and m are the same as in Definition 4 satisfying, in addition:*

(IS1) *For each $a \in AC$, $\bigcup\{m(a, v) : v \in VC_a\} = W$.*
(IS2) *For each $a \in AC$, $m(a, v) \cap m(a, v') = \emptyset$, for $v \neq v'$.*
(IS3) $R_\emptyset = W \times W$.
(IS4) $R_B \subseteq R_C$ *for* $C \subseteq B \subseteq AC$.
(IS5) *For $B \subseteq AC$ and $b \in AC$, $R_B \cap R_b \subseteq R_{B \cup \{b\}}$.*
(IS6) *For $b \in AC$, $(w, w') \in R_b$ if and only if there exists $v \in VC_b$ such that $w, w' \in m(b, v)$*

Note that $R_B = \bigcap_{b \in B} R_b$, and so $R_{B \cup C} = R_B \cap R_C$. Each R_B is an equivalence relation. So the tuple $(W, \{R_B\}_{B \subseteq AC})$ in $\mathfrak{F}$ forms an information structure (cf. Section 3).

Also note that in the definition of IS-structure, one can replace the condition (IS5) by

(IS5') *For $B \subseteq AC$ and $b \in AC$, if $(w, w') \in R_B$ and there exists $v \in VC_b$ such that $w, w' \in m(b, v)$, then $(w, w') \in R_{B \cup \{b\}}$.*

(IS5') is useful for getting the axiomatization of the logic for IS-structures, as we shall see in the next section.

LIS-models based on IS-structures are called IS-models.

Given an information system $\mathcal{S} := (W, AC, \bigcup_{a \in AC} VC_a, f)$, the structure $(W, \{Ind_\mathcal{S}(B)\}_{B \subseteq AC}, m^\mathcal{S})$, where $m^\mathcal{S}(a, v) := \{w \in W : f(w, a) = v\}$, is an IS-structure. We shall call it the *standard IS-structure* generated by $\mathcal{S}$, following Vakarelov [12]. It is not difficult to show that every IS-structure is a standard IS-structure generated by some information system. Let *standard IS-models* be the IS-models based on standard IS-structures. We shall write $\models_{IS} \alpha$ and $\models_{SIS} \alpha$ if α is valid in all IS-models and all standard IS-models respectively. From the above remark, we obtain,

Proposition 1. $\models_{IS} \alpha$ *if and only if* $\models_{SIS} \alpha$ *for all* $\alpha \in \mathfrak{L}$.

Let $\emptyset \neq B := \{b_1, b_2, \ldots, b_n\} \subseteq \mathcal{AC}$. Let $\mathcal{D}_B$ be the set of all wffs of the form $(b_1, v_1) \wedge (b_2, v_2) \wedge \ldots \wedge (b_n, v_n)$, $v_i \in \mathcal{VC}_{b_i}$, $i = 1, 2, \ldots n$. In the case when $B = \emptyset$, we define $\mathcal{D}_B := \{\top\}$. Then each element of the set $\mathcal{D}_B$ represents the empty set or an equivalence class with respect to the equivalence relation $Ind(B)$. In fact, $\bigwedge_{i=1}^{n}(b_i, v_i)$ represents the equivalence class of objects which take the value v_i for the attribute b_i, $i = 1, 2, \ldots, n$. More formally, we have the following proposition.

Proposition 2. *Let* $\mathfrak{M} := (W, \{R_B\}_{B \subseteq \mathcal{AC}}, m, V)$ *be an IS-model. Then*

$$\{\mathfrak{M}(\alpha) : \alpha \in \mathcal{D}_B\} \setminus \{\emptyset\} = \{[w]_{R_B} : w \in W\}, \text{ for any } B \subseteq \mathcal{AC}.$$

So if $\mathfrak{M}' := (W, \{R_B\}_{B \subseteq \mathcal{AC}}, m, V')$ *then* $\mathfrak{M}(\alpha) = \mathfrak{M}'(\alpha)$ *for all* $\alpha \in \mathcal{D}_B$.

The next propositions show how the language of LIS may be used to express the concepts presented in Section 2.

Let $\mathcal{S} := (W, \mathcal{AC}, \bigcup_{a \in \mathcal{AC}} \mathcal{VC}_a, f)$ be an information system and consider the corresponding standard IS-structure $\mathfrak{F}_\mathcal{S} = (W, \{Ind_\mathcal{S}(B)\}_{B \subseteq \mathcal{AC}}, m^\mathcal{S})$.

Proposition 3. *Let* $P, Q, S \subseteq \mathcal{AC}$, *and* p, q *be distinct propositional variables. Then the following hold.*

1. $P \Rightarrow Q$ *if and only if* $[Q]p \to [P]p$ *is valid in* $\mathfrak{F}_\mathcal{S}$, *i.e.* $\mathfrak{M} \models [Q]p \to [P]p$, *for all models* $\mathfrak{M}$ *based on* $\mathfrak{F}_\mathcal{S}$.
2. $P \not\equiv Q$ *if and only if* $\neg[\emptyset]([Q]p \to [P]p) \wedge \neg[\emptyset]([P]q \to [Q]q)$ *is satisfiable in* $\mathfrak{F}_\mathcal{S}$, *i.e. there is* $\mathfrak{M}$ *based on* $\mathfrak{F}_\mathcal{S}$, *and* $w \in W$ *where the wff is satisfied.*
3. $b \in P$ *is dispensable in* P *if and only if* $[P]p \leftrightarrow [P \setminus \{b\}]p$ *is valid in* $\mathfrak{F}_\mathcal{S}$.
4. P *is dependent if and only if* $\bigvee_{b \in P}([P]p_b \leftrightarrow [P \setminus \{b\}]p_b)$ *is valid in* $\mathfrak{F}_\mathcal{S}$, *where* $\{p_b : b \in P\}$ *is a set of distinct propositional variables.*
5. $Q \subseteq P$ *is a reduct of* P *if and only if* $\bigwedge_{b \in Q}[\emptyset]\neg([Q]q_b \leftrightarrow [Q \setminus \{b\}]q_b)$ *is satisfiable in* $\mathfrak{F}_\mathcal{S}$ *and* $[Q]p \leftrightarrow [P]p$ *is valid in* $\mathfrak{F}_\mathcal{S}$.

Proposition 4. *Let* $\mathfrak{M} := (W, \{Ind_\mathcal{S}(B)\}_{B \subseteq \mathcal{AC}}, m^\mathcal{S}, V)$ *be the standard IS-model on* $\mathfrak{F}_\mathcal{S}$, *for some valuation function* V. *Let* $P, Q, S \subseteq \mathcal{AC}$. *Then the following hold.*

1. $\mathfrak{M}(\bigvee_{\alpha \in \mathcal{D}_Q}[P]\alpha) = POS_P(Q)$.
2. $b \in P$ *is* Q*-dispensable in* P *if and only if* $\bigvee_{\alpha \in \mathcal{D}_Q}[P]\alpha \leftrightarrow \bigvee_{\alpha \in \mathcal{D}_Q}[P \setminus \{b\}]\alpha$ *is valid in* $\mathfrak{M}$.
3. $b \in P$ *is* Q*-indispensable in* P *if and only if* $\langle \emptyset \rangle \neg(\bigvee_{\alpha \in \mathcal{D}_Q}[P]\alpha \leftrightarrow \bigvee_{\alpha \in \mathcal{D}_Q}[P \setminus \{b\}]\alpha)$ *is valid in* $\mathfrak{M}$.
4. P *is* Q*-independent in* P *if and only if* $\bigwedge_{b \in P}\langle \emptyset \rangle \neg(\bigvee_{\alpha \in \mathcal{D}_Q}[P]\alpha \leftrightarrow \bigvee_{\alpha \in \mathcal{D}_Q}[P \setminus \{b\}]\alpha)$ *is valid in* $\mathfrak{M}$.

5. $S \subseteq P$ is a $Q-$reduct of P if and only if
$$[\emptyset](\bigwedge_{b \in S} \langle \emptyset \rangle \neg (\bigvee_{\alpha \in \mathcal{D}_Q} [S]\alpha \leftrightarrow \bigvee_{\alpha \in \mathcal{D}_Q} [S \setminus \{b\}]\alpha) \wedge (\bigvee_{\alpha \in \mathcal{D}_Q} [P]\alpha \leftrightarrow \bigvee_{\alpha \in \mathcal{D}_Q} [S]\alpha))$$ is valid in $\mathfrak{M}$.

We next see that the expressive power of LIS will not be affected even if only one modal operator, viz. $[\emptyset]$, is taken in the language.

Proposition 5. Let $B \subseteq \mathcal{AC}$. Then $[B]\alpha \leftrightarrow \bigwedge_{\beta \in \mathcal{D}_B}(\beta \to [\emptyset](\beta \to \alpha))$ is valid in all IS-models.

So Proposition 5 shows that for every wff α, there exists a wff α' such that $\alpha \leftrightarrow \alpha'$ is valid in all IS-models, and α' does not involve any modal operator $[B]$, where $B(\neq \emptyset) \subseteq \mathcal{AC}$. However, the complexity of α', denoted as $|\alpha'|$, could be very large compared to α. For instance, if α is of the form $\underbrace{[B][B]\ldots[B]}_{n-times} \beta$, where $|\mathcal{D}_B| = m$, then $|\alpha'| > m^n|\beta|$.

One may think of strengthening this result by requiring α' to be such that it does not even involve the modal operator $[\emptyset]$. This is not possible as shown by the following example.

Example 2. Let $B \subseteq \mathcal{AC}$. Choose $a \in \mathcal{AC}$ such that $a \notin B$. For each $b(\neq a) \in \mathcal{AC}$, choose a $v_b \in \mathcal{VC}_b$, and $v_a^1, v_a^2 \in \mathcal{VC}_a$. Let $W := \{x, y\}$, and consider the IS-structures $\mathfrak{F} := (W, \{R_C\}_{C \subseteq \mathcal{AC}}, m)$ and $\mathfrak{F}' := (W, \{R'_C\}_{C \subseteq \mathcal{AC}}, m')$, where

- $m(b, v_b) = m'(b, v_b) := \{x, y\}$, $m(b, v) = m'(b, v) := \emptyset$ for all $b(\neq a) \in \mathcal{AC}$ and $v(\neq v_b) \in \mathcal{VC}_b$,
- $m(a, v_a^1) := \{x, y\}$, $m'(a, v_a^1) := \{x\}$, $m'(a, v_a^2) := \{y\}$,
- $m(a, v) := \emptyset$ for all $v(\neq v_a^1) \in \mathcal{VC}_a$,
- $m'(a, v) := \emptyset$ for all $v \in \mathcal{VC}_a \setminus \{v_a^1, v_a^2\}$,
- $R_C := W \times W$ for all $C \subseteq \mathcal{AC}$,
- for $a \notin C \subseteq \mathcal{AC}$, $R'_C := W \times W$ and for $a \in C$, $R'_C := Id_W$.

Let us consider the models $\mathfrak{M} := (\mathfrak{F}, V)$ and $\mathfrak{M}' := (\mathfrak{F}', V)$, for any V. We see that $\mathfrak{M}, x \models [B](a, v_a^1)$, while $\mathfrak{M}', x \not\models [B](a, v_a^1)$. But one can show that for any α which does not involve *any* modal operator,
$$\mathfrak{M}, x \models \alpha \text{ if and only if } \mathfrak{M}', x \models \alpha.$$
So the wff $[B](a, v_a^1)$ cannot be logically equivalent to a wff that does not contain any modal operator.

We end this section by giving some wffs which are satisfiable/valid in both the class of IS and standard IS-structures.

Proposition 6

1. $[B]p \leftrightarrow ((a_1, v_1) \wedge (a_2, v_2) \wedge \cdots (a_n, v_n))$ is satisfiable.
2. $\neg[B]p \leftrightarrow (\bigwedge_{\alpha \in \mathcal{D}_B}(\alpha \to \langle \emptyset \rangle(\alpha \wedge \neg p)))$ is valid.
3. $\langle \emptyset \rangle(b, v)$ is satisfiable.
4. $[\emptyset](\bigwedge_{i \in \{1,2,\ldots,n\}}(a_i, v_i) \to (a, v))$ is satisfiable.

5. $\bigwedge_{i\in\{1,2,\ldots,n\}}(b_i,v_i) \leftrightarrow [B](\bigwedge_{i\in\{1,2,\ldots,n\}}(b_i,v_i))$, $B = \{b_1, b_2, \ldots, b_n\}$ *is valid.*
6. $\bigwedge_{i\in\{1,2,\ldots,n+1\}}\langle\emptyset\rangle(p_i \wedge (b,v)) \to \bigvee_{i\neq j}\langle\emptyset\rangle(p_i \wedge p_j)$ *is satisfiable.*
7. $[\emptyset][B](a,v) \leftrightarrow [\emptyset](a,v)$ *is valid.*
8. $\langle B\rangle\delta \to [B]\delta$ *is valid, where δ is the wff obtained by applying only Boolean connectives on descriptors (b,v), $b \in B$.*

If the wff in (1) is valid in an IS-model $\mathfrak{M}$, an object x is in the lower approximation of the set represented by p with respect to indiscernibility corresponding to the attribute set B, if and only if x takes the value v_i for the attribute a_i, $1 \leq i \leq n$. Validity of the wff in (2) guarantees that an object x is not a positive element of a set X with respect to a attribute set, say B if and only if there exists an object y which takes the same attribute value as x for each attribute of B but $y \notin X$. The wff in (3) says that there is an object that takes the value v for the attribute b. The wff in (4) represents a situation where there is an attribute set $\{a_1, a_2, \ldots, a_n, a\}$ and attribute values $v \in \mathcal{VC}_a$, $v_i \in \mathcal{VC}_{a_i}, 1 \leq i \leq n$ such that any object which takes the value v_i for the attribute a_i, $1 \leq i \leq n$, will take the value v for the attribute a. The wff in (5) represents the fact that if any object, say x takes the value v_i for the attribute b_i, $1 \leq i \leq n$, then every $R_{\{b_1,b_2,\ldots,n\}}$ related objects of x will also take the value v_i for the attribute b_i, $1 \leq i \leq n$. The wff in (6) is valid only in an IS-frame where there are at most n elements taking the value v for the attribute b. The wff in (8) represents the fact that any property defined using only Boolean connectives and attributes from the set B is definable with respect to the partition induced by $Ind(B)$.

5   Soundness and Completeness Theorems for LIS

In this section we present an axiomatic system for LIS and get the soundness and completeness theorems with respect to the IS-models. Note that in case of the IS-model, the modal operator $[\emptyset]$ is interpreted as the global modal operator [4]. Let $B, C \subseteq \mathcal{AC}$.

Axiom schema:

Ax1. All axioms of classical propositional logic (PL).
Ax2. $[B](\alpha \to \beta) \to ([B]\alpha \to [B]\beta)$.
Ax3. $[\emptyset]\alpha \to \alpha$.
Ax4. $\alpha \to [\emptyset]\langle\emptyset\rangle\alpha$.
Ax5. $\langle\emptyset\rangle\langle\emptyset\rangle\alpha \to \langle\emptyset\rangle\alpha$.
Ax6. $[C]\alpha \to [B]\alpha$ for $C \subseteq B \subseteq \mathcal{AC}$.
Ax7. $(a,v) \to \neg(a,v')$, for $v \neq v'$.
Ax8. $\bigvee_{v\in\mathcal{VC}_a}(a,v)$.
Ax9. $(a,v) \to [a](a,v)$.
Ax10. $((b,v) \wedge [B\cup b]\alpha) \to [B]((b,v) \to \alpha)$.

Rules of inference:

$$N. \quad \frac{\alpha}{[B]\alpha} \qquad MP. \quad \frac{\alpha \quad \alpha \to \beta}{\beta}$$
$$\textit{where } B \subseteq \mathcal{AC}$$

Ax3-Ax5 are the usual modal axioms. So $[\emptyset]$ satisfies all the $S5$–axioms. The notion of theoremhood is defined in the usual way, with notation $\vdash \alpha$ to say 'α is a theorem'. We note that it is not necessary to write the $S5$–axioms for the operators $[B]$, $B \neq \emptyset$, as these are theorems. Ax6-Ax8 correspond to (IS4),(IS2) and (IS1) respectively of Definition 6. Ax9 and Ax10 establish the relationship between the indiscernibility relation and attribute, attribute value pairs. Ax10 is the syntactic counterpart for the condition (IS5'). Note that $[B \cup b]\alpha \rightarrow [B]\alpha \wedge [b]\alpha$ would appear to be the counterpart of the condition (S5), but in fact, there are IS-models in which it is not valid. For instance, consider any IS-model $\mathfrak{M} := (W := \{w_1, w_2, w_3\}, \{R_B\}_{B \subseteq \mathcal{AC}}, m, V)$ such that for $a, b \in \mathcal{AC}$, $W/R_a := \{\{w_1, w_2\}, \{w_3\}\}$, $W/R_b := \{\{w_1, w_3\}, \{w_2\}\}$ and $V(p) = \{w_1\}$. Then $\mathfrak{M}, w_1 \not\models [\{a, b\}]p \rightarrow [a]p \wedge [b]p$.

We observe that the n-agent epistemic logic $S5_n^D$ [6] with knowledge operators K_i ($i = 1, \ldots, n$) and distributed knowledge operators D_G for groups G of agents, is embeddable in LIS with $|\mathcal{AC}| \geq n$. Suppose $\mathcal{AC} := \{a_1, a_2, \ldots, a_m\}$, $m \geq n$. Then the embedding Ψ fixes propositional variables, and takes $K_i\alpha$ to $[a_i]\,\Psi(\alpha)$ and $D_{\{i_1, i_2, \ldots, i_s\}}\alpha$ to $[\{a_{i_1}, a_{i_2}, \ldots, a_{i_s}\}]\,\Psi(\alpha)$. On the other hand, LIS is more expressive than $S5_n^D$, having the extra feature of the descriptors. Any study of indiscernibility relations induced by information systems naturally involves descriptors, as these determine both what value an object will take for an attribute, and also the indiscernibility relation itself (shown by Ax9 and Ax10).

To illustrate the proof system, let us give a Hilbert-style proof of a LIS-wff.

Proposition 7. $\vdash (a, v_a) \wedge (b, v_b) \rightarrow [\{a, b\}]((a, v_a) \wedge (b, v_b))$.

Proof

(1) $\vdash (a, v_a) \wedge (b, v_b) \rightarrow [a](a, v_a) \wedge [b](b, v_b)$ (Ax9 and PL).
(2) $\vdash [a](a, v_a) \wedge [b](b, v_b) \rightarrow [\{a, b\}](a, v_a) \wedge [\{a, b\}](b, v_b)$ (Ax6 and PL).
(3) $\vdash [\{a, b\}](a, v_a) \wedge [\{a, b\}](b, v_b) \rightarrow [\{a, b\}]((a, v_a) \wedge (b, v_b))$
$\qquad\qquad\qquad\qquad\qquad\qquad\qquad$ (Modal (K-)theorem).
(4) $\vdash (a, v_a) \wedge (b, v_b) \rightarrow [\{a, b\}]((a, v_a) \wedge (b, v_b))$ ((1), (2), (3) and PL). $\quad\square$

It is not difficult to obtain

Theorem 1 (Soundness). *If $\vdash \alpha$, then $\models_{IS} \alpha$.*

5.1 Completeness

The completeness theorem is proved for any LIS-wff α, following the standard modal logic technique [4]. As in normal modal logic, we have the following result.

Proposition 8. *Every consistent set of LIS-wffs has a maximally consistent extension.*

We now describe the *canonical model* $\mathfrak{M}^C$ for LIS.

Definition 7 (Canonical model). $\mathfrak{M}^C := (W^C, \{R_B^C\}_{B \subseteq AC}, m^C, V^C)$, *where*

- $W^C := \{w : w$ *is a* maximally consistent set$\}$,
- *for each* $B \subseteq AC$, $(w, w') \in R_B^C$ *if and only if for all wffs* α,
$$[B]\alpha \in w \text{ implies } \alpha \in w',$$
- $m^C(a, v) := \{w \in W^C : (a, v) \in w\}$,
- $V^C(p) := \{w \in W^C : p \in w\}$.

By giving the same argument as in normal modal logic, we obtain

Proposition 9 (Truth Lemma). *For any wff* β *and* $w \in W^C$,
$$\beta \in w \text{ if and only if } \mathfrak{M}^C, w \models \beta.$$

Using the Truth Lemma and Proposition 8 we have

Proposition 10. *If* α *is consistent then there exists a maximal consistent set* Σ *such that* $\mathfrak{M}^C, \Sigma \models \alpha$.

Proposition 11. *The canonical model* $\mathfrak{M}^C$ *satisfies* (IS1), (IS2), (IS4), (IS5′) *and* **(IS6a)** *for* $b \in AC$, *if* $(w, w') \in R_b$, *there is* $v \in VC_b$ *such that* $w, w' \in m(b, v)$.

Proof. We only prove (IS5′) and (IS6a).
(IS5′) Let $(w, w') \in R_B$ and let there exist a $v \in VC_b$ such that $w, w' \in m^C(b, v)$. Further, suppose $[B \cup \{b\}]\alpha \in w$. We need to prove $\alpha \in w'$. Using Ax10 and the fact that $(b, v) \wedge [B \cup \{b\}]\alpha \in w$, we obtain $[B]((b, v) \rightarrow \alpha) \in w$. This gives $(b, v) \rightarrow \alpha \in w'$ and hence $\alpha \in w'$ as $(b, v) \in w'$.

(IS6a) Let $(w, w') \in R_b$, and $v \in VC_b$ be such that $w \in m^C(b, v)$, i.e. $(b, v) \in w$. Then by Ax9, we obtain $[b](b, v) \in w$. So $(b, v) \in w'$, i.e $w' \in m^C(b, v)$. □

Note that we still have not proved (IS3) and the other direction of (IS6). In order to get these properties, we construct a new model from $\mathfrak{M}^C$.
    Let $\mathfrak{M}^g := (W^g, \{R_B^g\}_{B \subseteq AC}, m^g, V^g)$ be the sub-model of $\mathfrak{M}^C$ generated by Σ using the equivalence relation $R_\emptyset^C$, i.e.

- W^g is the equivalence class of Σ with respect to the equivalence relation $R_\emptyset^C$,
- R_B^g, m^g, V^g are the restrictions of R_B^C, m^C, V^C to W^g respectively.

Proposition 12. $\mathfrak{M}^g := (W^g, \{R_B^g\}_{B \subseteq AC}, m^g, V^g)$ *is an IS model.*

Proof. Clearly (IS1)-(IS4), (IS5′) and (IS6a) are satisfied. We only prove the other direction of (IS6). Let there exist $v \in VC_b$ such that $w, w' \in m^g(b, v)$. Let $[b]\alpha \in w$. We want to show $\alpha \in w'$. Here we have $(b, v) \wedge [b]\alpha \in w$ and hence from Ax10 with $B = \emptyset$, we obtain $[\emptyset]((b, v) \rightarrow \alpha) \in w$. Since $wR_\emptyset^g w'$, we obtain $(b, v) \rightarrow \alpha \in w'$. This together with $(b, v) \in w'$ gives $\alpha \in w'$. □

Since $R_B^C \subseteq R_\emptyset^C$ for all $B \subseteq AC$, $\mathfrak{M}^g$ is also a generated sub-model of $\mathfrak{M}^C$ with respect to R_B^C. An easy induction on the complexity of the wff α gives us

Proposition 13. *For each wff α and $w \in W^g$, we have*
$$\mathfrak{M}^C, w \models \alpha, \text{ if and only if } \mathfrak{M}^g, w \models \alpha.$$

Thus we get the completeness theorem with respect to the class of all IS-models.

Theorem 2 (Completeness). *For any LIS-wff α, if $\models_{IS} \alpha$, then $\vdash \alpha$.*

Note that due to Proposition 1, this also gives the completeness theorem with respect to the class of all standard IS-models.

6   Decidability

In this section, our aim is to prove the following result.

Theorem 3. *We can decide for a given $\alpha \in \mathfrak{L}$, whether $\models_{IS} \alpha$ ($\models_{SIS} \alpha$).*

For this, we prove that LIS has the finite model property (Proposition 17 below). The standard filtration technique [4] is used, with natural modifications to the definitions. Let Σ denote a finite sub-formula closed set of LIS-wffs.

Let $\mathfrak{M} := (W, \{R_B\}_{B \subseteq AC}, m, V)$ be a LIS-model. We define a binary relation $\equiv_\Sigma$ on W as follows:

$w \equiv_\Sigma w'$, if and only if for all $\beta \in \Sigma \cup \mathcal{D}$, $\mathfrak{M}, w \models \beta$ if and only if $\mathfrak{M}, w' \models \beta$.

Definition 8 (Filtration model). *Given a model $\mathfrak{M} = (W, \{R_B\}_{B \subseteq AC}, m, V)$ and Σ as above, we define a model $\mathfrak{M}^f = (W^f, \{R_B^f\}_{B \subseteq AC}, m^f, V^f)$, where*

- *$W^f := \{[w] : w \in W\}$, $[w]$ is the equivalence class of w with respect to the equivalence relation $\equiv_\Sigma$;*
- *$R_B^f \subseteq W^f \times W^f$ is defined as:*
  $[w]R_B^f[u]$ if and only if there exist $w' \in [w]$ and $u' \in [u]$ such that $w'R_Bu'$;
- *$V^f : PV \to 2^{W^f}$ is defined as: $V^f(p) := \{[w] \in W^f : w \in V(p)\}$;*
- *$m^f(a, v) := \{[w] \in W^f : w \in m(a, v)\}$.*

$\mathfrak{M}^f$ is the filtration of $\mathfrak{M}$ through the sub-formula closed set Σ.

Proposition 14. *For any model $\mathfrak{M}$, if $\mathfrak{M}^f$ is a filtration of $\mathfrak{M}$ through Σ, then the domain of $\mathfrak{M}^f$ contains at most 2^n elements, where $|\Sigma \cup \mathcal{D}| = n$.*

Proof. Define the map $g : W^f \to 2^{\Sigma \cup \mathcal{D}}$ where $g([w]) := \{\beta \in \Sigma \cup \mathcal{D} : \mathfrak{M}, w \models \beta\}$. Since g is injective, W^f contains at most 2^n elements. $\qquad\square$

If $(W, \{R_B\}_{B \subseteq AC}, m)$ is an IS-structure, so is $(W^f, \{R_B^f\}_{B \subseteq AC}, m^f)$. This gives

Proposition 15. *If the model $\mathfrak{M}$ is an IS-model then $\mathfrak{M}^f$ is also an IS-model.*

By induction on the complexity of the wff α, we therefore have

Proposition 16 (Filtration Theorem). *Let Σ be a finite sub-formula closed set of LIS-wffs. For all wffs $\beta \in \Sigma \cup \mathcal{D}$, all models $\mathfrak{M}$, and all objects $w \in W$,*
$$\mathfrak{M}, w \models \beta \text{ if and only if } \mathfrak{M}^f, [w] \models \beta.$$

So we get

Proposition 17 (Finite model property). *Let α be a wff and Σ be the set of all sub-wffs of α. If α is satisfiable, then it is satisfiable in a finite model with at most $2^{|\Sigma \cup \mathcal{D}|}$ elements.*

7   Conclusions

A logic LIS for the study of information systems is proposed. It is shown that the wffs of LIS can express many important properties related to rough set theory. Questions of axiomatization and decidability of the logic are also addressed.

The language of LIS contains finite sets of attributes and attribute values. Although in practical problems, we usually consider finite sets, it would be interesting to consider infinite sets of attributes and attribute values. Another issue of interest to us is to see whether LIS may be suitably extended to capture the situation when the information system changes with time. This could be the result of an inflow of information, due to which there is a change in the attributes or attribute values.

References

1. Balbiani, P.: Axiomatization of logics based on Kripke models with relative accessibility relations. In: Orłowska, E. (ed.) Incomplete Information: Rough Set Analysis, pp. 553–578. Physica Verlag, Heidelberg (1998)
2. Balbiani, P., Orłowska, E.: A hierarchy of modal logics with relative accessibility relations. Journal of Applied Non-Classical Logics 9(2-3), 303–328 (1999)
3. Banerjee, M., Khan, M.A.: Propositional logics from rough set theory. In: Peters, J.F., Skowron, A., Düntsch, I., Grzymała-Busse, J.W., Orłowska, E., Polkowski, L. (eds.) Transactions on Rough Sets VI. LNCS, vol. 4374, pp. 1–25. Springer, Heidelberg (2007)
4. Blackburn, P., de Rijke, M., Venema, Y.: Modal Logic. Cambridge University Press, Cambridge (2001)
5. Demri, S., Orłowska, E.: Incomplete Information: Structure, Inference, Complexity. Springer, Heidelberg (2002)
6. Fagin, R., Halpern, J.Y., Moses, Y., Vardi, M.Y.: Reasoning About Knowledge. MIT Press, Cambridge (1995)
7. Orłowska, E.: Dynamic information systems. Fundamenta Informaticae 5, 101–118 (1982)
8. Orłowska, E.: Logic of nondeterministic information. Studia Logica 44(1), 91–100 (1985)
9. Orłowska, E.: Logic of indiscernibility relations. In: Goos, G., Hartmanis, J. (eds.) SCT 1984. LNCS, vol. 208, pp. 177–186. Springer, Heidelberg (1985)
10. Orłowska, E.: Kripke semantics for knowledge representation logics. Studia Logica 49(2), 255–272 (1990)
11. Pawlak, Z.: Rough Sets. Theoretical Aspects of Reasoning about Data. Kluwer Academic Publishers, Dordrecht (1991)
12. Vakarelov, D.: Abstract characterization of some knowledge representation systems and the logic NIL of nondeterministic information. In: Jorrand, Ph., Sgurev, V. (eds.) Artificial Intelligence II, pp. 255–260. North-Holland, Amsterdam (1987)

An Uncertainty-Based Belief Selection Method for POMDP Value Iteration

Qi Feng, Xuezhong Zhou, Houkuan Huang, and Xiaoping Zhang

School of Computer and Information Technology, Beijing Jiaotong University,
Beijing 100044, China
fengqi2008@gmail.com, xzzhou@bjtu.edu.cn, hkhuang@bjtu.edu.cn

Abstract. Partially Observable Markov Decision Process (POMDP) provides a probabilistic model for decision making under uncertainty. Point-based value iteration algorithms are effective approximate algorithms to solve POMDP problems. Belief selection is a key step of point-based algorithm. In this paper we provide a belief selection method based on the uncertainty of belief point. The algorithm first computes the uncertainties of the belief points that could be reached, and then selects the belief points that have lower uncertainties and whose distances to the current belief set are larger than a threshold. The experimental results indicate that this method is effective to gain an approximate long-term discounted reward using fewer belief states than the other point-based algorithms.

Keywords: POMDP, value iteration, point-based algorithm, belief selection, uncertainty.

1 Introduction

Partially Observable Markov Decision Process (POMDP) constitutes a powerful mathematical model for planning under uncertainty environment [1, 2].

Value iteration algorithm [3] for POMDP is a well-known method. The optimal policy of POMDP could be computed using the optimal value function over the belief space. But traditional exact value iteration needs to update the value function over the entire belief space which makes it infeasible to solve the real-world POMDP problems. This motivated approximate algorithms for POMDP which have been proven to scale efficiently. Recently, point-based [4-10] approximate algorithms are known to be the promising approaches for computing value functions, such as PBVI [5], HSVI [6], Perseus [7], Breadth first belief selection (BFBS) [8], Distance-based belief expansion (DBBE) [9] and FSVI [10].

The selection of belief points is a crucial step in point-based algorithm. The algorithms mentioned above use different heuristic methods for searching through belief space. For example, PBVI [5] chooses the belief point that is farthest away from the points already existing in the belief set, but PBVI may ignore some belief points that are important to update the value function. BFBS algorithm [8] expands the belief set to include all the beliefs that are reachable in the next time step. And the drawback of this algorithm is the size of the belief set may be expanded exponentially. DBBE [9]

C. Sossai and G. Chemello (Eds.): ECSQARU 2009, LNAI 5590, pp. 841–849, 2009.
© Springer-Verlag Berlin Heidelberg 2009

improves BFBS algorithm by adding these beliefs that are farther than a threshold, but it is not feasible for large-scale domains.

In this paper, we propose a belief selection algorithm based on the uncertainty of belief state (Uncertainty-based belief selection, UBBS). There exists the fact that the optimal value function over the belief space is piecewise linear and convex [1]. Agent could gain more long-term reward at the belief state that has lower uncertainty [2]. This property of convexity motivates that belief states that have lower uncertainties would be more helpful to improve value function. Our algorithm first computes the uncertainties of belief points that could be reached, and then adds to belief set the belief states that have lower uncertainties and whose distances to the current set are larger than a threshold. We give two different methods to describe uncertainty of a belief point: one uses entropy of a belief state, the other is based on the gap between the maximal and the minimal elements of a belief vector.

The experimental results show that UBBS method for belief set expansion is effective: it could gain an approximate long-term discounted reward using fewer belief states compared with other point-based algorithms.

This paper is organized as follows: in section 2, we introduce some basic concepts of POMDP and value iteration. In section 3, we present several related point-based value iteration algorithms. In section 4, we propose the UBBS algorithm and give two methods to describe the uncertainty of a belief point. In section 5, we describe the experimental results on POMDP benchmark problems. And we draw conclusions in section 6.

2   POMDP and Value Iteration

A POMDP framework is represented as a tuple $<S, A, \Omega, R, T, O, \gamma>$, where S is a finite set of discrete world states, A is a set of discrete actions, Ω is a set of observations that provide incomplete information about world states, R is the reward function where $R(s, a)$ is the reward received when taking action a at state s, T is the transition probability distribution where $T(s,a,s') = P(s'|s,a)$ represents the probability of transition from state s to state s' when taking action a, O is the observation function where $O(s',a,o) = P(o|a,s')$ is the distribution describing the probability of observing o from state s' after using action a, and γ is discount factor [3].

An important assumption of POMDP is that the states are not completely observable, so the agent maintains a belief state, denoted as $|S|$-dimension vector b, to represent the probability distribution over states. A belief state is a sufficient statistic for the given history. And the transition from a belief state to another is still Markovian, that is, the next belief state is depended on the current belief state and the current action.

Given an action a and an observation o, the belief state is updated at each time step by Bayesian rule:

$$b_a^o(s) = \tau(b,a,o) = \frac{O(a,s',o)\sum_{s \in S} b(s)T(s,a,s')}{p(o \mid b,a)} \qquad (1)$$

where,

$$p(o \mid b, a) = \sum_{s \in S} b(s) \sum_{s' \in S} T(s, a, s') O(a, s', o) \quad. \tag{2}$$

The goal of POMDP is to find a optimal sequence of actions $\{a_0, a_1, \ldots, a_t\}$, denoted as the optimal policy π, that maximizes the expected sum of long-term discounted reward $E\left[\sum_t \gamma^t R(s_t, a_t)\right]$. If the state of agent is partially observable, the goal function above is changed to maximize expected reward for a belief state as follows:

$$V(b) = \max_{a \in A} \left[R(b, a) + \gamma \sum_{b' \in B} T(b, a, b_a^o) V(b_a^o) \right] \quad. \tag{3}$$

The optimal policy could be computed using value iteration algorithm which updates the value function by applying backup operator on the previous value.

It is well known that the value function over belief states is piecewise linear and convex [1]. Then the optimal value function over the belief space is represented as a maximum of the inner product of the belief state and the α-vectors: $V(b) = \max_{\{\alpha\}} b \cdot \alpha$. After each iteration, the value function is constructed by a collection of hyperplanes. Each hyperplane is related to an action. And the α-vector is a vector of coefficients of the related hyperplane. The maximal value over the belief space is represented by the upper surface of hyperplanes (Fig. 1).

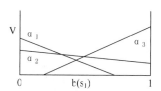

Fig. 1. Vector representation of value function

Rewriting the value function (equation 3) using vector representation, we could write the value update as:

$$V_{n+1}(b) = \max_a \left[\sum_{s \in S} b(s) R(s, a) + \gamma \sum_{o \in \Omega} P(o \mid b, a) V_n(b_a^o) \right] \tag{4}$$

$$= \max_a \left[\sum_{s \in S} b(s) R(s, a) + \gamma \sum_{o \in \Omega} P(o \mid b, a) \max_{\{\alpha_n\}} \sum_{s' \in S} b_a^o(s') \alpha_n(s') \right] \tag{5}$$

$$= \max_a \left[\sum_{s \in S} b(s) R(s, a) + \gamma \sum_{o \in \Omega} \max_{\{\alpha_n\}} \sum_{s' \in S} O(a, s', o) \sum_{s \in S} T(s, a, s') b(s) \alpha_n(s') \right] \tag{6}$$

$$= \max_a \sum_{s \in S} b(s) \left[R(s, a) + \gamma \sum_{o \in \Omega} \arg\max_{\{g_{a,o}\}} b \cdot g_{a,o} \right], \tag{7}$$

where,

$$g_{a,o}(s) = \sum_{s' \in S} O(a, s', o)T(s, a, s')\alpha_n(s')$$ (8)

Value iteration algorithm implements backup operator directly on α-vectors, which is:

$$backup(b) = \arg\max_{\{g_a^b\}_{a \in A}} b \cdot g_a^b$$ (9)

where,

$$g_a^b = R_a + \gamma \sum_{o \in \Omega} \arg\max_{\{g_{a,o}\}} b \cdot g_{a,o}$$ (10)

3 Point-Based Value Iteration

The exact value iteration algorithms for solving POMDP problems attempt to update the value function over the entire belief space, hence they are believed not to be able to solve large-scale POMDP problems. In recent years researchers have proposed lots of approximate algorithms, among which point-based methods [5-10] are most promising. Point-based value iteration is based on the following fact: for most of POMDP problems, even given arbitrary action and arbitrary observation, agent would only reach a small part of the belief points in the belief space [5]. Once the value of a belief is updated, the values of nearby belief points are likely to be improved. Thus point-based value iteration need not to compute the value function over the belief space, it solves the POMDP on a finite set of belief points that are more probable to be reached.

PBVI algorithm [5] is a typical point-based method. It solves the POMDP on a finite set of belief points $B = \{b_0, b_1, ..., b_m\}$. PBVI starts with an initial belief set B_0 which usually contains only the initial belief state, updates the values using backup operator for the belief set, and then expands the belief set. In PBVI algorithm, steps of belief set expansion and steps of value iteration are applied iteratively until some stopping criteria is reached. Belief selection in expansion of B_t is based on the fact that the approximate value is better when the belief points are uniformly distributed in the belief simplex as many as possible. So for each belief point of the set, PBVI finds a successive belief point that is farthest away from the points already existing in B_t.

Because PBVI attempts to improve the density of belief set in the step of belief selection, it may ignore some belief points that are important to update the value function. Izadi et al. proposed BFBS algorithm [8]. BFBS expands the belief set to include all the beliefs that are reachable in the next time step from the existing points of the set. Although BFBS is likely to provide the best approximation of value function, the size of the belief set is expanded exponentially. So it is not feasible for real-world problems. DBBE [9] improves BFBS algorithm, adding not all the beliefs that are reachable but a subset of these belief points that are farther from B_t than a given threshold. A reachable belief point b' is added to the current set B if it satisfies the following formula:

$$D(b',B) = \min_{b \in B} \left\| b - b' \right\|_1 > \varepsilon .$$ (11)

But it is showed in our experiments that the belief set is still too large for value iteration. FSVI algorithm [10] is a trial-based algorithm. It selects a sequence of belief points using the action that is optimal for a sampled state of the underlying MDP, and then updates the value in reversed order. FSVI dose not perform very well in the problems that need many steps of information-gathering.

4 Uncertainty-Based Belief Selection

Belief selection is a key step in point-based value iteration. We propose a belief selection algorithm based on the uncertainty of a belief point (Uncertainty-based belief selection, UBBS). This algorithm is motivated by the fact that the optimal value function has a property of convexity over the entire belief space [2]. It could be illustrated in Fig. 1 that the values at the belief points in the middle of space are lower than that in the corner. Belief states in the middle of belief space have high uncertainties about the real-world state, so the agent could not take appropriate actions to gain higher reward. On the contrary, agent could gain higher reward in less uncertain belief states which are in the corner of the belief space.

This fact indicates that the less uncertain belief points in the belief simplex may be more worthwhile for improving the value function of the belief space. UBBS algorithm modifies the step of belief expansion in point-based value iteration, adding belief points that have lower uncertainties, and shares the same main framework with PBVI algorithm. The main framework of the algorithm is shown in Fig. 2.

Algorithm PBVI
    Initialize B_0 and value function
    while stopping criteria have not been reached
        Value iteration;
        Belief set expansion;
    End while.

Fig. 2. Framework of the algorithm

In this paper, we proposed two different methods to represent uncertainty of a belief point, named Entropy-based belief selection (EBBS) and Gap-based belief selection (GBBS) respectively. The former is based on the entropy of the belief state; the latter is based on the gap between the maximal and the minimal probabilities of the belief state.

Entropy is a well known scalar to describe the uncertainty of a random variable. EBBS algorithm measures the uncertainty of a belief state by entropy. Given a belief point $b = (b(s_1), b(s_1), \ldots b(s_{|S|}))$ describing the probabilities of the world states, the entropy could be computed by equation (12):

$$Entropy(b) = -\sum_{s \in S} b(s) \log b(s) .$$ (12)

The lower the entropy of a belief point, the less uncertainty it has, and the more long-term reward would be gained at the belief state.

GBBS algorithm describes the uncertainty of a belief point using the gap between the maximal element and the minimal element of the belief vector. The value of each element $b(s)$ in the belief state shows a probability that agent is at state s. If agent is sure that it is at a certain state s_i by observation, so the element $b(s_i)$ of the belief vector is set to be 1, the other elements are set as 0, and the uncertainty of this belief state is the least. The gap of this belief point is highest. Another extreme case is when agent is so ambiguous that it could not distinct the states at all, the elements in the belief state are set as the same probabilities. Then the uncertainty of this belief is the highest, and the gap is 0. The wider of the gap of a belief point, the less uncertainty the belief point contains. GBBS algorithm uses equation (13) to measure uncertainty of belief state b:

$$Gap(b) = \max(b) - \min(b).$$ (13)

Algorithm UBBS(threshold ε)
    for all $b \in B$ do
        for all $a \in A$ do
            Sample current state s from b;
            Sample next state s' from $T(s,a,s^*)$;
            for all $o \in \Omega$ do
                Compute next belief b' reachable from b;
                Insert into a sorted list L according to uncertainty;
            end for
        end for
        while Get the first b' from the sorted list L do
            if $D(b',B) > \varepsilon$
                then return $B = B \cup \{b'\}$;
        end while
    end for

Fig. 3. Algorithm of Uncertainty-based belief selection

These two methods (EBBS and GBBS) proposed above differ in the computing of uncertainty, whichever could be used in the framework of UBBS algorithm. They are both tested in the experiments in the next section.

The nearby belief points have approximate values, so the values of the belief points in the neighborhood are improved at the same time when updating the value of a belief state. This fact suggests that two selected belief points should not be too close in distance. In our UBBS algorithm, we share the same criteria with DBBE which is represented by formula (11).

Uncertainty-based belief selection (EBBS or GBBS) algorithm is a part of point-based value iteration illustrated in Figure 2, which improves the step of belief expansion. EBBS and GBBS algorithms use the same process described in Fig. 3, which differ in the computing of uncertainty of a belief point.

5 Experiments

We conduct the comparative experiments on 4 POMDP problems (Hallway, Hallway2, Tag Avoid and Rock Sample 4,4) with EBBS and GBBS algorithms, and the other three point-based algorithms. Table 1 shows the description of these 4 problems, where $|S|$ represents the number of states, $|A|$ is the number of actions and $|\Omega|$ represents the number of observations. Average discounted reward (ADR) is used as a target value that an agent is expected to gain in the problem.

Table 1. Problem description

Problem	$|S|$	$|A|$	$|\Omega|$	Target ADR
Hallway	61	5	21	0.51
Hallway2	93	5	17	0.33
Tag Avoid	870	5	30	-9.3
Rock Sample 4,4	257	9	2	18

In all domains, the discount factor is 0.95. The threshold value of the minimal 1-norm distance is set to 0.9. For each problem we execute these algorithms for 10 times with different random seeds, but DBBE algorithm on Hallway2 is run only once because it is time-consuming. To test the policy for each run of the algorithms on these problems, we do 100 trials of random exploration of 100 time steps and compute the average reward. And the results reported here are averaged over 10 runs.

Table 2 presents the results for the problems, comparing the average discounted reward, the computation time for backup, the size of the belief set and the number of α-vectors.

UBBS algorithm improves the step of belief set expansion of PBVI algorithm. Experiment results indicate that EBBS and GBBS algorithms work more effectively than PBVI since they need fewer number of belief points than PBVI to get the similar long-term discounted reward, especially on problem Rock Sample 4,4. EBBS and GBBS algorithms could gain higher reward than PBVI on Tag Avoid problem using fewer belief states.

EBBS and GBBS algorithms are more powerful than DBBE algorithm. The size of belief set grows exponentially during each expansion which makes DBBE method not feasible for real-world POMDP problems, although this method can gain good reward. As PBVI, UBBS algorithm doubles the size of belief set at most for a belief set expansion. UBBS method needs not much unnecessary belief points and value updates.

Our algorithm outperforms FSVI on Hallway and Hallway2. But FSVI does better compared with UBBS algorithm especially on Rock Sample 4,4. A reason for the failure of UBBS algorithm is that our method considers only the uncertainty of a belief state but not the reward which the belief state could provide. FSVI simulates underlying MDP to search heuristically in the belief space in order to get higher reward. Taking Rock Sample 4,4 for example, the immediate reward at a certain state may be 1, -1 or -100. Agent should avoid the state at which agent would get a reward like -100 by taking some actions, because this kind of actions seems fatal. Although

Table 2. Experiment results

Method	ADR	Time	$|B|$	$|V|$
Hallway(61s 5a 21o)				
PBVI	0.513	18.7	74.8	71.1
DBBE	0.519	598	1747	178.9
FSVI	0.512	44.8	289.7	136.4
EBBS	0.517	19.7	86.2	80.5
GBBS	0.513	8.7	63	59.9
Hallway2(93s 5a 17o)				
PBVI	0.336	1303.7	304.8	254.6
DBBE*	0.325	79762	6673	2127
FSVI	0.329	523.4	642.4	556.5
EBBS	0.327	539.9	189.9	165.6
GBBS	0.334	224.4	128.9	118.3
Tag Avoid(870s 5a 30o)				
PBVI	-9.112	304.6	983.5	224.1
DBBE	-7.909	65.8	1035.3	186.5
FSVI	-11.50	15.6	547.5	29.9
EBBS	-8.760	361.5	527.3	279.3
GBBS	-8.511	280.8	518.8	308.7
Rock Sample 4,4(257s 9a 2o)				
PBVI	17.958	37260	32897.6	617
DBBE	18.020	975.6	6376.9	306.9
FSVI	17.666	1.3	107	50
EBBS	17.916	191.8	628.3	438
GBBS	17.956	197.2	650.2	438.8

the belief state keeps low uncertainty about this kind of real state, it should not be selected to expand the belief set. We claim that these beliefs would not be much helpful to improve the long-term discounted reward. But UBBS algorithm would not avoid this case which could be avoided by FSVI. We will consider it in our future work. Note that the ADR of FSVI on Tag Avoid in [10] is -6.612, but we do not achieve such a good result in this experiment.

Usually, the number of the belief states $|B|$ being backed up and the number of α-vectors satisfy the following inequation [11]: $|B| \geq |V|$. But if the ratio of $|B| / |V|$ is large, it means lots of belief points share the same optimal policy. But a small ratio may suggest us that a better approximate value function could be gotten if we add more belief states to complete the value iteration. The experimental results indicate that this ratio of EBBS and GBBS algorithm ranges from 1.05 to 1.88, which outperforms the other algorithms in these 4 domains.

6   Conclusion

In this paper we propose a belief selection algorithm (UBBS) based on the uncertainty of belief state in point-based value iteration for POMDP. UBBS algorithm chooses the belief state that has low uncertainty and the minimal 1-norm distance between selected belief state and the existing points already in the belief set should be larger than a threshold. We use two different methods to represent uncertainty of a belief point: one is EBBS, and the other is GBBS algorithm. EBBS uses entropy to describe the uncertainty of a belief point, and GBBS is based on the gap between the maximal and minimal elements of a belief state to compute the uncertainty.

Uncertainty-based belief selection method improves the belief expansion in PBVI and DBBE. The experimental results show that UBBS algorithm could gain the approximate long-term discounted reward using fewer belief states compared with other point-based algorithms. Our future work is to consider the reward during belief selection in order to gain higher reward as quickly as possible.

Acknowledgments

This work is partially supported by Scientific Breakthrough Program of Beijing Municipal Science & Technology Commission, China (H020920010130), China Key Technologies R & D Programme (2007BA110B06-01) and China 973 project (2006CB504601). Thank Professor Shani for sharing the POMDP software written in JAVA.

References

1. Sondik, E.J.: The optimal control of partially observable Markov processes over the infinite horizon: Discounted costs. Operations Research 26(2), 282–304 (1978)
2. Kaelbling, L.P., Littman, M.L., Cassandra, A.R.: Planning and acting in partially observable stochastic domains. Artificial Intelligence 101, 99–134 (1998)
3. Hu, Q.Y., Liu, J.Y.: An Introduction to Markov Decision Processes. Xi Dian University Press, Xi'an (2000) (in Chinese)
4. Zhang, N.L., Zhang, W.: Speeding up the Convergence of Value Iteration in Partially Observable Markov Decision Processes. Journal of Artificial Intelligence Research (JAIR) 14, 29–51 (2001)
5. Pineau, J., Gordon, G., Thrun, S.: Point-based value iteration: An anytime algorithm for POMDPs. In: Int. Joint Conf. on Artificial Intelligence (IJCAI), Acapulco, Mexico, pp. 1025–1030 (2003)
6. Smith, T., Simmons, R.: Heuristic search value iteration for POMDPs. In: Proc. of Uncertainty in Artificial Intelligence (UAI) (2004)
7. Spaan, M.T.J., Vlassis, N.: Perseus: Randomized point-based value iteration for POMDPs. Journal of Artificial Intelligence Research (JAIR) 24, 195–220 (2005)
8. Izadi, M.T., Precup, D., Azar, D.: Belief selection in point-based planning algorithms for POMDPs. In: Proceedings of Canadian Conference on Artificial Intelligence (AI), Quebec City, Canada, pp. 383–394 (2006)
9. Izadi, M.T., Precup, D.: Exploration in POMDP belief space and its impact on value iteration approximation. In: European Conference on Artificial Intelligence (ECAI), Riva del Garda, Italy (2006)
10. Shani, G., Brafman, R.I., Shimony, S.E.: Forward search value iteration for POMDPs. In: Proc. Int. Joint Conf. on Artificial Intelligence (IJCAI), pp. 2619–2624 (2007)
11. Pineau, J., Gordon, G., Thrun, S.: Point-based approximations for fast POMDP solving. Technical Report, SOCS-TR-2005.4, School of Computer Science, McGill University (2005)

Optimal Threshold Policies for Multivariate Stopping-Time POMDPs

Vikram Krishnamurthy

Department of Electrical and Computer Engineering, University of British Columbia,
Vancouver, V6T 1Z4, Canada
vikramk@ece.ubc.ca

Abstract. This paper deals with the solving multivariate partially observed Markov decision process (POMDPs). We give sufficient conditions on the cost function, dynamics of the Markov chain target and observation probabilities so that the optimal scheduling policy has a threshold structure with respect to the multivariate TP2 ordering. We present stochastic approximation algorithms to estimate the parameterized threshold policy.

1 Introduction

This paper deals with multivariate POMDPs with two actions, where one of the actions is a stop action. Such POMDPs arise in radar resource management of sophisticated military radar systems. Consider L dynamical targets tracked by an agile beam multi-function radar. How should the radar manager decide which target to track with high priority during the time slot and for how long? Given Bayesian track estimates of the underlying targets, the goal is to devise a sensor management strategy that at each time dynamically decides how much radar resource to invest in each target. Several recent works in statistical signal processing [1,2,3], study the problem as a multivariate partially observed Markov Decision Process (POMDP) in the context of radar and sensor management. A major concern with the POMDP formulation is that in most realistic cases, POMDPs are numerically intractable as they are PSPACE hard problems. The main aim of this paper is to show that by introducing structural assumptions on multivariate POMDPs, the optimal scheduling policy can be characterized by a simple structure and computed efficiently. We formulate a two level optimization framework. The inner level optimization termed *sensor micro-management* deals with how long to maintain a given priority allocation of targets. It is formulated as a multivariate partially observed Markov Decision Process (POMDP). The main goal of this paper is to prove that under reasonable conditions, the multivariate POMDP has a remarkable structure: the optimal scheduling policy is a simple threshold. Therefore, the optimal policy can be computed efficiently. Showing this result requires using the TP2 (totally positive of order 2) multivariate stochastic ordering. We give a novel necessary and sufficient condition for the optimal threshold policy to be approximated by the best linear hyperplane. We present stochastic approximation algorithms to compute these parameterized thresholds.

C. Sossai and G. Chemello (Eds.): ECSQARU 2009, LNAI 5590, pp. 850–862, 2009.

This paper generalizes [4,1] which dealt with structural results for scalar POMDPs. These papers use the univariate monotone likelihood ratio (MLR) ordering. For multivariate POMDPs one needs to use the TP2 multivariate stochastic ordering of Bayesian estimate which is not necessarily reflexive (unlike the univariate monotone likelihood ratio ordering). The results of this paper are also related to [5] where conditions are given for a POMDP to have a TP2 monotone increasing policy. However, to make those results useful from a practical point of view, one needs to translate monotonicity of the policy to the existence of a threshold policy. A major contribution of the current paper is to develop the properties of a specialized version of the TP2 stochastic ordering on multilinear curves and lines. We show that this specialized TP2 order requires less restrictive conditions on the costs compared to [5]. Moreover, we present necessary and sufficient conditions can be given for the best linear and multi-linear approximation to the optimal threshold policy. This allows us to estimate the optimal linear and multi-linear estimate to the threshold policy via stochastic approximation algorithms. We also refer to [6,7] for applications of POMDPs in sensor scheduling and multi-armed bandits.

2   Model and Dynamic Programming Formulation

We motivate our multivariate POMDP problem in terms of a radar resource management problem. Consider L targets evolving over the *fast time scale* denoted $n = 1, 2, \ldots,$. The *slow time scale* denoted by $t = 1, 2 \ldots$ indexes random length intervals of time in the fast time scale. These intervals of length denoted τ_t are called *scheduling intervals*. We use $k = 1, \ldots, \tau_t$ to denote the time within a scheduling interval.

2.1   Target Dynamics

Consider a radar with an agile beam, tracking L moving targets (e.g., aircraft). Each target l is modeled as a finite state random process $X_n^{(l)}$ indexed by $l \in \{1, \ldots, L\}$ evolving over discrete time $n = 0, 1, \ldots$. To simplify notation, assume each process $X_k^{(l)}$ has the same finite state space $\mathbf{S} = \{1, \ldots, S\}$. Each process $X_n^{(l)}$ models a specific attribute of target l. For example in [1], it models the distance of the target to the base-station. The radar resource manager uses this information to micro-manage the radar by adapting the target dwell time. Finally denote the composite process $X_n = (X_n^{(1)}, \ldots, X_n^{(L)})$ with state space $\mathbf{S}^{\text{comp}} = \mathbf{S} \times \cdots \times \mathbf{S} = \{1, \ldots, S^L\}$, where $\times$ denotes Cartesian product. We index the states of X_k by the vector index $\mathbf{i}$ or $\mathbf{j}$, where $\mathbf{i} = (i_1, \ldots, i_L) \in \mathbf{S}^{\text{comp}}$ with generic element $i_l \in \mathbf{S}$, $l = 1, 2, \ldots, L$. Assume X_k evolves according to a S^L state Markov chain, with transition matrix

$$P = [p_{\mathbf{ij}}]_{S^L \times S^L}, \ p_{\mathbf{ij}} = \mathbb{P}(X_n = \mathbf{j} | X_{n-1} = \mathbf{i}); \text{ with } \pi_0(\mathbf{i}) = \mathbb{P}(X_0 = \mathbf{i}). \quad (1)$$

Macro-management: Target Selection a_t: At each instant t on the slow time scale, the sensor manager picks one target denoted $a_t \in \{1, \ldots, L\}$, to track/estimate with high priority, while the other $L - 1$ targets are tracked/estimated with lower priority.

Micro-management: Scheduling Control u_k: Once the action a_t is chosen, the micro-manager is initiated for the t-th scheduling interval. The clock on the fast time scale k is reset to $k = 0$ and commences ticking. At this fast time scale, $k = 0, 1, \ldots$, the L targets are estimated by a Bayesian tracker. The target a_t is given highest priority and allocated the best quality sensors (or more time within the scheduling interval) for measuring its state. The remaining $L - 1$ targets are given lower priority and tracked using lower quality sensors (or given less time per scheduling interval). Micro-management is the main focus of this paper. *How long should the micro-manager track target a_t with high priority before returning control to the macro-manager to pick a new high priority target?*

2.2   Formulation of Micro-management as a Multivariate POMDP

Below we formulate the micro-management of L targets as a multivariate POMDP.

Markov Chain: X_k defined in Sec.2.1 models the L evolving targets.

Action Space: At the $(k+1)$th time instant within the t-th scheduling interval, the micro-manager picks action u_{k+1} as a function μ_{a_t} of the Bayesian estimates π_k (defined in (4)) of all L targets as $u_{k+1} = \mu_{a_t}(\pi_k) \in \{\text{continue} = 1, \text{stop} = 2\}$ where $\mu_a \in A_a$, and A_a denotes stationary scheduling policies. If $u_k = \text{continue} = 1$, the micro-manager continues with the current target priority allocation a_t. So k increments to $k+1$ and the L targets are tracked with target a_t given the highest priority. If $u_k = \text{stop} = 2$, the micro-manager stops the current scheduling interval t, and returns control to the macro-manager to determine a new target a_{t+1}.

Multivariate Target Measurements: Given the state X_k of the L targets, measurement vector $Y_k = (Y_k^{(1)}, \ldots, Y_k^{(L)})$ is obtained at time k from the multivariate distribution $\mathbb{P}_{a_t}(Y_k|X_k, u_k)$. Assume each target's observation $Y_k^{(l)}, l = 1, \ldots, L$, is finite valued,

$$Y_k^{(l)} \in \mathbf{y} = \{O_1, O_2, \ldots, O_M\}, \text{ so } Y_k \in \mathbf{Y} \overset{\triangle}{=} \mathbf{y} \times \cdots \times \mathbf{y} \qquad (2)$$

Multi-target Bayesian Estimation: In scheduling interval t, with priority allocation a_t, at time k, denote the history of past observations and actions as

$$Z_k^{(l)} = \{a_t, \pi_{t-1}, u_1, Y_1^{(l)}, \ldots, u_k, Y_k^{(l)}\}$$

where $Z_k = (Z_k^{(1)}, \ldots, Z_k^{(L)})$. Here π_{t-1} is the *a posteriori* distribution of the L targets from the macro-manager at the end of the $(t-1)$th scheduling interval. Based on Z_{k+1}, the Bayesian tracker computes the posterior state distribution π_{k+1} of the L targets

$$\pi_{k+1} = (\pi_{k+1}(\mathbf{i}), \mathbf{i} \in \mathbf{S}^{\text{comp}}\}, \quad \pi_{k+1}(\mathbf{i}) = \mathbb{P}_{a_t}(X_{k+1} = \mathbf{i}|Z_{k+1}) \qquad (3)$$

The S^L-dimensional vector π_k is computed via the Hidden Markov Bayesian filter: $\pi_{k+1} = T_{a_t}(\pi_k, u_{k+1}, Y_{k+1})$ where

$$T_a(\pi, u, Y) = \frac{B_a(u, Y)P'\pi}{\sigma_a(\pi, u, Y)}, \quad \sigma_a(\pi, u, Y) = \mathbf{1}'_{S^L} B_a(u, Y)P'\pi$$

$$\text{and } B_a(u, Y) = \text{diag}(\mathbb{P}_a(Y|1, u), \ldots, \mathbb{P}_a(Y|S^L, u)). \qquad (4)$$

Here $\mathbf{1}_{S^L}$ is the S^L dimension vector of ones. π_k is referred to as the *information state*, since (see [4]) it is a sufficient statistic to describe the history Z_k. The composite Bayesian estimate π_k in (4) lives in an $S^L - 1$ dimensional unit simplex

$$\Pi^{comp} \triangleq \left\{ \pi \in \mathbb{R}^{S^L} : \mathbf{1}'_{S^L}\pi = 1, \quad 0 \le \pi(\mathbf{i}) \le 1 \text{ for all } \mathbf{i} \in \mathbf{S}^{comp} \right\} \tag{5}$$

The Bayesian posterior distribution of each target l, defined as $\pi_k^{(l)} = \mathbb{P}_{a_t}(X_k^{(l)}|Z_k^{(l)})$, can be computed by marginalizing the joint distribution π_k. $\pi_k^{(l)}$ lives in an $S-1$ dimensional unit simplex $\Pi \triangleq \{\pi^{(l)} \in \mathbb{R}^S : \mathbf{1}'_S\pi^{(l)} = 1, 0 \le \pi^{(l)}(i) \le 1, i \in \{1\dots,S\}\}$.

Tracking Cost: At time k, with given current composite state X_k of the L targets, if action $u_{k+1} = \mu_{a_t}(\pi_k) \in \{\text{continue} = 1, \text{stop} = 2\}$ is taken, then the micro-manager accrues an instantaneous cost $c_{a_t}^{\mathbf{P}}(X_k, u_{k+1})$. Here $c_a^{\mathbf{P}}(X, u) \ge 0$ and

$$c_a^{\mathbf{P}}(X, u) = \begin{cases} u = 1 & \text{cost of continuing with current allocation } a \text{ given state } X \\ u = 2 & \text{cost of terminating current allocation } a \text{ given state } X. \end{cases} \tag{6}$$

In (6), the non-negative L dimensional vector $\mathbf{p} = (\mathbf{p}(1), \dots, \mathbf{p}(L))'$, denotes target priority allocations and is set by the macro-manager. $\mathbf{p}$ links the micro and macro-management. The costs $c_a^{\mathbf{P}}(X, u)$ are chosen as decreasing functions (elementwise) of $\mathbf{p}$ since higher priority targets should incur lower tracking costs. The cost $c_a^{\mathbf{P}}(X, 2)$ can also be viewed as a *switching cost* incurred by the micro-manager. If $u = 2$ is chosen, control reverts back to the macro-manager to determine a new target priority allocation. Let τ_t denote the time k (in the t-th scheduling interval) at which action $u_k = \text{stop} = 2$ is chosen. The random variable τ_t is a stopping time, i.e., the event $\{\tau_t \le k\}$ for any positive integer k is a measurable function of the sigma algebra generated by history Z_k. Let $0 \le \rho < 1$ denote a user defined economic discount factor. Then the sample path cost incurred during this interval is

$$\hat{J}^{\mathbf{P}}(\mu_{a_t}, \pi_{t-1}) = \sum_{k=1}^{\tau_t} \rho^{k-1} C_{a_t}^{\mathbf{P}}(\pi_k, u_{k+1}) \text{ where } u_{k+1} = \mu_{a_t}(\pi_k) \in \{1, 2\} \tag{7}$$

$$C_a^{\mathbf{P}}(\pi_k, u_{k+1}) = c_a^{\mathbf{P}'}(u_{k+1})\pi_k, \quad u_{k+1} = \mu_a(\pi_k), \quad c_a^{\mathbf{P}}(u) \triangleq \left[c_a^{\mathbf{P}}(1, u) \cdots c_a^{\mathbf{P}}(S^L, u) \right]'$$

Discounted Cost Stopping Time Problem Formulation: The objective is to compute the optimal policy $\mu_{a_t}^*(\pi_k) \in \{\text{continue} = 1, \text{stop} = 2\}$ to minimize the expected discounted cost $J(\mu_{a_t}, \mathbf{p})$ over the set of admissible control laws $\mathcal{A}$. That is, compute

$$\inf_{\mu \in \mathcal{A}} J^{\mathbf{P}}(\mu_{a_t}, \pi_{t-1}) \text{ where } J^{\mathbf{P}}(\mu_{a_t}, \pi_{t-1}) \triangleq \mathbb{E}\left\{ \sum_{k=1}^{\tau_t} \rho^{k-1} C_{a_t}^{\mathbf{P}}(\pi_k, \mu_a(\pi_k)) | \pi_{t-1} \right\}. \tag{8}$$

(1), (2), (4), (6), (8) form a multivariate POMDP for sensor micro-management.

Remark: Special case. Independent targets with Independent Observations: If each target $X_k^{(l)}$ $l \in \{1, \dots, L\}$, evolves as an independent Markov chain with $S \times S$ transition matrix $P^{(l)} = [p_{ij}^{(l)}]_{S \times S}$, (for $i, j \in \{1, \dots, S\}$), where $p_{ij}^{(l)} = \mathbb{P}(X_k^{(l)} = j | X_{k-1}^{(l)} = i)$. The initial distribution is then denoted as $\pi_0^{(l)} = [\pi_0^{(l)}(i)]_{S \times 1}$ where $\pi_0^{(l)}(i) = \mathbb{P}(X_0^{(l)} = i)$. If the measurements of the targets are also mutually independent, then we observe $Y_k = (Y_k^{(1)}, \dots, Y_k^{(L)})$ from product distribution $\mathbb{P}_{a_t}(Y_k^{(1)} | X_k^{(1)}, u_k) \times \cdots \times \mathbb{P}_{a_t}(Y_k^{(L)} | X_k^{(L)}, u_k)$. Note that the individual distributions $\mathbb{P}_a(Y^{(l)} | X^{(l)}, u)$ can be multivariate. The posterior distribution $\pi_k^{(l)} \triangleq \mathbb{P}_{a_t}(X_k^{(l)} | Z_k^{(l)})$ for target $l = 1, \dots, L$ is computed as $\pi_{k+1}^{(l)} = T_{a_t}^{(l)}(\pi_k^{(l)}, u_{k+1}, Y_{k+1}^{(l)})$ where

$$B_a^{(l)}(u, Y) = \text{diag}(\mathbb{P}_a^{(l)}(Y | 1, u), \dots, \mathbb{P}_a^{(l)}(Y | S, u))$$

$$T_a^{(l)}(\pi^{(l)}, u, Y^{(l)}) = \frac{B_a^{(l)}(u, Y^{(l)}) P^{(l)\prime} \pi^{(l)}}{\sigma_a^{(l)}(\pi^{(l)}, u, Y^{(l)})} \qquad (9)$$

$$\sigma_a^{(l)}(\pi^{(l)}, u, Y^{(l)}) = 1_S' B_a^{(l)}(u, Y^{(l)}) P^{(l)\prime} \pi^{(l)}.$$

The joint state π is the Kronecker product of the individual information states: $\pi = \pi^{(1)} \otimes \cdots \otimes \pi^{(L)} \in \Pi^{\text{prod}}$, where $\Pi^{\text{prod}} \triangleq \{\pi \in \Pi^{\text{comp}} : \pi = \pi^{(1)} \otimes \pi^{(2)} \cdots \otimes \pi^{(L)}\}$.

3   Micro-management: Multivariate POMDP with Threshold Policy

Consider the micro-management POMDP problem with objective function (8). For fixed priority allocation vector $\mathbf{p}$, the optimal stationary policy $\mu_a^{\mathbf{P},*} : \Pi^{\text{comp}} \to \{1, 2\}$ and associated optimal cost $J^{\mathbf{P}}(\mu_a^*, \pi)$ are the solution to "Bellman's equation" for $V_a^{\mathbf{P}}(\pi)$

$$J^{\mathbf{P}}(\mu_a^*, \pi) = V_a^{\mathbf{P}}(\pi) = \min_{u \in \{1,2\}} Q_a^{\mathbf{P}}(\pi, u), \quad \mu_a^{\mathbf{P},*}(\pi) = \arg \min_{u \in \{1,2\}} Q_a^{\mathbf{P}}(\pi, u) \qquad (10)$$

$$Q_a^{\mathbf{P}}(\pi, 1) = C_a^{\mathbf{P}}(\pi, 1) + \rho \sum_{Y \in \mathbf{Y}} V_a^{\mathbf{P}}(T(\pi, 1, Y)) \, \sigma(\pi, 1, Y), \quad Q_a^{\mathbf{P}}(\pi, 2) = C_a^{\mathbf{P}}(\pi, 2)$$

Recall $C_a^{\mathbf{P}}(\pi, u)$ is defined in (7). Since the information state space Π^{comp} of a POMDP is an uncountable set, the dynamic programming equations (10) do not translate into practical solution methodologies as $V(\pi)$ needs to be evaluated at each $\pi \in \Pi^{\text{comp}}$, an uncountable set. In our multivariate POMDP, the state space dimension is S^L (exponential in the number of targets) and so applying value iteration is completely intractable. The rest of this section focuses on the structure of the POMDP. Theorem 1, shows that under suitable conditions, the optimal scheduling policy is a simple threshold policy. We then develop novel parameterizations of this threshold curve and compute them efficiently.

3.1   Main Result: Existence of Threshold Policy for Multivariate POMDP

We list the assumptions for correlated targets. Assume for any fixed $a \in \{1, \dots, L\}$, the following conditions hold for the multivariate POMDP (8).

(A1): The cost $c_a^P(X, u) \geq c_a^P(S^L, u)$ for all $X \in \mathbf{S}^{\text{comp}}$.

(A2): The transition matrix P in (1) of the L targets is MTP2 (see appendix for definition).

(A3): For a, u, L-variate observation probabilities $\mathbb{P}_a(Y|X, u)$ is MTP2 in (Y, X).

(S): $c_a^P(X, 2) - c_a^P(X, 1) \geq c_a^P(S^L, 2) - c_a^P(S^L, 1)$. (submodular costs)

A special case of the above assumptions involving independent targets is:

(A1'): $c_a^P(X, u)$ is a separable cost function for the L targets of the form $c_a^P(X, u) = \sum_{l=1}^{L} c_a^{P^{(l)}}(X^{(l)}, u)$ where $c_a^{P^{(l)}}(X^{(l)}, u)$ denotes the cost for individual target l. Assume $c_a^{P^{(l)}}(S, u) \leq c_a^{P^{(l)}}(i, u)\ i = 1, 2, \ldots S-1,\ u \in \{1, 2\},\ \ l = 1, \ldots, L$.

(A2'): The transition probability matrix $P^{(l)}$ of each target l is MTP2 (see Definition 1).

(A3'): $\mathbb{P}_a^{(l)}(Y^{(l)}|X^{(l)}, u)$ is is MTP2 in $Y^{(l)}, X^{(l)}$.

(S'): $c_a^{P^{(l)}}(i, 2) - c_a^{P^{(l)}}(i, 1) \geq c_a^{P^{(l)}}(S, 2) - c_a^{P^{(l)}}(S, 1)$ for each target l.

Examples of the above conditions in radar management are discussed in Sec.3.4. The main result below shows that the optimal micro-management policy has a threshold structure, see Appendix 4 for definitions.

Theorem 1 (Existence of Threshold Policy for Sensor Micro-management). *Consider the multivariate POMDP and a fixed target priority allocation $a \in \{1, \ldots, L\}$. Then:*

(i) Dependent Targets: Under (A1), (A2), (A3), (S), the optimal policy $\mu_a^*(\pi)$ is TP2 increasing on lines in Π^{comp}. That is, $\pi \underset{TP2L}{\geq} \tilde{\pi}$, implies $\mu_a^*(\pi) \geq \mu_a^*(\tilde{\pi})$.

(ii) Independent Targets: Under (A1'), (A2'), (A3'), (S'), the optimal policy $\mu_a^*(\pi)$ is TP2 increasing on curves in Π^{comp}. That is, $\pi \underset{TP2C}{\geq} \tilde{\pi}$, implies $\mu_a^*(\pi) \geq \mu_a^*(\tilde{\pi})$.

(iii) There exists a curve Γ (which we call a "threshold curve") that partitions information state space Π^{comp} into two individually connected regions $\mathcal{R}_1, \mathcal{R}_2$, such that:

$$\text{Optimal scheduling policy } \mu_a^*(\pi) = \begin{cases} \text{continue} = 1 & \text{if } \pi \in \mathcal{R}_1 \\ \text{stop} = 2 & \text{if } \pi \in \mathcal{R}_2 \end{cases} \quad (11)$$

Also region $\mathcal{R}_2$ is convex. So Γ is continuous and differentiable almost everywhere. $\square$

Under the conditions of Theorem 1, the optimal scheduling policy for the multivariate POMDP is a threshold policy with a threshold curve Γ that partitions the information state space Π^{comp}. Note that without these conditions, the optimal policy of the multivariate POMDP can be an arbitrarily complex partition of the simplex Π^{comp} – and solving such a multivariate POMDP is computationally intractable. The convexity of region $\mathcal{R}_2$ (statement (iii) of the theorem) follows from [8, Lemma 1].

3.2   Characterization of Best Linear and Multi-linear Threshold

Due to the existence of a threshold curve Γ, computing the optimal policy μ_a^* reduces to estimating this threshold curve. In this section, we derive *linear* and *multi-linear* approximations to Γ. Such linear/multi-linear thresholds have two attractive properties:

(i) Estimating them is computationally efficient. (ii) We give novel conditions on the threshold coefficients that are *necessary and sufficient* for the resulting linear/multi-linear threshold policy to be TP2 increasing on lines. Due to the necessity and sufficiency of the condition, optimizing over the space of linear/multi-linear thresholds yields the "best" linear/multi-linear approximation to the threshold curve Γ.

Dependent Targets. We start with the following definition of a linear threshold policy: For fixed target priority $a \in \{1, \ldots, L\}$, define the linear threshold policy $\mu_{\theta_a}(\pi)$ as

$$\mu_{\theta_a}(\pi) = \begin{cases} \text{continue} = 1 & \text{if } \theta_a'\pi < 1 \\ \text{stop} = 2 & \text{if } \theta_a'\pi \geq 1, \end{cases} \quad \pi \in \Pi^{\text{comp}}. \tag{12}$$

Here $\theta_a \in \mathbb{R}_+^{S^L}$ denotes the vector of coefficients of the linear threshold policy.

Theorem 2 (Dependent Targets). *Assume conditions (A1), (A2), (A3), (S) hold for the multivariate POMDP (8). Then for any fixed target priority $a \in \{1, \ldots, L\}$:*

(i) The linear threshold policy $\mu_{\theta_a}(\pi)$ defined in (12) is TP2 increasing on lines iff $\theta_a(S^L) \geq \theta_a(i)$, $i = 1, \ldots, S^L - 1$. (ii) Therefore, the optimal linear threshold approximation to threshold curve Γ of Theorem 1 is the solution of the following constrained optimization problem:

$$\theta_a^* = \arg\min_{\theta_a \in \mathbb{R}_+^{S^L}} J^P(\mu_{\theta_a}, \pi), \quad \text{subject to } \theta_a(S^L) \geq \theta_a(i), \quad i = 1, \ldots, S^L - 1.$$

$$\tag{13}$$

where $J^P(\mu_{\theta_a}, \pi)$ is obtained as in (8) by applying threshold policy μ_{θ_a} in (12). □

Remark: To make the threshold vector parametrization θ_a^* unique, we have incorporated the following steps: The term '1' on the right hand side of (12) (and also in (14) below), is without loss of generality; otherwise one could scale both sides of these equations resulting in non-uniqueness. The requirement that θ is a non-negative vector is without loss of generality since a positive vector with identical elements can always be added.

Independent Targets. The main point in Theorem 3 below is that for L independent but *non-identical* targets, we can construct a SL dimension threshold as the best multilinear approximation of Γ. Define SL dimension vector $\theta_a' = (\theta_a^{(1)'}, \theta_a^{(2)'}, \ldots, \theta_a^{(L)'})$, where each sub-vector $\theta_a^{(l)} \in \mathbb{R}_+^S$, $l = 1, \ldots, L$. The elements of each sub-vector $\theta_a^{(l)}$ are denoted $\theta_a^{(l)}(i)$ and are associated with target l. The dimension SL of θ_a here is in contrast to the S^L dimension threshold for dependent targets in Theorem 2 above. Then for any fixed $a \in \{1, \ldots, L\}$, define the multi-linear threshold policy

$$\mu_{\theta_a}(\pi) = \begin{cases} \text{continue} = 1 & \text{if } \prod_{l=1}^L \theta_a^{(l)'}\pi^{(l)} < 1 \\ \text{stop} = 2 & \text{if } \prod_{l=1}^L \theta_a^{(l)'}\pi^{(l)} \geq 1, \end{cases} \quad \pi \in \Pi^{\text{prod}} \tag{14}$$

To make the threshold parameterization θ_a unique, we need to disallow scaling $\theta_a^{(l)'}\pi^{(l)}$ by a constant for one target l and dividing by the same constant for another target l'. So we assume that $\max_i \theta^{(1)}(i) = \max_i \theta^{(2)}(i) = \cdots = \max_i \theta^{(L)}(i)$.

Theorem 3 (Independent targets). *Assume conditions (A1'), (A2'), (A3') and (S') hold for the multivariate POMDP (8). Then for any fixed $a \in \{1, \ldots, L\}$:*

(i) The multi-linear policy $\mu_{\theta_a}(\pi)$ in (14) is TP2 increasing on lines iff $\theta_a(S) \geq \theta_a^{(l)}(i)$.

(ii) Therefore, the optimal multi-linear threshold approximation to threshold Γ is the SL dimension threshold vector θ_a^* which is the solution to the optimization problem:

$$\theta_a^* = \arg \min_{\theta \in \mathbb{R}_+^{SL}} J^{\mathrm{P}}(\mu_\theta, \pi), \quad \text{with} \quad \begin{cases} \theta_a^{(l)}(S) = constant \; indpt \; of \; l \; denoted \; as \; \theta_a(S) \\ \theta_a(S) \geq \theta_a^{(l)}(i), i = 1, \ldots, S-1, \; l = 1, \ldots, L \end{cases}$$
(15)

In (15), $J^{\mathrm{P}}(\mu_\theta, \pi)$ is obtained as in (8) by applying policy μ_θ in (14). $\qquad \square$

Remark: The multi-linear threshold policy coefficients $\theta_a^{(l)}$ in (14) are the same, it suffices to pick θ_a as a S dimension threshold vector: $\theta_a \triangleq \theta_a^{(1)} = \theta_a^{(2)} = \cdots = \theta_a^{(L)}$.

3.3   Algorithm to Compute the Optimal Multi-linear Threshold Policy

We focus on the independent identical targets case of Theorem 3, the algorithms for dependent targets is similar and omitted. We resort to sample-path based simulation optimization to estimate $\theta_a^* \in \mathbb{R}_+^S$: For batches indexed by $n = 1, 2, \ldots$, evaluate the sample path cost $\hat{J}_n^{\mathrm{P}}(\mu_{\theta_a}, \pi_{t-1})$ by simulating the multivariate POMDP. The aim is:

$$\text{Compute } \theta_a^* = \arg \min_{\theta \in \Theta} \mathbb{E}\{\hat{J}_n^{\mathrm{P}}(\mu_{\theta_a}, \pi_{t-1})\} \text{ subject to constraints in (15).} \qquad (16)$$

Consider the unconstrained S dimensional vector ϕ, with component vectors ϕ having dimensions identical to θ. Set $\theta_a(S) = [\phi_a(S)]^2$, $\theta_a(i) = [\phi_a(S)]^2 \sin^2 \phi_a(i)$. Since the square of a number is non-negative, $[\phi_a]^2 \geq 0$. Also $\sin^2(\cdot) \in [0, 1]$. Thus θ_a automatically satisfies constraints. This equivalent unconstrained optimization problem in solved via a stochastic approximation algorithm. For iterations $n = 0, 1, 2, \ldots$:

1. Evaluate sample cost $\hat{J}_n^{\mathrm{P}}(\mu_{\phi_a}, \pi_{t-1})$. Compute gradient estimate $\widehat{\nabla}_\phi \hat{J}_n^{\mathrm{P}}(\mu_{\phi_a}, \pi_{t-1})$ as: (we denote $\hat{J}_n^{\mathrm{P}}(\mu_{\phi_a}, \pi_{t-1})$ as $\hat{J}_n(\hat{\phi})$ to simplify notation):

$$\widehat{\nabla}_\phi J_n = \frac{J_n(\hat{\phi}_n + \mu_n \omega_n) - J_n(\hat{\phi}_n - \mu_n \omega_n)}{2\mu_n} \omega_n, \quad \omega_n(i) = \begin{cases} -1 & \text{with prob } 0.5 \\ +1 & \text{with prob } 0.5. \end{cases}$$

where $\mu_n = \frac{\mu}{(n+1)^\gamma}$.

2. Update threshold coefficients $\hat{\phi}_n$ via (where ϵ_n below denotes step size)

$$\hat{\phi}_{n+1} = \hat{\phi}_n - \epsilon_{n+1} \widehat{\nabla}_\phi \hat{J}_n^{\mathrm{P}}(\mu_{\phi_a}, \pi_{t-1}), \quad \epsilon_n = \epsilon/(n+1+s)^\kappa, \; 0.5 < \kappa \leq 1, \text{ and } \epsilon, s > 0.$$
(17)

In Step 2, the initial value $\pi_{t-1} \in \Pi^{\mathrm{prod}}$ can be chosen arbitrarily, since by definition any stationary policy does not depend on the initial condition (but of course, the cost does). The simultaneous perturbation stochastic approximation (SPSA) algorithm [9] picks a single random direction ω_n along which direction the derivative is evaluated at each batch n. So to evaluate the gradient estimate $\widehat{\nabla}_\phi J_n$ in (17) requires only 2 POMDP simulations, i.e., the number of evaluations is independent of dimension of parameter ϕ. Because the stochastic gradient algorithm (17) converges with probability one to local optima, it is necessary to try several initial conditions $\hat{\phi}_0$.

3.4   Discussion of Assumptions

(A1), (A1'): Suppose states $1, 2, \ldots, S$ denotes decreasing distance of the target to a base-station. Then the closer the target, the higher the threat, and more incentive to track it. (A1) means that the reward (negative of cost) for tracking the target is smallest when it is at maximum distance. This is natural since the further away the target the lower the threat. Similarly, if states $1, 2, \ldots, S$ denote increasing covariances of the target estimate, then the larger the covariance, the higher the incentive to track the target.

(A2) and (A2'): For independent targets, if the target is at state i, $1 \leq i \leq S$, at time k then at time $k + 1$, it is reasonable to assume that it is either still in state i or with a lesser probability in the neighboring states $i + 1$ or $i - 1$. Each target can then be modeled as a S state Markov chain with tridiagonal transition probability matrix P. As shown in [10, pp.99–100], a necessary and sufficient condition for tridiagonal A to be TP2 is that $p_{i,i}p_{i+1,i+1} \geq p_{i,i+1}p_{i+1,i}$. Such a diagonally dominant tridiagonal matrix satisfies (A2'). (A2) can model correlated convoy behavior of targets.

(A3), (A3'): Several observation models satisfy the TP2 ordering (A3), see [11]. Suppose sensor l measures the target in quantized Gaussian noise. The observation proba-

bilities are $P_a^{(l)}(Y|i, u) = \frac{\bar{b}_{iY}(u)}{\sum_{m=1}^{M} \bar{b}_{iY}(u)}$ where $\bar{b}_{iY}(u) = \frac{1}{\sqrt{2\pi\Sigma_u^{(l)}}} \exp\left(-\frac{1}{2}\frac{(Y - g'e_i)^2}{\Sigma_u^{(l)}}\right)$

with $O_1 > O_2 > \cdots > O_M$. Here $\Sigma_u^{(l)} > 0$ denotes the noise variance of the sensor u and thus reflects the quality of its measurements. Here $g_1 > g_2 > \ldots > g_S$ denotes the quantized distance of the target to the base-station. It is easily verified that (A3') holds. The ordering $O_1 > O_2 > \cdots > O_M$ is consistent with our discussion in (A1') where state 1 was the farthest distance and S the closest.

(S), (S'): The difference in rewards between deploying an accurate estimator and a less accurate estimator should increase as the threat level goes up. This gives economic incentive to pick the more accurate action when the target is close or threat is high.

4   Numerical Example

We consider $L = 3$ independent Markovian targets, each with $S = 30$ states corresponding to quantized distance. The composite state space of 30^3 is enormous. Without structural results, the POMDP is intractable. We construct a POMDP to satisfy assumptions (A1'), (A2'), (A3') and (S'). All targets have the same tridiagonal $S \times S$ transition matrix P with $p_{i,i} = 0.8$, $p_{i,i+1} = p_{i,i-1} = 0.1$. This satisfies (A2'). The target observation probabilities are $\mathbb{P}_a(Y^{(l)}|X^{(l)}, u) = 0.99$ for $l = a$ and 0.6 for $l \neq a$. Then (A3') holds. For fixed target priority allocation a, we chose the tracking cost for target a as $c_a(X^{(a)} = x, u = 1) = e^{-x/10}$, $c_a(X^{(a)} = x, u = 2) = 2c_a(x, 1)$, $x = 1, \ldots, S$. For the remaining $L-1$ targets, $c_a(X^{(l)}, u) = 0.1c_a(X^{(a)}, u), l \neq a$. Thus $c_a(X^{(l)} = x, u)$ decreases with x and is submodular so (A1') and (S') hold.

Since the POMDP satisfies (A1'), (A2'), (A3'), (S'), Theorem 1 implies the existence of an optimal threshold policy. Theorem 3 implies that the best multi-linear policy approximation of dimension $S = 30$ to the optimal threshold curve can be constructed.

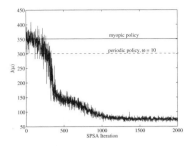

Fig. 1. Performance of the policy gradient for the multivariate POMDP comprising of 30^3 states. This policy is compared with a heuristic myopic policy and periodic policy with period $\omega = 10$.

The sample path cost of the POMDP was evaluated according to (7). The SPSA algorithm parameters in (17) were chosen as $\mu = 8.0, \epsilon = 0.05, \gamma = 0.8$. As shown in Fig.1, the SPSA algorithm converges to the optimal multi-linear threshold. We compared the performance with a simple myopic policy and a periodic policy. Fig.1 shows the performance of the optimal multi-linear policy is significantly better than the myopic and periodic policies.

References

1. Krishnamurthy, V., Djonin, D.: Structured threshold policies for dynamic sensor scheduling–a partially observed Markov decision process approach. IEEE Trans. Signal Proc. 55(10), 4938–4957 (2007)
2. Moran, W., Suvorova, S., Howard, S.: Application of sensor scheduling concepts to radar. In: Hero, A., Castanon, D., Cochran, D., Kastella, K. (eds.) Foundations and Applications for Sensor Management, pp. 221–256. Springer, Heidelberg (2006)
3. Evans, R., Krishnamurthy, V., Nair, G.: Networked sensor management and data rate control for tracking maneuvering targets. IEEE Trans. Signal Proc. 53(6), 1979–(1991)
4. Lovejoy, W.: Some monotonicity results for partially observed Markov decision processes. Operations Research 35(5), 736–743 (1987)
5. Rieder, U.: Structural results for partially observed control models. Methods and Models of Operations Research 35, 473–490 (1991)
6. Krishnamurthy, V.: Algorithms for optimal scheduling and management of hidden Markov model sensors. IEEE Trans. Signal Proc. 50(6), 1382–1397 (2002)
7. Krishnamurthy, V., Wahlberg, B.: POMDP multiarmed bandits – structural results. Mathematics of Operations Research (May 2009)
8. Lovejoy, W.: On the convexity of policy regions in partially observed systems. Operations Research 35(4), 619–621 (1987)
9. Spall, J.: Introduction to Stochastic Search and Optimization. Wiley, Chichester (2003)
10. Gantmacher, F.: Matrix Theory, vol. 2. Chelsea Publishing Company, New York (1960)
11. Karlin, S., Rinott, Y.: Classes of orderings of measures and related correlation inequalities. I. Multivariate totally positive distributions. Journal of Multivariate Analysis 10, 467–498 (1980)
12. Topkis, D.: Supermodularity and Complementarity. Princeton University Press, Princeton (1998)

Appendix 1: TP2 Stochastic Ordering, and Submodularity

To compare multivariate information states π and $\tilde{\pi}$, we use the multivariate totally positive of order 2 (TP2) stochastic partial ordering. It is ideal for information states since it is preserved after conditioning on any information [4,11]. Let $\mathbf{i} = (i_1, \ldots, i_L)$ and $\mathbf{j} = (j_1, \ldots, j_L)$ denote indices of L-variate probability mass functions (pmfs) Denote $\mathbf{i} \wedge \mathbf{j} = [\min(i_1, j_1), \ldots, \min(i_L, j_L)]'$, $\mathbf{i} \vee \mathbf{j} = [\max(i_1, j_1), \ldots, \max(i_L, j_L)]'$.

Definition 1 (TP2 ordering and MLR ordering). *Let P and Q denote L-variate pmfs. Then $P \underset{TP2}{\geq} Q$ if $P(\mathbf{i})Q(\mathbf{j}) \leq P(\mathbf{i} \vee \mathbf{j})Q(\mathbf{i} \wedge \mathbf{j})$. For univariate P and Q this is equivalent to the MLR ordering denoted as $P \geq_r Q$. A multivariate distribution P is said to be multivariate TP2 (MTP2) if $P \underset{TP2}{\geq} P$ holds, i.e., $P(\mathbf{i})P(\mathbf{j}) \leq P(\mathbf{i} \vee \mathbf{j})P(\mathbf{i} \wedge \mathbf{j})$.*

Denote $\Pi^{\mathrm{TP2}} \overset{\triangle}{=} \{\pi \in \Pi^{\mathrm{comp}} : \pi \text{ is TP2 reflexive}\}$. Because every product state is TP2 reflexive, $\Pi^{\mathrm{prod}} \subset \Pi^{\mathrm{TP2}} \subset \Pi^{\mathrm{comp}}$.

Lemma 1
(i) For all $\pi \in \Pi^{\mathrm{comp}}$, $e_1 \underset{TP2}{\leq} \pi \underset{TP2}{\leq} e_{S^L}$.

(ii) If $\pi_0 \in \Pi^{\mathrm{TP2}}$, then under (A2), (A3), the information state trajectory π_k, $k = 1, 2, \ldots$ computed via the Bayesian estimator (4), satisfies $\pi_k \in \Pi^{\mathrm{TP2}}$.

(iii) $\pi \in \Pi^{\mathrm{TP2}}$, implies for $\epsilon \in [0,1]$, $\tilde{\pi} \overset{\triangle}{=} \epsilon e_{S^L} + (1-\epsilon)\pi$ is reflexive and $\pi \underset{TP2}{\leq} \tilde{\pi} \underset{TP2}{\leq} e_{S^L}$.

(iv) If $\pi^{(l)} \geq_r \tilde{\pi}^{(l)}$, $l = 1, \ldots, L$, then $\pi^{(1)} \otimes \cdots \otimes \pi^{(L)} \underset{TP2}{\geq} \tilde{\pi}^{(1)} \otimes \cdots \otimes \tilde{\pi}^{(L)}$.

TP2 Ordering over lines: Although the TP2 ordering over Π^{comp} is used in [5], it is a stronger condition than we require and it does not yield a constructive procedure to implement a threshold scheduling policy for a multivariate POMDP. We define a novel TP2 ordering over lines. Define the $S^L - 2$ dimensional simplex $\mathcal{H} \in \Pi^{\mathrm{comp}}$ comprising of $\pi \in \Pi^{\mathrm{prod}}$ with last element $\pi(S^L) = 0$. That is, $\mathcal{H} = \text{convex hull}(e_s, \ldots, e_{S^L-1}) = \{\bar{\pi} : \bar{\pi} \in \Pi^{\mathrm{comp}} \text{ and } \bar{\pi}(S^L) = 0\}$. For each $\bar{\pi} \in \mathcal{H}$, construct the line $\mathcal{L}(e_{S^L}, \bar{\pi})$ that connects $\bar{\pi}$ to e_{S^L}. Thus $\mathcal{L}(e_{S^L}, \bar{\pi})$ comprises of information states π of the form:

$$\mathcal{L}(e_{S^L}, \bar{\pi}) = \{\pi \in \Pi^{\mathrm{comp}} : \pi = (1-\epsilon)\bar{\pi} + \epsilon e_{S^L}, \ 0 \leq \epsilon \leq 1\}, \bar{\pi} \in \mathcal{H}. \qquad (18)$$

For notational simplicity, we denote a generic line $\mathcal{L}(e_{S^L}, \bar{\pi})$ as $\mathcal{L}(e_{S^L})$.

Definition 2 (TP2 ordering on lines). *π_1 is greater than π_2 with respect to the TP2 ordering on the line $\mathcal{L}(e_{S^L})$ – denoted as $\pi_1 \underset{TP2L}{\geq} \pi_2$, if $\pi_1, \pi_2 \in \mathcal{L}(e_S^L, \bar{\pi})$ for some $\bar{\pi} \in \mathcal{H}$, i.e., π_1, π_2 are on the same line connected to e_S^L, and $\pi_1 \underset{TP2}{\geq} \pi_2$.*

A nice property of $\underset{TP2L}{\geq}$ is that if $\pi \in \mathcal{L}(e_S^L, \bar{\pi})$ is TP2 reflexive, then all points in the line $\mathcal{L}(e_S^L, \bar{\pi})$ between e_S^L and π are TP2 orderable and TP2 reflexive; see Lemma 1.

Result 1 (Submodular function, [12]). *A function $f(\pi)$ is TP2 increasing on lines in Π^{comp} if $\pi_1 \underset{TP2L}{\geq} \pi_2$ implies $f(\pi_1) \geq f(\pi_2)$. $f : \mathcal{L}_r(e_S, \bar{\pi}) \times \{1, 2\} \to \mathbb{R}$ is submodular if $f(\pi, u) - f(\pi, \bar{u}) \leq f(\bar{\pi}, u) - f(\bar{\pi}, \bar{u})$, for $\bar{u} \leq u$, $\bar{\pi} \underset{TP2L}{\leq} \pi$. If $f : \mathcal{L}(e_S, \bar{\pi}) \times \{1, 2\} \to \mathbb{R}$ is submodular, then $u^*(\pi) = \mathrm{argmin}_{u \in \{1,2\}} f(\pi, u)$ is TP2 increasing on Π^{comp}.*

Appendix 2: Proof of Theorems

Theorem 4. *The following properties hold for the multi-variate POMDP.*

1. Under (A1), $C(\pi, u)$ is TP2 decreasing on lines $\mathcal{L}(e_S, \bar{\pi})$..
2. Under (A1), (A2), (A3), $Q(\pi, u)$ is TP2 decreasing on lines $\mathcal{L}(e_S, \bar{\pi})$..
3. Under (A1), (A2), (A3), (S), $Q(\pi, u)$ is submodular wrt $\underset{TP2L}{\geq}$.
Thus the optimal policy $\mu^*(\pi)$ is TP2 increasing on lines $\mathcal{L}(e_S, \bar{\pi})$.

Proof of Part 1: Let $\pi \underset{TP2L}{\geq} \tilde{\pi}$. Then $\pi = (1 - \epsilon)\bar{\pi} + \epsilon e_{SL}$ and $\tilde{\pi} = (1 - \tilde{\epsilon})\bar{\pi} + \tilde{\epsilon} e_{SL}$,
where $1 \geq \epsilon \geq \tilde{\epsilon} \geq 0$. But $C(\pi, u) - C(\tilde{\pi}, u) = (\epsilon - \tilde{\epsilon})\left[c(S^L, u) - c'\bar{\pi}\right]$. Therefore
$c(s^L, u) \leq c(X, u)$, i.e., (A1), is a sufficient condition for $C(\pi, u) < C(\tilde{\pi}, u)$.

Proof of Part 2: The proof is by mathematical induction on the value iteration:

$$V_{k+1}(\pi) = \min_{u \in \{1,2\}} Q_{k+1}(\pi, u), \quad \mu_{k+1}^*(\pi) = \text{argmin}_{u \in \{1,2\}} Q_{k+1}(\pi, u) \qquad (19)$$

$$Q_{k+1}(\pi, u) = C(\pi, u) + \rho \sum_{Y \in \mathbf{Y}} V_k\left(T(\pi, u, Y)\right)\sigma(\pi, u, Y), \ \pi \in \Pi, u \in \{1, 2\}.$$

The value iteration algorithm converges uniformly in π [4], i.e., $\lim_{k \to \infty} V_k(\pi) \to V(\pi)$ and $\lim_{k \to \infty} \mu_k^*(\pi) \to \mu^*(\pi)$ uniformly in π. Choose $V_0(\pi)$ as an arbitrary TP2 decreasing function of π in (19). Consider (19) at any stage k. Assume $V_k(\pi)$ is TP2 decreasing in π. Consider $\pi \underset{TP2L}{\geq} \tilde{\pi}$. Denote optimal actions for $\pi, \tilde{\pi}$ as $\mu^*(\pi)$ and $\mu^*(\tilde{\pi})$. From [5, Theorem 4.2] under (A2) and (A3), $\sum_{Y \in \mathbf{Y}} V_k\left(T(\pi, u, Y)\right)\sigma(\pi, u, Y)$ is TP2 decreasing in π. From Part 1, under (A1), $C(\pi, u)$ is TP2 decreasing on lines. Since the sum of decreasing functions is decreasing, the result follows.

Proof of Part 3: To show that $Q(\pi, u)$ is submodular, requires showing that $Q(\pi, 1) - Q(\pi, 2)$ is TP2 decreasing on lines. From Part 2 $V_k(\pi)$ is TP2 decreasing over lines if (A1), (A2), (A3) hold. So to prove $Q(\pi, u)$ is submodular, we show $C(\pi, 1) - C(\pi, 2)$ is TP2 decreasing over lines. Similar to proof of Part 1, $C(\pi, 1) - C(\pi, 2)$ is decreasing over lines if (S) holds. Then, Result 1, implies $\mu^*(\pi)$ is TP2 increasing on lines.

Proof of Theorem 1: Part 3 in the above proof establishes the first claim of Theorem 1. To prove the second claim, for each $\bar{\pi} \in \mathcal{H}$ construct the line segment $\mathcal{L}(e_S, \bar{\pi})$ connecting $\mathcal{H}$ to e_S as in (18). Part 2 of Theorem 4 says that $\mu^*(\pi)$ is monotone for $\pi \in \mathcal{L}(e_S, \bar{\pi})$. There are two possibilities: (i) There is at least one reflexive information state on line information $\mathcal{L}(e_{SL}, \bar{\pi})$ apart from e_{SL}. In this case, pick the reflexive state with the smallest $\epsilon \in [0, 1]$ – call this state $\underline{\pi}$. Then by Lemma 1, on the line segment connecting $(1 - \epsilon)\underline{\pi} + \epsilon e_{SL}$, all information states are TP2 orderable and reflexive. Moving along this line segment towards e_{SL}, pick the largest ϵ for which the $\mu(\pi) = 1$. The information state corresponding to this ϵ is the threshold information state – denote it by $\Gamma(\bar{\pi}) = \pi^{\epsilon^*, \bar{\pi}} \in \mathcal{L}(e_1, \bar{\pi})$ where $\epsilon^* = \max\{\epsilon \in [0, 1] : \mu^*(\pi^{\epsilon, \bar{\pi}}) = 1\}$. (ii) There is no reflexive state on $\mathcal{L}(e_{SL}, \bar{\pi})$ apart from e_{SL}. In this case, define the threshold information state arbitrarily. It is irrelevant since from Lemma 1, the trajectory of all information states is TP2 reflexive. The above construction implies that on $\mathcal{L}(e_S, \bar{\pi})$, there is a unique threshold point $\Gamma(\bar{\pi})$. The entire simplex can be covered by considering all pairs of lines $\mathcal{L}(e_s, \bar{\pi})$, for $\bar{\pi} \in \mathcal{H}$, i.e., $\Pi = \cup_{\bar{\pi} \in \mathcal{H}} \mathcal{L}(e_s, \bar{\pi})$. Combining all points $\Gamma(\bar{\pi})$ for all pairs of lines $\mathcal{L}(e_s, \bar{\pi})$, $\bar{\pi} \in \mathcal{H}$, yields the threshold curve $\Gamma = \cup_{\bar{\pi} \in \mathcal{H}} \Gamma(\bar{\pi})$.

Proof of Theorem 2: Given $\pi_1, \pi_2 \in \mathcal{L}(\bar{\pi})$ with $\pi_1 \underset{TP2L}{\leq} \pi_2$, we need to prove the linear threshold policy satisfies $\mu_\theta(\pi_1) \leq \mu_\theta(\pi_2)$ iff $\theta(S^L) \geq \theta(i)$, $i = 1, 2, \ldots, L-1$. Also $\pi \underset{TP2L}{\leq} \pi_2$ means that $\pi_1 = \epsilon_1 e_{S^L} + (1 - \epsilon_1)\bar{\pi}$, $\pi_2 = \epsilon_2 e_{S^L} + (1 - \epsilon_2)\bar{\pi}$ and $\epsilon_1 \leq \epsilon_2$.

Necessity: We show that if $\theta(S^L) \geq \theta^{(l)}(i)$ for $i \neq S^L$, then $\mu_\theta(\pi)$ is TP2 increasing on lines $\mathcal{L}(\bar{\pi})$. Note that from (12), $\theta_a'\pi_2 - \theta_a'\pi_1 = (\epsilon_2 - \epsilon_1)(\theta(S^L) - \theta'\bar{\pi})$, is of the same sign as $(\epsilon_2 - \epsilon_1)$ for all $\bar{\pi} \in \mathcal{H}$. Therefore $\epsilon_2 \geq \epsilon_1$ implies $\mu_\theta(\pi_2) > \mu_\theta(\pi_1)$. That is, $\pi_2 \underset{TP2L}{\geq} \pi_1$ implies $\mu_\theta(\pi_2) \geq \mu_\theta(\pi_1)$. This implies $\mu_\theta(\pi)$ is TP2 increasing on $\mathcal{L}(\bar{\pi})$.

Sufficiency: Suppose $\mu_\theta(\pi)$ is TP2 increasing on lines. We need to prove $\theta(S^L) \geq \theta(i)$. From (12), for $\pi_1 \underset{TP2L}{\leq} \pi_2$, since $\mu_\theta(\pi)$ is TP2 increasing, it follows that $\mu_\theta(\pi_1) \leq \mu_\theta(\pi_2)$. This is equivalent to $(\epsilon_2 - \epsilon_1)(\theta(S^L) - \theta'\bar{\pi}) \geq 0$ for all $\bar{\pi} \in \mathcal{H}$. Since $\epsilon_2 \geq \epsilon_2$, the expression is positive iff $\theta(S^L) > \theta'\bar{\pi} \ \forall \bar{\pi} \in \mathcal{H}$. This implies $\theta(S^L) \geq \theta(i)$ for $i \neq S^L$.

An Evidential Measure of Risk in Evidential Markov Chains

Hélène Soubaras

Thales Research & Technology France
Campus Polytechnique - 1 av. Augustin Fresnel, F-91767 Palaiseau cedex, France
helene.soubaras@thalesgroup.com

Abstract. This paper provides a new method for computing a risk criterion for decision-making in systems modelled by an Evidential Markov Chain (EMC), which is a generalization to the Dempster-Shafer's Theory of Evidence [1]: it is a Markov chain manipulating sets of states instead of the states themselves. A cost is associated to each state. An evidential risk measurement derived from the statistical ones will be proposed. The vector of the costs of the states, the transition matrix of the Markov model, and the gauge matrix describing the repartition of the sets will be used to construct matrix calculations in order to provide an upper and a lower bound of the estimated risk. The former is a Choquet integral following the belief function, and the latter is established from the plausibility function.

1 Introduction

On applications such as crisis management, the decision-maker's role is crucial. Indeed, if he performs a good and early choice of what actions to do, taking into account his limited available resources, he can avoid an important part of human or financial losses. He can manage at two levels: the survey before the crisis, when one must decide whether to intervene or not, and during the ongoing crisis, when one must decide what to do. In both cases, the objective is to provide him with an aid by computing an evaluation of the future risk as decision criterion. The proposed approach focuses on a Markov modelling of the system (the interacting phenomena), which is well suited in many situations involving propagation (e.g. fire). Risk measurement requires also a model for the cost of the system states, which leads us to aim at defining a simplified Markov Decision Process (MDP) [2]. This is a sequential decision process consisting in a collection of possible actions, with a Markov and a cost models depending on each one. Solving entirely a MDP is finding a policy, i.e. which action to do for each state of the system. In this paper we focus on one single action (the fact to wait) to plot the risk value as a function of future time.

Methods have been proposed in the literature for pure probabilistic Markov models [3] [4] [5] [6].

But the main difficulty in applications such as early crisis management is that the data are missing for a probabilistic model, due to the uncertainty about

C. Sossai and G. Chemello (Eds.): ECSQARU 2009, LNAI 5590, pp. 863–874, 2009.

what may happen, the lack of observations at the very beginning of a crisis, and their imprecision (e.g. text). This is why we shall consider a generalization of the Markov chain called Evidential Markov Chain (EMC) [7], to the Dempster-Shafer's Theory of Evidence [1]. We shall use this generalization in order to propose a new measure of risk.

There already exist works on generalized MDPs with imprecise parameters [8] (a POMDP with a given model of parameter intervals) and [9] (some conclusions about POMDP with Imprecise Probabilities). Harmanec [10] proposes also to delimitate the utility function (which is a generalization of the risk criterion) by two bounds, as in the proposed approach, which also includes a calculus from a Choquet integral. But Harmanec's objective [10] is different, since he aims at provinding ideas for solving a whole MDP.

In the proposed approach the cost is modelled by a vector (and not a reward matrix as in MDPs). This vector-based model for the cost allows to perform the matrix computation that will be described in this paper.

2 Preliminary

2.1 Reminder about Belief Functions

One calls *frame of discernment* a set Ω of all possible mutually exclusive hypotheses; It can be discrete or continuous. A *mass function*, also called BBA (Basic Belief Assignment) [1], is a normalized mapping $m : 2^{\Omega} \to [0, 1]$ (2^{Ω} is the set of all subsets of Ω). A subset $A \subseteq \Omega$ whose mass is nonzero is called a *focal set*. In this paper, we suppose $m(\emptyset) = 0$. The mass function becomes a classical probability when the focal sets are singletons. $\mathcal{F} \subseteq 2^{\Omega}$ will denote the set of all focal sets. A mass function allows to define the classical *belief function* *Bel* and *plausibility function* *Pl* [11]. Several interpretations of Shafer's model of belief functions [1] have been proposed, like the upper and lower probability [11]; let Pr be a probability with its associated σ-algebra on Ω. Pr is compatible with the belief mass m if and only if for any subset $S \subseteq \Omega$ belonging to the σ-algebra, one has $Pr(S) \geq Bel(S)$. (Thus $Bel(S)$ is the lower probability). One has the consequent relationship:

$$Bel(S) \leq Pr(S) \leq Pl(S) \tag{1}$$

Thus a given mass allows a family of compatible probabilities. They are said *imprecise probabilities* since they belong to an interval. A geometric model illustrates this in [12].

2.2 Choquet Integral

A *capacity* [13] [14] on Ω is a set function $\mu : 2^{\Omega} \to [0, 1]$ satisfying $\mu(\emptyset) = 0$, $\mu(\Omega) = 1$, and for subsets A and B of Ω, $A \subseteq B \Rightarrow \mu(A) \leq \mu(B)$. Note that a belief function *Bel* is a capacity.

Definition 1. *Let μ be a capacity on Ω defined on a set of subsets $\mathcal{F} \subseteq 2^{\Omega}$. Let be a function $f : \Omega \rightarrow \mathbb{R}$ which is measurable w.r.t. μ, i.e. $f^{-1}(y) \in \mathcal{F}$ for all $y \in \mathbb{R}$. Suppose that $f(x)$ can take N possible values $f_1 < f_2 < ... < f_N$. The Choquet integral [13] fo f w.r.t. μ is defined by:*

$$C_{\mu}(f) = \sum_{n=1}^{N} f_n \left(\mu(A_n) - \mu(A_{n+1}) \right)$$

where $A_n = \{x : f(x) \geq f_n\}$ and $A_{N+1} = \emptyset$.

2.3 Matrix Tools

We consider a BBA on a finite discrete frame Ω containing N elements. N_f is the number of focal sets. One will define the mass vector M by its coordinates: $M(j) = m(A_j) = m_j$ for all focal set A_j, $1 \leq j \leq N_f$.

Matrix Representations for Belief Functions. It is known [14] that the relation between a BBA m and the corresponding belief function Bel is a bijection. A mass function m can be deduced from a belief function Bel thanks to the so-called *Möbius transform*:

$$m(A) = \sum_{B \subseteq A} (-1)^{|A \setminus B|} Bel(B) \tag{2}$$

The column vector Bel containing all the values of the belief function on the nonempty subsets will also be of size 2^N, and it can be calculated from the mass vector M thanks to a matrix product: $Bel = \mathbf{BfrM}.M$, and the Möbius transform is then performed by the inverse matrix $\mathbf{BfrM}^{-1}$. $\mathbf{BfrM}$ is a generalization matrix defined by $BfrM(A, B) = 1$ if $B \subseteq A$ and 0 otherwise [15]. Similarly, the plausibility function will be computed the gauge (pattern) matrix:

Definition 2. *The* gauge *matrix of a collection of subets A_i is defined by:*

$$G_a(i, j) = \begin{cases} 1 \text{ if } & A_i \cap A_j \neq \emptyset \\ 0 \text{ otherwise} \end{cases}$$

The 2^N-size column vector Pl of the plausibility function is then defined by: $Pl = G_a M$. Note that G_a can also be deduced from $\mathbf{BfrM}$ because of the relationship $Pl(A) = 1 - Bel(A)$.

Markov Kernel Matrix. Let be X and Y two discrete random variables. A Markov kernel is a matrix of the conditional probabilities $p(i|k)$ of the occurrence $Y = y_i$ given $X = x_k$ has occurred. (In Markov chains, the Markov kernel is the state transition matrix).

Now, let be a frame Ω with a mass function m and let be $\mathcal{F}$ the set of ts focal sets. Let be a finite partition $\mathcal{H} = \{X_i \ / \ 1 \leq i \leq N_x\} \subseteq 2^{\Omega}$ on Ω. Each one of the N_x subsets X_i can be called a *class*. They may be for example the singletons of Ω. One would like to estimate which class contains the truth. Classes and focal sets can be viewed as random sets X and S, taking values in $\mathcal{H}$ and $\mathcal{F}$ respectively. Each focal set A_k can occur with a probability m_k.

One assumes that there exists a *Markov kernel* K, transforming masses into compatible probabilities. K is defined by $K(i,k) = p(i|k)$ such that

$$p_i = \sum_{k=1}^{N_f} p(i|k) m_k \tag{3}$$

where $p_i = Pr(X_i)$ is a compatible probability and $m_k = m(A_k)$ is a mass. This can also be written with P, the vector of the probabilities of the classes:

$$P = K.M \tag{4}$$

Note that K is zero where G_a is zero. As $0 \le p(i|k) \le 1$ for all (i,k), one can verify from the definitions of the belief and the plausibility functions that for any compatible kernel K, the relation (1) is satisfied.

Matrix Representations for Classes. One can still define the gauge matrix G_a of size $N_x \times N_f$ by $G_a(i,j) = 1$ if $X_i \cap A_j \ne \emptyset$, and 0 otherwise, for all classes X_i and for all focal sets A_j. Any Markov kernel K compatible with the BBA is zero where G_a is zero. The lines of the transposed matrix G_a^T can be seen as base-2 representations of the focal sets. One can describe entirely a belief mass by its gauge matrix G_a and its mass vector M. When the classes are not singletons, the cardinality $|A|$ of a focal set A is defined as the number of classes it hits. This number is obtained by

$$(11...1).G_a = \begin{pmatrix} |A_1| \\ |A_2| \\ \vdots \\ |A_{N_f}| \end{pmatrix}$$

The computation of the belief and the plausibility functions with matrix products is still possible, as it was shown by Smets [15] and at paragraph 2.3, for the 2^{N_x} subsets of Ω that are unions of subsets X_i:

$$Bel = \mathbf{BfrM}.M \qquad\qquad Pl = G_a.M$$

where $\mathbf{BfrM}$ was defined by Smets [15] ($\mathbf{BfrM}$ is a generalization matrix whose non null elements are 1); G_a is the gauge matrix (2). They are all $(2^{N_x} \times N_f)$-sized matrices.

3 Proposed Evidential Risk Measures

Risk analysis consists first in listing qualitatively the risks in all possible sequences of events, then determining the cost of each of these risks, and finally in measuring the associated corresponding level of risk, taking into account the probability of occurrence of each risk. A review of probabilistic measures for the level of risk can be found at [3] [6]. A generic expression will be reminded here. Extensions to belief functions will be proposed in this section. Let Ω be the set of all possible hypotheses, defined as the frame of discernment in the belief functions approach. The classical probabilistic risk measures (average cost, variance, worst case, partial momentum) can be written as the average of a function of the cost $f(c)$: $R =< f(c) >$.

Evidential Approach. The generalization of the calculus of averaging, which is performed through an integration in the probabilistic case, will be achieved with a Choquet integral [16]. The cost is no more a random variable but an uncertain number C whose probability (for each value of cost) is bounded by Bel and Pl (1), these bounds will be used in the calculation, to define an interval of risk, thus two values R_0 and R_1 to propose to the decision-maker:

$$R_0 \leq R \leq R_1$$

The details of the calculus of R_0 and R_1 will be described in this section. Then we shall verify that these values are bounds for the probabilistic expected cost risk.

Lower Choquet Risk Based on the Belief Function. As said at paragraph 2.2, a belief function allows to compute a Choquet integral. Let's define the cumulative cost belief: $\mu(A_n) = Bel(C > c_n)$. An example is shown at figure 2 for the BBA shown at figure 1.

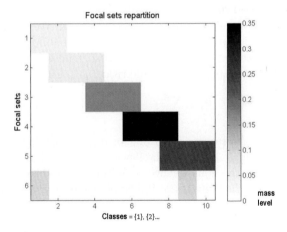

Fig. 1. A basic belief assignment for $\Omega = \{1, 2, 3, 4, 5, 6, 7, 8, 9, 10\}$. Here the clases are the singletons of Ω. The focal sets are represented with a colour corresponding to their mass.

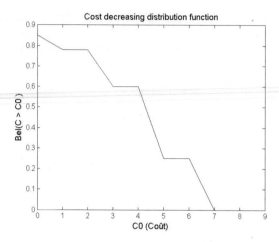

Fig. 2. Cumulative belief corresponding to the BBA of figure 1 for the calculus of the Choquet integral

Thus, one can extend the calculus of the probabilistic risk to the belief function by the Choquet integral providing the *Choquet risk*:

$$R_0 = \mathcal{C}_{Bel}(f(C))) = \sum_n f(c_n)\,(Bel(C > c_n) - Bel(C > c_{n+1})) \qquad (5)$$

where f(c) is the cost function. It is the quantity which is averaged in the probabilistic risk computation.

Upper Bound Based on the Plausibility Function. The plausibility of the classes allows to calculate an upper bound for the risk. We propose to integrate the plausibility function (as a upper bound of the probability) in a Riemann way on all the sets defined by the value of their cost. One will obtain the following expression for the upper bound R_1 of the risk:

$$R_1 = \sum_n f(c_n)Pl(C = c_n)$$

If $f(c) \geq 0$ is a non-decreasing function, for any probability distribution compatible with a given mass function, the lower Choquet risk is less than the probabilistic risk. This is the consequence of Schmeidler's theorem upon expected utility [17]. On the other hand, the upper plausibility risk is greater than the probabilistic risk, since $Pr(c_n) = Pr(C = c_n) \leq Pl(C = c_n)$ for all n.

4   Use in an Evidential Markov Chain

An Evidential Markov Chain (EMC) [18] [7] [19] is a Markov chain involving subsets of Ω instead of its elements themselves. A generalization of Hidden Markov

Models (HMMs) to EMCs has even been proposed by Ramasso [20] [21] and applied to human motion in video sequences. The states of the system are the classes. At time t, the state transition matrix Q operates on the vector M_t of the masses of the focal sets, instead of the vector of probabilities of classical Markov chains:

$$M_t = QM_{t-1}$$

The focal sets do not change in this model. The gauge matrix G_a is thus fixed. Note that the $(N_f \times N_f)$-sized matrix Q must not be confused with the Markov kernels K which relates the masses to compatible probabilities (4).

The objective is to perform the evidential computation of the risk in such a model. For that purpose one considers the costs of the classes; their values are sorted in a vector C in a strictly increasing order. The size of C is thus $N_c \leq N_x$.

4.1   Choquet Risk Calculus

The Choquet risk (5) is computable with matrix computation in EMCs. One must first compute a matrix D_{Ch} which provides the decreasing distribution function of the cost $\underline{F}$ as a function of the mass vector M. D_{Ch} depends only on the gauge G_a. It has N_c lines and N_f columns and is calculated as follows:

$$D_{Ch}(i, j) = \begin{cases} 1 \text{ if} & A_j \subseteq B(c_i) \\ 0 \text{ otherwise} \end{cases}$$

where A_j is a focal set. The cost decreasing distribution is then obtained by:

$$\underline{F} = D_{Ch}.M$$

Then from D_{Ch} one computes the matrix M_{Ch} which performs the Choquet integration. It has the same dimensions as D_{Ch} and is obtained by:

$$\forall j, \qquad M_{Ch}(1, j) = 1 - D_{Ch}(1, j)$$

and

$$\forall i > 1, \forall j \qquad M_{Ch}(i, j) = D_{Ch}(i - 1, j) - D_{Ch}(i, j)$$

The Choquet risk calculus is then performed through the matrix product: $R_0 = f(C)^T M_{Ch}.M$ where $f(C)$ is the vector of coordinates $|C - c_m|^\alpha$ whose coordinates outside $[c_0, c_1]$ are assigned to zero. This is the function to be averaged to obtain the probabilistic risk.

4.2   Plausibility Risk Calculus

This time, the computation matrix for the plausibility of each cost will be named M_{Pl}; it is also calculated from the gauge G_a. For a given cost value $c = c_i$, the plausibility will involve all the focal sets hitting the set

$$D(c_i) = \{x \in \Omega / c(x) = c_i\}$$

In consequence, the computation matrix is:

$$M_{Pl}(i,j) = \begin{cases} 1 \text{ if} & A_j \cap D(c_i) \neq \emptyset \\ 0 \text{ otherwise} \end{cases}$$

The upper bound of the risk R_1 is obtained by: $R_1 = f(C)^T M_{Pl} M$.

4.3   Mid-Term and Long-Term Risk

For an initial mass vector M_0, we have at time t:

$$R_0(t) = f(C)^T M_{Ch} Q^t M_0 \qquad\qquad R_1(t) = f(C)^T M_{Pl} Q^t M_0$$

These expressions are the estimate of the future risk when the mass allocation is known at the present time $t = 0$.

5   Application Example

An earthquake has just started, or a hurricane has just arrived, in a region covering several towns. We are in the very first moments of the crisis. Local rescue teams (brigades...) have already started to intervene. It is crucial that the decision-makers take arrangements very rapidly. To help them to decide when and where to send rescuers, the EMC-based tool will propose them an interval of risk evaluation expressed as a *number of possible new casualties by time unit* as a forecast for the next hours of the crisis.

In the numerical example proposed here, we focus on 8 values for the cost: 0, 10, 20, 50, 100, 200, 500, and 10000 new casualties per time unit. These values define 8 classes. The considered focal sets are described in the table 5.

In this example, the gauge matrix G_a, which is also illustrated at figure 3, and the vector of the costs C are:

$$G_a = \begin{pmatrix} 1\,0\,0\,0\,0\,0 \\ 0\,1\,0\,0\,1\,0 \\ 0\,1\,0\,0\,1\,1 \\ 0\,1\,0\,0\,0\,1 \\ 0\,1\,1\,1\,0\,1 \\ 0\,0\,1\,1\,0\,1 \\ 0\,0\,1\,1\,0\,0 \\ 0\,0\,1\,0\,0\,0 \end{pmatrix} \qquad C = \begin{pmatrix} 0 \\ 10 \\ 20 \\ 50 \\ 100 \\ 200 \\ 500 \\ 1000 \end{pmatrix}$$

and one chooses the following transition matrix for this EMC:

$$Q = \begin{pmatrix} 0.9 & 0 & 0 & 0 & 0.5 & 0 \\ 0.06 & 0.06 & 0.06 & 0.06 & 0.06 & 0.06 \\ 0.04 & 0.04 & 0.04 & 0.04 & 0.04 & 0.04 \\ 0 & 0.3 & 0.4 & 0.4 & 0 & 0 \\ 0 & 0.5 & 0.1 & 0.1 & 0.3 & 0.1 \\ 0 & 0.1 & 0.4 & 0.2 & 0.1 & 0.8 \end{pmatrix}$$

Table 1. Focal sets in the crisis management example at the beginning of an earthquake

No.	Focal set description	Covered classes
1	nothing happens (ex : false alarm)	0 casualties
2	new incident (ex : building collpases...)	10 to 100 casualties
3	new incident, more serious	100 to 1000 casualties
4	consequence (ex : wall falls on evacuating people)	100 to 500 casualties
5	efficient local intervention	10 to 20 casualties
6	need for more intervention	20 to 200 casualties

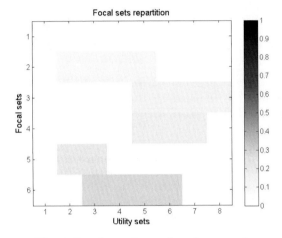

Fig. 3. Result on the earthquake example

This transition matrix expresses that a minor incident is likely to occur at any moment; it is the same with a greater incident, but less probably; the over-accident probability appears in the transition from A_2 to A_4 and A_3 to A_4. When the incident is serious, there is a greater probability that the local rescuer resources will not be enough.

At present, the date is $t = 0$, and some information about what is going on or has just occurred are available (from phone calls, for example). They allow to establish belief masses for the focal sets. For example, if we know that something has just occurred and we think that it is rather serious; we have some belief in the hypothesis that the local rescuers will be able to cope with the problem. The corresponding initial mass vector M_0 is then:

$$M_0 = \begin{pmatrix} 0.3 \\ 0.6 \\ 0 \\ 0 \\ 0.1 \\ 0 \end{pmatrix}$$

With those data, the computation of the risk bounds (Choquet risk as lower bound, and plausibility risk as upper bound), has been performed. We used the function $f(c) = c$, so this is the expected cost risk. We obtained the results illustrated at figure 4.

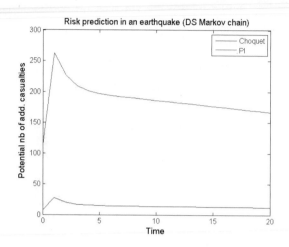

Fig. 4. Result on the earthquake example

These curves show that the forecasted risk jumps at relatively short term, and afterwards it decreases. This is not surprising, because the initial state indicates that an incident has likely occurred, which may provoke consequent incidents within the next moments.

6   Conclusion

Until now, Evidential Markov chains (EMCs) had been proposed only in a very different context (airborne image segmentation for remote sensing [18] [7]). But in fact EMC models have potentially interesting applications in the field of uncertain systems, particularly those involving human behaviors or imprecise data such as text. An example was given in crisis management of a earthquake or a hurricane. Validated by numerical results, this approach has the advantage to address an operational need: to provide a fast decision support tool for the very beginning of a large-scale crisis, in spite of uncertainty in the knowledge about the ongoing situation. The proposed algorithm may be extended to other models such as hidden Markov models (HMMs); it can also be used in an algorithm for planning or for proposing alternatives to the decision-maker, particularly in a generalization of the Markov Decision Processes (MDPs) [2], to belief functions. The EMCs themselves may also be useful to perform simulations (e.g. events due to the tenseness between two conflicting countries).

References

1. Shafer, G.: A Mathematical Theory of Evidence. Princeton University Press, Princeton (1976)
2. Puterman, M.L.: Markov Decision Processes. Wiley, New York (1994)
3. Soubaras, H., Mattioli, J.: Une approche markovienne pour la prévision du risque. In: Proc. of 7th Congrès int. Pluridisciplinaire Qualité et Sûreté de Fonctionnement QUALITA 2007, Tanger, Maroc, March 20-22, pp. 64–71 (2007)
4. Longchampt, J.: Méthode matricielle de quantification des risques associés à la maintenance d'un composant par un modèle semi-markovien. In: Actes du 6e Congrès int. pluridisciplinaire Qualité et Sûreté de Fonctionnement QUALITA 2005, Bordeaux, France (March 2005)
5. Soubaras, H.: Risk prediction in a space-time markov model applied to propagating contamination. In: Information Processing and Management of Uncertainty in knowledge-based systems (IPMU 2008) Conf., Malaga, Spain, June 22-27 (2008)
6. Soubaras, H., Mattioli, J.: Prévision de l'évolution future du risque dans les systèmes markoviens. In: Diagnostic des Systèmes Complexes (QUALITA 2007), pp. 157–181 (2008)
7. Lanchantin, P., Pieczynski, W.: Chaînes et arbres de markov évidentiels avec applications à la segmentation des processus non stationnaires. Revue Traitement du Signal 22 (2005)
8. Itoh, H., Nakamuraa, K.: Partially observable markov decision processes with imprecise parameters. Artificial Intelligence 171, 453–490 (2007)
9. Trevizan, F.W., Cozman, F.G., de Barros, L.N.: Unifying nondeterministic and probabilistic planning through imprecise markov decision processes (2006)
10. Harmanec, D.: A generalization of the concept of markov decision process to imprecise probabilities. In: Proc. of ISIPTA'99, 1st Int. Symposium on Imprecise Probabilities and their applications, Ghent, Belgium, June 30-July 2 (1999)
11. Smets, P., Kennes, R.: The transferable belief model. Artificial Intelligence 66, 191–234 (1994)
12. Cuzzolin, F.: A geometric approach to the theory of evidence. IEEE Tansactions on Systems, Man and Cybernetics - part C: Applications and reviews 38, 522–534 (2008)
13. Kojadinovic, I.: Multi-attribute utility theory based on choquet integral: a theoretical and practical overview. In: MOPGP 2006, 7th Int. Conf. on Multi-Objective Programming and Goal Programming, Tours, France, June 12-14 (2006)
14. Grabisch, M., Murofushi, T., Sugeno, M.: Fuzzy Measures and Integrals. Physica-Verlag (2000)
15. Smets, P.: The application of the matrix calculus to belief functions. Int. J. Approximate Reasoning 31, 1–30 (2002)
16. Peters, J., Ramanna, S.: Application of the choquet integral in software cost estimation. In: Proc. of 5th IEEE Conf. on Fuzzy Systems, September 8-11, vol. 2, pp. 862–866 (1996)
17. Schmeidler, D.: Subjective probability and expected utility without additivity. Econometrica 57, 571–587 (1989)
18. Pieczynski, W.: Hidden evidential markov trees and image segmentation. In: Proc. of Int. Conf. on Image Processing ICIP 1999, vol. 1, pp. 338–342 (1999)

19. Soubaras, H.: On evidential markov chains. In: Information Processing and Management of Uncertainty in knowledge-based systems (IPMU 2008) conf., Malaga, Spain, June 22-27 (2008)
20. Ramasso, E., Rombaut, M., Pellerin, D.: Un filtre temporel crédibiliste pour la reconnaissance d'actions humaines dans les vidéos. In: Proc. of LFA 2006 (Rencontres Francophones sur la Logique Floue et ses applications), Toulouse, France, October 19-20 (2006)
21. Ramasso, E.: Reconnaissance de séquences d'états par le modéle des Croyances Transférables. Application á l'analyse de vidéos d'athlétisme. Ph.d. thesis, Université Joseph Fourier de Grenoble (2007)

Algebras of Fuzzy Sets in Logics Based on Continuous Triangular Norms

Stefano Aguzzoli[1], Brunella Gerla[2], and Vincenzo Marra[3]

[1] Università degli Studi di Milano, Dipartimento di Scienze dell'Informazione
via Comelico 39–41, I-20135 Milano, Italy
`aguzzoli@dsi.unimi.it`
[2] Università dell'Insubria, Dipartimento di Informazione e Comunicazione
via Mazzini 5, I-21100 Varese, Italy
`brunella.gerla@uninsubria.it`
[3] Università degli Studi di Milano, Dipartimento di Informazione e Comunicazione
via Comelico 39–41, I-20135 Milano, Italy
`marra@dico.unimi.it`

Abstract. Associated with any $[0, 1]$-valued propositional logic with a complete algebraic semantics, one can consider algebras of families of fuzzy sets over a classical universe, endowed with the appropriate operations. For the three most important schematic extensions of Hájek's Basic (Fuzzy) Logic, we investigate the existence and the structure of such algebras of fuzzy sets in the corresponding algebraic varieties. In the general case of Basic Logic itself, and in sharp contrast to the three aforementioned extensions, we show that there actually exist different, incomparable notions of algebras of fuzzy sets.

1 Introduction

By a *fuzzy set* we shall mean, as usual, a function $f\colon X \to [0, 1]$ from a set X to the real unit interval $[0, 1]$. We write $[0, 1]^X$ to denote the family of all fuzzy sets over X. Already in his first paper on fuzzy sets, Zadeh introduced an algebraic structure on $[0, 1]^X$ by considering operations between fuzzy sets that generalise classical union, intersection, and complement [23, § II]. Specifically, given $f, g\colon X \to [0, 1]$, Zadeh defined union as $(f \vee g)(x) = \max\{f(x), g(x)\}$, intersection as $(f \wedge g)(x) = \min\{f(x), g(x)\}$, and complement as $(\neg f)(x) = 1 - f(x)$, where $x \in X$. In passing, he remarked on the connection between the nascent theory of fuzzy sets, and many-valued logic [23, Footnote 3, and Comment on pp. 341–342]. Later, several other possible families of fundamental operations on fuzzy sets have been considered. Each such family induces a notion of *algebra of fuzzy sets*. Specifically, any subset of $[0, 1]^X$ that is closed under the chosen operations is such an algebra. Algebras of fuzzy sets are also known as *bold algebras* [4], and are related to *clans of fuzzy sets*, see e.g. [6, Ch. I] and [20, 12.4].

A *triangular norm* (*t-norm*, for short) is an operation $T\colon [0, 1] \times [0, 1] \to [0, 1]$ that is associative, commutative, has 1 as identity element, and is monotone,

C. Sossai and G. Chemello (Eds.): ECSQARU 2009, LNAI 5590, pp. 875–886, 2009.

meaning that $T(x, y) \leq T(x', y')$ whenever $x \leq x'$ and $y \leq y'$, for any $x, y, x', y' \in [0, 1]$. For background, we refer to [20]. Now, t-norms that are left-continuous can be used as the $[0, 1]$-valued semantics of conjunction in certain many-valued logics. This is because each such t-norm T has an associated *residuum* R defined by the condition

$$R(x, y) = \max \{ z \mid T(z, x) \leq y \} .$$

The residuum R provides the $[0, 1]$-valued semantics for an implication connective associated with a conjunction whose semantics is given by T. If absolute falsity is interpreted as 0, negation can then be interpreted as $R(x, 0)$, for each $x \in [0, 1]$. This leads to a framework of *monoidal t-norm-based logics*, first introduced in [11]. Throughout, we shall focus on the important special case of t-norms that are continuous functions; hence, for the rest of this paper, 't-norm' means 'continuous t-norm'. In this paper we assume familiarity with Hájek's treatment [16] of *Basic Logic* (BL, for short) — the logic all t-norms and their residua —, and with its most important schematic extensions, namely, *Łukasiewicz, Gödel,* and *Product logic*.

In this paper we address a problem that can be informally stated as follows.

Does each one of Łukasiewicz, Gödel, Product, and Basic logic admit a well-defined notion of algebra of fuzzy sets?

In order to make this question precise, it is convenient to work in an algebraic setting. We recall that there is a variety $\mathbb{BL}$ of algebras, called *BL-algebras*, corresponding to BL. This variety provides the (complete) algebraic semantics for Hájek's Basic Logic, and each BL-algebra arises as the Tarski-Lindenbaum algebra of a theory in BL, provided the language is sufficiently large. Schematic extensions of BL are in one-one correspondence with subvarieties of $\mathbb{BL}$. The algebras thus associated with Product and Gödel logic are just called *Product* and *Gödel algebras*, respectively. Those associated with Łukasiewicz logic are known as *MV-algebras*[1] [8]. The corresponding varieties are denoted $\mathbb{P}$, $\mathbb{G}$, and $\mathbb{MV}$, respectively. Let $\mathbb{S}$ be a subvariety of $\mathbb{BL}$ associated with the schematic extension $\mathscr{E}$ of BL. If a t-norm T and its residuum R satisfy the additional equations satisfied by $\mathbb{S}$, equivalently, if they satisfy the corresponding additional axiom schemata for $\mathscr{E}$, then the structure $S = \langle [0, 1], T, R, 0 \rangle$ is an algebra of $\mathbb{S}$, called a *standard algebra* (for $\mathbb{S}$ or $\mathscr{E}$). If $\mathscr{E}$ is complete with respect to $[0, 1]$-valued assignments to atomic formulæ extended to non-atomic formulæ through the use of a fixed t-norm T and its residuum, then $\mathscr{E}$ is said to satisfy *standard completeness with respect to the t-norm T*. In algebraic language, this is equivalent to saying that the variety $\mathbb{S}$ is generated[2] by S or, equivalently, that S is *generic* for $\mathbb{S}$. (In general $\mathscr{E}$ satisfies standard completeness if the corresponding variety $\mathbb{S}$ is generated

---

[1] Traditionally, MV-algebras are presented on a different signature, see [8, 1.1.1]. Here, we regard them as a subvariety of BL-algebras, and thus adopt the corresponding signature.

[2] Recall that this means that any equation in the language of $\mathbb{S}$ that fails for some evaluation into some algebra in $\mathbb{S}$, must already fail for some evaluation into the algebra S.

by a class of standard algebras.) Any generic standard algebra S can be used to define a notion of algebras of fuzzy sets (for $\mathbb{S}$ or $\mathscr{E}$). For, we note first that S induces a structure of S-algebra on the family of functions $[0,1]^X$, for any set X, by defining operations pointwise. Then, we shall say that an S-algebra A is an *S-algebra of fuzzy sets* if and only if A is isomorphic to a subalgebra of $[0,1]^X$. However, this notion leaves much to be desired when $\mathbb{S}$ is not generated by S. Indeed, in this case a standard argument shows that there exists a formula φ in the logic $\mathscr{E}$ that is not provable, and yet no countermodel for φ can be produced in any S-algebra of fuzzy sets. Here, the connection between S-algebras of fuzzy sets and the underlying logic is too tenuous to be defensible. We therefore always ask that $\mathbb{S}$ *be generated by* S. Under this assumption, let us write $\mathscr{F}(\mathbb{S}, S)$ for the class of all S-algebras of fuzzy sets in $\mathbb{S}$. Now the mathematical counterpart of the question above is tackled by looking at how $\mathscr{F}(\mathbb{S}, S)$ depends on S. Specifically, given another standard algebra $S' = \langle [0,1], T', R', 0 \rangle$ that generates $\mathbb{S}$, let us write

$$\mathscr{F}(\mathbb{S}, S) \sqsubseteq \mathscr{F}(\mathbb{S}, S')$$

if each algebra in $\mathscr{F}(\mathbb{S}, S)$ is isomorphic to some algebra in $\mathscr{F}(\mathbb{S}, S')$, and

$$\mathscr{F}(\mathbb{S}, S) \equiv \mathscr{F}(\mathbb{S}, S') \tag{1}$$

when the converse also holds. The latter is then a happy situation, for *the logic $\mathscr{E}$ induces an essentially unique notion of algebras of fuzzy sets*, and the dependence of the notion of fuzzy sets on the choice of S becomes immaterial. In particular, (1) certainly holds when $\mathbb{S}$ has, up to isomorphism, just one standard generic algebra. This is known to be the case for $\mathbb{MV}$, $\mathbb{P}$, and $\mathbb{G}$ — see below for details and references. In these cases, then, the question arises whether one can characterize those algebras that are isomorphic to some algebra of fuzzy sets. Again, in some cases the answer is available via the literature, as explained below. Our contribution here is Theorem 1, where we characterize product algebras of fuzzy sets. Zadeh's fuzzy logic — a case not encompassed by Hájek's framework — has the interesting feature that all its algebras are algebras of fuzzy sets; see §4.2.

The last part of our paper deals with BL-logic itself. Our second main result, Theorem 2, shows that in this case there indeed are incomparable notions of algebras of fuzzy sets. More generally, as an anonymous referee pointed out to us, from the results in [13] an algorithm can be extracted that settles the question whether a given subvariety of $\mathbb{BL}$ generated by a single standard algebra has the property that all its generating standard algebras are isomorphic. Such an algorithm semi-decides whether (1) holds. A further concluding discussion is in the final §6.

2   Łukasiewicz Logic

Let $x \odot y = \max\{x + y - 1, 0\}$ denote the *Łukasiewicz t-norm*, and let $x \to_L y$ be its associated residuum, that is, $x \to_L y = 1$ if $x \le y$, $x \to_L y = 1 - x + y$, otherwise.

2.1   Standard Generic MV-Algebras

Up to isomorphism, the algebra $[0,1]_L = \langle [0,1], \odot, \to_L, 0 \rangle$ is the only standard MV-algebra: see [16, 2.1.22–23]. It is also generic for MV, by Chang's Completeness Theorem [8, 2.5.3]. Therefore, there is a unique notion of Łukasiewicz algebras of fuzzy sets.

2.2   Łukasiewicz Algebras of Fuzzy Sets

In accordance with general universal algebra, an MV-algebra is called *simple* if it has no non-trivial congruences. Up to an isomorphism, simple MV-algebras are the subalgebras of $[0,1]_L$ [8, 3.5.1]. An MV-algebra is *semisimple* if the intersection of all its maximal ideals is the singleton ideal. Using the correspondence between congruences and ideals, this is the same thing as saying that the MV-algebra is a subdirect product of simple MV-algebras. (Semisimplicity is also equivalent to the absence of "infinitesimal elements"; for a precise statement and proof, see [8, 3.6.3–4].) From this, and the above-mentioned characterization of simple MV-algebras, it follows at once [8, 3.6.1] that *an MV-algebra embeds into* $[0,1]_L^X$, *for some set X, if and only if it is semisimple*. This result is essentially due to Chang [7, 4.9]; see also also [4, Theorem 4].

3   Product Logic

Let $x \cdot y = xy$ denote the usual multiplication of real numbers, and let $x \to_\Pi y$ be its associated residuum, that is, $x \to_\Pi y = 1$ if $x \leq y$, $x \to_\Pi y = y/x$, otherwise.

3.1   Standard Generic Product Algebras

Up to isomorphism, the algebra $[0,1]_\Pi = \langle [0,1], \cdot, \to_\Pi, 0 \rangle$ is the only standard product algebra: see [16, 2.1.22]. It is also generic for $\mathbb{P}$, by [16, 4.1.13]. Therefore, as for Łukasiewicz logic, there is a unique notion of product algebras of fuzzy sets. We now turn to the problem of characterizing them.

3.2   Product Algebras of Fuzzy Sets

Our analysis of product algebras of fuzzy sets requires some background on hoops. We recall that a *hoop* is an algebra $\langle H, *, \Rightarrow, \top \rangle$ such that $\langle H, *, \top \rangle$ is a commutative monoid, and for all $x, y, z \in H$,

1. $x \Rightarrow x = \top$,
2. $x * (x \Rightarrow y) = y * (y \Rightarrow x)$,
3. $x \Rightarrow (y \Rightarrow z) = (x * y) \Rightarrow z$.

Derived operations are $x \wedge y = x * (x \Rightarrow y)$ and $x \vee y = ((x \Rightarrow y) \Rightarrow y) \wedge ((y \Rightarrow x) \Rightarrow x)$. These make H into a lattice, and we shall feel free to use the associated lattice order $\leq$ on H. A *bounded* hoop is an algebra $\langle H, *, \Rightarrow, \bot, \top \rangle$ such that $\langle H, *, \Rightarrow, \top \rangle$ is a hoop, and $\bot \leq x$ for all $x \in H$.

A *basic hoop* is a hoop satisfying $(x \Rightarrow y) \vee (y \Rightarrow x) = \top$. A *Wajsberg hoop* is a basic hoop satisfying $(x \Rightarrow y) \Rightarrow y = (y \Rightarrow x) \Rightarrow x$. A *product hoop* is a basic hoop satisfying $(y \Rightarrow z) \vee ((y \Rightarrow (x * y)) \Rightarrow x) = \top$. A basic hoop is *cancellative* if it satisfies $x \Rightarrow (x * y) = y$. For more details on hoops, we refer the reader to [1].

BL-algebras are the same thing as bounded basic hoops, and basic hoops are the hoop subreducts of BL-algebras. Similarly, the classes of MV-algebras and product algebras respectively coincide with the class of bounded Wajsberg hoops and bounded product hoops; Wajsberg hoops and product hoops are respectively the hoop-subreducts of MV-algebras and product algebras.

Note that each cancellative hoop is a Wajsberg hoop. In general a product hoop is not cancellative. However, totally ordered product hoops are cancellative by [9, Cor. 2.6].

If H is a hoop, a subset $F \subseteq H$ is a *filter* of H if and only if it is closed under $*$ ($x, y \in F$ implies $x * y \in F$) and is an upper set ($x \in F$ and $x \leq y \in H$ imply $y \in F$). Filters of a BL-algebra B are defined as the filters of its hoop reduct.

The hoop subreduct of the standard product algebra $\langle (0, 1]_\Pi, \cdot, \rightarrow, 1 \rangle$ is a cancellative hoop, commonly called *the standard cancellative hoop*.

We shall also need to use lattice-groups as tools. We refer the reader to [15] for all the additional background needed. Let $\langle G, +, -, 0, \leq \rangle$ be an *abelian lattice-ordered group* — for short, just an *ℓ-group*. That is, $\langle G, +, -, 0 \rangle$ is an abelian group with identity 0 and unary inverse operation $-$, $\langle G, \leq \rangle$ is a lattice order, and $x \leq y$ implies $x + z \leq y + z$ for all $x, y, z \in G$. The *negative cone* of such an ℓ-group is

$$G^- = \{x \in G \mid x \leq 0\}.$$

Now, G^- is made into a hoop by endowing it with the operation

$$x \ominus y = (y - x) \wedge 0.$$

Lemma 1. *Each cancellative hoop $\langle H, \cdot, \rightarrow, 1 \rangle$ is isomorphic (as a hoop) to the hoop $\langle G^-, +, \ominus, 0 \rangle$ obtained from the negative cone of an ℓ-group $\langle G, +, -, 0, \leq \rangle$. Further, the latter ℓ-group is unique, to within an isomorphism of ℓ-groups.*

Proof. This is proved in [3]. See also [10, Thm. 2], [9, Thm. 2.5].   □

The ℓ-group $G \equiv \langle G, +, -, 0, \leq \rangle$ is *archimedean* if for all $0 < x, y \in G$ there exists a positive integer n such that $nx \not< y$. If G is totally ordered then the above condition reduces to the following.

For all $0 < x, y \in G$ there exists $0 < n \in \mathbb{Z}$ such that $nx \geq y$.   (2)

Following usual algebraic terminology, a hoop $H \equiv \langle H, \cdot, \rightarrow, 1 \rangle$ is called *simple* if the set of its filters is $\{\{1\}, H\}$. Notice that *if H is simple then it is totally ordered*. This is proved by a standard argument.

Lemma 2. *Let $H \equiv \langle H, \cdot, \rightarrow, 1 \rangle$ be a totally ordered hoop and let $G^- \equiv \langle G^-, +, \ominus, 0 \rangle$ be the negative cone of the ℓ-group $G \equiv \langle G, +, -, 0, \leq \rangle$ such that $H \cong G^-$, as in Lemma 1. Then H is a simple hoop if and only if the group G is archimedean.*

880 S. Aguzzoli, B. Gerla, and V. Marra

Proof. By (2), it suffices to show that H is simple if and only if for all $x, y < 1$ there exists a positive integer n such that $x^n \leq y$. If H is simple then for each $x \in H$, with $x < 1$, the filter $\langle x \rangle$ generated by x coincides with the whole of H. Since $\langle x \rangle = \{y \mid x^n \leq y$ for some $0 < n \in \mathbb{Z}\}$, it follows that for all $x, y < 1$ there exists a positive integer n such that $x^n \leq y$. For the converse implication, assume H is not simple, and let F be a proper non-trivial filter of H and y an element of $H \setminus F$. Note that $y < 1$. Then for all $x \in F$ and for all positive integers n we have $x^n > y$, otherwise $y \in F$, a contradiction. □

In [9] it is observed that the only simple product algebra is the two-element chain. Simplicity is thus a very strong notion for product algebras. Our analysis hinges upon a weaker notion of simplicity that will turn out to be more useful.

A product algebra $A \equiv \langle A, \cdot, \to, 0 \rangle$ is *hoop-simple* if its hoop-subreduct $H(A) = \langle A \setminus \{0\}, \cdot, \to, 1 \rangle$ is a simple hoop. Here, if A is totally ordered, by Lemma 2 it follows that A is hoop-simple if and only if $H(A)$ is isomorphic to the hoop obtained from the negative cone of an archimedean ℓ-group. Furthermore, $H(A)$ is cancellative. By direct inspection, the standard product algebra $[0, 1]_\Pi$ is hoop-simple.

A product algebra is *hoop-semisimple* if it is isomorphic to a subdirect product of a family of hoop-simple product algebras. For example, the free n-generated product algebra is hoop-semisimple, for each integer $n \geq 0$.

In the next proof we shall use a classical result, Hölder's Theorem: *A totally ordered group is archimedean if and only if it is isomorphic (as an ordered group) to a subgroup of the additive group of reals* $\mathbb{R}$. For a proof, see [15, 4.A].

Lemma 3. *A product algebra* $A \equiv \langle A, \cdot, \to, 0 \rangle$ *is hoop-simple if and only if it is isomorphic to a subalgebra of the standard product algebra* $[0, 1]_\Pi$.

Proof. First assume A is a subalgebra of $[0, 1]_\Pi$. Pick $0 < x, y < 1 \in A$. By the elementary properties of real multiplication there exists a positive integer n such that $x^n \leq y$. Then $x^n \in A$, because A is a subalgebra. This means that the only filters of the cancellative hoop $A \setminus \{0\}$ are $\{1\}$ and $A \setminus \{0\}$, that is, A is hoop-simple.

Conversely, assume now A is hoop-simple. Then A is of the form $\{\bot\} \cup G^-$, where, by Lemma 2, G^- is the negative cone of an archimedean ℓ-group G. Note that G is necessarily totally ordered. By Hölder's Theorem, G embeds into the additive group of reals $\mathbb{R}$. Say $\varphi \colon G \to \mathbb{R}$ is this embedding. Next observe that there is an isomorphism of product algebras $\psi \colon \{-\infty\} \cup \mathbb{R}^- \to [0, 1]_\Pi$, where $-\infty$ is a new element such that $-\infty < x$ for all $x \in \mathbb{R}^-$, and the operations are extended in the obvious manner. To wit, the isomorphism $\psi \colon \{-\infty\} \cup \mathbb{R}^- \to [0, 1]_\Pi$ is given by the map $\psi(x) = e^x$ for all $x \in \mathbb{R}^-$, and $\psi(-\infty) = 0$. Then the restriction of $\psi \circ \varphi$ to G^- is a hoop embedding of G^- into the hoop subreduct $(0, 1]$ of $[0, 1]_\Pi$. This proves that A is a product subalgebra of $[0, 1]_\Pi$. □

We are now ready to characterize product algebras of fuzzy sets.

Theorem 1. *A product algebra is isomorphic to a subalgebra of* $[0, 1]_\Pi^X$ *for some set X if and only if it is hoop-semisimple.*

Proof. By Lemma 3, if such an algebra A is hoop-semisimple, then it embeds into a direct product of subalgebras of $[0,1]_\Pi$, say

$$A \hookrightarrow \prod_{x \in X} A_x \hookrightarrow \prod_{x \in X} [0,1]_\Pi \,,$$

where each A_x is isomorphic to a subalgebra of $[0,1]_\Pi$. We conclude $A \cong B$ for some $B \subseteq [0,1]_\Pi^X$. The converse is obvious.   □

Just like MV-algebras are not all semisimple, so there are product-algebras that are not hoop-semisimple — Theorem 1 is not void. As an example, consider the ordered set $Z = \{\bot\} \cup (\mathbb{Z} \overset{\rightarrow}{\times} \mathbb{Z})^-$, where $\mathbb{Z}$ is the additive ordered group of integers, $\overset{\rightarrow}{\times}$ denotes the lexicographic product (please see [15, 1.3.25] for details), and $\bot < (h,k)$ for all $(h,k) \in (\mathbb{Z} \overset{\rightarrow}{\times} \mathbb{Z})^-$. It is easy to see that Z can be endowed with the structure of a totally ordered product algebra, and $(\mathbb{Z} \overset{\rightarrow}{\times} \mathbb{Z})^-$ is a cancellative hoop. By construction, the ℓ-group $(\mathbb{Z} \overset{\rightarrow}{\times} \mathbb{Z})$ is not archimedean, and hence Z is not hoop-simple. The ordered set of filters of Z contains a minimal proper element, namely $\{(0,h) \mid h \in \mathbb{Z}^-\}$. Hence, Z is subdirectly irreducible. We conclude that Z is not hoop-semisimple, and thus, by Theorem 1, cannot be represented as a product algebra of fuzzy sets.

4 Gödel and Zadeh-Kleene Logics

4.1 Gödel Logic

We next consider *Gödel logic*, the logic of the minimum t-norm and its residuum $x \to_G y = 1$ if $x \le y$, $x \to_G y = y$ otherwise. Its algebraic semantics is the subvariety $\mathbb{G}$ of those BL-algebras that have an idempotent monoidal operation, called *Gödel algebras*. We write $[0,1]_G$ for the standard Gödel algebra $\langle [0,1], \min, \to_G, 0 \rangle$.

As is well known, Gödel logic coincides with the extension of the intuitionistic propositional calculus by the *prelinearity* axiom scheme $(\varphi \to \psi) \vee (\psi \to \varphi)$; see [16, 4.2.8]. Thus, Gödel algebras are the same thing as the subvariety of Heyting algebras satisfying prelinearity. Background on Heyting algebras can be found e.g. in [18].

Standard Generic Gödel Algebras. Observe that the isomorphism type of a Gödel algebra G is entirely determined by its bounded lattice reduct. Indeed, regarding G as a BL-algebra, the monoidal operation coincides with the lattice meet operation, and the implication is uniquely determined by the lattice structure via residuation. It follows immediately that the only standard Gödel algebra is $[0,1]_G$, to within an isomorphism. Moreover, $[0,1]_G$ is generic for $\mathbb{G}$ by [16, 4.2.17].

Gödel Algebras of Fuzzy Sets. Some Gödel algebras are not algebras of fuzzy sets. To explain this, let us recall a well-known fact. If H is any Heyting algebra, and $t \notin H$, then one constructs another Heyting algebra $H \cup \{t\}$ with $t > h$ for all $h \in H$, and having the same order as H otherwise. Since, by a standard universal-algebraic theorem [5, 8.4], an algebra is subdirectly irreducible if and only if its lattice of congruences has a unique atom, one verifies that $H \cup \{t\}$ is subdirectly irreducible. Further, it is easy to check that $H \cup \{t\}$ is a linearly ordered Gödel algebra if H is. Now consider a linearly ordered Gödel algebra C of cardinality $> 2^{\aleph_0}$ (there obviously exist Gödel chains of any cardinality), then the Gödel algebra $C \cup \{t\}$ is not an algebra of fuzzy sets. The argument needed to prove this assertion is analogous — *mutatis mutandis* — to the one we shall give below in the proof of Theorem 2; due to space constraints, we omit details.

Quite unlike the case of $\mathbb{MV}$ and $\mathbb{P}$, it is doubtful that Gödel algebras of fuzzy sets enjoy significant structural properties. It is clear that a Gödel algebra is isomorphic to a subalgebra of $[0,1]_G^X$ for some set X if and only if it is a subdirect product of Gödel chains, each of which embeds into $[0,1]_G$. However, the latter chains do not appear to admit a non-trivial characterization. Indeed, the problem reduces to characterizing the suborders of $(0,1)$ (equivalently, of $\mathbb{R}$). It is known that there are at least continuosly many non-isomorphic such suborders [22, 2.25], but to the best of our knowledge no complete characterization is available.

4.2   Zadeh-Kleene Logic

By this we mean the equational logic of *Kleene algebras*. The latter are algebras $\langle A, \wedge, \vee, \neg, \bot, \top \rangle$ such that $\langle A, \wedge, \vee, \bot, \top \rangle$ is a bounded distributive lattice, $\neg$ is an involution satisfying De Morgan's laws, i.e. $\neg\neg x = x$ and $\neg(x \wedge y) = \neg x \vee \neg y$, and, moreover, $x \wedge \neg x \leq y \vee \neg y$. Let us stress that the logic of Kleene algebras is not an extension of BL. As a matter of fact, the involutive negation $\neg$ is not definable as the pseudo-complement with respect to $\bot$ of the residuum of $\wedge$. In the literature, when an implication operator is added to the signature of Kleene algebras, is not the residual one, usually. Logics comprising both residual implication of the t-norm and involutive negation have been studied in [12,14].

Standard Generic Kleene Algebras. Kalman proved[3] [19, Theorem 2] that the only non-trivial subdirectly irreducible Kleene algebras are the two-element Boolean algebra, and the three-element chain $K = \{\top, m, \bot\}$ (where m satisfies $\neg\neg m = m$). Since the former is a Kleene subalgebra of the latter, it follows that the variety $\mathbb{K}$ of Kleene algebras is generated by K. Now consider the standard Kleene algebra $[0,1]_K = \langle [0,1], \min, \max, \neg, 0, 1 \rangle$, where $\neg x = 1 - x$. Since K embeds into $[0,1]_K$ via the map $m \mapsto 1/2$, it follows that $[0,1]_K$ is generic for $\mathbb{K}$. Any other standard Kleene algebra is isomorphic to $[0,1]_K$ (see [21]). In summary, in the variety $\mathbb{K}$ there is, up to an isomorphism, just one standard generic algebra, namely $[0,1]_K$.

---

[3] Kalman called Kleene algebras *normal i-lattices*, and did not assume that they are bounded.

Kleene Algebras of Fuzzy Sets. As mentioned in the Introduction, *every* Kleene algebra A is an algebra of fuzzy sets. This follows at once from Kalman's [19, Theorem 2]. The latter says that A is a subdirect power of K over some index set X. Thus, we have an embedding $A \hookrightarrow K^X$. But since $K \hookrightarrow [0,1]_K$, we also have an embedding $K^X \hookrightarrow [0,1]_K^X$. By composition, we get the desired representation of A as an algebra of fuzzy sets.

5   Hájek's Basic Logic

5.1   Standard Generic BL-Algebras

Let $\langle I, \leq \rangle$ be a linearly ordered set and for any $i \in I$ let A_i be a hoop with top element 1 such that $A_i \cap A_j = \{1\}$ for $i \neq j$. The *ordinal sum* of the family $\{A_i\}_{i \in I}$ is the structure

$$\bigoplus_{i \in I} A_i = \left(\bigcup_{i \in I} A_i, \cdot, \rightarrow, 1 \right)$$

where

$$x \cdot y = \begin{cases} x \cdot^{A_i} y & \text{if } x, y \in A_i \\ y & \text{if there exists } j < i, x \in A_i, 1 \neq y \in A_j \\ x & \text{otherwise,} \end{cases}$$

and

$$x \rightarrow y = \begin{cases} x \rightarrow^{A_i} y & \text{if } x, y \in A_i \\ y & \text{if there exists } j < i, x \in A_i, y \in A_j \\ 1 & \text{otherwise.} \end{cases}$$

The ordinal sum of I copies of the same hoop A will be denoted by IA. Thus, for instance, the standard Gödel algebra $[0,1]_G$ is (isomorphic to) the ordinal sum $[0,1]\{0,1\}$.

Up to isomorphism, for every integer $n \geq 1$ there exists a unique linearly ordered Wajsberg hoop with $n+1$ elements that we shall denote by $\mathbf{L}_n$.

Lemma 4. *An ordinal sum $\bigoplus_{i \in I} A_i$, where each A_i is a linearly ordered Wajsberg hoop, i_0 is the minimum element of I, and A_{i_0} is bounded, is generic for the variety of BL-algebras if and only if*

- *for every $n \geq 1$, $\mathbf{L}_n$ embeds into A_{i_0}, and*
- *for every $n \geq 1$, the set $\{i \in I \mid \mathbf{L}_n$ embeds into $A_i\}$ is infinite.*

Proof. This is proved in [17] (see also [2]).   □

We shall presently use Lemma 4 to deduce our main lemma to Theorem 2. To this end, we need a few preliminaries. We denote by ω the set of natural numbers equipped with the usual order. If $(P, \leq_P)$ and $(Q, \leq_Q)$ are posets, we denote by $P + Q$ the poset obtained by taking the disjoint union of P and Q with the order given by $x \leq y$ if either $x \leq_P y$, or $x \leq_Q y$, or $x \in P$ and $y \in Q$. By 1 we denote an arbitrarily fixed poset with one element. By P^∂ we denote the set P equipped with the order $\leq'$ such that $p \leq' q$ if and only if $q \leq p$.

Lemma 5. *The BL-algebras* $\omega[0,1]_{\text{Ł}}$, $(\omega+1)[0,1]_{\text{Ł}}$, *and* $(1+\omega^{\partial})[0,1]_{\text{Ł}}$ *are standard and generic for the variety of BL-algebras. Moreover, the latter two are subdirectly irreducible, while the first one is not.*

Proof. By Lemma 4, to establish the first statement we only have to prove that the lattice reduct of each of the above algebras is order isomorphic to $[0,1]$.

1. Let f_1 be the function that, for every $n \in \omega$, linearly maps the n-th copy of $[0,1]_{\text{Ł}}$ in the ordinal sum $\bigoplus_{n \in \omega}[0,1]_{\text{Ł}}$ into the subinterval $[n/(n+1),(n+1)/(n+2)]$ of $[0,1]$. The function f_1 is an order isomorphism between $\omega[0,1]_{\text{Ł}}$ and $[0,1]$.

2. Let f_2 be the function that, for every $n \in \omega$, linearly maps the n-th copy of $[0,1]_{\text{Ł}}$ in the ordinal sum $\bigoplus_{x \in \omega+1}[0,1]_{\text{Ł}}$ into the subinterval $[n/(2(n+1)),(n+1)/(2(n+2))]$ of $[0,1]$ and the last copy of $[0,1]_{\text{Ł}}$ to $[1/2,1]$. Then f_2 is an order isomorphism between $(\omega+1)[0,1]_{\text{Ł}}$ and $[0,1]$.

3. Let f_3 be the function that linearly maps the first copy of $[0,1]_{\text{Ł}}$ in the ordinal sum $\bigoplus_{x \in 1+\omega^{\partial}}[0,1]_{\text{Ł}}$ to $[0,1/2]$ and for every $n \in \omega$, linearly maps the n-th copy of $[0,1]_{\text{Ł}}$ into the subinterval $[1/2+1/(2(n+2)),1/2+1/(2(n+1))]$ of $[0,1]$. Then f_3 is an order isomorphism between $(1+\omega^{\partial})[0,1]_{\text{Ł}}$ and $[0,1]$.

As to the second statement, by a general theorem in universal algebra [5, 8.4], an algebra is subdirectly irreducible if and only if its lattice of congruences has a unique atom. Direct inspection shows that both $(\omega+1)[0,1]_{\text{Ł}}$ and $(1+\omega^{\partial})[0,1]_{\text{Ł}}$ do have a unique non-trivial minimal filter, namely, the last summand of the ordinal sum, while $\omega[0,1]_{\text{Ł}}$ fails this property. □

5.2   Incomparable Notions of Algebras of Fuzzy Sets for BL

We can now exhibit two standard generic BL-algebras S and S' such that

$$\mathscr{F}(\mathbb{BL},S) \not\subseteq \mathscr{F}(\mathbb{BL},S') \qquad \text{and} \qquad \mathscr{F}(\mathbb{BL},S') \not\subseteq \mathscr{F}(\mathbb{BL},S). \qquad (3)$$

Theorem 2. *The logic BL does not induce a unique notion of algebra of fuzzy sets. Specifically, (3) holds for the standard generic BL-algebras* $S = (\omega+1)[0,1]_{\text{Ł}}$ *and* $S' = (1+\omega^{\partial})[0,1]_{\text{Ł}}$ *introduced above.*

Proof. Clearly, $T \in \mathscr{F}(\mathbb{BL},T)$ for each BL-algebra T, because the T-algebra of those fuzzy sets having as domain a singleton is obviously isomorphic to T. Let us show that S is not isomorphic to a subalgebra of S'^X, whatever the choice of the set X. Suppose by way of contradiction that $e \colon S \to \prod_X S'$ is an embedding of BL-algebras. Writing A_x for the range of the map $p_x \circ e \colon S \to S'$, where $p_x \colon \prod_X S' \to S'$ is the xth projection map, we obtain a subdirect embedding $e^* \colon S \hookrightarrow \prod_{x \in X} A_x$ by setting $e^*(s) = e(s)$ for $s \in S$. But since S is subdirectly irreducible by Lemma 5, it follows that S is isomorphic to A_x for some choice of $x \in X$. This is a contradiction, as we will show that S does not embed into S'. To see this, for each BL-algebra B, write $\mathcal{I}(B)$ for the poset of idempotent elements of B. Now note that $\mathcal{I}(S)$ is order-isomorphic to $\omega+2$, while

$\mathcal{I}(S')$ is order-isomorphic to $1 + \omega^{\partial}$. Since any homomorphism of BL-algebras carries idempotents to idempotents and preserves order, from any BL-algebraic embedding $S \hookrightarrow S'$ we would obtain an order-embedding $\omega + 2 \hookrightarrow 1 + \omega^{\partial}$, which is impossible. It can be similarly shown that S' is not isomorphic to a subalgebra of S^X, whatever the set X. $\qquad\qquad\qquad\qquad\qquad\qquad\qquad\qquad\qquad\qquad\qquad$ $\square$

6   Conclusions

As we have seen in this paper, each of the varietis $\mathbb{MV}$, $\mathbb{P}$, and $\mathbb{G}$ induces a unique notion of algebra of fuzzy sets. Following an analogy with the classical representation theory for MV-algebras, we have characterized product algebras of fuzzy sets as hoop-semisimple algebras (Theorem 1). As to Gödel algebras, we have explained in §4.1 why a satisfactory structure theory along similar lines appears problematic. For Kleene algebras, the algebraic semantics of Zadeh's fuzzy logic, we have indicated in §4.2 that there is just one possible notion of algebras of fuzzy sets, and, interestingly, all Kleene algebras are representable as algebras of fuzzy sets.

For the general case of BL-logic, we have proved as our second main result (Theorem 2) that there are incomparable notions of algebras of fuzzy sets in $\mathbb{BL}$. It is thus clear that more research on BL-algebras is needed to clarify the situation — for instance, is there a *most general* notion of BL-algebra of fuzzy set, i.e., is there a standard generic BL-algebra S^* such that for any other standard generic BL-algebra S one has $\mathscr{F}(\mathbb{BL}, S) \sqsubseteq \mathscr{F}(\mathbb{BL}, S^*)$? This and other related directions for future research require investigation of the structure of subalgebras of standard generic algebras. By way of conclusion, we give a result along these lines.

Let B be a BL-algebra with top element 1, and let $\mathcal{F}(B)$ be the collection of its filters (including the improper filter B). For each $\mathfrak{p} \in \mathcal{F}(B)$, let $\mathfrak{p}^*$ be the subset of B defined by

$$\mathfrak{p}^* = \left(\mathfrak{p} \setminus \bigcup \{ \mathfrak{q} \in \mathcal{F}(B) \mid \mathfrak{q} \subsetneq \mathfrak{p} \} \right) \cup \{1\}.$$

Theorem 3. *For any BL-chain B the following are equivalent.*

1. *There exists a standard generic BL-algebra A such that B is a subalgebra of A.*
2. *The set $\mathcal{F}(B)$ of filters of B is such that the set $\mathcal{K} = \{ \mathfrak{p} \in \mathcal{F}(B) \mid |\mathfrak{p}^*| > 2 \}$ satisfies $|\mathcal{K}| \leq \aleph_0$, and $\mathcal{F}(B) \setminus \mathcal{K}$ is the union of at most denumerably many intervals of $\mathcal{F}(B)$ (linearly ordered by inclusion), each of which order-embeds into $[0, 1]$. Moreover, $\mathfrak{p}^*$ is either a bounded or a cancellative Wajsberg subhoop of B for each filter $\mathfrak{p} \in \mathcal{F}(B)$, and B^* is an MV-subalgebra of B.*

We cannot include a proof of Theorem 3 here for lack of space, but we plan to publish an extended account elsewhere.

Acknowledgements

We thank the anonymous reviewers for their helpful suggestions.

References

1. Aglianò, P., Ferreirim, I.M.A., Montagna, F.: Basic hoops: an algebraic study of continuous t-norms. Studia Logica 87(1), 73–98 (2007)
2. Aglianò, P., Montagna, F.: Varieties of BL-algebras I: general properties. Journal of Pure and Applied Algebra 181, 105–129 (2003)
3. Amer, K.: Equationally complete classes of commutative monoids with monus. Algebra Universalis 18(1), 129–131 (1984)
4. Belluce, L.P.: Semisimple algebras of infinite valued logic and bold fuzzy set theory. Canad. J. Math. 38(6), 1356–1379 (1986)
5. Burris, S., Sankappanavar, H.P.: A course in universal algebra. Graduate Texts in Mathematics, vol. 78. Springer, New York (1981)
6. Butnariu, D., Klement, E.P.: Triangular norm-based measures and games with fuzzy coalitions. Kluwer Academic Publishers, Dordrecht (1993)
7. Chang, C.C.: Algebraic analysis of many valued logics. Trans. Amer. Math. Soc. 88, 467–490 (1958)
8. Cignoli, R.L.O., D'Ottaviano, I.M.L., Mundici, D.: Algebraic foundations of many-valued reasoning. Trends in Logic—Studia Logica Library, vol. 7. Kluwer Academic Publishers, Dordrecht (2000)
9. Cignoli, R., Torrens, A.: An algebraic analysis of product logic. Mult.-Valued Log. 5(1), 45–65 (2000)
10. Cignoli, R., Torrens, A.: Free cancellative hoops. Algebra Universalis 43(2-3), 213–216 (2000)
11. Esteva, F., Godo, L.: Monoidal t-norm based logic: towards a logic for left-continuous t-norms. Fuzzy Sets and Systems 124(3), 271–288 (2001)
12. Esteva, F., Godo, L., Hájek, P., Navara, M.: Residuated fuzzy logics with an involutive negation. Arch. Math. Logic 39(2), 103–124 (2000)
13. Esteva, F., Godo, L., Montagna, F.: Equational characterization of the subvarieties of BL generated by t-norm algebras. Studia Logica 76(2), 161–200 (2004)
14. Flaminio, T., Marchioni, E.: T-norm-based logics with an independent involutive negation. Fuzzy Sets and Systems 157(24), 3125–3144 (2006)
15. Glass, A.M.W.: Partially ordered groups. Series in Algebra, vol. 7. World Scientific Publishing Co. Inc., Singapore (1999)
16. Hájek, P.: Metamathematics of fuzzy logic. Trends in Logic—Studia Logica Library, vol. 4. Kluwer Academic Publishers, Dordrecht (1998)
17. Montagna, F.: Generating the variety of BL-algebras. Soft Computing 9, 869–874 (2005)
18. Johnstone, P.T.: Stone spaces. Cambridge Studies in Advanced Mathematics, vol. 3. Cambridge University Press, Cambridge (1982)
19. Kalman, J.: Lattices with involution. Trans. Amer. Math. Soc. 87, 485–491 (1958)
20. Klement, E.P., Mesiar, R., Pap, E.: Triangular norms. Trends in Logic—Studia Logica Library, vol. 8. Kluwer Academic Publishers, Dordrecht (2000)
21. Trillas, E.: Negation functions in the theory of fuzzy sets. Stochastica 3(1), 47–60 (1979)
22. Rosenstein, J.G.: Linear orderings. Pure and Applied Mathematics, vol. 98. Academic Press Inc.[Harcourt Brace Jovanovich Publishers], London (1982)
23. Zadeh, L.A.: Fuzzy sets. Information and Control 8, 338–353 (1965)

Soft Constraints Processing
over Divisible Residuated Lattices

Simone Bova

Department of Computer Science, University of Milan
Via Comelico 39, 20135 Milan, Italy
bova@dico.unimi.it

Abstract. We claim that divisible residuated lattices (DRLs) can act
as a unifying evaluation framework for soft constraint satisfaction problems (soft CSPs). DRLs form the algebraic semantics of a large family
of substructural and fuzzy logics [13,15], and are therefore natural candidates for this role. As a preliminary evidence in support to our claim,
along the lines of Cooper et al. and Larrosa et al. [11,18], we describe a
polynomial-time algorithm that enforces k-hyperarc consistency on soft
CSPs evaluated over DRLs. Observed that, in general, DRLs are neither
idempotent nor totally ordered, this algorithm accounts as a generalization of available enforcing algorithms over commutative idempotent
semirings and fair valuation structures [4,11].

1 Introduction

A constraint satisfaction problem (CSP) is the problem of deciding, given a
collection of constraints on variables, whether or not there is an assignment to the
variables satisfying all the constraints. In the *crisp* setting [19], any assignment
satisfying all the constraints provides a solution, and any solution is as good
as any other. In the *soft* setting [5], more generally, each constraint maps the
assignments to a *valuation structure*, which is a bounded poset equipped with a
suitable *combination* operator; the task is to find an assignment such that the
combination of its images under all the constraints is maximal in the order of
the valuation structure. Formal definitions are given in Section 2.

In its general formulation, the soft CSP is NP-complete, so that research
effort is currently aimed to characterize tractable cases [7,6], and investigating
constraints processing heuristics; amongst the latter, *enforcing* algorithms are
of the foremost importance.[1] A typical enforcing algorithm takes in input a soft
CSP and enforces a *local consistency* property over it, producing two possible
outcomes: either the input problem is found locally inconsistent, implying its
global inconsistency; or else, the input problem is transformed into an *equivalent*
problem (called *closure*), possibly inconsistent but *easier*, that is, with a smaller

---

[1] For further background on constraint processing, we refer the reader to [12].

C. Sossai and G. Chemello (Eds.): ECSQARU 2009, LNAI 5590, pp. 887–898, 2009.
© Springer-Verlag Berlin Heidelberg 2009

solution space. Despite their incompleteness as inconsistency tests, enforcing algorithms are useful as subprocedures in the exhaustive search for an optimal solution, for instance in *branch and bound* search. The generalization of local consistency notions and techniques from the crisp to the soft setting plays a central role in the algorithmic investigation of the soft CSP: for this reason, any class of structures that allows for an easy migration of local consistency techniques in the soft setting deserves consideration [4,18,11].

Not surprisingly, the *weaker* the properties of the valuation structure are, the harder it is to migrate a local consistency technique from the crisp to the soft setting. Indeed, a crisp CSP is equivalent to a soft CSP over a valuation structure with very *strong* properties: the algebra $(\{0,1\}, \leq, \odot, \bot, \top)$, where $\bot = 0 \leq 1 = \top$ and $x \odot y = 1$ if and only if $x = y = 1$. At the other extreme, the *weakest* possible valuation structure has to be a bounded poset, with top element $\top$ and bottom element $\bot$, equipped with a commutative, associative operation $x \odot y$ which is monotone over the order ($x \leq y$ implies $x \odot z \leq y \odot z$), has $\top$ as identity ($x \odot \top = x$) and $\bot$ as annihilator ($x \odot \bot = \bot$). Intuitively, an assignment mapped to $\top$ by a constraint is entirely satisfactory, and an assignment mapped to $\bot$ is entirely unsatisfactory; if two assignments are mapped to x and y, in case $x \leq y$, the latter is preferred to the former, and in case $x \parallel y$, none is preferred over the other; the operator $\odot$ combines constraints in such a way that adding constraints shrinks the solution space (as boundary cases, $\top$ does not shrink the solution space, and $\bot$ empties the solution space). In this setting, two options arise: whether or not to allow incomparability (formally, whether or not to admit *non-totally* ordered valuation structures); and, whether or not to keep into account repetitions (formally, whether or not to allow for valuation structures with *nonidempotent* combination operators). [2] The aforementioned algebra $(\{0,1\}, \leq, \odot, \bot, \top)$ is strong in the sense that it is totally ordered and idempotent.

In this paper, we propose *(commutative bounded) divisible residuated lattices* (in short, *DRLs*) as a unifying evaluation framework for soft constraints. Despite DRLs form an intensively studied algebraic variety since [22], they have never been proposed as an evaluation framework for soft constraints. However, there are robust motivations for considering DRLs in the soft CSP setting, coming from logic and algebra. As already mentioned, the soft CSP is a generalization of the crisp CSP. Conversely, the crisp CSP can be seen as a particular soft CSP, evaluated over the algebra $(\{0,1\}, \leq, \odot, \bot, \top)$, that is, a reduct of the familiar Boolean algebra **2** (taking $\odot$ as $\wedge$). Since **2** and the meet operation in **2** form the algebraic counterparts of Boolean logic and Boolean conjunction respectively, it is natural to intend the combination operator $\odot$ in a valuation structure as a generalization of the meet operation in **2**, and to investigate the algebraic counterparts of logics that generalize Boolean conjunction as candidate valuation structures for soft CSPs. Intriguingly, a central approach in the area of *mathematical fuzzy logic*, popularized by Hájek [15], relies on the idea of generalizing Boolean logic starting from a generalization of Boolean conjunction by means of a class of

---

[2] In the idempotent case $x \odot x = x$, so that repetitions do not matter.

functions called *(continuous) triangular norms* [17]. The idea is the following. A triangular norm $*$ is an associative, commutative, continuous binary function over the real interval $[0, 1]$; moreover, $*$ is monotone over the (total, dense and complete) order of reals in $[0, 1]$, has 1 as identity and 0 as annihilator. Given a (continuous) triangular norm $*$, there exists a unique binary function $\rightarrow_*$ on $[0, 1]$ satisfying the *residuation* equivalence $x * z \leq y$ if and only if $z \leq x \rightarrow_* y$, namely, $x \rightarrow_* y = \max\{z \mid x * z \leq y\}$. This function is called *residuum*, and is a generalization of Boolean implication. Corresponding to any triangular norm $*$, a propositional fuzzy logic $L_* = ([0, 1], \wedge, \vee, \odot, \rightarrow, \neg, \bot, \top)$ is obtained by interpreting propositional variables over $[0, 1]$, $\bot$ over 0, $\odot$ over $*$, $\rightarrow$ over $\rightarrow_*$, and eventually by defining $\neg x = x \rightarrow \bot$, $\top = \neg\bot = 1$, $x \wedge y = x \odot (x \rightarrow y) = \min(x, y)$, and $x \vee y = ((x \rightarrow y) \rightarrow y) \wedge ((y \rightarrow x) \rightarrow x) = \max(x, y)$. Boolean logic can be readily recovered from L_* by restricting the domain and the connectives to $\{0, 1\}$. As much as Boolean algebras form the equivalent algebraic semantics of Boolean logic, the variety of *BL-algebras* (defined in Section 3) forms the algebraic semantics of *Hájek's basic logic*, the logic of all continuous triangular norms and their residua [15,10].

Therefore, BL-algebras can be regarded as a first candidate evaluation framework for soft CSPs. We shall see that, as far as we are concerned with the implementation of local consistency techniques, for instance the k-hyperarc consistency enforcing algorithm presented in Section 4, *prelinearity* turns out to be redundant. Since prelinearity is exactly the property that specializes BL-algebras inside the class of DRLs [16], we are led to the latter as a defensible level of generality for our unifying evaluation framework. On the logical side, the variety of DRLs forms the algebraic semantics of an intersecting common fragment of basic logic and intuitionistic logic, called *generalized basic logic* [3].

We shall observe that DRLs, in general lattice ordered and nonidempotent, "subsume" preeminent valuation structures where local consistency techniques succeeded, namely, commutative idempotent semirings, lattice ordered and idempotent [4], and fair valuation structures, totally ordered and nonidempotent [11]. Compare Proposition 1 and Proposition 2 in Section 3.

As a preliminary, initial evidence in support of the proposal of DRLs as valuation structures for soft CSPs, we shall prove that DRLs readily host a polynomial-time algorithm that enforces a useful local consistency property, called k-*hyperarc consistency* (compare Definition 5 and Theorem 5 in Section 4). This property guarantees that any *consistent* assignment to a variable i extends to an assignment to any other $\leq k-1$ variables constrained by i, without producing additional costs. We insist that our algorithm works *uniformly* over every DRL, including the aforementioned, previously investigated structures as special cases.

DRLs allow for an extensive, smooth migration of constraint processing techniques from the crisp to the soft setting, far beyond the technical result presented in Section 4, which is intended as a first, concrete example of this new research line. For instance, we reasonably expect that the problem of finding efficiently *optimal* closures (in a suitable sense, required to embed enforcing algorithms into branch and bound exhaustive search) can be formalized in purely algebraic and logical terms in the setting proposed in this paper.[3]

We remark that analogous local consistency techniques have been investigated by Bistarelli and Gadducci over tropical residuated semirings [2], and we encourage a future comparison of the two settings in terms of unifying potential, structural insight, and computational viability. We also remark that the idea of formalizing soft constraints consistency techniques as many-valued logics refutations appears in the work of Ansótegui et al. [1].

Outline. The paper is organized as follows. In Section 2, we define soft CSPs and valuation structures. In Section 3, we define divisible residuated lattices, and we list a number of properties qualifying DRLs as suitable and natural valuation structures for soft constraints. Then, we describe the relation between evaluation frameworks such as commutative idempotent semirings and fair valuation structures, and DRLs. In Section 4 we present the main technical contribution of this paper, that is a uniform polynomial-time algorithm for k-hyperarc consistency enforcing on soft CSPs evaluated over DRLs. For background on partial orders and universal algebra, we refer the reader to any standard reference.

2   Soft Constraint Satisfaction Problems

In this section, we define formally the notions of soft CSPs, valuation structure, and optimal solution to a soft CSP.

A *(soft) constraint satisfaction problem* (in short, *CSP*) is a tuple $\mathbf{P} = (X, D, P, \mathbf{A})$ specified as follows. $X = \{1, \ldots, n\} = [n]$ is a set of *variables*, and $D = \{D_i\}_{i \in [n]}$ is a set of finite *domains* over which variables are assigned, variable i being assigned over domain D_i. Let $Y \subseteq X$. We let

$$l(Y) = \prod_{i \in Y} D_i$$

denote all the assignments of variables in Y onto the corresponding domains (*tuples*). If $Y = \emptyset$, then $l(Y)$ contains only the empty tuple. For any $Z \subseteq Y$, we denote by $t|_Z$ the *projection* of t onto the variables in Z. For every $i \in Y$, $a \in D_i$ and $t \in l(Y \setminus \{i\})$, we let $t \cdot a$ denote the tuple t' in $l(Y)$ such that $t'|_{\{i\}} = a$ and $t'|_{Y \setminus \{i\}} = t$ (if $Y = \{i\}$, then $t \cdot a = a$).

$\mathbf{A}$ is an algebra with domain A and signature including a binary relation $\leq$, a binary operation $\odot$ and constants $\top$, $\bot$, such that the reduct $(A, \leq, \top, \bot)$ is a bounded poset (that is, $\leq$ is a partial order with greatest element $\top$ and least element $\bot$), and the reduct $(A, \odot, \top)$ is a commutative monoid (that is, $\odot$ is commutative and associative and has identity $\top$) where $\odot$ is *monotone* over $\leq$, that is $x \leq y$ implies $x \odot z \leq y \odot z$. $\mathbf{A}$ is called the *valuation structure* of $\mathbf{P}$, and $\odot$ is called the *combination* operator over $\mathbf{A}$.

---

[3] This key problem has been recently solved over fair valuation structures [8]. Another, weaker consistency property, to be investigated in the DRLs setting, is *virtual consistency* [9].

P is a finite multiset[4] of *constraints*. Each constraint $C_Y \in P$ is defined over a subset $Y \subseteq X$ as a map

$$C_Y : \prod_{i \in Y} D_i \to A.$$

A constraint C_Y has *scope* Y and *arity* $|Y|$.

Let $(C_{Y_1}, \ldots, C_{Y_m})$ be an m-tuple of constraints in P, and let f be an m-ary operation on A. Then, $f(C_{Y_1}, \ldots, C_{Y_m})$ is the constraint with scope $Y_1 \cup \cdots \cup Y_m$ defined by putting, for every $t \in l(Y_1 \cup \cdots \cup Y_m)$:

$$f(C_{Y_1}, \ldots, C_{Y_m})(t) = f(C_{Y_1}(t|_{Y_1}), \ldots, C_{Y_m}(t|_{Y_m})).$$

The set $S(\mathbf{P})$ of *(optimal) solutions* to $\mathbf{P}$ is equal to the set of $t \in l(X)$ such that $\bigodot_{C_Y \in P} C_Y(t|_Y)$ is *maximal* in the poset:

$$\left\{ \bigodot_{C_Y \in P} C_Y(t|_Y) \;\middle|\; t \in l(X) \right\} \subseteq A,$$

where an element x is maximal in a poset if there is no element $y > x$ in the poset (notice that maximal elements in a poset form an antichain). If $S(\mathbf{P}) = \{\bot\}$, we say that $\mathbf{P}$ is *inconsistent*.

Let $\mathbf{P} = (X, D, P, \mathbf{A})$ and $\mathbf{P}' = (X, D, P', \mathbf{A})$ be CSPs. We say that $\mathbf{P}$ and $\mathbf{P}'$ are *equivalent* (in short, $\mathbf{P} \equiv \mathbf{P}'$) if and only if for every $t \in l(X)$,

$$\bigodot_{C_Y \in P} C_Y(t|_Y) = \bigodot_{C_Y \in P'} C_Y(t|_Y).$$

In particular, if $\mathbf{P} \equiv \mathbf{P}'$, then $S(\mathbf{P}) = S(\mathbf{P}')$.

In the sequel we shall assume the following, without loss of generality: P contains at most one constraint with scope $Y \neq \emptyset$ for every $Y \subseteq X$ (otherwise, we replace any pair of constraints C_Y', C_Y'' by the constraint C_Y defined by $C_Y(t|_Y) = C_Y'(t|_Y) \odot C_Y''(t|_Y)$ for every $t \in l(Y)$); P contains all the constraints $C_{\{i\}}$ for $i = 1, \ldots, n$ (otherwise, we add the constraint $C_{\{i\}}$ stipulating that $C_{\{i\}}(a) = \top$ for every $a \in D_i$); $C_{\{i\}}(a) > \bot$ for every $a \in D_i$ (otherwise, we remove a from D_i, declaring the problem inconsistent if D_i becomes empty). Moreover, we shall assume that constraints are implemented as tables, such that entries can be both retrieved and modified, and that algebraic operations over the valuation structure are polynomial-time computable in the size of their inputs.

3   Divisible Residuated Lattices

In this section, we introduce the variety of DRLs and some of its subvarieties, which are interesting with respect to soft CSPs. We give the logical interpretation of each mentioned algebraic variety, and we formalize the relation between DRLs and, commutative idempotent semirings and fair valuation structures.

---

[4] Multisets are necessary to support nonidempotent combinations of constraints.

Definition 1 (Divisible Residuated Lattice, DRL). *A divisible residuated lattice* [5] *is an algebra* $(A, \vee, \wedge, \odot, \rightarrow, \top, \bot)$ *such that: (i)* $(A, \odot, \top)$ *is a commutative monoid; (ii)* $(A, \vee, \wedge, \top, \bot)$ *is a bounded lattice (we write* $x \leq y$ *if and only if* $x \wedge y = x$*); (iii) residuation holds, that is,* $x \odot z \leq y$ *if and only if* $z \leq x \rightarrow y$*; (iv) divisibility holds, that is,* $x \wedge y = x \odot (x \rightarrow y)$*. A DRL is called a DRL-chain if its lattice reduct is totally ordered.*

We remark that residuation can be readily rephrased in equational terms, so that DRLs form a variety. Notice that divisible residuated chains are not closed under direct products, thus they do not form a variety. As a matter of fact, the lattice reduct of a DRL is *distributive*, that is, $x \wedge (y \vee z) = (x \wedge y) \vee (x \wedge z)$.

The monoidal operation of a DRL matches the minimal requirements imposed over the combination operator of a valuation structure in Section 2, as summarized by the following fact [14].

Fact 1 (DRLs Basic Properties). *Let* **A** *be a DRL. For every* $x, y, z \in A$*:* (i) $x \odot (y \odot z) = (x \odot y) \odot z$*,* $x \odot y = y \odot x$*,* $x \odot \top = x$*, and* $x \odot \bot = \bot$*; (ii)* $x \leq y$ *implies* $x \odot z \leq y \odot z$ *(in particular,* $x \odot x \leq x$*).*

We exploit the following facts from [14].

Fact 2 (DRLs Extra Properties). *Let* **A** *be a DRL. For every* $x, y, z \in A$*:* (i) $x \leq y$ *if and only if* $x \rightarrow y = \top$*; (ii)* $y \leq x$ *implies* $x \odot (x \rightarrow y) = y$*; (iii)* $y \leq z$ *implies* $(x \odot z) \odot (z \rightarrow y) = x \odot y$*; (iv)* $x \odot (y \vee z) = (x \odot y) \vee (x \odot z)$*.*

Fact 3. *[20] Let* $(A, \vee, \wedge, \top, \bot)$ *be a complete bounded lattice, and let* $\odot$ *be a commutative monotone* [6] *operation over* A *such that* $\odot$ *distributes over* $\vee$*. There exists a unique operation* $x \rightarrow y$ *satisfying residuation, namely,* $x \rightarrow y = \bigvee \{z \mid x \odot z \leq y\}$*.*

In the rest of this section, we discuss the relation between commutative idempotent semirings and fair valuation structures, and DRLs. We first introduce some subvarieties of DRLs.

Definition 2 (DRLs Subvarieties). *A BL-algebra is a DRL satisfying prelinearity, that is,* $(x \rightarrow y) \vee (y \rightarrow x) = \top$*. A Heyting algebra is a DRL satisfying idempotency, that is,* $x \odot x = x$*. A Gödel algebra is an idempotent BL-algebra. A Heyting algebra (or a Gödel algebra) is a Boolean algebra if it satisfies involutiveness, that is,* $\neg \neg x = x$ *where* $\neg x = x \rightarrow \bot$*.*

As we mentioned in the introduction, the variety of BL-algebras form the equivalent algebraic semantics of Hájek's basic logic. Analogously, the varieties of Heyting algebras, Gödel algebras, and Boolean algebras respectively, form the equivalent algebraic semantics of *intuitionistic logic, Gödel logic*, and *classical logic* [21,15].

---

[5] To our aims, we can restrict to commutative and bounded residuated lattices. We refer the reader to [16] for a general definition.

[6] Monotonicity of $\odot$ on both arguments is sufficient.

We consider first commutative idempotent semirings. The restriction to the idempotent case is motivated in this context, since local consistency techniques succeed only over idempotent semirings [5].

Definition 3 (Commutative Idempotent Semiring, CIS). *A commutative idempotent semiring is an algebra* $(A, \vee, \odot, \top, \bot)$ *such that: (i)* $\vee$ *is commutative, associative, idempotent,* $x \vee \bot = x$ *and* $x \vee \top = \top$*; (ii)* $\odot$ *is commutative, associative, idempotent,* $x \odot \top = x$ *and* $x \odot \bot = \bot$*; (iii)* $\odot$ *distributes over* $\vee$*, that is* $x \odot (y \vee z) = (x \odot y) \vee (x \odot z)$*.*

Fact 4. *[4, Theorem 2.9, Theorem 2.10] Let* $\mathbf{A} = (A, \vee, \odot, \top, \bot)$ *be a CIS. Then,* $(A, \vee, \wedge, \top, \bot)$*, where* $x \wedge y = x \odot y$*, is a complete bounded distributive lattice.*

Proposition 1. *Let* $\mathbf{A} = (A, \vee, \odot, \top, \bot)$ *be a CIS. Then, the expansion* $\mathbf{A}' = (A, \vee, \wedge, \odot, \rightarrow, \top, \bot)$ *of* $\mathbf{A}$*, where* $x \wedge y = x \odot y$ *and* $x \rightarrow y = \bigvee \{z \mid x \odot z \leq y\}$*, is a Heyting algebra.*

Proof. It is sufficient to prove that $(A, \vee, \wedge, \top, \bot)$ is a bounded distributive lattice, and that $\rightarrow$ is the residuum of $\wedge$. The first part is given by Fact 4. The second part is given by Fact 3: indeed, $(A, \vee, \odot, \top, \bot)$ is complete by Fact 4, $\odot$ is monotone by [4, Theorem 2.4], and $\odot$ distributes over $\vee$ by Definition 3(*iii*), hence the operation $\rightarrow$ is the uniquely determined residuum of $\wedge$.   □

Next we consider fair valuation structures. According to [11], a fair valuation structure is a structure $(A, \leq, \oplus, \ominus, \top, \bot)$ such that $(A, \leq, \top, \bot)$ is a bounded chain, the combination operator $\oplus$ is commutative, associative, monotone, with identity $\bot$ and annihilator $\top$, and the structure is *fair*, that is, for every $x \leq y \in A$ there exists a maximum $z \in A$, denoted by $y \ominus x$, such that $x \oplus z = y$. The fairness property is crafted ad hoc to preserve the soundness of constraints processing inside the adopted nonidempotent framework [11, Section 4]. Technically, the authors have to guarantee that $z \leq y$ implies $x \oplus y = (x \oplus z) \oplus (y \ominus z)$. We propose here a different, dual definition of a fair valuation structure.

Definition 4 (Dual Fair Valuation Structure, FVS). *A (dual) fair valuation structure is an algebra* $\mathbf{A} = (A, \vee, \wedge, \odot, \rightarrow, \top, \bot)$ *such that* $(A, \vee, \wedge, \top, \bot)$ *is a bounded chain,* $(A, \odot, \top)$ *is a commutative monoid, and* $\mathbf{A}$ *satisfies residuation and divisibility.*

Remarkably, the aforementioned technical condition, which in our setting becomes $y \leq z$ implies $x \odot y = (x \odot z) \odot (z \rightarrow y)$, holds by divisibility. The proposed dualization is defensible in logical terms, since the operation of combining soft constraints is intended as a *conjunction*, and the monoidal operation of a DRL is in fact a generalization of Boolean conjunction. In [11], the authors explicitly relate their combination operator with *triangular conorms*. The latter operations, dual to triangular norms, are customarily intended as generalizations of Boolean *disjunction*.

Proposition 2. *A FVS is a DRL-chain.*

We conclude this section mentioning that the soft CSP evaluation framework known as *fuzzy CSP* [4], which has the form $([0,1], \vee, \wedge, \top, \bot)$, can be extended to the Gödel chain $([0,1], \vee, \wedge, \odot, \rightarrow, \top, \bot)$ putting $x \odot y = x \wedge y$ and $x \rightarrow y$ equal to y if $y > x$ and to $\top$ otherwise. This chain singly generates the variety of Gödel algebras.

4   Enforcing k-Hyperarc Consistency on Soft CSPs over Divisible Residuated Lattices

In this section, we define a property of local consistency, called k-hyperarc consistency, and we describe a polynomial-time algorithm that enforces k-hyperarc consistency on soft CSPs evaluated over DRLs. Syntactically, the pseudocode is almost identical to that presented in [11,18]; the important and nontrivial point here is to show that it is sound over weaker structures, namely, DRLs (Lemma 2).[7]

Definition 5 (k-Hyperarc Consistency). *Let* $\mathbf{P} = (X, D, P, \mathbf{A})$ *be a CSP. Let* $Y \subseteq X$ *such that* $2 \leq |Y| \leq k$ *and* $C_Y \in P$. *We say that* Y *is* k-*hyperarc consistent if for each* $i \in Y$ *and each* $a \in D_i$ *such that* $C_{\{i\}}(a) > \bot$, *there exists* $t \in l(Y \setminus \{i\})$ *such that,*

$$C_{\{i\}}(a) = C_{\{i\}}(a) \odot C_Y(t \cdot a). \tag{1}$$

We say that $\mathbf{P}$ *is* k-*hyperarc consistent if every* $Y \subseteq X$ *such that* $2 \leq |Y| \leq k$ *and* $C_Y \in P$ *is* k-*hyperarc consistent.*

Notice that equation (1) holds if $C_Y(t \cdot a) = \top$. In words, Y is k-hyperarc consistency if each assignment $a \in D_i$ of variable $i \in Y$ such that $C_{\{i\}}(a) > \bot$, extends to an assignment $t \in l(Y \setminus \{i\})$ of variables $Y \setminus \{i\}$ without producing additional costs.

The idea beyond enforcing algorithms is to explicitate implicit constraints induced by the problem over certain subsets of variables, thus possibly discovering a *local* unsatisfiability at the level of these variables. As a specialization of this strategy, our algorithm shifts costs from constraints of arity greater than one to constraints of arity one, thus it possibly reveals the unsatisfiability of the subproblem induced over a single variable (or else, it possibly shrinks the domain of that variable). Such a local unsatisfiability implies the unsatisfiability of the whole problem, as the following proposition shows.

Proposition 3. *Let* $\mathbf{P} = (X, D, P, \mathbf{A})$ *be a CSP and let* $i \in [n]$ *be such that* $C_{\{i\}} \in P$ *and* $C_{\{i\}}(a) = \bot$ *for every* $a \in D_i$. *Then,* $\mathbf{P}$ *is inconsistent.*

---

[7] A technical advance of the present procedure, compared with the analogous procedure presented by Bistarelli and Gadducci over tropical residuated semirings [2], is termination.

Proof. First recall that for every $x \in A$ it holds that $x \odot \bot = \bot$. But then, $C_{\{i\}}(t|_{\{i\}}) = \bot$ for every $t \in l(X)$, so that $\bigodot_{C_Y \in P} C_Y(t|_Y) = \bot$. Therefore, $S(\mathbf{P}) = \{\bot\}$ and $\mathbf{P}$ is inconsistent.   □

Algorithm: k-HYPERARCCONSISTENCY
Input: A CSP $\mathbf{P} = (X, D, P, \mathbf{A})$.
Output: $\bot$ or a k-hyperarc consistent CSP $\mathbf{P}' = (X, D, P', \mathbf{A})$ equivalent to $\mathbf{P}$.

k-HYPERARCCONSISTENCY$((X, D, P, \mathbf{A}))$
1    $Q \leftarrow \{1, \ldots, n\}$
2   **while** $Q \neq \emptyset$ **do**
3        $i \leftarrow \text{POP}(Q)$
4   **foreach** $Y \subseteq X$ such that $2 \leq |Y| \leq k$, $i \in Y$ and $C_Y \in P$ **do**
5   domainShrinks $\leftarrow \text{PROJECT}(Y, i)$
6   **if** $C_{\{i\}}(a) = \bot$ for each $a \in D_i$ **then**
7   **return** $\bot$
8   **else if** domainShrinks **then**
9                $\text{PUSH}(Q, i)$
10   **endif**
11   **endforeach**
12   **endwhile**
13   **return** $(X, D, P', \mathbf{A})$

PROJECT(Y, i)
14   domainShrinks $\leftarrow$ **false**
15   **foreach** $a \in D_i$ such that $C_{\{i\}}(a) > \bot$ **do**
16       $x \leftarrow$ a maximal element in $\{C_Y(t \cdot a) \mid t \in l(Y \setminus \{i\})\}$
17       $C_{\{i\}}(a) \leftarrow C_{\{i\}}(a) \odot x$
18   **if** $C_{\{i\}}(a) = \bot$ **then**
19   domainShrinks $\leftarrow$ **true**
20   **endif**
21   **foreach** $t \in l(Y \setminus \{i\})$ **do**
22           $C_Y(t \cdot a) \leftarrow (x \rightarrow C_Y(t \cdot a))$
23   **endforeach**
24   **endforeach**
25   **return** domainShrinks

As already discussed in the introduction, enforcing k-hyperarc consistency over the k-hyperarc inconsistent problem $\mathbf{P}$ may return in output several distinct k-hyperarc consistent problems, depending on the choices made on Lines 1, 3, 4 and 16.

In the rest of this section, we prove that the algorithm runs in polynomial-time (Lemma 1) and is sound (Lemma 2), leading to our main technical result (Theorem 5).

Lemma 1 (Complexity). *Let* $\mathbf{P} = (X, D, P, \mathbf{A})$ *be a CSP, where* $X = [n]$, $d = \max_{i \in [n]} |D_i|$ *and* $e = |P|$. *Then,* k-HYPERARCCONSISTENCY *terminates in at most* $O(e^2 \cdot d^{k+1})$ *time.*

Proof. The main loop in Lines 2-12 iterates at most $n(d+1) \leq e(d+1)$ times, since $n \leq e$ without loss of generality and each $i \in [n]$ is added to Q once on Line 1 and at most d times on Line 9 (once for each shrink of domain D_i of size $\leq d$). Each iteration of the main loop involves at most e iterations of the loop nested in Lines 4-11, since there are at most e constraints satisfying the condition in Line 4 with respect to any given $i \in [n]$. Each such nested iteration amounts to an invocation of PROJECT and an iteration over domain D_i of size $\leq d$. Any invocation of PROJECT amounts to an iteration over domain D_i of size $\leq d$ on Line 15, and for each such iteration, two iterations over all the $\leq d^{k-1}$ tuples $t \in l(Y \setminus \{i\})$, observing that $1 \leq |Y \setminus \{i\}| \leq k-1$ (Line 16 and Lines 21-23). Summarizing, the algorithm executes at most

$$(e(d+1))e(d+d(2d^{k-1}))$$

many iterations, so that it terminates in at most $O(e^2 \cdot d^{k+1})$ time. □

Lemma 2 (Soundness). *Let* $\mathbf{P} = (X, D, P, \mathbf{A})$ *be a CSP, and consider the output of* k-HYPERARCCONSISTENCY$(\mathbf{P})$:

(i) *if it is* $\perp$, *then* $\mathbf{P}$ *is inconsistent;*
(ii) *otherwise, it is a* k-*hyperarc consistent CSP equivalent to* $\mathbf{P}$.

Proof. First we show that the subprocedure PROJECT preserves equivalence, in the following sense. Let R' be the multiset of constraints before the jth invocation of PROJECT in Line 5, let Y and i be the parameters of such invocation, and let R'' be the multiset of constraints computed by the jth execution of Lines 14-25. We aim to show that for every $t \in l(X)$,

$$\bigodot_{C_Y \in R'} C_Y(t|_Y) = \bigodot_{C_Y \in R''} C_Y(t|_Y), \qquad (2)$$

that is, problems $(X, D, R', \mathbf{A})$ and $(X, D, R'', \mathbf{A})$ are equivalent.

Let $t \in l(X)$ and let $t|_{\{i\}} = a \in D_i$ such that $C_{\{i\}}(a) > \perp$ (Line 15). Clearly, $t|_{Y \setminus \{i\}} \in l(Y \setminus \{i\})$. In Line 16, x is settled to a maximal element in the poset

$$\{C_Y(t|_{Y \setminus \{i\}} \cdot t|_{\{i\}}) \mid t|_{Y \setminus \{i\}} \in l(Y \setminus \{i\})\},$$

so that by construction $C_Y(t|_Y) \leq x$. By Line 17, the constraint $C_{\{i\}}(t|_{\{i\}})$ in R' becomes

$$C_{\{i\}}(t|_{\{i\}}) \odot x$$

in R'', and by Line 22, at some iteration of the loop in Lines 21-23, the constraint $C_Y(t|_Y)$ in R' becomes

$$x \to C_Y(t|_Y)$$

in R''. Now, we claim that:

$$C_{\{i\}}(t|_{\{i\}}) \odot C_Y(t|_Y) = (C_{\{i\}}(t|_{\{i\}}) \odot x) \odot (x \to C_Y(t|_Y)).$$

Indeed, in light of Fact 2(iii) and the aforementioned fact that $C_Y(t|_Y) \leq x$,

$$(C_{\{i\}}(t|_{\{i\}}) \odot x) \odot (x \to C_Y(t|_Y)) = C_{\{i\}}(t|_{\{i\}}) \odot (x \odot (x \to C_Y(t|_Y)))$$
$$= C_{\{i\}}(t|_{\{i\}}) \odot (x \wedge C_Y(t|_Y))$$
$$= C_{\{i\}}(t|_{\{i\}}) \odot C_Y(t|_Y).$$

Eventually, PROJECT does not modify constraints $C_Z \in R'$ such that $Z \neq \{i\}$ and $Z \neq Y$, so that,

$$\bigodot_{C_Z \in R', Z \neq \{i\}, Z \neq Y} C_Z(t|_Z) = \bigodot_{C_Z \in R'', Z \neq \{i\}, Z \neq Y} C_Z(t|_Z).$$

Thus, since $z' = z''$ implies $z \odot z' = z \odot z''$ in $\mathbf{A}$ for every $z, z', z'' \in A$ by Fact 1(ii), we conclude that (2) holds.

Now suppose that the algorithm outputs $\bot$ in Line 7. We claim that the input problem $\mathbf{P} = (X, D, P, \mathbf{A})$ is inconsistent. Indeed, let j be such that after the jth execution of PROJECT, say over parameters Y and i, it holds that $C_{\{i\}}(a) = \bot$ for each $a \in D_i$. Let P' be the multiset of constraints computed by such jth execution. Since PROJECT preserves equivalence, $\mathbf{P}' = (X, D, P', \mathbf{A})$ is equivalent to $\mathbf{P}$. But, by Proposition 3, $\mathbf{P}'$ is inconsistent, so that $\mathbf{P}$ is inconsistent too.

Next suppose that the algorithm outputs $\mathbf{P}' = (X, D, P', \mathbf{A})$ in Line 13. We claim that the output problem is k-hyperarc consistent and equivalent to the input problem $\mathbf{P} = (X, D, P, \mathbf{A})$. For equivalence, simply note that PROJECT preserves equivalence. For k-hyperarc consistency, first note that every $i \in [n]$ is such that $C_{\{i\}}(a) > \bot$ for some $a \in D_i$. Indeed, this holds in the input problem $\mathbf{P}$ without loss of generality, and each execution of PROJECT, which possibly pushes some $C_{\{i\}}(a)$ down to $\bot$ in Line 17, is followed by the check of Lines 18-20.

Now, let $Y \subseteq X$ be such that $2 \leq |Y| \leq k$, $i \in Y$ and $C_Y \in P'$, and let $a \in D_i$ be such that $C_{\{i\}}(a) > \bot$. We claim that there exists $t \in l(Y \setminus \{i\})$ such that

$$C_{\{i\}}(a) = C_{\{i\}}(a) \odot C_Y(t \cdot a).$$

Note that, by Fact 1(i), equality holds if $C_Y(t \cdot a) = \top$. Let R' and R'' be respectively the multisets of constraints before and after the last execution of PROJECT on input Y and i. Let $t \in l(Y \setminus \{i\})$ be such that $C_Y(t \cdot a)$ is the maximal element in $\{C_Y(t \cdot a) \mid t \in l(Y \setminus \{i\})\}$ assigned to x in Line 16. Thus, at some iteration of loop in Lines 21-23, we have that the constraint $C_Y(t \cdot a)$ in R' is updated to $x \to C_Y(t \cdot a)$ in R''. But, by Fact 2(i),

$$x \to C_Y(t \cdot a) = C_Y(t \cdot a) \to C_Y(t \cdot a) = \top,$$

therefore, $C_Y(t \cdot a) = \top$ in R''. Noticing that subsequent assignments to $C_Y(t \cdot a)$ during the main loop have the form $x \to \top$, which is equal to $\top$ by Fact 2(i), the claim is settled. $\square$

Theorem 5. *Let $\mathbf{P}$ be a CSP, and let $\mathbf{P}' = k$-HYPERARC-CONSISTENCY$(\mathbf{P})$. Then, $\mathbf{P}'$ is a k-hyperarc consistent CSP equivalent to $\mathbf{P}$, computed in polynomial time in the size of $\mathbf{P}$.*

Acknowledgments. The author thanks the anonymous referees for careful comments on the preliminar version of this paper.

References

1. Ansótegui, C., Bonet, M.L., Levy, J., Manyà, F.: The Logic Behind Weighted CSP. In: 20th International Joint Conference on Artificial Intelligence, pp. 32–37 (2007)
2. Bistarelli, S., Gadducci, F.: Enhancing Constraints Manipulation in Semiring-Based Formalisms. In: 17th European Conference on Artificial Intelligence, pp. 63–67. IOS Press, Amsterdam (2006)
3. Bova, S., Montagna, F.: The Consequence Relation in the Logic of Commutative GBL-Algebras is PSPACE-complete. Theor. Comput. Sci. 410(12-13), 1143–1158 (2009)
4. Bistarelli, S., Montanari, U., Rossi, F.: Semiring-Based Constraint Satisfaction and Optimization. J. ACM 44(2), 201–236 (1997)
5. Bistarelli, S., Montanari, U., Rossi, F., Schiex, T., Verfaillie, G., Fargier, H.: Semiring-Based CSPs and Valued CSPs: Frameworks, Properties, and Comparison. Constraints 4(3), 199–240 (1999)
6. Cohen, D.A., Cooper, M.C., Jeavons, P.: An Algebraic Characterisation of Complexity for Valued Constraint. In: Benhamou, F. (ed.) CP 2006. LNCS, vol. 4204, pp. 107–121. Springer, Heidelberg (2006)
7. Cohen, D.A., Cooper, M.C., Jeavons, P., Krokhin, A.A.: The Complexity of Soft Constraint Satisfaction. Artif. Intell. 170, 983–1016 (2006)
8. Cooper, M.C., de Givry, S., Schiex, T.: Optimal Soft Arc Consistency. In: 20th International Joint Conference on Artificial Intelligence, pp. 68–73 (2007)
9. Cooper, M.C., de Givry, S., Schiex, T., Sánchez, M., Zytnicki, M.: Virtual Arc Consistency for Weighted CSP. In: 23rd AAAI Conference on Artificial Intelligence, pp. 253–258 (2008)
10. Cignoli, R., Esteva, F., Godo, L., Torrens, A.: Basic Fuzzy Logic is the Logic of Continuous t-Norms and their Residua. Soft Comput. 4(2), 106–112 (2000)
11. Cooper, M.C., Schiex, T.: Arc Consistency for Soft Constraints. Artif. Intell. 154(1-2), 199–227 (2004)
12. Dechter, R.: Constraint Processing. Morgan Kaufmann, San Francisco (2003)
13. Galatos, N., Jipsen, P., Kowalski, T., Ono, H.: Residuated Lattices: An Algebraic Glimpse at Substructural Logics. Elsevier, Amsterdam (2007)
14. Gottwald, S.: A Treatise on Many-Valued Logics. Research Studies Press (2000)
15. Hájek, P.: Metamathematics of Fuzzy Logic. Kluwer, Dordrecht (1998)
16. Jipsen, P., Montagna, F.: On the Structure of Generalized BL-Algebras. Algebra Univ. 55, 226–237 (2006)
17. Klement, E.P., Mesiar, R., Pap, E.: Triangular Norms. Kluwer, Dordrecht (2000)
18. Larrosa, J., Schiex, T.: Solving Weighted CSP by Maintaining Arc Consistency. Artif. Intell. 159(1-2), 1–26 (2004)
19. Montanari, U.: Networks of Constraints: Fundamental Properties and Applications to Picture Processing. Inform. Sciences 7, 95–132 (1974)
20. Pavelka, J.: On Fuzzy Logic I, II, III. Zeitschr. f. Math. Logik und Grundlagen der Math. 25, 45–52, 119–134, 447–464 (1979)
21. Rasiowa, H.: An Algebraic Approach to Non-Classical Logics. North-Holland, Amsterdam (1974)
22. Ward, M., Dilworth, R.P.: Residuated Lattices. T. Am. Math. Soc. 45, 335–354 (1939)

On the Convergence with Fixed Regulator in Residuated Structures

Lavinia Corina Ciungu

Polytechnical University of Bucharest
Splaiul Independenței 313, Bucharest, Romania
lavinia_ciungu@math.pub.ro

Abstract. The convergence with a fixed regulator has been studied for lattice ordered groups and MV-algebras. In this paper we present some particular results for the case of perfect MV-algebras using Di Nola-Lettieri functors and we extend the notion of convergence with a fixed regulator for residuated lattices. The main results consist of proving that any locally Archimedean MV-algebra has a unique v-Cauchy completion and that in an Archimedean residuated lattice the v-limit is unique.

1 Preliminaries

In this section we recall some definitions and results regarding the convergence with fixed regulator in ℓ-groups. For more details on the subject we refer the reader to [2,3,4]. On an MV-algebra A, the *distance function* $d : A \times A \to A$ is defined by:

$$d(x,y) = (x \odot y^-) \oplus (x^- \odot y).$$

Among the properties of the distance function (see [8]), we will use the following:

(1) $d(x,y) = 0$ iff $x = y$, (2) $d(x,y) = d(y,x)$, (3) $d(x,0) = x$, (4) $d(x,z) \leq d(x,y) \oplus d(y,z)$, (5) $x \leq y$ implies $y = x \oplus d(x,y)$.

An element x in an MV-algebra is said to be *infinitely small* or *infinitesimal* if $x \neq 0$ and $nx \leq x^-$ for all $n \in \mathbb{N}$. The set of all infinitesimals in A is denoted by $Infinit(A)$.

The *radical* $Rad(A)$ of an MV-algebra A is the intersection of all maximal ideals of A. The MV-algebra A is called *perfect* if $A = Rad(A) \cup (Rad(A))^-$, where $(Rad(A))^- = \{x^- \mid x \in Rad(A)\}$. For any MV-algebra A, $Rad(A) = Infinit(A) \cup \{0\}$.

An MV-algebra A is said to be *Archimedean* or *semisimple* if $nx \leq x^-$ for all $n \in \mathbb{N}$ implies $x = 0$ (see [5]).

According to [11], a perfect MV-algebra A is called *locally Archimedean* whenever $x, y \in Rad(A)$ are such that $nx \leq y$ for all $n \in \mathbb{N}$, it follows that $x = 0$.

Mundici proved in [13] that for any MV-algebra A there is an abelian ℓ-group $(G, +, u)$ with strong unit u such that A is isomorphic to $\Gamma(G, u) = [0, u]$ endowed with a canonical structure of MV-algebra:

$$x \oplus y = (x + y) \wedge u, \quad x^- = u - x, \quad x \odot y = (x + y - u) \vee 0.$$

The Mundici functor Γ is a categorical equivalence between the category of abelian ℓ-groups with strong unit and the category of MV-algebras.

C. Sossai and G. Chemello (Eds.): ECSQARU 2009, LNAI 5590, pp. 899–910, 2009.

In the case of perfect MV-algebras a crucial result is the categorical equivalence between the category of perfect MV-algebras and the category of abelian ℓ-groups established by A. Di Nola and A. Lettieri ([10]).

Let G be an ℓ-group and $g \in G$. We denote by $g^+ = g \vee \{0\}$, $g^- = (-g) \vee \{0\}$ and we remind that $g = g^+ - g^-$ and $| g | = g \vee (-g) = g^+ + g^-$.

For each abelian ℓ-group $(G, +)$, consider the lexicographic product $\mathbb{Z} \times_{lex} G$ and define the perfect MV-algebra $\Delta(G) = \Gamma(\mathbb{Z} \times_{lex} G, (1,0))$ with the operations:

$$(x, y) \oplus (u, v) = (1, 0) \wedge (x + u, y + v)$$
$$(x, y)^- = (1, 0) - (x, y) = (1 - x, -y)$$
$$(x, y) \odot (u, v) = (0, 0) \vee (x + u - 1, y + v).$$

An element of $\Delta(G)$ has either the form $(0, g)$ with $g \geq 0$ or the form $(1, g)$ with $g \leq 0$ ($g \in G$). With the above definitions, the distance function on $\Delta(G)$ becomes:

$$d((x, y), (u, v)) = (1, 0) \wedge ((0, 0) \vee (x - u, y - v) + (0, 0) \vee (u - x, v - y)) =$$
$$= (1, 0) \wedge (|x - u|, |y - v|) = \begin{cases} (0, |y - v|) \text{ if } x = u \\ (1, 0) \text{ otherwise.} \end{cases}$$

According to [8]) we have:

(1) $(1, 0)$ is a strong unit of $\mathbb{Z} \times_{lex} G$;
(2) If A is a perfect MV-algebra, then $(Rad(A), \oplus, 0)$ is a cancellative abelian monoid;
(3) $Rad(\Delta(G)) = \{(0, x) \mid x \geq 0\}$; $(Rad(\Delta(G)))^- = \{(1, x) \mid x \leq 0\}$.
   On $Rad(A) \times Rad(A)$ we define the congruence $\approx$ by

$$(x, y) \approx (u, v) \text{ iff } x \oplus v = y \oplus u$$

and denote by $[x, y]$ the congruence class of $(x, y) \in Rad(A) \times Rad(A)$. Denote $\mathcal{D}(A) = Rad(A) \times Rad(A)/ \approx$ and define:

$$[x, y] + [u, v] = [x \oplus u, y \oplus v]$$
$$[x, y] \leq [u, v] \text{ iff } x \oplus v \leq y \oplus u.$$

With these operations $\mathcal{D}(A)$ becomes an abelian ℓ-group such that:

$$[x, y] \wedge [u, v] = [(x \oplus v) \wedge (y \oplus u), y \oplus v]$$
$$[x, y] \vee [u, v] = [x \oplus u, (x \oplus v) \wedge (y \oplus u)].$$

Di Nola-Lettieri functors $\mathcal{D} : \mathcal{P} \rightarrow \mathcal{A}$ and $\Delta : \mathcal{A} \rightarrow \mathcal{P}$ realize a categorical equivalence between the category $\mathcal{P}$ of perfect MV-algebras and the category $\mathcal{A}$ of abelian ℓ-groups([10]).

Proposition 1. *([8]) In $\mathcal{D}(A)$ we have:*

(1) $\mathcal{D}(A)^+ = \{[x, 0] \mid x \in Rad(A)\}$; (2) $-[x, y] = [y, x]$;
(3) $[x, y]^+ = [x \odot y^-, 0], [x, y]^- = [x^- \odot y, 0]$; (4) $|[x, y]| = [d(x, y), 0]$.

We recall some notions regarding the v-convergence with a fixed regulator in ℓ-groups presented in [2] and [3]. Let G be an abelian ℓ-group and $0 < v \in G$.

The sequence $(x_n)_n$ in G is said to be v-*convergent* to the element $x \in G$ if for each $p \in \mathbb{N}$ there is $n_0 \in \mathbb{N}$ such that $p|x_n - x| \leq v$ for each $n \in \mathbb{N}$, $n \geq n_0$. In this case we denote $x_n \to_v x$ and we say that x is a v-*limit* of $(x_n)_n$. The element v is called *convergence regulator* in G.

The sequence $(x_n)_n$ in G is said to be v-*fundamental* or v-*Cauchy* if for each $p \in \mathbb{N}$ there is $n_0 \in \mathbb{N}$ such that $p|x_n - x_m| \le v$ for each $n, m \in \mathbb{N}$, $m \ge n \ge n_0$.

If every v-Cauchy sequence is convergent in G, then G is said to be v-*Cauchy complete*.

Definition 2. *([2]) If G is Archimedean, then an Archimedean ℓ-group H is called a v-Cauchy completion of G if the following conditions are satisfied:*

(1) G is an ℓ-subgroup of H;
(2) H is v-Cauchy complete;
(3) Every element of H is a v-limit of some sequence in G.

The v-Cauchy completion for an arbitrary ℓ-group G is constructed in [2].

Residuated lattices are algebraic counterparts of logics without the contraction rule and their properties have been studied by many authors, such as Ward and Dilworth ([16]) and Kowalski and Ono ([12], [14]).

Definition 3. *A residuated lattice is an algebra $\mathcal{L} = (L, \wedge, \vee, \odot, \to, 0, 1)$ of the type $(2, 2, 2, 2, 0, 0)$ satisfying the following conditions:*

(L1) $(L, \wedge, \vee, 0, 1)$ is a bounded lattice;
(L2) $(L, \odot, 1)$ is a commutative monoid and the binary operation $\odot$ is isotone with respect the lattice order;
(L3) $x \odot y \le z$ iff $x \le y \to z$ for any $x, y, z \in L$.

A totally ordered residuated lattice is called *chain* or *linearly ordered* residuated lattice. In the sequel we will agree that the operations $\wedge, \vee, \odot$ have higher priority than the operation $\to$. In a residuated lattice $\mathcal{L} = (L, \wedge, \vee, \odot, \to, 0, 1)$ we define for all $x \in L$: $x^- = x \to 0$. We will refer to $\mathcal{L}$ by its universe L.

In a residuated lattice L we define the *distance function* $d : L \times L \longrightarrow L$ by

$$d(x, y) = (x \to y) \wedge (y \to x).$$

Proposition 4. *([9]) The distance function satisfies the following properties:*

(1) $d(x, y) = d(y, x)$; (2) $d(x, y) = 1$ iff $x = y$; (3) $d(x, 1) = x$;
(4) $d(x, 0) = x^-$; (5) $d(x, z) \odot d(z, y) \le d(x, y)$; (6) $d(x, y) \le d(x \odot u, y \odot u)$;
(7) $d(x, u) \odot d(y, v) \le d(x \odot y, u \odot v)$; (8) $d(x, u) \odot d(y, v) \le d(y \to x, v \to u)$;
(9) $d(x, u) \wedge d(y, v) \le d(x \wedge y, u \wedge v)$; (10) $d(x, u) \wedge d(y, v) \le d(x \vee y, u \vee v)$;
(11) if $x, y \in [x', y']$ then $d(x', y') \le d(x, y)$.

For any $n \in N, x \in L$ we put $x^0 = 1$ and $x^{n+1} = x^n \odot x = x \odot x^n$. The order of $x \in L$, denoted $ord(x)$ is the smallest $n \in \mathbb{N}$ such that $x^n = 0$. If there is not such an n, then $ord(x) = \infty$.

Lemma 5. *Let L be a linearly ordered residuated lattice. If $x \in L$ such that $ord(x) = \infty$, then $x^n > x^-$ for any $n \in \mathbb{N}$.*

Proof. First, note that $x > 0$ (if $x = 0$, then $x^2 = 0$), so $x^- < 1$. For $n = 0$, $x^0 = 1 > x^-$. For $n \geq 1$, if $x^n \leq x^- = x \to 0$, then $x^{n+1} = 0$, a contradiction. Hence, $x^n > x^-$. Thus, $x^n > x^-$ for any $n \in \mathbb{N}$.   □

Lemma 6. *In any linearly ordered residuated lattice L, if $x, y \in L$ such that $x^n \odot y > 0$ for some $n \in \mathbb{N}$, then $x^n > y^-$.*

Proof. If we suppose $x^n \leq y^- = y \to 0$, then $x^n \odot y = 0$, contradiction.
Therefore, $x^n > y^-$.   □

Proposition 7. *In any residuated lattice the following properties are equivalent:*

(i) $x^n \geq x^-$ *for any $n \in \mathbb{N}$ implies $x = 1$;*
(ii) $x^n \geq y^-$ *for any $n \in \mathbb{N}$ implies $x \vee y = 1$;*
(iii) $x^n \geq y^-$ *for any $n \in \mathbb{N}$ implies $x \to y = y$ and $y \to x = x$.*

Proof. $(i) \Rightarrow (ii)$ Let $x, y \in L$ such that $x^n \geq y^-$ for any $n \in N$. We have:
$$(x \vee y)^- = x^- \wedge y^- \leq x^n \leq (x \vee y)^n,$$
hence $(x \vee y)^n \geq (x \vee y)^-$ for any $n \in N$. Thus, by the hypothesis we get $x \vee y = 1$.
$(ii) \Rightarrow (i)$ Consider $x \in L$ such that $x^n \geq y^-$ for any $n \in N$.
By (ii), taking $y = x$ we get $x \vee x = 1$, hence $x = 1$.
$(i) \Rightarrow (iii)$ Let $x, y \in L$ such that $x^n \geq y^-$ for any $n \in N$.
Similarly with $(i) \Rightarrow (ii)$, if $x, y \in L$ we have $(x \vee y)^n \geq (x \vee y)^-$ for any $n \in N$, hence, by the hypothesis, we get $x \vee y = 1$.
In any resituated lattice we have $x \vee y \leq [(x \to y) \to y] \wedge [(y \to x) \to x]$ (see [6]).
Since $x \vee y = 1$, it follows that $[(x \to y) \to y] \wedge [(y \to x) \to x] = 1$,
hence $(x \to y) \to y = 1$ and $(y \to x) \to x = 1$.
From $(x \to y) \to y = 1$, we have $x \to y \leq y$ and considering that $y \leq x \to y$ we obtain $x \to y = y$. Similarly, $y \to x = x$.
$(iii) \Rightarrow (i)$. Consider $x \in L$ such that $x^n \geq x^-$, for any $n \in N$.
By the hypothesis we obtain $x \to x = x$, hence $x = 1$.   □

Definition 8. *A residuated lattice is called Archimedean if one of the equivalent conditions from the above proposition is satisfied.*

Proposition 9. *In any Archimedean linearly ordered residuated lattice L the following properties hold:*

(1) if $x^n > 0$ for any $n \in \mathbb{N}, n \geq 2$, then $x = 1$;
(2) if $x^n \odot y > 0$ for any $n \in \mathbb{N}$, then $x \vee y = 1$;
(3) if $x^n \odot y > 0$ for any $n \in \mathbb{N}$, then $x \to y = y$ and $y \to x = x$.

Proof. It follows from Lemmas 5, 6 and Proposition 7.   □

For a residuated lattice L let denote:
$$L^* = \{x \in L \mid x \geq x^-\} \text{ and } G(L) = \{x \in L \mid x^2 = x\}.$$

Remark 10. *Let L, L_1 be two residuated lattices and $h : L \longrightarrow L_1$ a RL-morphism. Then:*

(1) $0, 1 \in G(L)$;
(2) if $x \in G(L)$, then $x^n = x$ for all $n \in \mathbb{N}, n \geq 2$;
(3) if $x \in G(L)$, then $h(x) \in G(L_1)$.

Proposition 11. *Let* L, L_1 *be two residuated lattices and* $h : L \longrightarrow L_1$ *a RL-morphism. Then:*

(1) $1 \in L^*$ *and* $0 \notin L^*$; (2) $x \leq y$ *and* $x \in L^*$, *then* $y \in L^*$;
(3) *if* $x, y \in L^*$, *then* $x \vee y \in L^*$; (4) *if* $x \in L^*$, *then* $h(x) \in L_1^*$.

Proof

(1) $1^- = 0 < 1; 0^- = 1 > 0$;
(2) Since $x \leq y$, we have $y^- \leq x^- \leq x \leq y$, hence $y \in L^*$;
(3) We have $(x \vee y)^- = x^- \wedge y^- \leq x^- \leq x \vee y$, thus $x \vee y \in L^*$;
(4) We have $x \in L^*$ iff $x \geq x^-$ iff $x^- \to x = 1$ iff $h(x^- \to x) = h(1) = 1$ iff $h(x^-) \to h(x) = 1$ iff $h(x^-) \leq h(x)$ iff $h(x)^- \leq h(x)$ iff $h(x) \in L_1^*$.   □

Corollary 12. *Let* L, L_1 *be two residuated lattices and* $h : L \longrightarrow L_1$ *a RL-morphism. Then:*
    (1) $L^* \cap G(L) \neq \emptyset$; (2) *if* $x \in L^* \cap G(L)$, *then* $h(x) \in L_1^* \cap G(L_1)$.

Proposition 13. *Let* L *be an Archimedean residuated lattice and* $x \in L, y \in L^*$. *Then:*

(1) $x^n \geq y$ *for any* $n \in N$ *implies* $x \vee y = 1$;
(2) $x^n \geq y$ *for any* $n \in N$ *implies* $x \to y = y$ *and* $y \to x = x$.

Proof. By the definition of an Archimedean residuated lattice and L^*.   □

Example 14. *([15]) Consider the residuated lattice* $(L, \wedge, \vee, \odot, \to, 0, 1)$ *defined on the unit interval* $L = [0, 1]$ *with the usual order and the operations:*

$$x \odot y = \begin{cases} 0, & \text{if } x + y \leq \frac{1}{2} \\ x \wedge y, & \text{otherwise} \end{cases} \qquad x \to y = \begin{cases} 1, & \text{if } x \leq y \\ max(\frac{1}{2} - x, y), & \text{otherwise.} \end{cases}$$

Since $(\frac{1}{3})^n = \frac{1}{3} > \frac{1}{6} = (\frac{1}{3})^-$, *it follows that* L *is not Archimedean. One can easily check that* $L^* = [\frac{1}{4}, 1]$ *and* $G(L) = \{0\} \cup (\frac{1}{4}, 1]$.

2   Convergence with a Fixed Regulator in Perfect MV-Algebras

The functor Γ was used in [4] to obtain the v-convergence for MV-algebras from the theory of v-convergence in ℓ-groups. Using the functors $\mathcal{D}$ and Δ we will investigate the v-convergence in perfect MV-algebras.

Definition 15. *([4]) Let* A *be an arbitrary MV-algebra and* $0 < v \in A$. *The sequence* $(x_n)_n$ *in* A *v-converges to an element* $x \in A$ *(or* x *is a v-limit of* $(x_n)_n$), *denoted* $x_n \to_v x$, *if for every* $p \in N$ *there is* $n_0 \in N$ *such that* $pd(x_n, x) \leq v$ *for each* $n \in N$, $n \geq n_0$.

Proposition 16. *([4]) If* $(x_n)_n$ *and* $(y_n)_n$ *are sequences in an arbitrary MV-algebra* A *and* $x, y \in A$ *such that* $x_n \to_v x$ *and* $y_n \to_v y$, *then:* $x_n^- \to_v x^-, x_n \oplus y_n \to_v x \oplus y, x_n \odot y_n \to_v x \odot y, x_n \vee y_n \to_v x \vee y, x_n \wedge y_n \to_v x \wedge y$.

Proposition 17. *In an arbitrary MV-algebra A the following hold:*

(1) *If $(x_n)_n \subseteq Rad(A)$, $0 < v \in Rad(A)$ and $x_n \to_v x$, then $x \in Rad(A)$;*
(2) *If $(x_n)_n \subseteq (Rad(A))^-$, $v \in (Rad(A))^-$, $v < 1$ and $x_n \to_{v^-} x$, then $x \in (Rad(A))^-$.*

Proof
(1) Since $x_n \to_v x$, for each $p \in \mathbb{N}$ there is $n_0 \in \mathbb{N}$ such that $pd(x_n, x) \leq v$ for each $n \in \mathbb{N}$, $n \geq n_0$. Using the properties of the distance function on A we have:
$$x = d(x, 0) \leq d(x, x_n) \oplus d(x_n, 0) = d(x_n, x) \oplus x_n \leq v \oplus x_n$$
Because $Rad(A)$ is an ideal and $v, x_n \in Rad(A)$ it follows that $v \oplus x_n \in Rad(A)$ and then $x \in Rad(A)$.
(2) We have $(x_n^-)_n \subseteq Rad(A)$, $0 < v^- \in Rad(A)$ and apply (1). □

Proposition 18. *Let A be a locally Archimedean MV-algebra. Then:*
(1) *A sequence $(x_n)_n \subseteq Rad(A)$ has a unique v-limit for any $0 < v \in Rad(A)$;*
(2) *If $(x_n)_n, (y_n)_n \subseteq Rad(A)$ and $0 < v \in Rad(A)$ such that $x_n \to_v x$, $y_n \to_v y$ and $x_n \leq y_n$ for any $n \in \mathbb{N}$, then $x \leq y$.*

Proof.
(1) Consider $x_1, x_2 \in A$ such that $x_n \to_v x_1$ and $x_n \to_v x_2$. Then, by the above proposition we have $x_1, x_2 \in Rad(A)$ and by the properties of distance:
$$pd(x_1, x_2) \leq pd(x_1, x_n) \oplus d(x_n, x_2) \leq 2v \text{ for all } p \in \mathbb{N}.$$
Since A is locally Archimedean, we get $d(x_1, x_2) = 0$, hence $x_1 = x_2$.
(2) Since $x_n \leq y_n$, we have $x_n^- \oplus y_n = 1 \to_v 1$. By Proposition 16 it follows that $x_n^- \oplus y_n \to_v x^- \oplus y$ and by (1) we get $x^- \oplus y = 1$. Thus $x \leq y$. □

Proposition 19. *([1]) If A is a perfect MV-algebra, then the following are equivalent:*
(i) A is locally Archimedean ; (ii) $\mathcal{D}(A)$ is an Archimedean ℓ-group.

Proposition 20. *If A is a perfect MV-algebra, $(x_n)_n \subseteq Rad(A)$ and $0 < v \in Rad(A)$ then the following are equivalent:*
(i) $x_n \to_v x$ in A; (ii) $[x_n, 0] \to_{[v,0]} [x, 0]$ in $\mathcal{D}(A)$.

Proof. (i)$\Rightarrow$(ii) Assume that for each $p \in \mathbb{N}$ there is $n_0 \in \mathbb{N}$ such that $pd(x_n, x) \leq v$ for each $n \in \mathbb{N}$, $n \geq n_0$. Then, for each $p \in \mathbb{N}$ and $n \in \mathbb{N}$, $n \geq n_0$ we have

$$p|[x_n, 0] - [x, 0]| = p|[x_n, x]| = p[d(x_n, x), 0] =$$
$$= [pd(x_n, x), 0] \leq [v, 0]$$

Thus $[x_n, 0] \to_{[v,0]} [x, 0]$ in $\mathcal{D}(A)$.
(ii)$\Rightarrow$(i) is proved similarly. □

Definition 21. *([4]) Let A be an arbitrary MV-algebra and $0 < v \in A$. The sequence $(x_n)_n$ in A is said to be v-fundamental or v-Cauchy if for each $p \in \mathbb{N}$ there is $n_0 \in \mathbb{N}$ such that $pd(x_n, x_m) \leq v$ for each $m, n \in \mathbb{N}$, $m \geq n \geq n_0$.*

Proposition 22. *([4]) Let A be an arbitrary MV-algebra and $0 < v \in A$. If the sequence $(x_n)_n$ is v-convergent in A, then $(x_n)_n$ is v-Cauchy in A.*

Proposition 23. *([4]) Let A be an arbitrary MV-algebra and $0 < v \in A$. If the sequence $(x_n)_n$ is v-Cauchy in A, then the sequences $x_n \oplus y_n$, $x_n \odot y_n$, $x_n \vee y_n$, $x_n \wedge y_n$, x_n^- are v-Cauchy in A.*

Corollary 24. *Let A be a perfect MV-algebra, $(x_n)_n \subseteq Rad(A)$ and $0 < v \in Rad(A)$. If $([x_n, y_n])_n$ is a $[v, 0]$-Cauchy sequence in $\mathcal{D}(A)$, then $([x_n, y_n]^+)_n$ and $([x_n, y_n]^-)_n$ are also $[v, 0]$-Cauchy sequences in $\mathcal{D}(A)$.*

Proposition 25. *Let $(x_n)_n$ be a v-Cauchy sequence in the perfect MV-algebra A with $0 < v \in Rad(A)$. Then there is $n_0 \in \mathbb{N}$ such that $\{x_n \mid n \geq n_0\} \subseteq Rad(A)$ or $\{x_n \mid n \geq n_0\} \subseteq (Rad(A))^-$.*

Proof. Because $(x_n)_n$ is a v-Cauchy sequence, for each $p \in \mathbb{N}$ there is $n_0 \in \mathbb{N}$ such that $pd(x_n, x_{n+k}) \leq v$ for each $n, k \in \mathbb{N}$, $n \geq n_0$. Thus $d(x_n, x_{n+k}) \in Rad(A)$. Assume there are $n \in \mathbb{N}$ and $k \in \mathbb{N}$ such that $x_n \in (Rad(A))^-$ and $x_{n+k} \in Rad(A)$, so $x_{n+k} \leq x_n$. It follows that $x_n = x_{n+k} \oplus d(x_n, x_{n+k})$, with $x_{n+k}, d(x_n, x_{n+k}) \in Rad(A)$. It follows that $x_n \in Rad(A)$, which is a contradiction. Similarly, if $x_n \in Rad(A)$ and $x_{n+k} \in (Rad(A))^-$, then $x_n \leq x_{n+k}$ and $x_{n+k} = x_n \oplus d(x_n, x_{n+k})$, with $x_n, d(x_n, x_{n+k}) \in Rad(A)$. It follows that $x_{n+k} \in Rad(A)$, which is again a contradiction. $\square$

Generaly, a v-Cauchy sequence in A is not convergent (see [4]). If every v-Cauchy sequence in A is convergent, then A is said to be v-Cauchy complete.

Similar to the proof of Proposition 20 we can prove the following result.

Proposition 26. *If A is perfect MV-algebra, $(x_n)_n \subseteq Rad(A)$ and $0 < v \in Rad(A)$, then the following are equivalent:*

(i) $(x_n)_n$ is a v-Cauchy sequence in A;
(ii) $([x_n, 0])_n$ is a $[v, 0]$-Cauchy sequence in $\mathcal{D}(A)$.

Theorem 27. *If A is a perfect MV-algebra and $0 < v \in Rad(A)$, then the following are equivalent:*

(i) A is v-Cauchy complete MV-algebra;
(ii) $\mathcal{D}(A)$ is $[v, 0]$-Cauchy complete ℓ-group.

Proof. (i)$\Rightarrow$(ii) Suppose that $([x_n, y_n])_n$ is a $[v, 0]$-Cauchy sequence in $\mathcal{D}(A)$. It follows that $([x_n, y_n]^+)_n$ and $([x_n, y_n]^-)_n$ are also $[v, 0]$-Cauchy sequences. By Proposition 1 and Proposition 26, $x_n \odot y_n^-$ and $x_n^- \odot y_n$ are v-Cauchy sequences in A. Since A is v-Cauchy complete, it follows that $x_n \odot y_n^- \to_v z_1$ and $x_n^- \odot y_n \to_v z_2$, with $z_1, z_2 \in A$. By Proposition 17 we have $z_1, z_2 \in Rad(A)$. By Proposition 1 and Proposition 20 we get $[x_n, y_n]^+ \to_{[v,0]} [z_1, 0]$ and $[x_n, y_n]^- \to_{[v,0]} [z_2, 0]$. Since $[z_1, z_2] = [z_1, 0] - [z_2, 0]$ we get $[x_n, y_n] \to_{[v,0]} [z_1, z_2]$. Thus $\mathcal{D}(A)$ is a $[v, 0]$-Cauchy complete ℓ-group.

(ii)$\Rightarrow$(i) Consider the v-Cauchy sequence $(x_n)_n$ in A. According to Proposition 25 we can assume $(x_n)_n \subseteq Rad(A)$ or $(x_n)_n \subseteq (Rad(A))^-$.

If $(x_n)_n \subseteq Rad(A)$, applying Proposition 26, the sequence $([x_n, 0])_n$ is $[v, 0]$-Cauchy in $\mathcal{D}(A)$. Therefore there is $x \in Rad(A)$ such that $[x_n, 0] \to_{[v,0]} [x, 0]$. From Proposition 26 it follows that $x_n \to_v x$ in A.

If $(x_n)_n \subseteq (Rad(A))^-$, then $(x_n^-)_n \subseteq Rad(A)$ and we get similarly that $x_n^- \to_v x^-$.

By Proposition 16 it follows that $x_n \to_v x$. Thus, A is v-Cauchy complete. □

Definition 28. *Let A be a locally Archimedean MV-algebra and $0 < v \in Rad(A)$. A locally Archimedean MV-algebra B is called v-Cauchy completion of A if the following are satisfied:*

(1) A is a subalgebra of B; (2) B is v-Cauchy complete;
(3) Every element of $Rad(B)$ is a v-limit of some sequence in $Rad(A)$.

Theorem 29. *Let A, B be two locally Archimedean MV-algebras, $A \subseteq B$ and $0 < v \in Rad(A)$. The following are equivalent:*

(i) B is a v-Cauchy completion of A;
(ii) $\mathcal{D}(B)$ is a $[v, 0]$-Cauchy completion of $\mathcal{D}(A)$.

Proof. (i)⇒(ii) We prove the conditions (1)-(3) from Definition 2:

(1) $A \subseteq B \Rightarrow \mathcal{D}(A) \subseteq \mathcal{D}(B)$;
(2) follows by Theorem 27;
(3) Take $[x, y] \in \mathcal{D}(B)$. Then there are two sequences $(x_n)_n$, $(y_n)_n \subseteq Rad(A)$ such that $x_n \to_v x$ and $y_n \to_v y$. Thus, $[x_n, y_n] \to_{[v,0]} [x, y]$, hence $\mathcal{D}(B)$ is a $[v, 0]$-Cauchy completion of $\mathcal{D}(A)$.

(ii)⇒(i) We show that conditions (1)-(3) from Definition 28 hold:
(1) holds by hypothesis and (2) holds by Theorem 27;
(3) Take $x \in Rad(B)$. There is a sequence $([x_n, 0])_n$ in $\mathcal{D}(A)$ such that $[x_n, 0] \to_{[v,0]} [x, 0]$. Thus $x_n \to_v x$ and therefore B is a v-Cauchy completion of A. □

Theorem 30. *Any locally Archimedean MV-algebra has a unique v-Cauchy completion.*

Proof. Let A be a locally Archimedean MV-algebra. By Proposition 19, $\mathcal{D}(A)$ is an Archimedean ℓ-group. By Theorems 3.16 and 3.17 from [2], there is a unique v-Cauchy completion G of the abelian ℓ-group $\mathcal{D}(A)$. But $G = \mathcal{D}(B)$ for some $B = \Delta(G)$, so $\mathcal{D}(B)$ is the unique v-Cauchy completion of $\mathcal{D}(A)$. By Theorem 29 it follows that B is the unique v-Cauchy completion of $\mathcal{D}(A)$. □

3 Convergence with a Fixed Regulator in Residuated Lattices

Definition 31. *Let L be a residuated lattice and $v \in G(L) \setminus \{0\}$. The sequence $(x_n)_n$ is said to be v-convergent to an element $x \in L$ (or x is v-limit of $(x_n)_n$) denoted by $x_n \to_v x$, if for every $p \in \mathbb{N}$ there is $n_0 \in \mathbb{N}$ such that $d(x_n, x)^p \geq v$ for all $n \in \mathbb{N}$, $n \geq n_0$. The element v is called convergence regulator in L.*

Example 32
a) The constant sequence $(x, x, x, ..., x, ...)$ v-converges to x for any $v \in G(L)$;
b) Consider the residuated lattice L in Example 14, $v = \frac{1}{2} \in G(L)$ and the sequence $(x_n)_n$ with $x_n = \frac{n-1}{n}$ for all $n \in \mathbb{N}$. Then $x_n \to_{\frac{1}{2}} 1$. Indeed, for any $p \in \mathbb{N}$, we have $d(x_n, 1)^p = x_n^p = (\frac{n-1}{n})^p = \frac{n-1}{n} \geq \frac{1}{2}$ for all $n \geq 2$. Thus, for any p, $n_0 = 2$.

Remarks 33

(1) *Let* $u \in G(L), u < v$. *If* $x_n \to_v x$, *then* $x_n \to_u x$.

(2) *If* $G(L) = \{0, 1\}$, *then the constant sequences are the only v-convergent sequences.*

Proposition 34. *Let* $(x_n)_n, (y_n)_n$ *be two sequences in* L *and* $x, y \in L$ *such that* $x_n \to_v$
x *and* $y_n \to_v y$. *Then the following properties hold:*

(1) $x_n \wedge y_n \to_v x \wedge y$; (2) $x_n \vee y_n \to_v x \vee y$; (3) $x_n \odot y_n \to_v x \odot y$;
(4) $(x_n \to y_n) \to_v (x \to y)$; (5) $x_n^- \to_v x^-$; (6) $a \odot x_n \to_v a \odot x$ *for any* $a \in L$.

Proof. By the hypothesis, for every $p \in \mathbb{N}$ there are $n_1, n_2 \in \mathbb{N}$ such that $d(x_n, x)^p \geq v$
for all $n \in \mathbb{N}, n \geq n_1$ and $d(y_n, y)^p \geq v$ for all $n \in \mathbb{N}, n \geq n_2$.

Obviously, $d(x_n, x) \geq d(x_n, x)^p \geq v$ for all $n \geq n_1$ and $d(y_n, y) \geq d(y_n, y)^p \geq$
v for all $n \geq n_2$. We will apply the properties of the distance function asserted in
Proposition 4.

(1) For every $p \in \mathbb{N}$ we have $d(x_n \wedge y_n, x \wedge y)^p \geq (d(x_n, x) \wedge d(y_n, y))^p \geq (v \wedge v)^p =$
$v^p = v$, for all $n \in \mathbb{N}, n \geq n_0 = max\{n_1, n_2\}$. Thus, $x_n \wedge y_n \to_v x \wedge y$;
(2), (3) and (4) can be proved in a similar manner as (1);
(5) We have $d(x_n^-, x^-) = d(x_n \to 0, x \to 0) \geq d(x_n, x) \odot d(0, 0) = d(x_n, x)$. Thus,
$x_n^- \to_v \overline{x}$;
(6) Since $d(a \odot x_n, a \odot x) \geq d(a, a) \odot d(x_n, x) = d(x_n, x)$, it follows that, for all
$p \in \mathbb{N} \, d(a \odot x_n, a \odot x)^p \geq d(x_n, x)^p \geq v$ for all $n \in \mathbb{N}, n \geq n_1$, so $a \odot x_n \to_v a \odot x$.
$\square$

Proposition 35. *Let* L, L_1 *be two residuated lattices,* $h : L \to L_1$ *a RL-morphism and*
$v \in G(L)$. *If* $x_n \to_v x$ *in* L, *then* $h(x_n) \to_{h(v)} h(x)$ *in* L_1.

Proof. According to Remark 10 and Proposition 11, it follows that $h(v) \in G(L_1)$.
Since $x_n \to_v x$, then for every $p \in \mathbb{N}$ there is $n_0 \in \mathbb{N}$ such that $d(x_n, x)^p \geq v$ for all
$n \in \mathbb{N}, n \geq n_0$. It follows that for every $p \in \mathbb{N}$ we have:

$$d(h(x_n), h(x))^p = h(d(x_n, x))^p = h(d(x_n, x)^p) \geq h(v) \text{ for all } n \in \mathbb{N}, n \geq n_0.$$
Thus, $h(x_n) \to_{h(v)} h(x)$.
$\square$

Theorem 36. *The v-convergence in a residuated lattice* L *defines a topology on* L.

Proof. We define a *closed set* X to be a subset of L such that, if $(x_n)_n \subseteq X$ with
$x_n \to_v x$, then $x \in X$. We show that the set T of all closed sets of L defines a topology
on L. Obviously, $\emptyset, L \in T$. Also, it is easy to check that if $X_i \in T$, then $\cap_{i \in I} X_i \in T$.
Consider $X_1, X_2 \in T$ and $(x_n)_n \subseteq X \cup Y$ such that $x_n \to_v x$ (that is, for every
$p \in \mathbb{N}$ there is $n_0 \in \mathbb{N}$ such that $d(x_n, x)^p \geq v$ for all $n \in \mathbb{N}, n \geq n_0$). Each of the
elements of $\{x_n, n \in \mathbb{N}\}$ is in X or Y, hence if we define $I_1 = \{n \in \mathbb{N} : x_n \in X\}$
and $I_2 = \{n \in \mathbb{N} : x_n \in Y\}$, then I_1 or I_2 is infinite. Suppose that I_1 is infinite. This
means that there exists a subsequence $(y_n)_n = (x_{k(n)})_n$ of $(x_n)_n$ such that $(y_n)_n \subseteq X$
($k : \mathbb{N} \longrightarrow \mathbb{N}$ is an arbitrary increasing function). It follows that for every $p \in \mathbb{N}$ there
is $n_0 \in \mathbb{N}$ such that $d(y_n, x)^p \geq v$ for all $n \in \mathbb{N}, n \geq n_0$, so $x \in X$, since X is closed.
We conclude that $x \in X \cup Y$, hence $X \cup Y$ is closed. Thus, T is a topology on L. $\square$

The above constructed topology is called *v-topology*.

As an immediate consequence of Proposition 34, the operations $\wedge, \vee, *, \rightarrow$ of the residuated lattice L are continuous w.r.t. the v-topology.

Proposition 37. *Let L be an Archimedean residuated lattice. Then:*

(1) *The v-limit is unique determined for each $v \in L^* \cap G(L)$;*
(2) *If $(x_n)_n, (y_n)_n \subseteq L$ such that $x_n \rightarrow_v x, y_n \rightarrow_v y$ and $x_n \leq y_n$ for all $n \in \mathbb{N}$, then $x \leq y$.*

Proof
(1) Suppose there exist $x_1, x_2 \in L$ such that $x_n \rightarrow_v x_1$ and $x_n \rightarrow_v x_2$.
It follows that for every $p \in \mathbb{N}$ there is $n_1, n_2 \in \mathbb{N}$ such that $d(x_n, x_1)^p \geq v$ for all $n \in \mathbb{N}, n \geq n_1$ and $d(x_n, x_2)^p \geq v$ for all $n \in \mathbb{N}, n \geq n_2$.
Let's consider $p \in \mathbb{N}$ and $m \geq n_1, n_2$. Then we have:

$$d(x_1, x_2)^p \geq d(x_1, x_m)^p \odot d(x_m, x_2)^p \geq v^2 = v \geq v^-.$$

Because L is Archimedean, we obtain $d(x_1, x_2) = 1$.
Thus, by Proposition 4(2) we have $x_1 = x_2$.
(2) Since $x_n \leq y_n$ for all $n \in N$, we have $x_n \rightarrow y_n = 1$, so $(x_n \rightarrow y_n) \rightarrow_v 1$.
On the other hand $(x_n \rightarrow y_n) \rightarrow_v (x \rightarrow y)$ and applying (1) we get $x \rightarrow y = 1$, hence $x \leq y$. $\qquad\qquad\square$

Remark 38. *If the residuated lattice L is not Archimedean, then the v-limit of a sequence $(x_n)_n$ is not always unique.*
   Indeed, let L be the non Archimedean residuated lattice from Example 14. Then for the sequence $(x_n)_n = (\frac{1}{3}, 1, \frac{1}{3}, 1, ...)$ we have:

$$d(x_n, \frac{1}{3}) = \begin{cases} 1 \text{ if } n \text{ is odd} \\ \frac{1}{3} \text{ if } n \text{ is even} \end{cases} \quad \text{and} \quad d(x_n, 1) = \begin{cases} \frac{1}{3} \text{ if } n \text{ is odd} \\ 1 \text{ if } n \text{ is even.} \end{cases}$$

It follows that $d(x_n, \frac{1}{3})^p \geq (\frac{1}{3})^p = \frac{1}{3}$ and $d(x_n, 1)^p \geq (\frac{1}{3})^p = \frac{1}{3}$ for all $p \in \mathbb{N}$, $p \geq 1$.
   Taking $v = \frac{1}{3} \in L^* \cap G(L)$, we get $x_n \rightarrow_v \frac{1}{3}$ and $x_n \rightarrow_v 1$.

Definition 39. *Let $v \in G(L)$. The sequence $(x_n)_n$ is said to be v-Cauchy or v-fundamental sequence if for every $p \in \mathbb{N}$ there is $n_0 \in \mathbb{N}$ such that $d(x_n, x_m)^p \geq v$ for all $n, m \in \mathbb{N}, m \geq n \geq n_0$.*

Proposition 40. *Let $v \in G(L)$. If the sequence $(x_n)_n$ is v-convergent, then $(x_n)_n$ is v-Cauchy.*

Proof. Suppose that $x_n \rightarrow_v x$, that is for every $p \in \mathbb{N}$ there is $n_0 \in \mathbb{N}$ such that $d(x_n, x)^p \geq v$ for all $n \in \mathbb{N}, n \geq n_0$. Let $m, n \in N, m \geq n \geq n_0$.
   By Proposition 4(5) we have $d(x_n, x_m)^p \geq d(x_n, x)^p \odot d(x_m, x)^p \geq v^2 = v$.
   Hence, the sequence $(x_n)_n$ is v-Cauchy. $\qquad\qquad\square$

Proposition 41. *Let $(x_n)_n, (y_n)_n$ be v-Cauchy sequences in L and $a \in L$. Then, the sequences $(x_n \wedge y_n)_n, (x_n \vee y_n)_n, (x_n \odot y_n)_n, (x_n \rightarrow y_n)_n, (x_n)_n^-, (a \odot x_n)_n$ are v-Cauchy.*

Proof. By the hypothesis, for every $p \in \mathbb{N}$ there is $n_1, n_2 \in \mathbb{N}$ such that $d(x_n, x_m)^p \geq v$ for all $n, m \in \mathbb{N}, m \geq n \geq n_1$ and $d(y_n, y_m)^p \geq v$ for all $n, m \in \mathbb{N}, m \geq n \geq n_2$.

By Proposition 4(9), for any $p \in N$ and $m \geq n \geq n_0 = max\{n_1, n_2\}$ we have

$$d(x_n \wedge y_n, x_m \wedge y_m)^p \geq (d(x_n, x_m) \wedge d(y_n, y_n))^p \geq (v \wedge v)^p = v^p = v.$$

Thus, $(x_n \wedge y_n)_n$ is a v-Cauchy sequence. Similarly for the other sequences. □

Corollary 42. $C(L) = \{(x_n)_n \subseteq L^\mathbb{N} \mid (x_n)_n \ v\text{-Cauchy}\}$ *is a residuated lattice with component-wise operations.*

Proof. By the above proposition, $C(L)$ is a stable part of $L^\mathbb{N}$.

The conditions in the definition of a residuated lattice are verified taking into consideration the component-wise defined operations. $C(L)$ has the bounds $(0)_n$ and $(1)_n$. □

On $C(L)$ we define the relation $\equiv$ by $(x_n)_n \equiv (y_n)_n$ iff $(d(x_n, y_n))_n \rightarrow_v 1$.

By $(x)_n$ we denote the constant x sequence.

Proposition 43. *The relation $\equiv$ is a congruence on $C(L)$.*

Proof. First we show that $\equiv$ is an equivalence:

Reflexivity: $(x_n)_n \equiv (x_n)_n$ because $(d(x_n, x_n))_n = (1)_n \rightarrow_v 1$.

Symmetry: $(x_n)_n \equiv (y_n)_n$ implies $(y_n)_n \equiv (x_n)_n$ because $d(x_n, y_n) = d(y_n, x_n)$.

Transitivity: Let's consider $(x_n)_n \equiv (y_n)_n$ and $(y_n)_n \equiv (z_n)_n$, that is $(d(x_n, y_n))_n \rightarrow_v 1$ and $(d(y_n, z_n))_n \rightarrow_v 1$. It follows that for every $p \in \mathbb{N}$ there exist $n_1, n_2 \in \mathbb{N}$ such that $d((d(x_n, y_n)), 1)^p \geq v$ for all $n \geq n_1$ and $d((d(y_n, z_n)), 1)^p \geq v$ for all $n \geq n_2$.

Taking under consideration that $d(x, 1) = x$ we get $d(x_n, y_n)^p \geq v$ for all $n \geq n_1$ and $d(y_n, z_n)^p \geq v$ for all $n \geq n_2$. We have: $d(x_n, z_n)^p \geq d(x_n, y_n)^p \odot d(y_n, z_n)^p \geq v^2 = v$.

Taking $n_0 = max\{n_1, n_2\}$, we get $d(x_n, z_n)^p \geq v$ for all $n \geq n_0$.

Hence, $(d(d(x_n, z_n)), 1)^p \geq v$ for all $n \geq n_0$, so $(d(x_n, z_n))_n \rightarrow_v 1$, that is $(x_n)_n \equiv (z_n)_n$.

For the compatibility of the relation $\equiv$ with the operations we apply Propoposition 4(6-9). Thus, $\equiv$ is a congruence on $C(L)$. □

Corollary 44. $L_1 = C(L)/\equiv$ *is a residuated lattice.*

Denote by $[(x_n)_n]$ the equivalence class of the sequence $(x_n)_n$.

Proposition 45. *If L is Archimedean, then the map $i : L \rightarrow L_1, i(x) = [(x)_n]$ is an embedding of residuated lattices.*

Proof. Consider $x_1, x_2 \in L$ such that $(x_1)_n \equiv (x_2)_n$, that is $(d(x_1, x_2))_n \rightarrow_v 1$.

On the other hand, since $(d(x_1, x_2))_n$ is a constant sequence, it follows that $(d(x_1, x_2))_n \rightarrow_v d(x_1, x_2)$. Hence, according to Proposition 37 we get $d(x_1, x_2) = 1$, that is $x_1 = x_2$. Thus, the map $x \rightarrow [(x_n)_n]$ is injective. Applying Proposition 4 and Proposition 34 it is easy to check that this map is morphism. Therefore, the map i is an embedding of residuated lattices. □

Definition 46. *A residuated lattice L is said to be v-Cauchy complete if every v-Cauchy sequence in L is v-convergent.*

A residuated lattice L is a complete if is complete as a lattice, that is, if every subset A of L has both a greatest lower bound (infimum, meet) and a least upper bound (supremum, join) in L. These are denoted by $\bigvee A$ and respectively $\bigwedge A$.

Theorem 47. *If the residuated lattice L is complete, then L is v-Cauchy complete.*

Proof. Let $(x_n)_n$ be a v-Cauchy sequence in L, that is for every $p \in \mathbb{N}$ there is $n_{0_p} \in \mathbb{N}$ such that $d(x_n, x_m)^p \geq v$ for all $n, m \in \mathbb{N}$, $m \geq n \geq n_{0_p}$. Since L is a complete lattice, there exists $x = \bigwedge_{p \in \mathbb{N}} \bigwedge_{n \geq n_{0_p}} x_n$. Obviously, $[x_n, x] \subseteq [x_n, x_m]$.

Hence, according to Proposition 4(11), for each $p \in \mathbb{N}$ we have $d(x_n, x)^p \geq d(x_n, x_m)^p \geq v$ for all $n \in \mathbb{N}$, $n \geq n_{0_p}$.

Thus, $x_n \to_v x$ and it follows that L is a v-Cauchy complete lattice. □

References

1. Belluce, L.P., Di Nola, A.: Yosida type representation for perfect MV-algebras. Math. Log. Quart. 42, 551–563 (1996)
2. Černák, Š., Liihová, J.: Convergence with a regulator in lattice ordered groups. Tatra Mountains Mathematical Publications 30, 35–45 (2005)
3. Černák, Š.: Convergence with a fixed regulator in Archimedean lattice ordered groups. Mathematica Slovaca 2, 167–180 (2006)
4. Černák, Š.: Convergence with a fixed regulator in lattice ordered groups and applications to MV-algebras. Soft Computing 12(5), 453–462 (2008)
5. Cignoli, R., D'Ottaviano, I.M.L., Mundici, D.: Algebraic Foundation of Many-valued Reasoning. Kluver academic Publishers, Dordrecht (2000)
6. Ciungu, L.C.: Classes of residuated lattices. An. Univ. Craiova Ser. Mat. Inform. 33, 189–207 (2006)
7. Georgescu, G., Liguori, F., Martini, G.: Convergence in MV-algebras. Soft Computing 4, 41–52 (1997)
8. Georgescu, G., Leuştean, I.: Convergence in perfect MV-algebras. Journal of Mathematical Analysis and Applications 228(1), 96–111 (1998)
9. Georgescu, G., Popescu, A.: Similarity Convergence in Residuated Structures. Logic Journal of IGPL 13(4), 389–413 (2005)
10. Di Nola, A., Letieri, A.: Perfect MV-algebras are categoricaly equivalent to abelian ℓ-groups. Studia Logica 53, 417–432 (1994)
11. Di Nola, A.: Algebraic analysis of Łukasiewicz logic, ESSLLI, Summer school, Utrecht (August 1999)
12. Kowalski, T., Ono, H.: Residuated lattices: An algebraic glimpse at logics without contraction, Monograph (2001)
13. Mundici, D.: Interpretation of AF C^*-algebras in Lukasiewicz calculus. J. Funct. Analysis 65, 15–63 (1986)
14. Ono, H.: Substructural logics and residuated lattices-an introduction. In: 50 Years of Studia Logica, Trends in Logic, vol. 21, pp. 193–228. Kluwer Academic Publisher, Dordrecht (2003)
15. Turunen, E.: Mathematics behind fuzzy logic. In: Advances in Soft Computing. Physica-Verlag, Heildelberg (1999)
16. Ward, M., Dilworth, R.P.: Residuated lattices. Transactions of the American Mathematical Society 45, 335–354 (1939)

Open Partitions and Probability Assignments in Gödel Logic

Pietro Codara, Ottavio M. D'Antona, and Vincenzo Marra

Università degli Studi di Milano
Dipartimento di Informazione e Comunicazione
via Comelico 39–41
I-20135 Milano, Italy
{codara,dantona,marra}@dico.unimi.it

Abstract. In the elementary case of finitely many events, we generalise to *Gödel (propositional infinite-valued) logic* — one of the fundamental fuzzy logics in the sense of Hájek — the classical correspondence between partitions, quotient measure spaces, and push-forward measures. To achieve this end, appropriate Gödelian analogues of the Boolean notions of probability assignment and partition are needed. Concerning the former, we use a notion of probability assignment introduced in the literature by the third-named author et al. Concerning the latter, we introduce and use *open partitions*, whose definition is justified by independent considerations on the relational semantics of Gödel logic (or, more generally, of the finite slice of intuitionistic logic). Our main result yields a construction of finite quotient measure spaces in the Gödelian setting that closely parallels its classical counterpart.

1 Introduction

We assume familiarity with Gödel (propositional infinite-valued) logic, one of the fundamental fuzzy logics in the sense of Hájek [4]; we recall definitions in Section 2.1 below. The problem of generalising elementary probability theory to such fuzzy logics has recently attracted considerable attention; let us mention e.g. [9] for Łukasiewicz logic, [8] for $[0,1]$-valued logics with continuous connectives, and [2], [1] for Gödel logic. This paper falls into the same general research field.

Consider a finite set of (classical, yes/no) events $\mathscr{E}$, along with the finite Boolean algebras $B(\mathscr{E})$ that they generate. A *probability assignment* to $B(\mathscr{E})$ is a function $P \colon B(\mathscr{E}) \to [0,1]$ satisfying *Kolmogorov's axioms*, namely,

(B1) $P(\top) = 1$, and
(B2) $P(X) + P(Y) = P(X \vee Y) + P(X \wedge Y)$ for all $X, Y \in B(\mathscr{E})$.

Here, $\top$ is the top element of $B(\mathscr{E})$, and $\vee$ and $\wedge$ denote the join and meet operation of $B(\mathscr{E})$, respectively. The assignment P is uniquely determined by its values on the atoms $A = \{a_1, \ldots, a_m\}$ of $B(\mathscr{E})$. In fact, there is a bijection between probability assignments to $B(\mathscr{E})$, and *probability distributions* on the set A, that is, functions $p \colon A \to [0,1]$ such that

C. Sossai and G. Chemello (Eds.): ECSQARU 2009, LNAI 5590, pp. 911–922, 2009.

(BD) $\sum_{i=1}^{m} p(a_i) = 1$.

In one direction, one obtains such a p from a probability assignment P to $B(\mathcal{E})$ just as the restriction of P to A. Conversely, from a probability distribution p on A one obtains a probability assignment P to $B(\mathcal{E})$ by setting $P(X) = \sum_{a_i \leq X} p(a_i)$ for any event $X \in B(\mathcal{E})$. If one represents $B(\mathcal{E})$ as the Boolean algebra of subsets of A, this means that, for any $X \subseteq A$, one has $P(X) = \sum_{a_i \in X} p(a_i)$. In probabilistic parlance, one calls the set A a *sample space*, and its singleton subsets *elementary events*.

In several contexts related to probability theory, *partitioning* the sample space A is a process of fundamental importance. A *partition* of A is a collection of non-empty, pairwise disjoint subsets of A — often called *blocks* — whose union is A. A partition π of A can be regarded as a quotient object obtained from A. Indeed, there is a natural projection map $A \to \pi$ given by

$$a \in A \mapsto [a]_\pi \in \pi ,$$

where $[a]_\pi$ denotes the unique block of π that a belongs to. (Thus, $[a]_\pi$ is the equivalence class of a under the equivalence relation on A uniquely associated with π.) When, as is the present case, A carries a probability distribution, one would like each such quotient set π to inherit a unique probability distribution, too. This is indeed the case. Define a function $p_\pi \colon \pi \to [0, 1]$ by

$$p_\pi([a]_\pi) = \sum_{a_i \in [a]_\pi} p(a_i) = P([a]_\pi) . \tag{1}$$

Then p_π is a probability distribution on the set π. To close a circle of ideas, let us return from the distribution p_π to a probability assignment to an appropriate algebra of events. For this, it suffices to observe that the partition $\pi = \{[a]_\pi \mid a \in A\}$ determines the unique subalgebra S_π of $B(\mathcal{E})$ whose atoms are given by $\bigvee [a]_\pi$, for $a \in A$. We can then define a probability assignment $P_{S_\pi} \colon S_\pi \to [0, 1]$ starting from P via

$$P_{S_\pi}(X) = \sum_{[a]_\pi \leq X} P([a]_\pi) .$$

In other words, in the light of (1), P_{S_π} is the unique probability assignment to S_π that is associated with the probability distribution p_π on the atoms of S_π.

Although our current setting is restricted to finitely many events, and is thus elementary, generalisations of the standard ideas above play an important rôle in parts of measure theory. In particular, in certain contexts one constructs quotient measure spaces using as a key tool the push-forward measure along the natural projection map.[1] For our purposes, let $f \colon A \to B$ be a function between the finite sets A and B, and let $p \colon A \to [0, 1]$ be a probability distribution on A. Define a map $p_f \colon B \to \mathbb{R}$ by setting

$$p_f(b) = \sum_{a \in f^{-1}(b)} p(a)$$

---

[1] For an influential account of these ideas, please see [10].

for any $b \in B$. We call p_f the *push-forward* of p along f. (Here, as usual, $p_f(b) = 0$ when the index set is empty.) One checks that p_f is again a probability distribution. If, moreover, f is a surjection, and thus induces a partition of A by taking fibres (=inverse images of elements in the codomain), then the following fact is easily verified.

Fact. *Let A be a finite set, π a partition of A, and $q \colon A \to \pi$ the natural projection map. Then, for every probability distribution $p \colon A \to [0,1]$ on A, the push-forward probability distribution p_q of p along q coincides with p_π in* (1).

Summing up, the fact above provides the desired construction of quotient measure spaces in the elementary case case of finitely many events. Our main result, Theorem 2 below, affords a generalisation of this construction to Gödel logic. To achieve this end, we need appropriate Gödelian analogues of the Boolean notions of probability assignment and partition. Concerning the former, we use a notion of probability assignment recently introduced in [2]; the needed background is in Subsection 2.3. Concerning the latter, in Section 3 we introduce *open partitions*, whose definition is justified by independent considerations on the relational semantics of Gödel logic. As a key tool for the proof of Theorem 2, we obtain in Theorem 1 a useful characterisation of open partitions.

2 Preliminary Results, and Background

2.1 Gödel Logic

Equip the real unit interval $[0,1]$ with the operations $\wedge$, $\to$, and $\perp$ defined by

$$x \wedge y = \min(x, y), \qquad x \to y = \begin{cases} 1 & \text{if } x \leq y, \\ y & \text{otherwise}, \end{cases} \qquad \perp = 0 .$$

The tautologies of Gödel logics are exactly the formulas $\varphi(X_1, \dots, X_n)$ built from connectives $\{\wedge, \to, \perp\}$ that evaluate constantly to 1 under any $[0,1]$-valued assignment to the propositional variables X_i, where each connective is interpreted as the operation denoted by the same symbol. As derived connectives, one has $\neg\varphi = \varphi \to \perp$, $\top = \neg\perp$, $\varphi \vee \psi = ((\varphi \to \psi) \to \psi) \wedge ((\psi \to \varphi) \to \varphi)$. Thus, $\top$ is interpreted by 1, $\vee$ by maximum, and negation by

$$\neg x = \begin{cases} 1 & \text{if } x = 0, \\ 0 & \text{otherwise}. \end{cases}$$

Gödel logic can be axiomatised in the style of Hilbert with *modus ponens* as only deduction rule. In fact, completeness with respect to the many-valued semantics above can be shown to hold for arbitrary theories. For details, we refer to [4].

Gödel logic also coincides with the extension of the intuitionistic propositional calculus by the *prelinearity* axiom scheme $(\varphi \to \psi) \vee (\psi \to \varphi)$ (see again [4]). Thus, the algebraic semantics of Gödel logic is the well-known subvariety of Heyting algebras[2] satisfying prelinearity, which we shall call *Gödel algebras*. By

---

[2] For background on Heyting algebras, see e.g. [7].

[6, Thm. 1], and in analogy with Boolean algebras, a finitely generated Gödel algebra is finite. Throughout, the operations of a Gödel algebra are always denoted by $\wedge$, $\vee$, $\rightarrow$, $\neg$, $\top$ (top element), and $\bot$ (bottom element).

2.2   Posets and Open Maps

For the rest of this paper, *poset* is short for partially ordered set, and all posets are assumed to be finite. If P is a poset (under the relation $\leq$) and $S \subseteq P$, the *lower set* generated by S is

$$\downarrow S = \{p \in P \mid p \leq s \text{ for some } s \in S\} \ .$$

(When S is a singleton $\{s\}$, we shall write $\downarrow s$ for $\downarrow \{s\}$.) A subposet $S \subseteq P$ is a *lower set* if $\downarrow S = S$. *Upper sets* and $\uparrow S$ are defined analogously. We write Min P and Max P for the set of minimal and maximal elements of P, respectively.

As we already mentioned, Gödel algebras are the same thing as Heyting algebras satisfying the prelinearity axiom. From any finite poset P one reconstructs a Heyting algebra, as follows. Let Sub P be the family of all lower sets of P. When partially ordered by inclusion, Sub P is a finite distributive lattice, and thus carries a unique Heyting implication adjoint to the lattice meet operation via *residuation*. Explicitly, if L is a finite distributive lattice, then its Heyting implication is given by

$$x \rightarrow y = \bigvee \{z \in L \mid z \wedge x \leq y\}$$

for all $x, y \in L$. Accordingly, we regard Sub P as a Heyting algebra.

Conversely, one can obtain a finite poset from any finite Heyting algebra H, by considering the poset Spec H of prime filters of H, ordered by reverse inclusion. Equivalently, one can think of Spec H as the poset of join-irreducible elements of H, with the order they inherit from H. (Let us recall that a *filter* of H is an upper set of H closed under meets; it is *prime* if it does not contain the bottom element of H, and contains either y or z whenever it contains $y \vee z$. We further recall that $x \in H$ is *join-irreducible* if it is not the bottom element of H, and whenever $x = y \vee z$ for $y, z \in H$, then either $x = y$ or $x = z$.)

The constructions of the two preceding paragraphs are inverse to each other, in the sense that for any finite Heyting algebra H one has an isomorphism of Heyting algebras

$$\text{Sub Spec } H \cong H \ . \tag{2}$$

In fact, the isomorphism (2) is natural. To explain this, let us recall that an order-preserving function $f : P \rightarrow Q$ between posets is called *open* if whenever $f(u) \geq v'$ for $u \in P$ and $v' \in Q$, there is $v \in P$ such that $u \geq v$ and $f(v) = v'$. From a logical point of view, if one regards P and Q as finite Kripke frames, then open maps are known as *p-morphisms*; cf. e.g. [3]. It is a folklore result that there is a categorical duality between finite Heyting algebras and their homomorphisms, and finite posets and open order-preserving maps between them. Given a homomorphism of finite Heyting algebras $h: A \to B$, the map $\operatorname{Spec} h \colon \operatorname{Spec} B \to \operatorname{Spec} A$ given by $\mathfrak{p} \mapsto h^{-1}(\mathfrak{p})$ (where $\mathfrak{p}$ is a prime filter of B) is open and order-preserving. Conversely, given an open order-preserving map $f \colon P \to Q$, the function $\operatorname{Sub} f \colon \operatorname{Sub} Q \to \operatorname{Sub} P$ given by $Q' \mapsto f^{-1}(Q')$ is a homomorphism of Heyting algebras. Specifically, the order-preserving property of f is equivalent to $\operatorname{Sub} f$ being a lattice homomorphism; and the additional assumption that f be open insures that $\operatorname{Sub} f$ preserves the Heyting implication, too. It can be checked that Spec and Sub (now regarded as functors) yield the aforementioned categorical duality.

Let us now restrict attention to finite *Gödel* algebras. A *forest* is a poset F such that $\downarrow x$ is totally ordered for any $x \in F$. In this case, it is customary to call $\operatorname{Min} F$ and $\operatorname{Max} F$ the sets of *roots* and *leaves* of F, respectively. Further, a lower set of the form $\downarrow x$, for $x \in F$, is called a *branch* of F. Note that any lower set of a forest F is itself a forest, and we shall call it a *subforest* of F. Horn proved [5, 2.4] that a Heyting algebra H is a Gödel algebra if and only if its prime filters are a forest under reverse inclusion, i.e. if $\operatorname{Spec} H$ is a forest. Using this fact, one sees that the categorical duality of the preceding paragraph restricts to a categorical duality between finite Gödel algebras with their homomorphisms, and forests with open maps between them. Since Gödel logic is a generalisation of classical logic, this duality has a Boolean counterpart as a special case. Namely, finite Boolean algebra and their homomorphisms are dually equivalent to finite sets and functions between them. To obtain this result starting from Gödel algebras, one just observes that a finite Gödel algebra G is Boolean if and only if $\operatorname{Spec} G$ is a forest consisting of roots only, that is, a finite set, and that an open map between such forests is just a set-theoretic function. Observe that the folklore duality between finite Boolean algebra and finite sets underlies the correspondence between probability assignments and distributions illustrated in the Introduction. Similarly, the folklore duality between forests and open maps will underlie the analogous correspondence for Gödel logic, which we will state in Proposition 1 below.

Example 1. If $G = \{\top, \bot\}$, then $\operatorname{Spec} G$ is a single point, the prime filter $\{\top\}$. If, even more trivially, G is the degenerate singleton algebra $G = \{\top = \bot\}$, then G has no prime filter at all, and thus $\operatorname{Spec} F$ is the empty forest. On the other hand, there is no such thing as a Gödel algebra with empty underlying set, because the signature contains the constant $\bot$. Next suppose G is the Gödel algebra whose Hasse diagram is depicted in Fig. 1. The join-irreducible elements of G are those labeled by X, $\neg X$, and $\neg\neg X$. Therefore, $\operatorname{Spec} G$ is the forest depicted in Fig. 2. To recover G from $\operatorname{Spec} F$, consider the collection $\operatorname{Sub} \operatorname{Spec} G$ of all lower sets of $\operatorname{Spec} G$ ordered by inclusion. This is depicted in Fig. 3. Ordering $\operatorname{Sub} \operatorname{Spec} G$ by inclusion, we get back (an algebra naturally isomorphic to) G. (Let us note that here, using algebraic terminology, G is the Gödel algebra freely generated by the generator X.)

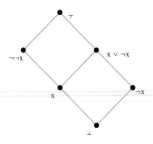

Fig. 1. A Gödel algebra G

Fig. 2. The forest $\operatorname{Spec} G$ (cf. Fig. 1)

Fig. 3. The elements of $\operatorname{Sub} \operatorname{Spec} G$ (cf. Fig. 1 and 2)

2.3 Probability Assignments

Let G be a finite Gödel algebra. By a *probability assignment* to G we mean a function $P\colon G \to [0,1]$ such that, for any $X, Y, Z \in G$,

(G1) $P(\top) = 1$ and $P(\bot) = 0$,
(G2) $X \le Y$ implies $P(X) \le P(Y)$,
(G3) $P(X) + P(Y) = P(X \vee Y) + P(X \wedge Y)$ for all $X, Y \in G$, and
(G4) if X is covered[3] by Y that is covered by Z, and each X, Y, Z is either join-irreducible or coincides with $\bot$, then $P(X) = P(Y)$ implies $P(Y) = P(Z)$.

These axioms were first put forth in [2]. Clearly, (G1–3) are just standard properties of Boolean probability assignments: the first is normalisation; the second, monotonicity; the third, finite additivity. On the other hand, (G4) is characteristic of the Gödel case. If G happens to be a Boolean algebra, (G4) holds trivially, for in this case any two join-irreducible elements (=atoms) are incomparable. When G is not Boolean, then (G4) is an actual constraint on admissible distributions of values — a constraint arising from the nature of implication in Gödel logic. A further discussion of (G4) can be found in [2, Section III].

---

[3] This means that $X < Y$, and there is no element lying properly between X and Y.

As in the Boolean case, there is a notion of probability distribution corresponding to (G1–G4). To define it, consider the forest $F = \operatorname{Spec} G$. A *probability distribution* on F is a function $p\colon F \to [0,1]$ such that

(GD1) $\sum_{x \in F} p(x) = 1$, and
(GD2) for all $x \leq y \in F$, $p(x) = 0$ implies $p(y) = 0$.

Axiom (GD2) is equivalent to the condition that $p^{-1}(0)$ be an upper set of F. Now, the following correspondence result holds.

Proposition 1. *Let G be a finite Gödel algebra. Without loss of generality, let us assume $G = \operatorname{Sub} F$ for a finite forest F. Let $\mathscr{P}$ be the family of all probability assignments to G, and $\mathscr{D}$ the family of all probability distributions on F. With each $P \in \mathscr{P}$, let us associate the map $p\colon F \to [0,1]$ such that, for each $x \in F$,*

$$p(x) = P(\downarrow x) - P(\downarrow x^{\triangleleft}) \,, \tag{3}$$

where $x^{\triangleleft}$ is the unique element of F that is covered by x, if $x \notin \operatorname{Min} F$, and $x^{\triangleleft} = \perp$, otherwise. Then the correspondence $p \mapsto P$ is a bijection between $\mathscr{P}$ and $\mathscr{D}$. Its inverse is given by

$$P(X) = \sum_{x \in X} p(x) \,, \tag{4}$$

for each $X \in G = \operatorname{Sub} F$.

For reasons of space, we shall omit the proof of Proposition 1. In the rest of the paper we shall find it expedient to work with distributions rather than assignments. Using Proposition 1 in a straightforward manner, it is possible to obtain a version of Theorem 2 below for probability assignments. Details are omitted, again due to space limitations.

3   Open Partitions

We next introduce a key tool for the statement and proof of Theorem 2, namely, a notion of partition for forests.

Remark 1. It turns out that our results in this section apply equally well to (always finite) posets, with no additional complications. Therefore, here we shall work with posets and open maps between them. As mentioned in Subsection 2.2, in logical terms this amounts to working with finite Kripke frames and their p-morphisms, that is, with the relational semantics of the finite slice of intuitionistic logic. On the other hand, both in Proposition 1 and in Theorem 2 we restrict attention to forests only. Indeed, while the notion of open partition can be justified for the whole finite slice of intuitionistic logic, the notion of probability distribution given in (GD1–2) is intimately related to a complete $[0,1]$-valued semantics for the logic at hand — and, as is well known, no such complete semantics is available for full intuitionistic logic. We reserve a thorough discussion of these points for a future occasion.

In the Boolean case, a partition of a set A is the same thing as the collection of fibres of an appropriate surjection $f: A \rightarrow B$. Accordingly, we define as follows.

Definition 1. *An* open partition *of a poset P is a set-theoretic partition $\pi = \{B_1, \ldots, B_m\}$ of P that is induced by some surjective open map $f: P \rightarrow Q$ onto a poset Q. That is, for each $i = 1, \ldots, m$ there is $y \in Q$ such that*

$$B_i = f^{-1}(y) = \{x \in P \mid f(x) = y\} \ .$$

It follows that an open partition π of P carries an *underlying partial order*: namely, define

$$B_i \preceq B_j$$

if and only if

$$f(B_i) \leq f(B_j) \ \text{in} \ Q \ .$$

It is easily verified that the order $\preceq$ does not depend on the choice of f and Q. Also note that π, regarded as a poset under $\preceq$, is order-isomorphic to (any choice of) Q.

Definition 1 has the advantage that it can be recast in quite general category-theoretic terms. However, it is also apparent that, in practice, it is quite inconvenient to work with — one needs to refer to f and Q, whereas in the Boolean case one has at hand the usual definition in terms of non-empty, pairwise disjoint subsets. This initial drawback is fully remedied by our main result on open partitions.

Theorem 1. *Let P be a poset, and let $\pi = \{B_1, \ldots, B_m\}$ be a set-theoretic partition of P. Then π is an open partition of P if and only if for each $B_i \in \pi$ there exist $i_1, i_2, \ldots, i_t \in \{1, \ldots, m\}$ such that*

$$\uparrow B_i = B_{i_1} \cup B_{i_2} \cup \cdots \cup B_{i_t} \ . \tag{5}$$

In this case, the underlying order $\preceq$ of π is uniquely determined by

$$B_i \preceq B_j \ \text{iff} \ B_j \subseteq \uparrow B_i \ \text{iff there are} \ x \in B_i, y \in B_j \ \text{with} \ x \leq y \ ,$$

for each $B_i, B_j \in \pi$.

Proof. Suppose π is an open partition of P. By Definition 1 there exists a surjective open map f from P onto a poset Q whose set of fibres is π. Suppose, by way of contradiction, that (5) does not hold. Thus, there exist $p, q \in B_j$ such that $p \in \uparrow B_i$, but $q \notin \uparrow B_i$, for some $B_i, B_j \in \pi$. Let $f(B_i) = y$. Since f is order-preserving, $y \in \downarrow f(p)$. Since f is open, $y \notin \downarrow f(q)$, for else we would find $x \in B_i$ with $x \leq q$. But $f(q) = f(p)$ and we have a contradiction.

Suppose now that π satisfies (5). Endow π with the relation $\preceq$, defined as in Theorem 1. Observe that under the condition (5), for each $B_i, B_j \in \pi$, $B_j \subseteq \uparrow B_i$ if and only if there are $x \in B_i, y \in B_j$ with $x \leq y$. Indeed, whenever $x \leq y$, the block B_j intersects the upper set of the block B_i. By (5), B_j must be entirely contained in $\uparrow B_i$. The converse is trivial.

We show that $\preceq$ is a partial order on π. One can immediately check that $\preceq$ is reflexive and transitive. Let $B_i, B_j \in \pi$ be such that $B_i \preceq B_j$ and $B_j \preceq B_i$. Let $x \in B_i$. Since $B_j \preceq B_i$ there exists $y \in B_j$ such that $y \leq x$. Since $B_i \preceq B_j$ there exists $z \in B_i$ such that $z \leq y \leq x$. Iterating, since P is finite, we will find $p \in B_i$ and $q \in B_j$ satisfying $p \leq q \leq p$. Since π is a partition, we obtain $B_i = B_j$. Thus, the relation $\preceq$ is antisymmetric, and it is a partial order on π.

Let us consider now the projection map $f : P \to \pi$ which sends each element of P to its block. Let $x \in B_i$, $y \in B_j$, for $B_i, B_j \in \pi$. If $x \leq y$ then $f(x) = B_i \preceq f(y) = B_j$ and f is order-preserving. By construction, since π does not have empty blocks, f is surjective. To show f is open, we consider $u \in P$, $f(u) = B_t$, and $B_s \preceq B_t$, for some $B_s \in \pi$. Since $B_t \subseteq \uparrow B_s$, there exists $v \in B_s$ such that $v \leq u$. Since $f(v) = B_s$, f is open.

It remains to show that the last statement holds. Endow π with a partial order $\preceq'$ different from $\preceq$ and consider the map $f' : P \to \pi$ that sends each element of P to its own block. We consider two cases.

(Case 1). There exist $B_i, B_j \in \pi$ such that $B_i \preceq B_j$, but $B_i \npreceq' B_j$. Since there are $x \in B_i, y \in B_j$ with $x \leq y$, f' is not order-preserving.

(Case 2). There exist $B_i, B_j \in \pi$ such that $B_i \preceq' B_j$, but $B_i \npreceq B_j$. Let $y \in B_j$. By the definition of $\preceq$, for every $x \in B_i$, $x \nleq y$. Thus, f' is not an open map.

Thus, if one endows π with an order different from $\preceq$, one cannot find any surjective open map from P to π which induces the partition π. We therefore conclude that the order on π is uniquely determined, and the proof is complete. $\square$

Fig. 4. Two set-theoretic partitions of a forest

Example 2. We consider two different set-theoretic partitions $\pi = \{\{x, y\}, \{z\}\}$, and $\pi' = \{\{x, z\}, \{y\}\}$ of the same forest F. The partitions are depicted in Figure 4. It is immediate to check, using Condition (5) in Theorem 1, that π is an open partition of F, while π' is not.

4   Main Result

In Table 1, we summarise the correspondence between the fragments of the probability theory for Gödel logic sketched in the above, and the classical elementary theory. To state and prove our main result, we need one more definition to generalise the push-forward construction from finite sets to forests. Let $f : F_1 \to F_2$ be

Table 1. Gödelian analogues of Boolean concepts

Concept	Boolean model	Gödelian model
Sample space	Set	Forest
Event	Subset	Subforest
Elementary event	Singleton	Branch
Partition	Set-theoretic partition	Open partition
Structure of events	Boolean algebra of sets	Gödel algebra of forests
Probability assignment	Function satisfying (B1–2)	Function satisfying (G1–4)
Probability distribution	Function satisfying (BD2)	Function satisfying (GD1–2)

an open map between forests, and let $p\colon F_1 \to [0,1]$ be a probability distribution. The *push-forward* of p along f is the function $p_f\colon F_2 \to \mathbb{R}$ defined by setting

$$p_f(y) = \sum_{x \in f^{-1}(y)} p(x)$$

for any $y \in F_2$.

Theorem 2. *Let F be a forest, π an open partition of F, and $q\colon F \to \pi$ the natural projection map, Then, for any probability distribution $p\colon F \to [0,1]$, the push-forward p_q of p along q is again a probability distribution on π.*

Proof. It is clear that p_q takes values in the non-negative real numbers, because p does. Thus we need only prove that p_q satisfies (GD1–2).

We first prove that (GD1) holds. Let us display the given open partition as $\pi = \{B_1, \ldots, B_m\}$, and let us write $\preceq$ for its underlying order, and $\prec$ for the corresponding strict order. By definition, we have

$$p_q(B_i) = \sum_{x \in q^{-1}(B_i)} p(x) \tag{6}$$

for each $B_i \in \pi$. Since $q\colon F \to \pi$ is the natural projection map onto π, (6) can be rewritten as

$$p_q(B_i) = \sum_{x \in B_i} p(x) \tag{7}$$

Summing (7) over $i = 1, \ldots, m$, we obtain

$$\sum_{i=1}^{m} p_q(B_i) = \sum_{i=1}^{m} \sum_{x \in B_i} p(x) . \tag{8}$$

Since π is, in particular, a set-theoretic partition of F, from (8) we infer

$$\sum_{i=1}^{m} p_q(B_i) = \sum_{x \in F} p(x) = 1 , \tag{9}$$

with the latter equality following from the fact that p satisfies (GD1). This proves that p_q satisfies (GD1), too.

To prove (GD2), suppose, by way of contradiction, that $p_q^{-1}(0)$ is not an upper set — in particular, it is not empty. Then there exist $B_i \neq B_j \in \pi$ with

$$p_q(B_i) = 0 \ , \tag{10}$$

but

$$B_i \prec B_j \tag{11}$$

and

$$p_q(B_j) > 0 \ . \tag{12}$$

From (11), together with Theorem 1 and the fact that $B_i \cap B_j = \emptyset$, we know that there exist $x_i \in B_i$ and $x_j \in B_j$ such that

$$x_i < x_j \ .$$

From (10), along with (7) and the fact that p has non-negative range, we obtain

$$p_q(x) = 0 \text{ for all } x \in B_i \ . \tag{13}$$

By precisely the same token, from (12) we obtain that there exists an element $x_j' \in B_j$ such that

$$p_q(x_j') > 0 \ . \tag{14}$$

In (14) we possibly have $x_j' \neq x_j$. However, we make the following

Claim. There exists $x_i' \in B_i$ with $x_i' < x_j'$.

Proof. By way of contradiction, suppose not. Then, writing $B_i = \{x_{i_1}, \ldots, x_{i_u}\}$, we have

$$x_j' \notin (\uparrow x_{i_1}) \cup \cdots \cup (\uparrow x_{i_u})$$

But, clearly,

$$(\uparrow x_{i_1}) \cup \cdots \cup (\uparrow x_{i_u}) = \uparrow B_i \ ,$$

so that

$$x_j' \notin \uparrow B_i \ .$$

Since, however, $x_j' \in B_j$, the latter statement immediately implies

$$B_j \nsubseteq \uparrow B_i \ .$$

Since, moreover, $B_i \prec B_j$ by (11), this contradicts Theorem 1. The Claim is settled. $\square$

Now the Claim, together with (13–14), amounts to saying that $p^{-1}(0)$ is not an upper set, contradicting the assumption that p satisfies (GD2). Thus, p_q satisfies (GD2), too. This completes the proof. $\square$

Example 3. We refer to the forest F and its open partition $\pi = \{\{x, y\}, \{z\}\}$ depicted in Figure 4. Consider the probability distribution $p : F \to [0, 1]$ such that $f(x) = 1$, and $f(y) = f(z) = 0$. Let $q : F \to \pi$ be the natural projection map.

The push-forward p_q of p along q is again a probability distribution on π. Indeed, $p_q(\{x, y\}) = 1$ and $p_q(\{z\}) = 0$, and thus p_q satisfies (GD1-2).

References

1. Aguzzoli, S., Gerla, B., Marra, V.: De Finetti's No-Dutch-Book Criterion for Gödel logic. Studia Logica 90(1), 25–41 (2008)
2. Aguzzoli, S., Gerla, B., Marra, V.: Defuzzifying formulas in Gödel logic through finitely additive measures. In: IEEE International Conference on Fuzzy Systems, 2008. FUZZ-IEEE 2008 (IEEE World Congress on Computational Intelligence), June 2008, pp. 1886–1893 (2008)
3. Chagrov, A., Zakharyaschev, M.: Modal logic. Oxford Logic Guides, vol. 35. The Clarendon Press/Oxford University Press, New York (1997)
4. Hájek, P.: Metamathematics of fuzzy logic. Trends in Logic—Studia Logica Library, vol. 4. Kluwer Academic Publishers, Dordrecht (1998)
5. Horn, A.: Logic with truth values in a linearly ordered Heyting algebra. J. Symbolic Logic 34, 395–408 (1969)
6. Horn, A.: Free L-algebras. J. Symbolic Logic 34, 475–480 (1969)
7. Johnstone, P.T.: Stone spaces. Cambridge Studies in Advanced Mathematics, vol. 3. Cambridge University Press, Cambridge (1982)
8. Kühr, J., Mundici, D.: De Finetti theorem and Borel states in [0, 1]-valued algebraic logic. Internat. J. Approx. Reason. 46(3), 605–616 (2007)
9. Mundici, D.: Bookmaking over infinite-valued events. Internat. J. Approx. Reason. 43(3), 223–240 (2006)
10. Rohlin, V.A.: On the fundamental ideas of measure theory. Amer. Math. Soc. Translation 1952(71), 55 (1952)

Exploring Extensions of Possibilistic Logic over Gödel Logic

Pilar Dellunde[1,2], Lluís Godo[2], and Enrico Marchioni[2]

[1] Universitat Autònoma de Barcelona,
08193 Bellaterra, Spain
pilar.dellunde@uab.cat
[2] IIIA – CSIC,
08193 Bellaterra, Spain
{enrico,pilar,godo}@iiia.csic.es

Abstract. In this paper we present completeness results of several fuzzy logics trying to capture different notions of necessity (in the sense of Possibility theory) for Gödel logic formulas. In a first attempt, based on different characterizations of necessity measures on fuzzy sets, a group of logics, with Kripke style semantics, are built over a restricted language, indeed a two level language composed of non-modal and modal formulas, the latter moreover not allowing for nested applications of the modal operator N. Besides, a full fuzzy modal logic for graded necessity over Gödel logic is also introduced together with an algebraic semantics, the class of NG-algebras.

1 Introduction

The most general notion of uncertainty is captured by monotone set functions with two natural boundary conditions. In the literature, these functions have received several names, like *Sugeno measures* [24] or *plausibility measures* [20]. Many popular uncertainty measures, like probabilities, upper and lower probabilities, Dempster-Shafer plausibility and belief functions, or possibility and necessity measures, can be therefore seen as particular classes of Sugeno measures.

In this paper, we specially focus on possibilistic models of uncertainty. A *possibility measure* on a complete Boolean algebra of events $\mathcal{U} = (U, \wedge, \vee, \neg, \overline{0}^{\mathcal{U}}, \overline{1}^{\mathcal{U}})$ is a Sugeno measure μ^* satisfiying the following $\vee$-decomposition property for any countable set of indices I

$$\mu^*(\vee_{i \in I} u_i) = \sup_{i \in I} \mu^*(u_i),$$

while a *necessity measure* is a Sugeno measure μ_* satisfying the $\wedge$-decomposition property

$$\mu_*(\wedge_{i \in I} u_i) = \inf_{i \in I} \mu_*(u_i).$$

Possibility and necessity are *dual* classes of measures, in the sense that if μ^* is a possibility measure, then the mapping $\mu_*(u) = 1 - \mu^*(\neg u)$ is a necessity

C. Sossai and G. Chemello (Eds.): ECSQARU 2009, LNAI 5590, pp. 923–934, 2009.
© Springer-Verlag Berlin Heidelberg 2009

measure, and vice versa. If $\mathcal{U}$ is the power set of a set X, then any dual pair of measures (μ^*, μ_*) on $\mathcal{U}$ is induced by a *normalized possibility distribution*, i.e. a mapping $\pi : X \to [0,1]$ such that, $\sup_{x \in X} \pi(x) = 1$, and, for any $A \subseteq X$,

$$\mu^*(A) = \sup\{\pi(x) \mid x \in A\} \text{ and } \mu_*(A) = \inf\{1 - \pi(x) \mid x \notin A\}.$$

Appropriate extensions of uncertainty measures on algebras of events more general than Boolean algebras need to be considered in order to represent and reason about the uncertainty of non-classical events. For instance, the notion of (finitely additive) probability has been generalized in the setting of MV-algebras by means of the notion of *state* [22]. In particular, the well-known Zadeh's notion of probability for fuzzy sets (as the expected value of the membership function) defines a state over an MV-algebra of fuzzy sets. States on MV-algebras have been used in [12] to provide a logical framework for reasoning about the probability of (finitely-valued) fuzzy events. Another generalization of the notion of probability measure has been recently studied in depth by defining probabilistic states over Gödel algebras [1].

On the other hand, extensions of the notions of possibility and necessity measures for fuzzy sets have been proposed under different forms and used in different logical systems extending the well-known Dubois-Lang-Prade's possibilistic logic to fuzzy events, see e.g. [7,9,16,3,2,4]. All the notions of necessity for fuzzy sets considered in the literature turn out to be of the form

$$N(A) = \inf_{x \in U} \pi(x) \Rightarrow A(x) \qquad (*)$$

where A is a fuzzy set in some domain U, $\pi : U \to [0,1]$ is a possibility distribution on U and $\Rightarrow$ is some suitable many-valued implication function. In particular, the following notions of necessity have been discussed:

(1) $x \Rightarrow_{KD} y = \max(1 - x, y)$ (Kleene-Dienes);
(2) $x \Rightarrow_{RG} y = 1$ if $x \le y$, and $x \Rightarrow_{RG} y = 1 - x$ otherwise (reciprocal of Gödel);
(3) $x \Rightarrow_{\text{L}} y = \min(1, 1 - x + y)$ (Łukasiewicz).

All these definitions actually extend the above definition over classical sets or events.

In the literature different logical formalizations to reason about such extensions of the necessity of fuzzy events can be found. In [19], and later in [17], a full many-valued modal approach is developed over the finitely-valued Łukasiewicz logic in order to capture the notion of necessity defined using $\Rightarrow_{KD}$. A logic programming approach over Gödel logic is investigated in [3] and in [2] by relying on $\Rightarrow_{KD}$ and $\Rightarrow_{RG}$, respectively. More recently, following the approach of [12], modal-like logics to reason about the necessity of fuzzy events in the framework of MV-algebras have been defined in [13], in order to capture the notion of necessity defined by $\Rightarrow_{KD}$ and $\Rightarrow_{\text{L}}$.

The purpose of this paper is to explore different logical approaches to reason about the necessity of fuzzy events over Gödel algebras. In more concrete terms, our ultimate aim is to study a full modal expansion of the $[0, 1]$-valued Gödel logic with a modality N such that the truth-value of a formula $N\varphi$ (in $[0, 1]$) can be interpreted as the degree of necessity of φ, according to some suitable semantics. In this context, although this does not extend the classical possibilistic logic, it seems also interesting to investigate the notion of necessity definable from Gödel implication, which is the standard fuzzy interpretation of the implication connective in Gödel logic:

(4) $x \Rightarrow_G y = 1$ if $x \leq y$, and $x \Rightarrow_G y = y$ otherwise (Gödel);

This work is structured as follows. After this introduction, in Section 2 we recall a characterization of necessity measures on fuzzy sets defined by implications $\Rightarrow_{KD}$ and $\Rightarrow_{RG}$ and provide a (new) characterization of those defined by $\Rightarrow_G$. These characterizations are the basis for the completeness results of several logics introduced in Section 3 capturing the corresponding notions of necessity for Gödel logic formulas. These logics, with Kripke style semantics, are built over a two-level language composed of modal and non-modal formulas, moreover the latter not allowing nested applications of the modal operator. In Section 4 a full fuzzy modal logic for graded necessity over Gödel logic is introduced together with an algebraic semantics. Finally, in Section 5 we mention some open problems and new research goals we plan to address in the near future.

Due to lack of space, we cannot include preliminaries on basic notions regarding Gödel logic and its expansions with truth-constants, with Monteiro-Baaz's operator Δ and with an involutive negation, that will be used throughout the paper. Instead, the reader is referred to [17,10,11] for the necessary background.

2   Some Necessity Measures over Gödel Algebras of Fuzzy Sets and Their Characterizations

Let X be a (finite) set and let $F(X) = [0,1]^X$ be the set of fuzzy sets over X, i.e. the set of functions $f : X \rightarrow [0,1]$. $F(X)$ can be regarded as a Gödel algebra equipped with the pointwise extensions of the operations of the standard Gödel algebra $[0,1]_G$. In the following, for each $r \in [0,1]$, we will denote by $\bar{r}$ the constant function $\bar{r}(x) = r$ for all $x \in X$.

Definition 1. *A mapping $N : F(X) \rightarrow [0,1]$ satisfying*

(N1) $N(\wedge_{i \in I} f_i) = \inf_{i \in I} N(f_i)$
(N2) $N(\bar{r}) = r$, *for all $r \in [0,1]$*

is called a basic necessity.

If $N : F(X) \rightarrow [0,1]$ is a basic necessity then it is easy to check that it also satisfies the following properties:

(i) $\min(N(f), N(\neg_G f)) = 0$
(ii) $N(f \Rightarrow_G g) \leq N(f) \Rightarrow_G N(g)$

The classes of necessity measures based on the Kleene-Dienes implication and the reciprocal of Gödel implication have been already characterized in the literature. We do not consider here the one based on Łukasiewicz implication.

Lemma 2 ([3,2]). *Let $N : F(X) \to [0,1]$ be a basic necessity. Consider the following properties:*

(N_{KD}) $N(\bar{r} \Rightarrow_{KD} f) = r \Rightarrow_{KD} N(f)$, *for all* $r \in [0,1]$
(N_{RG}) $N(\bar{r} \Rightarrow_{RG} f) = r \Rightarrow_{RG} N(f)$, *for all* $r \in [0,1]$

Then, we have:

(1) N satisfies (N_{KD}) iff $N(f) = \inf_{x \in X} \pi(x) \Rightarrow_{KD} f(x)$
(2) N satisfies (N_{RG}) iff $N(f) = \inf_{x \in X} \pi(x) \Rightarrow_{RG} f(x)$

for some possibility distribution $\pi : X \to [0,1]$ such that $\sup_{x \in X} \pi(x) = 1$.

The characterization of the necessity measures based on Gödel implication is somewhat more complex since it needs to consider also an associated class of possibility measures which are not dual in the usual strong sense.

Definition 3. *A mapping $\Pi : F(X) \to [0,1]$ satisfying*

$(\Pi 1)$ $\Pi(\vee_{i \in I} f_i) = \sup_{i \in I} \Pi(f_i)$
$(\Pi 2)$ $\Pi(\bar{r}) = r$, *for all* $r \in [0,1]$

is called a basic possibility.

Note that if $\Pi : F(X) \to [0,1]$ is a basic possibility then it also satisfies $\max(\Pi(\neg f), \Pi(\neg\neg f)) = 1$.

For each $x \in X$, let us denote by $\mathbf{x}$ its characteristic function, i.e. the function from $F(X)$ such that $\mathbf{x}(y) = 1$ if $y = x$ and $\mathbf{x}(y) = 0$ otherwise. Observe that each $f \in F(X)$ can be written as

$$f = \bigwedge_{x \in X} \mathbf{x} \Rightarrow_G \overline{f(x)} = \bigvee_{x \in X} \mathbf{x} \wedge \overline{f(x)}.$$

Therefore, if N and Π are a pair of basic necessity and possibility on $F(X)$ respectively, by $(N1)$ and $(\Pi 1)$ we have

$$N(f) = \inf_{x \in X} N(\mathbf{x} \Rightarrow_G \overline{f(x)}) \quad \text{and} \quad \Pi(f) = \sup_{x \in X} \Pi(\mathbf{x} \wedge \overline{f(x)}).$$

Then we obtain the following characterizations.

Proposition 4. *Let $\Pi : F(X) \to [0,1]$ be a basic possibility. Π further satisfies*

$(\Pi 3)$ $\Pi(f \wedge \bar{r}) = \min(\Pi(f), r)$, *for all* $r \in [0,1]$

iff there exists $\pi : X \to [0,1]$ such that $\sup_{x \in X} \pi(x) = 1$ and, for all $f \in F(X)$, $\Pi(f) = \sup_{x \in X} \min(\pi(x), f(x))$.

Proof. One direction is easy. Conversely, assume that $\Pi : F(X) \to [0,1]$ satisfies $(\Pi 1)$ and $(\Pi 3)$. Then, taking into account the above observations, we have

$$\Pi(f) = \sup_{x \in X} \Pi(\mathbf{x} \wedge \overline{f(x)}) = \sup_{x \in X} \min(\Pi(\mathbf{x}), f(x)).$$

Hence, the claim easily follows by defining $\pi(x) = \Pi(\mathbf{x})$. $\square$

Proposition 5. *Let $N : F(X) \rightarrow [0,1]$ be a basic necessity and $\Pi : F(X) \rightarrow [0,1]$ be a basic possibility satisfying ($\Pi 3$). N and Π further satisfy*

(ΠN) $N(f \Rightarrow_G \bar{r}) = \Pi(f) \Rightarrow_G r$, *for all $r \in [0,1]$*

iff there exists $\pi : X \rightarrow [0,1]$ such that $\sup_{x \in X} \pi(x) = 1$ and

$$N(f) = \inf_{x \in X} \pi(x) \Rightarrow_G f(x) \text{ and } \Pi(f) = \sup_{x \in X} \min(\pi(x), f(x)).$$

Proof. As for the possibility Π, this is already shown above in Proposition 4. Let N be defined as $N(f) = \inf_{x \in X} \pi(x) \Rightarrow_G f(x)$ for the possibility distribution $\pi : F(X) \rightarrow [0,1]$ determined by Π. We have $N(f \Rightarrow_G \bar{r}) = \inf_{x \in X}(\pi(x) \Rightarrow_G (f(x) \Rightarrow_G r)) = \inf_{x \in X}((\pi(x) \wedge f(x)) \Rightarrow_G r) = (\sup_{x \in X} \pi(x) \wedge f(x)) \Rightarrow_G r = \Pi(f) \Rightarrow_G r$. Hence, Π and N satisfy (ΠN).

Conversely, suppose that N and Π satisfy (ΠN). Then, using the fact that $\Pi(\mathbf{x}) = \pi(x)$ for each $x \in X$, we have $N(f) = \inf_{x \in X} N(\mathbf{x} \Rightarrow_G \overline{f(x)}) = \inf_{x \in X} \Pi(\mathbf{x}) \Rightarrow_G f(x) = \inf_{x \in X} \pi(x) \Rightarrow_G f(x)$. $\square$

3   Four Complete Logics: The Two-Level Language Approach

The language of the logics we are going to consider in this section consists of two classes of formulas:

(i) The set $Fm(V)$ of non-modal formulas $\varphi, \psi \dots$, which are formulas of $G_\Delta(\mathbb{Q})$ (Gödel logic G expanded with Baaz's projection connective Δ and truth constants $\bar{r}$ for each rational $r \in [0,1]$) built from the set of propositional variables $V = \{p_1, p_2, \dots\}$;

(ii) And the set $MFm(V)$ of modal formulas $\Phi, \Psi \dots$, built from atomic modal formulas $N\varphi$, with $\varphi \in Fm(V)$, where N denotes the modality *necessity*, using the connectives from G_Δ and truth constants $\bar{r}$ for each rational $r \in [0,1]$. Notice that nested modalities are not allowed.

The axioms of the logic NG^0 of basic necessity are the axioms of $G_\Delta(\mathbb{Q})$ for non-modal and modal formulas plus the following necessity related modal axioms:

$(N1)$ $N(\varphi \rightarrow \psi) \rightarrow (N\varphi \rightarrow N\psi)$
$(N2)$ $N(\bar{r}) \leftrightarrow \bar{r}$,   for each $r \in [0,1] \cap \mathbb{Q}$.

The rules of inference of NG^0 are *modus ponens* (for modal and non-modal formulas) and *necessitation*: from $\vdash \varphi$ infer $\vdash N\varphi$.

It is worth noting that NG^0 proves the formula $N(\varphi \wedge \psi) \leftrightarrow (N\varphi \wedge N\psi)$, which encodes a characteristic property of necessity measures.

As for the semantics we consider several classes of *possibilistic* Kripke models. A *basic necessity Kripke model* is a system $\mathcal{M} = \langle W, e, I \rangle$ where:

- W is a non-empty set whose elements are called *nodes* or *worlds*,
- $e : W \times V \to [0,1]$ is such that, for each $w \in W$, $e(w, \cdot) : V \to [0,1]$ is an evaluation of propositional variables which is extended to a $G_\Delta(\mathbb{Q})$-evaluation of (non-modal) formulas of $Fm(V)$ in the usual way.
- For each $\varphi \in Fm(V)$ we define its associated function $\hat{\varphi}_W : W \to [0,1]$, where $\hat{\varphi}_W(w) = e(w, \varphi)$. Let $\widehat{Fm} = \{\hat{\varphi} \mid \varphi \in Fm(V)\}$
- $I : \widehat{Fm} \to [0,1]$ is a basic necessity over $\widehat{Fm}$ (as a G-algebra), i.e. it satisfies
  (i) $I(\hat{\bar{r}}_W) = r$, for all $r \in [0,1] \cap \mathbb{Q}$
  (ii) $I(\wedge_{i \in I} \hat{\varphi}_{iW}) = \inf_{i \in I} I(\hat{\varphi}_{iW})$.

Now, given a modal formula Φ, the truth value of Φ in $\mathcal{M} = \langle W, e, I \rangle$, denoted $\|\Phi\|_{\mathcal{M}}$, is inductively defined as follows:

- If Φ is an atomic modal formula $N\varphi$, then $\|N\varphi\|_{\mathcal{M}} = I(\hat{\varphi}_W)$
- If Φ is a non-atomic modal formula, then its truth value is computed by evaluating its atomic modal subformulas, and then by using the truth functions associated to the $G_\Delta(\mathbb{Q})$ connectives occurring in Φ.

We will denote by $\mathcal{N}$ the class of basic necessity Kripke models. Then, taking into account that $G_\Delta(Q)$-algebras are locally finite, following the same approach of [13] with the necessary modifications, one can prove the following result.

Theorem 6. NG^0 *is sound and complete for modal theories w.r.t. the class* $\mathcal{N}$ *of basic necessity structures.*

Now our aim is to consider extensions of NG^0 which faithfully capture the three different notions of necessity measure considered in the previous section. We start by considering the following additional axiom:

$$(N_{KD}) \quad N(\bar{r} \vee \varphi) \leftrightarrow (\bar{r} \vee N\varphi), \text{ for each } r \in [0,1] \cap \mathbb{Q}.$$

Let NG_{KD} be the axiomatic extension of NG^0 with (N_{KD}). Then, using Lemma 2, it is easy to prove that indeed NG_{KD} captures the reasoning about KD-necessity measures.

Theorem 7. N_{KD} *is sound and complete for modal theories w.r.t. the subclass* $\mathcal{N}_{KD}$ *of necessity structures* $\mathcal{M} = \langle W, e, I \rangle$ *such that the necessity measure* I *is defined as, for every* $\varphi \in Fm(V)$, $I(\hat{\varphi}_W) = \inf_{w \in W} \pi(w) \Rightarrow_{KD} \hat{\varphi}_W(w)$ *for some possibility distribution* $\pi : W \to [0,1]$ *on the set of possible worlds* W.

To capture RG-necessities, we need to expand the base logic $G_\Delta(\mathbb{Q})$ with an involutive negation $\sim$. This corresponds to the logic $G_\sim(\mathbb{Q})$, as defined in [10]. So we define NG_{RG} as the axiomatic extension of NG^0 over $G_\sim(\mathbb{Q})$ (instead of over $G_\Delta(\mathbb{Q})$) with the following axiom:

$$(N_{RG}) \quad N(\sim\varphi \to \overline{1-r}) \leftrightarrow (\sim N\varphi \to \overline{1-r}), \text{ for each } r \in [0,1] \cap \mathbb{Q}.$$

Then, using again Lemma 2 and the fact that also $G_\sim(\mathbb{Q})$-algebras are locally finite, one can also prove the following result.

Theorem 8. NG_{RG} *is sound and complete for modal theories w.r.t. the subclass* $\mathcal{N}_{RG}$ *of necessity structures*[1] $\mathcal{M} = \langle W, e, I \rangle$ *such that the necessity measure* I *is defined as, for every* $\varphi \in Fm(V)$, $I(\hat{\varphi}_W) = \inf_{w \in W} \pi(w) \Rightarrow_{RG} \hat{\varphi}_W(w)$ *for some possibility distribution* $\pi : W \to [0,1]$ *on the set of possible worlds* W.

It is worth pointing out that if we added the Boolean axiom $\varphi \vee \neg\varphi$ to the logics N_{KD} and N_{RG}, both extensions would basically collapse into the classical possibilistic logic.

Finally, to define a logic capturing N_G-necessities, we need to expand the language of NG^0 with an additional operator Π to capture the associated possibility measures according to Proposition 5. Therefore we consider the extended set $MFm(V)^+$ of modal formulas $\Phi, \Psi \ldots$ as those built from atomic modal formulas $N\varphi$ and $\Pi\varphi$, with $\varphi \in Fm(V)$, truth-constants $\overline{r}$ for each $r \in [0,1] \cap \mathbb{Q}$ and G_Δ connectives. Then the axioms of the logic $N\Pi_G$ are those of $G_\Delta(\mathbb{Q})$ for non-modal and modal formulas, plus the following necessity related modal axioms:

(N1) $N(\varphi \to \psi) \to (N\varphi \to N\psi)$
(N2) $N(\overline{r}) \leftrightarrow \overline{r}$,
(Π1) $\Pi(\varphi \vee \psi) \leftrightarrow (\Pi\varphi \vee \Pi\psi)$
(Π2) $\Pi(\overline{r}) \leftrightarrow \overline{r}$,
(Π3) $\Pi(\varphi \wedge \overline{r}) \leftrightarrow (\Pi\varphi \wedge \overline{r})$
(NΠ) $N(\varphi \to \overline{r}) \leftrightarrow (\Pi\varphi \to \overline{r})$

where $(N2), (\Pi 2), (\Pi 3)$ and $(N\Pi)$ hold for each $r \in [0,1] \cap \mathbb{Q}$. Inference rules of $N\Pi_G$ are those of $G_\Delta(\mathbb{Q})$ and necessitation for N and Π.

Now, we also need to consider expanded Kripke structures of the form $\mathcal{M} = \langle W, e, I, P \rangle$, where W and e are as above and the mappings $I, P :\to [0,1]$ are such that, for every $\varphi \in Fm(V)$, $I(\hat{\varphi}_W) = \inf_{w \in W} \pi(w) \Rightarrow_G \hat{\varphi}_W(w)$ and $P(\hat{\varphi}_W) = \sup_{w \in W} \min(\pi(w), \hat{\varphi}_W(w))$, for some possibility distribution $\pi : W \to [0,1]$. Call $\mathcal{NP}_G$ the class for such structures. Then, using Proposition 5 we get the following result.

Theorem 9. $N\Pi_G$ *is sound and complete for modal theories w.r.t. the class* $\mathcal{NP}_G$ *of structures.*

4   Possibilistic Necessity Gödel Logic and Its Algebraic Semantics: The Full Modal Approach

The logics defined in the previous section are not proper modal logics since the notion of well-formed formula excludes those formulas with occurrences of

---

[1] With the proviso that the evaluations e of propositional variables extend to $G_\sim(\mathbb{Q})$-evaluations for non-modal formulas and not simply to $G_\Delta(\mathbb{Q})$-evaluations.

nested modalities. Our aim in this section is then to explore a full (fuzzy) modal approach.

We start as simple as possible by defining a fuzzy modal logic over Gödel propositional logic G to reason about the necessity degree of G-propositions. The language of *Possibilistic Necessity Gödel logic*, PNG, is defined as follows: formulas of PNG are built from the set of G-formulas using G-connectives and the operator N. Axioms of PNG are those of Gödel logic plus the following modal axioms:

1. $N(\varphi \to \psi) \to (N\varphi \to N\psi)$.
2. $N\psi \leftrightarrow NN\psi$.
3. $\neg N\bar{0}$.

Deduction rules for PNG are Modus Ponens and Necessitation for N (from ψ derive $N\psi$). These axioms and rules define a notion of proof $\vdash_{PNG}$ in the usual way.

Notice that in PNG the *Congruence Rule* "from $\varphi \leftrightarrow \psi$ derive $N\varphi \leftrightarrow N\psi$" as well as the theorems $N\bar{1}$ and $N(\varphi \wedge \psi) \leftrightarrow (N\varphi \wedge N\psi)$ are derivable. Also observe that, if we had restricted the Necessitation Rule only to theorems, we would have obtained a local consequence relation (instead of the global one we have introduced here). For this weaker version of the logic, the Deduction Theorem in its usual form would holds, nevertheless this logic turns out not to be algebraizable.

Theorem 10. [Deduction Theorem] *If $T \cup \{\varphi, \phi\}$ is any set of PNG-formulas, then $T \cup \{\varphi\} \vdash_{PNG} \phi$ iff $T \vdash_{PNG} (\varphi \wedge N\varphi) \to \phi$.*

Kripke style semantics based on possibilistic structures (W, e, I) could be also defined as in Section 3, but now the situation is more complex due to the fact that we are dealing with a full modal language. Moreover, it seems even more complex to try to get some completeness results with respect to this semantics so this is left for future research. This is the reason why in the rest of the paper we will turn our attention to the study of an algebraic semantics, following the ideas developed in [15,14] for the case of a probabilistic logic over Łukasiewicz logic, and see how far we can go.

We start by defining a suitable class of algebras which are expansions of Gödel algebras with a new unary operator trying to capture the notion of necessity.

Definition 11. *An NG-algebra is a structure (A, N) where A is a G-algebra and $N : A \to A$ is a monadic operator such that:*

1. $N(x \Rightarrow y) \Rightarrow (Nx \Rightarrow Ny) = 1$
2. $Nx = NNx$
3. $N1 = 1$

The function N is called an internal possibilistic state *on the G-algebra A.*

Observe that, so defined, the class of NG-algebras is a variety. Examples of internal possibilistic states are the identity function Id, the Δ operator and the

$\neg\neg$ operator. The variety of G-algebras can be considered as a subvariety of NG-algebras, namely the subvariety obtained by adding the equation $N(x) = x$. It is easy to check, using the definition of NG-algebra that, for every NG-algebra (A, N) such that $N(A) = A$ we have $N = Id$, and that, given $a, b \in A$, $a \le b$ implies $Na \le Nb$.

Definition 12. *An* NG*-filter F on an NG-algebra (A, N) is a filter on the G-algebra A with the following property: if $a \in F$, then $Na \in F$.*

By an argument analogous to the one in Lemma 2.3.14 of [17], if $\sim_F$ is the relation defined by: for every $a, b \in A$, $a \sim_F b$ iff $(a \Rightarrow b) \in F$ and $(b \Rightarrow a) \in F$, then $\sim_F$ is a congruence on (A, N) and the quotient algebra $(A, N)/\sim_F$ is an NG-algebra.

Lemma 13. *Let F be an NG-filter on an NG-algebra (A, N). Then, the least NG-filter containing F as a subset and a given $a \in A$ is*

$$F' = \{u \in A : \exists v \in F \ such \ that \ u \ge v * a * Na\}$$

By Corollary 4.8 of [5], it is easy to check that PNG is finitely algebraizable and that the equivalent algebraic semantics of PNG is the variety of NG-algebras. As a corollary we obtain the following general completeness result.

Theorem 14. *The logic* PNG *is strongly complete with respect to the variety of NG-algebras. This means that for any set of formulas $\Gamma \cup \{\Phi\}$, $\Gamma \vdash_{\text{PNG}} \Phi$ iff, for all NG-algebra A and for all evaluation e on A, if $e(\Psi) = 1^A$ for all $\Psi \in \Gamma$, then $e(\Phi) = 1^A$.*

Observe that it is not possible to prove completeness with respect to linearly ordered NG-algebras. Otherwise $N(\Phi \vee \Psi) \leftrightarrow (N\Phi \vee N\Psi)$ would be a theorem. Now we prove some satisfiability results of formulas of the logic *PNG*.

Formulas of the language of PNG can be seen also as terms of the language of NG-algebras. Therefore for the sake of clarity, in the following proofs we work with first-order formulas of the language of NG-algebras proving that they are satisfiable, if the corresponding formulas of the language of PNG are satisfiable.

Proposition 15. *Let $\phi(x_1, \ldots, x_n)$ be a PNG-formula. If ϕ is satisfiable, then $\phi = \bar{1}$ is satisfiable in an NG-algebra (B, Ω), by a sequence $(b_1, \ldots, b_n)$ of elements of B such that, for every $0 < i \le n$, we have either $b_i = 1$ or $\Omega(b_i) = 0$.*

Proof. Let (A, N) be an NG-algebra such that $\phi = \bar{1}$ is satisfiable in (A, N) by $(a_1, \ldots, a_n)$. Without loss of generality we assume that there is $k \le n$ such that for every $0 < i \le k$, $N(a_i) \ne 0$ and for $i > k$, $N(a_i) = 0$.

Now we build a finite sequence of NG-algebras $(B_1, \ldots, B_k)$ and homomorphisms $(h_1, \ldots, h_k)$ such that for every $0 < i \le k$, ϕ is satisfied in B_i by

$$(c_1, \ldots, c_{i-1}, h_i \circ h_{i-1} \circ \cdots \circ h_1(a_i), \ldots, h_i \circ h_{i-1} \circ \cdots \circ h_1(a_n))$$

where each $c_i \in \{0, 1\}$. We define only the first homomorphism, the others can be introduced analogously. Let $F = \{x \in A : Nx \ge Na_1\}$. So defined, it is easy to check that F is an NG-filter. And since, by a previous assumption, $Na_1 \neq 0$, the filter F is proper. Thus, $(A, N)/\sim_F$ is an NG-algebra. Now let h_1 be the canonical homomorphism from (A, N) to $(A, N)/\sim_F$, and let $B_1 = (A, N)/\sim_F$. It is easy to check that $\phi = \bar{1}$ is satisfied in B_1 by $(h_1(a_1), \ldots, h_1(a_n))$, that $h_1(a_1) = 1$ and that for $i > k$, $N(h_1(a_i)) = 0$. Finally, take $(B, \Omega) = (B_k, h_k \circ \cdots \circ h_1 \circ N)$. $\qquad\square$

Definition 16. *An* unnested atomic formula *of the language of NG-algebras, is an atomic formula of one of the following four forms: $x = y$, $c = y$ (where c is a constant $c \in \{\bar{0}, \bar{1}\}$), $Nx = y$ or $F(\bar{x}) = y$ (for some function symbol F of the language of the Gödel algebras).*

Lemma 17. *Let ϕ be a term of the language of NG-algebras. Then there is a set Γ^ϕ of unnested atomic formulas such that, for every NG-algebra (A, N):*

$$\phi = \bar{1} \text{ is satisfiable in } (A, N) \text{ iff } \Gamma^\phi \text{ is satisfiable in } (A, N).$$

Proof. It is a direct consequence of Theorem 2.6.1 of [21]. $\qquad\square$

Example: Let ϕ be the term $x_1 \vee N(x_2 \Rightarrow N(x_3 \Rightarrow \bar{0}))$, take Γ^ϕ to be the following set of unnested atomic formulas:

$$\{x_1 \vee y = z, \bar{1} = z, Nw = y, (x_2 \Rightarrow v) = w, Nq = v, (x_3 \Rightarrow p) = q, \bar{0} = p\}$$

Theorem 18. *Let $\phi(x_1, \ldots, x_n)$ be a PNG-formula. If ϕ is satisfiable, then $\phi = \bar{1}$ is satisfiable in the NG-algebra $([0, 1]_G, \Delta)$ by a sequence of rational numbers.*

Proof. Let (A, Ω) be an NG-algebra in which $\phi(x_1, \ldots, x_n) = \bar{1}$ is satisfiable by an n-tuple $(a_1, \ldots, a_n)$. Without loss of generality we may assume that:

- ϕ is a conjunction of unnested atomic formulas (by using Lemma 17);
- for every $0 < i \leq n$, $a_i \neq 0$ and $a_i \neq 1$ (otherwise we can work with the formulas $\phi(x_i/\bar{1})$ or $\phi(x_i/\bar{0})$, by substituting the corresponding variables by the constants $\bar{0}$ or $\bar{1}$);
- for every i, we have $\Omega(a_i) = 0$ (by Proposition 15).

Now we consider the unnested conjuncts of ϕ. For the sake of simplicity, assume that there is $k \leq n$ such that only in case that $0 < i \leq k$, the variable x_i has an occurrence in an unnested atomic formula of the form $Nx_i = y$. We work now with the formula $\gamma = \phi(Nx_i/\bar{0})$, obtained by substituting in ϕ all the occurrences of Nx_i by the constant $\bar{0}$, for every $0 < i \leq k$.

Observe that, so defined, γ is a conjunction of unnested atomic formulas in the language of the G-algebras which is satisfied in (A, Ω) by $(a_1, \ldots, a_n)$. Therefore, the conjunction $\gamma_0 = \gamma \wedge \bigwedge_{0 < i \leq k} (x_i \neq 1)$ is also satisfied in (A, Ω) by $(a_1, \ldots, a_n)$ (by our assumptions at the beginning of this proof). Finally, since γ_0 is a formula in the language of the G-algebras, it is satisfied in $[0, 1]_G$ by a sequence of rational numbers, and thus, by definition of γ_0, it is easy to check that ϕ is also satisfied in $([0, 1]_G, \Delta)$. $\qquad\square$

5   Future Work

Several issues related to the logic PNG deserve further investigation. A topic that is worth studying in depth is the relation between the algebraic semantics for the logic PNG (and of some meaningful axiomatic extensions) and the Kripke style semantics of the kind used in Section 3. This is crucial if one wants to keep as the intended graded semantics of the N operator one of the possibilistic necessities of the families described in Section 2. Actually, the PNG logic might only capture the logic of basic necessities, and so, different axioms (and possibly operators as well) must be added in order to capture other more specific families of necessities, somehow related to axioms (N_{KD}), (N_{RG}), $(\Pi 3)$ or the axiom $(N\Pi)$.

Also as a future task, we aim at studying the complexity of the sets of satisfiable formulas for both NG_{KD}, NG_{RG} and $N\Pi_G$. Given the results in [18], we conjecture that the problem of checking satisfiability for those logics is in PSPACE. As for PNG, notice that from the results in the above section and the fact that satisfiability for G_Δ is an NP-complete problem (easily derivable from [17]), we immediately obtain that the set of satisfiable PNG-formulas is in NP.

Acknowledgments. The authors are grateful to the anonymous referees for their valuable comments for improving the final version of this paper. They also acknowledge partial support from the Spanish projects MULOG2 (TIN2007-68005-C04) and *Agreement Technologies* (CONSOLIDER CSD2007-0022, INGENIO 2010), as well as the ESF Eurocores-LogICCC/MICINN project (FFI2008-03126-E/FILO). Marchioni also acknowledges partial support of the Juan de la Cierva Program of the Spanish MICINN.

References

1. Aguzzoli, S., Gerla, B., Marra, V.: De Finetti's no-Dutch-book criterion for Gödel logic. Studia Logica 90, 25–41 (2008)
2. Alsinet, T.: Logic Programming with Fuzzy Unification and Imprecise Constants: Possibilistic Semantics and Automated Deduction. Monografies de l'Institut d'Investigació en Intel·ligència Artificial, CSIC, Barcelona (2003)
3. Alsinet, T., Godo, L., Sandri, S.: On the Semantics and Automated Deduction for PLFC: a logic of possibilistic uncertainty and fuzziness. In: Proc. of 15th Conference on Uncertainty in Artificial Intelligence Conference UAI 1999, Stokholm, Sweden, pp. 3–12 (1999)
4. Alsinet, T., Godo, L., Sandri, S.: Two formalisms of extended possibilistic logic programming with context-dependent fuzzy unification: a comparative description. Electr. Notes Theor. Comput. Sci. 66(5) (2002)
5. Blok, W.J., Pigozzi, D.: Algebraizable logics. Memoirs of the American Mathematical Society A.M.S. 396 (1989)
6. Burris, S., Sankappanavar, H.P.: A Course in Universal Algebra. Graduate texts in mathematics, vol. 78. Springer, Heidelberg (1981)
7. Dubois, D., Prade, H.: Possibility theory. Plenum Press, New York (1988)
8. Dubois, D., Prade, H.: Resolution principles in possibilistic logic. International Journal of Approximate Reasoning 4(1), 1–21 (1990)

9. Dubois, D., Lang, J., Prade, H.: Possibilistic logic. In: Gabbay, et al. (eds.) Handbook of Logic in Artificial Intelligence and Logic Programming. Nonmonotonic Reasoning and Uncertain Reasoning, vol. 3, pp. 439–513. Oxford University Press, Oxford (1994)

10. Esteva, F., Godo, L., Hájek, P., Navara, M.: Residuated fuzzy logics with an involutive negation. Archive for Mathematical Logic 39(2), 103–124 (2000)

11. Esteva, F., Gispert, J., Godo, L., Noguera, C.: Adding truth-constants to logics of continuous t-norms: Axiomatization and completeness results. Fuzzy Sets and Systems 158, 597–618 (2007)

12. Flaminio, T., Godo, L.: A logic for reasoning about the probability of fuzzy events. Fuzzy Sets and Systems 158, 625–638 (2007)

13. Flaminio, T., Godo, L., Marchioni, E.: On the Logical Formalization of Possibilistic Counterparts of States over n-Valued Lukasiewicz Events. Journal of Logic and Computation (in press), doi:10.1093/logcom/exp012

14. Flaminio, T., Montagna, F.: MV-algebras with internal states and probabilistic fuzzy logics. International Journal of Approximate Reasoning 50, 138–152 (2009)

15. Flaminio, T., Montagna, F.: An algebraic approach to states on MV-algebras. In: Proc. of EUSFLAT 2007, Ostrava, Czech Republic, vol. II, pp. 201–206 (2007)

16. Godo, L., Vila, L.: Possibilistic temporal reasoning based on fuzzy temporal constraints. In: Proceedings of IJCAI 1995, pp. 1916–1922. Morgan Kaufmann, San Francisco (1995)

17. Hájek, P.: Metamathematics of Fuzzy Logic. Trends in Logic, vol. 4. Kluwer Academic Publishers, Dordrecht (1998)

18. Hájek, P.: Complexity of fuzzy probability logics II. Fuzzy Sets and Systems 158, 2605–2611 (2007)

19. Hájek, P., Harmancová, D., Esteva, F., Garcia, P., Godo, L.: On Modal Logics for Qualitative Possibility in a Fuzzy Setting. In: Proc. of the 94 Uncertainty in Artificial Intelligence Conference (UAI 1994), pp. 278–285. Morgan Kaufmann, San Francisco (1994)

20. Halpern, J.Y.: Reasoning about uncertainty. MIT Press, Cambridge (2003)

21. Hodges, W.: Model Theory. Encyclopedia of Mathematics and its Applications, vol. 42. Cambridge University Press, Cambridge (1993)

22. Mundici, D.: Averaging the truth-value in Lukasiewicz logic. Studia Logica 55(1), 113–127 (1995)

23. Shafer, G.: A mathematical theory of evidence. Princeton University Press, Princeton (1976)

24. Sugeno, M.: Theory of Fuzzy Integrals and its Applications. PhD thesis, Tokyo Institute of Technology, Tokio, Japan (1974)

Author Index